Your gateway to the most widely-used online biology self-study materials and more...

With your purchase of a new copy of Freeman's *Biological Science,* Third Edition, you should have received a **Student Access Kit for MasteringBiology™**. This kit contains instructions and a code for accessing this dynamic website. Your Student Access Kit looks like this:

DON'T THROW YOUR ACCESS KIT AWAY!

If you did not purchase a new textbook or cannot locate the Student Access Kit, you can purchase your subscription to MasteringBiology™ online with a major credit card by selecting Buy Now. To do so, go to www.masteringbio.com. Identify the exact title and edition of your textbook when prompted.

INSTRUCTORS:

Look for your "Instructor Access Kit" that you received with your book. Contact your Pearson Benjamin Cummings Sales Representative if you need assistance.

MasteringBIOLOGY What is MasteringBiology™?

MasteringBiology™ offers two valuable learning systems:

1. The **Study Area** for studying on your own or in a study group
2. **MasteringBiology™ Assignments** that your instructor may require

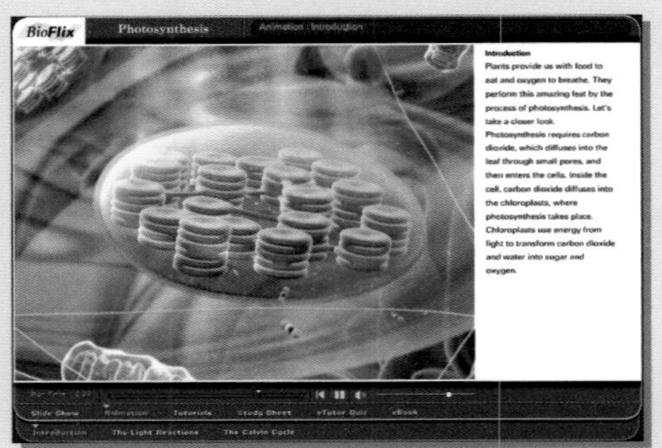

The **Study Area** provides state-of-the-art BioFlix™ 3-D Animations, Discovery videos, Web Animations, Cell Biology videos, interactive eBook, answers to textbook questions and much more (see other side).

MasteringBiology™ Assignments may include Tutorials and other Multiple Choice Questions required by your instructor. MasteringBiology™ tutorials coach you with feedback specific to your needs and offer hints when you get stuck, optimizing your study time and maximizing your learning.

Features of the MasteringBiology™ Study Area for Freeman's *Biological Science*, Third Edition

CHAPTER GUIDE

Prepare for the test by reviewing important information outlined in the Chapter Guide. The Chapter Guide provides a quick overview of key concepts, practice quizzes, and review questions to help students study for the exam. Link directly to the BioFlix™ 3-D animations, Discovery Videos, BioSkills, Cell Biology Animations, and interactive eBook to help students master biology's most difficult topics.

BIOFLIX™ 3-D ANIMATIONS

BioFlix BioFlix™ are 3-D movie-quality animations with carefully constructed student tutorials, labeled slide shows, study sheets, and quizzes which help bring biology to life. Topics include tour of an animal cell, tour of a plant cell, cellular respiration, photosynthesis, mitosis, meiosis, protein synthesis, water transport in plants, how neurons work, and muscle contraction.

VIDEOS

Watch brief Discovery Channel™ video clips on 29 different topics including antibiotic resistance, fighting cancer, and introduced species.

WEB ANIMATIONS

Web Animation Learn about biological concepts in a visual format with online web animations. Animations include pre-quizzes and post-quizzes to test students' understanding of dynamic concepts and processes. Additional Cell Biology Animations provide vivid images of the functions and processes of the cell.

GRAPHIT!

Graphing tutorials show students how to plot, interpret, and critically evaluate real data.

CHAPTER QUIZZES

Students can assess their comprehension of material with 20 multiple-choice quiz questions for each chapter.

ANSWER KEY

Find answers to the book's figure and table caption questions and exercises, "you should be able to..." activities, and end-of-chapter questions.

ART

View and print artwork from the textbook to follow along in class or use as a study tool in review sessions.

AUDIO GLOSSARY

Build your biology vocabulary with our audio glossary. Hear the correct pronunciation and learn the meaning of all key terms.

E-BOOK

Refer to a convenient online version of the book while you study.

BIOLOGICAL SCIENCE

VOLUME 2 **EVOLUTION, DIVERSITY, AND ECOLOGY**

Actual Size: 4 cm

ABOUT THE PHOTO:
He'e, *Octopus* sp. (juvenile), photographed at Midway Atoll National Wildlife Refuge, located in the Northwestern Hawaiian Islands, 29 March 2003, by David Liittschwager and Susan Middleton, authors of *Archipelago: Portraits of Life in the World's Most Remote Island Sanctuary* (National Geographic Society, 2005).

BIOLOGICAL SCIENCE

THIRD EDITION

SCOTT FREEMAN
University of Washington

BENJAMIN CUMMINGS
PEARSON
Benjamin Cummings

San Francisco Boston New York
Cape Town Hong Kong London Madrid Mexico City Montreal Munich Paris Singapore Sydney Tokyo Toronto

Editor-in-Chief, Biology: Beth Wilbur
Sponsoring Editors: Andrew Gilfillan, Susan Winslow, Becky Ruden
Editorial Project Managers: Ann Heath, Sonia DiVittorio
Development Editors: Erin Mulligan, Mary Catherine Hager, Susan Weisberg
Senior Production Supervisor: Shannon Tozier
Associate Editor: Mercedes Grandin
Assistant Editor: Anna Amato
Editorial Assistant: Lisa Tarabokjia
Media Producer: Ericka O'Benar
Executive Directors of Development: Carol Trueheart, Deborah Gale
Executive Marketing Manager: Lauren Harp
Director of Marketing: Christy Lawrence
Managing Editor: Mike Early
Executive Managing Editor: Erin Gregg
Production Service and Composition: Pre-Press PMG
Illustrations: Kim Quillin, Imagineering Media Services, Inc., Greg Williams
Interior Design: Marilyn Perry
Cover Design: Kim Quillin, Marilyn Perry
Cover Production: Laura Wieglab and Side by Side Studios
Manufacturing Buyer: Mike Early and Michael Pene
Director, Image Resource Center: Melinda Patelli
Manager, Rights and Permissions: Zina Arabia
Visual Research Coordinator: Elaine Soares
Image Permission Coordinator: Debbie Latronica
Photo Researcher: Yvonne Gerin
Cover Printer: Phoenix Color
Printer and Binder: R.R. Donnelley, Willard

Cover Image: Octopus–he'e, *Octopus* sp. (juvenile): David Liittschwager/Susan Middleton

Library of Congress Cataloging-in-Publication Data

Freeman, Scott
Biological science / Scott Freeman.—3rd ed.
p. cm.
Includes index.
ISBN 0-13-224950-2 (student edition)—ISBN 0-13-224985-5 (professional copy)—ISBN 0-13-225308-9 (volume 1)—ISBN 0-13-232543-8 (volume 2)—ISBN 0-13-156816-7 (volume 3)
1. Biology. I. Title.
QH308. 2. F73 2007
570—dc22

2007036866

ISBN 10-digit 0-13-224950-2; 13-digit 978-0-13-224950-8 (Student edition)
ISBN 10-digit 0-13-224985-5; 13-digit 978-0-13-224985-0 (Professional copy)
ISBN 10-digit 0-13-225308-9; 13-digit 978-0-13-225308-6 (Volume 1)
ISBN 10-digit 0-13-232543-8; 13-digit 978-0-13-232543-1 (Volume 2)
ISBN 10-digit 0-13-156816-7; 13-digit 978-0-13-156816-7 (Volume 3)

3 4 5 6 7 8 9 10—DOW—11 10 09 08
www.aw-bc.com

Brief Contents: Volume 2

VOLUME 1: Chapters 1–23 • VOLUME 2: Chapters 1, 24–35, 50–55 • VOLUME 3: Chapters 1, 36–49

About the Author

Scott Freeman received his PhD. in Zoology from the University of Washington and was subsequently awarded an Alfred P. Sloan Postdoctoral Fellowship in Molecular Evolution at Princeton University. His current research focuses on the scholarship of teaching and learning—specifically, (1) how active learning and peer teaching techniques increase student learning and improve performance in introductory biology, and (2) how the levels of exam questions vary among introductory biology courses, standardized post-graduate entrance exams, and professional school courses. He has also done research in evolutionary biology on topics ranging from nest parasitism to the molecular systematics of the blackbird family. Scott teaches introductory biology for majors at the University of Washington and is co-author, with Jon Herron, of the standard-setting undergraduate text *Evolutionary Analysis*.

UNIT ADVISORS

An elite group of eleven content experts and star teachers worked with Scott and Kim on every aspect of the third edition. These advisors came to be treasured as they read and interpreted reviews, provided recommendations on outstanding recent papers to check, answered questions, and provided advice on an array of specific issues. The quality and accuracy of this book are a tribute to their efforts and skills.

Ross Feldberg, *Tufts University* (Unit 1)

David Wilson, *Parkland College* (Unit 1)

Paula Lemons, *Duke University* (Unit 2)

Greg Podgorski, *Utah State University* (Units 3 and 4)

George Gilchrist, *College of William and Mary* (Unit 5)

Brianna Timmerman, *University of South Carolina* (Unit 6)

Marc Perkins, *Orange Coast College* (Unit 6)

Michael Black, *California Polytechnic State University* (Units 6 and 8)

Diane Marshall, *University of New Mexico* (Unit 7)

James M. Ryan, *Hobart and William Smith Colleges* (Unit 8)

Alan Molumby, *University of Illinois, Chicago* (Unit 9)

ILLUSTRATOR

Kim Quillin combines expertise in biology and information design to create pedagogically effective and scientifically accurate visual representations of biological principles. She received her B.A. in Biology at Oberlin College and her Ph.D. in Integrative Biology (as a National Science Foundation Graduate Fellow) from the University of California, Berkeley, and has taught undergraduate biology at both schools. Students and instructors alike have praised Kim's illustration programs for *Biological Science*, as well as *Biology: A Guide to the Natural World*, by David Krogh, and *Biology: Science for Life*, by Colleen Belk and Virginia Borden, for their success at applying core principles of information design to convey complex biological ideas in a visually appealing manner.

Preface to Instructors

☞STUDENTS, There is also a preface for you, located right before Chapter 1. It's called "Using this Book as a Tool for Learning." Please read it—it should help you get organized and be successful in this course. ***Scott Freeman***

This book is for instructors who want to help their students learn how to think like a biologist. The content knowledge, problem-solving ability, and analytical skills that this requires can help students become better human beings, in addition to preparing them for success in clinical medicine, scientific research, conservation, law, teaching, journalism, and other careers.

A course goal that focuses on thinking and learning, rather than simply memorizing, is important because today's biology students are going to be tomorrow's problem solvers. This is biology's century, not only because of the breathtaking pace of research but also because many of the most profound challenges we face today—resource shortages, overpopulation, species extinctions, drug resistance, global warming—are biological in nature. The world needs our students.

A Student-Centered Textbook

The first edition of *Biological Science* focused on offering a new approach to teaching biology—one that emphasized higher-order thinking skills over an encyclopedic grasp of what is known about biology. The second edition stayed true to this vision, but added topics and features that made the book easier for professors to use. The third edition also went through several rounds of revision driven by expert reviewers and advisors. Based on input from hundreds of professors around the world, I made thousands of changes to make the book even more accurate, current, and easy to use. But fundamentally, this edition is all about students. For the past three years, the book team and I have used insights from research on student learning—as well as direct feedback from students—as a way to create a better teaching tool.

I plunged into the literature on student learning, and members of the book team and I conducted two dozen focus groups with over 130 students. We asked students the same questions, over and over: What topics are most difficult in this chapter? What helped you "get it"? What tripped you up? How can this figure, or table, or passage of text teach better? Our goal was to create a text that is innovative, engaging, filled with the excitement that drives research, and inspired by data on how students learn.

Fairly quickly, our reading and research identified two fundamental problems:

1. *Novice learners have trouble picking out important information.* This is one of the most striking findings emerging from research on how people learn. It's also an issue that we hear about all the time as instructors. "Do we have to know X?" "Will Y be on the test?" We also see it when students come to office hours and open up their text—they've highlighted everything. Students need help figuring out which material is really important and which offers supporting details. This is crucial to their success, because if we're doing our job right, we're going to test them on the important stuff—not details that they'll forget five minutes after the exam.
2. *Novice learners have a terrible time with self-assessment.* How many times have students told you, "I understood the concepts *so well*, but just did badly on the exam"? Novice learners struggle to understand when they don't understand something. They'll sit in class or read the text and tell themselves, "Yeah, I get this, I get this," but then they crash on the exam. Experts are much more skeptical—they make themselves *use* information and ideas before they're confident they understand what's going on.

Fundamentally, our task is simple: We need to help our students become better students. Textbooks should help learners acquire the skills they need to make the novice-to-expert transition.

Supporting Novice Learners

The Gold Thread—"Learn It"

Our response to the "can't pick out the important points" problem is a battery of tools, highlighted in gold throughout the text.

- **Key Concepts** are listed at the start of the chapter. When material related to these key concepts is presented in the chapter itself, it is flagged with a gold bullet.
- **Chapter Summaries** revisit each of the key concepts. In this way, the big ideas in each chapter are laid out at the start, developed in detail, and then summarized.
- **Check Your Understanding** boxes appear at the ends of key sections within each chapter. Each box briefly summarizes

one or two fundamental points—the key ideas that students ought to have mastered before they move on.

- **Highlighted Passages** help students focus on particularly important information throughout the text.
- **Summary Tables** pull information together into a compact format that is easy to review.

In effect, the gold thread offers students expert guidance on picking out and focusing on the important, unifying points in an information-laden discipline. In addition, and in the same spirit,

- **"Pointing Hands"** in illustrations act like your hand at the whiteboard, guiding students' attention to a figure's central teaching points.

The Blue Thread—"Practice It"

Our response to the self-assessment problem is an array of questions and exercises, highlighted in blue throughout the text.

- **In-text "You should be able to's"** offer exercises on topics that professors and students have identified as the most difficult concepts in each chapter.
- **Caption Questions and Exercises** challenge students to critically examine the information in a figure or table—not just absorb it.
- **Check Your Understanding** boxes present two to three tasks that students should be able to complete in order to demonstrate a mastery of specified key ideas.
- **Chapter Summaries** include "You should be able to" problems or exercises related to each of the key concepts declared in the gold, "Learn It" thread.
- **End-of-Chapter Questions** are organized around Bloom's taxonomy of learning, so students can test their understanding at the knowledge, comprehension, and application levels.

The fundamental idea is that if students really understand a piece of information or a concept, they should be able to do something with it.

Supporting Visual Learners

Kim Quillin—this book's illustrator—is a 1-in-6.6-billion talent. She combines superb academic training in biology (a Ph.D. from the University of California, Berkeley), insights from Edward Tufte's research on information architecture, impressive artistic talent, and a teacher's sensitivity to students.

Kim's goals in the first edition were to build a visual narrative that was (1) a direct extension of the text narrative and (2) guided by rigorous attention to the latest advances in information design. For example, figures that illustrate stepwise processes in biology were presented in a standardized, linear, and intuitive format, to spare students the "speed bumps" of convoluted compositions. In all of the figures, the drawings, colors, and labels were placed with surgical precision to maximize cleanliness and present a clear hierarchy of information. And to help bring a narrative voice to the figures, Kim integrated the "pointing hand" feature—inspired by National Academy of Sciences member M.A.R. Koehl—as a quiet and friendly way of highlighting important or challenging information such as data trends in graphs, or implications of phylogenetic trees.

The second edition stayed faithful to the original art philosophy, adding details requested by professors, formalizing experiments into boxes that made the experimental process consistent and explicit, and increasing the vibrancy and color of the drawings. Refinements in the third edition have focused on bringing the content of many individual figures up-to-date and honing the clarity, impact, and student-friendliness of every figure.

Supporting Skill Building

When instructors write learning objectives for introductory biology courses, they are paying close attention to skill building, in addition to the traditional emphasis on mastering content and concepts and the newer focus on developing higher-order thinking skills.

To aid skill building in introductory biology, I've added a set of nine new appendixes, called BioSkills, to the end matter of the book—just before the glossary. The BioSkills are meant to provide background on skills and techniques that are used throughout biological science. In addition, some of the BioSkills should help shore up problems with student preparation in mathematics and chemistry. They are called out at relevant points in the text. Please point them out to students who are having trouble with specific skills.

1. Reading Graphs
2. Reading a Phylogenetic Tree
3. Using Statistical Tests and Interpreting Standard Error Bars
4. Reading Chemical Structures
5. Using Logarithms
6. Making Concept Maps
7. Using Electrophoresis to Separate Molecules
8. Observing Microscopic Structures and Processes
9. Combining Probabilities

Serving a Community of Teachers

There is nothing that inspires me more than watching a passionate teacher work with motivated students. Teaching and learning is the essence of humanity, of who we are.

I write *Biological Science* because it's my way of being part of that work. As a textbook author, my greatest reward comes from interacting with inspired instructors and students from around the world—through e-mails, phone calls, focus groups, reviews, seminars, and teaching workshops.

Introductory biology courses are undergoing a slow and steady, yet remarkable, change: from a memorization-driven, largely passive exercise to a dynamic, active interchange that emphasizes higher-order thinking skills. The change is good. Instructors are stimulated and having more fun. They're approaching teaching problems in a hypothesis-testing framework, collecting data, and changing course designs based on empirical evidence. Students are performing at a higher level and should be better prepared for graduate school, professional school, and careers related to biology. My hope is that *Biological Science* can support this change.

Thank you for considering this text, for your passion about biology, and for your work on behalf of your students. What you do is *important*.

Scott Freeman
University of Washington

This book is dedicated to the world's greatest profession: teaching. Teachers are like Johnny Appleseed—they sow seeds, but seldom get to see the trees or fruit. This edition reflects in particular four teachers who had an exceptional impact on me: Vern Bailey, Owen Jenkins, Sievert Rohwer, and Barbara Wakimoto. Thank you.

Detailed Contents

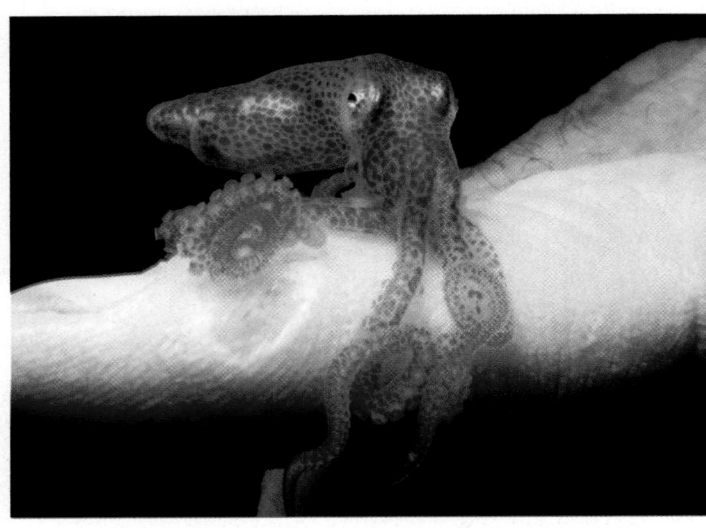

Acknowledgments

Expert Feedback

Work on this edition was organized around two pillars: feedback from our advisory board and peer reviewers, and feedback from students.

Advisory Board

A cadre of highly select advisors analyzed reviews, provided citations for particularly important recent papers, and worked with me and Kim Quillin to brainstorm ideas for improving each chapter and unit. It was a joy to work with them, and their expertise and perspective were vital to this edition.

Unit 1 Ross Feldberg, *Tufts University*
David Wilson, *Parkland College*
Unit 2 Paula Lemons, *Duke University*
Unit 3 Greg Podgorski, *Utah State University*
Unit 4 Greg Podgorski, *Utah State University*
Unit 5 George Gilchrist, *William & Mary College*
Unit 6 Brianna Timmerman, *University of South Carolina*
Marc Perkins, *Orange Coast College*
Michael Black, *California Polytechnic State University, San Luis Obispo*
Unit 7 Diane Marshall, *University of New Mexico*
Unit 8 Jim Ryan, *Hobart and William Smith Colleges*
Michael Black, *California Polytechnic State University, San Luis Obispo*
Unit 9 Alan Molumby, *University of Illinois, Chicago*

Reviewers

Peer review is the backbone of scientific publication. Acting as a reviewer is a fundamental component of our service as professional biologists, and reviewing chapters from an introductory text is one of the most important things we can do to improve the training of the next generation of biologists. I am deeply grateful for the insights offered by the following reviewers, who drew on their extensive content expertise, teaching experience, and research practice.

Marc Albrecht, *University of Nebraska, Kearney*
David Asch, *Youngstown State University*
Mariette Baxendale, *University of Missouri, St. Louis*
Greg Beaulieu, *University of Victoria*
Christopher Beck, *Emory University*
Peter Berget, *Carnegie Mellon University*
Janet Bester-Meredith, *Seattle Pacific University*
Cynthia Bishop, *Seattle Pacific University*
Michael Black, *California Polytechnic State University, San Luis Obispo*
Anthony Bledsoe, *University of Pittsburgh*
Patrice Boily, *University of New Orleans*
Scott Bowling, *Auburn University*
Maureen Brandon, *Idaho State University*
John Briggs, *Arizona State University*
Art Buikema, *Virginia Tech University*
Kim Caldwell, *University of Alabama*
Jeff Carmichael, *University of North Dakota*
Patrick Carter, *Washington State University*
John Caruso, *University of New Orleans*
Mary Lynn Casem, *California State University, Fullerton*
Cynthia Church, *Metropolitan State College*
Alison Cleveland, *University of South Florida*
Anita Davelos Baines, *University of Texas, Pan-American*
Jeff Demuth, *Indiana University*
Todd Duncan, *University of Colorado, Denver*
Johnny El Rady, *University of South Florida*
Peter Facchini, *University of Calgary*
Zen Faulkes, *University of Texas, Pan-American*
Ross Feldberg, *Tufts University*
Lewis Feldman, *University of California, Berkeley*
Jonathan Fisher, *St. Louis University*
Steve Frankel, *Northeastern University*
Amy Frary, *Mount Holyoke College*
Jed Fuhrman, *University of Southern California*
Caitlin Gabor, *Texas State University, San Marcos*
Michael Gaines, *University of Miami*
John R. Geiser, *Western Washington University*
D. Timothy Gerber, *University of Wisconsin, La Crosse*
Lisa Gerheart, *University of California, Davis*
Kathy Gillen, *Kenyon College*
Florence K. Gleason, *University of Minnesota, Twin Cities*
John Godwin, *North Carolina State University*
Reuben Goforth, *Michigan State University*
Linda Green, *University of Virginia*
Joe Harsh, *University of North Carolina, Charlotte*
Clare Hays, *Metropolitan State University*
Kerry Heafner, *University of Louisiana, Monroe*
Harold Heatwole, *North Carolina State University*
Brian Helmuth, *University of Southern California*
Susan Hengeveld, *Indiana University*
Mark Hens, *University of North Carolina, Greensboro*
Albert Herrera, *University of Southern California*
Malcolm Hill, *University of Richmond*
Ron Hoham, *Colgate University*
Kelly Howe, *University of New Mexico*
Cindy Johnson-Groh, *Gustavus Adolphus College*

Walter Judd, *University of Florida*
Nancy Kaufmann, *University of Pittsburgh*
Loren Knapp, *University of South Carolina*
Scott Knight, *Montclair State University*
Paul Lagos, *University of Mississippi*
Paula Lemons, *Duke University*
Vicky Lentz, *SUNY, College at Oneonta*
Georgia Lind, *Kingsborough Community College*
Chris Little, *University of Texas, Pan-American*
Andrea Lloyd, *Middlebury College*
Christopher Loretz, *University of Buffalo*
Cindy Martinez Wedig, *University of Texas, Pan-American*
Andrew McCubbin, *Washington State University*
Kelly McLaughlin, *Tufts University*
Victoria McMillan, *Colgate University*
Jennifer Miskowski, *University of Wisconsin, La Crosse*
Alan Molumby, *University of Illinois, Chicago*
Daniel Moon, *University of North Florida*
Mike Muller, *University of Illinois, Chicago*
Dana Nayduch, *Georgia Southern University*
Jacalyn S. Newman, *University of Pittsburgh*
Harry Nickla, *Creighton University*
Mary Jane Niles, *University of San Francisco*
Shawn Nordell, *St. Louis University*
Celia Norman, *Arapahoe Community College*
Nicole Obert, *University of Illinois, Urbana-Champaign*
John Osterman, *University of Nebraska, Lincoln*
John Palisano, *University of the South, Sewanee*
Glenn Parsons, *University of Mississippi, Oxford*
Andrew Pease, *Villa Julie College*
Deborah Pelli, *University of North Carolina, Greensboro*
Shelley A. Phelan, *Fairfield University*
Debra Pires, *University of California, Los Angeles*
Peggy Pollak, *Northern Arizona University*
Harvey Pough, *Rochester Institute of Technology*
Colin Purrington, *Swarthmore College*
Margaret Qazi, *Gustavus Adolphus College*
Rajinder Ranu, *Colorado State University*
Pamela C. Rasmussen, *Michigan State University*
Ann E. Rushing, *Baylor University*
James Ryan, *Hobart and William Smith Colleges*
Adam Ryburn, *SUNY, College at Oneonta*
Margaret Saha, *College of William and Mary*
Mark Sandheinrich, *University of Wisconsin, La Crosse*
Glenn Sauer, *Fairfield University*
Stephen G. Saupe, *St. John's University*
Andrew Scala, *Dutchess Community College*
Richard Showman, *University of South Carolina, Columbia*
Walter Shriner, *Mt. Hood Community College*
Sue Simon-Westendorf, *Ohio University*
Mark Spiro, *Bucknell University*
Paul Stapp, *California State University, Fullerton*
Scott Steinmaus, *California Polytechnic State University, San Luis Obispo*
John Stiller, *Eastern Carolina University*
John Stolz, *Duquesne University*
Kirk A. Stowe, *University of South Carolina*
Brianna Timmerman, *University of South Carolina*
Martin Tracey, *Florida International University*
Ashok Updhyaya, *University of South Florida*
Ann Vogel, *Illinois State University*
Fred Wasserman, *Boston University*
Elizabeth Weiss, *University of Texas, Austin*
Susan Whittemore, *Keene State College*
Ted Zerucha, *Appalachian State University*

Accuracy Reviewers

Once several rounds of peer review are completed and revised chapters and figures are produced, we rely on accuracy reviewers to check every page for errors. This work is demanding intellectually and has to be done under time pressure. The accuracy reviewers for this edition were exceptionally talented and timely.

Unit 1 Wayne Becker, *University of Wisconsin, Madison*
Unit 2 James Manser, *Harvey Mudd College (formerly)*
Unit 3 Peter Berget, *Carnegie Mellon University*
Mary Rose Lamb, *University of Puget Sound*
Unit 4 James Manser, *Harvey Mudd College (formerly)*
Unit 5 Jeffrey Feder, *University of Notre Dame*
Andrew Forbes, *University of Notre Dame*
Andrew Michel, *University of Notre Dame*
Tom Powell, *University of Notre Dame*
Unit 6 Laura Baumgartner, *University of Colorado, Boulder*
Michael Black, *California Polytechnic State University, San Luis Obispo*
Kimberly Erickson, *University of Colorado, Boulder*
Steve Trudell, *University of Washington, Seattle*
Unit 7 Susan Waaland, *University of Washington, Seattle*
Unit 8 Warren Burggren, *University of North Texas*
Susan Whittemore, *Keene State College*
Unit 9 Mark Johnston, *Dalhousie University*

BioSkills Julie Aires, *Florida Community College at Jacksonville;* Ross Feldberg, *Tufts University;* George Gilchrist, *William and Mary College;* Doug Luckie, *Michigan State University;* Greg Podgorski, *Utah State University*

Correspondents

I am grateful to colleagues who take the initiative to contact me directly or through my publisher to make suggestions on how to improve the text and figures. Please never hesitate to do this—I take your comments to heart, in the spirit of a shared commitment to improved student learning. This list also includes friends and colleagues who were kind enough to respond to emails or calls from me, asking for ideas on how to clarify specific topics.

Julie Aires, *Florida Community College, Jacksonville*
Gerald Borgia, *University of Maryland*
Scott Bowling, *Auburn University*
Elizabeth Cowles, *Rice University*
Fred Delcomyn, *University of Illinois, Urbana-Champaign*
Leslie Dendy, *University of New Mexico, Los Alamos*
John Dudley, *University of Illinois, Urbana-Champaign*
Larry Forney, *University of Idaho*
Arthur Gibson, *University of California, Los Angeles*
Matt Gilg, *University of Northern Florida*
Jean Heitz, *University of Wisconsin, Madison*
Jack Hogg, *University of Montana*
Johnathan Kupferer, *University of Illinois, Chicago*
Hans Landel, *Edmonds Community College*
Frederick Lanni, *Carnegie Mellon University*
Andi Lloyd, *Middlebury College*
Carmen Mannella, *Wadsworth Center, SUNY Albany*
Andrew McCubbin, *Washington State University*
Tim Nelson, *Seattle Pacific University*
Shawn Nordell and students, *University of St. Louis*
Carol Pollock, *University of British Columbia*
Joelle Presson, *University of Maryland*
William Saunders, *LaGuardia Community College*
David Senseman, *University of Texas, San Antonio*
Bryan Spohn, *Florida Community College, Jacksonville*
Scott Steinmaus, *California Polytechnic State University, San Luis Obispo*
Judy Stone, *Colby College*
Dean Wendt, *California Polytechnic State University, San Luis Obispo*

Student Feedback

The second pillar of this edition—in addition to the role played by advisors and reviewers—was an extensive series of focus groups with students who were currently taking introductory biology or who had just completed the course.

Student Focus Group Coordinators

Planning and implementing the student focus groups would have been impossible without the support of key faculty members, who went out of their way to provide opportunities for their students to be heard.

Julie Aires, *Florida Community College, Jacksonville*
Frank Cantelmo, *St. Johns University, New York*
Matt Gilg, *University of Northern Florida*
Bill Hoese, *California State University, Fullerton*
John Nagey, *Scottsdale Community College*
Debra Pires, *University of California, Los Angeles*
Emily Taylor, *California Polytechnic State University, San Luis Obispo*
John Weser, *Scottsdale Community College*

Student Focus Group Participants

The students who attended focus groups were asked three questions about chapters they were assigned to read: (1) what were the most difficult concepts, (2) why were they hard, and (3) what helped you finally get it? We usually had them work in groups, and when they reported back to us, we never failed to be deeply impressed by the quality of their ideas and their ability to articulate them. It is not possible to overstate how important student feedback was to this edition. Combined with the superb advice I was getting from advisors, reviewers, and other colleagues, I had a wealth of ideas on how to make each chapter work better for both instructors and learners. These students were *inspiring*.

California Polytechnic State University, San Luis Obispo
Jenna Arruda, Katie Camfield, Benjamin Capper, Mandsa Chandra, Rebekah Clarke, Annalisa Conn, Marisa Crawford, Katie Duffield, T.J. Eames, Megan Fay, Margaret Hackney, Steffani Hall, Gemma Hill, John Kong, Taylor Lindblom, Adam Marre, Vik Mediratta, Serena Moen, Sunil Patel, Corinne Ross, Teresa Sais, Jessie Singer, Stephanie Szeto, Kelsey Tallon, Gregory Thurston, Greg Vidovic, Melody Wilkinson, Taiga Young

California State University, Fullerton
Redieat Assefa, Josemari Feliciano, Civon Gewelber, Sarah Harpst, Jeff Kuhnlein, Linda Ong, T. Richard Parenteau, Robert Tran, Nicole Bournival

Fairfield University
Sally Casper, Tamika Dickens, Pryce Gaynor, Cindi Munden

Florida Community College, Jacksonville
Algen Albritten III, Danielle Boss, Chantel Callier, Eugenia Cruz, Lauren Faulkner, Jonathan Hopkins, Chantae Knight, David Lambert, Amber McCurdy, Tara Pladsen, Lauren Spruiell, Courtney Torgeon, Theresa Tran

LaGuardia Community College
Kristine Azzoli, Felicita Gonzalez, Pedro Granados, Mike Levine, Kris Ragoonath, Maria Reyes

Scottsdale Community College
Tatum Arthur, Shadi Asayesh, Angela Bikl, Abrey Britt, Drew Bryck, Jason Butler, Cindy Clifton, Dean Doty, Tannaz Farahani, Bethany Garcia, Jeff Godfrey, Troy Graziadei, Dina Habhab, Loreley Hall, Crista Jackson, Paul Krueger, David Levine, Chad Massena, Jessica Massena, Brian Martinez, Sam Mohmand, Esther Morantz, Jill Patel, Staci Puckett, Rebecca Rees, David Rosenbaum, Samantha Schrepel, Chris Schroeder, Kelsey Thomsen, Chris Volpe, Jamie Wagner, Lianne Wharton

St. Johns University, New York
Diana Carroccia, Milea Emmons, Tunc Ersoy, Blayre Linker, Zain Mirza, Mohammed Sheikh, Richardson Talarera, Michael Weinberg, Win Aung Yeni

University of California, Los Angeles
Farhan Banani, Stephanie Davis, Krystal De La Rosa, Samantha Hammer, Jennifer Okuda Hein, Neha Jashi, Marissa Lee, Calvin Leung, Venkat Mocherla, David Nguyen, Isabella Niu, Aya Obara

University of Maryland
Megan Berg, Lauren Fitzgerald, Megan Janssen, Avita Jones, Deidre Robinson

University of Northern Florida
Elysia Brennan, Christopher Ferrara, Lindsay Googe, Samantha Grogan, Marie Haagensen, Crystal Harris, Madeline Parhalo, Sherline Pierre, Stacy Pohlman, Nichole Polito, Sarah Lynn Redding, Megan Richardson, Megan Smart, Frank Snyder

Supplements Contributors

Our goal for the supplements package to accompany the Third Edition was to create learning tools that incorporate principles of active learning. Research shows that students do better in class when they are asked to use the material they are learning about. From a new workbook that encourages students to practice biology, to interactive web animations that test their knowledge, our supplements ask students to work with information, not just memorize it. My sincere thanks to the following people for their important contribution to the book's core teaching values.

Media Supplements

Marc Albrecht, *University of Nebraska, Kearney*
John Bell, *Brigham Young University*
Michael Black, *California Polytechnic State University, San Luis Obispo*
Warren Burggren, *University of North Texas*
Fannie Chen
Carol Chihara, *University of San Francisco*
Clarissa Dirks, *University of Washington, Seattle*
Kimberly Erickson, *University of Colorado, Boulder*
Zen Faulkes, *University of Texas, Pan-American*
Kathy Gillen, *Kenyon College*
Mary Catherine Hager
Susan Hengeveld, *Indiana University, Bloomington*
Loren Knapp, *University of South Carolina*
Jonathan Lochamy, *Georgia Perimeter College*
James Manser, *Harvey Mudd College (formerly)*
Cynthia Martinez-Wedig, *University of Texas, Pan-American*
Victoria McMillan, *Colgate University*
Andrew Pease, *Villa Julie College*
Debra Pires, *University of California, Los Angeles*
Pamela Rasmussen, *Michigan State University*
Susan Rouse, *Brenau University*
Christina Russin, *Northwestern University*
William Russin, *Northwestern University*
Cheryl Ingram Smith, *Clemson University*
Ellen M. Smith
Mark Spiro, *Bucknell University*
Eric Stavney, *DeVry University*
Michael Wenzel, *California State University, Sacramento*

Print Supplements

Marc Albrecht, *University of Nebraska, Kearney*
Charles Austerberry, *Creighton University*
Brian Bagatto, *University of Akron*
Jay Brewster, *Pepperdine University*
Warren Burggren, *University of North Texas*
Cynthia Giffen, *University of Wisconsin, Madison*
Jean Heitz, *University of Wisconsin, Madison*
Laurel Hester, *Cornell University*
Cynthia Martinez-Wedig, *University of Texas, Pan-American*
Jenny McFarland, *Edmonds Community College*
Greg Podgorski, *Utah State University*
Carol Pollock, *University of British Columbia*
Susan Rouse, *Southern Wesleyan University*
Elena Shpak, *University of Tennessee*
Sally Sommers Smith, *Boston University*
Briana Timmerman, *University of South Carolina*
David Wilson, *Parkland College*

Book Team

Finally, this edition would not have been published without the encouragement and support of our publishing partners at Pearson Arts & Science. I would like to acknowledge those individuals in the Pearson Science Group who helped make the Third Edition possible.

Prentice Hall

This edition was launched by Prentice Hall and then transferred—along with all other Prentice Hall biology titles—to their sister company, Benjamin Cummings. The editorial team at Prentice Hall was responsible for establishing the vision that directed this edition. In addition, they recruited the first advisors and media and supplement contributors, and implemented the initial set of student focus groups, prior to turning the project over to their Benjamin Cummings colleagues. I am grateful for their talent, energy, and friendship, and for the extraordinary efforts they made to make the management transition as smooth as possible. These people are Andrew Gilfillan (Sponsoring Editor), Ann Heath (Executive Project Manager), Erin Mulligan (Development Editor), Lisa Tarabokjia (Editorial Assistant), and Carol Trueheart (VP, Executive Director of Development).

Benjamin Cummings

The Benjamin Cummings editorial and production team welcomed *Biological Science* into their publishing house and moved the project forward through the final critical stages of development and into production. The team was initially led by Sponsoring Editor Susan Winslow, who brought a fresh perspective to the project. Then, Market Development Manager Becky Ruden took over the Sponsoring Editor reins with focused energy and verve. She is a bright, young star. Special thanks go to Project Manager Sonia DiVittorio, who has proven herself one of the sharpest talents in textbook publishing. Sonia's tireless pursuit of quality is evident on every page. In tribute, I have given her a new title at the company: Goddess of Bookmaking.

Thanks also go to Senior Production Supervisor Shannon Tozier, who led the production team with enormous skill and perseverance, and to Design Manager Marilyn Perry, who helped create the striking interior and cover designs. Others on the book team deserving acknowledgement include: Mary Catherine Hager and Susan Weisberg (Development Editors), Anna Amato (Assistant Editor), Mercedes Grandin (Associate Editor), Ericka O'Benar (Media Producer), Yvonne Gerin, Elaine Soares, and Debbie Latronica (photo research team), Christy Lawrence, Lauren Harp, Lillian Carr, and Mansour Bethany (marketing team), Josh Frost (market development), and Deborah Gale (Executive Director of Development).

Additionally, I'd like to thank Chris Thillen (copyeditor); Frank Purcell, Ellen Sanders, Pete Shanks (proofreaders); production editor Katy Faria and her production colleagues at Pre-Press PMG; and the illustration team at Imagineering for all the hours and energy put into making this edition the best it can be.

Finally, I'd like to extend my appreciation to the people whose vantage point allows them to assess the bigger picture. I am grateful to Linda Davis, President of Pearson Arts and Sciences, for helping me understand Benjamin Cummings' overarching publishing goals and how *Biological Science* fits into them. Thank you, Beth Wilbur, Editor-in-Chief of Biology, for providing editorial leadership and helping the book take root in its new home and allowing room for its personality to flourish. A very special thank-you goes to Paul Corey, President of the Pearson Science Group. Paul has been an advocate for the core values of *Biological Science* from its beginning over a decade ago. I am grateful for his continuing friendship and professional guidance.

Supplements

The Freeman *Biological Science* supplements package offers a robust suite of print and electronic tools designed to help instructors make the most of their limited time and to help students study efficiently.

INSTRUCTOR RESOURCES

- The entire textbook illustration program is available in JPEG format. All tables, photos, and line drawings with (and without) labels are individually enhanced for optimal in-class projection and are pre-loaded into chapter-correlated PowerPoint presentations.
- A second set of PowerPoint presentations consists of lecture outlines for each chapter, augmented by key text illustrations and hyperlinks to Web Animations.
- A third set of PowerPoint presentations allows select key figures to be presented in a step-by-step manner. In-text figure caption exercises with illustrated answers are included in this step-edited set.
- In-class active lecture questions correlated by chapter can be used with any classroom response system and are available in PowerPoint format.
- The Instructor Guide includes lecture outlines, active learning lecture activities, answers to end-of-chapter questions, and innovative material to help motivate and engage students.
- The Printed Test Bank and Computerized Test Bank have been peer reviewed and student tested. Test questions are ranked according to Bloom's taxonomy and the improved TestGen® software makes assembling tests much easier. The Test Bank is also available in Course Management systems and in Word® format on the Instructor Resources DVD.
- Four-color Transparency Acetates with every illustration from the text are available. Labels and images have been enlarged and modified to ensure optimal projection in a large lecture hall.

STUDENT RESOURCES

- **NEW!** *Practicing Biology: A Student Workbook* offers a variety of activities such as modeling, mapping, and graphing to help students with different learning styles visualize and understand biological processes.
- **NEW!** The eBook addresses the changing needs of students and instructors in the majors biology course by offering an electronic version of the text that links directly to animations, quizzes, and videos.
- **NEW!** Complimentary access to Pearson Tutor Services provides highly interactive one-on-one biology tutoring by qualified instructors seven nights a week during peak study hours. Students will be able to "drop in" for live online help, submit questions to an e-structor anytime, or pre-schedule a tutoring session with an e-structor to receive help at their convenience.
- The Study Guide presents a breakdown of key biological concepts, and helps students focus on the fundamentals of each chapter. It is designed in two parts to help students study more effectively. Part I is intended as a "survival guide," and Part II explores the material in the textbook, chapter by chapter.

MULTIMEDIA RESOURCES

- **NEW!** MasteringBIOLOGY MasteringBiology™ offers in-depth online tutorials on biology's toughest topics. These tutorials provide hints and feedback specific to each student's misconceptions. MasteringBiology also includes diagnostic test questions and is useful for "just-in-time" teaching.
- **NEW!** ***BioFlix*** BioFlix cover the most difficult biology topics with 3-D, movie-quality animations, labeled slide shows, carefully constructed student tutorials, study sheets, and quizzes that support all types of learners. Topics include Tour of an Animal Cell, Tour of a Plant Cell, Membrane Transport, Cellular Respiration, Photosynthesis, Mitosis, Meiosis, DNA Replication, Protein Synthesis, Water Transport in Plants, How Neurons Work, Synapses, and Muscle Contraction.
- **Web Animation** Web Animations add depth and visual clarity to the most important topics and processes described in the text. Animations include pre-quizzes and post-quizzes to help students prepare for exams.
- **NEW!** Discovery Channel video clips on 29 different topics include antibiotic resistance, fighting cancer, and introduced species. Additional Cell Biology animations provide vivid images of the functions and processes of the cell.
- **NEW!** Access to BioForum, a new online community forum created just for biology educators. Learn from your peers about how they're teaching difficult topics or what resources they use to help them teach. Post activities, handouts, Web links, or other tools you would like to share with forum participants.
- **NEW!** Video demonstration shows biology educators using active learning techniques you can incorporate in your classroom immediately.
- Course Management content for *Biological Science,* 3rd Edition, is available for institutions using **WebCT** or **Blackboard** and is also available in our nationally hosted **Course Compass** course management system. If desired, WebCT and Blackboard cartridges containing only the Test Bank are available for download.

Preface to Students Using this Book as a Tool for Learning

FOCUS ON THE GOLD THREAD ...

The gold thread helps you pick out important information.

Start with *Key Concepts* on the first page of every chapter. Read these gold bullet points first to familiarize yourself with the chapter's big ideas.

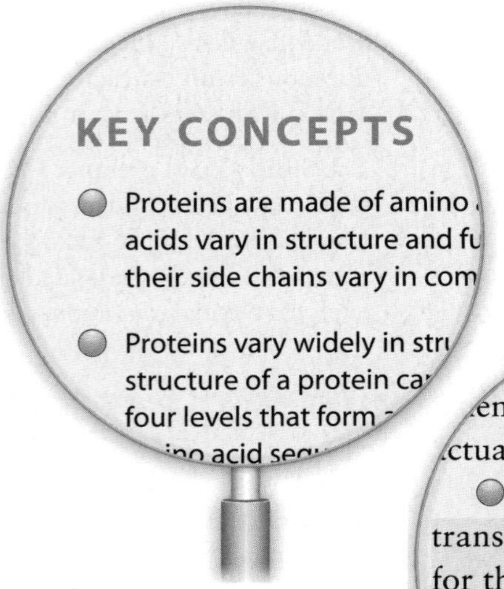

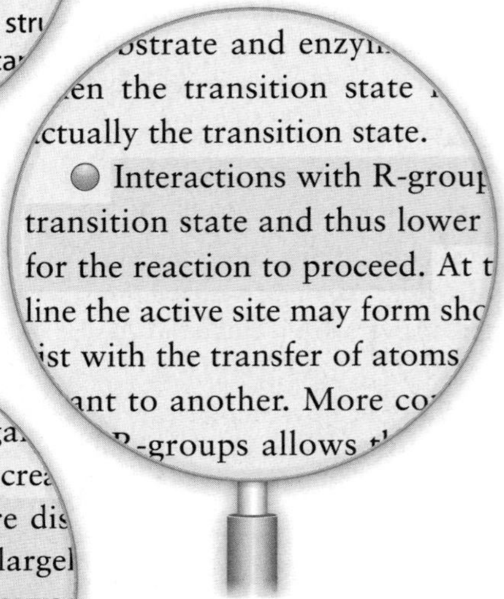

Watch for material related to Key Concepts inside chapters; it will also be flagged with a gold bullet. Slow down and pay close attention to these highlighted passages.

Other particularly important information is highlighted in gold. Gold highlighting is always a signal to slow down and pay special attention.

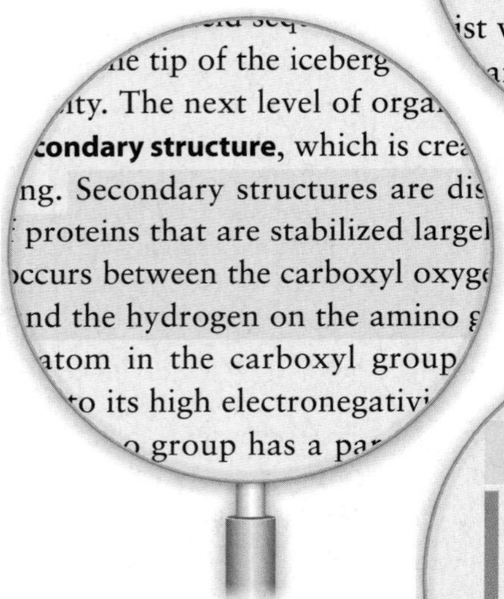

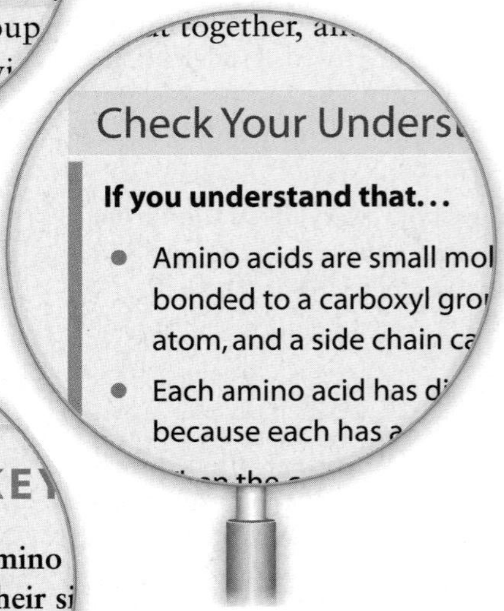

The gold half of *Check Your Understanding* boxes summarizes important information from the section you just read. Stop and ask yourself, Do I *really* understand every bullet point?

Summary of Key Concepts at the ends of chapters is a good place to start reviewing when it's time to study for an exam. Key Concepts are revisited here in detail.

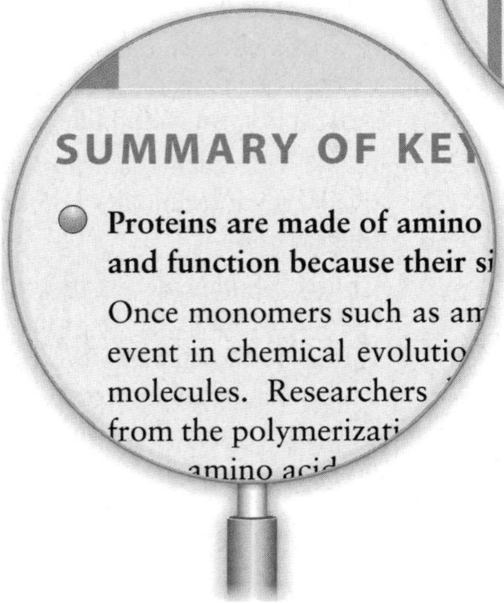

... PRACTICE WITH THE BLUE THREAD.

The blue thread helps you practice what you've learned.

The best way to succeed on exams is to *practice*. If you really understand a piece of information or a concept, **you should be able to** *do something with it.* If you can't, you haven't mastered the material.

Practice It activities are always flagged with blue bullets and blue type. Answers to all ***Practice It*** activities are available at **www.masteringbio.com** under the Study Area button.

Text passages flagged with a blue bullet, blue type, and the words ***"you should be able to"*** offer exercises on topics that professors and students have identified as most difficult. These are the topics most students struggle with on exams.

Many figures and some tables include blue ***Questions*** or ***Exercises*** to help you check your understanding of the material they present.

The blue half of ***Check Your Understanding*** boxes asks you to do something with the information in the gold half. If you can't complete these exercises, go back and re-read that section of the chapter.

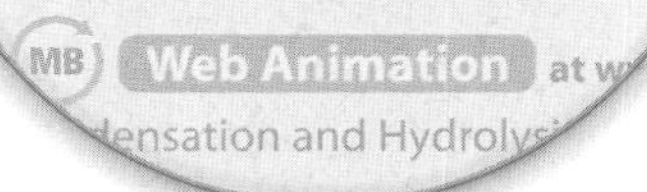

Blue ***"you should be able to"*** exercises also let you test your understanding of each item in the ***Summary of Key Concepts*** at the ends of chapters.

Half of this book is text. The other half is figures and tables. Figures always focus on important concepts. Tables present the raw material of science; they are data rich.

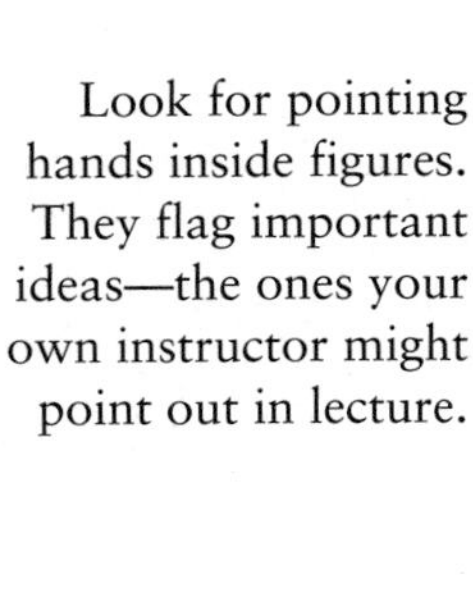

Look for pointing hands inside figures. They flag important ideas—the ones your own instructor might point out in lecture.

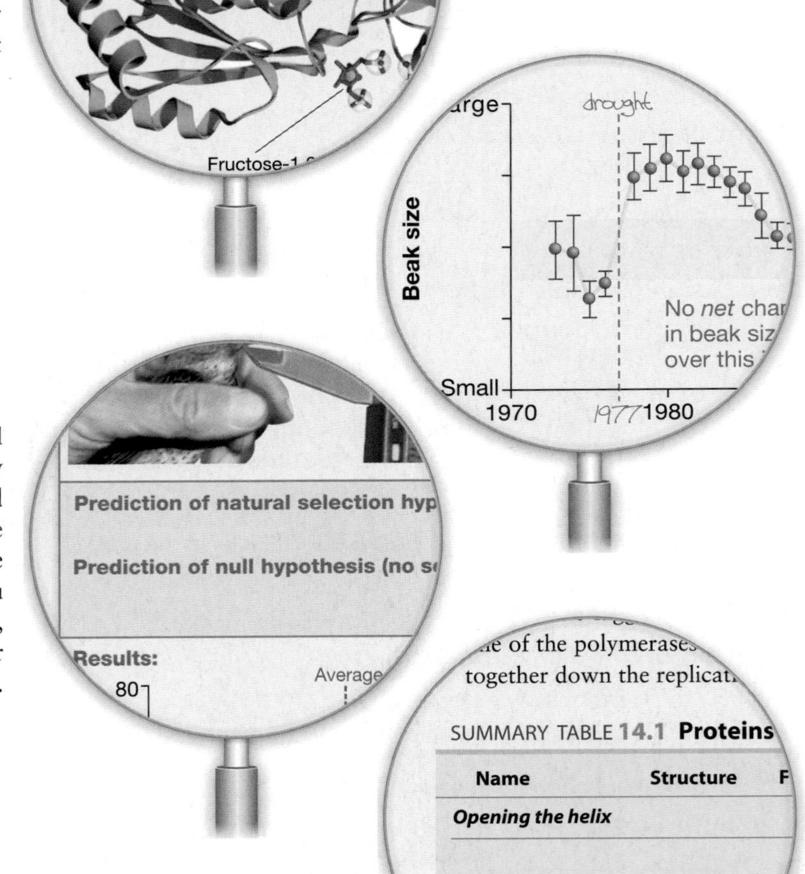

Some figures and tables ask you to to analyze data and fill in information. Take the time to do these exercises; they will help you master important concepts.

Experiment Boxes will help you understand how experiments are designed and give you practice interpreting data. Some leave space open for you to fill in null hypotheses, predicted outcomes, or the conclusion.

Summary Tables pull important information together in a format that's easy to review. A complete list of Summary Tables is provided inside the back cover.

Part I: ***The Study Area*** includes Web Animations and BioFlix.

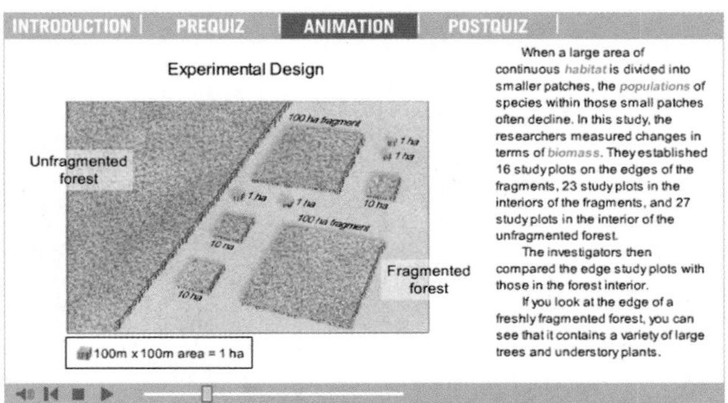

Web Animations **Web Animation** help you review important textbook topics. Pre- and post-quizzes let you test your proficiency.

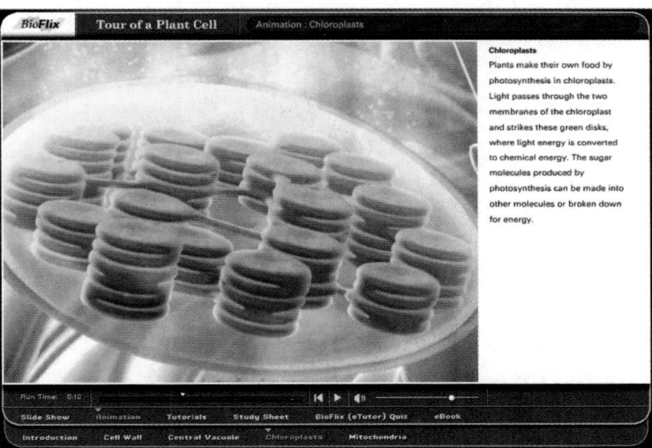

BioFlix **BioFlix** cover the most difficult biology topics with 3-D, movie-quality animations, labeled slide shows, carefully constructed tutorials, study sheets, and quizzes that support all types of learners. Topics include: Tour of an Animal Cell, Tour of a Plant Cell, Membrane Transport, Cellular Respiration, Photosynthesis, Mitosis, Meiosis, DNA Replication, Protein Synthesis, Water Transport in Plants, How Neurons Work, Synapses, and Muscle Contraction.

Part 2: ***MasteringBiology Instructor Assignments*** include in-depth tutorials on biology's toughest topics.

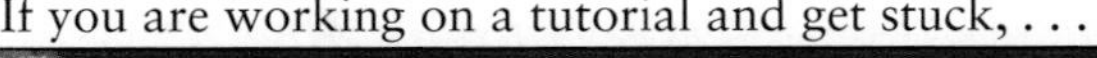

If you are working on a tutorial and get stuck, . . .

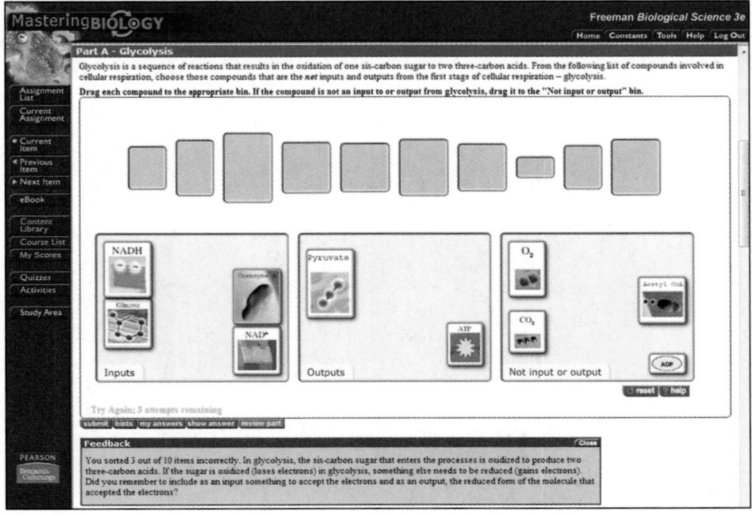

. . . you will receive instant feedback specific to your error.

You can also ask for hints . . .

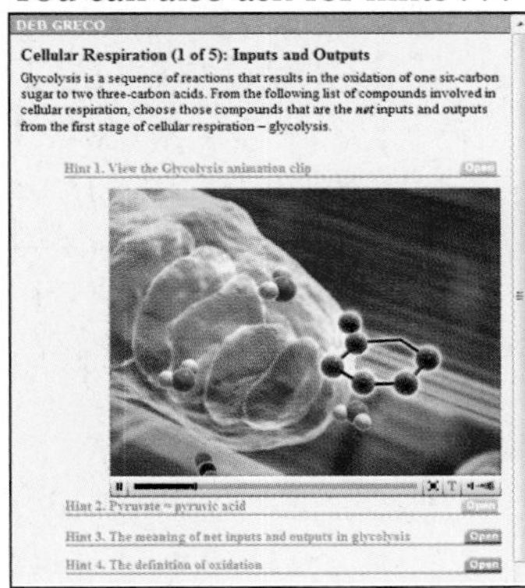

. . . to receive background information and Socratic questions and exercises designed to guide you to mastery of the topic.

Biology and the Tree of Life

1

A young reef octopus, from Hawaii, also appears on the cover of this book. It is one of perhaps 10 million species alive today.

KEY CONCEPTS

- Biological science was founded with the development of (1) the cell theory, which proposes that all organisms are made of cells and that all cells come from preexisting cells, and (2) the theory of evolution by natural selection, which maintains that species change through time because individuals with certain heritable traits produce more offspring than other individuals do.
- A phylogenetic tree is a graphical representation of the evolutionary relationships among species. These relationships can be estimated by analyzing similarities and differences in traits. Species that share distinctive traits are closely related and are placed close to each other on the tree of life.
- Biologists ask questions, generate hypotheses to answer them, and design experiments that test the predictions made by competing hypotheses.

In essence, biology is a search for ideas and observations that unify our understanding of the diversity of life, from bacteria living in rocks a mile underground to octopuses and humans. Chapter 1 is an introduction to this search.

The goals of this chapter are to introduce the amazing variety of life-forms alive today, consider some fundamental traits shared by all organisms, and explore how biologists go about answering questions about life. The chapter also introduces themes that will resonate throughout this book: (1) analyzing how organisms work at the molecular level, (2) understanding why organisms have the traits they do in terms of their evolutionary history, and (3) helping you learn how to think like a biologist.

We begin by examining two of the greatest unifying ideas in all of science: the cell theory and the theory of evolution by natural selection. When these concepts emerged in the mid-1800s, they revolutionized the way that biologists understand the world. The cell theory proposed that all organisms are made of cells and that all cells come from preexisting cells. The theory of evolution by natural selection maintained that species have changed through time and that all species are related to one another through common ancestry. The theory of evolution by natural selection established that bacteria, mushrooms, roses, robins, and humans are all part of a family tree, similar to the genealogies or family trees that connect individual people.

A **theory** is an explanation for a very general class of phenomena or observations. The cell theory and the theory of evolution provide a foundation for the development of modern biology because they focus on two of the most general questions possible: What are organisms made of? Where did they come from? Let's begin by tackling the first of these two questions.

Key Concept | Important Information | Practice It

1.1 The Cell Theory

The initial conceptual breakthrough in biology—the cell theory—emerged after some 200 years of work. In 1665 Robert Hooke used a crude microscope to examine the structure of cork (a bark tissue) from an oak tree. The instrument magnified objects to just 30 times (30×) their normal size, but it allowed Hooke to see something extraordinary. In the cork he observed small, pore-like compartments that were invisible to the naked eye (**Figure 1.1a**). These structures came to be called cells.

Soon after Hooke published his results, Anton van Leeuwenhoek succeeded in developing much more powerful microscopes, some capable of magnifications up to 300×. With these instruments, Leeuwenhoek inspected samples of pond water and made the first observations of single-celled organisms like the *Paramecium* in **Figure 1.1b**. He also observed and described the structure of human blood cells and sperm cells.

In the 1670s a researcher who was studying the leaves and stems of plants with a microscope concluded that these large, complex structures are composed of many individual cells. By the early 1800s enough data had accumulated for a biologist to claim that *all* organisms consist of cells. But between then and now, biologists have developed microscopes that are tens of thousands of times more powerful than Leeuwenhoek's and have described well over a million new species. Did the biologist's claim hold up?

(a) The first view of cells: Robert Hooke's drawing from 1665

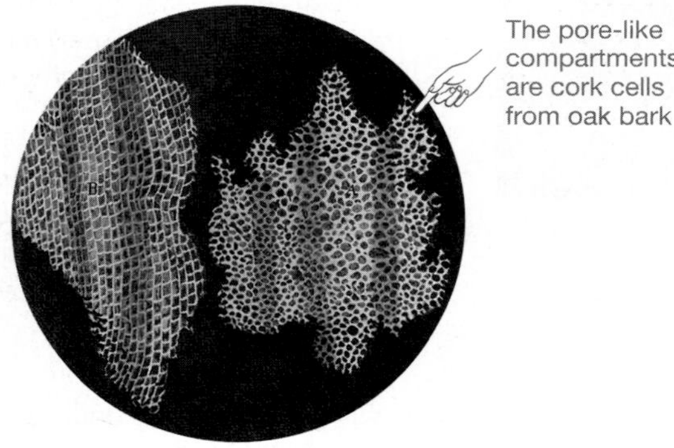

(b) Anton van Leeuwenhoek was the first to view single-celled "animalcules" in pond water.

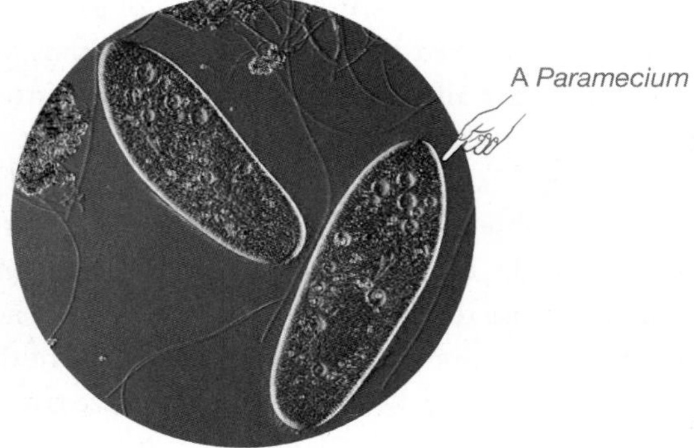

FIGURE 1.1 The Discovery of Cells.

Are *All* Organisms Made of Cells?

The smallest organisms known today are bacteria that are barely 200 nanometers wide, or 200 *billionths* of a meter. (See the endpapers of this book to review the metric system and its prefixes.) It would take 5000 of these organisms lined up end to end to span a millimeter. This is the distance between the smallest hash marks on a metric ruler. In contrast, sequoia trees can be over 100 meters tall. This is the equivalent of a 20-story building. Bacteria and sequoias are composed of the same fundamental building block, however—the cell. Bacteria consist of a single cell; sequoias are made up of many cells.

Biologists have become increasingly dazzled by the diversity and complexity of cells as advances in microscopy have made it possible to examine cells at higher magnifications. The basic conclusion made in the 1800s is intact, however: As far as is known, all organisms are made of cells. Today, a **cell** is defined as a highly organized compartment that is bounded by a thin, flexible structure called a plasma membrane and that contains concentrated chemicals in an aqueous (watery) solution. The chemical reactions that sustain life take place inside cells. Most cells are also capable of reproducing by dividing—in effect, by making a copy of themselves.

The realization that all organisms are made of cells was fundamentally important, but it formed only the first part of the cell theory. In addition to understanding what organisms are made of, scientists wanted to understand how cells come to be.

Where Do Cells Come From?

Most scientific theories have two components: The first describes a pattern in the natural world, while the second identifies a mechanism or process that is responsible for creating that pattern. Hooke and his fellow scientists had articulated the pattern component of the cell theory. In 1858 Rudolph Virchow added the process component by stating that all cells arise from preexisting cells. The complete cell theory, then, can be stated as follows: All organisms are made of cells, and all cells come from preexisting cells.

This claim was a direct challenge to the prevailing explanation, called spontaneous generation. At the time, most biologists believed that organisms arise spontaneously under certain conditions. For example, the bacteria and fungi that spoil foods such as milk and wine were thought to appear in these nutrient-rich media of their own accord—they spring to life from nonliving materials. Spontaneous generation was a **hypothesis**: a proposed explanation. The all-cells-from-cells hypothesis, in contrast, maintained that cells do not spring to life spontaneously but are produced only when preexisting cells grow and divide. Biologists usually use *theory* to refer to proposed explanations for broad patterns in nature and *hypothesis* to refer to explanations for more tightly focused questions.

Soon after the all-cells-from-cells hypothesis appeared in print, Louis Pasteur set out to test its predictions experimentally. A **prediction** is something that can be measured and that must be correct if a hypothesis is valid. Pasteur wanted to determine whether microorganisms could arise spontaneously in a nutrient broth or whether they appear only when a broth is exposed to a source of preexisting cells. To address the question, he created two treatment groups: a broth that was not exposed to a source of preexisting cells and a broth that was. The spontaneous generation hypothesis predicted that cells would appear in both treatments. The all-cells-from-cells hypothesis predicted that cells would appear only in the treatment exposed to a source of preexisting cells.

Figure 1.2 shows Pasteur's experimental setup. Note that the two treatments are identical in every respect but one. Both used glass flasks filled with the same amount of the same nutrient broth. Both were boiled for the same amount of time to kill any existing organisms such as bacteria or fungi. But because the flask pictured in **Figure 1.2a** had a straight neck, it was exposed to preexisting cells after sterilization by the heat treatment. These preexisting cells are the bacteria and fungi that cling to dust particles in the air. They could drop into the nutrient broth because the neck of the flask was straight. In contrast, the flask

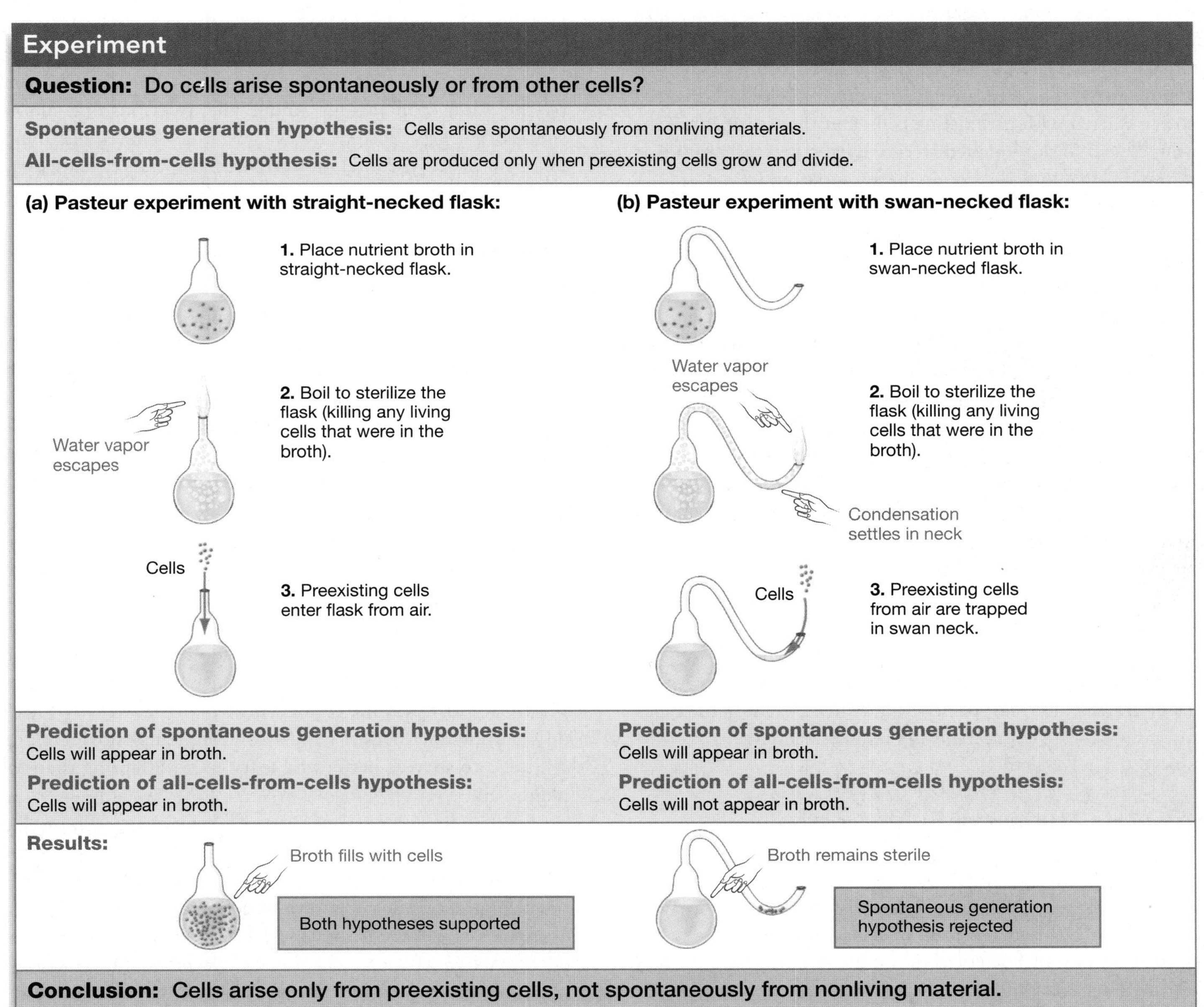

FIGURE 1.2 The Spontaneous Generation Hypothesis Was Tested Experimentally.

drawn in **Figure 1.2b** had a long swan neck. Pasteur knew that water would condense in the crook of the swan neck after the boiling treatment and that this pool of water would trap any bacteria or fungi that entered on dust particles. Thus, the swan-necked flask was isolated from any source of preexisting cells even though it was still open to the air.

Pasteur's experimental setup was effective because there was only one difference between the two treatments and because that difference was the factor being tested—in this case, a broth's exposure to preexisting cells. ● If you understand this concept, you should be able to identify problems that would arise if he had put different types of broth in the two treatments, heated them for different lengths of time, or used a ceramic flask for one treatment and a glass flask for the other.

And Pasteur's results? As Figure 1.2 shows, the treatment exposed to preexisting cells quickly filled with bacteria and fungi. This observation was important because it showed that the heat sterilization step had not altered the nutrient broth's capacity to support growth, and because it supported the hypothesis that growth started with preexisting cells. But the treatment in the swan-necked flask remained sterile. Even when the broth was left standing for months, no organisms appeared in it.

Because Pasteur's data were in direct opposition to the predictions made by the spontaneous generation hypothesis, the results persuaded most biologists that the all-cells-from-cells hypothesis was correct.

The success of the cell theory's process component had an important implication: If all cells come from preexisting cells, it follows that all individuals in a population of single-celled organisms are related by common ancestry. Similarly, in a multicellular individual such as you, all of the cells present are descended from preexisting cells, tracing back to a fertilized egg. A fertilized egg is a cell created by the fusion of sperm and egg—cells that formed in individuals of the previous generation. In this way, all of the cells in a multicellular organism are connected by common ancestry.

The second great founding idea in biology is similar, in spirit, to the cell theory. It also happened to be published the same year as the all-cells-from-cells hypothesis. This was the realization, made independently by Charles Darwin and Alfred Russel Wallace, that all *species*—all distinct, identifiable types of organisms—are connected by common ancestry.

1.2 The Theory of Evolution by Natural Selection

In 1858 short papers written separately by Darwin and Wallace were read to a small group of scientists attending a meeting of the Linnean Society of London. A year later, Darwin published a book that expanded on the idea summarized in those brief papers. The book was called *The Origin of Species*. The first edition sold out in a day.

What Is Evolution?

Like the cell theory, the theory of evolution by natural selection has a pattern and a process component. Darwin and Wallace's theory made two important claims concerning patterns that exist in the natural world. The first claim was that species are related by common ancestry. This contrasted with the prevailing view in science at the time, which was that species represent independent entities created separately by a divine being. The second claim was equally novel. Instead of accepting the popular hypothesis that species remain unchanged through time, Darwin and Wallace proposed that the characteristics of species can be modified from generation to generation. Darwin called this process "descent with modification."

Evolution, then, means that species are not independent and unchanging entities, but are related to one another and can change through time. This part of the theory of evolution—the pattern component—was actually not original to Darwin and Wallace. Several scientists had already come to the same conclusions about the relationships among species. The great insight by Darwin and Wallace was in proposing a process, called **natural selection**, that explains *how* evolution occurs.

What Is Natural Selection?

Natural selection occurs whenever two conditions are met. The first is that individuals within a population vary in characteristics that are **heritable**—meaning, traits that can be passed on to offspring. A **population** is defined as a group of individuals of the same species living in the same area at the same time. Darwin and Wallace had studied natural populations long enough to realize that variation among individuals is almost universal. In wheat, for example, some individuals are taller than others. As a result of work by wheat breeders, Darwin and Wallace knew that short parents tend to have short offspring. Subsequent research has shown that heritable variation exists in most traits and populations. The second condition of natural selection is that in a particular environment, certain versions of these heritable traits help individuals survive better or reproduce more than do other versions. For example, if tall wheat plants are easily blown down by wind, then in windy environments shorter plants will tend to survive better and leave more offspring than tall plants will.

If certain heritable traits lead to increased success in producing offspring, then those traits become more common in the population over time. In this way, the population's characteristics change as a result of natural selection acting on individuals. This is a key insight: Natural selection acts on individuals, but evolutionary change affects only populations. In this example, populations of wheat that grow in windy environments tend to become shorter from generation to generation. But in any given generation, none of the individual wheat plants get taller or shorter as a result of natural selection. This sort of change in the characteristics of a population, over time, is evolution. Evolution occurs when heritable variation leads to differential success in reproduction. ● If you understand this concept, you should be able to graph how average stem height will change over time in a population that occupies a windy environment versus a non-windy environment—one where tall plants have an advantage due to better access to light. (For help with reading and making graphs, see **BioSkills 1** in the back of this book.)

Darwin also introduced some new terminology to identify what is happening during natural selection. For example, in everyday English the word *fitness* means health and well-being. But in biology, **fitness** means the ability of an individual to produce offspring. Individuals with high fitness produce many offspring. Similarly, the word *adaptation* in everyday English means that an individual is adjusting and changing to function in new circumstances. But in biology, an **adaptation** is a trait that increases the fitness of an individual in a particular environment. Once again, consider wheat: In windswept habitats, wheat plants with short stalks have higher fitness than do individuals with long stalks. Short stalks are an adaptation to windy environments.

To clarify further how natural selection works, consider the origin of the vegetables called the cabbage family plants. Broccoli, cauliflower, Brussels sprouts, cabbage, kale, savoy, and collard greens descended from the same species—the wild plant in the mustard family pictured in **Figure 1.3a**. To create the plant called broccoli, horticulturists selected individuals of the wild mustard species with particularly large and compact flowering stalks. In mustards, the size and shape of the flowering stalk is a heritable trait. When the selected individuals were mated with one another, their offspring turned out to have larger and more compact flowering stalks, on average, than the original population (**Figure 1.3b**). By repeating this process over many generations, horticulturists produced a population with extraordinarily large and compact flowering stalks. The derived population has been artificially selected for the size and shape of the flowering stalk; as **Figure 1.3c** shows, it barely resembles the ancestral form. Note that during this process, the size and shape of the flowering stalk in each individual plant did not change within its lifetime—the change occurred in the characteristics of the population over time. Darwin's great insight was that natural selection changes the characteristics of a wild population over time, just as the deliberate manipulation of "artificial selection" changes the characteristics of a domesticated population over time.

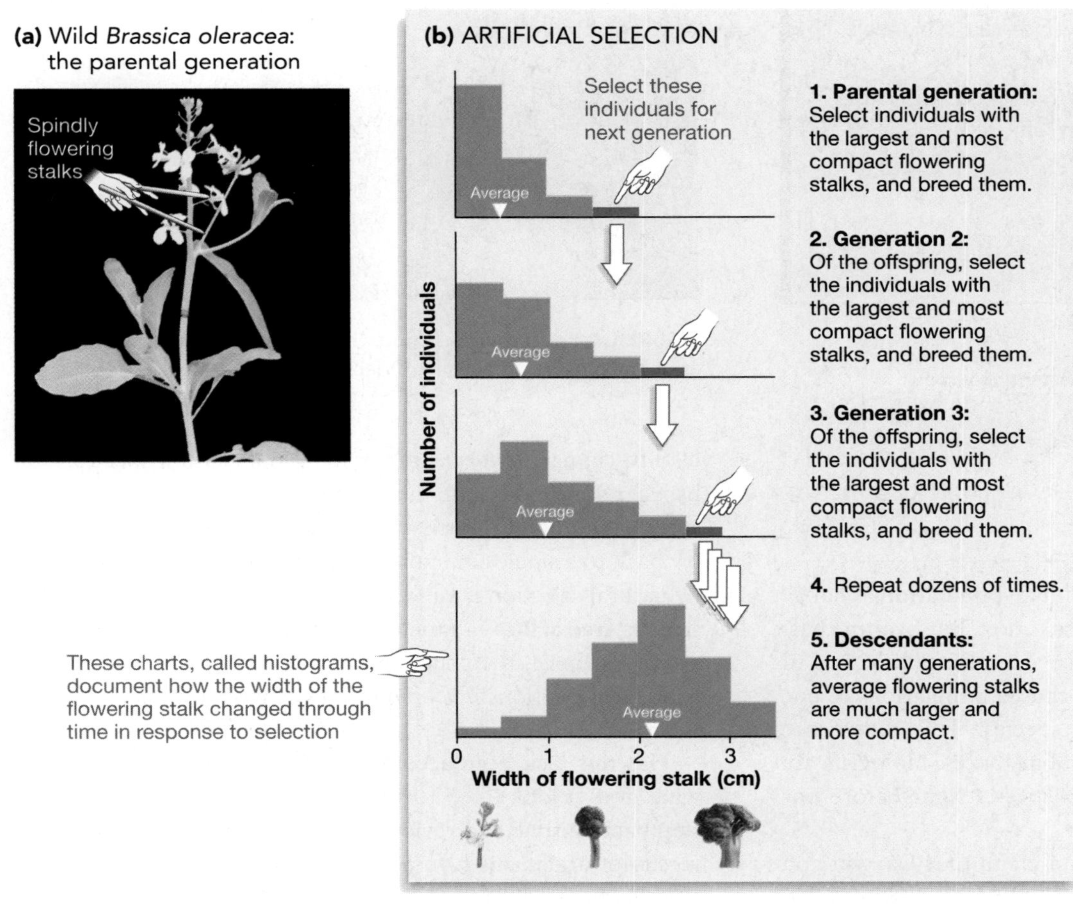

FIGURE 1.3 Artificial Selection Can Produce Dramatic Changes in Organisms.

Since Darwin and Wallace published their work, biologists have succeeded in measuring hundreds of examples of natural selection in wild populations. They have accumulated a massive body of evidence documenting that species have changed through time.

Together, the cell theory and the theory of evolution provided the young science of biology with two central, unifying ideas:

1. The cell is the fundamental structural unit in all organisms.
2. All species are related by common ancestry and have changed over time in response to natural selection.

Check Your Understanding

If you understand that...

- Natural selection occurs when heritable variation in certain traits leads to improved success in reproduction. Because individuals with these traits produce many offspring with the same traits, the traits increase in frequency and evolution occurs.
- Evolution is simply a change in the characteristics of a population over time.

You should be able to...

Explain why each of the following common misconceptions about evolution by natural selection is incorrect, using the example of selection on the height of wheat stalks.

1) evolution is progressive, meaning that species always get larger, more complex, or "better" in some sense;
2) individuals as well as populations change when natural selection occurs; or
3) individuals with high levels of fitness are stronger or bigger or "more dominant."

at www.masteringbio.com
Artificial Selection

1.3 The Tree of Life

Section 1.2 focused on how individual populations change through time in response to natural selection. But over the past several decades, biologists have also documented dozens of cases in which natural selection has caused populations of one species to diverge and form new species. This divergence process is called **speciation**. In several instances, biologists are documenting the formation of new species right before our eyes (**Figure 1.4**).

Research on speciation supports a claim that Darwin and Wallace made over a century ago—that natural selection can lead to change *between* species as well as within species. The broader conclusions are that all species come from preexisting species and that all species, past and present, trace their ancestry back to a single common ancestor. If the theory of evolution by natural selection is valid, biologists should be able to reconstruct a **tree of life**—a family tree of organisms. If life on Earth arose just once, then such a diagram would describe the genealogical relationships among species with a single, ancestral species at its base.

Has this task been accomplished? If the tree of life exists, what does it look like? To answer these questions, we need to step back in time and review how biologists organized the diversity of organisms *before* the development of the cell theory and the theory of evolution.

(a) *Tragopogon mirus* (left) evolved from *Tragopogon dubius* (right).

(b) Different soapberry bug species feed on native (left) or introduced (right) plants.

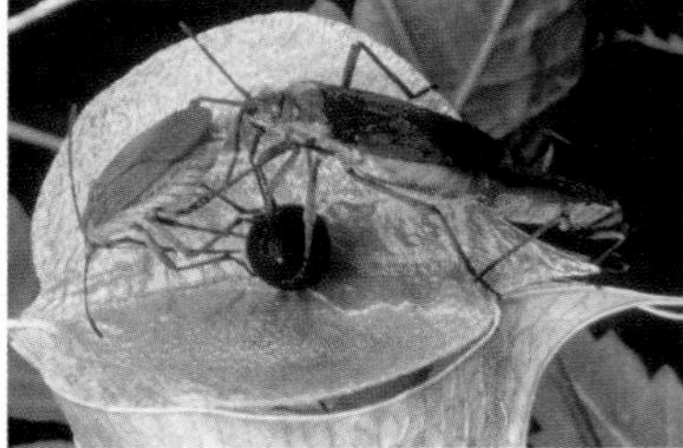

(c) Different maggot fly species feed on hawthorn (left) versus apple (right) fruits.

FIGURE 1.4 Speciation in Action. The pairs of organisms shown here are in the process of becoming independent species (see Chapter 26).

Linnaean Taxonomy

In science, the effort to name and classify organisms is called **taxonomy**. This branch of biology began to flourish in 1735 when a botanist named Carolus Linnaeus set out to bring order to the bewildering diversity of organisms that were then being discovered.

The building block of Linnaeus' system is a two-part name unique to each type of organism. The first part indicates the organism's **genus** (plural: **genera**). A genus is made up of a closely related group of species. For example, Linnaeus put humans in the genus *Homo*. Although humans are the only living species in this genus, several extinct organisms, all of which walked upright and made extensive use of tools, were later also assigned to *Homo*. The second term in the two-part name identifies the organism's species. Section 1.1 defined a species as a distinct, identifiable type of organism. More formally, a **species** is made up of individuals that regularly breed together or have characteristics that are distinct from those of other species. Linnaeus gave humans the species name *sapiens*.

An organism's genus and species designation is called its scientific name or Latin name. Scientific names are always italicized. Genus names are always capitalized, but species names are not—for instance, *Homo sapiens*. Scientific names are based on Latin or Greek word roots or on words "Latinized" from other languages (see **Box 1.1**). Linnaeus gave a scientific name to every species then known. (He also Latinized his own name—from Karl von Linné to Carolus Linnaeus.)

Linnaeus maintained that different types of organisms should not be given the same genus and species names. Other species may be assigned to the genus *Homo*, and members of other genera may be named *sapiens*, but only humans are named *Homo sapiens*. Each scientific name is unique.

Linnaeus' system has stood the test of time. His two-part naming system, or **binomial nomenclature**, is still the standard in biological science.

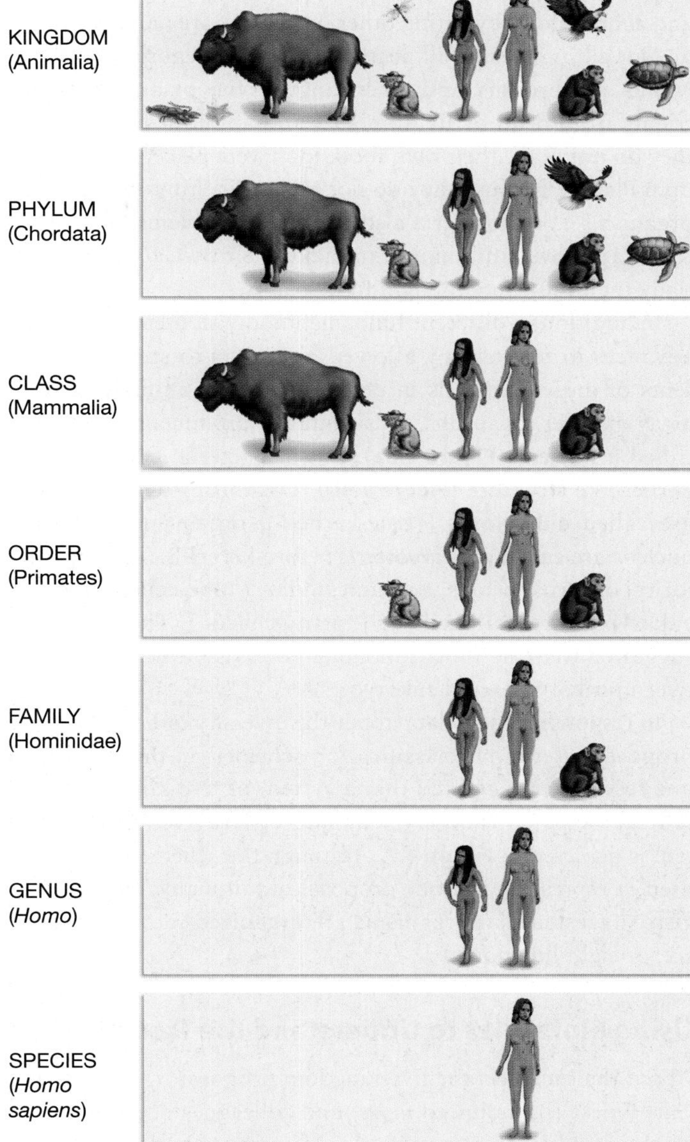

FIGURE 1.5 Linnaeus Defined Taxonomic Levels. In the Linnaean system, each animal species is placed in a taxonomic hierarchy with seven levels. Lower levels are nested within higher levels.

Taxonomic Levels To organize and classify the tremendous diversity of species being discovered in the 1700s, Linnaeus created a hierarchy of taxonomic groups: From the most specific grouping to the least specific, the levels are **species, genus, family, order, class, phylum** (plural: **phyla**), and **kingdom**. **Figure 1.5** shows how this nested, or hierarchical, classification scheme works, using humans as an example. Although our species is the sole living member of the genus *Homo*, humans are now grouped with the orangutan, gorilla, common chimpanzee, and bonobo in a family called Hominidae. Linnaeus grouped members of this family with gibbons, monkeys, and lemurs in an order called Primates. The Primates are grouped in the class Mammalia with rodents, bison, and other organisms that have fur and produce milk. Mammals, in turn, join other animals with structures called notochords in the phylum Chordata, and they join all other animals in the kingdom Animalia. Each of these named groups—primates, mammals, or *Homo sapiens*—can be referred to as a **taxon** (plural: **taxa**). The essence of Linnaeus's system is that lower-level taxa are nested within higher-level taxa.

Aspects of this hierarchical scheme are still in use. As biological science matured, however, several problems with Linnaeus's original proposal emerged.

How Many Kingdoms Are There? Linnaeus proposed that species could be organized into two kingdoms—plants and animals. According to Linnaeus, organisms that do not move and that produce their own food are plants; organisms that move and acquire food by eating other organisms are animals.

Not all organisms fall neatly into these categories, however. Molds, mushrooms, and other fungi survive by absorbing nutrients from dead or living plants and animals. Even though they do not make their own food, they were placed in the kingdom Plantae because they do not move. The tiny, single-celled organisms called bacteria also presented problems. Some bacteria can move, and many can make their own food. But initially they, too, were thought to be plants.

In addition, a different fundamental division emerged when advances in microscopy allowed biologists to study the contents of individual cells in detail. In plants, animals, and an array of other organisms, cells contain a prominent component called a nucleus (**Figure 1.6a**). But in bacteria, cells lack this kernel-like structure (**Figure 1.6b**). Organisms with a nucleus are called **eukaryotes** ("true-kernel"); organisms without a nucleus are called **prokaryotes** ("before-kernel"). The vast majority of prokaryotes are unicellular ("one-celled"); many eukaryotes are multicellular ("many-celled"). These findings suggested that the most fundamental division in life was between prokaryotes and eukaryotes.

In response to new data about the diversity of life, biologists proposed alternative classification schemes. In the late 1960s one researcher suggested that a system of five kingdoms best reflects the patterns observed in nature. This five-kingdom system is depicted in **Figure 1.7**. Although the scheme was widely used, it represents just one proposal out of many. Other biologists suggested that organisms are organized into three, four, six, or eight kingdoms.

(a) Eukaryotic cells have a membrane-bound nucleus.

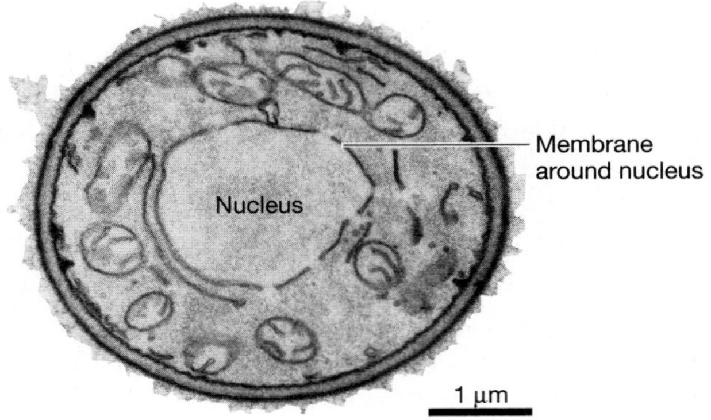

(b) Prokaryotic cells do *not* have a membrane-bound nucleus.

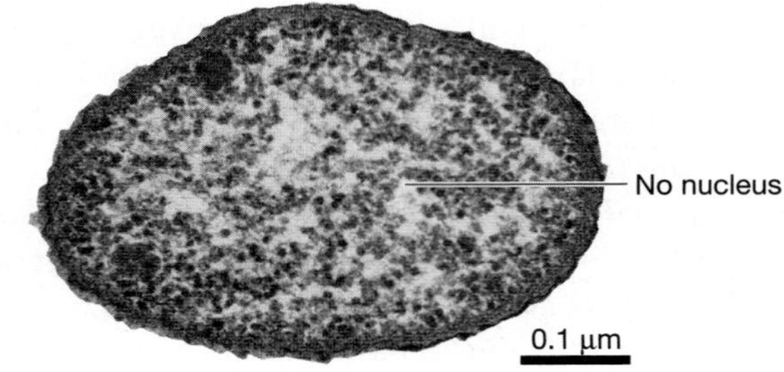

FIGURE 1.6 Eukaryotes and Prokaryotes.

EXERCISE Study the scale bars; then draw two ovals that accurately represent the relative sizes of a eukaryotic cell and a prokaryotic cell.

Using Molecules to Understand the Tree of Life

About the time that the five-kingdom proposal was published, Carl Woese (pronounced *woes*) and colleagues began working on the problem from a radically different angle. Instead of assigning organisms to kingdoms based on characteristics such as the presence of a nucleus or the ability to move or to manufacture food, they attempted to understand the relationships among organisms by analyzing their chemical components. Their goal was to understand the **phylogeny** (meaning "tribe-source") of all organisms—their actual genealogical relationships.

To understand which organisms are closely versus distantly related, Woese and co-workers needed to study a molecule that is found in all organisms. The molecule they selected is called small subunit rRNA. It is an essential part of the machinery that all cells use to grow and reproduce.

Although rRNA is a large and complex molecule, its underlying structure is simple. The rRNA molecule is made up of sequences of four smaller chemical components called ribonucleotides. These ribonucleotides are symbolized by the letters

BOX 1.1 Scientific Names and Terms

Scientific names and terms are often based on Latin or Greek word roots that are descriptive. For example, *Homo sapiens* is derived from the Latin *homo* for "man" and *sapiens* for "wise" or "knowing." The yeast that bakers use to produce bread and that brewers use to brew beer is called *Saccharomyces cerevisiae*. The Greek root *saccharo* means "sugar," and *myces* refers to a fungus. *Saccharomyces* is aptly named "sugar fungus" because yeast is a fungus and because the domesticated strains of yeast used in commercial baking and brewing are often fed sugar. The specific name of this organism, *cerevisiae*, is Latin for "beer." Loosely translated, then, the scientific name of brewer's yeast means "sugar fungus for beer."

Most biologists find it extremely helpful to memorize some of the common Latin and Greek roots. To aid you in this process, new terms in this text are often accompanied by a reference to their Latin or Greek word roots in parentheses.

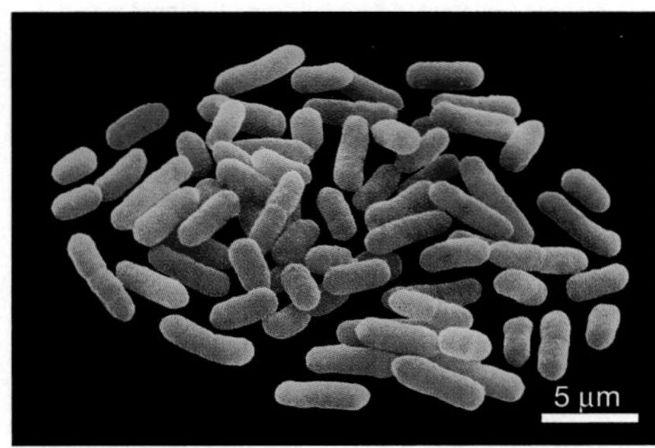

KINGDOM MONERA (includes all prokaryotes)

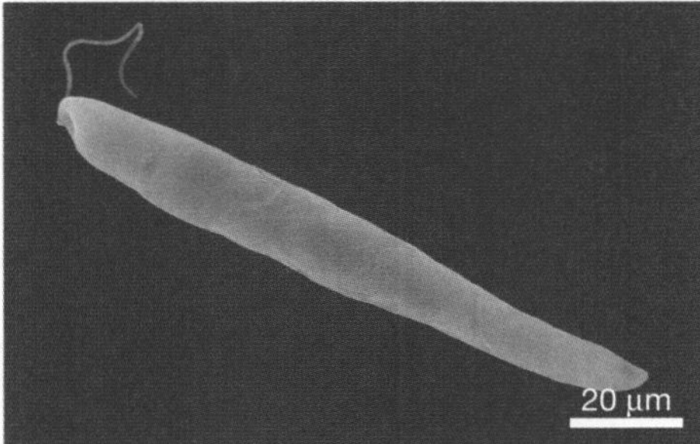

KINGDOM PROTISTA (includes several groups of unicellular eukaryotes)

KINGDOM PLANTAE

KINGDOM FUNGI

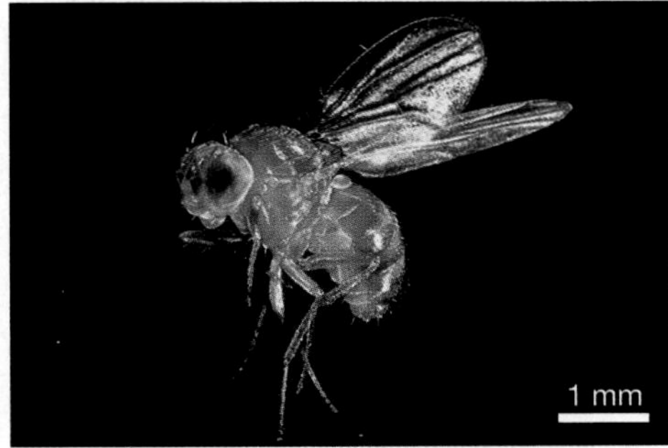

KINGDOM ANIMALIA

FIGURE 1.7 The Five-Kingdom Scheme. For decades, most biologists accepted the hypothesis that organisms naturally fall into the five kingdoms depicted here.

QUESTION How many times bigger is the fruit fly in the bottom panel than one of the prokaryotic cells in the top panel?

A, U, C, and G. In rRNA, ribonucleotides are connected to one another linearly, like boxcars of a freight train (**Figure 1.8**).

Why might rRNA be useful for understanding the relationships among organisms? The answer is that the ribonucleotide sequence in rRNA is a trait, similar to the height of wheat stalks or the size of flowering stalks of broccoli, that can change during the course of evolution. Although rRNA performs the same function in all organisms, the sequence of ribonucleotide building blocks in this molecule is not identical among species. In land plants, for example, the molecule might start with the sequence A-U-A-U-C-G-A-G. In green algae, which are closely related to land plants, the same section of the molecule might contain A-U-A-U-G-G-A-G. But in brown algae, which are not closely related to green algae or to land plants, the same part of the molecule might consist of A-A-A-U-G-G-A-G.

The research program that Woese and co-workers pursued was based on a simple premise: If the theory of evolution is correct, then rRNA sequences should be very similar in closely related organisms but less similar in organisms that are less closely related. Groups that are closely related, like the plants, should share certain changes in rRNA that no other species have.

To test this premise, the researchers determined the sequence of ribonucleotides in the rRNA of a wide array of species. Then they considered what the similarities and differences in the sequences implied about relationships among the species. The goal was to produce a diagram that described the phylogeny of the organisms in the study. A diagram that depicts evolutionary history in this way is called a **phylogenetic tree**. Just as a family tree shows relationships among individuals, a phylogenetic tree shows relationships among species. On a phylogenetic tree, branches that are close to one another represent species that are closely related; branches that are farther apart represent species that are more distantly related.

FIGURE 1.8 RNA Molecules Are Made Up of Smaller Molecules. The complete small subunit rRNA molecule contains about 2000 ribonucleotides; just 8 are shown in this comparison.

QUESTION Suppose that in the same portion of rRNA, molds and other fungi have the sequence A-U-A-U-G-G-A-C. According to these data, are fungi more closely related to green algae or to land plants? Explain your logic.

The Tree of Life Estimated from an Array of Genes To construct a phylogenetic tree, researchers use a computer to find the arrangement of branches that is most consistent with the similarities and differences observed in the data. Although the initial work was based only on the sequences of ribonucleotides observed in rRNA, biologists now use data sets that include sequences from a wide array of genes. A recent tree produced by comparing these sequences is shown in **Figure 1.9**. Because this tree includes species from many different kingdoms and phyla, it is often called the universal tree, or the tree of life. For help in learning how to read a phylogenetic tree, see **BioSkills 2**.

The tree of life implied by rRNA and other genetic data astonished biologists. For example:

- The fundamental division in organisms is not between plants and animals or even between prokaryotes and eukaryotes. Rather, *three* major groups occur: (1) the Bacteria; (2) another group of prokaryotic, single-celled organisms called the Archaea; and (3) the eukaryotes. To accommodate this new perspective on the diversity of organisms, Woese created a new taxonomic level called the **domain**. As Figure 1.9 indicates, the three domains of life are now called the Bacteria, Archaea, and Eukarya.
- Some of the kingdoms that had been defined earlier do not reflect how evolution actually occurred. For example, recall that Linnaeus grouped the multicellular eukaryotes known as fungi with plants. But the genetic data indicate that fungi are much more closely related to animals than they are to plants.
- Bacteria and Archaea are much more diverse than anyone had imagined. If the differences among animals, fungi, and plants warrant placing them in separate kingdoms, then dozens of kingdoms exist among the prokaryotes.

The Tree of Life Is a Work in Progress Just as researching your family tree can help you understand who you are and where you came from, so the tree of life helps biologists understand the relationships among organisms and the history of species. For example, the discovery of the Archaea and the placement of lineages such as the fungi qualify as exciting breakthroughs in our understanding of life's diversity. Work on the tree of life continues at a furious pace, however, and the location of certain branches on the tree is hotly debated. As databases expand and as techniques for analyzing data improve, the shape of the tree of life presented in Figure 1.9 will undoubtedly change.

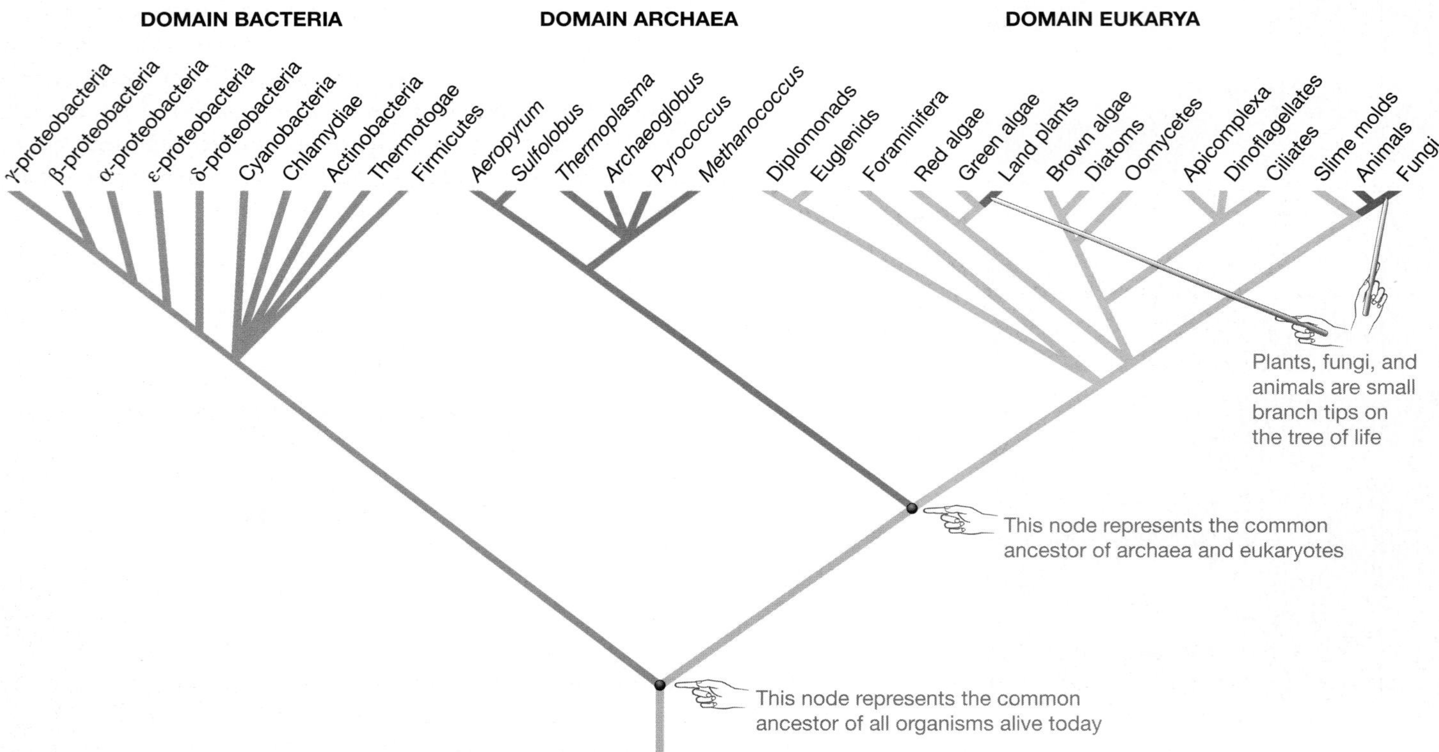

FIGURE 1.9 The Tree of Life. "Universal tree" estimated from a large amount of gene sequence data. The three domains of life revealed by the analysis are labeled. Common names are given for most lineages in the domains Bacteria and Eukarya. Genus names are given for members of the domain Archaea, because most of these organisms have no common names.

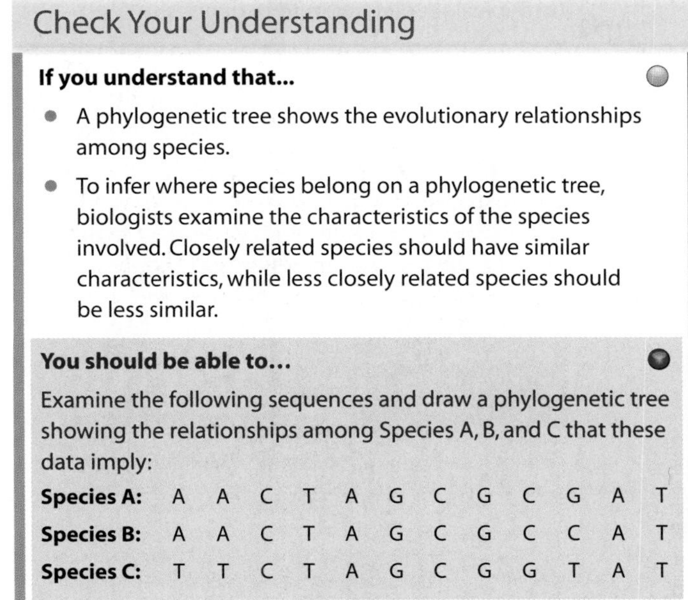

Check Your Understanding

If you understand that...

- A phylogenetic tree shows the evolutionary relationships among species.
- To infer where species belong on a phylogenetic tree, biologists examine the characteristics of the species involved. Closely related species should have similar characteristics, while less closely related species should be less similar.

You should be able to...

Examine the following sequences and draw a phylogenetic tree showing the relationships among Species A, B, and C that these data imply:

Species A:	A	A	C	T	A	G	C	G	C	G	A	T
Species B:	A	A	C	T	A	G	C	G	C	C	A	T
Species C:	T	T	C	T	A	G	C	G	G	T	A	T

1.4 Doing Biology

This chapter has introduced some of the great ideas in biology. The development of the cell theory and the theory of evolution by natural selection provided cornerstones when the science was young; the tree of life is a relatively recent insight that has revolutionized the way researchers understand the diversity of life on Earth.

These theories are considered great because they explain fundamental aspects of nature, and because they have consistently been shown to be correct. They are considered correct because they have withstood extensive testing. How do biologists test ideas about the way the natural world works? The answer is that they test the predictions made by alternative hypotheses, often by setting up carefully designed experiments. To illustrate how this approach works, let's consider two questions currently being addressed by researchers.

Why Do Giraffes Have Long Necks? An Introduction to Hypothesis Testing

If you were asked why giraffes have long necks, you might say that long necks enable giraffes to reach food that is unavailable to other mammals. This hypothesis is expressed in African folktales and has traditionally been accepted by many biologists. The food competition hypothesis is so plausible, in fact, that for decades no one thought to test it. Recently, however, Robert Simmons and Lue Scheepers assembled data suggesting that the food competition hypothesis is only part of the story. Their analysis supports an alternative hypothesis—that long necks allow giraffes to use their heads as effective weapons for battering their opponents.

How did biologists test the food competition hypothesis? What data support their alternative explanation? Before attempting to answer these questions, it's important to recognize that hypothesis testing is a two-step process. The first step is to state the hypothesis as precisely as possible and list the predictions it makes. The second step is to design an observational or experimental study that is capable of testing those predictions. If the predictions are accurate, then the hypothesis is supported. If the predictions are not met, then researchers do further tests, modify the original hypothesis, or search for alternative explanations.

The Food Competition Hypothesis: Predictions and Tests Stated precisely, the food competition hypothesis claims that giraffes compete for food with other species of mammals. When food is scarce, as it is during the dry season, giraffes with longer necks can reach food that is unavailable to other species and to giraffes with shorter necks. As a result, the longest-necked individuals in a giraffe population survive better and produce more young than do shorter-necked individuals, and average neck length of the population increases with each generation. To use the terms introduced earlier, long necks are adaptations that increase the fitness of individual giraffes during competition for food. This type of natural selection has gone on so long that the population has become extremely long necked.

The food competition hypothesis makes several explicit predictions. For example, the food competition hypothesis predicts that (1) neck length is variable among giraffes; (2) neck length in giraffes is heritable; and (3) giraffes feed high in trees, especially during the dry season, when food is scarce and the threat of starvation is high.

The first prediction is correct. Studies in zoos and natural populations confirm that neck length is variable among individuals.

The researchers were unable to test the second prediction, however, because they studied giraffes in a natural population and were unable to do breeding experiments. As a result, they simply had to accept this prediction as an assumption. In general, though, biologists prefer to test every assumption behind a hypothesis.

What about the prediction regarding feeding high in trees? According to Simmons and Scheepers, this is where the food competition hypothesis breaks down. Consider, for example, data collected by a different research team about the amount of time that giraffes spend feeding in vegetation of different heights. **Figure 1.10a** shows that in a population from Kenya, both male and female giraffes spend most of their feeding time eating vegetation that averages just 60 percent of their full height. Studies on other populations of giraffes, during

(a) Most feeding is done below neck height.

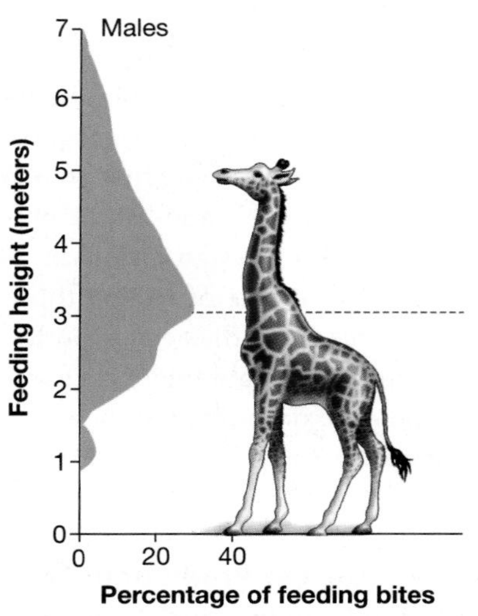

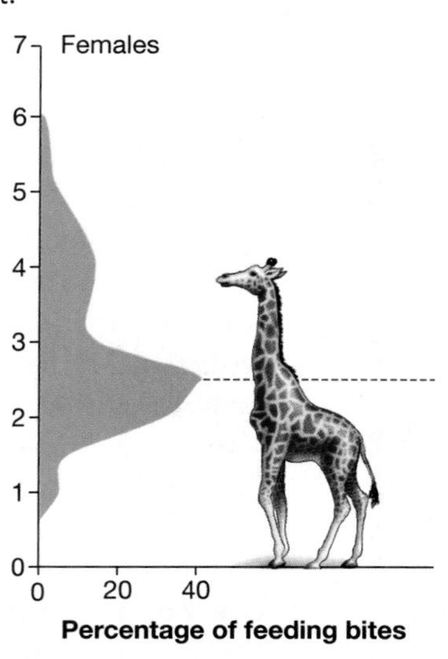

(b) Typical feeding posture in giraffes

FIGURE 1.10 Giraffes Do Not Usually Extend Their Necks to Feed.

both the wet and dry seasons, are consistent with these data. Giraffes usually feed with their necks bent **(Figure 1.10b)**.

These data cast doubt on the food competition hypothesis, because one of its predictions does not appear to hold. Biologists have not abandoned this hypothesis completely, though, because feeding high in trees may be particularly valuable during extreme droughts, when a giraffe's ability to reach leaves far above the ground could mean the difference between life and death. Still, Simmons and Scheepers have offered an alternative explanation for why giraffes have long necks. The new hypothesis is based on the mating system of giraffes.

The Sexual Competition Hypothesis: Predictions and Tests Giraffes have an unusual mating system. Breeding occurs year round rather than seasonally. To determine when females are coming into estrus (or "heat") and are thus receptive to mating, the males nuzzle the rumps of females. In response, the females urinate into the males' mouths. The males then tip their heads back and pull their lips to and fro, as if tasting the liquid. Biologists who have witnessed this behavior have proposed that the males taste the females' urine to detect whether estrus has begun.

Once a female giraffe enters estrus, males fight among themselves for the opportunity to mate. Combat is spectacular. The bulls stand next to one another, swing their necks, and strike thunderous blows with their heads. Researchers have seen males knocked unconscious for 20 minutes after being hit and have cataloged numerous instances in which the loser died. Giraffes are the only animals known to fight in this way.

These observations inspired a new explanation for why giraffes have long necks. The sexual competition hypothesis is based on the idea that longer-necked giraffes are able to strike harder blows during combat than can shorter-necked giraffes. In engineering terms, longer necks provide a longer moment arm. A long moment arm increases the force of the impact. (Think about the type of sledge hammer you'd use to bash down a concrete wall—one with a short handle or one with a long handle?) Thus, longer-necked males should win more fights and, as a result, father more offspring than do shorter-necked males. If neck length in giraffes is inherited, then the average neck length in the population should increase over time. Under the sexual competition hypothesis, long necks are adaptations that increase the fitness of males during competition for females.

Although several studies have shown that long-necked males are more successful in fighting and that the winners of fights gain access to estrous females, the question of why giraffes have long necks is not closed. With the data collected to date, most biologists would probably concede that the food competition hypothesis needs further testing and refinement and that the sexual selection hypothesis appears promising. It could also be true that both hypotheses are correct. For our purposes, the important take-home message is that all hypotheses must be tested rigorously.

In many cases in biological science, testing hypotheses rigorously involves experimentation. Experimenting on giraffes is difficult. But in the case study considered next, biologists were able to test an interesting hypothesis experimentally.

Why Are Chili Peppers Hot? An Introduction to Experimental Design

Experiments are a powerful scientific tool because they allow researchers to test the effect of a single, well-defined factor on a particular phenomenon. Because experiments testing the effect of neck length on food and sexual competition in giraffes have yet to be done, let's consider a different question: Why do chili peppers taste so spicy?

The jalapeño, Anaheim, and cayenne peppers used in cooking descended, via artificial selection, from a wild shrub that is native to the deserts of the American Southwest. As **Figure 1.11a** shows, wild chilies produce fleshy fruits with seeds inside, just like their domesticated descendants. In both wild chilies and the cultivated varieties, the "heat" or pungent flavor of the fruit and seeds is due to a molecule called capsaicin. In humans and other mammals, capsaicin binds to heat-sensitive cells in the tongue and mouth. In response to this binding, signals are sent to the brain that produce the sensation of burning. Similar signals would be transmitted if you drank boiling water. Asking why chilies are hot, then, is the same as asking why chilies contain capsaicin.

Josh Tewksbury and Gary Nabhan proposed that the presence of capsaicin is an adaptation that protects chili fruits from being eaten by animals that destroy the seeds inside. To understand this hypothesis, it's important to realize that the seeds inside a fruit have one of two fates when the fruit is eaten. If the seeds are destroyed in the animal's mouth or digestive system, then they never germinate (sprout). In this case, "seed predation" has occurred. But if seeds can travel undamaged through the animal, then they are eventually "planted" in a new location along with a valuable supply of fertilizer. In this case, seeds are dispersed. Here's the key idea: Natural selection should favor fruits that taste bad to animal species that act as seed predators. But these same fruits should not deter species that act as seed dispersers. This proposal is called the directed dispersal hypothesis.

Does capsaicin deter seed predators, as the directed dispersal hypothesis predicts? To answer this question, the researchers captured some cactus mice (**Figure 1.11b**) and birds called curve-billed thrashers (**Figure 1.11c**). These species are among the most important fruit- and seed-eating animals in habitats where chilies grow. Based on earlier observations, the biologists predicted that cactus mice destroy chili seeds but that curve-billed thrashers disperse them effectively.

To test the directed dispersal hypothesis, the biologists offered both cactus mice and curve-billed thrashers three kinds of fruit: hackberries, fruits from a strain of chilies that can't synthesize capsaicin, and pungent chilies that have lots of capsaicin. The non-pungent chilies are about the same size and color as normal chilies and have similar nutritional value. The hackberries look like chilies, except that they are not quite as red and contain no capsaicin. The three fruits were present in equal amounts. For each animal tested, the researchers recorded the percentage of hackberry, non-pungent chili, and pungent chili that was eaten during a specific time interval. Then they calculated the average amount of each fruit that was eaten by five test individuals from each species.

The directed dispersal hypothesis predicts that seed dispersers will eat the pungent chilies readily but that seed predators won't. Recall that a prediction specifies what we should observe if a hypothesis is correct. Good scientific hypotheses make testable predictions—predictions that can be supported or rejected by collecting and analyzing data. If the directed dispersal hypothesis is wrong, however, then there shouldn't be any difference in what various animals eat. This latter possibility is called a **null hypothesis**. A null hypothesis specifies what we should observe when the hypothesis being tested

(a) Wild chilies produce fruits that contain seeds.

(b) Cactus mice are seed eaters.

(c) Curve-billed thrashers are fruit eaters.

FIGURE 1.11 Chilies . . . and Chili Eaters?

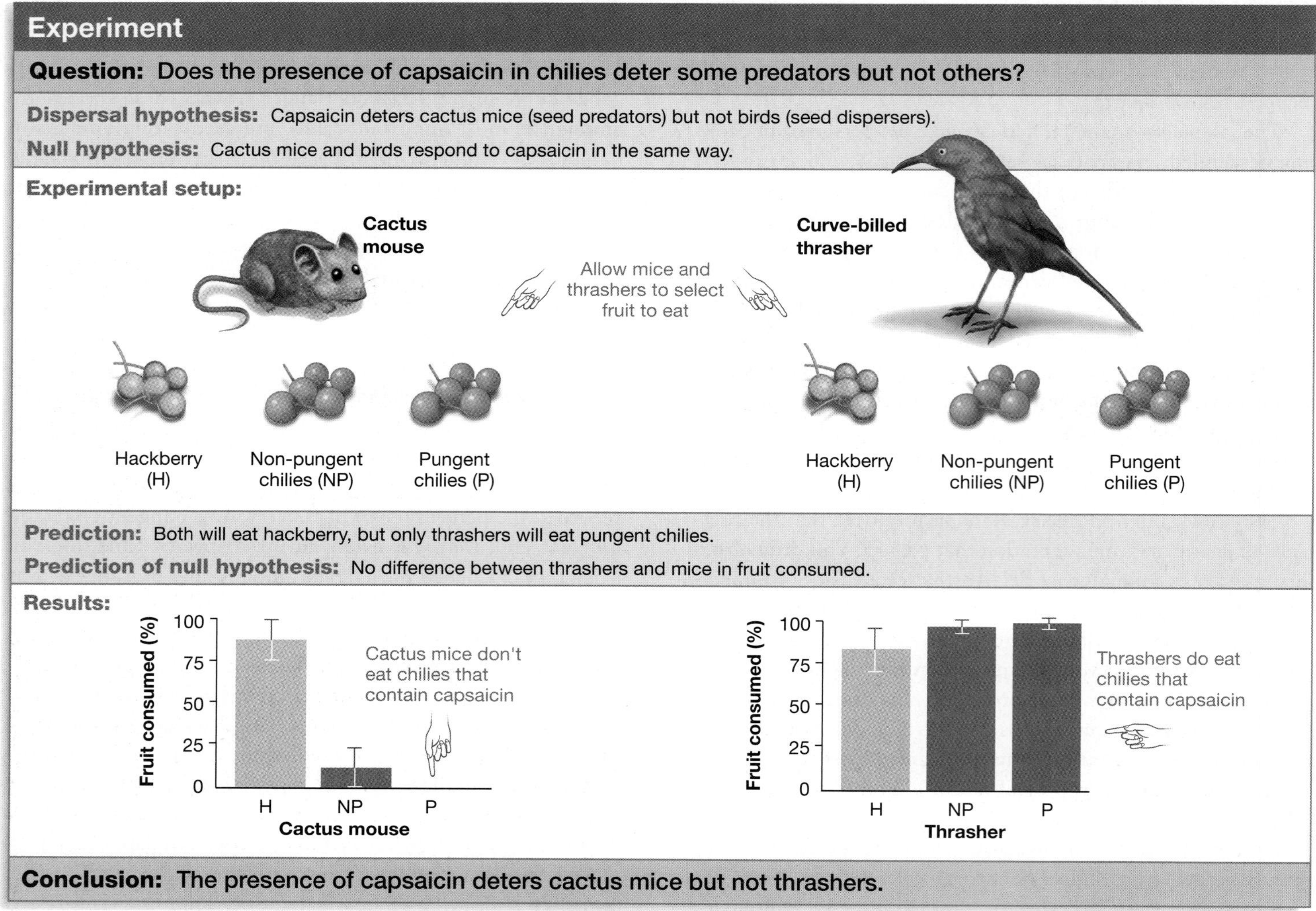

FIGURE 1.12 An Experimental Test: Does Capsaicin Deter Some Fruit Eaters? The graphs in the Results section indicate the average percentage of fruit consumed by the animals that were tested. The thin vertical lines indicate the standard error associated with each average. The standard error is a measure of variability and uncertainty in the data. For more details, see **BioSkills 3**.

doesn't hold. These predictions are listed in **Figure 1.12.** Do the predictions of the directed dispersal hypothesis hold? To answer this question, look at the results plotted in Figure 1.12. View the data for each type of fruit to see if different amounts were eaten by different types of animals, and ask yourself whether the two animals ate the same or different amounts of each type of fruit. Based on your analysis of the data, decide whether the results support the directed dispersal hypothesis or the null hypothesis. Use the conclusion stated in Figure 1.12 to check your answer.

In relation to designing effective experiments, this study illustrates several important points:

- It is critical to include control groups. A control checks for factors, other than the one being tested, that might influence the experiment's outcome. For example, if hackberries had not been included as a control, it would have been possible to claim that the cactus mice in the experiment didn't eat pungent chilies simply because they weren't hungry. But the not-hungry hypothesis can be rejected because all of the animals ate hackberries.

- The experimental conditions must be carefully controlled. The investigators used the same feeding choice setup, the same time interval, and the same definition of what "fruit consumed" meant in each test. Controlling all of the variables except one—the types of fruits presented—is crucial because it eliminates alternative explanations for the results. For example, what types of problems could arise if the cactus mice were given less time to eat than the thrashers, or if the test animals were always presented with hackberries first and pungent chilies last?

- Repeating the test is essential. It is almost universally true that larger sample sizes in experiments are better. For example, suppose that the experimenters had used just one cactus mouse instead of five, and that this mouse was unlike other cactus mice because it ate almost anything. If so, the resulting data would be badly distorted. By testing many individuals, the amount of distortion or "noise" in the data caused by unusual individuals or circumstances is reduced.

MB Web Animation at www.masteringbio.com
Introduction to Experimental Design

● If you understand these concepts, you should be able to design an experiment that tests the directed dispersal hypothesis by adding capsaicin to small pieces of apple—a fruit that both mice and thrashers eat readily.

To test the assumption that cactus mice are seed predators and curve-billed thrashers are seed dispersers, the researchers did a follow-up experiment. They fed fruits of the non-pungent chili to each type of predator. When the seeds had passed through the animals' digestive systems and were excreted, the researchers collected and planted the seeds—along with 14 uneaten seeds. Planting uneaten seeds served as a control treatment, because it tested the hypothesis that the seeds were viable and would germinate if they were not eaten. About 50 percent of the uneaten seeds germinated, and almost 60 percent of the seeds eaten by a thrasher germinated. But none of the seeds eaten by cactus mice germinated. The data indicate that seeds pass through curve-billed thrashers unharmed but are destroyed when eaten by cactus mice.

Based on the outcomes of these two experiments, the researchers concluded that curve-billed thrashers are efficient seed dispersers and are not deterred by capsaicin. The cactus mice, in contrast, refuse to eat chilies. If they ate chilies, the mice would kill the seeds. These are exactly the results predicted by the directed dispersal hypothesis. The biologists concluded that the presence of capsaicin in chilies is an adaptation that keeps their seeds from being destroyed by mice. In habitats that contain cactus mice, the production of capsaicin increases the fitness of individual chili plants.

These experiments are a taste of things to come. In this text you will encounter hypotheses and experiments on questions ranging from how water gets to the top of 100-meter-tall sequoia trees to why the bacterium that causes tuberculosis has become resistant to antibiotics. A commitment to tough-minded hypothesis testing and sound experimental design is a hallmark of biological science. Understanding their value is an important first step in becoming a biologist.

Check Your Understanding

If you understand that...

- Hypotheses are proposed explanations that make testable predictions.
- Predictions are observable outcomes of particular conditions.
- Well-designed experiments alter just one condition—a condition relevant to the hypothesis being tested.

You should be able to...

1) Design an experiment to test the hypothesis that the use of capsaicin as cooking spice is an adaptation—specifically, that the presence of capsaicin in food kills disease-causing bacteria.
2) State the predictions of the adaptation hypothesis and the null hypothesis in your experiment.
3) Answer the following questions about your experimental design:

   How does the presence of a control group in your experiment allow you to test the null hypothesis?

   Why isn't the experiment valid without the presence of a control group?

   How are experimental conditions controlled or standardized in a way that precludes alternative explanations of the data?

   Why do you propose to repeat the experiment many times?

Chapter Review

SUMMARY OF KEY CONCEPTS

For over two hundred years, biologists have been discovering traits that unify the spectacular diversity of living organisms.

Biological science was founded with the development of (1) the cell theory, which proposes that all organisms are made of cells and that all cells come from preexisting cells, and (2) the theory of evolution by natural selection, which maintains that the characteristics of species change through time—primarily because individuals with certain heritable traits produce more offspring than do individuals without those traits.

The cell theory is an important unifying principle in biology, because it identified the fundamental structural unit common to all life. The theory of evolution by natural selection is another key unifying principle, because it states that all organisms are related by common ancestry. It also offered a robust explanation for why species change through time and why they are so well adapted to their habitats.

You should be able to describe the evidence that supported the cell theory. You should also be able to explain why a population of wild *Brassica oleracea* will evolve by natural selection, if in response to global warming individuals with large leaves begin producing the most offspring.

MB Web Animation at www.masteringbio.com
Artificial Selection

A phylogenetic tree is a graphical representation of the evolutionary relationships among species. Phylogenies can be established by analyzing similarities and differences in traits. Species that share many traits are closely related and are placed close to each other on the tree of life.

The cell theory and the theory of evolution predict that all organisms are part of a genealogy of species, and that all species trace their ancestry back to a single common ancestor. To reconstruct this phylogeny, biologists have analyzed the sequence of components in rRNA and other molecules found in all cells. A tree of life, based on similarities and differences in these molecules has three major lineages: the Bacteria, Archaea, and Eukarya.

You should be able to explain why biologists can determine whether newly discovered species are members of the Bacteria, Archaea, or Eukarya by analyzing their rRNA or other molecules.

Biologists ask questions, generate hypotheses to answer them, and design experiments that test the predictions made by competing hypotheses.

Another unifying theme in biology is a commitment to hypothesis testing and to sound experimental design. Analyses of neck length in giraffes and the capsaicin found in chilies are case studies in the value of testing alternative hypotheses and conducting experiments. Biology is a hypothesis-driven, experimental science.

You should be able to explain (1) the relationship between a hypothesis and a prediction, and (2) why experiments are convincing ways to test predictions.

MB Web Animation at www.masteringbio.com
Introduction to Experimental Design

QUESTIONS

Test Your Knowledge

1. Anton van Leeuwenhoek made an important contribution to the development of the cell theory. How?
   a. He articulated the pattern component of the theory—that all organisms are made of cells.
   b. He articulated the process component of the theory—that all cells come from preexisting cells.
   c. He invented the first microscope and saw the first cell.
   d. He invented more powerful microscopes and was the first to describe the diversity of cells.

2. Suppose that a proponent of the spontaneous generation hypothesis claimed that cells would appear in Pasteur's swan-necked flask eventually. According to this view, Pasteur did not allow enough time to pass before concluding that life does not originate spontaneously. Which of the following is the best response?
   a. The spontaneous generation proponent is correct: Spontaneous generation would probably happen eventually.
   b. Both the all-cells-from-cells hypothesis and the spontaneous generation hypothesis could be correct.
   c. If spontaneous generation happens only rarely, it is not important.
   d. If spontaneous generation did not occur after weeks or months, it is not reasonable to claim that it would occur later.

3. What does the term *evolution* mean?
   a. The strongest individuals produce the most offspring.
   b. The characteristics of an individual change through the course of its life, in response to natural selection.
   c. The characteristics of populations change through time.
   d. The characteristics of species become more complex over time.

4. What does it mean to say that a characteristic of an organism is heritable?
   a. The characteristic evolves.
   b. The characteristic can be passed on to offspring.
   c. The characteristic is advantageous to the organism.
   d. The characteristic does not vary in the population.

5. In biology, what does the term *fitness* mean?
   a. how well trained and muscular an individual is, relative to others in the same population
   b. how slim an individual is, relative to others in the same population
   c. how long a particular individual lives
   d. the ability to survive and reproduce

6. Could *both* the food competition hypothesis and the sexual selection hypothesis explain why giraffes have long necks? Why or why not?
   a. No. In science, only one hypothesis can be correct.
   b. No. Observations have shown that the food competition hypothesis cannot be correct.
   c. Yes. Long necks could be advantageous for more than one reason.
   d. Yes. All giraffes have been shown to feed at the highest possible height and fight for mates.

Test Your Knowledge answers: 1. d; 2. d; 3. c; 4. b; 5. d; 6. c

Test Your Understanding

Answers are available at www.masteringbio.com

1. The Greek roots of the term taxonomy can be translated as "arranging rules." Explain why these roots were an appropriate choice for this term.

2. It was once thought that the deepest split among life-forms was between two groups: prokaryotes and eukaryotes. Draw and label a phylogenetic tree that represents this hypothesis. Then draw and

label a phylogenetic tree that shows the actual relationships among the three domains of organisms.

3. Why was it important for Linnaeus to establish the rule that only one type of organism can have a particular genus and species name?
4. What does it mean to say that an organism is adapted to a particular habitat?
5. Compare and contrast natural selection with the process that led to the divergence of a wild mustard plant into cabbage, broccoli, and Brussels sprouts.
6. The following two statements explain the logic behind the use of molecular sequence data to estimate evolutionary relationships:

   "If the theory of evolution is true, then rRNA sequences should be very similar in closely related organisms but less similar in organisms that are less closely related."

   "On a phylogenetic tree, branches that are close to one another represent species that are closely related; branches that are farther apart represent species that are more distantly related."

   Is the logic of these statements sound? Why or why not?

Applying Concepts to New Situations

Answers are available at www.masteringbio.com

1. A scientific theory is a set of propositions that defines and explains some aspect of the world. This definition contrasts sharply with the everyday usage of the word theory, which often carries meanings such as "speculation" or "guess." Explain the difference between the two definitions, using the cell theory and the theory of evolution by natural selection as examples.
2. Turn back to the tree of life shown in Figure 1.9. Note that Bacteria and Archaea are prokaryotes, while Eukarya are eukaryotes. On the simplified tree below, draw an arrow that points to the branch where the structure called the nucleus originated. Explain your reasoning.

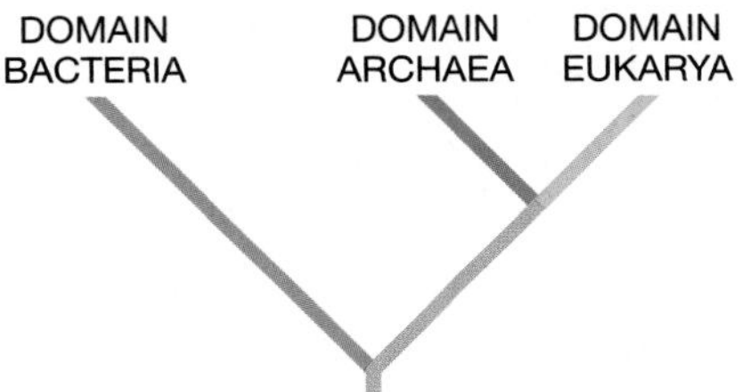

3. The proponents of the cell theory could not "prove" that it was correct in the sense of providing incontrovertible evidence that all organisms are made up of cells. They could state only that all organisms examined to date were made of cells. Why was it reasonable for them to conclude that the theory was valid?
4. How do the tree of life and the taxonomic categories created by Linnaeus (kingdom, phylum, class, order, family, genus, and species) relate to one another?

Evolution by Natural Selection

24

The color difference between these two peppered moths is largely due to the action of different alleles at a single gene. Biologists have documented changes in the frequencies of these alleles in several different populations around the world, due to natural selection.

KEY CONCEPTS

- Populations and species evolve, meaning that their heritable characteristics change through time. More precisely, evolution is defined as changes in allele frequencies over time.
- Natural selection occurs when individuals with certain alleles produce the most surviving offspring in a population. An adaptation is a genetically based trait that increases an individual's ability to produce offspring in a particular environment.
- Evolution by natural selection is not progressive, and it does not change the characteristics of the individuals that are selected—it changes only the characteristics of the population. Animals do not do things for the good of the species, and not all traits are adaptive. All adaptations are constrained by trade-offs and genetic and historical factors.

This chapter is about one of the great ideas in science. The theory of evolution by natural selection, formulated independently by Charles Darwin and Alfred Russel Wallace, explains how organisms have come to be adapted to environments ranging from arctic tundra to tropical wet forest. As an example of a revolutionary breakthrough in our understanding of the world, the theory of evolution by natural selection ranks alongside Copernicus's theory of the Sun as the center of our solar system, Newton's laws of motion and theory of gravitation, the germ theory of disease, the theory of plate tectonics, and Einstein's general theory of relativity. These ideas are the foundation stones of modern science; all are accepted on the basis of overwhelming evidence.

Evolution by natural selection has become one of the best-supported and most important theories in the history of scientific research. But like most scientific breakthroughs, this one did not come easily. When Darwin published his theory in 1859 in a book called *On the Origin of Species by Means of Natural Selection*, it unleashed a firestorm of protest throughout Europe. At that time, the leading explanation for the diversity of organisms was a theory called special creation. This theory held that all species were created independently, by God, perhaps as recently as 6000 years ago. The theory of special creation also maintained that species were immutable, or incapable of change, and thus had been unchanged since the moment of their creation. Darwin's ideas were radically different. He proposed that life on Earth was ancient and that species change through time.

To understand the contrast between the theory of special creation and the theory of evolution by natural selection more thoroughly, recall from Chapter 1 that scientific theories usually have two components: a pattern and a process. The first

Key Concept Important Information Practice It

component is either a claim about a pattern that exists in nature or a statement that summarizes a series of observations about the natural world. In short, the pattern component is about facts—about how things *are* in nature. The second component of a scientific theory is a process that produces that pattern or set of observations. For example, the pattern component in the theory of special creation was that species were created independently of one another and that they do not change through time. The process that explained this pattern was the instantaneous and independent creation of living organisms by a supernatural being.

To help you understand the pattern and process components of evolution by natural selection, the chapter begins by examining the ideas that led to Darwin's breakthrough. Section 24.2 analyzes the pattern component of the theory—specifically, the evidence behind the claim that species are not independent but are instead related, and that they are not immutable but instead have changed through time. Darwin proposed that natural selection, introduced in Section 24.3, explains this pattern. The concluding sections review two recent studies of evolution by natural selection. These case studies illustrate how biologists test Darwin's theory by studying evolution in action and help clarify some common misunderstandings about how evolution by natural selection works.

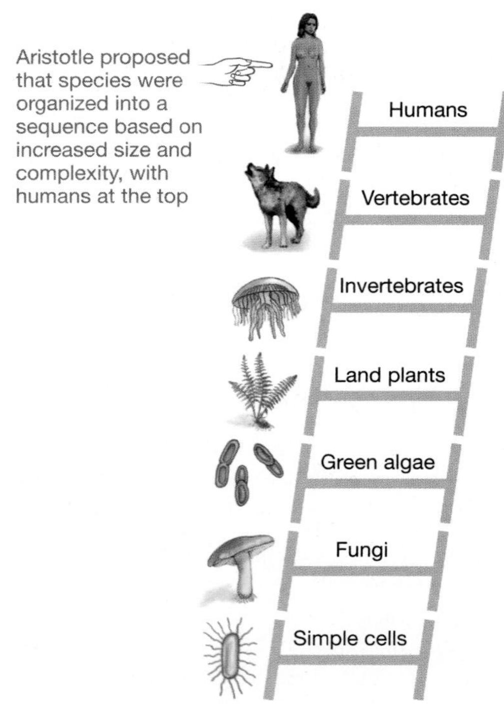

FIGURE 24.1 The Great Chain of Being Proposed by Aristotle.

24.1 The Evolution of Evolutionary Thought

People often use the word *revolutionary* to describe the theory of evolution by natural selection. Revolutions overturn things—they replace an existing entity with something new and often radically different. A political revolution removes the ruling class or group and replaces it with another. The industrial revolution replaced small shops for manufacturing goods by hand with huge, mechanized assembly lines. A scientific revolution, in contrast, overturns an existing idea about how nature works and replaces it with another, radically different, idea.

The idea that Darwin and Wallace overturned had dominated thinking about the nature of organisms for over 2000 years. The Greek philosopher Plato claimed that every organism was an example of a perfect essence, or type, created by God, and that these types were unchanging. Plato acknowledged that the individual organisms present on Earth might deviate slightly from the perfect type, but he said this was similar to seeing shadows created by the perfect type on a wall. The key to understanding life, in Plato's mind, was to ignore the shadows and focus on understanding each type of unchanging, perfect essence.

Not long after Plato developed his ideas, Aristotle ordered the types of organisms known at the time into a linear scheme called the great chain of being, also called the scale of nature or the ladder of life (**Figure 24.1**). Aristotle proposed that species were organized into a sequence based on increased size and complexity, with humans at the top.

Today, philosophers and biologists refer to ideas like these as typological thinking. Typological thinking is based on the idea that species are unchanging types and that variations within species are unimportant or even misleading. Typological thinking occurs in the Bible's book of Genesis, where God creates one of each type of organism, and inspired Linneaus's effort in the 1700s to classify each type of organism in the hierarchy of kingdoms, classes, orders, families, and genera introduced in Chapter 1.

The idea that types never change eventually began to break down, however. In 1809 the biologist Jean-Baptiste de Lamarck proposed that species are not static, but have changed through time. Lamarck was the first to propose a formal theory of **evolution**—the idea that species change through time.

The pattern component of Lamarck's theory was initially based on the great chain of being, however. When he started his work on evolution, Lamarck claimed that simple organisms originate at the base of the chain by spontaneous generation (see Chapter 1) and then evolve by moving up the chain over time. Thus, Lamarckian evolution is progressive in the sense of always producing larger and more complex, or "better," species. To capture this point, biologists like to say that Lamarck turned the ladder of life into an escalator.

Lamarck also contended that species change through time via the inheritance of acquired characters. The idea here is that individuals change as they develop in response to challenges posed by the environment, and they pass on these phenotypic changes to offspring. A classic scenario is that giraffes develop long necks as they stretch to reach leaves high in treetops, and they then produce offspring with elongated necks.

In contrast, Darwin and Wallace proposed that evolution does not follow this linear, progressive pattern. (**Box 24.1** discusses why the theory of evolution by natural selection became associated primarily with Darwin's name.) What is more important, they emphasized that the process responsible for change through time is based on variation among individuals in populations. This was a radical break from the typological thinking that had dominated scientific thought since Plato. Darwin claimed that instead of being unimportant or an illusion, variation among individuals in a population was the key to understanding the nature of species. Biologists refer to this view as population thinking. A **population** consists of individuals of the same species that are living in the same area at the same time. Darwin and Wallace proposed that evolution occurs because traits vary among the individuals in a population, and

BOX 24.1 Why Darwin Gets Most of the Credit

Although Charles Darwin and Alfred Russel Wallace formulated the same explanation for how species change over time, Darwin's name is much more prominently associated with the theory of evolution because he developed the idea more thoroughly and provided massive evidence for it in *On the Origin of Species*. But historians of science speculate about whether Darwin would have published his theory at all had Wallace not threatened to scoop him (**Figure 24.2**).

Darwin wrote a paper explaining evolution by natural selection in 1842—a full 17 years before the first edition of *On the Origin of Species* came out. He never submitted the work for publication, however. Why? Darwin claimed that he needed time to document all of the arguments for and against the theory and to examine its many implications. There is probably an element of truth in this—Darwin was a remarkably thorough thinker and writer. But many historians of science argue that he held off largely out of fear. Because his theory was inconsistent with the creation story in the Bible's Book of Genesis, Darwin knew that he would be exposed to scathing criticism from religious and scientific leaders. He was also an extremely private person, had a strong religious upbringing, and was frequently in poor health. He responded to stress or personal attacks by suffering long bouts of debilitating illness. The prospect of fighting for his ideas against the most powerful men in Europe was daunting. But Wallace forced Darwin's hand.

(a) Charles Darwin

(b) Alfred Russel Wallace

FIGURE 24.2 The Codiscoverers of Evolution by Natural Selection. (a) Charles Darwin in 1840, four years after he returned from the voyage of the *Beagle* and two years before he drafted his first paper explaining evolution by natural selection. **(b)** Alfred Russel Wallace, who in 1858 independently formulated the theory of natural selection.

[(b) Alfred Russel Wallace by unknown artist, after a photograph by Thomas Sims, fl. 1860s. Reg. No.: 1765. National Portrait Gallery, London.]

Wallace was also a native of England, but he had been making a living by collecting butterflies and other natural history specimens in Malaysia and selling them to private collectors. While recuperating from a bout of malaria there in 1858, he wrote a brief article outlining the logic of evolution by natural selection. He sent a copy to Darwin, who immediately recognized that they had formulated the same explanation for how populations change through time. The two had their papers read together before the Linnean Society of London, and Darwin then rushed *On the Origin of Species* into publication a year later. The first edition sold out in a day.

Fortunately for Darwin's health, a fellow biologist and friend named Thomas Huxley publicly defended the theory against criticism, which came from both scientific and religious quarters. Darwin continued to live quietly on his estate in Down, England, and actively continued a brilliant research career.

because individuals with certain traits leave more offspring than others do.

The theory of evolution by natural selection was revolutionary because it overturned the idea that species are static and unchanging, and because it replaced typological thinking with population thinking. It also proposed a mechanism that could account for change through time and be tested through observation and experimentation. Plato and his followers emphasized the existence of fixed types; evolution by natural selection is all about change and diversity.

Now the question is, what evidence backs the claim that species are not fixed types? What data convinced biologists that the theory of evolution by natural selection is correct?

24.2 The Pattern of Evolution: Have Species Changed through Time?

In *On the Origin of Species*, Darwin repeatedly used the phrase **descent with modification** to describe evolution. By this he meant that the species existing today have descended from other, preexisting species and that species are modified, or change, through time. This view was a radical departure from the pattern of independently created and immutable species as embodied in Plato's work and in the theory of special creation. In essence, the pattern component of the theory of evolution by natural selection makes two claims about the nature of species: (1) They change through time, and (2) they are related by common ancestry. Let's consider the evidence for each of these claims in turn.

Evidence for Change through Time

When Darwin began his work, biologists and geologists had just begun to assemble and interpret the fossil record. A **fossil** is any trace of an organism that lived in the past. These traces range from bones and branches to shells, tracks or impressions, and dung (**Figure 24.3**). The **fossil record** consists of all the fossils that have been found and described in the scientific literature.

Initially, fossils were organized according to their relative ages. This was possible because most fossils are found in **sedimentary rocks**, which form from sand or mud or other materials deposited at locations such as beaches or river mouths, and because sedimentary rocks are known to form in layers. Fossils from rocks underneath other rocks were judged to be older than the fossils found above them. In this way, researchers began putting fossils in an older-to-younger sequence. They also began naming different periods of geologic time, creating the sequence of eons, epochs, and periods called the **geologic time scale**. After the discovery of radioactivity in the late 1800s, researchers used radiometric dating techniques to assign absolute ages to the relative ages in the geologic time scale. According to data from radiometric dating, Earth is about 4.6 billion years old, and the earliest signs of life appear in rocks that formed 3.4–3.8 billion years ago. Instead of being 6000 years old as some proponents of the theory of special creation claimed, life on Earth is ancient.

The fossil record continues to expand in size and quality. Several observations about this data set convinced biologists that species have indeed changed through time.

Extinction In the early nineteenth century, researchers began discovering fossil bones, leaves, and shells that were unlike structures from any known animal or plant. At first, many scientists insisted that living examples of these species would be found in unexplored regions of the globe. But as research continued and the number and diversity of fossil collections grew, the argument became less and less plausible. After Baron Georges Cuvier published a detailed analysis of an **extinct**

(a) 110-million-year-old ammonite shell

(b) 50-million-year-old bird tracks

(c) 20,000-year-old sloth dung

FIGURE 24.3 A Fossil Is *Any* Trace of an Organism That Lived in the Past. In addition to **(a)** body parts such as shells or bones or branches, fossils may consist of **(b)** tracks or impressions, or even **(c)** pieces of dung.

species—that is, a species that no longer exists—called the Irish "elk" in 1812, most scientists accepted extinction as a reality. This gigantic deer was judged to be too large to have escaped discovery and too distinctive to be classified as a large-bodied population of an existing species.

Advocates of the theory of special creation argued that the fossil species were victims of the flood at the time of Noah. Darwin, in contrast, interpreted them as evidence that species are not static, immutable entities, unchanged since the moment of special creation. His reasoning was that if species have gone extinct, then the array of species living on Earth has changed through time. Recent analyses of the fossil record support the claim that more species have gone extinct than exist today. The data also indicate that species have gone extinct continuously throughout Earth's history—not just in one or even a few catastrophic events.

Transitional Forms Long before Darwin published his theory, researchers reported striking resemblances between the fossils found in the rocks underlying certain regions and the living species found in the same geographic areas. The pattern was so widespread that it became known as the "law of succession." The general observation was that extinct species in the fossil record were succeeded, in the same region, by similar species (**Figure 24.4a**). Early in the nineteenth century, the pattern was simply reported and not interpreted. But later, Darwin pointed out that it provided strong evidence in favor of the hypothesis that species had changed through time. His idea was that the extinct forms and living forms were related—that they represented ancestors and descendants.

As the fossil record improved, researchers discovered transitional forms that broadened the scope of the law of succession. A **transitional form** is a fossil species with traits that are intermediate between those of older and younger species. For example, intensive work over the past several decades has yielded fossils that document a gradual change over time from land-dwelling mammals that had limbs to ocean-dwelling mammals that had reduced limbs or no limbs (**Figure 24.4b**). All of the species in this sequence have distinctive types of ear bones that researchers use to identify whales. The oldest whale fossils found to date are from fox-sized animals that had eyes located at the tops of their heads. Based on this observation and the fact that the fossils were found in rocks that form only in ocean deposits, biologists suggest that the earliest whales were semiaquatic animals not unlike hippopotamuses. Over the subsequent 12 million years, the fossil record shows that the limbs of whale species became more reduced. These observations support the hypothesis that whales gradually became more strictly aquatic and more like today's whales in appearance and lifestyle. Whale species have clearly changed through time.

(a) Living species "succeed" fossil species in the same region.

Fossil sloth from South America

Present-day sloth from South America

(b) Transitional forms during the evolution of whales

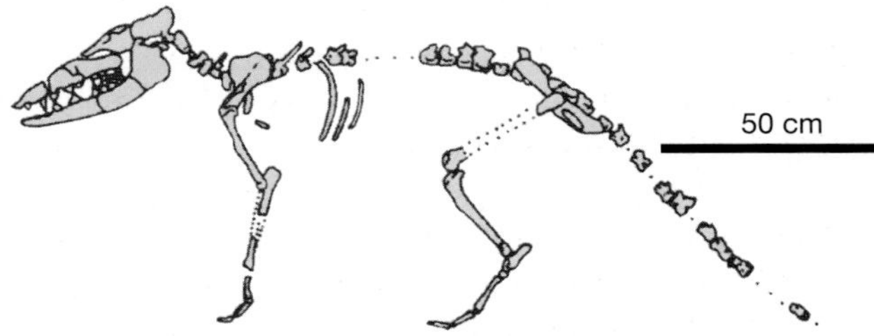

Pakicetus, about 50 million years old

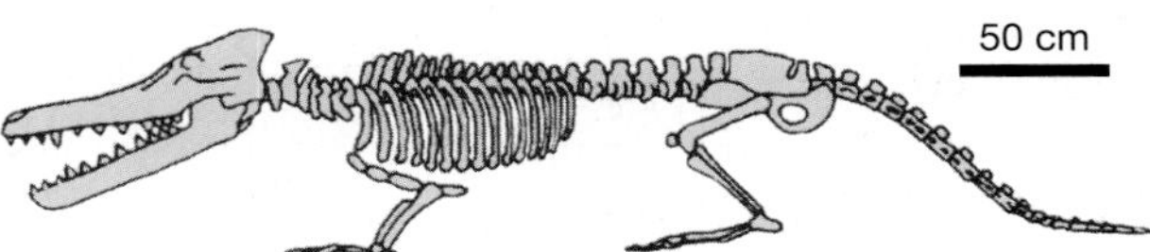

Ambulocetus, about 49 million years old

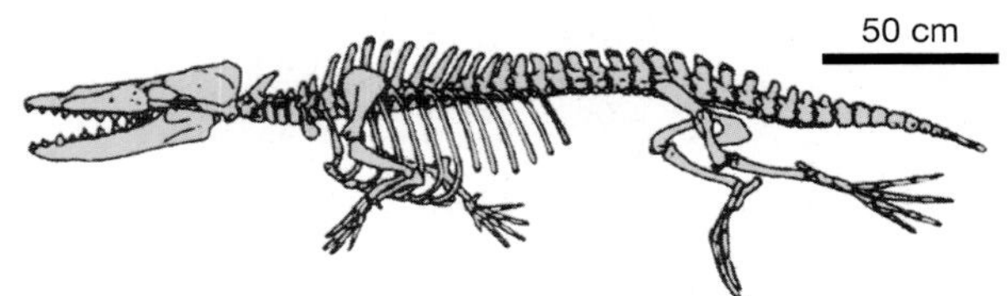

Rhodocetus, about 47 million years old

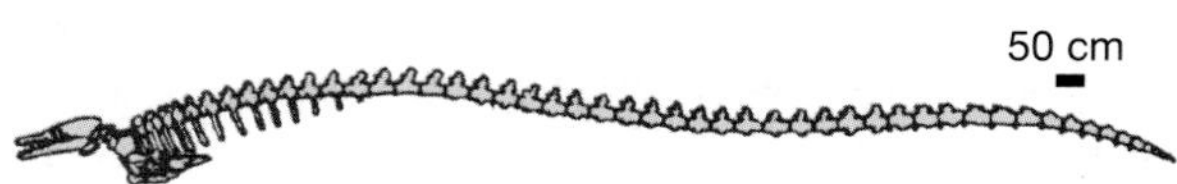

Basilosaurus, about 38 million years old

FIGURE 24.4 Evidence That Species Have Changed through Time. **(a)** Fossil and living sloths are found only in Central America and South America. Darwin argued that living species are descended from ancestors that lived in the same area. **(b)** Transitional forms document the changes that occurred as whales evolved from terrestrial mammals to the aquatic mammals of today.

● **QUESTION** How would these observations be explained under the theory of special creation?

Similar sequences of transitional forms document changes that led to the evolution of feathers and flight in birds, stomata and vascular tissue in plants, upright posture and large brains in humans, jaws in vertebrates (animals with backbones), limbs in amphibians and other vertebrates, the loss of limbs in snakes, and other traits. Each of these transitional forms provides strong evidence for change through time.

Vestigial Traits Darwin was the first to provide a widely accepted interpretation of vestigial traits. A **vestigial trait** is a reduced or incompletely developed structure that has no function or reduced function, but it is clearly similar to functioning organs or structures in closely related species.

Biologists have documented thousands of examples of vestigial traits. The genomes of humans and other organisms contain hundreds of pseudogenes—the functionless DNA sequences introduced in Chapter 20. Bowhead whales and rubber boas have tiny hip and leg bones that do not help them swim or slither; ostriches and kiwis have reduced wings and cannot fly; blind cave-dwelling fish still have eye sockets. The human appendix is a reduced version of the cecum—an organ found in other vertebrates that functions in digestion. Monkeys and many other primates have long tails; but our coccyx, illustrated in **Figure 24.5a**, is too tiny to help us maintain balance. Many mammals, including primates, are able to erect their hair when they are cold or excited. But our sparse fur does little to keep us warm, and goose bumps are largely ineffective in signaling our emotional state (**Figure 24.5b**).

The existence of vestigial traits is inconsistent with the theory of special creation, which maintains that species were perfectly designed by a supernatural being and that the characteristics of species are static. Instead, vestigial traits are evidence that the characteristics of species have changed over time.

Current Examples Biologists have documented hundreds of contemporary populations that are changing in response to changes in their environment. Bacteria have become resistant to drugs; insects have become resistant to pesticides; weedy plants have become resistant to herbicides. Section 24.4 provides a detailed analysis of research on two examples of evolution in action.

Biologists have also studied dozens of cases where new species are forming, right before our eyes. You'll be able to analyze data from one of these research projects in Chapter 26.

To summarize, change through time continues and can be measured directly. Evidence from the fossil record and living species indicates that life is ancient, that species have changed through the course of Earth's history, and that species continue to change. The take-home message is that species are dynamic—not static and unchanging, as claimed by the theory of special creation.

(a) The human tailbone is a vestigial trait.

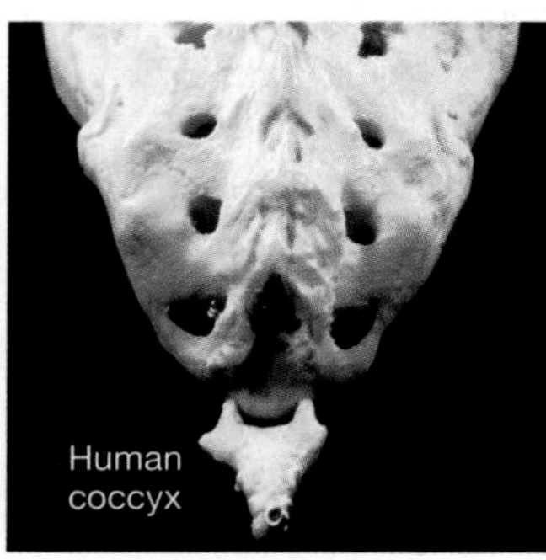

(b) Goose bumps are a vestigial trait.

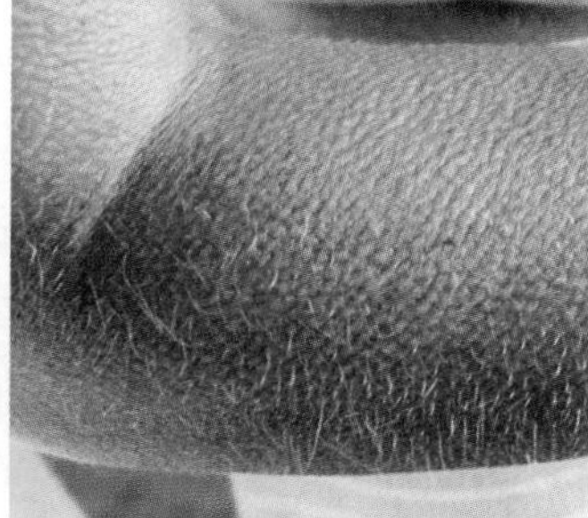

FIGURE 24.5 Vestigial Traits Are Reduced Versions of Traits in Other Species. (a) The tailbone and **(b)** goose bumps are human traits that have reduced function. They are similar to larger, fully functional structures in other species.

QUESTION How would these observations be explained if evolution occurred via inheritance of acquired characters?

Evidence That Species Are Related

Data from the fossil record and contemporary species refute the hypothesis that species are immutable. What about the claim that species were created independently—meaning that they are unrelated to each other?

Geographic Relationships Charles Darwin began to realize that species are related by common ancestry, just as individuals within a family are, during a five-year voyage he took aboard the English naval ship HMS *Beagle*. While fulfilling its mission to explore and map the coast of South America, the *Beagle* spent considerable time in the Galápagos Islands off the coast of present-day Ecuador. Darwin had taken over the role of ship's naturalist and gathered extensive collections of the plants and animals found in these islands. Among the birds he collected were what came to be known as the Galápagos mockingbirds, pictured in **Figure 24.6a**.

Several years after Darwin had returned to England, a naturalist friend in London pointed out that the mockingbirds Dar-

(a) Four mockingbird species on the Galápagos islands

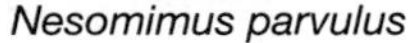

Nesomimus parvulus

Nesomimus melanotis

Nesomimus trifasciatus

Nesomimus macdonaldi

(b) Darwin reasoned that they share a common ancestor.

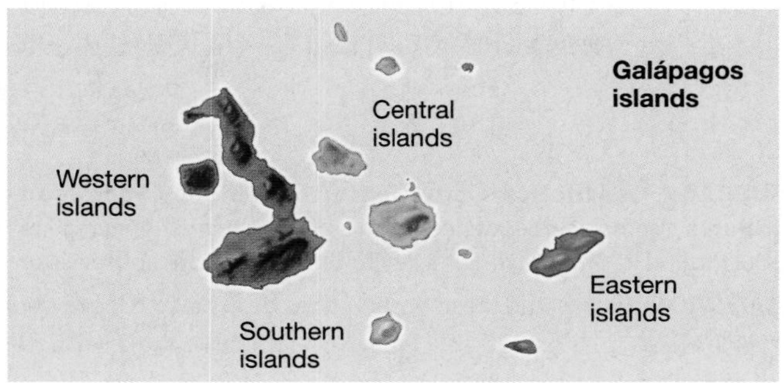

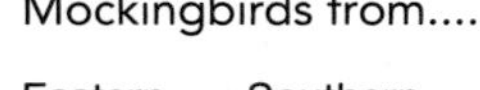

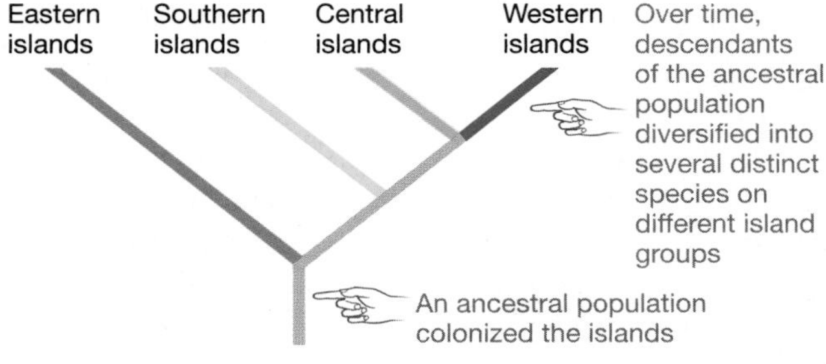

FIGURE 24.6 Close Relationships among Island Forms Argue for Shared Ancestry. (a) Darwin collected mockingbirds from several islands of the Galápagos. **(b)** Phylogeny illustrating Darwin's explanation for why mockingbirds from different islands are similar yet distinct.

QUESTION How would these observations be explained under the theory of special creation?

win had collected on different islands were distinct species, based on differences in coloration and beak size and shape. This struck Darwin as remarkable. Why would species that inhabit neighboring islands be so similar, yet clearly distinct? This turns out to be a very general pattern: In island groups across the globe, it is routine to find similar species on neighboring islands.

Darwin realized that this pattern—puzzling when examined as a product of special creation—made perfect sense when interpreted in the context of evolution, or descent with modification. He proposed that the mockingbirds were similar because they had descended from the same common ancestor. If so, then the mockingbird species are part of a phylogeny—a family tree of populations or species. Further, the mockingbirds can be placed on a phylogenetic tree, a branching diagram that describes the ancestor–descendant relationships among species, much as a genealogy describes the ancestor–descendant relationships among individual humans. **Figure 24.6b** illustrates this idea. Darwin's hypothesis was that instead of being created independently, mockingbird populations that colonized different islands had changed through time and formed new species. The presence of similar species in the same geographic area is still considered strong evidence that species are related by common ancestry—specifically, that their common ancestor lived in the same region.

Homology Translated literally, *homology* means "the study of likeness." When biologists first began to study the anatomy of humans and other vertebrates, they were struck by the remarkable similarity of their skeletons, muscles, and other structures. But because the biologists who did these early studies were advocates of the theory of special creation, they could not explain why striking similarities existed among certain organisms but not others. Today, biologists recognize that **homology** is a similarity that exists in species descended from a common ancestor. Human hair and dog fur are homologous. Humans have hair and dogs have hair because they share a common ancestor that was a mammal and also had hair.

Homology can be recognized and studied at three interacting levels. The most fundamental of these levels is **genetic homology**—a similarity in the DNA sequences of different species. As an example, consider the *eyeless* gene in fruit flies and the *Aniridia* gene in humans. Both genes act in determining where eyes will develop. The genes are so similar in DNA sequence that they code for proteins that are nearly identical in

Gene:	Amino acid sequence (single-letter abbreviations):
Aniridia (Human)	LQRNRTSFTQEQIEALEKEFERTHYPDVFARERLAAKIDLPEARIQVWFSNRRAKWRREE
eyeless (Fruit fly)	LQRNRTSFTNDQIDSLEKEFERTHYPDVFARERLAGKIGLPEARIQVWFSNRRAKWRREE

Only six of the 60 amino acids in these sequences are different. The two sequences are 90% identical.

FIGURE 24.7 Genetic Homology: Genes from Different Species May Be Similar in DNA Sequence or Other Attributes. Amino acid sequences from a portion of the *Aniridia* gene product found in humans and the *eyeless* gene product found in *Drosophila*. For a key to the single-letter abbreviations used for the amino acids, see Chapter 3.

● **QUESTION** How would these observations be explained if evolution occurred via inheritance of acquired characters?

amino acid sequence (**Figure 24.7**). This observation is interesting because eye structure is so different in the two species—fruit flies have a compound eye with many lenses, and humans have a camera eye with a single lens. To explain this observation, biologists propose that fruit flies and humans descended from a common ancestor that had a gene similar to *eyeless* and *Aniridia*, and that this gene was involved in the formation of a simple light-gathering organ. The structure of the organ diverged as insects and mammals evolved, but the same gene remained responsible for where eyes are located.

Although Chapter 18 and Chapter 20 considered other examples of genetic homologies, the most remarkable of all genetic homologies is the genetic code introduced in Chapter 15. Except for one or two codons in a handful of species, the same 64 mRNA codons specify the same amino acids in all organisms that have been studied. To explain the existence of the universal genetic code, biologists hypothesize that today's code also existed in the common ancestor of all organisms alive today. Similarly, all organisms living today have a plasma membrane consisting of a phospholipid bilayer with interspersed proteins, transcribe the information coded in DNA to RNA via RNA polymerase, use ribosomes to synthesize proteins, employ ATP as an energy currency, and make copies of their genome via DNA polymerase. Like the genetic code, these traits undoubtedly existed in the cell that gave rise to all the species alive today.

The second level where biologists analyze homology was introduced in Chapters 21–23. A **developmental homology** is a similarity in embryonic traits. Developmental homologies are routinely observed in the overall **morphology,** or form, of embryos and in the fate of particular embryonic tissues.

Figure 24.8 illustrates the strong general resemblance among the embryos of vertebrates. Early in development, structures called gill pouches and tails that extend past the anus form in chicks, humans, and cats. Later in development, gill pouches are lost in all three species and tails are lost in humans. In fish, however, the gill pouches stay intact and give rise to functioning gills in adults. To explain this observation, biologists hypothesize that gill pouches and tails exist in chicks, humans, and cats because they existed in the fishlike species that was the common ancestor of today's fish, birds, and mammals. Embryonic gill pouches are a vestigial trait in chicks, humans, and cats; embryonic tails are a vestigial trait in humans.

Developmental homologies are also observed at the level of specific tissues. Even though the structure of the adult jaw is different in fish and mammals, the same group of embryonic cells develops into the jaw structure in both groups. This observation is logical if fish and mammals descended from the same common ancestor, and if this ancestor also had a jaw that developed from the same population of embryonic cells.

Developmental homologies are due to homologous genes and give rise to **structural homology**—similarities in adult morphology. A classic example is the common structural plan observed in the limbs of vertebrates (**Figure 24.9**). In Darwin's own words, "What could be more curious than that the hand of a man, formed for grasping, that of a mole for digging, the leg of the horse, the paddle of the porpoise, and the wing of the bat, should all be constructed on the same pattern, and should include the same bones, in the same relative positions?" Darwin raised the question because an engineer would

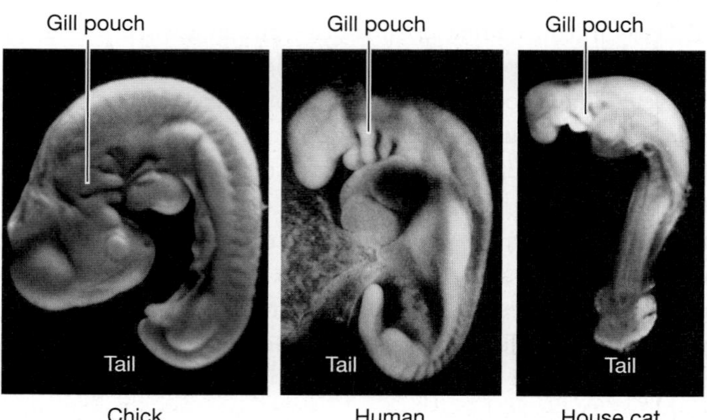

FIGURE 24.8 Developmental Homology: Structures That Appear Early in Development Are Similar. The early embryonic stages of a chick, a human, and a cat, showing a strong resemblance.

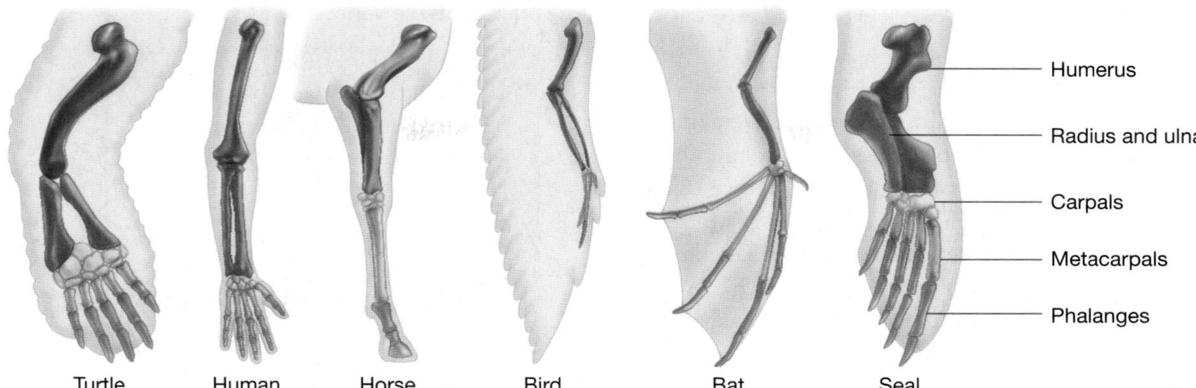

FIGURE 24.9 Structural Homology: Limbs with Different Functions Have the Same Underlying Structure. Even though their function varies, all vertebrate limbs are modifications of the same number and arrangement of bones. Darwin interpreted structural homologies like these as a product of descent with modification. (These limbs are not drawn to scale.)

QUESTION What is the relationship among genetic homologies, developmental homologies, and structural homologies?

never use the same underlying pattern to design the structure of a grasping tool, a digging implement, a walking device, a propeller, and a wing. But if all mammals descended from a common ancestor, and if that ancestor had a limb with the same basic plan shown in Figure 24.9, then it would be logical to observe that its descendants had modified forms of the same design.

The general point is that, in many cases, traits are similar in different species because the species in question are related to each other by common descent. If species were created independently of one another, these types of similarities would not occur.

Evolution Is Change through Time

Biologists draw upon data from several sources to challenge the hypothesis that species are immutable and were created independently. The data support the idea that species have descended, with modification, from a common ancestor. **Table 24.1** summarizes this evidence.

By the late 1800s, the vast majority of biologists were convinced that the pattern component of the theory of evolution was valid and that the theory of special creation was incorrect. This was long before genetic homologies had been described, before many of the most important transitional forms were discovered, and before contemporary examples of populations changing through time and new species being created had been documented. Among biologists, controversy over the fact of evolution ended more than 120 years ago.

As you evaluate the evidence supporting the pattern component of the theory of evolution, it's important to recognize that no single observation or experiment instantly "proved" the fact of evolution and swept aside belief in special creation. Rather, Darwin and others argued that the pattern called evolution was much more consistent with the data than was the pattern predicted by special creation. Descent with modification was a more successful and powerful scientific theory because it explained observations—such as vestigial traits and the close relationships among species on neighboring islands—that special creation could not.

What about the process component of the theory of evolution by natural selection? If the limbs of bats and humans were not created independently and recently, how did they come to be?

TABLE **24.1** **Evidence for Evolution**

Prediction 1: Species Are Not Static, but Change through Time

- Most species have gone extinct.
- Fossil (extinct) species frequently resemble living species found in the same area.
- Transitional forms document change in traits through time.
- Earth is ancient.
- Vestigial traits are common.
- Populations and species can be observed changing today.

Prediction 2: Species Are Related, Not Independent

- Closely related species often live in the same geographic area.
- Homologous traits are common and exist at three levels:
  1. genetic (gene structure and the genetic code)
  2. developmental (embryonic structures and processes)
  3. structural (morphological traits)

24.3 The Process of Evolution: How Does Natural Selection Work?

Darwin's greatest contribution did not lie in recognizing the fact of evolution. Lamarck and other researchers had already proposed evolution as a pattern in nature long before Darwin began his work. Instead, Darwin's crucial insight lay in recognizing a process, called **natural selection**, that could explain the pattern of descent with modification.

In his original formulation, Darwin broke the process of evolution by natural selection into four simple postulates, or steps in a logical sequence:

1. The individual organisms that make up a population vary in the traits they possess, such as their size and shape.
2. Some of the trait differences are heritable, meaning that they are passed on to offspring genetically. For example, tall parents may tend to have tall offspring.
3. In each generation, many more offspring are produced than can possibly survive. Thus, only some individuals in the population survive long enough to produce offspring; and among the individuals that produce offspring, some will produce more than others.
4. The subset of individuals that survive best and produce the most offspring is not a random sample of the population. Instead, individuals with certain heritable traits are more likely to survive and reproduce. Natural selection occurs when individuals with certain characteristics produce more offspring than do individuals without those characteristics.

Because the selected traits are passed on to offspring, the frequency of the selected traits increases from one generation to the next. **Evolution**—a change in the genetic characteristics of a population over time—is simply an outcome of these four steps.

In studying these steps, you should realize that variation among individuals is essential if evolution is to occur. You should also recognize that Darwin had to introduce population thinking into biology because it is populations that change over time when evolution occurs. To come up with these postulates and understand their consequences, Darwin had to think in a revolutionary way.

Today, biologists usually condense Darwin's four postulates into two statements that communicate the essence of evolution by natural selection more forcefully: Evolution by natural selection occurs when (1) heritable variation leads to (2) differential success in survival and reproduction.

To illustrate this condensed version of Darwin's logic, consider a population of 10 moths where differences in wing coloration are due primarily to two alleles of a single gene (**Figure 24.10a**). These alleles are called A_1 and A_2. The moths are diploid so there are a total of 20 alleles present in the population, before selection starts. The A gene product is a protein

(a) If heritable variation...

Color varies among individuals primarily because of differences in their genotype

(b) ... leads to differential success...

Birds find and eat many more dark-winged moths than light-winged moths

(c) ... then evolution results.

	Frequency of A_1 allele	Frequency of A_2 allele
Start	0.5 (10 of 20 alleles present)	0.5 (10 of 20 alleles present)
End	0.625 (5 of 8 alleles present)	0.375 (3 of 8 alleles present)

Allele frequencies have changed in the population of surviving moths

FIGURE 24.10 Evolution by Natural Selection Occurs When Heritable Variation Leads to Differential Success. See text for explanation. Note that this example is not hypothetical. The allele frequency changes diagrammed here have occurred independently in populations of peppered moths (*Biston betularia*) in England and several locations in North America.

QUESTION Why is it important that similar changes in allele frequencies have been observed in different populations of this species, independently?

that is involved in the synthesis or deposition of a dark pigment called melanin. Individuals with the genotype A_1A_1 have a light gray coloration, while individuals with the A_1A_2 or A_2A_2 genotype are black. Now, it's important to recognize that some of the variation in wing color in this population is due to differences in the environment that the individuals experience, and some is due to differences in the alleles they carry. For example, not all individuals with the A_2A_2 genotype will be exactly the same shade of black, because each individual was exposed to different temperatures and had different amounts of nutrients available when its wings were developing and the A gene product was active. But because the A_1 and A_2 alleles have such different effects on wing color, there is heritable variation in the trait within the population.

These moths are active at night and spend the day resting on tree trunks and branches, where they are hunted by birds. In an environment where trees with light-colored bark are common, birds can find and eat dark-winged individuals much more readily than they can find and eat light-winged individuals. Because predation by birds causes natural selection on wing color, there is differential success: In this environment, light-winged individuals survive better than dark-winged individuals. In our example, only 4 individuals survived—meaning that only 8 alleles are now present in the population (**Figure 24.10b**). As a result, evolution occurs. ● Evolution is defined as a change in allele frequencies in a population over time. In this case, the frequency of the A_1 allele increases in this population over time (**Figure 24.10c**).

To explain the process of natural selection, Darwin referred to successful individuals as "more fit" than other individuals. He gave the word *fitness* a definition different from its everyday English usage. **Darwinian fitness** is the ability of an individual to produce offspring, relative to that ability in other individuals in the population. This is a measurable quantity. Researchers study populations in the lab or in the field and estimate the relative fitness of each individual by counting how many offspring it produces relative to other individuals. In environments where most trees have light gray bark, light-winged moths have higher fitness than dark-winged moths. But in environments dominated by trees with dark bark, black-winged moths have higher fitness than light-winged moths.

The concept of fitness, in turn, provides a compact way of formally defining adaptation. ● The biological meaning of adaptation, like the biological meaning of fitness, is quite different from its normal English usage. In biology, an **adaptation** is a heritable trait that increases the fitness of an individual in a particular environment relative to individuals lacking the trait. Adaptations increase fitness—the ability to produce offspring. Light-colored wings are an adaptation in environments where most trees have gray bark.

To summarize, evolution by natural selection occurs when heritable variation in traits leads to differential success in survival and reproduction. ● If you understand this concept, you should be able to make a figure analogous to Figure 24.10, illustrating what would happen to allele frequencies in the population if dark-barked trees were much more abundant than light-barked trees. You should also be able to explain what would happen if all of the variation in wing coloration were due to differences in nutrition or temperature that the individuals experienced—meaning that they all had the same genotype.

To help you get a thorough understanding of how evolution by natural selection works, Section 24.4 is devoted to data—specifically, to recent studies of how natural selection works in actual populations. Biologists accept Darwin's theory not only because of its explanatory power but also because evolution by natural selection has been observed directly.

24.4 Evolution in Action: Recent Research on Natural Selection

Darwin's theory of evolution by natural selection is testable. If the theory is correct, biologists should be able to test the validity of each of Darwin's postulates and actually observe evolution in natural populations.

This section summarizes two examples in which evolution by natural selection has been, or is being, observed in nature. Literally hundreds of other case studies are available, involving a wide variety of traits and organisms. To begin, let's explore the evolution of drug resistance—one of the great challenges facing today's biomedical researchers and physicians.

How Did *Mycobacterium tuberculosis* Become Resistant to Antibiotics?

Mycobacterium tuberculosis, the bacterium that causes tuberculosis, or TB, has long been a scourge of humankind. TB was responsible for almost 25 percent of all deaths in New York City in 1804; in nineteenth-century Paris, the figure was closer to 33 percent. To put these numbers in perspective, consider that all types of cancer, combined, currently account for about 30 percent of the deaths that occur in the United States. TB was once as great a public health issue as cancer is now.

Although tuberculosis still kills more adults than any other viral or bacterial disease in the world, TB attracted relatively little attention in the industrialized nations between about 1950 and 1990. During that time, TB was primarily a disease of developing nations.

The decline of tuberculosis in Western Europe, North America, Japan, Korea, and Australia is one of the great triumphs of modern medicine. In these countries, sanitation, nutrition, and general living conditions began to improve dramatically in the early twentieth century. When people are healthy and well nourished, their immune systems work well enough to stop most *M. tuberculosis* infections quickly—before the infection can harm the individual and before the bacteria can be transmitted to a new host. In addition, antibiotics such as rifampin started to become available in the industrialized countries in the early 1950s. These drugs allowed physicians to stop even advanced infections and saved millions of lives.

In the late 1980s, however, rates of *M. tuberculosis* infection surged in many countries, and in 1993 the World Health Organization (WHO) declared TB a global health emergency. Physicians were particularly alarmed because the strains of *M. tuberculosis* responsible for the increase were largely or completely resistant to rifampin and other antibiotics that were once extremely effective. How and why did the evolution of drug resistance occur? The case of a single patient—a young man who lived in Baltimore—will illustrate what is happening all over the world.

The story begins when the individual was admitted to the hospital with fever and coughing. Chest X-rays, followed by bacterial cultures of fluid ejected from the lungs, showed that he had an active TB infection. He was given several antibiotics for 6 weeks, followed by twice-weekly doses of rifampin and isoniazid for an additional 33 weeks. Ten months after therapy started, bacterial cultures from his chest fluid indicated no *M. tuberculosis* cells. His chest X-rays were also normal. The antibiotics seemed to have cleared the infection.

Just two months after the TB tests proved normal, however, the young man was readmitted to the hospital with a fever, severe cough, and labored breathing. Despite being treated with a variety of antibiotics, including rifampin, he died of respiratory failure 10 days later. Samples of material from his lungs showed that *M. tuberculosis* was again growing actively there. But this time the bacterial cells were completely resistant to rifampin.

Drug-resistant bacteria had killed this patient. Where did they come from? Is it possible that a strain that was resistant to antibiotic treatment evolved *within* him? To answer this question, a research team analyzed DNA from the drug-resistant strain and compared it with stored DNA from *M. tuberculosis* cells that had been isolated a year earlier from the same patient. After examining extensive stretches from each genome, the biologists were able to find only one difference: a point mutation in a gene called *rpoB*. This gene codes for a component of the enzyme RNA polymerase. Recall from Chapter 16 that RNA polymerase transcribes DNA to mRNA, and that a point mutation is a single base change in DNA. In this case, the mutation changed a cytosine to a thymine, altering the normal codon TCG to a mutant one, TTG. As a result, the RNA polymerase produced by the drug-resistant strain had leucine instead of serine at the 153rd amino acid in the polypeptide chain.

This result is meaningful. The drug that was being used to treat the patient works by binding to the RNA polymerase of *M. tuberculosis*. When the drug enters an *M. tuberculosis* cell and binds to RNA polymerase, it interferes with transcription. If sufficient quantities of rifampin are present for long enough and if the drug binds tightly, bacterial cells will not be able to make proteins efficiently and produce few offspring. But apparently the substitution of a leucine for a serine prevents rifampin from binding efficiently. Consequently, cells with the C → T mutation continue to produce offspring efficiently even in the presence of the drug.

These results suggest that a chain of events led to this patient's death (**Figure 24.11**). The researchers hypothesized that by chance, one or a few of the cells present early in the course of the infection happened to have an *rpoB* gene with the C → T mutation. Under normal conditions, mutant forms of RNA polymerase do not work as well as the more common form, so cells with the C → T mutation would not produce many offspring and would stay at low frequency—even while the overall population grew to the point of inducing symptoms that sent the young man to the hospital.

At that point, therapy with rifampin began. In response, cells in the population with normal RNA polymerase began to grow much more slowly or to die outright. As a result, the overall bacterial population declined in size so drastically that the patient appeared to be cured—his symptoms began to disappear. But cells with the C → T mutation had an advantage in the new environment. They began to grow more rapidly than the normal cells and continued to increase in number after therapy ended. Eventually the *M. tuberculosis* population regained its former abundance, and the patient's symptoms reappeared. However, drug-resistant cells now dominated the population. This is why the second round of rifampin therapy was futile. If health-care workers or the patient's family had contracted TB from him, rifampin therapy would have been useless on them, too, and the disease would have continued to spread.

Does this sequence of events mean that evolution by natural selection occurred? One way of answering this question is to review Darwin's four postulates and determine whether each was tested and verified:

1. *Did variation exist in the population?* The answer is yes. Due to mutation, both resistant and nonresistant strains of TB were present prior to administration of the drug. Most *M. tuberculosis* populations, in fact, exhibit variation for the trait; studies on cultured *M. tuberculosis* show that a mutation conferring resistance to rifampin is present in one out of every 10^7 to 10^8 cells.

2. *Was this variation heritable?* The answer is yes. The researchers showed that the variation in the phenotypes of the two strains—from drug susceptibility to drug resistance—was due to variation in their genotypes. Because the mutant *rpoB* gene is passed on to daughter cells when a *Mycobacterium* replicates, the allele and the phenotype it produces—drug resistance—are passed on to offspring.

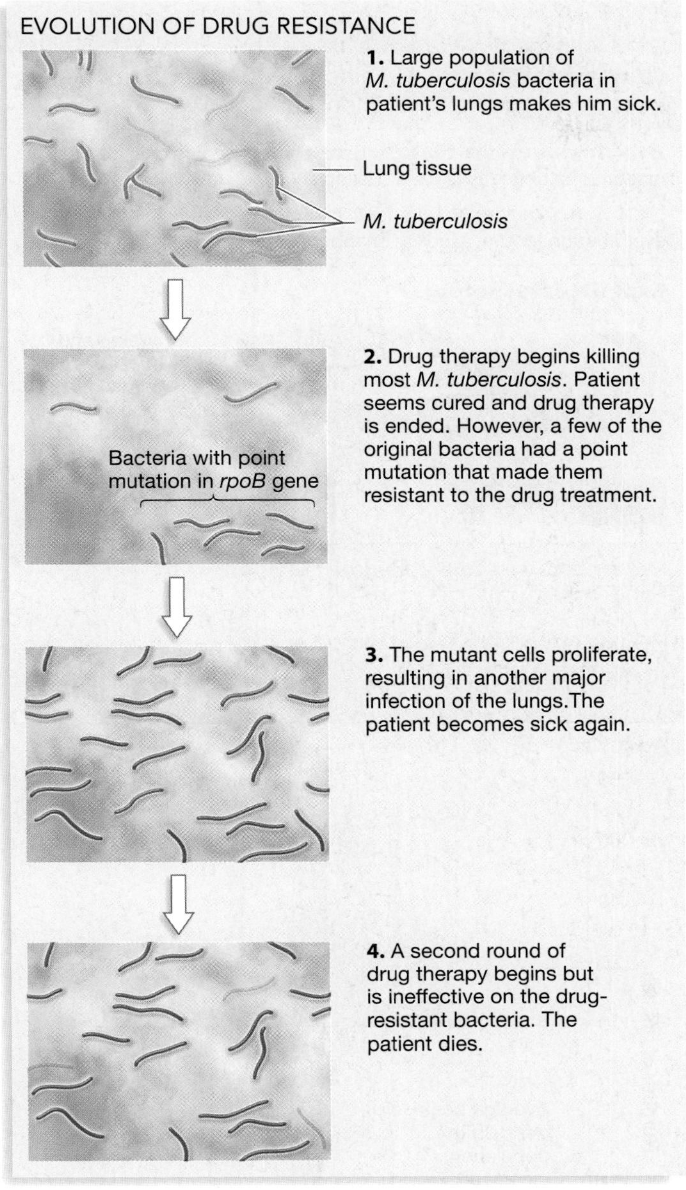

FIGURE 24.11 Alleles That Confer Drug Resistance Increase in Frequency When Drugs Are Used.

QUESTION In most individuals, the immune system is able to eliminate the few bacteria that remain at step 2. This individual had AIDS, however, so his immune system was compromised. Why did step 3 occur? If he had transmitted the infection to another person at step 3 or step 4, would they respond to drug therapy?

3. *Was there variation in reproductive success?* That is, did some *M. tuberculosis* individuals survive better and leave more offspring than other *M. tuberculosis* individuals? The answer is yes. Only a tiny fraction of M. tuberculosis cells in the patient survived the first round of antibiotics long enough to reproduce—so few that, after the initial therapy, his chest X-ray was normal and his fluid sample contained no *M. tuberculosis* cells. This happened because a small number of drug-resistant bacterial cells were able to survive and keep reproducing after the onset of drug treatment.

4. *Did selection occur?* That is, did a nonrandom subset of the population produce the most offspring? The answer is yes. The *M. tuberculosis* population present early in the infection was different from the *M. tuberculosis* population present at the end. This could have occurred only if cells with the drug-resistant allele had higher reproductive success when rifampin was present than did cells with the normal allele. *M. tuberculosis* individuals with the mutant *rpoB* gene had higher fitness in an environment where rifampin was present. The mutant allele produces a protein that is an adaptation when the cell's environment contains the antibiotic.

This study verified all four postulates and confirmed that evolution by natural selection had occurred. The *M. tuberculosis* population evolved because the mutant *rpoB* allele increased in frequency. The individual cells themselves did not evolve, however. When natural selection occurred, the individual cells did not change through time; they simply survived or died, or produced more or fewer offspring. This is a fundamentally important point: Natural selection acts on individuals, because individuals experience differential success. But only populations evolve. Allele frequencies change in populations, not in individuals. Understanding evolution by natural selection requires population thinking—not typological thinking.

The events just reviewed have occurred many times in other patients. Recent surveys indicate that drug-resistant strains now account for about 10 percent of the *M. tuberculosis*–causing infections throughout the world. And the emergence of drug resistance in TB is far from unusual. Resistance to a wide variety of insecticides, fungicides, antibiotics, antiviral drugs, and herbicides has evolved in hundreds of insects, fungi, bacteria, viruses, and plants. In many cases, the specific mutations that lead to a fitness advantage and the spread of the resistance alleles are known.

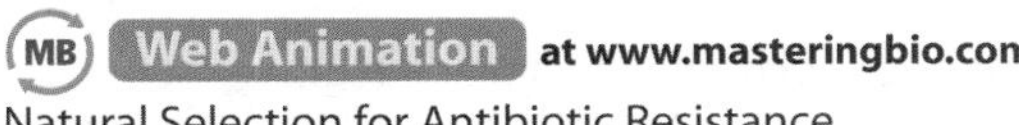

at www.masteringbio.com
Natural Selection for Antibiotic Resistance

Why Are Beak Size, Beak Shape, and Body Size Changing in Galápagos Finches?

The TB example is particularly satisfying, because the molecular basis of both heritable variation and differential success is understood. It is one of many examples of natural selection and rapid evolution induced by drugs, pesticides, herbicides, or other human-caused changes in the environment. But can biologists study evolution in response to natural environmental change—when humans are not involved? The answer is yes. As an example of how this work is done, let's review research led by Peter and Rosemary Grant. These biologists have been

FIGURE 24.12 The Medium Ground Finch Is a Seed-Eater.

investigating changes in beak size, beak shape, and body size that have occurred in finches native to the Galápagos Islands.

The medium ground finch, pictured in **Figure 24.12**, makes its living by eating seeds. Finches crack seeds with their beaks. For over three decades, the population of medium ground finches on Isle Daphne Major of the Galápagos has been studied intensively by the Grants' team. Because Daphne Major is small—about the size of 80 football fields—the researchers have been able to catch, weigh, and measure all individuals and mark each one with a unique combination of colored leg bands.

Early studies of the finch population established that beak size and shape and body size vary among individuals, and that beak morphology and body size are heritable. Stated another way, parents with particularly deep beaks tend to have offspring with deep beaks. Large parents also tend to have large offspring. Beak size and shape and body size are traits with heritable variation.

Not long after the team had established these results, a dramatic selection event occurred. In the annual wet season of 1977, Daphne Major received just 24 mm of rain instead of the 130 mm that normally falls. During the drought, few plants were able to produce seeds, and 84 percent (about 660 individuals) of the medium ground finch population disappeared.

Two observations support the hypothesis that most or all of these individuals died of starvation. The researchers found a total of 38 dead birds, and all were emaciated. Further, none of the missing individuals were spotted on nearby islands, and none reappeared once the drought had ended and food supplies returned to normal.

The research team realized that the die-off presented an opportunity to study natural selection. The change in the environment produced what biologists call a **natural experiment**. Instead of comparing groups created by direct manipulation under controlled conditions, natural experiments allow researchers to compare treatment groups created by an unplanned, natural change in conditions. In this case, the Grant's team could compare the population before and after the drought. Were the survivors different from nonsurvivors? When the biologists analyzed the characteristics of each group, they found that survivors tended to have much deeper beaks than did the birds that died (**Figure 24.13**). This was an impor-

Experiment

Question: Did natural selection on ground finches occur when the environment changed?

Hypothesis to be tested: Beak characteristics changed in response to changes in food availability.

Null hypothesis: No changes in beak characteristics occurred, even though food availability changed.

Experimental setup:

Weigh and measure all birds in the population before and after the drought.

Prediction of natural selection hypothesis:

Prediction of null hypothesis (no selection):

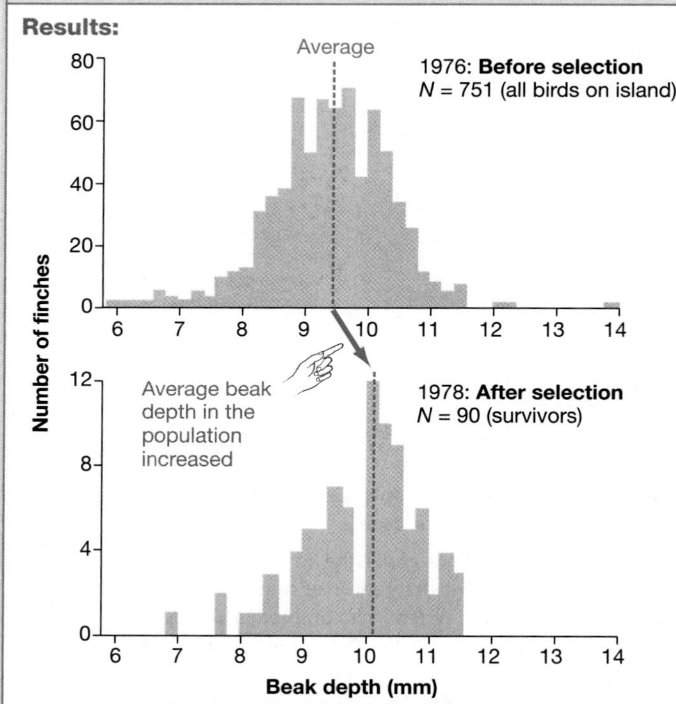

Conclusion: Natural selection occurred. The characteristics of the population have changed.

FIGURE 24.13 A Natural Experiment: Changes in a Medium Ground Finch Population in Response to a Change in the Environment (a Drought). The histograms in the Results section show the distribution of beak depth in medium ground finches on Daphne Major before and after the drought of 1977. *N* is the sample size.

● **QUESTION** Why was the sample size so much smaller in 1978?

● **EXERCISE** Fill in the predictions made by the two hypotheses.

tant finding, because the type of seeds available to the finches had changed dramatically as the drought continued. At the drought's peak, the tough fruits of a plant called *Tribulus cistoides* served as the finches' primary food source. These fruits are so difficult to crack that they are ignored in years when food supplies are normal. Grant's group hypothesized that individuals with particularly large and deep beaks were more likely to crack these fruits efficiently enough to survive.

At this point, the Grants had shown that natural selection led to an increase in average beak depth in the population. When breeding resumed in 1978, the offspring that were produced had beaks that were half a millimeter deeper, on average, than those in the population that existed before the drought. This result confirmed that evolution had occurred. In only one generation, natural selection had led to a measurable change in the characteristics of the population. Alleles that led to the development of deep beaks must have increased in frequency. Large, deep beaks were an adaptation for cracking large fruits and seeds. More recent work, reviewed in **Box 24.2** on page 499, has identified one of the genes in which allele frequencies may have changed during this event.

In 1983, however, the environment changed again. Over a seven-month period, a total of 1359 mm of rain fell. Plant growth was luxuriant, and finches fed primarily on small, soft seeds that were being produced in abundance. During this interval, small individuals with small, pointed beaks had exceptionally high reproductive success. As a result, the characteristics of the population changed again. In fact, the Grants have documented continued evolution in response to continued changes in the environment. **Figure 24.14** documents changes that have occurred in average body size, beak size, and beak shape over the past 30 years. Although beak size showed no net change when the start and end of this interval are compared, beak shape changed dramatically. On average, finch beaks got much pointier over the course of the study. In addition, overall body size got smaller. Long-term studies such as this have been powerful, because they have succeeded in documenting natural selection in response to changes in the environment.

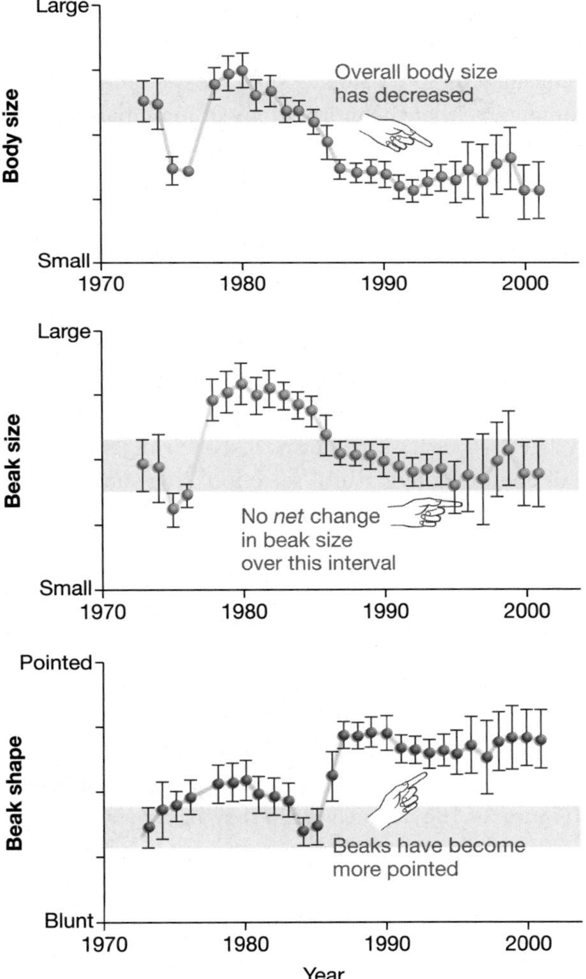

FIGURE 24.14 Body Size, Beak Size, and Beak Shape in Finches Changed over a 30-Year Interval.

EXERCISE Label the drought in 1977 and the wet year in 1983. Circle years when (1) average body size increased, (2) average beak size declined, and (3) beaks became pointier.

EXERCISE The Grants recently published more data on beak size (the middle graph). In 2005 and 2006, average beak size dropped to about the first hash mark above "Small" on the vertical axis. Place these data points on the graph. Do the new data change your interpretation?

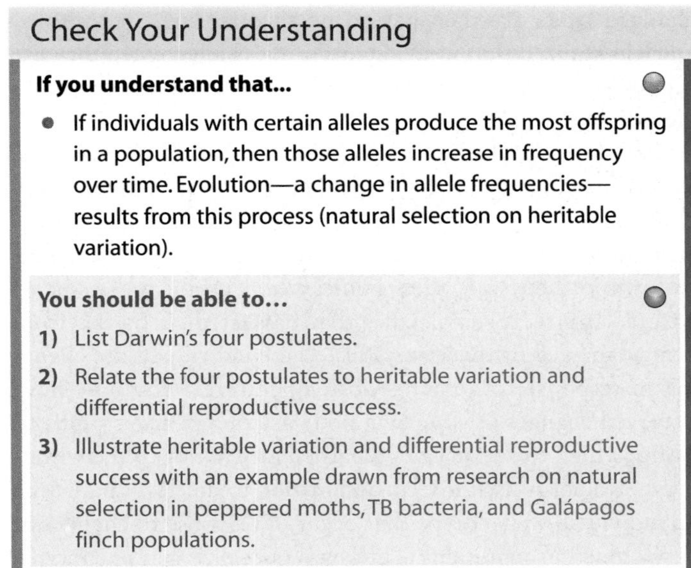

Check Your Understanding

If you understand that...

- If individuals with certain alleles produce the most offspring in a population, then those alleles increase in frequency over time. Evolution—a change in allele frequencies—results from this process (natural selection on heritable variation).

You should be able to...

1) List Darwin's four postulates.
2) Relate the four postulates to heritable variation and differential reproductive success.
3) Illustrate heritable variation and differential reproductive success with an example drawn from research on natural selection in peppered moths, TB bacteria, and Galápagos finch populations.

24.5 The Nature of Natural Selection and Adaptation

Natural selection appears to be a simple process, but appearances can be deceiving. Research has shown that it is often misunderstood. To help clarify how natural selection works, let's consider some of the more common misconceptions about natural selection in light of data on drug resistance in the TB bacterium and changes in finch populations.

Selection Acts on Individuals, but Evolutionary Change Occurs in Populations

Perhaps the most important point to clarify about natural selection is that during the process, individuals do not change—only the population does. During the drought, the beaks of individual finches did not become deeper. Rather, the average beak depth in the population increased over time, because deep-beaked individuals produced more offspring than shallow-beaked individuals did. Natural selection acted on individuals, but the evolutionary change occurred in the characteristics of the population.

In the same way, individual bacterial cells did not change when rifampin was introduced to their environment. Each *M. tuberculosis* cell had the same polymerase alleles all its life. But because the mutant allele increased in frequency over time, the characteristics of the bacterial population changed.

This point should make sense, given that evolution is defined as changes in allele frequencies. An individual's allele frequencies cannot change over time—it has the alleles it was born with all its life.

This point also highlights a sharp contrast between evolution by natural selection and evolution by the inheritance of acquired characters—the hypothesis promoted by Jean-Baptiste de Lamarck. Lamarck proposed that individuals change in response to challenges posed by the environment and that the changed traits are then passed on to offspring. In contrast, Darwin realized that individuals do not change when they are selected—they simply produce more offspring than other individuals do.

The issue is tricky because individuals often *do* change in response to changes in the environment. For example, wood frogs native to northern North America are exposed to extremely cold temperatures as they overwinter. When ice begins to form in their skin, their bodies begin producing a sort of natural antifreeze—molecules that protect their tissues from being damaged by the ice crystals. These individuals are changing in response to a change in temperature.[1] You may have observed changes in your own body as you got accustomed to living at high elevation or in a particularly hot or cold environment. Biologists use the term **acclimation** to describe changes in an individual's phenotype that occur in response to changes in environmental conditions. The key is to realize that phenotypic changes due to acclimation are not passed on to offspring. As a result, they cannot cause evolution. ● If you understand this concept, you should be able to explain the difference between the biological definition of adaptation and its use in everyday English, and then explain the difference between acclimation and adaptation.

[1]In some species of frogs, so much extracellular fluid freezes during cold snaps that individuals appear to be frozen solid. Their hearts also stop beating. When temperatures warm in the spring, their hearts start beating again, their tissues thaw, and they resume normal activities.

Evolution Is Not Goal-Directed or Progressive

It is tempting to think that evolution by natural selection is goal directed. For example, you might hear a fellow student say that *M. tuberculosis* cells "wanted" or "needed" the mutant, drug-resistant allele so that they could survive and continue to reproduce in an environment that included rifampin. This does not happen. The mutation that created the mutant allele occurred randomly, due to an error during DNA synthesis, and it just happened to be advantageous when the environment changed. Adaptations do not occur because organisms want or need them.

It is also tempting to think that evolution by natural selection is progressive—meaning organisms have gotten "better" over time. (In this context, better usually means more complex.) It is true that the groups appearing later in the fossil record are often more morphologically complex or "advanced" than closely related groups that appeared earlier. Flowering plants are considered more complex than mosses, and most biologists would agree that the morphology of mammals is more complex than that of amphioxus. But there is nothing predetermined or absolute about this tendency. In fact, complex traits are routinely lost or simplified over time as a result of evolution by natural selection. Populations that become parasitic are particularly prone to this trend. Tapeworms, for example, lack a mouth and digestive system. As parasites that live in the intestines of humans and other mammals, they simply absorb nutrients directly from their environment, across their plasma membranes. But tapeworms evolved from species with a sophisticated digestive tract. Tapeworms lost their digestive tract as a result of evolution by natural selection. In a similar vein, snakes evolved from lizard-like ancestors that had legs (see Chapter 21).

The nonprogressive nature of evolution by natural selection contrasts sharply with Lamarck's conception of the evolutionary process, in which organisms progress over time to higher and higher levels on a chain of being. Under Aristotle's and Lamarck's hypothesis, it is sensible to refer to "higher" and "lower" organisms (**Figure 24.15a**). But under evolution by natural selection, there is no such thing as a higher or lower organism (**Figure 24.15b**). Green algae may be a more ancient group than flowering plants, but neither group is higher or lower than the other. Green algae simply have a different suite of adaptations than do flowering plants, so they thrive in different types of environments.

The general message here is that all populations have evolved by natural selection based on their ability to gather resources and produce offspring. All organisms are adapted to their environment. Evolution by natural selection is not progressive, so no organism is any "higher" than any other organism.

To drive this point home, recall what happened when torrential rains fell on Daphne Major. Instead of continuing to

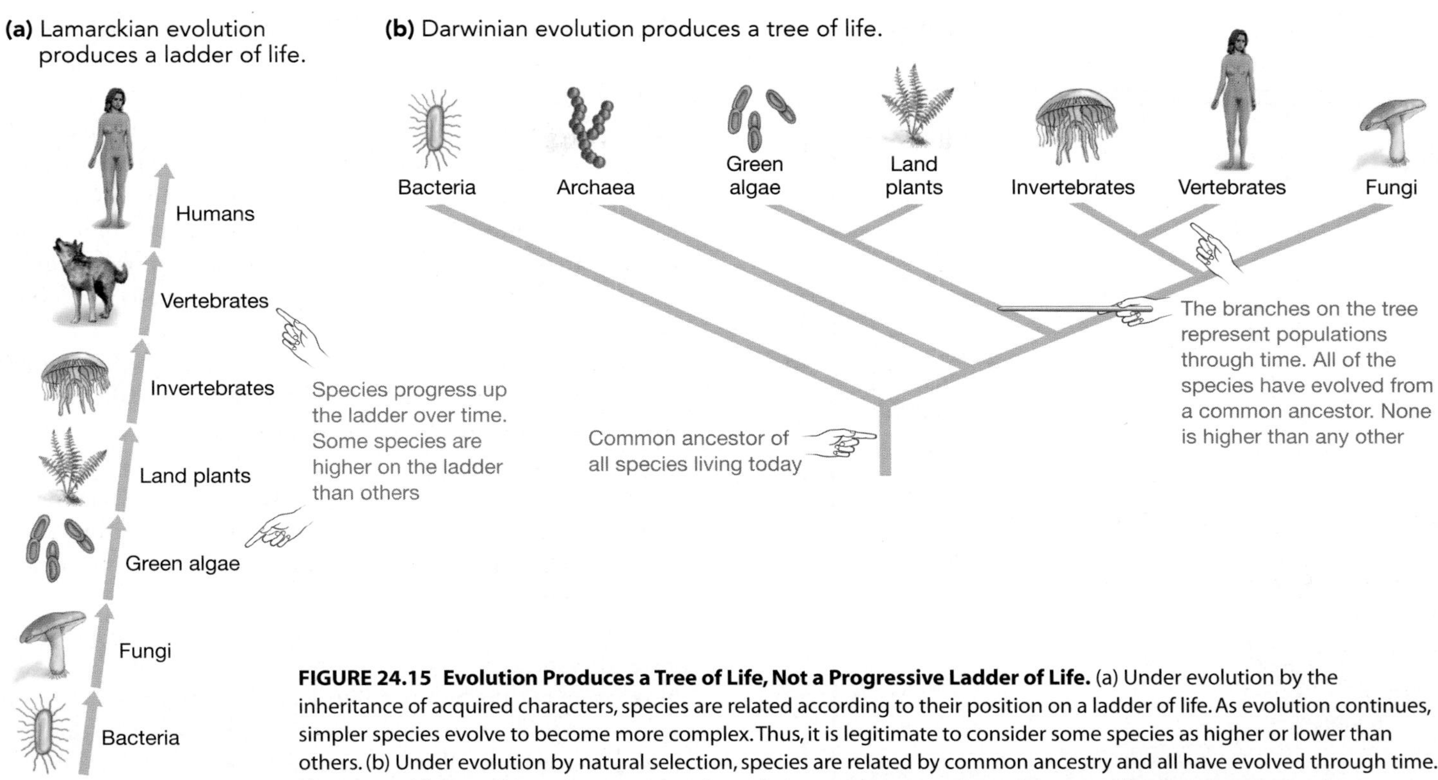

FIGURE 24.15 Evolution Produces a Tree of Life, Not a Progressive Ladder of Life. (a) Under evolution by the inheritance of acquired characters, species are related according to their position on a ladder of life. As evolution continues, simpler species evolve to become more complex. Thus, it is legitimate to consider some species as higher or lower than others. (b) Under evolution by natural selection, species are related by common ancestry and all have evolved through time. None is any higher or lower than any other. As evolution continues, species may become simpler or more complex, depending on what traits are favored by the environment. (For help in reading evolutionary trees, see **BioSkills 2**.)

increase in size, the average beak size of the finch population declined. Natural selection did not produce a progression of ever-larger beaks. Instead, the population simply responded to whatever change happened to occur in the environment. In many bacterial and viral populations, the frequency of drug-resistant individuals has declined dramatically when the drug in question was discontinued.

Natural selection is not goal directed or progressive. It simply favors individuals that happen to be better adapted to the environment existing at the time.

Animals Do Not Do Things for the Good of the Species

Consider the widely circulated story that rodents called lemmings sacrifice themselves for the good of their species. The story claims that when lemming populations are high, overgrazing is so extensive that the entire species is threatened with starvation and extinction. In response, some individuals throw themselves into the sea and drown. This lowers the overall population size and allows the vegetation to recover enough to save the species. Even though individuals suffer, the good-of-the-species hypothesis maintains that the behavior evolved because the group benefits.

The lemming suicide story is false. Although lemmings do disperse from areas of high population density in order to find habitats with higher food availability, they do not throw themselves into the sea. The individual wearing the inner tube in **Figure 24.16** represents the reason lemmings do not kill themselves for the good of the species. The cartoon assumes that certain alleles predispose lemmings to sacrifice themselves for others. But the inner tube represents what biologists call a "cheater," or "selfish," allele. Individuals with self-sacrificing alleles die and do not produce offspring. But individuals with selfish, cheater alleles survive and produce offspring. As a result, selfish alleles increase in frequency while self-sacrificing alleles decrease in frequency. Thus, it is not possible for individuals to sacrifice themselves for the good of the species. No instance of purely self-sacrificing behavior—where the individual received no fitness benefit in return—has ever been recorded in nature. Chapter 52 provides additional details on this point.

Not All Traits Are Adaptive

Although organisms are often exquisitely adapted to their environment, adaptation is far from perfect. Vestigial traits such as the human coccyx (tailbone), goose bumps, and appendix do not increase the fitness of individuals with those traits. The structures are not adaptive. They exist simply because they were present in the ancestral population.

FIGURE 24.16 Self-Sacrificing Behavior Cannot Evolve if "Cheater" Alleles Exist. Most individuals in this population have alleles that lead to self-sacrificing behavior and result in death; the individual with the inner tube has alleles that prevent self-sacrificing behavior.

QUESTION Why is the individual with the inner tube smiling?

Vestigial traits are not the only types of structures with no function. Some adult traits exist as holdovers from structures that appear early in development. For example, human males have rudimentary mammary glands. The structures are not adaptive. They exist only because nipples form in the human embryo before sex hormones begin directing the development of male organs instead of female organs.

Perhaps the best example of nonadaptive traits involves evolutionary changes in DNA sequences. Recall from Chapter 16 that mutation may change a base in the third position of a codon without changing the amino acid sequence of the protein encoded by that gene. Changes such as these are said to be neutral, or silent. They occur because of the redundancy of the genetic code (see Chapter 16). Neutral changes in DNA sequences are extremely common, yet not adaptive.

The general point here is that not all traits are adaptive. Evolution by natural selection does not lead to "perfection." Besides carrying an array of traits that have no function, the adaptations that organisms have are constrained in a variety of important ways.

Genetic Constraints The Grants' team analyzed data on the characteristics of finches that survived the 1977 drought, and the team made an interesting observation: Although individuals with deep beaks survived better than individuals with shallow beaks, birds with particularly narrow beaks survived better than individuals with wider beaks. This observation made sense because finches crack *Tribulus* fruits by twisting them. Narrow beaks concentrate the twisting force more efficiently than wider beaks, so they are especially useful for cracking the fruits. But narrower beaks did not evolve in the population. To explain why, the biologists noted that parents with deep beaks tend to have offspring with beaks that are both deep and wide. This is a common pattern. Many alleles that affect body size have an effect on all aspects of size—not just one structure or dimension. As a result, selection for increased beak depth overrode selection for narrow beaks, even though a deep and narrow beak would have been more advantageous.

The general point here is that selection was not able to optimize all aspects of a trait. In the case of the finches, wider beaks were not the best possible beak shape for individuals living in an arid habitat. Wider beaks evolved anyway, due to a type of constraint called a **genetic correlation.** Genetic correlations occur because of pleiotropy (see Chapter 13). In this case, selection on alleles for one trait (increased beak depth) caused a correlated, though suboptimal, increase in another trait (beak width).

Genetic correlations are not the only genetic constraint on adaptation. Lack of genetic variation is also important. Consider that salamanders have the ability to regrow severed limbs. Some eels and sharks can sense electric fields. Birds can sense magnetic fields and see ultraviolet light. Even though it is possible that these traits would confer increased reproductive success in humans, they do not exist—because the requisite genes are lacking.

Fitness Trade-offs In everyday English, the term *trade-off* refers to a compromise between competing goals. It is difficult to design a car that is both large and fuel efficient, a bicycle that is both rugged and light, or a plane that is both fast and maneuverable.

In nature, selection occurs in the context of fitness trade-offs. A **fitness trade-off** is a compromise between traits, in terms of how those traits perform in the environment. During the drought in the Galápagos, for example, medium ground finches with large bodies had an advantage because they won fights over the few remaining sources of seeds. But individuals with large bodies also require large amounts of food to maintain their mass; they also tend to be slower and less nimble than smaller individuals. When food is short, large individuals are more prone to starvation. Even if large size is advantageous in an environment, there is always counteracting selection that prevents individuals from getting even bigger.

BOX 24.2 Which Allele Frequencies Changed When Finch Beaks Changed?

Researchers investigating the evolution of *M. tuberculosis* populations have strong evidence that alleles of the *rpoB* gene change in frequency when the antibiotic rifampin is introduced into the environment. In this and many other cases of evolution in response to drugs or herbicides or pesticides, biologists know exactly which base pairs in specific genes are favored by natural selection.

In cases like the evolution of finch beaks, however, it is much more difficult to understand the exact genes and alleles involved. Characteristics like body size and body shape are polygenic, meaning that many genes—each one exerting a relatively small effect—influence the trait (see Chapter 13). And because most work in molecular genetics has been done on organisms like *E. coli* and *Drosophila melanogaster*, we know relatively little about which genes might be causing evolutionary change in organisms like birds, fish, and mammals.

The situation is starting to change rapidly, however, thanks to advances in the field called evolution and development, or evo-devo (see Chapter 21). As an example, consider recent work done in Cliff Tabin's lab. These researchers normally work on topics like the genetic control of limb development in chickens and other vertebrates. They focus on understanding the cell-cell signals and regulatory transcription factors that regulate development. But developmental biologists are aware that different alleles exist for each of the cell-cell signals and regulatory transcription factors they identify. These alleles represent heritable variation that can influence the evolution of traits like limb size and shape.

Interesting things began to happen when Tabin's group started interacting with Peter and Rosemary Grant. The researchers began studying beak development in an array of Galápagos finch species. More specifically, they looked for variation in the pattern of expression of cell-cell signals that had already been identified as important in the development of chickens. They struck pay dirt when they did in situ hybridizations showing where a cell-cell signal called *Bmp4* is expressed. As **Figure 24.17a** shows, there is a strong correlation between the amount of *Bmp4* expression when beaks are developing in young Galápagos finches and the width and depth of adult beaks. And when the researchers experimentally increased *Bmp4* expression in young chickens, they found that beaks got wider and deeper (**Figure 24.17b**). Based on these data, the researchers suggest that alleles that increase *Bmp4* expression were selected for during the drought of 1977. *Bmp4* may have been one of several genes in which allele frequencies changed and caused evolution.

(a) Natural variation in *Bmp4* expression: Finches

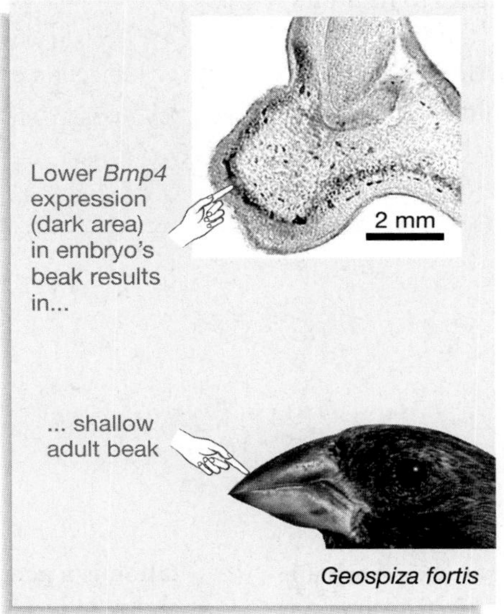

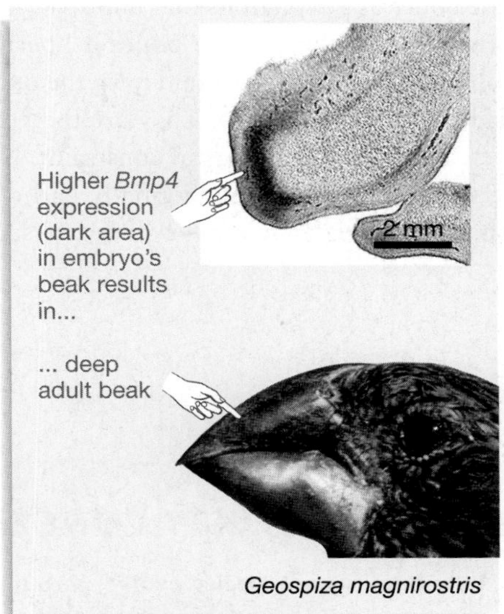

(b) Experimental variation in *Bmp4* expression: Chickens

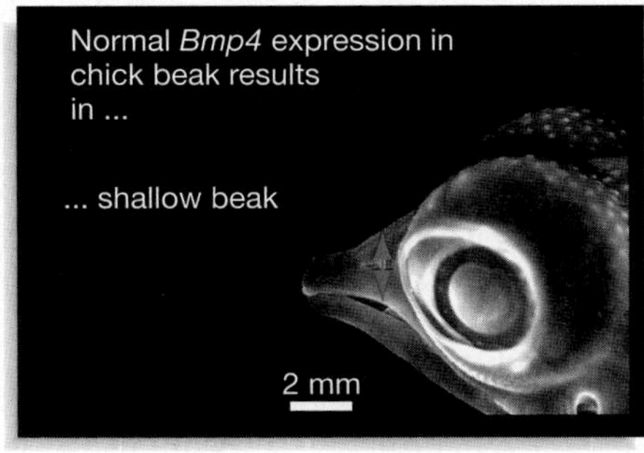

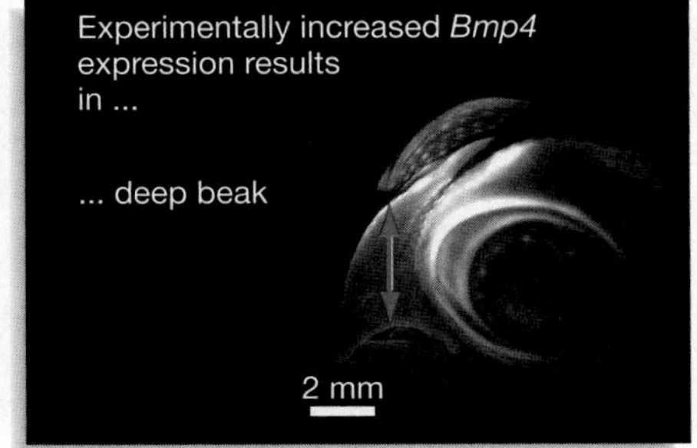

FIGURE 24.17 Changes in *Bmp4* Expression Change Beak Depth and Width. (a) The micrographs are in situ hybridizations (see Chapter 21) showing the location and extent of *Bmp4* expression in young *Geospiza fortis* and *G. magnirostris*. In these and four other species that were investigated, the amount of Bmp4 protein produced correlates with the depth and width of the adult beak. **(b)** When *Bmp4* expression in developing chickens is increased experimentally, beak depth and width increase.

Biologists have documented trade-offs between the size of eggs or seeds that an individual makes and the number of offspring it can produce, between rapid growth and long lifespan, and between bright coloration and tendency to attract predators. The message of this research is simple: Because selection acts on many traits at once, every adaptation is a compromise.

Historical Constraints In addition to being constrained by genetic correlations, lack of genetic variation, and fitness trade-offs, adaptations are constrained by history. The reason is simple: All traits have evolved from previously existing traits. For example, the tiny hammer, anvil, and stirrup bones found in your middle ear evolved from bones that were part of the jaw and braincase in the ancestors of mammals. These bones now function in the transmission and amplification of sound from your outer ear to your inner ear. Biologists routinely interpret these bones as adaptations that improve your ability to hear air-borne sounds. But are the bones a "perfect" solution to the problem of transmitting sound from the outside of the ear to the inside? The answer is no. They are the best solution possible, given an important historical constraint. Natural selection was acting on structures that originally had a very different function. Other vertebrates have different structures involved in transmitting sound to the ear. In at least some cases, those structures may be more efficient than our hammer, anvil, and stirrup.

To summarize, not all traits are adaptive, and even adaptive traits are constrained by genetic and historical factors. In addition, natural selection is not the only process that causes evolutionary change. Chapter 25 introduces three other processes that change allele frequencies over time. Compared with natural selection, these processes have very different consequences.

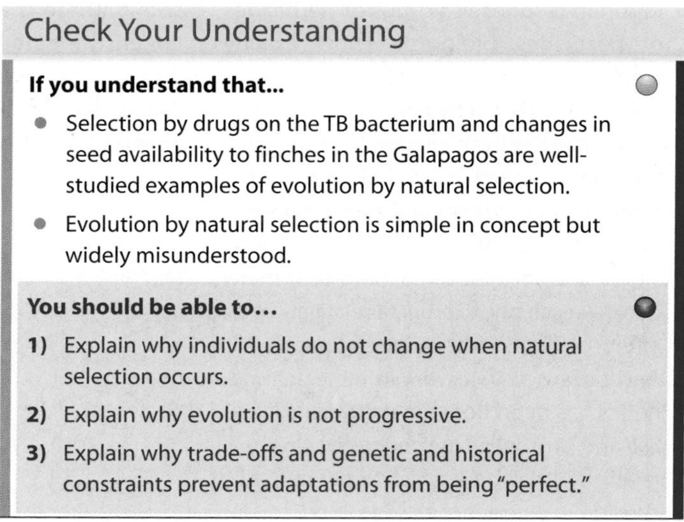
Check Your Understanding

If you understand that...

- Selection by drugs on the TB bacterium and changes in seed availability to finches in the Galapagos are well-studied examples of evolution by natural selection.
- Evolution by natural selection is simple in concept but widely misunderstood.

You should be able to...

1) Explain why individuals do not change when natural selection occurs.
2) Explain why evolution is not progressive.
3) Explain why trade-offs and genetic and historical constraints prevent adaptations from being "perfect."

Chapter Review

SUMMARY OF KEY CONCEPTS

Populations and species evolve, meaning that their heritable characteristics change through time. More precisely, evolution is defined as changes in allele frequencies over time.

Evidence for the fact of evolution—that species are related through shared ancestry and have changed over time—has accumulated over the past years. An array of observations is inconsistent with the alternative theory that species were formed instantaneously, recently, and independently, and have remained unchanged through time. These observations include the geographic proximity of closely related species such as the Galápagos mockingbirds; the existence of structural, developmental, and genetic homologies; the near universality of the genetic code; resemblances of modern to fossil forms; transitional fossils; the fact of extinction; and the presence of vestigial traits. These data support the hypothesis that populations and species change through time.

You should be able to predict how changes in finch populations and *Mycobacterium* populations would be explained under the theory of special creation and under evolution by inheritance of acquired characters.

Natural selection occurs when individuals with certain alleles produce the most surviving offspring in a population. An adaptation is a genetically based trait that increases an individual's ability to produce offspring in a particular environment.

Natural selection occurs whenever genetically based differences among individuals lead to differences in their ability to reproduce—when heritable variation leads to reproductive success in survival and reproduction. Alleles or traits that increase the reproductive success of an individual are said to increase the individual's fitness. A trait that leads to higher fitness, relative to individuals without the trait, is an adaptation. If a particular allele increases fitness and leads to adaptation, the allele will increase in frequency in the population.

Evolution is an outcome of natural selection. It is defined as changes in allele frequencies that occur in populations from one generation to the next. Evolution by natural selection has been confirmed by a wide variety of studies and has long been considered to be the central organizing principle of biology.

You should be able to explain the difference between the biological and everyday English definitions of the words *fitness* and *adaptation*.

MB Web Animation at www.masteringbio.com
Natural Selection for Antibiotic Resistance

Evolution by natural selection is not progressive, and it does not change the characteristics of the individuals that are selected—it changes only the characteristics of the population. Animals do not do things for the good of the species, and not all traits are adaptive. All adaptations are constrained by trade-offs and genetic and historical factors.

Individuals that are naturally selected are not changed by the process—they simply produce more offspring than other individuals do. Traits that increase in frequency under natural selection do so because they improve fitness, not because they are necessarily larger or more complex. Because natural selection acts only on existing traits, it does not lead to perfection.

You should be able to give an example of an adaptation—like the large brains of *Homo sapiens* or the ability of falcons to fly very fast—and discuss how it is constrained. **You should also be able to** explain why self-sacrificing alleles cannot increase in frequency and why individuals do not change when selection acts on them.

QUESTIONS

Test Your Knowledge

1. How can Darwinian fitness be estimated?
   a. Document how long different individuals in a population survive.
   b. Count the number of offspring produced by different individuals in a population.
   c. Determine which individuals are strongest.
   d. Determine which phenotype is the most common one in a given population.
2. Why are some traits considered vestigial?
   a. They improve the fitness of an individual who bears them, compared with the fitness of individuals without those traits.
   b. They change in response to environmental influences.
   c. They existed a long time in the past.
   d. They are reduced in size, complexity, and function compared with traits in related species.
3. What is an adaptation?
   a. a trait that improves the fitness of its bearer, compared with individuals without the trait
   b. a trait that changes in response to environmental influences within the individual's lifetime
   c. an ancestral trait—one that was modified to form the trait observed today
   d. the ability to produce offspring
4. Why does the presence of extinct and transitional forms in the fossil record support the pattern component of the theory of evolution by natural selection?
   a. It supports the hypothesis that individuals change over time.
   b. It supports the hypothesis that weaker species are eliminated by natural selection.
   c. It supports the hypothesis that species evolve to become more complex and better adapted over time.
   d. It supports the hypothesis that species have changed through time.
5. Why are homologous traits similar?
   a. They are derived from a common ancestor.
   b. They are derived from different ancestors.
   c. They result from convergent evolution.
   d. Their appearance, structure, or development is similar.
6. Which of the following statements is correct?
   a. When individuals change in response to challenges from the environment, their altered traits are passed on to offspring.
   b. Species are created independently of each other and do not change over time.
   c. Populations—not individuals—change when natural selection occurs.
   d. The Earth is young, and most of today's landforms were created during the floods at the time of Noah.

Test Your Knowledge answers: 1. b; 2. d; 3. a; 4. d; 5. a; 6. c

Test Your Understanding

Answers are available at www.masteringbio.com

1. Compare and contrast the theory of evolution by natural selection and the theory of special creation and evolution by inheritance of acquired characters. What are the central claims of each theory? What testable predictions does each make?
2. Some biologists encapsulate evolution by natural selection with the phrase "mutation proposes, selection disposes." Explain what they mean, using the formal terms introduced in this chapter.
3. Review the section on the evolution of drug resistance in *Mycobacterium tuberculosis*.
   - In *M. tuberculosis*, how does heritable variation arise for the trait of drug resistance?
   - What evidence do researchers have that a drug-resistant strain evolved in the patient analyzed in their study, instead of having been transmitted from another infected individual?
   - If the antibiotic rifampin were banned, would the mutant *rpoB* gene have lower or higher fitness in the new environment? Would strains carrying the mutation continue to increase in frequency in *M. tuberculosis* populations?
4. Compare and contrast typological thinking with population thinking. Why was Darwin's emphasis on the importance of variation among individuals so crucial to his theory, and why was it a revolutionary idea in Western science?
5. The evidence supporting the pattern component of the theory of evolution can be criticized on the grounds that it is indirect. For example, no one has directly observed the formation of a vestigial trait over time. Due to the indirect nature of the evidence, it could be argued that structural and genetic homologies are coincidental and do not result from common ancestry. Is indirect evidence for a scientific theory legitimate? Are you persuaded that descent with modification is the best explanation available for the data reviewed in Section 24.2? Why or why not?
6. Why isn't evolution by natural selection progressive? Why don't the strongest individuals in a population always produce the most offspring?

Applying Concepts to New Situations

Answers are available at www.masteringbio.com

1. The geneticist James Crow wrote that successful scientific theories have the following characteristics: (1) They explain otherwise puzzling observations; (2) they provide connections between otherwise disparate observations; (3) they make predictions that can be tested; and (4) they are heuristic, meaning that they open up new avenues of theory and experimentation. Crow added two other elements that he considered important on a personal, emotional level: (5) They should be elegant, in the sense of being simple and powerful; and (6) they should have an element of surprise.

   How well does the theory of evolution by natural selection fulfill these six criteria? Think of a theory you've been introduced to in another science course—for example, the atomic theory or the germ theory of disease—and evaluate it by using this list.
2. The average height of humans has increased steadily for the past 100 years in industrialized nations. This trait has clearly changed over time. Most physicians and human geneticists believe that the change is due to better nutrition and a reduced incidence of disease. Has human height evolved?
3. Genome sequencing projects may dramatically affect how biologists analyze evolutionary changes in quantitative traits. For example, suppose that the genomes of many living humans are sequenced and that genomes could be sequenced from many people who lived 100 years ago. (That might be possible with preserved tissue.) If 20 genes have been shown to influence height, how could you use the sequence data from these genes to test the hypothesis that human height has evolved in response to natural selection?
4. In some human populations, individuals tan in response to exposure to sunlight. Tanning is an acclimation response to a short-term change in the environment. It is adaptive because it prevents sunburn. The ability to tan varies among individuals in these populations, however. Is the ability to tan an acclimation or an adaptation? Explain your logic.

Evolutionary Processes

25

A male raggiana bird of paradise, left, displays for a female, right. His long, colorful feathers and dramatic behavior result from sexual selection—a process introduced in this chapter.

KEY CONCEPTS

- The Hardy-Weinberg principle acts as a null hypothesis when researchers want to test whether evolution or nonrandom mating is occurring at a particular gene.
- Each of the four evolutionary mechanisms has different consequences. Only natural selection produces adaptation. Genetic drift causes random fluctuations in allele frequencies. Gene flow equalizes allele frequencies between populations. Mutation introduces new alleles.
- Inbreeding changes genotype frequencies but does not change allele frequencies.
- Sexual selection leads to the evolution of traits that help individuals attract mates. It is usually stronger on males than on females.

Chapter 24 defined evolution as a change in allele frequencies. One of the key concepts from that chapter was that even though natural selection acts on individuals, evolutionary change occurs in **populations**. A population is a group of individuals from the same species that live in the same area and regularly interbreed.

Natural selection is not the only process that causes evolution, however. There are actually four mechanisms that shift allele frequencies in populations:

1. *Natural selection* increases the frequency of certain alleles—the ones that contribute to success in survival and reproduction.
2. *Genetic drift* causes allele frequencies to change randomly. In some cases, drift may even cause alleles that decrease fitness to increase in frequency.
3. *Gene flow* occurs when individuals leave one population, join another, and breed. Allele frequencies may change when gene flow occurs, because arriving individuals introduce alleles to their new population and departing individuals remove alleles from their old population.
4. *Mutation* modifies allele frequencies by continually introducing new alleles. The alleles created by mutation may be beneficial or detrimental or have no effect on fitness.

This chapter has two fundamental messages: Natural selection is not the only agent responsible for evolution, and each of the four evolutionary processes has different consequences. Natural selection is the only mechanism that acting alone can result in adaptation. Mutation, gene flow, and genetic drift do not favor certain alleles over others. Mutation and drift introduce a nonadaptive component into evolution.

Key Concept Important Information Practice It

Let's take a closer look at the four evolutionary processes by examining a null hypothesis—what happens to allele frequencies when the evolutionary mechanisms are *not* operating.

25.1 Analyzing Change in Allele Frequencies: The Hardy-Weinberg Principle

To study how the four evolutionary processes affect populations, biologists take a three-pronged approach. First they create mathematical models that track the fate of alleles over time. Then they collect data to test predictions made by the models' equations. Finally, they apply the results to solve problems in human genetics, conservation of endangered species, or other fields.

This research strategy began in 1908, when both G. H. Hardy and Wilhelm Weinberg published a major result independently. At the time, it was commonly believed that changes in allele frequency occur simply as a result of sexual reproduction—meiosis followed by the random fusion of gametes (eggs and sperm) to form offspring. Some biologists claimed that dominant alleles inevitably increase in frequency. Others predicted that two alleles of the same gene inevitably reach a frequency of 0.5.

To test these hypotheses, Hardy and Weinberg analyzed what happens to the frequencies of alleles when many individuals in a population mate and produce offspring. Instead of thinking about the consequences of a mating between two parents with a specific pair of genotypes, as we did with Punnett squares in Chapter 13, Hardy and Weinberg wanted to know what happened in an entire population, when *all* of the individuals—and thus all possible genotypes—bred. Like Darwin, Hardy and Weinberg were engaged in population thinking.

To analyze the consequences of matings among all of the individuals in a population, Hardy and Weinberg invented a novel approach: They imagined that all of the gametes produced in each generation go into a single group called the **gene pool** and then combine at random to form offspring. Something very much like this happens in species like clams and sea stars and sea urchins, which release their gametes into the water, where they mix randomly with gametes from other individuals in the population and combine to form zygotes.

To determine which genotypes would be present in the next generation and in what frequency, Hardy and Weinberg simply had to calculate what happened when two gametes were plucked at random out of the gene pool, many times, and each of these gamete pairs was then combined to form offspring. These calculations would predict the genotypes of the offspring that would be produced, as well as the frequency of each genotype.

The researchers began by analyzing the simplest situation possible—that just two alleles of a particular gene exist in a population. Let's call these alleles A_1 and A_2. We'll use p to symbolize the frequency of A_1 alleles in the gene pool and q to symbolize the frequency of A_2 alleles in the same gene pool. Because there are only two alleles, the two frequencies must add up to 1; that is, $p + q = 1$. Although p and q can have any value between 0 and 1, let's suppose that the initial frequency of A_1 is 0.7 and that of A_2 is 0.3 (**Figure 25.1**, step 1). In this case, 70 percent of the gametes in the gene pool carry A_1 and 30 percent carry A_2 (Figure 25.1, step 2).

Because only two alleles are present, three genotypes are possible: A_1A_1, A_1A_2, and A_2A_2 (Figure 25.1, step 3). What will the frequency of these three genotypes be in the next generation? Figure 25.1, step 4, explains the logic of Hardy's and Weinberg's result:

- The frequency of the A_1A_1 genotype is p^2.
- The frequency of the A_1A_2 genotype is $2pq$.
- The frequency of the A_2A_2 genotype is q^2.

The genotype frequencies in the offspring generation must add up to 1, which means that $p^2 + 2pq + q^2 = 1$. In our numerical example, $0.49 + 0.42 + 0.09 = 1$. Figure 25.1, step 5, shows how the frequencies of alleles A_1 and A_2 are calculated from these genotype frequencies. In our example, the frequency of allele A_1 is still 0.7 and the frequency of allele A_2 is still 0.3. Thus, the frequency of allele A_1 in the next generation is still p and the frequency of allele A_2 is still q. No allele frequency change occurred. Even if A_1 is dominant to A_2, it does not increase in frequency (Figure 25.1, step 6). And there is no trend toward both alleles reaching a frequency of 0.5. **Figure 25.2** illustrates the same result a little differently. The figure uses a Punnett square in a novel way: to predict the outcome of random mating— meaning, random combinations of all gametes in a population. The outcome is the same as in Figure 25.1.

This result is called the **Hardy-Weinberg principle.** It makes two fundamental claims:

1. If the frequencies of alleles A_1 and A_2 in a population are given by p and q, then the frequencies of genotypes A_1A_1, A_1A_2, and A_2A_2 will be given by p^2, $2pq$, and q^2 for generation after generation.
2. When alleles are transmitted according to the rules of Mendelian inheritance, their frequencies do not change over time. For evolution to occur, some other factor or factors must come into play.

What are these other factors?

The Hardy-Weinberg Model Makes Important Assumptions

The Hardy-Weinberg model is based on important assumptions about how populations and alleles behave. Specifically, for a population to conform to the Hardy-Weinberg principle, none of the four mechanisms of evolution can be acting on the population. In addition, the model assumes that mating is random with respect

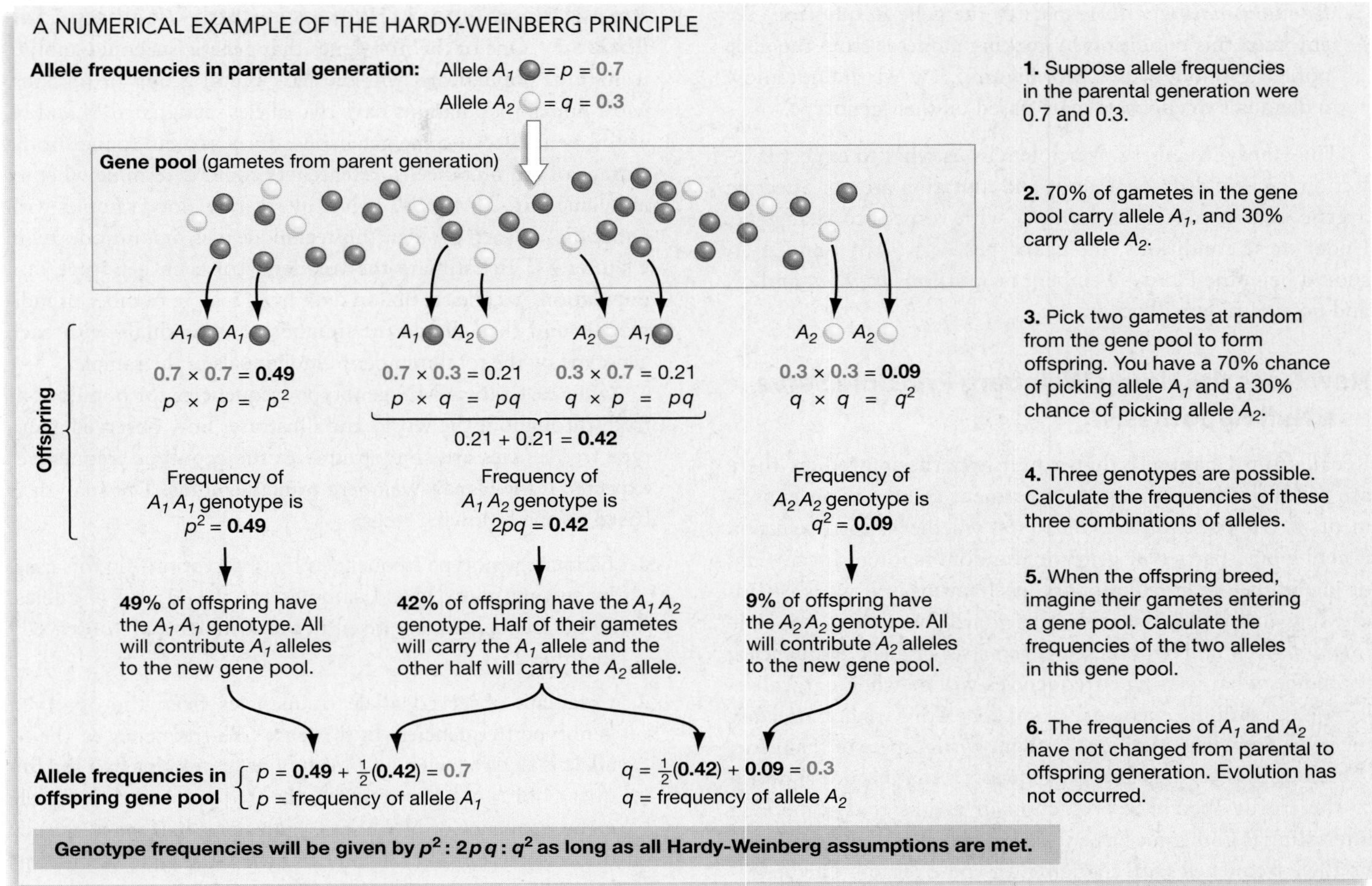

FIGURE 25.1 A Numerical Example of the Hardy-Weinberg Principle. To understand the logic behind calculating the frequency of A_1A_2 genotypes in step 4, see **BioSkills 9.**

to the gene in question. Thus, here are the five conditions that must be met:

1. *No natural selection* at the gene in question. In step 2 of Figure 25.1, the model assumed that all members of the parental generation survived and contributed equal numbers of gametes to the gene pool, no matter what their genotype.

2. *No genetic drift, or random allele frequency changes*, affecting the gene in question. We avoided this type of allele frequency change in step 4 of Figure 25.1 by assuming that we drew alleles in their exact frequencies p and q, and not at some different values caused by chance. For example, allele A_1 did not "get lucky" and get drawn more than 70 percent of the time. No random changes due to luck occurred.

3. *No gene flow*. No new alleles were added by immigration or lost through emigration anywhere in Figure 25.1. As a result, all of the alleles in the offspring population came from the original population's gene pool.

4. *No mutation*. We didn't consider that new A_1s or A_2s or other, new alleles might be introduced into the gene pool in step 2 or step 5 of Figure 25.1.

Allele frequencies in parental generation:

A_1 = p = 0.7 A_2 = q = 0.3

All eggs in gene pool

All sperm in gene pool	0.7 A_1	0.3 A_2
0.7 A_1	A_1A_1 0.49	A_1A_2 0.21
0.3 A_2	A_2A_1 0.21	A_2A_2 0.09

Genotype frequencies in offspring generation:

$A_1A_1 = p^2 = \mathbf{0.49}$

$A_1A_2 = 2pq = \mathbf{0.42}$

$A_2A_2 = q^2 = \mathbf{0.09}$

Allele frequencies in offspring generation:

$A_1 = p = \mathbf{0.49} + \frac{1}{2}\mathbf{(0.42)} = 0.70$

$A_2 = q = \frac{1}{2}\mathbf{(0.42)} + \mathbf{0.09} = 0.30$

Allele frequencies have not changed

FIGURE 25.2 A Punnett Square Illustrates the Hardy-Weinberg Principle.

5. *Random mating* with respect to the gene in question. We enforced this condition by picking gametes from the gene pool at random in step 3 of Figure 25.1. We did not allow individuals to choose a mate based on their genotype.

The Hardy-Weinberg principle tells us what to expect if selection, genetic drift, gene flow, and mutation are not affecting a gene, *and* if mating is random with respect to that gene. Under these conditions, the genotypes A_1A_1, A_1A_2, and A_2A_2 should be in the Hardy-Weinberg proportions p^2, $2pq$, and q^2, and no evolution will occur.

How Does the Hardy-Weinberg Principle Serve as a Null Hypothesis?

Recall from Chapter 1 that a null hypothesis predicts there are no differences among the treatment groups in an experiment. Biologists often want to test whether natural selection is acting on a particular gene, nonrandom mating is occurring, or one of the other evolutionary mechanisms is at work. In addressing questions like these, the Hardy-Weinberg principle functions as a null hypothesis. Given a set of allele frequencies, it predicts what genotype frequencies will be when natural selection, mutation, genetic drift, and gene flow are not affecting the gene; and when mating is random with respect to that gene. If biologists observe genotype frequencies that do not conform to the Hardy-Weinberg prediction, it means that something interesting is going on: Either nonrandom mating is occurring, or allele frequencies are changing for some reason. Further research is needed to determine which of the five Hardy-Weinberg conditions is being violated.

Let's consider two examples to illustrate how the Hardy-Weinberg principle is used as a null hypothesis: MN blood types and *HLA* genes, both in humans.

Are MN Blood Types in Humans in Hardy-Weinberg Equilibrium? One of the first genes that geneticists could analyze in natural populations was the MN blood group of humans. Most human populations have two alleles, designated *M* and *N*, at this gene. Because the gene codes for a protein found on the surface of red blood cells, researchers could determine whether individuals are *MM*, *MN*, or *NN* by treating blood samples with antibodies to each protein (this technique was first introduced in Chapter 8). To estimate the frequency of each genotype in a population, geneticists obtain data from a large number of individuals and then divide the number of individuals with each genotype by the total number of individuals in the sample.

Table 25.1 shows MN genotype frequencies for populations from throughout the world and illustrates how observed genotype frequencies are compared with the genotype frequencies expected if the Hardy-Weinberg principle holds. The analysis is based on the following steps:

1. Estimate genotype frequencies by observation—in this case, by testing many blood samples for the *M* and *N* alleles. These frequencies are given in the rows labeled "observed" in Table 25.1.
2. Calculate observed allele frequencies from the observed genotype frequencies. In this case, the frequency of the *M* allele is the frequency of *MM* homozygotes plus half the frequency of *MN* heterozygotes; the frequency of the *N* allele is the frequency of *NN* homozygotes plus half the frequency of *MN* heterozygotes. (You can review the logic behind this calculation in steps 5 and 6 of Figure 25.1.)
3. Use the observed allele frequencies to calculate the genotypes expected according to the Hardy-Weinberg principle. Under the null hypothesis of no evolution and random mating, the expected genotype frequencies are p^2: $2pq$: q^2.

TABLE **25.1** **The MN Blood Group of Humans: Observed and Expected Genotype Frequencies**

The expected genotype frequencies are calculated from the observed allele frequencies, using the Hardy-Weinberg principle.

		Genotype Frequencies			Allele Frequencies	
Population and Location		***MM***	***MN***	***NN***	***M***	***N***
Inuit (Greenland)	Observed	0.835	0.156	0.009	0.913	0.087
	Expected	0.834	0.159	0.008		
Native Americans (U.S.)	Observed	0.600	0.351	0.049	0.776	0.224
	Expected	0.602	0.348	0.050		
Caucasians (U.S.)	Observed	0.292	0.494	0.213	0.540	0.460
	Expected	0.290	0.497	0.212		
Aborigines (Australia)	Observed	0.025	0.304	0.672	0.178	0.825
	Expected	0.031	0.290	0.679		
Ainu (Japan)	Observed — Step 1 →	0.179	0.502	0.319 — Step 2 →		
	Expected			← Step 3		

EXERCISE Fill in the values for allele frequencies and expected genotype frequencies for the Ainu people of Japan.

4. Compare the observed and expected values. Researchers must use statistical tests to determine whether the differences between the observed and expected genotype frequencies are small enough to be due to chance or large enough to reject the null hypothesis of no evolution and random mating.

Although using statistical testing is beyond the scope of this text (see **BioSkills 3** for a brief introduction to the topic), you should be able to inspect the numbers and comment on them. In these populations, for example, the observed and expected *MN* genotype frequencies are almost identical. (A statistical test shows that the small differences observed are probably due to chance.) For every population surveyed, genotypes at the *MN* locus are in Hardy-Weinberg proportions. As a result, biologists conclude that the assumptions of the Hardy-Weinberg model are valid for this locus. The results imply that when these data were collected, the *M* and *N* alleles in these populations were not being affected by the four evolutionary mechanisms and that mating was random with respect to this gene—meaning that humans were not choosing mates on the basis of their *MN* genotype.

Before moving on, however, it is important to note that a study such as this does not mean that the *MN* gene has never been under selection or subject to nonrandom mating or genetic drift. Even if selection has been very strong for many generations, one generation of no evolutionary forces and of random mating will result in genotype frequencies that conform to Hardy-Weinberg expectations. The Hardy-Weinberg principle is used to test the hypothesis that currently no evolution is occurring at a particular gene and that in the previous generation, mating was random with respect to the gene in question.

Are *HLA* Genes in Humans in Hardy-Weinberg Equilibrium? A research team recently collected data on the genotypes of 125 individuals from the Havasupai tribe native to Arizona. These biologists were studying two genes that are important in the functioning of the human immune system. More specifically, the genes that they analyzed code for proteins that help immune system cells recognize and destroy invading bacteria and viruses. Previous work had shown that different alleles exist at both the *HLA-A* and *HLA-B* genes, and that the alleles at each gene code for proteins that recognize slightly different disease-causing organisms. Like the *M* and *N* alleles, *HLA* alleles are *codominant*—meaning that both are expressed and create the phenotype (see Chapter 13). As a result, the research group hypothesized that individuals who are heterozygous at one or both of these genes may have a strong fitness advantage. The logic is that heterozygous people have a wider variety of HLA proteins, so their immune systems can recognize and destroy more types of bacteria and viruses. They should be healthier and have more offspring than homozygous people do.

To test this hypothesis, the researchers used their data on observed genotype frequencies to determine the frequency of each allele present. When they used these allele frequencies to calculate the expected number of each genotype according to the Hardy-Weinberg principle, they found the observed and expected values reported in **Table 25.2**. When you inspect these data, notice that there are many more heterozygotes and many fewer homozygotes than expected under Hardy-Weinberg conditions. Statistical tests show it is extremely unlikely that the difference between the observed and expected numbers could occur purely by chance.

TABLE **25.2 HLA Genes of Humans: Observed and Expected Genotypes**

	Observed Number	Expected Number under Hardy-Weinberg
HLA-A		
Homozygotes	38	48
Heterozygotes	84	74
HLA-B		
Homozygotes	21	30
Heterozygotes	101	92

SOURCE: T. Markow et al., *HLA* polymorphism in the Havasupai: Evidence for balancing selection. *American Journal of Human Genetics* 53 (1993): 943–952. Published by the University of Chicago Press.

These results supported the team's prediction and indicated that one of the assumptions behind the Hardy-Weinberg principle was being violated. But which? The researchers argued that mutation, migration, and drift are negligible in this case and offered two competing explanations for their data:

1. *Mating may not be random with respect to the* HLA *genotype*. Specifically, people may subconsciously prefer mates with *HLA* genotypes unlike their own and thus produce an excess of heterozygous offspring. This hypothesis is plausible. For example, experiments have shown that college students can distinguish each others' genotypes at genes related to *HLA* on the basis of body odor. Individuals in this study were more attracted to the smell of genotypes unlike their own. If this is true among the Havasupai, then nonrandom mating would lead to an excess of heterozygotes compared with the proportion expected under Hardy-Weinberg.

2. *Heterozygous individuals may have higher fitness*. This hypothesis is supported by data collected by a different research team, studying Hutterite people living in South Dakota. In this population, married women who have the same *HLA*-related alleles as their husbands have more trouble getting pregnant and experience higher rates of spontaneous abortion than do women with *HLA*-related alleles different from those of their husbands. The data suggest that homozygous fetuses have lower fitness than do fetuses heterozygous at these genes. If this were true among the Havasupai, selection would lead to an excess of heterozygotes relative to Hardy-Weinberg expectations.

Which explanation is correct? It is possible that both are. But the fact is, no one knows. Using the Hardy-Weinberg principle as a null hypothesis allowed biologists to detect an interesting pattern in a natural population. Research continues on the question of why the pattern exists.

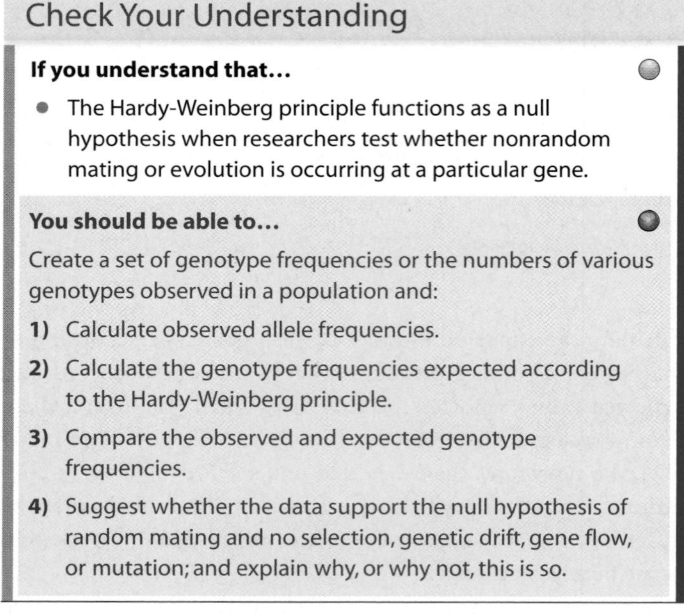

Check Your Understanding

If you understand that...

- The Hardy-Weinberg principle functions as a null hypothesis when researchers test whether nonrandom mating or evolution is occurring at a particular gene.

You should be able to...

Create a set of genotype frequencies or the numbers of various genotypes observed in a population and:

1) Calculate observed allele frequencies.
2) Calculate the genotype frequencies expected according to the Hardy-Weinberg principle.
3) Compare the observed and expected genotype frequencies.
4) Suggest whether the data support the null hypothesis of random mating and no selection, genetic drift, gene flow, or mutation; and explain why, or why not, this is so.

MB **Web Animation** at www.masteringbio.com
The Hardy-Weinberg Principle

25.2 Types of Natural Selection

Natural selection occurs when individuals with certain phenotypes produce more offspring than individuals with other phenotypes do. If certain alleles are associated with the favored phenotypes, they increase in frequency while other alleles decrease in frequency. The result is evolution. To use the language introduced in Chapter 24, evolution by natural selection occurs when heritable variation leads to differential success in survival and reproduction.

Although you should have a solid understanding of why evolution by natural selection occurs, it is important to recognize that natural selection occurs in a wide variety of patterns. Each of these patterns has different causes and consequences. For example, the second explanation for the data in Table 25.2 is a pattern of natural selection called **balancing selection** or **heterozygote advantage**. When balancing selection occurs, heterozygous individuals have higher fitness than homozygous individuals do. The consequence of this pattern is that genetic variation is maintained in populations. **Genetic variation** refers to the number and relative frequency of alleles that are present in a particular population.

When biologists analyze the consequences of different patterns of selection, they often focus on genetic variation. The reason is simple: Lack of genetic variation in a population is usually a bad thing. To understand why, recall from Chapter 24 that selection can occur only if heritable variation exists in a population. If genetic variation is low and the environment changes—perhaps due to the emergence of a new disease-causing virus, a rapid change in climate, or a reduction in the availability of a particular food source—it is unlikely that any alleles will be present that have high fitness under the new conditions. As a result, the average fitness of the population will decline. If the environmental change is severe enough, the population may even be faced with extinction.

Let's examine some of the different types of natural selection with this question in mind: How do they affect the level of genetic variation in the population?

Directional Selection

According to the data introduced in Chapter 24, natural selection has increased the frequency of drug-resistant strains of the tuberculosis bacterium and caused changes in beak shape and body size in medium ground finches. This type of natural selection is called **directional selection**, because the average phenotype of the populations changed in one direction.

Figure 25.3a illustrates how directional selection works when the trait in question has a bell-shaped, normal distribution in a population. Recall from Chapter 13 that when many different genes influence a trait, the distribution of phenotypes in the population tends to form a bell-shaped curve. In such cases, directional selection is acting on many different genes at once. In the case of selection on drug resistance in the TB bacterium, however, selection was acting on a single gene.

Most often, directional selection tends to reduce the genetic diversity of populations. If directional selection continues over time, the favored alleles will eventually reach a frequency of 1.0 while disadvantageous alleles will reach a frequency of 0.0. Alleles that reach a frequency of 1.0 are said to be fixed; those that reach a frequency of 0.0 are said to be lost. When disadvantageous alleles decline in frequency **purifying selection** is said to occur.

Fixation and loss may not always occur under directional selection, however. To appreciate why, consider recent data on the body size of cliff swallows native to the Great Plains of North America. In 1996 a population of cliff swallows endured a six-day period of exceptionally cold, rainy weather. Cliff swallows feed by catching mosquitoes and other insects in flight. Insects disappeared during this cold snap, however, and the biologists recovered the bodies of 1853 swallows that died of starvation. As soon as the weather improved, the researchers caught and measured the body size of 1027 survivors from the same population. As the histograms in **Figure 25.3b** show, survivors were much larger on average than the birds that died. Directional selection, favoring large body size, had occurred. To explain this observation, the investigators suggest that larger birds survived because they had larger fat stores and did not get as cold as the

(a) Directional selection changes the average value of a trait.

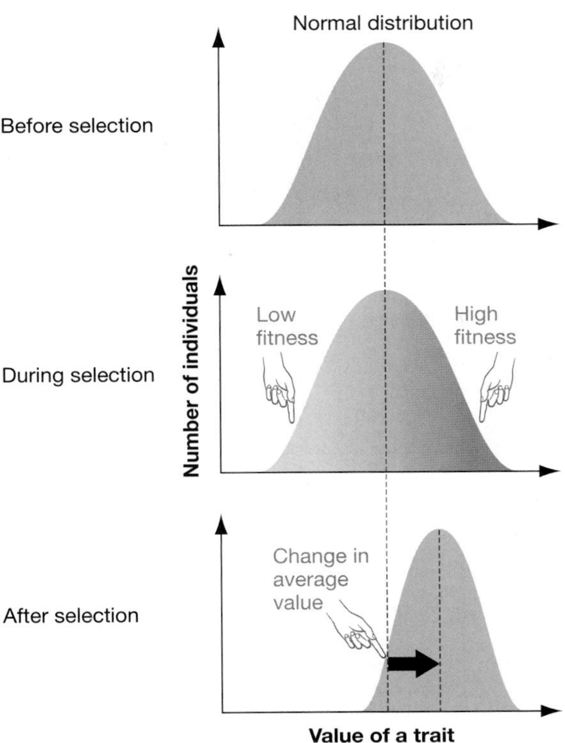

(b) For example, directional selection caused average body size to increase in a cliff swallow population.

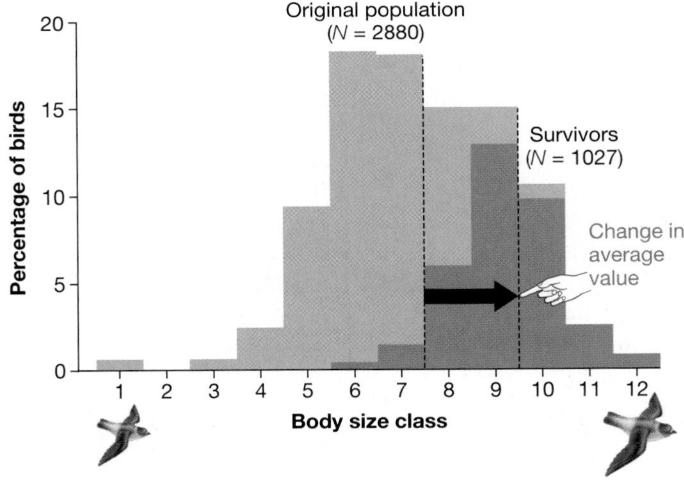

FIGURE 25.3 Directional Selection. (a) When directional selection acts on traits that have a normal distribution, individuals with one set of extreme values experience poor reproductive success. **(b)** The light green histogram shows the distribution of overall body size in cliff swallows prior to an extended cold snap that killed many individuals. (The various body-size classes were calculated from measurements of wing length, tail length, leg length, and beak size.) The dark green histogram shows the size distribution of individuals from the same population that survived the cold spell. Here *N* indicates the sample size.

smaller birds. As a result, the larger birds were less likely to die of exposure to cold and more likely to avoid starvation until the weather warmed up and insects were again available.

If the differences in body size among individuals were due in part to differences in their genotypes—meaning that heritable variation in body size existed—then the population evolved. If so, and if this type of directional selection continued, then alleles that contribute to small body size would quickly be eliminated from the cliff swallow population. It is not clear that this will be the case, however, because directional selection is rarely constant throughout a species' range and through time. By examining weather records, the researchers established that cold spells as severe as the one that occurred in 1996 are rare. Further, research on other swallow species suggests that smaller birds are more maneuverable in flight and thus more efficient when they feed. If so, then selection for feeding efficiency could counteract selection by cold weather. When this is the case, individuals with intermediate body size should be favored. Opposing patterns of directional selection will help maintain genetic variation for this trait.

When studies on a wide array of populations and species are considered, it is common to find that one cause of directional selection on a trait is counterbalanced by a different factor that causes selection in the opposite direction. This concept, known as a fitness trade-off, was introduced in Chapter 24. In such cases, the optimal phenotype is intermediate. The same pattern can result from an entirely different type of natural selection, called stabilizing selection.

Stabilizing Selection

When cliff swallows were exposed to cold weather, selection greatly reduced one extreme in the range of phenotypes and resulted in a directional change in the average characteristics of the population. But selection can also reduce both extremes in a population, as illustrated in **Figure 25.4a**. This pattern of selection is called **stabilizing selection**. It has two important consequences: There is no change in the average value of a trait over time, and genetic variation in the population is reduced.

Figure 25.4b shows a classical data set in humans illustrating stabilizing selection. Biologists who analyzed birth weights and mortality in 13,730 babies born in British hospitals in the 1950s found that babies of average size (slightly over 7 pounds) survived best. Mortality was high for very small babies and very large babies. This is persuasive evidence that birth weight was under strong stabilizing selection in this population. Alleles associated with high birthweight or low birthweight were subject to purifying selection.

Disruptive Selection

Disruptive selection has the opposite effect of stabilizing selection. Instead of favoring phenotypes near the average value and

(a) Stabilizing selection reduces the amount of variation in a trait.

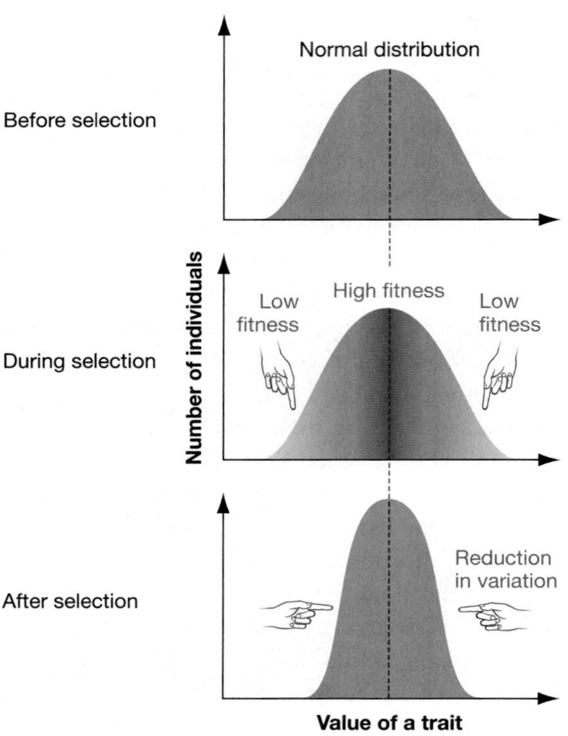

(b) For example, very small and very large babies are the most likely to die, leaving a narrower distribution of birth weights.

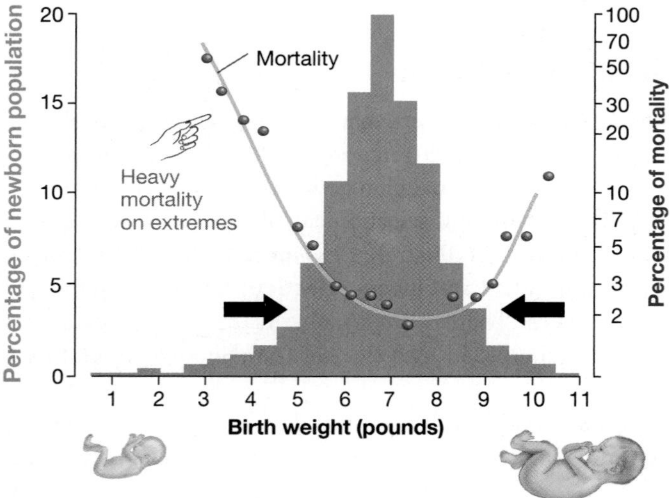

FIGURE 25.4 Stabilizing Selection. (a) When stabilizing selection acts on normally distributed traits, individuals with extreme phenotypes experience poor reproductive success. **(b)** Histogram showing the percentages of newborns with various birth weights on the left-hand axis. The purple dots indicate the percentage of newborns in each weight class that died, plotted on the logarithmic scale shown on the right.

(a) Disruptive selection increases the amount of variation in a trait.

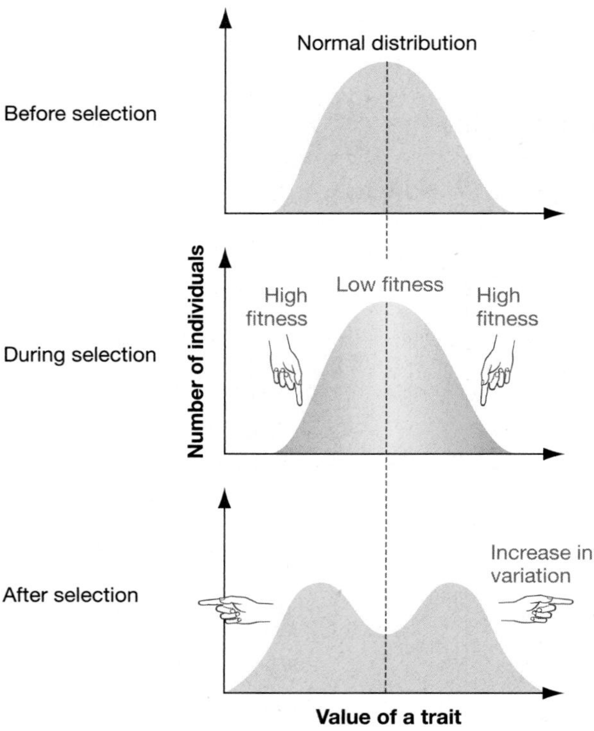

(b) For example, only juvenile black-bellied seedcrackers that had very long or very short beaks survived long enough to breed.

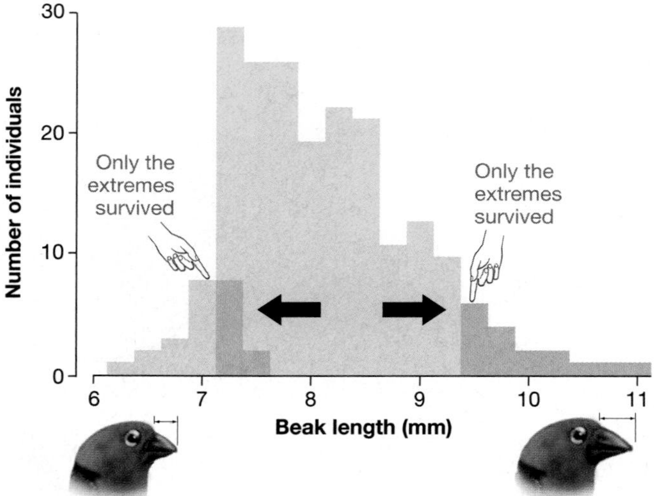

FIGURE 25.5 Disruptive Selection. (a) When disruptive selection occurs on traits with a normal distribution, individuals with extreme phenotypes experience high reproductive success. **(b)** Histogram showing the distribution of beak length in a population of black-bellied seedcrackers. The light orange bars represent all juveniles; the dark orange bars, juveniles that survived to adulthood.

eliminating extreme phenotypes, it eliminates phenotypes near the average value and favors extreme phenotypes (**Figure 25.5a**). When disruptive selection occurs, the overall amount of genetic variation in the population is maintained.

Recent research has shown that disruptive selection is responsible for the striking bills of black-bellied seedcrackers, pictured in **Figure 25.5b**. The data plotted in the figure show that individuals with either very short or very long beaks survive best and that birds with intermediate phenotypes are at a disadvantage. In this case, the agent that causes natural selection is food. At a study site in south-central Cameroon, West Africa, a researcher found that only two sizes of seed are available to the seedcrackers: large and small. Birds with small beaks crack and eat small seeds efficiently. Birds with large beaks handle large seeds efficiently. But birds with intermediate beaks have trouble with both, so alleles associated with medium-sized beaks are subject to purifying selection. Disruptive selection maintains high overall variation in this population.

Disruptive selection is important because it sometimes plays a part in **speciation**, or the formation of new species. If small-beaked seedcrackers began mating with other small-beaked individuals, their offspring would tend to be small beaked and would feed on small seeds. Similarly, if large-beaked individuals chose only other large-beaked individuals as mates, they would tend to produce large-beaked offspring that would feed on large seeds. In this way, selection would result in two distinct populations. Under some conditions, the populations may eventually form two new species. The process of species formation, based on disruptive selection and other mechanisms, is explored in detail in Chapter 26.

MB **Web Animation** at www.masteringbio.com
Three Modes of Natural Selection

25.3 Genetic Drift

Natural selection is not random. It is directed by the environment and results in adaptation. **Genetic drift**, in contrast, is undirected. It is defined as any change in allele frequencies in a population that is due to chance. The process is aptly named, because it causes allele frequencies to drift up and down randomly over time. When drift occurs, allele frequencies change due to blind luck—what is formally known as **sampling error**.

To understand why genetic drift occurs, consider the group of people who founded the present population of Pitcairn Island in the South Pacific. The founding event occurred in 1789, when a small band of mutineers led by Fletcher Christian took over the British warship HMS *Bounty* and fled to Pitcairn. The six sailors were joined by two Tahitian men and six Tahitian women. Suppose that six couples formed and raised two children each. How will allele frequencies compare in the founding population that bred versus the next generation?

To answer this question, let's focus on a hypothetical gene A with alleles A_1, A_2, and A_3. Suppose that the six males and six females in the original couples have the genotypes given in **Table 25.3**. The gametes formed by each of these parents have an equal chance of carrying either allele. Further, eggs and sperm combine at random when fertilization occurs—meaning that each type of egg and sperm has an equal probability of combining, irrespective of its genotype. To simulate the fertilization process, you can flip a coin to decide which gametes combine. Let heads stand for the first allele listed in the parent's genotype, while tails represents the second allele listed. Doing these coin flips resulted in the offspring genotypes given in the "Child 1" and "Child 2" columns of Table 25.3.

TABLE **25.3 A Thought Experiment on Genetic Drift**

The genotypes of the children in this table were generated by flipping a coin to simulate which alleles combined during fertilization. The frequencies of alleles A_1, A_2, and A_3 differ in the two generations, due to genetic drift.

	Father	Mother	Child 1	Child 2
Couple 1	A_3A_3	A_2A_3	A_3A_3	A_3A_3
Couple 2	A_2A_3	A_2A_3	A_2A_3	A_3A_2
Couple 3	A_1A_2	A_2A_3	A_2A_3	A_1A_2
Couple 4	A_1A_1	A_1A_2	A_1A_2	A_1A_1
Couple 5	A_1A_2	A_3A_3	A_1A_3	A_1A_3
Couple 6	A_1A_2	A_1A_3	A_2A_3	A_2A_1

Now let's count up alleles and calculate allele frequencies. Because there are 12 breeding individuals and 12 offspring, the total number of alleles present in each generation (ignoring the two founding individuals who did not breed) is 24. From the data in Table 25.3, the allele frequencies are as follows:

	A_1	A_2	A_3
Allele frequencies in the parents	$\frac{7}{24} = 29.2\%$	$\frac{8}{24} = 33.3\%$	$\frac{9}{24} = 37.5\%$
Allele frequencies in the offspring	$\frac{7}{24} = 29.2\%$	$\frac{7}{24} = 29.2\%$	$\frac{10}{24} = 41.6\%$

Although the frequency of allele A_1 did not change, the frequency of A_2 declined by over 4 percent and the frequency of A_3 increased by over 4 percent, purely by chance. Allele A_3 did not confer higher fitness. Instead, it just got lucky. Random chance caused evolution—a change in allele frequencies in a population. Instead of each allele being sampled in exactly its original frequency when offspring formed, as the Hardy-Weinberg principle assumes, a chance sampling error occurred.

This exercise helps to illustrate several important points:

- Genetic drift is random with respect to fitness. The allele frequency changes it produces are not adaptive.
- Genetic drift is most pronounced in small populations. If the original population had included only couples 1 and 2, then the frequency of allele A_2 would have declined by 12.5 percent and the frequency of A_3 would have increased by 12.5 percent. The smaller the sample, the larger the sampling error. Conversely, if the population had consisted of 500 couples who produced 1000 offspring (necessitat-

ing 2000 coin flips), it is extremely unlikely that drift would produce a change in allele frequency as large as 4 percent.

- Over time, genetic drift can lead to the random loss or fixation of alleles. If allele A_1 continued to be unlucky generation after generation, it would eventually drift to a frequency of 0.0 and be lost. If allele A_3 continued to be lucky for several more generations, it might drift to a frequency of 1.0. When random loss or fixation occurs, genetic variation in the population declines.

Figure 25.6 provides an example of these points. The graphs show the output from computer simulations that combine the alleles in a gene pool at random to create an offspring generation, calculate the allele frequencies in the offspring generation, and use those allele frequencies to create a new gene pool; this process was continued for 100 generations. The top graph shows eight replicates of this process with a population size of 4; the bottom graph shows eight replicates with a population of 400. Notice the striking differences between the effects of drift in the small versus large population and the consequences for genetic variation when alleles drift to fixation or loss.

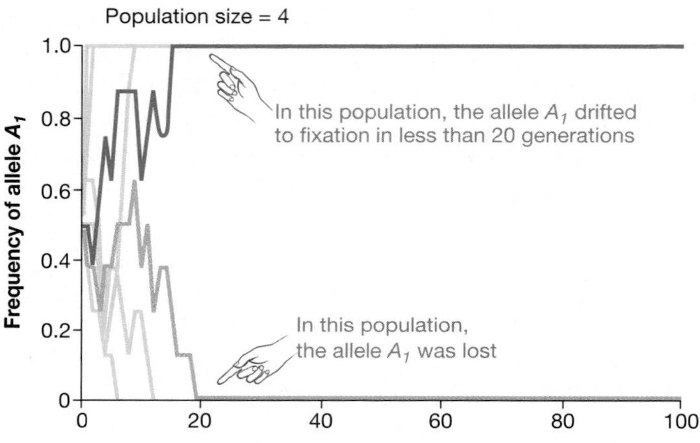

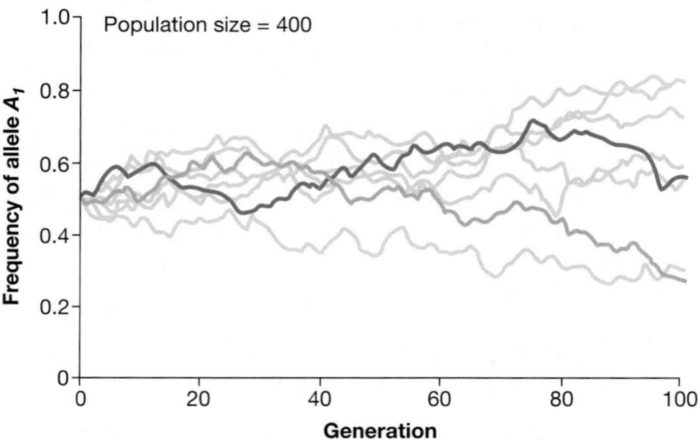

FIGURE 25.6 Genetic Drift Is More Pronounced in Small Populations than Large Populations.

EXERCISE Draw graphs predicting what these graphs would look like for a population size of 40 and a population size of 4000.

Given enough time, drift can be an important factor even in large populations. To drive this point home, consider two types of alleles that were introduced in earlier chapters and that have no effect on fitness. Recall from Chapter 16 that alleles containing silent mutations, usually in the third position of a codon, do not change the gene product. As a result, they have no effect on the phenotype. Yet these alleles routinely drift to high frequency or even fixation over time. Similarly, recall from Chapter 20 that pseudogenes do not code for a product. Although their presence does not affect an individual's phenotype, dozens of pseudogenes in the human genome have reached fixation, due to drift.

Experimental Studies of Genetic Drift

Research on genetic drift began with theoretical work in the 1930s and 1940s, which used mathematical models to predict the effect of genetic drift on allele frequencies and genetic variation. In the mid-1950s, Warwick Kerr and Sewall Wright did an experiment to show how drift works in practice. The biologists started with a large laboratory population of fruit flies that contained a **genetic marker**—a specific allele that causes a distinctive phenotype. In this case, the marker was the morphology of bristles. Fruit flies have bristles that can be either straight or bent. This difference in bristle phenotype depends on a single gene. Kerr and Wright's lab population contained just two alleles—normal (straight) and "forked" (bent).

To begin the experiment, the researchers set up 96 cages in their lab. Then they placed four adult females and four adult males of the fruit fly *Drosophila melanogaster* in each. They chose individual flies to begin these experimental populations so that the frequency of the normal and forked alleles in each of the 96 starting populations was 0.5. The two alleles do not affect the fitness of flies in the lab environment, so Kerr and Wright could be confident that if changes in the frequency of normal and forked phenotypes occurred, they would not be due to natural selection.

After these first-generation adults bred, Kerr and Wright reared their offspring. Then they randomly chose four males and four females—meaning that they simply grabbed individuals without regard to whether their bristles were normal or forked—from each of the 96 offspring populations and allowed them to breed and produce the next generation. The researchers repeated this procedure until all 96 populations had undergone a total of 16 generations. During the entire course of the experiment, no migration from one population to another occurred. Previous studies had shown that mutations from normal to forked are rare. Thus, the only evolutionary process operating during the experiment was genetic drift.

Their result? After 16 generations, the 96 populations fell into three groups (**Figure 25.7**). Forked bristles were found on all of the individuals in 29 of the experimental populations. Due to drift, the forked allele had been fixed in these 29 populations and the normal allele had been lost. In 41 other populations, however, the opposite was true: All individuals had normal bristles. In these populations, the forked allele had been lost due to chance. Both alleles were still present in 26 of the populations. The message of the study is startling: In 73 percent of the experimental populations (70 out of the 96), genetic drift had reduced allelic diversity at this gene to zero. As predicted, genetic drift decreased genetic variation within populations and increased genetic differences between populations. Is drift important in natural populations as well?

Genetic Drift in Natural Populations

The sampling process that occurs during fertilization and that caused changes in allele frequencies in Pitcairn islanders and fruit flies occurs in every population in every generation in every species that reproduces sexually. It is particularly important in small populations, where sampling error tends to be high. This is of enormous concern to conservation biologists, because populations all over the planet are being drastically reduced in size by habitat destruction and other human activities. Small populations that occupy nature reserves or zoos are particularly susceptible to genetic drift. If drift leads to a loss of genetic diversity, it could darken the already bleak outlook for some endangered species.

It is important to realize that because drift is caused by sampling error, it can occur by *any* process or event that involves sampling—not just the sampling of gametes that occurs during fertilization. Let's consider two such examples, called founder effects and bottlenecks.

How Do Founder Effects Cause Drift? When a group of individuals emigrates to a new geographic area and establishes a new population, a founder event is said to occur. If the group is small enough, the allele frequencies in the new population are almost guaranteed to be different from those in the source population, due to sampling error. A change in allele frequencies that occurs when a new population is established is called a **founder effect**. As soon as the mutineers and Tahitians set foot on Pitcairn Island and took up residence, evolution had occurred due to drift, via a founder effect. The allele frequencies in the new population were different from those in England or in Tahiti. (Once breeding took place, drift continued to act via sampling errors that occurred during fertilization.)

Fishermen on the island of Anguilla in the Caribbean recently witnessed a founder event involving green iguanas. A few weeks after Hurricane Luis and Hurricane Marilyn swept

(a) Bristle shape is a useful genetic marker in fruit flies

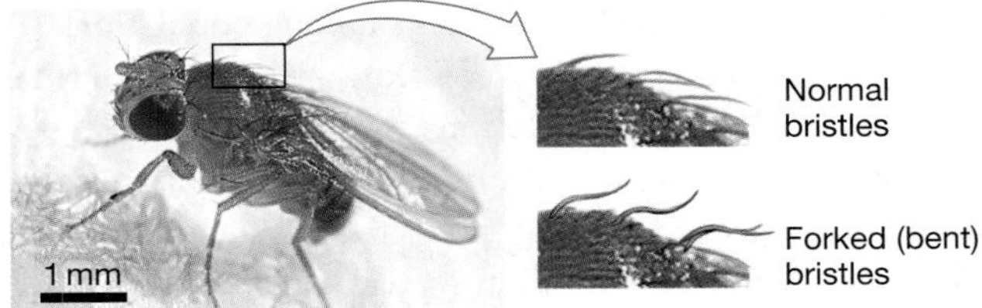

(b) In most experimental populations, the forked bristle allele drifted to fixation or loss

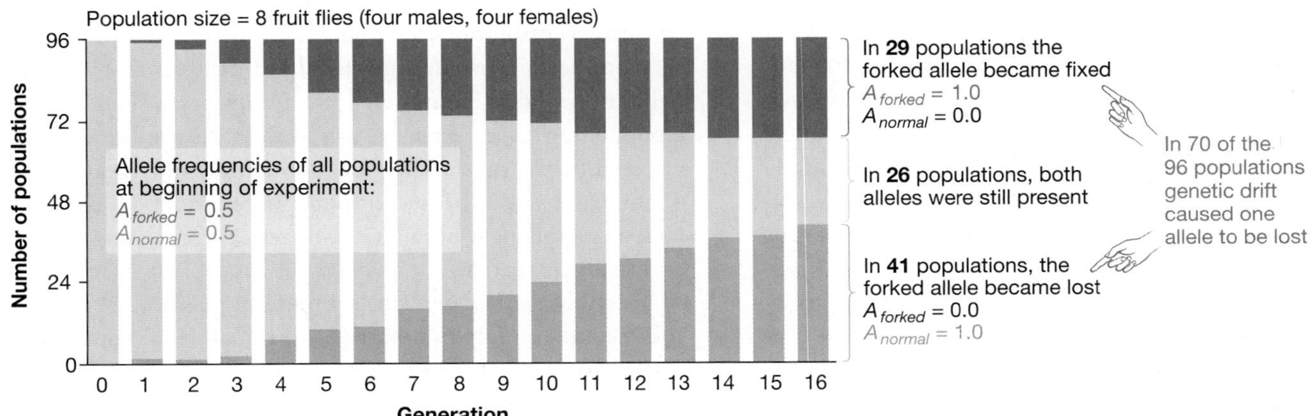

FIGURE 25.7 An Experiment on the Effects of Genetic Drift in Small Populations. In this experiment on fruit flies, population size in 96 separate populations was kept at eight individuals in each generation for 16 generations. The only evolutionary force acting in these populations was genetic drift. At the start of the experiment, the frequencies of the forked-bristle alleles and wild-type bristle alleles were each 50 percent.

● **QUESTION** Predict what the results of this experiment would be if the allele for forked-bristles was originally at a frequency of 90% instead of 50%.

through the region in September 1995, a large raft composed of downed logs tangled with other debris floated onto a beach on Anguilla. The fishermen noticed green iguanas on the raft and several on shore. Because green iguanas had not previously been found on Anguilla, the fishermen notified biologists. The researchers were able to document that at least 15 individuals had arrived; two years later they were able to confirm that at least some of the individuals were breeding. During this founder event, it is extremely unlikely that allele frequencies in the new Anguilla population of green iguanas exactly matched those of the source population, thought to be on the islands of Guadeloupe.

Founder events like these have been the major source of populations that occupy islands all over the world, as well as island-like habitats such as mountain meadows, caves, and ponds. Each time a founder event occurs, a founder effect is likely to accompany it, changing allele frequencies via genetic drift.

How Do Population Bottlenecks Cause Drift? If a large population experiences a sudden reduction in size, a population bottleneck is said to occur. The term comes from the metaphor of a few individuals passing through the neck of a bottle, by chance. Disease outbreaks, natural catastrophes such as floods or fires or storms, or other events can cause population bottlenecks.

Genetic bottlenecks follow population bottlenecks, just as founder effects follow founder events. A **genetic bottleneck** is a sudden reduction in the number of alleles in a population. Drift occurs during genetic bottlenecks and causes a change in allele frequencies. As an example of a genetic bottleneck, consider the human population of Pingelap Atoll in the South Pacific. On this island, only about 20 people out of a population of several thousand managed to survive the effects of a typhoon and a subsequent famine that occurred around 1775.

The survivors apparently included a person who carried a loss-of-function allele at a gene called *CNGB3*, which codes for a protein involved in color vision. The allele is recessive, and when it is homozygous it causes a serious vision deficit called achromatopsia. The condition is extremely rare in most populations, with the homozygous genotype and the affliction occurring in about 0.005 percent of the population. If genotypes at this locus are in Hardy-Weinberg proportions, then $q^2 = 0.00005$ and the frequency of the loss-of-function allele in most populations is about $\sqrt{0.00005}$, or 0.7 percent. In the population that survived the Pingelap Atoll disaster, however, the loss-of-function allele was at a frequency of about 1/40, or 2.5 percent. If the allele was at the typical frequency of 0.7 percent prior to the population bottleneck, then a huge frequency change occurred during the bottleneck, due to drift.

In today's population on Pingelap Atoll, about 1 in 20 people is afflicted with achromatopsia and the allele is at a frequency of over 20 percent. Because it is extremely unlikely that the loss-of-function allele is favored by directional selection or heterozygote advantage, researchers hypothesize that its frequency in this small population has continued to increase over the past 230 years due to drift.

Check Your Understanding

If you understand that...

- Genetic drift occurs any time allele frequencies change due to chance.
- Drift violates the assumptions of the Hardy-Weinberg principle and occurs during many different types of events, including random fusion of gametes at fertilization, founder events, and population bottlenecks.

You should be able to...

1) Explain why drift leads to a random loss or fixation of alleles.
2) Explain why drift is particularly important as an evolutionary force in small populations.

25.4 Gene Flow

Gene flow is the movement of alleles from one population to another. It occurs when individuals leave one population, join another, and breed. As an evolutionary mechanism, gene flow usually has one outcome: It equalizes allele frequencies between the source population and the recipient population. When alleles move from one population to another, the populations tend to become more alike. To capture this point, biologists say that gene flow homogenizes allele frequencies among populations.

To see the consequences of gene flow in action, consider data from prairie lupine populations on Mount St. Helens in the state of Washington. Prairie lupine is a perennial, meaning that individuals grow from the same root system year after year. They are fairly short lived, however, with each individual rarely living for more than 5 years. Prairie lupines are most commonly found living at high altitude on volcanoes, in habitats like that shown in **Figure 25.8a**.

The explosion of Mount St. Helens in 1980 created thousands of hectares of mudflows and ash deposits that are ideal habitat for prairie lupines. In 1981 a single individual founded a population on an extensive ash plain in the area most highly affected by the blast. This new population was located over 4 km from the nearest surviving population. In the blast zone near this new population, dozens of other populations eventually appeared.

To study genetic diversity in these populations, biologists collected tissue samples from 532 individuals in 32 populations. The age of each of the study populations was known. The researchers made sure they analyzed tissues from the oldest individuals in populations that had been established just 1 to 3 years prior to the start of the study. These individuals were of

(a) Lupines colonize sites and form populations.

(b) Gene flow reduces genetic differences among populations.

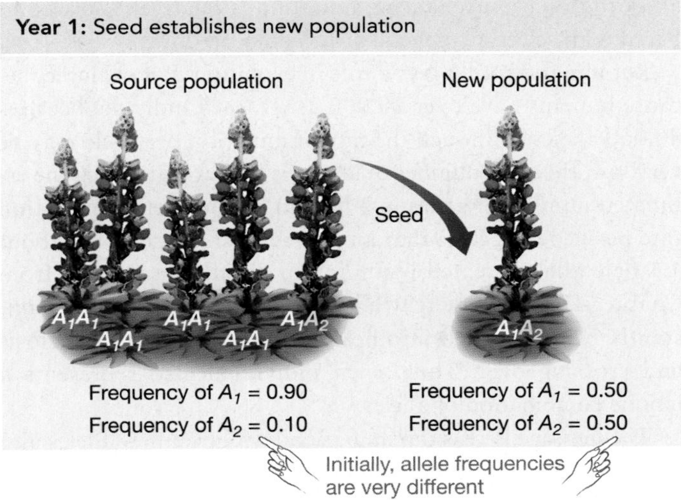

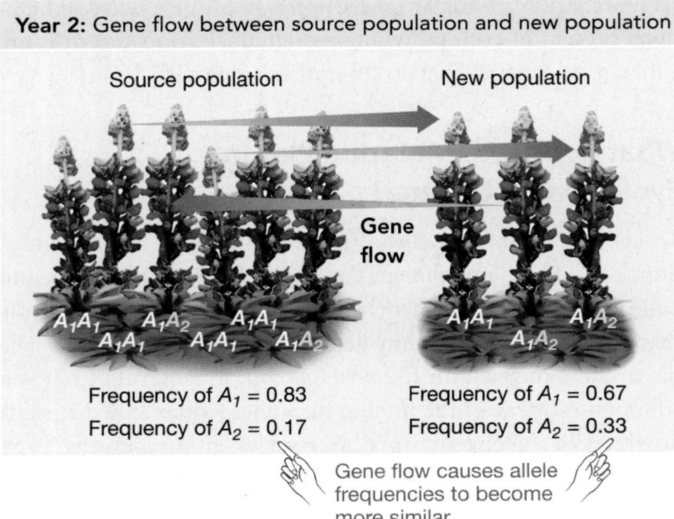

FIGURE 25.8 Gene Flow Equalizes Allele Frequencies between Populations. (a) Several species of lupine grow on the slopes of volcanoes. **(b)** The movement of alleles between populations tends to equalize allele frequencies over time.

special interest, because they were undoubtedly the founders of the new populations. Once tissues from all 532 individuals had been collected, the investigators used the polymerase chain reaction (see Chapter 19) to amplify the alleles of two genes in each individual. These data allowed them to estimate the number of alleles present in each population, as well as the frequency of each allele.

Their results? The data showed that the oldest individuals in newly established populations had allele frequencies that were very different from allele frequencies in the source populations. To interpret this observation, the biologists hypothesize that strong founder effects occurred each time a new lupine population became established. But their data also showed that as populations got older, allele frequencies became progressively more like those in the oldest populations—specifically, like the populations that had survived the eruption. Because selection due to differences in soil and moisture conditions, diseases, and predators present in each population would tend to make populations less similar over time, the researchers doubted that natural selection played a large role in equalizing allele frequencies. Instead, they propose that a continuous flow of alleles between populations, primarily via pollen carried by bees, was responsible for making the populations genetically more alike over time (**Figure 25.8b**).

What impact did gene flow have on the average fitness of individuals in these populations? The answer is not known, but theory suggests that it might vary. If a population had lost alleles due to genetic drift, then the arrival of new alleles via gene flow should increase genetic diversity. In this way, gene flow might increase the average fitness of individuals. But if natural selection had resulted in a population that was highly adapted to a specific habitat, then gene flow from other populations might introduce alleles that have low fitness in that particular environment. In this case, gene flow would have a negative effect on average fitness in the population.

To summarize, gene flow is random with respect to fitness—the arrival or departure of alleles can increase or decrease average fitness, depending on the situation. But a movement of alleles between populations tends to reduce genetic differences between them. This latter generalization is particularly important in our own species right now. Large numbers of people from Africa, the Middle East, Mexico, Central America, and Asia are emigrating to the countries of the European Union and the United States. Because individuals from different cultural and ethnic groups are intermarrying frequently and having offspring, allele frequencies in human populations are becoming more similar.

25.5 Mutation

To appreciate the role of mutation as an evolutionary force, let's return to one of the central questions that biologists ask about an evolutionary mechanism: How does it affect genetic variation in a population? Gene flow can increase genetic diversity in a recipient population if new alleles arrive with immigrating individuals. But gene flow can instead decrease genetic variation in the source population if alleles leave with emigrating individuals. The impact of genetic drift on genetic variation is much more clear cut than the impact of gene flow. Genetic drift tends to decrease genetic diversity over time, as alleles are randomly lost or fixed. Similarly, most forms of selection favor certain alleles and lead to a decrease in overall genetic variation.

If most of the evolutionary mechanisms lead to a loss of genetic diversity over time, what restores it? In particular, where do entirely new alleles come from? The answer to both of these questions is **mutation**.

As Chapter 16 noted, mutations occur when DNA polymerase makes an error as it copies a DNA molecule, resulting in a change in the sequence of deoxyribonucleotides. If a mutation occurs in a stretch of DNA that codes for a protein, the changed codon may result in a polypeptide with a novel amino acid sequence. Because errors are inevitable, mutation constantly introduces new alleles into populations in every generation. Mutation is an evolutionary mechanism that increases genetic diversity in populations.

The second important point to recognize about mutation is that it is random with respect to the fitness of the affected allele. Mutation just happens. Changes in DNA do not occur in a way that tends to increase fitness or decrease fitness. Because most organisms are well adapted to their current habitat, random changes in genes usually result in products that do not work as well as the alleles that currently exist. Stated another way, most mutations in sequences that code for a functional protein or RNA result in **deleterious** alleles—alleles that lower fitness. On rare occasions, however, mutation in these types of sequences produces a beneficial allele—an allele that allows individuals to produce more offspring. Beneficial alleles should increase in frequency in the population due to natural selection.

Because mutation produces new alleles, it can in principle change the frequencies of alleles through time. But does mutation occur often enough to make it an important factor in changing allele frequencies of a particular gene? The short answer is no.

Mutation as an Evolutionary Mechanism

To understand why mutation is not a significant mechanism of evolutionary change by itself, consider that the highest mutation rates that have been recorded at individual genes in humans are on the order of 1 mutation in every 10,000 gametes produced by an individual. This rate means that for every 10,000 alleles produced, on average one will have a mutation. When two gametes combine to form an offspring, then, at most about 1 in every 5000 offspring will carry a mutation at a particular gene. Will mutation affect allele frequencies in a species such as ours? To answer this question, suppose that 195,000 humans live in a population; that 5000 offspring are born one year; and that at the end of that year, the population numbers 200,000. Humans are diploid, so in a population this size, there is a total of 400,000 copies of each gene. Only one of them is a new allele created by mutation, however. Over the course of a year, the allele frequency change introduced by mutation is 1/400,000, or 0.0000025 (2.5×10^{-6}). At this rate, it would take 4000 years for mutation to produce a change in allele frequency of 1 percent.

These calculations support the conclusion that mutation does little to change allele frequencies on its own. Although mutation can be a significant evolutionary force in bacteria and archaea, which have extremely short generation times, mutation in eukaryotes rarely causes a change from the genotype frequencies expected under the Hardy-Weinberg principle. As an evolutionary mechanism, mutation is relatively slow compared with selection, genetic drift, and gene flow.

But mutation still plays a role in evolution. For example, because humans have over 20,000 genes, each individual carries 40,000 alleles. Although the rate of mutation per allele may be very low, the total number of alleles is high. Multiplying the estimated number of genes in a human by the average mutation rate per gene suggests that an average person contains about 1.1 new alleles created by mutation. In addition, recall from Chapter 12 that these new alleles and existing alleles are constantly being shuffled into new combinations during meiosis and crossing over. Thus, each individual also represents a unique combination of alleles.

The message here is that mutation introduces new alleles into every individual in every population in every generation. And in species that undergo sexual reproduction, meiosis and genetic recombination create variation in terms of the allele combinations present in each individual. Even if selection and drift are eliminating genetic diversity, mutation renews it.

What Role Does Mutation Play in Evolutionary Change?

To get a better appreciation for how mutation affects evolution, consider an experiment conducted by Richard Lenski and colleagues. The experiment was designed to evaluate the role that mutation plays in many genes over many generations. The researchers focused on *Escherichia coli*, a bacterium that is a common resident of the human intestine. **Figure 25.9** diagrams how Lenski's group set up a large series of populations, each founded with a single cell. Half of the cells contained an allele that allowed them to metabolize the sugar arabinose. The populations founded by these cells were designated Ara^+. The other cells carried an allele that did not allow them to metabolize arabinose. The populations founded by these cells were designated Ara^-. When Ara^+ cells are grown on special plates, they produce colonies that look red. When Ara^- colonies are

Experiment

Question: How does average fitness in a population change over time?

Hypothesis: Average fitness increases over time.

Null hypothesis: Average fitness does not increase over time.

Experimental setup:

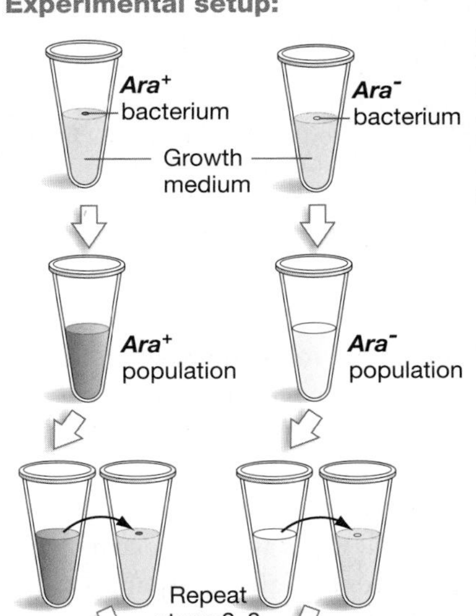

1. Place 10 mL of identical growth medium into many replicate tubes with one bacterium in each, some *Ara*⁺ and some *Ara*⁻ so that they can be distinguished by color when grown on special plates.

2. Incubate overnight. Average population in each tube is now 5×10^8 cells. (Cells are red or white when grown on special plates.)

3. Remove 0.1 mL from each tube and move to 10 mL of fresh medium. Freeze remaining cells for later analysis.

Repeat steps 2–3 1500 times

Ara⁺ bacteria from a later generation

Ara⁻ bacteria from first generation

4. Put an equal number of cells from generation 1 and a later generation in fresh growth medium (competition experiment).

5. Incubate overnight and count the cells. Which are more numerous?

Prediction: Descendant populations have higher average fitness. (There will be more individuals from descendant than ancestral populations on the plates.)

Prediction of null hypothesis: There will be no difference in fitness between descendant and ancestral populations.

Results:

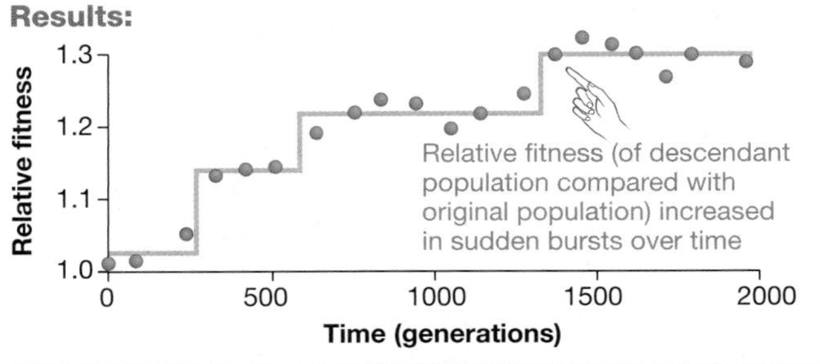

Conclusion: Descendant populations have higher fitness than do ancestral populations.

FIGURE 25.9 An Experiment to Test Changes in Fitness over Time.

grown on the same plates, they produce colonies that look white. Under the temperature and nutrient conditions used in the experiment, however, neither *Ara*⁻ nor *Ara*⁺ cells had a selective advantage. The presence of the *Ara* allele simply gave the researchers a way to identify particular populations by color. What happened to the descendants of these founding cells over time?

To answer this question, the researchers transferred a small number of cells from each of the populations into a new batch of the same growth medium, under the same light and temperature conditions, every day for over four years. In this way, each population grew continuously. Over the course of the experiment, the researchers estimated that each population underwent a total of 10,000 generations. This is the equivalent of over 200,000 years of human evolution. In addition, the biologists saved a sample of cells from each population at regular intervals and stored them in a freezer. The frozen cells served as a fossil record of cells that existed over the 10,000 generation time interval. But because frozen *E. coli* cells resume growth when they are thawed, the "fossil" individuals could be brought back to life.

The strain of *E. coli* used in the experiment is completely asexual and reproduces by cell division. Thus, mutation was the only source of genetic variation in these populations. Although no gene flow occurred, both selection and genetic drift were operating in each population.

Were cells from the older and newer generations of each population different? This question was addressed via competition experiments. *Ara*⁺ cells from one generation and *Ara*⁻ cells from a different generation were put together in the same flask. After allowing the cells to grow and compete, the researchers grew them on special plates and counted the number of red versus white colonies. At the end of these competition experiments, the population of *Ara*⁺ or *Ara*⁻ cells that was more numerous had grown the fastest, meaning that it was better adapted to the experimental environment. In this way the researchers could measure the fitness of descendant populations relative to ancestral populations. If relative fitness was greater than 1, it meant that recent-generation cells outnumbered older-generation cells when the competition was over.

The data from a series of competition experiments are graphed at the bottom of Figure 25.9. Notice that relative fitness increased dramatically—almost 30 percent—over time. But notice also that fitness increased in fits and starts. This pattern is emphasized by the solid line on the graph, which represents a mathematical function fitted to the data points.

What caused this stair-step pattern? Lenski's group hypothesizes that genetic drift was relatively unimportant in this experiment because population sizes were so large. Instead, they propose that each jump was caused by a novel mutation that conferred a fitness benefit. Their interpretation is that cells that happened to have the beneficial mutation grew rapidly and came to dominate the population. After a beneficial

SUMMARY TABLE 25.4 **Evolutionary Mechanisms**

Process	Definition and Notes	Effect on Genetic Variation	Effect on Average Fitness
Selection	Certain alleles are favored	Can lead to maintenance, increase, or reduction of genetic variation	Can produce adaptation
Genetic drift	Random changes in allele frequencies: most important in small populations	Tends to reduce genetic variation, via loss or fixation of alleles	Usually reduces average fitness
Gene flow	Movement of alleles between populations; reduces differences between populations	May increase genetic variation by introducing new alleles; may decrease it by removing alleles	May increase average fitness by introducing high-fitness alleles or decrease it by introducing low-fitness alleles
Mutation	Production of new alleles	Increases genetic variation by introducing new alleles	Random with respect to fitness; most mutations in coding sequences lower fitness

mutation occurred, the fitness of the population stabilized—sometimes for hundreds of generations—until another random but beneficial mutation occurred and produced another jump in fitness.

The experiment makes an important point: Mutation is the ultimate source of genetic variation. If mutation did not occur, evolution would eventually stop. Without mutation, there is no variation for natural selection to act on.

To summarize, mutation alone is usually inconsequential in changing allele frequencies at a particular gene. When considered across the genome and when combined with natural selection, however, it becomes an important evolutionary mechanism. Mutation is the ultimate source of the heritable variation that makes evolution possible.

Table 25.4 summarizes the four evolutionary forces and their consequences. If one or more of these processes affects a gene, then genotypes will not be in Hardy-Weinberg proportions. But we have yet to consider the effects of another assumption in the Hardy-Weinberg model—that mating takes place at random with respect to the gene in question. What happens when this assumption is violated?

25.6 Nonrandom Mating

In the Hardy-Weinberg model, gametes were picked from the gene pool at random and paired to create offspring genotypes. In nature, however, matings between individuals may not be random with respect to the gene in question. Even in species like clams that simply broadcast their gametes into the surrounding water, gametes from individuals who live close to each other are more likely to combine than gametes from individuals that live farther apart. In insects, vertebrates, and many other animals, females don't mate at random but actively choose certain males.

Two mechanisms that violate the Hardy-Weinberg assumption of random mating have been studied intensively. One is the phenomenon known as inbreeding; the other is the process called sexual selection.

Inbreeding

Mating between relatives is called **inbreeding**. By definition, relatives share a recent common ancestor. Individuals that inbreed are likely to share alleles they inherited from their common ancestor.

To understand how inbreeding affects populations, let's follow the fate of alleles and genotypes when it occurs. We'll again focus on a single locus with two alleles, A_1 and A_2. Suppose that these alleles initially have equal frequencies of 0.5. But now let's imagine that the individuals in the population don't produce gametes that go into a gene pool. Instead, they self-fertilize. Many flowering plants, for example, contain both male and female organs and routinely self-pollinate. Homozygous parents that self-fertilize produce all homozygous offspring. Heterozygous parents, in contrast, produce homozygous and heterozygous offspring in a 1:2:1 ratio (**Figure 25.10a**). **Figure 25.10b** shows the outcome for the population as a whole. In this figure, the width of the boxes represents the frequency of the three genotypes, which start out at the Hardy-Weinberg ratio of $p^2 : 2pq : q^2$. Notice that the homozygous proportion of the population increases each generation, while the heterozygous proportion is halved. At the end of the four generations illustrated, heterozygotes are rare.

This simple exercise demonstrates two fundamental points about inbreeding. First, inbreeding increases homozygosity. When inbreeding occurs, the frequency of homozygotes increases and the frequency of heterozygotes decreases. In essence, inbreeding takes alleles from heterozygotes and puts them into homozygotes. Second, inbreeding does not cause evolution, because allele frequencies do not change in the population as a whole. Inbreeding changes genotype frequencies—not allele frequencies. Self-fertilization, or selfing, is the most extreme form of inbreeding; but the same outcomes occur, more slowly, with less-extreme forms of inbreeding. If you understand this concept, you should be able to predict how observed genotype frequencies should differ from those expected under the Hardy-Weinberg principle, when inbreeding is occurring.

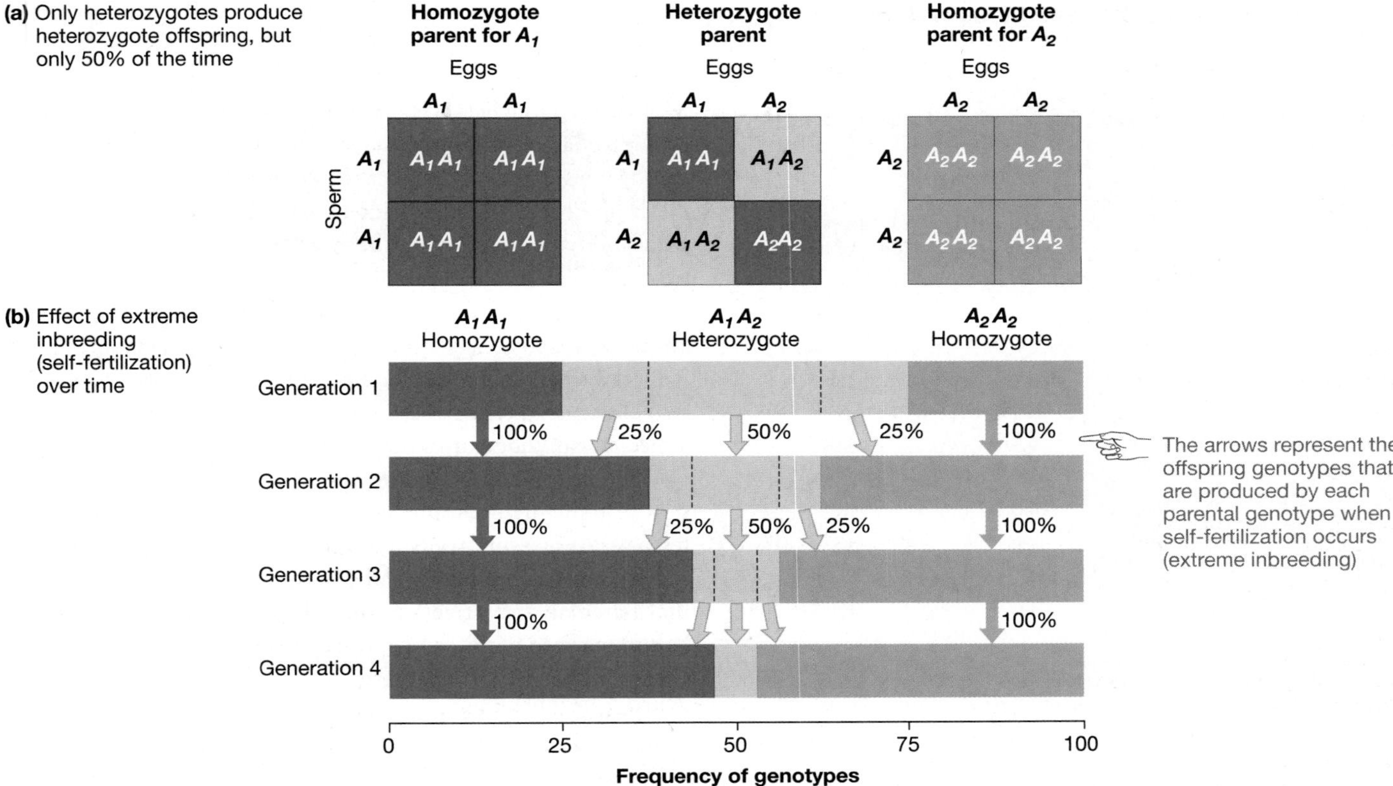

FIGURE 25.10 Inbreeding Increases Homozygosity and Decreases Heterozygosity. (a) Heterozygous parents produce homozygous and heterozygous offspring in a 1:2:1 ratio. **(b)** The width of the boxes corresponds to the frequency of each genotype.

QUESTION What proportion of this population is homozygous in generation 1? In generation 4?

Inbreeding has attracted a great deal of attention because of a phenomenon called inbreeding depression. Inbreeding depression is a decline in average fitness that takes place when homozygosity increases and heterozygosity decreases in a population. Inbreeding depression results from two processes:

1. Many recessive alleles represent loss-of-function mutations. Because these alleles are usually rare, there are normally very few homozygous recessive individuals in a population. Instead, most loss-of-function alleles exist in heterozygous individuals. The alleles have little or no effect when they occur in heterozygotes, because one normal allele usually produces enough functional protein to support a normal phenotype. But inbreeding increases the frequency of homozygous recessive individuals. Loss-of-function mutations are usually deleterious or even lethal when they are homozygous.
2. Many genes—especially those involved in fighting disease—are under intense selection for heterozygote advantage. If an individual is homozygous at these genes, then fitness declines.

The upshot here is that the offspring of inbred matings are expected to have lower fitness than the progeny of outcrossed matings. This prediction has been verified in a wide variety of species, often through laboratory or greenhouse studies that compare the fitnesses of offspring from controlled matings (**Figure 25.11**). As **Table 25.5** shows, inbreeding depression is also

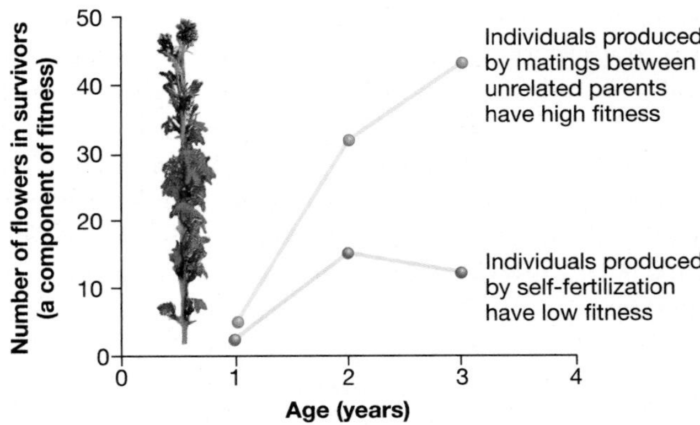

FIGURE 25.11 Inbreeding Depression Occurs in *Lobelia cardinalis*. Inbreeding depression is the fitness difference between non-inbred and inbred individuals.

EXERCISE Label the parts on this graph that indicate inbreeding depression. Does inbreeding depression increase with age in this species or remain constant throughout life?

TABLE 25.5 **Inbreeding Reduces Fitness in Humans**

The percentages reported here give the mortality rate of children produced by first-cousin marriages versus marriages between nonrelatives. In every study, children of first-cousin marriages have a higher mortality rate.

		Deaths	
Age	**Period**	**Children of First Cousins (%)**	**Children of Nonrelatives (%)**
Children under 20 (U.S.)	18th–19th century	17.0	12.0
Children under 10 (U.S.)	1920–1956	8.1	2.4
At/before birth (France)	1919–1950	9.3	3.9
Children (France)	1919–1950	14.0	10.0
Children under 1 (Japan)	1948–1954	5.8	3.5
Children 1–8 (Japan)	1948–1954	4.6	1.5

SOURCE: From *Principles of Human Genetics,* by Curt Stern: © 1973 by W. H. Freeman. Used with permission.

pronounced in humans. Because inbreeding has such deleterious consequences in humans, it is not surprising that many contemporary human societies have laws forbidding marriages between individuals who are related as first cousins or closer. ● If you understand inbreeding depression, you should be able to explain why it does not occur in species like garden peas, where self-fertilization has occurred routinely for many generations.

The trickiest point to grasp about inbreeding is that even though it does not cause evolution directly—because it does not change allele frequencies—it can speed the rate of evolutionary change. More specifically, it increases the rate at which purifying selection eliminates deleterious recessive alleles from a population. A moment's thought should convince you why this is so. Deleterious recessives are usually very rare in populations, because they lower fitness. Rare alleles are usually found in heterozygotes because it is more likely for individuals to have one copy of a rare allele than two. When no inbreeding is occurring, the deleterious recessives found in heterozygotes cannot be eliminated by natural selection. But when inbreeding occurs, most deleterious recessives are found in homozygotes and are quickly eliminated by selection. ● If you understand this concept, you should be able to explain why inbreeding helps "purge" deleterious recessive alleles.

Sexual Selection

If female peacocks choose males with the longest and most iridescent tails as mates, then nonrandom mating is occurring. Due to this type of nonrandom mating, the frequency of alleles that contribute to long, iridescent tails will increase in the population.

Charles Darwin was the first biologist to recognize that selection based on success in courtship is a mechanism of evolutionary change. The process is called **sexual selection**, and it can be considered a special case of natural selection. ● Sexual selection occurs when individuals within a population differ in their ability to attract mates. It favors individuals with heritable traits that enhance their ability to obtain mates.

In 1948 A. J. Bateman contributed a fundamental insight about how sexual selection works. His idea was elaborated by Robert Trivers in 1972. The Bateman-Trivers theory contains two elements: a claim about a pattern in the natural world and a mechanism that causes the pattern.

The pattern component of their theory is that sexual selection usually acts on males much more strongly than on females. As a result, traits that attract members of the opposite sex are much more highly elaborated in males. The mechanism that Bateman and Trivers proposed to explain this pattern can be summarized with a quip: "Eggs are expensive, but sperm are cheap." That is, the energetic cost of creating a large egg is enormous, whereas a sperm contains few energetic resources. Thus, in most species, females invest much more in their offspring than do males. This phenomenon is called the fundamental asymmetry of sex. It is characteristic of almost all sexual species and has two important consequences:

1. Because eggs are large and energetically expensive, females produce relatively few young over the course of a lifetime. A female's fitness is limited primarily by her ability to gain the resources needed to produce more eggs and healthier young—not by the ability to find a mate.

2. Sperm are so simple to produce that a male can father an almost limitless number of offspring. For males, fitness is limited not by the ability to acquire the resources needed to produce sperm, but by the number of females they can mate with.

The theory of sexual selection makes strong predictions. If females invest a great deal in each offspring, then they should protect that investment by being choosy about their mates. Conversely, if males invest little in each offspring, then they should be willing to mate with almost any female. If there are an equal number of males and females in the population, and if males are trying to mate with any female possible, then males will have to compete with each other for mates. As a result, sexual selection should act more strongly on males than on females. Traits that evolve due to sexual selection—meaning traits that are useful only in courtship or in competition for mates—should be found primarily in males.

Do data from experimental or observational studies agree with these predictions? Let's consider each of them in turn.

Sexual Selection via Female Choice If females are choosy about which males they mate with, what criteria do females use to make their choice? Recent experiments have shown that in several bird species, females prefer to mate with males that are well fed and in good health. These experiments were motivated by three key observations: (1) In many bird species, the existence of colorful feathers or a colorful beak is due to the presence of the red and yellow pigments called carotenoids. (2) Carotenoids protect tissues and stimulate the immune system to fight disease more effectively. (3) Animals cannot synthesize their own carotenoids, but plants can. Animals have to obtain carotenoids by eating carotenoid-rich plant tissues.

These observations suggest that the healthiest and best-nourished birds in a population have the most colorful beaks and feathers. Sick birds have dull coloration because they are using all of their carotenoids to stimulate their immune system. Poorly fed birds have dull coloration because they have few carotenoids available. By choosing a colorful male as the father of her offspring, a female is likely to have offspring with alleles that will help the offspring fight disease effectively and feed efficiently.

To test the hypothesis that females prefer to mate with colorful males, a team of researchers experimented with zebra finches (**Figure 25.12a**). They fed one group of male zebra finches a diet that was heavily supplemented with carotenoids, and they fed a second group of male zebra finches (the control group) a diet that was similar in every way except for the additional carot-enoids. To control for other differences between the groups, the researchers used individuals that had been raised and maintained in environments that were as similar as possible, identified pairs of brothers, and randomly assigned one brother to the control group and one brother to the treatment group.

As predicted, the males eating the carotenoid-supplemented diet developed more colorful beaks than did the males fed the carotenoid-poor diet (**Figure 25.12b**). When given a choice of mating with either of the two brothers, 9 out of 10 females tested preferred the more-colorful male. These results are strong evidence that females of this species are choosy about their mates and that they prefer to mate with healthy, well-fed males.

Enough experiments have been done on other bird species to support a general conclusion: Colorful beaks and feathers, along with songs and dances and other types of courtship displays, carry the message "I'm healthy and well fed because I have good alleles. Mate with me."

Choosing "good alleles" is not the entire story in sexual selection via female choice, however. In many species, females prefer to mate with males that care for young or that provide the resources required to produce eggs. Brown kiwi females make an enormous initial investment in their offspring—their eggs routinely represent over 15 percent of the mother's total body weight (**Figure 25.13**)—but choose to mate with males that take over all of the incubation and other care of the offspring. It is common to find that female fish prefer to mate

(a) Male zebra finches have orange beaks.

(b) Males fed carotenoids get more colorful beaks.

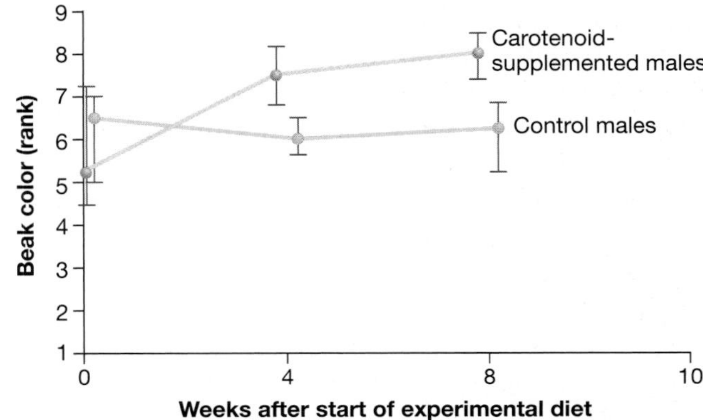

FIGURE 25.12 If Male Zebra Finches Are Fed Carotenoids, Their Beaks Get More Colorful.

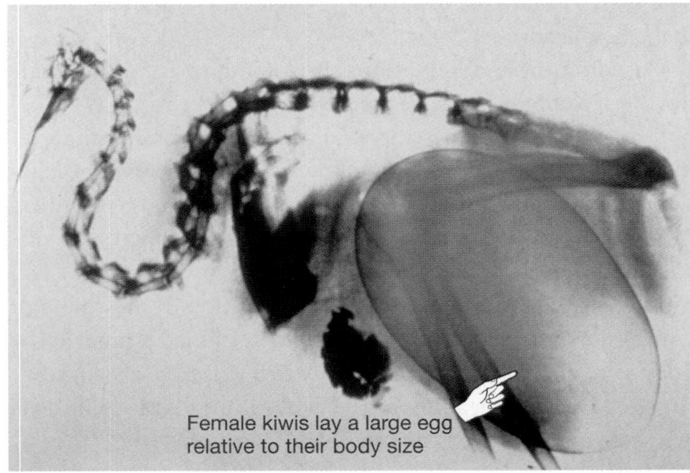

FIGURE 25.13 In Many Species, Females Make a Large Investment in Each Offspring. X-ray of a female kiwi, ready to lay an egg.

● **QUESTION** Suppose you are examining the gametes of a newly discovered species. How do you know which are sperm and which are eggs?

with males that protect a nest site and care for the eggs until they hatch. In humans and many species of birds, males provide food, protection, and other resources required for rearing young.

To summarize, females may choose mates on the basis of (1) physical characteristics that signal male genetic quality, (2) resources or parental care provided by males, or (3) both. In some species, however, females do not have the luxury of choosing a male. Instead, competition among males is the primary cause of sexual selection.

Sexual Selection via Male-Male Competition As an example of research on how males compete for mates, consider data from a long-term study of a northern elephant seal population breeding on Año Nuevo Island, off the coast of California. Elephant seals feed on marine fish and spend most of the year in the water. But when females are ready to mate and give birth, they haul themselves out of the water onto land. Females prefer to give birth on islands, where newborn pups are protected from terrestrial and marine predators. Because elephant seals have flippers that are ill suited for walking, females can haul themselves out of the water only on the few beaches that have gentle slopes. As a result, large numbers of females congregate in tiny areas to breed.

Male elephant seals establish territories on breeding beaches by fighting (**Figure 25.14a**). A **territory** is an area that is actively defended and that provides exclusive use by the owner. Males that win battles with other males monopolize matings with the females residing in their territories. Females can't choose among males—they simply have to mate with the winning male. Males that lose battles are relegated to territories with few females or are excluded from the beach. Fights are essentially slugging contests and are usually won by the larger male. The males stand face to face, bite each other, and land blows with their heads.

Based on these observations, it is not surprising that male northern elephant seals frequently weigh three tons (2700 kg) and are over four times more massive, on average, than females. The logic here runs as follows: Males that own beaches with large congregations of females will father large numbers of offspring. Males that lose fights will father few or no offspring. As a result, the alleles of territory-owning males will rapidly increase in frequency in the population. If the ability to win fights and produce offspring is determined primarily by body size, then alleles for large body size will have a significant fitness advantage, leading to the evolution of large male size. The fitness advantage is due to sexual selection.

Figure 25.14b provides evidence for intense sexual selection in males. Biologists have marked many of the individuals in the seal population on Año Nuevo to track the lifetime reproductive success of a large number of individuals. As the data show, in this population a few males father a large number of offspring, while most males father few or none. Among females, variation in reproductive success is also high; but it is much lower than in males (**Figure 25.14c**). In this species, most sexual selection is driven by male-male competition rather than female choice.

(a) Males compete for the opportunity to mate with females.

(b) Variation in reproductive success is high in males.

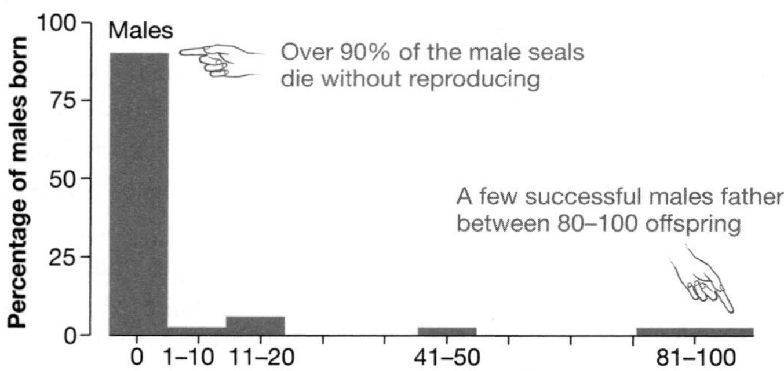

(c) Variation in reproductive success is relatively low in females.

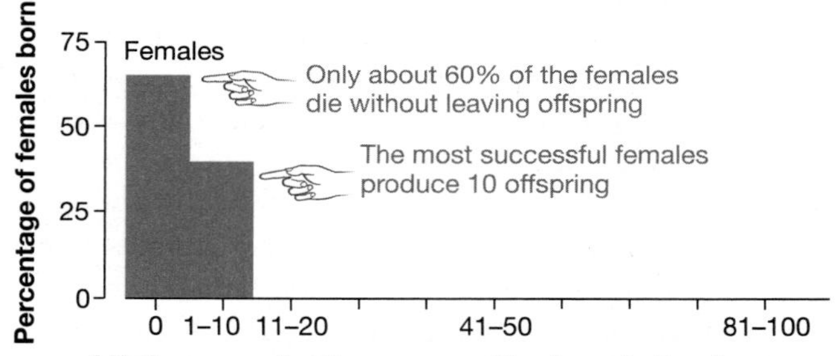

FIGURE 25.14 Intense Sexual Selection on Male Elephant Seals.
(a) Two bull northern elephant seals compete for the opportunity to mate with females that will breed on this beach. The histograms show that variation in lifetime reproductive success is even higher **(b)** in male northern elephant seals than it is **(c)** in females.

● **QUESTION** Consider an allele that increases reproductive success in males versus an allele that increases reproductive success in females. Which allele will increase in frequency faster, and why?

During the breeding season, males of the beetle *Dynastes granti* use their elongated horns to fight over females.

Male scarlet tanagers use their bright coloration in territorial and courtship displays.

Male lions are larger than female lions and have an elaborate ruff of fur called a mane.

FIGURE 25.15 Sexually Selected Traits Are Used to Compete for Mates. Males often have exaggerated traits that they use in fighting or courtship. In many species, females lack these traits.
[(b) top ©B. Schorre/VIREO; (b) bottom ©R. & A. Simpson/VIREO]

QUESTION How would you set up an experiment to test the hypothesis that the exaggerated traits shown here help males of these species attract mates?

What Are the Consequences of Sexual Selection? In elephant seals and most other animals studied, most females that survive to adulthood get a mate. In contrast, many males do not. Because sexual selection tends to be much more intense in males than females, males tend to have many more traits that function only in courtship or male-male competition. Stated another way, sexually selected traits often differ sharply between the sexes.

Sexual dimorphism (literally "two-forms") refers to any trait that differs between males and females. **Figure 25.15** illustrates sexually dimorphic traits. They range from weapons that males use to fight over females, such as antlers and horns, to the elaborate ornamentation and behavior used in courtship displays. Humans are sexually dimorphic in size, distribution of body hair, and many other traits.

Like inbreeding and other forms of nonrandom mating, sexual selection violates the assumptions of the Hardy-Weinberg principle. Unlike inbreeding, however, it causes certain alleles to increase or decrease in frequency and results in evolution.

Check Your Understanding

If you understand that...

- Inbreeding and sexual selection are forms of nonrandom mating.
- Inbreeding is mating between relatives. It is not an evolutionary process, because it changes genotype frequencies—not allele frequencies. Inbreeding increases homozygosity and may lead to inbreeding depression.
- Sexual selection is based on differential success in obtaining mates. It causes evolution by increasing the frequency of alleles associated with successful courtship.

You should be able to...

1) Define the fundamental asymmetry of sex.
2) Explain why males are usually the sex with exaggerated traits used in courtship.
3) Explain how male-male competition for mates differs from female choice of mates.

Chapter Review

SUMMARY OF KEY CONCEPTS

The Hardy-Weinberg principle acts as a null hypothesis when researchers want to test whether evolution or nonrandom mating is occurring at a particular gene.

Biologists study the consequences of the different evolutionary mechanisms through a combination of mathematical modeling and experimental or observational research. The Hardy-Weinberg principle can serve as a null hypothesis in evolutionary studies because it predicts what genotype and allele frequencies are expected to be if mating is random with respect to the gene in question and none of the four evolutionary processes is operating on that gene.

You should be able to predict how genotype frequencies differ from Hardy-Weinberg proportions under selection for heterozygote advantage, and directional selection.

MB Web Animation at www.masteringbio.com
The Hardy-Weinberg Principle

Each of the four evolutionary mechanisms has different consequences. Only natural selection produces adaptation. Genetic drift causes random fluctuations in allele frequencies. Gene flow equalizes allele frequencies between populations. Mutation introduces new alleles.

Natural selection occurs in a wide variety of patterns and a wide variety of intensities. Directional selection may lead to certain alleles becoming fixed—and thus reduces allelic diversity in populations. Stabilizing selection eliminates phenotypes with extreme characteristics and decreases allelic diversity in populations. The rate of evolution under natural selection depends on both the intensity of selection and the amount of genetic variation available.

Genetic drift results from sampling error and is an important evolutionary force in small populations. Drift leads to the random fixation of alleles and tends to reduce overall allelic diversity.

Gene flow is the movement of alleles between populations. Gene flow tends to homogenize allele frequencies and decrease differentiation among populations, but it can also serve as an important source of new variation in populations.

Mutation is too infrequent to be a major cause of allele frequency change. But because mutation continually introduces new alleles at all genes, it is essential to evolution. Without mutation, natural selection and genetic drift would eventually eliminate genetic variation, and evolution would cease.

You should be able to explain how the evolutionary forces affect the management of endangered species. What effect will drift have? How will gene flow from captive populations affect wild populations?

MB Web Animation at www.masteringbio.com
Three Modes of Natural Selection

Inbreeding changes genotype frequencies but does not change allele frequencies.

Inbreeding, or mating among relatives, is a form of nonrandom mating. Inbreeding does not change allele frequencies, so it is not an evolutionary mechanism. It does, however, change genotype frequencies by leading to an increase in homozygosity and a decrease in heterozygosity. These patterns can accelerate natural selection and can cause inbreeding depression.

You should be able to predict how extensive inbreeding during the 1700s and 1800s affected the royal families of Europe.

Sexual selection leads to the evolution of traits that help individuals attract mates. It is usually stronger on males than on females.

Sexual selection is a form of nonrandom mating. It occurs when certain traits help males succeed in contests over mates or when certain traits are attractive to females. It is responsible for the evolution of phenotypic differences between males and females.

You should be able to predict the pattern of sexual dimorphism observed in animal species where males make a much larger investment in offspring than females do.

QUESTIONS

Test Your Knowledge

1. Why isn't inbreeding considered an evolutionary mechanism?
   a. It does not change genotype frequencies.
   b. It does not change allele frequencies.
   c. It does not occur often enough to be important in evolution.
   d. It does not violate the assumptions of the Hardy-Weinberg principle.
2. Why is genetic drift aptly named?
   a. It causes allele frequencies to drift up or down randomly.
   b. It is the ultimate source of genetic variability.
   c. It is an especially important mechanism in small populations.
   d. It occurs when populations drift into new habitats.
3. How do sexual selection and inbreeding differ?
   a. Unlike inbreeding, sexual selection changes allele frequencies and affects only genes involved in attracting mates.
   b. Unlike sexual selection, inbreeding changes allele frequencies and involves any mating between relatives—not just self-fertilization.
   c. Unlike inbreeding, sexual selection results from the random fusion of gametes during fertilization. It is particularly important in small populations, where few mates are available.
   d. Inbreeding occurs only in small populations, while sexual selection can occur in any size population.

4. What does it mean when an allele reaches "fixation"?
   a. It is eliminated from the population.
   b. It has a frequency of 1.0.
   c. It is dominant to all other alleles.
   d. It is adaptively advantageous.

5. In what sense is the Hardy-Weinberg principle a null hypothesis, similar to the control treatment in an experiment?
   a. It defines what genotype frequencies should be if nonrandom mating is occurring.
   b. Expected genotype frequencies can be calculated from observed allele frequencies and then compared with the observed genotype frequencies.
   c. It defines what genotype frequencies should be if natural selection, genetic drift, gene flow, or mutation is occurring *and* if mating is random.
   d. It defines what genotype frequencies should be if evolutionary mechanisms are *not* occurring.

6. Mutation is the ultimate source of genetic variability. Why is this statement correct?
   a. DNA polymerase (the enzyme that copies DNA) is remarkably accurate.
   b. "Mutation proposes and selection disposes."
   c. Mutation is the only source of new alleles.
   d. Mutation occurs in response to natural selection. It generates the alleles that are required for a population to adapt to a particular habitat.

Test Your Knowledge answers: 1.b; 2.a; 3.a; 4.b; 5.d; 6.c

Test Your Understanding

Answers are available at www.masteringbio.com

1. Create concept map summarizing the effects of selection, drift, gene flow, mutation, and inbreeding on genetic variation and fitness. In addition to indicating whether each of the five processes increases or decreases genetic variation and average fitness, your diagram should indicate why.
2. Directional selection can lead to the fixation of favored alleles. When this occurs, genetic variation is zero and evolution stops. Explain why this rarely occurs.
3. Why does sexual selection often lead to sexual dimorphism? Why are males usually the sex that exhibits exaggerated characteristics?
4. Is it possible for one gene in a population *not* to have genotype frequencies that are in Hardy-Weinberg proportions, even though all other genes in the population are in Hardy-Weinberg proportions? Explain why or why not.
5. Explain why small populations become inbred.
6. Explain why genetic drift is much more important in small populations than large populations.

Applying Concepts to New Situations

Answers are available at www.masteringbio.com

1. In humans, albinism is caused by loss-of-function mutations in genes involved in the synthesis of melanin, the dark pigment in skin. Only people homozygous for a loss-of-function allele have the relevant phenotype. In Americans of northern European ancestry, albino individuals are present at a frequency of about 1 in 10,000 (or 0.0001). Knowing this genotype frequency, we can calculate the frequency of the loss-of-function alleles. If we let p_2 stand for this frequency, we know that $p_2^2 = 0.0001$; therefore $p_2 = \sqrt{0.0001} = 0.01$. By subtraction, the frequency of normal alleles is 0.99. If the genes responsible for albinism conform to the conditions required by the Hardy-Weinberg principle, what is the frequency of "carriers"—or people who are heterozygous for this condition? Your answer indicates the percentage of Caucasians in the United States who carry an allele for albinism.
2. Conservation managers frequently use gene flow, in the form of transporting individuals or releasing captive-bred young, to counteract the effects of drift on small, endangered populations. Explain how gene flow can also mitigate the effects of inbreeding.
3. Suppose you were studying several species of human. In one, males never lifted a finger to help females raise children. In another, males did just as much parental care as females except for actually carrying the baby during pregnancy. How does the fundamental asymmetry of sex compare in the two species? How would you expect sexual dimorphism to compare between the two species?
4. You are a conservation biologist charged with creating a recovery plan for an endangered species of turtle. The turtle's habitat has been fragmented into small, isolated but protected areas by suburbanization and highway construction. Some evidence indicates that certain turtle populations are adapted to marshes that are normal, whereas others are adapted to acidic wetlands or salty habitats. Further, some turtle populations number less than 25 breeding adults, making genetic drift and inbreeding a major concern. In creating a recovery plan, the tools at your disposal are captive breeding, the capture and transfer of adults to create gene flow, or the creation of habitat corridors between wetlands to make migration possible. Write a two-paragraph essay outlining the major features of your proposal.

wwww.masteringbio.com is also your resource for • Answers to text, table, and figure caption questions and exercises • Answers to *Check Your Understanding* boxes • Online study guides and quizzes • Additional study tools including the *E-Book for Biological Science* 3rd ed., textbook art, animations, and videos.

26 Speciation

KEY CONCEPTS

- Speciation occurs when populations of the same species become genetically isolated by lack of gene flow and then diverge from each other due to selection, genetic drift, or mutation.
- Populations can be recognized as distinct species if they are reproductively isolated from each other, if they have distinct morphological characteristics, or if they form independent branches on a phylogenetic tree.
- Populations can become genetically isolated from each other if they occupy different geographic areas, if they use different habitats or resources within the same area, or if one population is polyploid and cannot breed with the other.
- When populations that have diverged come back into contact, they may fuse, continue to diverge, stay partially differentiated, or have offspring that form a new species.

The Bullock's oriole was once considered a member of the same species as the Baltimore oriole, because the two populations interbreed in certain regions. Recent analyses, based on the phylogenetic species concept introduced in this chapter, have shown that Bullock's orioles are an independent species.

Although Darwin called his masterwork *On the Origin of Species by Means of Natural Selection*, he actually had little to say about how new species arise. Instead, his data and analyses focused on the process of natural selection and the changes that occur within populations over time. He spent much less time considering changes that occur *between* populations.

Since Darwin, however, biologists have realized that populations of the same species may diverge from each other when they are isolated in terms of gene flow. Recall from Chapter 25 that gene flow equalizes allele frequencies between populations. When gene flow ends, allele frequencies in the isolated populations are free to diverge—meaning that the populations begin to evolve independently of each other. If mutation, selection, and genetic drift cause isolated populations to diverge sufficiently, distinct types, or species, form—that is, the process of **speciation** takes place. Speciation is a splitting event that creates two or more distinct species from a single ancestral group (**Figure 26.1**). When speciation is complete, a new branch has been added to the tree of life.

In essence, then, speciation results from genetic isolation and genetic divergence. Isolation results from lack of gene flow, and divergence occurs because selection, genetic drift, and mutation proceed independently in the isolated populations. How does genetic isolation come about? And how do selection, drift, and mutation cause divergence?

This chapter is devoted to exploring these questions. Our first task is to examine how species are defined and identified. Subsequent sections focus on how speciation occurs in two contrasting situations: when populations are separated into

Key Concept | Important Information | Practice It

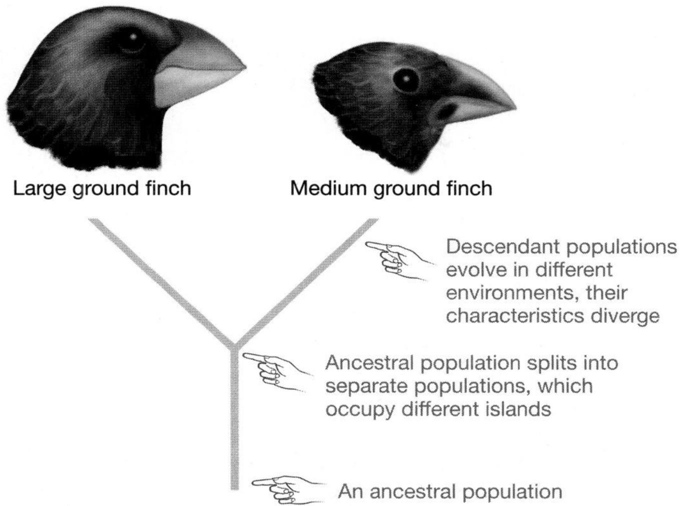

FIGURE 26.1 Speciation Creates Evolutionarily Independent Populations. The large ground finch and medium ground finch are derived from the same ancestral population. This ancestral population split into two populations isolated by lack of gene flow. Because the populations began evolving independently, they acquired the distinctive characteristics observed today.

distinct regions and when they occupy the same geographic area. The chapter concludes with a look at a classical question in speciation research: What happens when populations that have been isolated from one another come back into contact? Do they interbreed and merge back into the same species, or do they remain independent and form new species?

26.1 How Are Species Defined and Identified?

Species are distinct types of organisms and represent evolutionarily independent groups. Like the Galápagos finches in Figure 26.1, species are distinct from one another in appearance, behavior, habitat use, or other traits. These characteristics differ among species because their genetic characteristics differ. Genetic distinctions occur because mutation, selection, and drift act on each species independently of what is happening in other populations.

What makes one species "evolutionarily independent" of other species? The answer begins with *lack* of gene flow. As Chapter 25 explained, gene flow eliminates genetic differences among populations. Allele frequencies in populations, and thus the populations' characteristics, become more alike when gene flow occurs between them. If gene flow between populations is extensive and continues over time, it eventually causes even highly distinct populations to coalesce into the unit known as a species. Conversely, if gene flow between populations stops, then mutation, selection, and drift begin to act on the populations independently. If a new mutation creates an allele that changes the phenotype of individuals in one population, there is no longer any way for that allele to appear in the other population. As a result, allele frequencies and other characteristics in the populations diverge. When allele frequencies change sufficiently over time, populations become distinct species.

Formally, then, a **species** is defined as an evolutionarily independent population or group of populations. Even though this definition sounds straightforward, it can be exceedingly difficult to put into practice. How can evolutionarily independent populations be identified in the field and in the fossil record? There is no single, universal answer. Even though biologists agree on the definition of a species, they frequently have to use different sets of criteria to identify them. Three criteria for identifying species are in common use: (1) the biological species concept, (2) the morphospecies concept, and (3) the phylogenetic species concept.

The Biological Species Concept

According to the **biological species concept**, the critical criterion for identifying species is reproductive isolation. This is a logical yardstick because no gene flow occurs between populations that are reproductively isolated from each other. Specifically, if two different populations do not interbreed in nature, or if they fail to produce viable and fertile offspring when matings take place, then they are considered distinct species. Groups that naturally or potentially interbreed, and that are reproductively isolated from other groups, belong to the same species. Biologists can be confident that reproductively isolated populations are evolutionarily independent.

Reproductive isolation can result from a wide variety of events and processes. To organize the various mechanisms that stop gene flow between populations, biologists distinguish (1) **prezygotic** (literally, "before-zygote") **isolation**, which prevents individuals of different species from mating, and (2) **postzygotic** (literally, "after-zygote") **isolation**, in which the offspring of matings between members of different species do not survive or reproduce. In prezygotic isolation, reproductive isolation occurs before mating can occur. In postzygotic isolation, interspecies mating does occur, but any hybrid offspring produced have low fitness. **Table 26.1** summarizes some of the more important mechanisms of prezygotic and postzygotic isolation.

Although the biological species concept has a strong theoretical foundation, it has disadvantages. The criterion of reproductive isolation cannot be evaluated in fossils or in species that reproduce asexually. In addition, it is difficult to apply when closely related populations do not happen to overlap with each other geographically. In this case, biologists are left to guess whether interbreeding and gene flow would occur if the populations happened to come into contact.

The Morphospecies Concept

How do biologists identify species when the criterion of repro ductive isolation cannot be applied? Under the **morphospe** ("form-species") **concept**, researchers identify evoluti

TABLE **26.1** **Mechanisms of Reproductive Isolation**

Type	Description	Example
Prezygotic Isolation		
Temporal	Populations are isolated because they breed at different times.	Bishop pines and Monterey pines release their pollen at different times of the year.
Habitat	Populations are isolated because they breed in different habitats.	Parasites that begin to exploit new host species are isolated from their original population.
Behavioral	Populations do not interbreed because their courtship displays differ.	To attract male fireflies, female fireflies give a species-specific sequence of flashes.
Gametic barrier	Matings fail because eggs and sperm are incompatible.	In sea urchins, a protein called bindin allows sperm to penetrate eggs. Differences in the amino acid sequence of bindin cause matings to fail between closely related populations.
Mechanical	Matings fail because male and female genitalia are incompatible.	In many insects, the male copulatory organ and female reproductive canal fit like a "lock and key." Changes in either organ initiate reproductive isolation.
Postzygotic Isolation		
Hybrid viability	Hybrid offspring do not develop normally and die as embryos.	When ring-necked doves mate with rock doves, less than 6% of eggs hatch.
Hybrid sterility	Hybrid offspring mature but are sterile as adults.	Eastern meadowlarks and western meadowlarks are almost identical morphologically, but hybrid offspring are largely infertile.

independent lineages by differences in size, shape, or other morphological features. The logic behind the morphospecies concept is that distinguishing features are most likely to arise if populations are independent and isolated from gene flow.

The morphospecies concept is compelling simply because it is so widely applicable. It is a useful criterion when biologists have no data on the extent of gene flow, and it is equally applicable to sexual, asexual, or fossil species. Its disadvantage is that the features used to distinguish species are subjective. In the worst case, different researchers working on the same populations disagree on the characters that distinguish species. For example, some researchers who work on the fossil record of humans argue that the specimens currently named *Homo habilis* and *Homo rudolfensis* (**Figure 26.2**) actually belong to the same species. Disagreements like these often end in a stalemate, because no independent criteria exist for resolving the conflict.

The Phylogenetic Species Concept

The **phylogenetic species concept** is a recent addition to the tools available for identifying evolutionarily independent lineages and is based on reconstructing the evolutionary history of populations. Proponents of this approach argue that it is widely applicable and precise.

The reasoning behind the phylogenetic species concept begins with Darwin's claim that all species are related by common ancestry. Chapter 1 and Chapter 25 introduced this claim and the phylogenetic trees that are used to represent the

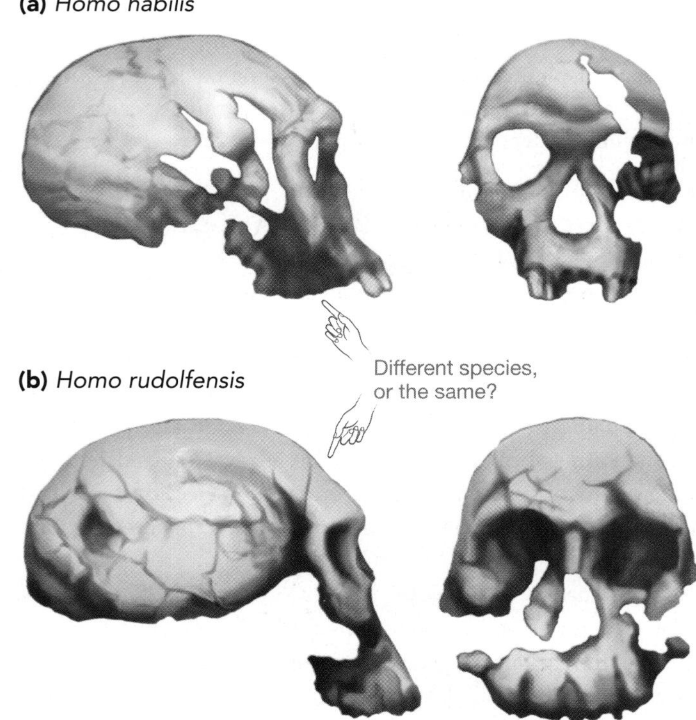

FIGURE 26.2 Morphospecies May Be Difficult to Distinguish from Each Other. Fossils from **(a)** *Homo habilis* and **(b)** *Homo rudolfensis* have been recovered from the same region in Africa, in rocks of the same age. Biologists argue over whether the two populations were distinct enough morphologically to be considered separate species or whether they should be considered the same species.

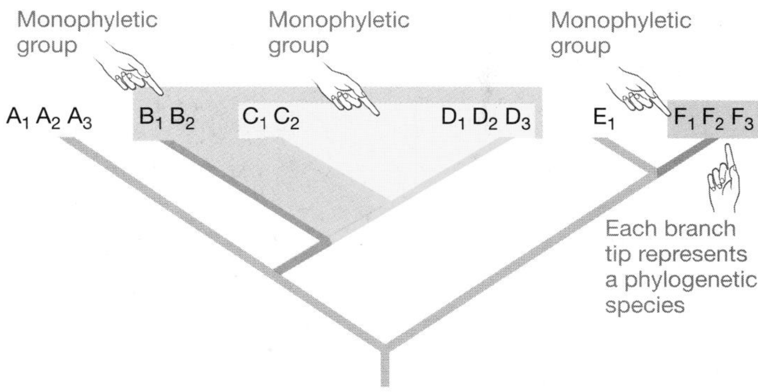

FIGURE 26.3 Monophyletic Groups. The color-coded lineages on this phylogenetic tree are monophyletic because they contain a common ancestor and all its descendants.

EXERCISE Circle the six monophyletic groups that represent phylogenetic species.

genealogical relationships among populations. (For help with interpreting phylogenetic trees, see **BioSkills 2**.) **Figure 26.3** shows such a tree. On this and all other phylogenetic trees, branches represent populations that are changing through time. Notice that each branch ends in a tip. On this tree, there are six tips.

Each of the tips on the tree in Figure 26.3 represents a phylogenetic species. To understand why, it is essential to understand the concept of a monophyletic ("one-tribe") group. A **monophyletic group**, also called a **clade** or **lineage**, consists of an ancestral population, all of its descendants, and *only* those descendants. Three monophyletic groups are color-coded on the tree in Figure 26.3, including one (containing species B, C, and D) that overlaps another monophyletic group (the one containing just species C and D). On any given evolutionary tree, there are many monophyletic groups. Under the phylogenetic species concept, a species is defined as the smallest monophyletic group in a phylogenetic tree that compares populations—as opposed to larger groups such as "pine trees" or "mammals." On a tree of populations, each tip is a phylogenetic species.

The tree in Figure 26.3, for example, is based on data from an array of populations; but many of the populations are not distinctive enough to represent separate tips. The phylogenetic species on the tree are labeled A, B, C, D, E, and F. These are the smallest monophyletic groups on the tree. The clusters at some of the tips (A_1, A_2, A_3, etc.) represent populations within those species. These may be separated geographically, but their characteristics are so similar that they do not form independent tips on the tree. They are simply part of the same monophyletic group containing other populations.

If you understand the phylogenetic species concept, you should be able to draw a tree showing that the relationships among gorillas, common chimps, and humans are like species B, C, and D in Figure 26.3. On this tree, add labels to represent populations of gorillas from Rwanda and Congo, chimps from East Africa versus West Africa, and humans from Siberia, Australia, and North America.

The phylogenetic species concept has two distinct advantages: (1) It can be applied to any population (fossil, asexual, or sexual), and (2) it is logical because populations are distinct enough to be monophyletic only if they are isolated from gene flow and have evolved independently. The approach has a distinct disadvantage, however: Carefully estimated phylogenies are available only for a tiny (though growing) subset of populations on the tree of life. Critics of this approach also point out that it would probably lead to recognition of many more species than either the morphospecies or biological species concept. Proponents counter that, far from being a disadvantage, the recognition of increased numbers of species might better reflect the extent of life's diversity.

In actual practice, researchers use all three species concepts summarized here **(Table 26.2)**. Conflicts have occurred, however, when different species concepts are applied to the real world. To appreciate this point, let's consider the case of the dusky seaside sparrow.

Species Definitions in Action: The Case of the Dusky Seaside Sparrow

Seaside sparrows live in salt marshes along the Atlantic and Gulf Coasts of the United States. Recall from Chapter 1 that

SUMMARY TABLE 26.2 **Species Concepts**

Species Concept	Criterion for Recognizing Species	Advantages	Disadvantages
Biological	Reproductive isolation between populations (they don't breed and don't produce viable offspring)	Reproductive isolation = evolutionary independence	Not applicable to asexual or fossil species; difficult to assess if populations do not overlap geographically
Morphospecies	Morphologically distinct populations	Widely applicable	Subjective (researchers often disagree about how much or what kinds of morphological distinction = speciation)
Phylogenetic	Smallest monophyletic group on phylogenetic tree	Widely applicable; based on testable criteria	Relatively few well-estimated phylogen[illegible] are currently available

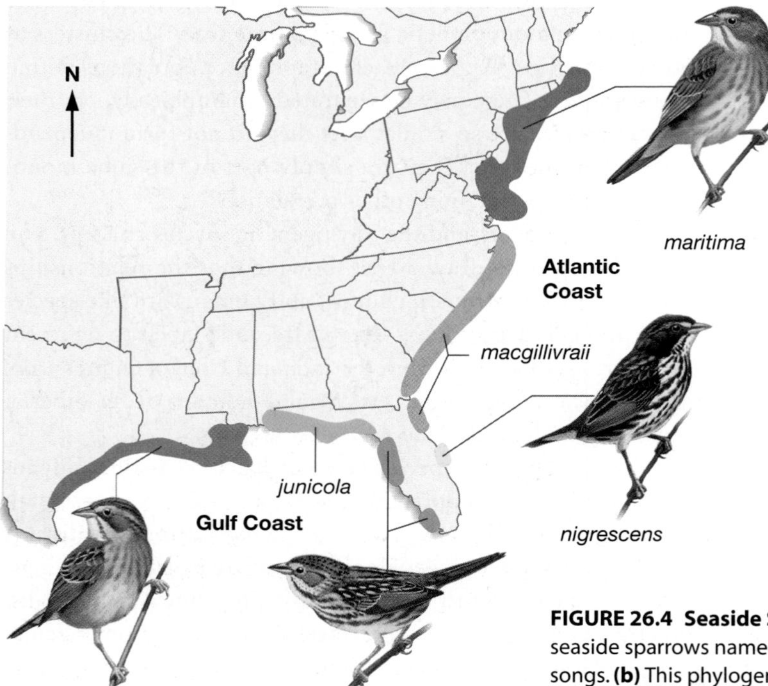

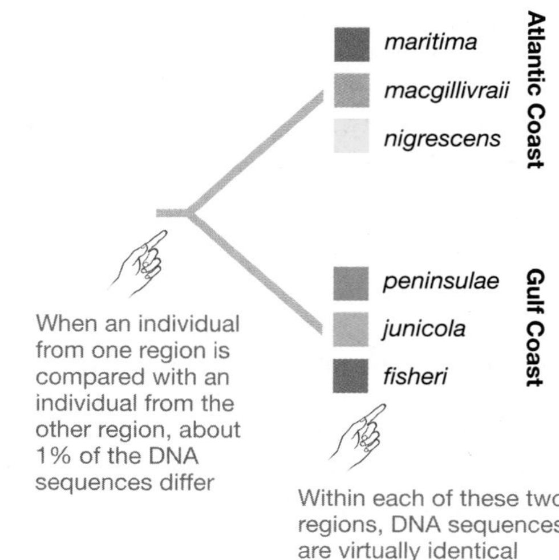

FIGURE 26.4 Seaside Sparrows Form Two Monophyletic Groups. (a) The "subspecies" of seaside sparrows named on this map are distinguished by their distinctive coloration and songs. **(b)** This phylogenetic tree was constructed by comparing DNA sequences. The tree shows that seaside sparrows represent two distinct monophyletic groups, one native to the Atlantic Coast and the other native to the Gulf Coast.

QUESTION If you were a conservation biologist and could save only two subspecies of seaside sparrows from extinction, would you choose two subspecies from the Atlantic Coast, two from the Gulf Coast, or one from the Atlantic and one from the Gulf? Explain why.

scientific names consist of a genus name followed by a species name; the scientific name of this species is *Ammodramus maritimus*. Researchers had traditionally named a variety of seaside sparrow "subspecies" under the morphospecies concept. **Subspecies** are populations that live in discrete geographic areas and have distinguishing features, such as coloration or calls, but are not considered distinct enough to be called separate species.

Because salt marshes are often destroyed for agriculture or oceanfront housing, by the late 1960s biologists began to be concerned about the future of some seaside sparrow populations. A subspecies called the dusky seaside sparrow (*Ammodramus maritimus nigrescens*) was in particular trouble; by 1980 only six individuals from this population remained. All were males.

At this point government and private conservation agencies sprang into action under the auspices of the Endangered Species Act, a law whose goal is to prevent the extinction of species. The law uses the biological species concept to identify species and calls for the rescue of endangered species through active management. Because current populations of seaside sparrows are physically isolated from one another, and because young seaside sparrows tend to breed near where they hatched, researchers believed that little to no gene flow occurred among populations. Under the biological species concept and morphospecies concept, there may be as many as six species of seaside sparrow (**Figure 26.4a**). The dusky seaside sparrow subspecies became a priority for conservation efforts because it was reproductively isolated.

To launch the rescue program, the remaining male dusky seaside sparrows were taken into captivity and bred with females from a nearby subspecies: *A. maritimus peninsulae*. Officials planned to use these hybrid offspring as breeding stock for a reintroduction program. The goal was to preserve as much genetic diversity as possible by reestablishing a healthy population of dusky-like birds. The plan was thrown into turmoil, however, when a different group of biologists estimated the phylogeny of the seaside sparrows by comparing gene sequences. This tree, shown in **Figure 26.4b**, shows that seaside sparrows represent just two distinct monophyletic groups: one native to the Atlantic Coast and the other native to the Gulf Coast. Under the phylogenetic species concept, only two species of seaside sparrow exist. Far from being an important, reproductively isolated population, the phylogeny showed that the dusky sparrow is part of the same monophyletic group that includes the other Atlantic Coast sparrows. Further, officials had unwittingly crossed the dusky males with females from the Gulf Coast lineage. Because the goal of the conservation effort was to preserve existing genetic diversity, this was the wrong population to use.

The researchers who did the phylogenetic analysis maintained that the biological and morphospecies concepts had misled a well-intentioned conservation program. Under the phylogenetic species concept, they claimed that officials should have allowed the dusky sparrow to go extinct and then concentrated their efforts on simply preserving one or more populations from each coast. In this way, the two monophyletic groups of sparrows—and the most genetic diversity—would be preserved. Under the morphospecies concept, however, officials did the right thing by preserving distinct types. They argue that dusky seaside sparrows had distinctive, heritable traits like coloration and songs that are now lost forever.

When conservation funding is scarce, life-and-death decisions like these are crucial. Now our task is to consider an even more fundamental question: How do isolation and divergence produce the event called speciation?

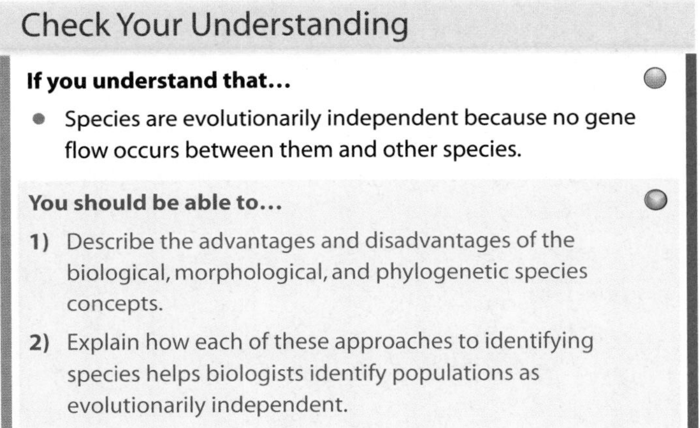
Check Your Understanding

If you understand that...

- Species are evolutionarily independent because no gene flow occurs between them and other species.

You should be able to...

1) Describe the advantages and disadvantages of the biological, morphological, and phylogenetic species concepts.
2) Explain how each of these approaches to identifying species helps biologists identify populations as evolutionarily independent.

26.2 Isolation and Divergence in Allopatry

Speciation begins when gene flow between populations is reduced or eliminated. Genetic isolation happens routinely when populations become physically separated. Physical isolation, in turn, occurs in one of two ways: dispersal or vicariance. As **Figure 26.5a** illustrates, a population can disperse to a new habitat, colonize it, and found a new population. Alternatively, a new physical barrier can split a widespread population into two or more subgroups that are physically isolated from each other (**Figure 26.5b**). A physical splitting of habitat is called **vicariance.** Speciation that begins with physical isolation via either dispersal or vicariance is known as **allopatric** ("different-homeland") **speciation**. Populations that live in different areas are said to be in **allopatry**.

The case studies that follow address two questions: How do colonization and range-splitting events occur? Answering this question takes us into the field of **biogeography**—the study of how species and populations are distributed geographically. Once populations are physically isolated, how do genetic drift and selection produce divergence?

Dispersal and Colonization Isolate Populations

Peter Grant and Rosemary Grant witnessed a colonization event while working in the Galápagos Islands off the coast of South America. Recall from Chapter 24 that they had been studying medium ground finches on the island of Daphne Major since 1971. In 1982 five members of a new species, called the large ground finch, arrived and began nesting. These colonists had apparently dispersed from a population that lived on a nearby island in the Galápagos. Because finches normally stay on the same island year-round, the colonists represented a new population, allopatric with their source population.

The large ground finches' arrival gave the researchers a chance to test a long-standing hypothesis about dispersal and colonization. Decades ago Ernst Mayr suggested that colonization events are likely to trigger speciation, for two reasons: (1) The physical separation between populations reduces or eliminates gene flow, and (2) genetic drift will cause the old and new populations to diverge rapidly. Drift occurs during the colonization event itself via the founder effect introduced in Chapter 25. And if the number of individuals in the new population remains small for several generations, genetic drift will continue to alter allele frequencies. In addition, natural selection may cause divergence if the newly colonized environment is different from the original habitat.

To evaluate whether genetic drift occurred when large ground finches colonized Daphne Major, Grant and Grant caught, weighed, and measured most of the parents and offspring produced on Daphne Major over the succeeding 12 years. When they compared these data with measurements of large ground finches in other populations, they discovered that the average bill size in the new population was much larger. As predicted, the colonists represented a nonrandom sample of the original population. Genetic drift produced a colonizing population with characteristics significantly different from those of the source population.

The novel environment experienced by the colonizers will also expose them to new forms of natural selection. In many cases, the agent of selection is changes in available food. In finches, the size and shape of the beak are closely correlated with the types of seeds or insects that individuals eat. Based on this observation, it is logical to predict that if large seeds are particularly common on Daphne Major, then large ground finches with large beaks will survive and reproduce well new large-beaked species will evolve.

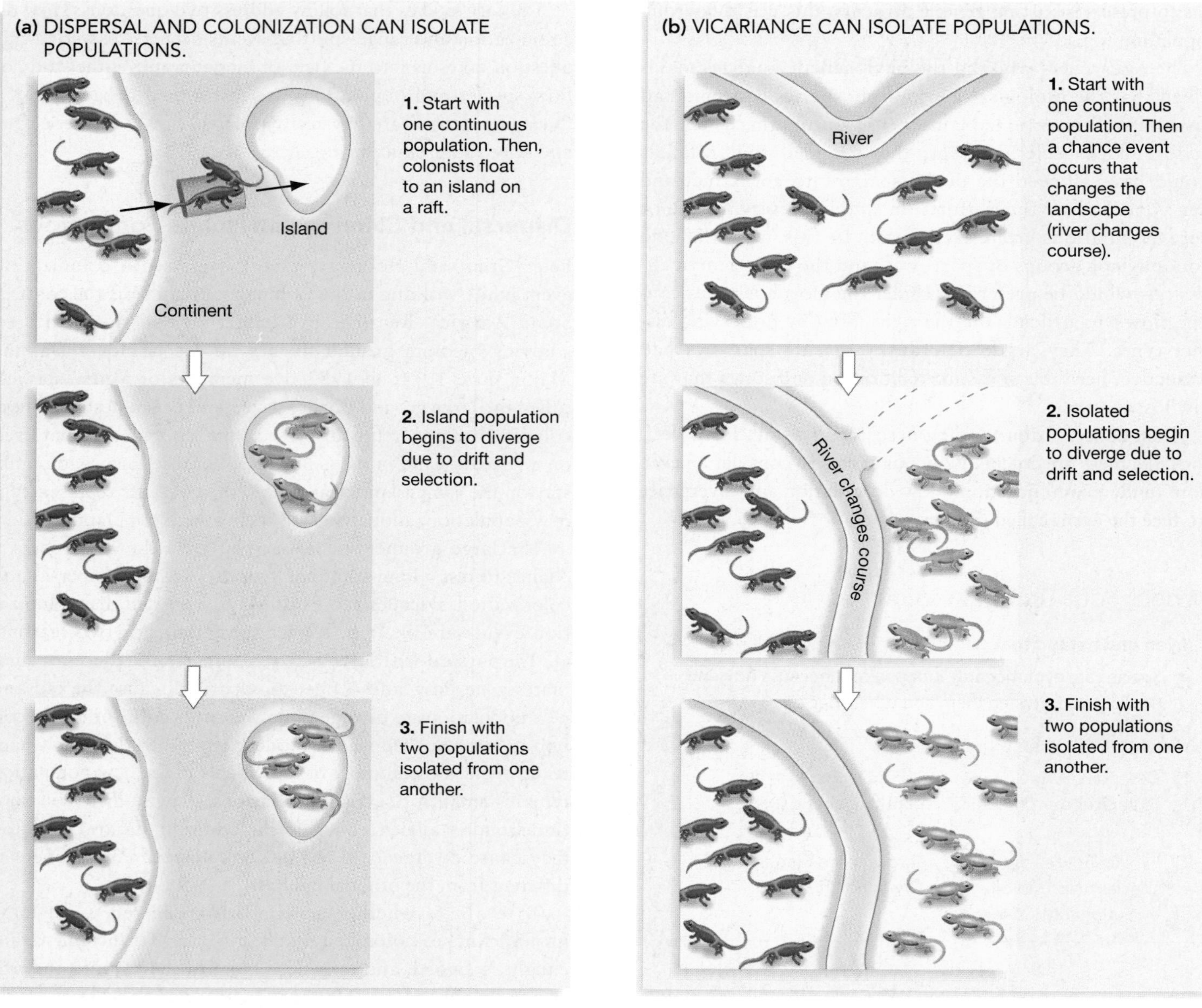

FIGURE 26.5 Allopatric Speciation Begins via Dispersal or Vicariance. (a) When dispersal occurs, colonists establish a new population in a novel location. **(b)** In vicariance, a widespread population becomes fragmented into isolated subgroups.

The general message here is that the characteristics of a colonizing population are likely to be different from the characteristics of the source population due to founder effects. Subsequent natural selection may extend the rapid divergence that begins with genetic drift. Colonization, followed by genetic drift and natural selection, is thought to be responsible for speciation in Galápagos finches and many other island groups.

Vicariance Isolates Populations

If a new physical barrier such as a mountain range or river splits the geographic range of a species, vicariance has taken place. Vicariance events during the most recent ice age are thought to be responsible for the origin of many of the species observed today. Over the past several million years, glaciers that covered large regions of the northern continents advanced and retreated repeatedly. During most of the glacial advances, the growing ice fields fragmented existing forest and grassland habitats into smaller regions that were isolated from each other by expanses of ice. If populations of the same species had occupied these isolated regions, then an inability of individuals to migrate over the ice fields would have left the populations genetically isolated. The advancing ice would have split turtle, flowering plant, insect, and fish species into geographically separated populations that then might have undergone speciation.

Another example of speciation by vicariance involves the group of large, flightless birds called the ratites. You probably are most familiar with the ostrich; kiwis, emus, rheas, and cassowaries are also ratites. Today, ratites are found in South America, Africa, Australia, and New Zealand. Unfortunately, habitat destruction by humans and hunting recently extinguished ratites called elephant birds that lived on the island of Madagascar, off the southeast coast of Africa, as well as 11 species of moas that were native to New Zealand. The elephant bird is the largest bird species ever recorded, with a maximum height of 3.5 m (11 ft), a maximum mass of 454 kg (1000 lbs), and eggs with a volume of up to 7 L (2 gallons). Moas may have been the tallest birds that ever lived, with some species possibly reaching 4 m (13 ft) when they held their necks erect.

The earliest ratites in the fossil record lived about 150 million years ago, on a landmass called Gondwana. As **Figure 26.6a** shows, Gondwana was made up of a number of physically distinct landmasses. The theory of plate tectonics holds that Earth's entire crust—the layer at its surface—is made up of moving blocks of rock called plates. Landmasses, including Gondwana, make up continental plates. The term continental drift is used to describe the ongoing motion of continental plates through time.

Although the "supercontinent" Gondwana existed for tens of millions of years, the continents in it began to drift apart about 140 million years ago (**Figure 26.6b**). Geologic data indicate that first, a landmass consisting of today's South America and Africa split off from a slightly smaller landmass composed of Antarctica, Madagascar, India, and Australia. Later, each of these large landmasses split up to form today's configuration of islands and continents. Ratites are flightless, so each vicariance event isolated distinct populations. Initially, each ratite population would have diverged via genetic drift. These differences would have been extended by natural selection as environments changed on each plate, leading to the evolution of the species observed today and their current distributions (**Figure 26.6c**). In this way, a series of vicariance events triggered a series of speciation events.

To summarize, physical isolation of populations via dispersal or vicariance produces genetic isolation—the first requirement of speciation. When genetic isolation is accompanied by genetic divergence due to mutation, selection, and genetic drift, speciation results.

MB Web Animation at www.masteringbio.com
Allopatric Speciation

(a) Gondwana was the original home of ratites.

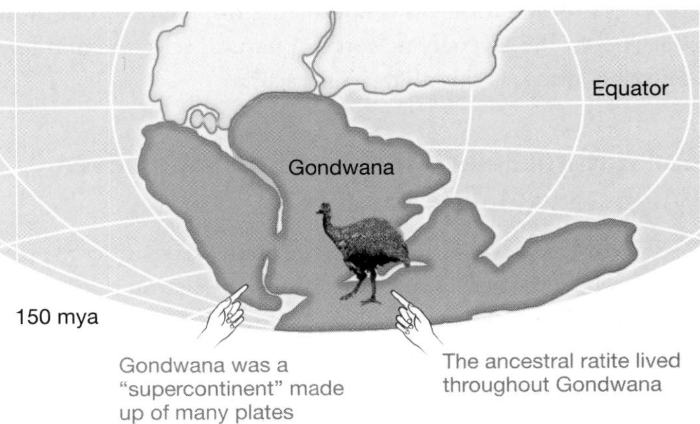

(b) Gondwana began to break up into separate continents.

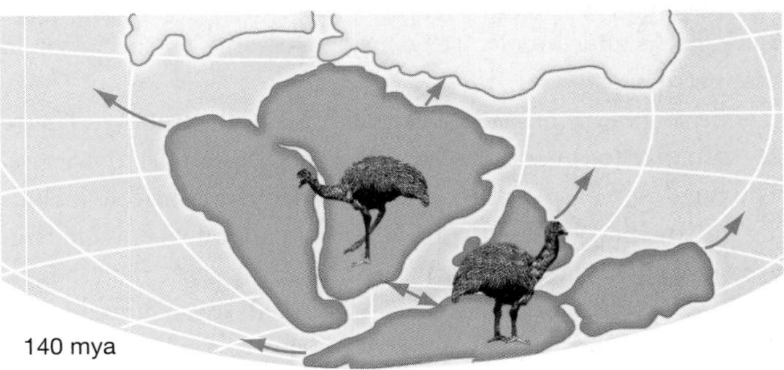

(c) Ratites speciated as the continents moved apart.

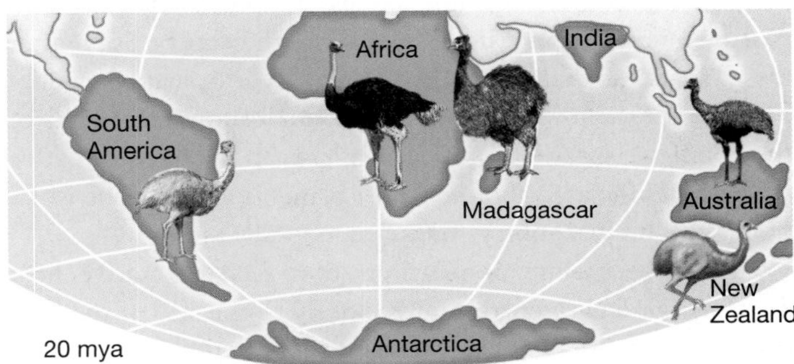

FIGURE 26.6 Continental Drift Caused Vicariance in Ratite Birds. **(a)** The ancestors of today's ratites lived on the supercontinent of Gondwana. **(b)** Continental drift led to the breakup of Gondwana starting about 140 million years ago. This was a vicariance event that isolated ratite populations. **(c)** Continued continental drift brought the continents and islands into their present positions about 20 million years ago (though they continue to move). Ratites in each area diverged in response to mutation, selection, and drift.

QUESTION Did speciation in ratites take 120 million years? Explain your answer.

26.3 Isolation and Divergence in Sympatry

When populations or species live in the same geographic area, or at least close enough to one another to make interbreeding possible, biologists say that they live in **sympatry** ("together-homeland"). Traditionally, researchers have predicted that speciation could not occur among sympatric populations, because gene flow is possible. The prediction was that gene flow would easily overwhelm any differences among populations cre- by genetic drift and natural selection. As Chapter 2- gene flow can homogenize gene frequencies even populations that are allopatric, such as an island population close to a population on a continent. In general, gene flow overwhelms the diversifying force of natural selection and prevents speciation. Is this always the case?

Can Natural Selection Cause Speciation Even When Gene Flow Is Possible?

Recently, several well-documented examples have upset the traditional view that **sympatric speciation**—speciation that occurs even though gene flow is possible—is rare or nonexistent. These studies are fueling a growing awareness that under certain circumstances, natural selection that causes populations to diverge can overcome gene flow and cause speciation. The key realization is that even though sympatric populations are not physically isolated, they may be isolated by preferences for different habitats. As an example, let's consider research on speciation in soapberry bugs.

The soapberry bug is a species of insect, illustrated in **Figure 26.7a**, native to the south-central and southeastern United States. The bugs feed on plants in a family called Sapindaceae, including the soapberry tree, serjania vine, and balloon vine. As the figure shows, the bugs feed by piercing fruits with their beaks, reaching in to penetrate the coats of the seeds located deep inside the fruit, and then sucking up the contents of the seeds through their beaks. The bugs also mate on their host plants.

The soapberry bug's story began to get interesting when horticulturists brought three new species of sapindaceous plants to North America from Asia in the twentieth century. Soon after these plants were introduced to the New World, soapberry bugs began using them as food. As **Figure 26.7b** shows, the fruits of the nonnative species are much smaller than the fruits of native species. Did the arrival of new host plants lead to genetic isolation? If so, have soapberry bugs begun to diverge?

In soapberry bug populations that feed on native host plants, beak length corresponds closely to the size of the host fruit. For example, bugs that feed on species with big fruit tend to have long beaks. The correlation between fruit size and beak length is logical, because it should allow individuals to reach the seeds inside the fruit efficiently. It also prompted a biologist to ask a simple question: In populations of soapberry bugs that exploit the introduced plant species, have beak lengths evolved to match the sizes of the new fruits? If so, it would imply that populations have become genetically isolated on different host plants and that natural selection is currently causing soapberry bug populations to diverge.

To answer this question, researchers measured large samples of bugs found on both native and nonnative hosts. Some of the resulting data are shown in **Figure 26.7c**. The histogram shows that bugs collected on native plants growing in south Florida have much longer beaks than those collected on nonnative plants growing in central Florida. The data support the argument that soapberry bug populations exploiting exotic species have indeed changed—presumably in response to natural selection for efficient use of host fruits. More specifically, the data support the hypothesis that disruptive selection has occurred on beak length ever since some soapberry bug populations switched to new host plant species.

(a) Soapberry bugs use their beaks to reach seeds inside fruits.

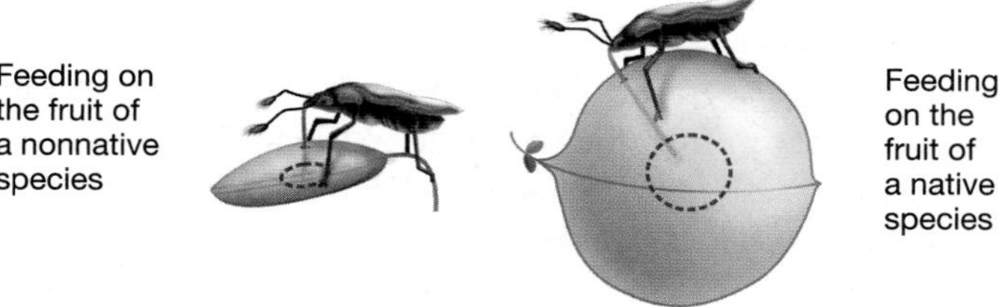

(b) Nonnative fruits are much smaller than native fruits.

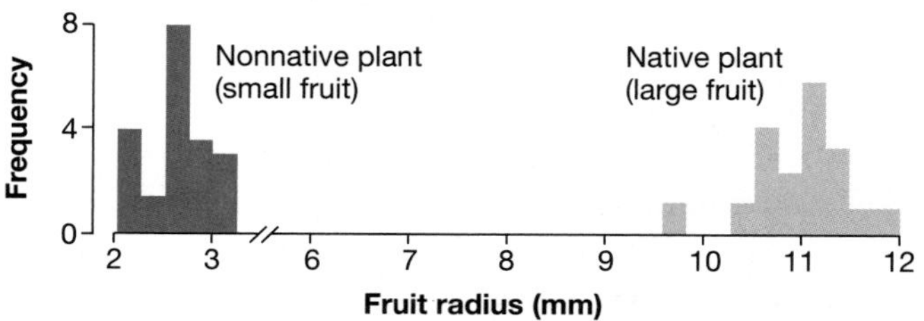

(c) Evidence for disruptive selection on beak length

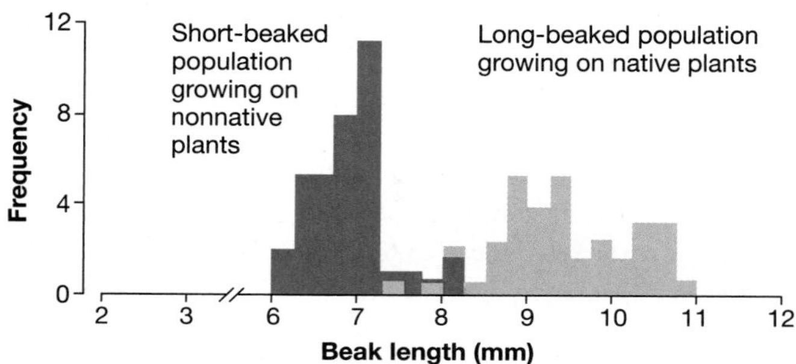

FIGURE 26.7 Disruptive Selection on Beak Length in Soapberry Bugs. (a) A soapberry bug feeding on a nonnative plant called the flat-podded golden rain tree (left) and one feeding on a native plant called the balloon vine (right). **(b)** Nonnative species of soapberry, with small fruits, were introduced to areas where soapberry bugs live in the 1940s and 1950s. **(c)** Soapberry bug populations that feed on these two host plants have very different beak lengths, which correspond to differences in the size of the host fruit.

QUESTION To promote divergence, why is it important for soapberry bugs to mate on their food plants?

The researchers who are following the story expect that soapberry bugs adapted to nonnative species with small fruit will continue to diverge from populations that exploit native species with large fruits, and that they will eventually form distinct species. Because soapberry bugs mate on or near their host plants, switching to a new host species should reduce gene flow among populations at the same time that it sets up disruptive selection. As a result, natural selection may be able to overwhelm gene flow and cause speciation even when populations are sympatric—meaning that they are close enough geographically to make mating physically possible.

Although the soapberry bug's story might seem localized and specific, the events may be common. Biologists currently estimate that over 3 million insect species exist. Most of these species are associated with specific host plants. Based on these observations, it is reasonable to hypothesize that switching host plants, as soapberry bugs have done, has been a major trigger for speciation throughout the course of insect evolution.

How Can Polyploidy Lead to Speciation?

Based on the theory and data reviewed thus far, it is clear that gene flow, genetic drift, and natural selection play important roles in speciation. Can the fourth evolutionary process—mutation—influence speciation as well? The answer might appear to be no. Chapter 25 emphasized that even though mutation is the ultimate source of genetic variation in populations, it is an inefficient mechanism of evolutionary change. If populations become isolated, it is unlikely that mutation, on its own, could cause them to diverge appreciably.

There is a particular type of mutation, though, that turns out to be extremely important in speciation—particularly in plants. The key is that the mutation reduces gene flow between mutant and normal, or wild-type, individuals. It does so because mutant individuals have more than two sets of chromosomes. This condition is known as **polyploidy**.

Polyploidy occurs when an error in meiosis or mitosis results in a doubling of the chromosome number. For example, chromosomes in a diploid ($2n$) species may fail to pull apart during anaphase of mitosis, resulting in a tetraploid cell ($4n$) instead of a diploid cell.

To understand why polyploid individuals are genetically isolated from wild-type individuals, consider what happens when that tetraploid cell undergoes meiosis to form gametes and mates with a diploid individual. By meiosis, the normal individuals produce haploid gametes, while the mutant individuals produce diploid gametes. These gametes unite to form a triploid ($3n$) zygote. Even if this offspring develops normally and reaches sexual maturity, it is rare that it will be able to form functional gametes. The sketch at the bottom of **Figure 26.8** illustrates why: When meiosis occurs in a triploid individual, homologous chromosomes cannot synapse and separate correctly. Thus, they are not distributed to daughter cells evenly, and virtually all of the gametes produced by the triploid individual end up with an uneven number of chromosomes. Because its gametes contain a dysfunctional set of chromosomes, the triploid individual is virtually sterile. As a result, the tetraploid and diploid individuals rarely produce fertile offspring when they mate. Tetraploid and diploid populations are reproductively isolated.

How do the polyploid individuals involved in speciation form? There are two general mechanisms:

1. **Autopolyploid** ("same-many-form") individuals are produced when a mutation results in a doubling of chromosome number and the chromosomes all come from the same species.

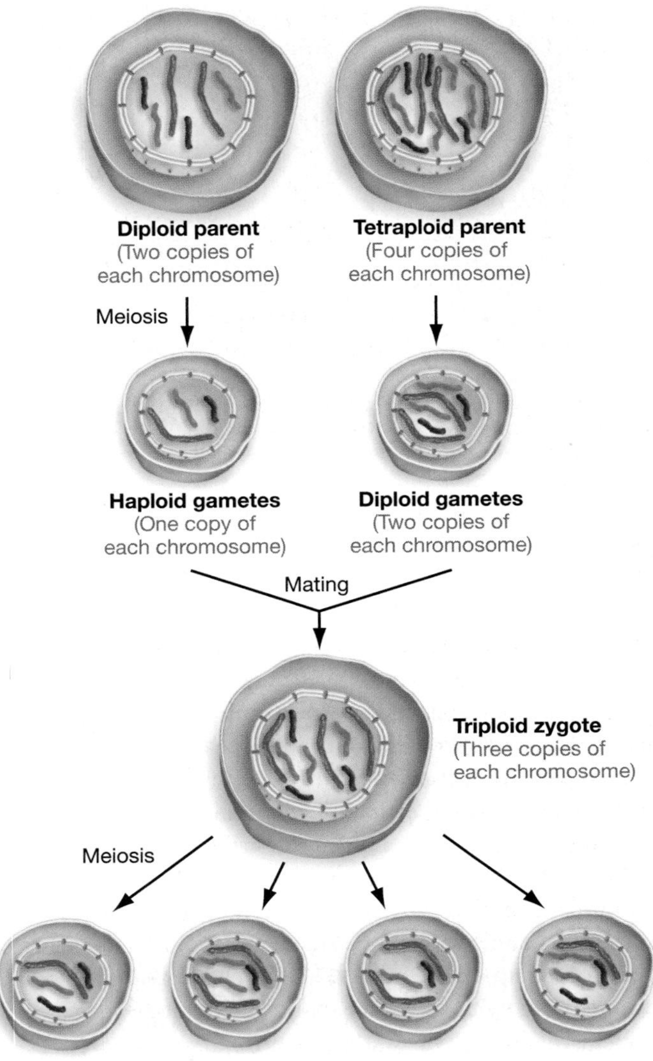

FIGURE 26.8 Polyploidy Can Lead to Reproductive Isolation. The mating diagrammed here illustrates why tetraploid individuals are reproductively isolated from diploid individuals.

EXERCISE Identify gamete combinations that *would* result in viable offspring. Would the likelihood of triploid individuals producing viable gametes increase or decrease if the number of chromosomes was 10 instead of 4?

2. **Allopolyploid** ("different-many-form") individuals are created when parents that belong to different species mate and produce an offspring where chromosome number doubles. Allopolyploid individuals have chromosome sets from different species.

Let's consider specific examples to illustrate how speciation by polyploidy occurs.

Autopolyploidy Although autopolyploidy is thou much less common than allopolyploidy, biologists

FIGURE 26.9 Maidenhair Ferns. This is the diploid form of the maidenhair fern; a recently discovered population of tetraploid individuals is indistinguishable morphologically.

documented autopolyploidy in the maidenhair fern. This plant inhabits woodlands across North America (**Figure 26.9**). During the normal life cycle of a fern, individuals alternate between a haploid (n) stage and a diploid ($2n$) stage. Biologists initially set out to do a routine survey of allelic diversity in a population of these ferns. They happened to be examining individuals in the haploid stage and found several individuals that had *two* versions of each gene instead of just one. These individuals were diploid even though they had the "haploid" growth form. The biologists followed these individuals through their life cycle and confirmed that when the ferns mated, they produced offspring that were tetraploid ($4n$). The researchers had stumbled upon polyploid mutants within a normal population.

To follow up on the observation, they located the parent of the mutant individuals. The parent turned out to have a defect in meiosis. Instead of producing normal, haploid cells as a result of meiosis, the mutant individual produced diploid cells. These diploid cells eventually led to the production of diploid gametes. Because maidenhair ferns can self-fertilize, the diploid gametes could combine to form tetraploid offspring. The tetraploid offspring could then self-fertilize or mate with their tetraploid parent. If the process continued, a polyploid population of maidenhair ferns would be established. The polyploid individuals would be genetically isolated from the original population and thus evolutionarily independent, because tetraploid individuals can breed with other tetraploids but not with diploids. If genetic drift and selection then caused the two populations to diverge, speciation would be under way.

● If you understand how autopolyploidy works, you should be able to create a scenario explaining how the process gave rise to a tetraploid grape with extra-large fruit, from a diploid population with smaller fruit. (You've probably seen both types of fruit in the supermarket.)

This autopolyploidy study documented the critical first step in speciation—the establishment of genetic isolation. In this population of maidenhair ferns, as in soapberry bugs, speciation is under way right before our eyes.

Allopolyploidy New tetraploid species may be created when two diploid species hybridize. **Figure 26.10** illustrates the sequence of events involved. If a diploid offspring that forms from a mating between two different species has chromosomes that do not pair normally during meiosis, the offspring is sterile. But if a mutation occurs that doubles the chromosome number, then homologs synapse, meiosis can proceed, and diploid gametes are produced. When diploid gametes fuse during self-fertilization, a tetraploid individual results.

Exactly this chain of events occurred repeatedly after three European species of weedy plants in the genus *Tragopogon* were introduced to western North America in the early 1900s. In 1950 a biologist described the first of two tetraploid species

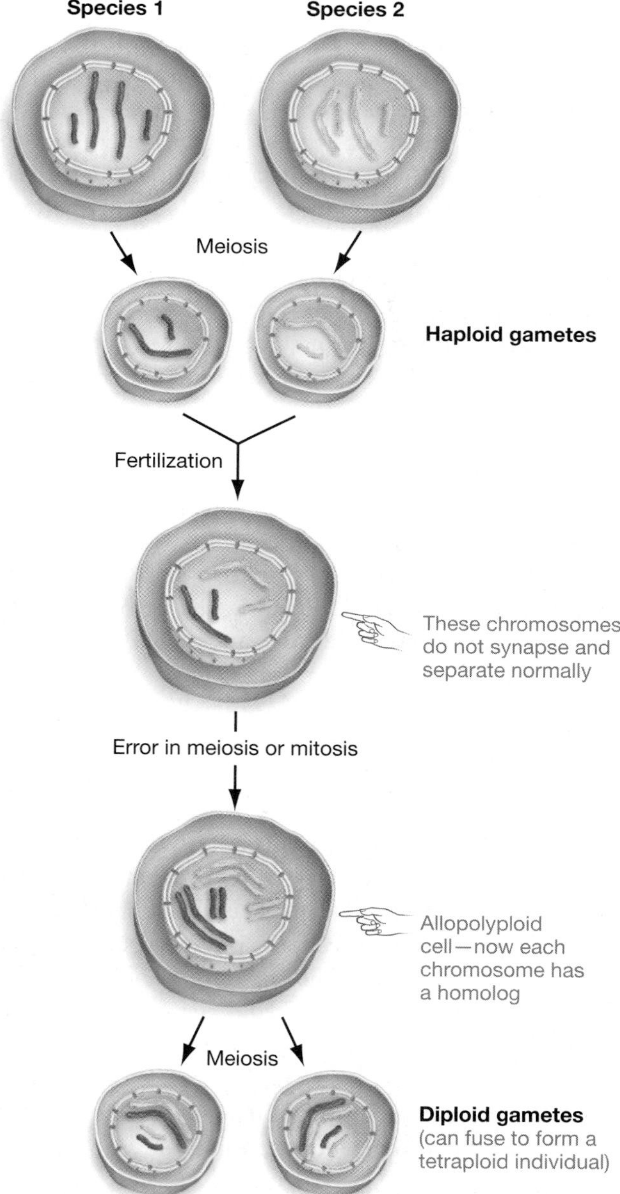

FIGURE 26.10 Allopolyploidy May Occur after Two Species Hybridize. Allopolyploid individuals contain chromosomes from two different species.

that have been discovered. Both were clearly the descendants of the introduced diploids. Further, the nature and amount of genetic variation in the tetraploid species suggests that the allopolyploid sequence occurred repeatedly—meaning that each tetraploid species of *Tragopogon* originated independently multiple times. Follow-up work has shown that at least one of the new tetraploid species is expanding its geographic range.

● If you understand how alloploidy works, you should be able to create a scenario explaining how a cross between a tetraploid population called Emmer wheat and a wild, diploid wheat gave rise to the hexaploid bread wheat grown throughout the world today.

The claim that speciation by polyploidization has been particularly important in plants is backed by the observation that many diploid species have close relatives that are polyploid. Three properties of plants have been noteworthy in making this mode of speciation possible: (1) Reproductive cells and somatic cells are not separated early in development, as they are in animals (see Chapter 23). Instead, plant somatic cells that have undergone many rounds of mitosis can undergo meiosis and produce gametes. If sister chromatids separate during anaphase of one of these mitotic divisions but do not migrate to opposite poles, the result can be a tetraploid daughter cell that later undergoes meiosis to form diploid gametes. (2) The ability of some plant species to self-fertilize makes it possible for diploid gametes to fuse and create genetically isolated tetraploid populations. (3) Hybridization between species is common, creating opportunities for speciation via formation of allopolyploids.

To summarize, speciation by polyploidization is driven by chromosome-level mutations and occurs in sympatry (**Table 26.3**). Compared to the gradual process of speciation by geographic isolation or by disruptive selection in sympatry, speciation by polyploidy is virtually instantaneous. It is fast, sympatric, and common.

Check Your Understanding

If you understand that...

- Speciation occurs when populations become isolated genetically and then diverge due to selection, genetic drift, or mutation.

You should be able to...

1) Give an example of at least three different types of events that lead to the genetic isolation of populations.
2) Explain why selection and drift cause the populations in each example to diverge.

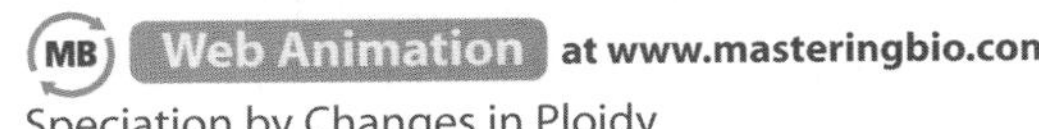

Speciation by Changes in Ploidy

26.4 What Happens When Isolated Populations Come into Contact?

Suppose two populations that have been isolated come into contact again. If divergence has taken place and if divergence has affected when, where, or how individuals in the populations mate, then it is unlikely that interbreeding will take place. In cases such as this, prezygotic isolation exists. When it does, mating between the populations is rare, gene flow is minimal, and the populations continue to diverge.

But what if prezygotic isolation does not exist, and the populations begin interbreeding? The simplest outcome is that the populations fuse over time, as gene flow erases any distinctions between them. Several other possibilities exist, however. Let's explore three of them: reinforcement, hybrid zones, and speciation by hybridization.

SUMMARY TABLE **26.3** **Mechanisms of Sympatric Speciation**

	Process	Notes and/or Example
Disruptive selection	Natural selection for different habitats or resources causes divergence.	Must be accompanied by some mechanism of genetic isolation (e.g., soapberry bugs selected to feed on different-sized fruits mate on those fruits, so little to no interbreeding occurs between small- and large-beaked populations).
Polyploidization	Genetic isolation is created by formation of polyploid individuals that can breed only with each other.	Particularly common in plants, because frequent hybridization occurs between species, and many mitotic divisions occur prior to meiosis.
Autopolyploidy	Polyploids have duplicate chromosome sets from same species (e.g., a chromosome doubling to produce a tetraploid).	Maidenhair fern individual became tetraploid due to an error in meiosis.
Allopolyploidy	Polyploids have chromosome sets from different species (they originate with a hybridization event, followed by chromosome doubling).	*Tragopogon* species introduced to North America hybridized and formed offspring that became tetraploid and formed new species.

Reinforcement

If two populations have diverged extensively and are distinct genetically, it is reasonable to expect that their hybrid offspring will have lower fitness than their parents. The logic here is that if organisms have evolved distinctive developmental sequences or reproductive systems, then a hybrid offspring will not be able to develop or reproduce normally. Recall from Table 26.1 that hybrid offspring may die early in development or survive to sexual maturity but be infertile. In such cases, postzygotic isolation exists. When it occurs, there should be strong natural selection against interbreeding. The hypothesis is that hybrid offspring represent a wasted effort on the part of parents. Individuals that do not interbreed, due to a different courtship ritual or pollination system or other form of prezygotic isolation, should be favored because they produce more viable offspring.

Natural selection for traits that isolate populations in this way is called **reinforcement**. The name is descriptive because the selected traits reinforce differences that developed while the populations were isolated from one another.

Some of the best data on reinforcement come from laboratory studies of closely related fruit fly species in the genus *Drosophila*. Researchers analyzed a large series of experiments that tested whether members of closely related fly species are willing to mate with one another. The biologists found an interesting pattern. If closely related species are sympatric—meaning that they live in the same area—individuals from the two species are seldom willing to mate with one another. But if the species are allopatric—meaning that they live in different areas—then individuals are often willing to mate with one another. This is exactly the pattern that is expected if reinforcement is occurring. The pattern is logical because natural selection can act to reduce mating between species only if their ranges overlap. Thus, it is reasonable to find that sympatric species exhibit prezygotic isolation but that allopatric species do not. There is a long-standing debate, however, over just how important reinforcement is in groups other than the genus *Drosophila*.

Hybrid Zones

Hybrid offspring are not always dysfunctional. In some cases they are capable of mating and producing offspring and have features that are intermediate between those of the two parental populations. When this is the case, hybrid zones can form. A **hybrid zone** is a geographic area where interbreeding occurs and hybrid offspring are common. Depending on the fitness of hybrid offspring and the extent of breeding between parental species, hybrid zones can be narrow or wide, and long or short lived. As an example of how researchers analyze the dynamics of hybrid zones, let's consider recent work on two bird species.

Townsend's warblers and hermit warblers live in the coniferous forests of North America's Pacific Northwest. In southern Washington State, where their ranges overlap, the two species hybridize extensively. As **Figure 26.11a** shows, hybrid offspring have characteristics that are intermediate relative to the two parental species. To explore the dynamics of this hybrid zone, a team of biologists examined gene sequences in the mitochondrial DNA (mtDNA) of a large number of Townsend's, hermit, and hybrid warblers collected from forests throughout the region. The team found that each of the parental species has certain species-specific mtDNA sequences. This result allowed the researchers to infer how hybridization was occurring. To grasp the reasoning here, it is critical to realize that mtDNA is maternally inherited in most animals and plants. If a hybrid individual has Townsend's mtDNA, its mother had to be a Townsend's warbler while its father had to be a hermit warbler. In this way, identifying mtDNA types allowed the research team to infer whether Townsend's females were mating with hermit males, or vice versa, or both.

Their data presented a clear pattern: Most hybrids form when Townsend's males mate with hermit warbler females. One of the investigators followed up on this result with experiments showing that Townsend's males are extremely aggressive in establishing territories and that they readily attack hermit warbler males. Hermit males, in contrast, do not challenge Townsend's males. The hypothesis, then, is that Townsend's males invade hermit territories, drive off the hermit males, and mate with hermit females.

The team also found something completely unexpected. When they analyzed the distribution of mtDNA types along the Pacific Coast and in the northern Rocky Mountains, they found that many Townsend's warblers actually had hermit mtDNA. **Figure 26.11b** shows that in some regions—such as the larger islands off the coast of British Columbia—*all* of the warblers had hermit mtDNA, even though they looked like full-blooded Townsend's warblers. To explain this result, the team hypothesized that hermit warblers were once found as far north as Alaska and that Townsend's warblers have gradually taken over their range. Their logic is that repeated mating with Townsend's warblers over time made the hybrid offspring look more and more like Townsend's, even while maternally inherited mtDNA kept the genetic record of the original hybridization event intact.

If this hypothesis is correct, then the hybrid zone should continue moving south. If it does so, hermit warblers may eventually become extinct. In many cases, however, hybridization does not lead to extinction but rather leads to the opposite—the creation of new species.

New Species through Hybridization

A team of researchers recently examined the relationships of three sunflower species native to the American West: *Helianthus annuus*, *H. petiolaris*, and *H. anomalus*. The first two of these species are known to hybridize in regions where their ranges overlap. The third species, *H. anomalus*, resembles these hybrids. In fact, because some gene sequences in *H. anomalus* are remarkably similar to those found in *H. annuus*, while other gene sequences are almost identical to those found in *H. petiolaris*, biologists have suggested that *H. anomalus* originated in hybridization between *H. annuus* and *H. petiolaris* (**Figure 26.12**). All three species have the same number of

(a) Hybrids have intermediate characteristics.

Townsend's warbler

Townsend's-hermit hybrid

Hermit warbler

(b) Hybrids inherit species-specific mtDNA sequences from their mothers.

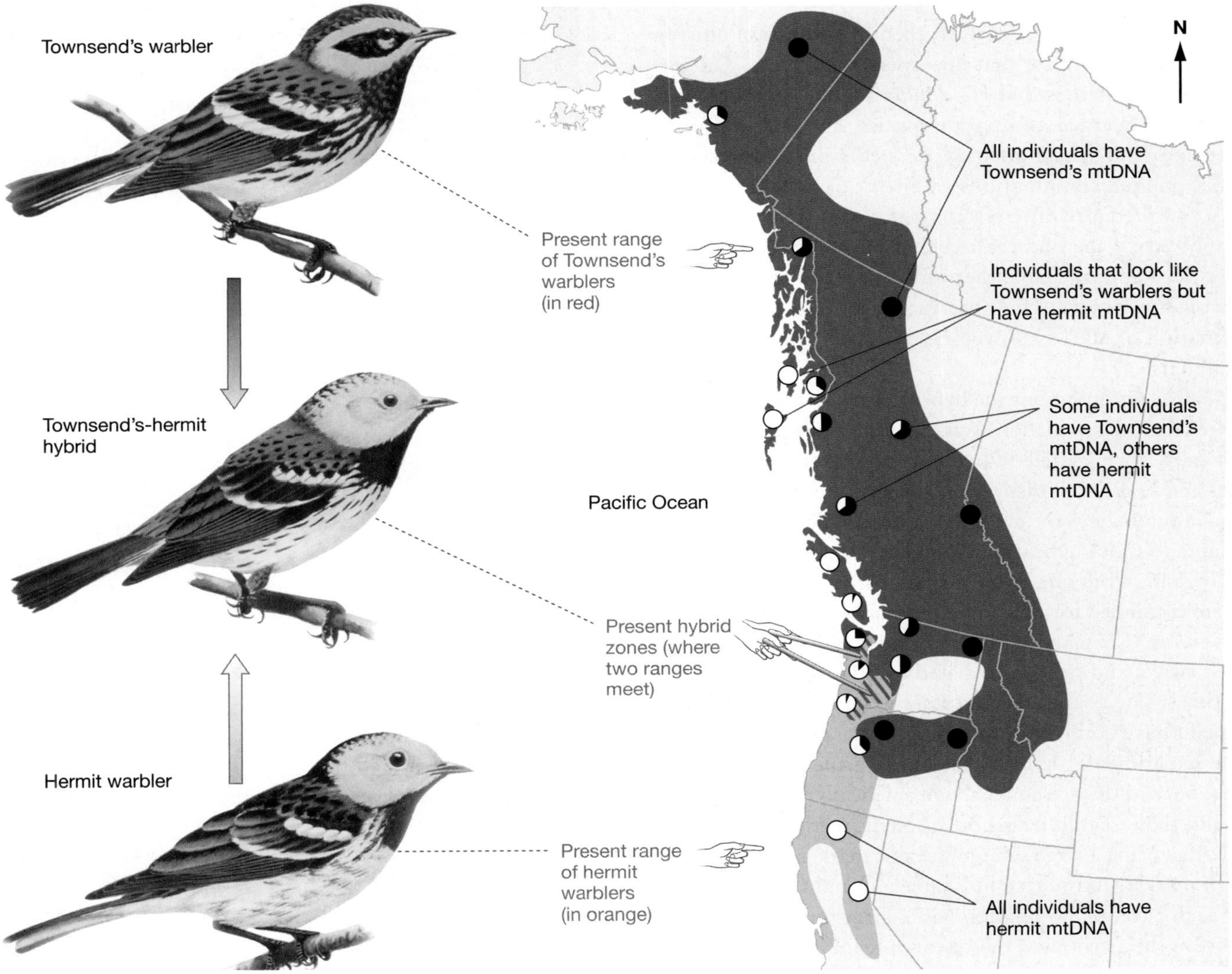

N

All individuals have Townsend's mtDNA

Present range of Townsend's warblers (in red)

Individuals that look like Townsend's warblers but have hermit mtDNA

Some individuals have Townsend's mtDNA, others have hermit mtDNA

Pacific Ocean

Present hybrid zones (where two ranges meet)

Present range of hermit warblers (in orange)

All individuals have hermit mtDNA

FIGURE 26.11 Analyzing a Hybrid Zone. (a) When Townsend's warblers and hermit warblers hybridize, the offspring have intermediate characteristics. **(b)** Map showing the current range of Townsend's and hermit warblers. The small pie charts show the percentage of individuals with Townsend's warbler mtDNA (in black) and hermit warbler mtDNA (in white).

FIGURE 26.12 Hybridization Occurs between Sunflower Species. The sunflower *Helianthus anomalus* may have originated in hybridization events between *H. annuus* and *H. petiolarus*.

chromosomes, so neither allopolyploidy nor autopolyploidy was involved.

If this interpretation is correct, then hybridization must be added to the list of ways that new species can form. The specific hypothesis here is that *H. annuus* and *H. petiolaris* were isolated and diverged as separate species, and later began interbreeding. The hybrid offspring created a third, new species that had unique combinations of alleles from each parental species and therefore different characteristics. This hypothesis is supported by the observation that *H. anomalus* grows in much drier habitats than either of the parental species and is distinct in appearance. This observation suggests that a unique combination of alleles allowed *H. anomalus* to thrive in certain habitats.

Biologists set out to test the hybridization hypothesis by trying to re-create the speciation event experimentally (**Figure 26.13**). Specifically, they mated individuals from the two parental species and raised the offspring in the greenhouse. When these hybrid individuals were mature, the researchers either mated the plants to other hybrid individuals or "backcrossed" them to individuals from one of the parental species. This breeding program continued for four more generations before the experiment ended. Ultimately, the experimental lines were backcrossed twice, and they were mated to other hybrid offspring three times. The goal of these crosses was to simulate matings that might have occurred naturally.

The experimental hybrids looked like the natural hybrid species, but did they resemble them genetically? To answer this question, the research team constructed genetic maps of each population, using a large series of genetic markers similar to the types of markers introduced in Chapter 19 and Chapter 20. Because each parental population had a large number of unique markers in their genomes, the research team hoped to identify which genes found in the experimental hybrids came from which parental species.

Their results are diagrammed in the "Results" section of Figure 26.13. The bottom bar in the illustration represents a region called S in the genome of the naturally occurring species *Helianthus anomalus*. As the legend indicates, this region contains three sections of sequences that are also found in *H. petiolaris* (indicated with the color orange) and two that are also found in *H. annuus* (indicated with the color red). The top bar shows the composition of this same region in the genome of the experimental hybrid lines. The key observation here is that the genetic composition of the synthesized hybrids matches that of the naturally occurring hybrid species.

In effect, the researchers had succeeded in re-creating a speciation event. Their results provide strong support for the hybridization hypothesis for the origin of *H. anomalus*. The data also highlight the dynamic range of possible outcomes as a result of secondary contact of two populations: fusion of the populations, reinforcement of divergence, founding of stable hybrid zones, extinction of one population, or the creation of new species (**Table 26.4**).

Experiment

Question: Can new species arise by hybridization between existing species?

Hypothesis: *Helianthus anomalus* originated by hybridization between *H. annuus* and *H. petiolaris*.

Null hypothesis: *Helianthus anomalus* did not originate by hybridization between *H. annuus* and *H. petiolaris*.

Experimental setup:

H. annuus X *H. petiolaris*

Hybrid X Hybrid

Hybrid Hybrid Hybrid

1. Mate *H. annuus* and *H. petiolaris* and raise offspring.

2. Mate F_1 hybrids or backcross F_1s to parental species; raise offspring.

3. Repeat for four more generations.

Prediction: Experimental hybrids will have the same mix of *H. annuus* and *H. petiolaris* genes as natural *H. anomalus*.

Prediction of null hypothesis: Experimental hybrids will not have the same mix of *H. annuus* and *H. petiolaris* genes as natural *H. anomalus*.

Results:

DNA comparison of a chromosomal region called S:

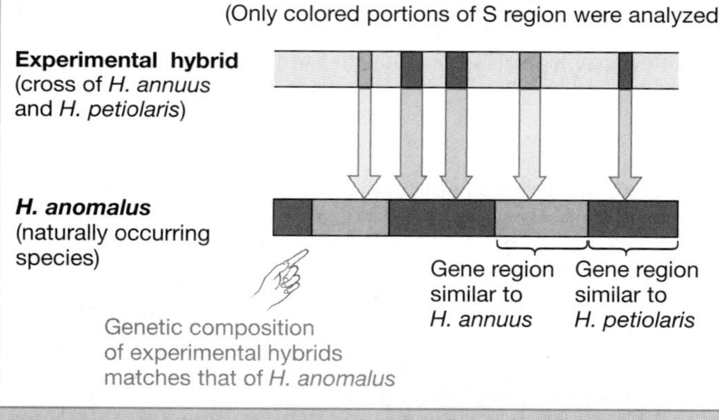

Conclusion: New species may arise via hybridization between existing species.

FIGURE 26.13 Experimental Evidence That New Species Can Originate in Hybridization Events.

SUMMARY TABLE **26.4** **Outcomes of Secondary Contact**

Possible Outcome	Description	Example
Fusion of the populations	The two populations freely interbreed.	Occurs whenever populations of the same species come into contact.
Reinforcement	If hybrid offspring have low fitness, natural selection favors the evolution of traits that prevent interbreeding between the populations.	Appears to be common in fruit fly species that occupy the same geographic areas.
Hybrid zone formation	There is a well-defined geographic area where hybridization occurs. This area may move over time or be stable.	Many stable hybrid zones have been described; the hybrid zone between hermit and Townsend's warblers appears to have moved over time.
Extinction of one population	If one population or species is a better competitor for shared resources, then the poorer competitor may be driven to extinction.	Townsend's warblers may be driving hermit warblers to extinction.
Creation of new species	If the combination of genes in hybrid offspring allows them to occupy distinct habitats or use novel resources, they may form a new species.	Hybridization between sunflowers gave rise to a new species with unique characteristics.

Chapter Review

SUMMARY OF KEY CONCEPTS

Speciation occurs when populations of the same species become genetically isolated by lack of gene flow and then diverge from each other due to selection, genetic drift, or mutation.

When gene flow does not occur between populations, natural selection, genetic drift, and mutation act on the populations independently. As a result, the genetic and physical characteristics of the populations change over time. Eventually, the populations become so different that they are recognized as distinct species—evolutionarily independent populations. Speciation is a splitting event in which one lineage gives rise to two or more independent descendant lineages.

You should be able to design an experiment that would, given enough time, result in the production of two species from a single ancestral population.

Populations can be recognized as distinct species if they are reproductively isolated from each other, if they have distinct morphological characteristics, or if they form independent branches on a phylogenetic tree.

Researchers use several criteria to test whether populations represent distinct species. The biological species concept focuses on the degree of hybridization between species to determine whether gene flow is occurring. The morphospecies concept infers that speciation has occurred if populations have distinctive morphological traits. The phylogenetic species concept defines *species* as the smallest monophyletic groups on evolutionary trees containing populations.

You should be able to explain whether human populations would be considered separate species under the biological, morphological, and phylogenetic species concepts.

Populations can become genetically isolated from each other if they occupy different geographic areas, if they use different habitats or resources within the same area, or if one population is polyploid and cannot breed with the other.

Speciation often begins when small groups of individuals colonize a new habitat or when a large, continuous population becomes fragmented into isolated habitats. Colonization is thought to be a major mode of speciation on islands, while range splitting is thought to be a major mode of speciation in glaciated areas and in species that occupied the supercontinent Gondwana during its breakup.

Contrary to traditional expectations, biologists have found that speciation occurs frequently due to disruptive selection when populations are sympatric. Similarly, mutations that produce polyploidy can trigger rapid speciation in sympatry because they lead to reproductive isolation between diploid and tetraploid populations.

You should be able to evaluate whether your experiment on speciation (see above) represents a case of vicariance, dispersal, different habitat use, or polyploidy.

MB **Web Animation** at www.masteringbio.com
Allopatric Speciation; Speciation by Changes in Ploidy

When populations that have diverged come back into contact, they may fuse, continue to diverge, stay partially differentiated, or have offspring that form a new species.

If prezygotic isolation exists, populations that come back into contact will probably continue to diverge. Alternatively, interbreeding can cause diverged populations to fuse into the same species. Researchers have also documented that secondary contact can lead to reinforcement and complete reproductive isolation, the formation of hybrid zones, or the creation of a new hybrid species.

You should be able to predict the outcome of hybridization when hybrid offspring have higher, lower, or equal fitness to the parental populations.

QUESTIONS

Test Your Knowledge

1. What distinguishes a morphospecies?
   a. It has distinctive characteristics, such as size, shape, or coloration.
   b. It represents a distinct twig in a phylogeny of populations.
   c. It is reproductively isolated from other species.
   d. It is a fossil from a distinct time in Earth history.
2. When does vicariance occur?
   a. small populations coalesce into one large population
   b. a population is fragmented into isolated subpopulations
   c. individuals colonize a novel habitat
   d. individuals disperse and found a new population
3. Why is "reinforcement" an appropriate name for the concept that natural selection should favor divergence and genetic isolation if populations experience postzygotic isolation?
   a. Selection should reinforce high fitness for hybrid offspring.
   b. Selection should reinforce the fact that they are "good species" under the morphological species concept.
   c. Selection acts because hybrid offspring do not develop at all or are sterile when mature.
   d. It reinforces selection for divergence that began when the species were geographically isolated.
4. The biological species concept can be applied only to which of the following groups?
   a. bird species living today **b.** dinosaurs
   c. bacteria **d.** archaea
5. Why are genetic isolation and genetic divergence occurring in soapberry bugs, even though populations occupy the same geographic area?
   a. Different populations feed and mate on different types of fruit.
   b. One population is tetraploid; others are diploid.
   c. Nonnative host plants caused a vicariance event.
   d. Beak length has changed due to disruptive selection.
6. When the ranges of different species meet, a stable "hybrid zone" occupied by hybrid individuals may form. How is this possible?
   a. Hybrid individuals may have intermediate characteristics that are advantageous in a given region.
   b. Hybrid individuals are always allopolyploid and are thus unable to mate with either of the original species.
   c. Hybrid individuals may have reduced fitness and thus be strongly selected against.
   d. One species has a selective advantage, so as hybridization continues, the other species will go extinct.

Test Your Knowledge answers: 1. a; 2. b; 3. d; 4. a; 5. a; 6. a

Test Your Understanding

Answers are available at www.masteringbio.com

1. Make an outline listing the sections and subsections in this chapter. Fill it in with notes on the experimental and analytical approaches used in the case studies provided for each topic. Which studies represent direct observation of speciation, and which are indirect studies of historical events?
2. In the case of the seaside sparrow, how did the species identified by the biological species concept, the morphospecies concept, and the phylogenetic species concept conflict?
3. Explain why genetic drift occurs during colonization events. Explain why natural selection occurs after colonization events.
4. Explain how isolation and divergence are occurring in soapberry bugs. Of the four evolutionary processes (mutation, gene flow, drift, and selection), which two are most important in causing this event?
5. Unlike animal gametes, plant reproductive cells do not differentiate until late in life. Explain why plants are much more likely to produce diploid gametes and produce polyploid offspring than are animals.
6. Summarize the possible outcomes when populations that have been separated for some time come back into contact and hybridize. Explain why the fitness of hybrid offspring determines each outcome.

Applying Concepts to New Situations

Answers are available at www.masteringbio.com

1. A large amount of gene flow is now occurring among human populations due to intermarriage among people from different ethnic groups and regions of the world. Is this phenomenon increasing or decreasing racial differences in our species? Explain.
2. Humans have introduced thousands of species to new locations around the globe. Few, if any, of these colonization events have resulted in speciation. Use these data to evaluate the hypothesis that founder events trigger speciation.
3. Recall from Chapter 25 that 15 green iguanas recently colonized the island of Anguilla. Iguanas were not found on Anguilla previously. It is likely that the 15 iguanas originally lived on the island of Guadalupe.
   - Outline a short-term study designed to test the hypothesis that genetic drift produced allele frequency differences in the two populations (the "old" green iguana population of Guadalupe and the new population on Anguilla).
   - Outline a long-term study designed to test the hypothesis that natural selection will produce changes in the characteristics of the two populations over time.
4. All over the world, natural habitats are being fragmented into tiny islands as suburbs, ranches, and farms expand. Predict how allele frequencies will change in these fragmented populations, and then predict the long-term outcome of the global fragmentation of habitats. In each case, explain your logic.

wwww.masteringbio.com is also your resource for • Answers to text, table, and figure caption questions and exercises • Answers to *Check Your Understanding* boxes • Online study guides and quizzes • Additional study tools including the *E-Book for Biological Science* 3rd ed., textbook art, animations, and videos.

Phylogenies and the History of Life

27

A fossilized shell from an ammonite. Ammonites were large, ocean-dwelling animals related to today's squid. The last ammonites disappeared during a mass extinction event analyzed in section 27.5.

KEY CONCEPTS

- Phylogenies and the fossil record are the major tools that biologists use to study the history of life.
- The Cambrian explosion was the rapid morphological and ecological diversification of animals that occurred during the Cambrian period.
- Adaptive radiations are a major pattern in the history of life. They are instances of rapid diversification associated with new ecological opportunities and new morphological innovations.
- Mass extinctions have occurred repeatedly throughout the history of life. They rapidly eliminate most of the species alive in a more or less random manner.

This chapter is about time and change. More specifically, it's about vast amounts of time and profound change in organisms. Both of these topics can be difficult for humans to grasp. Our lifetimes are measured in decades, and our knowledge of history is usually measured in centuries or millennia. But this chapter analyzes events that occurred over millions and even billions of years. A million years is completely beyond our experience and almost beyond our imagination.

It takes practice to get comfortable analyzing the profound changes that occur in organisms over deep time. To help you get started, the chapter begins by introducing the two major analytical tools that biologists use to reconstruct the history of life: phylogenetic trees and the fossil record. In the remaining three sections of the chapter, we'll explore three of the great events in the history of life: the initial diversification of animals, the phenomenon called adaptive radiation, and the phenomenon known as mass extinction.

27.1 Tools for Studying History: Phylogenetic Trees

The evolutionary history of a group of organisms is called its **phylogeny**. Phylogenies are usually summarized and depicted in the form of a phylogenetic tree. A **phylogenetic tree** shows the ancestor-descendant relationships among populations or species, and clarifies who is related to whom. In a phylogenetic tree, a **branch** represents a population through time; the point where two branches diverge, called a **node** (or fork) represents the point in time when an ancestral group split into two or more descendant groups; and a **tip** (or terminal node), the endpoint of a branch, represents a group (a species or larger taxon) living today or one that ended in extinction.

Evolutionary trees have been introduced at various points in earlier chapters—often with just enough information to help you understand a new concept. Now let's focus on the trees themselves. How do you read a finished tree, and how are trees put together in the first place? We'll consider each question in turn.

BioSkills 2 introduces the parts of a phylogenetic tree and how to read one. (It would be a *very* good idea to review BioSkills 2 right now!) Here let's focus on how biologists go about building them.

How Do Researchers Estimate Phylogenies?

Phylogenetic trees are an extremely effective way of summarizing data on the evolutionary history of a group of organisms. But like any other pattern or measurement in nature, from the average height of a person in a particular human population to the speed of a passing airplane, the genealogical relationships among species cannot be known with absolute certainty. Instead, the relationships depicted in an evolutionary tree are estimated from data.

To infer the historical relationships among species, researchers analyze the species' morphological or genetic characteristics, or both. For example, to reconstruct relationships among fossil species of humans, scientists analyze aspects of tooth, jaw, and skull structure. To reconstruct relationships among contemporary human populations, investigators usually compare the sequences of bases in a particular gene.

The fundamental idea in phylogeny inference is that closely related species should share many of their characteristics, while distantly related species should share fewer characteristics. But there are two general strategies for using data to estimate trees: the phenetic approach and the cladistic approach.

The **phenetic approach** to estimating trees is based on computing a statistic that summarizes the overall similarity among populations, based on the data. For example, researchers might use gene sequences to compute an overall "genetic distance" between two populations. A genetic distance summarizes the average percentage of bases in a DNA sequence that differ between two populations. A computer program then builds a tree that clusters the most similar populations and places more-divergent populations on more-distant branches.

The **cladistic approach** to inferring trees is based on the realization that relationships among species can be reconstructed by identifying shared derived characters, or **synapomorphies** (literally, "union-forms"), in the species being studied. A synapomorphy is a trait that certain groups of organisms have that exists in no others. Synapomorphies allow biologists to recognize **monophyletic groups**—also called **clades** or **lineages**.

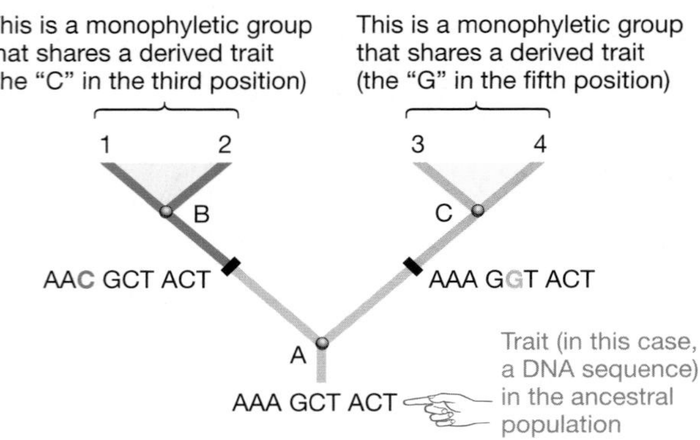

FIGURE 27.1 Synapomorphies Identify Monophyletic Groups.

(See Chapter 26 and BioSkills 2.) For example, fur and lactation are synapomorphies that identify mammals as a monophyletic group. Synapomorphies are characteristics that are shared because they are derived from traits that existed in their common ancestor.

Figure 27.1 illustrates the logic behind a cladistic analysis. When the ancestral population at the bottom of the figure splits into two descendant lineages at node A, each descendant group begins evolving independently and acquires unique traits. These traits are derived from their common ancestor via mutation, selection, and genetic drift. As an example, the traits might be a particular region of DNA that changes as follows:

AAA GCT ACT	ancestral population
AAC GCT ACT	a descendant population
AAA GGT ACT	another descendant population

When the two lineages themselves split at nodes B and C, the species that result share the derived characteristics. In this way, the ancestor at node B and species 1 and 2 can be recognized as a monophyletic group. Similarly, the ancestor at node C and species 3 and 4 can be recognized as a different monophyletic group. When many such traits have been measured, a computer program can be used to identify which traits are unique to each monophyletic group and then place the groups in a tree in the correct relationship to each other.

How Can Biologists Distinguish Homology from Homoplasy? Although the logic behind phenetic and cladistic analyses is elegant, problems arise. The issue is that traits can be similar in two species not because those traits were present in a common ancestor, but because similar traits evolved independently in two distantly related groups. In the example given in Figure 27.1, it is possible that species 2 is not at all closely related to species 1. Its ancestors may have had the sequence TAT GGT AGT, which happened to change to AAC GCT ACT due to mutation, selection, and drift that took place independently of the changes that took place in the ancestors of species 1.

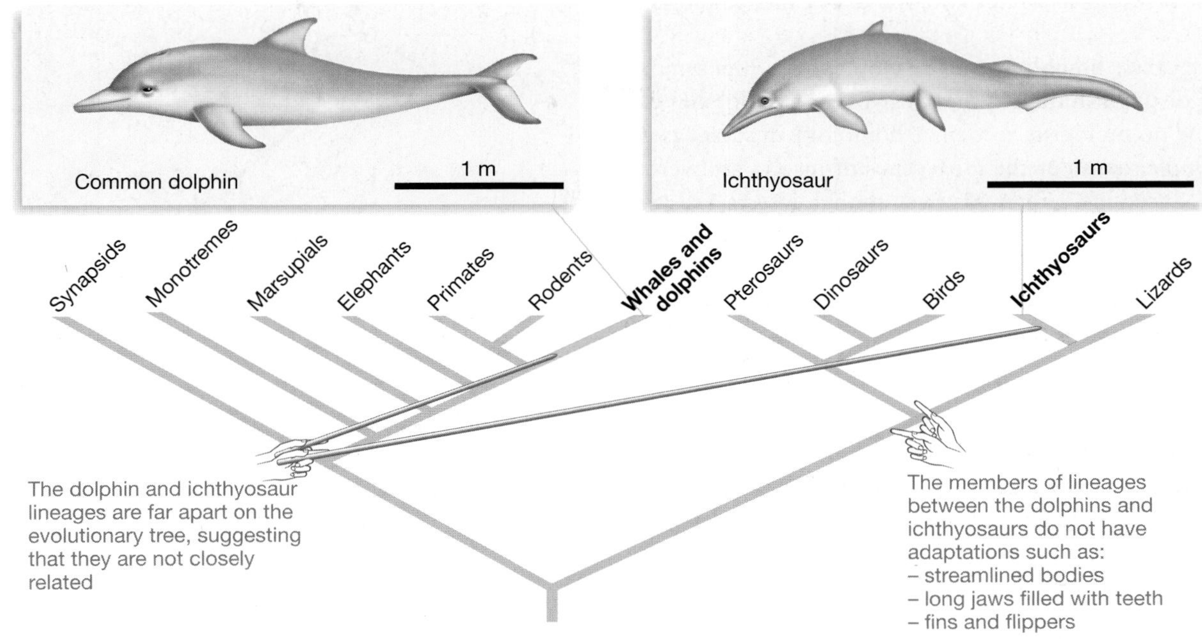

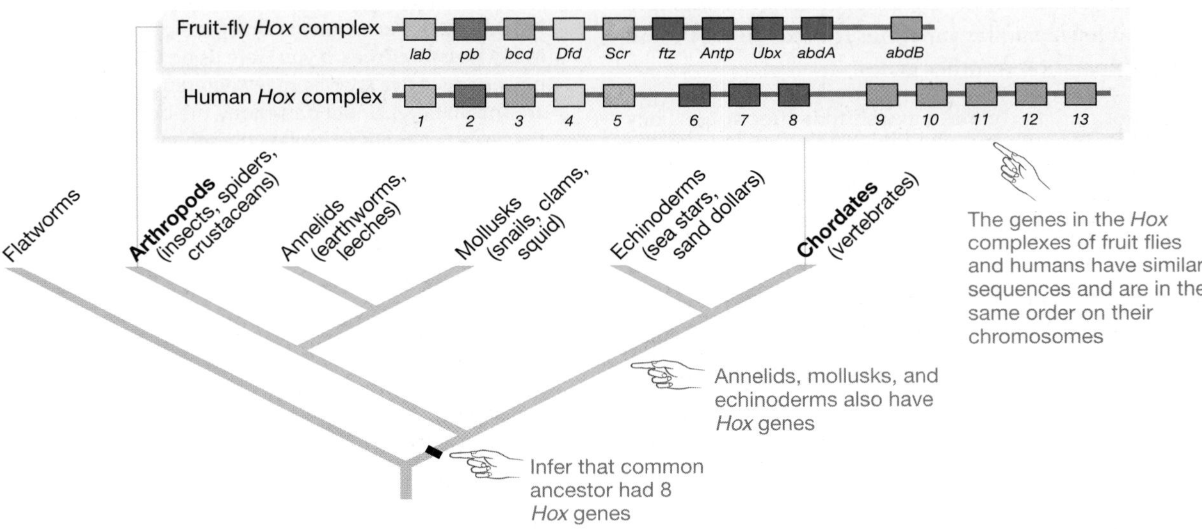

FIGURE 27.2 Homoplasy and Homology: Different Causes of Similarity in Traits. (a) Dolphins and ichthyosaurs look similar but are not closely related—dolphins are mammals; ichthyosaurs are reptiles. Because their similar traits did not exist in the common ancestor of mammals and reptiles, biologists infer that traits such as streamlined bodies, sharp teeth, and flippers evolved independently in these two groups. **(b)** All of the animal groups illustrated on this phylogeny have *Hox* complexes that are similar to those illustrated for fruit flies and humans.

Homology ("same-source") occurs when traits are similar due to shared ancestry (see Chapter 24); **homoplasy** ("same-form") occurs when traits are similar for reasons other than common ancestry. For example, the aquatic reptiles called ichthyosaurs were strikingly similar to modern dolphins (**Figure 27.2a**). Both are large marine animals with streamlined bodies and large dorsal fins. Both chase down fish and capture them between elongated jaws filled with dagger-like teeth. But no one would argue that ichthyosaurs and dolphins are similar because the traits they share existed in a common ancestor. As the phylogeny in Figure 27.2a shows, analyses of other traits show that ichthyosaurs are reptiles whereas dolphins are mammals.

Based on these data, it is logical to argue that the similarities between ichthyosaurs and dolphins result from convergent evolution. **Convergent evolution** occurs when natural selection favors similar solutions to the problems posed by a similar way of making a living. But convergent traits do not occur in the common ancestor of the similar species. Streamlined bodies and elongated jaws filled with sharp teeth help *any* species—whether it is a reptile or a mammal—chase down fish in open water. Convergent evolution is a common cause of homoplasy; it results in what biologists once called analogous traits.

In many cases, homology and homoplasy are much more difficult to distinguish than in the ichthyosaur and dolphin example. How do biologists recognize homology in such cases? As an example, consider the *Hox* genes of insects and vertebrates introduced in Chapter 21. Even though insects and vertebrates last shared a common ancestor some 600–700 million years ago, biologists argue that their *Hox* genes are derived from the same ancestral sequences. There are several lines of evidence to support this hypothesis:

- The genes are organized in a similar way. **Figure 27.2b** shows that these genes in both insects and vertebrates are found in gene complexes, with similar genes found adjacent to one another on the chromosome. Recall from Chapter 20 that genes with these characteristics are called gene families. The organization of the gene families is nearly identical among insect and vertebrate species.
- All of the *Hox* genes share a 180-base-pair sequence called the *homeobox*, introduced in Chapter 21. The polypeptide encoded by the homeobox is almost identical in insects and vertebrates and has a similar function: It binds to DNA and regulates the expression of other genes.
- The products of the *Hox* genes have similar functions: They identify the locations of cells in embryos. They are also expressed in similar patterns in time and space.

In addition, many other animals, on lineages that branched off between insects and mammals, have similar genes. This is a crucial observation: If similar traits found in distantly related lineages are indeed similar due to common ancestry, then similar traits should be found in many intervening lineages on the tree of life. (Examine the phylogeny in Figure 27.2a and ask yourself whether the analogous traits found in ichthyosaurs and dolphins fulfill this criterion.)

Now suppose that a researcher set out to infer a phylogenetic tree from a set of morphological traits or DNA sequences. Without already having a tree in hand, it can be difficult or impossible to tell which traits are homologous and qualify as synapomorphies. Just as any data set contains unavoidable errors, or "noise," the data sets used to infer phylogenies inevitably include homoplasy. To reduce the chance that homoplasy will lead to erroneous conclusions about which species are most closely related, biologists who are using cladistic approaches invoke the logical principle of **parsimony**. Under parsimony, the most likely explanation or pattern is the one that implies the least amount of change. For example, a biologist might compare all of the branching patterns that are theoretically possible and count the number of changes in DNA sequences that would be required to produce each pattern. For example, the tree in **Figure 27.3a** requires two changes in base sequence; the tree in **Figure 27.3b** requires four. Convergent evolution and other causes of homoplasy should be rare compared with similarity due to shared descent, so the tree that implies the fewest overall evolutionary changes should be the one that most accurately reflects what really happened during evolution. If the branching pattern in Figure 27.3a is the most parsimonious of all possible trees, then biologists conclude that it is the most likely representation of actual phylogeny based on the data in hand.

(a) Two changes

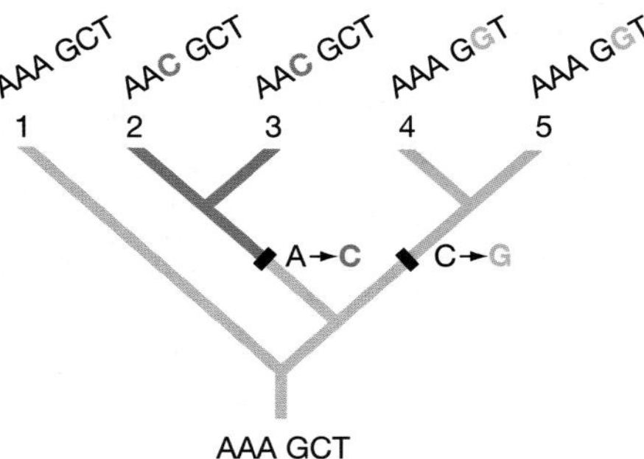

(b) Four changes

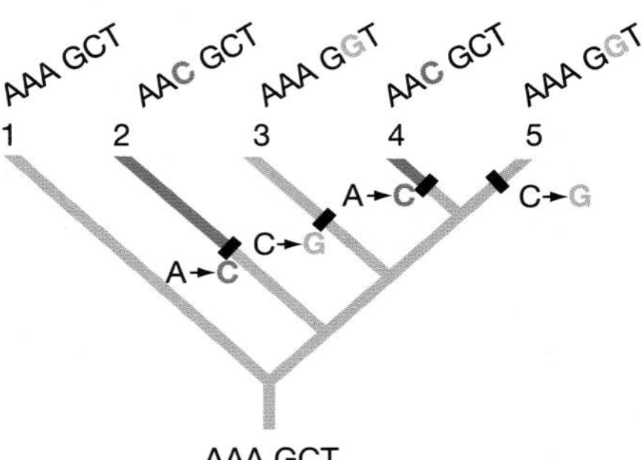

FIGURE 27.3 Parsimony Is One Method for Choosing among the Many Possible Trees. If you were using DNA sequence data to infer the phylogeny of five species, many different trees are possible. (Only two are shown here). Under parsimony, the best tree is the one that requires the fewest changes to explain the sequences observed in the five species.

Whale Evolution: A Case History As an example of how a cladistic approach works, consider the evolutionary relationships of the whales and the lineage of mammals called the Artiodactyla. Cows, deer, and hippos are artiodactyls. Members of this group have hooves and an even number of toes. They also share another feature: the unusual pulley shape of an ankle bone called the astragalus. Along with having feet with hooves and an even number of toes, the shape of the astragalus is a synapomorphy that identifies the artiodactyls as a monophyletic group. These data support the tree shown in **Figure 27.4a**. Whales do not have an astragalus and are shown as an outgroup on this tree—that is, a species or group that is closely related to the monophyletic group but not part of it.

(a) The astragalus is a synapomorphy that identifies artiodactyls as a monophyletic group.

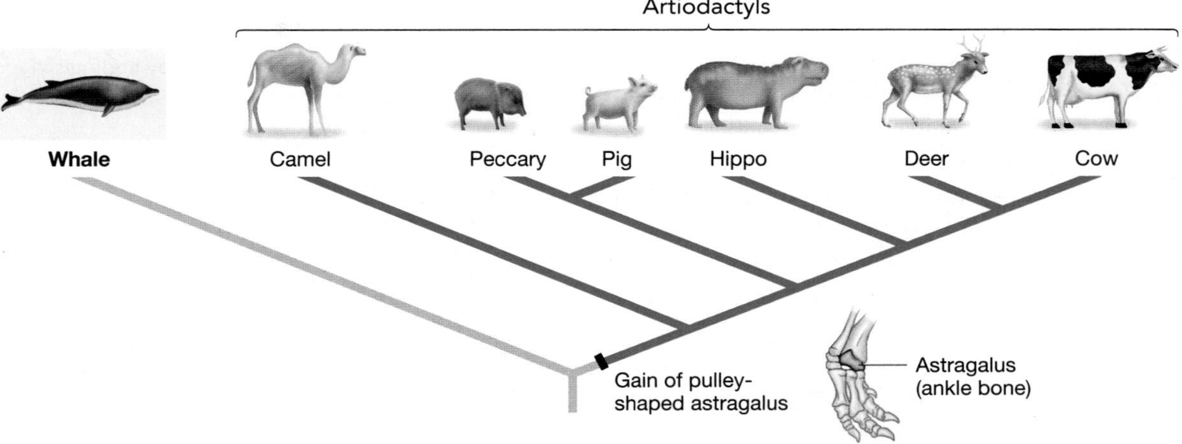

(b) If whales are related to hippos, then two changes occurred in the astragalus.

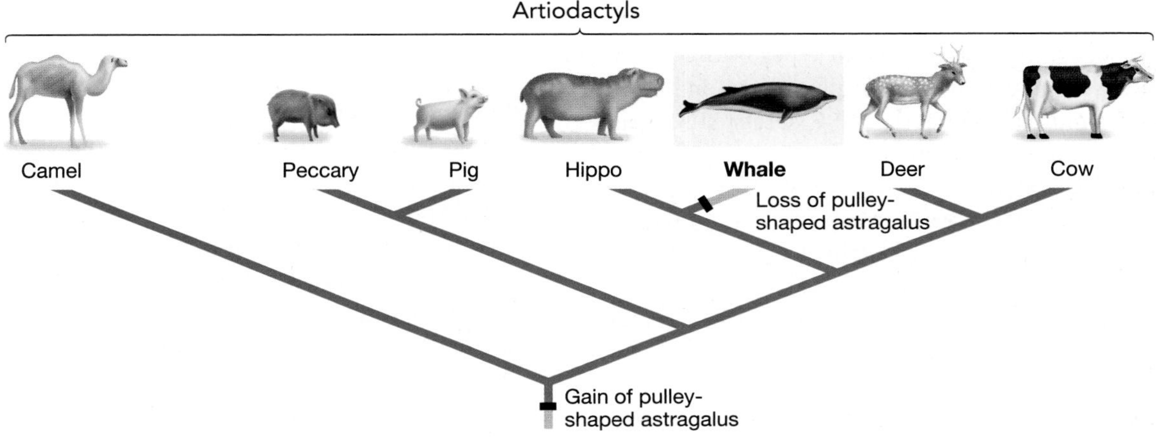

(c) Data on the presence and absence of SINE genes support the close relationship between whales and hippos.

Locus	1	2	3	4	5	6	7	8	9	10	11	12	13	14	15	16	17	18	19	20
Cow	0	0	0	0	0	0	0	**1**	**1**	**1**	**1**	**1**	**1**	**1**	**1**	**1**	**1**	**1**	0	0
Deer	0	0	0	0	0	0	0	**1**	?	**1**	**1**	**1**	**1**	**1**	**1**	?	**1**	**1**	0	0
Whale	**1**	**1**	**1**	**1**	**1**	**1**	**1**	0	?	**1**	0	**1**	**1**	0	0	0	?	**1**	0	0
Hippo	0	?	0	**1**	**1**	**1**	**1**	0	**1**	**1**	0	**1**	**1**	0	0	0	?	**1**	0	0
Pig	0	0	0	?	0	0	0	0	?	0	0	0	?	?	0	0	0	**1**	**1**	**1**
Peccary	?	?	?	?	?	?	?	?	?	?	?	?	?	?	?	?	?	?	**1**	**1**
Camel	0	0	0	0	0	0	0	0	0	0	0	0	0	0	0	0	0	0	0	0

1 = gene present
0 = gene absent
? = still undetermined

Whales and hippos share four *unique* SINE genes (4, 5, 6, and 7)

FIGURE 27.4 Evidence That Whales and Hippos Form a Monophyletic Group. Based on a parsimony analysis of the pulley-shaped astragalus, biologists favored **(a)** a tree excluding whales from the Artiodactyla over **(b)** the hypothesis that whales are artiodactyls and are closely related to hippos. **(c)** Data on the presence and absence of SINE genes support the whales + hippo hypothesis.

EXERCISE The presence of SINE genes 4, 5, 6, 7 identifies hippos and whales as part of a monophyletic group. The presence of SINE genes 8, 11, 14, 15, 17 identifies deer and cows as part of a monophyletic group. SINE genes 10, 12, 13 identify hippos, whales, deer, and cows as part of a monophyletic group. Map the origin of these SINE genes on the tree in part (b). Where did SINE genes 19 and 20 first insert themselves into the genomes of artiodactyls?

When researchers began comparing DNA sequences of artiodactyls and other species of mammals, however, the data showed that whales share many similarities with hippos. These results supported the tree shown in **Figure 27.4b**. The tree supported by the DNA data conflicts with the tree supported by morphological data because it implies that the pulley-shaped astragalus evolved in artiodactyls and then was lost during whale evolution. The tree in Figure 27.4b implies two changes in the astragalus (a gain *and* a loss of the astragalus), while the tree in Figure 27.4a implies just one (a gain only). In terms of the evolution of the astragalus, the "whale + hippo" tree is less parsimonious than the "whales-are-not-artiodactyls" tree.

The conflict between the data sets was resolved when researchers analyzed the distribution of the parasitic gene sequences called **SINEs (short interspersed nuclear elements),** which occasionally insert themselves into the genomes of mammals. SINEs are transposable elements, similar to the LINEs introduced in Chapter 20. As the data in **Figure 27.4c** show, whales and hippos share several types of SINES that are not found in other groups. Specifically, whales and hippos share the SINEs numbered 4, 5, 6, and 7. Other SINE genes are present in some artiodactyls but not in others; camels have no SINE genes at all. To explain these data, biologists hypothesize that no SINEs were present in the population that is ancestral to all of the species in the study. After the branching event that led to the split between the camels and all the other artiodactyls however, different SINEs became inserted into the genomes of descendant populations. As a result, the presence of a particular SINE represents a derived character. Because whales and hippos share four of these derived characters, it is logical to conclude that these animals are closely related. ● If you understand this concept, you should be able to explain why SINES numbered 4–7 are synapomorphies that identify whales and hippos as a monophyletic group, and why the similarity in these SINES is unlikely to represent homoplasy.

Based on these data, most biologists accepted the phylogeny shown in Figure 27.4b as the most accurate estimate of evolutionary history. According to this phylogeny, whales are artiodactyls and share a relatively recent common ancestor with hippos. This observation inspired the hypothesis that both whales and dolphins are descended from a population of artiodactyls that spent most of their time feeding in shallow water, much as hippos do today.

Recently this hypothesis was supported in spectacular fashion. In 2001 two teams of researchers announced the independent discoveries of fossil artiodactyls that were clearly related to whales and yet had a pulley-shaped astragalus. One of these species is illustrated at the top of Figure 24.4b. The combination of phylogenetic data and data from the fossil record has clarified how a particularly interesting group of mammals evolved. What else do fossils have to say?

Check Your Understanding

If you understand that...

- Phylogenies can be estimated by finding synapomorphies that identify monophyletic groups.

You should be able to...

1) Explain whether the following traits represent homoplasy or homology: hair in humans and whales; extensive hair loss in humans and whales; limbs in humans and whales; social behavior in certain whales (e.g., dolphins) and humans.
2) Explain why the presence and absence of certain SINE genes serve as synapomorphies supporting the hypothesis that species of cows and deer form a monophyletic group.

27.2 Tools for Studying History: The Fossil Record

Phylogenetic analyses are powerful ways to infer the order in which events occurred during evolution and to understand how particular groups of species are related. ● But only the fossil record provides direct evidence about what organisms that lived in the past looked like, where they lived, and when they existed. A **fossil** is a piece of physical evidence from an organism that lived in the past. The **fossil record** is the total collection of fossils that have been found throughout the world. The fossil record is housed in thousands of private and public collections.

Let's review how fossils form, analyze the strengths and weaknesses of the fossil record, and then summarize major events that have taken place in life's approximately 3.5-billion-year history. (Although most evidence suggests that life originated 3.4–3.6 billion years ago, research continues.)

How Do Fossils Form? Most of the processes that form fossils begin when part or all of an organism is buried in ash, sand, mud, or some other type of sediment. Consider a series of events that begins when a tree falls into a swamp. **Figure 27.5** illustrates the leaves of a tree falling onto a patch of mud, where they are buried by soil and debris before they decay. Pollen and seeds settle into the muck at the bottom of the swamp, where decomposition is slow. The stagnant water is too acidic and too oxygen poor to support large populations of bacteria and fungi, so much of this material is buried intact before it decomposes. The trunk and branches that sit above the water line rot fairly quickly, but as pieces break off they, too, sink to the bottom and are buried.

Once burial occurs, several things can happen. If decomposition does not occur, the organic remains can be preserved intact—like the fossil pollen in **Figure 27.6a**. Alternatively, if

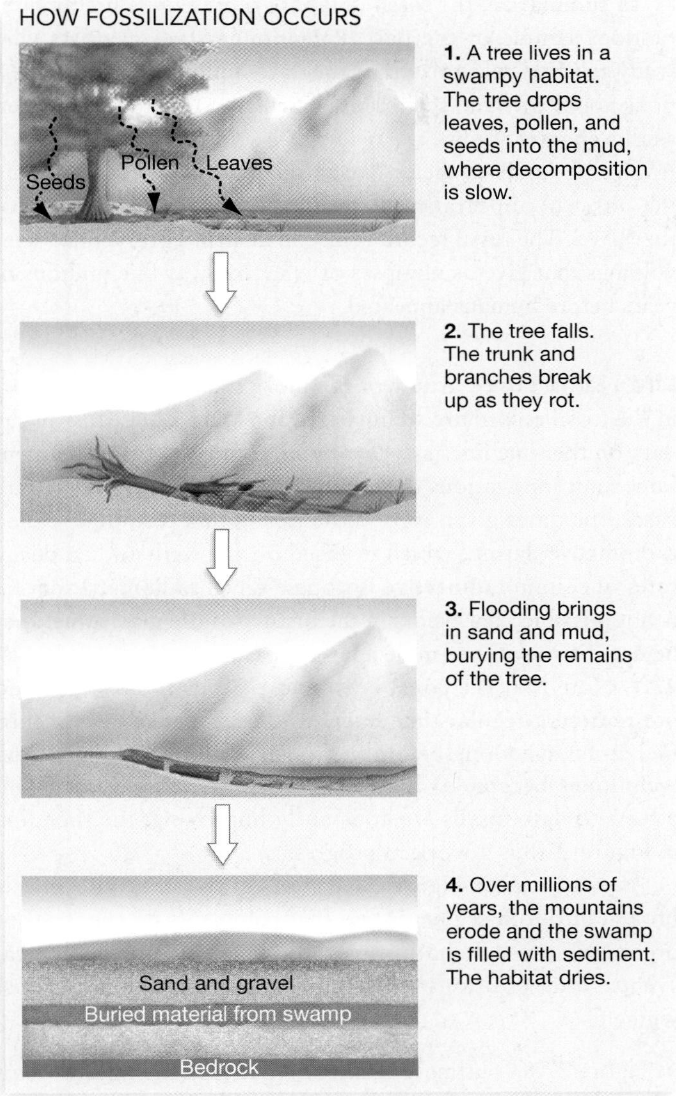

FIGURE 27.5 Fossilization Preserves Traces of Organisms That Lived in the Past. Fossilization occurs most readily when the remains of an organism are buried in sediments, where decay is slow.

sediments accumulate on top of the material and become cemented into rocks such as mudstone or shale, the sediments' weight can compress the organic material below into a thin, carbonaceous film. This happened to the leaf in **Figure 27.6b**. If the remains decompose *after* they are buried—as did the branch in **Figure 27.6c**—the hole that remains can fill with dissolved minerals and faithfully create a **cast** of the remains. If the remains rot extremely slowly, dissolved minerals can gradually infiltrate the interior of the cells and harden into stone, forming a permineralized fossil, such as petrified wood (**Figure 27.6d**).

After many centuries have passed, fossils can be exposed at the surface by many mechanisms, including erosion, a road cut, or quarrying. If researchers find a fossil, they can prepare it for study by painstakingly clearing away the surrounding rock. If the species represented is new, researchers describe its morphology in a scientific publication, name the species, estimate the fossil's age based on dates assigned to nearby rock layers, and add the specimen to a collection so that it is available for study by other researchers. It is now part of the fossil record. This is the information database that supports much of the research reviewed in this chapter.

The scenario just presented is based on conditions that are ideal for fossilization: The tree fell into an environment where decomposition was slow and burial was rapid. In most habitats the opposite situation occurs—decomposition is rapid and burial is slow. In reality, then, fossilization is an extremely rare event. To appreciate this point, consider that there are 10 specimens of the first bird to appear in the fossil record, *Archaeopteryx*. All were found at the same site in Germany where limestone is quarried for printmaking (the bird's specific name is *lithographica*). If you accept an estimate that crow-sized birds native to wetland habitats in northern Europe would have a population size of around 10,000 and a life span of 10 years, and if you accept the current estimate that the species existed for about 2 million years, then you can calculate that about 2 billion *Archaeopteryx*

(a) Intact fossil

The pollen was preserved intact because no decomposition occurred.

(b) Compression fossil

Sediments accumulated on top of the leaf and compressed it into a thin carbon-rich film.

(c) Cast fossil

The branch decomposed after it was buried. This left a hole that filled with dissolved minerals, faithfully creating a cast of the original.

(d) Permineralized fossil

The wood decayed very slowly, allowing dissolved minerals to infiltrate the cells gradually and then harden into stone.

FIGURE 27.6 Fossils Are Formed in Several Ways. Different preservation processes give rise to different types of fossils.

lived. But as far as researchers currently know, only 1 out of every 200,000,000 individuals fossilized. For this species, the odds of becoming a fossil were almost 40 times worse than your odds are of winning the grand prize in a state lottery.

Limitations of the Fossil Record Before looking at how the fossil record is used to answer questions about the history of life, it is essential to review the nature of this archive and recognize several features:

- *Habitat bias* Because burial in sediments is so crucial to fossilization, there is a strong habitat bias in the database. Organisms that live in areas where sediments are actively being deposited—including beaches, mudflats, and swamps—are much more likely to form fossils than are organisms that live in other habitats. Within these habitats, burrowing organisms such as clams are already underground—pre-buried—at death and are therefore much more likely to fossilize. Organisms that live aboveground in dry forests, grasslands, and deserts are much less likely to fossilize.

- *Taxonomic bias* Slow decay is almost always essential to fossilization, so organisms with hard parts such as bones or shells are most likely to leave fossil evidence. This requirement introduces a strong taxonomic bias into the record. Clams, snails, and other organisms with hard parts have a much higher tendency to be preserved than do worms. A similar bias exists for tissues within organisms. For instance, pollen grains are encased in a tough outer coat that resists decay, so they fossilize much more readily than do flowers. Teeth are the most common mammalian fossil, simply because they are so hard and decay resistant. Shark teeth are abundant in the fossil record; but shark bones, which are made of cartilage, are almost nonexistent.

- *Temporal bias* Recent fossils are much more common than ancient fossils. To understand why, consider that when two of Earth's tectonic plates converge, the edge of one plate usually sinks beneath the other plate. The rocks composing the edge of the descending plate are either melted or radically altered by the increased heat and pressure they encounter as they move downward into Earth's interior. These alterations obliterate any fossils in the rock. In addition, fossil-bearing rocks on land are constantly being broken apart and destroyed by wind and water erosion. The older a fossil is, the more likely it is to be demolished.

- *Abundance bias* Because fossilization is so improbable, the fossil record is weighted toward common species. Organisms that are abundant, widespread, and present on Earth for long periods of time leave evidence much more often than do species that are rare, local, or ephemeral.

To summarize, the fossil database represents a highly nonrandom sample of the past. **Paleontologists**—scientists who study fossils—recognize that they are limited to asking questions about tiny and scattered segments on the tree of life. Yet, as this chapter shows, the record is a scientific treasure trove. Analyzing fossils is the only way scientists have of examining the physical appearance of extinct forms and inferring how they lived. The fossil record is like an ancient library, filled with volumes that give us glimpses of what life was like millions of years before humans appeared.

Life's Time Line A few of the more significant data points in the fossil record are summarized in **Figure 27.8**. Most of the data on the time lines are "evolutionary firsts" that document important innovations during the history of life. In almost all cases, the dates given were estimated with a technique called radiometric dating, which is based on the well-studied decay rates of certain radioactive isotopes. When radiometric dating is not possible, key events in the history of life may sometimes be dated by using the molecular approaches introduced in **Box 27.1**. Note that the eons, eras, and periods in the figure do not represent regular time intervals, because they were identified and named long before radiometric and molecular dating techniques became available. Because the fossil record and efforts to date fossils are constantly improving, the time line in Figure 27.8 is a work in progress.

As you inspect Figure 27.8, notice that the time line is broken into four segments and that comments on the types of organisms present, nature of the climate, major geological events, and positions of the continents accompany most segments:

- Figure 27.8a outlines the interval between the formation of Earth about 4.6 billion years ago and the appearance of most animal groups about 542 million years ago (abbreviated mya). The entire interval is called the **Precambrian**; it is divided into the Hadean, Archaean, and Proterozoic eons. The important things to note about this time are that (1) life was exclusively unicellular for most of Earth's history and (2) oxygen was virtually absent from the oceans and atmosphere for almost 2 billion years after the origin of life.

- The interval between 542 mya and the present is called the Phanerozoic eon and is divided into three eras. Each of these eras is further divided into intervals called periods.

- Figure 27.8b summarizes the **Paleozoic** ("ancient life") **era**, which begins with the appearance of many animal lineages and ends with the obliteration of almost all multicellular life-forms at the end of the Permian period. The Paleozoic saw the origin and initial diversification of the animals, land plants, and fungi, as well as the appearance of land animals.

BOX 27.1 The Molecular Clock

Several researchers have proposed that the fossil record can be supplemented with information from molecules in living species. This hypothesis, called the **molecular clock**, is based on analyzing silent mutations (Chapter 16) and other changes in DNA that do not affect gene function. These changes are not affected by natural selection, but can increase to fixation by genetic drift. Research has shown that if mutation rates stay roughly constant over time, drift fixes mutations at a steady rate. For example, consider data on the protein hemoglobin. A biologist compared the amino acid sequence of this molecule in vertebrates ranging from sharks to humans. When he plotted the number of amino acid differences between species against the date that the species diverged, according to the fossil record, the graph shown in **Figure 27.7** resulted. The evolution of hemoglobin seems to proceed at a constant rate over time.

To see how researchers use molecular clocks to supplement the fossil record, let's consider some recent analyses of human evolution. Seven fossil species of *Homo* have been identified to date. The first fossils of our species, *H. sapiens*, appear about 100,000 years ago. But *sapiens* had to be around before that. The question is, when did *Homo sapiens* actually arise?

A team of investigators used a molecular clock to answer this question. The group compared the complete mitochondrial genomes sequenced from African, European, and Japanese individuals. These data allowed the team to estimate the number of bases in mitochondrial DNA (mtDNA) that had changed as human populations diverged. To estimate how long it took these genetic differences to develop, the researchers also sequenced the mtDNA of our close relatives, the orangutan and African apes. According to the fossil record, orangutans and African apes diverged 13 million years ago. By dividing the number of bases that differ in the mtDNA of orangutans and African apes by 13 million years, the researchers arrived at an estimate that mtDNA changes at the rate of 7×10^{-8} per base per year.

The total amount of sequence divergence observed between two species should be equal to the rate of sequence evolution $\times$ the time since divergence occurred $\times$ 2. When the researchers used this equation with their estimate of the number of bases in mtDNA that had changed as human populations diverged, they concluded that the most recent common ancestor of all living humans lived between 125,000 and 161,000 years ago. Large species of mammals typically last in the fossil record for about 1.5 million years prior to extinction. *H. sapiens*, then, is a relatively young species.

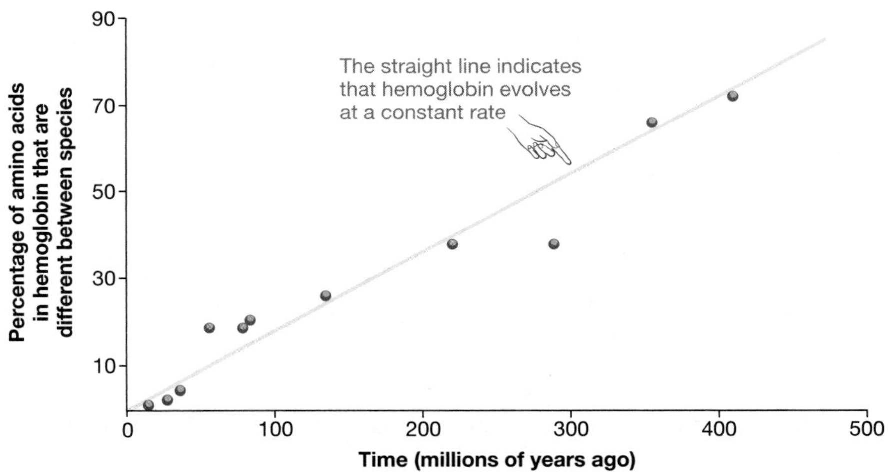

FIGURE 27.7 A Data Set That Supports a Molecular Clock. To interpret this graph, consider that the point at the upper right represents a comparison of hemoglobin proteins in sharks and mammals. According to the fossil record, sharks branched off from other vertebrates about 450 million years ago. The other data points represent comparisons between other species of vertebrates.

QUESTION How fast does this clock tick? That is, what percentage of amino acids in hemoglobin change per 100 million years, on average?

- The **Mesozoic** ("middle life") **era** is outlined in Figure 27.8c. This interval is nicknamed the Age of Reptiles. It begins with the end-Permian extinction events and ends with the extinction of the dinosaurs and other groups at the boundary between the Cretaceous period and Paleogene period. In terrestrial environments of the Mesozoic, gymnosperms were the dominant plants and dinosaurs were the dominant vertebrates.
- Figure 27.8d highlights the **Cenozoic** ("recent life") **era**, which is divided into the Paleogene period and the Neogene period. The Cenozoic is sometimes nicknamed the Age of Mammals, because mammals diversified after the disappearance of the dinosaurs. On land, angiosperms were the dominant plants and mammals were the dominant vertebrates. Events that occur today are considered to be part of the Cenozoic era.

(a) The **Precambrian** (Hadean, Archaean, and Proterozoic Eons) included the origin of life, photosynthesis, and the oxygen atmosphere.

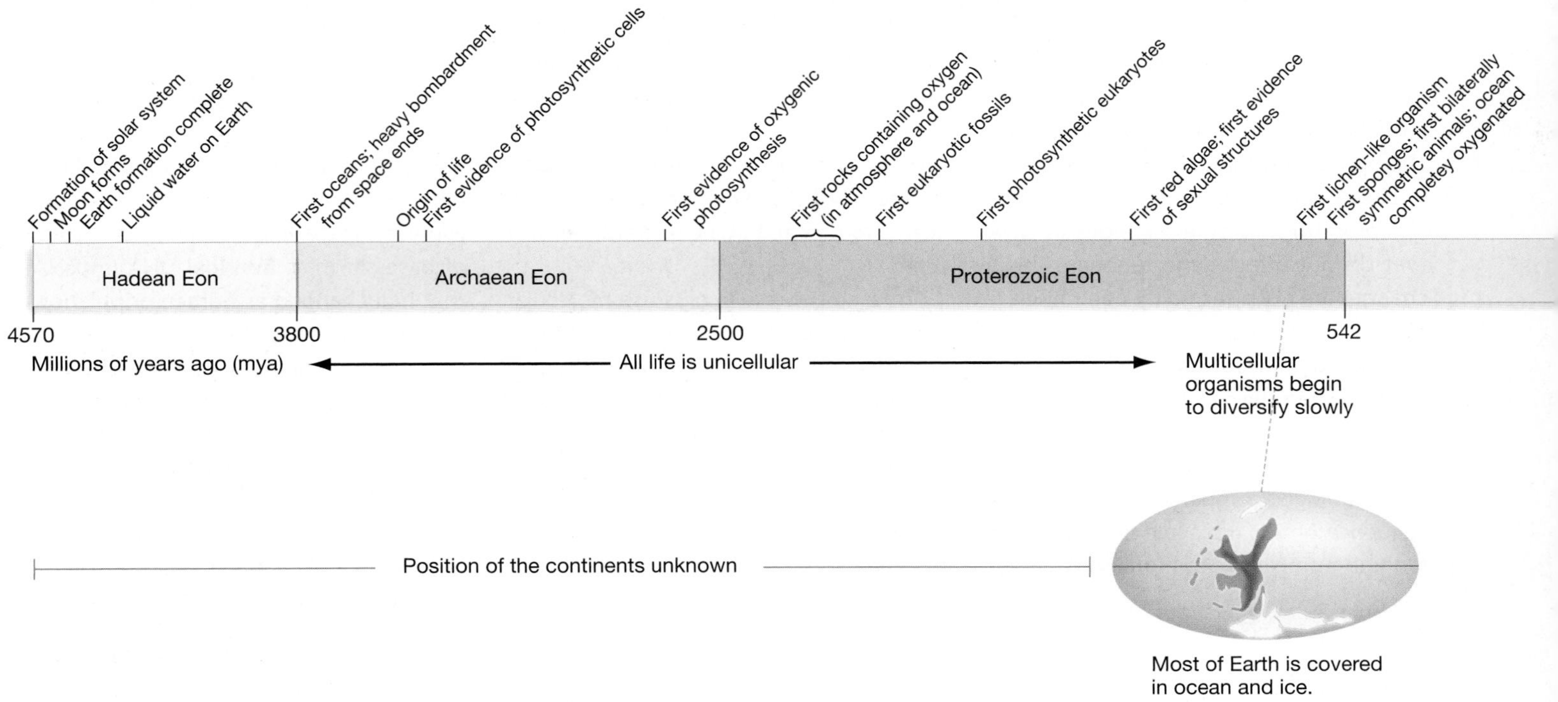

(c) Phanerozoic Eon: The **Mesozoic Era** is sometimes called the Age of Reptiles.

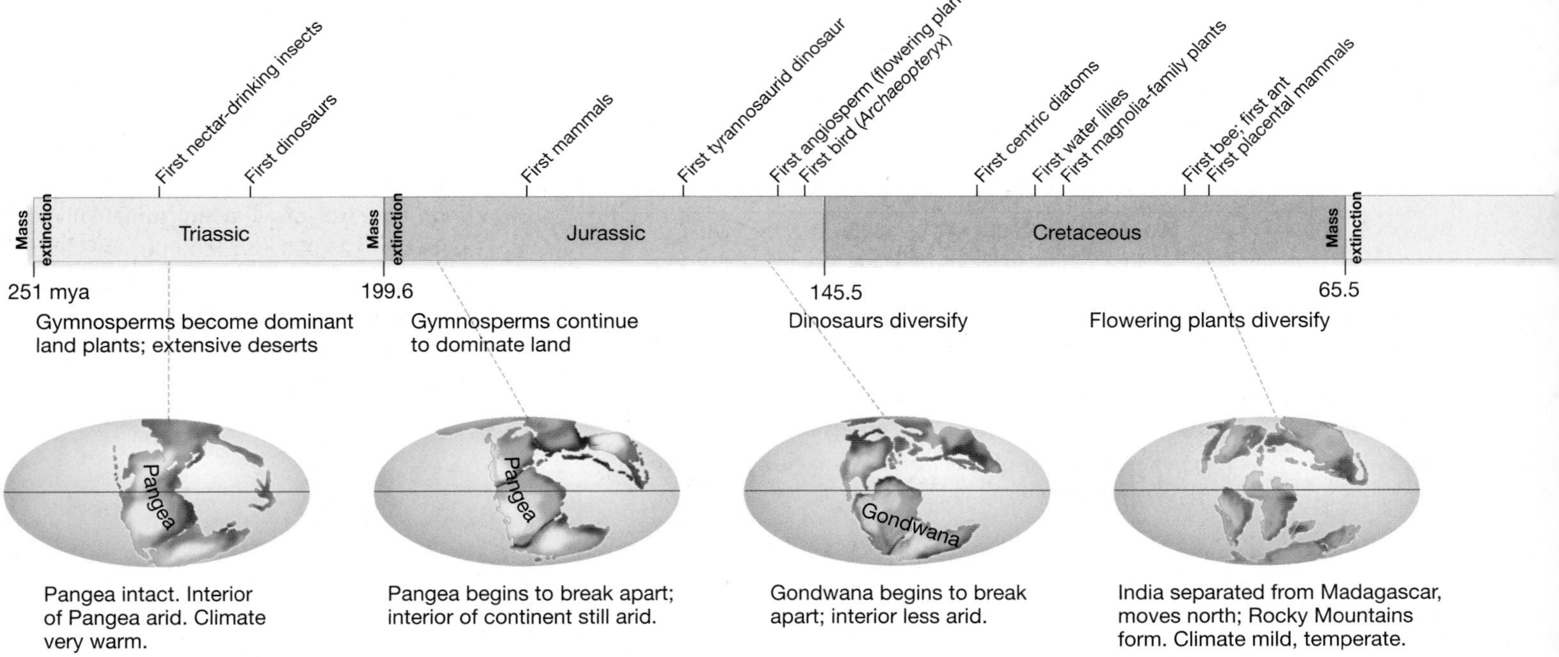

FIGURE 27.8 Life's Time Line. Significant events in Earth's history and the history of life are plotted for the **(a)** Precambrian, **(b)** Paleozoic era, **(c)** Mesozoic era, and **(d)** Cenozoic era (*mya* stands for "million years ago"). The Hadean, Archaean, and Proterozoic are vast stretches of time called eons. The Paleozoic, Mesozoic, and Cenozoic eras make up the Phanerozoic eon.

(b) Phanerozoic Eon: The **Paleozoic Era** included the origin and early diversification of animals, land plants, and fungi.

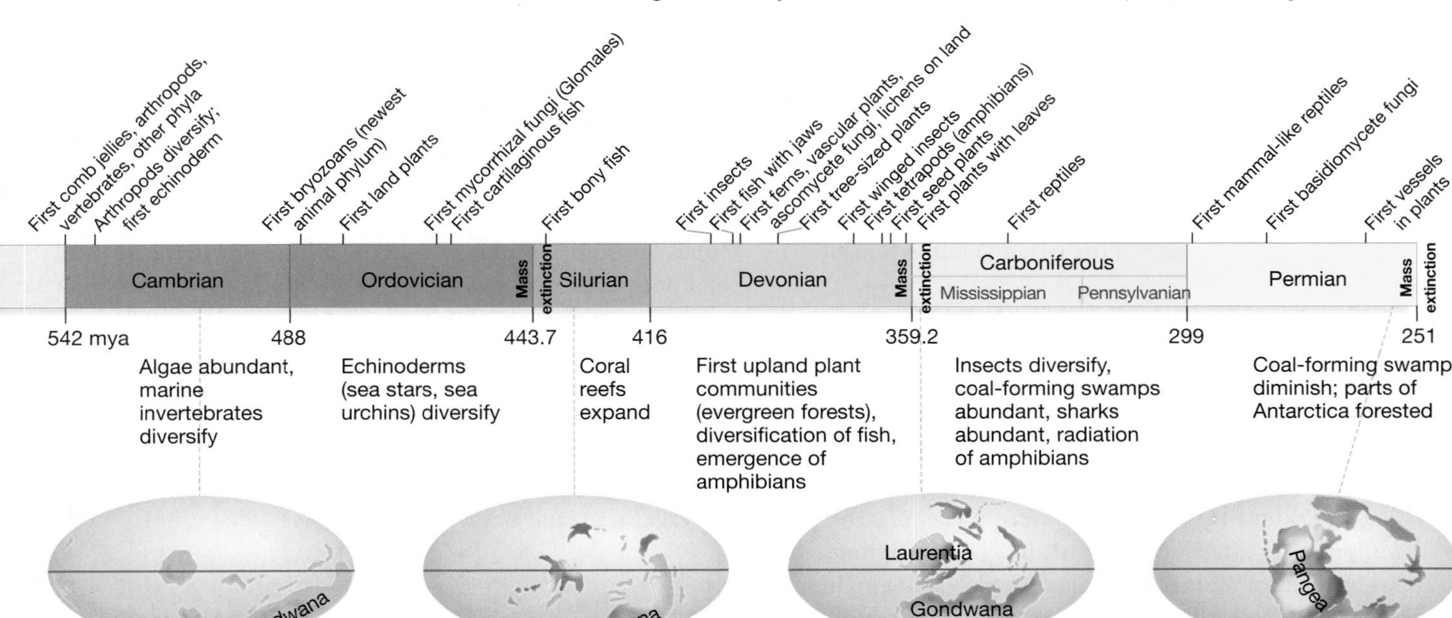

(d) Phanerozoic Eon: The **Cenozoic Era** is nicknamed the Age of Mammals.

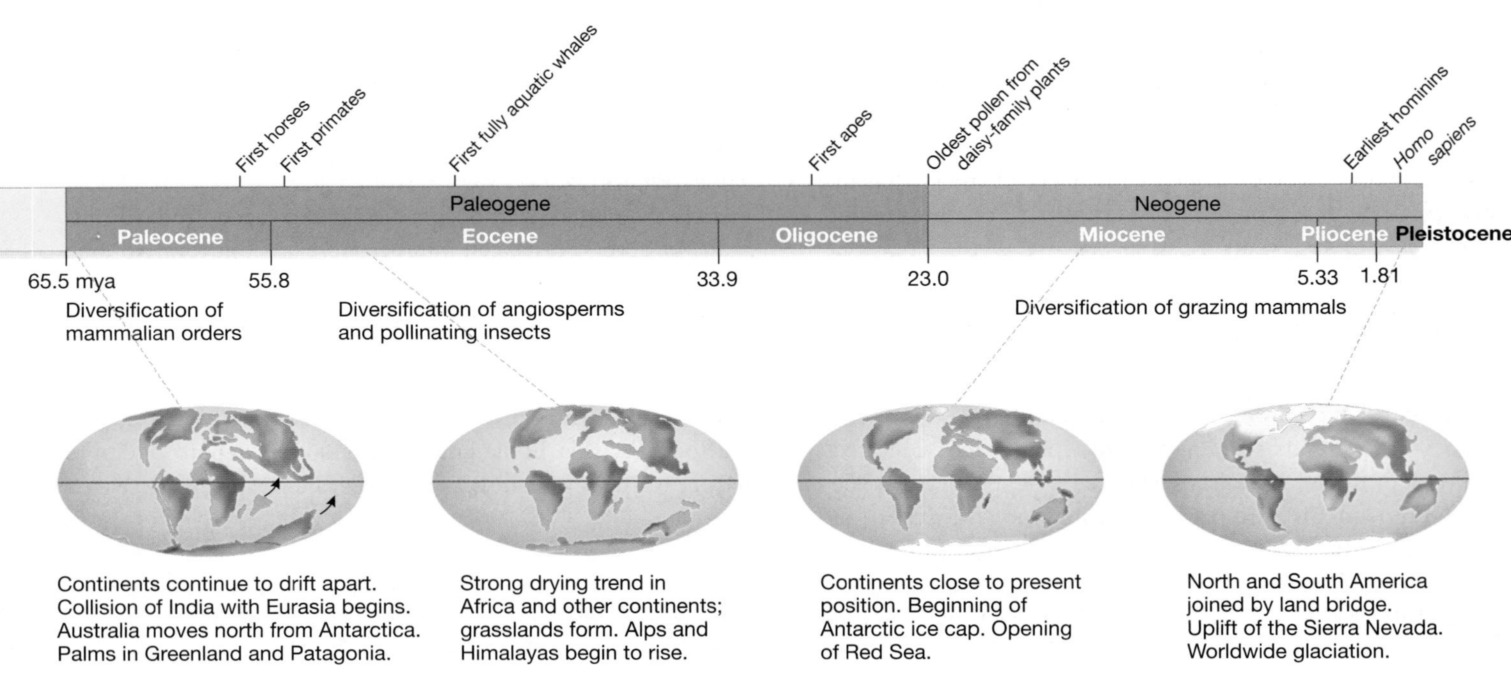

The data in Figure 27.8 show that the nature of life has changed radically since the first cell appeared on Earth. Although the changes that have taken place over the past 3.4 billion years are well documented in the fossil record, the sweep of time involved is difficult for the human mind to comprehend. A semester can seem long to a college student; but in relation to Earth's history, 100,000 years or even a million years is the blink of an eye.

27.3 The Cambrian Explosion

Phylogenetic trees and the fossil record give researchers powerful tools for understanding what life was like during the millions of years before biologists existed. They tell thousands of stories about the sweeping changes that have occurred throughout Earth's history. But here there is space to consider just a few of the most dramatic events. Let's begin with research on the origin and early diversification of animals.

The first animals appear in the fossil record about 570 million years ago. A burst of diversification occurred soon after that, at the start of the Cambrian period. This diversification happened so rapidly that it earned a nickname: the **Cambrian explosion**.

It is impossible to appreciate the drama of this event without recalling that almost all life-forms were unicellular for almost 3 *billion* years after the origin of life. The exceptions were several lineages of multicellular algae, which show up in the fossil record about 1 billion years ago. But the ancestors of animals were still unicellular, marine creatures with cilia or flagella—probably not too different from some of the aquatic eukaryotes living today. Then, about 565 million years ago, the first animals—sponges, jellyfish, and perhaps simple worms—appear in the fossil record. Just 50 million years later, virtually every major group of animals had appeared. In a relatively short time, creatures with shells, exoskeletons, internal skeletons, legs, heads, tails, eyes, antennae, jaw-like mandibles, segmented bodies, muscles, and brains had evolved. It was arguably the most spectacular period of evolutionary change in the history of life.

By combining evidence from the fossil record and phylogeny reconstruction, researchers have begun to clarify the timing and sequence of key events.

Cambrian Fossils: An Overview

The Cambrian explosion is documented by three major fossil assemblages that record the state of animal life at 570 mya, at 565 to 542 mya, and at 525 to 515 mya. The species collected from each of these intervals are referred to respectively as the Doushantuo fossils (from the Doushantuo formation in China), Ediacaran fossils (from Ediacara Hills, Australia), and Burgess Shale fossils (from British Columbia, Canada), as **Figure 27.9a** shows. Fossils from the Ediacaran interval and Burgess Shale interval have now been found at localities throughout the world.

Fortunately for biologists, the glimpses of life provided by the three groups of organisms are extraordinarily clear. These three assemblages all break one of the cardinal rules of fossil preservation. Soft-bodied animals, which usually do not fossilize efficiently, are well represented in all three groups. By sheer luck, during each of these time intervals there happened to be a few habitats in which burial occurred so rapidly and decomposition occurred so slowly that organisms without shells were able to fossilize.

To appreciate why this extraordinarily efficient preservation is so important, consider the Burgess Shale. In several localities around the world dated to 525–515 million years ago, fossilization processes were typical—only shelled organisms are found. But soft-bodied organisms also fossilized in the atypical conditions of the Burgess Shale. At this locality, over *five times* as many species are represented. The database is 500 percent better than usual.

The presence of these exceptionally rich deposits before, during, and after the Cambrian explosion makes the fossil record for this event extraordinarily complete. Let's take a brief look at the **faunas**—the animal types—found in the Doushantuo, Ediacaran, and Burgess Shale fossils.

The Doushantuo Microfossils

Two papers, published within days of each other in 1998, introduced the world to the fossils of the Doushantuo rock formation. Researchers were able to identify several dozen tiny sponges, ranging from 150 μm to 750 μm across, in samples dated at approximately 580 million years ago. In deposits dated to about 570 million years ago, a different team of biologists found clusters of cells that they interpreted as animal embryos. This conclusion was based on a simple observation: Their samples contained one-celled, two-celled, four-celled, and eight-celled fossils, along with individuals containing larger cell numbers whose overall size was the same (**Figure 27.9b**). Recall from Chapter 22 that this is exactly the pattern that occurs during cleavage in today's animals. In other words, cell number increases but total volume remains constant. What type of animal did such embryos develop into? The answer is still unknown. In addition, a different team of researchers has presented evidence that a large fraction of the Doushantuo "embryos" are actually fossils from large, sulfur-oxidizing bacteria that today live on the seafloor. As this book goes to press, the debate over the nature of the fossils in Figure 27.9b is unresolved. Research continues.

The sponges and possible embryos are scattered among abundant cyanobacterial cells as well as among multicellular algae that are the ancestors of today's seaweeds. These bacteria and algae were undoubtedly photosynthetic. The composite picture, then, is of a shallow-water marine habitat dominated by photosynthetic organisms. Scattered among other organisms in the Doushantuo formation were sponges and possibly other tiny creatures that probably made their living by filtering organic debris from the water. These were the first animals on Earth.

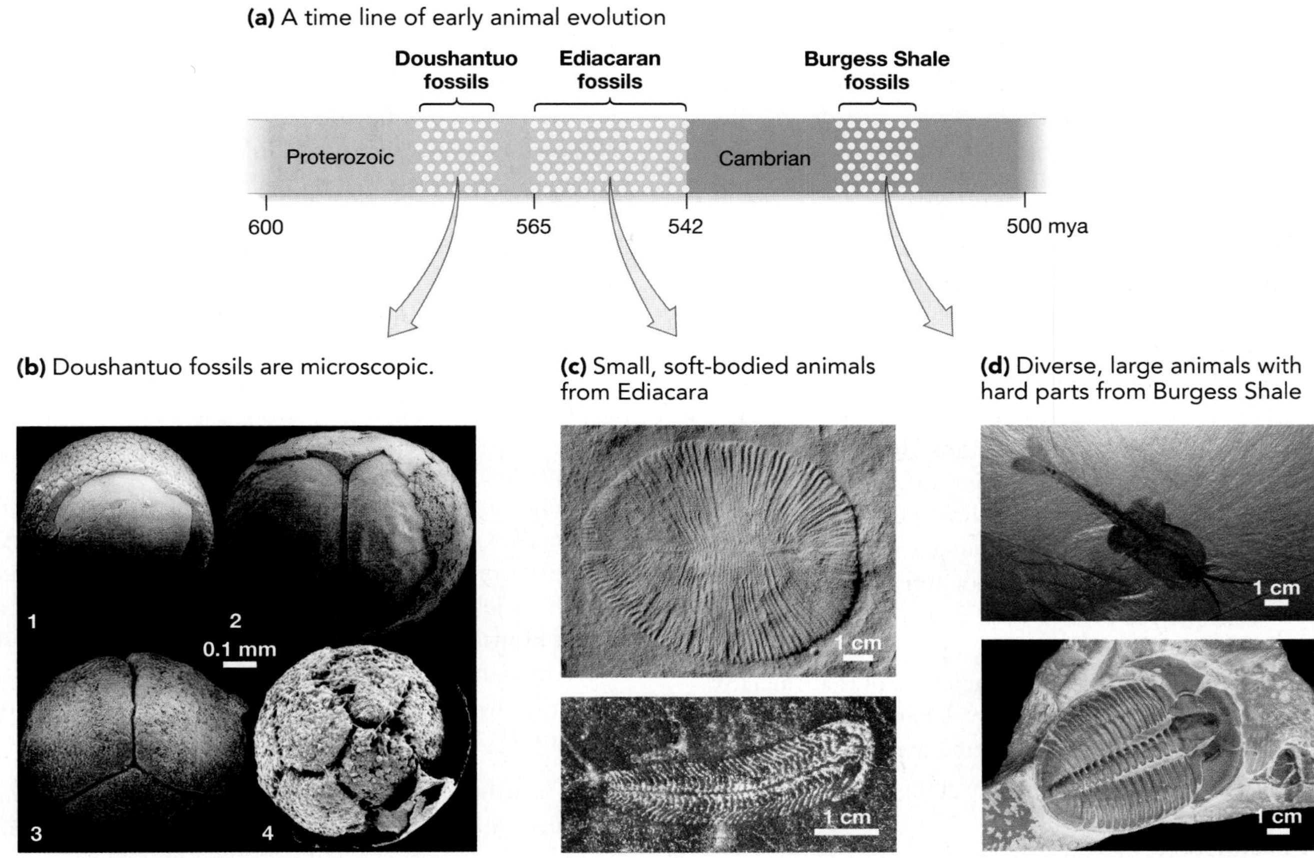

FIGURE 27.9 Fossils Document the Cambrian Explosion. (a) The origin of animals and their diversification during the Cambrian explosion is documented by three major fossil assemblages. **(b)** These fossilized cells from the Doushantuo deposits are arranged according to the hypothesis that they represent (1) a fertilized egg and then embryos at (2) the 2-cell, (3) the 4-cell, and (4) the 16-cell stages. **(c)** These 560-million-year-old fossils of small, soft-bodied animals were found in the Ediacara Hills of Australia. **(d)** Similar to this shrimplike arthropod (top) and trilobite (bottom), many of the animals fossilized in the Burgess Shale had heads, tails, shells (exoskeletons), and appendages.

The Ediacaran Faunas

Paleontologists discovered animal fossils in the Ediacara Hills of southern Australia in the 1940s. The specimens included the compressed bodies of sponges, jellyfish, and comb jellies as well as many burrows, tracks, and other traces from unidentified animal species (**Figure 27.9c**). In the decades since the initial discovery, similar faunas that are dated between 565 and 544 million years ago have been found at sites around the world. None of the organisms that fossilized during this period have shells or limbs, however, and none have heads or mouths or feeding appendages. As the scale bars in Figure 27.9c show, these organisms were also little in comparison to today's animals. These observations suggest that Ediacaran animals were small individuals that burrowed in sediments, sat immobile on the sea floor, or floated in the water. There is no evidence that they had structures associated with actively hunting and capturing food. Instead, it is likely that Ediacaran animals simply filtered or absorbed organic material from their surroundings.

The Burgess Shale Faunas

The discovery of fossils in the Burgess Shale formation of British Columbia, Canada, early in the twentieth century ranks among the most sensational additions ever made to the fossil record. Combined with the later unearthing of an extraordinary fossil assemblage in the Chengjiang deposits of China, the Burgess Shale gives researchers a compelling picture of life in the oceans 525–515 million years ago. Sponges, jellyfish, and comb jellies are abundant in these rocks; but entirely new lineages are present as well. Principal among these are the arthropods and mollusks. Today, the arthropods include the spiders, insects, and crustaceans (crabs, shrimp, and lobsters); mollusks include the clams, mussels, squid, and octopi. But echinoderms (sea stars and sea urchins), several types of worm and wormlike creatures, and even a chordate—the group that includes today's vertebrates—are found in these fossil faunas. In short, virtually every major animal lineage is documented there.

This tremendous increase in the size and morphological complexity of animals, illustrated in **Figure 27.9d**, was accompanied by diversification in how they made a living. The Cambrian seas were filled with animals that had eyes, mouths, limbs, and shells. They swam, burrowed, walked, ran, slithered, clung, or floated; there were predators, scavengers, filter feeders, and grazers. The diversification created and filled many of the ecological niches found in marine habitats today.

The Cambrian explosion still echoes. Animals that fill today's teeming tide pools, beaches, and mudflats trace their ancestry to species preserved in the Burgess Shale.

Did Gene Duplications Trigger the Cambrian Explosion?

The Doushantuo, Ediacaran, and Burgess Shale faunas document what happened during Cambrian explosion. Now the question is: *How* did all this morphological change come about? In particular, what genetic changes were responsible for the increased size and complexity of animals?

Chapter 21 introduced the homeotic genes that play a key role in laying out the three-dimensional pattern of multicellular organisms as they develop. You might recall that the *Hox* genes found in animals help specify the location and shape of limbs, antenna, and many other key structures. As soon as developmental biologists discovered how important these genes are in specifying morphological traits, they began asking about their role in evolution. Was the increase in morphological complexity during the Cambrian possible because of an increase in the number and complexity of homeotic genes?

To answer this question, biologists are working to determine the number and identity of homeotic genes found in different animal lineages. The idea is to look for correlations between the evolutionary history of animal lineages, their genetic makeup, and their morphology. Many researchers predicted there would be a strong association between the order in which animal lineages appeared during evolutionary history, the number of *Hox* genes present in each lineage, and each lineage's morphological complexity and body size. Some biologists even suggested that each major animal lineage would have unique homeotic genes associated with its unique body plans and appendages.

The logic behind this "new genes, new bodies" hypothesis was that gene duplication events could have occurred before and during the Cambrian explosion and produced new copies of existing homeotic genes. (Look back at Chapter 20 to review how gene duplication events occur.) These new genes would make possible the new body plans and appendages recorded in the Burgess Shale fauna. Again, the idea was that the number of homeotic genes present would correlate directly with morphological complexity.

To see if the predictions made by this hypothesis hold up, study the data in **Figure 27.10**. The phylogenetic tree represents the best current estimate of the evolutionary relationships among major animal lineages, and the colored boxes represent the *Hox* genes documented in each lineage. Note that *Hox* genes are found in clusters, with genes lined up on the chromosome one after the other. Note, too, that each gene in the *Hox* cluster has a distinct function as an embryo develops. Specifically, each *Hox* gene is involved in a different aspect of pattern formation—the events that organize cells in the space inside an embryo. To reflect these observations, the colors of the boxes indicate homology with other genes. If genes have the same color, it means their DNA sequences are so similar that researchers are confident the genes are related by common descent.

The following conclusions can be made from the data in the figure:

- As predicted by the new genes, new bodies hypothesis, the number and identity of *Hox* genes varies widely among animals. Groups that branched off early according to the fossil record and that have relatively small, simple bodies, such as jellyfish, have fewer *Hox* genes than do groups that branched off later—such as vertebrates. The number of genes in the *Hox* cluster appears to have expanded during the course of evolution (from top to bottom in Figure 27.10).
- It is sensible to argue that the new Hox genes were indeed created by gene duplication events, because genes within the cluster are similar in their structure and base sequence. For example, consider the genes that are colored green in Figure 27.10. These genes don't appear in jellyfish and probably originated when a mutation resulted in a duplication of the blue-colored gene, which is present in jellyfish. The two important points here are that (1) biologists can use the phylogeny to understand the order in which different genes within the cluster appeared, and (2) gene duplication provides a mechanism for the new genes, new bodies hypothesis.
- The entire *Hox* cluster was duplicated, and then duplicated again, in the lineage leading to vertebrates. All of the animals shown in Figure 27.10 have a single *Hox* cluster except for mice and zebra fish. Mice have 4 distinct clusters; zebra fish have 7. Because all vertebrates examined to date have more than one *Hox* cluster, the mutations that led to the duplication of the cluster probably occurred close to the origin of vertebrates. In addition, zebrafish and other ray-finned fishes have enough distinct *Hox* clusters to suggest that the entire cluster was duplicated yet again in their ancestors. Because vertebrates include some of the largest and most complex animals, and because ray-finned fishes are the most species rich and among the most morphologically diverse vertebrates, these observations also support the new genes, new bodies hypothesis. But it is important to note that in some cases, copies of genes within the duplicated clusters have been lost. Mice, for ex-

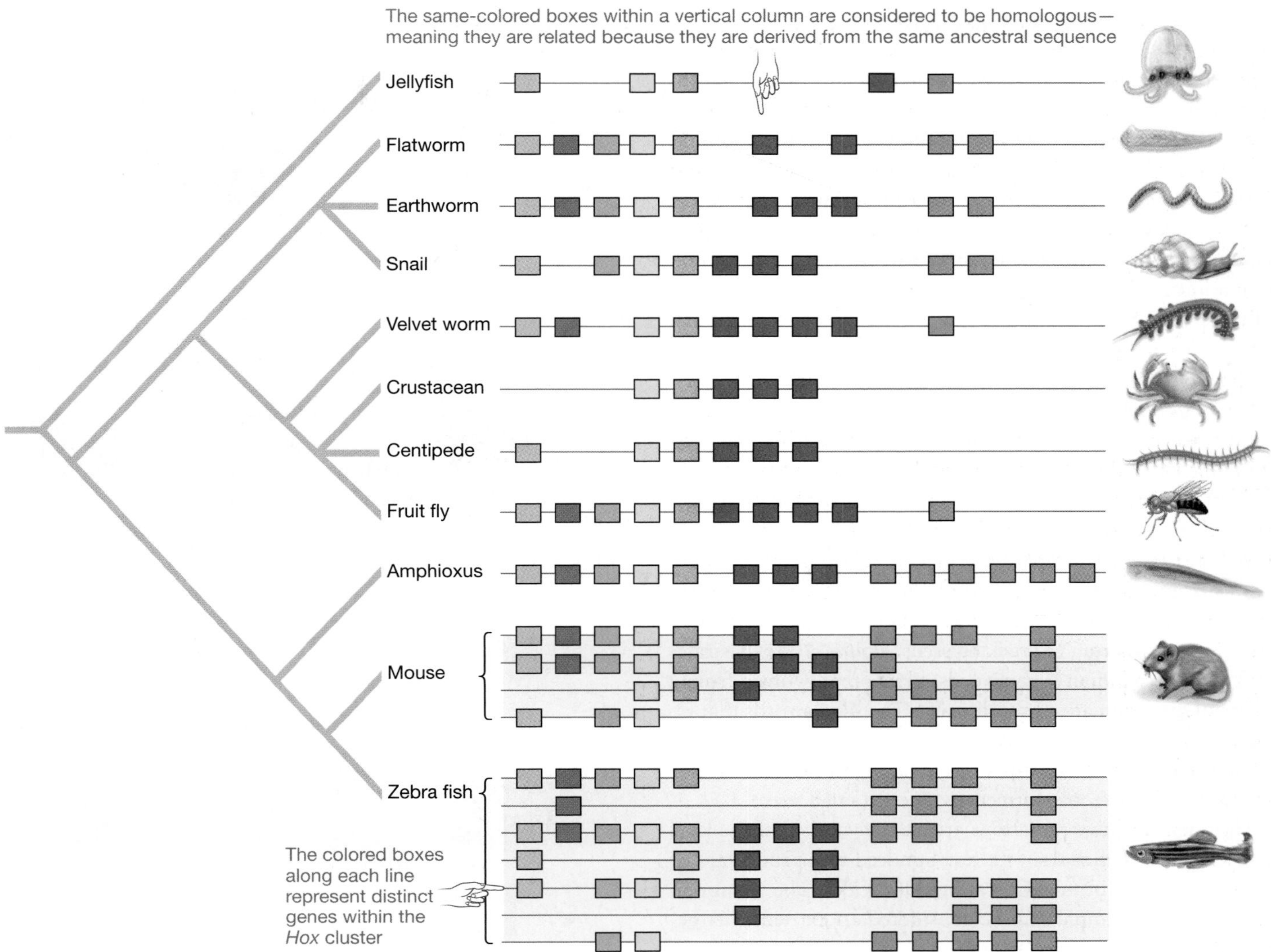

FIGURE 27.10 ***Hox*** **Genes in Animals.**

EXERCISE Next to the illustration of each animal on this evolutionary tree, write the number of Hox genes it has.

ample, have no blue-colored gene in their third set of *Hox* genes.

The data would certainly appear to support a simple-minded version of the new genes, new bodies hypothesis, except for one crucial observation: In arthropods (represented by the fruit fly, centipede, and crustacean), there is no correspondence at all between the number of *Hox* genes and the complexity of the resulting organisms. Crabs have just 5 *Hox* genes, while centipedes have 6 and the closely related but morphologically simpler velvet worms have 9.

Clearly, the situation is more involved than initially predicted. Duplication of *Hox* genes has undoubtedly been important in making the elaboration of animal body plans possible. But new genes are not the whole story: Changes in the expression and function of existing genes have been equally or even more important (see Chapter 22).

Check Your Understanding

If you understand that...

- The Cambrian explosion saw the rise of virtually every major animal lineage, and the evolution of a wide array of morphological innovations and food-getting strategies, in the relatively short time frame of 50 million years.

You should be able to...

1) Compare and contrast the animals documented in the Doushantuo, Ediacara, and Burgess Shale faunas.
2) Identify trends in early animal evolution—for example, whether animals tended to get larger or smaller, or less or more complex in their overall morphology.
3) Evaluate the hypothesis that the morphological evolution recorded in the Cambrian explosion was made possible by the duplication of *Hox* genes.

27.4 Adaptive Radiations

Suppose that biologists succeed in their goal of estimating the evolutionary relationships of all major lineages of organisms. The resulting phylogenetic tree would represent the complete tree of life. Now imagine that you could look at this tree from afar. One of the patterns that would jump out is that dense, bushy outgrowths are scattered among the branches. As **Figure 27.11a** shows, this shape results when many species branch off from a lineage in a short amount of time. Biologists sometimes call this pattern a star phylogeny, because of its starburst shape. Why does this pattern exist? One of the leading causes is a phenomenon known as an adaptive radiation.

An **adaptive radiation** occurs when a single lineage produces many descendant species that live in a wide diversity of habitats and find food in a variety of ways. **Figure 27.11b** shows a few of the Hawaiian honeycreepers—a diverse lineage of songbirds that evolved after a finchlike ancestor happened to colonize the Hawaiian islands. Considering how varied honeycreeper beaks are in size and shape, you should not be surprised that different types of honeycreepers obtain food in different ways—by eating insects, sucking nectar, or cracking seeds. **Figure 27.11c** illustrates a few of the Hawaiian silverswords, plants that evolved from a species of tarweed native to California. The silverswords that resulted from the adaptive radiation in Hawaii live in habitats ranging from lush rain forests to austere lava flows, and they range from mosslike mat-formers to vines to small trees.

Although adaptive radiations are usually studied at the level of species and time scales of a few hundred thousand years or less, they can also be analyzed at broader scales. The Cambrian explosion, for example, can be considered an extremely large-scale adaptive radiation. Another classical example is the diversification of mammals that took place between 65 and 50 million years ago. During this relatively short interval, the primates (monkeys and apes), bats, carnivores, deer, whales, horses, and rodents originated. The organisms resulting from this rapid divergence represent a remarkable array of adaptive forms. They swim, fly, glide, burrow, swing through trees, walk on four legs, or walk on two legs. They occupy habitats from the open ocean to mountaintops and from rain forests to deserts. They eat fruit, nuts, leaves, twigs, bark, insects, crustaceans, mollusks, fish, and other mammals.

The hallmark of an adaptive radiation is rapid speciation and ecological diversification within a single lineage. What makes adaptive radiations occur?

(a) Adaptive radiations produce star phylogenies.

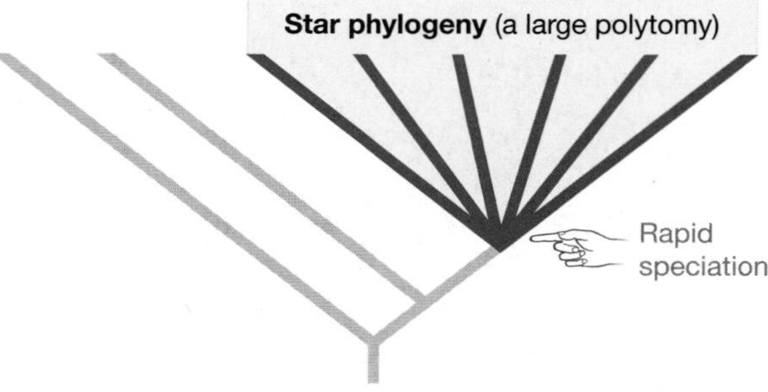

(b) Hawaiian honeycreepers underwent adaptive radiation.

(c) Hawaiian silverswords underwent adaptive radiation.

FIGURE 27.11 Adaptive Radiations Often Produce "Star Phylogenies." (a) A phylogeny has a starburst shape when speciation has been so rapid that the order of branching can't be resolved. If speciation is followed by divergence into many different adaptive forms, then an adaptive radiation has taken place. On the Hawaiian islands, adaptive radiations followed the arrival of both a finch from North America and tarweed seeds from California. These radiations produced the Hawaiian **(b)** finches and **(c)** silverswords.

EXERCISE On each tree, add a North American finch or a Californian tarweed as an outgroup. (To review what an outgroup is, see BioSkills 2.)

Ecological Opportunity as a Trigger

Whether adaptive radiations are analyzed at a large or small scale, biologists find that one of the most consistent themes in these events is ecological opportunity—meaning the availability of new or novel types of resources. The radiation of mammals, for example, occurred immediately after the extinction of dinosaurs. As mammals diversified, they took over the ecological roles formerly filled by dinosaurs and swimming reptiles. Adaptive radiations often occur when habitats are unoccupied by competitors.

Recently a group of biologists documented this process in detail. They did not study a radiation that followed an extinction event, however. Instead they analyzed radiations triggered by colonization events on islands that had distinct habitats and were free of competitors.

The study focused on the *Anolis* lizards of the Caribbean. Biologists have interpreted the history of this group of lizards as an adaptive radiation for two reasons: The lineage includes 150 species, and there is a strong correspondence between the size and shape of each species and the habitat it occupies. Most *Anolis* species that are twig-dwellers, for example, have relatively short legs and tails, while those that spend most of their time clinging to broad tree trunks or running along the ground tend to have long legs and tails (**Figure 27.12a**). These data suggest that lizard species have diversified in a way that allows them to occupy many different habitats.

Exactly how did the diversification occur? As the first step in answering this question, the biologists estimated the phylogeny of *Anolis* from DNA sequence data. Then they compared the habitats occupied by each species with their relationships on the phylogenetic tree. The results shown in **Figure 27.12b**, for species found on two different islands, are typical. Notice that the original colonist on each island belonged to a different ecological type. The initial species on Hispaniola lived on the trunks and crowns of trees, while the original colonist on Jamaica occupied twigs. From different evolutionary starting points, then, an adaptive radiation occurred on both islands. The key point is that on both islands, the same four ecological types eventually evolved. New species arose on each island independently, but because both islands had similar varieties of habitats, each island ended up with a complement of species that was similar in lifestyle and appearance. The researchers found the same pattern on many other islands, meaning that the same type of convergent evolution occurred repeatedly. In other words, a series of "miniature adaptive radiations" occurred—one on each island—within the overall *Anolis* radiation. The small-scale and large-scale radiations were triggered by two conditions: opportunity in the form of both available habitat when new islands were colonized, and lack of competitors.

Morphological Innovation as a Trigger

The other major trigger for adaptive radiations, in addition to ecological opportunity, is morphological innovation. Important new traits such as multicellularity, shells, exoskeletons, and limbs were a driving force behind the adaptive radiation called the Cambrian explosion. The evolution of the limb triggered

(a) Species of *Anolis* vary in leg length and tail length. Some species are ground dwelling; others live in distinct regions of shrubs or trees.

(b) The same adaptive radiation of *Anolis* has occurred on different islands, starting from different types of colonists.

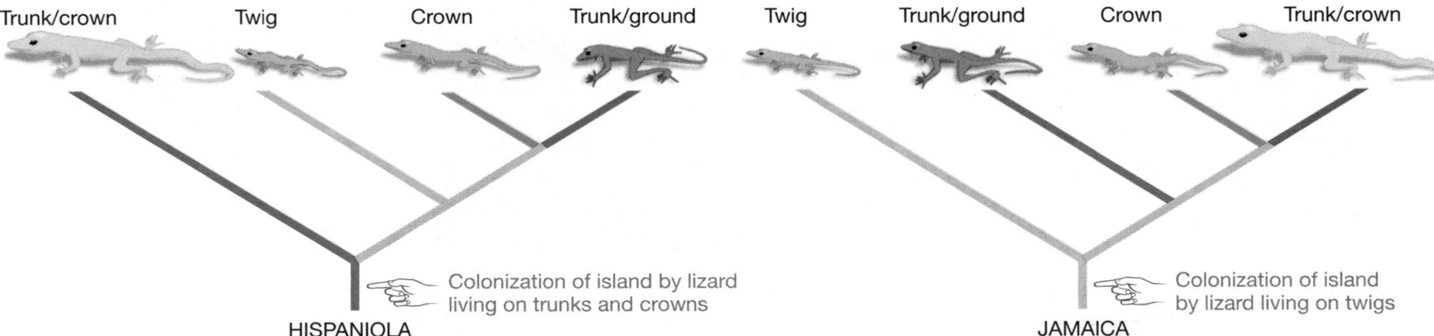

FIGURE 27.12 Adaptive Radiations of *Anolis* Lizards. (a) Short-legged lizard species (left, right) spend most of their time on the twigs of trees and bushes; long-legged species (center) live on tree trunks and the ground. **(b)** The evolutionary relationships among lizard species on the islands of Hispaniola and Jamaica. The initial colonist species was different on these islands; but in terms of how they look and where they live, a similar suite of four species evolved. According to these data, similar adaptive radiations took place independently on the two islands.

the diversification of land-dwelling vertebrates called the tetrapods. Similarly, many of the other important diversification events in the history of life started off with the evolution of a key morphological trait that allowed descendants to live in new areas, exploit new sources of food, or move in new ways:

- The evolution of wings, three pairs of legs, and a protective external skeleton helped make the insects the most diverse lineage on Earth, with perhaps well over 3 million species in existence today (**Figure 27.13a**).
- Flowers are a unique reproductive structure that helped trigger the diversification of angiosperms (flowering plants). Because flowers are particularly efficient at attracting pollinators, the evolution of the flower made angiosperms more efficient in reproduction. Today angiosperms are far and away the most species-rich lineage of land plants, with over 250,000 species known (**Figure 27.13b**).
- Cichlids are a lineage of fish that evolved a unique set of jaws in their throat. These second jaws make food processing extremely efficient. Different species have throat jaws specialized for crushing snail shells, shredding tissue from other fish, or mashing bits of algae. Over 300 species of cichlid live in Africa's Lake Victoria alone (**Figure 27.13c**).
- Feathers and wings gave some dinosaurs the ability to fly (**Figure 27.13d**). Today the lineage called birds contains about 10,000 species, with representatives that live in virtually every habitat on the planet.

In sum, adaptive radiation is a key pattern in the history of life and is usually associated with a new ecological opportunity or a morphological innovation. During an adaptive radiation, rapid speciation and morphological divergence are tightly linked.

MB **Web Animation** at www.masteringbio.com
Adaptive Radiation

27.5 Mass Extinctions

Mass extinction events are evolutionary hurricanes. They buffet the tree of life, snapping twigs and breaking branches. They are catastrophic episodes that wipe out huge numbers of species and lineages in a short time, giving the tree of life a drastic pruning. One mass extinction event, about 251 million years ago, nearly uprooted the tree entirely. The end-Permian extinction came close to ending multicellular life on Earth (**Box 27.2**).

Mass extinction events need to be distinguished from background extinctions. A **mass extinction** refers to the rapid extinction of a large number of lineages scattered throughout the tree of life. More specifically, a mass extinction occurs when at least 60 percent of the species present are wiped out within 1 million years. **Background extinction** refers to the lower, average rate of extinction observed when a mass extinction is not occurring. Although there is no hard-and-fast rule for distinguishing between the two extinction rates, paleontologists traditionally recognize and study five mass extinction events. **Figure 27.14**, for example,

(a) Insects have a distinctive body plan.

(b) Flowering plants have a unique reproductive structure (flower).

(c) Cichlids have "throat jaws" that can bite and process food.

(d) Feathers evolved in dinosaurs.

FIGURE 27.13 Some Adaptive Radiations Are Associated with Morphological Innovations. **(a)** The insect body includes several important innovations, including wings, compound eyes, three pairs of legs, and a segmented body organized into three general regions. **(b)** The flower was a morphological innovation that made pollen transfer and reproduction more efficient. **(c)** Cichlid species have distinctive mouthparts, including a pair of specialized jaws located in their throat. [©Don P. Northup, www.africancichlidphotos.com] **(d)** The evolution of feathers and flight in dinosaurs triggered the adaptive radiation of birds. This dinosaur, *Caudipteryx,* was covered with short downy body feathers and had longer feathers on its forelimbs and tail. Feathers originally evolved for display or insulation; later they were used in gliding and in powered flight.

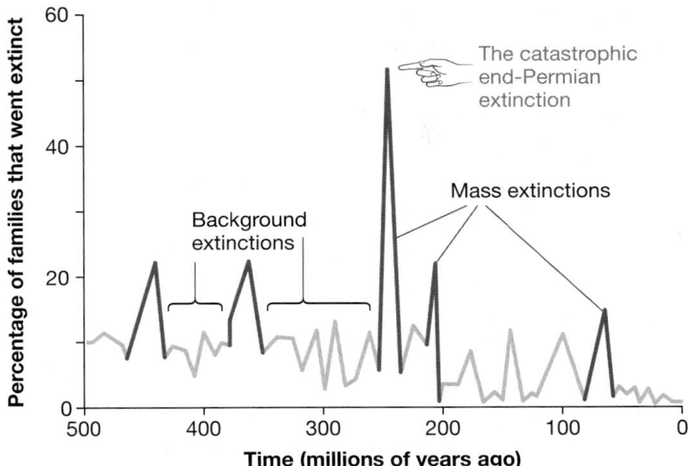

FIGURE 27.14 The Big Five Mass Extinction Events. This graph shows the percentage of families that went extinct over each interval in the fossil record since the Cambrian explosion. The five mass extinction events are drawn in dark red.

EXERCISE Circle the end-Cretaceous extinction event that occurred 65 million years ago, ending the era of the dinosaurs.

plots the percentage of plant and animal families that died out during each stage in the geologic time scale since the Cambrian explosion. (A plant or animal family is a grouping of similar genera in Linnean taxonomy—see Chapter 1.) Five spikes in the graph—denoting a large number of extinctions within a short time—are drawn in red. These are referred to as "The Big Five."

How Do Background and Mass Extinctions Differ?

Biologists are interested in distinguishing between background and mass extinctions because these events have contrasting causes and effects. Background extinctions are thought to occur when normal environmental change, emerging diseases, or competition with other species reduces certain populations to zero. Mass extinctions, in contrast, are thought to result from extraordinary, sudden, and temporary changes in the environment. During a mass extinction, species do not die out because individuals are poorly adapted to normal or gradually changing environmental conditions. Rather, species die out from exposure to exceptionally harsh, short-term conditions—such as huge volcanic

BOX 27.2 The End-Permian Extinction

The end-Permian has been called the Mother of Mass Extinctions. To appreciate the scale of what happened, imagine that you took a walk along a seashore and identified 100 different species of algae and animals living on the beach and tide pools and the shallow water offshore. Now imagine that you snapped your fingers and 90 of those species disappeared forever. Only 10 species are left. An area that was teeming with diverse forms of life would look barren. This is what happened, all over the world, during the end-Permian extinction. The event was a catastrophe of almost unimaginable proportions. On a personal level, it would be like nine of your ten best friends dying.

Although biologists have long appreciated the scale of the end-Permian extinction, research on its causes is ongoing. Consider the following:

- Flood basalts are outpourings of molten rock that flow across the Earth's surface. The largest flood basalts on Earth, called the Siberian traps, occurred during the end-Permian. They added enormous quantities of heat, CO_2, and sulfur dioxide to the atmosphere (**Figure 27.15**). The CO_2 led to intense global warming, and sulfur dioxide reacted with water to form sulfuric acid, which is toxic to most organisms.
- There is convincing evidence that sea level dropped dramatically during the extinction event, reducing the amount of habitat available for marine organisms, and that the oceans became completely or largely anoxic—meaning that they lacked oxygen.
- Terrestrial animals may have been restricted to low-elevation habitats, due to low oxygen concentrations and high CO_2 levels in the atmosphere.

In short, it is clear that both marine and terrestrial environments deteriorated dramatically for organisms that depend on oxygen to live. What biologists don't understand is *why* the environment changed so radically, and so quickly. Research continues.

FIGURE 27.15 The Siberian Traps Formed at the End-Permian. The largest flood basalt on Earth, the Siberian traps probably resembled this fissure eruption in Iceland, but on a much larger scale. The Siberian traps formed a layer of rock between 400-3,000 m thick that covered up to 7,000,000 km^2 of the end-Permian landscape, in what is now northeast Asia.

eruptions or catastrophic sea-level changes. In a general sense, background extinctions are thought to result primarily from natural selection. Mass extinctions, in contrast, function like genetic drift. The extinctions they cause are largely random with respect to the fitness of individuals under normal conditions.

To drive these points home, and to see what happens after a mass extinction has occurred, let's examine one of The Big Five in detail. The event we'll analyze, the mass extinction at the end of the Cretaceous period, was not the largest in history—the mass extinction at the end of the Permian period wiped out 90 percent of the multicellular organisms alive at the time. However, the end-Cretaceous extinction is the best understood and is among the most dramatic: It extinguished the dinosaurs and ushered in the diversification of mammals.

What Killed the Dinosaurs?

The end-Cretaceous extinction of 65 million years ago is as satisfying a murder mystery as you could hope for, but the butler didn't do it. The impact hypothesis for the extinction of the dinosaurs, first put forth in the early 1970s, proposed that an asteroid struck Earth and snuffed out an estimated 60–80 percent of the multicellular species alive.

The impact hypothesis was intensely controversial at first. As researchers set out to test its predictions, however, support began to grow:

- Worldwide, sedimentary rocks that formed at the Cretaceous-Paleogene (K–P)[1] boundary were found to contain extremely high quantities of the element iridium. Iridium is vanishingly rare in Earth rocks, but it is an abundant component of asteroids and meteorites (**Figure 27.16a**).
- Shocked quartz and microtektites are minerals that are found only at documented meteorite impact sites (**Figure 27.16b**). Shocked quartz forms when shock waves from an asteroid impact alter the structure of sand grains. Microtektites form when minerals are melted at an impact site and then cool and resolidify. In Haiti and an array of other locations, both shocked quartz and microtektites have been discovered in abundance in rock layers dated to 65 million years ago.
- A crater the size of Sicily was found just off the northwest coast of Mexico's Yucatán peninsula (**Figure 27.16c**). Microtektites are abundant in sediments from the crater's walls, and the crater dates to the K-P boundary.

[1]Geologists use *K* to abbreviate Cretaceous, because C refers to the Cambrian period.

Taken together, these data provided conclusive evidence in favor of the impact hypothesis. Researchers now agree that the mystery is solved.

Based on currently available data, astronomers and paleontologists estimate that the asteroid that struck Earth 65 million years ago was about 10 km across. To get a sense of this scale, consider that Mt. Everest is about 10 km above sea level and that planes cruise at an altitude of about 10 km. Imagine Earth being hit by a rock the size of Mt. Everest, or a rock that would fill the space between you and a jet in the sky.

The distribution of shocked quartz and microtektites dated to 65 mya indicates that the asteroid hit Earth at an angle and splashed material over much of southeastern North America. To understand the impact's consequences, consider the results of the Tunguska event. On June 30, 1908, a piece of a comet about 30 m across and about 2 megatons in mass exploded at least 5 km above Earth's surface near the Tunguska River in Siberia. The explosion released about 1000 times as much energy as the atomic bomb that destroyed Hiroshima in World War II. The event incinerated vegetation over hundreds of square kilometers, leveled trees over thousands of square kilometers, and significantly increased dust levels across the Northern Hemisphere. A deafening blast was heard 500 km (300 mi) away; people standing 60 km away were thrown to the ground or knocked unconscious. These events pale in comparison to what happened 65 million years ago. The energy the K-P asteroid unleashed was 37 *million* times greater than the Tunguska event.

According to both computer models and geologic data, the consequences of the K-P asteroid strike were nothing short of devastating. A tremendous fireball of hot gas would have spread from the impact site; large soot and ash deposits in sediments dated to 65 million years ago testify to catastrophic wildfires, worldwide. The largest tsunami in the last 3.5 billion years would have disrupted ocean sediments and circulation patterns. The impact site itself is underlain by a sulfate-containing rock called anhydrite. The SO_4^{2-} released by the impact would have reacted with water in the atmosphere to form sulfuric acid (H_2SO_4), triggering extensive acid rain. Massive quantities of dust, ash, and soot would have blocked the Sun for long periods, leading to rapid global cooling and a crash in plant productivity.

Selectivity The asteroid impact did not kill indiscriminately. Perhaps by chance, certain lineages escaped virtually unscathed while others vanished. Among vertebrates, for example, the dinosaurs, pterosaurs (flying reptiles), and all of the large-bodied marine reptiles (mosasaurs, ichthyosaurs, and plesiosaurs) expired; mammals, crocodilians, amphibians, and turtles survived.

Why? Answering this question has sparked intense debate. For years the leading hypothesis was that the K-P extinction event was size selective. The logic here was that the extended

(a) Iridium is present at high concentration in rocks formed 65 million years ago.

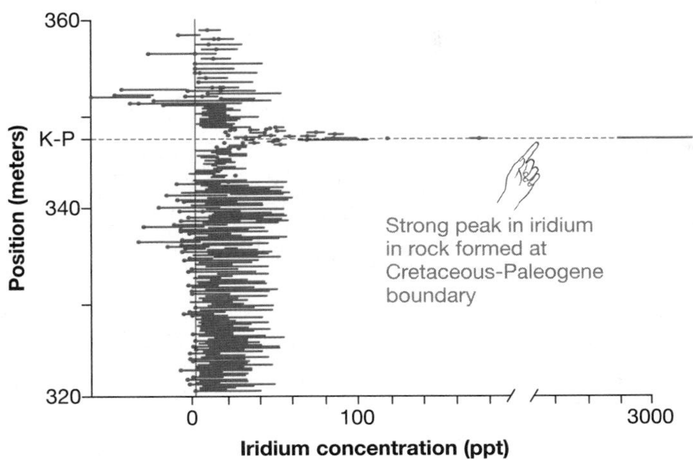

(b) Minerals that form during asteroid impacts

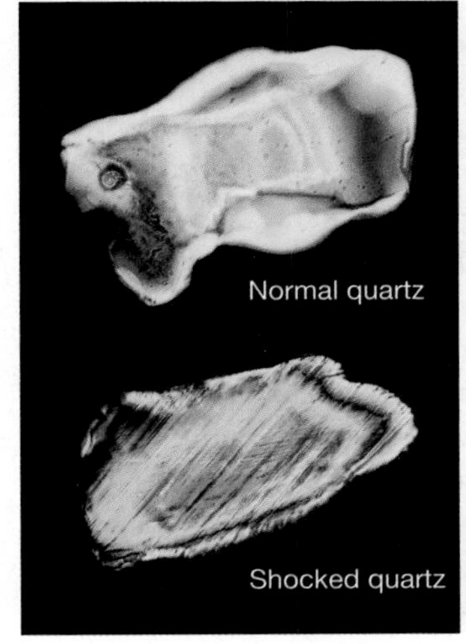

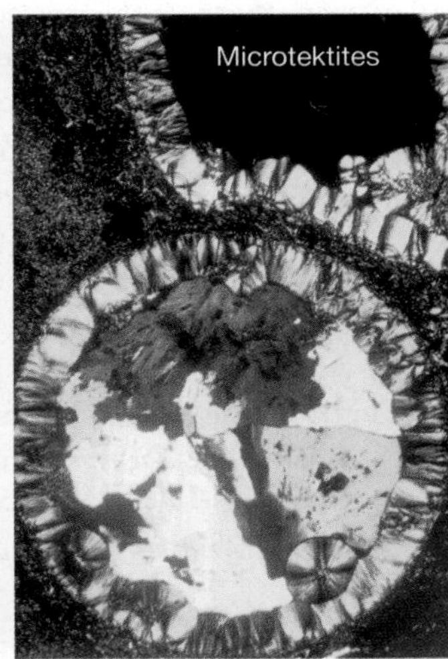

(c) The asteroid left a crater 180 km (112 miles) wide.

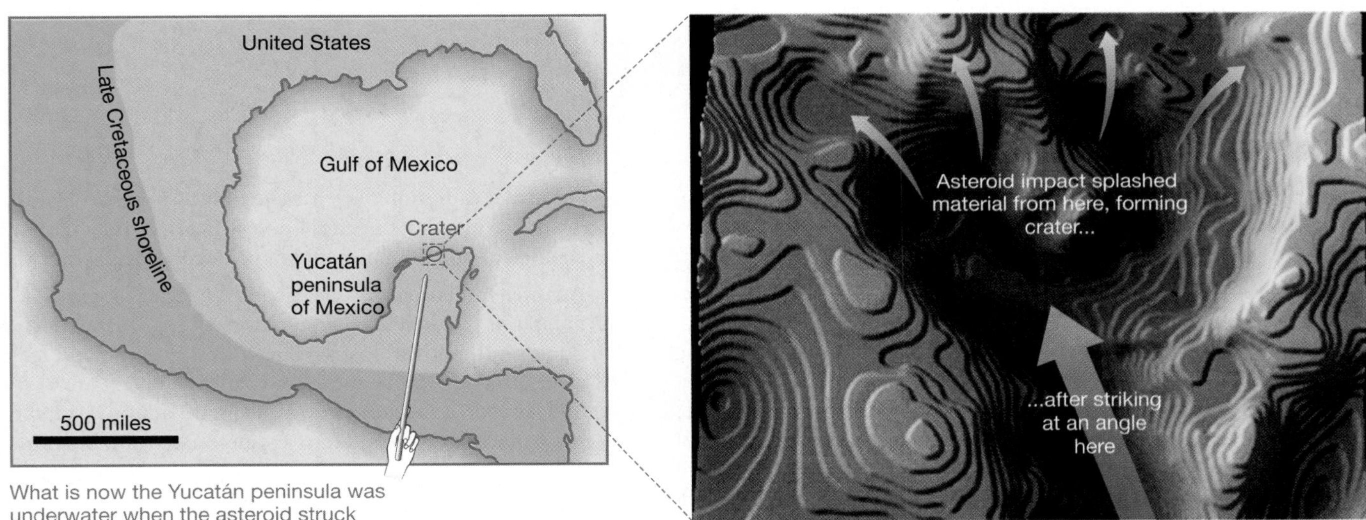

FIGURE 27.16 Evidence of an Asteroid Impact 65 Million Years Ago. (a) The concentration of iridium in rocks that formed on either side of the Cretaceous-Paleogene (K-P) boundary, in parts per trillion (ppt). **(b)** Normal quartz grains are markedly different from shocked quartz grains. The striations in the shocked quartz are caused by a sudden increase in pressure. Microtektites (right) are tiny glass particles formed when minerals melt at an impact site and then recrystallize. **(c)** Geologists have identified the walls of a crater off the northwest coast of the Yucatán Peninsula, dated to 65 million years ago.

EXERCISE On the map in part (c), label the "splash zone" where microtektites and debris from the impact would have landed. Label where the tsunami generated by the impact would have struck land.

darkness and cold would affect large organisms disproportionately, because they require more food than do small organisms. But extensive data on the survival and extinction of marine clams and snails have shown no hint of size selectivity, and small-bodied and juvenile dinosaurs perished along with large-bodied and adult forms. One hypothesis currently being tested is that organisms that were capable of inactivity for long periods—by hibernating or resting as long-lived seeds or spores—were able to survive the catastrophe. But this aspect of the mystery is still unsolved.

Recovery After the K-P extinction, fern fronds and fern spores dominate the plant fossil record from North America and Australia. These data suggest that extensive stands of ferns replaced diverse assemblages of cone-bearing and flowering plants after the impact. The fundamental message here is that terrestrial ecosystems around the world were radically simplified. In marine environments, some invertebrate groups do not exhibit normal levels of species diversity in the fossil record until 4–8 million years past the K-P boundary. Recovery was slow.

The organisms present in the Paleogene were markedly different from those of the preceding period. The lineage called Mammalia, which had consisted largely of rat-sized predators and scavengers in the heyday of the dinosaurs, exploded after the impact and took the place of the dinosaurs. Within 10–15 million years, all of the major mammalian orders observed today had appeared—from pigs to primates. Why? A major branch on the tree of life had disappeared. With competitors removed, mammals flourished.

The Mesozoic is sometimes called the Age of Reptiles, and the Cenozoic the Age of Mammals. The change was not due to a competitive superiority conferred by adaptations such as fur and lactation. Rather, it was due to a chance event: a once-in-a-billion-years collision with a massive rock from outer space.

Check Your Understanding

If you understand that...

- Mass extinctions have occurred repeatedly throughout the history of life.

You should be able to...

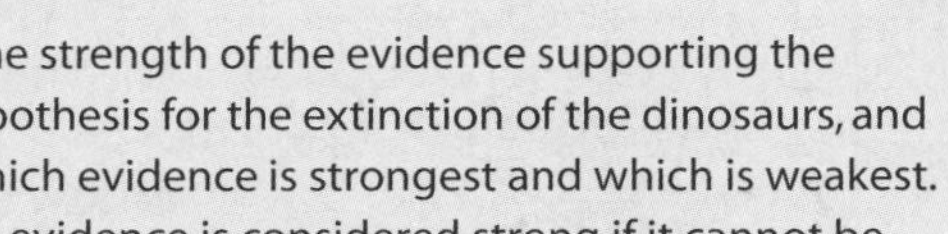

1) Evaluate the strength of the evidence supporting the impact hypothesis for the extinction of the dinosaurs, and identify which evidence is strongest and which is weakest. (In science, evidence is considered strong if it cannot be explained by alternative hypotheses.)
2) Explain why mass extinctions wipe out species more or less randomly—much the way genetic drift affects changes in allele frequencies.

Chapter Review

SUMMARY OF KEY CONCEPTS

Phylogenies and the fossil record are the major tools that biologists use to study the history of life.

Phylogenetic trees can be estimated by grouping species based on overall similarity in traits or by analyzing shared, derived characters (synapomorphies) that identify monophyletic groups. Phylogenetic trees document the evolutionary relationships among species and identify the order in which events occurred. To minimize the impact of homoplasy when inferring phylogenies, researchers use parsimony or other approaches to decide which of the many trees that are possible is most likely to reflect actual evolutionary history. The fossil record is used in conjunction with phylogenetic analyses because it is the only direct source of data about what extinct organisms looked like and where they lived.

You should be able to describe how biologists use phylogenetic trees and the fossil record to infer when traits—like upright posture in humans—evolved.

The Cambrian explosion was the rapid morphological and ecological diversification of animals that occurred during the Cambrian period.

The diversification of animals over a 5-million-year period, starting about 565 million years ago, is perhaps the best-studied event in the history of life. The animals that lived in marine environments just before and after the Cambrian explosion are documented in the Doushantuo, Ediacaran, and Burgess Shale faunas. During this period the first heads, tails, appendages, shells, exoskeletons, and segmented bodies evolved. Data on the number and identity of homeotic genes in different animal lineages suggest that at least some of the animal radiation was possible because gene duplication events created new copies of *Hox* genes.

You should be able to explain why the Cambrian explosion can be considered an example of an adaptive radiation.

Adaptive radiations are a major pattern in the history of life. They are instances of rapid diversification associated with new ecological opportunities and new morphological innovations.

Adaptive radiations can be triggered by the colonization of a new habitat or the demise of competitors after a mass extinction. Morphological innovations, such as limbs, flowers, and feathers can also initiate adaptive radiations. Speciation events and morphological change occur rapidly during an adaptive radiation, as a single lineage diversifies into a wide variety of ecological roles.

You should be able to to generate and defend a hypothesis for an ecological opportunity or morphological innovation that triggered the Cambrian explosion.

MB Web Animation at www.masteringbio.com
Adaptive Radiation

Mass extinctions have occurred repeatedly throughout the history of life. They rapidly eliminate most of the species alive in a more or less random manner.

Mass extinctions have altered the course of evolutionary history at least five times. They prune the tree of life more or less randomly and have marked the end of several prominent lineages and the rise of new branches. The end-Cretaceous extinction killed 60 to 80 percent of existing species and was caused by an asteroid impact. After the devastation of a mass extinction, it can take 10–15 million years for ecosystems to recover their former levels of diversity. Data introduced in Chapter 55 suggest that a sixth mass extinction event—caused by humans—is currently under way.

You should be able to evaluate whether environmental changes caused by humans are eliminating species in a random manner, as opposed to those species being poorly adapted to the environment.

QUESTIONS

Test Your Knowledge

1. Choose the best definition of a fossil.
   a. any trace of an organism that has been converted into rock
   b. a bone, tooth, shell, or other hard part of an organism that has been preserved
   c. any trace of an organism that lived in the past
   d. the process that leads to preservation of any body part from an organism that lived in the past
2. Why are the Doushantuo, Ediacaran, and Burgess Shale fossil deposits unusual?
   a. Soft-bodied animals are preserved in them.
   b. They are easily accessible to researchers.
   c. They are the only fossil-bearing rock deposits from their time period.
   d. They include terrestrial, instead of just marine, species.
3. Which of the following best characterizes an adaptive radiation?
   a. Speciation occurs extremely rapidly, and descendant populations occupy a large geographic area.
   b. A single lineage diversifies rapidly, and descendant populations occupy many habitats and ecological roles.
   c. Natural selection is particularly intense, because disruptive selection occurs.
   d. Species recover after a mass extinction.
4. Which of the following is most accurate?
   a. Mass extinctions are due to asteroid impacts; background extinctions may have a wide variety of causes.
   b. Mass extinctions focus on particularly prominent groups, such as dinosaurs; background extinctions affect species from throughout the tree of life.
   c. Only five mass extinctions have occurred, but hundreds of background extinctions have occurred.
   d. Mass extinctions extinguish groups rapidly and randomly; background extinctions are slower and often result from natural selection.
5. Why do molecular clocks exist?
   a. Natural selection is not important at the molecular level.
   b. Homologous genes have the same structure.
   c. They can be calibrated.
   d. Some DNA sequences, in some lineages, change at a steady rate over time.
6. Why is burial a key step in fossilization?
   a. It slows the process of decay by bacteria and fungi.
   b. It allows tissues to be preserved as casts or molds.
   c. It protects tissues from wind, rain, and other corrosive elements.
   d. All of the above.

Test Your Knowledge answers: 1. c; 2. a; 3. b; 4. d; 5. d; 6. d

Test Your Understanding

Answers are available at www.masteringbio.com

1. The text claims that the fossil record is biased in several ways. What are these biases? If the database is biased, is it still an effective tool to use in studying the diversification of life? Explain.
2. The initial diversification of animals took place over some 50 million years, at the start of the Cambrian period. Why is the diversification called an "explosion"?
3. What is the "new genes, new bodies" hypothesis? Explain whether the data summarized in this chapter support or contradict the hypothesis.
4. Give an example of an adaptive radiation that occurred after a colonization event, after a mass extinction, and after a morphological innovation. In each case, provide a hypothesis to explain why the adaptive radiation occurred.
5. Summarize the evidence that supports the impact hypothesis for the K-P extinction.
6. Why are monophyletic groups identified by shared, derived traits?

Applying Concepts to New Situations

Answers are available at www.masteringbio.com

1. Suppose that the dying wish of a famous eccentric was that his remains be fossilized. His family has come to you for expert advice. What steps would you recommend to maximize the chances that his wish will be fulfilled?
2. Using data from the molecular phylogenies presented in this section and data on the fossil record of whales presented in Chapter 24, summarize how whales evolved from the common ancestor they share with today's hippos.
3. Some researchers contend that the end-Permian extinction event was also caused by an impact with a large extraterrestrial object. List the evidence, ordered from least convincing to most convincing, that you would like to see before you accept this hypothesis. Explain your rankings.
4. One of the "triggers" proposed for the Cambrian explosion is a dramatic rise in oxygen concentrations that occurred in the oceans about 800 mya. Review material in Chapter 9 on how oxygen compares with other atoms or compounds as an electron acceptor during cellular respiration (look near the end of Section 9.6, where aerobic and anaerobic respiration are compared). Then explain the logic behind the "oxygen-trigger" hypothesis for the Cambrian explosion.

www.masteringbio.com is also your resource for • Answers to text, table, and figure caption questions and exercises • Answers to *Check Your Understanding* boxes • Online study guides and quizzes • Additional study tools including the *E-Book for Biological Science* 3rd ed., textbook art, animations, and videos.

28 Bacteria and Archaea

KEY CONCEPTS

- Bacteria and archaea have a profound impact on humans and global ecosystems. A few bacteria cause important infectious diseases; some bacterial and archaeal species are effective at cleaning up pollution; photosynthetic bacteria were responsible for the evolution of the oxygen atmosphere; bacteria and archaea cycle nutrients through every terrestrial and aquatic environment.
- Bacteria and archaea have been evolving for billions of years and are extremely sophisticated organisms. Although they are small and relatively simple morphologically, they live in virtually every habitat known and use remarkably diverse types of compounds in cellular respiration and fermentation.

Although this hot spring looks devoid of life, it is actually teeming with billions of bacterial and archaeal cells from a wide variety of species. As this chapter will show, bacteria and archaea occupy virtually every environment on Earth.

Bacteria and Archaea form two of the three largest branches on the tree of life (**Figure 28.1**). The third major branch or domain consists of eukaryotes and is called the Eukarya. Virtually all members of the Bacteria and Archaea domains are unicellular, and all are prokaryotic—meaning that they lack a membrane-bound nucleus.

Although their relatively simple morphology makes bacteria and archaea appear similar to the untrained eye, they are strikingly different at the molecular level (**Table 28.1**). Organisms in the Bacteria and Archaea domains are distinguished by the types of molecules that make up their plasma membranes and cell walls, and by the machinery they use to transcribe DNA and translate messenger RNA into proteins. Most notably, **Bacteria** have a unique compound called peptidoglycan in their cell walls, and **Archaea** have unique phospholipids—compounds containing hydrocarbons called isoprenes in their tails—in their plasma membranes (see Chapters 5 and 6). In addition, the structures of the RNA polymerases and ribosomes found in Archaea and Eukarya are distinct from those found in Bacteria and similar to each other. This is why antibiotics that poison bacterial ribosomes do not affect the ribosomes of archaea or eukaryotes. If you were unicellular, bacteria and archaea would look and act as different as mammals and fish do to you now.

The lineages in the domains Bacteria and Archaea are ancient, diverse, ubiquitous, and abundant. The oldest fossils of any type found to date are 3.4-billion-year-old carbon-rich deposits derived from bacteria. Because eukaryotes do not appear

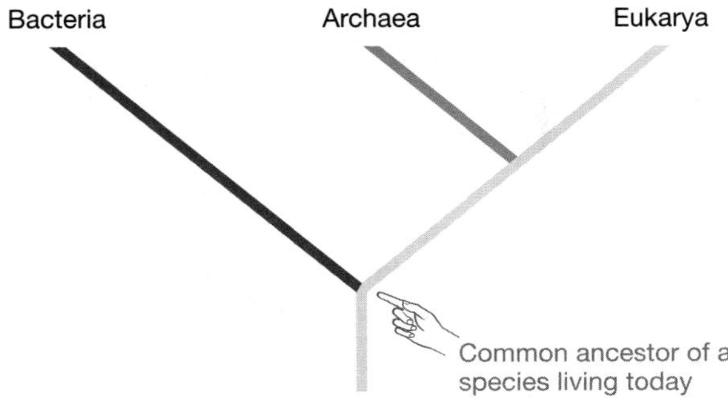

FIGURE 28.1 Bacteria, Archaea, and Eukarya Are the Three Domains of Life. Archaea are more closely related to eukaryotes than they are to bacteria.

● **EXERCISE** Circle the prokaryotic lineages on the tree of life.

● **QUESTION** Was the common ancestor of all species living today prokaryotic or eukaryotic? Explain your reasoning.

in the fossil record until 1.75 billion years ago, biologists infer that prokaryotes were the only form of life on Earth for almost 1.7 billion years. The Bacteria and Archaea have diversified into species that now blanket the planet.

Just how many bacteria and archaea are alive today? Although a mere 5000 species have been formally named and described to date—most by the morphological species concept introduced in Chapter 26—it is virtually certain that millions exist. Consider that over 400 species of prokaryotes are living in your gastrointestinal tract right now. Another 500 species live in your mouth, of which just 300 have been described and named. Norman Pace points out that there may be tens of millions of different insect species but notes, "If we squeeze out any one of these insects and examine its contents under the microscope, we find hundreds or thousands of distinct microbial species." Most of these **microbes** (microscopic organisms) are bacteria or archaea, and virtually all are unnamed and undescribed. If you want to discover and name new species, then study bacteria or archaea.

In addition to recognizing how diverse bacteria and archaea are in terms of numbers of species, it's critical to appreciate their abundance.

- The approximately 10^{13} (10 trillion) cells in your body are vastly outnumbered by the bacterial and archaeal cells living on and in you. An estimated 10^{12} bacterial cells live on your skin, and an additional 10^{14} bacterial and archaeal

SUMMARY TABLE **28.1 Characteristics of Bacteria, Archaea, and Eukarya**

	Bacteria	Archaea	Eukarya
Nuclear envelope?	No	No	Yes
Circular chromosome?	Yes (but linear in some species)	Yes	No
DNA associated with histone proteins? (see Chapter 18)	No	Yes	Yes
Organelles present?	Some in limited number of species	None described to date	Extensive in number and diversity
Flagella present?	Yes—spin like propeller	Yes; spin like bacterial flagella, but distinctive in molecular composition	Yes, but undulate back and forth and have completely different molecular composition compared with bacteria and archaea
Unicellular or multicellular?	Almost all unicellular	All unicellular	Many multicellular
Sexual reproduction?	No*	Not known	Common
Structure of lipids in plasma membrane	Glycerol bonded to straight-chain fatty acids via ester linkage	Glycerol bonded to branched fatty acids (synthesized from isoprene subunits) via ether linkage	Glycerol bonded to straight-chain fatty acids via ester linkage
Cell-wall material	Almost all include peptidoglycan, which contains muramic acid	Varies widely among species, but no peptidoglycan and no muramic acid	When present, usually made of cellulose or chitin
Transcription and translation machinery	One relatively simply RNA polymerase; translation begins with formylmethionine; translation poisoned by several antibiotics that do not affect archaea or eukaryotes	One relatively complex RNA polymerase; translation begins with methionine	Several relatively complex RNA polymerases; translation begins with methionine

*Sexual reproduction begins with meiosis and often involves the exchange of haploid genomes between individuals of the same species. In bacteria, meiosis does not occur. Small numbers of genes can be transferred from one bacterial cell to another, however, and genetic recombination may occur. For more detail, see Chapter 12.

● **EXERCISE** Using the data in this table, add labeled marks to Figure 28.1 indicating where the following traits evolved: peptidoglycan in cell wall, archaeal-type plasma membrane, archaeal and eukaryote-type ribosomes, nuclear envelope.

cells occupy your stomach and intestines. You, along with most other mammals, would not be able to digest food properly without these gut bacteria. You are a walking, talking habitat—one that is teeming with bacteria and archaea.

- A mere teaspoon of good-quality soil contains *billions* of microbial cells, most of which are bacteria and archaea.
- In sheer numbers, species in a lineage called the Group I marine archaea may be the most successful organisms on Earth. Biologists routinely find these cells at concentrations of 10,000 to 100,000 individuals per milliliter of seawater, at depths from 200 to 4000 or more meters below the surface, in most of the world's oceans. At these concentrations, a drop of seawater contains a population equivalent to that of a large human city. Yet this lineage was first described in the early 1990s.
- Biologists estimate the total number of individual bacteria and archaea alive today at 5×10^{30}. If they were lined up end to end, they would make a chain longer than the Milky Way galaxy. These cells contain 50 percent of all the carbon and 90 percent of all the nitrogen and phosphorus found in organisms. In terms of the total volume of living material on our planet, bacteria and archaea are dominant life-forms.

Bacteria and archaea are also found almost everywhere. They live in environments as unusual as oxygen-free mud, hot springs, and salt flats. They have been discovered living in bedrock to a depth of 1500 meters below Earth's surface. In the ocean they are found from the surface to depths of 10,000 m and at temperatures ranging from near 0°C in Antarctic sea ice to over 121°C near submarine volcanoes.

Although there are far more prokaryotes than eukaryotes, much more is known about eukaryotic diversity than about prokaryotic diversity. Researchers who study prokaryotic diversity are exploring one of the most wide-open frontiers in all of science. So little is known about the extent of these domains that recent collecting expeditions have turned up two entirely new **kingdoms** and numerous **phyla** (singular: **phylum**). These are names given to major lineages within each domain. To a biologist, this achievement is equivalent to the sudden discovery of a new group of eukaryotes as distinctive as the flowering plants or animals with backbones.

The physical world has been explored and mapped, and many of the larger plants and animals are named. But in **microbiology**—the study of organisms that can be seen only with the aid of a microscope—this is an age of exploration and discovery.

28.1 Why Do Biologists Study Bacteria and Archaea?

Biologists study bacteria and archaea for the same reasons they study any organisms. First, these organisms are intrinsically fascinating. Discoveries such as finding bacterial cells living a kilometer underground or in 95°C hot springs keep biologists awake at night, staring at the ceiling. They can't wait to get into the lab in the morning and start trying to figure out how those cells are staying alive. Second, researchers know how important understanding diversity is to the rest of biology. In fields from cell biology and genomics to ecology, the ability to compare characteristics in a diverse array of organisms is fundamental to increased understanding. Third, almost all biologists are aware that a mass extinction is currently under way (see Chapter 55). To preserve biodiversity, we have to understand it.

Those reasons aren't necessarily why governments, businesses, and foundations fund research on biological diversity, however. These institutions usually have more practical goals in mind. In many cases they want to know how studying organismal diversity improves human health and welfare. To answer this question, each chapter in this unit starts with a section on how the organisms in question affect humans and other species. The second section of each chapter focuses on methods—on *how* biologists study the group in question. The third section in each chapter is the heart of the matter: It delves into how the organisms make a living and analyzes the unique evolutionary innovations that were responsible for the group's origin and diversification. It also explains what's cool about the group, in a biological sense. The final section in each of this unit's chapters summarizes the key characteristics of prominent lineages within the group. It's written in a succinct, note-like style and is intended to give you an overview of who's who in the group.

And with that, let's plunge in. The ubiquity and abundance of bacteria and archaea make them exceptionally important in both human and natural economies. Let's consider their role in biomedicine, pollution control, industry, and global environmental change.

Some Bacteria Cause Disease

No archaea are known to cause disease in humans. But of the hundreds or thousands of bacterial species living in and on your body, a tiny fraction can disrupt normal body functions enough to cause illness. Bacteria that cause disease are said to be **pathogenic** (literally, "disease-producing"). Pathogenic bacteria have been responsible for some of the most devastating epidemics in human history.

Robert Koch was the first biologist to establish a link between a particular species of bacterium and a specific disease. When Koch began his work on the nature of disease in the late 1800s, microscopists had confirmed the existence of the particle-like organisms we now call bacteria, and Louis Pasteur had shown that bacteria and other microorganisms are responsible for spoiling milk, wine, broth, and other foods. Koch hypothesized that bacteria might also be responsible for causing infectious diseases, which spread by being passed from an infected individual to an uninfected individual.

Koch set out to test this hypothesis by identifying the organism that causes anthrax. Anthrax is a disease of cattle and other grazing mammals that can result in fatal blood poisoning. The disease also occurs infrequently in humans and mice.

To establish a causative link between a specific microbe and a specific disease, Koch proposed that four criteria had to be met:

1. *The microbe must be present in individuals suffering from the disease and absent from healthy individuals.* By careful microscopy, Koch was able to show that the bacterium *Bacillus anthracis* was always present in the blood of cattle suffering from anthrax, but absent from asymptomatic individuals.

2. *The organism must be isolated and grown in a pure culture away from the host organism.* Koch was able to grow pure colonies of *B. anthracis* in glass dishes on a nutrient medium, using gelatin as a substrate.

3. *If organisms from the pure culture are injected into a healthy experimental animal, the disease symptoms should appear.* Koch demonstrated this crucial causative link in mice injected with *B. anthracis*. The symptoms of anthrax infection appeared, and then the infected mice died.

4. *The organism should be isolated from the diseased experimental animal, again grown in pure culture, and demonstrated by its size, shape, and color to be the same as the original organism.* Koch did this by purifying *B. anthracis* from the blood of diseased experimental mice.

These criteria, now called **Koch's postulates**, are still used to confirm a causative link between new diseases and a suspected infectious agent. Koch's experimental results also became the basis for the **germ theory of disease**. This theory, which laid the foundation for modern medicine, holds that infectious diseases are caused by bacteria and viruses. (Viruses are acellular particles that parasitize cells and are analyzed in detail in Chapter 35.) Some of the bacteria that cause illness in human beings are listed in **Table 28.2**. The important things to note about the list are that pathogenic forms come from many lineages in the domain Bacteria and that pathogenic bacteria tend to affect tissues at the entry points to the body, such as wounds or pores in the skin, the respiratory and gastrointestinal tracts, and the urogenital canal.

In the industrialized countries, improvements in sanitation and nutrition have caused dramatic reductions in mortality rates due to infectious diseases (**Figure 28.2**), most of which are due to bacterial and viral infections. In addition, the discovery of antibiotics in 1928, their development over subsequent decades, and widespread use starting in the late 1940s gave physicians effective tools to combat most bacterial infections. **Antibiotics** are molecules that kill bacteria. Extensive use of antibiotics in the late twentieth century led to the evolution of drug-resistant strains of bacteria, however (see Chapter 24). Most of the bacterial species listed in Table 28.2 now include strains that are resistant to one or more of the commonly prescribed antibiotics. Coping with antibiotic resistance in pathogenic bacteria has become a great challenge of modern medicine.

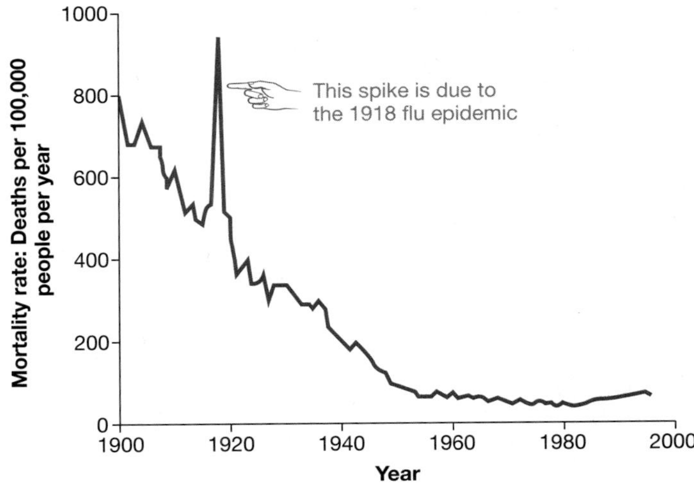

FIGURE 28.2 Deaths due to Bacterial Infections Have Declined Dramatically in Some Countries. This graph shows the death rate due to infectious diseases—meaning, bacterial or viral infections—in a country that industrialized in the late 1800s (the United States).

● **EXERCISE** The first widely prescribed antibiotic in this country was available in the late 1940s. Label this event on the graph.

● **QUESTION** Why was there a slight rise in the graph starting in the mid-1980s?

Although Pasteur, Koch, and other biologists began studying bacteria because of the role these prokaryotes play in disease, the scope of microbiology research has broadened enormously. Let's consider some topics that are currently inspiring studies of bacteria and archaea.

Bacteria Can Clean Up Pollution

Throughout the industrialized world, some of the most serious pollutants in soils, rivers, and ponds consist of organic compounds that were originally used as solvents or fuels but leaked or were spilled into the environment. Most of these compounds are highly hydrophobic. Because they do not dissolve in water, they tend to accumulate in sediments. If the compounds are subsequently ingested by burrowing worms or clams or other organisms, they can be passed along to fish, insects, humans, birds, and other species. Most of these compounds are toxic to eukaryotes in moderate to high concentrations. Petroleum from oil spills and compounds that contain ring structures and chlorine atoms, such as the family of compounds called dioxins, are particularly notorious because of their toxicity to humans.

Biologists who are responsible for cleaning up sites polluted with organic solvents and fuels are faced with a challenge, because at least some of the toxic compounds present are highly resistant to decomposition. Instead of being broken down into

TABLE 28.2 **Some Diseases Caused by Bacteria**

Lineage	Species	Tissues Affected	Disease
Firmicutes	*Clostridium botulinum*	Gastrointestinal tract, nervous system	Food poisoning (botulism)
	Clostridium tetani	Wounds, nervous system	Tetanus
	Staphylococcus aureus	Skin, urogenital canal	Acne, boils, impetigo, toxic shock syndrome
	Streptococcus pneumoniae	Respiratory tract	Pneumonia
	Streptococcus pyogenes	Respiratory tract	Strep throat, scarlet fever
Spirochaetes	*Borrelia burgdorferi*	Skin and nerves	Lyme disease
	Treponema pallidum	Urogenital canal	Syphilis
Actinomycetes	*Mycobacterium leprae*	Skin and nerves	Leprosy
	Mycobacterium tuberculosis	Respiratory tract	Tuberculosis
	Propionibacterium acnes	Skin	Acne
Chlamydiales	*Chlamydia trachomatis*	Urogenital canal	Genital tract infection
Proteobacteria (ε group)	*Helicobacter pylori*	Stomach	Ulcer
Proteobacteria (β group)	*Neisseria gonorrhoeae*	Urogenital canal	Gonorrhea
Proteobacteria (γ group)	*Haemophilus influenzae*	Ear canal, nervous system	Ear infections, meningitis
	Pseudomonas aeruginosa	Urogenital canal, eyes, ear canal	Infections of eye, ear, urinary tract
	Salmonella enterica	Gastrointestinal tract	Food poisoning
	Vibrio parahaemolyticus	Gastrointestinal tract	Food poisoning
	Yersinia pestis	Lymph and blood	Plague

harmless compounds, certain toxic molecules tend to just "sit there." These compounds pose a long-term threat to nearby fish, birds, people, and other organisms.

To clean up sites like these, researchers have begun to explore more extensive use of **bioremediation**, the use of bacteria and archaea to degrade pollutants. Bioremediation is often based on complementary strategies:

- *Fertilizing contaminated sites* to encourage the growth of existing bacteria and archaea that degrade toxic compounds. After several recent oil spills, researchers added nitrogen to affected sites as a fertilizer, but left nearby beaches untreated as controls. Dramatic increases occurred in the growth of bacteria and archaea that use hydrocarbons in cellular respiration, probably because the added nitrogen was used to synthesize enzymes and other key compounds. In at least some cases, the fertilized sediments cleaned up much faster than the unfertilized sites (**Figure 28.3**).
- *"Seeding," or adding, specific species of bacteria and archaea* to contaminated sites shows promise of alleviating pollution in some situations. For example, researchers have recently discovered bacteria that are able to render certain chlorinated, ring-containing compounds harmless. Instead of being poisoned by the pollutants, these bacteria use ring-containing, chlorinated compounds as electron acceptors during cellular respiration. In at least some cases, the by-product is dechlorinated and nontoxic to humans and other eukaryotes. To follow up on these discoveries, researchers are now growing the bacteria in quantity and testing them in the field, to test the hypothesis that seeding can speed the rate of decomposition in contaminated sediments. Initial reports suggest that seeding may help clean up at least some polluted sites.

Extremophiles

Bacteria or archaea that live in high-salt, high-temperature, low-temperature, or high-pressure habitats are called **extremophiles**

FIGURE 28.3 Bacteria and Archaea Can Play a Role in Cleaning Up Pollution. On the left is a rocky coast that was polluted by an oil spill but fertilized to promote the growth of oil-eating bacteria. The portion of the beach on the right was untreated.

("extreme-lovers"). Studying them has been extraordinarily fruitful for understanding the tree of life, developing industrial applications, and exploring the structure and function of enzymes.

As an example of these habitats, consider hot springs at the bottom of the ocean, where water as hot as 300°C emerges and mixes with 4°C seawater. At locations like these, archaea are abundant forms of life. Researchers recently discovered an archaean that grows so close to these hot springs that its surroundings are at 121°C—a record for life at high temperature. This organism can live and grow in water that is heated past its boiling point (100°C) and at pressures that would instantly destroy a human being.

Extremophiles have become a hot area of research. The genomes of a wide array of extremophiles have been sequenced, and expeditions regularly seek to characterize new species. Why? Based on models of conditions that prevailed early in Earth's history, it appears likely that the first forms of life were extremophiles. Thus, understanding extremophiles may help explain how life on Earth began. In a similar vein, astrobiologists ("space-biologists") use extremophiles as model organisms in the search for extraterrestrial life. The idea is that if bacteria and archaea can thrive in extreme habitats on Earth, it is likely that cells might be found in similar environments on other planets or moons of planets. And because enzymes that function at extreme temperatures and pressures are useful in many industrial processes, extremophiles are of commercial interest as well. Chapter 19 introduced *Taq* polymerase, which is a DNA polymerase that is stable up to 95°C. Recall that *Taq* polymerase is used to run the polymerase chain reaction (PCR) in research and commercial settings. This enzyme was isolated from a bacterium called *Thermus aquaticus* ("hot water"), which was discovered in hot springs in Yellowstone National Park.

How Do Small Cells Affect Global Change?

Certain bacteria and archaea can live in extreme environments and use toxic compounds as food because they produce extremely sophisticated enzymes. The complex chemistry that they carry out, combined with their numerical abundance, has made them potent forces for global change throughout Earth's history. Bacteria and archaea have altered the composition of the oceans, atmosphere, and terrestrial environments for billions of years. They continue to do so today.

The Oxygen Revolution Today, oxygen represents almost 21 percent of the molecules in Earth's atmosphere. But researchers who study the composition of the atmosphere are virtually certain that no free molecular oxygen (O_2) existed for the first 2.3 billion years of Earth's existence. This conclusion is based on two observations: (1) There was no plausible source of oxygen at the time the planet formed; and (2) the oldest Earth rocks indicate that, for many years afterward, any oxygen that formed reacted immediately with iron atoms to produce iron oxides, such as hematite (Fe_2O_3) and magnetite (Fe_3O_4) Early in Earth's history, the atmosphere was dominated by nitrogen and carbon dioxide. Where did the oxygen we breathe come from? The answer is cyanobacteria.

Cyanobacteria are a lineage of photosynthetic bacteria (**Figure 28.4**). According to the fossil record, species of cyanobacteria first became numerous in the oceans about 2.7–2.55 billion years ago. Their appearance was momentous, because cyanobacteria were the first organisms to perform oxygenic ("oxygen-producing") photosynthesis. Oxygenic photosynthesis depends on the proteins and pigments in photosystem II. You might recall from Chapter 10 that photosystem II includes enzymes capable of stripping electrons from water molecules. The reaction that "splits" water results in the production of oxygen as well as electrons. The electrons are required for photosynthesis to continue; the oxygen molecules are simply released as a waste product.

The fossil record and geological record indicate that oxygen concentrations in the oceans and atmosphere began to increase 2.3–2.1 billion years ago. Once oxygen was common in the oceans, cells could begin to use it as the final electron acceptor during cellular respiration. **Aerobic** respiration was now a possibility. Prior to this, organisms had to use compounds other than oxygen as a final electron acceptor—only **anaerobic** respiration was possible.

The evolution of aerobic respiration was a crucial event in the history of life. Because oxygen is extremely electronegative, it is an efficient electron acceptor. Much more energy is released as electrons move through electron transport chains with oxygen as the ultimate acceptor than is released with other substances as the electron acceptor (**Figure 28.5**). Once oxygen was available, much more ATP could be produced for each electron donated by NADH or $FADH_2$. As a result, the rate of energy production could rise dramatically.

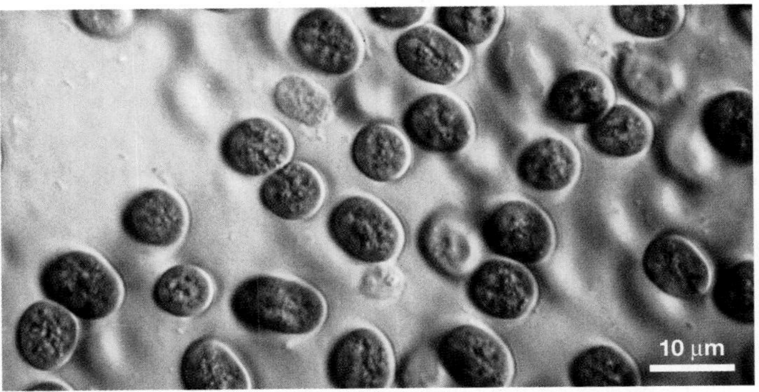

FIGURE 28.4 Cyanobacteria Were the First Organisms to Perform Oxygenic Photosynthesis. Life as we know it today would not have evolved if cyanobacteria had not begun producing oxygen as a by-product of photosynthesis about 2.7 billion years ago.

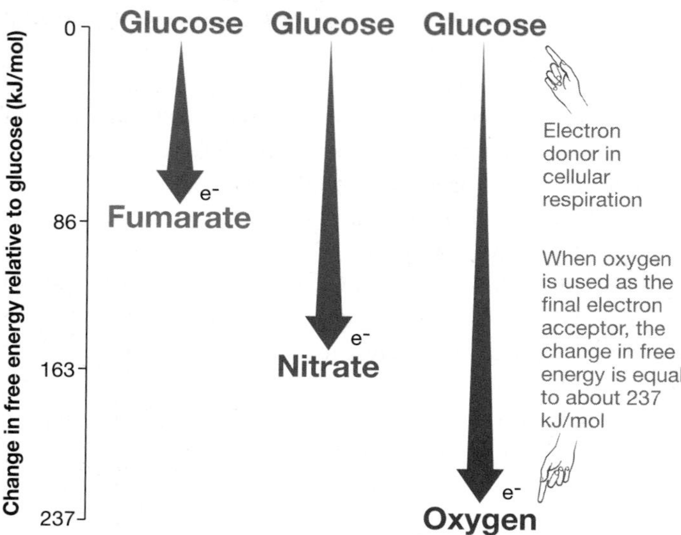

FIGURE 28.5 Cellular Respiration Can Produce More Energy When Oxygen Is the Final Electron Acceptor. Because oxygen has such high electronegativity, the potential energy of the electrons used in cellular respiration is much lower when oxygen is the final acceptor compared to other molecules or ions. As a result, a larger amount of free energy is released during cellular respiration with oxygen as the final electron acceptor.

QUESTION Which organisms grow faster—those using aerobic respiration or those using anaerobic respiration? Explain your reasoning.

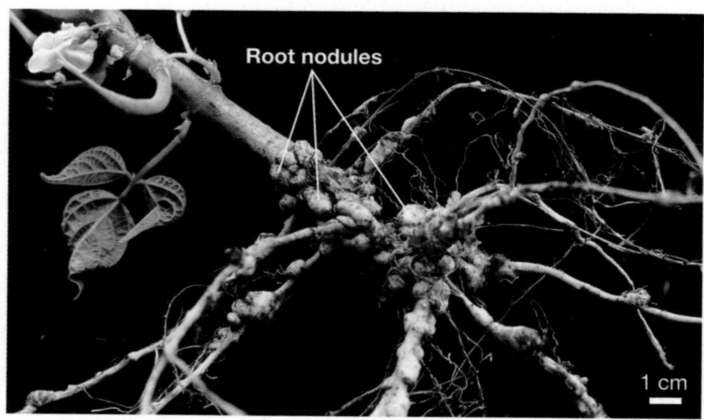

FIGURE 28.6 Some Nitrogen-Fixing Bacteria Live in Association with Plants. Root nodules form a protective structure for bacteria that fix nitrogen.

Once oxygen levels had built up to significant levels in the oceans and atmosphere, the first eukaryotes—which were macroscopic algae—appear in the fossil record. Biologists hypothesize that a causal link was involved with the availability of oxygen. The claim is that large cell size, then multicellularity, and eventually large body size were made possible by the high metabolic rates and rapid growth fueled by aerobic respiration.

To summarize, data indicate that cyanobacteria were responsible for a fundamental change in Earth's atmosphere—to one with a high concentration of oxygen. Never before, or since, have organisms done so much to alter the nature of our planet.

The Nitrogen Cycle In many environments, fertilizing forests or grasslands with nitrogen results in increased growth. Researchers infer from these results that plant growth is limited by the availability of nitrogen.

Organisms must have nitrogen to synthesize proteins and nucleic acids. Although molecular nitrogen (N_2) is extremely abundant in the atmosphere, most organisms cannot use it. To incorporate nitrogen atoms into amino acids and nucleotides, all eukaryotes and many bacteria and archaea have to obtain N in a form such as ammonia (NH_3) or nitrate (NO_3^-).

The only organisms that are capable of converting molecular nitrogen to ammonia are bacteria. The steps in the process, called **nitrogen fixation**, are complex and highly endergonic reduction-oxidation (redox) reactions (see Chapter 9). The enzymes required to accomplish nitrogen fixation are found only in selected bacterial lineages. Certain species of cyanobacteria that live in surface waters of the ocean or in association with water plants are capable of fixing nitrogen. In terrestrial environments, nitrogen-fixing bacteria that are not cyanobacteria live in close association with plants—often taking up residence in special root structures called nodules (**Figure 28.6**).

If bacteria could not fix nitrogen, it is virtually certain that only a tiny fraction of life on Earth would exist today. Large or multicellular organisms would probably be rare to nonexistent, because too little nitrogen would be available to make large quantities of proteins and build a large body.

Nitrate Pollution Corn, rice, wheat, and many other crop plants do not live in association with nitrogen-fixing bacteria. To increase yields of these crops, farmers use fertilizers that are high in nitrogen. In parts of the world, massive additions of nitrogen in the form of ammonia are causing serious pollution problems. **Figure 28.7** shows why. When ammonia is added to a cornfield—in midwestern North America, for example—much of it never reaches the growing corn plants. Instead, a significant fraction of the ammonia molecules is used as food by bacteria in the soil. Bacteria that use ammonia as an electron donor to fuel cellular respiration release nitrite (NO_2^-) as a waste product. Other bacteria use nitrite as an electron donor and release nitrate (NO_3^-). Nitrate molecules are extremely soluble in water and tend to be washed out of soils into groundwater or streams. From there they eventually reach the ocean, where they can cause pollution.

To understand why nitrates can pollute the oceans, consider the Gulf of Mexico. Nitrates carried by the Mississippi River are used as a nutrient by cyanobacteria and algae that live in the Gulf. These cells explode in numbers in response. When they die and sink to the bottom of the Gulf, bacteria and archaea and other decomposers use them as food. The decomposers use so much oxygen as an electron acceptor in cellular respiration that oxygen levels in the sediments and even in Gulf waters decline. Nitrate pollution has been so

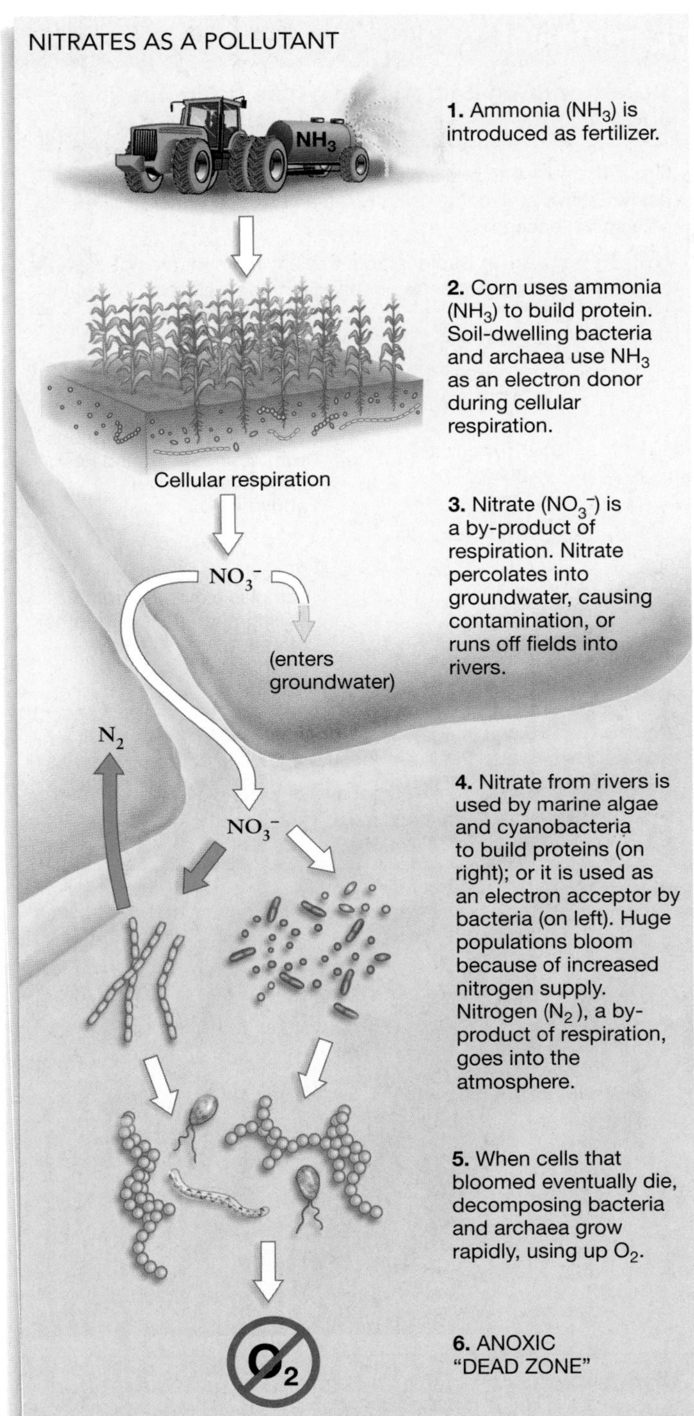

FIGURE 28.7 Nitrates Act as a Pollutant in Aquatic Ecosystems.

severe that large areas in the Gulf of Mexico are anoxic (lacking in oxygen). The oxygen-free "dead zone" in the Gulf of Mexico is devoid of fish, shrimp, and other organisms that require oxygen. Lately, the dead zone has encompassed about 18,000 km^2—roughly the size of New Jersey. Similar problems are cropping up in other parts of the world. Virtually every link in the chain of events leading to nitrate pollution involves bacteria and archaea.

The general message of this section is simple: Bacteria and Archaea may be small in size, but because of their abundance, ubiquity, and ability to do sophisticated chemistry, they have an enormous influence on the global environment.

28.2 How Do Biologists Study Bacteria and Archaea?

Biologists who study bacteria and archaea claim that they are investigating some of the most interesting and important organisms on Earth. Bacteria have been important model organisms in biological science for many decades (see **Box 28.1** on page 576). It is legitimate to state, however, that our understanding of the domains Bacteria and Archaea is advancing more rapidly right now than at any time during the past 100 years—and perhaps faster than our understanding of any other lineages on the tree of life.

As an introduction to the domains Bacteria and Archaea, let's examine a few of the techniques that biologists use to answer questions about them. Some of these research strategies have been used since bacteria were first discovered; some were invented less than 10 years ago.

Using Enrichment Cultures

Which species of bacteria and archaea are present at a particular location, and what do they use as food? To answer questions like these, biologists rely heavily on their ability to culture organisms in the lab. Of the 5000 species of bacteria and archaea that have been described to date, almost all were discovered when they were isolated from natural habitats and grown under controlled conditions in the laboratory.

One classical technique for isolating new types of bacteria and archaea is called **enrichment culture**. Enrichment cultures are based on establishing a specified set of growing conditions—temperature, lighting, substrate, types of available food, and so on. The idea is to sample cells from the environment and grow them under extremely specific conditions. Cells that thrive under the specified conditions will increase in numbers enough to be isolated and studied in detail.

To appreciate how this strategy works in practice, consider research on bacteria that live deep below Earth's surface. One recent study began with samples of rock and fluid from drilling operations in Virginia and Colorado. The samples came from sedimentary rocks at depths ranging from 860 to 2800 meters below the surface, where temperatures are between 42°C and 85°C. The questions posed in the study were simple: Is anything alive down there? If so, what do the organisms use to fuel cellular respiration?

The research team hypothesized that if organisms were living deep below the surface of the Earth, the cells might use hydrogen molecules (H_2) as an electron donor and the ferric ion (Fe^{3+}) as an electron acceptor (**Figure 28.8**). (Recall from Chapter 9 that most eukaryotes use sugars as electron donors and use oxygen as an electron acceptor during cellular respiration.) Fe^{3+} is the oxidized form of iron, and it is abundant in the rocks the biologists collected from great depths. It exists at great depths below the surface in the form of ferric oxyhydroxide. The researchers predicted that if an organism in the samples reduced the ferric ions during cellular respiration, a black, oxidized, and magnetic mineral called magnetite (Fe_3O_4) would start appearing in the cultures as a by-product of cellular respiration.

What did their enrichment cultures produce? In some culture tubes, a black compound began to appear within a week. Using a variety of tests, the biologists confirmed that the black substance was indeed magnetite. As the Results section of Figure 28.8 shows, microscopy revealed the organisms themselves—previously undiscovered bacteria. Because they grow only when incubated at between 45°C and 75°C, these organisms are considered **thermophiles** ("heat-lovers"). The discovery was spectacular, because it hinted that Earth's crust may be teeming with organisms to depths of over a mile below the surface. Enrichment culture continues to be a productive way to isolate and characterize new species of bacteria and archaea.

Experiment

Question: Can bacteria live a mile below Earth's surface?

Hypothesis: Bacteria are capable of cellular respiration deep below Earth's surface by using H_2 as an electron donor and Fe^{3+} as an electron acceptor.

Null hypothesis: Bacteria from this environment are not capable of using H_2 as an electron donor and Fe^{3+} as an electron acceptor.

Experimental setup:

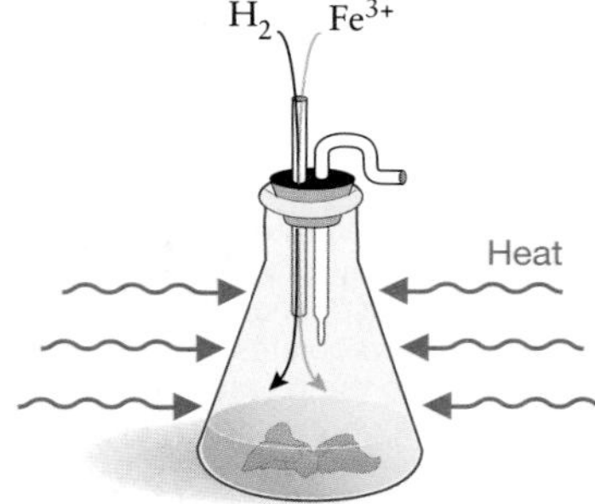

Enrichment culture method:

1. Create culture conditions with abundant H_2 and Fe^{3+}; raise temperatures above 45°C.

2. Add rock and fluid samples extracted from drilling operations at depths of about 1000 m below Earth's surface.

Prediction: Black, magnetic grains of magnetite (Fe_3O_4) will accumulate because Fe^{3+} is reduced by growing cells and shed as waste product. Cells will be visible.

Prediction of null hypothesis: No magnetite will appear. No cells will grow.

Results:

Magnetite is detectable, and cells are visible

Conclusion: At least one bacterial species that can live deep below Earth's surface grew in this enrichment culture. Different culture conditions might result in the enrichment of different species present in the same sample.

FIGURE 28.8 Enrichment Cultures Isolate Large Populations of Cells That Grow under Specific Conditions. Bacteria that grow underground can withstand high temperatures. The species enriched in this experiment needs hydrogen (H_2) and ferric ions (Fe^{3+}) to live.

● **QUESTION** How could the researchers determine to which bacterial lineage this newly discovered species belongs?

Using Direct Sequencing

Researchers estimate that of all the bacteria and archaea living today, less than 1 percent have been grown in culture. To augment research based on enrichment cultures, researchers are employing a technique called direct sequencing. **Direct sequencing** is a strategy for documenting the presence of bacteria and archaea that cannot be grown in culture. It is based on identifying phylogenetic species—populations that have enough distinctive characteristics to represent an independent twig on an evolutionary tree (see Chapter 26).

Direct sequencing allows biologists to identify and characterize organisms that have never been seen. The technique has revealed huge new branches on the tree of life and produced revolutionary data on the habitats where archaea are found.

Figure 28.9 outlines the steps performed in a direct sequencing study. To begin, researchers collect a sample from a habitat—often seawater or soil. Next, the cells in the sample are broken open (lysed), and their DNA is purified. Using the polymerase chain reaction introduced in Chapter 19, researchers can isolate specific genes from the species in the sample. After sequencing these genes, biologists compare the data with sequences in existing databases. If the sequences are markedly different, the sample probably contains previously undiscovered organisms.

Direct sequencing studies have produced new and sometimes startling results. For two decades after the discovery of the

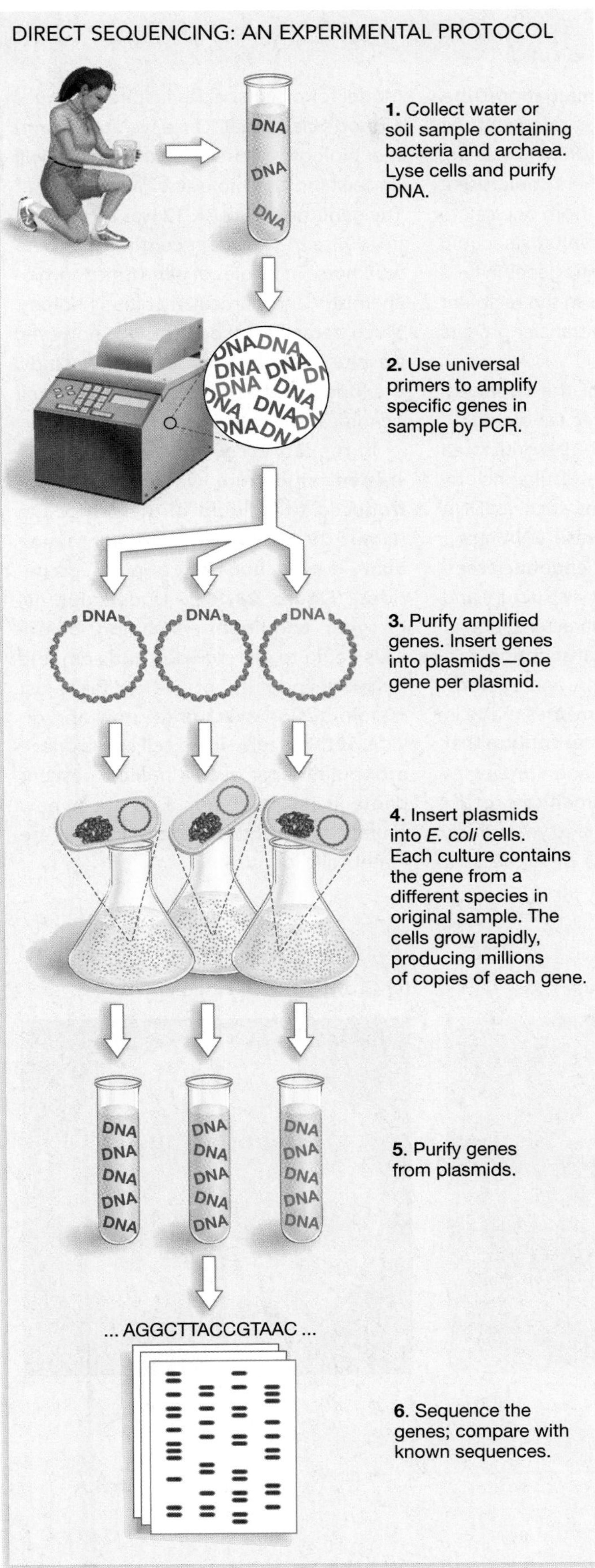

Archaea, for example, researchers thought that these organisms could be conveniently grouped into just four categories: extreme halophiles, sulfate-reducers, methanogens, and extreme thermophiles. Extreme **halophiles** ("salt-lovers") live in salt lakes, salt ponds, and salty soils. **Sulfate reducers** are species that produce hydrogen sulfide (H_2S) as a by-product of cellular respiration. (H_2S may be familiar because it smells like rotten eggs.) **Methanogens** produce methane (natural gas; CH_4) as a by-product of cellular respiration. Extreme **thermophiles** grow best at temperatures above 80°C. Based on these early data, researchers thought that archaea were restricted to extreme environments and that the four phenotypes corresponded to separate lineages within the domain Archaea. As a result of direct sequencing studies, however, these generalizations have been discarded. Beginning in the mid-1990s, direct sequencing revealed archaea in habitats as diverse as rice paddies and the Arctic Ocean. Some of these newly discovered organisms appear to belong to entirely new lineages, tentatively called the **Korarchaeota** and **Nanoarchaeota**. These species' DNA sequences are so distinctive that they might represent "kingdoms" analogous to plants or animals. Yet both groups were identified and named before any of their members had actually been observed. Follow-up research is focused on confirming these initial results. Thanks to studies like these, direct sequencing is revolutionizing our understanding of bacterial and archaeal diversity.

Evaluating Molecular Phylogenies

To put data from enrichment culture and direct sequencing studies into context, biologists depend on the accurate placement of species on phylogenetic trees. Recall from Chapter 1, Chapter 27, and **BioSkills 2** that phylogenetic trees illustrate the evolutionary relationships among species and lineages. They are a pictorial summary of which species are more closely or distantly related to others.

Some of the most useful phylogenetic trees for the Bacteria and the Archaea have been based on studies of the RNA molecule found in the small subunit of ribosomes, or what biologists call SSU RNA. (See Chapter 16 for more information on the structure and function of ribosomes.) In the late 1960s Carl Woese and colleagues began a massive effort to determine and compare the base sequences of SSU RNA molecules from a wide array of species. The result of their analysis was the **universal tree**, or the **tree of life**, illustrated in Figure 28.1.

FIGURE 28.9 Direct Sequencing Allows Researchers to Identify Species That Have Never Been Seen. Direct sequencing allows biologists to isolate specific genes from the organisms present in a sample. The polymerase chain reaction is used to generate enough copies of these genes so that the DNA from different species can be sequenced.

BOX 28.1 A Model Organism: *Escherichia coli*

Of all model organisms in biology, perhaps none has been more important than the bacterium *Escherichia coli*—a common inhabitant of the human gut. The strain that is most commonly worked on today, called K-12 (**Figure 28.10a**), was originally isolated from a hospital patient in 1922.

Escherichia coli K-12 did not begin to rise to its present prominence as a model organism until 1945, however. In that year, biologists observed that some K-12 cells were capable of a process called conjugation. Like all bacteria and archaea, K-12 and other strains of *E. coli* reproduce by **binary fission**—the splitting of a cell into two daughter cells. DNA replication precedes fission, so the daughter cells are genetically identical. But as Chapter 12 noted, some bacterial cells are also capable of transferring copies of the extracellular loops of DNA called **plasmids**. When **conjugation** occurs, a copy of a plasmid from one bacterial cell—sometimes joined by one or more genes from the main bacterial chromosome—is transferred to a recipient cell through a structure called a pilus or conjugation tube. Conjugation is sometimes referred to as bacterial sex. Its discovery in K-12 was important because researchers could use it to transfer specific alleles from one cell to another. For example, biologists could study the function of specific genes in K-12 by documenting changes in the recipient cell's phenotype once the transfer process was complete.

During the last half of the twentieth century, key results in molecular biology—introduced in Chapters 14–19—originated in studies of *E. coli*. These results include the discovery of enzymes such as DNA polymerase, RNA polymerase, DNA repair enzymes, and restriction endonucleases; the elucidation of ribosome structure and function; and the initial characterization of promoters, regulatory transcription factors, regulatory sites in DNA, and operons. In many cases, initial discoveries made in *E. coli* allowed researchers to confirm that homologous enzymes and processes existed in an array of organisms, often ranging from other bacteria to yeast, mice, and humans. The success of *E. coli* as a model for other species inspired Jacques Monod's claim that "Once we understand the biology of *Escherichia coli*, we will understand the biology of an elephant." The genome of *E. coli* K-12 was sequenced in 1997, and the strain continues to be a workhorse in studies of gene function, biochemistry, and, particularly, biotechnology. Much remains to be learned, however. Despite over 60 years of intensive study, the function of about a third of the *E. coli* genome is still unknown.

In the lab, *E. coli* is usually grown in **suspension culture**, where cells are introduced to a liquid nutrient medium (**Figure 28.10b**), or on plates containing **agar**—a gelatinous mix of polysacccharides (**Figure 28.10c**). Under optimal growing conditions—meaning before cells begin to get crowded and compete for space and nutrients—a cell takes just 30 minutes on average to grow and divide. At this rate, a single cell can produce a population of over a million descendants in just 10 hours. Except for new mutations, all of the descendant cells are genetically identical.

(a) *Escherichia coli*, strain K-12

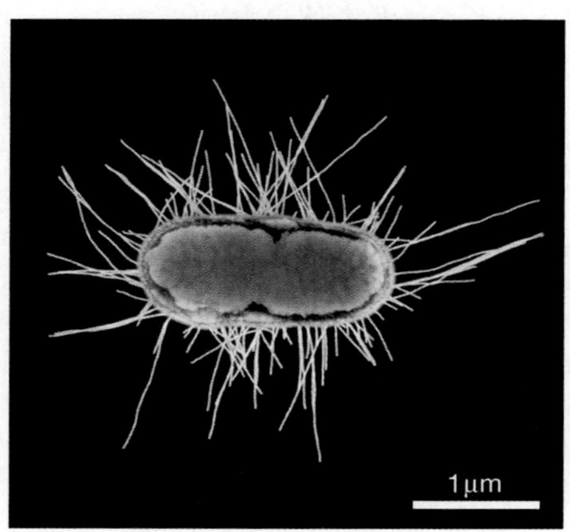

(b) Growth in liquid medium

(c) Growth on solid medium

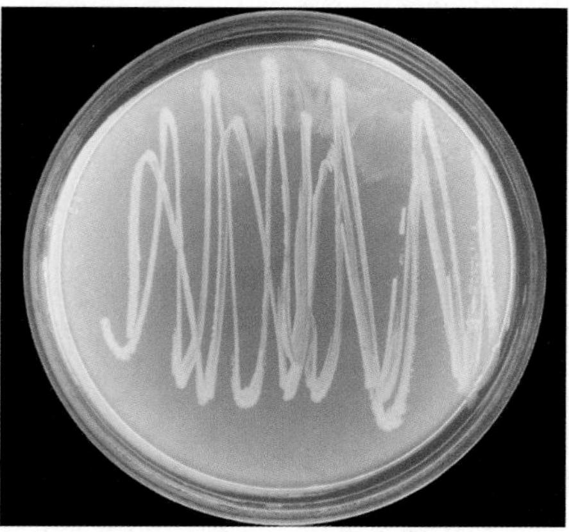

FIGURE 28.10 *E. coli* Is Readily Cultured in the Laboratory.

QUESTION *Escherichia coli* is grown at a temperature of 37°C. Why?

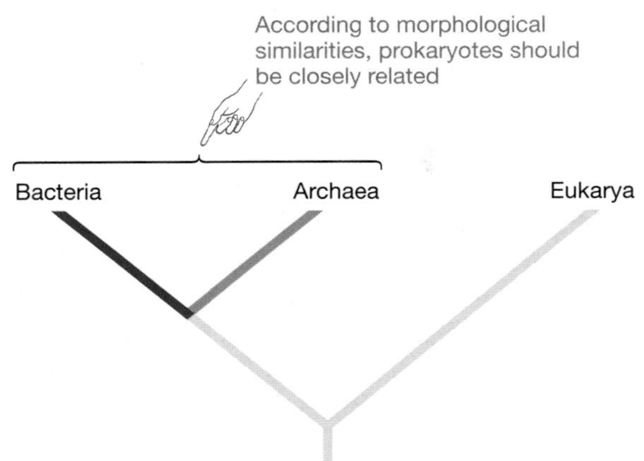

FIGURE 28.11 The Tree of Life Based on Morphology Was Incorrect. Until recently, the major division among organisms was thought to be between those without a membrane-bound nucleus (the prokaryotes) and those with a membrane-bound nucleus (the eukaryotes). Comparisons of RNA sequences have shown that this tree is not correct.

● **EXERCISE:** Draw the correct tree next to this one. Check your sketch against Figure 28.1. On the correct and incorrect trees, indicate where the nuclear envelope evolved.

Woese's tree is now considered a classic result. Prior to its publication, biologists thought that the major division among organisms was between prokaryotes and eukaryotes—between cells that lacked a membrane-bound nucleus and cells that possessed a membrane-bound nucleus (**Figure 28.11**). But based on data from the ribosomal RNA molecule, the major divisions of life-forms are the three groups that Woese named the Bacteria, Archaea, and Eukarya. Follow-up work documented that Bacteria were the first of the three lineages to diverge from the common ancestor of all living organisms. This result means that the Archaea and Eukarya are more closely related to each other than they are to the Bacteria.

Although virtually all biologists accept the three-domains system, understanding the relationships of major lineages within the Bacteria and Archaea has proven difficult. Analyses of both morphological and molecular characteristics have succeeded in identifying a large series of monophyletic groups within the domains. Recall from Chapter 27 that a **monophyletic group** consists of an ancestral population and all of its descendants. Monophyletic groups can also be called clades, or lineages. Rapid progress is being made on the question of how the major clades within Bacteria and Archaea are related to each other, and work continues. The phylogenetic tree in **Figure 28.12** summarizes recent results but is still considered highly provisional. As more data become available, it is almost certain that at least some of the branches on this tree will change position. In addition, recall that direct sequencing studies have recently led to the discovery of major new groups. If the existence of these groups is confirmed, entirely new branches will be added to this tree. Work on molecular phylogenies continues at a brisk pace.

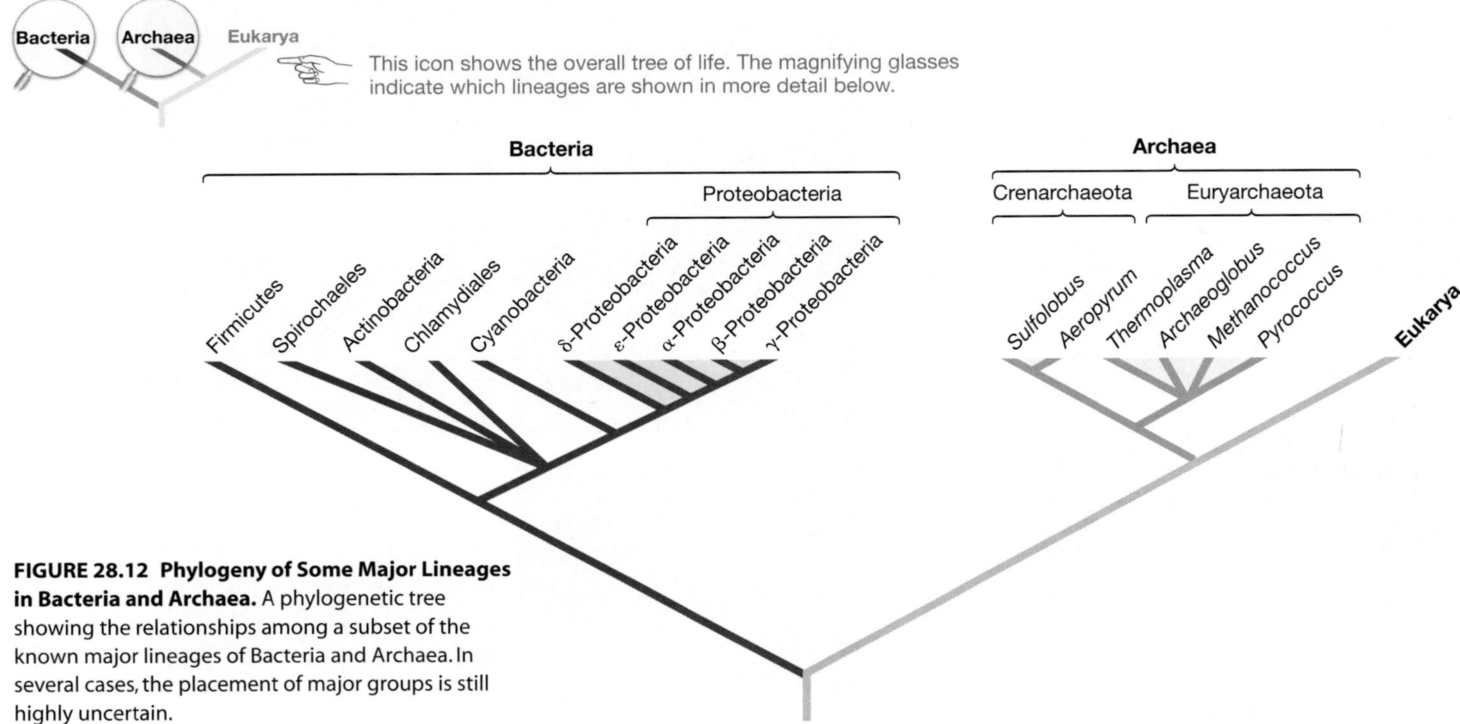

FIGURE 28.12 Phylogeny of Some Major Lineages in Bacteria and Archaea. A phylogenetic tree showing the relationships among a subset of the known major lineages of Bacteria and Archaea. In several cases, the placement of major groups is still highly uncertain.

Check Your Understanding

If you understand that...

- Enrichment cultures are based on setting up specified conditions in the laboratory and isolating the cells that grow rapidly in response. They create an abundant, pure sample of bacteria that thrive under particular conditions and can be studied further.
- Direct sequencing is based on isolating DNA from samples taken directly from the environment, purifying and sequencing specific genes, and then analyzing where those DNA sequences are found on the phylogenetic tree of Bacteria and Archaea.

You should be able to...

1) Design an enrichment culture that would isolate species that could be used to clean up oil spills.
2) Outline a study designed to identify the bacterial and archaeal species present in a soil sample near the biology building on your campus.

at www.masteringbio.com
The Tree of Life

28.3 What Themes Occur in the Diversification of Bacteria and Archaea?

Initially, the diversity of bacteria and archaea can seem almost overwhelming. To make sense of the variation among lineages and species, biologists focus on two themes: diversification in morphology and metabolism. Regarding metabolism, the key question is which molecules are used as food. Bacteria and archaea are capable of living in a wide array of environments because they vary in cell structure and in how they make a living.

Morphological Diversity

Because we humans are so large, it is hard for us to appreciate the morphological diversity that exists among bacteria and archaea. To us, they all look small and similar. But at the scale of a bacterium or archaean, different species are wildly diverse in morphology. For example, bacteria and archaea range in size from the smallest of all free-living cells—bacteria called mycoplasmas with volumes as small as 0.03 μm^3—to the largest bacterium known, *Thiomargarita namibiensis*, with volumes as large as 200×10^6 μm^3. Over a billion *Mycoplasma* cells could fit inside an individual *Thiomargarita* (**Figure 28.13a**). Bacteria

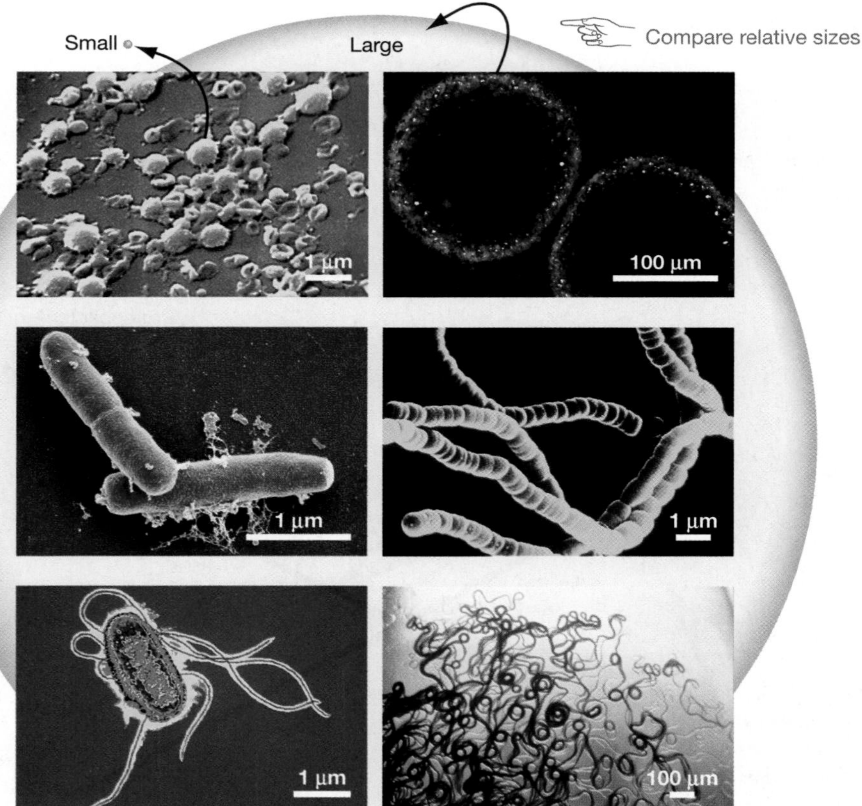

(a) Size varies
The sizes of bacteria and archaea vary. *Mycoplasma* cells (left) are about 0.5 μm in diameter, while *Thiomargarita namibiensis* cells (right) are about 150 μm in diameter.

(b) Shape varies
The shapes of bacteria and archaea vary from rods such as *Bacillus anthracis* (left) and spheres to filaments or spirals such as *Rhodospirillum*. In some species, such as *Streptococcus faecalis* (right), cells attach to one another and form chains.

(c) Mobility varies
A wide variety of bacteria and archaea use flagella (left) to power swimming movements. These cyanobacterial cells (right) move by gliding across a substrate.

FIGURE 28.13 Morphological Diversity among Bacteria and Archaea Is Extensive.

and archaea exhibit a variety of shapes as well, including filaments, spheres, rods, chains, and spirals (**Figure 28.13b**). Many cells are motile, with swimming movements powered by flagella (**Figure 28.13c**). Some cells can swim 10 more body lengths per second—much faster than any human can sprint. Although gliding movement occurs in several groups, the molecular mechanism responsible for this form of motility is still unknown.

At this scale, the composition of the plasma membrane and cell wall are particularly important. The introduction to this chapter highlighted the dramatic differences between the plasma membranes and cells walls of bacteria versus archaea. And within bacteria, two general types of cell wall exist. These types can be readily distinguished because they react differently when treated with a dye called the **Gram stain**. As **Figure 28.14a** shows, Gram-positive cells look purple but Gram-negative cells look pink. At the molecular level, cells that are **Gram-positive** have a plasma membrane surrounded by a cell wall with extensive peptidoglycan (**Figure 28.14b**). You might recall from Chapter 5 that peptidoglycan is a complex substance composed of cross-linked carbohydrate strands. Cells that are **Gram-negative**, in contrast, have a plasma membrane surrounded by a cell wall that has two components—a thin gelatinous layer containing peptidoglycan, and an outer phospholipid bilayer (**Figure 28.14c**).

To summarize, members of the Bacteria and the Archaea are remarkably diverse in their overall size, shape, and motility, as well as in the composition of their cell walls and plasma membranes. But when asked to name the innovations that were most responsible for the diversification of these two domains, biologists do not point to their morphological diversity. Instead, they point to metabolic diversity—variation in the chemical reactions that go on inside these cells. The most important thing to remember about bacteria and archaea is how diverse they are in the types of compounds they can use as food.

Metabolic Diversity

Bacteria and archaea are the masters of metabolism. Taken together, they can use almost anything as food—from hydrogen molecules to crude oil. Bacteria and archaea look small and relatively simple to us in their morphology, but their biochemical capabilities are dazzling.

Just how varied are bacteria and archaea when it comes to making a living? To appreciate the answer, recall from Chapters 9 and 10 that organisms have two fundamental nutritional needs—acquiring chemical energy in the form of adenosine triphosphate (ATP) and obtaining molecules with carbon-carbon bonds that can be used as building blocks for the synthesis of fatty acids, proteins, DNA, RNA, and other large, complex compounds required by the cell.

(a) Gram-positive cells retain Gram stain more than Gram-negative cells do.

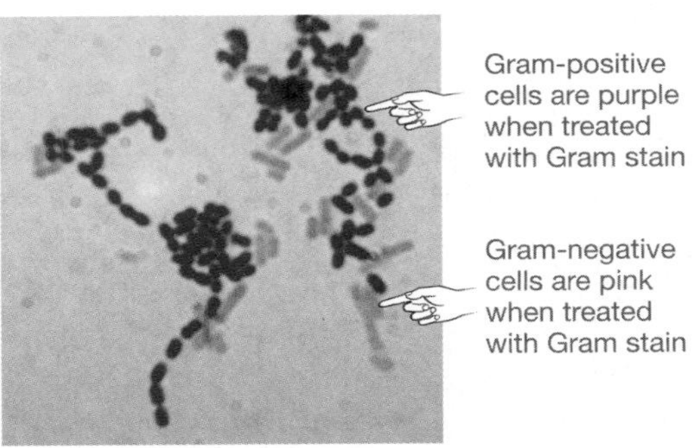

(b) Cell walls in Gram-positive bacteria have extensive peptidoglycan.

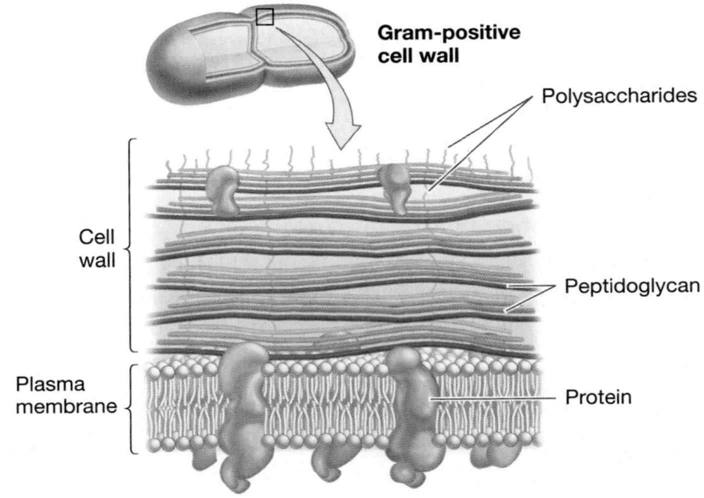

(c) Cell walls in Gram-negative bacteria have some peptidoglycan and an outer membrane.

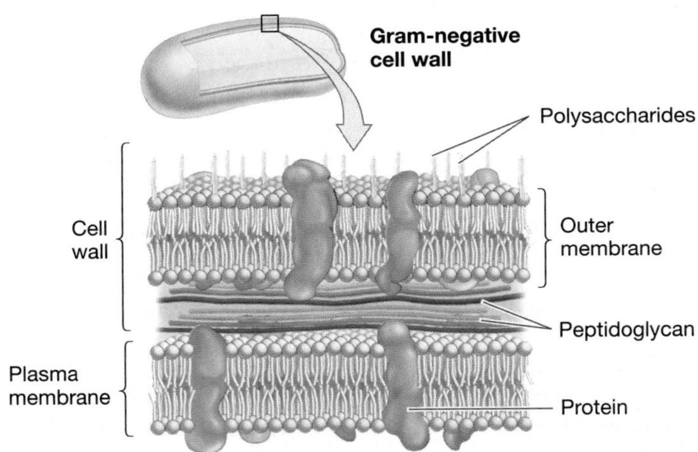

FIGURE 28.14 Gram Staining Distinguishes Two Types of Cell Walls in Bacteria. (a) When treated with the Gram stain, some cells retain a large amount of stain and look purple while other cells retain little stain and look pink. The amount of Gram stain retained is high in cells with walls that contain a large amount of peptidoglycan **(b)** and small in cells that contain little peptidoglycan but have an outer membrane layer **(c)**.

SUMMARY TABLE 28.3 **Six General Methods for Obtaining Energy and Carbon**

Source of Energy (for ATP production)	Source of Building-Block Compounds with Carbon-Carbon Bonds (for synthesis of complex organic compounds): **Self-Synthesized** (*autotrophs* synthesize building-block compounds from CO_2, CH_4, or other simple molecules)	**Obtained from Others** (*heterotrophs* acquire building-block compounds from other organisms)
Light (*phototrophs*)	*photoautotrophs*	*photoheterotrophs*
Organic Molecules with High Potential Energy (*organotrophs*)	*chemoorganotrophs*	*chemoorganoheterotrophs*
Inorganic Molecules with High Potential Energy (*lithotrophs*)	*chemolithotrophs*	*chemolithotrophic heterotrophs*

Bacteria and archaea produce ATP in three ways:

1. **Phototrophs** ("light-feeders") use light energy to promote electrons to the top of electron transport chains. ATP is produced by photophosphorylation (see Chapter 10).
2. **Chemoorganotrophs** oxidize organic molecules with high potential energy, such as sugars. ATP may be produced by cellular respiration—with sugars serving as electron donors—or via fermentation pathways (see Chapter 9).
3. **Chemolithotrophs** ("rock-feeders") oxidize inorganic molecules with high potential energy, such as ammonia (NH_3) or methane (CH_4). ATP is produced by cellular respiration, with inorganic compounds serving as the electron donor.

Bacteria and archaea fulfill their second nutritional need—obtaining building block compounds with carbon-carbon bonds—in two ways: (1) by synthesizing their own from simple starting materials such CO_2 and CH_4, or (2) by absorbing ready-to-use organic compounds from their environment. Organisms that manufacture their own building-block compounds are termed **autotrophs** ("self-feeders"). Organisms that acquire building-block compounds from other organisms are called **heterotrophs** ("other-feeders").

Because there are three distinct ways of producing ATP and two general mechanisms for obtaining carbon, there are a total of six methods for producing ATP and obtaining carbon. The names that biologists use for organisms that use these six "feeding strategies" are given in **Table 28.3**. Of the six possible ways of producing ATP and obtaining carbon, just two are observed in eukaryotes. But bacteria and archaea do them all. In their metabolism, eukaryotes are simple compared with bacteria and archaea. ● If you understand the essence of metabolic diversity in bacteria and archaea, you should be able to match the six example species described in **Table 28.4** to the appropriate category in Table 28.3.

What makes this remarkable diversity possible? Bacteria and archaea have evolved dozens of variations on the basic processes you learned about in Chapters 9 and 10. They use compounds with high potential energy to produce ATP via cellular respiration (electron transport chains) or fermentation, they use light to produce high-energy electrons, and they reduce carbon from CO_2 or other sources to produce sugars or

TABLE 28.4 **Examples of Metabolic Diversity in Bacteria and Archaea**

Bacteria/Archaea	How ATP Is Produced	How Building-Block Molecules Are Synthesized
Cyanobacteria	via photosynthesis	from CO_2 via the Calvin cycle
Clostridium aceticum	fermentation of glucose	from CO_2 via reactions called the acetyl-CoA pathway
Nitrifying bacteria (e.g., *Nitrosomonas* sp.)	via cellular respiration, using ammonia (NH_3) as an electron donor	from CO_2 via the Calvin cycle
Heliobacteria	via photosynthesis	absorb carbon-containing building-block molecules from the environment
Escherichia coli	fermentation of organic compounds or cellular respiration, using organic compounds as electron donors	absorb carbon-containing building-block molecules from the environment
Beggiatoa	via cellular respiration, using hydrogen sulfide (H_2S) as an electron donor	absorb carbon-containing building-block molecules from the environment

other building-block molecules with carbon-carbon bonds. The story of bacteria and archaea can be boiled down to two sentences: The basic chemistry required for photosynthesis, cellular respiration, and fermentation originated in these lineages. Then the evolution of variations on each of these processes allowed prokaryotes to diversify into millions of species that occupy diverse habitats. Let's take a closer look.

Producing ATP via Photosynthesis: Variation in Electron Sources and Pigments Instead of using molecules as a source of high-energy electrons, phototrophs pursue a radically different strategy: **photosynthesis**. Phototrophs use the energy in light to raise electrons to high-energy states. As these electrons are stepped down to lower energy states by electron transport chains, the energy released is used to generate ATP.

Chapter 10 introduced an important feature of photosynthesis: The process requires a source of electrons. Recall that in cyanobacteria and plants, the required electrons come from water. When these organisms "split" water molecules apart to obtain electrons, they generate oxygen as a by-product. Species that use water as a source of electrons for photosynthesis are said to complete **oxygenic** photosynthesis. In contrast, many phototrophic bacteria use a molecule other than water as the source of electrons. In many cases, the electron donor is hydrogen sulfide (H_2S); a few species can use the ion known as ferrous iron (Fe^{2+}). Instead of producing oxygen as a by-product of photosynthesis, these cells produce elemental sulfur (S) or the ferric ion (Fe^{3+}). They are said to complete **anoxygenic** photosynthesis and live in habitats where oxygen is rare.

Chapter 10 also introduced the photosynthetic pigments found in plants and explored the light-absorbing properties of chlorophylls *a* and *b*. Cyanobacteria have these two pigments. But researchers have isolated seven additional chlorophylls from bacterial phototrophs. Each major group of photosynthetic bacteria has one or more of these distinctive chlorophylls, and each type of chlorophyll absorbs light best at a different wavelength.

Why are bacterial chlorophylls so diverse? The leading hypothesis is that photosynthetic species with different absorption spectra are able to live together without competing for light. If so, then the diversity of photosynthetic pigments observed in bacteria has been an important mechanism for generating species diversity among phototrophs.

Producing ATP via Cellular Respiration: Variation in Electron Donors and Acceptors Millions of bacterial, archaeal, and eukaryotic species—including animals and plants—are organotrophs. These organisms obtain the energy required to make ATP by oxidizing organic compounds such as sugars, starch, or fatty acids. As Chapter 9 showed, cellular enzymes can strip electrons from organic molecules that have high potential energy and then transfer these high-energy electrons to the electron carriers NADH and $FADH_2$. These compounds feed electrons to an electron transport chain (ETC), where electrons are stepped down from a high-energy state to a low-energy state (**Figure 28.15a**). The energy that is released allows components of the ETC to generate a proton gradient across the plasma membrane (**Figure 28.15b**). The resulting flow of protons through the enzyme ATP synthase results in the production of ATP, via the process called chemiosmosis.

(a) Electrons pass from an electron donor through an electron transport chain, to a final electron acceptor.

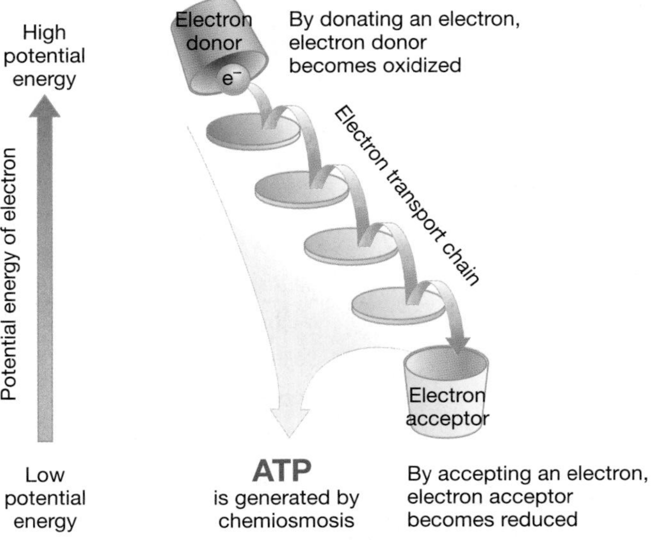

(b) The electron transport chain is organized in a series of multi-molecular complexes in the plasma membrane.

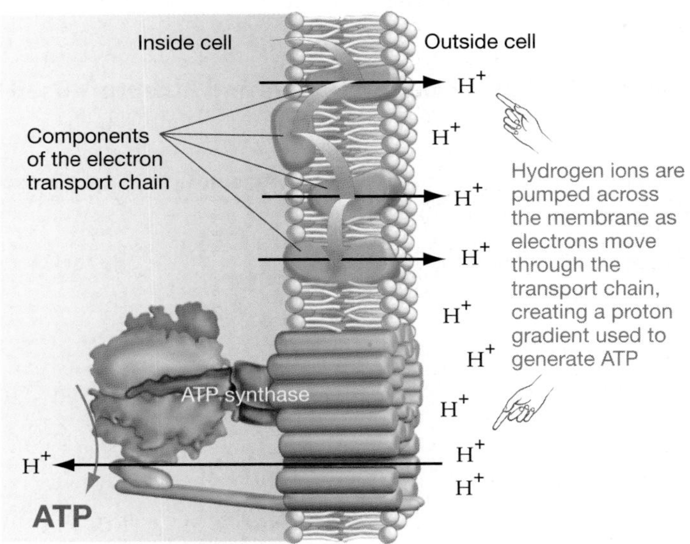

FIGURE 28.15 Cellular Respiration Is Based on Electron Transport Chains.

● **EXERCISE** In part (a), add the chemical formula for a specific electron donor, electron acceptor, and reduced by-product for a species of bacteria or archaea. Then write in the electron donor, electron acceptor, and reduced by-product observed in humans.

The essence of this process, called **cellular respiration**, is that a molecule with high potential energy serves as an original electron donor and is oxidized, while a molecule with low potential energy serves as a final electron acceptor and becomes reduced. Much of the potential energy difference between the electron donor and electron acceptor is transformed into chemical energy in the form of ATP.

In introducing respiration, Chapter 9 focused on the role of organic compounds with high potential energy—such as glucose—as the original electron donor and oxygen as the final electron acceptor (see Figure 9.22). Many bacteria and archaea, as well as all eukaryotes, rely on these molecules. When cellular respiration is complete, glucose is completely oxidized to CO_2, which is given off as a by-product. When oxygen acts as the final electron acceptor, water is also produced as a by-product.

However, many other bacteria and archaea employ an electron donor other than sugars and an electron acceptor other than oxygen during cellular respiration, and they produce by-products other than carbon dioxide and water. As **Table 28.5** shows, many bacteria and archaea are lithotrophs—they use inorganic ions or molecules as electron donors. The substances used as electron donors range from hydrogen molecules (H_2) and hydrogen sulfide (H_2S) to ammonia (NH_3) and methane (CH_4). And instead of using oxygen, some organisms use compounds with low potential energy—such as sulfate (SO_4^{2-}), nitrate (NO_3^-), carbon dioxide (CO_2), or ferric ions (Fe^{3+})—as electron acceptors. Instead of producing CO_2 and H_2O as by-products, these organisms might produce CH_4 or NO_3^-. It is only a slight exaggeration to claim that researchers have found bacterial and archaeal species that can use almost any compound with relatively high potential energy as an electron donor and almost any compound with relatively low potential energy as an electron acceptor.

Because the electron donors and electron acceptors used by bacteria and archaea are so diverse, one of the first questions biologists ask about a species is how it accomplishes cellular respiration. The best way to answer this question is through the enrichment culture technique introduced in Section 28.2. Recall that in an enrichment culture, researchers supply specific electron donors and electron acceptors in the medium and try to isolate cells that can use those compounds to support growth.

The remarkable metabolic diversity of bacteria and archaea is important. First, it explains their ecological diversity. Bacteria and archaea are found almost everywhere because they exploit an almost endless variety of molecules as electron donors and electron acceptors. Second, it explains why they play such a key role in cleaning up some types of pollution. Species that use organic solvents or petroleum-based fuels as electron donors or electron acceptors may be effective agents in bioremediation efforts if the waste products are less toxic than the solvents or fuels. Finally, metabolic diversity is what makes bacteria and archaea major players in global change. Nitrogen, phosphorus, sulfur, carbon, and other crucial nutrients cycle from one organism to another because bacteria and archaea can use them in almost any molecular form. The nitrite (NO_2^-) that some bacteria produce as a by-product of respiration does not build up in the environment, because it is used as an electron acceptor by other species and converted to molecular nitrate (NO_3^-). This molecule, in turn, is converted to molecular nitrogen (N_2) by yet another suite of bacterial and archaeal species. Bacteria can convert N_2 in the atmosphere to NH_3 in proteins (by bacterial or archaeal fixation and decomposition). Similar types of interactions occur with molecules that contain phosphorus, sulfur, and carbon. In this way, bacteria and archaea play a key role in the cycling of nutrients (**Figure 28.16**).

TABLE **28.5** **Some Electron Donors and Acceptors Used by Bacteria and Archaea**

		By-Products		
Electron Donor	**Electron Acceptor**	**From Electron Donor**	**From Electron Acceptor**	**Category***
Sugars	O_2	CO_2	H_2O	Organotrophs
H_2 or organic compounds	SO_4^{2-}	H_2O or CO	H_2S	Sulfate reducers
H_2	CO_2	H_2O	CH_4	Methanogens
CH_4	O_2	CO_2	H_2O	Methanotrophs
S^{2-} or H_2S	O_2	SO_4^{2-}	H_2O	Sulfur bacteria
Organic compounds	Fe^{3+}	CO_2	Fe^{2+}	Iron reducers
NH_3	O_2	NO_2^-	H_2O	Nitrifiers
Organic compounds	NO_3^-	CO_2	N_2O, NO, or N_2	Denitrifiers (or nitrate reducers)
NO_2^-	O_2	NO_3^-	H_2O	Nitrosifiers

*The name biologists use to identify species that use a particular metabolic strategy.

QUESTION Explain why the terms *organotrophs*, *sulfate reducers*, and *methanogens* are appropriate. (The word root *–gen* means source or origin; *–troph* refers to feeding.)

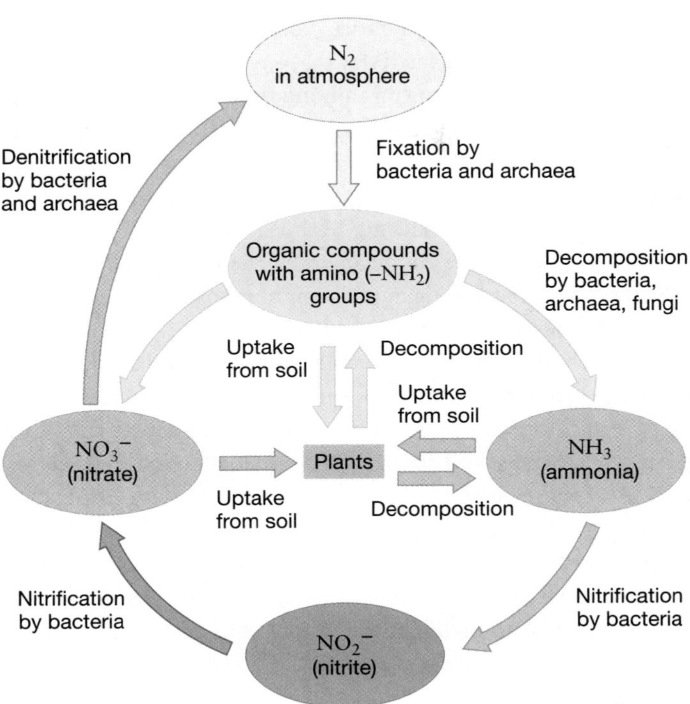

FIGURE 28.16 Nitrogen Atoms Cycle through Environments in Different Molecular Forms

EXERCISE Suppose that bacteria and archaea were no longer capable of fixing nitrogen. Draw an X through the part(s) of the cycle that would be most directly affected.

EXERCISE Add arrows and labels to indicate that animals ingest amino groups from plants or other animals and release amino groups or ammonia via excretion and decomposition.

Producing ATP via Fermentation: Variation in Substrates Chapter 9 introduced **fermentation** as a strategy for making ATP that does not involve electron transport chains. In fermentation, no outside electron acceptor is used, and the redox reactions required to produce ATP are internally balanced. Because fermentation is a much less efficient way to make ATP compared with cellular respiration, in many species it occurs as an alternative metabolic strategy when no electron acceptors are available to make cellular respiration possible. In other species, fermentation does not occur at all. In still other species, fermentation is the only way for cells to make ATP.

Although the presentation in Chapter 9 focused on how glucose is fermented to ethanol or lactic acid, some bacteria and archaea are capable of using other organic compounds as the starting point for fermentation. Bacteria and archaea that produce ATP via fermentation are still classified as organotrophs, but they are much more diverse in the substrates used. For example, the bacterium *Clostridium aceticum* can ferment ethanol, acetate, and fatty acids as well as glucose. Other species of *Clostridium* ferment complex carbohydrates (including cellulose or starch), proteins, amino acids, or even purines. Species that ferment amino acids produce by-products with names such as cadaverine and putrescine. These molecules are responsible for the odor of rotting flesh. Other bacteria can ferment lactose, a prominent component of milk. This fermentation has two end products: propionic acid and CO_2. Propionic acid is responsible for the taste of Swiss cheese; the CO_2 produced during fermentation creates the holes in cheese.

The diversity of enzymatic pathways observed in bacterial and archaeal fermentations extends the metabolic repertoire of these organisms and supports the claim that as a group, bacteria and archaea can use virtually any molecule with relatively high potential energy as a source of high-energy electrons for producing ATP. Given this diversity, it is no surprise that bacteria and archaea are found in such widely varying habitats. Different environments offer different energy-rich molecules. Various species of bacteria and archaea have evolved the biochemical machinery required to exploit most or all of these food sources.

Obtaining Building-Block Compounds: Variation in Pathways for Fixing Carbon In addition to acquiring energy, organisms must obtain building-block molecules that contain carbon-carbon bonds. Chapters 9 and 10 introduced the two mechanisms that organisms use to procure usable carbon—either making their own or getting it from other organisms. Autotrophs make their own building-block compounds; heterotrophs don't.

In many autotrophs, including cyanobacteria and plants, the enzymes of the Calvin cycle transform carbon dioxide (CO_2) to organic molecules that can be used in synthesizing cell material. The carbon atom in CO_2 is reduced during the process and is said to be "fixed." Animals and fungi, in contrast, obtain carbon from living plants or animals or by absorbing the organic compounds released as dead tissues decay.

Bacteria and archaea pursue these same two strategies. Some interesting twists occur among bacterial and archaeal autotrophs, however. Not all of them use the Calvin cycle to make building-block molecules, and not all start with CO_2 as a source of carbon atoms. For example,

- Several groups of bacteria fix CO_2 using pathways other than the Calvin cycle. Three of these distinctive pathways have been discovered to date.
- Some proteobacteria are called **methanotrophs** ("methane-eaters") because they use methane (CH_4) as their carbon source. (They also use CH_4 as an electron donor in cellular respiration.) Methanotrophs process CH_4 into more complex organic compounds via one of two enzymatic pathways, depending on the species.
- Some bacteria can use carbon monoxide (CO) or methanol (CH_3OH) as a starting material.

These observations drive home an important message from this chapter: Compared with eukaryotes, the metabolic capabilities of bacteria and archaea are remarkably sophisticated and complex.

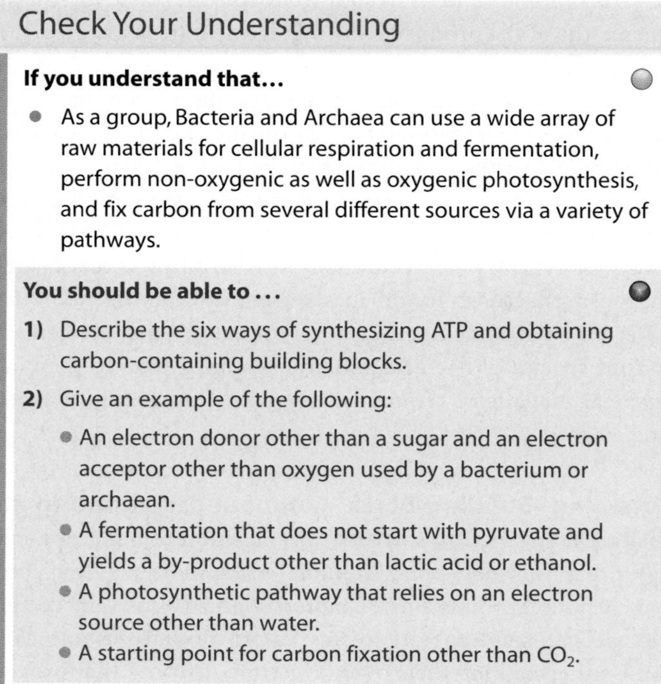

Check Your Understanding

If you understand that...

- As a group, Bacteria and Archaea can use a wide array of raw materials for cellular respiration and fermentation, perform non-oxygenic as well as oxygenic photosynthesis, and fix carbon from several different sources via a variety of pathways.

You should be able to ...

1) Describe the six ways of synthesizing ATP and obtaining carbon-containing building blocks.
2) Give an example of the following:
   - An electron donor other than a sugar and an electron acceptor other than oxygen used by a bacterium or archaean.
   - A fermentation that does not start with pyruvate and yields a by-product other than lactic acid or ethanol.
   - A photosynthetic pathway that relies on an electron source other than water.
   - A starting point for carbon fixation other than CO_2.

28.4 Key Lineages of Bacteria and Archaea

In the decades since the phylogenetic tree identifying the three domains of life was first published, dozens of studies have confirmed the result. It is now well established that all organisms alive today belong to one of the three domains, and that archaea and eukaryotes are more closely related to each other than either group is to bacteria.

Although the relationships among the major lineages within Bacteria and Archaea are still uncertain in some cases (see Figure 28.12), many of the lineages themselves are well studied. Let's survey the attributes of species from selected major lineages within the Bacteria and Archaea, with an emphasis on themes explored earlier in the chapter: their morphological and metabolic diversity, their impacts on humans, and their importance to other species and to the environment.

Each chapter in this unit will feature a similar "key lineages" section containing notes on what species in selected lineages look like, how they make a living, and where they live. Often you'll be asked to analyze a phylogenetic tree and add labels to indicate the origin of synapomorphies—traits that identify groups as monophyletic (see Chapter 27).

The overall goal of these "key lineages" sections is to give you a concise summary of who's who in major groups on the tree of life, and to highlight the innovations or features that allowed particular groups to arise and diversify.

Bacteria

The name *bacteria* comes from the Greek root *bacter*, meaning "rod" or "staff." The name was inspired by the first bacteria to be seen under a microscope, which were rod shaped. But as the following descriptions indicate, bacterial cells come in a wide variety of shapes. If the group were to be named now, biologists might use roots from the Latin words *diversus* or *abundantia*.

Biologists who study bacterial diversity currently recognize at least 16 major lineages, or phyla, within the domain. Some of these lineages were recognized by distinctive morphological characteristics; others by phylogenetic analyses of gene sequence data. The lineages reviewed here are just a sampling of bacterial diversity (**Figure 28.17**).

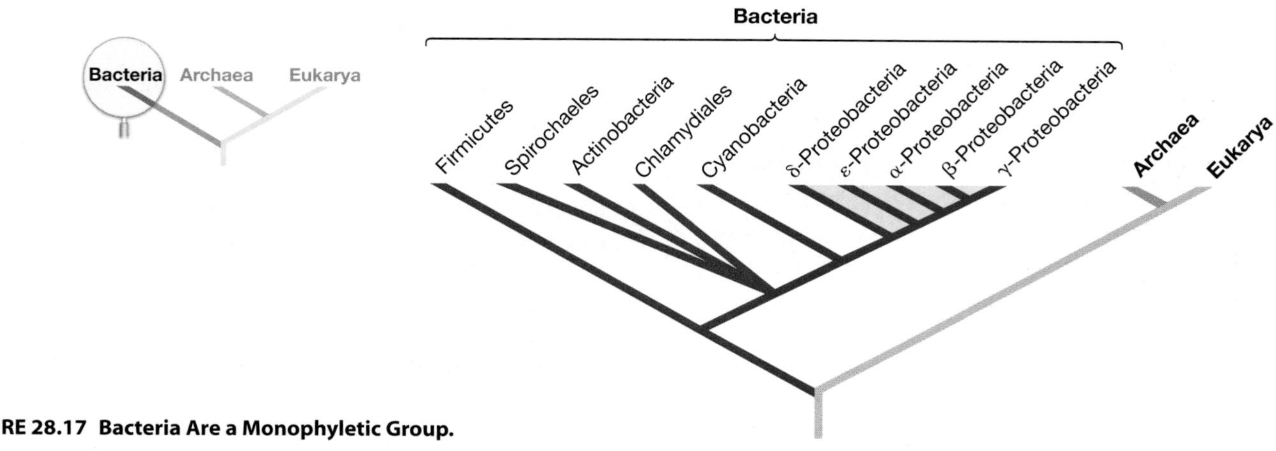

FIGURE 28.17 Bacteria Are a Monophyletic Group.

Bacteria > Firmicutes

The Firmicutes have also been called "low-GC Gram positives" because their cell walls react positively to the Gram stain—meaning that they lack a membrane outside their cell wall—and because they have a relatively low percentage of guanine and cytosine (G and C) in their DNA. In some species, G and C represent less than 25 percent of the bases present. There are over 1100 species. ● You should be able to mark the origin of the Gram-positive cell wall and low-GC genome on Figure 28.17 (only Firmicutes have a low-GC genome; Actinobacteria are the only other Gram-positive lineage).

Morphological diversity Most are rod shaped or spherical. Some of the spherical species form chains or tetrads (groups of four cells). A few form a durable resting stage called a **spore**. One subgroup lacks cell walls entirely; another synthesizes a cell wall made of cellulose.

Metabolic diversity Some species can fix nitrogen; some perform non-oxygenic photosynthesis. Others make all of their ATP via various fermentation pathways; still others perform cellular respiration, using hydrogen gas (H_2) as an electron donor.

Human and ecological impacts Species in this group cause a variety of diseases, including anthrax, botulism, tetanus, walking pneumonia, boils, gangrene, and strep throat. *Bacillus thuringiensis* produces a toxin that is one of the most important insecticides currently used in farming. Species in the genus *Lactobacillus* are used to ferment milk products into yogurt or cheese (**Figure 28.18**). Species in this group are important components of soil, where they speed the decomposition of dead plants, animals, and fungi.

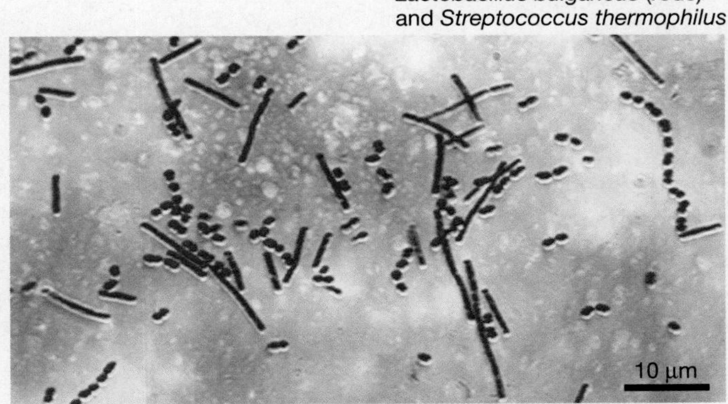

FIGURE 28.18 Firmicutes in Yogurt.

Bacteria > Spirochaetes (Spirochetes)

The spirochetes are one of the smaller bacterial phyla in terms of numbers of species: only 13 genera and a total of 62 species have been described to date. Recent analyses place spirochetes near the base of the bacterial phylogenetic tree (**Figure 28.19**).

Morphological diversity Spirochetes are distinguished by their unique corkscrew shape and unusual flagella. Instead of extending into the water surrounding the cell, spirochete flagella are contained within a structure called the outer sheath, which surrounds the cell. When the flagella beat, the cell lashes back and forth and swims forward. ● You should be able to mark the origin of the spirochete flagellum on Figure 28.17.

Metabolic diversity Most spirochetes manufacture ATP via fermentation. The substrate used in fermentation varies among species and may consist of sugars, amino acids, starch, or the pectin found in plant cell walls. A spirochete that lives only in the hindgut of termites can fix nitrogen.

Human and ecological impacts The sexually transmitted disease syphilis is caused by a spirochete. So is Lyme disease, which is transmitted to humans by deer ticks. Spirochetes are extremely common in freshwater and marine habitats; many live only under anaerobic conditions.

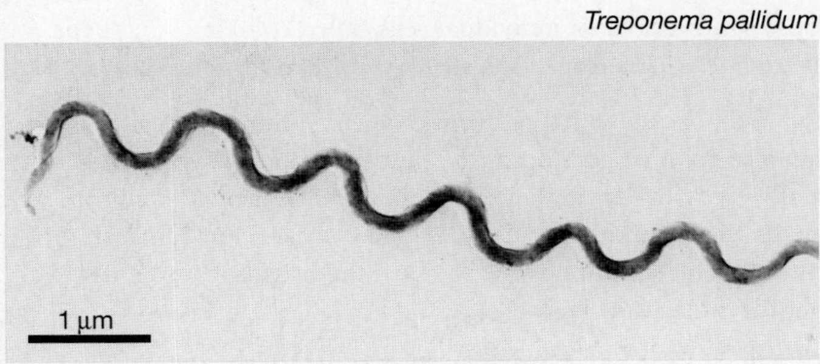

FIGURE 28.19 Spirochetes Are Corkscrew-Shaped Cells Inside an Outer Sheath.

Bacteria > Actinobacteria

Actinobacteria are sometimes called the "high-GC Gram positives" because (1) their cell-wall material appears purple when treated with the Gram stain—meaning that they have a peptidoglycan-rich cell wall and lack an outer membrane—and (2) their DNA contains a relatively high percentage of guanine and cytosine. In some species, G and C represent over 75 percent of the bases present. Over 1100 species have been described to date (**Figure 28.20**). ◉ You should be able to mark the origin of the high-GC genome in Actinobacteria on Figure 28.17.

Morphological diversity Cell shape varies from rods to filaments. Many of the soil-dwelling species are found as chains of cells that form extensive branching filaments called **mycelia**.

Metabolic diversity Many are heterotrophs that use an array of organic compounds as electron donors and oxygen as an electron acceptor. There are a handful of parasitic species. Like other parasites, they get most of their nutrition from host organisms.

Human and ecological impacts Over 500 distinct antibiotics have been isolated from species in the genus *Streptomyces*; 60 of these—including streptomycin, neomycin, tetracycline, and erythromycin—are now actively prescribed to treat diseases in humans or domestic livestock. Tuberculosis and leprosy are caused by members of this group. One species is critical to the manufacture of Swiss cheese. Species in the genus *Streptomyces* and *Arthrobacter* are abundant in soil and are vital as decomposers of dead plant and animal material. Some species in these genera live in association with plant roots and fix nitrogen; others can break down toxins such as herbicides, nicotine, and caffeine.

Streptomyces griseus

5 µm

FIGURE 28.20 A *Streptomyces* Species That Produces the Antibiotic Streptomycin.

Bacteria > Cyanobacteria

The cyanobacteria were formerly known as the "blue-green algae"—even though algae are eukaryotes. Only about 80 species of cyanobacteria have been described to date, but they are among the most abundant organisms on Earth. In terms of total mass, cyanobacteria dominate the surface waters in many marine and freshwater environments.

Morphological diversity Cyanobacteria may be found as independent cells, in chains that form filaments (**Figure 28.21**), or in the loose aggregations of individual cells called colonies. The shape of colonies varies from flat sheets to ball-like clusters of cells.

Metabolic diversity All perform oxygenic photosynthesis; many can also fix nitrogen. Because cyanobacteria can synthesize virtually every molecule they need, they can be grown in culture media that contain only CO_2, N_2, H_2O, and a few mineral nutrients. ◉ You should be able to mark the origin of oxygenic photosynthesis on Figure 28.17.

Human and ecological impacts If cyanobacteria are present in high numbers, their waste products can make drinking water smell bad. Some species release molecules called microcystins that are toxic to plants and animals. Cyanobacteria were responsible for the origin of the oxygen atmosphere on Earth. Today they still produce much of the oxygen and nitrogen and many of the organic compounds that feed other organisms in freshwater and marine environments. A few species live in association with fungi, forming lichens.

Nostoc species

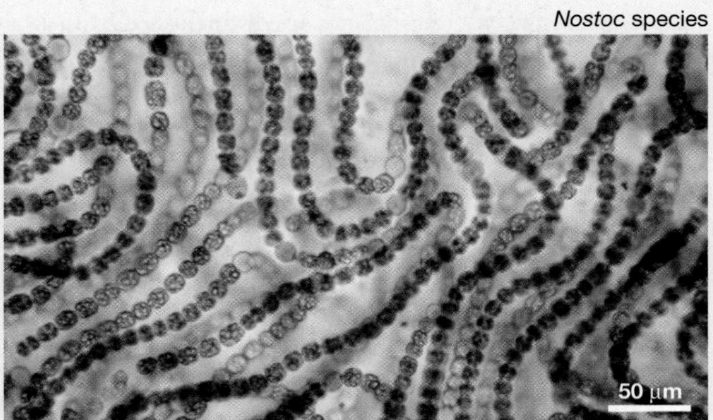

FIGURE 28.21 Cyanobacteria Contain Chlorophyll and Are Green.

Bacteria > Chlamydiales

In terms of numbers of species living today, Chlamydiales may be the smallest of all major bacterial lineages. Although chlamydiae are highly distinct phylogenetically, only four species in one genus (*Chlamydia*) are known.

Morphological diversity Chlamydiae are spherical. They are tiny, even by bacterial standards.

Metabolic diversity All known species live as parasites *inside* host cells and are termed **endosymbionts** ("inside-together-living"). Chlamydiae contain few enzymes of their own and get almost all of their nutrition from their hosts. In **Figure 28.22,** the chlamydiae are the pink-stained cells, which are living inside blue-stained animal cells. ● You should be able to mark the origin of the endosymbiotic lifestyle in this lineage on Figure 28.17. (The endosymbiotic lifestyle has also arisen in other bacterial lineages, independently of Chlamydiales.)

Human and ecological impacts *Chlamydia trachomatis* infections are the most common cause of blindness in humans. When the same organism is transmitted from person to person via sexual intercourse, it can cause serious urogenital tract infections. One species causes epidemics of a pneumonia-like disease in birds.

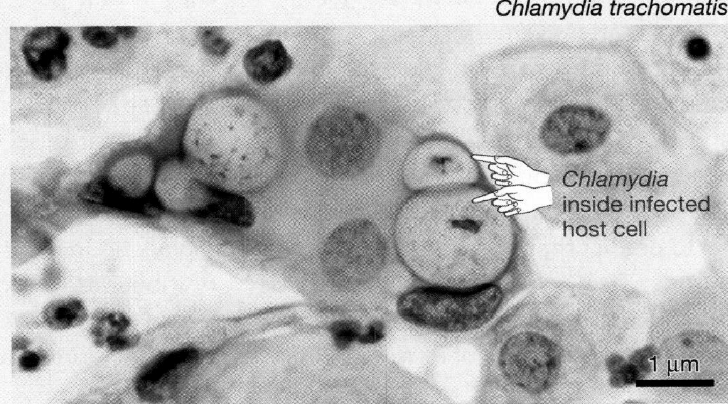

FIGURE 28.22 Chlamydiae Live Only inside Animal Cells.

Bacteria > Proteobacteria

The approximately 1200 species of proteobacteria form five major subgroups, designated by the Greek letters α (alpha), β (beta), γ (gamma), δ (delta), and ε (epsilon). The lineage is named after the Greek god Proteus, who could assume many shapes, because they are so diverse in their morphology and metabolism.

Morphological diversity Proteobacterial cells can be rods, spheres, or spirals. Some form stalks (**Figure 28.23a**). Some are motile. In one group, cells may move together to form colonies, which then transform into the specialized cell aggregate shown in **Figure 28.23b**. This structure is known as a **fruiting body**. Cells that are surrounded by a durable coating are produced at the tips of fruiting bodies. These spores sit until conditions improve, and then they resume growth.

Metabolic diversity Proteobacteria make a living in virtually every way known to bacteria—except that none perform oxygenic photosynthesis. Various species may perform cellular respiration by using organic compounds, nitrite, methane, hydrogen gas, sulfur, or ammonia as electron donors and oxygen, sulfate, or sulfur as an electron acceptor. Some perform non-oxygenic photosynthesis (**Box 28.2**).

Human and ecological impacts Pathogenic proteobacteria cause Legionnaire's disease, cholera, food poisoning, dysentery, gonorrhea, Rocky Mountain spotted fever, typhus, ulcers, and diarrhea. *Wolbachia* infections are common in insects and are often transmitted from mothers to offspring via eggs. Biologists use *Agrobacterium* cells to transfer new genes into crop plants. Certain acid-loving species of proteobacteria are used in the production of vinegars. Species in the genus *Rhizobium* (α-proteobacteria) live in association with plant roots and fix nitrogen. The bdellovibrios are a group in the δ-proteobacteria that are predators—they drill into bacterial cells and digest them. Proteobacteria are critical players in the cycling of nitrogen atoms through terrestrial and aquatic ecosystems.

(a) Stalked bacterium

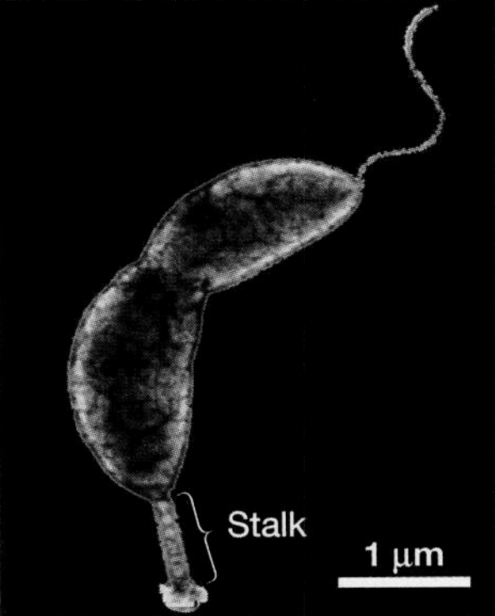

(b) Fruiting bodies

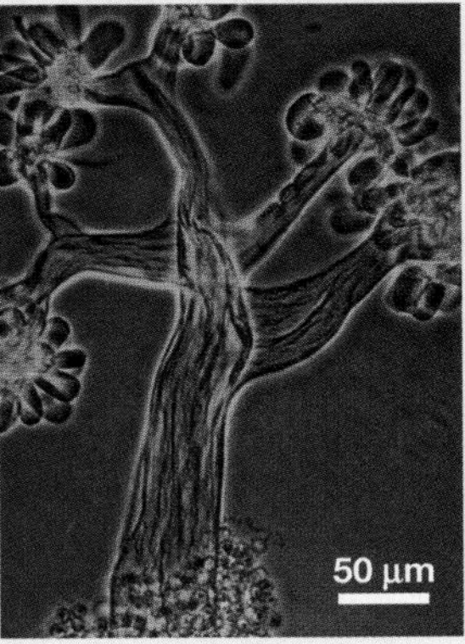

FIGURE 28.23 Some Proteobacteria Grow on Stalks or Form Fruiting Bodies.

BOX 28.2 Lateral Gene Transfer and Metabolic Diversity in Bacteria

If you read the notes on metabolic diversity in Section 28.4 carefully, you will realize that species capable of performing various types of photosynthesis are scattered among many bacterial lineages. The same is true for species that can fix nitrogen.

The "scattered among lineages" pattern is interesting because it is reasonable to predict that extremely complex structures and processes such as photosystem I, photosystem II, and nitrogen fixation evolved just once. If they did, then it would be logical to predict that photosynthesizers and nitrogen-fixers would form monophyletic groups, each group consisting of an ancestral species and all of its descendants. In contrast, a group that consists of a common ancestor and some *but not all* of its descendants is said to be **paraphyletic** ("beside-group"). If species that perform photosynthesis were monophyletic, then they would be arranged in the pattern shown in **Figure 28.24a**. But instead, the data in **Figure 28.24b** shows that photosynthetic species are paraphyletic. Nitrogen fixers are also paraphyletic.

How could the photosystems and nitrogen fixation evolve just once and yet be paraphyletic in their present phylogenetic distribution? This answer is a process called **lateral gene transfer**—the physical transfer of genes from species in one lineage to species in another lineage. Chapter 20 described several of the mechanisms responsible for lateral gene transfer. According to the lateral gene transfer hypothesis for the distribution of photosynthesis and nitrogen fixation between lineages, it is correct to claim that photosystem I, photosystem II, and the nitrogen-fixing enzymes evolved just once. But over the past 3.4 billion years, the genes responsible for these processes have been picked up by organisms from a wide array of bacterial lineages and incorporated into their genomes. Lateral gene transfer is thought to be an important mechanism for generating metabolic diversity in the Bacteria.

(a) Expected: Monophyletic distribution of photosynthetic groups

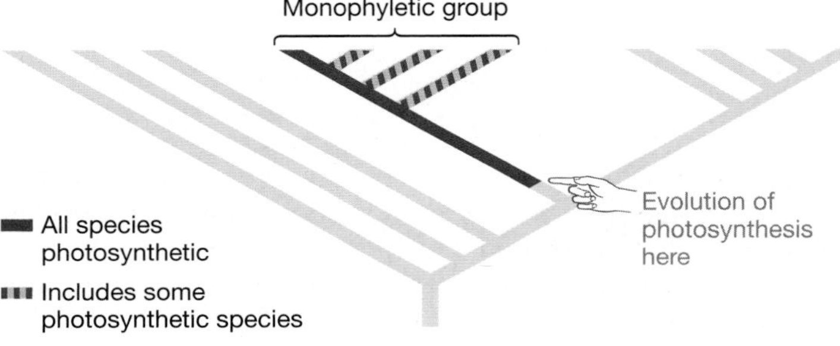

(b) Observed: Paraphyletic distribution of photosynthetic groups

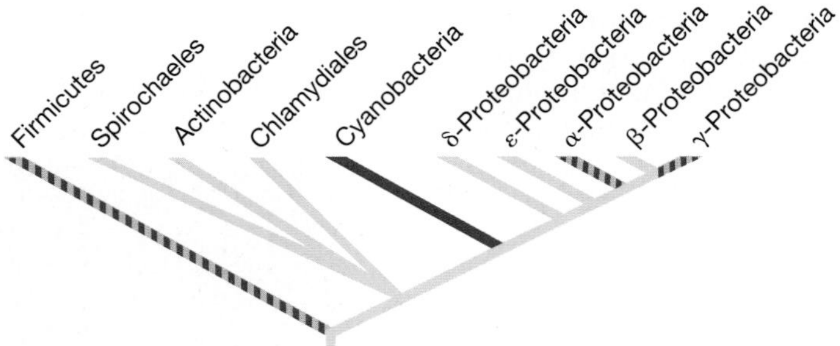

FIGURE 28.24 Photosynthetic Bacteria Are Paraphyletic. (a) If photosynthetic bacteria were monophyletic, their distribution on the phylogenetic tree of Bacteria would look something like this. **(b)** Species that can perform photosynthesis are scattered throughout the phylogenetic tree. Two bacterial lineages not shown—green sulfur bacteria and green nonsulfur bacteria—also perform a type of photosynthesis.

Archaea

The name *archaea* comes from the Greek root *archae*, for "ancient." The name was inspired by the hypothesis that this is a particularly ancient group, which turned out to be incorrect. Also incorrect was the hypothesis that archaeans are restricted to hot springs, salt ponds, and other extreme habitats. If the group were to be named now, biologists might start with the Latin root *ubiquit*, for "everywhere." Archaea live in virtually every habitat known. As far as biologists currently know, however, there are no parasitic archaea.

Phylogenies based on DNA sequence data have consistently shown that the domain is composed of at least two major phyla, called the Crenarchaeota and Euryarchaeota (**Figure 28.25**). Although it is clear that these two groups are highly differentiated at the DNA sequence level, biologists are still searching for shared, derived morphological traits that help define each lineage as a monophyletic group. In addition, the domain Archaea was discovered so recently that major groups are still being discovered and described. As mentioned earlier in this chapter, preliminary data indicate that two additional phyla may exist.

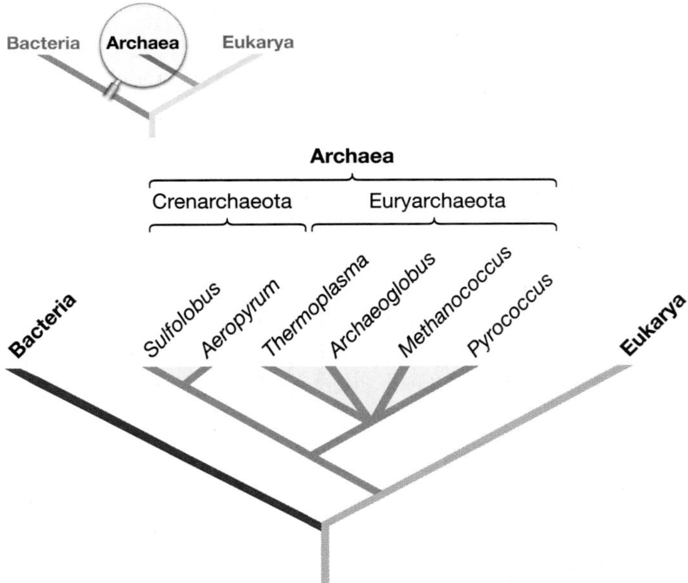

FIGURE 28.25 Archaea Are Monophyletic.

Archaea > Crenarchaeota

The Crenarchaeota got their name because they are considered similar to the oldest archaeans; the word root *cren-* refers to a source or fount. Although only 37 species have been named to date, it is virtually certain that thousands are yet to be discovered.

Morphological diversity Crenarchaeota cells can be shaped like filaments, rods, discs, or spheres. One species that lives in extremely hot habitats has a tough cell wall consisting solely of glycoprotein.

Metabolic diversity Depending on the species, cellular respiration can involve organic compounds, sulfur, hydrogen gas, ammonia, or Fe^{2+} ions as electron donors and oxygen, nitrate, sulfate, sulfur, carbon dioxide, or Fe^{3+} ions as electron acceptors. Some species make ATP exclusively through fermentation pathways.

Human and ecological impacts Crenarchaeota have yet to be used in the manufacture of commercial products. In certain extremely hot, high-pressure, cold, or acidic environments, crenarchaeota may be the only life-form present (**Figure 28.26**). Acid-loving species thrive in habitats with pH 1–5; some species are found in ocean sediments at depths ranging from 2500 to 4000 m below the surface.

FIGURE 28.26 Some Crenarchaeota Live in Sulfur-Rich Hot Springs.

Archaea > Euryarchaeota

The Euryarchaeota are aptly named, because the word root *eury-* means "broad." Members of this phylum live in every conceivable habitat. Some species are adapted to high-salt habitats with pH 11.5—almost as basic as household ammonia (**Figure 28.27**). Other species are adapted to acidic conditions with a pH as low as 0. Species in the genus *Methanopyrus* live near hot springs called black smokers that are 2000 m (over 1 mile) below sea level. About 170 species have been identified thus far, and more are being discovered each year.

Morphological diversity Euryarchaeota cells can be spherical, filamentous, rod shaped, disc shaped, or spiral. Rod-shaped cells may be short or long or arranged in chains. Spherical cells can be found in ball-like aggregations. Some species have several flagella. Some species lack a cell wall; others have a cell wall composed entirely of glycoproteins.

Metabolic diversity The group includes a variety of methane-producing species. These methanogens can use up to 11 different organic compounds as electron acceptors during cellular respiration; all produce CH_4 as a by-product of respiration. In other species of Euryarchaeota, cellular respiration is based on hydrogen gas or Fe^{2+} ions as electron donors and nitrate or sulfate as electron acceptors. Species that live in high-salt environments can use the molecule retinal—which is responsible for light reception in your eyes—to capture light energy and perform photosynthesis.

Human and ecological impacts Species in the genus *Ferroplasma* (literally, "hot iron") live in piles of waste rock near abandoned mines. As a by-product of metabolism, they produce acids that drain into streams and pollute them. Methanogens live in the soils of swamps and the guts of mammals (including yours). They are responsible for adding about 2 billion tons of methane to the atmosphere each year.

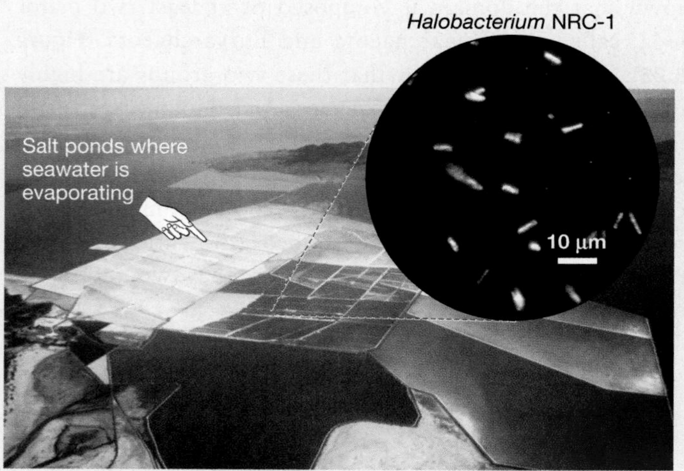

FIGURE 28.27 Some Euryarchaeota Live in High-Salt Habitats.

Chapter Review

SUMMARY OF KEY CONCEPTS

- **A few bacteria cause important infectious diseases; and collectively, bacteria and archaea play a key role in global ecosystems. Some species are effective at cleaning up pollution; photosynthetic bacteria were responsible for the evolution of the oxygen atmosphere; and bacteria and archaea cycle nutrients through terrestrial and aquatic environments.**

Bacteria and archaea may be tiny, but they have a huge impact on global ecosystems and human health. Cyanobacteria produce much of the oxygen in the oceans and atmosphere, and nitrogen-fixing bacteria and archaea keep the global nitrogen cycle running. Bacteria also cause some of the most dangerous human diseases, including plague, syphilis, botulism, cholera, and tuberculosis. Disease results when bacteria kill host cells or produce toxins that disrupt normal cell functions.

Enrichment cultures are used to grow large numbers of bacterial or archaeal cells that thrive under specified conditions, such as in the presence of certain electron donors and electron acceptors. To study bacteria and archaea that cannot be cultured, biologists frequently take advantage of direct sequencing. In this research strategy, DNA sequences are extracted directly from organisms in the environment—without first culturing the organisms in the lab. By analyzing where these sequences are placed on the tree of life, biologists can determine whether the branches represent organisms that are new to science. If so, then information on where the original sample was collected can expand knowledge about the types of habitats used by bacteria and archaea.

You should be able to explain what the composition of the atmosphere and what the nitrogen cycle would be like if bacteria and archaea did not exist.

- **Many bacterial and archaeal species are restricted in distribution and have a limited diet, but as a group they live in virtually every habitat known and use remarkably diverse types of compounds in cellular respiration and fermentation. Although their**

overall morphologies are small and relatively simple, the chemistry they can do is extremely sophisticated.

Metabolic diversity and complexity are the hallmarks of the bacteria and archaea, just as morphological diversity and complexity are the hallmarks of the eukaryotes. Like eukaryotes, many bacteria and archaea can extract energy from carbon-containing compounds with high potential energy, such as sugars. These molecules are processed through fermentation pathways or by transferring high-energy electrons to electron transport chains with oxygen as the final electron acceptor. But among the bacteria and archaea, many other inorganic or organic compounds with high potential energy serve as electron donors, and a wide variety of inorganic or organic molecules with low potential energy serve as electron acceptors. Dozens of distinct organic compounds are fermented, including proteins, purines, alcohols, and an assortment of carbohydrates.

Photosynthesis is also widespread among bacteria. In cyanobacteria, water is used as a source of electrons during photosynthesis, and oxygen gas is generated as a by-product. But in other species, the electron excited by photon capture comes from a reduced substance such as ferrous iron (Fe^{2+}) or hydrogen sulfide (H_2S) instead of water (H_2O); the oxidized by-product is the ferric ion (Fe^{3+}) or elemental sulfur (S) instead of oxygen (O_2). These organisms also contain chlorophylls not found in plants or cyanobacteria.

To acquire building-block molecules containing carbon-carbon bonds, some species use the enzymes of the Calvin cycle to reduce CO_2. But biologists have also discovered three additional biochemical pathways in bacteria and archaea that transform carbon dioxide (CO_2), methane (CH_4), or other sources of inorganic carbon into organic compounds such as sugars or carbohydrates.

You should be able to describe a habitat on Earth where bacteria and archaea would *not* be found, and explain why.

MB **Web Animation** at www.masteringbio.com
The Tree of Life

QUESTIONS

Test Your Knowledge

1. How do molecules that function as electron donors and those that function as electron acceptors differ?
   a. Electron donors are almost always organic molecules; electron acceptors are always inorganic.
   b. Electron donors are almost always inorganic molecules; electron acceptors are always organic.
   c. Electron donors have relatively high potential energy; electron acceptors have relatively low potential energy.
   d. Electron donors have relatively low potential energy; electron acceptors have relatively high potential energy.
2. What do some photosynthetic bacteria use as a source of electrons instead of water?
   a. oxygen (O_2)
   b. hydrogen sulfide (H_2S)
   c. organic compounds (e.g., CH_3COO^-)
   d. nitrate (NO_3^-)
3. What is distinctive about the chlorophylls found in different photosynthetic bacteria?
   a. their membranes
   b. their role in acquiring energy
   c. their role in carbon fixation
   d. their absorption spectra
4. What are organisms called that use inorganic compounds as electron donors in cellular respiration?
   a. phototrophs
   b. heterotrophs
   c. organotrophs
   d. lithotrophs
5. What has direct sequencing allowed researchers to do for the first time?
   a. identify important morphological differences among species
   b. study organisms that cannot be cultured (grown in the lab)
   c. It is based on sampling organisms from an environment and sequencing DNA from a particular gene.
   d. It is based on sampling organisms from an environment and sequencing the entire genomes present.
6. Koch's postulates outline the requirements for which of the following?
   a. showing that an organism is autotrophic
   b. showing that a bacterium's cell wall lacks an outer membrane and consists primarily of peptidoglycan
   c. showing that an organism causes a particular disease
   d. showing that an organism can use a particular electron donor and electron acceptor

Test Your Knowledge answers: 1. c; 2. b; 3. d; 4. d; 5. b; 6. c

Test Your Understanding

Answers are available at www.masteringbio.com

1. Biologists often use the term *energy source* as a synonym for "electron donor." Why?
2. The text claims that the tremendous ecological diversity of bacteria and archaea is possible because of their impressive metabolic diversity. Do you agree with this statement? Why or why not?
3. Suppose that universal PCR primers were available for genes involved in electron transport chains or for some of the different types of chlorophyll found in bacteria. Why would it be interesting to use these genes in a direct sequencing study?
4. The text claims that the evolution of an oxygen atmosphere paved the way for increasingly efficient cellular respiration and higher growth rates in organisms. Explain.
5. Look back at Table 28.5 and note that the by-products of respiration in some organisms are used as electron donors or acceptors by other organisms. In the table, draw lines between the dual-use molecules listed in the "Electron Donor," "Electron Acceptor," and "By-products" columns.
6. Explain the statement, "Prokaryotes are a paraphyletic group."

Applying Concepts to New Situations

Answers are available at www.masteringbio.com

1. The researchers who observed that magnetite was produced by bacterial cultures from the deep subsurface carried out a follow-up experiment. These biologists treated some of the cultures with a drug that poisons the enzymes involved in electron transport chains. In cultures where the drug was present, no more magnetite was produced. Does this result support or undermine their hypothesis that the bacteria in the cultures perform anaerobic respiration with Fe^{3+} serving as the electron acceptor? Explain your reasoning.
2. *Streptococcus mutans* obtains energy by oxidizing sucrose. This bacterium is abundant in the mouths of Western European and North American children and is a prominent cause of cavities. The organism is virtually absent in children from East Africa, where tooth decay is rare. Propose a hypothesis to explain this observation. Outline the design of a study that would test your hypothesis.
3. Suppose that you've been hired by a firm interested in using bacteria to clean up organic solvents found in toxic waste dumps. Your new employer is particularly interested in finding cells that are capable of breaking a molecule called benzene into less toxic compounds. Where would you go to look for bacteria that can metabolize benzene as an energy or carbon source? How would you design an enrichment culture capable of isolating benzene-metabolizing species?
4. Would you predict that disease-causing bacteria, such those listed in Table 28.2, obtain energy from light, organic molecules, or inorganic molecules? When they perform cellular respiration, which substance would you predict that they use as an electron acceptor? Explain your answer.

Protists

29

KEY CONCEPTS

- Protists are a paraphyletic grouping that includes all eukaryotes except the green plants, fungi, and animals. Biologists study protists to understand how eukaryotes evolved, because they are important in freshwater and marine ecosystems and global warming, and because some species cause debilitating diseases in plants, humans, and other organisms.
- Protists are diverse morphologically. They vary in the types of organelles they contain; they may be unicellular or multicellular, and they may have a cell wall or other external covering, or no such covering.
- Protists vary widely in terms of how they find food. Many species are photosynthetic, while others obtain carbon compounds by ingesting food or parasitizing other organisms.
- Protists vary widely in terms of how they reproduce. Sexual reproductive evolved in protists, and many protist species can reproduce both sexually and asexually.

Diatoms are single-celled protists that live inside a glassy case. They may be the most abundant of all eukaryotes found in aquatic environments.

This chapter introduces the third domain on the tree of life: the **Eukarya**. Eukaryotes range from single-celled organisms that are the size of bacteria to sequoia trees and blue whales. The largest and most morphologically complex organisms on the tree of life—algae, plants, fungi, and animals—are eukaryotes.

Although species in the Eukarya are astonishingly diverse, they share fundamental features that distinguish them from bacteria and archaea. Eukaryotes are defined by the presence of the shared, derived character called the nuclear envelope. Chapter 7 pointed out other key distinctions between eukaryotes and prokaryotes. You might recall that most eukaryotic cells are much larger than bacterial or archaeal cells. They also have many more organelles and have a much more extensive system of structural proteins called the cytoskeleton. Multicellularity is rare in bacteria and unknown in archaea, but has evolved several times in eukaryotes (see Chapter 8). Unlike bacteria and archaea, which reproduce by fission, eukaryotes undergo cell division via mitosis; many eukaryotes can also undergo meiosis. Like archaea, most eukaryotes have chromosomes where DNA is complexed with proteins called histones (see Chapter 18).

One of this chapter's fundamental goals is to explore how these morphological innovations—features like the nuclear envelope and organelles—evolved. Another goal is to analyze how morphological innovations allowed eukaryotes to pursue novel ways of performing basic life tasks such as feeding, moving, and reproducing.

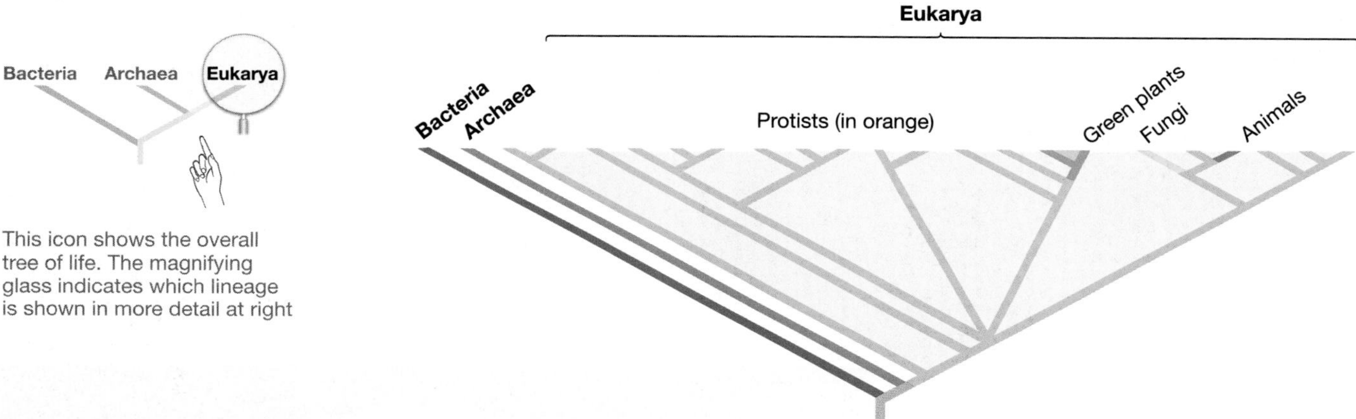

FIGURE 29.1 Protists Are Paraphyletic. The group called protists includes some, but not all, descendants of a single common ancestor.

● **EXERCISE** On the tree, indicate the common ancestor of all eukaryotes.

In introducing the Eukarya, this chapter focuses on an informal grab bag of lineages known as the protists. The term **protist** refers to all eukaryotes that are not green plants, fungi, or animals. Protist lineages are colored yellow-orange in **Figure 29.1**. ● As you study this tree, note that protists do not make up a monophyletic group. Instead, protists constitute a **paraphyletic group**—meaning that they represent some, but not all, of the descendants of a single common ancestor (see **Box 29.1**). To use the vocabulary introduced in Chapter 27, no synapomorphies define the protists. There is no trait that is found in protists but no other organisms.

By definition, then, the protists are a diverse lot. The story of the eukaryotes as a whole revolves around the evolution of complex new cell structures and methods for reproducing; the story of protists is how these new features diversified over time. The organelles and cytoskeletons found in protists vary widely in structure. Where and when meiosis occurs is highly variable among species. Many protists are microscopic single cells; others are multicellular organisms up to 60 meters long. Their lifestyles are just as diverse. Some are parasitic while others are predatory or photosynthetic. They may be stationary all their lives or in virtually constant motion. Their cells can change shape almost continuously or be encased in rigid, glassy shells. The common feature among protists is that they tend to live in environments where they are surrounded by water (**Figure 29.2**). Most plants, fungi, and animals are terrestrial, but protists are found in wet soils, aquatic habitats, or the bodies of other organisms.

BOX 29.1 How Should We Name the Major Branches on the Tree of Life?

Taxonomy is the branch of biology devoted to describing and naming new species and classifying groups of species. Carolus Linnaeus founded this field and published the first work in it in 1735. Linnaeus invented the system of Latin binomials, still being used today, in which each organism is given a unique name consisting of its genus and species. He also invented a hierarchy of more general taxonomic categories that included kingdoms, classes, orders, and families (see Chapter 1). Species were placed into different families, orders, classes, and kingdoms based on their morphological similarity.

Linnaeus worked long before Darwin had discovered the principle of evolution by natural selection, however. As a result, Linnaeus viewed the groups he described as entities that had been created separately and independently from one another. When biologists came to realize that all organisms are related by common descent, they had to reinterpret the categories that Linnaeus had established. Instead of representing neat "bins" holding species with similar characteristics, biologists recognized that any groupings that are named should actually be designated as twigs, branches, and stems on a tree of life.

According to the rules currently being adopted, most biologists assign names only to monophyletic groups—that is, to branches on the tree of life that include all the descendants of a common ancestor. Because it violates this rule, the name *protist* will probably be abandoned before long.

According to a Chinese proverb, "The first step in wisdom is to call things by their right name." In taxonomy, a massive effort is under way to do just that. The initial task is to produce accurate estimates of where each branch occurs on the tree of life. Then monophyletic groups can be named, with confidence that each represents a distinct stem, branch, or twig on the tree.

(a) Open ocean:

Surface waters teem with microscopic protists, such as these diatoms.

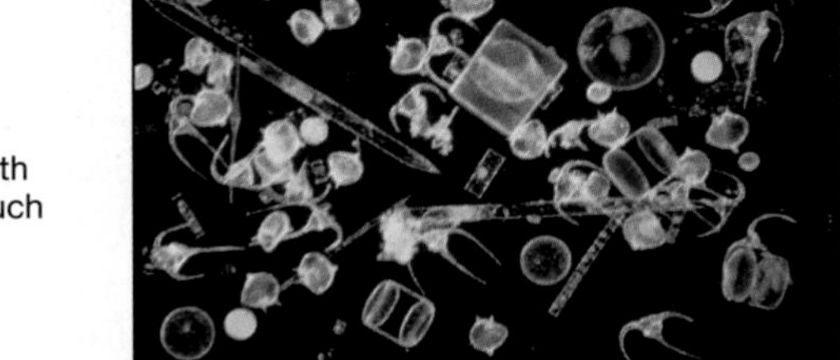

(b) Shallow coastal waters:

Gigantic protists, such as these kelp, form underwater forests.

(c) Intertidal habitats:

Protists such as these red algae are particularly abundant in tidal habitats.

FIGURE 29.2 Protists Are Particularly Abundant in Aquatic Environments.

To begin, let's ask why biologists are spending so much time and energy studying protists, and which techniques are proving most useful. Section 29.3 delves into the heart of the protist story by analyzing the innovations and subsequent diversification that took place in their morphology and their methods of feeding, moving, and reproducing. The chapter concludes with a survey of characteristics in key lineages—a who's who of protists that you are almost sure to encounter later on in your career.

29.1 Why Do Biologists Study Protists?

Biologists study protists in part because protists are intrinsically interesting, in part because they are so important medically and ecologically, and in part because they are critical to understanding the evolution of plants, fungi, and animals. The remainder of the chapter will focus on why protists are interesting in their own right and how they evolved; here let's consider their impact on human health and the environment.

Impacts on Human Health and Welfare

The most spectacular crop failure in history, the Irish potato famine, was caused by a protist. In 1845 most of the 3 million acres that had been planted to grow potatoes in Ireland became infested with *Phytophthora infestans*—a parasite that belongs to a lineage of protists called Oomycota. Potato tubers that were infected with *P. infestans* rotted in the fields or in storage.

As a result of crop failures in Ireland in two consecutive years, an estimated 1 million people out of a population of less than 9 million died of starvation or starvation-related illnesses. Several million others emigrated. Many people of Irish heritage living in North America, New Zealand, and Australia trace their ancestry to relatives who left Ireland to evade the famine. As devastating as the potato famine was, however, it does not begin to approach the misery caused by the protist *Plasmodium*.

Malaria Physicians and public health officials point to three major infectious diseases that are currently afflicting large numbers of people worldwide: tuberculosis, HIV, and malaria. Tuberculosis is caused by a bacterium and was introduced in Chapter 24; HIV is caused by a virus and is analyzed in Chapter 35. Here we consider malaria—a disease caused by several species in the eukaryotic lineage called Apicomplexa.

Malaria ranks as the world's most chronic public health problem. In India alone, over 30 million people each year suffer from debilitating fevers caused by malaria. At least 300 million people worldwide are sickened by it each year, and over 1 million die from the disease annually. The toll is equivalent to eight 747s, loaded with passengers, crashing every day. Most of the dead are children of preschool age.

Four species of the protist *Plasmodium* are capable of parasitizing humans and causing malaria. Infections start when *Plasmodium* cells enter a person's bloodstream during a mosquito bite. As **Figure 29.3** shows, *Plasmodium* initially infects liver cells; later, some cells change into a distinctive cell type that infects red blood cells. The *Plasmodium* cells multiply inside the host cells and kill them as they exit to infect additional liver cells or red blood cells. If cells released from infected red blood cells are transferred to a mosquito during a bite, they differentiate to form gametes. Inside the mosquito, gametes fuse to form a diploid cell called an oocyst, which undergoes meiosis. The haploid cells that result from meiosis can infect a human when the mosquito bites again.

Because each *Plasmodium* species spends part of its life cycle inside mosquitoes, most antimalaria campaigns have focused on controlling these insects. This strategy has become less and less effective over time, however, because natural selection has favored mosquito strains that are resistant to the insecticides that have been sprayed in their breeding habitats. Further, *Plasmodium* itself has evolved resistance to most of the drugs used to control its growth in infected people. Efforts to develop a vaccine against *Plasmodium* have also been fruitless to date, in part because the parasite evolves so quickly. (Chapter 35 explains why it has not been possible to develop effective vaccines against rapidly evolving viruses and organisms such as *Plasmodium*, cold and influenza viruses, and HIV.) Although *Plasmodium* is arguably the best studied of all protists, researchers have still not been able to devise effective and sustainable measures to control it.

Unfortunately, malaria is not the only important human disease caused by protists. **Table 29.1** lists protists that have been the cause of human suffering and economic losses. Parasitic protists affect hundreds of millions of people every year and are a major concern for physicians worldwide. Harmful protists also worry biologists who manage the fisheries that many people depend on for their food or livelihood.

Harmful Algal Blooms When a unicellular species experiences rapid population growth and reaches high densities in an aquatic environment, it is said to "bloom." Unfortunately, a handful of the many protist species involved in blooms can be harmful. Harmful algal blooms are usually due to photosynthetic protists called dinoflagellates. Certain dinoflagellates are harmful at high population density because they synthesize toxins to protect themselves from predation by small animals called copepods. Because toxin-producing dinoflagellates have high concentrations of red accessory pigments called xanthophylls, their blooms are known informally as red tides (**Figure 29.4**)

Algal blooms can be harmful to people because clams and other shellfish filter photosynthetic protists out of the water as food. During a bloom, high levels of toxins can build up in the flesh of these shellfish. Typically, the shellfish themselves are not harmed. But if a person eats contaminated shellfish, several

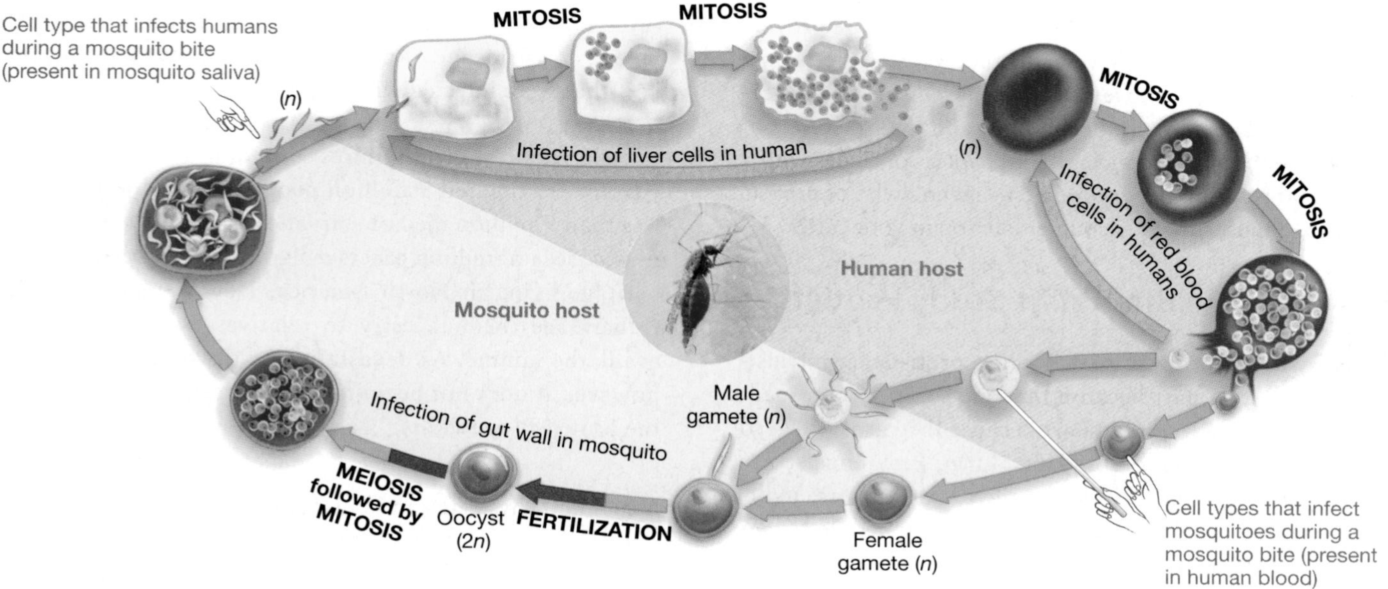

FIGURE 29.3 *Plasmodium* Lives in Mosquitoes and in Humans, where It Causes Malaria. Over the course of its life cycle, *Plasmodium falciparum* develops into a series of distinct cell types. Each type is specialized for infecting a different host cell in mosquitoes or humans. In mosquitoes, the protist lives in the gut and salivary glands. In humans, it infects and kills liver cells and red blood cells, contributing to anemia and high fever.

TABLE 29.1 **Human Health Problems Caused by Protists**

Species	Disease
Four species of *Plasmodium*, primarily *P. falciparum* and *P. vivax*	Malaria has the potential to affect 40 percent of the world's total population.
Toxoplasma	Toxoplasmosis may cause eye and brain damage in infants and in AIDS patients.
Many species of dinoflagellates	Toxins released during "red tides" accumulate in clams and mussels and poison people if eaten.
Giardia	Diarrhea due to giardiasis (beaver fever) can last for several weeks.
Trichomonas	Trichomoniasis is a reproductive tract infection and one of the most common sexually transmitted diseases. About 2 million young women are infected in the United States each year; some of them become infertile.
Leishmania	Leishmaniasis can cause skin sores or affect internal organs—particularly the spleen and liver.
Trypanosoma gambiense and *T. rhodesiense*	Trypanosomiasis ("sleeping sickness") is a potentially fatal disease transmitted through bites from tsetse flies. Occurs in Africa.
Trypanosoma cruzi	Chagas disease affects 16–18 million people and causes 50,000 deaths annually, primarily in South and Central America.
Entamoeba histolytica	Amoebic dysentery results from severe infections.
Phytophthora infestans	An outbreak of this protist wiped out potato crops in Ireland in 1845–1847, causing famine.

types of poisoning can result. For example, paralytic shellfish poisoning occurs when people eat shellfish that have fed heavily on protists that synthesize poisons called saxitoxins. Saxitoxins block ion channels that have to open for electrical signals to travel through nerve cells (see Chapter 45). In humans, high dosages of saxitoxins cause unpleasant symptoms such as prickling sensations in the mouth or even life-threatening symptoms such as muscle weakness and paralysis.

No antidote exists to the poisons secreted by protists during harmful blooms. As a result, biologists prevent poisonings by carefully monitoring protist populations in regions where shellfish are harvested for food. If harmful protist species begin to bloom, the shellfish beds are immediately closed to harvest until toxins are at lower levels.

FIGURE 29.4 Harmful Algal Blooms Are Sometimes Called Red Tides. Dinoflagellates with red coloration have "bloomed" along this beach.

Ecological Importance of Protists

As a whole, the protists represent just 10 percent of the total number of named eukaryote species. Although the species diversity of protists may be relatively low, their abundance is extraordinarily high. The number of individual protists found in some habitats is astonishing. One milliliter of pond water can contain well over 500 single-celled protists that swim with the aid of flagella. Under certain conditions, dinoflagellates can reach concentrations of 60 million cells per liter of seawater.

The great abundance of protists is important to an array of ecological events. For example, photosynthetic organisms take in carbon dioxide from the atmosphere and reduce, or "fix," it to form sugars or other organic compounds with high potential energy (Chapter 10). Photosynthesis transforms some of the energy in sunlight into chemical energy that organisms can use to grow and produce offspring. Species that produce chemical energy in this way are called **primary producers**. Diatoms, for example, are photosynthetic protists that rank among the leading primary producers in the oceans, simply because they are so abundant. Production of organic molecules in the world's oceans, in turn, represents almost half of the total carbon dioxide that is fixed on Earth. Why is this important?

Protists Play a Key Role in Aquatic Food Chains Small organisms that live near the surface of oceans or lakes and that drift along or swim only short distances are called **plankton**. The sugars and other organic compounds produced by **phytoplankton**—that is, photosynthetic species of plankton—are the basis of food chains in freshwater and marine environments. A **food chain** describes nutritional relationships

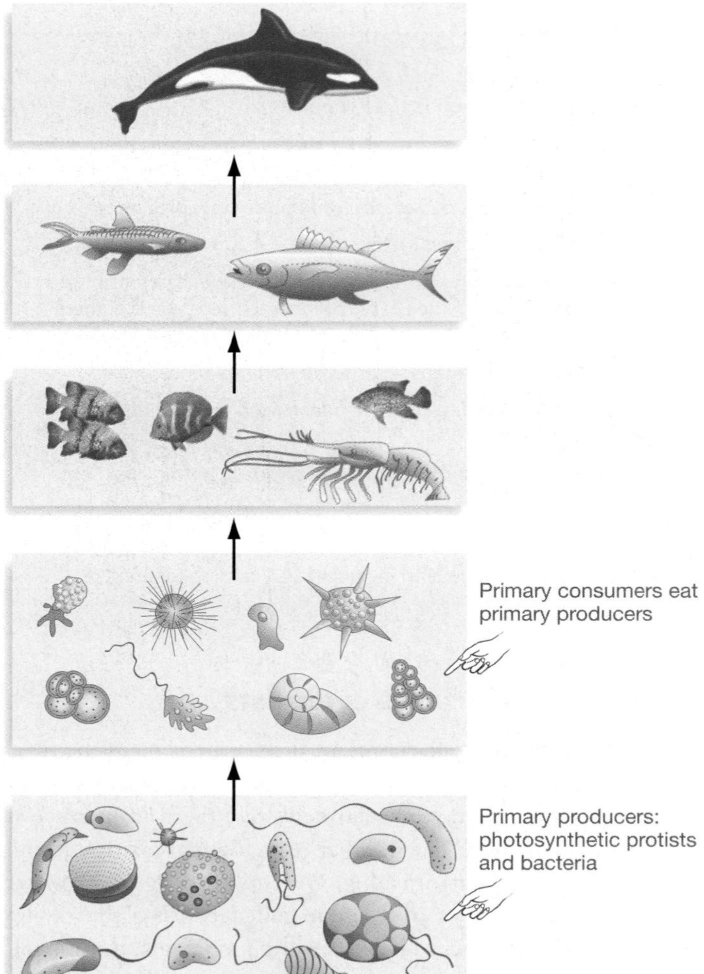

FIGURE 29.5 Protists Are Key Primary Producers in Marine and Freshwater Habitats. In aquatic habitats, photosynthetic protists form the base of the food chain along with photosynthetic bacteria.

● **EXERCISE** Indicate the levels at which humans feed

among organisms. Thus, it also describes how chemical energy flows within ecosystems.

Figure 29.5 shows a food chain for a marine environment. At the bottom of the chain are photosynthetic protists, photosynthetic bacteria, and other primary producers. These species are eaten by primary consumers, many of which are protists. Primary consumers are eaten by secondary consumers, which in turn are eaten by tertiary consumers, and so on. The fish and shellfish at the middle levels of the food chain are important sources of protein not only for whales, squid, and very large fish (such as tuna) at the uppermost levels but also for people.

Many of the species at the base of food chains in aquatic environments are protists. It is not an exaggeration to say that without protists, most food chains in freshwater and marine habitats would collapse.

Could Protists Help Reduce Global Warming? As Chapter 54 will explain in detail, carbon dioxide levels in the atmosphere are increasing rapidly due to human activities such as the burning of fossil fuels and forests. Because carbon dioxide traps heat that is radiating from Earth back out to space, high CO_2 levels in the atmosphere contribute to global warming—an issue that many observers consider today's most pressing environmental problem.

To understand global warming, it is critical to analyze how carbon atoms move among organisms and molecules in both terrestrial and aquatic habitats. The movement of carbon atoms from carbon dioxide molecules in the atmosphere to organisms in the soil or the ocean, and then back to the atmosphere is called the **global carbon cycle**. To reduce global warming, researchers are trying to figure out ways to decrease carbon dioxide concentrations in the atmosphere and increase the amount of carbon stored in terrestrial and marine environments.

Recall from Chapter 10 that in terrestrial environments, trees take CO_2 from the atmosphere and use some of the carbon atoms to make wood. Wood is a storage area, or "sink," for carbon, because carbon atoms tend to remain in wood for decades or centuries—until trees burn or decay. As a result, the growth of trees reduces the amount of carbon in the atmosphere and increases the amount of carbon stored in forests. Efforts to plant trees in deforested areas might play a role in efforts to reduce atmospheric CO_2 concentrations and slow the rate of global warming.

In the world's oceans, attempts to fertilize photosynthetic protists and bacteria might have a similar effect. To understand why, first consider the carbon cycle diagrammed in **Figure 29.6**. The cycle starts when CO_2 from the atmosphere dissolves in water and is taken up by phytoplankton. The phytoplankton are eaten by primary consumers, die and are consumed by decomposers or scavengers, or die and sink to the bottom of the ocean. There they may enter one of two long-lived repositories—sedimentary rocks or petroleum:

1. Several lineages of protists have shells made of calcium carbonate ($CaCO_3$) When these shells rain down from the ocean surface and settle in layers at the bottom, the deposits that result are compacted by the weight of the water and by sediments accumulating above them. Eventually they turn into rock. The limestone used to build the pyramids of Egypt consists of protist shells. Limestone and other carbon-containing rocks that form on the ocean floor can potentially lock up carbon atoms for tens of millions of years.

2. Although the process of petroleum (oil) formation is not well understood, it begins with accumulations of dead bacteria, archaea, and protists at the bottom of the ocean. CO_2 that passes from the atmosphere to the body of a photosynthetic protist and then to petroleum is removed from the atmosphere for millions of years—unless humans pump the petroleum out of the ground and burn it.

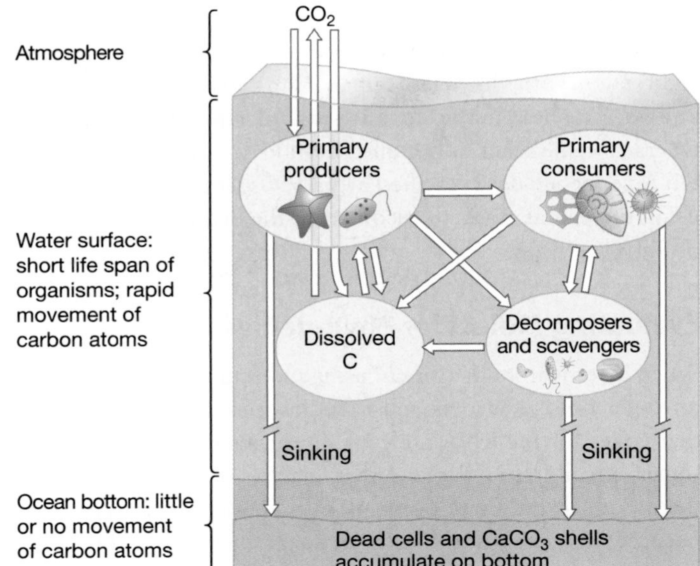

FIGURE 29.6 Protists Play a Key Role in the Marine Carbon Cycle. Arrows indicate the movement of carbon atoms. At the surface, carbon atoms tend to shuttle quickly among organisms. But if carbon atoms sink to the bottom of the ocean in the form of shells or dead cells, they may be locked up for long periods in carbon sinks. (Arrows "Sinking" are broken to indicate that the bottom may be miles below the surface.)

QUESTION Why are there arrows leading from primary producers and primary consumers to dissolved C?

When carbon atoms are removed from CO_2 in the atmosphere and tied up in limestone or petroleum, they have entered a particularly long-lived carbon sink.

How do these observations relate to global warming? Recent experiments have shown that dramatic things happen when habitats in the middle of the ocean are fertilized with iron. Iron is a critical component of the electron transport chains responsible for photosynthesis and respiration, but it is in particularly short supply in the open ocean. After iron is added to ocean waters, it is not uncommon to see populations of protists and other primary producers increase by a factor of 10. Some researchers hypothesize that when these blooms occur, the amount of carbon that rains down into carbon sinks in the form of shells and dead cells may increase. If so, then fertilizing the ocean to promote blooms might be an effective way to reduce CO_2 concentrations in the atmosphere.

The effectiveness of iron fertilization is hotly debated, however. As some researchers point out, fertilizing the ocean with iron might lead to large accumulations of dead organic matter and the formation of anaerobic dead zones like those described in Chapter 28. Consequently, many biologists are not convinced that adding iron and inducing massive blooms of protists, bacteria, and archaea would be a good idea. But if further research shows that iron fertilization is safe and effective, it could be added to the list of possible approaches for reducing carbon dioxide levels in the atmosphere.

Check Your Understanding

If you understand that...

- Biologists study protists because they cause disease and harmful algal blooms, and because they are key primary producers in aquatic environments.

You should be able to...

1) Explain why the World Health Organization and the World Bank are promoting the use of insecticide-treated sleeping nets as a way of reducing malaria.
2) Make a diagram of the carbon cycle that predicts the consequences of adding massive amounts of iron to marine plankton over many years.

29.2 How Do Biologists Study Protists?

Although biologists have made great strides in understanding pathogenic protists and the role that protists play in the global carbon cycle, it has been extremely difficult to gain any sort of solid insight into how the group as a whole diversified over time. The problem is that protists are so diverse that it has been difficult to find any overall patterns in the evolution and diversification of the group.

Recently, researchers have made dramatic progress in understanding protist diversity by combining data on the morphology of key groups and phylogenetic analyses of DNA sequence data. For the first time, a clear picture of how the Eukarya diversified may be within sight. Let's analyze how this work is being done, beginning with classical results on the morphological traits that distinguish major eukaryote groups.

Microscopy: Studying Cell Structure

Using light microscopy, biologists were able to identify and name many of the protist species known today. When transmission electron microscopes became available, a major breakthrough in understanding protist diversity occurred: Detailed studies of cell structure revealed that protists could be grouped according to characteristic overall form, according to organelles with distinctive features, or both. For example, both light and electron microscopy confirmed that the species that caused the Irish potato famine has reproductive cells with an unusual type of **flagellum**. Flagella are organelles that project from the cell and whip back and forth to produce swimming movements (Chapter 7). In reproductive cells of *Phytophthora infestans*, one of the two flagella present has tiny, hollow, hairlike projections. Biologists noted that kelp and other forms of brown algae also have cells with this type of flagellum.

To make sense of these results, researchers interpreted these types of distinctive morphological features as **synapomorphies**—shared, derived traits that distinguish major monophyletic

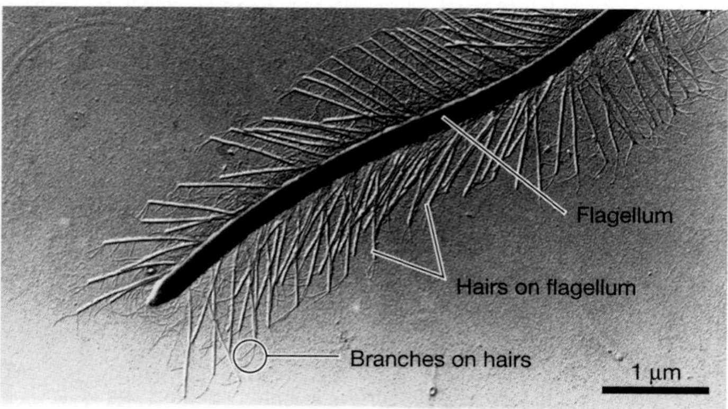

FIGURE 29.7 Species in the Lineage Called Stramenopiles Have a Distinctive Flagellum. The unusual, hollow "hairs" that decorate the flagella of stramenopiles often have three branches at the tip.

groups (see Chapter 27). Species that have a flagellum with hollow, hairlike projections became known as stramenopiles ("straw-hairs"); the hairs typically have three branches at the tip (**Figure 29.7**). In recognizing this group, investigators hypothesized that because an ancestor had evolved a distinctive flagellum, all or most of its descendants also had this trait. The qualifier *most* is important, because it is not unusual for certain subgroups to lose particular traits over the course of evolution, much as humans are gradually losing fur and tailbones.

Eventually, eight major groups of eukaryotes came to be identified on the basis of diagnostic morphological characteristics. These groups and the synapomorphies that identify them are listed in **Table 29.2**. Note that in almost every case, the synapomorphies listed in the table represent changes in structures that protect or support the cell or that influence the organism's ability to move or feed. Note also that the plants, fungi, and animals analyzed in Chapters 30 through 34 represent subgroups within two of the eight major eukaryotic lineages.

Although individual groups of protists were well characterized on the basis of morphology, the relationships among the eight eukaryotic lineages remained almost completely unknown. The next major advance in understanding eukaryotic diversity came when it became possible to obtain and analyze DNA sequence data. In the early 1990s, investigators began using molecular traits to reconstruct the evolutionary history of Eukarya.

Evaluating Molecular Phylogenies

When investigators first began using molecular data to infer the phylogeny of eukaryotes, most studies were based on the gene that codes for the RNA molecule in the small subunit of ribosomes. Recall from Chapter 1 that analyses of this gene revealed the nature of the three domains of life. When researchers sequenced this gene from an array of eukaryotes and used the data to infer a phylogeny of the domain Eukarya, the analysis suggested that the eight groups identified on the basis of distinctive morphological characteristics were indeed monophyletic groups. This was important support for the hypothesis that the distinctive morphological features were shared, derived characters that existed in a common ancestor of each lineage.

To understand the relationships among the eight lineages and their subgroups, biologists began analyzing sequence data from other genes and combining these results with data from morphological traits. Although estimating the phylogeny of Eukarya is still a work in progress, the phylogenetic tree in **Figure 29.8** is the current best estimate of the group's evolutionary history.

One of the emerging results is that the Amoebozoa and the Opisthokonta—which include fungi and animals—appear to form a monophyletic group recently given the provisional name Unikonta. Similarly, the Alveolata and Stramenopila appear to form a monophyletic group that biologists are call-

SUMMARY TABLE **29.2 Major Lineages of Eukaryotes**

Lineage	Distinguishing Morphological Features (synapomorphies)
Excavata	Cells have a pronounced "feeding groove" where prey or organic debris is ingested. No functioning mitochondria are present, although genes derived from mitochondria are found in the nucleus.
Discicristata	Cells have mitochondria with distinctive disc-shaped cristae.
Alveolate	Cells have sac-like structures called alveoli that form a continuous layer just under the plasma membrane. Alveoli are thought to provide support.
Stramenopila	If flagella are present, cells usually have two—one of which is covered with hairlike projections.
Rhizaria	Cells lack cell walls, although some produce an elaborate shell-like covering. When portions of the cell extend outward to move the cell, they are slender in shape.
Plantae	Cells have chloroplasts with a double membrane.
Opisthokonta	Reproductive cells have a single flagellum at their base. The cristae inside mitochondria are flat, not tube shaped as in other eukaryotes. (This lineage includes protists as well as the fungi and the animals. Fungi and animals are discussed in detail in Chapters 31 through 34.)
Amoebozoa	Cells lack cell walls. When portions of the cell extend outward to move the cell, they form large lobes.

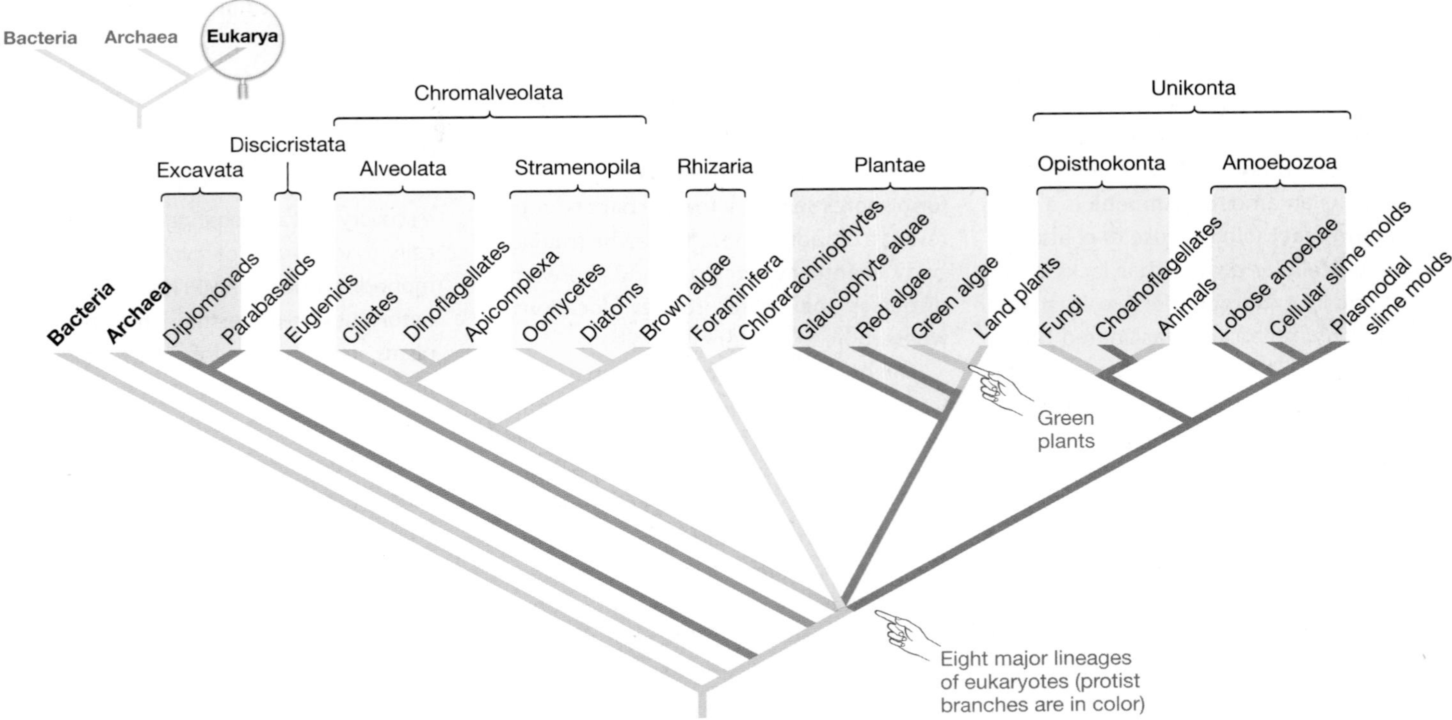

FIGURE 29.8 Phylogenetic Analyses Have Identified Eight Major Lineages of Eukaryotes. This tree shows selected subgroups from the eight major lineages discussed in this chapter. Many other lineages of protists have been identified.

EXERCISE Put a bar and label across the branch where the unusual flagellum of stramenopiles evolved. Put another bar and label to indicate where the distinctive mitochondria of the Discicristata evolved.

ing the Chromalveolata. Understanding where the root or base of the tree lies has been more problematic, however. The latest data suggest that a group of single-celled organisms called Excavates, which have a distinctive "feeding groove," forms the most basal lineage of the Eukarya. Groups with chloroplasts, specialized types of flagella, or unusual mitochondria evolved later. As more data become available, our understanding of eukaryote phylogeny will continue to improve.

Discovering New Lineages via Direct Sequencing

The effort to refine the phylogeny of the Eukarya is ongoing, and protists continue to be used as important model organisms in biology (see **Box 29.2**). But of all the research frontiers in eukaryotic diversity, the most exciting may be the one based on the technique called direct sequencing.

As Chapter 28 explained, **direct sequencing** is based on collecting organisms from a habitat and analyzing the DNA sequence of specific genes without growing larger populations of individuals in laboratory culture. The approach is based on using the polymerase chain reaction (PCR; see Chapter 19) to amplify certain genes in the organisms that have been collected. The resulting DNA sequence data are then used to place the organisms on a phylogenetic tree.

Recall that direct sequencing led to the discovery of previously unknown but major lineages of Archaea. To the amazement of biologists all over the world, the same thing happened when researchers used direct sequencing to survey eukaryotes.

The first direct sequencing studies that focused on eukaryotes were published in 2001 and were motivated by the hypothesis that direct sequencing could detect the presence of previously unknown species. To test this hypothesis, researchers sampled cells at various depths and locations in the oceans, used PCR primers to amplify genes that encode the RNA component of the ribosome's small subunit, sequenced the genes, and compared the data with the sequences of previously studied eukaryotes. One study sampled organisms at depths from 250 to 3000 m below the surface in waters off Antarctica; another focused on cells at depths of 75 m in the Pacific Ocean, near the equator. Both studies found a wide array of distinctive ribosomal RNA sequences—new species under the phylogenetic species concept introduced in Chapter 26.

Investigators who followed up on these results by examining the samples under the microscope were astonished to find that many of the newly discovered eukaryotes were tiny—from $0.2\mu m$ to $5\mu m$ in diameter. Other direct sequencing studies have now confirmed the existence of protists that are less than $0.2\mu m$ in diameter. These eukaryotic cells overlap in size with bacteria, which typically range from $0.5\mu m$ to $2\mu m$ in

BOX 29.2 A Model Organism: *Dictyostelium discoideum*

Several species of protists have served as model organisms in biology. Among the most important has been the cellular slime mold *Dictyostelium discoideum*. *Dictyostelium* is not always slimy, and it is not a mold—meaning a type of fungus. Instead, it is an amoeba. **Amoeba** is a general term that biologists use to characterize a unicellular protist that lacks a cell wall and is extremely flexible in shape. *Dictyostelium* has long fascinated biologists because it is a social organism. Independent cells sometimes aggregate to form a multicellular structure.

Figure 29.9 shows the *Dictyostelium discoideum* **life cycle**—the sequence of events that occurs over an individual's life span. Note that under most conditions, *D. discoideum* cells are haploid (n) and move about in decaying vegetation on forest floors or other habitats. They feed on bacteria by engulfing them whole. When these cells reproduce, they do so asexually, by mitosis. If food begins to run out, however, the cells begin to aggregate. In many cases, tens of thousands of cells cohere to form a 2-mm-long mass called a **slug**. (This is not the slug that is related to snails.) After migrating to a sunlit location, the slug stops and individual cells differentiate according to their position in the slug. Some form a stalk; others form a mass of spores at the tip of the stalk. A spore is a single cell that develops into an adult organism, but it is not formed from gamete fusion like a zygote. The entire structure, stalk plus mass of spores, is called a **fruiting body**. Cells that form spores secrete a tough coat and represent a durable resting stage. The fruiting body eventually dries out, and the wind disperses the spores to new locations, where more food might be available.

On occasion, *Dictyostelium* may also undergo sexual reproduction. When an aggregation is forming, two cells inside it may fuse to form a diploid ($2n$) zygote called a *giant cell*. The giant cell grows by feeding on the haploid amoebae in the aggregation. It then secretes a tough, protective coat. Later the zygote undergoes meiosis to form haploid offspring, which undergo mitosis. Eventually the haploid cells break out of the coating and begin to move about in search of food.

Dictyostelium discoideum has been an important model organism for investigating questions about eukaryotes:

- Cells in a slug are initially identical in morphology but then differentiate into distinctive stalk cells and spores. Studying this process helped biologists better understand how cells in plant and animal embryos differentiate into distinct cell types.
- The process of slug formation has helped biologists study how animal cells move and how they aggregate as they form specific types of tissues. In addition, the discovery that aggregating *D. discoideum* cells follow trails of cyclic adenosine monophosphate (cAMP) helped investigators better understand the mechanisms responsible for **chemotaxis**—movement in response to a chemical—in other organisms.
- When *D. discoideum* cells aggregate to form a slug, they stick to each other. The discovery of membrane proteins responsible for cell-cell adhesion helped biologists understand some of the general principles of multicellular life highlighted in Chapter 8.
- Cells that aggregate to form a slug are not genetically identical. About 20 percent of the cells help other individuals produce offspring by forming the stalk, but die without producing offspring themselves. Why? Researchers are using *D. discoideum* to answer questions about why some individuals appear to sacrifice themselves to help others.

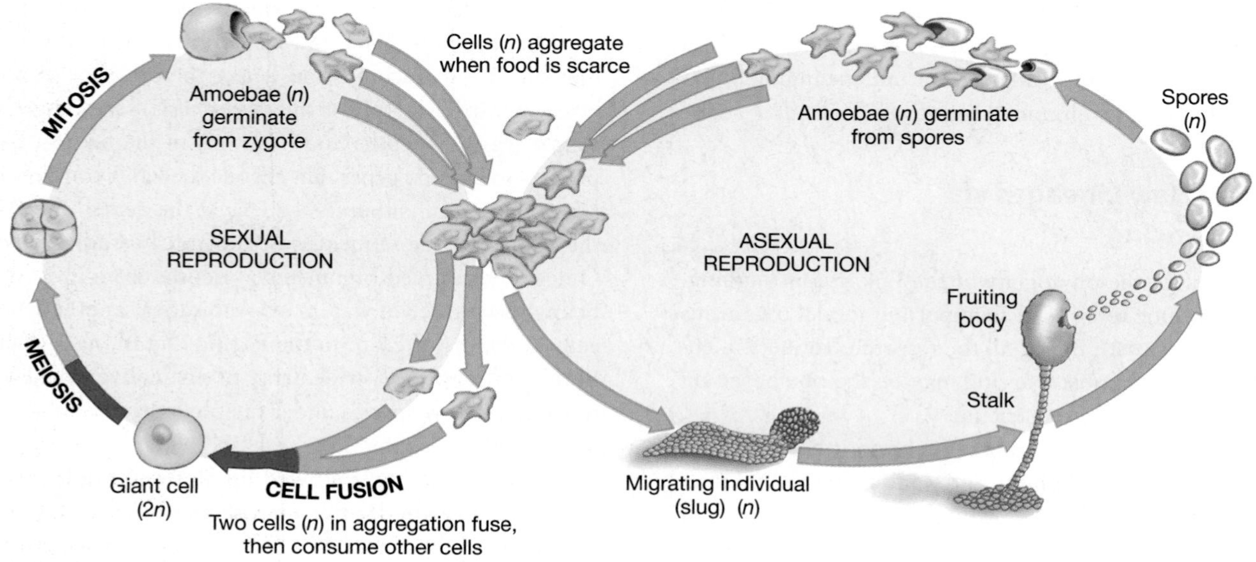

FIGURE 29.9 The Life Cycle of *Dictyostelium discoideum*.

QUESTION In slime molds, why isn't spore production considered a form of sexual reproduction?

QUESTION Most biologists consider the slime mold slug to be a simple form of multicellularity. Why?

diameter. The take-home message is that eukaryotic cells are much more variable in size than previously imagined. A whole new world of tiny protists has just been discovered.

The initial analyses suggest that at least some of the new species may represent important new lineages of dinoflagellates, stramenopiles, or alveolates. As additional direct sequencing studies are done, it is virtually certain that our understanding of eukaryotic diversity will change—perhaps radically—and improve.

Check Your Understanding

If you understand that...

- Biologists use data from microscopy and DNA sequencing to estimate phylogenetic trees and study the diversity of protists.
- According to the most recent analyses, the domain Eukarya comprises eight major lineages. Members of each lineage have distinctive aspects of cell structure.
- Direct sequencing has allowed investigators to recognize large numbers of previously undescribed eukaryotes, some of which are extremely small.

You should be able to...

1) Describe the synapomorphies that distinguish two of the eight major lineages of eukaryotes, and suggest how these traits might influence cell support or the ability to move and feed.
2) Outline the steps in a direct sequencing study that would allow you to characterize the protists present in the waters surrounding a coral reef.

29.3 What Themes Occur in the Diversification of Protists?

The protists range in size from bacteria-sized single cells to giant kelp. They live in habitats from the open oceans to dank forest floors. They are almost bewildering in their morphological and ecological diversity. Because they are a paraphyletic group, they do not share derived characteristics that set them apart from all other lineages on the tree of life.

Fortunately, one general theme helps tie protists together. Once an important new innovation arose in protists, it triggered the evolution of species that live in a wide array of habitats and make a living in diverse ways. Let's first consider novel morphological traits that arose in protists and then analyze new methods of feeding, moving, and reproducing that evolved in these lineages. Throughout, you should focus on understanding how each innovation helped make the fantastic diversity of protists possible.

What Morphological Innovations Evolved in Protists?

What did the earliest eukaryotes look like? Because the most ancient eukaryotic groups are unicellular, and because virtually all bacteria and all archaea are also unicellular, biologists infer that the first eukaryote was a single-celled organism. Further, all eukaryotes alive today have a nucleus and endomembrane system, mitochondria or genes that are normally found in mitochondria, and a cytoskeleton. Based on these observations, biologists conclude that their common ancestor also had these structures. And because the most ancient eukaryotic lineages lack cell walls, biologists suggest that the common ancestor of all eukaryotes living today also lacked this feature.

In sum, the earliest eukaryotes were probably single-celled organisms with a nucleus and endomembrane system, mitochondria, and a cytoskeleton, but no cell wall. It is also likely that these cells swam using a novel type of flagellum. Eukaryotic flagella are completely different structures from bacterial flagella and evolved independently. The flagella found in protists and other eukaryotes are made up of microtubules, and dynein is the major motor protein. An undulating motion occurs as dynein molecules walk down microtubules. The flagella of bacteria and archaea, in contrast, are composed primarily of a protein called flagellin (see Chapter 7). Instead of undulating, these flagella rotate to produce movement.

If you understand the synapomorphies that identify the eukaryotes as a monophyletic group, you should be able to map the origin of the nuclear envelope and the eukaryotic flagellum on Figure 29.8. Once you've done that, let's consider how several of these key new morphological features arose and influenced the subsequent diversification of protists, beginning with the trait that defines the Eukarya—the nuclear envelope.

The Nuclear Envelope Although researchers are still gathering evidence on how the nuclear envelope originated, the leading hypothesis is that it is derived from infoldings of the plasma membrane. As the drawings in **Figure 29.10** show, these infoldings are thought to have given rise to the nuclear envelope and the endoplasmic reticulum (ER) together. Two lines of evidence support this hypothesis: Infoldings of the plasma membrane occur in some bacteria living today, and the nuclear envelope and ER of today's eukaryotes are continuous (see Chapter 7). If you understand the infolding hypothesis, you should be able to explain why these observations support it.

According to current thinking, the evolution of the nuclear envelope was advantageous because it separated transcription and translation. RNA transcripts are processed inside the nucleus but translated outside the nucleus. In bacteria and archaea, transcription and translation occur together. Once a simple nuclear envelope was in place, alternative splicing and other forms of RNA processing could occur—giving the early

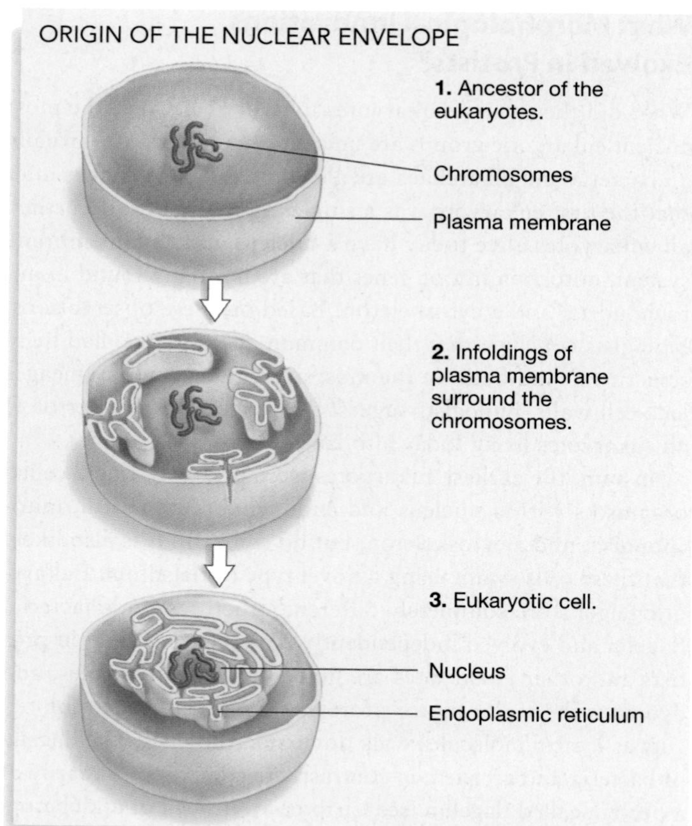

FIGURE 29.10 A Hypothesis for the Origin of the Nuclear Envelope. Infoldings of the plasma membrane, analogous to those shown here, have been observed in bacteria living today.

eukaryotes a novel way to control gene expression (see Chapter 18). The take-home message here is that an important morphological innovation gave the early eukaryotes a new way to manage and process genetic information.

Once a nucleus had evolved, it underwent diversification. In some cases, unique types of nuclei are associated with the founding of important lineages of protists. Ciliates, for example, have a diploid micronucleus that is involved only in reproduction and a polyploid macronucleus where transcription occurs. Diplomonads have two nuclei that look identical; it is not known how they interact. In foraminifera, red algae, and plasmodial slime molds, certain cells may contain many nuclei. Dinoflagellates have chromosomes that lack histones and attach to the nuclear envelope. In each of these groups the distinctive structure of the nucleus—and presumably, differences in how genetic information is processed—represent synapomorphies present in a common ancestor.

The Mitochondrion Mitochondria are organelles that generate ATP using pyruvate as an electron donor and oxygen as the ultimate electron acceptor (see Chapter 9). In 1981 Lynn Margulis expanded on a radical hypothesis—first proposed in the nineteenth century—to explain the origin of mitochondria. The **endosymbiosis theory** proposes that mitochondria originated when a bacterial cell took up residence inside a eukaryote about 2 billion years ago. Its name is inspired by the Greek word roots *endo*, *sym*, and *bio* (literally, "inside-together-living"). **Symbiosis** is said to occur when individuals of two different species live in physical contact; **endosymbiosis** occurs when an organism of one species lives inside an organism of another species.

In its current form, the endosymbiosis theory proposes that mitochondria evolved through a series of steps. As **Figure 29.11** shows, the process began when eukaryotic cells started to use their cytoskeletal elements to surround and engulf smaller prey.

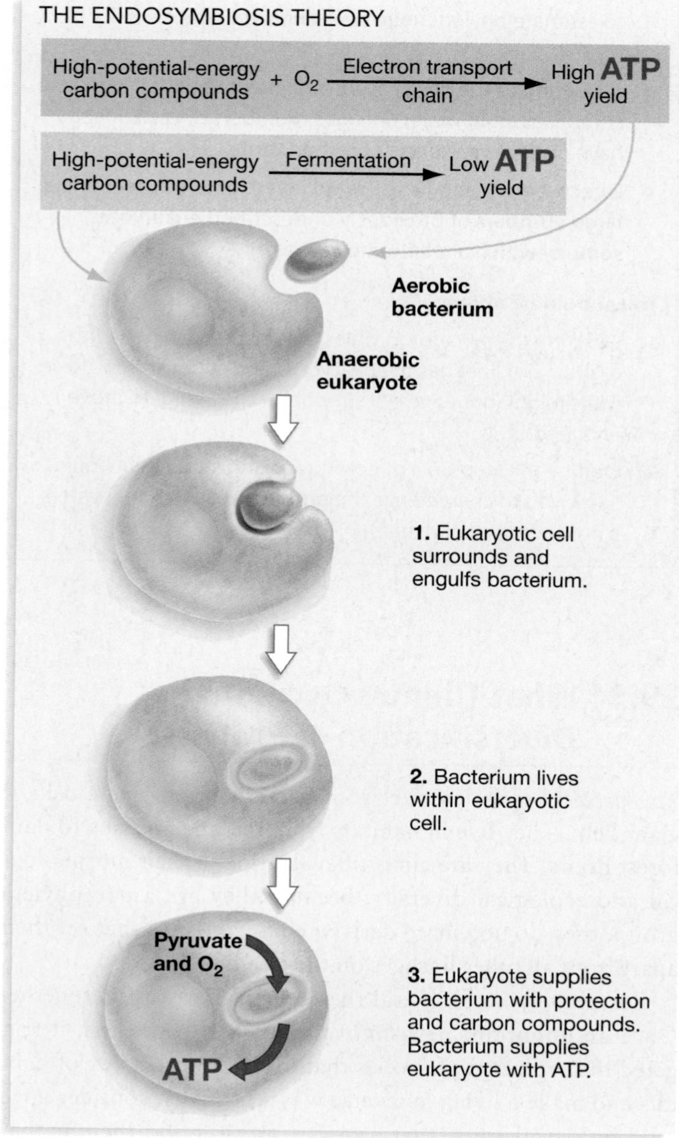

FIGURE 29.11 Proposed Initial Steps in the Evolution of the Mitochondrion.

QUESTION According to this hypothesis, how many membranes should surround a mitochondrion? Explain your logic.

The theory proposes that instead of being digested, an engulfed bacterium began to live inside its eukaryotic host. Specifically, the theory maintains that the engulfed cell survived by absorbing carbon molecules with high potential energy from its host and oxidizing them, using oxygen as a final electron acceptor. The host cell, in contrast, is proposed to be a predator capable only of anaerobic fermentation—meaning it could not use oxygen as an electron acceptor in cellular respiration. The relationship between the host and the engulfed cell was presumed to be stable because a mutual advantage existed between them: The host supplied the bacterium with protection and carbon compounds from its other prey, while the bacterium produced much more ATP than the host cell could synthesize on its own. Cells that can use oxygen during cellular respiration are able to produce much more ATP than are cells that cannot (see Chapter 28).

When Margulis first began promoting the theory, it met with a storm of criticism—largely because it seems slightly preposterous. But gradually biologists began to examine it rigorously. For example, endosymbiotic relationships between protists and bacteria exist today. Among the α-probacteria alone, three major groups are found *only* inside eukaryotic cells. Several observations about the structure of mitochondria are also consistent with the endosymbiosis theory:

- Mitochondria are about the size of an average bacterium and replicate by fission, as do bacterial cells. The duplication of mitochondria takes place independently of division by the host cell. When eukaryotic cells divide, each daughter cell receives some of the many mitochondria present.
- Mitochondria have their own ribosomes and manufacture their own proteins. Mitochondrial ribosomes closely resemble bacterial ribosomes in size and composition and are poisoned by antibiotics such as streptomycin that inhibit bacterial, but not eukaryotic, ribosomes.
- Mitochondria have double membranes, consistent with the engulfing mechanism of origin illustrated in Figure 29.11.
- Mitochondria have their own genomes, which are organized as circular molecules—much like a bacterial chromosome. Mitochondrial genes code for the enzymes needed to replicate and transcribe the mitochondrial genome.

Although these data are impressive, they are only consistent with the endosymbiosis theory. Stated another way, they do not exclude other explanations. This is a general principle in science: Evidence is considered strong when it cannot be explained by reasonable alternative hypotheses. In this case, the key was to find data that tested predictions made by Margulis's idea against predictions made by an alternative theory: that mitochondria evolved within eukaryotic cells, separately from bacteria.

A breakthrough occurred when researchers realized that according to the "within-eukaryotes" theory, the genes found in mitochondria had to have been derived from some of the nuclear genes of ancestral eukaryotes. Margulis's theory, in contrast, proposed that the genes found in mitochondria were bacterial in origin.

These predictions were tested by studies on the phylogenetic relationships of mitochondrial genes (**Figure 29.12**). For example, researchers compared gene sequences isolated from the

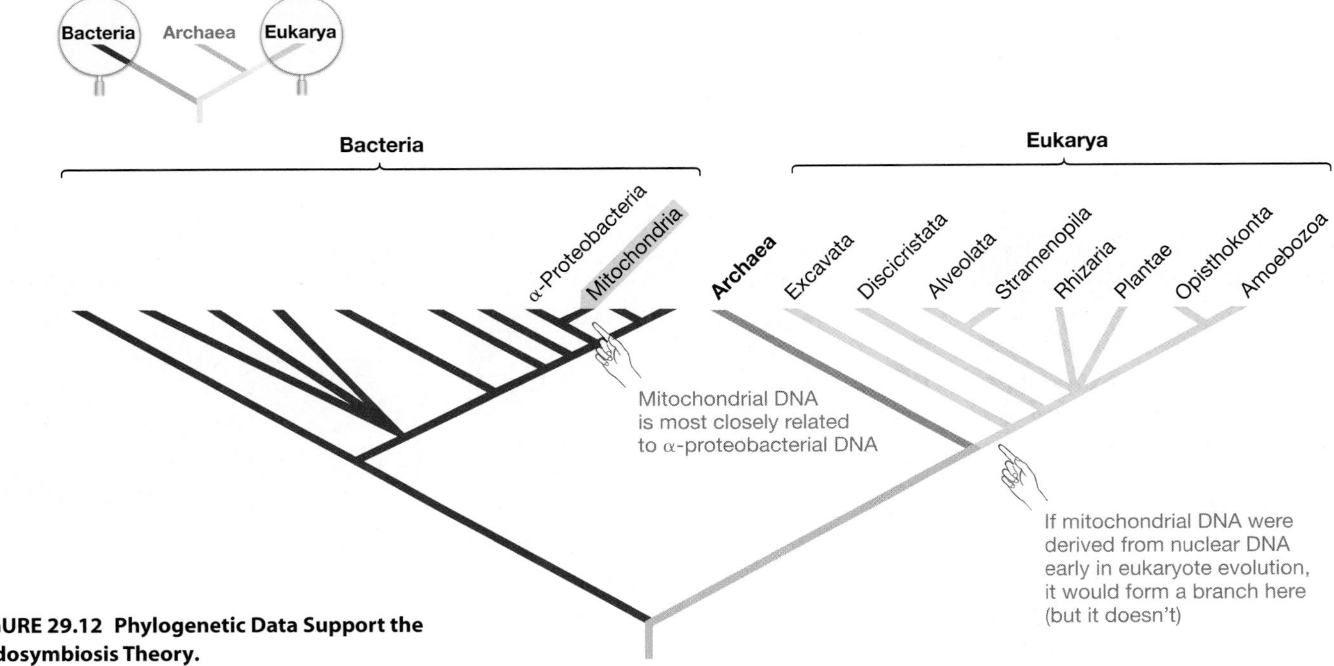

FIGURE 29.12 Phylogenetic Data Support the Endosymbiosis Theory.

nuclear DNA of eukaryotes, mitochondrial DNA from eukaryotes, and DNA from several species of bacteria. Exactly as the endosymbiosis theory predicted, the mitochondrial gene sequences turned out to be much more closely related to the sequences from the α-proteobacteria than to sequences from the nuclear DNA of eukaryotes. The result was considered overwhelming evidence that the mitochondrial genome came from an α-proteobacterium rather than from a eukaryote. The endosymbiosis theory was the only reasonable explanation for the data. The results were a stunning vindication of a theory that had once been intensely controversial. Mitochondria evolved via endosymbiosis. ● If you understand the endosymbiosis theory and the evidence for it, you should be able to describe how the chloroplast—the organelle where photosynthesis takes place in eukaryotes—could arise via endosymbiosis. You should also be able to list the types of evidence that would support an endosymbiotic origin for the chloroplast.

Once the mitochondrion was present in the ancestor of today's protists, it underwent diversification. In Diplomonads and Parabasalids, the organelle was lost entirely or is now a vestigial trait. In the Discicristata and Opisthokonta, the sac-like cristae inside the mitochondrion—where the electron transport chain and ATP synthase are located—have distinctive shapes (see Table 29.2). Presumably, variation in the presence or structure of mitochondria reflects variation in how ATP is produced by protists. In these cases, the mitochondrion's absence or structure qualifies as a synapomorphy that identifies a monophyletic group. The mitochondrion was a morphological innovation that subsequently diversified.

Structures for Support and Protection Bacterial and archaeal cells contain protein filaments that provide a rudimentary cytoskeleton. In contrast, the microfilaments, intermediate filaments, and microtubules found in eukaryotes form an extensive and dynamic internal skeleton (see Chapter 7). Understanding how these cytoskeletal elements evolved is a subject of current research. And as far as is known, the basic structure of the cytoskeleton does not vary much among protists. What does vary significantly is the presence and nature of other structures that provide support and protection for the cell. Many protists have cell walls outside their plasma membrane; others have hard external structures called a **test** or a **shell**; others have rigid structures inside the plasma membrane. In many cases, these novel structures represent synapomorphies that identify monophyletic groups among protists. For example:

- Diatoms are surrounded by a glass-like, silicon-oxide shell (**Figure 29.13a**). The shell is made up of two pieces that fit together in a box-and-lid arrangement, like the petri plates you may have seen in lab.
- Dinoflagelletes have a cell wall made up of cellulose plates (**Figure 29.13b**).
- Within Foraminifera, some lineages secrete an intricate, chambered test of calcium carbonate ($CaCO_3$; **Figure 29.13c**).
- Members of other Foraminifera lineages, and some amoebae, cover themselves with tiny pebbles.
- The parabasalids have a distinctive internal support rod, consisting of cross-linked microtubules that run the length of the cell.
- The euglenids have a collection of protein strips located just under the plasma membrane. The strips are supported by microtubules and stiffen the cell.
- The alveolates have distinctive sac-like structures called alveoli, located just under the plasma membrane, that help stiffen the cell.

In many cases, the diversification of protists has been associated with the evolution of innovative structures for support and protection.

(a) Diatom

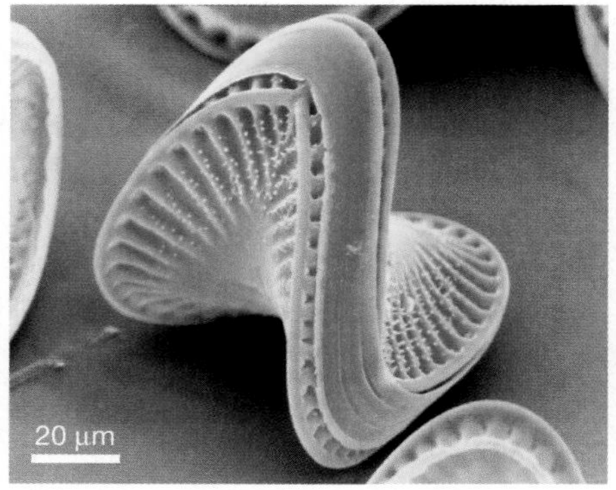

Test made of silicon oxides

(b) Dinoflagellate

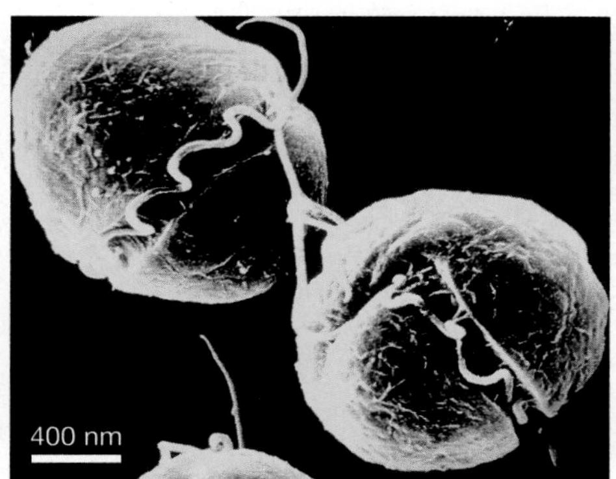

Plates made of cellulose

(c) Foraminiferan

Calcium carbonate test, with chambers

FIGURE 29.13 Hard Outer Coverings in Protists Vary in Composition.

Multicellularity The story of many protist lineages starts with the evolution of novel structures associated with managing genetic information, producing ATP, and supporting or protecting the cell. In other lineages of protists, though, the key morphological innovation was **multicellularity**. Multicellular individuals contain more than one cell. At least some of the cells are attached to each other, and some are specialized for different functions. In the simplest multicellular species, certain cells are specialized for producing or obtaining food while other cells are specialized for reproduction. The key point about multicellularity is that not all cells express the same genes.

A few species of bacteria are capable of aggregating and forming fruiting bodies similar to those observed in the protist *Dictyostelium discoideum* (see Box 29.2). Because cells in the fruiting bodies of these bacteria differentiate into specialized stalk cells and spore-forming cells, they are considered multicellular. But the vast majority of multicellular species are members of the Eukarya. Based on data presented in Figure 8.8, biologists are convinced that multicellularity evolved several times as protists diversified. The condition also evolved independently in the green plants, fungi, and animals. Multicellularity is a synapomorphy shared by all of the brown algae and all of the plasmodial and cellular slime molds. It also arose in some lineages of red algae.

To summarize, an array of novel morphological traits played a key role as protists diversified: the nucleus and endomembrane system, the mitochondrion, structures for protection and support, and multicellularity. Protists invented a variety of new ways to build and manage an individual. Once a new type of eukaryotic cell or multicellular individual existed, subsequent diversification was often triggered by novel ways of finding food, moving, or reproducing. Let's consider each of these life processes in turn.

How Do Protists Find Food?

According to Chapter 28, bacteria and archaea can use a wide array of molecules as electron donors and electron acceptors during cellular respiration. They get these molecules by absorbing them directly from the environment. Other bacteria don't absorb their nutrition—instead, they make their own food via photosynthesis.

Many groups of protists are similar to bacteria in the way they find food: They perform photosynthesis or absorb their food directly from the environment. But one of the most important stories in the diversification of protists was the evolution of a novel method for finding food. Many protists ingest their food—they eat bacteria, archaea, or even other protists whole. When ingestive feeding occurs, an individual takes in packets of food much larger than individual molecules. Thus, protists feed by (1) ingesting packets of food, (2) absorbing organic molecules directly from the environment, or (3) performing photosynthesis. Some ingest food as well as performing photosynthesis—meaning that they use a combination of feeding strategies.

Ingestive Feeding Ingestive lifestyles are based on eating live or dead organisms or on scavenging loose bits of organic debris. Protists such as the cellular slime mold *Dictyostelium discoideum* are large enough to engulf bacteria and archaea; many protists are large enough to surround and ingest other protists or microscopic animals. The engulfing process is possible in protists that lack a cell wall. A flexible membrane and dynamic cytoskeleton give these species the ability to surround and "swallow" prey using long, fingerlike projections called **pseudopodia** ("false-feet"). The engulfing process is illustrated in **Figure 29.14a**; the evolution of large cell size in eukaryotes is analyzed in **Box 29.3**.

(a) Pseudopodia engulf food

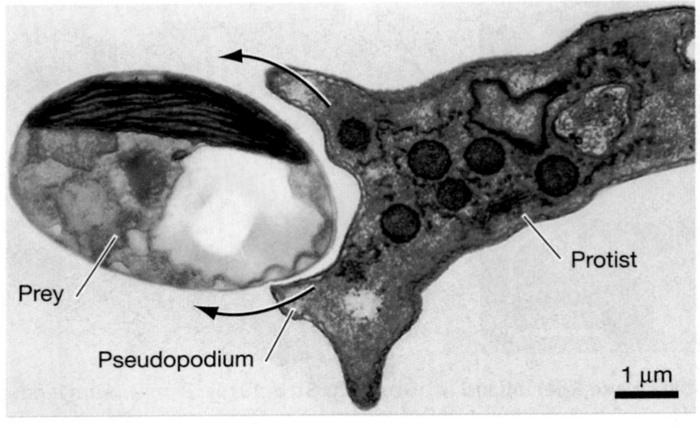

(b) Ciliary currents sweep food into gullet

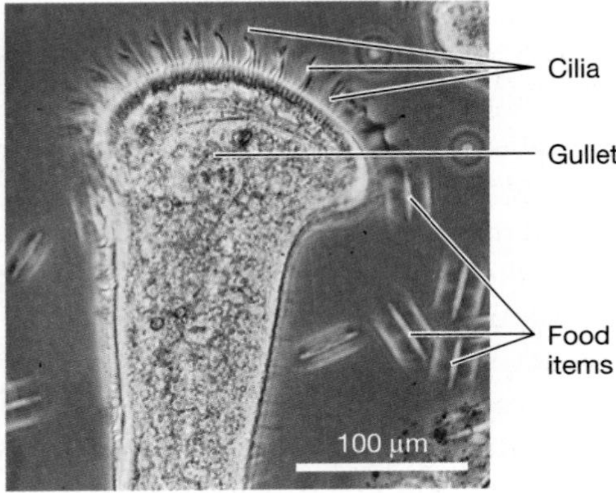

FIGURE 29.14 Ingestive Feeding. Methods of prey capture vary among ingestive protists. **(a)** Some predators engulf prey with pseudopodia; **(b)** other predators sweep them into their gullets with water currents set up by the beating of cilia.

Losing a cell wall and gaining the ability to move their plasma membrane around prey items or food particles was an important innovation during the evolution of protists. Recall from Chapter 28 that bacteria and archaea are abundant in wet soils and aquatic habitats, where protists also live. Instead of competing with bacteria and archaea for sunlight or food molecules, protists could eat them.

Although many ingestive feeders actively hunt down prey and engulf them, others do not. Instead of taking themselves to food, these species attach themselves to a surface. Protists that feed in this way have cilia that surround the mouth and beat in a coordinated way. The motion creates water currents that sweep food particles into the cell (**Figure 29.14b**).

Absorptive Feeding When nutrients are taken up directly from the environment, across the plasma membrane, absorptive feeding occurs. Absorptive feeding is common among protists. It's important to recognize, though, that protists are not nearly

BOX 29.3 Why Can Eukaryotes Have Such Large Cells?

Although eukaryotic cells vary widely in size, the domain Eukarya is distinguished by the evolution of the largest cells known. An average-sized eukaryotic cell is 10 times larger in diameter than an average-sized bacterial cell.

How could eukaryotic cells get so big? The answer to this question isn't obvious, because large cell size presents an important physical challenge: As cells become larger, their volume increases much more rapidly than does their surface area. This is because a cell's volume increases as the cube of a sphere's diameter (volume $\propto$ diameter^3) while its surface area increases as the square of a sphere's diameter (surface area $\propto$ diameter^2). As cells get larger, then, the proportion of surface area available gets smaller and smaller relative to the volume present (see Section 42.3). This creates a problem because food, gases, and waste molecules must diffuse across the cell's surface, while the volume is filled with biosynthetic machinery that requires raw materials and generates waste. As cell size increases, metabolism in the cell's interior can outstrip the transport and exchange processes at the surface—where raw materials enter and waste products leave.

Eukaryotic cells manage this dilemma in part because they are divided into membrane-bound compartments, each of which has a relatively high surface area and relatively low volume. As an example of how the compartmentalization of the eukaryotic cell works, consider the ciliate called *Paramecium* (**Figure 29.15**). This organism makes a living by eating bacteria, which it sweeps into an indentation known as the gullet and then into the cell mouth. After ingesting a bacterium, a *Paramecium* surrounds it with an internal membrane, forming a compartment called a **food vacuole**. Food vacuoles merge with membrane-bound structures called lysosomes, which hold digestive enzymes, and circulate around the cell. When the food has been digested and nutrients have diffused out of the food vacuole, the vacuole merges with the plasma membrane at a special organelle—the anal pore—and expels waste molecules. The food vacuoles provide a large area of internal membrane, allowing nutrients to be delivered efficiently throughout the volume of the cell and waste products to be expelled rapidly.

Although not all eukaryotic cells contain food vacuoles and anal pores, they all have many other internal compartments. Each of the internal structures introduced in Chapter 7—including the nucleus, lysosomes, peroxisomes, mitochondria, chloroplasts, central vacuole, Golgi apparatus, and rough and smooth endoplasmic reticulum (ER)—has a distinct function. Many are membrane bound and are devoted to the synthesis, transport, and distribution of molecules. These organelles were morphological innovations that made the evolution of large size possible. Large cell size was important because it made ingestive feeding possible. Without ingestive feeding, endosymbiosis and the evolution of mitochondria and chloroplasts never would have occurred.

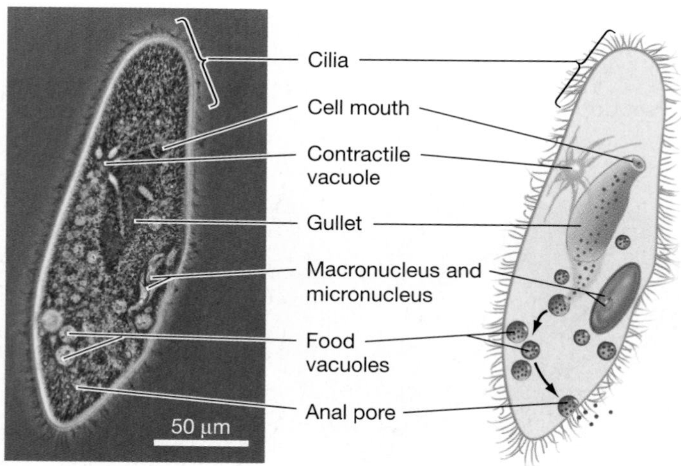

FIGURE 29.15 Protists Have Specialized Intracellular Structures. *Paramecium* feeds by sweeping bacteria into a subcellular structure called the gullet.

● **EXERCISE** Next to the drawing of *Paramecium,* list other organelles and intracellular structures that are found in protist cells.

as diverse as bacteria and archaea in terms of the electron donors and electron acceptors they absorb to use in cellular respiration. Most protists use glucose or other sugars as electron donors and oxygen as an electron acceptor. Unlike bacteria and archaea, which can ferment many different types of molecules, most protists can only ferment sugars.

Some protists that live by absorptive feeding are decomposers. A **decomposer** is an organism that feeds on dead organic matter, or **detritus**. But many of the protists that absorb their nutrition directly from the environment live inside other organisms. If they damage their host, the absorptive species is called a **parasite**. Most of the protists responsible for the diseases listed in Table 29.1 are parasites.

Photosynthesis You might recall from Chapter 10 and Chapter 28 that photosystems I and II evolved in bacteria, and that both photosystems occur in cyanobacteria. None of the basic machinery required for photosynthesis evolved in eukaryotes. Instead, they "stole" it via endosymbiosis.

The endosymbiosis theory contends that the organelle where photosynthesis takes place in eukaryotes originated when a protist engulfed a cyanobacterium. Once inside the protist, the photosynthetic bacterium provided its eukaryotic host with oxygen and glucose in exchange for protection and access to light. If the endosymbiosis theory is correct, today's chloroplasts trace their ancestry to cyanobacteria.

All of the photosynthetic eukaryotes have both a chloroplast and a mitochondrion, and the evidence for an endosymbiotic origin for the chloroplast is even more persuasive than for mitochondria. Chloroplasts have the same list of bacteria-like characteristics presented earlier for mitochondria. There are many examples of endosymbiotic cyanobacteria living inside protists or animals today, and the DNA sequences inside chloroplasts are extremely similar to cyanobacterial genes. In addition:

- The photosynthetic organelle of one group of protists, called the glaucophyte algae, has an outer layer containing the same constituent (peptidoglycan) found in the cell walls of cyanobacteria.
- Like chloroplasts, the cyanobacterium *Prochloron* contains chlorophylls *a* and *b* and has a system of internal membranes where the photosynthetic pigments and enzymes are located.

● If you understand the evidence for the endosymbiotic origin of the chloroplast, you should be able to explain why these latter two observations support the theory. You should also be able to add a branch indicating the location of chloroplast genes on the phylogenetic tree in Figure 29.12.

The acquisition of the chloroplast, like the acquisition of the mitochondrion, is called primary endosymbiosis. Because all of the species in the Plantae have chloroplasts with two membranes and similar molecular composition, biologists infer that primary endosymbiosis occurred in their common ancestor. Stated another way, the chloroplast is the synapomorphy that identifies the glaucophyte algae, red algae, and green plants (green algae and land plants) as part of the same monophyletic group.

The plot thickens, however, because chloroplasts also occur in three of the other major lineages of protists—the Discicristata, Chromalveolates, and Rhizaria. But in these species, the chloroplast is surrounded by more than two membranes—usually four. To explain this observation, researchers hypothesize that the ancestors of these groups acquired their chloroplasts by ingesting photosynthetic protists that already had chloroplasts. This process, called secondary endosymbiosis, occurs when an organism engulfs a photosynthetic eukaryotic cell and retains its chloroplasts as intracellular symbionts (**Figure 29.16**).

Figure 29.17 shows where primary and secondary endosymbiosis occurred on the phylogenetic tree of eukaryotes. Once protists obtained the chloroplast, it was "swapped around" to new lineages via secondary endosymbiosis. Many lineages of protists acquired the ability to photosynthesize independently of each other. In each case, the acquisition of a chloroplast triggered a radiation of photosynthetic species. ● If you under-

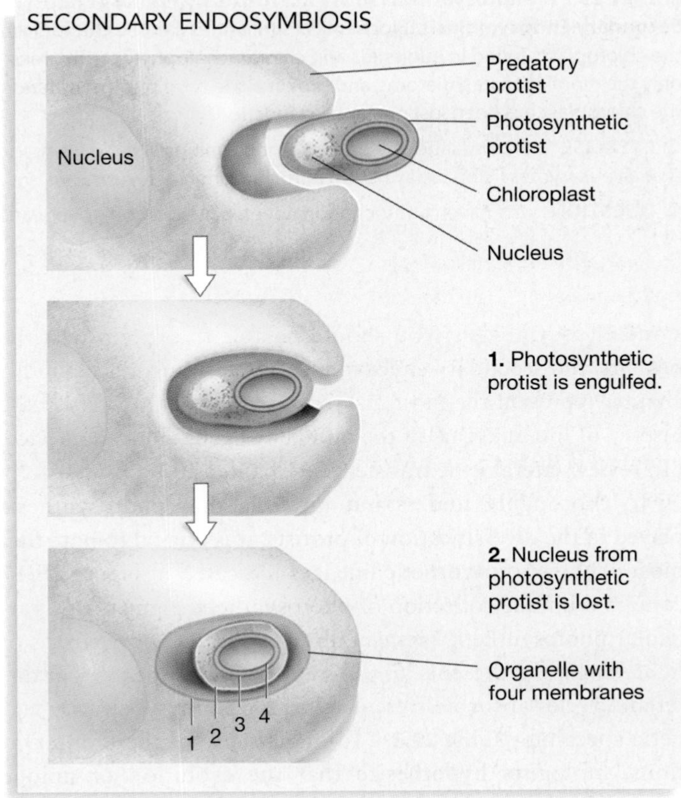

FIGURE 29.16 Secondary Endosymbiosis Leads to Organelles with Four Membranes. The chloroplasts found in some protists have four membranes and are hypothesized to be derived by secondary endosymbiosis. In species where chloroplasts have three membranes, biologists hypothesize that secondary endosymbiosis was followed by the loss of one membrane.

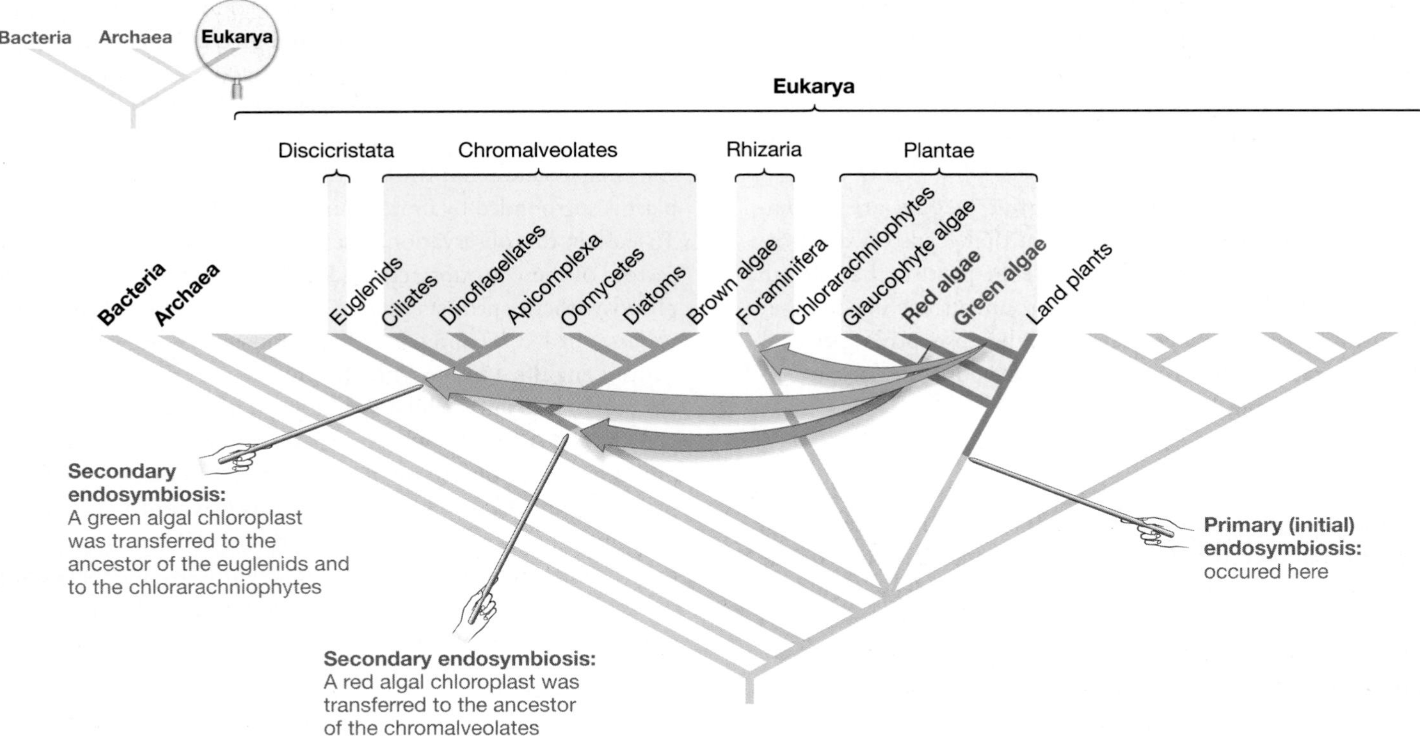

FIGURE 29.17 Photosynthesis Arose in Protists by Primary Endosymbiosis, Then Spread among Lineages via Secondary Endosymbiosis. Biochemical similarities link the chloroplasts found in chromalveolates with red algae and the chloroplasts found in euglenids and chlorarachniophytes with green algae. Note that among the chromalveolates, only the dinoflagellates, diatoms, and brown algae have photosynthetic species. In ciliates, apicomplexa, and oomycetes, the chloroplast has been lost or changed function.

● **EXERCISE** Mark and label the primary endosymbiosis that gave rise to the mitochondrion, based on the observation that all eukaryotes living today have a mitochondrion or a vestige of one.

● **QUESTION** Why haven't mitochondria been gained by secondary endosymbiosis?

stand endosymbiosis, you should be able to explain why the primary and secondary endosymbiosis events introduced in this chapter represent the most massive lateral gene transfers in the history of life, in terms of the number of genes moved at once. (To review lateral gene transfer, see Chapter 20.)

To thoroughly understand the role that photosynthesis played in the diversification of protists, it is critical to note that most of the photosynthetic lineages indicated in Figure 29.17 contain a unique collection of photosynthetic pigments. A particular photosynthetic pigment absorbs specific wavelengths of light. Thus, the presence of different pigments means that different species absorb different wavelengths in the electromagnetic spectrum (**Table 29.3**). To make sense of these observations, biologists hypothesize that the evolution of unique combinations of pigments gave photosynthetic eukaryotes the ability to harvest unique wavelengths of light and avoid competition. Chapter 28 pointed out that a similar phenomenon appears to have occurred among photosynthetic bacteria. Like the nuclear envelope, mitochondrion, multicellularity, ingestive feeding, and other traits that originated with protists, the acquisition of the chloroplast was an innovation that triggered subsequent diversification.

It is hard to overstate the importance of photosynthesis in the diversification of protists, and it is hard to overstate the importance of photosynthetic protists to other species. For example, many photosynthetic protists live symbiotically with animals or other protists—either inside or outside host cells. Although there are hundreds of examples, perhaps the most important involve the animals called corals. Virtually all reef-building corals harbor single-celled, photosynthetic protists in their skin or gut tissue. The concentrations of endosymbiotic cells can be impressive—up to a million protist cells per square centimeter of coral surface (**Figure 29.18**). The sheer abundance of these and other photosynthetic protists is significant. Plants may be the dominant producers of sugars and other high-energy organic compounds on land, but photosynthetic protists and bacteria are the dominant producers of carbon compounds in oceans and lakes.

Diversity in Lifestyles Before leaving the topic of obtaining food in protists, it's important to recognize that all three

TABLE 29.3 **Pigments in Photosynthetic Protists**

Lineage	Photosynthetic pigments	Accessory pigments
Glaucophytes	Chlorophyll *a*	Phycocyanin, allophycocyanin, β-carotene
Red algae	Chlorophyll *a*	Primarily phycoerythrin
Brown algae, diatoms, dinoflagellates	Chlorophyll *a* and *c*	Fucoxanthin
Green algae	Chlorophyll *a* and *b*	Fucoxanthin
Chlorarachniophytes	Chlorophyll *a* and *b*	Xanthophylls
Euglenids	Chlorophyll *a* and *b*	Xanthophylls, β-carotene

● **QUESTION** Are the pigments listed in this table consistent with the primary endosymbiosis and secondary endosymbiosis hypotheses described in Figure 29.17? Explain your answer.

lifestyles—ingestive, absorptive, and photosynthetic—occur in many different eukaryote lineages. All three types of food getting can occur within a single clade.

To drive this point home, consider the monophyletic group called the alveolates. There are three major subgroups of this lineage, called dinoflagellates, apicomplexa, and ciliates. Within each subgroup, species vary in how they make a living. About half of the dinoflagellates are photosynthetic, while many others are parasitic. Apicomplexa are parasitic. Ciliates include many species that ingest prey, but some ciliates live in the guts of cattle or the gills of fish and absorb nutrients from their hosts. Other ciliate species make a living by holding algae or other types of photosynthetic symbionts inside their cells.

FIGURE 29.18 Many Photosynthetic Protists Live Symbiotically. The coral pictured here contains thousands of symbiotic dinoflagellates.

Within each of the eight major lineages of eukaryotes, different methods for feeding helped trigger diversification.

How Do Protists Move?

Many protists actively move to find food. Predators such as *Dictyostelium discoideum* crawl over a substrate in search of prey. Most of the unicellular, photosynthetic species are capable of swimming to sunny locations. How are these crawling and swimming movements possible?

Amoeboid motion is a sliding movement observed in some protists. In the classic mode illustrated in **Figure 29.19a**, pseudopodia stream forward over a substrate, with the rest of the cytoplasm, organelles, and plasma membrane following. The motion requires ATP and involves interactions between proteins called actin and myosin inside the cytoplasm. The mechanism is related to muscle movement in animals, which is detailed in Chapter 47. But at the level of the whole cell, the precise sequence of events during amoeboid movement is still somewhat uncertain.

The other major mode of locomotion in protists involves flagella or cilia (**Figure 29.19b and c**). Both have identical structures. Recall from Chapter 7 that both consist of nine sets of doublet (paired) microtubules arranged around two central, single microtubules. Flagella, however, are long and are usually found alone or in pairs; cilia are short and usually occur in large numbers on any one cell. Flagella and cilia can also be distinguished by the types of structures associated with the basal bodies where they originate.

Even closely related protists can use radically different forms of locomotion. For example, consider again the lineage of protists called the alveolates and the subgroups known as ciliates, dinoflagellates, and apicomplexa. The ciliates swim

(a) Amoeboid motion via pseudopodia

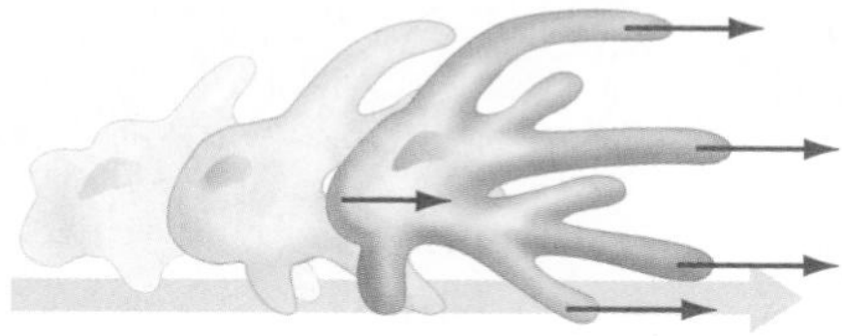

(b) Swimming via flagella

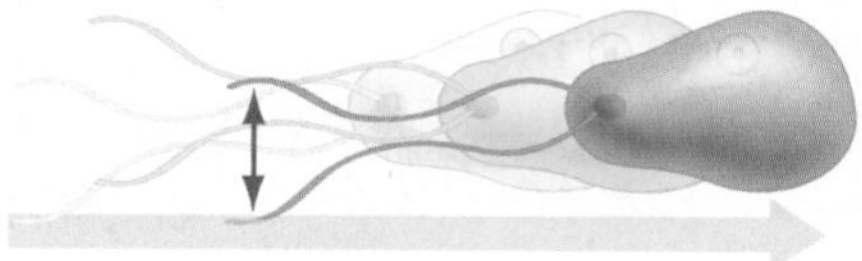

(c) Swimming via cilia

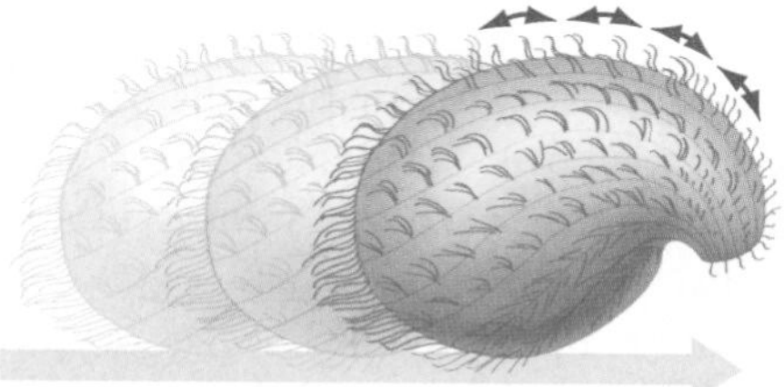

FIGURE 29.19 Modes of Locomotion in Protists Vary. (a) In amoeboid motion, long pseudopodia stream out from the cell. The rest of the cytoplasm, organelles, and external membrane follow. **(b)** Flagella are long and few in number, and they power swimming movements. **(c)** Cilia are short and numerous. In many cases they are used in swimming.

by beating their cilia. Dinoflagellates swim by whipping their flagella. Mature cells in apicomplexa move by amoeboid motion, but their gametes swim via flagella. Within the other major lineages of protists, it is common to observe variation in how cells move and common to find species that do not exhibit active movement but instead float passively in water currents. Movement is yet another example of the extensive diversification that occurred within each of the eight major eukaryote lineages.

How Do Protists Reproduce?

Sexual reproduction evolved in protists. As Chapter 12 pointed out, sexual reproduction can best be understood in contrast to asexual reproduction. Asexual reproduction is based on mitosis in eukaryotic organisms and on fission in bacteria and archaea, and it results in offspring that are genetically identical to the parent. Most protists undergo asexual reproduction routinely. When sexual reproduction occurs, in contrast, offspring are genetically different from their parents and from each other. Sexual reproduction is based on meiosis and fusion of gametes and occurs only intermittently in many protists—often at one particular time of year, or when individuals are crowded or food is scarce—and is based on meiosis. The evolution of sexual reproduction ranks among the most significant evolutionary innovations observed in eukaryotes.

Sexual versus Asexual Reproduction When reproduction occurs via meiosis, the resulting offspring are genetically variable—meaning their genotypes are different from those of other offspring and from those of their parents. The leading hypothesis to explain why meiosis evolved states that genetically variable offspring may be able to thrive if the environment changes. For example, offspring with genotypes different from those of their parents may be better able to withstand attacks by parasites that successfully attacked their parents (see Chapter 12). This is a key point because many types of parasites, including bacteria and viruses, have short generation times and evolve very quickly. Because the genotypes and phenotypes of parasites are constantly changing, natural selection is constantly favoring host individuals with new genotypes. The idea is that new offspring genotypes generated by meiosis may contain combinations of alleles that allow hosts to withstand attack by new strains of parasites. In short, many biologists view sexual reproduction as an adaptation to fight disease.

Variation in Life Cycles A life cycle describes the sequence of events that occur as individuals grow, mature, and reproduce. The evolution of meiosis introduced a new event in the life cycle of protist species; what's more, it created a distinction between haploid and diploid phases in the life of an individual. Recall from Chapter 12 that diploid individuals have two of each type of chromosome inside each cell, while haploid individuals have just one of each type of chromosome inside each cell. When meiosis occurs in diploid cells, it results in the production of haploid cells.

The life cycle of most bacteria and archaea is extremely simple: A cell divides, feeds, grows, and divides again. Bacteria and archaea do not undergo sexual reproduction, and they are always haploid. In contrast, among protists virtually every aspect of a life cycle is variable—whether meiosis occurs, whether asexual reproduction occurs, and whether the haploid or the diploid phase of the life cycle is the longer and more prominent phase.

Figure 29.20 illustrates some of the wide variation that occurs in life cycles of unicellular protists. Figure 29.20a depicts the haploid-dominated life cycle observed in many unicellular protists. The specific example given here is the dinoflagellate *Gymnodinium fuscum*. To analyze a life cycle, start with **fertilization**—the fusion of two gametes to form a diploid zygote. Then trace what happens to the zygote. In this case, the diploid zygote undergoes meiosis. The haploid products

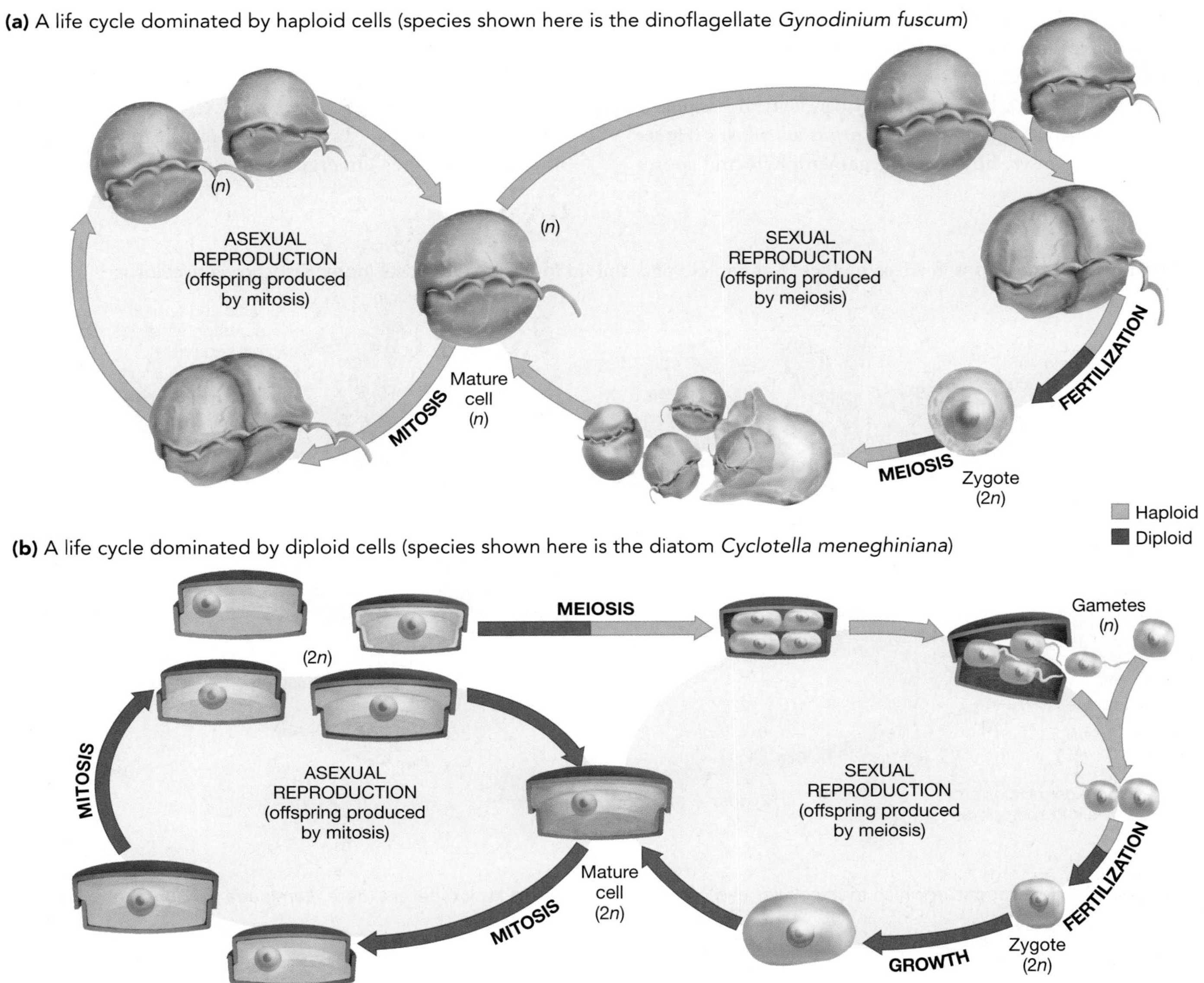

FIGURE 29.20 Life Cycles Vary Widely among Protists. Many protists can reproduce by both asexual reproduction and sexual reproduction. In unicellular species, the cell may be **(a)** haploid for most of its life or **(b)** diploid for most of its life.

of meiosis then grow into mature cells that eventually undergo asexual reproduction or produce gametes by mitosis. Contrast that cycle with the diploid-dominated life cycle of Figure 29.20b. The specific organism shown here is the diatom *Cyclotella meneghiniana*. To analyze this life cycle, start again with fertilization. In this case, the diploid zygote develops into a sexually mature, diploid adult cell. Meiosis occurs in the adult and results in the formation of haploid gametes, which then fuse to form a diploid zygote. The important contrasts are that meiosis occurs in the adult cell rather than in the zygote and that gametes are the only haploid cells in the life cycle.

In contrast to the relatively simple life cycles of Figure 29.20, many multicellular protists have one phase in their life cycle that is based on a multicellular haploid form and another phase that is based on a multicellular diploid form. This alternation of multicellular haploid and diploid forms is known as **alternation of generations**. The multicellular haploid form is called a **gametophyte**, because specialized cells in this individual produce gametes by mitosis. The multicellular diploid form is called a **sporophyte**, because it has specialized cells that undergo meiosis to produce haploid cells called spores. A **spore** is a single cell that develops into an adult organism but is not a product of fusion by gametes. When alternation of generation occurs, a spore divides by mitosis to form a haploid, multicellular gametophyte. The haploid gametes produced by the gametophyte then fuse to form a diploid zygote, which grows into the diploid, multicellular sporophyte. Among the protists, alternation of generations evolved independently in brown algae, red algae, and other groups.

Gametophytes and sporophytes may be identical in appearance, as in the brown alga called *Ectocarpus siliculosus* (**Figure 29.21a**). In many cases, however, the gametophyte and sporophyte look different, as in the brown alga called *Laminaria solidungula* (**Figure 29.21b**).

To get a better understanding of alternation of generations, compare and contrast the events described in Figure 29.20a and Figure 29.21a, starting with fertilization. Specifically, notice that zygotes undergoing alternation of generations do

(a) Alternation of generations in which multicellular haploid and diploid forms look identical (here, *Ectocarpus siliculosus*)

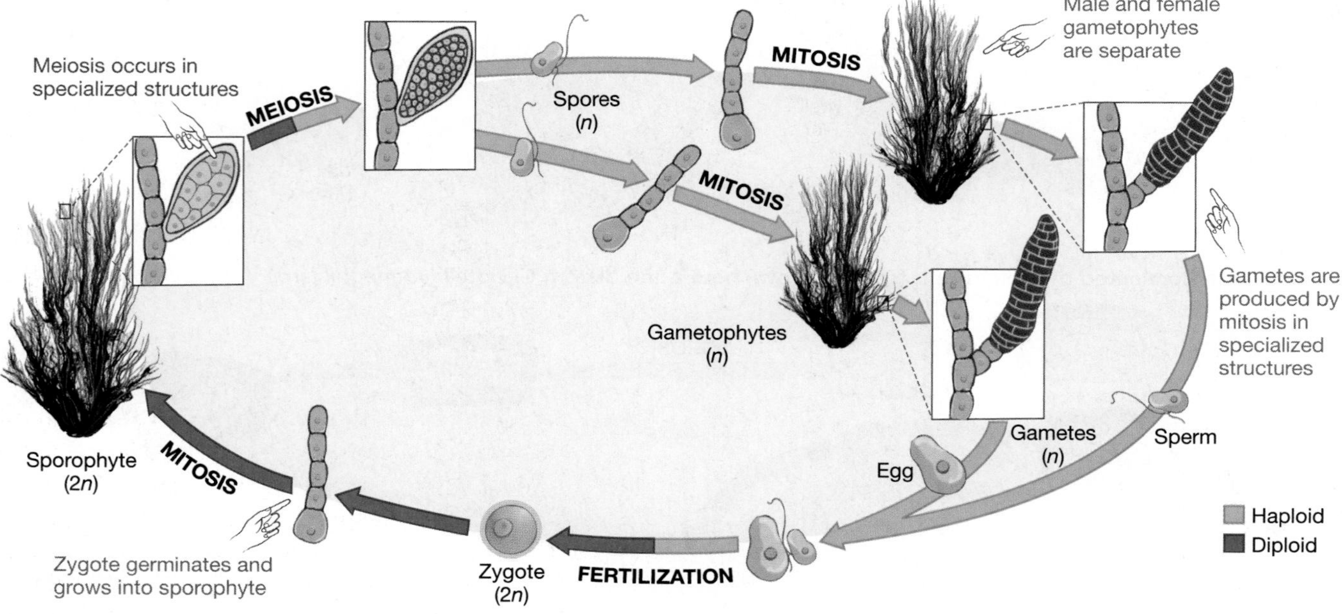

(b) Alternation of generations in which multicellular haploid and diploid forms look different (here, *Laminaria solidungula*)

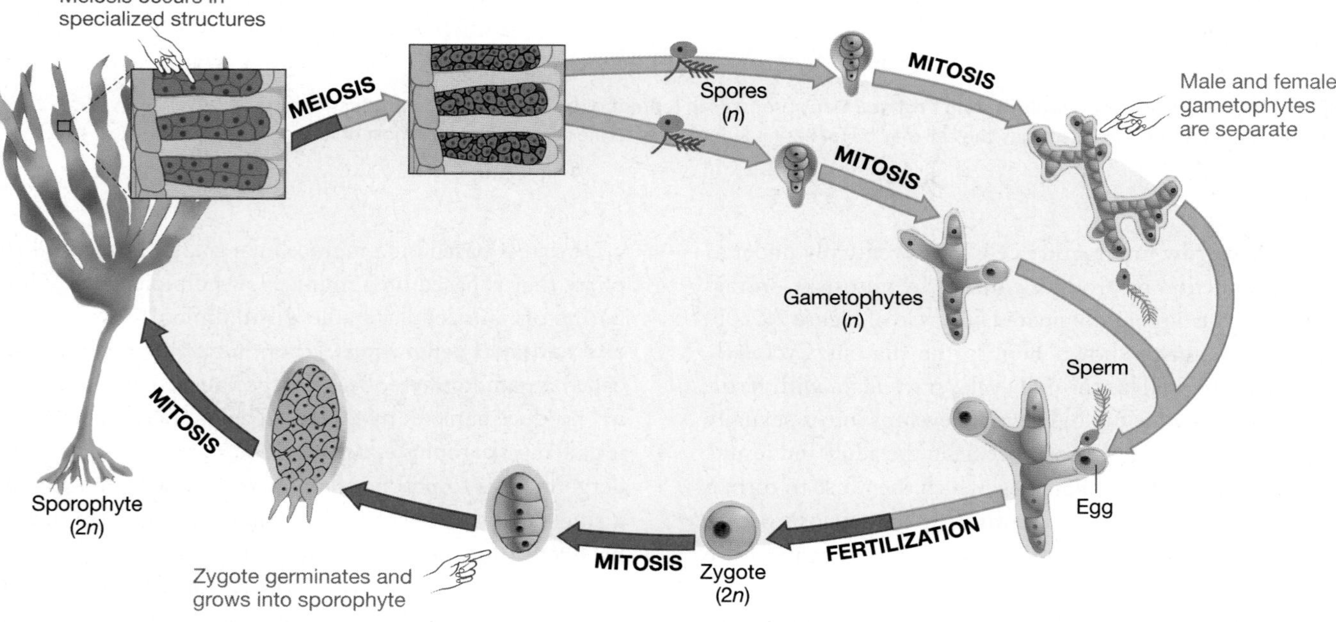

FIGURE 29.21 Alternation of Generations Occurs in an Array of Multicellular Protist Lineages.

not undergo meiosis immediately. Instead they germinate into diploid individuals that function as sporophytes. Cells in the sporophyte then undergo meiosis to form haploid spores, which germinate and grow into haploid gametophytes.

● If you understand how life cycles vary among multicellular protists, you should be able to define the terms *alternation of generations*, *gametophyte*, *sporophyte*, and *spore*. You should also be able to diagram a life cycle where alternation of generation occurs, without looking at Figure 29.21.

Why does so much variation occur in the types of life cycles observed among protists? The answer is not known. Variation in life cycles is a major theme in the diversification of protists. Explaining why that variation exists remains a topic for future research.

Check Your Understanding

If you understand that...

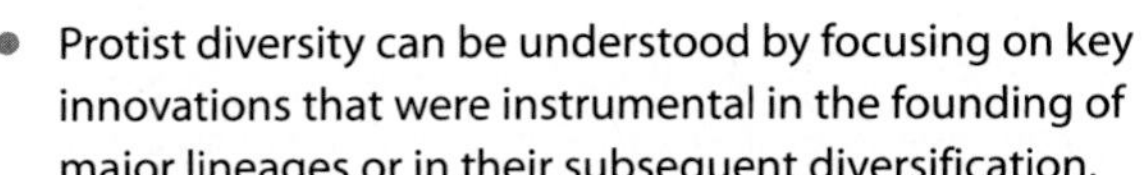

- Protist diversity can be understood by focusing on key innovations that were instrumental in the founding of major lineages or in their subsequent diversification.
- Important morphological innovations include the nucleus and endomembrane system, the mitochondrion, novel structures that provide protection or support, and multicellularity.
- Protists obtain chemical energy and building-block molecules via ingestive feeding, photosynthesis, or absorptive feeding (often, as parasites).
- Protists move via amoeboid movement, flagella, or cilia.
- Meiosis and sexual reproduction were crucial evolutionary innovations that originated in protists.

You should be able to...

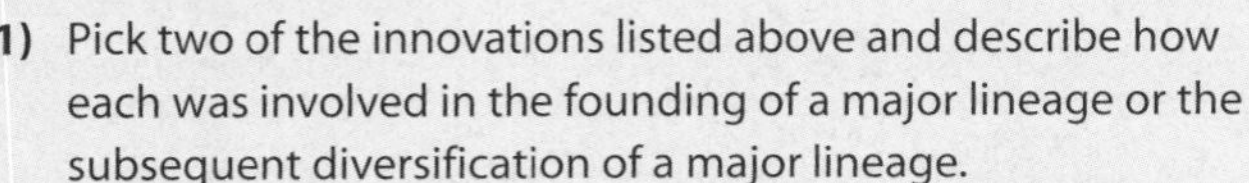

1) Pick two of the innovations listed above and describe how each was involved in the founding of a major lineage or the subsequent diversification of a major lineage.
2) Explain why organelles made the evolution of large cell size possible.
3) Provide evidence that traits such as multicellularity and photosynthesis arose repeatedly over the course of eukaryote evolution.

(MB) **Web Animation** at www.masteringbio.com
Alteration of Generations in a Protist

29.4 Key Lineages of Protists

An important generalization jumps out of the Eukarya phylogenetic tree and data on variation among protists in morphology, feeding method, locomotion, and mode of reproduction: Each of the eight major Eukarya lineages has at least one distinctive morphological characteristic. But once an ancestor evolved a set of distinctive characteristics, its descendants diversified into a wide array of lifestyles. For example, parasitic species evolved independently in all eight major lineages. Photosynthetic species exist in most of the eight, and multicellularity evolved independently in four. Similar statements could be made about the evolution of life cycles and modes of locomotion.

In effect, each of the eight lineages represents a similar radiation of species into a wide array of lifestyles. In each case, the radiation began with a morphological innovation—often a change in the structure or function of one or more organelles. Let's take a more detailed look at some representative taxa from seven of the eight major groups. The eighth group—the opisthokonts—includes fungi, animals, and several protist lineages. The fungi and animals are discussed in Chapters 31 through 34. The lineage called green plants is analyzed in Chapter 30.

Excavata

The unicellular species that form the Excavata are named for the morphological feature that distinguishes them—an "excavated" feeding groove found on one side of the cell. Because all excavates lack mitochondria, they were once thought to trace their ancestry to eukaryotes that existed prior to the origin of mitochondria. But researchers have found that excavates either have nuclear genomes containing genes that are normally found in mitochondria or unusual organelles that appear to be vestigial mitochondria. These observations support the hypothesis that the ancestors of excavates had mitochondria, but that these organelles were lost or reduced over time. There are several major lineages of Excavata, two of which are detailed here (**Figure 29.22**). ● You should be able to mark the loss of the mitochondrion on Figure 29.22.

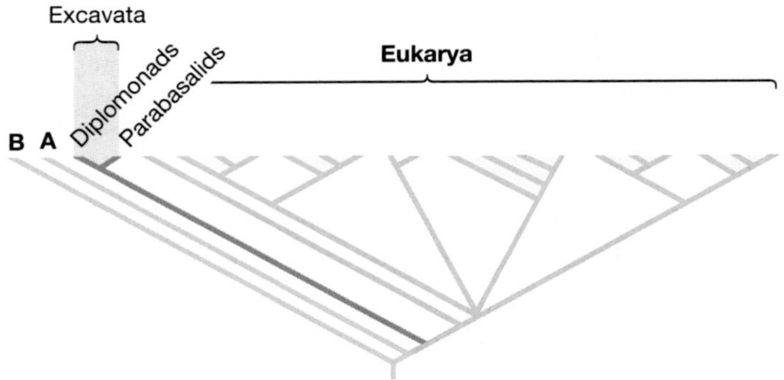

FIGURE 29.22 Excavates Are a Monophyletic Group.

Excavata > Diplomonadida

The first species of diplomonad known was described by the early microscopist Anton van Leeuwenhoek, who found *Giardia intestinalis* while examining samples of his own feces. About 100 species have been named to date. Many live in the guts of animal species without causing harm to their host; other species live in stagnant water habitats. In both types of environment, oxygen availability tends to be very low.

Morphology Each cell has two nuclei, which resemble eyes when viewed under the microscope (**Figure 29.23**). Each nucleus is associated with four flagella, for a total of eight flagella per cell. Some species lack the organelles called peroxisomes and lysosomes in addition to lacking functional mitochondria. All diplomonads lack a cell wall. ● You should be able to mark the origin of the double nucleus on Figure 29.22.

Feeding and locomotion Some are parasitic, but most ingest bacteria whole. They swim using their flagella.

Reproduction Only asexual reproduction occurs; meiosis has yet to be observed in members of this lineage.

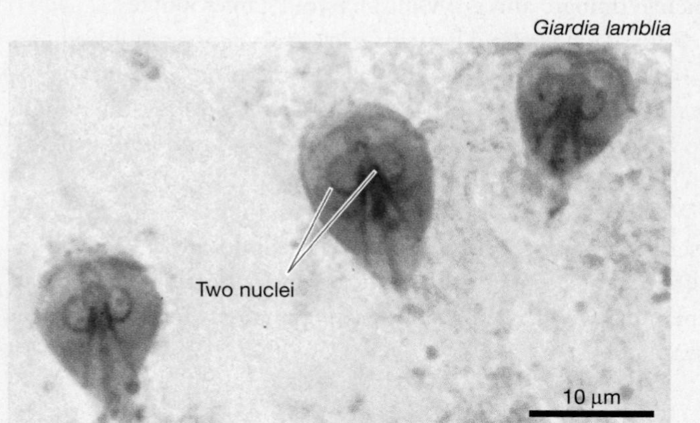

FIGURE 29.23 ***Giardia*** **Causes Intestinal Infections in Humans.**

Human and ecological impacts Both *Giardia intestinalis* and *G. lamblia* are common intestinal parasites in humans and cause giardiasis, or beaver fever. *Hexamida* is an agricultural pest that causes heavy losses in turkey farms each year.

Excavata > Parabasalida

No free-living parabasalids are known; all of the species described to date live inside animals. Of the 300 species of parabasalid, several live only in the guts of termites. Termites eat wood but cannot digest it themselves. Instead, the parabasalids produce enzymes that digest the cellulose in wood and release compounds that can be used by the termite host. The relationship between termites and parabasalids is considered mutualistic, because both parties benefit from the symbiosis.

Morphology Parabasalid cells lack a cell wall and mitochondria, and they have a single nucleus. A distinctive rod of cross-linked microtubules runs the length of the cell. The rod is attached to the basal bodies where a cluster of flagella arise (**Figure 29.24**). Although the number of flagella present varies widely, four or five is typical. ● You should be able to mark the origin of the parabasalid structure for supporting flagella on Figure 29.22.

Feeding and locomotion Parabasalids feed by engulfing bacteria, archaea, and organic matter. They swim using their flagella.

Reproduction All parabasalids reproduce asexually. Sexual reproduction also has been observed in a few species.

Human and ecological impacts *Trichomonas* infections can sometimes cause reproductive tract problems in humans, although members of this genus may also live in the gut or mouth of humans without causing harm. *Histomonas meleagridis* is an agricultural pest that causes disease outbreaks on chicken and turkey farms.

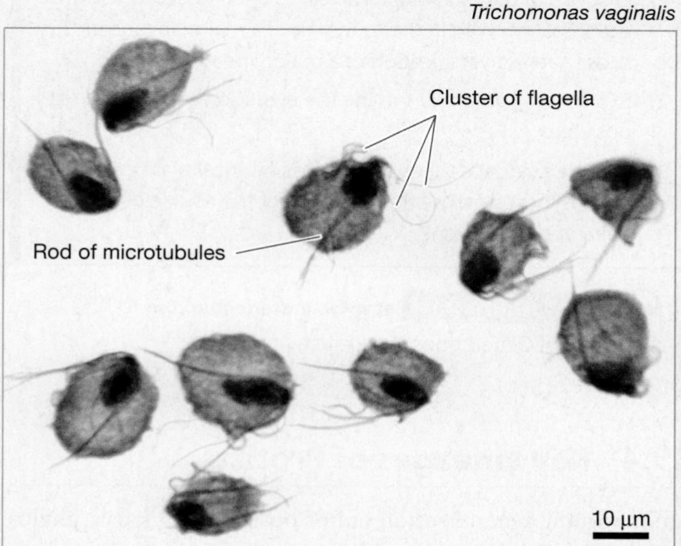

FIGURE 29.24 ***Trichomonas*** **Causes the Sexually Transmitted Disease Trichomoniasis.**

Discicristata

The discicristates are all unicellular and were named for the distinctive disc shape of the cristae within their mitochondria. Several major subgroups have been identified, such as amoeboid forms and lineages that include the species responsible for the diseases leishmaniasis and trypanosomiasis in humans (see Table 29.1). Because they are extremely common in ponds and lakes, the euglenids are the best-studied discicristates. ● You should be able to mark the origin of disc-shaped cristae in mitochondria on **Figure 29.25**.

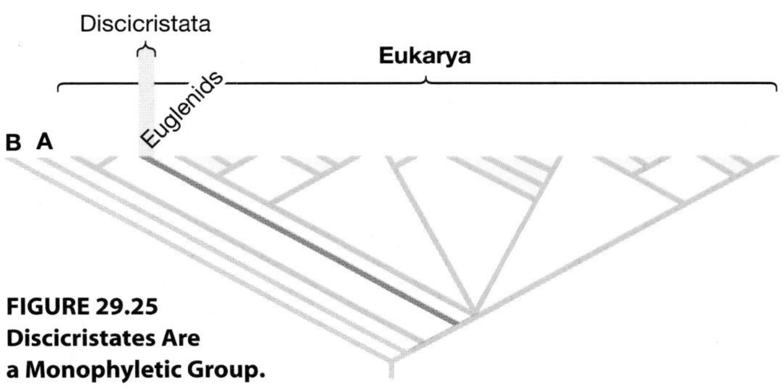

FIGURE 29.25 Discicristates Are a Monophyletic Group.

Discicristata > Euglenida

There are about 1000 known species of euglenid. Although most live in freshwater, a few are found in marine habitats. Fossil euglenids have been found in rocks over 410 million years old.

Morphology Most euglenids lack an external wall but have a unique system of interlocking protein molecules lying under the plasma membrane, which stiffen and support the cell.

Feeding and locomotion About one-third of the species have chloroplasts and perform photosynthesis, but most ingest bacteria and other small cells or particles. ● You should be able to explain how the observation of ingestive feeding in euglenids relates to the hypothesis that this lineage gained chloroplasts via secondary endosymbiosis. Instead of storing starch, the photosynthetic species synthesize a unique storage carbohydrate called paramylon. Some cells have a light-sensitive "eyespot" and use flagella to swim toward light (**Figure 29.26**). Other euglenids can inch along substrates using a unique form of movement based on extension of microtubules.

Reproduction Only asexual reproduction is known to occur among euglenids.

Human and ecological impacts Euglenids are important components of freshwater plankton and food chains.

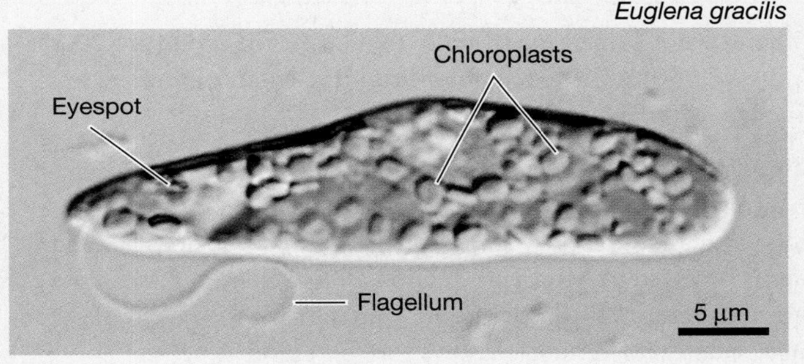

FIGURE 29.26 Euglenids Are Common in Ponds and Lakes.

Alveolata

Alveolates are distinguished by small sacs, called alveoli, that are located just under their plasma membranes. Although all members of this lineage are unicellular, the groups highlighted here—the Ciliata, the Dinoflagellata, and the Apicomplexa—are remarkably diverse in morphology and lifestyle. ● You should be able to mark the origin of alveoli on **Figure 29.27**.

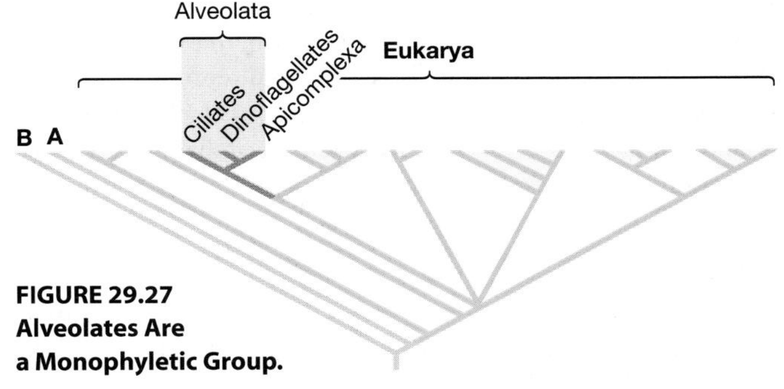

FIGURE 29.27 Alveolates Are a Monophyletic Group.

Alveolata > Apicomplexa

All of the 5000 known species of Apicomplexa are parasitic. Species in the genus *Plasmodium* are well studied because they cause malaria in humans and other vertebrates.

Morphology Apicomplexa cells have a system of organelles at one end, called the apical complex, that is unique to the group. The apical complex allows the apicomplexan to penetrate the plasma membrane of its host. ● You should be able to mark the origin of the apical complex on Figure 29.27. The cells have a chloroplast-derived, nonphotosynthetic organelle with four membranes, indicating that apicomplexans descended from an ancestor that gained a chloroplast via secondary endosymbiosis.

(continued on next page)

Alveolata > Apicomplexa *continued*

Feeding and locomotion All absorb nutrition directly from their host. They lack cilia or flagella, but some species can move by amoeboid motion.

Reproduction Apicomplexans can reproduce sexually or asexually. In some species, the life cycle involves two distinct hosts, and cells must be transmitted from one host to the next.

Human and ecological impacts Various species in the genus *Plasmodium* infect birds, reptiles, or mammals and cause malaria. *Toxoplasma* is an important pathogen in people infected with HIV (**Figure 29.28**), and chicken farmers estimate that they lose $600 million per year controlling the apicomplexan *Eimeria*. Apicomplexans that parasitize insects have been used to control pests.

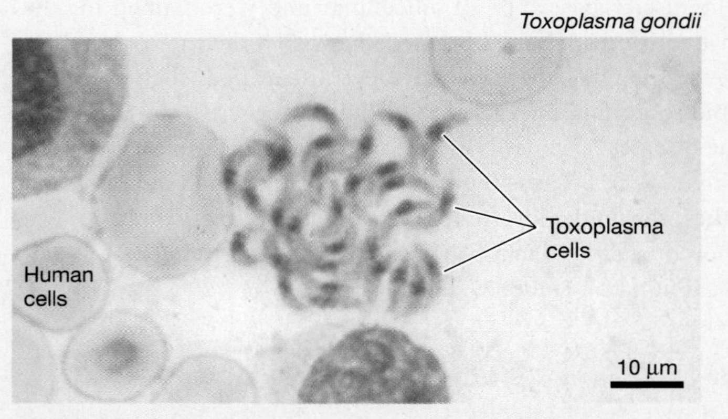

FIGURE 29.28 *Toxoplasma* Causes Infections in AIDS patients.

Alveolata > Ciliata

Ciliates were named for the cilia that cover them (**Figure 29.29**). Some 12,000 species are known from freshwater habitats, marine environments, and wet soils.

Morphology Ciliata cells have two distinctive nuclei: a large macronucleus and a small micronucleus. The macronucleus is polyploid and is actively transcribed. The micronucleus is diploid and is involved only in reproduction. A few ciliates secrete an external skeleton. ● You should be able to mark the origin of the macronucleus/micronucleus structure on Figure 29.27.

Feeding and locomotion Depending on the species, ciliates may be filter feeders, predators, or parasites. They use cilia to swim and have a mouth area where food is ingested.

Reproduction Ciliates divide to produce daughter cells asexually. They also undergo an unusual type of sexual reproduction called conjugation. During conjugation in ciliates, two cells line up side by side and physically connect. Micronuclei exchange between cells and fuse. The resulting nucleus eventually forms a new macronucleus and micronucleus.

Human and ecological impacts Ciliates are abundant in marine plankton and are important consumers. They are common in the digestive tracts of goats, sheep, cattle, and other grazers, where they feed on plant matter and help the host animal digest it.

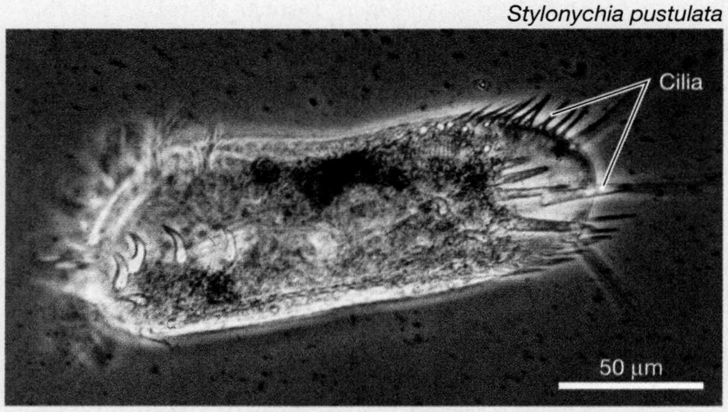

Stylonychia pustulata

Cilia

50 μm

FIGURE 29.29 Ciliates Are Abundant in Freshwater Plankton.

Alveolata > Dinoflagellata

Most of the 4000 known species of dinoflagellates are ocean-dwelling plankton, although they are abundant in freshwater as well. Some species are capable of **bioluminescence**, meaning they emit light via an enzyme-catalyzed reaction (**Figure 29.30**).

Morphology Most dinoflagellates are unicellular, although some live in the aggregations of individual cells called colonies. Each species has a distinct shape maintained by plates of cellulose inside the cell. Unlike chromosomes of other eukaryotes, chromosomes in dinoflagellates are attached to the nuclear envelope at all times and do not contain histones.

Feeding and locomotion About half of the dinoflagellates are photosynthetic. Some of the photosynthetic species live in association with corals or sea anemones, but most are planktonic. The other species are predatory or parasitic. Dinoflagellates are distinguished by the arrangement of their two flagella: One flagel-

(continued on next page)

Alveolata > Dinoflagellata *continued*

lum projects out from the cell while the other runs across the cell in a groove (see Figure 29.20a). The two flagella are perpendicular to each other. Cells swim in a spinning motion using the two flagella. ● You should be able to mark the origin of perpendicularly oriented flagella on Figure 29.27.

Reproduction Both asexual and sexual reproduction occur. Cells that result from sexual reproduction may form tough cysts. A cyst is a resistant structure that can remain dormant until environmental conditions improve.

Human and ecological impacts Photosynthetic dinoflagellates are important primary producers in marine ecosystems—second only to diatoms in the amount of carbon they fix per year. A few species are responsible for harmful algal blooms, or red tides (see Figure 29.4).

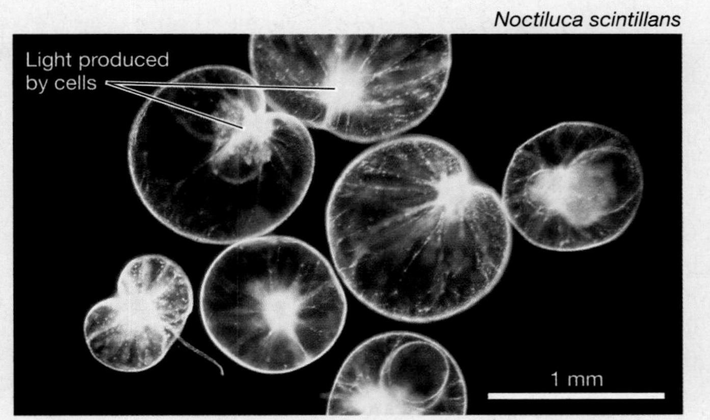

FIGURE 29.30 Some Dinoflagellates Species Are Bioluminescent.

Stramenopila (Heterokonta)

Stramenopiles are sometimes called *heterokonts*, which translates as "different hairs." At some stage of their life cycle, all stramenopiles have flagella that are covered with distinctive hollow "hairs." The structure of these flagella is unique to the stramenopiles (**Figure 29.31**). ● You should be able to mark the origin of the hairy flagellum on Figure 29.31. The lineage includes a large number of unicellular forms, although the brown algae are multicellular and include the world's tallest marine organisms, the kelp.

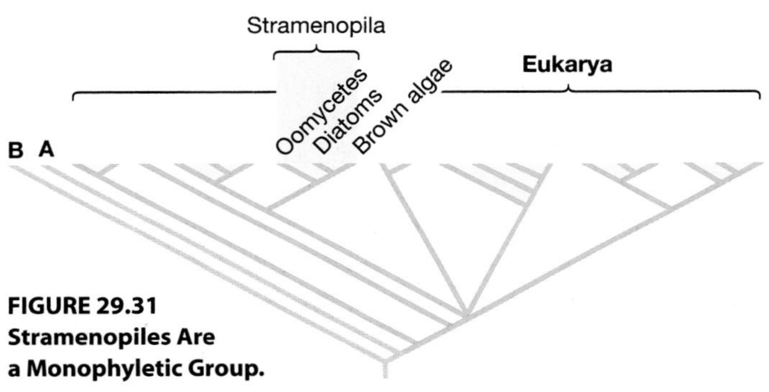

FIGURE 29.31 Stramenopiles Are a Monophyletic Group.

Stramenopila > Diatoms

Diatoms are unicellular or form chains of cells. About 10,000 species have been described to date, but some researchers claim that millions of species exist and have yet to be discovered.

Morphology Diatom cells are supported by external, silicon-rich, glassy shells that form a box-and-lid arrangement (**Figure 29.32** and Figure 29.20b). ● You should be able to mark the origin of the glassy test on Figure 29.31.

Feeding and locomotion Diatoms are photosynthetic. Only sperm cells have flagella and are capable of powered movement. In many species, the adult cells float in the water. But species living on a surface can glide via microtubules that project from their shells and that move in response to motor proteins.

Reproduction Diatoms divide by mitosis to reproduce asexually or by meiosis to form gametes that fuse to form a new individual. Many species can produce spores that are dormant during unfavorable growing conditions.

Human and ecological impacts In abundance, diatoms dominate the plankton of cold, nutrient-rich waters. They are found in virtually all aquatic habitats and are considered the most important producer of carbon compounds in fresh and salt water. Their shells settle into massive accumulations that are mined and sold commercially as diatomaceous earth, which is used in filtering applications and as an ingredient that adds bulk to polishes, paint, cosmetics, and other products.

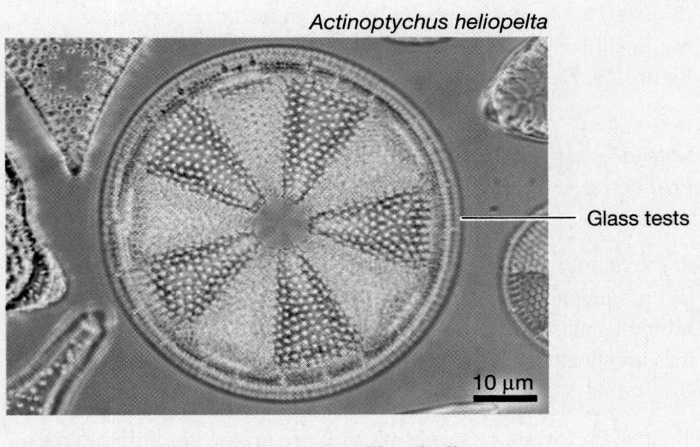

FIGURE 29.32 Diatoms Have Glass-like Shells.

Stramenopila > Oomycota (Water Molds)

Based on morphology, oomycetes were thought to be fungi and were given fungus-like names, such as downy mildew and water molds. They are readily distinguished from fungi at the DNA level, are primarily aquatic, have cellulose instead of chitin as the primary carbohydrate in the cell walls, and have gametes with stramenopile-like flagella. The morphological similarities between oomycetes and fungi result from convergent evolution, because both groups make their living absorbing nutrition from living or dead hosts. ● You should be able to mark the origin of the absorptive lifestyle in oomycetes on Figure 29.31.

Morphology Some oomycetes are unicellular and some form long, branching filaments called **hyphae**. Species that form hyphae often have multinucleate cells. Cells have walls containing cellulose.

Feeding and locomotion Most species feed on decaying organic material in freshwater environments; a few are parasitic. Mature individuals are sessile.

Reproduction Most species are diploid throughout the majority of their life cycle. In aquatic species, spores that are produced via asexual or sexual reproduction have flagella and swim to find new food sources; in terrestrial species, spores swim in rainwater or are dispersed by the wind in dry conditions. Spores form in special structures, like those shown in **Figure 29.33**.

Human and ecological impacts Oomycetes are extremely important decomposers in aquatic ecosystems. Along with certain bacteria and archaea, they are responsible for breaking down dead organisms and releasing nutrients for use by other species. The parasitic species can be harmful to humans, however. The organism that caused the Irish potato famine was an oomycete. An oomycete parasite almost wiped out the French wine industry in the 1870s, and other species are responsible for epidemic diseases of trees, including the diebacks currently occurring in oaks found in Europe and the western United States, and in eucalyptus in Australia.

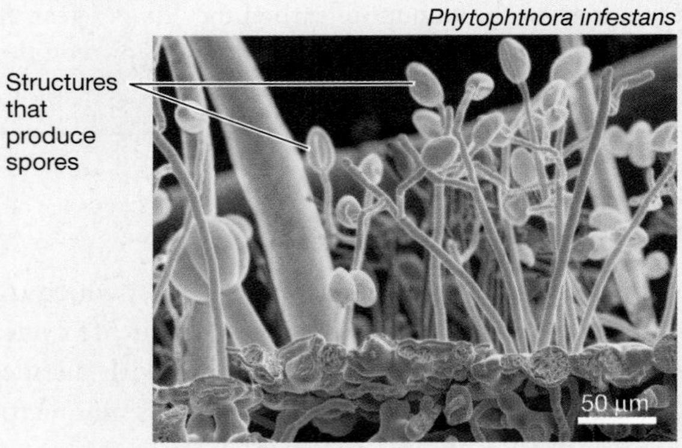

Phytophthora infestans

Structures that produce spores

50 µm

FIGURE 29.33 *Phytophthora infestans* Infects Potatoes.

Stramenopila > Phaeophyta (Brown Algae)

The color of brown algae is due to their unique suite of photosynthetic pigments. Over 1500 species have been described, most of them living in marine habitats.

Morphology The walls of Phaeophyta cells contain cellulose in addition to other complex polymers. Brown algae are unique among protists, because all species are multicellular. The body typically consists of leaflike blades, a stalk known as a stipe, and a rootlike holdfast, which attaches the individual to a substrate (**Figure 29.34**). ● You should be able to mark the origin of multicellularity and brown algal photosynthetic pigments on Figure 29.31.

Feeding and locomotion Brown algae are photosynthetic and **sessile**—that is, permanently fixed to a substrate—though reproductive cells may have flagella and be motile—capable of locomotion.

Reproduction Sexual reproduction occurs via the production of swimming gametes, which fuse to form a zygote. Most species exhibit alternation of generations. Depending on the species, the gametophyte and sporophyte stages may look similar or different.

Human and ecological impacts In many coastal areas, brown algae form forests or meadows that are important habitats for a wide variety of animals. In the Sargasso Sea, off Bermuda, floating brown algae form extensive rafts that harbor an abundance of animal species. The compound *algin* is purified from kelp and used in the manufacture of cosmetics and paint.

FIGURE 29.34 Many Brown Algae Have a Holdfast, Stalk, and Leaflike Blades.

Rhizaria

The rhizarians are single-celled amoebae that lack cell walls, though some species produce elaborate shell-like coverings. They move by amoeboid motion and produce long, slender pseudopodia. Over 11 major subgroups have been identified and named, including the planktonic organisms called actinopods, which synthesize glassy, silicon-rich skeletons, and the chlorarachniophytes, which obtained a chloroplast via secondary endosymbiosis and are photosynthetic. The best-studied and most abundant group is the Foraminifera (**Figure 29.35**). ● You should be able to mark the origin of the amoeboid form on Figure 29.35.

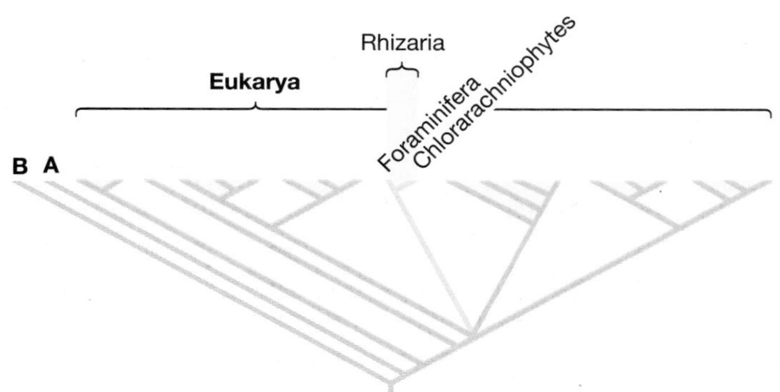

FIGURE 29.35 Rhizaria Are a Monophyletic Group.

Rhizaria > Foraminifera

Foraminifera, or forams, got their name from the Latin *foramen*, meaning "hole." Forams produce tests or shells that have holes through which pseudopodia emerge (**Figure 29.36**; see Figure 29.13c for a dead test on which holes are visible). ● You should be able to mark the origin of hard shells on Figure 29.35. Fossil tests of foraminifera are abundant in marine sediments—there is a continuous record of fossilized forams that dates back 530 million years. Although they are abundant in marine plankton as well as bottom habitats, their biology is relatively poorly known.

Morphology Foraminifera cells generally have multiple nuclei. The tests of forams are usually made of organic material stiffened with calcium carbonate ($CaCO_3$), and most species have several chambers. One species known from fossils was 12 cm long, but most species are much smaller. The size and shape of the test are traits that distinguish foram species from each other.

Feeding and locomotion Like other rhizarians, forams feed by extending their pseudopodia and using them to capture and engulf bacterial and archaeal cells or bits of organic debris, which are digested in food vacuoles. Some species have symbiotic algae that perform photosynthesis and contribute sugars to their host. Forams simply float in the water.

Reproduction Asexual reproduction occurs by mitosis. When meiosis occurs, the resulting gametes are released into the open water, where pairs fuse to form a new individual.

Human and ecological impacts The tests of dead forams commonly form extensive sediment deposits when they settle out of the water, producing layers that eventually solidify into chalk, limestone, or marble. Geologists use the presence of certain foram species to date rocks—particularly during petroleum exploration.

FIGURE 29.36 Forams Are Shelled Amoebae.

Plantae

Biologists are beginning to use the name **Plantae** to refer to the monophyletic group that includes red algae, green algae, land plants, and glaucophyte algae (**Figure 29.37**). All of these lineages are descended from a common ancestor that engulfed a cyanobacterium, beginning the endosymbiosis that led to the evolution of the chloroplast—their distinguishing morphological feature. This initial endosymbiosis probably occurred in an ancestor of today's glaucophyte algae. To support this hypothesis, biologists point to several important similarities between cyanobacterial cells and the chloroplasts that are found in the glaucophytes. Both have cell walls that contain peptidoglycan and that can be disrupted by molecules called lysozyme and penicillin. The glaucophyte chloroplast also has a membrane outside its wall that is similar to the membrane found in Gram-negative bacteria (see Chapter 28). Consistent with these observations, phylogenetic analyses of DNA sequence data place the glaucophytes as the most basal group of the Plantae. ● You should be able to mark the origin of the plant chloroplast on Figure 29.37.

The glaucophyte algae are unicellular or colonial. They live in plankton or attached to substrates in freshwater environments—particularly in bogs or swamps. Some glaucophyte species have flagella or produce flagellated spores, but sexual reproduction has never been observed in the group. The chloroplasts of glaucophytes have a distinct bright blue-green color. Chapter 30 introduces the green algae and land plants; here we consider just the red algae in detail.

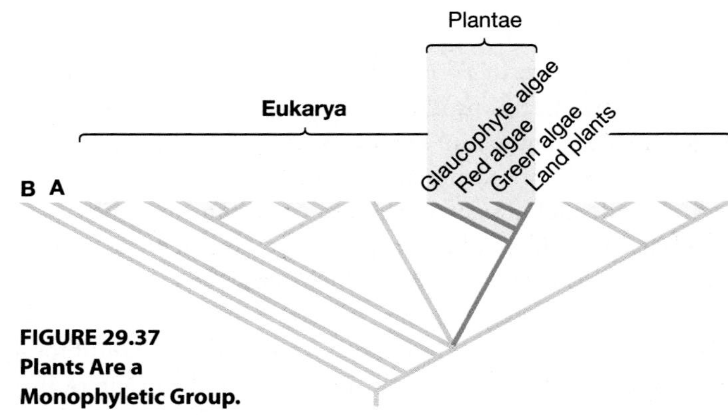

FIGURE 29.37 Plants Are a Monophyletic Group.

Plantae > Rhodophyta (Red Algae)

The 6000 species of red algae live primarily in marine habitats. One species lives over 200 m below the surface; another is the only eukaryote capable of living in acidic hot springs. Although their color varies, many species are red because their chloroplasts contain the accessory pigment phycoerythrin, which absorbs strongly in the blue and green portions of the visible spectrum. Because blue light penetrates water better than other wavelengths, red algae are able to live at considerable depth in the oceans. ● You should be able to mark the origin of red algal photosynthetic pigments on Figure 29.37.

Morphology Red algae cells have walls that are composed of cellulose and other polymers. A few species are unicellular, but most are multicellular. Many of the multicellular species are filamentous, but others grow as thin, hard crusts on rocks or coral (**Figure 29.38**). Some species have erect, leaf-like structures called thalli. Some species have cells with many nuclei.

Feeding and locomotion The vast majority of red algae are photosynthetic, though a few parasitic species have been identified. Red algae are the only type of algae that lack flagella.

Reproduction Asexual reproduction occurs through production of spores by mitosis. Alternation of generations is common, but the types of life cycles observed in red algae are extremely variable.

Human and ecological impacts On coral reefs, some red algae become encrusted with calcium carbonate. These species contribute to reef building and help stabilize the entire reef structure. Cultivation of *Porphyra* (or nori) for sushi and other foods is a billion-dollar-per-year industry in East Asia.

Lithothamnion species

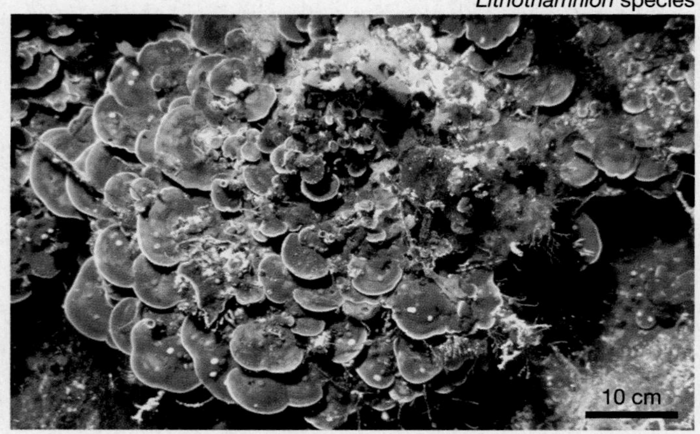

FIGURE 29.38 Red Algae Adopt an Array of Growth Forms.

Amoebozoa

Species in the Amoebozoa lack cell walls and take in food by engulfing it. They move via amoeboid motion and produce large, lobe-like pseudopodia. Major subgroups in the lineage are lobose amoebae, cellular slime molds, and plasmodial slime molds (**Figure 29.39**). ● You should be able to mark the origin of the amoeboid form on Figure 29.39 and explain whether it evolved independently of the amoeboid form in Rhizaria. The cellular slime mold *Dictyostelium discoideum* was described in detail in Box 29.2. Amoebae are abundant in freshwater habitats and in wet soils; some are parasites of humans and other animals. More details on plasmodial slime molds follow.

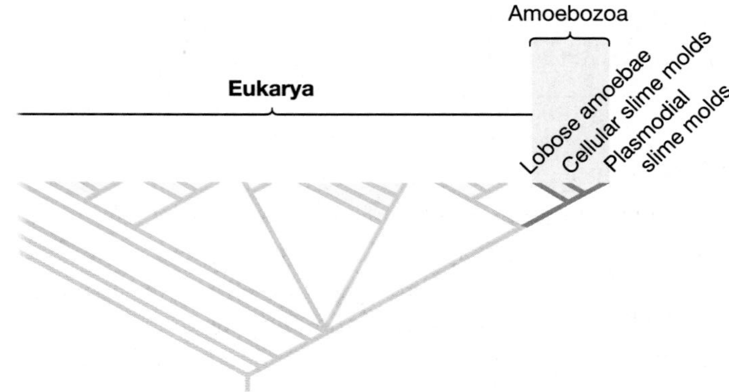

FIGURE 29.39 Amoebozoids Are a Monophyletic Group.

Amoebozoa > Myxogastrida (Plasmodial Slime Molds)

The plasmodial slime molds got their name because individuals form a large, weblike structure that consists of a single cell containing many diploid nuclei (**Figure 29.40**). Like oomycetes, the plasmodial slime molds were once considered fungi on the basis of their general morphological similarity (see Chapter 31).

Morphology The huge "supercell" form, with many nuclei in a single cell, occurs in few protists other than plasmodial slime molds. ● You should be able to mark the origin of the supercell on Figure 29.39.

Feeding and locomotion Myxogastrida cells feed on decaying vegetation and move by amoeboid motion.

Reproduction When food becomes scarce, part of the amoeba forms a stalk topped by a ball-like structure in which nuclei undergo meiosis and form spores. The spores are then dispersed to new habitats by the wind or small animals. After spores germinate to form amoebae, two amoebae fuse to form a diploid cell that begins to feed and eventually grows into a supercell.

Human and ecological impacts Like cellular slime molds, plasmodial slime molds are important decomposers in forests. They help break down leaves, branches, and other dead plant material, releasing nutrients that they and other organisms can use.

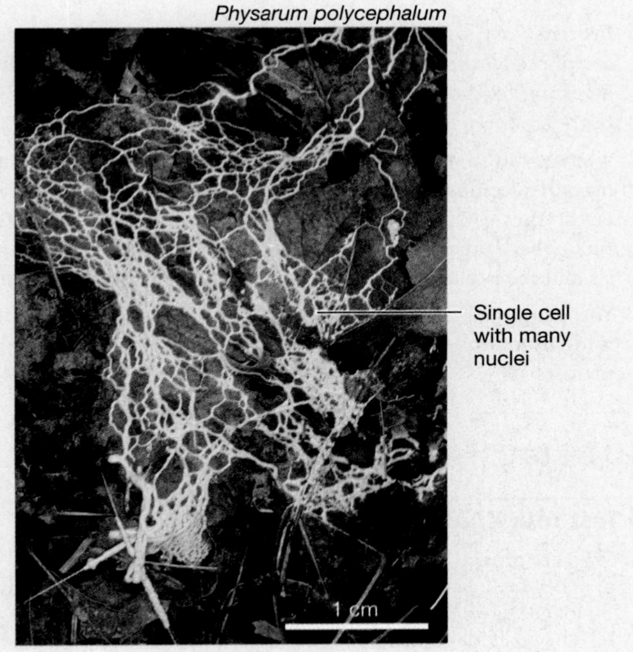

FIGURE 29.40 Plasmodial Slime Molds Are Important Decomposers in Forests.

Chapter Review

SUMMARY OF KEY CONCEPTS

- **Protists are a paraphyletic grouping that includes all eukaryotes except the green plants, fungi, and animals. Biologists study protists to understand how eukaryotes evolved, because they are important in freshwater and marine ecosystems and global warming, and because some species cause debilitating diseases in humans and other organisms.**

Protists are often tremendously abundant in marine and freshwater plankton and other habitats. As a result, protists provide food for many organisms in aquatic ecosystems and fix so much carbon that they have a large impact on the global carbon budget. Toxin-producing protists that grow to high densities result in a harmful algal bloom. Parasitic protists cause several important diseases in humans, including malaria.

You should be able to explain what terms biologists would substitute for *protists* if only monophyletic groups were named. ●

- **Protists are diverse morphologically. They vary in the types of organelles they contain; they may be unicellular or multicellular, and they may have a cell wall or other external covering or no such covering.**

Several morphological features are common to all or almost all eukaryotes, including the nucleus, endomembrane system, cytoskeleton, and flagellum. In addition to these common features, many aspects of morphology and lifestyle are extremely variable among the protist lineages. The protists include many unicellular organisms as well as multicellular slime molds, red algae, and brown algae. Multicellularity and structures that provide support or protection have evolved in many different protist groups independently.

Eukaryotes also contain mitochondria or have genes indicating that their ancestors once contained mitochondria. Several types of data support the hypothesis that mitochondria originated as endosymbiotic bacteria. The symbiosis is thought to have been successful because the endosymbiotic bacterium provided its host with larger amounts of ATP than it could produce on its own, while the host provided the bacterium with carbon compounds and protection. Similarly, the chloroplast's size, DNA structure, ribosomes, double membrane, and evolutionary relationships are consistent with the hypothesis that this organelle originated as an endosymbiotic cyanobacterium. After primary endosymbiosis occurred, chloroplasts were "passed around" to new lineages of protists via secondary endosymbiosis.

You should be able to make a drawing that shows how primary symbiosis would result in an organelle with two membranes containing genes that are closely related to bacteria, and how secondary endosymbiosis would result in an organelle with four membranes containing genes that are closely related to bacteria.

Protists vary widely in the way they find food. Many species are photosynthetic, while others obtain carbon compounds by ingesting food or parasitizing other organisms.

Protists exhibit predatory, parasitic, or photosynthetic lifestyles, which evolved in many groups independently. The evolution of ingestive feeding was important for two reasons: (1) It allowed eukaryotes to obtain resources in a new way—by eating bacteria, archaea, and other eukaryotes; and (2) it made endosymbiosis and the evolution of mitochondria and chloroplasts possible.

You should be able to propose a hypothesis to explain the wide diversity of photosynthetic pigments observed among protists.

Protists vary widely in the way they reproduce. Sexual reproduction evolved in protists, and many protist species can reproduce both sexually and asexually.

Protists undergo cell division based on mitosis and reproduce asexually. Many protists also undergo meiosis and sexual reproduction at some phase in their life cycle. Alternation of generations is common in multicellular species—meaning there are separate haploid and diploid forms of the same species. When alternation of generations occurs, haploid gametophytes produce gametes by mitosis; diploid sporophytes produce spores by meiosis.

You should be able to explain how you can tell a gametophyte from a sporophyte, and how you can distinguish a spore from a zygote.

MB **Web Animation** at www.masteringbio.com
Alternation of Generations in a Protist

QUESTIONS

Test Your Knowledge

1. Why are protists considered paraphyletic?
   a. They include many extinct forms, including lineages that no longer have any living representatives.
   b. They include some but not all, descendants of their most recent common ancestor.
   c. They represent all of the descendants of a single common ancestor.
   d. Not all protists have all of the synapomorphies that define the Eukarya, such as a nucleus.
2. What material is *not* used by protists to manufacture hard outer coverings?
   a. cellulose
   b. lignin
   c. glass-like compounds that contain silicon
   d. mineral-like compounds such as calcium carbonate ($CaCO_3$)
3. What does amoeboid motion result from?
   a. interactions among actin, myosin, and ATP
   b. coordinated beats of cilia
   c. the whiplike action of flagella
   d. action by the mitotic spindle, similar to what happens during mitosis and meiosis
4. According to the endosymbiosis theory, what type of organism is the original ancestor of the chloroplast?
   a. a photosynthetic archaean
   b. a cyanobacterium
   c. an algal-like, primitive photosynthetic eukaryote
   d. a modified mitochondrion
5. Multicellularity is defined in part by the presence of distinctive cell types. At the cellular level, what does this criterion imply?
   a. Individual cells must be extremely large.
   b. The organism must be able to reproduce sexually.
   c. Cells must be able to move.
   d. Different cell types express different genes.
6. Why are protists an important part of the global carbon cycle and marine food chains?
   a. They have high species diversity.
   b. They are numerically abundant.
   c. They have the ability to parasitize humans.
   d. They have the ability to undergo meiosis.

Test Your Knowledge answers: 1. b; 2. b; 3. a; 4. b; 5. d; 6. b

Test Your Understanding

Answers are available at www.masteringbio.com

1. What is the connection between the evolution of large cell size in protists and the evolution of ingestive feeding? Why is an advanced cytoskeleton and the lack of a cell wall required for ingestive modes of feeding? How does it relate to the acquisition of mitochondria and chloroplasts by endosymbiosis?
2. Is the evolution of novel structures for support and protection a synapomorphy that identifies the eukaryotes as a monophyletic group? Explain why or why not.
3. What is the relationship between meiosis and the alternation of generations? Why doesn't alternation of generations occur in unicellular species?
4. Outline the steps in the endosymbiosis theory for the origin of the mitochondrion. What did each partner provide the other, and what did each receive in return? Answer the same questions for the chloroplast.
5. Why was finding a close relationship between mitochondrial DNA and bacterial DNA considered particularly strong evidence in favor of the endosymbiosis theory? What evidence suggests that some protists acquired chloroplasts via secondary endosymbiosis?
6. The text claims that the evolutionary history of protists can be understood as a series of innovations that founded new lineages and/or triggered the diversification of existing lineages. Give an example that supports this claim.

Applying Concepts to New Situations

Answers are available at www.masteringbio.com

1. Consider the following:
   - All living eukaryotes have mitochondria or have evidence in their genomes that they once had these organelles. Thus it appears that eukaryotes acquired mitochondria very early in their history.
   - The first eukaryotic cells in the fossil record correlate with the first appearance of rocks formed in an oxygen-rich ocean and atmosphere.

   How are these observations connected? (HINT: Before answering, glance at Chapter 9 and remind yourself what happens in a mitochondrion.)
2. Consider the following:
   - *Plasmodium* has an unusual organelle called an apicoplast. Recent research has shown that apicoplasts are derived from chloroplasts via secondary endosymbiosis and have a large number of genes encoded by chloroplast DNA.
   - Glyphosate is one of the most widely used herbicides. It works by poisoning an enzyme encoded by a gene in chloroplast DNA.
   - Biologists are testing the hypothesis that glyphosate could be used as an antimalarial drug in humans.

   How are these observations connected?
3. Suppose a friend says that we don't need to worry about global warming. Her claim is that increased temperatures will make planktonic algae grow faster and that carbon dioxide (CO_2) will be removed from the atmosphere faster. According to her, this carbon will be buried at the bottom of the ocean in calcium carbonate tests. As a result, the amount of carbon dioxide in the atmosphere will decrease and global warming will decline. Comment.
4. Biologists are beginning to draw a distinction between "species trees" and "gene trees." A *species tree* is a phylogeny that describes the actual evolutionary history of a lineage. A *gene tree*, in contrast, describes the evolutionary history of one particular gene, such as a gene required for the synthesis chlorophyll *a*. In some cases, species trees and gene trees don't agree with each other. For example, the species tree for green algae indicates that their closest relatives are protists and plants. But the gene tree based on chlorophyll *a* from green algae suggests that this gene's closest relative is a bacterium, not a protist. What's going on? Why do these types of conflicts exist?

www.masteringbio.com is also your resource for • Answers to text, table, and figure caption questions and exercises • Answers to *Check Your Understanding* boxes • Online study guides and quizzes • Additional study tools including the *E-Book for Biological Science* 3rd ed., textbook art, animations, and videos.

30 Green Plants

KEY CONCEPTS

- The green plants include both the green algae and the land plants. Green algae are an important source of oxygen and provide food for aquatic organisms; land plants hold soil and water in place, build soil, moderate extreme temperatures and winds, and provide food for other organisms.
- Land plants were the first multicellular organisms that could live with most of their tissues exposed to the air. A series of key adaptations allowed them to survive on land. In terms of total mass, plants dominate today's terrestrial environments.
- Once plants were able to grow on land, a sequence of important evolutionary changes made it possible for them to reproduce efficiently—even in extremely dry environments.

Mosses are common in moist habitats and share many similarities with the earliest land plants. According to data reviewed in this chapter, the earliest land plants evolved from green algae that inhabited ponds, streams, and other freshwater habitats.

In terms of their total mass and their importance to other organisms, the **green plants** dominate terrestrial and freshwater habitats. When you walk through a forest or meadow, you are surrounded by green plants. If you look at pond or lake water under a microscope, green plants are everywhere.

The green plants comprise two major types of organisms: the green algae and the land plants. Green algae are important photosynthetic organisms in aquatic habitats—particularly lakes, ponds, and other freshwater settings—while land plants are the key photosynthesizers in terrestrial environments.

Although green algae have traditionally been considered protists, it is logical to study them along with land plants for two reasons: (1) They are the closest living relative to land plants and form a monophyletic group with them, and (2) the transition from aquatic to terrestrial life occurred when land plants evolved from green algae.

Land plants were the first organisms that could thrive with their tissues completely exposed to the air instead of being partially or completely submerged. When they evolved, multicellular organisms began to occupy dry terrestrial habitats that had been largely barren of visible life for over 3 billion years. Land plants made the Earth green. According to the fossil record, plants colonized the land in conjunction with fungi that grew in a mutually beneficial association. The fungi grew belowground and helped provide land plants with nutrients from the soil; in return, the plants provided the fungi with sugars and other products of photosynthesis. Not long after fungi and land plants evolved and began to diversify, animals also accom-

Key Concept Important Information Practice It

plished the feat of moving from aquatic to terrestrial habitats. But they could do so only because plants were there first and provided them something to eat. Close associations between fungi and land plants continue to this day and are detailed in Chapter 31; the importance of plants for humans and other animals is a major theme of this chapter.

Before land plants evolved, it is likely that the only life on the continents consisted of bacteria, archaea, and single-celled protists that thrive in wet soils. By colonizing the continents, plants transformed the nature of life on Earth. In the words of Karl Niklas, the movement of green plants from water to land ranks as "one of the greatest adaptive events in the history of life." They were the first multicellular organisms on land. Let's look at why so many biologists have devoted their lives to studying plants, then analyze how these organisms made the momentous transition to life on land.

30.1 Why Do Biologists Study the Green Plants?

Biologists study plants because people could not live without them and because they are fascinating. Along with most other animals and fungi, humans are almost completely dependent on plants for food. People rely on plants for other necessities of life as well—oxygen, fuel, building materials, and the fibers used in making clothing, paper, ropes, and baskets. But we also rely on land plants for important intangible values, such as the aesthetic appeal of landscaping and bouquets. To drive this point home, consider that the sale of cut flowers generates over $1 billion each year in the United States alone.

Based on these observations, it is not surprising that agriculture, forestry, and horticulture are among the most important endeavors supported by biological science. Tens of thousands of biologists are employed in research designed to increase the productivity of plants and to create new ways of using them in ways that benefit people. Research programs also focus on two types of land plants that cause problems for people: weeds that decrease the productivity of crop plants and newly introduced species that invade and then degrade natural areas.

Plants Provide Ecosystem Services

An **ecosystem** consists of all the organisms in a particular area, along with physical components of the environment such as the atmosphere, precipitation, surface water, sunlight, soil, and nutrients. Plants are said to provide **ecosystem services** because they add to the quality of the atmosphere, surface water, soil, and other physical components of an ecosystem. Stated another way, plants alter the landscape in ways that benefit other organisms:

- *Plants produce oxygen.* Recall from Chapter 10 that plants perform oxygenic (literally, "oxygen-producing") photosynthesis. In this process, electrons that are removed from water molecules are used to reduce carbon dioxide and produce sugars. In the process of stripping electrons from water, plants release oxygen molecules (O_2) as a by-product. As Chapter 28 noted, oxygenic photosynthesis evolved in cyanobacteria and was responsible for the origin of an oxygen-rich atmosphere. The evolutionary success of plants continued this trend because plants add huge amounts of oxygen to the atmosphere. Without the green plants, we and other terrestrial animals would be in danger of suffocating for lack of oxygen.
- *Plants build soil.* Leaves and roots and stems that are not eaten when they are alive fall to the ground and provide food for worms, fungi, bacteria, archaea, protists, and other decomposers in the soil. These organisms add organic matter to the soil, which improves soil structure and the ability of soils to hold nutrients and water.
- *Plants hold soil.* The extensive network of fine roots produced by trees, grasses, and other land plants helps hold soil particles in place. And by taking up nutrients in the soil, plants prevent the nutrients from being blown or washed away. When areas are devegetated by grazing, farming, logging, or suburbanization, large quantities of soil and nutrients are lost to erosion by wind and water (**Figure 30.1**).

(a) Wind erosion

(b) Water erosion

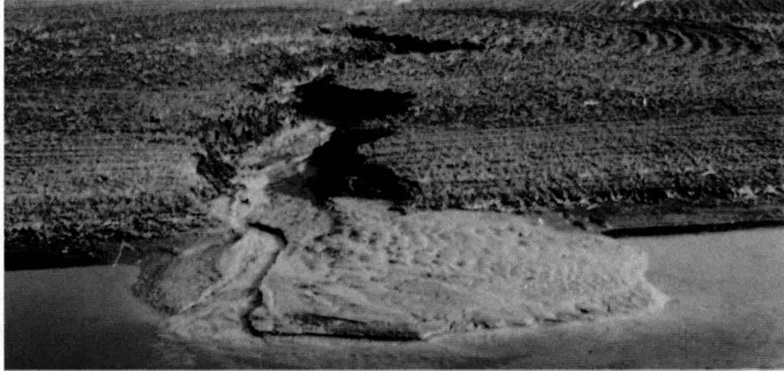

FIGURE 30.1 Plants Hold Soil in Place.

- *Plants hold water.* Plant tissues take up and retain water. Intact forests, prairies, and wetlands also prevent rain from quickly running off a landscape, for several reasons: Plant leaves soften the physical impact of rainfall on soil; plant roots hold soil particles in place during rainstorms; and plant organic matter builds the soil's water-holding capacity. When areas are devegetated, streams are more prone to flooding and groundwater is not replenished efficiently. It is common to observe streams alternately flooding and then drying up completely when the surrounding area is deforested.
- *Plants moderate the local climate.* By providing shade, plants reduce temperatures beneath them and increase relative humidity. They also reduce the impact of winds that dry out landscapes or make them colder. When plants are removed from landscapes to make way for farms or suburbs, habitats become much dryer and are subject to more extreme temperature swings.

Perhaps the most important ecosystem service provided by plants, however, involves food. Land plants are the dominant primary producers in terrestrial ecosystems. (As Chapter 29 indicated, primary producers convert energy in sunlight into chemical energy.) The sugars and oils that land plants produce by photosynthesis provide the base of the food chain in the vast majority of terrestrial habitats. As **Figure 30.2** shows, plants are eaten by **herbivores** ("plant-eaters"), which range in size from insects to elephants. These consumers are eaten by **carnivores** ("meat-eaters"), ranging in size from the tiniest spiders to polar bears. Humans are an example of **omnivores** ("all-eaters"), organisms that eat both plants and animals. Omnivores feed at several different levels in the terrestrial food chain. For example, people consume plants, herbivores such as chicken and cattle, and carnivores such as salmon and tuna.

Finally, just as photosynthetic protists and bacteria are the key to the carbon cycle in the oceans, green plants are the key to the carbon cycle on the continents. Plants take CO_2 from the atmosphere and reduce it to make sugars. Although both green algae and land plants also produce a great deal of CO_2 as a result of cellular respiration, they fix much more CO_2 than they release. The loss of plant-rich prairies and forests, due to fires or logging or suburbanization, has contributed to increased concentrations of CO_2 in the atmosphere. Higher carbon dioxide levels, in turn, are responsible for the rapid warming that is occurring worldwide (see Chapter 54).

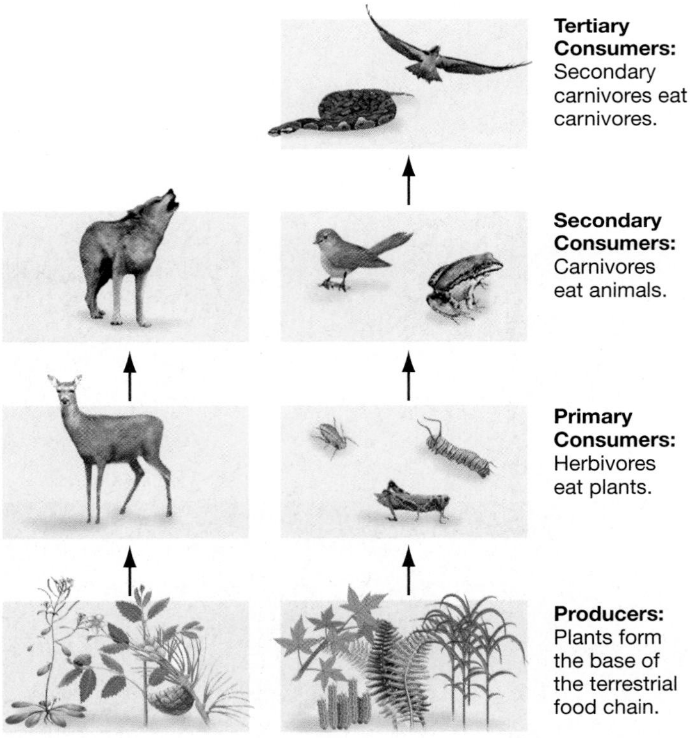

FIGURE 30.2 Plants Are the Basis of Food Chains in Terrestrial Environments. Virtually every organism that lives on land depends on plants for food, either directly or indirectly.

EXERCISE Label the levels in this food chain where humans feed.

Plants Provide Humans with Food, Fuel, Fiber, Building Materials, and Medicines

It is difficult to overstate the importance of plant research to the well-being of human societies. Plants provide our food supply as well as a significant percentage of the fuel, fibers, building materials, and medicines that we use.

- Agricultural research began with the initial domestication of crop plants, which occurred independently at several locations around the world between 10,000 and 2000 years ago. By actively selecting individuals with the largest and most nutritious seeds or leaves or stalks year after year, our ancestors gradually changed the characteristics of several wild species (**Figure 30.3a**). This process is called **artificial selection**, and it continues today (see Chapter 1). Over the past 100 years, for example, artificial selection has been responsible for dramatic increases in the oil content of corn kernels (**Figure 30.3b**). (Corn oil serves as a cooking oil and is used in other products.) Chapter 20 highlighted a current focus in agricultural research—the improvement of crop varieties through genetic engineering.
- For perhaps 100,000 years, wood burning was the primary source of energy used by all humans. As **Figure 30.4a** shows, however, wood has been replaced in the industrialized countries by other sources of energy. The first fuel to replace wood in many cases has been coal, which forms from partially decayed plant material that is compacted over time by overlying sediments and hardened into rock (**Figure 30.4b**). Starting in the mid-1800s, people in England, Germany, and the United States began to mine coal deposits that originally formed during the Carboniferous period some 350–275 million years ago. The coal fueled blast furnaces that smelted

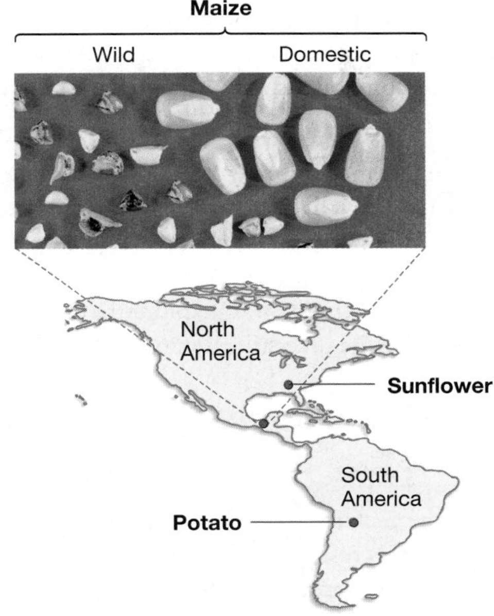

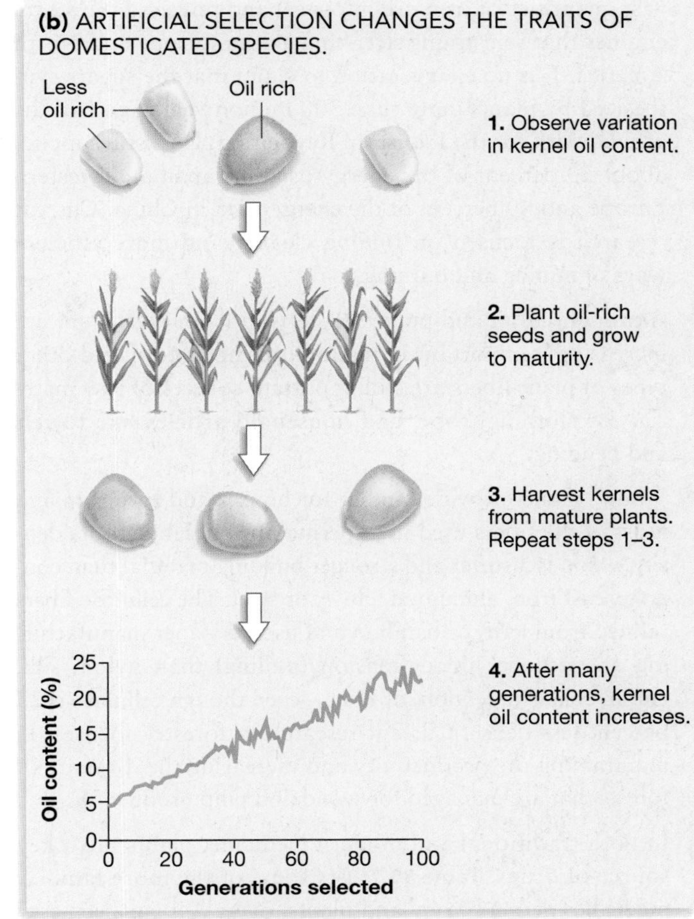

FIGURE 30.3 Crop Plants Are Derived from Wild Species via Artificial Selection. (a) Crop plants have originated on virtually every continent, including millet from Africa, wheat and barley from the Middle East, and rice and soybean from east Asia. **(b)** Artificial selection can lead to dramatic changes in plant characteristics.

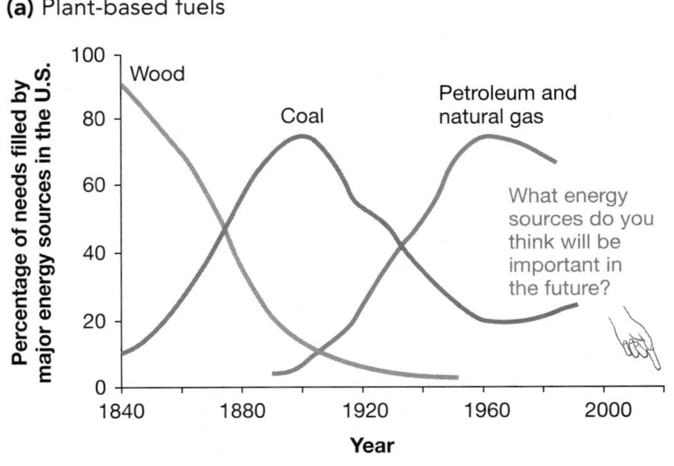

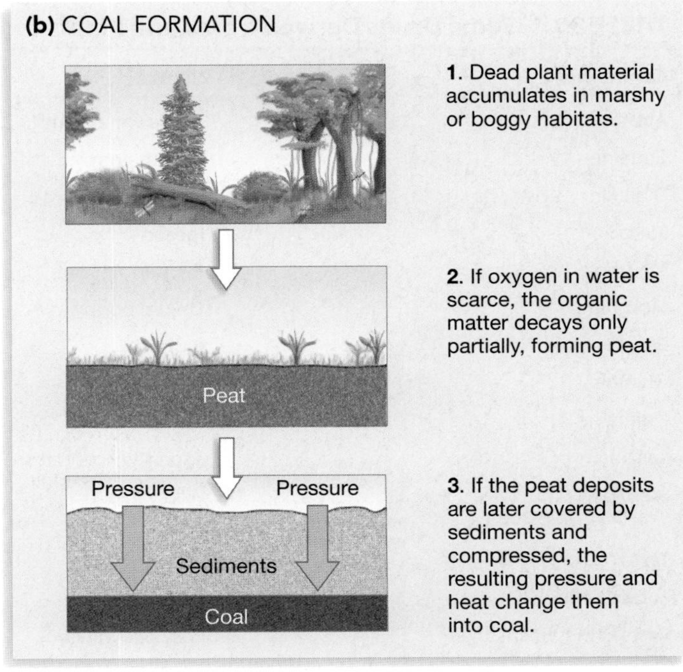

FIGURE 30.4 Humans Have Relied on Plant-Based Fuels. (a) Although wood has declined in importance in the industrialized countries, it is still the primary cooking and heating fuel in many areas of the world. **(b)** Coal is formed by terrestrial deposits of partially decayed organisms.

● **QUESTION** Is coal a renewable resource? Explain why or why not.

vast quantities of iron ore into steel and powered the steam engines that sent trains streaking across Europe and North America. It is no exaggeration to claim that the sugars synthesized by plants more than 300 million years ago laid the groundwork for the Industrial Revolution. Coal still supplies about 20 percent of the energy used in Japan and Western Europe and 80 percent of the energy used in China. Current research is focused on finding cleaner and more efficient ways of mining and burning coal.

- Although nylon and polyester derived from petroleum are increasingly important in manufacturing, cotton and other types of plant fibers are still important sources of raw material for clothing, rope, and household articles like towels and bedding.
- Woody plants provide lumber for houses and furniture and most of the fibers used in papermaking. Relative to its density, wood is a stiffer and stronger building material than concrete, cast iron, aluminum alloys, or steel. The cellulose fibers refined from trees or bamboo and used in paper manufacturing are stronger under tension (pulling) than nylon, silk, chitin, collagen, tendon, or bone—even though cellulose is 25 percent less dense. Current research in forestry focuses on maintaining the productivity and increasing the diversity of forests that are managed for wood and pulp production.
- In both traditional and modern medicine, plants are a key source of drugs. **Table 30.1** lists some of the more familiar medicines derived from land plants; overall, it has been estimated that about 25 percent of the prescriptions written in the United States each year include at least one molecule derived from plants. In most cases, plants synthesize these compounds in order to repel insects, deer, or other types of herbivores. For example, experiments have confirmed that morphine, cocaine, nicotine, caffeine, and other toxic compounds found in plants are effective deterrents to insect or mammalian consumers. Researchers continue to isolate and test new plant compounds for medicinal use in humans and domesticated animals.

Given the importance of plants to the planet in general and humans in particular, it is not surprising that understanding plant diversity is an important component of biological science. Let's first consider how biologists go about analyzing the diversity of green plants; then go on to explore the evolutionary innovations that made the diversification possible.

30.2 How Do Biologists Study Green Plants?

To understand the genetics and developmental biology of plants, researchers use thale cress, a weedy mustard, as a model organism (**Box 30.1**). To understand how green plants originated and diversified, biologists use three tools: They (1) compare the fundamental morphological features of various green algae and green plants; (2) analyze the fossil record of the lineage; and (3) assess similarities and differences in DNA sequences from homologous genes to estimate phylogenetic trees. The three approaches are complementary and have produced a remarkably clear picture of how land plants evolved from green algae and then diversified. Let's consider each of these research strategies.

Analyzing Morphological Traits

By comparing and contrasting the morphological traits of green plants, biologists have identified several distinct groups

TABLE 30.1 Some Drugs Derived from Land Plants

Compound	Source	Use
Atropine	Belladonna plant	Dilating pupils during eye exams
Codeine	Opium poppy	Pain relief, cough suppressant
Digitalin	Foxglove	Heart medication
Ipecac	Ipecac	Treating amoebic dysentery, poison control
Menthol	Peppermint tree	Cough suppressant, relief of stuffy nose
Morphine	Opium poppy	Pain relief
Papain	Papaya	Reduce inflammation, treat wounds
Quinine	Quinine tree	Malaria prevention
Quinidine	Quinine tree	Heart medication
Salicin	Aspen, willow trees	Pain relief (aspirin)
Steroids	Wild yams	Precursor compounds for manufacture of birth control pills and cortisone (to treat inflammation)
Taxol	Pacific yew	Ovarian cancer
Tubocurarine	Curare vine	Muscle relaxant used in surgery
Vinblastine, vincristine	Rosy periwinkle	Leukemia (cancer of blood)

BOX 30.1 A Model Organism: *Arabidopsis thaliana*

In the early days of biology, the best-studied plants were agricultural varieties such as maize (corn), rice, and garden peas. When biologists began to unravel the mechanisms responsible for oxygenic photosynthesis in the early to mid-1900s, they relied on green algae that were relatively easy to grow and manipulate in the lab—often the unicellular species *Chlamydomonas reinhardii*—as an experimental subject. Although crop plants and green algae continue to be the subject of considerable research, a new model organism emerged in the 1980s and now serves as the preeminent experimental subject in plant biology. That organism is *Arabidopsis thaliana*, commonly known as thale cress or wall cress (**Figure 30.5**).

FIGURE 30.5 *Arabidopsis thaliana* Is the Most Important Model Organism in Plant Biology. Like fruit flies and *Escherichia coli*, *A. thaliana* is small and short lived enough to grow easily in the laboratory, yet it is complex enough to be an interesting organism for study.

Arabidopsis is a member of the mustard family, or Brassicaceae, so it is closely related to radishes and broccoli. In nature it is a **weed**—meaning a species that is adapted to thrive in habitats where soils have been disturbed. In Europe, for example, *Arabidopsis* is common along roadsides and the edges of agricultural fields. It is also an **annual** plant, which means that individuals do not live from year to year but overwinter as seeds. (Plants that survive from one year to the next are said to be **perennial**.) Indeed, one of the most attractive aspects of working with *A. thaliana* is that individuals can grow from a seed into a mature, seed-producing plant in just four to six weeks. Several other attributes make it an effective subject for study: It has just five chromosomes, has a relatively small genome with limited numbers of repetitive sequences, can self-fertilize as well as undergo cross-fertilization, can be grown in a relatively small amount of space and with a minimum of care in the greenhouse, and produces up to 10,000 seeds per individual per generation.

Arabidopsis has been instrumental in a variety of studies in plant molecular genetics and development, and it is increasingly popular in ecological and evolutionary studies. In addition, the entire genome of the species has now been sequenced, and studies have benefited from the development of an international "*Arabidopsis* community"—a combination of informal and formal associations of investigators who work on *Arabidopsis* and use regular meetings, e-mail, and the Internet to share data, techniques, and seed stocks.

of green algae and a series of major lineages, or **phyla** (singular: **phylum**), of land plants.

The green algae have long been hypothesized to be closely related to land plants on the basis of several key morphological traits. Both green algae and land plants have chloroplasts that contain the photosynthetic pigments chlorophyll *a* and *b* and the accessory pigment β-carotene as well as similar arrangements of the internal, membrane-bound sacs called thylakoids (see Chapter 10). The cell walls of green algae and land plants are similar in composition, both groups synthesize starch as a storage product in their chloroplasts, and their sperm and peroxisomes are similar in structure and composition. (Recall from Chapter 7 that peroxisomes are organelles in which specialized oxidation reactions take place.)

The green algae include species that are unicellular, colonial, or multicellular and that live in marine, freshwater, or moist terrestrial habitats. Of all the green algal groups, the two most similar to land plants are the Coleochaetophyceae (coleochaetes) and Charaphyceae (stoneworts; **Figure 30.6**). Because the species that make up these groups are multicellular and live in ponds and other types of freshwater environments, biologists hypothesize that land plants evolved from multicellular green algae that lived in freshwater habitats.

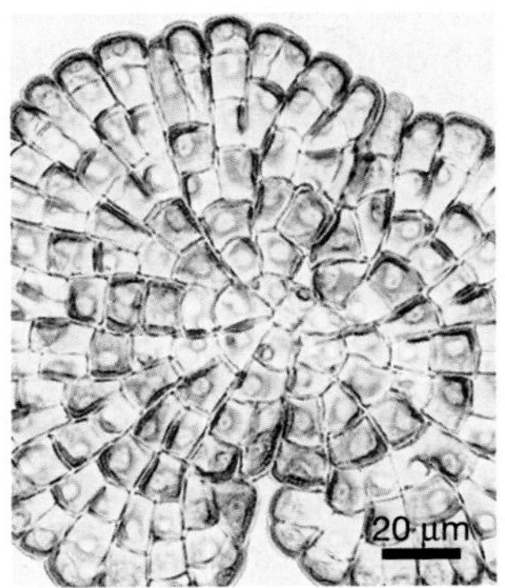

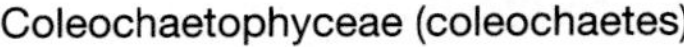

FIGURE 30.6 Most Green Algae Are Aquatic. Examples of species from the green algal lineages most closely related to the land plants.

Although some land plants live in ponds or lakes or rivers, the vast majority of species in this lineage live on land. Based on morphology, the most important phyla are traditionally

(a) Nonvascular plants do not have vascular tissue to conduct water and provide support.

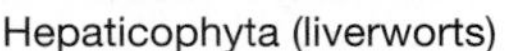
Hepaticophyta (liverworts)

Anthocerophyta (hornworts)

Bryophyta (mosses)

(b) Seedless vascular plants have vascular tissue but do not make seeds.

Lycophyta (lycophytes or club mosses)

Psilotophyta (whisk ferns)

Sphenophyta (horsetails)

Pteridophyta (ferns)

FIGURE 30.7 Morphological Diversity in Land Plants.

clustered into three broad categories:

1. **Nonvascular plants**, which include the groups called Hepaticophyta (liverworts), Anthocerophyta (hornworts), and Bryophyta (mosses). Liverworts and hornworts lack **vascular tissue**—meaning specialized groups of cells that conduct water or dissolved nutrients from one part of the plant body to another. Some moss species have specialized tissues that conduct water and food, but the cells that make up these tissues do not have the reinforced cell walls that define true vascular tissue. Nonvascular plants are extremely abundant in certain habitats but are usually small and grow close to the ground (**Figure 30.7a**).

2. **Seedless vascular plants**, which have well-developed vascular tissue but do not make seeds. A **seed** consists of an embryo and a store of nutritive tissue, surrounded by a tough protective layer. Lycophyta (lycophytes or club mosses), Psilophyta (whisk ferns), Sphenophyta (horsetails), and Pteridophyta (ferns) are among the major groups of seedless vascular plants (**Figure 30.7b**). Although most of the living representatives of these lineages are relatively small in stature, some fossil and some other living representatives of the seedless vascular plants are tree sized.

3. **Seed plants**, which have vascular tissue and make seeds. Although the members of this group vary a great deal in size and shape, seed plants encompass some of the world's largest organisms. Biologists recognize six major lineages in the group: Cycadophyta (cycads), Ginkgophyta (ginkgos), Gnetophyta (gnetophytes), Pinophyta (pines, spruces, firs), other conifers (redwoods, junipers, yews—species which bear seeds in cones, like the pinophytes, but represent a distinct lineage), and Anthophyta (angiosperms or flowering plants), as pictured in **Figure 30.7c.** (The group denoted "other conifers" was discovered so recently that it has not yet received a formal name.) The cycads, ginkgos, gnetophytes, pines, and other conifers are collectively known as **gymnosperms** ("naked-seeds"), because their seeds do not develop in an enclosed structure. In the flowering plants, or **angiosperms** ("encased-seeds"), seeds develop inside a protective structure called a carpel. Today, angiosperms are far and away the most important lineage of land plants in terms of species diversity. Almost 90 percent of the land plant species alive today, and virtually all of the domesticated forms, are angiosperms.

This quick overview of morphological diversity in land plants raises a number of questions:

(c) Seed plants have vascular tissue and make seeds.

Cycadophyta (cycads)

Ginkgophyta (ginkgo)

Other conifers (redwoods, junipers, yews)

Gnetophyta (gnetophytes)

Pinophyta (pines, spruces, firs)

Anthophyta (angiosperms or flowering plants)

FIGURE 30.7 *(continued)*

- Does other evidence support the hypothesis that green algae are ancestral to land plants, and that coleochaetes and stoneworts are the closest living relatives to land plants?
- Did the green algae evolve first, followed by nonvascular plants, then seedless vascular plants, and finally seed plants?
- Are both green algae and land plants monophyletic—meaning each is a distinct lineage tracing back to a single common ancestor?
- Are the nonvascular plants, seedless vascular plants, and seed plants monophyletic?

Answers are emerging from data in the fossil record and DNA sequences used to infer the green plant phylogeny.

Using the Fossil Record

The first green plants that appear in the fossil record are green algae in rocks that formed 700–725 million years ago. The first land plant fossils are much younger—they are found in rocks that are about 475 million years old. Because green algae appear long before land plants, the fossil record supports the hypothesis that land plants are derived from green algae.

The appearance and early diversification of green algae about 700 million years ago is also significant because at roughly the same time, the oceans and atmosphere were starting to become oxygen-rich—as never before in Earth's history. Based on this time correlation, it is reasonable to hypothesize that the evolution of green algae contributed to the rise of oxygen levels on Earth. The origin of the oxygen atmosphere occurred not long before the appearance of animals in the fossil record and may have played a role in their origin and early diversification.

The fossil record of the land plants themselves is massive. In an attempt to organize and synthesize the database, **Figure 30.8** breaks it into five time intervals—each encompassing a major event in the diversification of land plants.

The oldest interval begins 475 million years ago (mya), spans 30 million years, and documents the origin of the group. Most of the fossils dating from this period are microscopic. They consist of the reproductive cells called **spores** and sheets of a waxy coating called **cuticle**. Several observations support the hypothesis that these fossils came from green plants that were growing on land. First, cuticle is a watertight barrier that coats today's land plants and helps them resist drying. Second, the fossilized spores are surrounded by a sheetlike coating. Under the electron microscope, the coating material appears

FIGURE 30.8 The Fossil Record of Land Plants Can Be Broken into Five Major Intervals.

EXERCISE Add the following to the time line: (1) first terrestrial vertebrate animals (370 Ma); (2) first mammals (195 Ma).

almost identical in structure to a watertight material called **sporopollenin**, which encases spores and pollen from modern land plants and helps them resist drying. Third, fossilized spores that are 475 million years old have recently been found in association with spore-producing structures called **sporangia** (singular: **sporangium**). The fossilized sporangia are similar in appearance to the sporangia observed in some of today's liverworts.

The second major interval in the fossil record of land plants is called the "Silurian-Devonian explosion." In rocks dated 445–359 mya, biologists find fossils from most of the major plant lineages. Virtually all of the adaptations that allow plants to occupy dry, terrestrial habitats are present, including water-conducting tissue and roots.

The third interval in the fossil history of plants spans the aptly named Carboniferous period. In sediments dated from about 359 to 299 mya, biologists find extensive deposits of coal. Coal is a carbon-rich rock packed with fossil spores, branches, leaves, and tree trunks. Most of these fossils are derived from lycophytes, horsetails, and ferns. Although the only living lycophytes and horsetails are small, during the Carboniferous these groups were species rich and included a wide array of tree-sized forms. Because coal formation is thought to start only in the presence of water, the Carboniferous fossils indicate the presence of extensive forested swamps.

The fourth interval in land plant history is characterized by gymnosperms. Recall that gymnosperms include the cycads, ginkgos, gnetophytes, pines, and other cone-bearing trees. Because gymnosperms grow readily in dry habitats, biologists infer that both wet and dry environments on the continents became blanketed with green plants for the first time during this interval. Gymnosperms are particularly prominent in the fossil record from from 251 to 145 mya.

The fifth interval in the history of land plants is still under way. This is the age of flowering plants—the angiosperms. The first flowering plants in the fossil record appear about 150 mya. The plants that produced the first flowers are the ancestors of today's grasses, orchids, daisies, oaks, maples, and roses.

According to the fossil record, then, the green algae appear first, followed by the nonvascular plants, seedless vascular plants, and seed plants. Organisms that appear late in the fossil record are often much less dependent on moist habitats than are groups that appear earlier. For example, the sperm cells of mosses and ferns swim to accomplish fertilization, while gymnosperms and angiosperms produce pollen grains that are transported via wind or insects and that then produce sperm. These observations support the hypotheses that green plants evolved from green algae and that in terms of habitat use, the evolution of green plants occurred in a wet-to-dry trend.

To test the validity of these observations, biologists analyze data sets that are independent of the fossil record. Foremost among these are the DNA sequences and morphological data used to infer phylogenetic trees. Does the phylogeny of land plants confirm or contradict the patterns in the fossil record?

Evaluating Molecular Phylogenies

Understanding the phylogeny of green plants is a monumental challenge. The challenge is being met by the biologists who use both morphological traits and DNA sequence data to infer the evolutionary relationships among green plants.

The phylogenetic tree in **Figure 30.9** is a recent version of results coming out of laboratories from around the world. The black bars across some branches on this tree show when key innovations occurred, based on the fossil record and their oc-

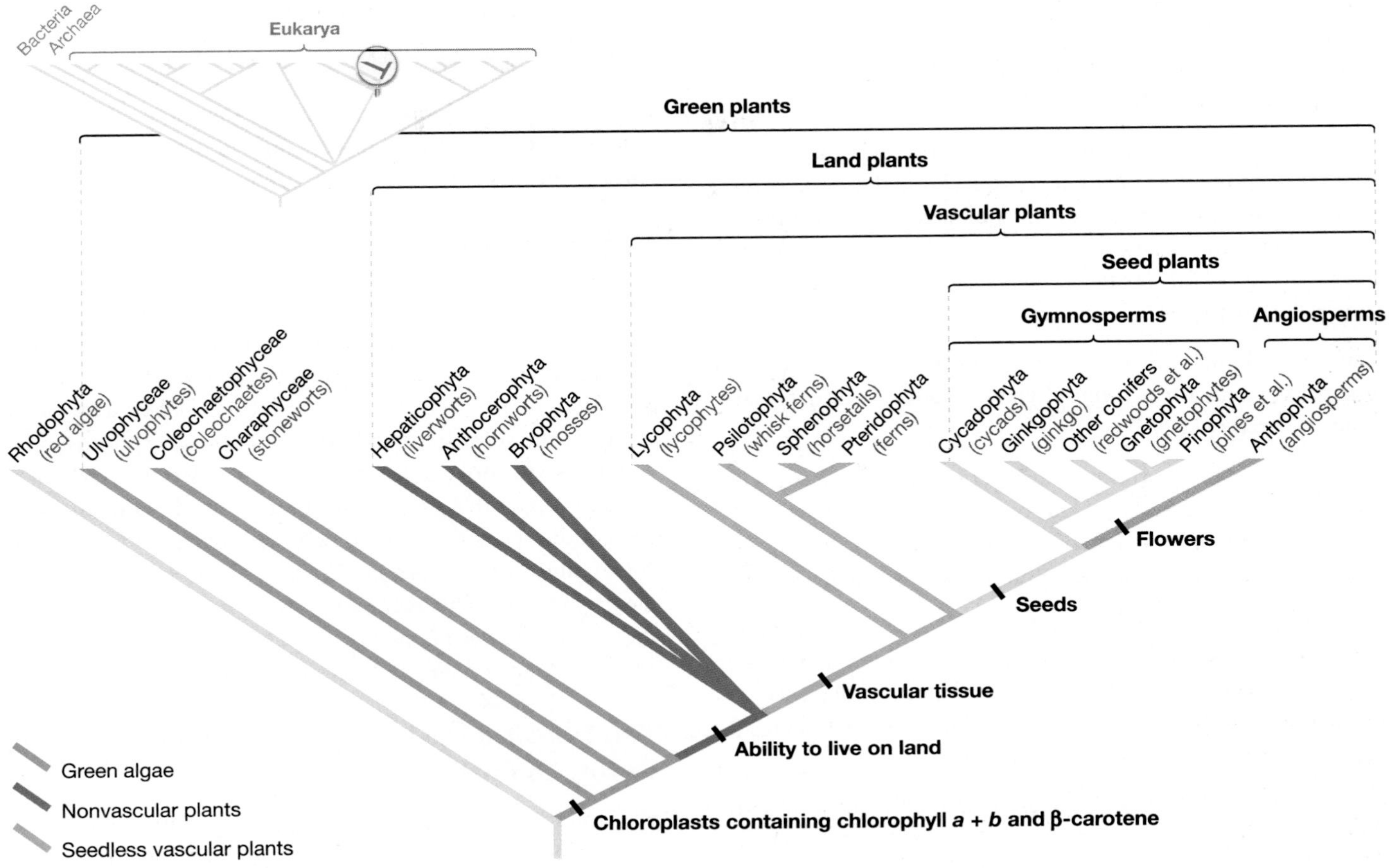

FIGURE 30.9 The Phylogeny of Green Plants. The blue branches on this figure indicate three of many major lineages collectively called green algae.

EXERCISE On the figure, identify why the groups called green algae and seedless vascular plants are paraphyletic.

currence in species living today. There are several important points to note about the relationships implied by this tree:

- The most basal branches on the tree, near the root, lead to coleochaetes, stoneworts, and other groups of green algae. This result supports the hypothesis that land plants evolved from green algae.
- The green algal group called Charaphyceae is the **sister group** to land plants—meaning that Charaphyceae are their closest living relative. Because living members of the Charaphyceae are multicellular and dwell in freshwater, the data support the hypothesis that land plants evolved from a multicellular ancestor that lived in ponds or lakes or other freshwater habitats.
- The green plants are monophyletic, meaning that a single common ancestor gave rise to all of the green algae and land plants. In contrast, the group called green algae is paraphyletic. Stated another way, the green algae include some but not all of the descendants of a single common ancestor.
- The land plants are monophyletic. This result supports the hypothesis that the transition from freshwater environments to land occurred just once. According to the phylogenetic tree, the population that made the transition then diversified into the array of land plants observed today.
- The nonvascular plants—that is, liverworts, hornworts, and mosses—are the earliest-branching, or most basal, groups among land plants. Although recent data confirm that each of these three lineages is monophyletic, the relationships among the three are still uncertain. For example, it is still not clear whether the nonvascular plants are monophyletic or paraphyletic.
- Some mosses have simple water-conducting tissues, and liverworts and hornworts have none at all. This finding suggests that the earliest land plants lacked water-conducting cells and vascular tissue, and that these traits evolved later. The phylogenetic tree supports the hypothesis that water-conducting cells and tissues evolved in a gradual fashion, with simpler structures preceding the evolution of more complex structures.
- The lycophytes are the sister group to all other seedless vascular plants. The whisk ferns, horsetails, and ferns form a monophyletic group; but as a whole the seedless vascular

plants form a **grade**—meaning a sequence of lineages that are not monophyletic. In contrast, the vascular plants are monophyletic—meaning that vascular tissue evolved once during the diversification of land plants.

- Because they lack roots and leaves, the whisk ferns were traditionally thought to be a basal group in the land plant radiation. Molecular phylogenies challenge this hypothesis and support an alternative hypothesis: The morphological simplicity of whisk ferns is a derived trait—meaning that complex structures have been lost in this lineage.
- The seed plants consist of the gymnosperms and the angiosperms and are a monophyletic group, meaning that seeds evolved once.
- The gymnosperms are monophyletic. Within the gymnosperms, the firs, spruces, and other species closely related to pines form an independent lineage to another group of cone-bearing plants (redwoods, junipers, yews, cypresses).
- The fossil record and the phylogenetic tree agree on the order in which groups appeared. Land plant evolution began with nonvascular plants, proceeded to seedless vascular plants, and continued with the evolution of seed plants.

The tree in Figure 30.9 will undoubtedly change and improve as additional data accumulate. Perhaps the most urgent need now is to clarify the relationships among the nonvascular plants—the liverworts, hornworts, and mosses. Because they are the most basal groups on the tree, understanding their relationships may help clarify how the earliest events in land plant evolution occurred.

Check Your Understanding

If you understand that...

- Biologists use the fossil record and phylogenetic analyses to study how green plants diversified.

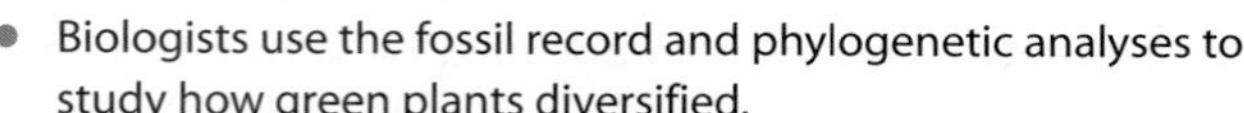

- The data analyzed to date support the hypotheses that green plants are monophyletic and that land plants evolved from multicellular green algae that inhabited freshwater.
- The fossil record and molecular analyses agree that the nonvascular plants evolved first, followed by the seedless vascular plants and the seed plants.

You should be able to...

1) Take a phylogenetic tree of green plants and label the groups called green plants, green algae, land plants, nonvascular plants, vascular plants, seed plants, and gymnosperms.
2) Explain which of these named groups are monophyletic and which are paraphyletic.
3) For monophyletic groups, name a shared, derived trait that distinguishes the lineage.

30.3 What Themes Occur in the Diversification of Green Plants?

Land plants have evolved from algae that grew on the muddy shores of ponds 475 million years ago to organisms that enrich the soil, produce much of the oxygen you breathe and most of the food you eat, and serve as symbols of health, love, and beauty. How did this happen?

Answering this question begins with recognizing the most striking trend in the phylogeny and fossil record of green plants: The most ancient groups in the lineage are dependent on wet habitats, while more recently evolved groups are tolerant of dry—or even desert—conditions. The story of land plants is the story of adaptations that allowed photosynthetic organisms to move from aquatic to terrestrial environments. Let's first consider adaptations that allowed plants to grow in dry conditions without drying out and dying, and then analyze the evolution of traits that allowed plants to reproduce efficiently on land. This section closes with a brief look at the radiation of flowering plants, which are the most important plants in many of today's terrestrial environments.

The Transition to Land, I: How Did Plants Adapt to Dry Conditions?

For aquatic green algae, terrestrial environments are deadly. Compared with a habitat in which the entire organism is bathed in fluid, in terrestrial environments only a portion, if any, of the plant's tissues are wet. Tissues that are exposed to air tend to dry out and die.

Once green plants made the transition to survive out of water, though, growth on land offered a bonanza of resources. Take light, for example. The water in ponds, lakes, and oceans absorbs and reflects light. As a result, the amount of light available to drive photosynthesis is drastically reduced even a meter or two below the water surface. In addition, the most important molecule required by photosynthetic organisms, carbon dioxide, is much more readily available in air than it is in water. Not only is it more abundant in the atmosphere, but it diffuses more readily there than it does in water.

Natural selection favored early land plants with adaptations that solved the water problem. These adaptations arose in two steps: (1) prevention of water loss from cells, which kept the cells from drying out and dying; and (2) transportation of water from tissues with direct access to water to tissues without access. Let's examine both of these steps in turn.

Preventing Water Loss: Cuticle and Stomata Section 30.2 pointed out that sheets of the waxy substance called cuticle are present early in the fossil record of land plants, along with encased spores. This observation is significant because the presence of cuticle in fossils is a diagnostic indicator of land plants. Cuticle is a waxy, watertight sealant that covers the aboveground parts of plants and gives them the ability to survive in dry environments (**Figure 30.10a**). If biologists had to point to one innovation that made the transition to land possible, it would be the production of cuticle.

Covering surfaces with wax creates a problem, however, regarding the exchange of gases across those surfaces. Plants need to take in carbon dioxide (CO_2) from the atmosphere in order to perform photosynthesis. But cuticle is almost as impervious to CO_2 as it is to water. Most modern plants solve this problem with a structure called a **stoma** ("mouth"; plural: **stomata**), consisting of an opening surrounded by specialized **guard cells** (**Figure 30.10b**). The opening, called a pore, opens or closes as the guard cells change shape. When guard cells become soft, they close the stomata. Pores are closed in this way to limit water loss from the plant. When guard cells become taut, in contrast, they open the pore, not only allowing CO_2 to diffuse into the interior of leaves and stems where cells are actively photosynthesizing but also allowing excess O_2 to diffuse out. (The mechanism behind guard-cell movement is explored in Chapter 39.)

Stomata are present in all land plants except the liverworts, which have pores but no guard cells. These data suggest that the earliest land plants evolved pores that allowed gas exchange to occur at breaks in the cuticle-covered surface. Later, the evolution of guard cells gave land plants the ability to regulate gas exchange—and control water loss—by opening and closing their pores.

Transporting Water: Vascular Tissue and Upright Growth Once cuticle and stomata had evolved, plants could keep from drying out and thus keep photosynthesizing while exposed to air. Cuticle and stomata allowed plants to grow on the saturated soils of lake or pond edges. The next challenge? Defying gravity.

Multicellular green algae can grow erect because they float. They float because the density of their cells is similar to water's density. But outside of water, the body of a multicellular green alga collapses. The water that fills its cells is 1000 times denser than air. Although the cell walls of green algae are strengthened by the presence of cellulose, their bodies lack the structural support to withstand the force of gravity and to keep an individual erect in air.

Based on these observations, biologists hypothesize that the first land plants were small or had a low, sprawling growth habit. Besides lacking rigidity, the early land plants would have had to obtain water through pores or through a few cells that lacked cuticle—meaning they would have had to grow in a way that kept many or most of their tissues in direct contact with moist soil.

The sprawling-growth hypothesis is supported by the observation that the most basal groups of land plants living today (liverworts, hornworts, and mosses) are all low-growing forms. If this hypothesis is correct, then competition for space and light would have become intense soon after the first plants began growing on land. To escape competition, plants would have to grow upright. Plants with adaptations that allowed some tissues to remain in contact with wet soil while other tissues grew erect would have much better access to sunlight compared with individuals that were incapable of growing erect.

For plants to adopt erect growth habits on land, though, two problems had to be overcome. The first is transporting

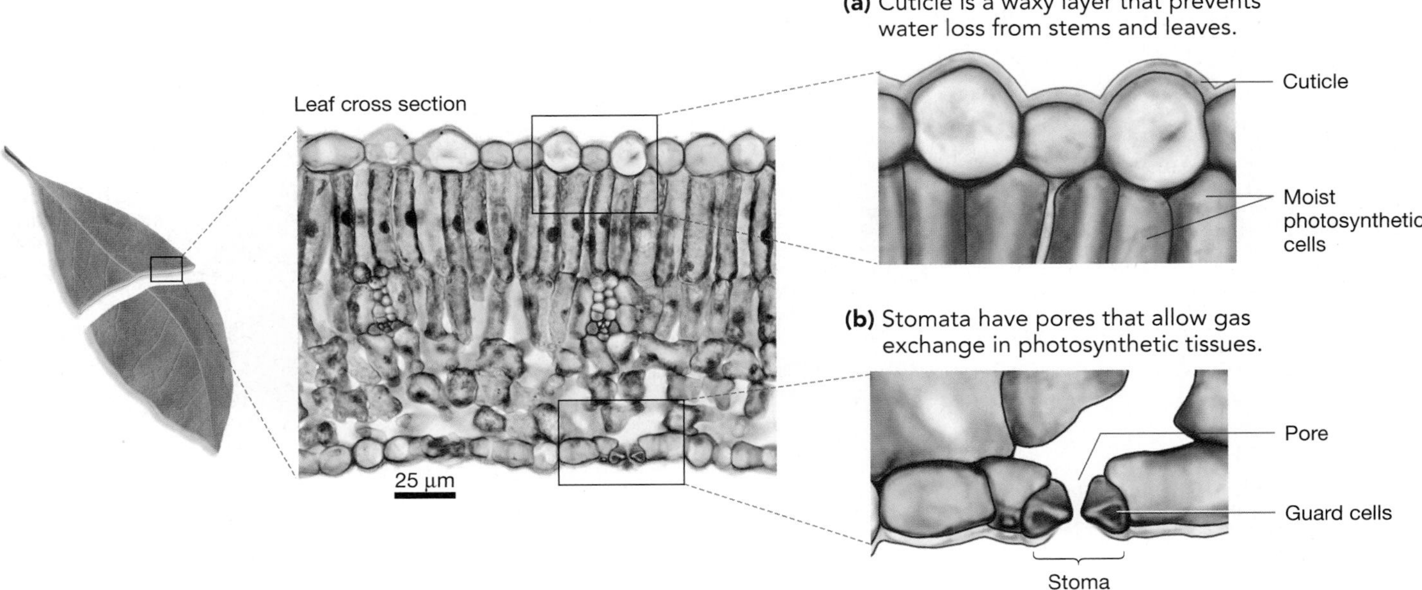

FIGURE 30.10 Cuticle and Stomata Are the Most Fundamental Plant Adaptations to Life on Land. In these micrographs, leaf cells have been stained blue to make their structure more visible. **(a)** The interior of plant leaves and stems is extremely moist; cuticle prevents water from evaporating away. **(b)** Stomata create pores to allow CO_2 to diffuse into the interior of leaves and stems where cells are actively photosynthesizing, and to allow excess O_2 to diffuse out.

● **QUESTION** Why was the evolution of guard cells important?

water from tissues that are in contact with wet soil to tissues that are in contact with dry air, against the force of gravity. The second is becoming rigid enough to avoid falling over in response to gravity and wind. As it turns out, vascular tissue helped to solve both problems.

Paul Kenrick and Peter Crane explored the origin of water-conducting cells and erect growth in plants by examining the extraordinary fossils found in a rock formation in Scotland called the Rhynie Chert. These rocks formed about 400 million years ago and contain some of the first large plant specimens in the fossil record—as opposed to the microscopic spores and cuticle found in older rocks. The Rhynie Chert also contains numerous plants that fossilized in an upright position. This indicates that many or most of the Rhynie plants grew erect. How did they stay vertical?

By examining fossils with the electron microscope, Kenrick and Crane established that species from the Rhynie Chert contained elongated cells that were organized into tissues along the length of the plant. Based on these data, the biologists hypothesized that the elongated cells were part of water-conducting tissue and that water could move from the base of the plants upward to erect portions through these specialized water-conducting cells. Some of the water-conducting cells had simple, cellulose-containing cell walls like the water-conducting cells found in today's mosses (**Figure 30.11a**). But in addition, some of the water-conducting cells present in the early fossils had cell walls with thickened rings containing a molecule called lignin (**Figure 30.11b**). **Lignin** is a complex polymer built from six-carbon rings. It is extraordinarily strong for its weight and is particularly effective in resisting compressing forces such as gravity.

These observations inspired the following hypothesis: The evolution of lignin rings gave stem tissues the strength to remain erect in the face of wind and gravity. Today, the presence of lignin in the cell walls of water-conducting cells is considered the defining feature of vascular tissue. The evolution of vascular tissue allowed early plants to support erect stems and transport water from roots to aboveground tissues.

Once simple water-conducting tissues evolved, evolution by natural selection elaborated them. In rocks that are about 380 million years old, biologists find the advanced water-conducting cells called tracheids. **Tracheids** are long, thin, tapering cells that have (1) a thickened, lignin-containing **secondary cell wall** in addition to a cellulose-based **primary cell wall**; and (2) gaps in the secondary cell wall, in the sides and ends of the cell, where water can flow efficiently from one tracheid to the next (**Figure 30.11c**). The secondary cell wall gave tracheids the ability to provide better structural support, but water could still move through the cells easily because of the gaps. Today, all vascular plants contain tracheids.

In fossils dated to 250–270 million years ago, biologists have documented the most advanced type of water-conducting cells observed in plants; these cells are known as vessel elements. **Vessel elements** are shorter and wider than tracheids. More importantly, the upper and lower ends of vessel elements have gaps in both the primary and secondary cell wall, which

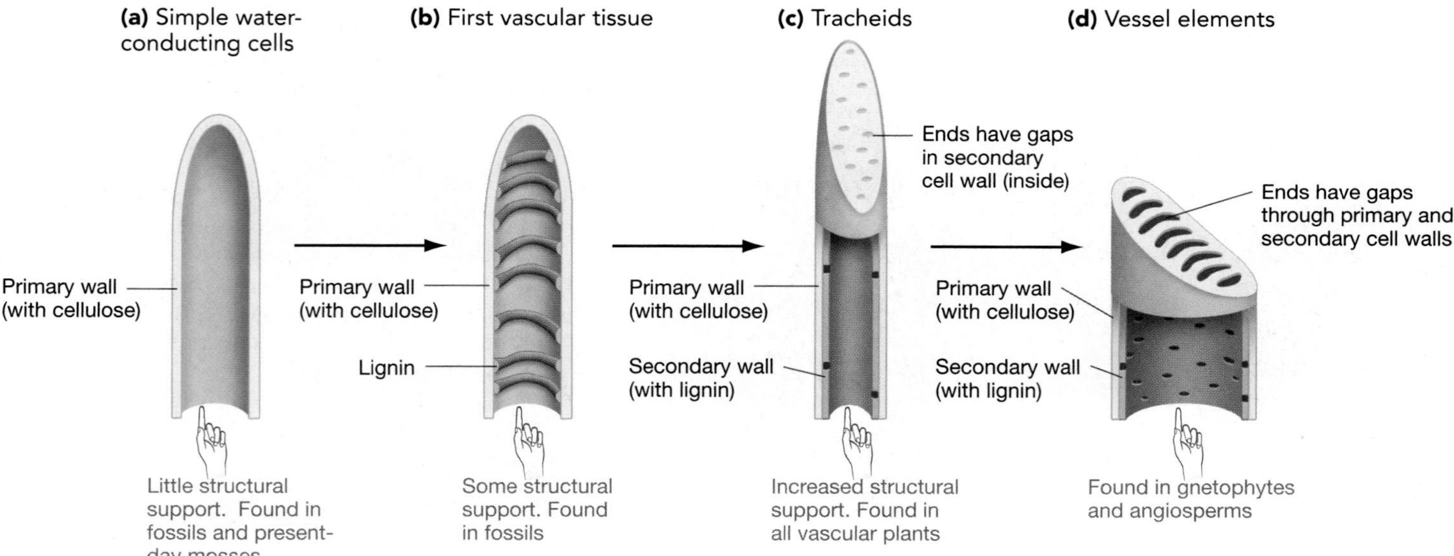

FIGURE 30.11 Evolutionary Sequence Observed in Water-Conducting Cells. According to the fossil record and the phylogeny of green plants, water-conducting cells became stronger over time due to the evolution of lignin and secondary cell walls. Efficient water transport was maintained through gaps in the secondary and/or primary cell wall.

QUESTION Biologists claim that vessels are more efficient than tracheids at transporting water, in part because vessels are shorter and wider than tracheids. Why does this claim make sense?

makes water movement extremely efficient (**Figure 30.11d**). In vascular tissue, vessel elements are lined up end to end to form a continuous pipelike structure.

In the stems and branches of some vascular plant species, tracheids or a combination of tracheids and vessels can form the extremely strong support material called **wood**. The anatomy of wood is explained in detail in Chapter 36.

All of the cell types shown in Figure 30.11 are dead when they mature, which means they lack cytoplasm. This feature allows water to move through the cells more efficiently. Taken together, the data summarized in the figure indicate that vascular tissue evolved in a series of gradual steps that provided increased structural support and increased efficiency in water transport.

Mapping Evolutionary Changes on the Phylogenetic Tree

Figure 30.12 summarizes how land plants adapted to dry conditions by mapping where major innovations occurred as the group diversified. Fundamentally important adaptations to dry conditions—such as the cuticle, pores, stomata, vascular tissue, and tracheids—evolved just once. In contrast, current analyses suggest that vessels evolved independently in gnetophytes and angiosperms.

The evolution of cuticle, stomata, and vascular tissue made it possible for plants to avoid drying out and to grow upright, while moving water from the base of the plant to its apex. Plants gained adaptations that allowed them not only to survive on land, but thrive. Now the question is, how did they reproduce?

at www.masteringbio.com
Plant Evolution and the Phylogenetic Tree

The Transition to Land, II: How Do Plants Reproduce in Dry Conditions?

Section 30.2 introduced one of the key adaptations for reproducing on land: spores that resist drying because they are encased in a tough coat of sporopollenin. Sporopollenin-like compounds are found in the walls of some green algal zygotes; thick-walled, sporopollenin-rich spores appear early in the fossil record of land plants and occur in all land plants living today. Based on these observations, biologists infer that sporopollenin-encased spores were one of the innovations that made the initial colonization of land possible. Two other innovations also occurred early in land plant evolution and were instrumental for efficient reproduction in a dry environment: (1) Gametes were produced in complex, multicellular structures; and (2) the embryo was retained on the parent (mother) plant and was nourished by it.

Retaining and Nourishing Offspring: Land Plants as Embryophytes The fossilized gametophytes of early land plants contain specialized reproductive organs called **gametangia** (singular: **gametangium**). Although members of the Charales also develop gametangia, the gametangia found in land plants are larger and more complex. The evolution of an elaborate gametangium was important because it protected gametes from drying and from mechanical damage. Gametangia are present in all land plants living today except angiosperms, where structures inside the flower perform the same functions.

In both the Charales and the land plants, individuals produce distinctive male and female gametangia. The sperm-producing structure is called an **antheridium** (plural:

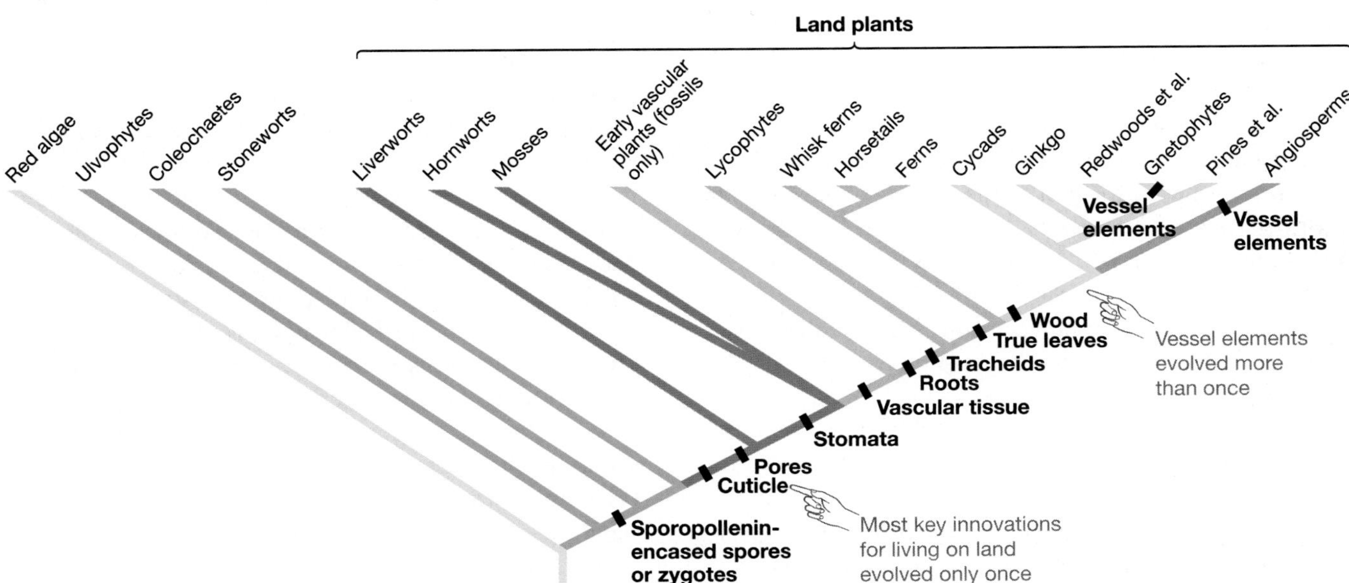

FIGURE 30.12 A Series of Evolutionary Innovations Allowed Plants to Adapt to Life on Land.

QUESTION Explain the logic that biologists used to map the location of each innovation indicated here.

(a) Sperm form in antheridia.

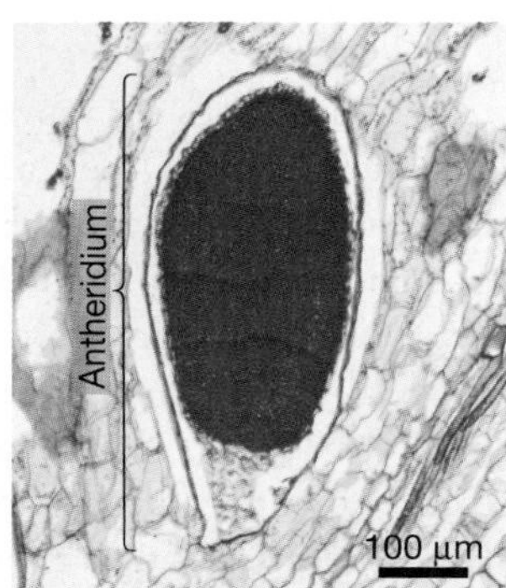

(b) Eggs form in archegonia.

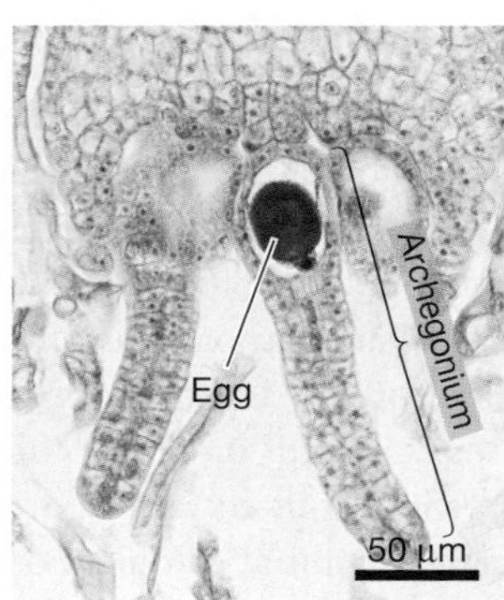

FIGURE 30.13 In All Land Plant Groups but Angiosperms, Gametes Are Produced in Gametangia. Gametangia are complex, multicellular structures that protect developing gametes from drying and mechanical damage. Note that there is one egg per archegonium.

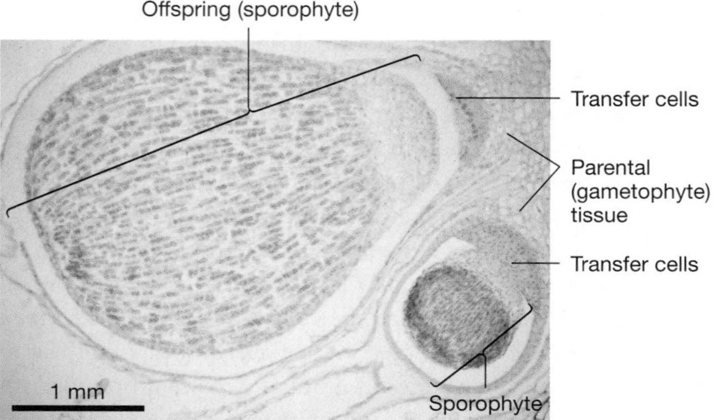

FIGURE 30.14 Land Plants Are Also Known as Embryophytes Because Parents Nourish Their Young. In land plants, fertilization and early development take place on the parent plant. As embryos develop, transfer cells carry nutrients from the mother to the offspring.

antheridia; Figure 30.13a, and the egg-producing structure is called an **archegonium** (plural: **archegonia; Figure 30.13b**). In terms of their function, the antheridium and archegonium are analogous to the testes and ovaries of animals.

The second innovation that occurred early in land plant evolution involved the eggs that formed inside archegonia. Instead of shedding their eggs into the water or soil, land plants retain them. Eggs are also retained in the green algal lineages that are most closely related to land plants: In Charales and other closely related groups, sperm swim to the egg, fertilization occurs, and the resulting zygote stays attached to the parent. Either before or after fertilization, the egg or zygote receives nutrients from the mother plant. But because these algae live in northern latitudes, the parent plant dies each autumn as the temperature drops. The zygote remains on the dead tissue from the parent, settles to the bottom of the lake or pond, and overwinters. In spring, meiosis occurs, and the resulting spores develop into haploid adult plants.

In land plants, the zygote is also retained on the parent plant after fertilization. But in contrast to the zygotes of most green algae, the zygotes of all land plants begin to develop on the parent plant, forming a multicellular embryo that remains attached to the parent and can be nourished by it. This is important because land plant embryos do not have to manufacture their own food early in life. Instead, they receive most or all of their nutrients from the parent plant.

The retention of the embryo was such a key event in land plant evolution that the formal name of the group is Embryophyta—literally, the "embryo-plants." The retention in these plants, commonly called **embryophytes**, is analogous to pregnancy in mammals, where offspring are retained by the mother and nourished through the initial stages of growth. Land plant embryos even have specialized **transfer cells**, which make physical contact with parental cells and facilitate the transfer of nutrients (**Figure 30.14**) much like the placenta that develops in a pregnant mammal.

Thick-walled spores, elaborate gametangia, and the embryophyte condition weren't the only key innovations associated with reproducing on land, though. In addition, all land plants undergo the phenomenon known as **alternation of generations**, introduced in Chapter 29.

Alternation of Generations When alternation of generations occurs, individuals have a multicellular haploid phase and a multicellular diploid phase. The multicellular haploid stage is called the **gametophyte**; the multicellular diploid stage is called the **sporophyte**. The two phases of the life cycle are connected by distinct types of reproductive cells—gametes and spores.

Although alternation of generations is observed in a wide array of protist lineages and in some groups of green algae, it does not occur in the algal groups most closely related to land plants. In the coleochaetes and stoneworts, the multicellular form is haploid. Only the zygote is diploid. As **Figure 30.15** shows, the zygote undergoes meiosis to form haploid spores. After dispersing with the aid of flagella, the spores begin dividing by mitosis and eventually grow into an adult, haploid individual. You might recall that this haploid-dominant life cycle is common in protists and was diagrammed in Figure 29.20a.

These data suggest that alternation of generations originated in land plants independently of its evolution in other groups of eukaryotes, and that it originated early in their history—soon after they evolved from green algae. The adaptive significance of alternation of generations and its role in the successful colonization of land is still being debated, however. Keep this in mind as you study alternation of generations in land plants: You are analyzing one of the great unsolved problems in contemporary biology.

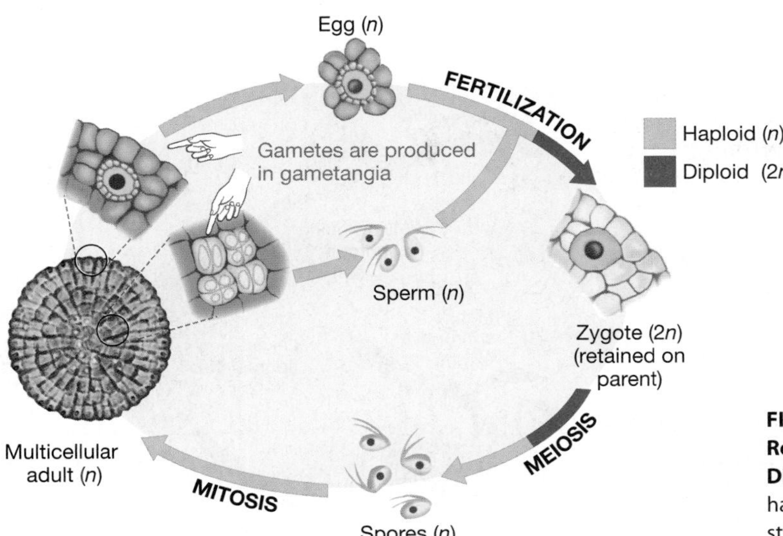

FIGURE 30.15 In Green Algae That Are Closely Related to Land Plants, Only the Zygote Is Diploid. The coleochaetes and stoneworts do not have alternation of generations. The multicellular stage is haploid.

Alternation of generations always involves the same basic sequence of events, illustrated in **Figure 30.16.** To review how this type of life cycle works, put your finger on the gametophyte in the figure and trace the cycle clockwise to find the following five key events:

1. Gametophytes produce gametes by mitosis. Both the gametophyte and the gametes are haploid.
2. Two gametes unite during fertilization to form a diploid zygote.
3. The zygote divides by mitosis and develops into a multicellular, diploid sporophyte.
4. The sporophyte produces spores by meiosis. Spores are haploid.
5. Spores divide by mitosis and develop into a haploid gametophyte.

Once you've traced the cycle successfully, take a moment to review the differences between a zygote and a spore. Both are single cells that divide by mitosis to form a multicellular individual. Zygotes, however, result from the fusion of two cells, such as a sperm and an egg. Spores are not formed by the fusion of two cells. In addition, zygotes produce sporophytes while spores produce gametophytes. Spores are produced inside structures called sporangia while gametes are produced inside gametangia.

Alternation of generations can be a difficult topic to master, for two reasons: (1) It is unfamiliar because it does not occur in humans or other animals, and (2) gamete formation begins with mitosis—not meiosis, as it does in animals (see Chapter 12). ● If you understand the basic principles of alternation of generations, you should be able to draw a life cycle starting with a gametophyte that produces gametes, which then fuse to form a zygote, which germinates to form a sporophyte, which produces spores via meiosis, which germinate to form a gametophyte. You should also be able to label the gametophyte, gametes, zygote, sporophyte, and spores, and indicate where meiosis takes place.

The Gametophyte-Dominant to Sporophyte-Dominant Trend in Life Cycles The five steps illustrated in Figure 30.16 occur in all species with alternation of generations. But in land plants, the relationship between the gametophyte and sporophyte is highly variable.

In nonvascular plants such as mosses, the sporophyte is small and short lived and is largely dependent on the gametophyte for nutrition (**Figure 30.17a**). When you see leafy-looking mosses growing on a tree trunk or on rocks like those in the photo at the start of this chapter, you are looking at gametophytes. Because

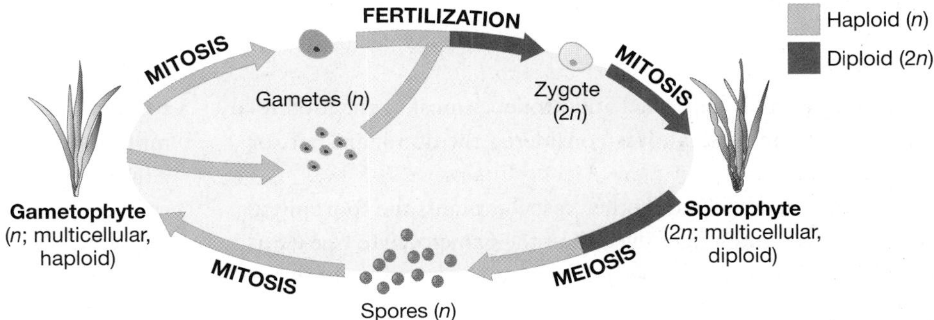

FIGURE 30.16 All Land Plants Undergo Alternation of Generations. Alternation of generations always involves the same sequence of five events: (1) Gametophytes produce haploid gametes by mitosis; (2) gametes fuse to form a diploid zygote; (3) zygotes develop into a diploid sporophyte; (4) sporophytes produce haploid spores by meiosis; and (5) spores develop into haploid gametophytes.

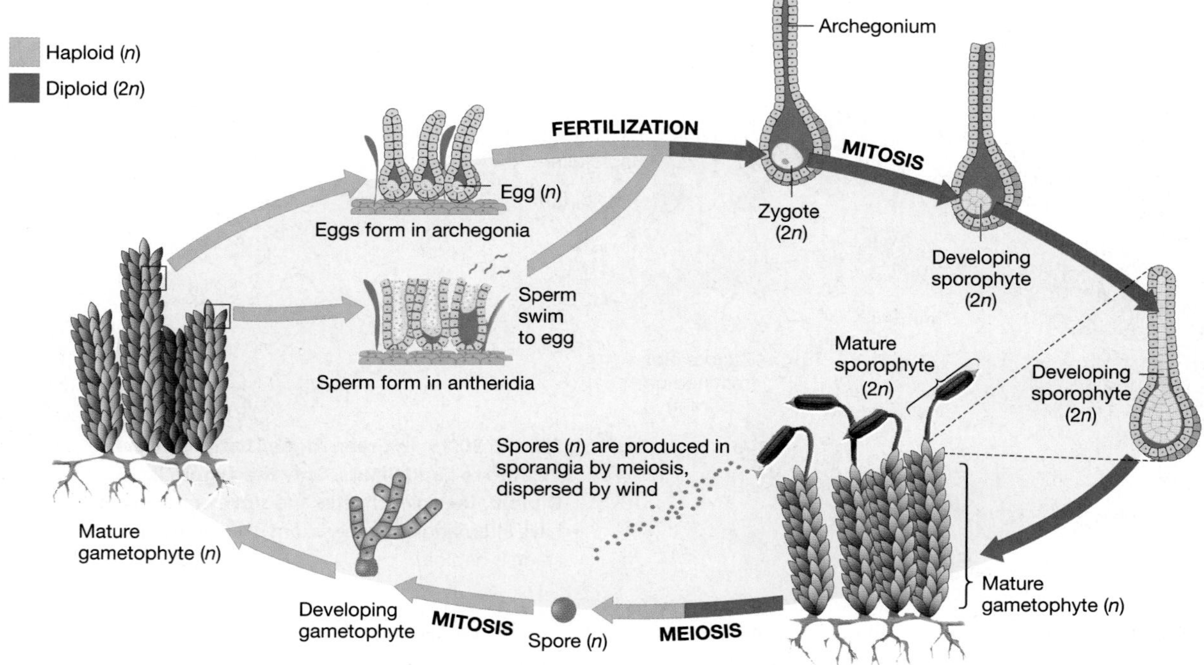

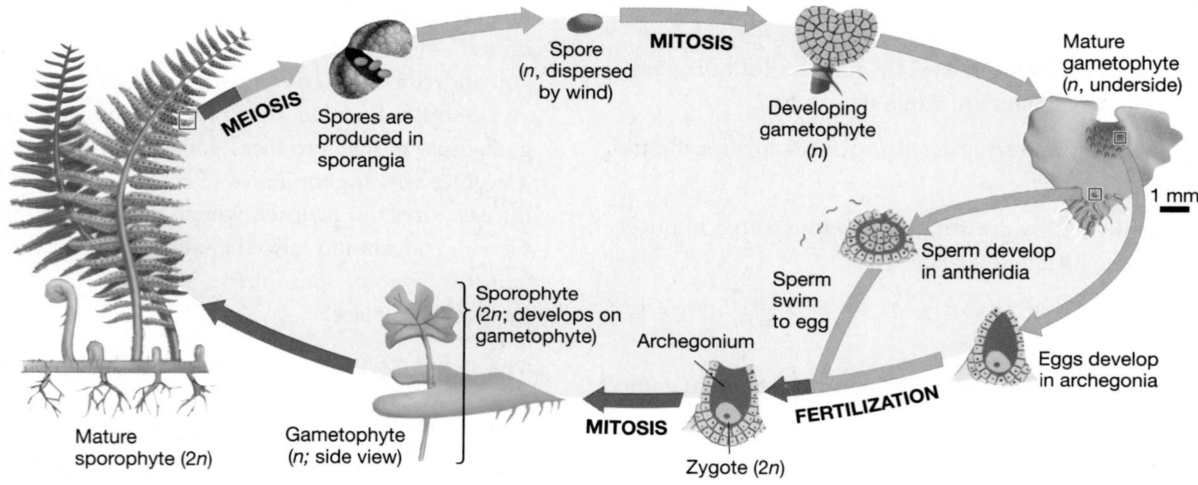

FIGURE 30.17 Gametophyte-Dominated Life Cycles Evolved Early; Sporophyte-Dominated Life Cycles Evolved Later. The most basal lineages of land plants, like mosses, have gametophytes that are much larger and longer lived than the sporophyte. In more derived lineages, such as ferns, the sporophyte is much larger and longer lived than the gametophyte.

● **QUESTION** How can you tell that alternation of generations occurs in these species?

● **QUESTION** Spores and zygotes are single cells that grow into an adult via mitosis. How do they differ?

the gametophyte is long lived and produces most of the food required by the individual, it is considered the dominant part of the life cycle.

In contrast, in ferns and other vascular plants the sporophyte is much larger and longer lived than the gametophyte (see **Figure 30.17b**). The ferns you see growing in gardens or forests are sporophytes. You'd have to hunt on your hands and knees to find their gametophytes, which are typically just a few millimeters in diameter. As you'll learn later in the chapter, the gametophytes of gymnosperms and angiosperms are even smaller—they are microscopic. Ferns and other vascular plants are said to have a sporophyte-dominant life cycle. ● If you understand the difference between gametophyte-dominant and sporophyte-dominant life cycles, you should be able to examine

(a) Hornwort gametophytes and sporophytes

(b) Horsetail gametophytes and sporophytes

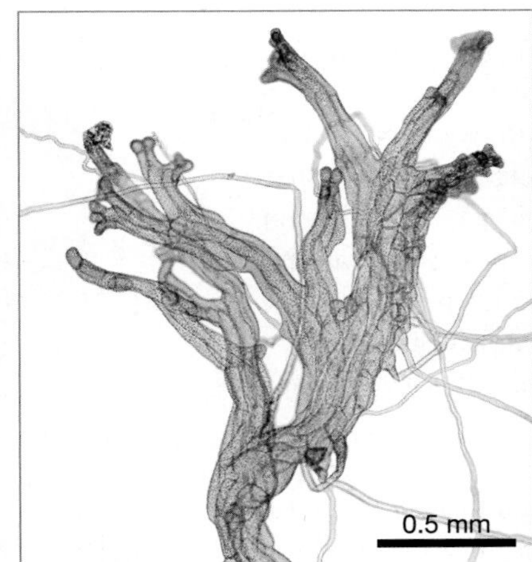

FIGURE 30.18 The Reduction of the Gametophyte Is One of the Strongest Trends in Land Plant Evolution. (a) A hornwort showing the sporophyte emerging from the gametophyte. **(b)** The same horsetail species as a tiny gametophyte and macroscopic sporophyte.

EXERCISE In parts (a) and (b), identify which is the gametophyte and which is the sporophyte.

the photos of hornworts (a nonvascular plant) and horsetails (a vascular plant) in **Figure 30.18**, and identify which is the gametophyte and which is the sporophyte.

The transition from gametophyte-dominated life cycles to sporophyte-dominated life cycles is one of the most striking of all trends in land plant evolution. To explain why it occurred, biologists hypothesize that sporophyte-dominated life cycles were advantageous because diploid cells can respond to varying environmental conditions more efficiently than haploid cells—particularly if the individual is heterozygous at many genes. But this idea has yet to be tested rigorously, and the trend is still considered something of a mystery. If you could come up with an explanation for why alternation of generations evolved in land plants and why the gametophyte-dominant to sporophyte-dominant trend occurred, and if your ideas stood up to intensive testing, the result would be celebrated worldwide.

Heterospory In addition to sporophyte-dominated life cycles, another important innovation found in seed plants is **heterospory**—the production of two distinct types of spore-producing structures and thus two distinct types of spores. Some lycophytes and a few ferns are also heterosporous, but all of the nonvascular plants and most of the seedless vascular plants are **homosporous**—meaning that they produce a single type of spore. Homosporous species produce spores that develop into bisexual gametophytes. Bisexual gametophytes produce both eggs and sperm (**Figure 30.19a**).

The two types of spore-producing structures found in heterosporous species are often found on the same individual. Microsporangia are spore-producing structures that produce microspores. **Microspores** develop into male gametophytes, which produce the small gametes called sperm. Megasporangia are spore-producing structures that produce megaspores. **Megaspores** develop into female gametophytes, which produce the large gametes called eggs (**Figure 30.19b**). Thus, the gametophytes of seed plants are either male or female, but never both.

The evolution of heterospory was a key event in land plant evolution because it made possible one of the most important adaptations for life in dry environments—pollen.

Pollen The nonvascular plants and the most ancient vascular plant lineages, such as ferns, have male gametes that swim to the egg to perform fertilization. For a sperm cell to swim to the egg and fertilize it, there has to be a continuous sheet of water between the male and female gametophyte, or a raindrop has to splash sperm onto a female gametophyte. In species that live in dry environments, these conditions are rare. The land plants made their final break with their aquatic origins and were able to reproduce efficiently in dry habitats when a structure evolved that could move their gametes without the aid of water.

(a) Non-vascular plants and most seedless vascular plants are homosporous.

Sporangium → Spores → Bisexual gametophyte → Sperm / Eggs

(b) Seed plants are heterosporous.

Microsporangia → Microspores → Male gametophyte → Sperm

Megasporangia → Megaspores → Female gametophyte → Eggs

FIGURE 30.19 Heterosporous Plants Produce Male and Female Spores That Are Morphologically Distinct.

In heterosporous seed plants, the microspore germinates to form a tiny male gametophyte that is surrounded by a tough, desiccation-resistant coat of sporopollenin. The resulting structure is called a **pollen grain**. Pollen grains can be exposed to the air for long periods of time without dying from dehydration. They are also tiny enough to be carried to female gametophytes by wind or animals. Upon landing near the egg, the male gametophyte releases the sperm cells that accomplish fertilization.

When pollen evolved, then, heterosporous plants lost their dependence on water to accomplish fertilization. Instead of swimming to the egg as a naked sperm cell, their tiny gametophytes took to the skies.

Seeds The evolution of large gametangia protected the eggs and sperm of land plants from drying. Embryo retention allowed offspring to be nourished directly by their parent, and pollen enabled fertilization to occur in the absence of water. Although retaining embryos ensures that off-spring are nourished, it has a downside: In ferns and horsetails, sporophytes have to live in the same place as their parent gametophyte. Seed plants overcome this limitation, however, because their embryos are portable and can disperse to new locations.

A **seed** is a structure that includes an embryo and a food supply surrounded by a tough coat (**Figure 30.20a**). Seeds allow embryos to be dispersed to a new habitat, away from the parent plant. Like a bird's egg, the seed provides a protective case for the embryo and a store of nutrients provided by the mother. In addition, seeds are often attached to structures that aid in dispersal by wind, water, or animals (**Figure 30.20b**).

The evolution of heterospory, pollen, and seeds triggered a dramatic radiation of seed plants starting about 290 million years ago. To make sure that you understand these key processes and structures, study the life cycle of the pine tree pictured in **Figure 30.21**. Starting with the sporophyte on the left, note that this and many other gymnosperm species have separate structures, called cones, where microsporangia and megasporangia develop. In this case, the two types of spores associated with heterospory develop in separate cones. The microsporangia contain a cell that divides by meiosis to form microspores, which then divide by mitosis to form pollen grains—tiny male gametophytes. Megasporangia are found inside protective structures called ovules, and they contain a mother cell that divides by meiosis to form a megaspore. The megaspore undergoes mitosis to form the female gametophyte, which contains egg cells. Notice that the female gametophyte stays attached to the sporophyte as pollen grains arrive and produce sperm that fertilize the eggs. Seeds mature as the embryo develops and cells derived from the female gametophyte become packed with nutrients provided by the sporophyte. When the seed disperses and germinates, the cycle of life begins anew.

Flowers Flowering plants, or angiosperms, are the most diverse land plants living today. About 250,000 species have been described, and more are discovered each year. Their success in terms of geographical distribution, number of individuals, and number of species revolves around a reproductive organ—the **flower**.

(a) Seeds package an embryo with a food supply.

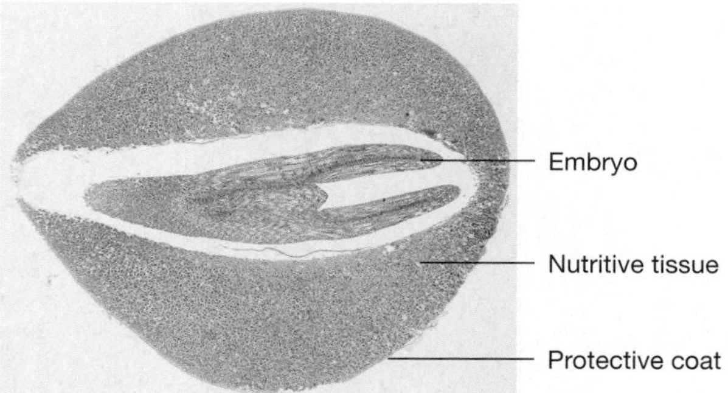

(b) Seeds are dispersed by wind, water, and animals.

FIGURE 30.20 Seeds Contain an Embryo and a Food Supply and Can Be Dispersed. (a) Seeds consist of an embryo, a store of carbohydrate or oil-rich nutritive tissue, and a protective coat. **(b)** Seeds are often found in structures that facilitate dispersal by wind, water, or animals.

Flowers contain two key reproductive structures: the stamens and carpels illustrated on the left-hand side of **Figure 30.22**. Stamens and carpels are responsible for heterospory. A **stamen** includes a structure called an anther, where microsporangia develop. Meiosis occurs inside the microsporangia, forming microspores. Microspores then divide by mitosis to form pollen grains. A flower's **carpel**, in contrast, contains a protective structure called an **ovary** where the **ovules** are found. The presence of enclosed ovules inspired the name *angiosperm* ("encased-seed") as opposed to *gymnosperm* ("naked-seed"). As in gymnosperms, ovules contain the megasporangia. A cell inside the megasporangium divides by meiosis to form the megaspore, which then divides by mitosis to form the female gametophyte. When a pollen grain lands on a carpel and produces sperm, fertilization takes place, as shown on the right-hand side of Figure 30.22. In angiosperms, though, fertilization involves two sperm cells. One sperm fuses with the egg to form the zygote, while a second

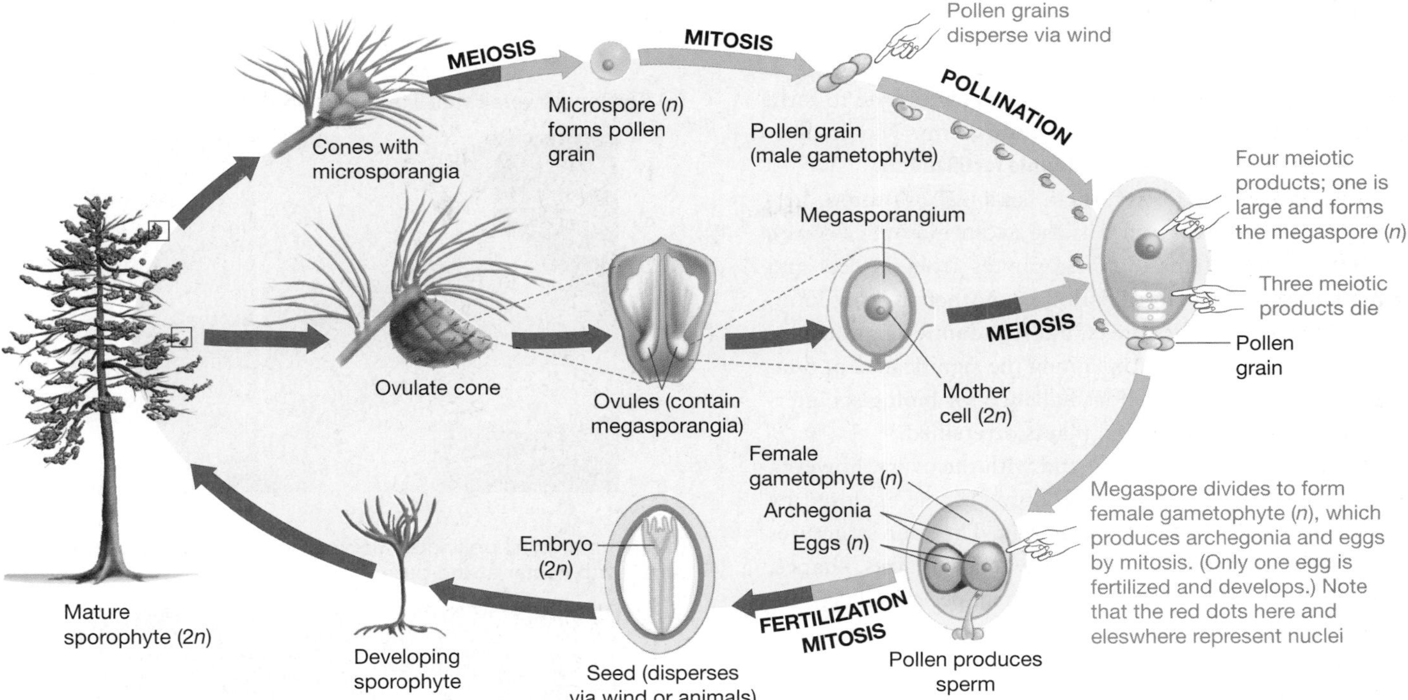

FIGURE 30.21 Heterospory in Gymnosperms: Microspores Produce Pollen Grains; Megaspores Produce Female Gametophytes.

● **QUESTION** Compare this life cycle with that of the fern pictured in Figure 30.17b. Is the gymnosperm gametophyte larger than, smaller than, or about the same size as a fern gametophyte? Compared with ferns, is the gymnosperm gametophyte more or less dependent on the sporophyte for nutrition?

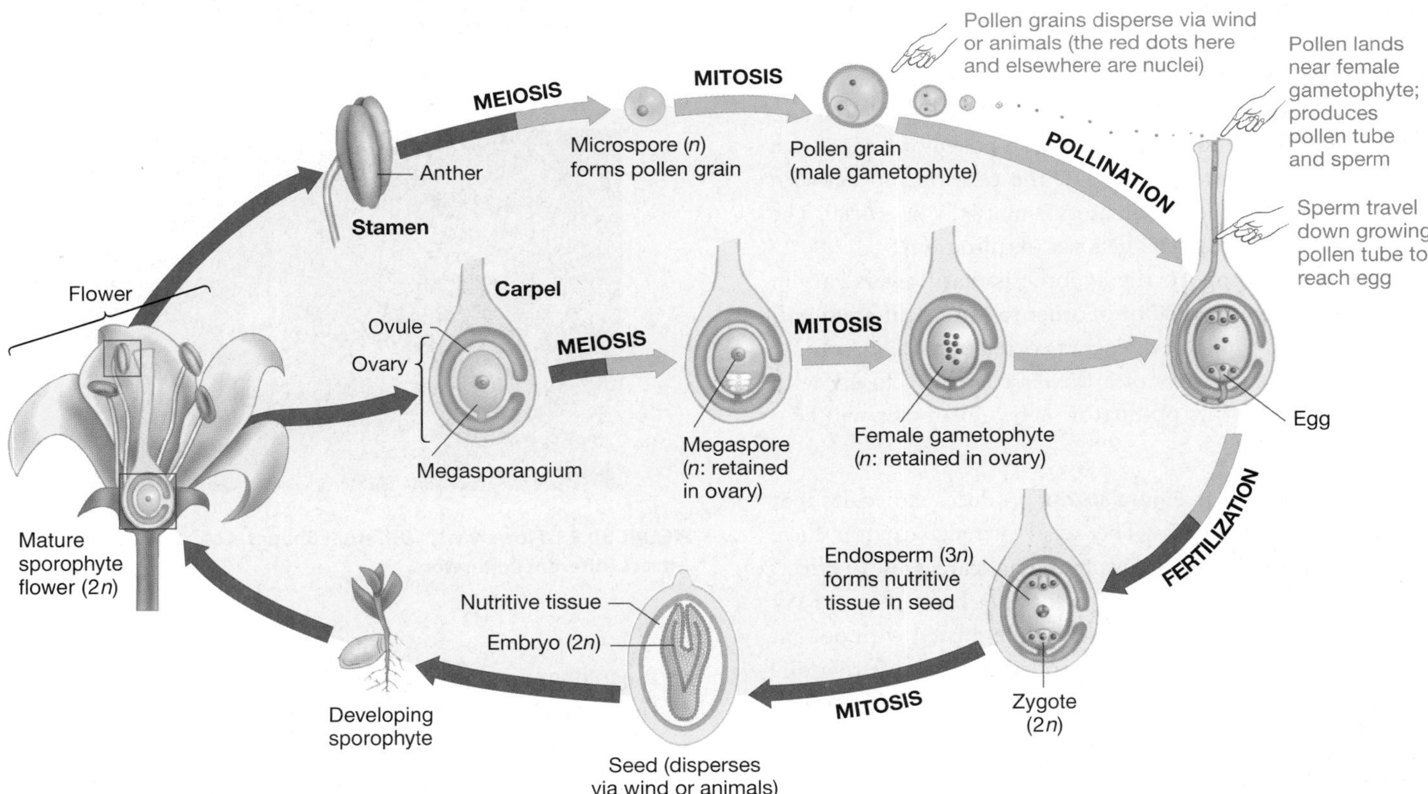

FIGURE 30.22 Heterospory in Angiosperms: Flowers Contain Microspores and Megaspores.

● **QUESTION** Gymnosperm pollen grains typically contain from 4 to 40 cells; mature angiosperm pollen grains contain three cells. Gymnosperm female gametophytes typically contain hundreds of cells; angiosperm female gametophytes typically contain seven. Do these observations conflict with the trend of reduced gametophytes during land plant evolution, or are they consistent with it? Explain your logic.

sperm fuses with two nuclei in the female gametophyte to form a triploid ($3n$) nutritive tissue called **endosperm**. The involvement of two sperm nuclei is called **double fertilization**.

The evolution of the flower, then, is an elaboration of heterospory. The key innovation was the evolution of the ovary, which helps protect female gametophytes from insects and other predators. Double fertilization is another striking innovation associated with the flower, but its adaptive significance is still not well understood. Explaining the significance of double fertilization is another major challenge for biologists interested in understanding how land plants diversified.

The story of the flower doesn't end with the ovary, however. Once stamens and carpels evolved, they became enclosed by modified leaves called **sepals** and **petals**. The four structures then diversified to produce a fantastic array of sizes, shapes, and colors—from red roses to blue violets. Specialized cells inside flowers also began producing a wide range of scents.

To explain these observations, biologists hypothesize that flowers are adaptations to increase the probability that an animal will perform **pollination**—the transfer of pollen from one individual's stamen to another individual's carpel. Instead of leaving pollination to an undirected agent such as wind, the hypothesis is that natural selection favored structures that reward an animal—usually an insect—for carrying pollen directly from one flower to another. Under the directed-pollination hypothesis, natural selection has favored flower colors and shapes and scents that are successful in attracting particular types of pollinators. A pollinator is an animal that disperses pollen. Pollinators are attracted to flowers because flowers provide the animals with food in the form of protein-rich pollen or a sugar-rich fluid known as **nectar.** In this way, the relationship between flowering plants and their pollinators is mutually beneficial. The pollinator gets food; the plant gets sex (fertilization).

What evidence supports the hypothesis that flowers vary in size, structure, scent, and color in order to attract different pollinators? The first type of evidence is correlational in nature. In general, the characteristics of a flower correlate closely with the characteristics of its pollinator. A few examples will help drive this point home:

- The carrion flower in **Figure 30.23a** produces molecules that smell like rotting flesh. The scent attracts carrion flies, which normally lay their eggs in animal carcasses. In effect, the plant tricks the flies. While looking for a place to lay their eggs on a flower, the flies become dusted with pollen. If the flies are already carrying pollen from a visit to a different carrion flower, they are likely to deposit pollen grains near the plant's female gametophyte. In this way, the carrion flies pollinate the plant.
- Flowers that are pollinated by hummingbirds typically have petals that form a long, tubelike structure corresponding to the size and shape of a hummingbird's beak (**Figure 30.23b**). Nectar-producing cells are located at the base of the tube. When hummingbirds visit the flower, they insert their beaks and harvest the nectar. In the process, th ey transfer pollen grains attached to their throats or faces.
- Hummingbird-pollinated flowers also tend to have red petals (Figure 30.23b), while bee-pollinated flowers tend to be purple or yellow (**Figure 30.23c**). Hummingbirds are attracted to red; bees have excellent vision in the purple and ultraviolet end of the spectrum.

(a) Carrion flowers smell like rotting flesh and attract carrion flies.

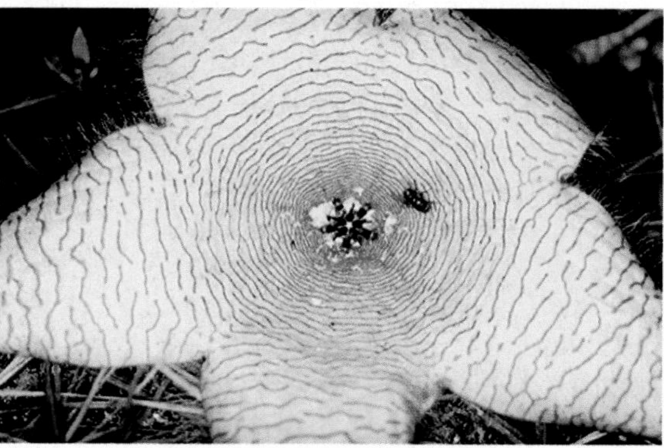

(b) Hummingbird-pollinated flowers are red and have long tubes with nectar at the base.

(c) Bumble-bee-pollinated flowers are often bright purple.

FIGURE 30.23 Flowers with Different Shapes, Colors, and Fragrances Attract Different Pollinators.

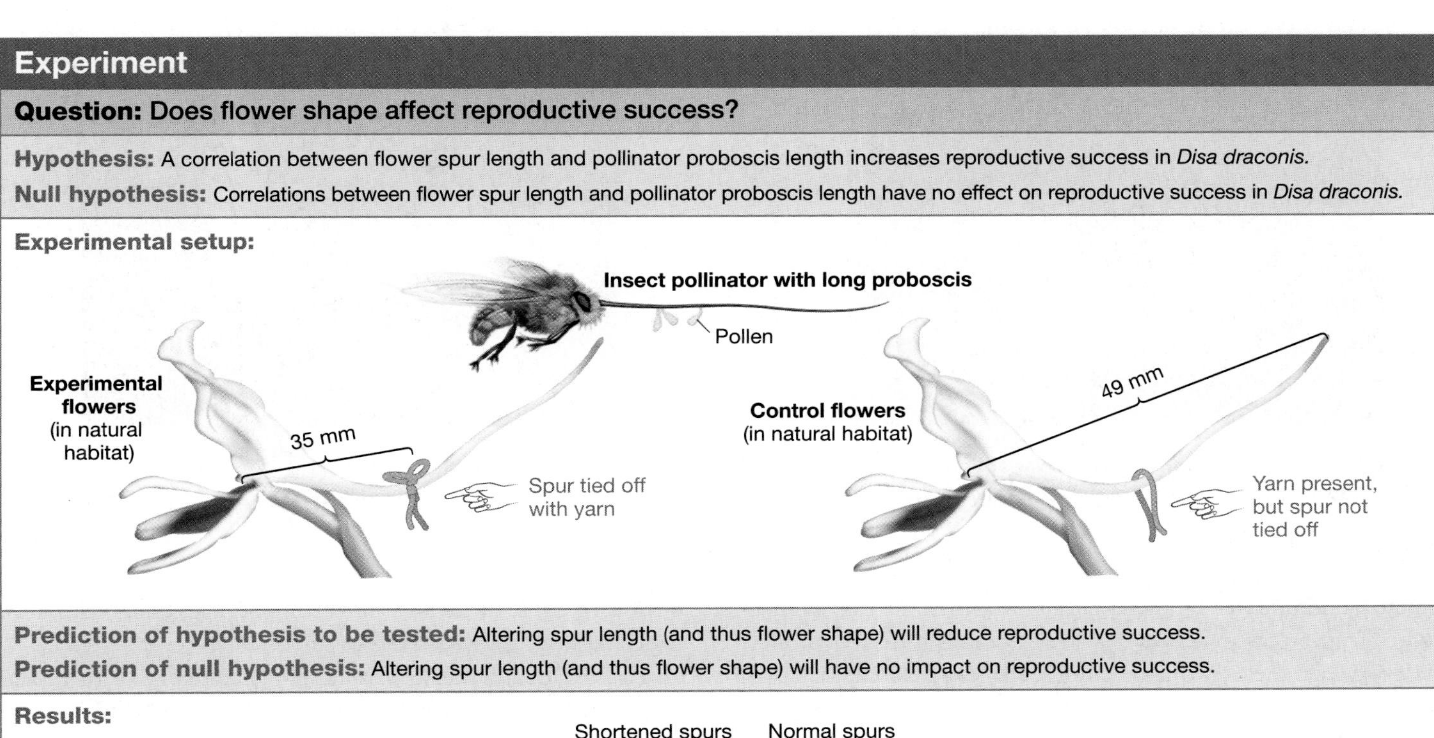

Results:

	Shortened spurs	Normal spurs
What fraction of flowers received pollen? (n = 59)	17%	46%
What fraction of flowers set seed? (n = 56)	18%	41%

Flowers with normal spurs have higher reproductive success

Conclusion: In this species, flower shape—specifically spur length—affects reproductive success.

FIGURE 30.24 The Adaptive Significance of Flower Shape: An Experimental Test. The spurs of experimental *D. draconis* flowers were tied off with yarn 35 mm from the opening; the spurs of control flowers had a piece of yarn loosely tied around them, too, but had normal spurs averaging 49 mm.

● **QUESTION** If spur length has no effect on reproductive success, what would the data in this table look like?

● **QUESTION** Why did the researchers bother to put yarn around the control flowers?

The directed-pollination hypothesis also has strong experimental support. Consider, for example, recent work on a South African orchid called *Disa draconis*. The length of the long tube, or spur, located at the back of this flower varies among populations of this species. As predicted by the directed-pollination hypothesis, each population is pollinated by a different insect. Short-spurred plants that grow in mountain habitats are pollinated by insect species that have a relatively short proboscis. (A *proboscis* is a specialized mouthpart found in some insects. When extended, it functions like a straw in sucking nectar or other fluids.) Long-spurred orchids that grow on low-lying sandplain habitats are pollinated by insects that have a particularly long proboscis.

To test the hypothesis that spur length affects pollination success, researchers artificially shortened the spurs of individuals in a long-spurred population (**Figure 30.24**). The biologists did this by tying off the spurs of randomly selected flowers with a piece of yarn, so that flies could not reach the end of the spur. The idea was that flies would not make contact with the flower's reproductive organs when they inserted their proboscis into the shortened tube. The biologists left nearby flowers alone but tied a piece of yarn near the spur. As the data in the "Results" box of Figure 30.24 indicate, flowers with short spurs received much less pollen and set much less seed than did flowers with normal-length spurs. These data strongly support the hypothesis that spur length is an adaptation that increases the frequency of pollination by particular insects. Based on results like this, biologists contend that the spectacular diversity of angiosperms resulted, at least in part, from natural selection exerted by the equally spectacular diversity of insect, mammal, and bird pollinators.

Fruits The evolution of the ovary was an important event in land plant diversification, but not only because it protected the

(a) Fruits are derived from ovaries and contain seeds.

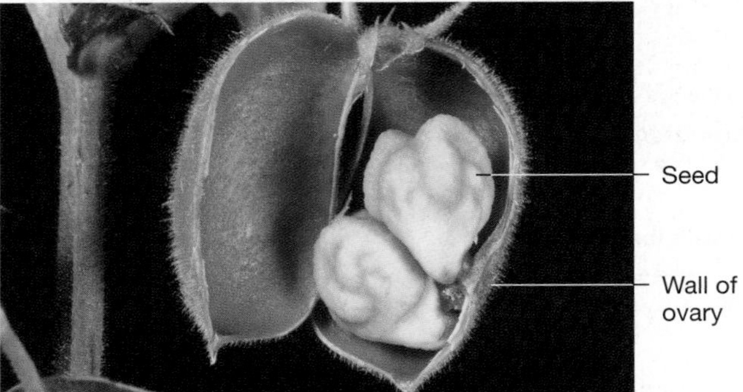

(b) Many fruits are dispersed by animals.

FIGURE 30.25 Fruits Are Derived from Ovaries Found in Angiosperms. (a) A pea pod is one of the simplest types of fruit. **(b)** The ovary wall often becomes thick, fleshy, and nutritious enough to attract animals that disperse the seeds inside.

female gametophytes of angiosperms. It also made the evolution of fruit possible. A **fruit** is a structure that is derived from the **ovary** and encloses one or more seeds (**Figure 30.25a**). Tissues derived from the ovary are often nutritious and brightly colored. Animals eat these types of fruits, digest the nutritious tissue around the seeds, and disperse seeds in their feces (**Figure 30.25b**). In other cases, the tissues derived from the ovary help fruits disperse via wind or water. The evolution of flowers made efficient pollination possible; the evolution of fruits made efficient seed dispersal possible.

The list of adaptations that allow land plants to reproduce in dry environments is impressive; **Figure 30.26** summarizes them. Once land plants had vascular tissue and could grow efficiently in dry habitats, the story of their diversification revolved around traits that allowed sperm cells to reach eggs efficiently and helped seeds disperse to new locations.

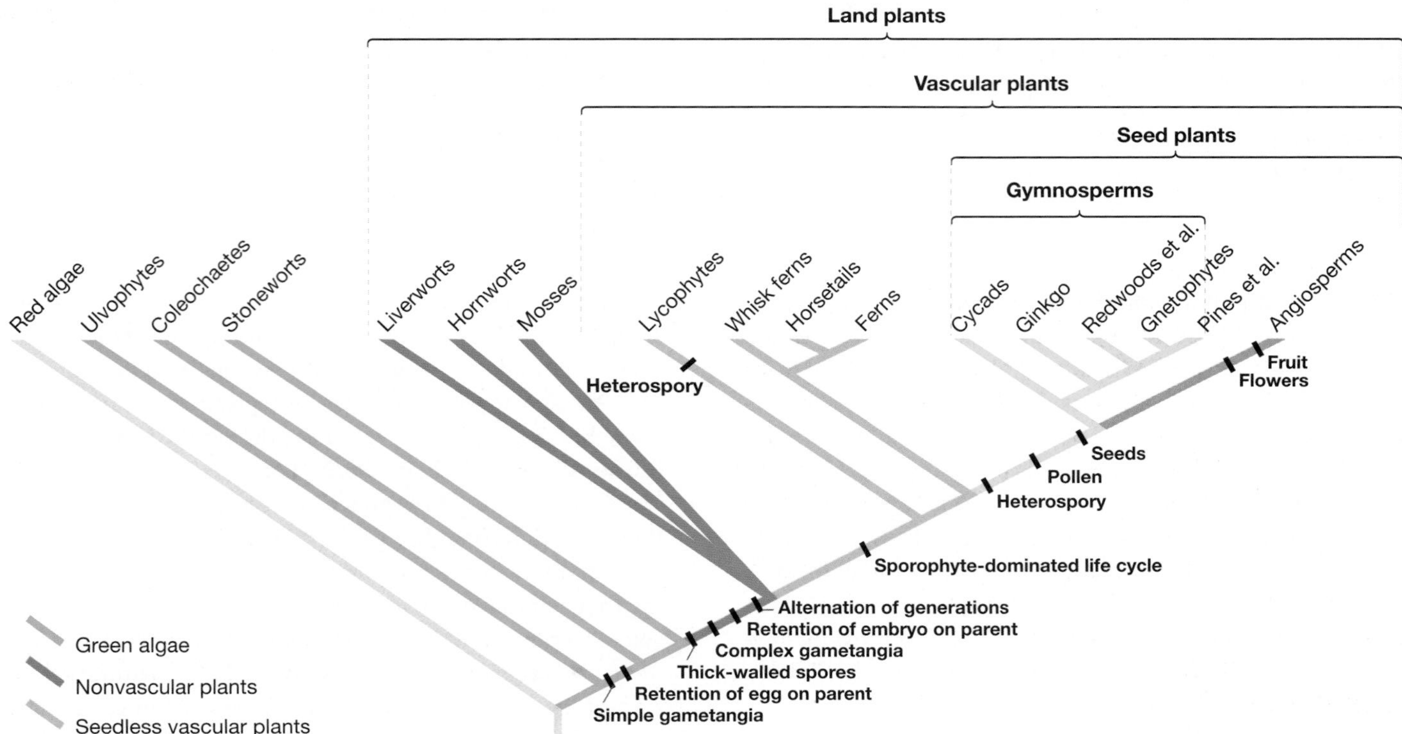

FIGURE 30.26 Evolutionary Innovations Allowed Plants to Reproduce Efficiently on Land.

EXERCISE Conifers, pines, gnetophytes, and angiosperms are the only land plants that do not have flagellated sperm that swim to the egg (at least a short distance). Mark the loss of flagellated sperm on the tree.

FIGURE 30.27 Four Morphological Differences between Monocots and Dicots.

The Angiosperm Radiation

For the past 125 million years, land plant diversification has really been about angiosperms. Angiosperms represent one of the great adaptive radiations in the history of life. As Chapter 27 noted, an **adaptive radiation** occurs when a single lineage produces a large number of descendant species that are adapted to a wide variety of habitats. The diversification of angiosperms is associated with three key adaptations: (1) vessels, (2) flowers, and (3) fruits. In combination, these traits allow angiosperms to transport water, pollen, and seeds efficiently. Based on these observations, it is not surprising that most land plants living today are angiosperms.

On the basis of morphological traits, the 250,000 species of angiosperms identified to date have traditionally been classified into two major groups: the monocotyledons, or **monocots**, and the dicotyledons, or **dicots**. Some familiar monocots include the grasses, orchids, palms, and lilies; familiar dicots include the roses, buttercups, daisies, oaks, and maples. The names of the two groups were inspired by differences in a structure called the cotyledon. A **cotyledon** is the first leaf that is formed in an embryonic plant. As **Figure 30.27** shows, monocots have a single cotyledon (visible inside the seed) while dicots have two cotyledons (visible in newly germinated plants). The figure also highlights other major morphological differences observed in monocots and dicots, concerning the arrangement of vascular tissue and leaf veins and the characteristics of flowers.

It would be misleading, however, to think that all species of flowering plants fall into one of these two groups—either monocots or dicots. Recent work has shown that dicots do not form a natural group consisting of a common ancestor and all of its descendants. To drive this point home, consider the phylogeny illustrated in **Figure 30.28**. These relationships were estimated by comparing the sequences of several genes that are shared by all angiosperms. Notice that species with dicot-like

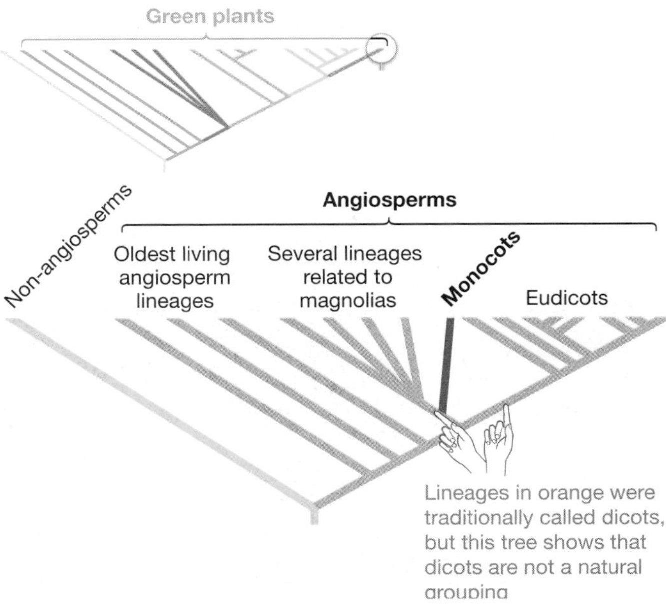

FIGURE 30.28 Monocots Are Monophyletic, but Dicots Are Paraphyletic.

characters are scattered around the angiosperm phylogenetic tree. Based on this analysis, biologists have concluded that although monocots are monophyletic, dicots are not. Instead, they are paraphyletic.

Biologists have adjusted the names assigned to angiosperm lineages to reflect this new knowledge of phylogeny. The most important of these changes was identifying the **eudicots** ("true dicots") as a lineage that includes roses, daisies, and maples. Plant systematists continue to work toward understanding relationships throughout the angiosperm phylogenetic tree, and there will undoubtedly be more name changes as knowledge grows.

Check Your Understanding

If you understand that...

- Land plants were able to make the transition to growing in terrestrial environments, where sunlight and carbon dioxide are abundant, based on a series of evolutionary innovations.
- Adaptations for growing on land included cuticle, stomata, and vascular tissue.
- Adaptations for effective reproduction on land included gametangia, the retention of embryos on the parent, pollen, seeds, flowers, and fruits.

You should be able to...

1) Explain why each of these adaptations was important in survival or reproduction.
2) Map where each of these evolutionary innovations occurred on the phylogenetic tree of green plants.

30.4 Key Lineages of Green Plants

The evolution of cuticle, pores, stomata, and water-conducting tissues allowed green plants to grow on land, where resources for photosynthesis are abundant. Once the green plants were on land, the evolution of gametangia, retained embryos, pollen, seeds, and flowers enabled them to reproduce efficiently even in very dry environments. The adaptations reviewed in Section 30.3 allowed the land plants to make the most important water-to-land transition in the history of life.

To explore green plant diversity in more detail, let's take a closer look at some major groups of green algae. Then we'll discuss the major phyla of land plants in the context of their broad morphological groupings: nonvascular plants, seedless vascular plants, and seed plants.

Green Algae

The **green algae** are a paraphyletic group that totals about 7000 species. Their bright green chloroplasts are similar to those found in land plants, with a double membrane and chlorophylls *a* and *b* but relatively few accessory pigments. And like land plants, green algae synthesize starch in the chloroplast as a storage product of photosynthesis and have a cell wall composed primarily of cellulose. Green algae are important primary producers in nearshore ocean environments and in all types of freshwater habitats. They are also found in several types of more exotic environments, including snowfields at high elevations, pack ice, and ice floes. These habitats are often splashed with bright colors due to large concentrations of unicellular green algae (**Figure 30.29a**). Although these cells live at near-freezing temperatures, they make all their own food via photosynthesis.

In addition, green algae live in close association with an array of other organisms:

- **Lichens** are stable associations between green algae and fungi or between cyanobacteria and fungi, and are often found in terrestrial environments that lack soil, such as tree bark or bare rock (**Figure 30.29b**). The algae or cyanobacteria are

(a) Green algae with red carotenoid pigments are responsible for pink snow.

(b) Most lichens are an association between fungi and green algae.

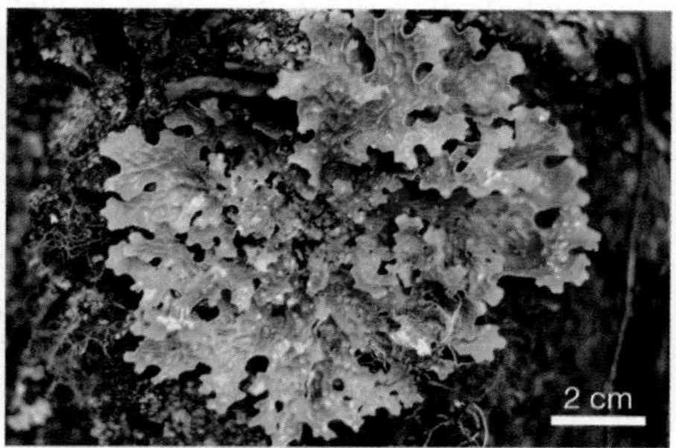

(c) Many unicellullar protists harbor green algae.

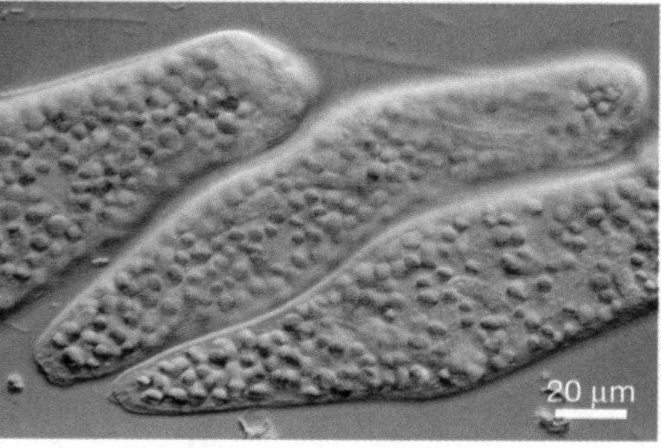

FIGURE 30.29 Some Green Algae Live in Unusual Environments.

protected from drying by the fungus; in return they provide sugars produced by photosynthesis. Of the 17,000 species of lichens described to date, about 85 percent involve green algae. The green algae that are involved are unicellular or grow in long filaments. Lichens are explored in more detail in Chapter 31.

- Unicellular green algae are common endosymbionts in planktonic protists that live in lakes and ponds (**Figure 30.29c**). The association is considered mutually beneficial: The algae supply the protists with food; the protists provide protection to the algae.

Green algae are a large and fascinating group of organisms. Let's take a closer look at just three of the many lineages (**Figure 30.30**).

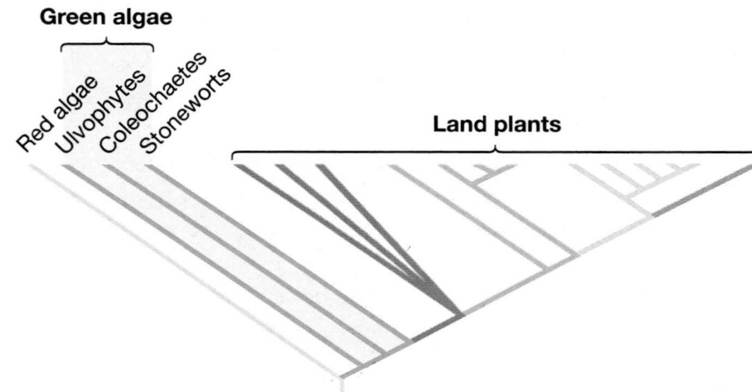

FIGURE 30.30 Green Algae Are Paraphyletic.

Green Algae > Coleochaetophyceae (Coleochaetes)

There are 19 species in this group. Most coleochaetes are barely visible to the unaided eye and grow as flat sheets of cells (**Figure 30.31**). They are considered multicellular because they have specialized photosynthetic and reproductive cells and because they contain **plasmodesmata**—structures introduced in Chapter 8 that connect adjacent cells.

The coleochaetes are strictly freshwater algae. They grow attached to aquatic plants such as water lilies and cattails or over submerged rocks in lakes and ponds. When they grow near beaches, they are often exposed to air when water levels drop in late summer.

Reproduction Asexual reproduction is common in coleochaetes and involves production of flagellated spores. During sexual reproduction, eggs are retained on the parent and are nourished after fertilization with the aid of transfer cells—a situation very similar to that observed in land plants. In some species certain individuals are male and produce only sperm, while other individuals are female and produce only eggs.

Life cycle Alternation of generations does not occur. Multicellular individuals are haploid; the only diploid stage in the life cycle is the zygote.

Human and ecological impacts Because they are very closely related to land plants, coleochaetes are studied intensively by researchers interested in how land plants made the water-to-land transition.

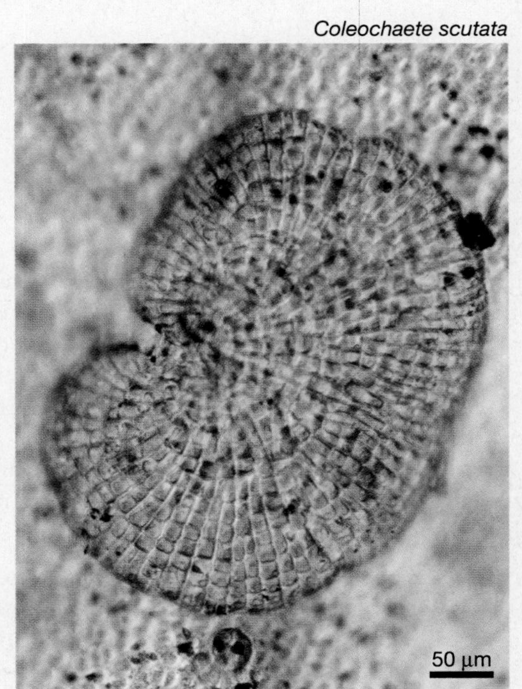

FIGURE 30.31 Coleochaetes Are Thin Sheets of Cells.

Green Algae > Ulvophyceae (Ulvophytes)

The Ulvophyceae are a monophyletic group composed of several diverse and important subgroups, with a total of about 4000 species. Members of this lineage range from unicellular to multicellular.

Many of the large green algae in habitats along ocean coastlines are members of the Ulvophyceae. *Ulva*, the sea lettuce (**Figure 30.32**), is a representative marine species. But there are also large numbers of unicellular or small multicellular species that inhabit the plankton of freshwater lakes and streams.

Reproduction Most ulvophytes reproduce both asexually and sexually. Asexual reproduction often involves production of spores that swim with the aid of flagella. Sexual reproduction usually results in production of a resting stage—a cell that is dormant in winter. In many species the gametes are not called eggs and sperm, because they are the same size and shape. In most species gametes are shed into the water, so fertilization takes place away from the parent plants.

Life cycle Many unicellular forms are diploid only as zygotes. Alternation of generations occurs in multicellular species. When alternation of generation occurs, gametophytes and sporophytes may look identical or different.

Human and ecological impacts Ulvophyceae are important primary producers in freshwater environments and in coastal areas of the oceans.

Ulva lactuca

FIGURE 30.32 Green Algae Are Important Primary Producers in Aquatic Environments.

Green Algae > Charaphyceae (Stoneworts)

There are several hundred species in this group. They are collectively known as stoneworts, because they commonly accumulate crusts of calcium carbonate ($CaCO_3$) over their surfaces. Like the coleochaetes, they have plasmodesmata and are multicellular. ● You should be able to mark the origin of plasmodesmata on Figure 30.30. (They do not occur in ulvophytes.) Some species of stonewort can be a meter or more in length.

The stoneworts are freshwater algae. Certain species are specialized for growing in relatively deep waters, though most live in shallow water near lake beaches or pond edges.

Reproduction Sexual reproduction is common and involves production of prominent, multicellular gametangia similar to those observed in early land plants. As in coleochaetes, in stoneworts the eggs are retained on the parent plant, which supplies eggs with nutrients prior to fertilization. ● You should be able to mark the origin of egg retention on Figure 30.30. (It does not occur in ulvophytes.)

Life cycle Alternation of generations does not occur. Multicellular individuals are haploid; the only diploid stage in the life cycle is the zygote.

Human and ecological impacts Some species form extensive beds in lake bottoms or ponds and provide food for ducks and geese as well as food and shelter for fish (**Figure 30.33**). They are a good indicator that water is not polluted.

Chara globularis (tall) and *C. fibrosa* (short)

FIGURE 30.33 Stoneworts Can Form Beds on Lake Bottoms.

Nonvascular Plants ("Bryophytes")

The most basal lineages of land plants are collectively known as the **bryophytes**. The evolutionary relationships among the three lineages with living species—liverworts, hornworts, and mosses—are still unclear (**Figure 30.34**).

All of the nonvascular plant species present today have a low, sprawling growth habit. In fact, it is unusual to find bryophytes more than 5 to 10 centimeters tall. Individuals are anchored to soil, rocks, or tree bark by structures called **rhizoids**. No bryophytes have vascular tissue with lignin-reinforced cell walls. In the lineages present today, simple water-conducting cells and tissues are found only in some mosses. All bryophytes have flagellated sperm that swim to eggs through raindrops or small puddles on the plant surface. Spores are dispersed by wind.

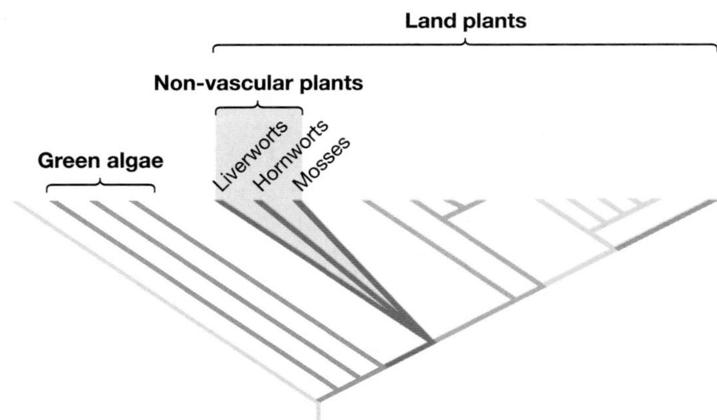

FIGURE 30.34 It Is Unclear Whether Bryophytes Are Monophyletic or Paraphyletic.

Nonvascular Plants > Bryophyta (Mosses)

Over 12,000 species of mosses have been named and described to date, and more are being discovered every year—particularly in the tropics. Mosses are informally grouped with other "bryophytes" (liverworts and hornworts) but are formally classified in their own monophyletic group: the phylum Bryophyta.

Although mosses are common in moist forests, they can also be abundant in more extreme environments, such as deserts and windy, treeless habitats in the Arctic, Antarctic, or mountaintops. In these severe conditions, mosses are able to thrive because their bodies can become extremely dry without dying. When the weather makes photosynthesis difficult, individuals dry out and become dormant or inactive (**Figure 30.35a**). Then when rains arrive or temperatures warm, the plants rehydrate and begin photosynthesis and reproduction (**Figure 30.35b**).

Adaptations to land One subgroup of mosses contains simple conducting tissues consisting of cells that are specialized for the transport of water or food. But because these cells do not have walls that are reinforced by lignin, they are not considered true vascular tissue. ● You should be able to mark the origin of the simple water-conducting cells and tissues in this moss sub-group on Figure 30.34. Because they lack true vascular tissue, most mosses are not able to grow much taller than a few centimeters.

Reproduction Asexual reproduction often occurs by fragmentation, meaning that pieces of gametophytes that are broken off by wind or a passing animal can begin growing independently. In many species, sexual reproduction cannot involve self-fertilization because the sexes are separate—meaning that an individual plant produces only eggs in archegonia or only sperm in antheridia. A typical sporophyte produces up to 50 million tiny spores. Spores are usually distributed by wind.

Life cycle The moss life cycle is similar to that of liverworts and hornworts: The sporophyte is retained on the much larger and longer-lived gametophyte and gets most of its nutrition from the gametophyte.

Human and ecological impacts Species in the genus *Sphagnum* are often the most abundant plant in wet habitats of northern environments. Because *Sphagnum*-rich environments account for 1 percent of Earth's total land area, equivalent to half the area of the United States, *Sphagnum* species are among the most abundant plants in the world. *Sphagnum*-rich habitats have an exceptionally short growing season, however, so the decomposition of dead mosses and other plants is slow. As a result, large deposits of semi-decayed organic matter, known as **peat**, accumulate. Researchers estimate that the world's peatlands store about 400 billion metric tons of carbon. If peatlands begin to burn or decay rapidly due to global warming, the CO_2 released will exacerbate the warming trend (see Chapter 54).

Peat is harvested as a traditional heating and cooking fuel in some countries. It is also widely used as a soil additive in gardening, because *Sphagnum* can absorb up to 20 times its dry weight in water. This high water-holding capacity is due to the presence of large numbers of dead cells in the leaves of these mosses, which readily fill with water via pores in their walls.

(a) Moss in dry weather

(b) Moss in wet weather

FIGURE 30.35 Many Moss Species Can Become Dormant When Conditions Are Dry.

Nonvascular Plants > Hepaticophyta (Liverworts)

Liverworts got their name because some species native to Europe have liver-shaped leaves. According to the medieval *Doctrine of Signatures*, God indicated how certain plants should be used by giving them a distinctive appearance. Thus, liverwort teas were hypothesized to be beneficial for liver ailments. (They are not.) About 6500 species are known. They are commonly found growing on damp forest floors or riverbanks, often in dense mats (**Figure 30.36**), or on the trunks or branches of tropical trees.

Adaptations to Land Liverworts are covered with cuticle. Some species have pores that allow gas exchange; in species that lack pores, the cuticle is very thin.

Reproduction Asexual reproduction occurs when fragments of a plant are broken off and begin growing independently. Some species also produce small structures called **gemmae** asexually, during the gametophyte phase. Mature gemmae are knocked off the parent plant by rain and grow into independent gametophytes. During sexual reproduction, sperm and eggs are produced in gametangia.

Life cycle The gametophyte is the largest and longest-lived phase in the life cycle. Sporophytes are small, grow directly from the gametophyte, and depend on the gametophyte for nutrition. Spores are shed from the sporophyte and are carried away by wind or rain.

Human and ecological impacts When liverworts grow on bare rock or tree bark, their dead and decaying body parts contribute to the initial stages of soil formation.

Marchantia bryophyta

1 cm

FIGURE 30.36 Liverworts Thrive in Moist Habitats.

Nonvascular Plants > Anthocerophyta (Hornworts)

Hornworts got their name because their sporophytes have a horn-like appearance (**Figure 30.37**) and because *wort* is the Anglo-Saxon word for *plant*. About 100 species have been described to date.

Adaptations to land Hornwort sporophytes have stomata. Research is under way to determine if they can open their pores to allow gas exchange or close their pores to avoid water loss during dry intervals.

Reproduction Depending on the species, gametophytes may contain only egg-producing archegonia or only sperm-producing antheridia, or both. Stated another way, individuals of some species are either female or male, while in other species each individual has both types of reproductive organs.

Life cycle The gametophyte is the longest-lived phase in the life cycle. Although sporophytes grow directly from the gametophyte, they are green because their cells contain chloroplasts. Sporophytes manufacture some of their own food but also get nutrition from the gametophyte. Spores disperse from the parent plant via wind or rain.

Human and ecological impacts Some species harbor symbiotic cyanobacteria that fix nitrogen.

Phaeocerus leavis

FIGURE 30.37 Hornworts Have Horn-Shaped Sporophytes.

Seedless Vascular Plants

The seedless vascular plants are a paraphyletic group that forms a grade between the nonvascular plants and the seed plants (**Figure 30.38**). All species of seedless vascular plants have conducting tissues with cells that are reinforced with lignin, forming vascular tissue. Tree-sized lycophytes and horsetails are abundant in the fossil record, and tree ferns are still common inhabitants of certain habitats, such as mountain slopes in the tropics.

The sporophyte is the larger and longer-lived phase of the life cycle in all of the seedless vascular plants. The gametophyte is physically independent of the sporophyte, however. Eggs are retained on the gametophyte, and sperm swim to the egg with the aid of flagella. Thus, seedless vascular plants depend on the presence of water for reproduction—they need enough water to form a continuous layer that "connects" gametophytes and allows sperm to swim to eggs. Sporophytes develop on the gametophyte and are nourished by the gametophyte when they are small.

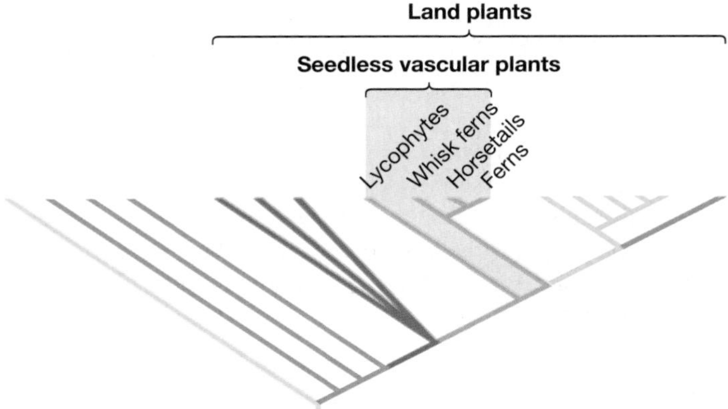

FIGURE 30.38 The Seedless Vascular Plants Are Paraphyletic.

Vascular Plants > Lycophyta (Lycophytes, or Club Mosses)

Although the fossil record documents lycophytes that were 2 m wide and 40 m tall, the 1000 species of lycophytes living today are all small in stature (**Figure 30.39**). Most live on the forest floor or on the branches or trunks of tropical trees. Because of their appearance, they are often called ground pines or **club mosses**—even though they are neither pines nor mosses.

Adaptations to land Lycophytes are the most ancient land plant lineage with **roots**—a belowground system of tissues and organs that anchors the plant and is responsible for absorbing water and mineral nutrients. Roots differ from the rhizoids observed in bryophytes, because roots contain vascular tissue and thus are capable of conducting water and nutrients from belowground to the upper reaches of the plant. Lycophytes are distinguished by small leaflike structures called microphylls that extend from the stems. ● You should be able to mark the origin of microphylls on Figure 30.38.

Reproduction Asexual reproduction can occur by fragmentation or gemmae. During sexual reproduction, spores of some species give rise to bisexual gametophytes—meaning that each gametophyte produces both eggs and sperm. In the genera called *Selaginella* and *Isoetes*, however, heterospory occurs and gametophytes are male or female. Self-fertilization is extremely rare in most of these species, however. Some club mosses have separate male and female gametophytes.

Lycopodium species

FIGURE 30.39 Lycophytes Living Today Are Small in Stature.

Life cycle The gametophytes of some species live entirely underground and get their nutrition from symbiotic fungi. In certain species, gametophytes live 6 to 15 years and give rise to a large number of sporophytes over time.

Human and ecological impacts Tree-sized lycophytes were abundant in the coal-forming forests of the Carboniferous period. In coal-fired power plants today, electricity is being generated by burning fossilized lycophyte trunks and leaves.

Vascular Plants > Psilotophyta (Whisk Ferns)

Only two genera of whisk ferns are living today, and there are perhaps six distinct species. Whisk ferns are restricted to tropical regions and have no fossil record. They are extremely simple morphologically, with aboveground parts consisting of branching stems that have tiny, scale-like outgrowths instead of leaves (**Figure 30.40**).

Adaptations to land Whisk ferns lack roots. Some species gain most of their nutrition from fungi that grow in association with the whisk ferns' extensive underground stems called **rhizomes**. Other species grow in rock crevices or are **epiphytes** ("upon-plants"), meaning that they grow on the trunks or branches of other plants—in this case, in the branches of tree ferns. ● After reviewing where leaves and roots originated during land plant evolution (see Figure 30.12), you should be able to mark the loss of leaves and roots in whisk ferns on Figure 30.38.

Tmesipteris species

FIGURE 30.40 Psilotophytes Are Extremely Simple Morphologically.

Reproduction Asexual reproduction is common in sporophytes via the extension of rhizomes and the production of new aboveground stems. When spores mature, they are dispersed by wind and germinate into gametophytes that contain both archegonia and antheridia.

Life cycle Sporophytes may be up to 30 cm tall, but gametophytes are less than 2 mm long and live under the soil surface. Gametophytes absorb nutrients directly from the surrounding soil and from symbiotic fungi. Fertilization takes place inside the archegonium, and the sporophyte develops directly on the gametophyte.

Human and ecological impacts Some whisk fern species are popular landscaping plants, particularly in Japan. The same species can be a serious pest in greenhouses.

Vascular Plants > Sphenophyta (or Equisetophyta) (Horsetails)

Although horsetails are prominent in the fossil record of land plants, just 15 species are known today. All 15 are in the genus *Equisetum*; the phylum as a whole is sometimes called Equisetophyta. Translated literally, *Equisetum* means "horse-bristle." Both the scientific name and the common name, horsetail, come from the brushy appearance of the stems and branches in some species (**Figure 30.41**). Horsetails may be locally abundant in wet habitats such as stream banks or marsh edges.

Adaptations to land Horsetails have an interesting adaptation that allows them to flourish in waterlogged, oxygen-poor soils. Horsetail stems are hollow, so oxygen readily diffuses down the stem to reach roots that cannot obtain oxygen from the surrounding soil. Horsetails are also distinguished by having whorled leaves and branches. ● You should be able to mark the origin of hollow stems, whorled branches, and whorled leaves on Figure 30.38.

Reproduction Asexual reproduction is common in sporophytes and occurs via fragmentation or the extension of rhizomes. From these rhizomes, two types of erect, specialized stems may grow—stems that contain tiny leaves and chloroplast-rich branches and that are specialized for photosynthesis, or stems that bear clusters of sporangia and produce huge numbers of spores by meiosis (Figure 30.41).

Life cycle Gametophytes perform photosynthesis but are small and short lived. They normally produce both antheridia and archegonia, but in most cases the sperm-producing structure matures first. This pattern is thought to be an adaptation that minimizes self-fertilization and maximizes cross-fertilization.

Human and ecological impacts Horsetail stems are rich in silica granules. The glass-like deposits not only strengthen the stem but also make these plants useful for scouring pots and pans—hence these plants are often called "scouring rushes."

Equisetum arvense

FIGURE 30.41 Horsetails Have Separate Reproductive and Vegetative Stalks.

Vascular Plants > Pteridophyta (Ferns)

With 12,000 species, ferns are by far the most species-rich group of seedless vascular plants. They are particularly abundant in the tropics. About a third of the tropical species are epiphytes, usually growing on the trunks or branches of trees. Species that can grow epiphytically live high above the forest floor, where competition for light is reduced, without making wood and growing tall themselves. The growth habits of ferns are highly variable among species, however, and ferns range in size from rosettes the size of your smallest fingernail to 20-meter-tall trees (**Figure 30.42a**).

Adaptations to land Ferns are the only seedless vascular plants that have large, well-developed leaves—commonly called **fronds**. Leaves give the plant a large surface area, allowing it to capture sunlight for photosynthesis efficiently.

Reproduction In a few species, gametophytes reproduce asexually via production of gemmae. Typically, species that can reproduce via gemmae never produce gametes or sporophytes. In most species, however, sexual reproduction is the norm.

Life cycle Although fern gametophytes contain chloroplasts and are photosynthetic, the sporophyte is typically the larger and longer-lived phase of the life cycle. In mature sporophytes, sporangia are usually found in clusters called **sori** on the undersides of leaves (**Figure 30.42b**). The structure of the sporangia is a distinctive feature of ferns: It arises from a single cell and has a wall composed of a single cell layer. ● You should be able to mark the evolution of the distinctive fern sporangium on Figure 30.38.

Human and ecological impacts In many parts of the world, people gather the young, unfolding fronds, or "fiddleheads," of ferns in spring as food. Ferns are also widely used as ornamental plants in landscaping.

(a) Ferns range from small to tree sized.

Gonocormus minutus

Dicksonia antarctica

(b) Fern sporangia are often located on the underside of fronds.

Polypodium vulgare

FIGURE 30.42 Ferns Are the Most Species-Rich Group of Seedless Vascular Plants.

Seed Plants

The seed plants are a monophyletic group consisting of the gymnosperms—cycads, ginkgo, conifers, gnetophytes, and pines—and the angiosperms (**Figure 30.43**). Seed plants are defined by the production of seeds and the production of pollen grains. Recall from Section 30.3 that seeds are a specialized structure for dispersing embryonic sporophytes to new locations. Seeds are the mature form of a fertilized ovule, the female reproductive structure that encloses the female gametophyte and egg cell. Also recall from Section 30.3 that pollen grains are tiny, sperm-producing gametophytes that are easily dispersed through air as opposed to water. The structure and function of seed plants is the focus of Chapters 36 through 40.

Seed plants are found in virtually every type of habitat, and they adopt every growth habit known in land plants. Their forms range from mosslike mats to shrubs and vines to 100-meter-tall trees. Seed plants can be annual or perennial, with life spans ranging from a few weeks to almost five thousand years.

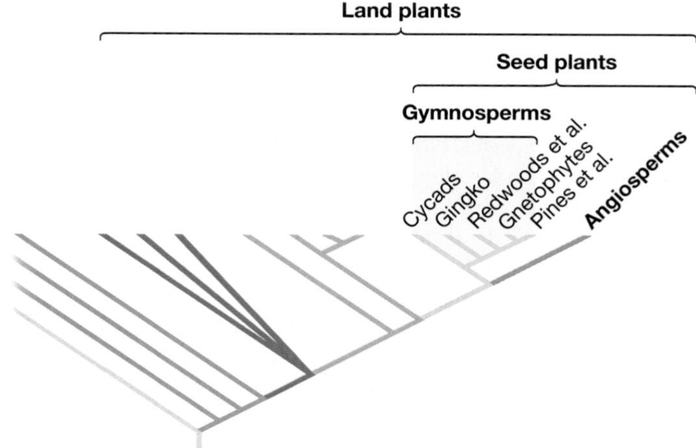

FIGURE 30.43 The Seed Plants Are a Monophyletic Group.

Seed Plants > Gymnosperms > Cycadophyta (Cycads)

The cycads are so similar in overall appearance to palm trees, which are angiosperms, that cycads are sometimes called "sago palms." Although cycads were extremely abundant when dinosaurs were present on Earth 150–65 million years ago, only about 140 species are living today. Most are found in the tropics (**Figure 30.44**).

FIGURE 30.44 Cycads Resemble Palms but Are Not Closely Related to Them.

Adaptations to land Cycads have woody stems; many are tree sized. They are unique among gymnosperms in having compound leaves—meaning that each leaf is divided into many smaller leaflets. ● You should be able to mark the origin of the distinctive cycad leaf on Figure 30.43.

Reproduction and life cycle Like other seed plants, cycads are heterosporous. Each sporophyte individual bears either microsporangia or megasporangia, but not both. Pollen is carried by insects (usually beetles or weevils) or, in some species, wind. Cycad seeds are large and often brightly colored. The colors attract birds and mammals, both of which eat and disperse the seeds.

Human and ecological impacts Cycads harbor large numbers of symbiotic cyanobacteria in specialized, aboveground root structures. The cyanobacteria are photosynthetic and fix nitrogen. The nitrogen acts as an important nutrient for nearby plants as well as the cycads themselves. Cycads are popular landscaping plants in some parts of the world.

Seed Plants > Gymnosperms > Ginkgophyta (Ginkgos)

Although ginkgos have an extensive fossil record, just one species is alive today. Leaves from the ginkgo, or maidenhair, tree are virtually identical in size and shape to those observed in fossil ginkgos that are 150 million years old (**Figure 30.45**).

(a) Fossil ginkgo *Ginkgo huttoni*

(b) Living ginkgo *Ginkgo biloba*

FIGURE 30.45 The Ginkgo Tree Is a "Living Fossil."

Adaptations to land Unlike most gymnosperms, the ginkgo is **deciduous**—meaning that it loses its leaves each autumn. This adaptation allows plants to be dormant during the winter, when photosynthesis and growth are difficult.

Reproduction and life cycle Sexes are separate—individuals are either male or female. Pollen is transported by wind. Sperm have flagella, however. Once pollen grains land near the female gametophyte and mature, the sperm cells leave the pollen grain and swim to the egg cells.

Human and ecological impacts Although today's ginkgo trees are native to southeast China, they are planted widely as an ornamental all over the world. They are especially popular in urban areas, as they are tolerant of air pollution. In some countries, the inside of the seed is eaten as a delicacy.

Seed Plants > Gymnosperms > Gnetophyta (Gnetophytes)

The gnetophytes comprise about 70 species in three genera. One genus consists of vines and trees from the tropics. A second is made up of desert-dwelling shrubs, including what may be the most familiar gnetophyte—the shrub called Mormon tea, which is common in the deserts of southwestern North America (see Figure 30.7c). The third genus contains a single species that probably qualifies as the world's most bizarre plant (**Figure 30.46**)—*Welwitschia mirabilis*, which is native to the deserts of southwest Africa. Although it has large belowground structures, the aboveground part consists of just two strap-like leaves, which grow continuously from the base and die at the tips. The leaves also split lengthwise as they grow and age.

Welwitschia mirabilis

FIGURE 30.46 *Welwitschia* Is an Unusual Plant.

Adaptations to land Gnetophytes have vessel elements in addition to tracheids. All of the living species make wood as a support structure.

Reproduction and life cycle The microsporangia and megasporangia are arranged in clusters at the end of stalks, similar to the way flowers are clustered in some angiosperms. Pollen is transferred by the wind or by insects. Double fertilization occurs in two of the three genera. As in other gymnosperms, seeds do not form inside an encapsulated structure.

Human and ecological impacts The drug ephedrine was originally isolated from species in the gnetophyte genus *Ephredra*, which is native to northern China and Mongolia. Ephedrine is used in the treatment of hay fever, colds, and asthma.

Seed Plants > Gymnosperms > Pinophyta (Pines, Spruces, Firs)

The gymnosperms include two major lineages of cone-bearing species: the pines and allies discussed here, and a group that includes redwoods, junipers, yews, and cypresses, featured below. Species in both lineages have a reproductive structure called the cone, in which microsporangia and megasporangia are produced (**Figure 30.47**).

The Pinophyta includes the familiar pines, spruces, firs, Douglas fir, tamaracks, and true cedars. These are among the largest and most abundant trees on the planet, as well as some of the most long-lived. One of the bristlecone pines native to southwestern North America is at least 4900 years old.

Adaptations to land Pinophyta have needle-like leaves, with a small surface area that allows them to thrive in habitats where water is scarce. Pines are common on sandy soils that have poor water-holding capacity, and spruces and firs are common in cold environments where water is often frozen. All of the living species make wood as a support structure.

Reproduction and life cycle Pollen is transferred by the wind.

Human and ecological impacts Pines, spruces, firs, and other species in this group dominate forests that grow at high latitudes and high elevations. Their seeds are key food sources for a variety of birds, squirrels, and mice, and their wood is the basis of the building products and paper industries in many parts of the world. The paper in this book was made from species in this group.

(a) Cones that produce microsporangia and pollen

Picea abies

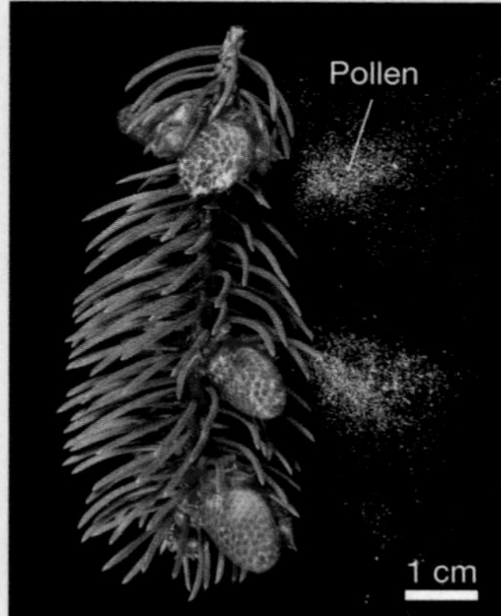

(b) Cones that produce macrosporangia and eggs

Picea abies

FIGURE 30.47 Pollen-Bearing Cones Produce Microsporangla; Ovulate Cones Produce Megasporangla.

Seed Plants > Gymnosperms > Other Conifers (Redwoods, Junipers, Yews)

The species in this lineage vary in growth form from sprawling juniper shrubs to the world's largest plants. Redwood trees growing along the Pacific Coast of North America can reach heights of up to 115 m (375 ft) and trunk diameters of over 11 m (36 ft).

Adaptations to land All of the species in this lineage are trees or large shrubs. Most have narrow leaves, which in many cases are arranged in overlapping scales (**Figure 30.48**). Narrow leaves have a small amount of surface area, which is not optimal for capturing sunlight and performing photosynthesis. But because the small surface area reduces water loss from leaves, many species in this lineage thrive in dry habitats or in cold environments where water is often frozen.

Reproduction and life cycle The species in this group are wind pollinated. As in all seed plants, the female gametophyte is retained on the parent. Thus, fertilization and seed development take place in the female cone. Like other gymnosperms, seeds do not form inside an encapsulated structure. Depending on the species, the seeds are dispersed by wind or by seed-eating birds or mammals.

Human and ecological impacts Redwoods, redcedar, whitecedar, and yellowcedar have wood that is highly rot-resistant and thus prized for making furniture, decks, house siding, or other applications where wood is exposed to the weather. Yew wood is often preferred for making traditional archery bows, and juniper berries are used to flavor gin.

Thuja plicata

FIGURE 30.48 Some Species in This Group Have Scale-like Leaves.

Seed Plants > Anthophyta (Angiosperms)

The flowering plants, or angiosperms, are far and away the most species-rich lineage of land plants. Over 250,000 species have already been described. They range in size from *Lemna gibba*—a floating, aquatic species that is less than half a millimeter wide—to massive oak trees. Angiosperms thrive in desert to freshwater to rain forest environments and are found in virtually every habitat except the deep oceans. They are the most common and abundant plants in terrestrial environments other than northern and high-elevation coniferous forests.

The defining adaptation of angiosperms is the flower. Flowers are reproductive structures that hold either pollen-producing microsporangia or the megasporangia that produce megaspores and eggs, or both. Nectar-producing cells are often present at the base of the flower, and the color of petals helps to attract insects, birds, or bats that carry pollen from one flower to another (**Figure 30.49a**). Some angiosperms are pollinated by wind, however. Wind-pollinated flowers lack both colorful petals and nectar-producing cells (**Figure 30.49b**).

Adaptations to land In addition to flowers, angiosperms evolved vessels, the conducting cells that make water transport particularly efficient. Most angiosperms contain both tracheids and vessels.

Reproduction and life cycle Unlike gymnosperms, angiosperms have a carpel, a structure within the flower that contains an ovary. The ovary encloses the ovule, which in turn encloses the female gametophyte. In most cases, male gametophytes are carried to female gametophytes by pollinators that are inadvertently dusted with pollen as they visit flowers to find food. Depending on the angiosperm species, self-fertilization may be common or absent. When the egg produced by the female gametophyte is fertilized, the ovule develops into a seed. When the ovary matures it forms a fruit, which contains the seed or seeds.

Human and ecological impacts It is almost impossible to overstate the importance of angiosperms to humans and other organisms. In most terrestrial habitats today, angiosperms supply the food that supports virtually every other species. For example, many insects eat flowering plants. Historically, the diversification of angiosperms correlated closely with the diversification of insects, which are by far the most species-rich lineage on the tree of life. It is not unusual for a single tropical tree to support dozens or even hundreds of insect species. Angiosperm seeds and fruits have also supplied the staple foods of virtually every human culture that has ever existed.

(a) Animal-pollinated flower (this species produces both pollen and eggs in the same flower)

Ornithogalum dubium

(b) Wind-pollinated flower (this species has separate male and female flowers)

Acer negundo

Acer negundo

FIGURE 30.49 Wind-Pollinated Flowers Lack Colorful Petals and Nectar.

Chapter Review

SUMMARY OF KEY CONCEPTS

The green plants include both the green algae and the land plants. Green algae are an important source of oxygen and provide food for aquatic organisms; land plants hold soil and water in place, build soil, moderate extreme temperatures and winds, and provide food for other organisms.

Plants improve the quality of the environment, and humans depend on them for food, fiber, and fuel. This dependence became extreme after the domestication of plant species beginning about 10,000 years ago. Artificial selection techniques have produced huge increases in the yields and dramatic changes in the characteristics of domesticated plants.

You should be able to predict how the current and massive loss of plant species and plant communities will affect the environment.

Land plants were the first multicellular organisms that could live with most of their tissues exposed to the air. A series of key adaptations allowed them to survive on land. In terms of total mass, plants dominate today's terrestrial environments.

The land plants evolved from green algae and colonized terrestrial environments in conjunction with fungi. The evolution of cuticle allowed plant tissues to be exposed to air without dying. The evolution of pores provided breaks in the cuticle and facilitated gas exchange, with CO_2 diffusing into leaves and O_2 diffusing out. Later, the evolution of guard cells allowed plants to control the opening and closing of pores in a way that maximizes gas exchange and minimizes water loss.

Vascular tissue evolved in a series of steps, beginning with simple water-conducting cells and tissues such as those observed in today's mosses. True vascular tissue has cells that are dead at maturity and that have secondary cell walls reinforced with lignin. As a result, vascular tissue conducts water and provides structural support that makes erect growth possible. Erect growth is important because it reduces competition for light. Tracheids are water-conducting cells found in all vascular plants; in addition, angiosperms and gnetophytes have water-conducting cells called vessels.

You should be able to describe plant adaptations that solved the following problems posed by terrestrial life: keeping tissues moist, holding the body erect, transporting water, and exchanging gases. Based on your answer, explain why it is logical that algae made the transition to terrestrial life just once.

MB Web Animation at www.masteringbio.com

Plant Evolution and the Phylogenetic Tree

Once plants were able to grow on land, a sequence of important evolutionary changes made it possible for them to reproduce efficiently—even in extremely dry environments.

All land plants are embryophytes, meaning that eggs and embryos are retained on the parent plant. Consequently, the developing embryo can be nourished by its mother. Seed plant embryos are then dispersed from the parent plant to a new location, encased in a protective housing, and supplied with a store of nutrients.

All land plants have alternation of generations. Over the course of land plant evolution, the gametophyte phase became reduced in size and life span and the sporophyte phase became more prominent. In seed plants, male gametophytes are reduced to pollen grains and female gametophytes are reduced to tiny structures that produce an egg. The evolution of pollen was an important breakthrough in the history of life, because sperm no longer needed to swim to the egg—tiny male gametophytes could be transported through the air via wind or insects.

You should be able to describe plant adaptations that solved the following problems posed by terrestrial life: transporting sperm, nourishing embryos, and dispersing embryos. Explain the disadvantage of having spores serve as the dispersal stage in a plant life cycle.

QUESTIONS

Test Your Knowledge

1. Which of the following groups is definitely monophyletic?
   a. nonvascular plants
   b. green algae
   c. green plants
   d. seedless vascular plants

2. What is a difference between tracheids and vessels?
   a. Tracheids are dead at maturity; vessels are alive and are filled with cytoplasm.
   b. Vessels have gaps in the primary and secondary cell wall; tracheids have gaps only in the secondary cell wall.
   c. Only tracheids have a thick secondary cell wall containing lignin.
   d. Only vessels have a thick secondary cell wall containing lignin.

3. Which of the following statements is *not* true?
   a. Green algae in the lineage called Charales are the closest living relatives of land plants.
   b. "Bryophytes" is a name given to the land plant lineages that do not have vascular tissue.
   c. The horsetails and the ferns form a distinct clade, or lineage. They have vascular tissue but reproduce via spores, not seeds.
   d. According to the fossil record and phylogenetic analyses, angiosperms evolved before the gymnosperms. Angiosperms are the only land plants with vessels.

4. The appearance of cuticle and stomata correlated with what event in the evolution of land plants?
   a. the first erect growth forms
   b. the first woody tissues
   c. growth on land
   d. the evolution of the first water-conducting tissues
5. What do seeds contain?
   a. male gametophyte and nutritive tissue
   b. female gametophyte and nutritive tissue
   c. embryo and nutritive tissue
   d. mature sporophyte and nutritive tissue
6. What is a pollen grain?
   a. male gametophyte
   b. female gametophyte
   c. male sporophyte
   d. sperm

Test Your Knowledge answers: 1. c; 2. b; 3. d; 4. c; 5. c; 6. a

Test Your Understanding

Answers are available at www.masteringbio.com

1. Soils, water, and the atmosphere are major components of the abiotic (nonliving) environment. Describe how green plants affect the abiotic environment in ways that are advantageous to humans.
2. The evolution of cuticle presented land plants with a challenge that threatened their ability to live on land. Describe this challenge and explain why stomata represent a solution. Compare and contrast stomata with the pores found in liverworts. Explain why it is logical to observe that liverworts that lack pores have extremely thin cuticle.
3. Diagram four steps in the evolution of vascular tissue. Why was the evolution of lignin-reinforced cell walls significant? In dry habitats, why are vascular plants more common than nonvascular plants, and why are most of them taller?
4. Land plants may have reproductive structures that (1) protect gametes as they develop; (2) nourish developing embryos, (3) allow sperm to be transported in the absence of water, (4) provide stored nutrients and a protective coat so that offspring can be dispersed away from the parent plant, and (5) provide nutritious tissue around seeds that facilitates dispersal by animals. Name each of these five structures, and state which land plant group or groups have each structure.
5. What does it mean to say that a life cycle is gametophyte-dominant versus sporophyte-dominant? Give an example of each type of life cycle.
6. Explain the difference between homosporous and heterosporous plants. Where are the microsporangium and a megasporangium found in a tulip? What happens to the spores that are produced by these structures?

Applying Concepts to New Situations

Answers are available at www.masteringbio.com

1. What is the significance of the observation that some members of the Coleochaetales and Charales synthesize sporopollenin and/or lignin?
2. Vessel elements transport water much more efficiently than tracheids, but are much more susceptible than tracheids to being blocked by air bubbles. Suggest a hypothesis to explain why the vascular tissue of angiosperms consists of a combination of vessel elements and tracheids.
3. Angiosperms such as grasses, oaks, and maples are wind pollinated. The ancestors of these subgroups were probably pollinated by insects, however. As an adaptive advantage, why might a species "revert" to wind pollination? (Hint: Think about the costs and benefits of being pollinated by insects versus wind.) Why is it logical to observe that wind-pollinated species usually grow in dense stands containing many individuals of the same species? Why is it logical to observe that in wind-pollinated deciduous trees, flowers form very early in spring—before leaves form?
4. You have been hired as a field assistant for a researcher interested in the evolution of flower characteristics in orchids. Design an experiment to determine whether color, size, shape, scent, or amount of nectar is the most important factor in attracting pollinators to a particular species. Assume that you can change any flower's color with a dye and that you can remove petals or nectar stores, add particular scents, add nectar by injection, or switch parts among species by cutting and gluing.

www.masteringbio.com is also your resource for • Answers to text, table, and figure caption questions and exercises • Answers to *Check Your Understanding* boxes • Online study guides and quizzes • Additional study tools, including the *E-Book for Biological Science* 3rd ed., textbook art, animations, and videos.

31 Fungi

KEY CONCEPTS

- Fungi are important in part because many species live in close association with land plants. They supply plants with key nutrients and decompose dead wood. They are the master recyclers of nutrients in terrestrial environments.
- All fungi make their living by absorbing nutrients from living or dead organisms. Fungi secrete enzymes so that digestion takes place outside their cells. Their morphology provides a large amount of surface area for efficient absorption.
- Many fungi have unusual life cycles. It is common for species to have a long-lived heterokaryotic stage, in which cells contain haploid nuclei from two different individuals. Although most species reproduce sexually, very few species produce gametes.

Amanita muscaria lives in association with living tree roots. The mushrooms you see are only reproductive structures—the vast majority of this individual exists underground. *Amanita muscaria* mushrooms contain molecules that can induce hallucinations in mammals. This species has been used in religious ceremonies by some native cultures but is considered highly toxic.

Fungi are eukaryotes that grow as single cells or as large, branching networks of multicellular filaments. Familiar fungi include the mushrooms you've encountered in woods or lawns, the molds and mildews in your home, the organism that causes athlete's foot, and the yeasts used in baking and brewing.

Along with the land plants and animals, the **fungi** are one of three major lineages of large, multicellular eukaryotes that occupy terrestrial environments. When it comes to making a living, the species in these three groups use radically different strategies. Land plants make their own food through photosynthesis. Animals eat plants, protists, fungi, or each other. Fungi absorb their nutrition from other organisms—dead or alive.

Fungi that absorb nutrients from dead organisms are the world's most important decomposers. Although a few types of organisms are capable of digesting the cellulose in plant cell walls, fungi and a handful of bacterial species are the only organisms capable of completely digesting both the lignin and cellulose that make up wood. Without fungi, Earth's surface would be piled so high with dead tree trunks and branches that there would be almost no room for animals to move or plants to grow.

Other fungi specialize in absorbing nutrients from living organisms. When fungi absorb these nutrients without providing any benefit in return, they lower the fitness of their host organism and act as parasites. If you've ever had athlete's foot or a vaginal yeast infection, you've hosted a parasitic fungus.

Key Concept Important Information Practice It

The vast majority of fungi that live in association with other organisms benefit their hosts, however. In these cases, fungi are not parasites but **mutualists**. The roots of virtually every land plant in the world are colonized by an array of mutualistic fungi. In exchange for sugars that are synthesized by the host plant, the fungi provide the plant with key nutrients such as nitrogen and phosphorus. Without these nutrients, the host plants grow much more slowly or even starve. It is not possible to overstate the importance of these relationships between living land plants and the fungi that live in their roots. In the soils beneath every prairie, forest, and desert, an underground economy is flourishing. Plants are trading the sugar they manufacture for nitrogen or phosphorus atoms that are available from fungi. These plant-fungal associations are the world's most extensive bartering system. The soil around you is alive with an enormous network of fungi that are fertilizing the plants you see above ground.

In short, fungi are the master traders and recyclers in terrestrial ecosystems. Some fungi release nutrients from dead plants and animals; others transfer nutrients they obtain to living plants. Because they recycle key elements such as carbon, nitrogen, and phosphorus and because they transfer key nutrients to plants, fungi have a profound influence on productivity and biodiversity. In terms of nutrient cycling on the continents, fungi make the world go around.

31.1 Why Do Biologists Study Fungi?

Given their importance to life on land and their intricate relationships to other organisms, it's no surprise that fungi are fascinating to biologists. But there are important practical reasons for humans to study fungi as well. They nourish the plants that nourish us. They affect global warming, because they are critical to the carbon cycle on land. Unfortunately, a handful of species can cause debilitating diseases in humans and crop plants. Let's take a closer look at some of the ways that fungi affect human health and welfare.

Fungi Provide Nutrients for Land Plants

Fungi that live in close association with plant roots are said to be **mycorrhizal** (literally, "fungal-root"). When biologists first discovered how extensive these fungal-plant associations are, they asked an obvious question: Does plant growth suffer if mycorrhizal fungi are absent? **Figure 31.1** shows a result typical of many experiments and provides a convincing answer. In this case, potato seedlings were grown in the presence and absence of the mycorrhizal fungi normally found on their roots. The photographs document that this species grows three to four times faster in the presence of its normal fungal associates than it does without them. For farmers, foresters, and ranchers, the presence of normal mycorrhizal fungi can mean the difference between profit and loss. Fungi are critical to the productivity of forests, croplands, and rangelands.

Fungi Speed the Carbon Cycle on Land

The introduction to this chapter claims that fungi are master decomposers and recyclers. To back up this assertion, consider two particularly dramatic events documented in the fossil record. One episode was based on a lack of fungi, the other on an abundance of fungi:

1. Researchers who examine the fossils present in coal from the Carboniferous period find remarkably few fungi that are capable of degrading dead plant material. Fungi that

FIGURE 31.1 Plants Grow Better in the Presence of Mycorrhizal Fungi. (Left) Typical experimental results when plants are grown with and without their normal mycorrhizal fungi (fungi are not visible in the photo). (Right) Root system of a larch tree seedling, with the mycelium from a mycorrhizal fungus visible.

make their living by digesting dead plant material are called **saprophytes** ("rotten-plants"). Noting the "dip" in the fossil record of saprophytic fungi, researchers hypothesized that their absence was responsible for the enormous buildup of dead plant material that occurred during the Carboniferous period. Recall from Chapter 30 that deposits of compressed, partially decayed plant material are called peat and that coal formed when the peat produced during the Carboniferous period was buried under other sediments and subjected to heat and pressure. Because the peatlands existing today are water-logged and highly acidic, biologists hypothesize that fungi did not grow in the coal-forming swamps of the Carboniferous period because the pH was too low. The message is that coal exists today because conditions were too acidic for fungi to do their job.

2. At the end of the Permian period, 250 million years ago, the greatest mass extinction in the history of life occurred. It is estimated that in substantially less than a million years, over 90 percent of multicellular species were wiped out. When the end-Permian extinction was first recognized, biologists thought it had affected primarily marine organisms. But then other researchers documented a huge "fungal spike" in rocks that formed in terrestrial environments. As **Figure 31.2** shows, the spike is a dramatic but short-lived increase in the number of fungal fossils that coincided with the mass extinction event. The explosion of fungal fossils was the first indication that land plants were also devastated during this interval. The hypothesis is that a massive die-off in trees and shrubs produced gigantic quantities of rotting wood and led to the explosion of fungal abundance documented in the spike.

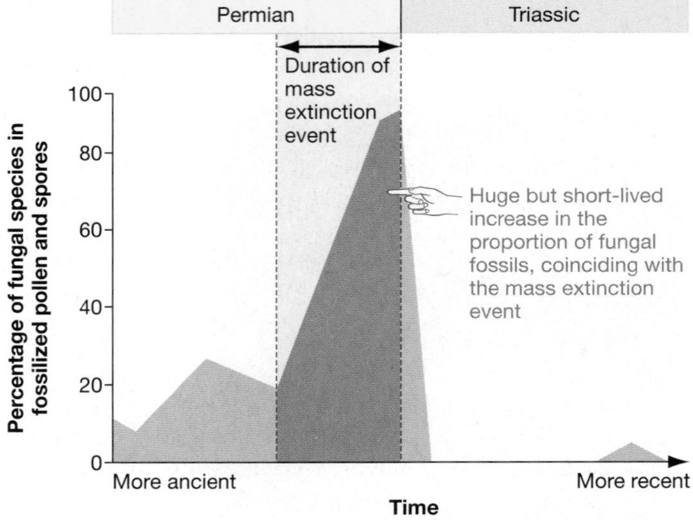

FIGURE 31.2 Fungal Fossils Spike during the End-Permian Extinction. The time interval documents changes in the abundance of fungi before, during, and after the mass extinction event that occurred at the boundary of the Permian and Triassic periods about 250 million years ago.

Saprophytic fungi play a key role in today's terrestrial environments as well. To understand why, recall from Chapter 30 that cells in the vascular tissues of land plants have secondary cell walls containing both lignin and cellulose. Wood forms when stems grow in girth by adding layers of lignin-rich vascular tissue. When trees die, fungi are the organisms that break down wood into sugars and other small organic compounds that they and other organisms can use as food.

Figure 31.3 illustrates the consequences of these facts by highlighting the role that fungi play as carbon atoms cycle through today's terrestrial environments. Note that there are two basic components of the **carbon cycle** on land: (1) the fixation of carbon by land plants—meaning that carbon in atmospheric CO_2 is converted to cellulose, lignin, and other complex organic compounds in the bodies of plants; and (2) the release of CO_2 from plants, animals, and fungi as the result of cellular respiration—meaning the oxidation of glucose and production of the ATP that sustains life. The fundamental point is that, for most carbon atoms, fungi connect the two components. If fungi had not evolved the ability to digest lignin and cellulose soon after land plants evolved the ability to make these compounds, carbon atoms would have been

FIGURE 31.3 Fungi Speed Up the Cycling of Carbon Atoms on Land. Carbon atoms cycle through terrestrial ecosystems. If fungi could not degrade lignin to CO_2 and cellulose to glucose, most carbon would eventually be tied up in indigestible woody tissues. As a result, the cycle would slow dramatically.

EXERCISE Draw an X through the arrow(s) that would not exist if fungi could not digest lignin and cellulose.

sequestered in wood for millennia instead of being rapidly recycled into glucose molecules and CO_2. Terrestrial environments would be radically different than they are today and probably much less productive. On land, fungi make the carbon cycle turn much more rapidly than it would without fungi. The nutrients that fungi release feed a host of other organisms.

Fungi Have Important Economic Impacts

In humans, parasitic fungi cause athlete's foot, vaginitis, diaper rash, ringworm, pneumonia, and thrush, among other miseries. But even though these maladies can be serious, in reality only about 31 species of fungi—out of the hundreds of thousands of existing species—regularly cause illness in humans. Compared with the frequency of diseases caused by bacteria, viruses, and protists, the incidence of fungal infections in humans is low. In addition, soil-dwelling fungi have been the source of many of the most important antibiotics currently being prescribed against bacterial infections. On balance, fungi have been much more helpful than harmful in human and veterinary medicine.

The major destructive impact that fungi have on people is through the food supply. Fungi known as rusts, smuts, mildews, wilts, and blights cause annual crop losses computed in the billions of dollars. These fungi are particularly troublesome in wheat, corn, barley, and other grain crops (**Figure 31.4a**). Saprophytic fungi are also responsible for enormous losses due to spoilage—particularly for fruit and vegetable growers (**Figure 31.4b**).

(a) Parasitic fungi infect corn and other crop plants.

(b) Saprophytic fungi rot fruits and vegetables.

FIGURE 31.4 Fungi Cause Problems with Crop Production and Storage. (a) A wide variety of grain crops are parasitized by fungi. Corn smut is a serious disease in sweet corn, although in Mexico the smut fungus is eaten as a delicacy. **(b)** Fungi decompose fruits and vegetables as well as leaves and tree trunks.

Fungi also have important positive impacts on the human food supply. Mushrooms are consumed in many cultures; in the industrialized nations they are commonly used in sauces, salads, and pizza. The yeast *Saccharomyces cerevisiae* was domesticated thousands of years ago; today it and other fungi are essential to the manufacture of bread, soy sauce, tofu, cheese, beer, wine, whiskey, and other products. In most cases, domesticated fungi are used by food and beverage producers in conditions where the cells grow via fermentation, creating by-products like the CO_2 that causes bread to rise and beer and champagne to fizz. In addition, enzymes derived from fungi are used to improve the characteristics of foods ranging from fruit juice and candy to meat.

In nature, recent epidemics caused by fungi have killed 4 billion chestnut trees and tens of millions of American elm trees in North America. The fungal species responsible for these epidemics were accidentally imported on species of chestnut and elm native to other regions of the world. When the fungi arrived in North America and began growing in chestnuts and elms native to North America, the results were catastrophic. The local chestnut and elm populations had virtually no genetic resistance to the pathogens and quickly succumbed. The epidemics radically altered the composition of upland and floodplain forests in the eastern United States. Before these fungal epidemics occurred, chestnuts and elms dominated these habitats.

Fungi Are Key Model Organisms in Eukaryotic Genetics

When biologists want to answer basic questions about how eukaryotic cells work, they usually turn to fungi. The filamentous fungus called *Neurospora crassa* was introduced in Chapter 15 because it was the study organism in classic experiments that supported the one-gene, one-enzyme hypothesis. The yeast *Saccharomyces cerevisiae* has been even more important in basic research on cell biology and molecular genetics.

S. cerevisiae is unicellular and relatively easy to culture and manipulate in the lab. In good conditions, yeast cells grow and divide almost as rapidly as bacteria. As a result, the species has become the organism of choice for experiments on control of the cell cycle and regulation of gene expression in eukaryotes. The morphology of this organism is so simple that it provides an example of a "pure" eukaryotic cell type—one that is suitable for experiments on how cell division occurs and how particular genes are turned on and off. Just as *Escherichia coli* serves as the model bacterial cell, *S. cerevisiae* serves as the model eukaryotic cell. For example, research has confirmed that several of the genes controlling cell division and DNA repair in yeast have homologs in humans; when mutated, these genes contribute to cancer. Strains of yeast that carry these mutations are now being used to test drugs that might be effective against cancer.

S. cerevisiae has become even more important as an increasing number of eukaryotic genomes are being sequenced (see Chapter 20). To begin interpreting the genomes of organisms like rice, mice, zebrafish, and humans, researchers turn to yeast. It is much easier to investigate the function of particular genes in *S. cerevisiae* by creating mutants or transferring specific alleles among individuals than it is to do the same experiments in mice or zebrafish. Once the function of a gene has been established in yeast, biologists can look for the homologous gene in other eukaryotes. If such a gene exists, they can usually infer that it has a function similar to its role in *S. cerevisiae*. In this way, yeast is serving as a key resource in the field called proteomics and functional genomics. *S. cerevisiae* was also the first eukaryote with a completely sequenced genome.

To summarize, biologists study fungi because they provide a window for understanding eukaryotic cells and because they affect a wide range of species in nature, including humans. How do biologists go about studying them? More specifically, what tools are helping researchers understand the diversity of fungi?

31.2 How Do Biologists Study Fungi?

About 80,000 species of fungi have been described and named to date, and about 1000 more are discovered each year. But the fungi are so poorly studied that the known species are widely regarded as a tiny fraction of the actual total. To predict the authentic number of species alive today, David Hawksworth looked at the ratio of vascular plant species to fungal species in the British Isles—the area where the two groups are the most thoroughly studied. According to Hawksworth's analysis, there is an average of six species of fungus for every species of vascular plant on these islands. If this ratio holds worldwide, then the estimated total of 275,000 vascular plant species implies that there are 1.65 million species of fungi.

Although this estimate sounds large, recent data on fungal diversity suggest that it may be an underestimate. Consider what researchers found recently when they analyzed fungi growing on Barro Colorado Island, Panama: Living on the healthy leaves of just two tropical tree species were a total of 418 distinct morphospecies of fungi. (Recall from Chapter 26 that morphospecies are distinguished from each other by some aspect of morphology.) Because over 310 species of trees and shrubs grow on Barro Colorado, the data suggest that tens of thousands of fungi may be native to this island alone. If further work on fungal diversity in the tropics supports these conclusions, there may turn out to be many millions of fungal species.

This viewpoint of fungal diversity was reinforced by a recent analysis of the fungi living in conjunction with the roots of a single species of grass native to Eurasia. In this study, researchers used the direct sequencing approach, introduced in Chapter 28, to analyze the gene that codes for the RNA molecule in the small subunit of fungal ribosomes. The data showed that a total of 49 phylogenetic species were living in conjunction with the grass roots. (The phylogenetic species concept was introduced in Chapter 26.) Most of the species had never before been described, and several represented completely new lineages of fungi. Biologists are only beginning to realize the extent of species diversity in fungi.

Let's consider how biologists are working to make sense of all this diversity, beginning with an overview of fungal morphology.

Analyzing Morphological Traits

Compared with animals and land plants, fungi have very simple bodies. Only two growth forms occur among them:

1. Single-celled forms called **yeasts** (**Figure 31.5a**), and
2. Multicellular, filamentous structures called **mycelia** (singular: **mycelium**; **Figure 31.5b**).

Many species of fungus grow only as a yeast or as a mycelium, but some regularly adopt both growth forms.

(a) Single-celled fungi are called yeasts.

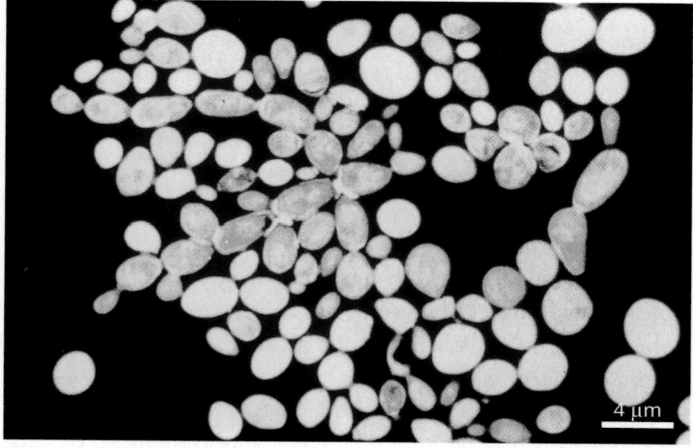

(b) Multicellular fungi have weblike bodies called mycelia.

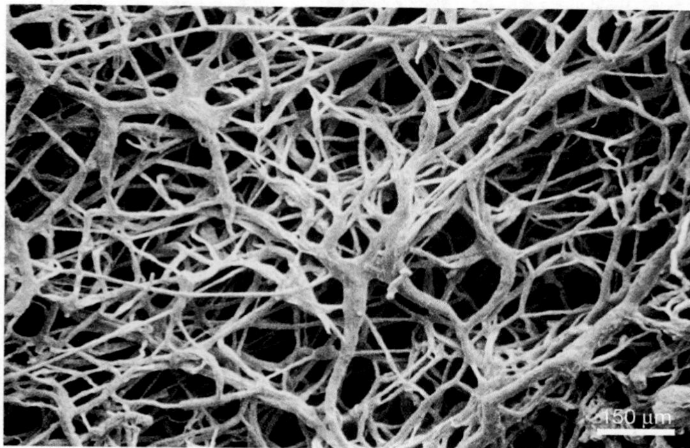

FIGURE 31.5 Fungi Have Just Two Growth Forms. Fungi grow **(a)** as single-celled yeasts and/or **(b)** as multicellular mycelia made up of long, thin, highly branched filaments.

Because most fungi form mycelia and because this body type is so fundamental to the absorptive mode of life, most studies of fungal morphology have focused on them. Let's take a closer look at the structure and function of a fungal mycelium.

The Nature of the Fungal Mycelium If food sources are plentiful, mycelia can be long lived and grow to be extremely large. Researchers recently discovered a mycelium growing across 1310 acres (6.5 km^2) in Oregon. This is an area substantially larger than most college campuses. The biologists estimated the individual's weight at hundreds of tons and its age at thousands of years, making it one of the largest and oldest organisms known.

Although most mycelia are much smaller and shorter lived than the individual in Oregon, all mycelia are dynamic. Mycelia constantly grow in the direction of food sources and die back in areas where food is running out. The body shape of a fungus can change almost continuously throughout its life.

The individual filaments that make up a mycelium are called **hyphae** (singular: **hypha**). Most hyphae are haploid though some are **heterokaryotic** ("different-kernel"), meaning that each cell contains two haploid nuclei—one from each parent. As **Figure 31.6a** shows, hyphae are long, narrow filaments that branch frequently. In most fungi, each filament is broken into cell-like compartments by cross-walls called **septa** (singular: **septum; Figure 31.6b**). Septa do not close off segments of hyphae completely. Instead, gaps called pores enable a wide variety of materials, even organelles and nuclei, to flow from one compartment to the next. Septa may have single large openings or a series of small gaps that give the septum a sieve-like appearance. Because nutrients, mitochondria, and even genes can flow though the entire mycelium—at least to a degree—the fungal mycelium is intermediate between a multicellular land plant or animal and an enormous single-celled organism. Some fungal species are even **coenocytic** ("common-celled"; pronounced *see-no-SIT-ick*)—meaning that they lack septa entirely. Coenocytic fungi have many nuclei scattered throughout the mycelium. In effect, they are a single, gigantic cell.

It's also important to appreciate just how thin hyphae are. Plant root tips range from 100 to 500 μm across, but fungal hyphae are typically less than 10 μm in diameter. This is equivalent to comparing the width of a piece of spaghetti to a railroad boxcar. Fungal mycelia can penetrate tiny fissures in soil and absorb nutrients that are inaccessible to plant roots.

Perhaps the most important aspect of mycelia and hyphae, however, is their shape. Because mycelia are composed of complex, branching networks of extremely thin hyphae, the body of a fungus has the highest surface-area-to-volume ratio possible in a multicellular organism. To drive this point home, consider that the hyphae found in any fist-sized ball of rich soil typically have a surface area equivalent to half a page of this book. This surface area is important because it makes absorption extremely efficient, and because fungi make their living via absorption.

The extraordinarily high surface area in a mycelium has a downside, however: Fungi are prone to drying out, because the amount of water that evaporates from an organism is a function of its surface area. Due to the high surface area of mycelia, fungi are most abundant in moist habitats. Fungi often endure dry conditions in the form of tough, watertight spores.

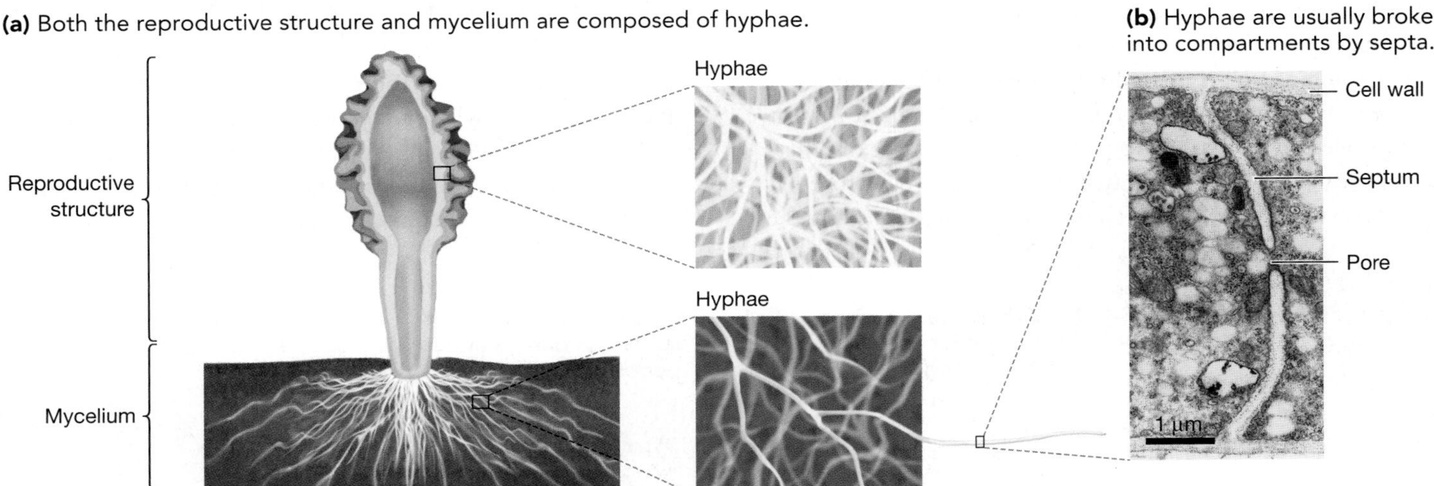

FIGURE 31.6 Multicellular Fungi Have Unusual Bodies. (a) The feeding structure of a fungus is a mycelium, which is made up of hyphae. In some species, hyphae come together to form multicellular structures such as mushrooms, brackets, or morels that emerge from the ground. **(b)** Hyphae are often divided into cell-like compartments by partitions called septa, which are broken by pores. As a result, the cytoplasm of different compartments is continuous.

EXERCISE The reproductive structure in part (a) is actually a tiny proportion of the entire fungal body. To give an idea of the correct scale, draw hyphae extending from below the reproduction structure across both pages of this text.

Mycelia are an adaptation that supports the absorptive lifestyle of fungi. The only thick, fleshy structures that fungi produce are reproductive organs—not feeding structures. Mushrooms, puffballs, and other dense, multicellular structures that arise from mycelia do not absorb food. Instead, they function in reproduction. Typically they are the only part of a fungus that is exposed to the atmosphere, where drying is a problem. The mass of filaments on the inside of mushrooms is protected from drying by the densely packed hyphae forming the surface. However, few species of fungi make the reproductive structures called mushrooms. Instead, each fungal species produces one of four types of distinctive reproductive structures—only one of which is found inside mushrooms.

Reproductive Structures When biologists study diversity in a lineage, they normally begin by comparing the morphologies of various species. Because fungal mycelia are so simple and are so similar among species, researchers have focused on the distinctive morphological structures that fungi produce during sexual reproduction. On the basis of these reproductive structures, most fungi fall into four major groups:

1. Members of the Chytridiomycota, or **chytrids**, live primarily in water or wet soils, and they are the only fungi that have motile cells. The spores that chytrids produce during asexual reproduction have flagella, as do the gametes produced by members of this group during sexual reproduction. These swimming cells are reproductive structures that distinguish chytrids from other fungi (**Figure 31.7a**). Structurally, chytrid flagella are similar to the flagella in the sperm cells of animals.
2. The hyphae belonging to the Zygomycota, are haploid and come in several mating types. Instead of having morphologically distinct males and females that produce sperm and eggs, hyphae of different mating types look identical but will not combine unless the individuals have different alleles of one or more genes involved in mating. If chemical messengers released by two hyphae indicate that they are of different mating types, the individuals may become yoked together as shown in **Figure 31.7b**. (The Greek root *zygos* means to be yoked together like oxen. Translated literally, *Zygomycota* means "yoked-together fungi.") Cells from the yoked hyphae fuse to form a spore-producing structure called a **zygosporangium**. Yoked hyphae that form a zygosporangium are the reproductive structure unique to this group.
3. Mushrooms, bracket fungi, and puffballs are among the complex reproductive structures produced by members of the Basidiomycota, or **club fungi**. Inside these structures, specialized cells called **basidia** ("little-pedestals") form at the ends of hyphae and produce spores (**Figure 31.7c**). Only members of the Basidiomycota produce basidia. It is common for multiple mating types to occur in these species; the mushroom-forming *Schizophyllum commune* is estimated to have 28,000 mating types.
4. Members of the Ascomycota, also called **sac fungi**, produce complex reproductive structures—the largest of which are often cup shaped. The tips of hyphae inside these structures produce distinctive sac-like cells called **asci** (singular: **ascus**; **Figure 31.7d**). An ascus is a spore-producing structure found only in Ascomycota.

In sum, morphological studies allowed biologists to describe and interpret the growth habit of mycelia as an adaptation that makes absorption extremely efficient. Careful analyses of morphological features also allowed researchers to identify four major lineages of fungi, based on the nature of their reproductive structures. Do the Chytridiomycota, Zygomycota, Basidiomycota, and Ascomycota each represent a monophyletic group? If so, how are these four major groups related? And how are fungi related to other eukaryotes?

Evaluating Molecular Phylogenies

Researchers have sequenced and analyzed an array of genes to establish where fungi fit on the tree of life. The top part of **Figure 31.8** shows where fungi occur among the many eukaryote lineages; the close-up in the bottom part of the figure emphasizes that fungi are much more closely related to animals than they are to land plants. In addition to the DNA sequence

(a) Chytridiomycota make swimming gametes and spores.

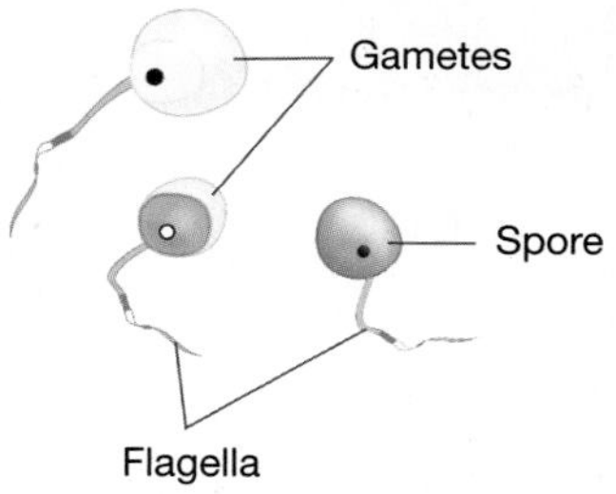

(b) Zygomycota hyphae yoke together and form a zygosporangium.

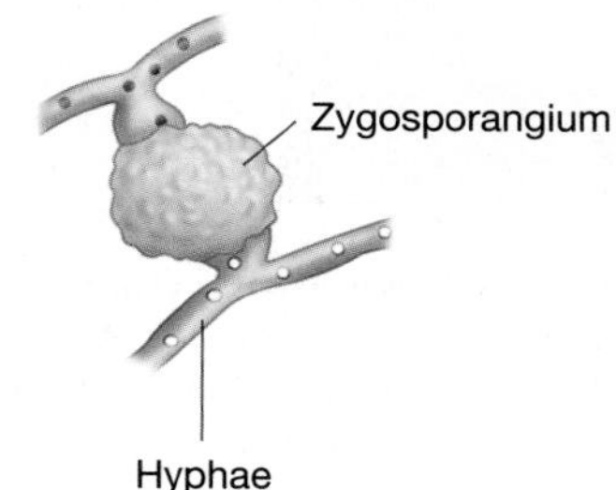

(c) Basidiomycota form spores on basidia (little pedestals).

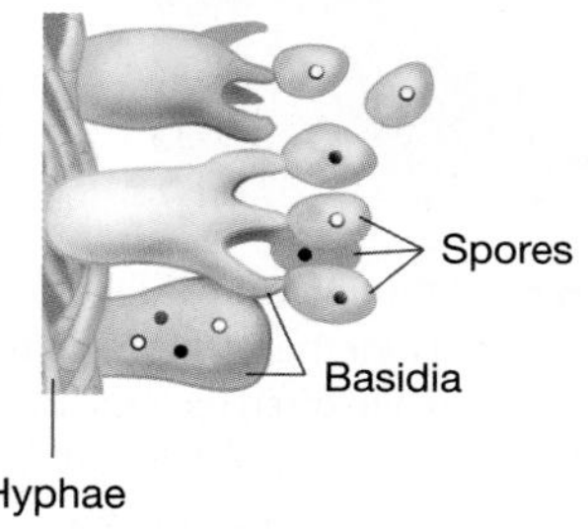

(d) Ascomycota form spores in asci (sacs).

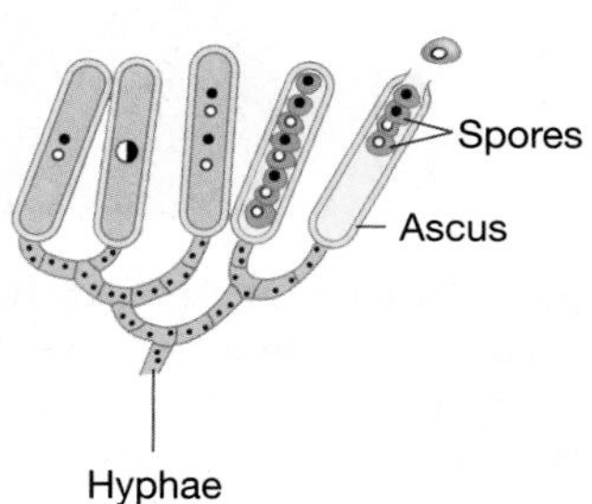

FIGURE 31.7 Four Distinct Reproductive Structures Are Observed in Fungi.

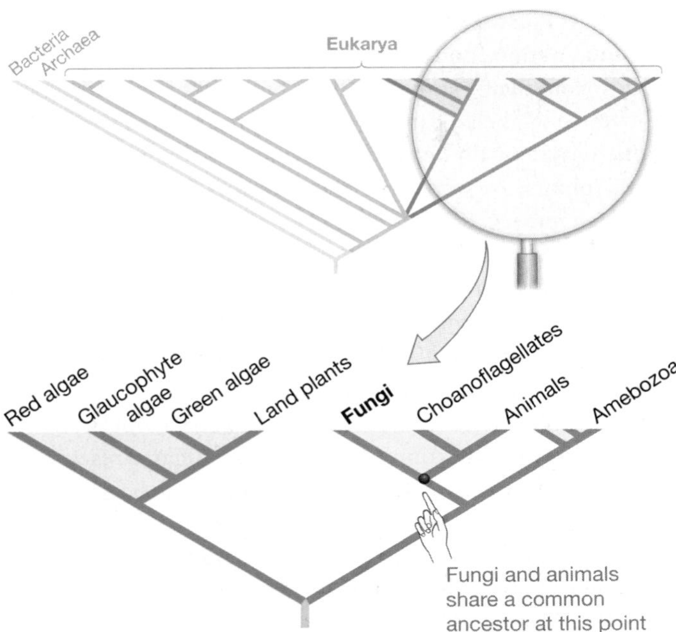

FIGURE 31.8 Fungi Are More Closely Related to Animals than to Land Plants. Phylogenetic tree showing the evolutionary relationships among the green plants, animals, fungi, and some groups of protists. (Choanoflagellates are solitary or colonial protists found in freshwater; they are introduced in Chapter 32.)

data, several morphological traits link animals and fungi. For example, most animals and fungi synthesize the tough structural material called chitin (see Chapter 5). **Chitin** is a prominent component of the cell walls of fungi. Also, the flagella that develop in chytrid spores and in chytrid gametes are very similar to those observed in animals: The flagella are single, are located at the back of the body, and move in a whiplash manner. Further, both animals and fungi store food by synthesizing the polysaccharide glycogen. Green plants, in contrast, synthesize starch as their storage product.

To understand the relationships among chytrids, zygomycetes, basidiomycetes, and ascomycetes, biologists have sequenced a series of genes from an array of fungal species and used the data to estimate the phylogeny of the group. The results, shown in **Figure 31.9**, support several important conclusions:

- Chytrids include the most basal groups of fungi. This result is consistent with the hypothesis that fungi evolved from aquatic ancestors. It also suggests that fungi made the transition to land early in their evolution.
- The Chytridiomycota and Zygomycota are paraphyletic. Neither grouping represents a single common ancestor and all of its descendants. This result means that either swimming gametes or yoked hyphae evolved more than once, or both were present in a common ancestor but then were lost in certain lineages.
- Although early work on molecular phylogenies suggested that organisms called microsporidians are the closest living relatives of fungi, subsequent analyses indicate that they are actually *within* the monophyletic group called fungi. This

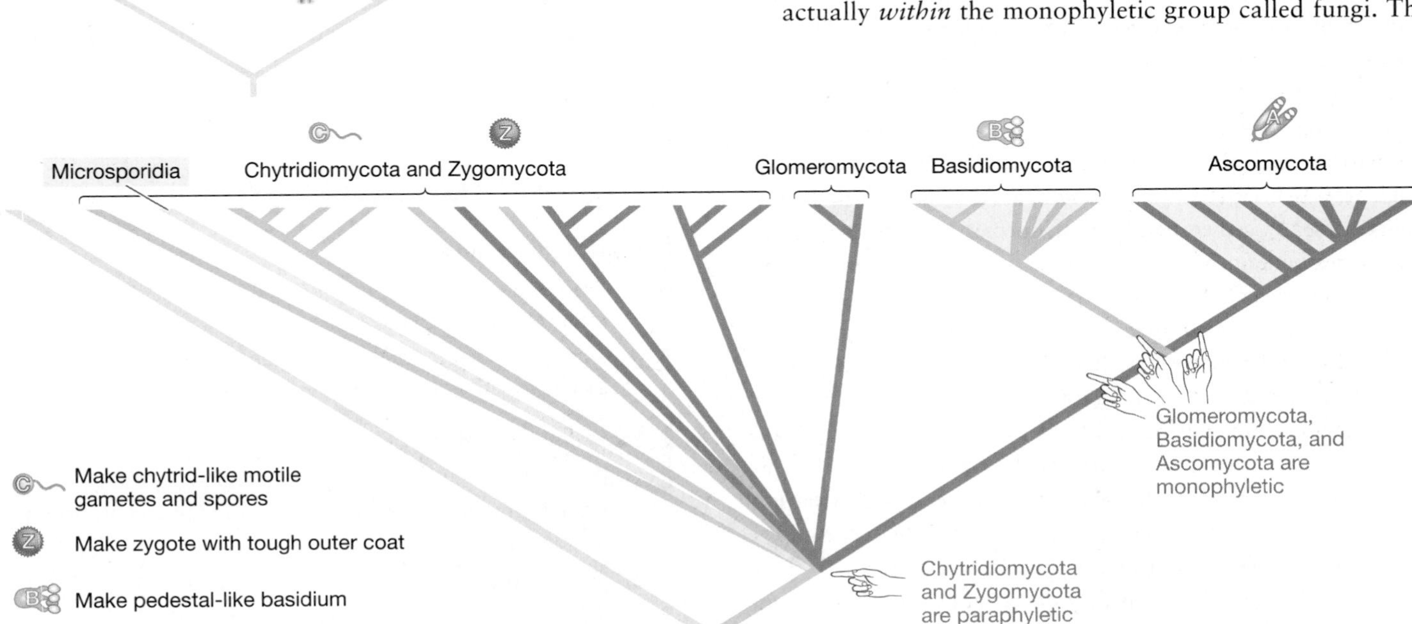

FIGURE 31.9 Phylogeny of the Fungi. A recent phylogenetic tree based on analyses of DNA sequence data. The key indicates the types of sexual reproductive structures observed in each major lineage.

was an important result because microsporidians cause serious disease in bee colonies, silkworm colonies, and people with AIDS. Based on the molecular phylogeny, researchers are testing the hypothesis that **fungicides**—meaning molecules that are lethal to fungi—may prove to be effective in combating microsporidian infections.

- An important group called the Glomeromycota is monophyletic. Sexual reproduction has yet to be observed in this lineage, so it is not known if they ever make spores via meiosis inside a zygosporangium.
- The Basidiomycota and Ascomycota are monophyletic and are the most highly derived groups of fungi.

Although progress on understanding the evolutionary history of fungi has been rapid, the phylogenetic tree in Figure 31.9 is still a work in progress. For example, it is not yet clear where microsporidians are placed relative to several lineages of chytrids and zygomycetes. (Microsporidians lack both swimming gametes and yoked hyphae.) In addition, efforts to understand the relationships of subgroups within the Basidiomycota and Ascomycota are in their infancy. Future work should clarify exactly how the masters of the absorptive lifestyle diversified and how the diversification of fungi relates to the diversification of land plants.

Experimental Studies of Mutualism

It is estimated that 90 percent of land plants live in close physical association with fungi. Stated another way, fungi and land plants often have a **symbiotic** ("together-living") relationship. Although some species of fungi live in association with an array of different land plant species, many documented fungal-plant associations are specific. It is not unusual for fungi to live in only a particular type of tissue, in one plant species. Based on this observation, biologists hypothesize that the evolution of symbiotic associations has played a large role in the diversification of fungi. Scientists categorize these symbiotic relationships as **mutualistic**, meaning they benefit both species; **parasitic**, meaning one species benefits at the expense of the other; or **commensal**, meaning one species benefits while the other is unaffected.

To understand the nature of the association, biologists turn to experimental approaches. Recall from Section 31.1 that in early experiments, researchers grew potatoes, trees, or other types of land plants with or without their normal mycorrhizal fungi. In these experiments, treatments lacking fungal symbionts are created by sterilizing soils with heat or by treating soils and seeds with fungicides. Presence-absence experiments have generally shown that plants grow much larger with their normal symbiotic fungi than they do without. Similarly, fungi that are typically symbiotic are usually unable to grow and reproduce if their regular host plant is absent.

To explore the nature of fungi-plant symbioses in more detail, researchers have used isotopes as tracers for specific elements (**Figure 31.10**). For example, to test the hypothesis that fungi obtain food in the form of carbon-containing compounds from their plant associates, biologists have introduced radioactively labeled carbon dioxide into the air surrounding plants that do or do not contain symbiotic fungi. The experimental plants are usually grown in pots inside laboratory growth chambers. The labeled CO_2 molecules are incorporated into the sugars produced during photosynthesis, and the location of the radioactive atoms can then be followed over time by means of a device that detects radioactivity. If plants feed their fungal symbionts, then labeled carbon compounds should be transferred from the plant to the fungi.

To test the hypothesis that plants are receiving nutrients from their symbiotic fungi in return for sugars, researchers have added radioactive phosphorus atoms or the heavy isotope of nitrogen (^{15}N) to potted plants that do or do not contain symbiotic fungi. If fungi facilitate the transfer of nutrients from soil to plants, then plants grown in the presence of their symbiotic fungi should receive much more of the radioactive phosphorus or heavy nitrogen than do plants grown in the absence of fungi.

As the "Results" section of Figure 31.10 shows, experiments with isotopes used as tracers have shown that sugars and other carbon-containing compounds produced by plants via photosynthesis are transferred to their fungal symbionts. In some cases, as much as 20 percent of the sugars produced by a plant end up in their symbiotic fungi. In exchange, the symbiotic fungi facilitate the transfer of phosphorus or nitrogen or both from soil to the plant. Because phosphorus and nitrogen are in extremely short supply in most environments, the nutrients supplied by symbiotic fungi are critical to the success of the plant. In this way, studies with isotopes have supported the hypothesis that most relationships between fungi and land plants are mutually beneficial.

Check Your Understanding

If you understand that...

- The bodies of fungi are either single-celled yeasts or multicellular mycelia.
- During sexual reproduction, different groups of fungi produce distinct reproductive structures.

You should be able to...

1) Describe the nature of a mycelium and explain why mycelia are interpreted as an adaptation to an absorptive lifestyle.
2) Identify the four types of reproductive structures observed in fungi.
3) Explain whether the Chytridiomycota, Zygomycota, Basidiomycota, and Ascomycota are paraphyletic or monophyletic.

Experiment

Question: Are mycorrhizal fungi mutualistic?

Hypothesis: Host plants provide mycorrhizal fungi with sugars and other photosynthetic products. Mycorrhizal fungi provide host plants with phosphorus and/or nitrogen from the soil.

Null hypothesis: No exchange of food or nutrients occurs between plants and mycorrhizal fungi. The relationship is not mutualistic.

Experimental setup:

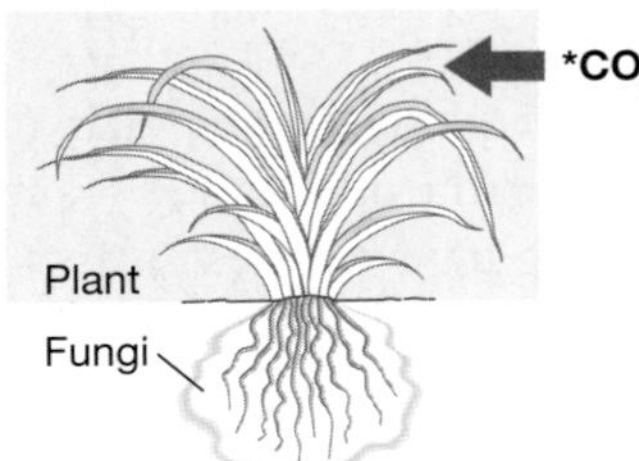

Labeled carbon treatment: Plant leaves are exposed to radioactive CO_2. Mycorrhizal fungi present.

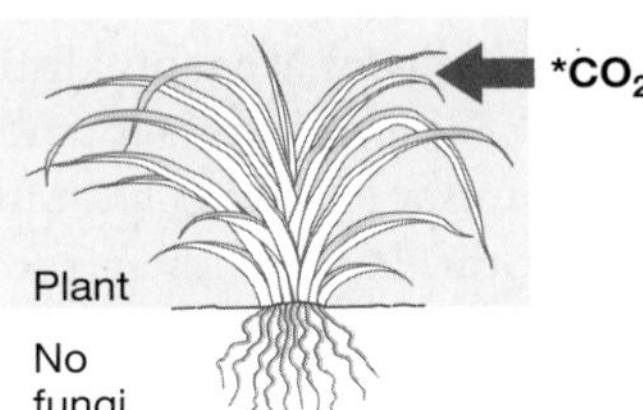

Labeled carbon control: Plant leaves are exposed to radioactive CO_2. Mycorrhizal fungi absent.

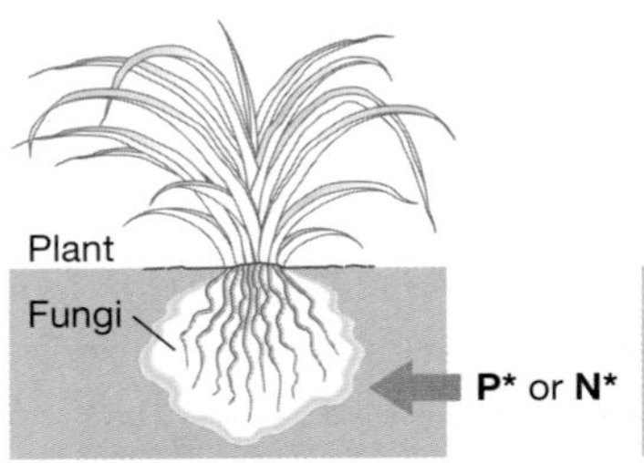

Labeled P or N treatment: Plant roots are exposed to radioactive P or heavy isotope of N. Mycorrhizal fungi present.

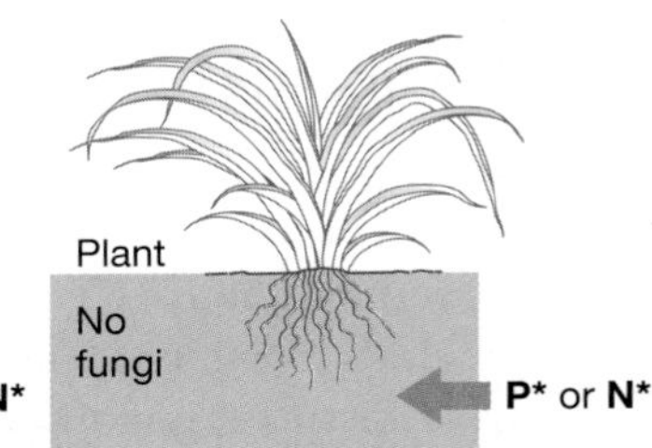

Labeled P or N control: Plant roots are exposed to radioactive P or heavy isotope of N. Mycorrhizal fungi absent.

Prediction for labeled carbon: A large percentage of the labeled carbon taken up by the plant will be transferred to mycorrhizal fungi. In the control, little labeled carbon will be present in the soil surrounding the roots.

Prediction of null hypothesis, labeled carbon: There will be no difference between amounts of labeled carbon in mycorrhizal fungi versus in soil when fungi are absent.

Prediction for labeled P or N: A large percentage of the labeled P or N taken up by the fungi will be transferred to the plant. In the control, little or no labeled P or N will be taken up by the plant.

Prediction of null hypothesis, labeled P or N: There will be no difference between amounts of labeled P or N found in plant in presence or absence of fungi.

Results:

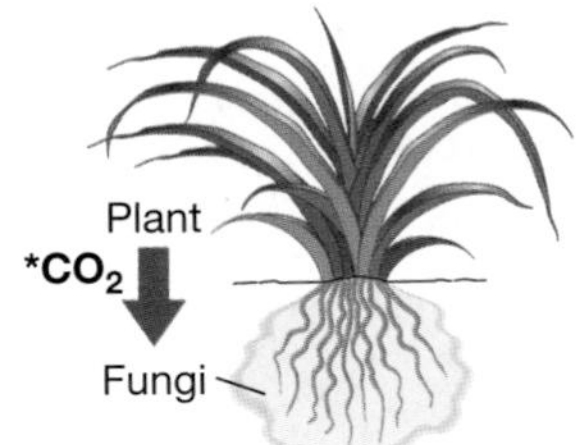

Labeled carbon treatment: Up to 20% of labeled carbon taken up by plant is transferred to mycorrhizal fungus.

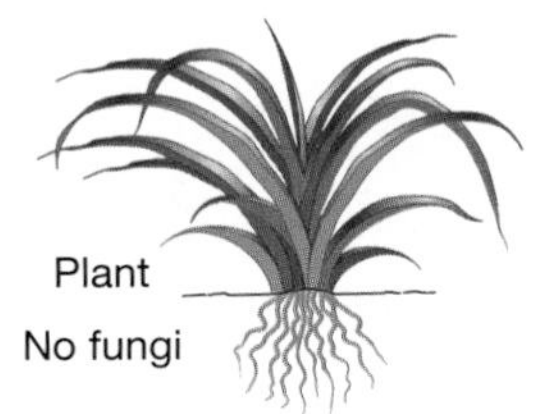

Labeled carbon control: Little to no labeled carbon is found in soil surrounding plant roots.

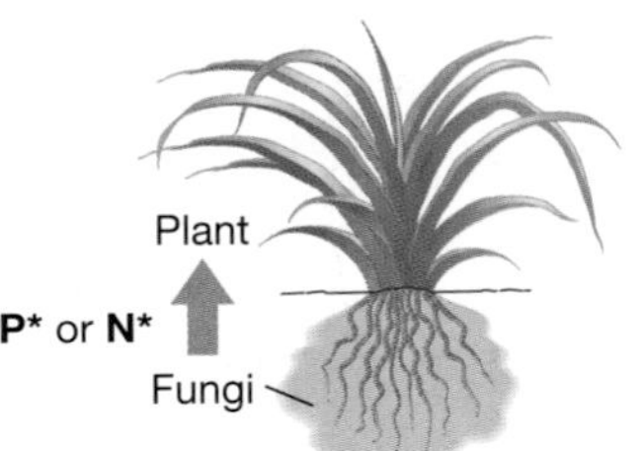

Labeled P or N treatment: Large amount of labeled P or N is found in host plant.

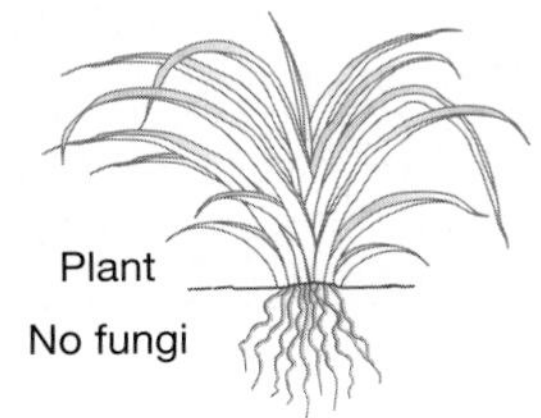

Labeled P or N control: Little labeled P or N is found in host plant.

Conclusion: The relationship between plants and mycorrhizal fungi is mutualistic. Plants provide mycorrhizal fungi with carbohydrates. Mycorrhizal fungi supply host plants with nutrients.

FIGURE 31.10 Experimental Evidence That Mycorrhizal Fungi and Plants Are Mutualistic. Nutrient transfer experiments indicate that sugars flow from plants to mycorrhizal fungi and that key nutrients flow from mycorrhizal fungi to plants.

31.3 What Themes Occur in the Diversification of Fungi?

Why are there so many different species of fungi? This question is particularly puzzling given that fungi share a common attribute: They all make their living by absorbing food directly from their surroundings. In contrast to the diversity of food-getting strategies observed in bacteria, archaea, and protists, all fungi make their living in the same basic way. In this respect, fungi are like plants—virtually all of which make their own food via photosynthesis.)

Chapter 30 showed that the diversification of land plants was driven not by novel ways of obtaining food, but by adaptations that allowed plants to grow and reproduce in a diverse array of terrestrial habitats. What drove the diversification of fungi? The answer is the evolution of novel methods for absorbing nutrients from a diverse array of food sources.

This section introduces a few of the ways that fungi go about absorbing nutrients from different food sources, as well as how they produce offspring. Let's explore the diversity of ways that fungi do what they do.

Fungi Participate in Several Types of Mutualisms

Not long after associations between fungi and the roots of land plants were discovered and shown to be mutualistic, researchers found that two types of mycorrhizal interactions are particularly common. The two major types of mycorrhizae have distinctive morphologies, geographic distributions, and functions. One type involves species from the Glomeromycota; the other usually involves species from the Basidiomycota, though some ascomycetes participate. But mycorrhizae are not the only type of symbiotic fungi found in plants. Researchers have also become interested in fungi that live in close association with the aboveground tissues of land plants—their leaves and stems. Fungi that live in the aboveground parts of plants are said to be **endophytic** ("inside-plants"). Recent research has shown that endophytic fungi are much more common and diverse than previously suspected. Further, data indicate that at least some species of endophytes are mutualistic.

These results support the general realization that most plants are covered with fungi—from the tips of their branches to the base of their roots. Many or even most plants are involved in several distinct types of mutualistic relationships with fungi.

Ectomycorrhizal Fungi The type of mycorrhizal fungus illustrated in **Figure 31.11a** is found on many of the tree species growing in the temperate latitudes of both hemispheres, as well

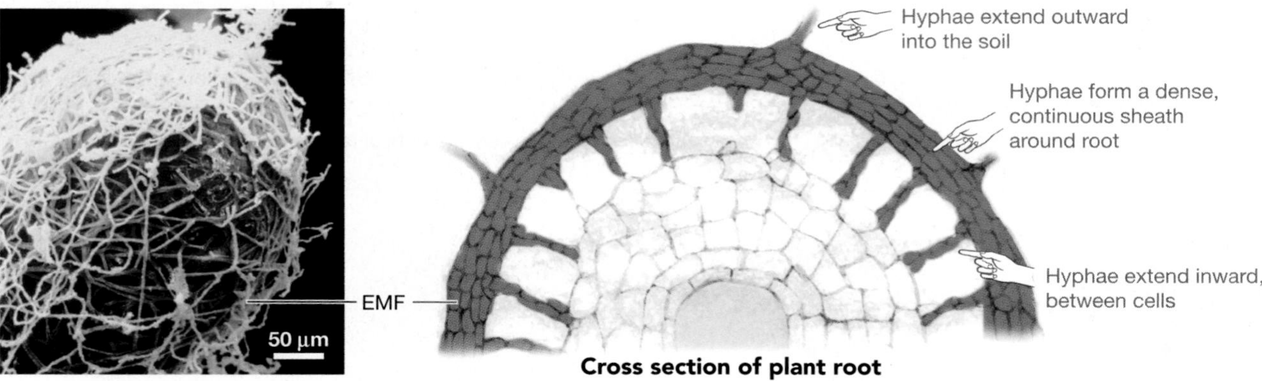

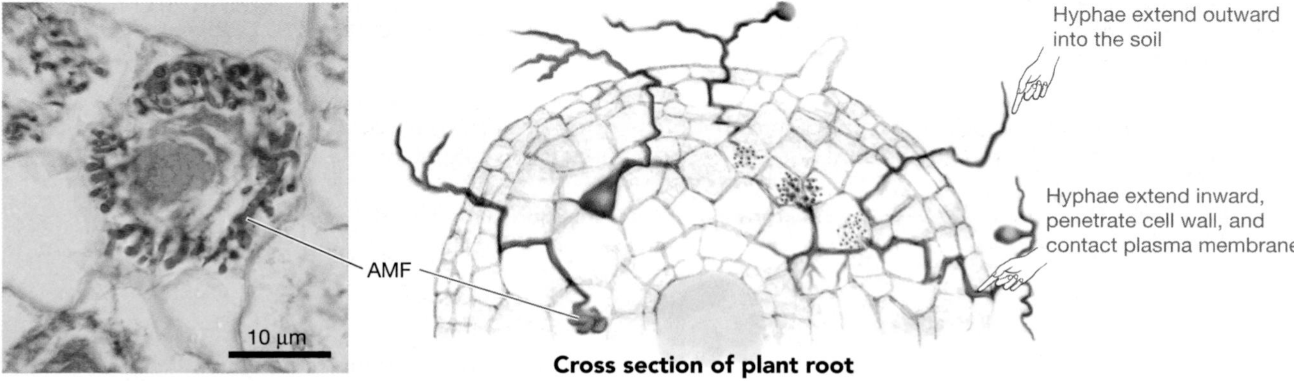

FIGURE 31.11 Mutualistic Fungi Interact with the Roots of Plants in Two Distinct Ways. (a) Ectomycorrhizal fungi (EMF) form a dense network around the roots of plants. Their hyphae penetrate the intercellular spaces of the root but do not enter the root cells. **(b)** The hyphae of arbuscular mycorrhizal fungi (AMF) penetrate the walls of root cells, where they branch into bushy structures or balloon-like vesicles that contact the root cell's plasma membrane.

EXERCISE In the cross sections, add arrows and labels showing the direction of movement of N (nitrogen-containing compounds), P (phosphorus-containing compounds), and C (carbon-containing compounds).

as on tree species in northern coniferous forests. In this type of association, hyphae form a dense network that covers a plant's root tips. As the cross section in Figure 31.11a shows, individual hyphae penetrate between cells in the outer layer of the root, but hyphae do not enter the root cells. Fungi with this growth form are called **ectomycorrhizal fungi** (**EMF**). The Greek root *ecto*, which refers to "outer," is appropriate because the fungi form an outer sheath that is often 0.1 mm thick. Hyphae also extend out from the sheath-like portion of the mycelium into the soil. Most EMF are basidiomycetes; a few are ascomycetes.

How and why do these trees and fungi interact? In the cold, northern habitats where EMF are abundant, the growing season is so short that the decomposition of needles, leaves, twigs, and trunks can be extremely sluggish. As a result, nitrogen atoms tend to remain tied up in amino acids and nucleic acids inside dead tissues instead of being available in the soil. Fortunately, the hyphae of EMF penetrate decaying material and release enzymes called peptidases that cleave the peptide bonds between amino acids in the dead tissues. The nitrogen released by this reaction is absorbed by the hyphae and then transported to the spaces between the root cells of trees, where it can be absorbed by the plant. EMF are also able to acquire phosphate ions that are bound to soil particles and transfer the ions to host plants. In return, the fungi receive sugars and other complex carbon compounds from the tree.

Researchers have found that when northern species such as birch tree seedlings are grown with and without their normal EMF in pots filled with forest soil, only the seedlings with EMF are able to acquire significant quantities of nitrogen and phosphorus. Inspired by such data, a biologist has referred to EMF as the "dominant nutrient-gathering organs in most temperate forest ecosystems." The hyphae of EMF are like an army of miners that discover, excavate, and deliver precious nuggets of nitrogen to the trees of northern forests.

Arbuscular Mycorrhizal Fungi (AMF) In contrast to the hyphae of EMF, the hyphae of **arbuscular mycorrhizal fungi** (**AMF**) grow *into* the cells of root tissue. The name *arbuscular* ("little-tree") was inspired by the bushy, highly branched hyphae, shown in **Figure 31.11b**, that form inside root cells. AMF are also called endomycorrhizal fungi, because they penetrate the interior of root cell walls, or vesicular-arbuscular mycorrhizae (VAM), because the hyphae of some species form large, balloon-like vesicles inside root cells.

The key point is that the hyphae of AMF penetrate the cell wall and contact the plasma membrane of root cells directly. The highly branched hyphae inside the plant cell wall are thought to be an adaptation that increases the surface area available for exchange of molecules between the fungus and its host. However, AMF do not form a tight sheath around roots, as do EMF. Instead, they form a pipeline extending from inside plant cells in the root to the soil well beyond the root.

Most AMF species belong to the lineage called Glomeromycota (see Figure 31.9). AMF are found in a whopping 80 percent of all land plant species, and are particularly common in grasslands and in the forests of warm or tropical habitats. Just as EMF are the dominant type of mycorrhizal association in cool, high-latitude environments, AMF are the dominant type in grasslands and in the tropics.

Besides being extremely common, AMF are ancient. Researchers have found AMF in fossilized root cells that are 400 million years old. This discovery confirms that mycorrhizal associations existed in the most ancient of all land plants. It also supports the hypothesis that plants and fungi colonized terrestrial environments together—meaning that fungi have been nourishing land plants since land plants first evolved.

What do AMF do? Plant tissues decompose quickly in the grasslands and tropical forests where AMF flourish because the growing season is long and warm. As a result, nitrogen is often readily available to plants. Phosphorus is often in short supply, though, because it is present as negatively charged phosphate ions that cling tightly to mineral particles. Based on these observations, biologists hypothesized that AMF transfer phosphorus atoms from the soil to the host plant. Experiments with radioactive atoms confirmed the phosphate-transfer hypothesis by showing that while AMF supply host plants with phosphorus, host plants supply fungi with reduced carbon. EMF mine nitrogen and some phosphorus in the north; AMF mine phosphorus in the south.

Are Endophytes Mutualists? Although endophytic fungi are relatively new to science, they are turning out to be both extremely common and highly diverse. Biologists in Brazil who examine tree leaves for the presence of fungi routinely find several previously undiscovered species of endophytes. Recall from Section 31.2 that a study on fungi in Panama found hundreds of fungal morphospecies living in the leaves of just two tree species. These newly discovered species are endophytes.

Recent research has shown that the endophytes found in some grasses produce compounds that deter or even kill herbivores in exchange for absorbing sugars from the plant. Based on these results, biologists have concluded that the relationship between endophytes and grasses is mutualistic. In other types of plants, however, researchers have not been able to document benefits for the plant host. The current consensus is that at least some endophytic fungi may be commensals—meaning the fungi and the plants simply coexist with no observable effect, either deleterious or beneficial, on the host plant. Although research on endophytes continues, it is already clear that most land plants are colonized by a wide array of fungi and that fungi have evolved an array of relationships with their plant symbionts.

What Adaptations Make Fungi Such Effective Decomposers?

The saprophytic fungi are master recyclers. Although bacteria and archaea are also important decomposers in terrestrial environments, fungi and a few bacterial species are the only organisms that can digest wood completely. Given enough time, fungi can turn even the hardest, most massive trees into soft soils that nourish an array of plants (**Figure 31.12**). You've already analyzed how extremely thin hyphae and the large surface area of a mycelium make absorption exceptionally efficient. Saprophytic fungi are also capable of growing toward the dead tissues that supply their food. What other adaptations help fungi decompose plant tissues?

Extracellular Digestion Large molecules such as starch, lignin, cellulose, proteins, and RNA cannot diffuse across the plasma membranes of hyphae. Only sugars, amino acids, nucleic acids, and other small molecules can enter the cytoplasm. As a result, fungi have to digest their food before they can absorb it.

Instead of digesting food inside a stomach or food vacuole, as most animals and some protists do, respectively, fungi synthesize digestive enzymes and then secrete them outside their hyphae, into their food. Fungi perform **extracellular digestion**—digestion that takes place outside the organism. The simple compounds that result from enzymatic action are then absorbed by the hyphae.

FIGURE 31.12 Fungi Recycle Nutrients. The fungi that are decomposing this section of tree trunk are breaking up its proteins, nucleic acids, lignin, and cellulose. In doing so, the fungi release nitrogen, phosphorus, and other nutrients that can be used by other organisms.

EXERCISE At least two species of basidiomycetes inhabit the area in this photo. Circle their reproductive structures.

As a case study in how this process occurs, consider the enzymes responsible for digesting lignin and cellulose. These are the two most abundant organic molecules on Earth. The term **lignin** refers to a family of extremely strong, complex polymers built from monomers that are six-carbon rings. Recall from Chapter 30 that lignin is found in the secondary cell walls of plant vascular tissues, where it furnishes structural support. **Cellulose** is a polymer of glucose and is found in the primary and secondary cell walls of plant cells. Basidiomycetes can degrade lignin completely—to CO_2—as well as digest cellulose. Let's take a closer look at how they do it.

Lignin Degradation Biologists have been keenly interested in understanding how basidiomycetes digest lignin. Paper manufacturers are also interested in this process because they need safe, efficient ways to degrade lignin in order to make soft, absorbent paper products.

To find out how lignin-digesting fungi do it, biologists began analyzing the proteins that these species secrete into extracellular space. After purifying these molecules, the investigators tested each protein for the ability to degrade lignin. Using this approach, investigators from two labs independently discovered an enzyme called lignin peroxidase.

The researchers who followed up on the discovery of lignin peroxidase found that the enzyme catalyzes the removal of a single electron from an atom in the ring structures of lignin. This oxidation step creates a free radical—an atom with an unpaired electron (see Chapter 2). This extremely unstable electron configuration leads to a series of uncontrolled and unpredictable reactions that end up splitting the polymer into smaller units.

Biologists have referred to this mechanism of lignin degradation as enzymatic combustion. The phrase is apt because the uncontrolled oxidation reactions triggered by lignin peroxidase are analogous to the uncontrolled oxidation reactions that occur when gasoline burns in a car engine. The nonspecific nature of the reaction is remarkable because virtually all of the other reactions catalyzed by enzymes are extremely specific. The lack of specificity in the lignin degradation reaction actually makes sense, however, given the nature of the compound. Unlike proteins, nucleic acids, and most other polymers with a regular and predictable structure, lignin is extremely heterogeneous. Over 10 types of covalent linkages are routinely found between the monomers that make up lignin. But once lignin peroxidase has created a free radical in the aromatic ring, any of these linkages can be broken.

The uncontrolled nature of the reactions has an important consequence. Instead of being stepped down an orderly electron transport chain as described in Chapter 9, the electrons involved in these reactions lose their potential energy in large, unpredictable jumps. As a result, the oxidation of lignin cannot be harnessed to drive the production of ATP and fuel the growth of hyphae. This conclusion is supported by an experimental observation: Fungi cannot grow with lignin as their sole source of food.

If wood-rotting fungi don't use lignin as food, why do they produce enzymes to digest it? The answer is simple. In wood, lignin forms a dense matrix around long strands of cellulose. Degrading the lignin matrix gives hyphae access to huge supplies of energy-rich cellulose. Saprophytic fungi are like miners. But instead of seeking out rare, gem-like nitrogen or phosphorus atoms as do EMF and AMF, the saprophytes use lignin peroxidase to blast away enormous lignin molecules, exposing rich veins of cellulose that can fuel growth and reproduction.

Cellulose Digestion Once lignin peroxidase has softened wood by stripping away its lignin matrix, the long strands of cellulose that remain can be attacked by enzymes called cellulases. Like lignin peroxidase, cellulases are secreted into the extracellular environment by fungi. But unlike lignin peroxidase, cellulases are extremely specific in their action. For example, biologists have purified seven different cellulases from the fungus *Trichoderma reesei*. Two of these enzymes catalyze a critical early step in digestion—they cleave long strands of cellulose into a disaccharide called cellobiose. The other cellulases are equally specific and also catalyze hydrolysis reactions. In combination, the suite of seven enzymes in *T. reesei* transforms long strands of cellulose into a simple monomer—glucose—that the fungus uses as a source of food.

Now that the enzymes responsible for lignin and cellulose degradation have been characterized, researchers would like to pursue a practical question: Can lignin peroxidase and cellulases be used to make paper production more efficient or perhaps speed efforts to clean up waste from old sawmill and paper mill operations? If the answer turns out to be yes, fungi could be used along with bacteria and archaea in the bioremediation efforts introduced in Chapter 29.

Variation in Life Cycles

Fungi evolved an array of ways to absorb nutrition as they diversified. But in addition, fungi evolved an array of ways to reproduce. Recall from Section 31.2 that fungi may produce swimming gametes and spores, yoked hyphae in which nuclei from different individuals fuse to form a zygote inside a protective structure, or specialized spore-producing cells called basidia and asci.

The **spore** is, in fact, the most fundamental reproductive cell in fungi. Spores are the dispersal stage in the fungal life cycle and are produced during both asexual and sexual reproduction. (Recall from Chapter 12 that asexual reproduction is based on mitosis, while sexual reproduction is based on meiosis.) Fungi produce spores in such prodigious quantities that it is not unusual for them to outnumber pollen grains in air samples. If a spore falls on a food source and is able to germinate, a mycelium begins to form. As the fungus expands, hyphae grow in the direction in which food is most abundant. If food begins to run out, mycelia respond by making spores, which are dispersed by wind or animals. Why would mycelia reproduce when food is low? To answer this question, biologists hypothesize that spore production is favored by natural selection when individuals are under nutritional stress. The logic is that spore production allows starving mycelia to disperse offspring to new habitats where more food might be available.

Unique Aspects of Fungal Life Cycles Compared with the life cycles of green plants, protists, and animals, the life cycles of fungi have important unique features. ● For example, only members of the Chytridiomycota produce gametes, and no species of fungi produce gametes that are so different in size that they are called sperm and egg. In addition, fertilization occurs in two distinct steps in many fungi: (1) fusion of cells and (2) fusion of nuclei from the fused cells. These two steps can be separated by long time spans and even long distances.

In many fungi, the process of sexual reproduction begins when hyphae from two individuals fuse to form a hybrid hypha. When the cytoplasm of two individuals fuses in this way, **plasmogamy** is said to occur (**Figure 31.13a**). But in some cases, the nuclei from the two or more individuals stay independent and grow into a **heterokaryotic** mycelium. The distinct nuclei that are present in heterokaryotic hyphae function independently, even though gene expression must be coordinated in order for growth and development to occur. For example, the two nuclei divide as the hyphae expand, so each compartment that is divided by a septum contains one of each of the two nuclei. How the activities of the two nuclei are coordinated is almost a complete mystery. Eventually, however, one or more pairs of unlike nuclei may fuse to form a diploid zygote. The fusion of nuclei is called **karyogamy**. The nuclei that are produced by karyogamy then divide by meiosis to form haploid spores.

As Figure 29.20 showed, most eukaryotes have life cycles dominated by haploid cells or diploid forms. Many fungi, in contrast, have a life cycle dominated by heterokaryotic cells. ● If you understand the relationship between plasmogamy, heterokaryosis, and karyogamy, you should be able to compare and contrast these events with the life cycles of other eukaryotes. For example, which human cells undergo plasmogamy and karyogamy? Does heterokaryosis occur?

Most fungal species can reproduce asexually as well as sexually. There are, in fact, large numbers of ascomycetes that have never been observed to reproduce sexually. During asexual reproduction, spore-forming structures are produced by a haploid mycelium, and spores are generated by mitosis. As a result, offspring are genetically identical to their parent.

Among the sexually reproducing species of fungi, the presence of a heterokaryotic stage and the morphology of the spore-producing structure vary. Let's take a closer look at each of the four major types of life cycles that have been observed in fungi:

1. The Chytridiomycota are the only fungal group with any species that exhibit alternation of generations. **Figure 31.13b** shows how alternation of generations occurs in the well-studied chytrid *Allomyces*. The key points are that swimming gametes are produced in haploid adults by mitosis and that gametes from the same individual or different individuals then fuse to form a diploid zygote, which grows into a sporophyte. The life cycle is completed when meiosis occurs in the sporophyte, inside a structure called a **sporangium**. The haploid spores produced by meiosis disperse by swimming.

2. In Zygomycota, sexual reproduction starts when hyphae from different individuals fuse, as shown in **Figure 31.13c**. Plasmogamy forms a zygosporangium that develops a tough, resistant coat and is followed by karyogamy.

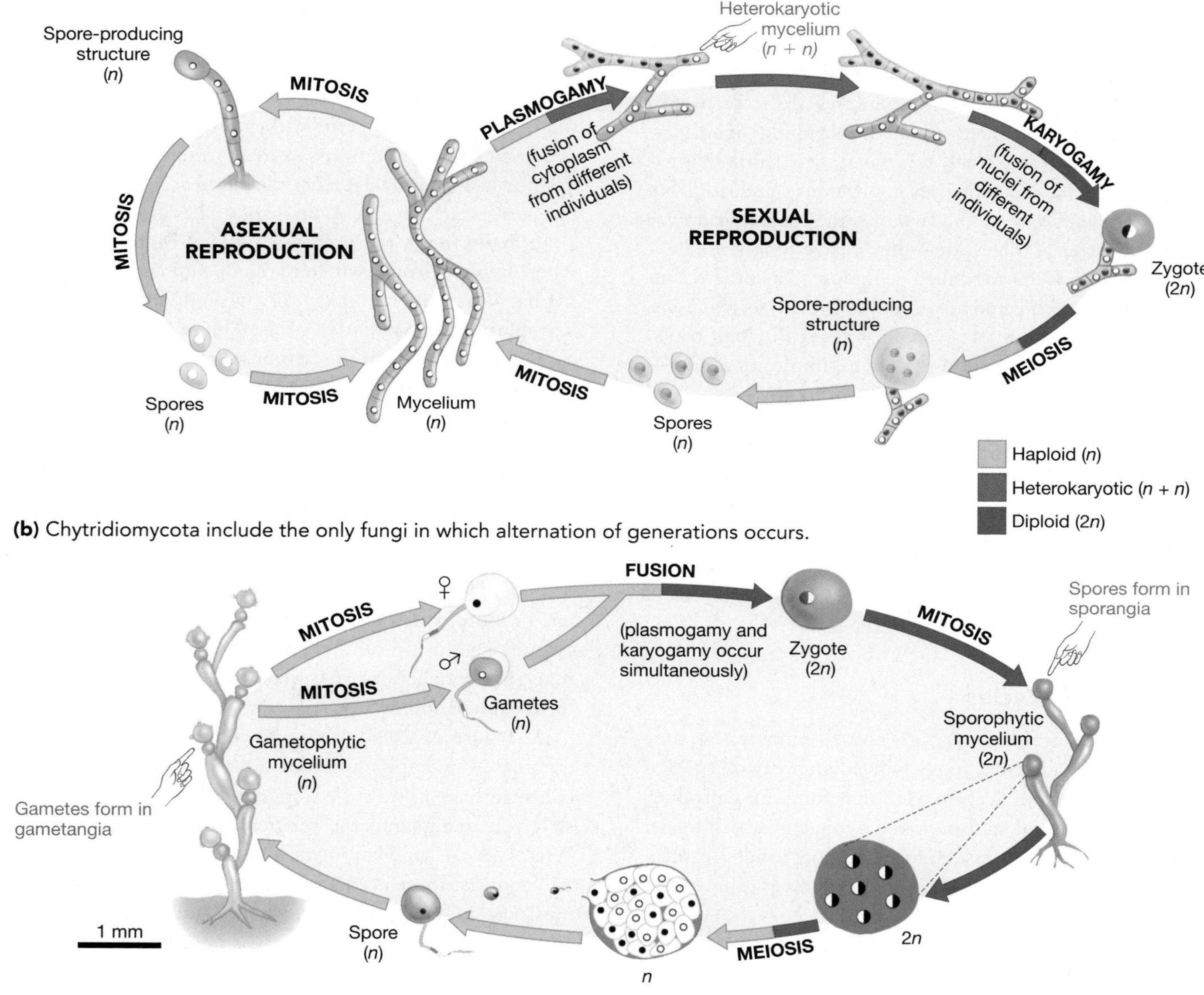

FIGURE 31.13 Fungi Have Unusual Life Cycles. **(a)** A generalized fungal life cycle, showing both asexual and sexual reproduction. **(b–e)** The sexual part of the life cycle in the four major groups of fungi: **(b)** Chytridiomycota, **(c)** Zygomycota, **(d)** Basidiomycota, and **(e)** Ascomycota. Asexual reproduction is particularly common in Zygomycota and Ascomycota, and is the only form of reproduction currently known in Glomeromycota.

EXERCISE Fungi spend most of their lives feeding. On each drawing, indicate the longest-lived component of the life cycle. In parts (c) and (e), add a loop labeled "Asexual reproduction" similar to that in part (a). This loop should start at the mycelium and return to it, via the growth of asexually produced spores.

(c) Zygomycota form yoked hyphae that produce a spore-forming structure (zygosporangium).

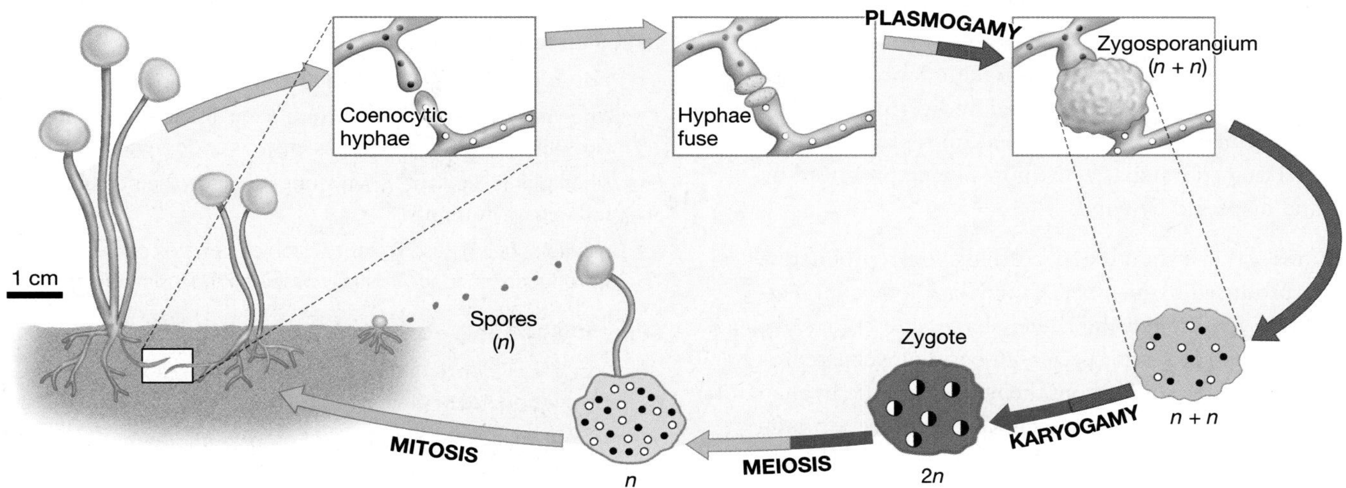

(d) Basidiomycota have reproductive structures with many spore-producing basidia.

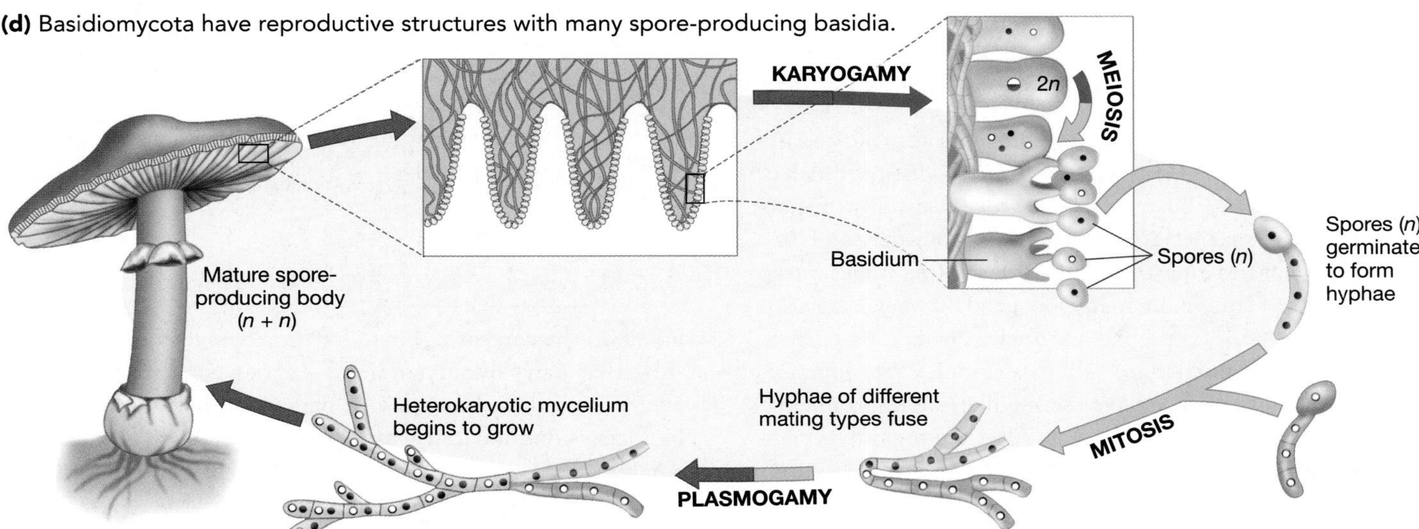

(e) Ascomycota have reproductive structures with many spore-producing asci.

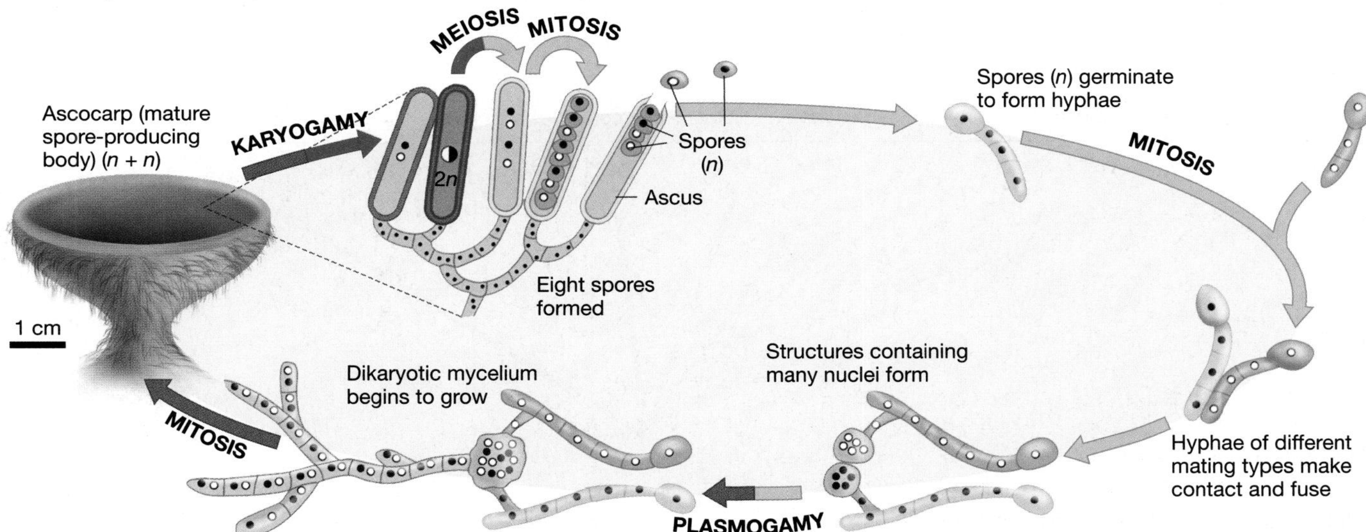

FIGURE 31.13 ***(continued)***

The zygosporangium can persist if conditions become too cold or dry to support growth. When temperature and moisture conditions are again favorable, meiosis occurs. The meiotic products within the sporangium produce haploid spores. When spores are released and germinate, they grow into new mycelia. These mycelia can reproduce asexually by making sporangia, with spores being produced by mitosis and dispersed by wind.

3. Mushrooms, bracket fungi, and puffballs are reproductive structures produced by members of the Basidiomycota (**Figure 31.13d**). Even though their size, shape, and color vary enormously from species to species, all basidiomycete reproductive structures originate from the heterokaryotic hyphae of mated individuals. Inside a mushroom or bracket or puffball, the pedestal-like, spore-producing cells called basidia form at the ends of hyphae. Karyogamy occurs within the basidia. The diploid nucleus that results undergoes meiosis, and haploid spores mature. Spores are ejected from the end of the basidia and are dispersed by the wind. It is not unusual for a single puffball or mushroom to produce a billion spores.

4. The reproductive structure in Ascomycota is produced by a dikaryotic hypha—one containing two distinct nuclei. As **Figure 31.13e** illustrates, the process usually begins when hyphae or specialized structures from the same ascomycete species but from different mating types fuse, forming a cell containing many independent nuclei. A short dikaryotic hypha, containing one nucleus from each parent, emerges and eventually grows into a complex reproductive structure whose hyphae have the sac-like, spore-producing structures called asci at their tips. After karyogamy occurs inside each ascus, meiosis takes place and haploid spores are produced. When the ascus matures, the spores inside are forcibly ejected. The spores are often picked up by the wind and dispersed.

Check Your Understanding

If you understand that...

- Most fungi live in close association with land plants.
- When plants are alive, mycorrhizal fungi associate with their roots and many fungal species grow as endophytes.
- When plants die, saprophytic fungi degrade their tissues and release nutrients.
- Instead of being based on the fusion of gametes, sexual reproduction in fungi is usually based on the fusion of hyphae.

You should be able to...

1) Describe evidence that mutualism occurs in EMF, AMF, and the endophytes of grasses.
2) Explain how extracellular digestion works, using the decomposition of lignin and cellulose as an example.
3) Draw a generalized fungal life cycle.
4) Explain the difference between plasmogamy and karyogamy, and relate them to the process of fertilization in plants and animals.
5) Explain what a heterokaryotic mycelium is.

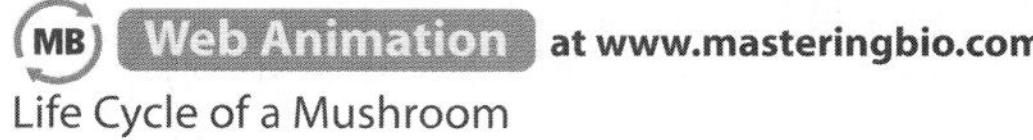

at www.masteringbio.com
Life Cycle of a Mushroom

31.4 Key Lineages of Fungi

Based on the current estimate for the phylogeny of Fungi, it is clear that Chytridiomycota and Zygomycota are not monophyletic and thus should not be named as phyla (**Figure 31.14**). Biologists continue to use the old names, though, because new naming schemes for these groups are still being developed. Although the following summaries treat the chytrids and zygomycetes as coherent groups, keep in mind that they are not and that they will be renamed soon.

FIGURE 31.14 Major Lineages of Fungi.

EXERCISE Mark the origin of the mycelia growth habit, ascus, and EMF association.

Chytridiomycota (Chytrids)

Chytrids are largely aquatic and are particularly common in freshwater environments. Some species live in wet soils, however, and a few have been found in desert soils that are wet only during a short rainy season. Species that are found in dry soils have resting spores that allow them to endure harsh conditions. Resting spores from chytrids have been shown to germinate normally after a resting period of 31 years. Members of this group are the only fungi that produce motile cells. In each case, the motile cells use flagella to swim.

Absorptive lifestyle Many species of chytrids have enzymes that allow them to digest cellulose. As a result, these species are important decomposers of plant material in wet soils, ponds, and lakes. Many of the freshwater species are parasitic, and on occasion parasitic chytrids cause disease epidemics in algae or aquatic insects (including mosquitoes). Other species parasitize mosses, ferns, or flowering plants. Mutualistic chytrids are among the most important of the many organisms living in the guts of deer, cows, elk, and other mammalian herbivores, because the chytrids help these mammals digest their food (**Figure 31.15**). Chytrids produce cellulases that degrade the cell walls of grasses and other plants ingested by the animal, releasing sugars that are used by both the chytrids and their hosts.

Life cycle During asexual reproduction, most chytrid species produce spores that swim to new habitats via a flagellum. A few species reproduce sexually as well as asexually and exhibit alternation of generations. (Recall from Section 31.3 that these chytrids are the only fungal species to do so.) In species that reproduce sexually, plasmogamy and karyogamy may occur in a variety of ways: through fusion of hyphae, gamete-forming structures, or gametes.

Human and ecological impacts Chytrids are important decomposers in aquatic habitats and key mutualists in the guts of large, plant-eating mammals. Parasitic chytrids are also common. Biologists are investigating the possibility of using a chytrid that parasitizes mosquito larvae as a biological control agent; potato tubers are sometimes invaded and spoiled by a parasitic chytrid that causes black wart disease; and chytrids may be responsible for catastrophic declines that have been occurring recently in frog populations all over the world.

Neocallimastix species

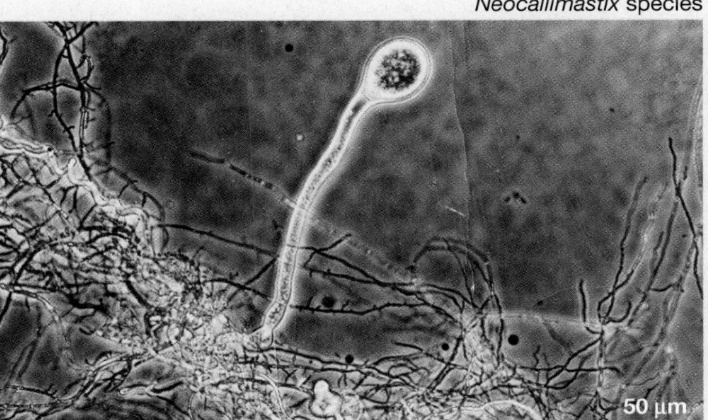

FIGURE 31.15 A Chytrid That Lives in the Gut of Cows and Digests the Cellulose in Grass Leaves.

Microsporidia

All of the estimated 1200 species of microsporidians are single celled and parasitic. They are distinguished by a unique structure called the polar tube (**Figure 31.16**), which allows them to enter the interior of the cells they parasitize. The tube shoots out from the microsporidian, penetrates the membrane of the host cell, and acts as a conduit for the contents of the microsporidian cell to enter the host cell. Once inside, the microsporidian replicates and produces a generation of daughter cells, which go on to infect other host cells. ● You should be able to mark the origin of the polar tube on Figure 31.14.

Microsporidians have a dramatically reduced genome—the smallest known among eukaryotes—and lack functioning mitochondria. Like other fungi, the polysaccharide chitin is a prominent component of their cell wall.

Absorptive lifestyle Most microsporidians parasitize insects or fish. Because they enter the interior of host cells, they are called intracellular parasites.

Nosema tractabile

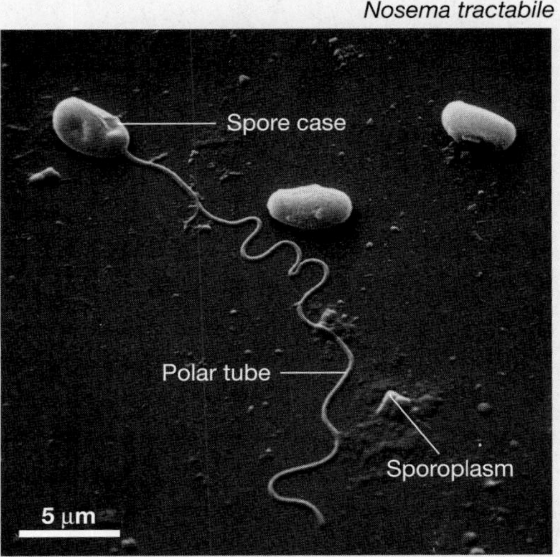

FIGURE 31.16 Microsporidia Infect Other Cells via a Polar Tube.

(Continued on next page)

Microsporidia *continued*

Life cycle Variation in life cycles is extensive. Some species appear to reproduce only asexually. Others produce several different types of sexual or asexual spores, and some must successfully infect several different host species to complete their life cycle.

Human and ecological impacts Species from eight different genera can infect humans. In most cases, however, microsporidians cause serious infections only in AIDS patients and other individuals whose immune systems are not functioning well. Some microsporidians are serious pests in honeybee and silkworm colonies, while others cause severe infections in grasshoppers and are marketed as a biological control agent.

Zygomycota

The Zygomycota ("yoked-fungi") are primarily soil-dwellers. Their hyphae yoke together and fuse during sexual reproduction and then form the durable, thick-walled zygosporangium that is characteristic of the group.

Absorptive lifestyle Many members of zygomycete lineages are saprophytes and live in plant debris. Some parasitize other fungi, however, or are important parasites of insects and spiders.

Life cycle Asexual reproduction is extremely common. Ball-like sporangia are produced at the tips of hyphae that form stalks, and mitosis results in the production of spores, which are dispersed by wind. During sexual reproduction, fusion of hyphae occurs only between individuals of different mating types. Fusion of hyphae from the same individual mycelium does not occur.

Human and ecological impacts The common bread mold *Rhizopus stolonifer* is a frequent household pest—probably the zygomycete that is most familiar to you. Saprophytic and parasitic members of these lineages are responsible for rotting strawberries and other fruits and vegetables and causing large losses in the fruit and vegetable industries. The black bread mold pictured in **Figure 31.17** is a familiar example of a saprophytic zygomycete. Some species of *Mucor* are used in the production of steroids for medical use. Species of *Rhizopus* and *Mucor* are used in the commercial production of organic acids, pigments, alcohols, and fermented foods.

(a) Black bread mold growing. A mycelium of tangled hyphae is visible as well as sporangia.

Rhizopus nigricans

100 µm

(b) A sporangium bearing spores.

Rhizopus nigricans

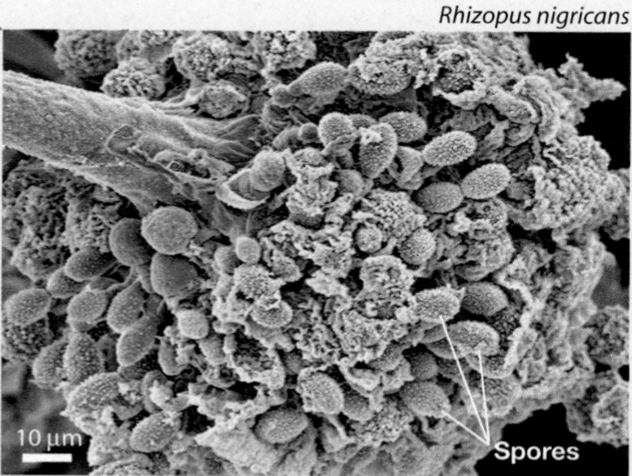

FIGURE 31.17 Black Bread Molds Are Zygomycetes.

Glomeromycota

Recent phylogenetic analyses have shown that the arbuscular mycorrhizal fungi (AMF) form a distinct phylum, indicating that they are a major monophyletic group. Before phylogenies based on molecular characters were available, the lineage now called Glomeromycota was called Glomales and considered part of the Zygomycota. ● You should be able to mark the origin of the AMF association on Figure 31.14.

Absorptive lifestyle Recall from Section 31.3 that AMF absorb phosphorus-containing ions or molecules in the soil and transfer them into the roots of trees, grasses, and shrubs in grassland and tropical habitats (**Figure 31.18**). In exchange, the host plant provides the symbiotic fungi with sugars and other organic compounds.

Life cycle Most species form spores underground. Glomeromycetes are difficult to grow in the laboratory, so their life cycle is not well known. No one has yet discovered a sexual phase in these species.

Human and ecological impacts Because grasslands and tropical forests are among the most productive habitats on Earth, the AMF are enormously important to both human and natural economies.

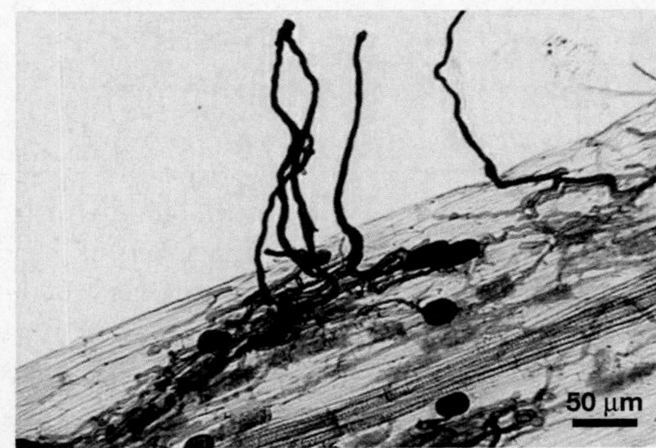

FIGURE 31.18 AMF Penetrate the Walls of Plant Root Cells.

Basidiomycota (Club Fungi)

Although most basidiomycetes form mycelia and produce multicellular reproductive structures, some species have the unicellular growth form. The group is named for basidia, the club-like or pedestal-like cells where meiosis and spore formation occur. ● You should be able to mark the origin of the basidium on Figure 31.14. About 31,000 species have already been described, and more are being discovered each year.

Absorptive lifestyle Basidiomycetes are important saprophytes. Along with a few soil-dwelling bacteria, they are the only organisms capable of synthesizing lignin peroxidase and completely digesting wood. Some basidiomycetes are ectomycorrhizal fungi (EMF) that associate with trees in temperate and northern forests. One subgroup consists entirely of parasitic forms called rusts, including species that cause serious infections in wheat and rye fields. The plant parasites called smut fungi are also basidiomycetes. Smuts specialize in infecting grasses; a few infect other fungi. Thus, the entire array of absorptive lifestyles found in fungi—saprophytic, mutualistic, and parasitic—is found within Basidiomycota.

Life cycle Asexual reproduction through production of spores is common in the Basidiomycota, although not as prevalent as in other groups of fungi. Asexual reproduction also occurs through growth and fragmentation of mycelia in the soil or in rotting wood, resulting in genetically identical individuals that are physically independent. During sexual reproduction, all basidiomycetes—even unicellular ones—produce basidia. In the largest subgroup in this lineage, basidia form in large, aboveground reproductive structures called mushrooms, brackets, earthstars, or puffballs (**Figure 31.19**).

Human and ecological impacts Because temperate and northern forests provide most of the hardwoods and softwoods used in furniture-making, building construction, and papermaking, the EMF are enormously important in forest management. Throughout the world, mushrooms are cultivated or collected from the wild as a source of food. The white button, crimini, and portabella mushrooms you may have seen in grocery stores are all varieties of the same species, *Agaricus bisporus*. Some of the toxins found in poisonous mushrooms are used in biological research; others have hallucinogenic effects on people and are used and traded illegally.

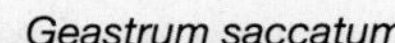

FIGURE 31.19 Earthstars Can Produce Billions of Spores.

Ascomycota (Sac Fungi)

Over half of all known fungi belong to the phylum Ascomycota. This group is named for the sac-like cells, or asci, where meiosis and spore formation take place.

The Ascomycota is extremely large and diverse, and the phylogenetic relationships among major subgroups are still being worked out. As a result, the discussion of "key lineages" that follows is based on two lifestyles observed in ascomycetes and not on distinct monophyletic groups. Work on the phylogenetic relationships within Ascomycota continues.

Ascomycota > Lichen-Formers

About half of the ascomycetes grow in symbiotic association with cyanobacteria and/or single-celled members of the green algae, forming the structures called **lichens**. Over 15,000 different lichens have been described to date; in most, the fungus involved is an ascomycete (although a few basidiomycetes participate as well). To name a lichen, biologists use the genus and species name assigned to the fungus that participates in the association.

Absorptive lifestyle It is not yet clear whether all of the relationships observed in lichens are mutualistic. In habitats where lichens are common, neither partner can exist as a free-living organism. The fungus in lichens appears to protect the photosynthetic bacterial or algal cells. The fungal hyphae form a dense protective layer that shields the photosynthetic species and reduces water loss. In return, the cyanobacterium or alga provides carbohydrates that the fungus uses as a source of carbon and energy. However, the hyphae of some lichen-forming fungi have been observed to invade algal cells and kill them. This observation suggests a partially parasitic relationship in at least some lichens. The nature of lichen-forming associations is the subject of ongoing research.

Life cycle As **Figure 31.20a** shows, many lichens reproduce asexually via the production of small "mini-lichen" structures called **soredia** that contain both symbionts. Soredia disperse to a new location via wind or water and then develop into a new, mature individual via the growth of the algal or bacterial and fungal symbionts. In addition, the fungal partner may form asci. Spores that are shed from asci germinate to form a small mycelium. If the growing hyphae encounter enough algal cells, a new lichen can form.

Human and ecological impacts In terms of their abundance and diversity, lichens dominate the Arctic and Antarctic tundras and are extremely common in boreal forests (see Chapter 50). They are the major food of caribou, as well as the most prevalent colonizers of bare rock surfaces throughout the world (**Figure 31.20b**). Rock-dwelling lichens are significant, because they break off mineral particles from the rock surface as they grow—launching the first step in soil formation. About 10,000 tons of lichens are processed annually and used in perfume production—either as a source of fragrant molecules or as a source of molecules that keep the fragrant components of perfumes from evaporating too rapidly.

(a) Cross section of a lichen, showing three layers

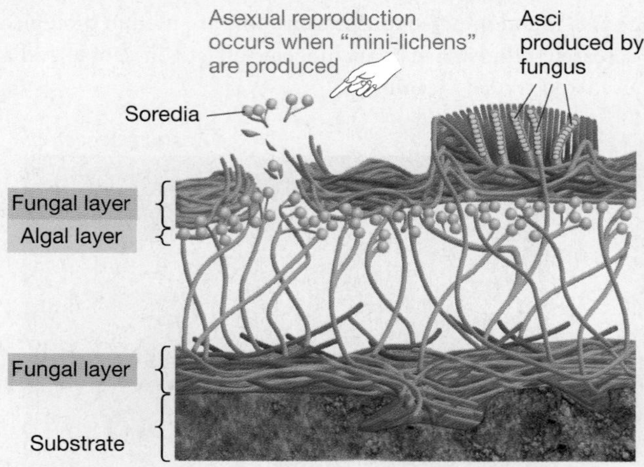

(b) Top view of lichens on a rock

FIGURE 31.20 Lichens Are Associations between a Fungus and a Cyanobacterium or Green Alga. (a) In a lichen, cyanobacteria or green algae are enmeshed in a dense network of fungal hyphae. **(b)** Lichens often colonize surfaces, such as tree bark or bare rock, where other organisms are rare.

Ascomycota > Non-Lichen-Formers

The ascomycetes that do not form lichens are found in virtually every terrestrial habitat, as well as some freshwater and marine environments. Although most ascomycetes form mycelia, many are single-celled yeasts.

Absorptive lifestyle A few members of the Ascomycota form mutualistic ectomycorrhizal fungi (EMF) associations with tree roots. Ascomycetes are also the most common endophytic fungi on aboveground tissues. Large numbers are saprophytic and are abundant in forest floors and in grassland soils. Parasitic forms are common as well. The entire array of absorptive lifestyles found in fungi has evolved within Ascomycota as well as within the Basidiomycota.

Life cycle The aboveground, ascus-bearing reproductive structures of these fungi may be a cup or saucer shape called an **ascocarp** (**Figure 31.21**). For that reason, these ascomycetes are known as **cup fungi**. It is also routine for spores called conidia to be produced asexually, at the ends of specialized hyphae called conidiophores.

Human and ecological impacts Some saprophytic ascomycetes can grow on jet fuel or paint and are used to help clean up contaminated sites. *Penicillium* is an important source of antibiotics, and *Aspergillus* produces citric acid used to flavor soft drinks and candy. Truffles and morels are so highly prized that they can fetch $1000 per pound, and the multibillion-dollar brewing, baking, and wine-making industries would collapse without the yeast *Saccharomyces cerevisiae*. A few parasitic ascomycetes cause infections in humans and other animals. In land plants, parasites from this group cause diseases including Dutch elm disease and chestnut blight.

Sarcoscypha coccinea

FIGURE 31.21 Ascomycetes Are Sometimes Called Cup Fungi.

Chapter Review

SUMMARY OF KEY CONCEPTS

- **Fungi are important in part because many species live in close association with land plants. They supply plants with key nutrients and decompose dead wood. They are the master recyclers of nutrients in terrestrial environments.**

Living plants are colonized by fungi. The roots of grasses and trees that grow in warm or tropical habitats are infiltrated by symbiotic glomeromycetes that supply the plant with phosphorus. Similar mycorrhizal associations, based on an exchange of nitrogen for sugars and other products of photosynthesis, are found between basidiomycetes and the roots of trees in cold, northern habitats. Many fungi also live in leaves and stems; in some grasses, these endophytic fungi secrete toxins that discourage herbivores.

Although mutualistic associations between fungi and living plants are common, so are parasitic relationships. Parasitic fungi are responsible for devastating blights in crops and other plants.

Once plants die, saprophytic fungi degrade the lignin and cellulose in wood and release nutrients from decaying plant material. Because they free up carbon atoms that would otherwise be locked up in wood, fungi speed up the carbon cycle in terrestrial habitats.

You should be able to predict the results of using fungicides to experimentally exclude fungi from 10m × 10m plots in a northern forest, a tropical forest, and a grassland.

- **All fungi make their living by absorbing nutrients from living or dead organisms. Fungi secrete enzymes so that digestion takes place outside their cells. Their morphology provides a large amount of surface area for efficient absorption.**

Several adaptations make fungi exceptionally effective at absorbing nutrients from the environment. Thin hyphae can penetrate tiny openings in soil, and the two growth habits found among Fungi—single-celled "yeasts" or mycelia composed of long, filamentous hyphae—give fungal cells extremely high surface-area-to-volume ratios. Extracellular digestion, in which enzymes are secreted into food sources, is another important adaptation in fungi. Perhaps the most important reactions catalyzed by these enzymes are the degradations of lignin and cellulose found in wood. Lignin decomposes through a series of uncontrolled oxidation reactions triggered by the enzyme lignin peroxidase. This "enzymatic combustion" contrasts sharply with the breakdown of cellulose, which occurs in a carefully regulated series of steps, each catalyzed by a specific cellulase.

You should be able to draw a mycelium penetrating soil or rotting wood and label the site of extracellular digestion and nutrient absorption.

- **Many fungi have unusual life cycles. It is common for species to have a long-lived heterokaryotic stage, in which cells contain haploid nuclei from two different individuals. Although**

most species reproduce sexually, very few species produce gametes.

Many species of fungi have never been observed to reproduce sexually, though most species can produce haploid spores either sexually or asexually. Sexual reproduction usually starts when hyphae from different individuals fuse—an event called plasmogamy. If the fusion of nuclei, or karyogamy, does not occur immediately, a heterokaryotic mycelium forms. Later, heterokaryotic cells produce spore-forming structures where karyogamy and meiosis take place.

The four traditional groupings of fungi are based on distinct reproductive structures: (1) the Chytridiomycota, which are aquatic fungi with motile gametes; (2) the Zygomycota, which are soil-dwelling fungi with tough sporangia; (3) the Basidiomycota, which have club-like, spore-forming structures; and (4) the Ascomycota, which have sac-like, spore-forming structures. Recent analyses of DNA sequence data have revealed that only the Basidiomycota and Ascomycota are monophyletic, however. Researchers are still trying to understand how the various groups of chytrids and zygomycetes are related to each other and to the microsporidians.

You should be able to explain why most fungi don't need to have gametes to accomplish sexual reproduction.

MB Web Animation at www.masteringbio.com
Life Cycle of a Mushroom

QUESTIONS

Test Your Knowledge

1. The mycelial growth habit leads to a body with a high surface-area-to-volume ratio. Why is this important?
   a. Mycelia have a large surface area for absorption.
   b. The hyphae that make up mycelia are long, thin tubes.
   c. Most hyphae are broken up into compartments by walls called septa, although some exist as single, gigantic cells.
   d. Hyphae can infiltrate living or dead tissues.

2. What do researchers hypothesize to be the cause of the "fungal spike" at the end of the Permian period?
   a. unusual soil conditions, which made it more likely that fungal spores and hyphae could fossilize
   b. wet, acidic environments, which slowed the rate of decomposition and allowed peat to form
   c. a switch to saprophytic lifestyles
   d. a mass extinction event that created a huge quantity of dead plant material

3. The Greek root *ecto* means "outer." Why are ectomycorrhizal fungi, or EMF, aptly named?
   a. Their hyphae form tree-like branching structures inside plant cell walls.
   b. They are mutualistic.
   c. Their hyphae form dense mats that envelop roots but do not penetrate the walls of cells inside the root.
   d. They transfer nitrogen from outside their plant hosts to the interior.

4. The hyphae of AMF form bushy or balloon-like structures after making contact with the plasma membrane of a root cell. Why?
   a. They anchor the fungus inside the root, so the association is more permanent.
   b. They increase the surface area available for the transfer of nutrients.
   c. They produce toxins that protect the plant cells against herbivores.
   d. They break down cellulose and lignin in the plant cell wall.

5. What does it mean to say that a hypha is dikaryotic or heterokaryotic?
   a. Two nuclei fuse during sexual reproduction to form a zygote.
   b. Two or more independent nuclei, derived from different individuals, are present.
   c. The nucleus is diploid or polyploid—not haploid.
   d. It is extremely highly branched, which increases its surface area and thus absorptive capacity.

6. Very few organisms besides fungi have a heterokaryotic stage in their life cycle. Which of the following is another unusual aspect of the fungal life cycle?
   a. Some fungi exhibit alternation of generations—meaning that there is a multicellular diploid stage and a multicellular haploid stage.
   b. They produce eggs and sperm in approximately equal numbers, instead of many sperm and a few eggs.
   c. Spores have to fuse with each other before they develop into a new mycelium.
   d. Most varieties undergo sexual reproduction without producing eggs or sperm.

Test Your Knowledge answers: 1. a; 2. d; 3. c; 4. b; 5. b; 6. d

Test Your Understanding

Answers are available at www.masteringbio.com

1. Explain why fungi that degrade dead plant materials are important to the global carbon cycle. Do you accept the text's statement that, without these fungi, "Terrestrial environments would be radically different than they are today and probably much less productive"? Why or why not?

2. Lignin and cellulose provide rigidity to the cell walls of plants. But in most fungi, chitin performs this role. Why is it logical that most fungi don't have lignin or cellulose in their cell walls?

3. Biologists claim that EMF and AMF species are better than plants at acquiring nutrients because they have a higher surface area and because they are more effective at acquiring phosphorus (P) and/or nitrogen (N). Compare and contrast the surface area of mutualistic fungi and plant roots. Explain why fungi are particularly efficient at acquiring P and N, compared to plants.

4. Using information from Chapters 3 through 6, list three key macromolecules found in plants that contain phosphorus. Explain why plant growth might be limited by access to P.

5. Compare and contrast the way that fungi degrade lignin with the way that they digest cellulose.

6. In most eukaryotes that reproduce sexually, only two sexes exist. But in some fungal species, thousands of different mating types exist. Explain why this contrast can occur.

Applying Concepts to New Situations

Answers are available at www.masteringbio.com

1. The box on chytrids in section 31.4 mentions that they may be responsible for massive die-offs currently occurring in frogs. Review Koch's postulates in Chapter 28, then design a study showing how you would use Koch's postulates to test the hypothesis that chytrid infections are responsible for the frog deaths.
2. Some biologists contend that the ratio of plant species to fungus species worldwide is on the order of 1:6. Explain why you agree or disagree with this claim. In doing so, consider the analyses of endophytic, parasitic, lichen-forming, mycorrhizal, and saprophytic strategies presented in this chapter. Also, consider the diversity of tissues and organs available in plants.
3. Experiments indicate that cellulase genes are transcribed and translated together. If cells are selected to be extremely efficient at digesting cellulose, is this result logical? Would you predict that the gene that codes for lignin peroxidase is transcribed along with the cellulase genes? How would you test your prediction?
4. Many mushrooms are extremely colorful. Fungi do not see, so it is unlikely that colorful mushrooms are communicating with one another. One hypothesis is that the colors serve as a warning to animals that eat mushrooms, much like the bright yellow and black stripes on wasps. Design an experiment capable of testing this hypothesis.

www.masteringbio.com is also your resource for • Answers to text, table, and figure caption questions and exercises • Answers to *Check Your Understanding* boxes • Online study guides and quizzes • Additional study tools including the *E-Book for Biological Science* 3rd ed., textbook art, animations, and videos.

32 An Introduction to Animals

KEY CONCEPTS

- Animals are a particularly species-rich and morphologically diverse lineage of multicellular organisms.
- Major groups of animals are recognized by their basic body plan, which differs in the number of tissues observed in embryos, symmetry, the presence or absence of a body cavity, and the way in which early events in embryonic development proceed.
- Recent phylogenetic analyses of animals have shown that there were three fundamental splits as animals diversified, resulting in two protostome groups (Lophotrochozoa and Ecdysozoa) and the deuterostomes.
- Within major groups of animals, evolutionary diversification was based on innovative ways of feeding and moving. Most animals get nutrients by eating other organisms, and most animals move under their own power at some point in their life cycle.
- Methods of sexual reproduction vary widely among animal groups, and many species can reproduce asexually. It is common for individuals to undergo metamorphosis during their life cycle.

Jellyfish are among the most ancient of all animals—they appear in the fossil record over 560 million years ago. Compared with most animals living today, they have relatively simple bodies. But like most other animals, they make their living by eating other organisms and are able to move under their own power.

The **animals** are a monophyletic group of eukaryotes that can be recognized by three traits: They are multicellular, they ingest their food, and they move under their own power at some point in their life cycle. Many unicellular protists also ingest other organisms or dead organic material (detritus) but are small, so they are limited to eating microscopic prey. Animals, in contrast, are the largest and most abundant predators, herbivores, and detritivores in virtually every ecosystem—from the deep ocean to alpine ice fields and from tropical forests to arctic tundras. Animals find food by tunneling, swimming, filtering, crawling, creeping, slithering, walking, running, or flying. They eat nearly every organism on the tree of life.

Over 1.2 million species of animals have been described and given scientific names to date, and biologists predict that tens of millions more have yet to be discovered. To analyze the almost overwhelming number and diversity of animals, this chapter presents a broad overview of how they diversified. It also provides information on the characteristics of the first groups of animals that evolved. The next two chapters provide a more

Key Concept | Important Information | Practice It

detailed exploration of two major phylogenetic groups in animals: protostomes and deuterostomes. Chapter 33 explores the protostomes, which include familiar organisms such as the insects, crustaceans (crabs and shrimp), and mollusks (clams and snails). Chapter 34 features the deuterostomes, which range from the sea stars to the vertebrates, including humans.

32.1 Why Do Biologists Study Animals?

If you ask biologists why they study animals, the first answer they'll give is, "Because they're fascinating." It's hard to argue with this statement. Consider ants. Ants live in colonies that routinely number millions of individuals. But colony-mates cooperate so closely in tasks such as food-getting, colony defense, and rearing young that each ant seems like a cell in a multicellular organism instead of an individual. Other species of ant parasitize this cooperative behavior, however. Parasitic ants look and smell like their host species but enslave them, forcing the hosts to rear the young of the parasitic species instead of their own. Ant colonies also vary widely in habitat. Some species live in trees and protect their host plants by attacking giraffes and other grazing animals a million times their size. Other species are ranchers or farmers. Rancher ants protect the plant-sucking insects called aphids and eat the sugar-rich honeydew that aphids secrete from their anus (**Figure 32.1a**). Farmer ants eat fungi that they plant, fertilize, and harvest in underground gardens (**Figure 32.1b**). New ant species are discovered every year.

Based on observations like these, most people would agree that ants—and by extension, other animals—are indeed fascinating. But beyond pure intellectual interest, there are other compelling reasons that biologists study animals:

- Animals are **heterotrophs**—meaning they obtain the carbon compounds they need from other organisms. Recall from earlier chapters that photosynthetic protists and bacteria are **primary producers**, which form the base of the food chain in most marine environments. Land plants play the same role in most terrestrial habitats. Heterotrophs eat producers and other organisms and are called **consumers**. Animals are consumers that occupy the upper levels of food chains in both marine and terrestrial regions. As a result, it is not possible to understand or preserve ecosystems without understanding and preserving animals.
- Animals are a particularly species rich and morphologically diverse lineage of multicellular organisms on the tree of life. Current estimates suggest that there are between 10 million and 50 million species of animals, although only a fraction of these have been formally described and named. Animals range in size and complexity from tiny sponges, which attach to a substrate and contain just a few cell types and simple tissues, to blue whales, which migrate tens of thousands of kilometers each year in search of food and contain trillions of cells, dozens of distinct tissues, an elaborate skeleton, and highly sophisticated sensory and nervous systems. A great deal of diversification has occurred in this lineage. To understand the history of life, it is important to understand how animals came to be so diverse.
- Humans in every country depend on wild and domesticated animals for food. Horses, donkeys, oxen, and other domesticated animals also provide most of the transportation and power used in preindustrial societies.
- Efforts to understand human biology depend on advances in animal biology. For example, most drug testing and other types of biomedical research are done on mice, rats, or primates.

Given that studying animals is interesting and valuable, let's get started. What makes an animal an animal, and how do biologists go about studying them?

(a) "Rancher ants" tend aphids and eat their sugary secretions.

(b) "Farmer ants" cultivate fungi in gardens.

FIGURE 32.1 Examples of Sophisticated Behavior in Ants.

32.2 How Do Biologists Study Animals?

Animals are distinguished by several traits other than multicellularity, eating, and moving. The cells of animals lack walls but have an extensive extracellular matrix, which includes proteins specialized for cell-cell adhesion and communication (see Chapter 8). Animals are the only lineage on the tree of life with species that have muscle tissue and nervous tissue. Although many animals reproduce both sexually and asexually, no animals undergo alternation of generations. During an animal's life cycle, adults of most species are diploid; the only haploid cells are gametes produced during sexual reproduction.

Beyond these shared characteristics, animals are almost overwhelmingly diverse—particularly in morphology. Biologists currently recognize about 34 **phyla**, or major lineages, of animals—including those listed in **Table 32.1**. Each animal phylum has distinct morphological features.

Analyzing Comparative Morphology

In essence, animals are moving and eating machines. A quick glance at the diversity of ways that animals find and capture food, like those illustrated in **Figure 32.2**, should convince you that evolution by natural selection has indeed produced a wide array of ways to move and eat. This diversity is possible because of extensive variation in appendages used in movement and in mouthparts or other organs used to capture and process food. Limbs and mouths are specialized structures that make particular ways of moving and eating possible.

In contrast to the spectacular diversity observed among animal limbs and mouthparts, the basic architecture of the animal body has been highly conserved throughout evolution. Just as there are only a few basic ways to frame a house—with posts and beams, stud walls, or cement blocks, for example—evolution has produced just a handful of ways to organize an animal body. Once these basic body plans evolved, an extraordinary radiation of species ensued—based on elaborations of limbs and mouthparts or other structures for moving and capturing food.

The most fundamental groups of animals are defined by variation in the basic body plan; the phyla and sub-phyla within these groups are usually defined by variation in appendages and mouthparts. A **body plan** is an animal's architecture—the major features of its structural and functional design. Four features define the basic elements of an animal's body plan: (1) the number of tissue layers found in embryos, (2) the type of body

TABLE **32.1** **An Overview of Major Animal Phyla**

Group and Phylum	Common Name or Example Taxa	Estimated Number of Species
Porifera	Sponges	7,000
Cnidaria	Jellyfish, corals, anemones, hydroids, sea fans	10,000
Ctenophora	Comb jellies	100
Acoelomorpha	Acoelomate worms	10
Protostomes (basal group)		
Chaetognatha	Arrow worms	100
Protostomes: Lophotrochozoa		
Rotifera	Rotifers	1,800
Platyhelminthes	Flatworms	20,000
Nemertea	Ribbon worms	900
Gastrotricha	Gastrotrichs	450
Acanthocephala	Acanthocephalans	1,100
Entoprocta	Entroprocts	150
Gnathostomulida	Gnathostomulids	80
Sipuncula	Peanut worms	320
Echiura	Spoon worms	135
Annelida	Segmented worms	16,500
Mollusca	Molluscs (clams, snails, octopuses)	94,000
Phoronida	Horseshoe worms	20
Ectoprocta	Ectoprocts	4,500
Brachiopoda	Brachiopods; lamp shells	335
Protostomes: Ecdysozoa		
Nematoda	Roundworms	25,000
Kinorhyncha	Kinorhynchs	150
Nematomorpha	Hair worms	320
Priapula	Priapulans	16
Onychophora	Velvet worms	110
Tardigrada	Water bears	800
Arthropoda	Arthropods (spiders, insects, crustaceans)	1,100,000
Deuterostomes		
Echinodermata	Echinoderms (sea stars,sea urchins, sea cucumbers)	7000
Hemichordata	Acorn worms	85
Chordata	Chordates (tunicates, lancelets, sharks, bony fish, amphibians, reptiles, mammals)	50,000

(a) Caterpillar mandibles harvest leaves.

(b) Feather worm tentacles filter debris.

(c) Shark jaws and teeth capture prey.

FIGURE 32.2 Animals Move and Eat in Diverse Ways. Variation in limbs and mouthparts allows animals to move and harvest food in a wide variety of ways.

symmetry and degree of cephalization (informally, this means the formation of a head region), (3) the presence or absence of a fluid-filled body cavity, and (4) the way in which the earliest events in the development of an embryo proceed. The origin and early evolution of animals was based on the origin and elaboration of these four features. Let's consider each in detail.

The Evolution of Tissues All animals have groups of similar cells that are organized into the tightly integrated structural and functional units called **tissues**. Although sponges lack many types of tissues found in other animals, they do have **epithelium**—a layer of tightly joined cells that covers the surface. Other animals have an array of other tissue types as well as epithelium.

In animals other than sponges, the number of tissue layers that exist in an embryo is a key trait. Animals whose embryos have two types of tissue layers are called **diploblasts** (literally, "two-buds"); animals whose embryos have three types are called **triploblasts** ("three-buds"). You might recall from Chapter 22 that these embryonic tissues are organized in layers, called **germ layers**. In diploblasts these germ layers are called **ectoderm** and **endoderm**. The Greek roots *ecto* and *endo* refer to *outer* and *inner*, respectively; the root *derm* means "skin." In most cases the outer and inner "skins" of diploblast embryos are connected by a gelatinous material that may contain some cells. In triploblasts, however, there is a tissue layer called **mesoderm** between the ectoderm and endoderm. (The Greek root *meso* refers to *middle*.)

The embryonic tissues found in animals develop into distinct adult tissues, organs, and organ systems (see Figure 22.10). In triploblasts, for example, ectoderm gives rise to skin and the nervous system. Endoderm gives rise to the lining of the digestive tract. The circulatory system, muscle, and internal structures such as bone and most organs are derived from mesoderm. In general, then, ectoderm produces the covering of the animal and endoderm generates the digestive tract. Mesoderm gives rise to the tissues in between. The same pattern holds in diploblasts, except that (1) muscle is simpler in organization and is derived from ectoderm, and (2) reproductive tissues are derived from endoderm.

Traditionally, two groups of animals have been recognized as diploblasts: the cnidarians and the ctenophorans (**Figure 32.3a**). The Cnidaria (pronounced *ni-DARE-ee-uh*) include jellyfish, corals, anemones, sea pens (**Figure 32.3b**), and hydra—which

(a) Cnidarians and ctenophores are diploblastic.

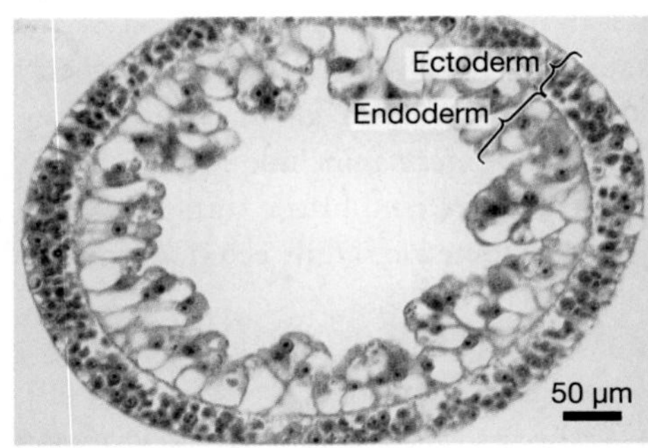

(b) Cnidaria include hydra, jellyfish, corals, and sea pens (shown).

(c) Ctenophora are the comb jellies.

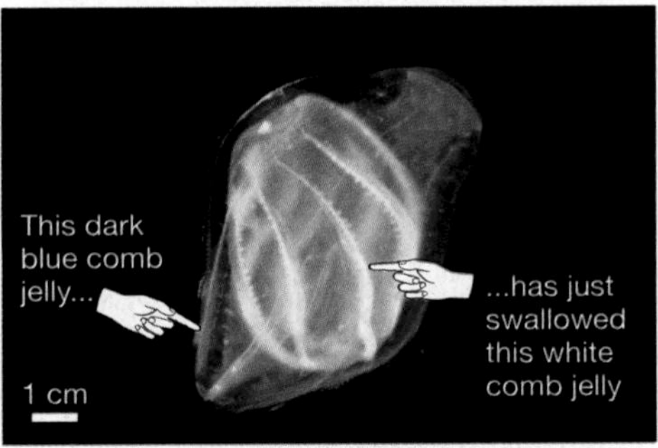

FIGURE 32.3 Diploblastic Animals Have Bodies Built from Ectoderm and Endoderm. (a) Diploblasts have just two embryonic tissue types. **(b)** Like most members of the Cnidaria, this sea pen lives in marine environments. **(c)** Comb jellies, belonging to the Ctenophora, are a major component of planktonic communities in the open ocean.

BOX 32.1 A Model Organism: *Hydra*

Cnidarians are of particular interest to biologists because they are the most ancient living lineage of animals with complex tissues. The ectoderm and endoderm present in their embryos give rise to a number of cell types and tissues in the adults. Cnidarian tissues may be composed of sensory cells that initiate electrical signals in response to environmental stimuli, nerve cells that process those electrical signals and conduct them throughout the body, or muscle cells that contract or relax in response to electrical signals. Cnidarians also have epithelium which consists of a tightly joined layer of cells that is attached to an extensive extracellular matrix. In animals, epithelium covers the outside of the body and lines the surfaces of internal organs.

To understand why the presence of these tissues is interesting, consider the freshwater cnidarian called hydra (**Figure 32.4**). Most species in the genus *Hydra* are about half a centimeter long, live attached to rocks or other firm substrates, and make their living by catching small prey or pieces of organic debris with a cluster of long tentacles. An adult hydra has three major body regions: (1) a basal disk, which attaches the individual to a rock; (2) a tubular section that makes up the bulk of the body; and (3) a "head" that contains the mouth and tentacles.

For over 100 years, biologists have been doing experiments based on cutting hydra bodies apart in various ways and studying how missing tissues and body regions **regenerate**—that is, re-form. This work has led to a deeper understanding of how nerve cells, muscles, epithelia, and other specialized cells and tissues arise from unspecialized cells called stem cells (see Chapter 21). In particular, hydra experiments provided fundamental insights into how cell-to-cell signals organize cells into tissues and, later in development, trigger the differentiation of cells into specialized cell types.

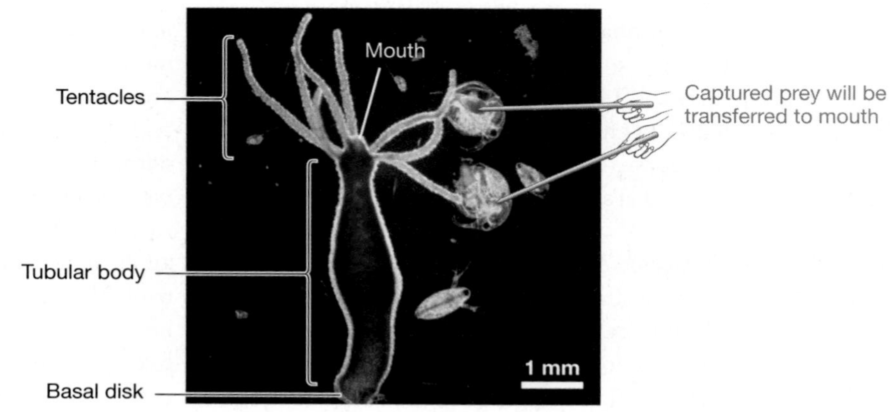

FIGURE 32.4 Hydra Is a Model Organism in Biology. Hydra grow quickly and are relatively easy to maintain in the lab. If an adult is cut into pieces, missing body parts can regenerate in some fragments to form complete adults.

have long been an important source of model organisms in developmental biology (**Box 32.1**). The Ctenophora (pronounced *ten-AH-for-ah*) are the comb jellies (**Figure 32.3c**). All other animals, from leeches to humans, are triploblastic.

Symmetry and Cephalization A basic feature of a multicellular body is the presence or absence of a plane of symmetry. An animal's body is symmetrical if it can be divided by a plane such that the resulting pieces are nearly identical. Animal bodies can have 0, 1, 2, or more planes of symmetry. Most sponges, including the one illustrated in **Figure 32.5a**, are asymmetrical—that is, having no planes of symmetry. They cannot be sectioned in a way that produces similar sides.

All other animals exhibit radial ("spoke") symmetry or bilateral ("two-sides") symmetry. Organisms with **radial symmetry** have at least two planes of symmetry (**Figure 32.5b**). Most of the radially symmetric animals living today either float in water or live attached to a substrate.

Organisms with **bilateral symmetry**, in contrast, have one plane of symmetry and tend to have a long, narrow body. The evolution of bilateral symmetry was a critical step in animal evolution because it triggered a series of associated changes that are collectively known as **cephalization**: the evolution of a head, or anterior region, where structures for feeding, sensing the environment, and processing information are concentrated (**Figure 32.5c**). Feeding and sensory structures on the head face the environment, while posterior regions, at the opposite end of the organism, are specialized for locomotion. All triploblastic animals have bilateral symmetry except for species in the phylum Echinodermata—a group where radial symmetry evolved independently of the diploblastic groups. The echinoderms (pronounced *ee-KINE-oh-derms*) include species such as sea stars, sea urchins, feather stars, and brittle stars. Although their larvae are bilaterally symmetric, adult echinoderms are radially symmetric.

To explain the pervasiveness of bilateral symmetry, biologists point out that locating and capturing food is particularly efficient when movement is directed by a distinctive head region and powered by a long posterior region. In combination with the origin of mesoderm, which made the evolution of extensive musculature possible, a bilaterally symmetric body plan enabled rapid, directed movement and hunting. Lineages with a

(a) Asymmetry

(b) Radial symmetry

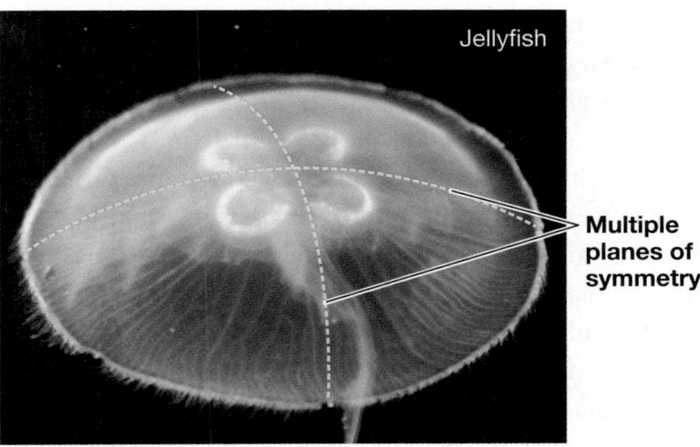

(c) Bilateral symmetry

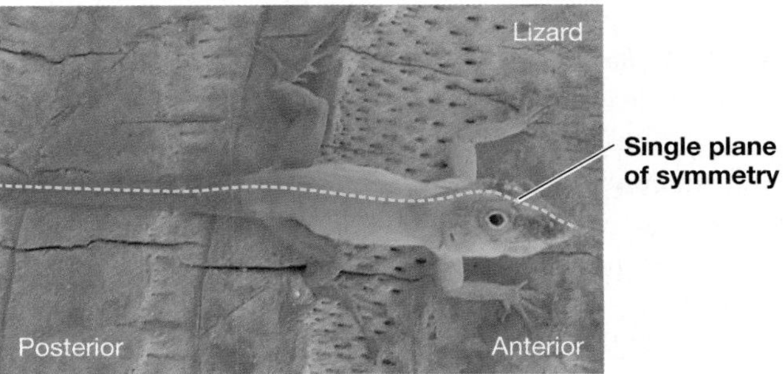

FIGURE 32.5 There Are Three Types of Body Symmetry in Animals.

QUESTION Are humans, sea stars, the comb jellies in Figure 32.3c, and the hydra in Figure 32.4 asymmetric, radially symmetric, or bilaterally symmetric?

triploblastic, bilaterally symmetric body had the potential to diversify into an array of formidable eating and moving machines.

Why Was the Evolution of a Body Cavity Important? A third architectural element that distinguishes animal phyla is the presence of an enclosed, fluid-filled cavity called a **coelom** (pronounced *SEE-lum*). Although diploblasts have a central canal that functions in digestion and circulation, they do not have a coelom.

Biologists were able to determine the nature of the body cavity in various triploblastic phyla through careful observation and

(a) Acoelomates have no enclosed body cavity.

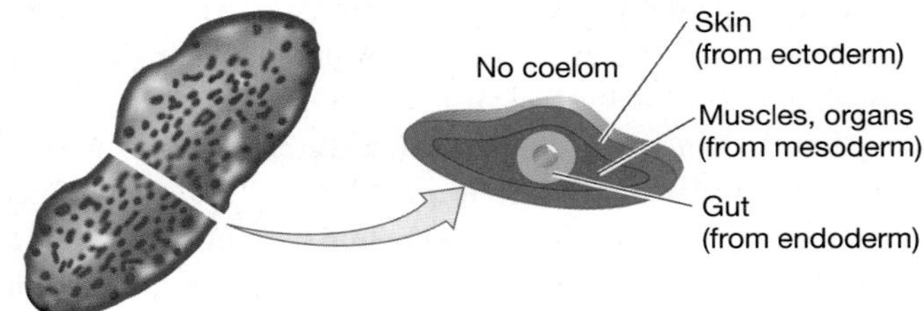

(b) Pseudocoelomates have an enclosed body cavity partially lined with mesoderm.

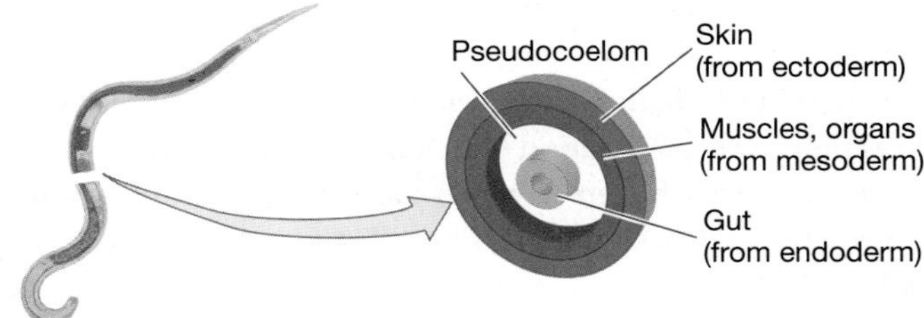

(c) Coelomates have an enclosed body cavity completely lined with mesoderm.

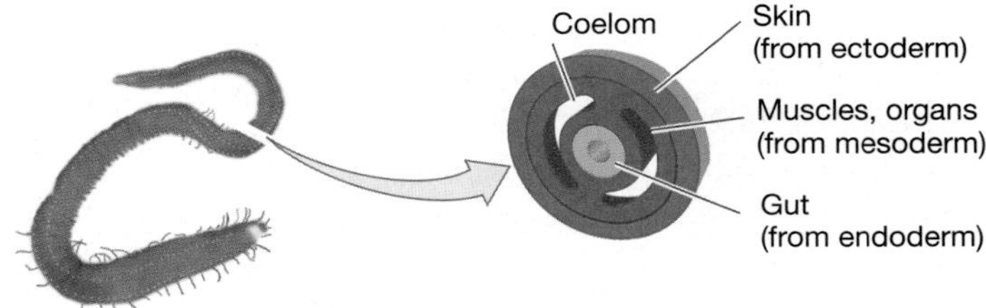

FIGURE 32.6 Animals May or May Not Have a Body Cavity.

dissection of developing embryos and adults. Triploblasts that do not have a coelom are called **acoelomates** ("no-cavity-form"; see **Figure 32.6a**); those that possess a coelom are known as **coelomates.** In a few coelomate groups, such as the roundworms and rotifers, the enclosed cavity forms between the endoderm and mesoderm layers in the embryo. This design is called a **pseudocoelom**, meaning "false-hollow" (**Figure 32.6b**). The term is unfortunate, because there is nothing false about the fluid-filled cavity—it exists. It is simply structured in a different way than a "true" coelom, which forms from within the mesoderm itself and thus is lined on both sides with cells from the mesoderm (**Figure 32.6c**). A pseudocoelom has mesoderm on the outer side of the fluid-filled cavity; a coelom has mesoderm on both the inner and outer sides. As a result, muscle and blood vessels can form on either side of the coelomate's body cavity. In this respect, the coelom represents a more sophisticated design than the pseudocoelom.

The coelom creates a container for the circulation of oxygen and nutrients, along with space where internal organs can move independently of each other. But the coelom is considered a critically important innovation during animal evolution because an enclosed, fluid-filled chamber can act as an efficient **hydrostatic skeleton**. Soft-bodied animals with hydrostatic skeletons can move even if they do not have fins or limbs. Movement is possible because the fluid inside the coelom stretches the body wall—much like a water balloon—meaning that it is under tension (**Figure 32.7a**). This force exerts pressure on the fluid inside the coelom. When muscles in the body wall contract against the pressurized fluid, the fluid moves because the water cannot be compressed.

Using a nematode (roundworm) as an example, **Figure 32.7b** shows how coordinated muscle contractions and relaxations produce changes in the shape of a hydrostatic skeleton that make writhing or swimming movements possible. ● If you understand how a hydrostatic skeleton works, you should be able to demonstrate its function with a long, tube-shaped water balloon, using your fingers to pinch each side and simulate muscle contractions. You should also be able to explain whether your balloon example models a coelom or a pseudocoelom. By providing a hydrostatic skeleton, the evolution of the coelom gave bilaterally symmetric organisms the ability to move efficiently in search of food.

(a) Hydrostatic skeleton of a nematode

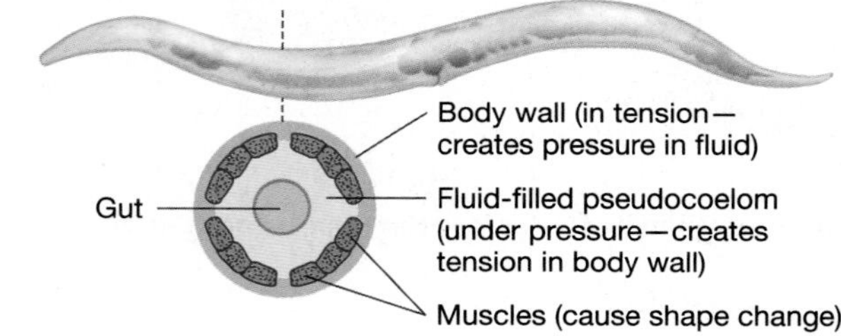

(b) Coordinated muscle contractions result in locomotion.

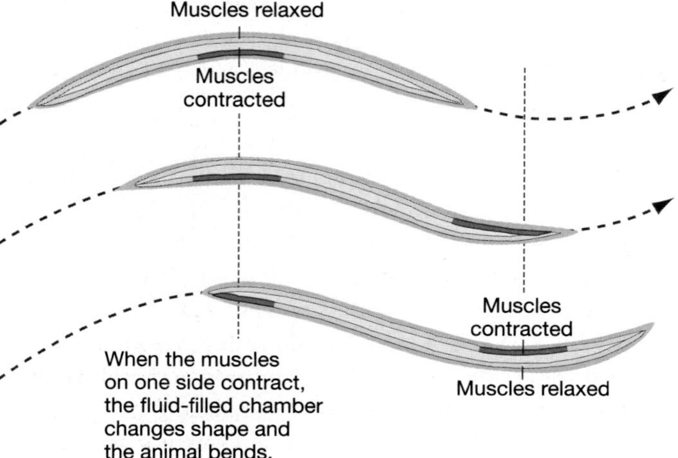

FIGURE 32.7 Hydrostatic Skeletons Allow Limbless Animals to Move. The nematode (roundworm) moves with the aid of its hydrostatic skeleton.

● **QUESTION** Suppose muscles on both sides of this nematode contracted at the same time. What would happen?

What Are the Protostome and Deuterostome Patterns of Development? Except for adult echinoderms, all of the coelomates—including juvenile forms of echinoderms—are bilaterally symmetric and have three embryonic tissue layers. This huge group of organisms is formally called the **Bilateria**, because they are bilaterally symmetric at some point in their life cycle. The bilaterians, in turn, can be split into two subgroups based on distinctive events that occur early in the development of the embryo. The two groups are the protostomes and the deuterostomes. The vast majority of animal species, including the arthropods, molluscs, and annelids, are protostomes; chordates and echinoderms are deuterostomes. (The phylum Arthropoda includes insects, spiders, and crustaceans; Annelida are the segmented worms; Chordata includes bony fishes, amphibians, and mammals.)

To understand the contrasts in how protostome and deuterostome embryos develop, recall from Chapter 22 that the development of an animal embryo begins with cleavage. **Cleavage** is a rapid series of mitotic divisions that occurs in the absence of growth. Cleavage divides the egg cytoplasm and results in a mass of cells. In many protostomes, these cell divisions take place in a pattern known as **spiral cleavage**. When spiral cleavage occurs, the mitotic spindles of dividing cells orient at an angle to the main axis of the cells and result in a helical arrangement of cells. In many deuterostomes, the mitotic spindles of dividing cells orient parallel or perpendicular to the main axes of the cells, resulting in cells that stack directly on top of each other in a pattern called **radial cleavage** (**Figure 32.8a**).

After cleavage has created a ball of cells, the process called gastrulation occurs. As you might recall from Chapter 22, **gastrulation** is a series of cell movements that forms the three embryonic tissue layers. In both protostomes and deuterostomes, gastrulation begins when cells move into the center of the embryo. The migration of the cells creates a pore that opens to the outside (**Figure 32.8b**). In **protostomes**, this pore becomes the mouth. The other end of the gut, the **anus**, forms later. In **deuterostomes**, however, this initial pore becomes the anus; the mouth forms later. Translated literally, *protostome* means "first-mouth" and *deuterostome* means "second-mouth."

The final developmental difference between the groups arises as gastrulation proceeds and the coelom begins to form. Protostome embryos have two blocks of mesoderm beside the gut. As the left side of **Figure 32.8c** indicates, their coelom begins to form when cavities open within each of the two blocks of mesoderm. In contrast, deuterostome embryos have layers of mesodermal cells located on either side of the gut. As the right side of Figure 32.8c shows, their coelom begins to form when these layers bulge out and pinch off to form fluid-filled pockets lined with mesoderm.

To summarize, the protostome and deuterostome patterns of development result from differences in three processes: cleavage, gastrulation, and coelom formation. In essence, the

PROTOSTOMES | DEUTEROSTOMES

(a) Cleavage
(zygote undergoes rapid divisions, eventually forming a mass of cells)

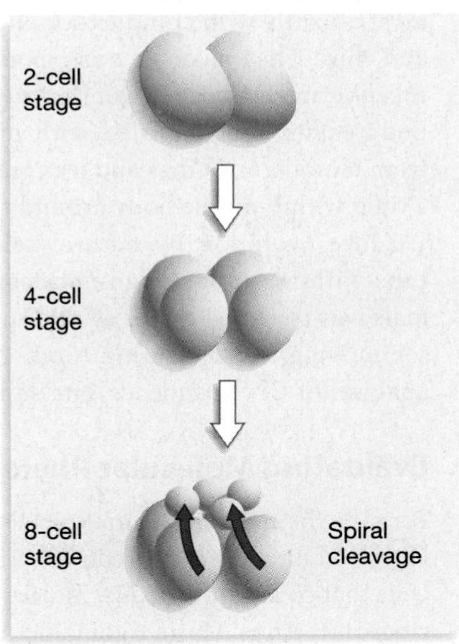

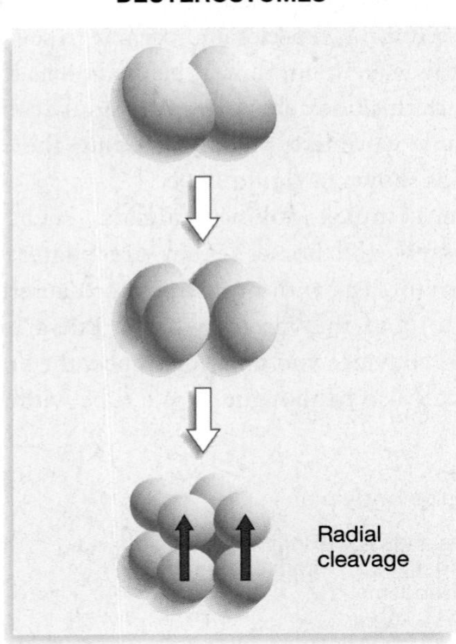

(b) Gastrulation
(mass of cells formed by cleavage is rearranged to form gut and embryonic tissue layers)

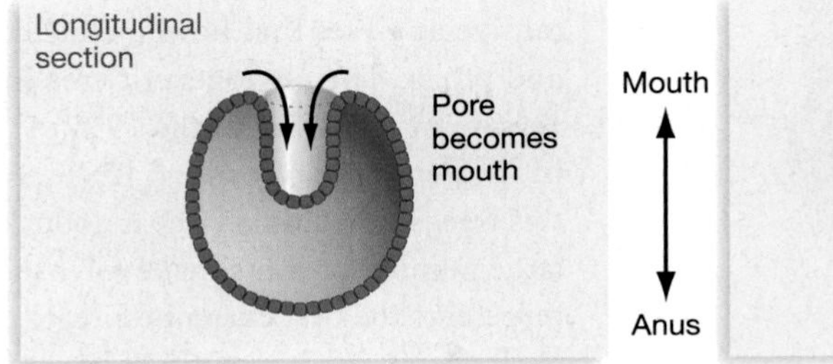

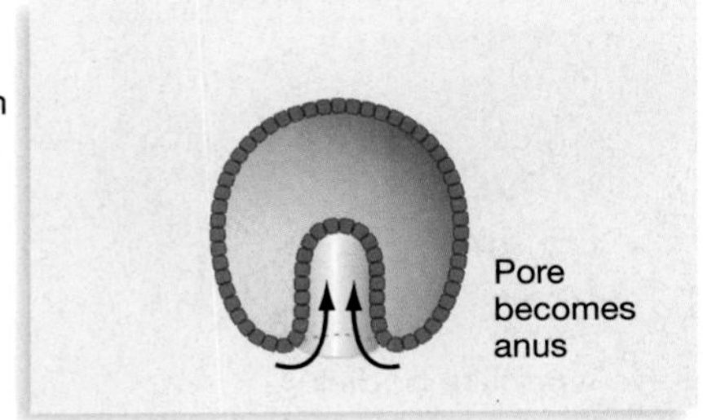

(c) Coelom formation
(body cavity lined with mesoderm develops)

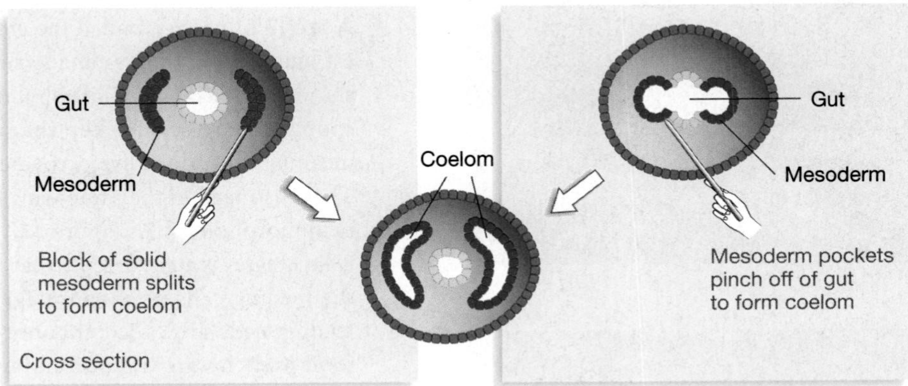

FIGURE 32.8 In Protostomes and Deuterostomes, Three Events in Early Development Differ. The differences between protostomes and deuterostomes show that there is more than one way to build a bilaterally symmetric, coelomate body plan.

protostome and deuterostome patterns of development represent two distinct ways of achieving the same end—the construction of a bilaterally symmetric body that contains a cavity lined with mesoderm.

The Tube-within-a-Tube Design Over 99 percent of the animal species alive today are bilaterally symmetric triploblasts that have coeloms and follow either the protostome or deuterostome pattern of development. This combination of features turned out to be a spectacularly successful way to organize a moving and eating machine.

Although it sounds complex to call a certain animal a "bilaterally symmetric, coelomic triploblast with protostome [or deuterostome] development," the bodies of most animals are actually extremely simple in form. The basic animal body is a tube within a tube. The inner tube is the individual's gut, and the outer tube forms the body wall, as illustrated in **Figure 32.9a**. The mesoderm in between forms muscles and organs. In several animal phyla, individuals have long, thin, tubelike bodies that lack limbs. Animals with this body shape are commonly called **worms**. There are many wormlike phyla, including the nemerteans and sipunculids shown in **Figure 32.9b**.

What about more complex-looking animals, such as grasshoppers and lobsters and horses? They are bilaterally symmetric, coelomic triploblasts with protostome or deuterostome development, too, and they aren't worms. But a moment's reflection should convince you that, in essence, the body plan of these animals can also be thought of as a tube within a tube, except that the tube is mounted on legs. Consider that most animals with complex-looking bodies are relatively long and thin. They have an outer body wall that is more or less tubelike and an internal gut that runs from mouth to anus. The body cavity itself is filled with muscles and organs derived from mesoderm. Wings and legs are just efficient ways to move a tube-within-a-tube body around the environment.

Once evolution by natural selection produced the basic tube-within-a-tube design, most of the diversification of animals was triggered by the evolution of novel types of structures for moving and capturing food. Do data from phylogenetic analyses of DNA sequence data support this view?

(a) The tube-within-a-tube body plan

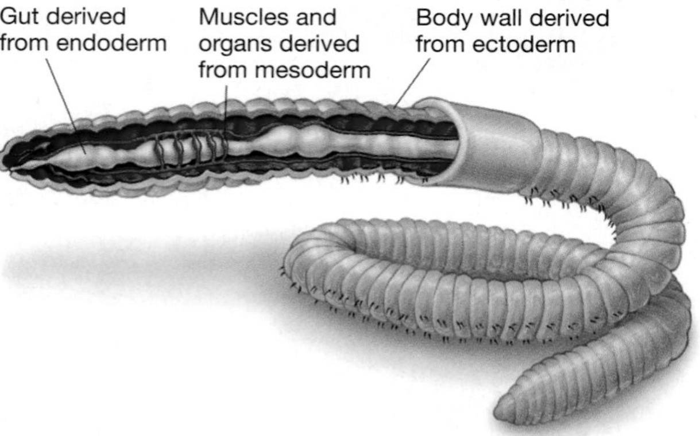

(b) Many animal phyla have wormlike bodies.

Nemertean (ribbon worm)

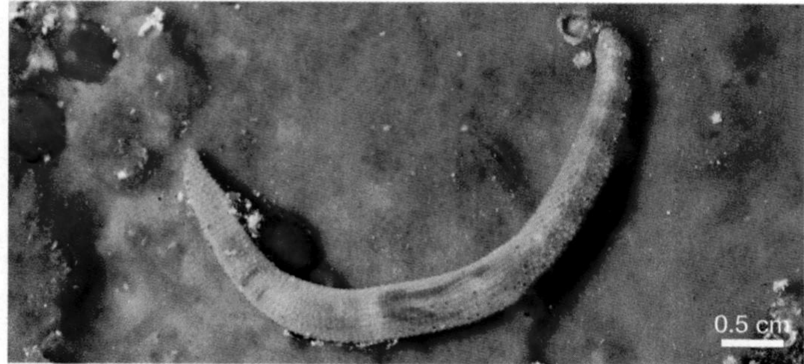

Sipunculid (peanut worm)

FIGURE 32.9 The Tube-within-a-Tube Body Plan Is Common in Animals.

● **QUESTION** How is the term *worm* similar to the term *yeast* (see Chapter 31)?

Evaluating Molecular Phylogenies

Perhaps the most influential paper ever published on the phylogeny of animals appeared in 1997. Using sequences from the gene that codes for the RNA molecule in the small subunit of the ribosome, Anna Marie Aguinaldo and colleagues estimated the phylogeny of species from 14 animal phyla. The results were revolutionary, and they have been verified by more recent and extensive analyses that have included data from additional genes and phyla. The phylogenetic tree in **Figure 32.10** is an updated version of the result of the 1997 study, based on further studies of the genes for ribosomal RNA and several proteins. Because this tree is based on a large amount of sequence data, and thus a large number of traits that evolve independently of each other, it represents the best current estimate of animal phylogeny.

To begin analyzing this tree, start from the root and work your way up. You should note several key points:

- A group of protists called the choanoflagellates are the closest living relatives of animals, and the Porifera (sponges) are the most basal animal phylum. Choanoflagellates and sponges share several key characteristics. Both are **sessile**, meaning that adults live permanently attached to a substrate. They also feed in the same way, using cells with nearly identical morphology. As **Figure 32.11** shows, the beating of flagella creates water currents that bring organic debris toward the feeding cells of choanoflagellates and sponges. Sponge feeding cells are called **choanocytes**. In these feeding cells, food particles are trapped and ingested. The key distinction between the two groups is that sponges are multicellular.
- Sponges are paraphyletic. Stated another way, not all sponges are derived from the same common ancestor. Instead, sponges represent several independent lineages.
- The radially symmetric, diploblastic phyla are placed just up from sponges on the tree, meaning they split off from other animals slightly later than sponges. This conclusion is bolstered by the fossil record of animals, because sponges appear prior to cnidarians and ctenophores. The placement of diploblasts at the base of the tree implies that endoderm and ectoderm were the first tissue types to evolve, and that radial symmetry evolved before bilateral symmetry.

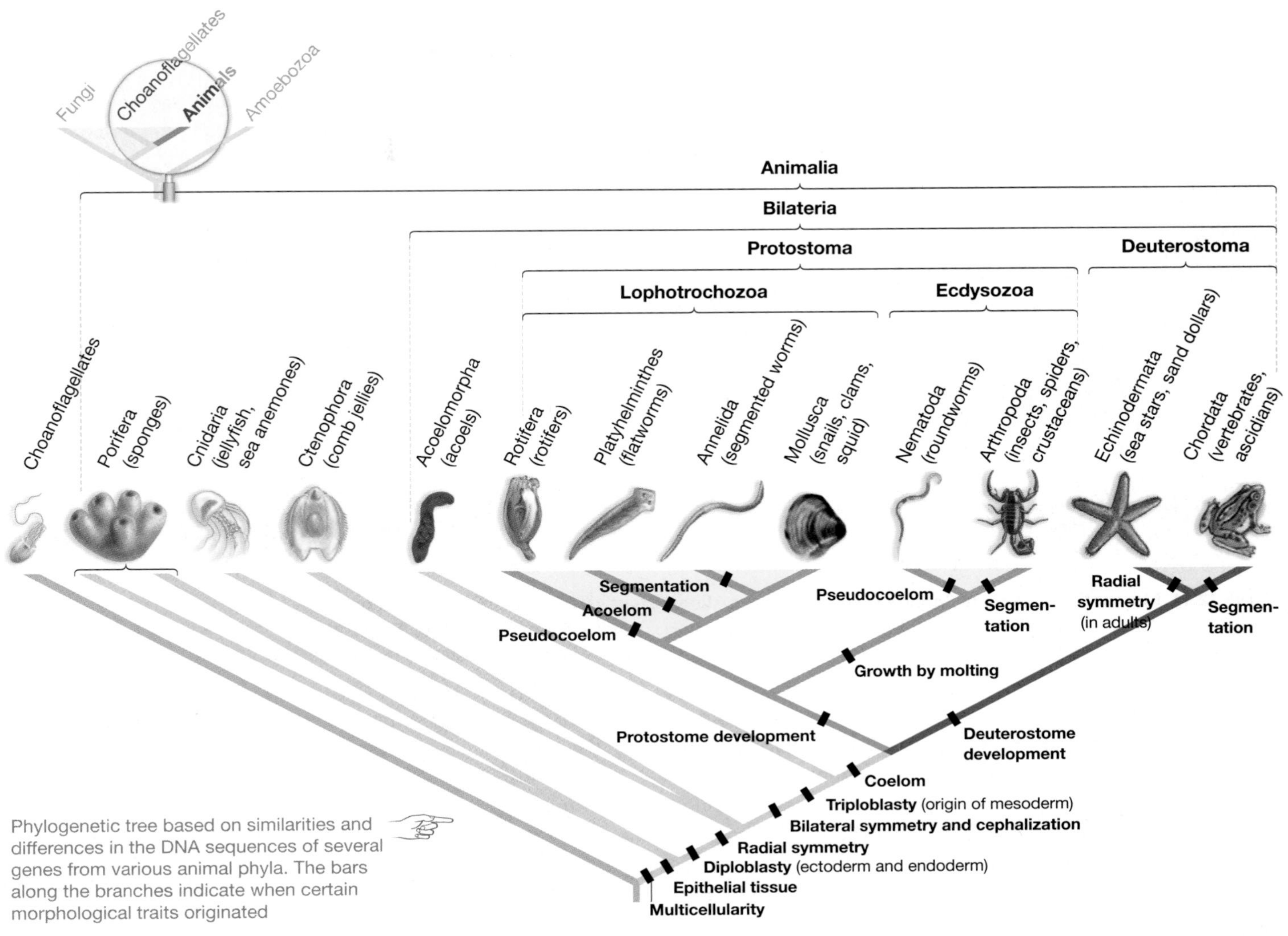

FIGURE 32.10 A Phylogeny of Animal Phyla Based on DNA Sequence Data. Phylogenetic tree based on similarities and differences in the DNA sequences of several genes from various animal phyla. The bars along the branches indicate when certain morphological traits originated.

EXERCISE Recent work suggests that segmentation may also have arisen independently within molluscs. Add a "Segmentation?" label to the tree to indicate this possibility.

- The most ancient groups of bilaterally symmetric triploblasts, the Acoelomorpha, lack a coelom. This result supports an important hypothesis—that animal bodies usually evolved from simpler to more complex forms.

- Based on morphology, biologists had thought that the major event in the evolution of the Bilateria was the split between the protostomes and the deuterostomes. The molecular data concur but show that an additional, equally fundamental split occurred within protostomes, forming two major subgroups with protostome development: (1) The Ecdysozoa (pronounced *eck-die-so-ZOH-ah*) includes the arthropods and the nematodes; (2) the Lophotrochozoa (pronounced *low-foe-tro-ko-ZOH-ah*) includes the molluscs and the annelids. **Ecdysozoans** grow by shedding their external skeletons and expanding their bodies, while **lophotrochozoans** grow by extending the size of their skeletons.

- Species in the phylum Platyhelminthes (flatworms) do not have a coelom but are lophotrochozoans. To interpret this result, biologists point out that platyhelminths had to have evolved from an ancestor that had a coelom. Stated another way, the acoelomate condition in these species is a derived condition. It represents the *loss* of a complex trait, indicated by the "Acoelom" label on the tree. In this case, animals did not evolve from simpler to more complex forms.

- Twice during the course of evolution, bodies with pseudocoeloms arose from ancestors that had "true" coeloms. A change from coelom to pseudocoelom occurred in the ancestors of today's (1) nematodes (roundworms) and (2) rotifers.

- When a body is divided into a series of repeated structures, such as an earthworm's segments or a fish's vertebral column and ribs, it is said to be segmented. Segmentation evolved

(a) Choanoflagellates are sessile protists; some are colonial.

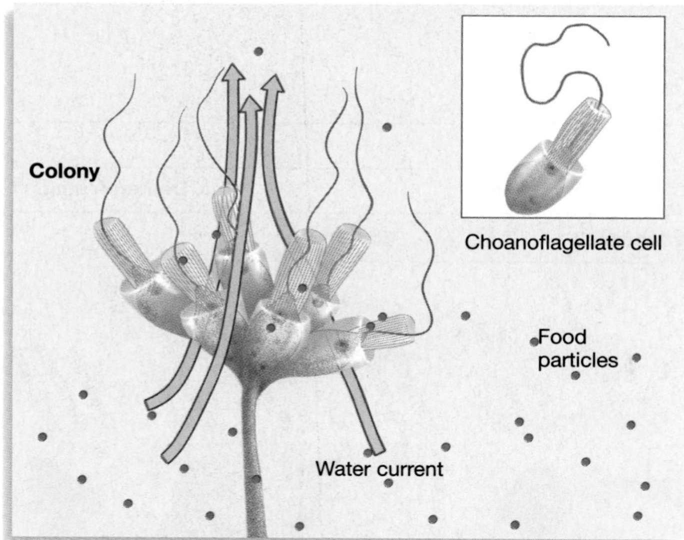

(b) Sponges are multicellular, sessile animals.

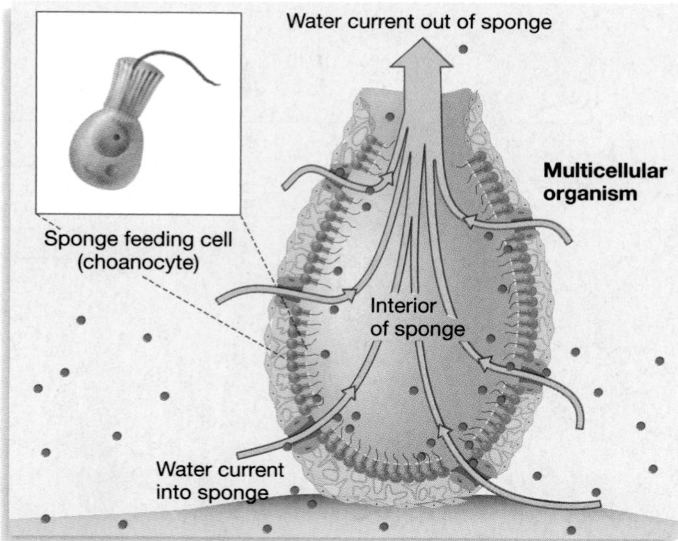

FIGURE 32.11 Choanoflagellates and Sponge Feeding Cells Are Almost Identical in Structure and Function.
(a) Choanoflagellates are suspension feeders. **(b)** A cross section of a simple sponge as it suspension feeds. The beating of flagella produces a water current that brings food into the body of the sponge, where it can be ingested by feeding cells.

independently in annelids (earthworms and other segmented worms) and arthropods (insect, spiders, and crustaceans), as well as in vertebrates. **Vertebrates** are a monophyletic lineage within the Chordata that is defined by the presence of a skull; many vertebrate species also have a backbone (see Chapter 34). The group called the **invertebrates**, which is defined as all animals that are not vertebrates, is paraphyletic—meaning that it includes some, but not all, of the descendants of a common ancestor. Recent evidence suggests that segmentation also evolved independently in molluscs (snails, clams, squid).

Although biologists are increasingly confident that most or all of these conclusions are correct, the phylogeny of animals is still very much a work in progress. As data sets expand, it is likely that new analyses will not only confirm or challenge these results but also contribute other important insights into how the most species-rich lineage on the tree of life originated and diversified. Stay tuned.

MB Web Animation at www.masteringbio.com
The Architecture of Animals

Check Your Understanding

If you understand that . . .

The origin and early diversification of animals was marked by changes in four fundamental features: body symmetry, the number of embryonic tissues present, the evolution of a body cavity, and protostome versus deuterostome patterns of development.

You should be able to . . .

1) Explain why bilateral symmetry in combination with triploblasty and a coelom is responsible for the "tube-within-a-tube" design observed in most animals living today.
2) Explain why cephalization was important.
3) Make a rough sketch of the phylogeny of animals, showing choanoflagellates as an out-group, sponges and jellyfish as basal groups, Acoelomorpha as the most ancient members of the Bilateria, and then three branches showing the relationships among the Lophotrochozoa, Ecdysozoa, and deuterostomes.
4) On the tree you sketched, mark the origin of multicellularity, triploblasty, bilateral symmetry, protostome development, and deuterostome development.

32.3 What Themes Occur in the Diversification of Animals?

Within each animal phylum, the basic features of the body plan do not vary from species to species. For example, molluscs are triploblastic, bilaterally symmetric protostomes with a reduced coelom; their body plan features a muscular foot, a cavity called the visceral mass, and a structure called a mantle. But there are over 100,000 species of mollusc. If the major animal lineages are defined by a particular body plan, what triggered the diversification of species within each lineage?

In most cases, the answer to this question is the evolution of innovative methods for feeding and moving. Recall that most animals get their food by ingesting other organisms. Animals are diverse because there are thousands of ways to find and eat the millions of different organisms that exist.

(a) Krill filter feed using their legs.

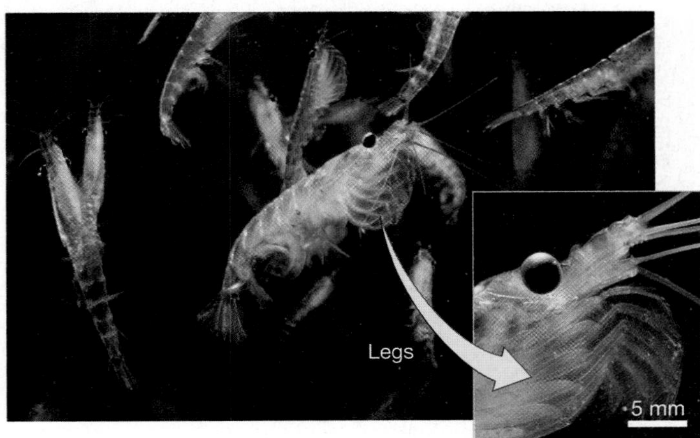

(b) Baleen whales filter feed using their baleen.

FIGURE 32.12 Suspension Feeders Harvest Suspended Food. (a) Krill and **(b)** baleen whales filter food particles from water, using the trapping structures shown in the close-ups.

Feeding

To organize the diversity of ways that animals find food, biologists distinguish *how* individuals eat from *what* they eat. In terms of how they eat, animals obtain food in four general ways: suspension feeding, deposit feeding, fluid feeding, and feeding on food masses. In terms of what they eat, animals have three general sources of food: plants or algae, other animals, and detritus—dead organic material. In many cases, species that pursue different food-getting strategies and food sources are found within a single lineage. Let's review the diversity of ways that animals eat, then consider the diversity of ways that animals move around to find food.

How Animals Feed: Four General Tactics Two simple ideas are key to understanding how animals from the same lineage can have the same basic body plan but feed in radically different ways: (1) Their mouthparts vary, and (2) the structure of an animal's mouthparts correlates closely with its method of feeding. Keep these concepts in mind as you review the general tactics that animals use to obtain food.

- **Suspension feeders**, also known as **filter feeders**, capture food by filtering out or concentrating particles suspended in water or air. **Figure 32.12a** shows the small marine animals called krill, which suspension feed as they swim. As individuals move forward, their legs wave in and out. Projections on their legs trap food particles that flow past. The food particles are then moved up the body to the mouth, where they are ingested. Krill, in turn, are eaten by suspension feeders called the baleen whales **(Figure 32.12b)**. These whales have a series of long plates, made from a horny material called baleen, hanging from their jaws. Baleen whales feed by gulping water that contains krill, squeezing the water out between their baleen plates, and trapping the krill inside their mouths. Suspension feeders employ a wide array of structures to trap suspended particles—usually small algae or animals or bits of detritus—and bring them to their mouths.
- **Deposit feeders** eat their way through a substrate. Earthworms, for example, are annelids that swallow soil as they tunnel through it **(Figure 32.13a)**. Many deposit feeders digest organic matter in the soil; their food consists of soil-dwelling

(a) Earthworms (Annelida) eat their way through soil.

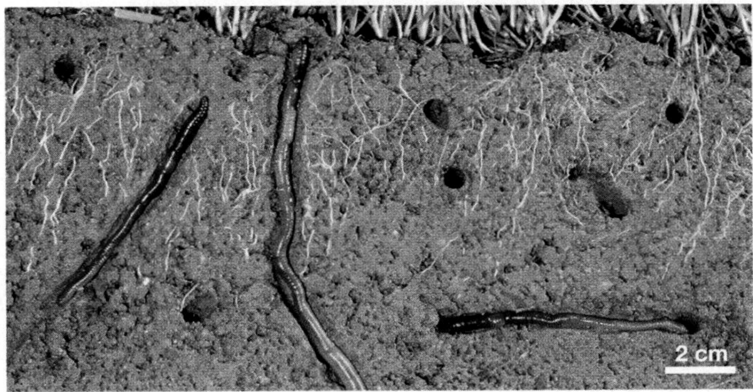

(b) Sea cucumbers eat their way through organic deposits.

FIGURE 32.13 Deposit Feeders Eat Their Way through a Substrate.

QUESTION Why is it logical that deposit feeders tend to have tubelike bodies?

(a) Butterflies have an extensible, hollow proboscis.

(b) Blowflies have an extensible, sponge-like mouthpart.

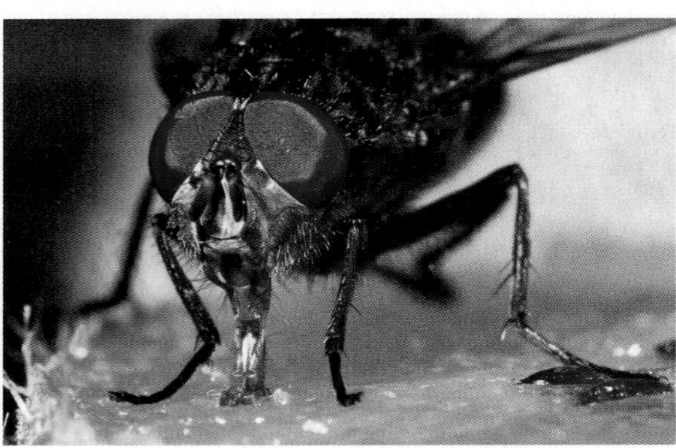

FIGURE 32.14 Fluid Feeders Drink Liquids.

● **QUESTION** Hummingbirds, ticks, and mosquitoes are also fluid feeders. How does the structure of their mouthparts correlate with function in feeding?

bacteria, archaea, protists, and fungi, along with detritus. Insects that burrow through plant leaves and stems, bore through piles of feces, or mine the carcasses of dead animals or plants can also be considered deposit feeders because they eat through a substrate. Perhaps the most common and important deposit feeders, though, are the echinoderms called sea cucumbers **(Figure 32.13b)**. The floor of the sea is rich in organic matter that rains down from the surface and collects in food-rich deposits, which sea cucumbers exploit.

Unlike suspension feeders, which are diverse in size and shape and use various trapping or filtering systems, deposit feeders are similar in appearance. They usually have simple mouthparts if they eat soft substrates, and their body shape is wormlike. Like suspension feeders, however, deposit feeders occur in a wide variety of lineages including roundworms (Nematoda), molluscs (Mollusca), peanut worms (Sipunculida), and chordates such as hagfish.

- **Fluid feeders** suck or mop up liquids like nectar, plant sap, blood, or fruit juice. They range from butterflies that feed on nectar with a straw-like proboscis (**Figure 32.14a**) to blowflies that feed on rotting fruit using a sponge-like mouthpart (**Figure 32.14b**). They are found in a wide array of lineages and often have mouthparts that allow them to pierce seeds, stems, skin, or other structures in order to withdraw the fluids inside.
- **Mass feeders** take chunks of food into their mouths. In these species, the structure of the mouthparts correlates with the type of food pieces that are harvested and ingested. Horses, for example, have sharp teeth in the front of the jaw for biting off grass stems and broad, flat molar teeth in the back of the jaw for mashing the coarse stems into a soft wad that can be swallowed (**Figure 32.15a**). In many snails, a feeding structure called a **radula** functions like a rasp or a file. The sharp plates on the radula move

(a) Horses feed on grass stems.

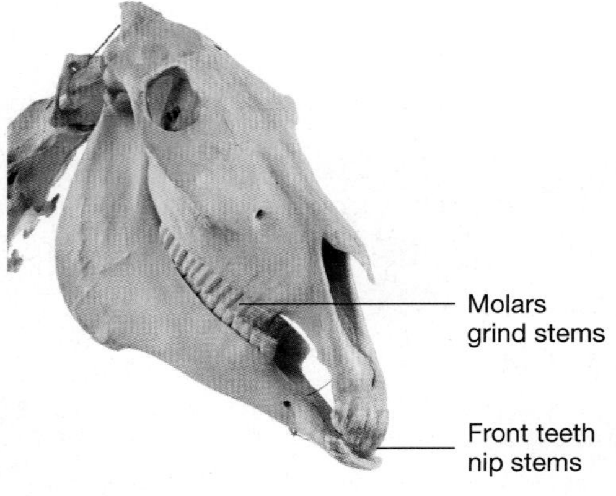

(b) Terrestrial snails feed on leaves.

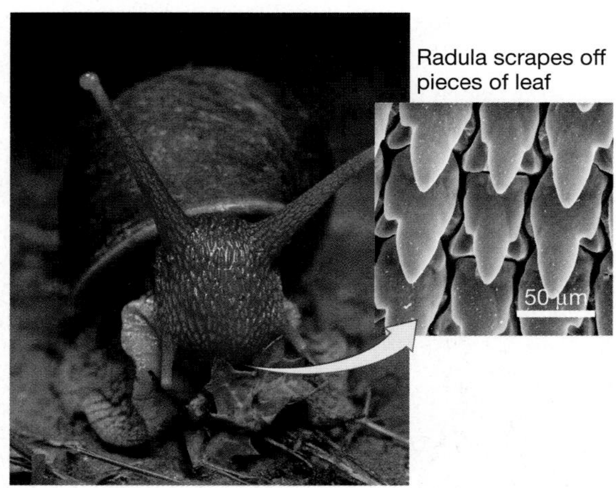

FIGURE 32.15 Food-Mass Feeders Ingest Bites or Lumps of Food.

● **QUESTION** Humans, dogs, and eagles are also food-mass feeders. How does the structure of their mouthparts correlate with function in feeding?

(a) Many frogs (Chordata) sit and wait for prey.

(b) Wolves (Chordata) chase prey.

FIGURE 32.16 Predators Kill and Eat Organisms.

back and forth to scrape material away from a plant or alga so that it can be ingested (Figure 32.15b).

What Animals Eat: Three General Sources Whether they feed by filtering, eating through deposits, or taking in fluids or masses, animals can be classified as **herbivores** that feed on plants or algae, **carnivores** that feed on animals, or **detritivores** that feed on dead organic matter. In addition, herbivores and carnivores can be sub-classified as predators or parasites; animals that eat both plants and animals are called **omnivores**.

Predators kill other organisms for food, using an array of mouthparts and hunting strategies. Many types of frogs, for example, are sit-and-wait predators. They sit still and wait for an insect or worm to move close, then capture it with a lightning-quick extension of their long, sticky tongue (**Figure 32.16a**). In contrast, wolves hunt by locating a prey organism and then running it down during an extended, long-distance chase (**Figure 32.16b**). Mountain lions stalk their prey slowly, then pounce on it or run it down in a short sprint. Although most predators kill other animals to live, species that eat seeds are commonly referred to as seed predators.

Unlike predators, **parasites** are usually much smaller than their victims and often harvest nutrients without causing death. **Endoparasites** live inside their hosts and usually have simple, wormlike bodies. Tapeworms, for example, are platyhelminths with no digestive system. Instead of a mouth, they have hooks or other structures on their head, called a scolex, that attach to their host's intestinal wall (**Figure 32.17a**). Instead of digesting food themselves, they absorb nutrients directly from their surroundings. **Ectoparasites,** in contrast, live outside their hosts. They usually have limbs or mouthparts that allow them to grasp the host and mouthparts that allow them to pierce their host's skin and suck the nutrient-rich fluids inside. The louse in **Figure 32.17b** is an example of an insect (Arthropoda) ectoparasite that afflicts humans.

(a) Tapeworms (Platyhelminthes) are endoparasites.

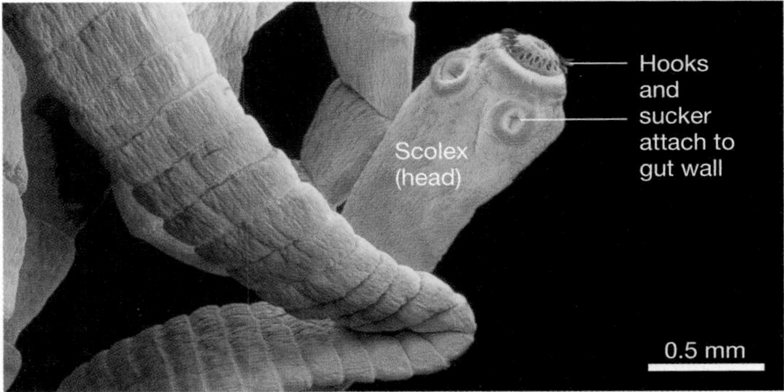

(b) Lice (Arthropoda) are ectoparasites.

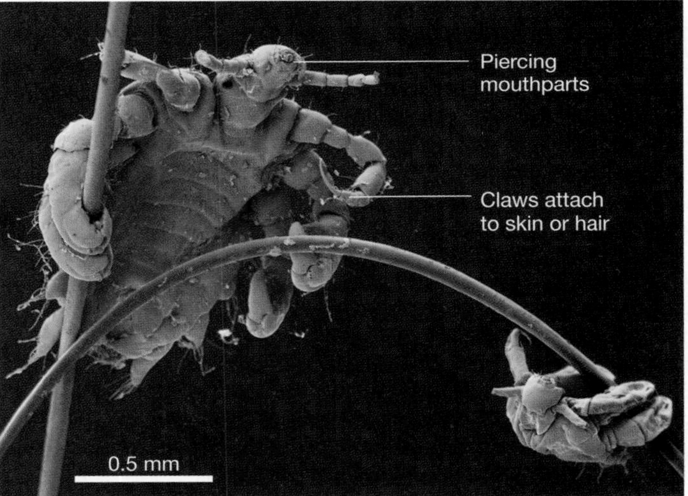

FIGURE 32.17 Parasites Take Nutrients from Living Animals.
(a) Tapeworms are common intestinal parasites of humans and other vertebrates. They attach to the wall of the digestive tract, using the barbed hooks and suckers on their anterior end, and absorb nutrition directly across their body wall. **(b)** Lice are insects that parasitize mammals and birds. This louse, *Phthirus pubis,* uses its clawlike legs to attach to the pubic region of humans. The animal pierces the host's skin with its mouthparts and feeds by sucking body fluids.

(a) Motile larval anemone

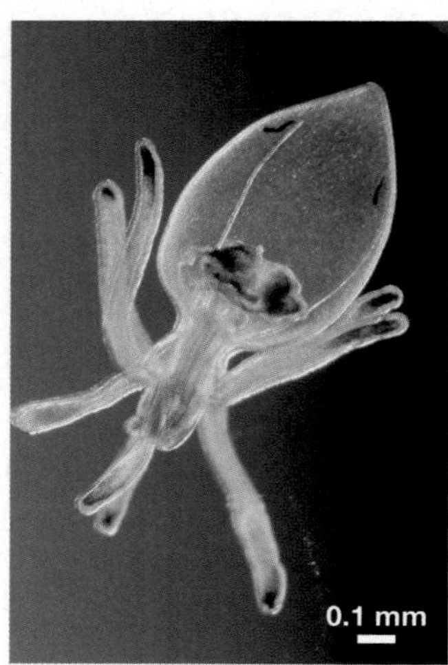

(b) Sessile adult anemone

FIGURE 32.18 If Adults Are Sessile, Then Larvae Disperse. (a) Larval anemones swim under their own power and can disperse to new habitats. **(b)** Anemones are sessile most of their lives.

Movement

Many animals are sit-and-wait predators, and some are sessile throughout their adult lives. But the vast majority of animals move under their own power either as a juvenile or as an adult. For example, the eggs of sea anemones hatch into larvae that swim (**Figure 32.18a**). Adult sea anemones, however, spend most of their lives attached to a rock or another substrate and make their living by capturing and eating fish or other organisms that pass by (**Figure 32.18b**). In species such as these, larvae function as a dispersal stage. They are a little like the seeds of land plants—a life stage that allows individuals to move to new habitats, where they will not compete with their parents for space and other resources.

In animals that move as adults, locomotion has three functions: (1) finding food, (2) finding mates, and (3) escaping from predators. The ways that animals move in search of food and sex are highly variable; they burrow, slither, swim, fly, crawl, walk, or run. The structures that power movement are equally variable—they include cilia, flagella, and muscles that attach to a hard skeleton or compress a hydrostatic skeleton, enabling wriggling movements. The hydrostatic skeleton is an evolutionary innovation unique to animals and is responsible for locomotion in the many animal phyla with wormlike bodies. Another major innovation occurred in animals, however, that made highly controlled, rapid movement possible: the limb.

Types of Limbs: Unjointed and Jointed Limbs are a prominent feature of species in many phyla and are particularly important in two major lineages: the ecdysozoans and the vertebrates. Some members of the Ecdysozoa, such as onychophorans (velvet worms), have unjointed, sac-like limbs (**Figure 32.19a**); others, such as crabs and other arthropods, have more complex, jointed limbs (**Figure 32.19b**). Jointed limbs make fast, precise movements possible and are a prominent type of limb in vertebrates and arthropods.

(a) Onychophorans are ecdysozoans with sac-like limbs.

(b) Crabs (Arthropoda) have jointed limbs.

(c) Polychaetes (Annelida) have parapodia.

(d) Sea urchins (Echinodermata) have tube feet.

FIGURE 32.19 Various Animal Appendages Function in Locomotion. Lineages within the Ecdysozoa have **(a)** sac-like legs or **(b)** jointed limbs. **(c)** Some species in the Lophotrochozoa have small projections called parapodia. **(d)** Echinoderms have unusual structures called tube feet.

Are All Animal Appendages Homologous? Chapter 24 introduced the concept of homology, which is defined as similarity in traits due to inheritance from a common ancestor. Traditionally, biologists have hypothesized that the major types of jointed and unjointed animal limbs were not homologous. To appreciate the logic behind this hypothesis, it's important to recognize just how diverse animal appendages are. Animals in a wide array of phyla have structures that stick out from the main body wall and function in locomotion. These appendages range from the human arm to bristle-like structures called parapodia in segmented worms (**Figure 32.19c**) and the soft, extensible tube feet found in echinoderms (**Figure 32.19d**). Because the structure of animal appendages is so diverse, it was logical to maintain that at least some appendages evolved independently of each other. As a result, biologists predicted that completely different genes are responsible for each major type of appendage.

Recent results have challenged this view, however. The experiments in question involve a gene called *Distal-less,* which was originally discovered in fruit flies. (*Distal* means "away from the body.") *Distal-less,* or *Dll,* is aptly named. In fruit flies that lack this gene's normal protein product, only the most rudimentary limb buds form. The mutant limbs are "distal-less." Based on the morphology of *Dll* mutants, the protein seems to deliver a simple message as a fruit-fly embryo develops: "Grow appendage out this way."

A group of biologists working in Sean Carroll's lab set out to test the hypothesis that *Dll* might be involved in limb or appendage formation in other animals. As **Figure 32.20** shows, they used a fluorescent marker that sticks to the *Dll* gene product to locate tissues where the gene is expressed. When they introduced the fluorescent marker into embryos from annelids, arthropods, echinoderms, chordates, and other phyla, they found that it bound to *Dll* in all of them. More important, the highest concentrations of *Dll* gene products were found in cells that form appendages—even in phyla with wormlike bodies that have extremely simple appendages. Other experiments have shown that *Dll* is also involved in limb formation in vertebrates.

Based on these findings, biologists are concluding that at least a few of the same genes are involved in the development of *all* appendages observed in animals. To use the vocabulary introduced in Chapter 24, the hypothesis is that all animal appendages have some degree of genetic homology and that they are all derived from appendages that were present in a common ancestor. The idea is that a simple appendage evolved early in the history of the Bilateria and that, subsequently, evolution by natural selection produced the diversity of limbs, antennae, and wings observed today. This hypothesis is controversial, however, and research continues at a brisk pace.

Reproduction and Life Cycles

An animal may be efficient at moving and eating, but if it does not reproduce, the alleles responsible for its effective locomotion

Experiment

Question: Is the gene *Dll* involved in limb formation in species other than insects?

Hypothesis: In all animals, *Dll* signals "grow appendage out here."

Null hypothesis: *Dll* is not involved in the development of appendages in species other than insects.

Experimental setup:

Add a stain to developing embryos that will attach to *Dll* gene products (proteins), revealing their location. The stain can be fluorescent green or dark brown.

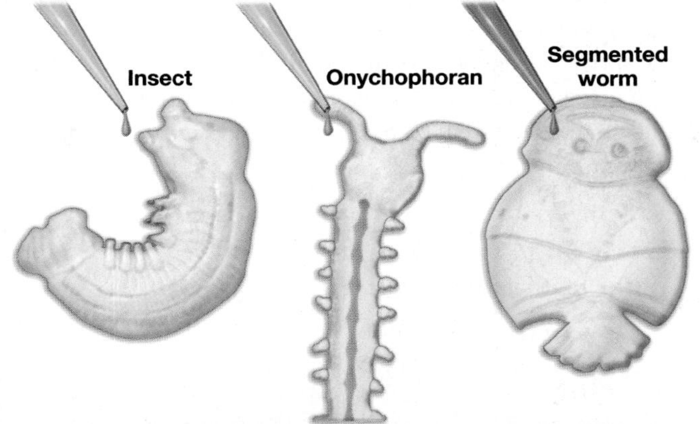

Prediction: In embryos from a wide array of species, stained *Dll* gene products will be localized to areas where appendages are forming.

Prediction of null hypothesis: Stained *Dll* gene products will be localized to areas where appendages are forming only in insects.

Results:

Insect: Developing legs

Onychophoran: Antennae; Developing legs

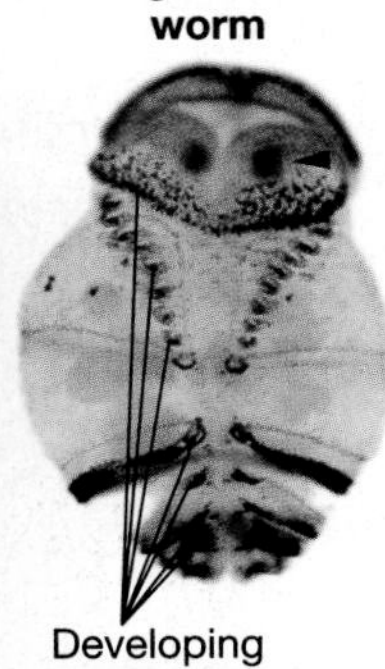

In species representing both Ecdysozoa and Lophotrochozoa, *Dll* is localized in areas of the embryo where appendages are forming.

Conclusion: The gene *Dll* is involved in limb formation in diverse species. The results suggest that all animal appendages may be homologous.

FIGURE 32.20 Experimental Evidence That All Animal Appendages Are Homologous.

QUESTION What results would have supported the null hypothesis?

and feeding will not increase in frequency in the population. As Chapter 24 emphasized, natural selection occurs when individuals with certain alleles produce more offspring than other individuals do. Organisms live to reproduce.

Given the array of habitats and lifestyles pursued by animals, it is not surprising that they exhibit a high degree of variation in how they reproduce. Although animal reproduction will be explored in detail in Chapter 48, a few examples will help drive home just how variable animal reproduction is:

- At least some species in most animal phyla can reproduce asexually, via mitosis, as well as sexually—via meiosis and fusion of gametes. In the lophotrochozoan phylum Rotifera, an entire lineage called the bdelloids (pronounced *DELL-oyds*) reproduces only asexually. Even certain fish, lizard, and snail species have never been observed to undergo sexual reproduction.
- When sexual reproduction does occur, fertilization may be internal or external. When internal fertilization takes place, males typically insert a sperm-transfer organ into the body of a female (**Figure 32.21a**). In some cases, males produce sperm in packets, which females then pick up and insert into their own bodies. But in seahorses, females insert eggs into the male's body, where they are fertilized. (The male is pregnant for a time and then gives birth to live young.) External fertilization is extremely common in aquatic species. Females lay eggs onto a substrate or into open water. Males shed sperm on or near the eggs (**Figure 32.21b**).
- Eggs or embryos may be retained in the female's body during development, or eggs may be laid outside to develop independently of the mother. Mammals and other species that nourish embryos inside the body and give birth to live young are said to be **viviparous** ("live-bearing"; **Figure 32.22a**); species that deposit fertilized eggs are **oviparous** ("egg-bearing"; **Figure 32.22b**); and some species are **ovoviviparous** ("egg-live-

(a) Internal fertilization

(b) External fertilization

FIGURE 32.21 Fertilization Can Be Internal or External in Animals. **(a)** Internal fertilization in dragonflies. **(b)** A giant clam emitting sperm that will fertilize a clutch of eggs released by a different individual.

(a) Viviparity ("live-bearing"): the birth of a shark

(b) Oviparity ("egg-bearing")

FIGURE 32.22 Some Animal Species Give Birth to Live Young, but Most Lay Eggs. **(a)** A lemon shark is viviparous, meaning its embryos develop for a period of time inside the female's body. **(b)** A corn snake is oviparous, meaning it lays fertilized eggs.

bearing"). In ovoviviparous species, the females retain eggs inside their body during early development; but the growing embryos are nourished by yolk inside the egg and not by nutrients transferred directly from the mother, as in viviparous species. Ovoviviparous females then give birth to well-developed young. Mammals and a few species of sea stars, onychophorans, fish, and lizards are viviparous; some snails, insects, reptiles, fishes, and sharks are ovoviviparous. But the vast majority of animals are oviparous.

Besides reproducing in a variety of ways, animal life cycles vary widely. Perhaps the most spectacular innovation in animal life cycles involves the phenomenon known as **metamorphosis** ("change-form")—a change from a juvenile to an adult body type. Juveniles are sexually immature, meaning their reproductive organs are undeveloped and the individual cannot breed. Adults are the reproductive stage in the life cycle.

Metamorphosis occurs in two basic ways. In **holometabolous** ("whole change") **metamorphosis**, the juvenile form is called a **larva** (plural: **larvae**) and looks substantially different from the adult form. Mosquito larvae, for example, live in quiet bodies of freshwater, where they suspension feed on bacteria, algae, and detritus (**Figure 32.23a,** top). When a larva has grown sufficiently, the individual stops feeding and moving and secretes a protective case. The individual is now known as a **pupa** (plural: **pupae**; Figure 32.23a, middle). During **pupation**, the pupa's body is completely remodeled into a new, adult form. In mosquitoes, the adult individual flies and gets its nutrition as a parasite—taking blood meals from mammals and sucking fluids from plants (Figure 32.23a, bottom). Holometabolous metamorphosis is also called complete metamorphosis. The transformation of butterfly and moth caterpillars to flying adults is another spectacular example of holometabolous metamorphosis.

In **hemimetabolous** ("half-change") **metamorphosis**, the juvenile form is called a **nymph** and looks like a miniature version of the adult (**Figure 32.23b**, top). For example, a grasshopper nymph sheds its external skeleton several times and grows—gradually changing from a wingless, sexually

(a) Mosquito: **Holo**metabolous metamorphosis

Larvae

Juveniles look substantially different from adults and eat different foods

Pupae

Adult

(b) Grasshopper: **Hemi**metabolous metamorphosis

FIGURE 32.23 During Metamorphosis, Individuals May or May Not Change Form Completely. (a) When holometabolous metamorphosis occurs, juvenile forms are called larvae. During pupation, the juvenile body is remodeled into the adult form. Adult forms are sexually mature. (b) When hemimetabolous metamorphosis occurs, juvenile forms are called nymphs. Adult forms are sexually mature.

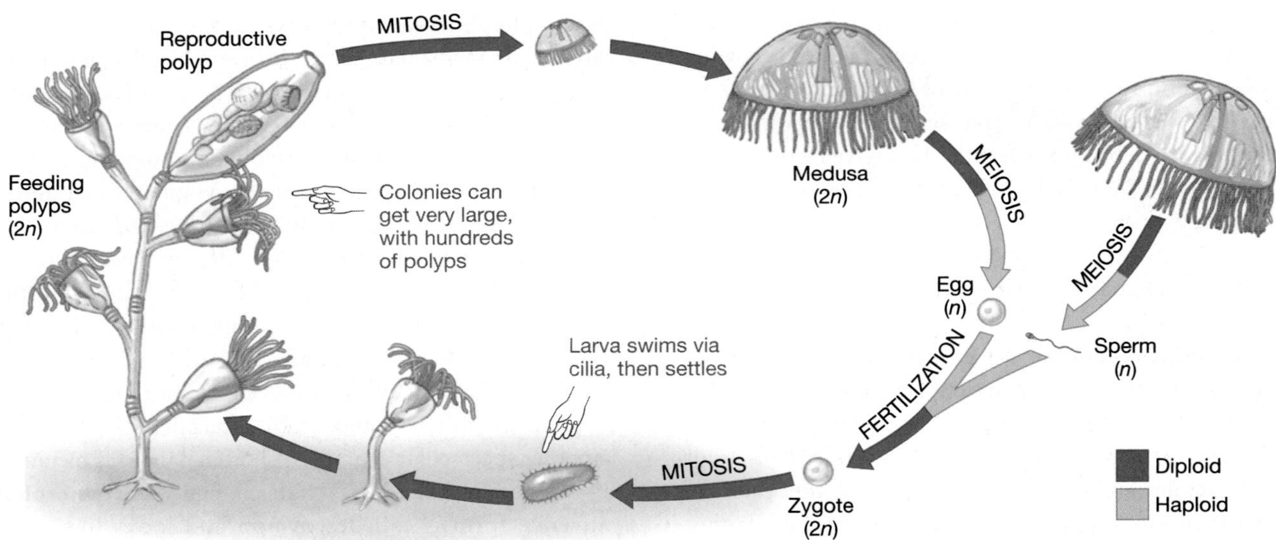

FIGURE 32.24 Cnidarian Life Cycles May Include a Polyp and Medusa Form. This is an example of a common life cycle in jellyfish.

● **QUESTION** Does alternation of generations occur in this species? Explain why or why not.

immature nymph to a sexually mature adult that is capable of flight (Figure 32.23b, bottom). Throughout this process, grasshoppers feed on the same food source in the same way: They chew leaves. Hemimetabolous metamorphosis is also called incomplete metamorphosis.

In insects, holometabolous metamorphosis is 10 times more common than hemimetabolous metamorphosis. One hypothesis to explain this observation is based on efficiency in feeding. Because juveniles and adults from holometabolous species feed on different materials in different ways and sometimes even in different habitats, they do not compete with each other. When complete metamorphosis is part of the life cycle, individuals of the same species but different ages often show dramatic variations in their mode of feeding. An alternative hypothesis to explain the evolutionary success of complete metamorphosis is based on specialization in feeding and mating. In many moths and butterflies, for example, larvae are specialized for feeding, whereas adults are specialized for mating. Larvae are largely sessile, whereas adults are highly mobile. If specialization leads to higher fitness, then complete metamorphosis would be advantageous. These hypotheses are not mutually exclusive, however—meaning that both could be correct—and they are still being tested.

Complete metamorphosis is also extremely common in marine animals. For example, some cnidarians have two distinct body types during their life cycle: (1) A largely sessile form called a **polyp** alternates with (2) a free-floating stage called a **medusa** (plural: **medusae**; **Figure 32.24**). Polyps usually live attached to a substrate, suspension feed on detritus, and frequently form large clusters of individuals called colonies. A **colony** is a group of identical individuals that are physically attached. Medusae, in contrast, float freely in the plankton and feed on crustaceans. Because polyps and medusae live in different habitats, the two stages of the life cycle exploit different food sources.

Check Your Understanding

If you understand that . . .

- The story of animal evolution is based on two themes:
  (1) the evolution of a small suite of basic body plans, and
  (2) a diversification of species with the same basic body plan, based on the evolution of innovative structures and methods for capturing food and moving.

You should be able to . . .

Give at least three examples of fundamental variations in the ways that animals find food, move, and reproduce.

32.4 Key Lineages of Animals: Basal Groups

The goal of this chapter is to provide a broad overview of how animals diversified into such a morphologically diverse lineage. According to recent phylogenetic analyses, the closest living relatives to the animals are the choanoflagellates, a group of protists. Phylogenetic analyses and the fossil record also indicate that the phyla Porifera (sponges), Cnidaria (jellyfish and others), Ctenophora, and Acoelomorpha are the most ancient of all animal groups (**Figure 32.25**). Let's explore the origins of animals by taking a more detailed look at each of the most basal lineages. The protostomes and deuterostomes are explored in Chapters 33 and 34, respectively.

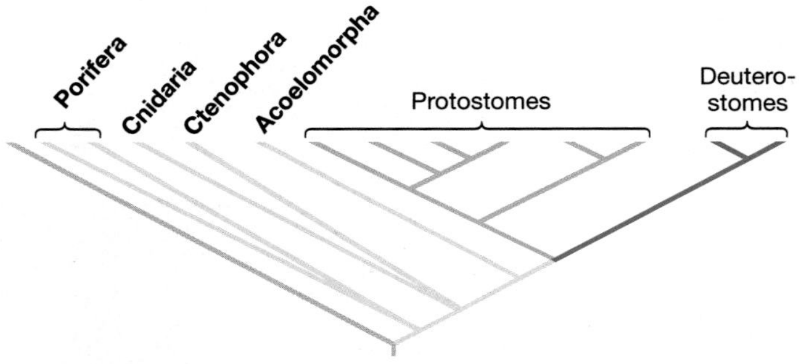

FIGURE 32.25 Animals Are a Monophyletic Group. Use this tree to map the origin of new features discussed in the following key lineage boxes.

● **EXERCISE** Write in the name of the out-group to animals.

Porifera (Sponges)

About 7000 species of sponges have been described to date. Although a few freshwater species are known, most are marine. All sponges are **benthic**, meaning that they live at the bottom of aquatic environments. Sponges are particularly common in rocky, shallow-water habitats of the world's oceans and in coastal areas of Antarctica.

The architecture of sponge bodies is built around a system of tubes and pores that create channels for water currents. Body symmetry varies among sponge species; most are asymmetrical, but some species are radially symmetric (**Figure 32.26**). Sponges have specialized cell types and well-organized epithelial tissue layers lining the inside and outside of the body, sealing a middle layer in between. ● You should be able to indicate the origin of multicellularity and epithelial tissue on Figure 32.25. They lack the other tissue types found in other animals, however. The "looseness" of sponge tissues is thought to be an adaptation that facilitates regeneration after wounding, the ability to remodel the body if water currents change, and the ability to move into and colonize new substrates as they become available. In many species, collagen fibers are augmented by **spicules**—stiff spikes of silica or calcium carbonate ($CaCO_3$)—to provide structural support for the body. One sponge species native to the Caribbean can grow to heights of 2 m.

Sponges have commercial and medical value to humans. The dried bodies of certain sponge species are able to hold large amounts of water and thus are prized for use in bathing and washing. In addition, researchers are increasingly interested in the array of toxins that sponges produce to defend themselves against predators and bacterial parasites. Some of these compounds have been shown to have antibacterial properties or to promote wound healing in humans.

Feeding Most sponges are suspension feeders. Their cells beat in a coordinated way to produce a water current that flows through small pores in the outer body wall, into chambers inside the body, and out through a single larger opening. As water passes by feeding cells, organic debris and bacteria, archaea, and small protists are filtered out of the current and then digested.

Movement Most adult sponges are sessile, though a few species are reported to move at rates of up to 4 mm per day. Most species produce larvae that swim with the aid of cilia.

Reproduction Asexual reproduction occurs in a variety of ways, depending on the species. Some sponge cells are totipotent, meaning that small groups of adult cells have the capacity to develop into a complete adult organism. Thus, a fragment that breaks off an adult sponge has the potential to grow into a new individual. Although individuals of most species produce both eggs and sperm, self-fertilization is rare because individuals release their male and female gametes at different times. Fertilization usually takes place in the water, but some ovoviviparous species retain their eggs and then release mature, swimming larvae after fertilization and early development have occurred.

Pseudoceratina crassa

FIGURE 32.26 Some Sponges Form Radially Symmetric Tubes.

Cnidaria (Jellyfish, Corals, Anemones, Hydroids)

Although a few species of Cnidaria inhabit freshwater, the vast majority of the 11,000 species are marine. They are found in all of the world's oceans, occupying habitats from the surface to the substrate, and are important predators. The phylum comprises four main lineages: Hydrozoa (hydroids—see Figure 32.4), Cubozoa (box jellyfish), Scyphozoa (jellyfish—see the chapter opening photo and Figure 32.24), and Anthozoa (anemones, corals, and sea pens—see Figure 32.3b).

Cnidarians are radially symmetric diploblasts consisting of ectoderm and endoderm layers that sandwich gelatinous material known as **mesoglea**, which contains a few scattered ectodermal cells. They have a gastrovascular cavity instead of a flow-through gut—meaning there is only one opening to the environment for both ingestion and elimination of wastes.

Many cnidarians have a life cycle that includes both a sessile polyp form (**Figure 32.27a**) and a free-floating medusa (**Figure 32.27b**). Anemones, and coral, however, exist only as polyps—never as medusae. Reef-building corals secrete outer skeletons of calcium carbonate that create the physical structure of a coral reef—one of the world's most productive habitats (see Chapter 54).

Feeding The morphological innovation that triggered the diversification of the cnidarians is a specialized cell, called a **cnidocyte**, which is used in prey capture. When cnidocytes brush up against a fish or other type of prey, the cells forcibly eject a barbed, spear-like structure called a cnidocyst, which is coated with toxins. The barbs hold the prey, and the toxins subdue it until it can be brought to the mouth and ingested. Cnidocytes are commonly located near the mouths of cnidarians or on elongated structures called tentacles. Cnidarian toxins can be deadly to humans as well as to prey organisms; in Australia, twice as many people die each year from stings by box jellyfish as from shark attacks ● You should be able to indicate the origin of cnidocytes on Figure 32.25. Besides capturing prey actively, most species of coral and many anemones host photosynthetic dinoflagellates. The relationship is mutually beneficial, because the protists supply the cnidarian host with food in exchange for protection.

Movement Both polyps and medusae have simple, muscle-like tissue derived from ectoderm or endoderm. In polyps, the gut cavity acts as a hydrostatic skeleton that works in conjunction with the muscle-like cells to contract or extend the body. Many polyps can also creep along a substrate, using muscle cells at their base. In medusae, the bottom of the bell structure is ringed with muscle-like cells. When these cells contract rhythmically, the bell pulses and the medusa moves by jet propulsion—meaning a forcible flow of water in the opposite direction of movement. Cnidarian larvae swim by means of cilia.

Reproduction Polyps may produce new individuals asexually by (1) budding, in which a new organism grows out from the body wall of an existing individual; (2) fission, in which an existing adult splits lengthwise to form two individuals; or (3) fragmentation, in which parts of an adult regenerate missing pieces to form a complete individual. During sexual reproduction, gametes are usually released from the mouth of a polyp or medusa and fertilization takes place in the open water. Eggs hatch into larvae that become part of the plankton before settling and developing into a polyp.

(a) Polyps attach to substrates.

Aurelia aurita

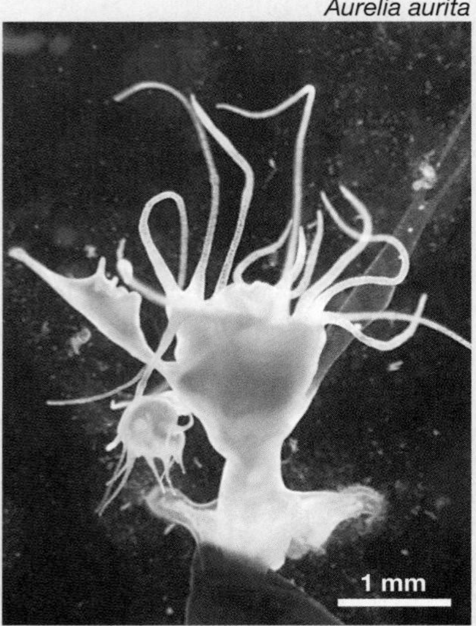

(b) Medusae float near the water surface.

Aurelia aurita

FIGURE 32.27 Most Cnidarians Have Polyp and Medusa Stages.

Ctenophora (Comb Jellies)

Ctenophores are transparent, ciliated, gelatinous diploblasts that live in marine habitats (**Figure 32.28**). Although a few species live on the ocean floor, most are planktonic—meaning that they live near the surface. Only about 100 species have been described to date, but some are abundant enough to represent a significant fraction of the total planktonic biomass.

Feeding Ctenophores are predators. Feeding occurs in several ways, depending on the species. Some comb jellies have long tentacles covered with cells that release an adhesive when they contact prey. These tentacles are periodically wiped across the mouth so that captured prey can be ingested. In other species, prey can stick to mucus on the body and be swept toward the mouth by cilia. Still other species ingest large prey whole. Figure 32.3c shows a comb jelly that has just swallowed a second comb jelly almost as large as itself.

Movement Adults move via the beating of cilia, which occur in comblike rows running the length of the body. Ctenophores are the largest animals known to use cilia for locomotion. ● You should be able to indicate the origin of swimming via rows of coordinated cilia on Figure 32.25.

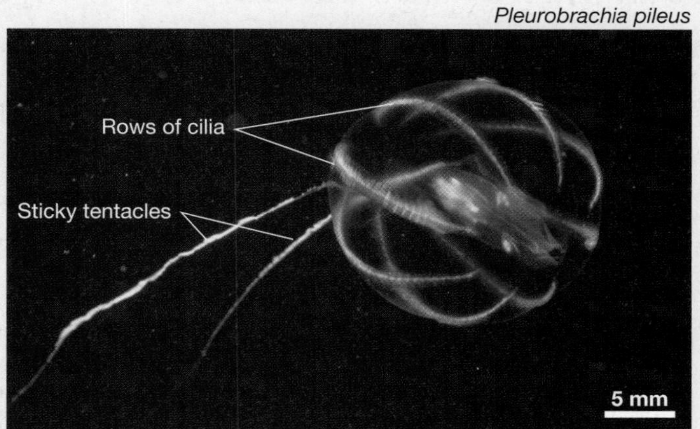

FIGURE 32.28 Ctenophores Are Planktonic Predators.

Reproduction Most species have both male and female organs and routinely self-fertilize, though fertilization is external. Larvae are free swimming. The few species that live on the ocean floor undergo internal fertilization and brood their embryos until they hatch into larvae.

Acoelomorpha

As their name implies, the acoelomorphs lack a coelom. They are bilaterally symmetric worms that have distinct anterior and posterior ends and are triploblastic. ● You should be able to indicate the origin of bilateral symmetry and mesoderm on Figure 32.25. They have simple digestive tracts or no digestive system at all. In some species with a digestive tract, the mouth is the only opening for the ingestion of food and excretion of waste products. Species that lack a digestive tract absorb nutrients across their body wall. Most acoelomorphs are only a couple of millimeters long and live in mud or sand in marine environments (**Figure 32.29**).

Feeding Acoelomorphs feed on detritus or prey on small animals or protists that live in mud or sand.

Movement Acoelomorphs swim, glide along the surface, or burrow through substrates with the aid of cilia that cover either the entire body or the ventral surface.

Reproduction Adults can reproduce asexually by fission (splitting in two) or by direct growth (budding) of a new individual from the parent's body. Individuals produce both sperm and eggs. Fertilization is internal, and fertilized eggs are laid outside the body.

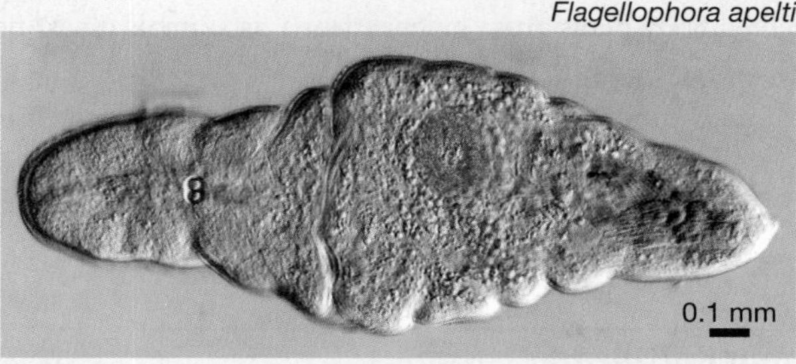

FIGURE 32.29 Acoelomorphs Are Small Worms That Live in Mud or Sand.

Chapter Review

SUMMARY OF KEY CONCEPTS

Animals are a particularly species rich and morphologically diverse lineage of multicellular organisms.

The animals consist of about 34 phyla and may number 10 million or more species. Biologists study animals because they are key consumers and because humans depend on them for transportation, power, or food.

You should be able to predict how a series of 10m × 10m plots in a forest would change if all animals were excluded.

Major groups of animals are recognized by nature of their basic body plan, which differs in the number of tissues observed in embryos, symmetry, the presence or absence of a body cavity, and the way in which early events in embryonic development proceed.

Sponges lack highly organized, complex tissues other than epithelium and may be asymmetric. The cnidarians and ctenophores have radial symmetry and two embryonic tissue layers. A handful of species have bilateral symmetry and three embryonic tissues but lack a body cavity, or coelom. The vast majority of animal species have bilateral symmetry, three embryonic tissues, and a coelom—features that gave rise to a "tube-within-a-tube" body plan. Depending on the species involved, the tube-within-a-tube design is built in one of two fundamental ways—via the protostome or deuterostome patterns of development.

You should be able to draw an asymmetric, radially symmetric, and bilaterally symmetric animal, and make a labeled drawing that shows how animals with tube-within-a-tube body plans move with limbs versus without limbs.

The Architecture of Animals

Recent phylogenetic analyses of animals have shown that there were three fundamental splits as animals diversified, resulting in two protostome groups (Lophotrochozoa and Ecdysozoa) and the deuterostomes.

Phylogenetic data support the hypothesis that triploblasty, bilateral symmetry, coeloms, and protostome and deuterostome development all evolved just once. However, phylogenetic data also suggest that coeloms were lost in the phylum Platyhelminthes; that pseudocoeloms evolved independently in the phyla Nematoda and Rotifera; and that at least three times, segmented body plans arose independently as animals diversified.

You should be able to describe how protostomes and deuterostomes differ, and how lophotrochozoans and ecdysozoans differ.

Within major groups of animals, evolutionary diversification was based on innovative ways of feeding and moving. Most animals get nutrients by eating other organisms, and most animals move under their own power at some point in their life cycle.

Among animals, there is a wide variety of feeding strategies. Individuals capture food in one of four ways: Suspension feeders filter organic material or small organisms from water; deposit feeders swallow soils or other materials and digest the food particles they contain; fluid feeders lap or suck up liquids; food-mass feeders harvest packets of material. In addition, the sources of food that animals exploit vary from plant and algal material to other animals to detritus.

The evolution of the coelom made efficient movement possible because an enclosed, fluid-filled body cavity could act as a hydrostatic skeleton. Limbs and other types of support structures (skeletons) evolved later. Recent research suggests that, even though the types of appendages used in animal locomotion range from simple sac-like limbs to complex lobster legs, all appendages may be homologous to some degree.

You should be able to (1) give an example—other than one provided in the text—of a suspension, fluid, deposit, and food-mass feeder; (2) explain how the animal's mouthparts or appendages make this mode of feeding efficient.

Methods of sexual reproduction vary widely among animal groups, and many species can reproduce asexually. It is common for individuals to undergo metamorphosis during their life cycle.

Sexual reproduction may involve laying eggs or giving birth to well-developed young. In most cases, metamorphosis involves a dramatic change between the larval and adult body form, as well as stark contrasts between larvae and adults in the habitat used and feeding strategy.

You should be able to (1) give an example—other than one provided in the text—of an animal that undergoes holometabolous metamorphosis; (2) explain how the food and habitats exploited by juveniles versus adults differ.

QUESTIONS

Test Your Knowledge

1. What synapomorphy distinguishes animals as a monophyletic group, distinct from choanoflagellates?
   a. Multicellularity and tissues.
   b. Movement via a hydrostatic skeleton.
   c. Growth by molting.
   d. Ingestive feeding.

2. Which of the following patterns in animal evolution is correct?
   a. All triploblasts have a coelom.
   b. All triploblasts evolved from an ancestor that had a coelom.
   c. Sponges have epithelial tissues that line an enclosed, fluid-filled body cavity.
   d. All bilaterally symmetric animals are triploblastic.

3. Which of the following patterns in animal evolution is correct?
   a. Segmentation evolved once.
   b. Pseudocoeloms evolved several times.
   c. Sponges lack tissues and are asymmetrical.
   d. All triploblasts are bilaterally symmetric as adults.
4. Why do some researchers maintain that the limbs of all animals are homologous?
   a. Homologous genes, such as *Dll*, are involved in their development.
   b. Their structure—particularly the number and arrangement of elements inside the limb—is the same.
   c. They all function in the same way—in locomotion.
   d. Animal appendages are too complex to have evolved more than once.
5. In a "tube-within-a-tube" body plan, what is the interior tube?
   a. ectoderm
   b. mesoderm
   c. either the coelom or the pseudocoelom
   d. the gut
6. What is the key difference between choanoflagellates and sponges?
   a. Sponges are multicellular.
   b. Sponges are asymmetrical and do not have tissues.
   c. Choanoflagellates have distinctive flagellated cells that function in suspension feeding.
   d. Choanoflagellates are strictly aquatic.

Test Your Knowledge answers: 1. a; 2. d; 3. b; 4. a; 5. d; 6. a

Test Your Understanding

Answers are available at www.masteringbio.com

1. Explain the difference between a diploblast and a triploblast. Why was the evolution of a third embryonic tissue layer important?
2. Explain how a hydrostatic skeleton works. Compare the types of movement that are possible with a hydrostatic skeleton to the types of movement observed in animals that lack a hydrostatic skeleton.
3. Draw a hypothetical phylogenetic tree of an animal lineage that radiated into a diverse array of species. Indicate where the group's basic body plan evolved. Indicate where adaptations for suspension, deposit, fluid, and food mass feeding may have evolved.
4. Compare and contrast the types of mouthparts and body shapes you would expect to find in herbivorous insect species that suck plant fluids from stems, bore through stems, bite leaves, and suck nectar from flowers.
5. Explain the differences in the protostome and deuterostome patterns of development. Do you agree with the text's claim that these are just different variations on the tube-within-a-tube body plan? Explain why or why not.
6. Why was the evolution of cephalization correlated with the evolution of bilateral symmetry? Why were both of these features significant?

Applying Concepts to New Situations

Answers are available at www.masteringbio.com

1. Would you expect internal fertilization to be more common in aquatic or terrestrial environments? Explain your answer. What types of differences would you expect to find in the structure of eggs that are laid in aquatic versus terrestrial environments?
2. Ticks are arachnids (along with spiders and mites); mosquitoes are insects. Both ticks and mosquitoes are ectoparasites that make their living by extracting blood meals from mammals. Ticks undergo incomplete metamorphosis, while mosquitoes undergo complete metamorphosis. Based on these observations, would you predict ticks or mosquitoes to be the more successful group in terms of number of species, number of individuals, and geographic distribution? Explain why. How could you test your prediction?
3. Suppose you are walking along an ocean beach at low tide and find an animal that is unlike any you have ever seen before. How would you go about determining how the animal feeds and how it moves? How would you go about determining the major features of its body plan?
4. Suspension feeding is extremely common in aquatic organisms but rare in terrestrial organisms. Generate a hypothesis to explain this observation.

www.masteringbio.com is also your resource for • Answers to text, table, and figure caption questions and exercises • Answers to *Check Your Understanding* boxes • Online study guides and quizzes • Additional study tools including the *E-Book for Biological Science* 3rd ed., textbook art, animations, and videos.

33 Protostome Animals

KEY CONCEPTS

- Molecular phylogenies support the hypothesis that protostomes are a monophyletic group divided into two major subgroups: the Lophotrochozoa and the Ecdysozoa.
- Although the members of many protostome phyla have limbless, wormlike bodies and live in marine sediments, the most diverse and species-rich lineages—Mollusca and Arthropoda—have body plans with a series of distinctive, complex features. Molluscs and arthropods inhabit a wide range of environments.
- Key events triggered the diversification of protostomes, including several lineages making the water-to-land transition, a diversification in appendages and mouthparts, and the evolution of metamorphosis in both marine and terrestrial forms.

A small sample of the insects collected from a single tree in the Amazonian wet forest. In numbers of individuals and species richness, protostomes are the most abundant and diverse of all animals.

Protostomes include some of the most familiar and abundant organisms on Earth. The phylum Arthropoda alone includes the insects, chelicerates (spiders and mites), crustaceans (shrimp, lobster, crabs, barnacles), and myriapods (millipedes, centipedes). Biologists estimate that there are over 10 million arthropod species—most of them unnamed and undescribed. Among insects alone, about 925,000 species have been formally identified to date. Well over a third of these insect species are beetles. In addition to being a species-rich clade, beetles are abundant. A typical acre of pasture in England is home to almost 18 *million* individual beetles. In tropical rain forests, beetles and other insects make up 40 percent of the total mass of organisms present. And beetles are not the only super-abundant type of arthropod. There may be over 6.5 billion humans on Earth, but the world population of ants is estimated to be 1 million billion individuals. The phylum Mollusca is almost as impressive. This lineage comprises over 93,000 snail, clam, chiton, and cephalopod (octopuses and squid) species.

It is important to recognize, though, that not all protostome lineages have been as spectacularly successful as the Arthropoda and Mollusca. The 22 phyla of protostomes include some of the most obscure lineages on the tree of life as well as some of the most prominent. There are just 135 species in the phylum Echiura (spoon worms), for example, and 150 species in the Kinorhyncha (mud dragons). The 80 species in the Gnathostomulida (pronounced *nath-oh-stoh-MEW-lida*) and the 450 species in the Gastrotricha (*GAS-troh-trika*) average less than 2 mm long.

Key Concept   Important Information   Practice It

BOX 33.1 Model Organisms: *Caenorhabditis elegans* and *Drosophila melanogaster*

Protostomes include two of the most important model organisms in the life sciences: the fruit fly *Drosophila melanogaster* and the roundworm *Caenorhabditis elegans*. (*Caenorhabditis* is pronounced *see-no-rab-DIE-tiss*.) If you walk into a biology building on any university campus around the world, you are almost certain to find at least one lab where fruit flies or roundworms are being studied.

Since the early 1900s, *Drosophila melanogaster* has been a key experimental subject in genetics. It was initially chosen as a focus for study by T. H. Morgan because it can be reared in the laboratory easily and inexpensively (**Figure 33.1a**), matings can be arranged, the life cycle is completed in less than two weeks, and females lay a large number of eggs. These traits made fruit flies valuable subjects for breeding experiments designed to test hypotheses about how traits are transmitted from parents to offspring (see Chapter 13). More recently, *D. melanogaster* has also become a key model organism in the field of developmental biology. The use of flies in developmental studies was inspired in large part by the work of Christianne Nüsslein-Volhard and Eric Wieschaus, who in the 1980s isolated flies with genetic defects in early embryonic development. By investigating the nature of these defects, researchers have gained valuable insights into how various gene products influence the development of eukaryotes (see Chapter 21).

Caenorhabditis elegans emerged as a model organism in developmental biology in the 1970s, due largely to work by Sydney Brenner and colleagues. This roundworm was chosen for three reasons: (1) Its cuticle (soft outer layer) is transparent, making individual cells relatively easy to observe (**Figure 33.1b**); (2) adults have exactly 959 nonreproductive cells; and, most important, (3) the fate of each cell in an embryo can be predicted because cell fates are invariant among individuals. For example, when researchers examine a 33-cell *C. elegans* embryo, they know exactly which of the 959 cells in the adult will be derived from each of those 33-embryonic cells. In addition, *C. elegans* are small (less than 1 mm long), are able to self-fertilize or cross-fertilize, and undergo early development in just 16 hours. The entire of genome of *C. elegans*, and that of *D. melanogaster*, has now been sequenced (see Chapter 20).

(a) Fruit flies can be reared in bottles.

(b) *Caenorhabditis elegans* is transparent and can be reared in petri dishes.

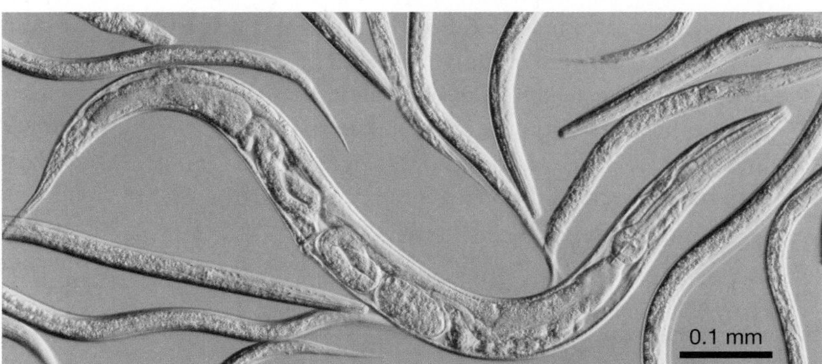

FIGURE 33.1 Model Organisms Can Be Readily Reared and Studied.

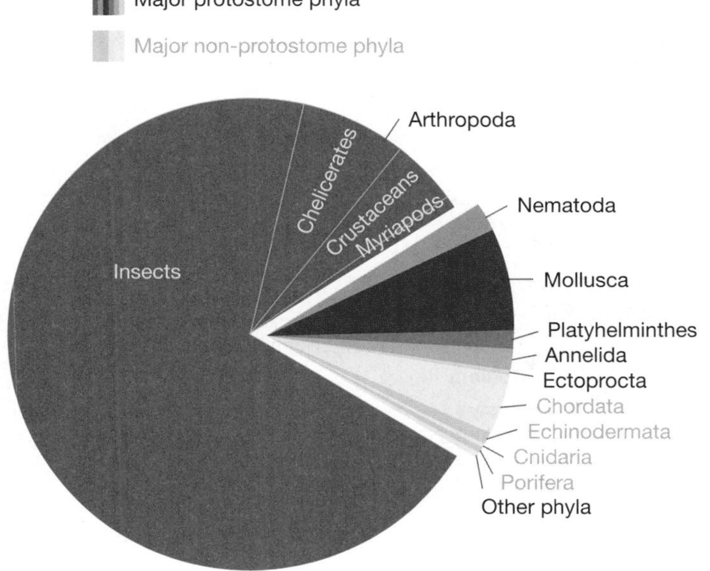

Figure 33.2 is a pie chart showing the relative numbers of species in various animal phyla. Except for the small wedges labeled Chordata, Cnidaria, Echinodermata, Porifera, and some of the "Other invertebrates," all animal species are protostomes.

As a group, protostomes live in just about every habitat that you might explore and include some of the most important model organisms in all of biological science (**Box 33.1**). If one of biology's most fundamental goals is to understand the diversity of life on Earth, then protostomes—particularly the arthropods and molluscs—demand our attention.

FIGURE 33.2 The Relative Abundances of Animal Lineages. Most animals are protostomes. About 70 percent of all known species of animals on Earth are insects, most of them beetles. (Humans and other vertebrates are deuterostomes, in the phylum Chordata.) This chapter focuses on the most species-rich lineages of protostomes.

● **QUESTION** About what percentage of known animal phyla are protostomes?

33.1 An Overview of Protostome Evolution

There are two major monophyletic groups of bilaterally symmetric, triploblastic, coelomate animals: the protostomes and deuterostomes. You might recall from Chapter 32 that protostomes and deuterostome embryos develop in dramatically different ways. The protostome and deuterostome patterns of early development represent distinctive pathways for building a bilaterally symmetric body with a coelom. In many protostomes, the pattern of cell divisions called spiral cleavage occurs after fertilization. During gastrulation, the initial pore that forms in the embryo becomes the mouth. If a **coelom** (a body cavity) forms later in development, it forms from openings that arise within blocks of mesodermal tissue.

Phylogenetic studies have long supported the hypothesis that protostomes are a monophyletic group. This result means that the protostome developmental sequence arose just once. But when recent analyses of DNA sequence data indicated that two major subgroups existed within the protostomes, it was considered a key insight into animal evolution. The two monophyletic groups of protostomes are called the Lophotrochozoa and Ecdysozoa, and they were initially identified by analyzing DNA sequences from the gene for the small subunit of ribosomal RNA. (The lineages are pronounced *low-foe-troe-koe-ZOE-an* and *eck-dye-so-ZOE-an*.) Newer work based on sequence data from several additional genes has supported this result. Although research continues, an emerging consensus holds that the divergence of protostomes into two major subgroups was an important event as protostomes diversified (**Figure 33.3**).

What Is a Lophotrochozoan?

The 14 phyla of lophotrochozoans include the molluscs, annelids, and flatworms (Platyhelminthes; pronounced *plah-tee-hell-MIN-theez*). Although **lophotrochozoans** can be distinguished from other protostomes because they grow by incremental additions to their body, the group's name was inspired by morphological traits that are found in some, but not all, of the phyla in the lineage: (1) a feeding structure called a lophophore, which is found in three phyla, and (2) a type of larva called a trochophore, which is common to many of the phyla in the lineage.

As **Figure 33.4a** shows, a **lophophore** (literally, "tuft-bearer") is a specialized structure that rings the mouth and functions in suspension feeding. Lophophores consist of tentacles that have ciliated cells along their surface. The beating of the cilia generates a water current that sweeps protists and other algae, detritus, and animal larvae into the region above the mouth, where they are trapped by the tentacles, swept into the mouth opening, and ingested. Lophophores are found in bryozoans ("moss animals"), brachiopods ("lamp shells"), and phoronids ("horseshoe worms").

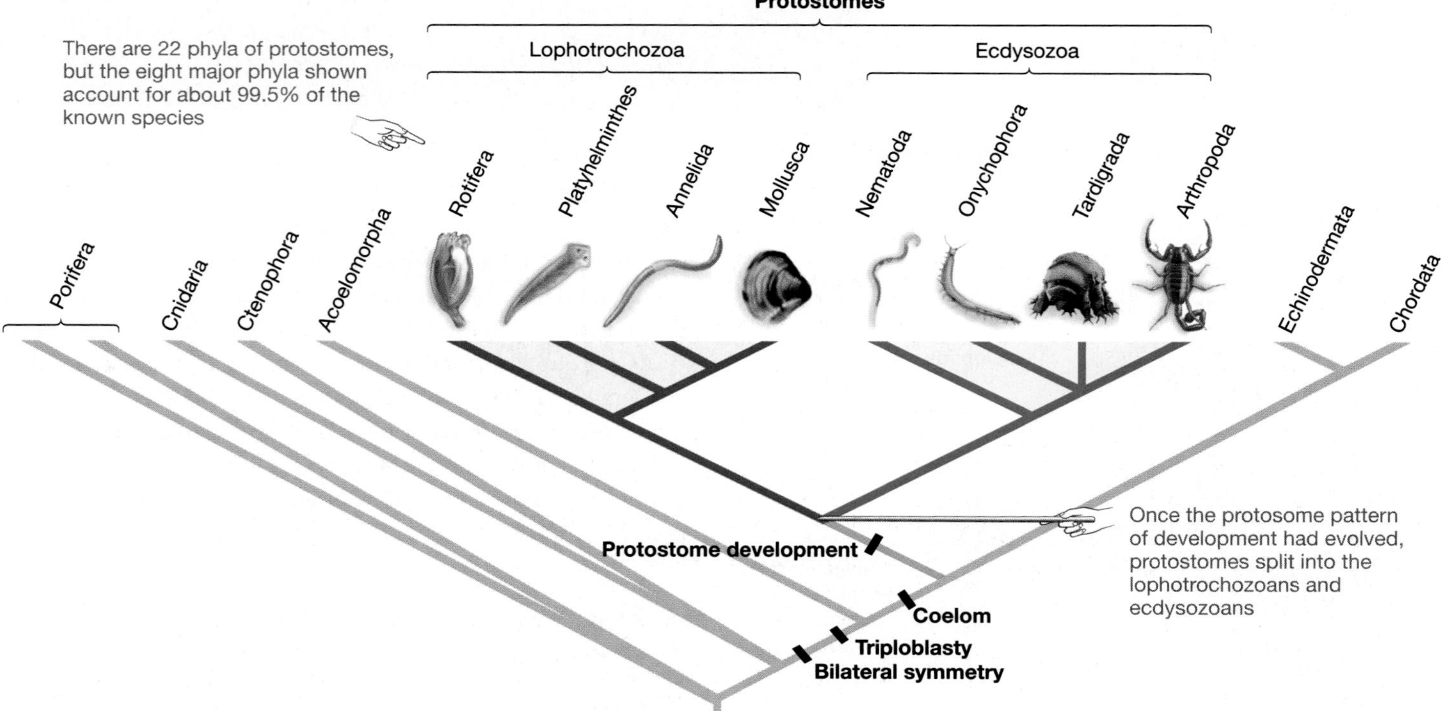

FIGURE 33.3 Protostomes Are a Monophyletic Group Comprising Two Major Lineages.

(a) Lophophores function in suspension feeding in adults.

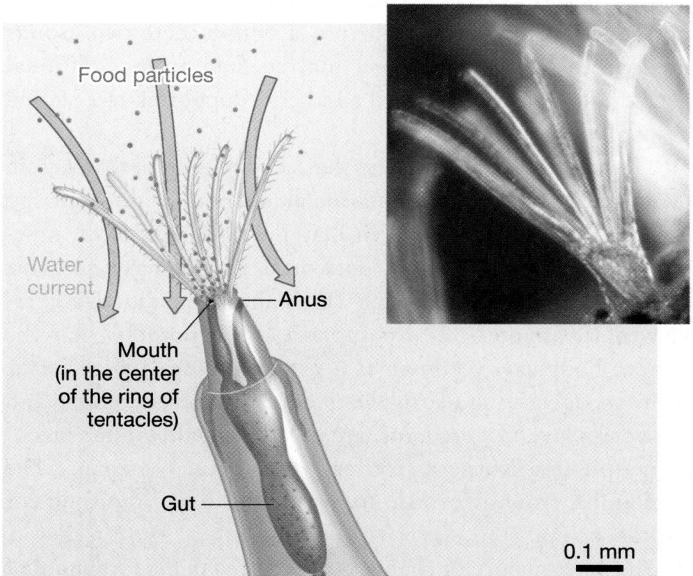

(b) Trochophore larvae swim and feed.

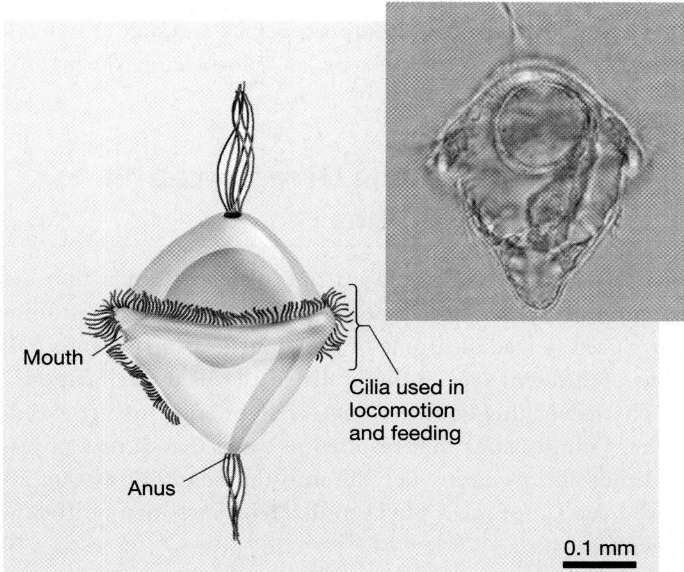

FIGURE 33.4 Lophotrochozoans Have Distinctive Traits. (a) Three phyla of lophotrochozoans have the feeding structure called a lophophore. **(b)** Many phyla of lophotrochozoans have the type of larva called a trochophore.

EXERCISE *Lopho–* means "tuft"; *trocho–* means *wheel.* Label the tuft and wheel in this figure.

Trochophores are a type of larvae common to marine molluscs, annelids that live in the ocean, and several other phyla in the Lophotrochozoa. As **Figure 33.4b** shows, a **trochophore** ("wheel-bearer") larva has a ring of cilia around its middle. These cilia aid in swimming. The beating of the cilia may also produce water currents that sweep food particles into the mouth, though many trochophores feed primarily on yolk provided by the mother. Thus, trochophores may represent a feeding stage as well as a dispersal stage in the life cycle. Recent analyses suggest that the trochophore may have originated early in the evolution of Lophotrochozoans, though different types of larva evolved later in some groups.

What Is an Ecdysozoan?

The primary contrast between lophotrochozoans and ecdysozoans involves the methods of growth used by organisms in each group. Instead of growing by steady and incremental additions to the body, **ecdysozoans** grow by **molting**—that is, by shedding an exoskeleton or external covering. The Greek root *ecdysis*, which means "to slip out or escape," is appropriate because during a molt, an individual sheds its outer layer—called a **cuticle** if it is soft or an **exoskeleton** if it is hard—and slips out of it (**Figure 33.5**). Once the old covering is gone, the body expands and a larger cuticle or exoskeleton then forms. As ecdysozoans grow and mature, they undergo a succession of molts. The most prominent of the seven ecdysozoan phyla are the roundworms (Nematoda) and arthropods (Arthropoda).

Growth by molting was required in ecdysozoans once they had evolved tough outer cuticles and thick exoskeletons. The leading hypothesis to explain the evolution of these stiff body coverings is based on protection from predators. To support this claim, biologists point to the secretive behavior of ecdysozoans while individuals are molting. When crabs and other crustaceans have shed an old exoskeleton, their new exoskeleton

FIGURE 33.5 Ecdysozoans Grow by Molting. Once an ecdysozoan has left its old exoskeleton, hours or days pass before the new exoskeleton is hardened. During this time, the individual is highly susceptible to predation or injury.

takes several hours to harden. During this interval, individuals hide and do not feed or move about. Experiments have shown that it is much easier for predators to attack and subdue individuals that are not protected by an exoskeleton during the intermolt period.

33.2 Themes in the Diversification of Protostomes

Protostomes have diverged into 22 different phyla that are recognized by distinctive body plans or specialized mouthparts used in feeding. Some of these phyla then split into millions of different species. What drove all this diversification?

To answer this question, let's analyze some of the evolutionary innovations that resulted in the origin of new phyla, and then follow up by delving into the adaptations that allowed certain of these phyla to diversify into many different species.

How Do Body Plans Vary among Phyla?

All protostomes are triploblastic and bilaterally symmetric, and all protostomes undergo embryonic development in a similar way. In contrast, radical changes occurred in coelom formation as protostomes diversified. **Figure 33.6** summarizes recent results on the relationships among protostome phyla and indicates where major changes occurred in the evolution of protostome body plans. Only a few of the 22 protostome phyla are included in this phylogenetic tree, however, because the relationships of most groups within the Lophotrochozoa and Ecdysozoa are still poorly resolved. Stated another way, it is not yet clear how most of the phyla within these two groups are related. As investigators analyze data from additional genes, a more highly resolved and better-supported tree should gradually emerge.

The most dramatic change that occurred after the coelom evolved was a reversion to an **acoelomate** body plan—meaning the lack of a body cavity—in Platyhelminthes. Flukes, tapeworms, and other types of flatworms do not have any sort of body cavity, even though their ancestors did. Data reviewed in Chapter 32 also support the hypothesis that the type of body cavity known as a **pseudocoelom**, which forms from an opening that originates between the endoderm and mesoderm layers of embryos, arose independently in the protostome phylum Rotifera (rotifers) and the ecdysozoans. The Nematoda (roundworms), for example, have a prominent pseudocoelom.

The other important change that occurred in the formation of body cavities in certain phyla was a drastic reduction of the coelom. To put this change in context, you might recall from Chapter 32 that most protostome phyla have wormlike bodies with a basic tube-within-a-tube design. The outside tube is the skin, which is derived from ectoderm; the inside tube is the gut, which is derived from endoderm. Muscles and organs derived from mesoderm are located between the two tubes. In the wormlike phyla, the coelom is well developed and functions as a hydrostatic skeleton that is the basis of movement. But in the most species-rich and morphologically complex protostome phyla—

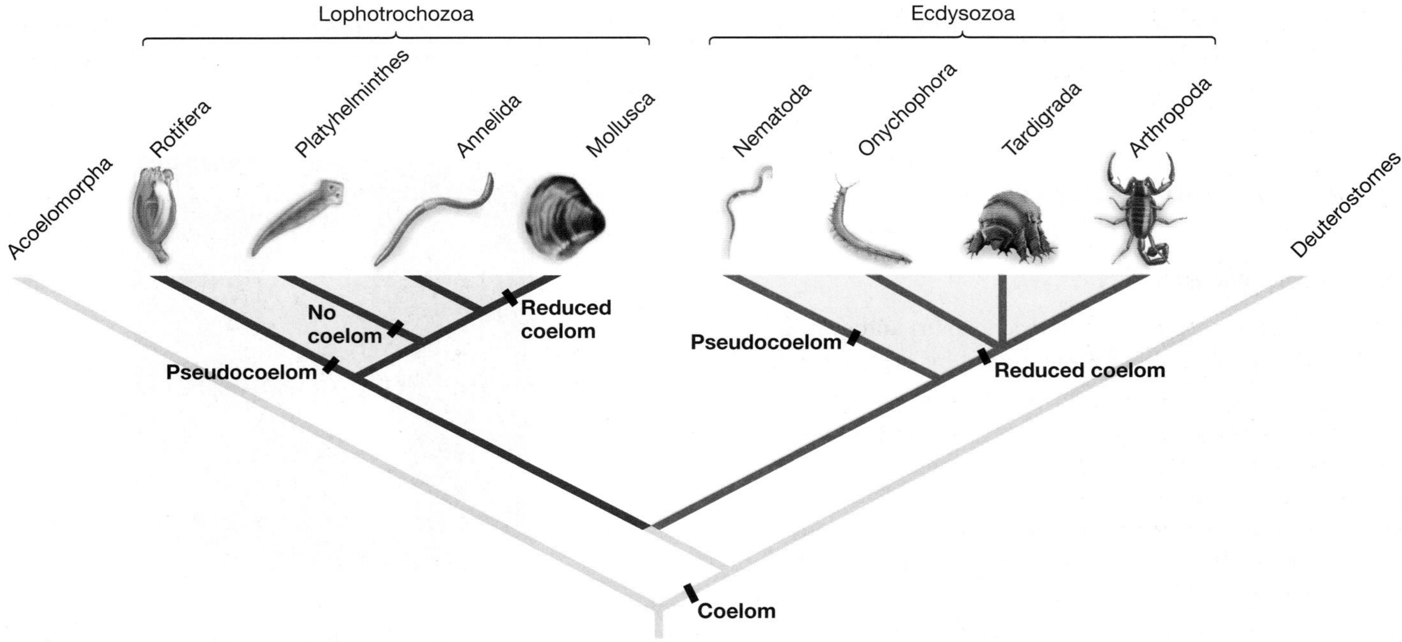

FIGURE 33.6 The Nature of the Coelom Varies Among Protostomes. This phylogenetic tree is based on analyses of DNA sequences from several genes. Most of the 14 protostome phyla not included in this tree have complete coeloms. The position of these phyla on the tree is still too uncertain to include them here, however.

the Arthropoda and the Mollusca (snails, clams, squid)—the coelom is drastically reduced. To use a term introduced in Chapter 24, the arthropod and mollusc coeloms are vestigial traits.

A fully functioning coelom has two roles: providing space for fluids to circulate among organs, and providing a hydrostatic skeleton for movement. In arthropods and molluscs, other structures fulfill these functions. Species in these phyla don't have a fully functioning coelom for a simple reason: They don't need one. Let's take a closer look.

The Arthropod Body Plan As **Figure 33.7a** shows, arthropods have segmented bodies that are organized into prominent regions called tagma. In the grasshopper shown, these regions are called the head, thorax, and abdomen. In addition, arthropods are distinguished by their jointed limbs and an exoskeleton made primarily of the polysaccharide chitin (see Chapter 5). In crustaceans, the exoskeleton is strengthened by the addition of calcium carbonate ($CaCO_3$).

Instead of being based on muscle contraction against a hydroskeleton, arthropod locomotion is based on muscles that apply force against the exoskeleton to move legs or wings. Species in the closely related phyla called Onychophora and Tardigrada (see Figure 33.3) also have limbs and reduced coeloms. Based on these observations, biologists suggest that the evolution of limbs made a hydrostatic skeleton unnecessary.

Besides their limbs and the other features labeled in Figure 33.7a, arthropods have a spacious body cavity called the **hemocoel** ("blood-hollow"; pronounced *HEE-mah-seal*) that provides space for internal organs and circulation of fluids. In caterpillars and other types of arthropod larva, the hemocoel also functions as a hydrostatic skeleton.

The Molluscan Body Plan Dramatic changes also occurred during the origin of molluscs, although the nature of the body plan that evolved is very different from that of arthropods. As **Figure 33.7b** shows, the mollusc body plan is based on three major components: (1) the **foot**, a large muscle located at the base of the animal and usually used in movement; (2) the **visceral mass**, the region containing most of the main internal organs and external gill; and (3) the **mantle**, a tissue layer that covers the visceral mass and that in many species secretes a shell made of calcium carbonate. Some mollusc species have a single shell; others have two, eight, or none at all.

In molluscs, the coelom's functions are replaced by the visceral mass and by the muscular foot. The visceral mass provides space for organs and the circulation of fluids. In some species the fluid enclosed by the visceral mass also functions as a hydrostatic skeleton. In other molluscs the muscular foot functions as a hydrostatic skeleton as it moves. Molluscs do not have limbs, but they did evolve ways of moving around that don't require a coelom.

Variation Among Body Plans of the Wormlike Phyla New types of body plans distinguish most of the phyla illustrated in Figure 33.6. In contrast, the general body plan is similar in most of the protostome phyla that have wormlike bodies. In many cases these lineages are distinguished by specialized mouthparts or feeding structures. To drive this point home, consider the three phyla illustrated in **Figure 33.8**.

- Echiurans (spoon worms) burrow into marine mud and suspension feed using an extended structure called a **proboscis**, which forms a gutter leading to the mouth (Figure 33.8a). Cells in the gutter secrete mucus, which is sticky enough to capture pieces of detritus. The combination of mucus and detritus is then swept into the mouth by cilia on cells in the gutter.

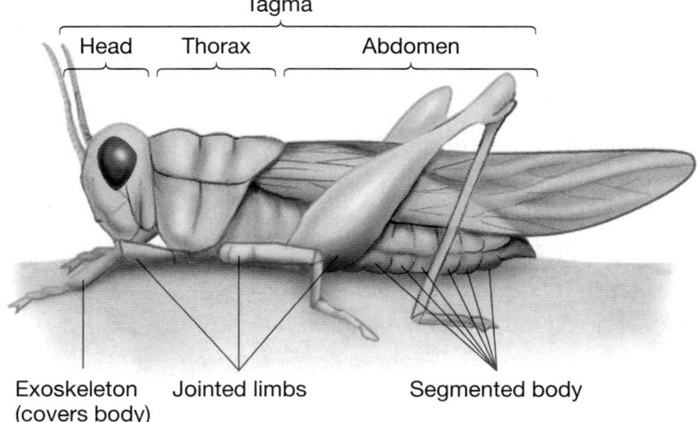

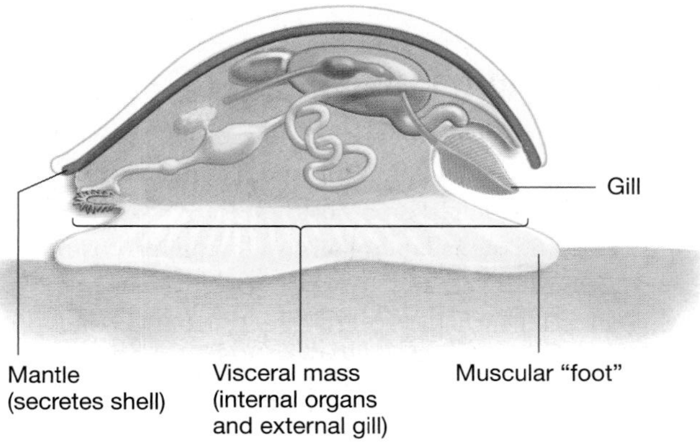

FIGURE 33.7 Arthropods and Molluscs Have Specialized Body Plans. (a) Arthropods have segmented bodies and jointed limbs, which enable these animals to move despite their hard outer covering, the exoskeleton. **(b)** The mollusc body plan is based on a foot, a visceral mass, and a mantle.

(a) Echiurans ("spoon worms")

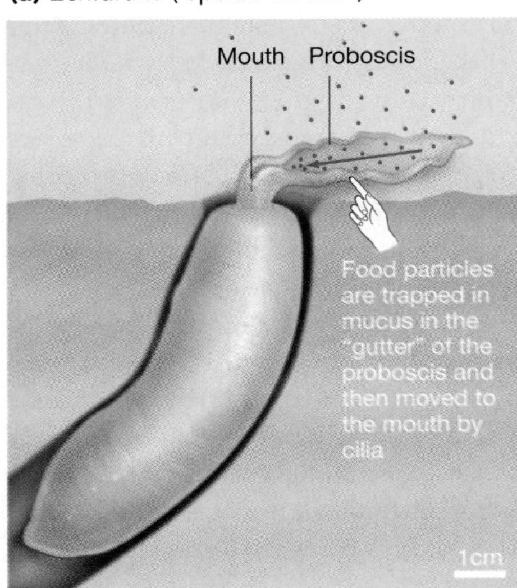

(b) Priapulids ("penis worms")

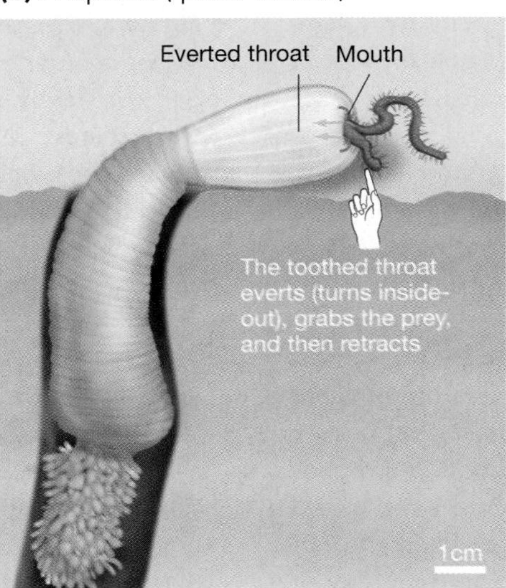

(c) Nemerteans ("ribbon worms")

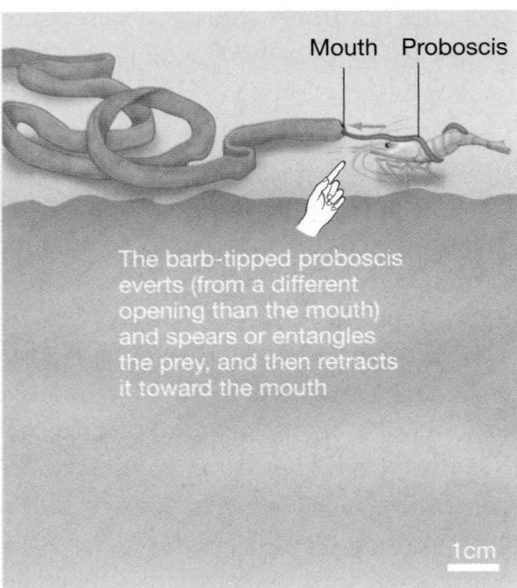

FIGURE 33.8 Some Protostome Phyla with Wormlike Bodies Have Specialized Mouthparts.

QUESTION It is common to find species from an array of wormlike phyla in the same habitat. Suggest a hypothesis to explain why this is possible.

- Priapulids (penis worms) also burrow into the substrate, but act as sit-and-wait predators. When a polychaete (Annelida) or other prey item approaches, the priapulid everts its toothed, cuticle-lined throat—meaning that it turns its throat inside out—grabs the prey item, and retracts the structure to take in the food (Figure 33.8b).
- The nemerteans (ribbon worms) are active predators that move around the ocean floor in search of food. Instead of everting their throat to capture prey, they have a proboscis that can extend or retract (Figure 33.8c). Nemerteans spear small animals with their proboscis or wrap the extended proboscis around prey. The food is then pulled into the mouth.

at www.masteringbio.com
Protostome Diversity

The Water-to-Land Transition

Protostomes are the most abundant animals in the surface waters and substrates of many marine and freshwater environments. But protostomes are common in virtually every terrestrial setting as well. Like the land plants and fungi, protostomes made the transition from aquatic to terrestrial environments. To help put this achievement in perspective, recall from Chapter 30 that green plants made the move from freshwater to land just once. From Chapter 31, recall that it is not yet clear whether fungi moved from aquatic habitats to land once or several times. Chapter 34 will show that only one lineage among deuterostomes moved onto land. But even given the current uncertainty in estimating the protostome phylogeny, it is clear that a water-to-land transition occurred multiple times as protostomes diversified. The ability to live in terrestrial environments evolved independently in arthropods (at least twice), molluscs, roundworms, and annelids.

Repeated water-to-land transitions are a prominent theme in the evolution of protostomes. In each case, the evidence for this claim is based on phylogenetic analyses, which support the hypothesis that the ancestors of the terrestrial lineages in each major subgroup were aquatic.

Why were so many different protostome groups able to move from water to land independently? The short answer is that it was easier for protostomes to accomplish this than for plants to do so. For example, land plants had to evolve roots and vascular tissue to transport water and support their stems. Similarly, the first terrestrial animals had to be able to support their body weight on land and move about. But the protostome groups that made the water-to-land transition already had hydrostatic skeletons, exoskeletons, appendages, or other adaptations for support and locomotion that happened to work on land as well as water.

To make the transition to land, new adaptations allowed protostomes to (1) exchange gases and (2) avoid drying out. As detailed in Chapter 44, land animals exchange gases with the atmosphere readily as long as a large surface area is exposed to the air. The bigger challenge is to prevent the gas-exchange surface and other parts of the body from drying out. Roundworms and earthworms solve this problem by living in wet soils or other moist environments. Arthropods and many molluscs have a watertight exoskeleton or a shell that minimizes water loss, and their respiratory structures are located inside the body. In insects, the openings to respiratory passages can be closed to minimize water loss. And unlike land plants and fungi, all land animals can move to moister habitats if the area they are in gets too dry.

Water-to-land transitions are important because they open up entirely new habitats and new types of resources to exploit. For example, plants moved onto land about 450 million years ago. But the first land-dwelling vertebrates don't appear in the fossil record until about 365 million years ago. For almost 90 million years, then, any protostome species that moved into terrestrial environments had an abundance of food available to eat, with a minimal number of competitors. Natural selection would then favor mutations that changed morphology in ways that allowed different species of protostomes to exploit different types of plant resources. Based on this reasoning, biologists claim that the ability to live in terrestrial environments was a key event in the diversification of several protostome phyla.

How Do Protostomes Feed, Move, and Reproduce?

Some of the protostome phyla have undergone dramatic amounts of speciation, while others have diversified hardly at all. Based on species counts, biologists can claim that certain body plans have proven much more successful than others at making efficient eating and moving machines.

Once the wormlike, arthropod, and mollusc body plans had evolved, subsequent diversification was largely driven by adaptations that allowed protostomes to feed, move, or reproduce in novel ways. Recall that an **adaptation** is a trait that increases the fitness of individuals relative to individuals without the trait.

Adaptations for Feeding Protostomes include suspension, deposit, liquid, and food-mass feeders. Besides exploiting detritus, they prey on or parasitize plants, algae, or other animals. Exploiting a diversity of foods is possible because protostomes have such a wide variety of mouthparts for capturing and processing food. Figure 33.8 highlighted mouthparts that distinguish some of the wormlike phyla. But within phyla, arthropods take the prize for mouthpart diversity. The mouthparts observed in this phylum vary in structure from tubes to pincers and allow the various species to pierce, suck, grind, bite, mop, chew, engulf, cut, or mash (**Figure 33.9**). All arthropods have the same basic body plan, but their mouthparts and food sources are highly diverse.

In many species of arthropods, the jointed limbs also play a key role in getting food. The crustaceans called krill use their legs to sweep food toward their mouths as they suspension feed; certain insects, crustaceans, spiders, and molluscs use their appendages to capture prey or hold food as it is being chewed or bitten by the mouthparts.

In most cases, juvenile and adult forms of the same species exploit different food sources. Metamorphosis is extremely common in protostomes and usually results in larvae and adults that live in different habitats and have different overall morphology and mouthparts.

Adaptations for Moving In protostomes, variation in movement depends on variation in two features: (1) the

(a) Leaf-cutter ants cut leaves.

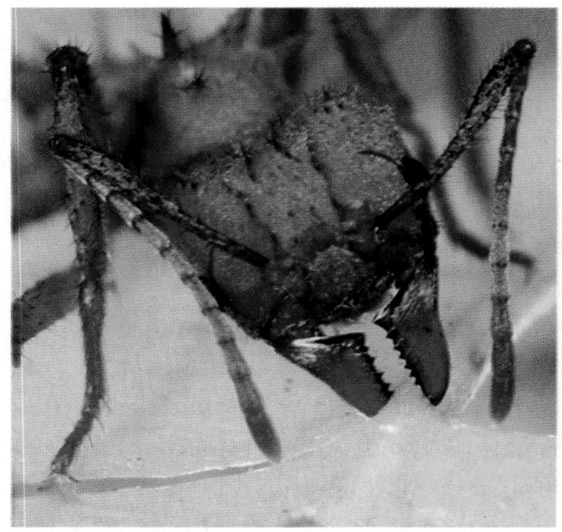

(b) Deer ticks pierce mammalian skin.

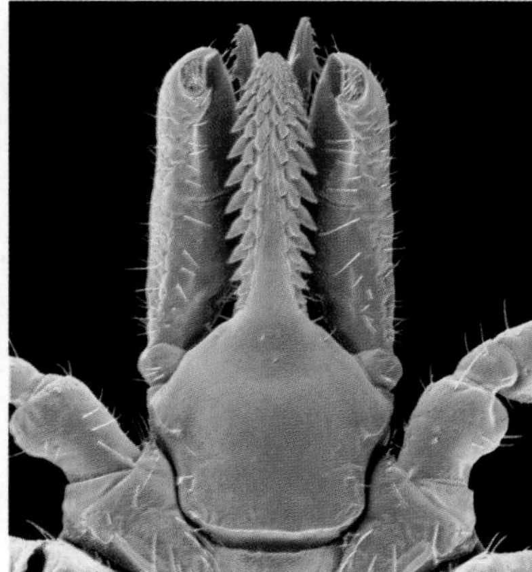

(c) Nut weevils bore into hard fruit.

FIGURE 33.9 Arthropod Feeding Structures Are Diverse. Even though their mouthparts and food sources are diverse, all of these arthropods have segmented bodies that are organized into tagma, with an exoskeleton and jointed appendages.

QUESTION How do the phyla with similar bodies but diverse mouthparts illustrated here compare with the wormlike phyla shown in Figure 33.8?

(a) Walking, running, and jumping

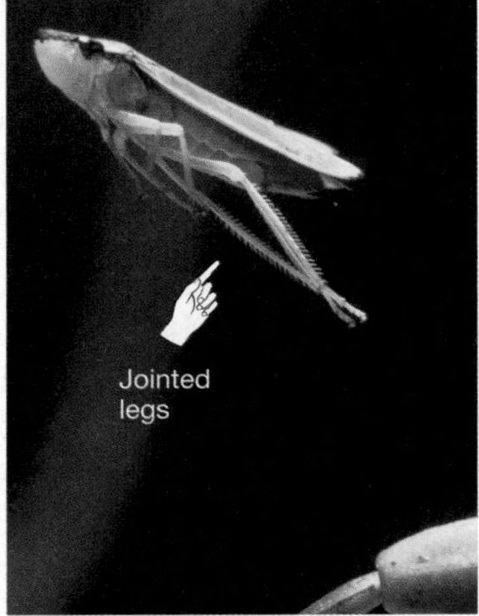

(b) Flying

(c) Gliding and crawling

(d) Jet propulsion

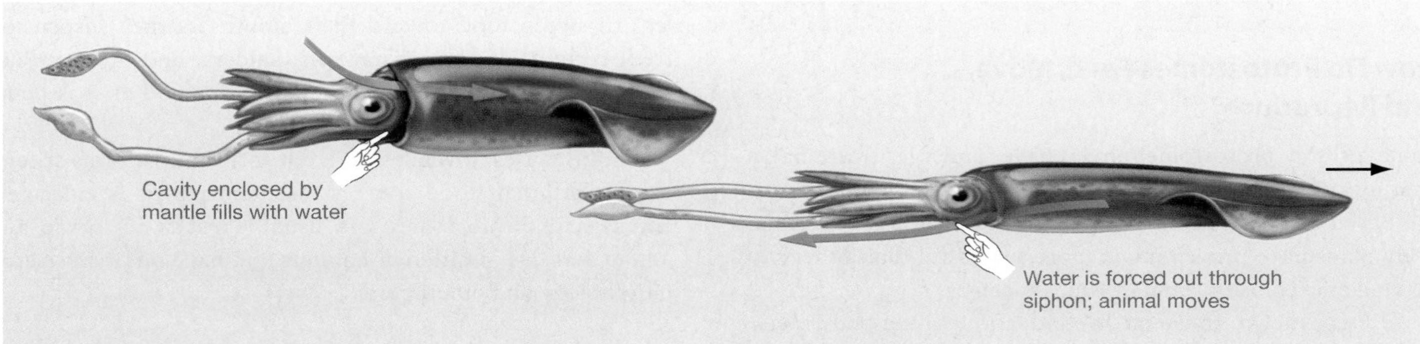

FIGURE 33.10 Protostome Locomotion Is Diverse. The evolution of **(a)** jointed limbs and **(b)** wings were key innovations in arthropod movement. **(c)** A wave of muscle contractions, down the length of the foot, allows molluscs to glide along a substrate. **(d)** In jet propulsion, muscular contractions force the water out through a moveable siphon. The siphon can aim outgoing water in a specific direction.

presence or absence of limbs and (2) the type of skeleton that is present. Wormlike protostomes that lack limbs move with the aid of a hydrostatic skeleton. This type of movement is also observed in caterpillars, grubs, maggots, and other types of insect larvae—even though adults of the same species have a hard exoskeleton and move with the aid of jointed limbs (**Figure 33.10a**). Although insect larvae have a highly reduced coelom, they do have an enclosed, fluid-filled body cavity that functions as a hydrostatic skeleton.

The evolution of jointed limbs made rapid, precise running movements possible and was a key innovation in the evolution of protostomes. The jointed limb is a major reason that arthropods have been so spectacularly successful in terms of species diversity, abundance of individuals, geographic range, and duration in the fossil record.

But the jointed limb is just one of several evolutionary innovations that allowed protostomes to move in unique ways:

- The insect wing is one of the most important adaptations in the history of life. About two-thirds of the multicellular species living today are winged insects. According to data in the fossil record, insects were the first organisms that had wings and could fly. Like most insects today, the earliest insects had two pairs of wings (**Figure 33.10b**). In most four-winged insects living today, however, the four wings function as two. Beetles fly with only their hindwings; butterflies, moths, bees, and wasps have hooked structures that make their two pairs of wings move together. Flies appear later in the fossil record. They have a single pair of large wings that power flight along with a pair of small, winglike structures called halteres that provide stability during flight.

- In molluscs, waves of muscle contractions sweep down the length of the large, muscular foot, allowing individuals to glide along a surface (**Figure 33.10c**). Cells in the foot secrete a layer of mucus, which reduces friction with the substrate and increases the efficiency of gliding.

- Squid are molluscs that have a mantle lined with muscle. When the cavity surrounded by the mantle fills with water and the mantle muscles contract, a stream of water is forced out of a tube called a **siphon**. The force of the water propels the squid forward (**Figure 33.10d**). This mechanism, jet propulsion, evolved in squid long before human engineers thought of using the same principle to power aircraft.

Adaptations in Reproduction When it comes to variation in reproduction and life cycles, protostomes do it all. Asexual reproduction by splitting the body lengthwise or by fragmenting the body is common in many of the wormlike phyla. Many crustacean and insect species reproduce asexually via **parthenogenesis** ("virgin-origin")—the production of unfertilized eggs that develop into offspring. Sexual reproduction is the predominant mode of producing offspring in most protostome groups, however. Sexual reproduction is often based on external fertilization in clams, bryozoans, brachiopods, and other groups. It often begins with copulation and internal fertilization in groups that are capable of movement—such as crustaceans, snails, and insects—because males and females can meet. Females of ovoviviparous insects and snails bear live young, although they do so by retaining fully formed eggs inside their bodies and then nourishing them via the nutrient-rich yolk that is inside the egg (see Chapter 32).

Two important reproductive innovations occurred during protostome diversification: (1) the evolution of metamorphosis and (2) an egg that would not dry out on land. Metamorphosis is common in marine protostomes. In these species, metamorphosis is hypothesized to be an adaptation that allows larvae to disperse to new habitats by floating or swimming in the plankton. Metamorphosis is also common in terrestrial insects, where it is hypothesized to be an adaptation that reduces competition for food between juveniles and adults. Regarding the ability to colonize terrestrial environments, though, the most critical adaptation was an egg that resists drying. Insect eggs have a thick membrane that keeps moisture in, and the eggs of slugs and snails have a thin calcium carbonate shell that helps retain water. Desiccation-resistant eggs evolved repeatedly in populations that made the transition to life on land. Recall from Chapter 30 that an analogous adaptation arose in land plants, when the evolution of the gametangia helped protect eggs and embryos from drying. And as Chapter 34 will show, a membrane-bound amniotic egg, analogous to the membrane-bound eggs of insects, evolved in the terrestrial vertebrates.

Check Your Understanding

If you understand that...

- In protostomes with wormlike bodies, phyla vary in mouthpart structure and mode of feeding.
- In protostomes that do not have wormlike bodies, phyla vary in the nature of the coelom—whether the coelom is complete, a pseudocoelom, vestigial, or absent.
- Arthropods and molluscs have vestigial coeloms but specialized body plans that perform the functions of a coelom.
- A water-to-land transition occurred in several lineages independently.

You should be able to...

1) Describe the major features of the arthropod and mollusc body plans, and explain which structures perform the functions of a coelom.
2) Give examples of variation in mouthpart structure and feeding, structures used in locomotion, and metamorphosis in protostomes.
3) Give two examples of adaptations that facilitated the water-to-land transition in protostomes.

33.3 Key Lineages: Lophotrochozoans

Although analyses of extensive DNA sequences and the presence of unique, derived features such as lophophores and trochophore larvae helped biologists identify the Lophotrochozoa, the lineages within this group (**Figure 33.11**) are highly diverse in morphology. To drive this point home, let's take a closer look at four of the key phyla in the group: (1) Rotifera, (2) Platyhelminthes, (3) Annelida, and (4) Mollusca.

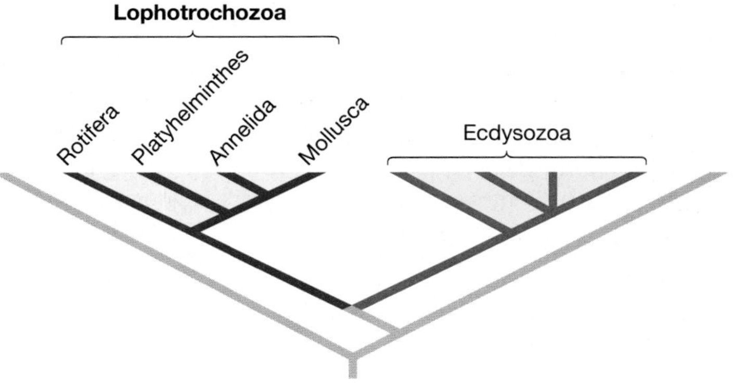

FIGURE 33.11 Lophotrochozoans Are a Monophyletic Group.

Lophotrochozoans > Rotifera (Rotifers)

The 1800 rotifer species that have been identified thus far live in damp soils as well as marine and freshwater environments. They are important components of the plankton in freshwater and in brackish waters, where rivers flow into the ocean and water is slightly salty. **Rotifers** have pseudocoeloms, and most species are less than 1 mm long. Although rotifers do not have a lophophore or a trochophore larval stage, extensive similarities in DNA sequence identify them as a member of the lophotrochozoan lineage.

Feeding Rotifers have a cluster of cilia at their anterior end called a **corona** (**Figure 33.12**). In many species, the beating of the cilia in the corona makes suspension feeding possible by creating a current that sweeps microscopic food particles into the mouth. The corona is the signature morphological feature of this group. ● You should be able to indicate the origin of this structure on Figure 33.11.

Movement Although a few species of rotifers are sessile, most swim via the beating of cilia in the corona.

Reproduction Females produce unfertilized eggs by mitosis; the eggs then hatch into new, asexually produced individuals. Recall that the production of offspring via unfertilized eggs is termed parthenogenesis. An entire group of rotifers—called the bdelloids—reproduces only asexually via parthenogenesis. Both sexual reproduction and asexual reproduction are observed in most species, however. Development is direct, meaning that fertilized eggs hatch and grow into adults without going through metamorphosis.

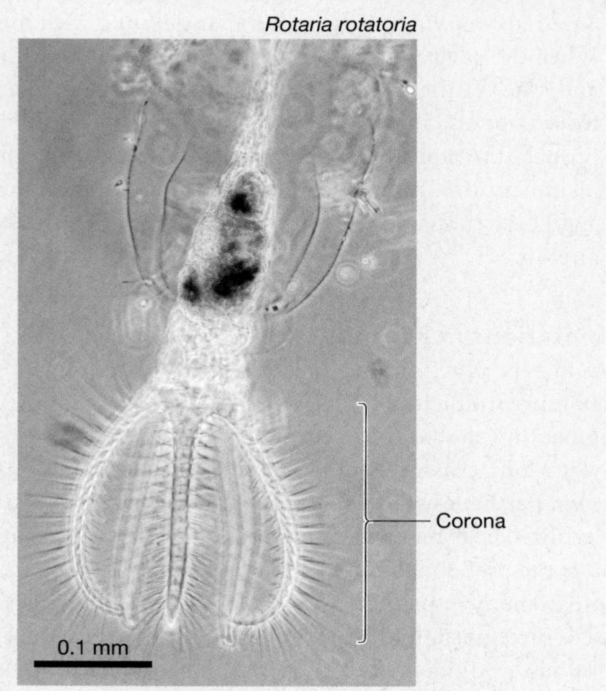

FIGURE 33.12 Rotifers Are Tiny, Aquatic Suspension Feeders.

Lophotrochozoans > Platyhelminthes (Flatworms)

The flatworms are a large and diverse phylum. More than 20,000 species have been described in three major subgroups. In traditional classification schemes like those described in Chapter 1, each of these lineages was referred to as a class. They are (1) the free-living species called Turbellaria (**Figure 33.13a**), (2) the endoparasitic tapeworms called Cestoda (**Figure 33.13b**), and (3) the endoparasitic or ectoparasitic flukes, called Trematoda (**Figure 33.13c**). Although a few turbellarian species are terrestrial, most live on the substrates of freshwater or marine environments. Tapeworms and other cestodes parasitize fish, mammals, or other vertebrates. Flukes parasitize vertebrates or molluscs.

Flatworms are named for the broad, flattened shape of their bodies. (The Greek roots *platy* and *helminth* mean "flatworm.") Species in the Platyhelminthes are unsegmented and lack a coelom. They also lack structures that are specialized for gas exchange—taking in oxygen and ridding the body of carbon dioxide. Further, they do not have blood vessels or any other type of system for circulating oxygen and nutrients to their cells. Based on these observations, biologists interpret the flattened bodies of these animals as an adaptation that gives flatworms an extremely high surface-area-to-volume ratio. Because they have so much surface area, a large amount of gas exchange can occur

(Continued on next page)

Lophotrochozoans > Platyhelminthes (Flatworms) *continued*

directly across their body wall, with oxygen diffusing into the body from the surrounding water and carbon dioxide diffusing out. Because the volume of the body is so small relative to the surface area available, nutrients and gases can diffuse efficiently to all of the cells inside the animal. This body plan has a downside, however: Because the body surface has to be moist for gas exchange to take place, flatworms are restricted to environments where they are surrounded by fluid. ● You should be able to indicate the origin of the flattened, acoelomate body plan on Figure 33.11.

Feeding Platyhelminthes lack a lophophore and have a digestive tract that is "blind"—meaning it has only one opening for ingestion of food and elimination of wastes. Most turbellarians are hunters that prey on protists or small animals; others scavenge dead animals. Tapeworms and flukes, in contrast, are strictly parasitic and feed on nutrients provided by hosts. Flukes gulp host tissues and fluids through a mouth and have a blind digestive tract. Tapeworms do not have a mouth, and they do not have a digestive tract. They obtain nutrients solely by diffusion through their body wall.

Movement Some turbellarians can swim a little by undulating their bodies, and most can creep along substrates with the aid of cilia on their ventral surface. Tapeworms and flukes move much less; adult cestodes have hooked attachment structures at their anterior end that permanently attach them to the interior of their host.

Reproduction Turbellarians can reproduce asexually by splitting themselves in half. If they are fragmented as a result of a predator attack, the body parts can regenerate into new individuals. Most turbellarians contain both male and female organs and reproduce sexually by aligning with another individual and engaging in mutual and simultaneous fertilization. Flukes (trematodes) and tapeworms (cestodes) also reproduce sexually and either cross-fertilize or self-fertilize. The reproductive systems and life cycles of flukes and tapeworms are extremely complex, in many cases involving two or even three distinct host species, with sexual reproduction occurring in the **definitive host** and asexual reproduction occurring in one or more **intermediate hosts**. For example, humans are the definitive host of the blood fluke *Schistosoma mansoni*. Fertilized eggs are shed in a human host's feces and enter aquatic habitats if sanitation systems are poor. The eggs develop into larvae that infect snails. Inside the snail, asexual reproduction results in the production of a different type of larva, which emerges from the snail and burrows into the skin of humans who wade in infested water. Once inside a human, the parasite lives in blood and develops into a sexually mature adult.

(a) Turbellarians are free living.
Pseudoceros ferrugineus

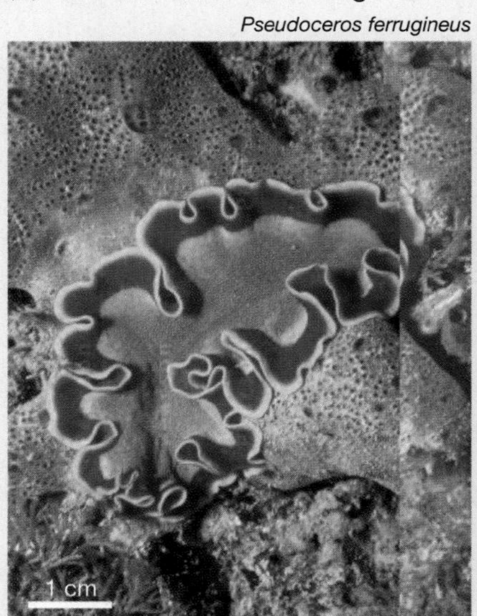

(b) Cestodes are endoparasitic.
Taenia species

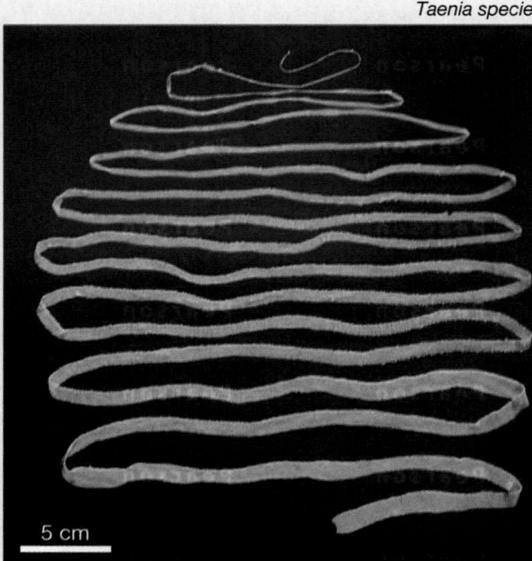

(c) Trematodes are endoparasitic.
Dicrocoelium dendriticum

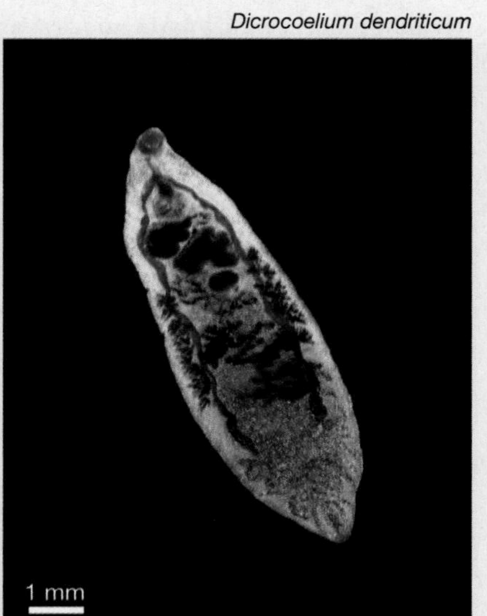

FIGURE 33.13 Flatworms Have Simple, Flattened Bodies.

Lophotrochozoans > Annelida (Segmented Worms)

All **annelids** have a segmented body and a coelom that functions as a hydrostatic skeleton. ● You should be able to indicate the origin of annelid segmentation on Figure 33.11. The 16,500 species that have been described thus far are traditionally divided into two major lineages, called Polychaeta and Clitellata:

1. The Polychaeta (*pol-ee-KEE-ta*), or polychaetes ("many-bristles"), are named for their numerous, bristle-like extensions called **chaetae** (**Figure 33.14a**). The chaetae extend from appendages called **parapodia**. Polychaetes are mostly marine, and they range in size from species that are less than 1 mm long to species that grow to lengths of 3.5 m. They include a large number of burrowers in mud or sand, sedentary forms that secrete chitinous tubes, and active, mobile species.
2. The Clitellata is composed of the oligochaetes (*oh-LIG-oh-keetes*) and leeches, or Oligochaeta and Hirudinea, respectively. Oligochaetes ("few-bristles") include the earthworms, which burrow in moist soils (**Figure 33.14b**); an array of freshwater species; and a few marine forms. Oligochaetes lack parapodia and, as their name implies, have many fewer chaetae than do polychaetes. The coelom of leeches is much reduced in size compared with that of other annelids and consists of a series of connected chambers. Leeches live in freshwater as well as marine habitats (**Figure 33.14c**).

Feeding Polychaetes have a wide variety of methods for feeding: The burrowing forms deposit feed or suspension feed using mucous-lined nets; the sedentary forms suspension feed with the aid of a dense crown of tentacles; and the active forms graze on algae ands hunt small animals, which they capture by everting their throats. Virtually all oligochaetes, in contrast, make their living by deposit feeding in soils. About half of the leeches are ectoparasites that attach themselves to fish or other hosts and suck blood and other body fluids. Hosts are usually unaware of the attack, because leech saliva typically contains an anesthetic. The host's blood remains liquid as the leech feeds, because the parasite's saliva also contains an anticoagulant. Parasitic leeches are still used by physicians to remove blood from particularly large bruises. The nonparasitic leech species are predators or scavengers.

Movement Polychaetes and oligochaetes crawl or burrow with the aid of their hydrostatic skeletons; the parapodia of polychaetes also act as paddles or tiny feet that aid in movement. Many polychaetes are excellent swimmers. Leeches can swim by using their hydrostatic skeletons to make undulating motions of the body.

Reproduction Asexual reproduction occurs in polychaetes and oligochaetes by transverse fission or fragmentation—meaning that body parts can regenerate a complete individual. Sexual reproduction in polychaetes may begin with internal or external fertilization, depending on the species. Polychaetes have separate sexes and usually release their eggs directly into the water. Some species produce eggs that hatch into trochophore larvae. In oligochaetes and leeches, individuals produce both sperm and eggs and engage in mutual, internal cross-fertilization. Eggs are enclosed in a mucus-rich, cocoon-like structure; after fertilization, offspring develop directly into miniature versions of their parents.

(a) Most polychaetes are marine.

Alvinella pompejana

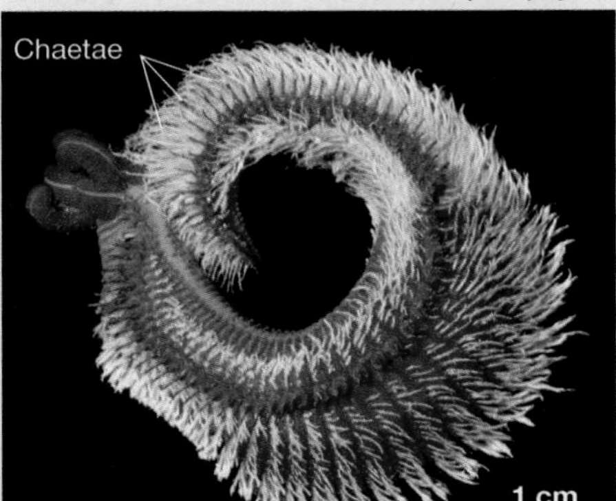

(b) Most oligochaetes are terrestrial.

Paranais litoralis

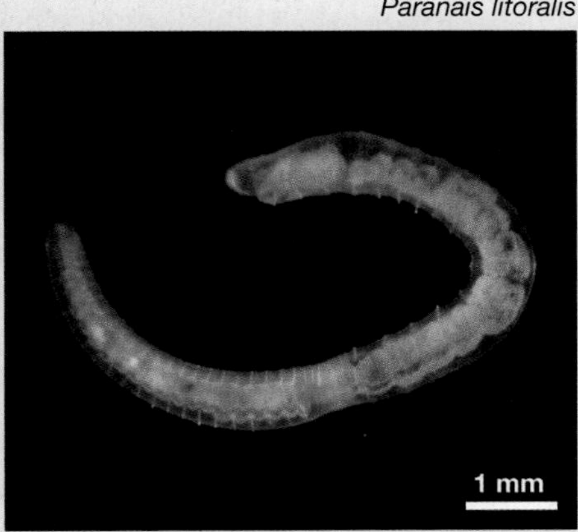

(c) Most leeches live in freshwater.

Hirudo medicinalis

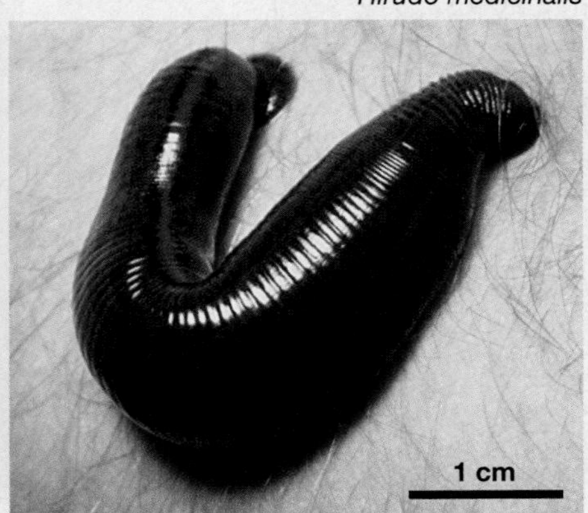

FIGURE 33.14 Annelids Are Segmented Worms.

Mollusca (Molluscs)

The **molluscs** are far and away the most species-rich and morphologically diverse group in the Lophotrochozoa. They have a specialized body plan based on a muscular foot, a visceral mass, and a mantle that may or may not secrete a calcium carbonate shell. The coelom is much reduced or absent. ● You should be able to indicate the origin of the molluscan body plan on Figure 33.11. Over 93,000 species have been described thus far. Although most molluscs live in marine environments, there are some terrestrial and freshwater forms.

Because molluscs are so diverse, let's consider each of the major subgroups or classes in the phylum separately and analyze the traits responsible for their diversification. The four most important lineages of molluscs are (1) **bivalves** (clams and mussels), (2) **gastropods** (slugs and snails), (3), **chitons,** and (4) **cephalopods** (squid and octopuses). The bivalves are suspension feeders; the other three groups of molluscs are herbivores or predators.

Mollusca > Bivalvia (Clams, Mussels, Scallops, Oysters)

The bivalves are so named because they have two separate shells made of calcium carbonate secreted by the mantle. The shells are hinged, and they open and close with the aid of muscles attached to them (**Figure 33.15a**). When the shell is closed, it protects the mantle, visceral mass, and foot. The bivalve shell is an adaptation that reduces predation.

Most bivalves live in the ocean, though there are many freshwater forms. Clams burrow into mud, sand, or other soft substrates and are sedentary as adults. Oysters and mussels are largely sessile as adults, but most of them live above the substrate, attached to rocks or other solid surfaces. Scallops are mobile and live on the surface of soft substrates. The smallest bivalves are freshwater clams that are less than 2 mm long; the largest bivalves are giant marine clams that may weigh more than 400 kg. All bivalves can sense gravity, touch, and certain chemicals, and scallops even have eyes.

Because most bivalves live on or under the ocean floor and because they are covered by a hard shell, their bodies are often buried in sediment after death. Bivalves thus fossilize readily, and the Bivalvia lineage has the most extensive fossil record of any animal, plant, or fungal group. This large database has allowed biologists to conduct thorough studies of the evolutionary history of bivalves.

Bivalves are important commercially. Clams, mussels, scallops, and oysters are farmed or harvested from the wild in many parts of the world and used as food by humans. Pearl oysters that are cultivated or collected from the wild are the most common source of natural pearls used in jewelry.

Feeding Bivalves are suspension feeders that take in any type of small animal or protist or detritus. Suspension feeding is based on a flow of water across gas-exchange structures called **gills**. The gills lie between the mantle and the visceral mass and consist of a series of thin membranes where particles are trapped. In many cases the water current flows through siphons, which are tubes formed by edges of the mantle, extending from the shell and form a plumbing system. The siphons conduct water into the gills and then back out of the body, powered by the beating of cilia on the gills (**Figure 33.15b**). Bivalves are the only major group of molluscs that lack the feeding structure called a radula (see Chapter 32).

(a) Scallops live on the surface of the substrate and suspension feed. *Lima scabra*

(b) Most clams burrow into soft subtrates and suspension feed

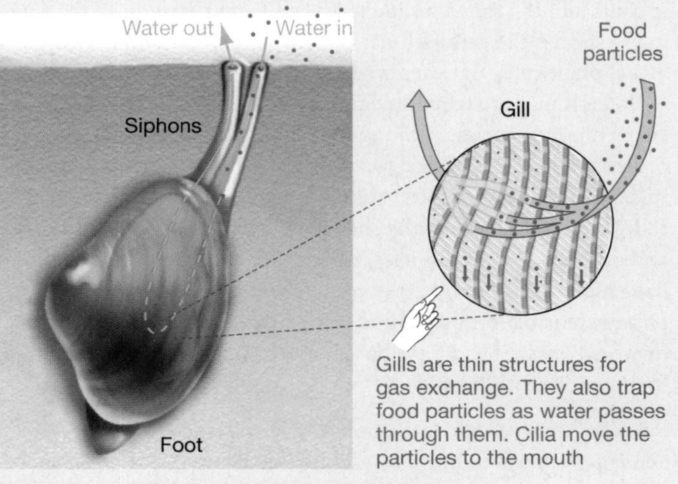

FIGURE 33.15 Bivalves Have Two Shells.

Movement Clams burrow with the aid of their muscular foot, which functions as a hydrostatic skeleton. But they are otherwise sedentary. Scallops are able to swim by clapping their shells together and forcing water to jet out, pushing them along.

(Continued on next page)

Mollusca > Bivalvia (Clams, Mussels, Scallops, Oysters) *continued*

Scallop locomotion is similar to the swimming of cnidarian medusae, which move when muscular contractions force water out of their bell (see Chapter 32). Bivalves produce a swimming trochophore larva that is responsible for dispersing individuals to new locations.

Reproduction Only sexual reproduction occurs in bivalves. Eggs and sperm are shed into the water, and fertilized eggs develop into trochophore larvae. Trochophores then metamorphose into a distinct type of larva called a **veliger**, which continues to feed and swim before settling to the substrate and metamorphosing into an adult form that secretes a shell.

Mollusca > Gastropoda (Snails, Slugs, Nudibranchs)

The gastropods ("belly-feet") are named for the large, muscular foot on their ventral side. Their most startling feature, though, is a developmental process called torsion. During gastropod development, the visceral mass usually rotates. The rotation twists the mantle cavity and digestive tract and results in the anus being located above the head.

Most snails can retract their foot and body into a shell when they are attacked or when their tissues begin to dry out (**Figure 33.16a**). Land slugs and nudibranchs (pronounced *NEW-da-branks*) lack shells but often contain toxins or foul-tasting chemicals to protect them from being eaten. The bright colors of nudibranchs, or sea slugs, are thought to act as a warning to potential predators (**Figure 33.16b**). Gastropods are used as food in some cultures and are important medically because they serve as a host for flukes that also infect humans before completing their life cycle. About 70,000 species of gastropods are known.

Feeding Gastropods and other molluscs have a unique structure in their mouths called a radula. Recall from Chapter 32 that in many species the **radula** functions like a rasp to scrape away algae, plant cells, or other types of food. It is usually covered with teeth that are made of chitin and that vary in size and shape among species. Although most gastropods are herbivores or detritivores, specialized types of teeth allow some gastropods to act as predators. Species called drills, for example, use their radula to bore a hole in the shells of oysters or other molluscs and expose the visceral mass, which they then eat. Cone snails have highly modified, harpoon-like "teeth" mounted at the tip of an extensible proboscis and armed with poison. When a fish or worm passes by, the proboscis shoots out. The prey is speared by the tooth, subdued by the poison, and consumed.

Movement Waves of contractions down the length of the foot allow gastropods to move by gliding (see Figure 33.10c). Sea butterflies are gastropods with a reduced or absent shell but a large, winglike foot that flaps and powers swimming movements.

Reproduction Females of some gastropod species can reproduce asexually by producing eggs parthenogenetically, but most reproduction is sexual. Sexual reproduction in some gastropods begins with internal fertilization. Some marine gastropods produce a trochophore larva that may disperse up to several hundred kilometers from the parent. But in most marine species and all terrestrial forms, larvae are not free living. Instead, offspring remain in an egg case while passing through several larval stages and then hatch as miniature versions of the adults.

(a) Snails have a single shell, which they use for protection.
Maxacteon flammea

(b) Land slugs and sea slugs (nudibranchs) lack shells.
Chromodoris geminus

FIGURE 33.16 Gastropods Have a Single Shell or Lack Shells.

Mollusca > Polyplacophora (Chitons)

The Greek word roots that inspired the name *Polyplacophora* mean "many-plate-bearing." The name is apt because chitons (pronounced *KITE-uns*) have eight calcium carbonate plates along their dorsal side (**Figure 33.17**). The plates form a protective shell. The approximately 1000 species of chitons are marine. They are usually found on rocky surfaces in the intertidal zone, where rocks are periodically exposed to air at low tides.

Feeding Chitons have a radula and use it to scrape algae and other organic matter off rocks.

Movement Chitons move by gliding on their broad, muscular foot—as gastropods do.

Reproduction Chiton sexes are separate, and fertilization is external. In some species, however, sperm that are shed into the water enter the female's mantle cavity and fertilize eggs inside the body. Depending on the species involved, eggs may be enclosed in a membrane and released or retained until hatching and early development are complete. Most species have trochophore larvae.

FIGURE 33.17 Chitons Have Eight Shell Plates.

Mollusca > Cephalopoda (Nautilus, Cuttlefish, Squid, Octopuses)

The cephalopods ("head-feet") have a well-developed head and a foot that is modified to form arms and/or tentacles. **Tentacles** are long, thin, muscular extensions that aid in movement and prey capture (**Figure 33.18**). Except for the nautilus, cephalopods have either highly reduced shells or none at all. They also have large brains and image-forming eyes with sophisticated lenses.

FIGURE 33.18 Cephalopods Have Highly Modified Bodies.

Feeding Cephalopods are highly intelligent predators that hunt by sight and use their arms or tentacles to capture prey—usually fish or crustaceans. They have a radula as well as a structure called a **beak**, which can exert powerful biting forces. Some cuttlefish and octopuses also inject poisons into their prey to subdue them.

Movement Cephalopods can swim by moving their fins to "fly" through the water, or by jet propulsion using the mantle cavity and siphon. They draw water into their mantle cavity and then force it out through a siphon (see Figure 33.10d). Squid are built for speed and hunt small fish by chasing them down. Octopuses, in contrast, crawl along the substrate by using their long, arm-like tentacles. They chase down crabs or other crustaceans, or they pry mussels or clams from the substrate and then use their beaks to crush the exoskeletons of their prey.

Reproduction Cephalopods have separate sexes, and some species have elaborate courtship rituals that involve color changes and interaction of tentacles. When a male is accepted by a female, he deposits sperm that are encased in a structure called a **spermatophore**. The spermatophore is transferred to the female, and fertilization is internal. Females lay eggs. When they hatch, juveniles develop directly into adults.

33.4 Key Lineages: Ecdysozoans

The ecdysozoans were first recognized as monophyletic when investigators began using DNA sequence data to estimate the phylogeny of protostomes (**Figure 33.19**). Seven phyla are currently recognized in the lineage (see Table 32.1), including the Onychophora (*on-ee-KOFF-er-uh*) and the Tardigrada (**Figure 33.20**). They are not species-rich phyla, but both are closely related to arthropods. Onychophorans and tardigrades are similar to arthropods in having a segmented body and limbs. Unlike arthropods, their limbs not jointed and they do not have an exoskeleton. The onychophorans, or velvet worms, are small, caterpillar-like organisms that live in moist leaf litter and prey on small invertebrates. Onychophorans have sac-like appendages and segmented bodies with a hemocoel.

The tardigrades, or water bears, are microscopic animals that live in benthic (bottom) habitats of marine or freshwater environments. Large numbers of water bears can also be found in the film of water that covers moss or other land plants in moist habitats. Tardigrades have a reduced coelom but a prominent hemocoel, and they walk on their clawed, sac-like legs. Most feed by sucking fluids from plants or animals; others are **detritivores**.

Let's take a closer look at the two most diverse and abundant phyla of the Ecdysozoa: (1) Nematoda and (2) Arthropoda.

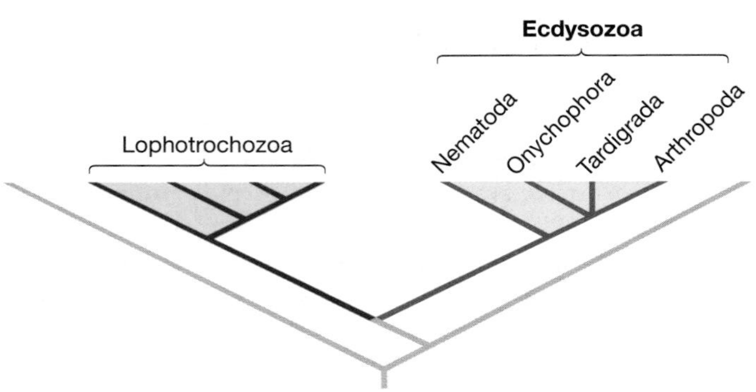

FIGURE 33.19 Ecdysozoans Are a Monophyletic Group.

(a) Onychophorans have lobe-like limbs.

Peripatus species

(b) Tardigrades have lobe-like limbs with claws.

Echiniscus species

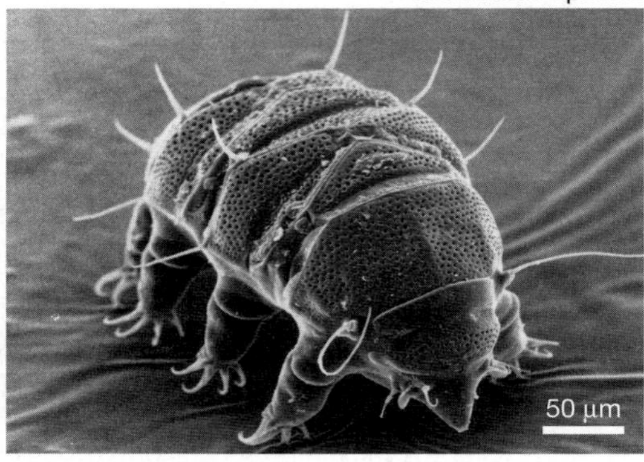

FIGURE 33.20 Onychophorans and Tardigrades Have Limbs and Are Closely Related to Arthropods. (a) The Onychophora are called velvet worms because they have wormlike bodies and a velvety appearance. **(b)** The Tardigrada are known as water bears because they live in moist habitats and, when viewed under the microscope, resemble bears.

Ecdysozoans > Nematoda (Roundworms)

Species in the phylum Nematoda are commonly called **nematodes** or **roundworms** (**Figure 33.21**). Roundworms are unsegmented worms with a pseudocoelom, a tube-within-a-tube body plan, and no appendages. They have no muscles that can change the diameter of the body—their body musculature consists solely of longitudinal muscles that shorten or lengthen the body upon contracting or relaxing, respectively. ● You should be able to indicate the evolution of a pseudocoelom and loss of circular muscles on Figure 33.19. Although some nematodes can grow to lengths of several meters, the vast majority of species are tiny—most are much less than 1 mm long. They lack specialized systems for exchanging gases and for circulating nutrients and wastes. Instead, gas exchange occurs across the body wall, and nutrients and wastes move by simple diffusion.

The nematode *Caenorhabditis elegans* is one of the most thoroughly studied model organisms in biology (see Box 33.1). Species that parasitize humans have also been studied intensively. Pin-

(Continued on next page)

Ecdysozoans > Nematoda (Roundworms) *continued*

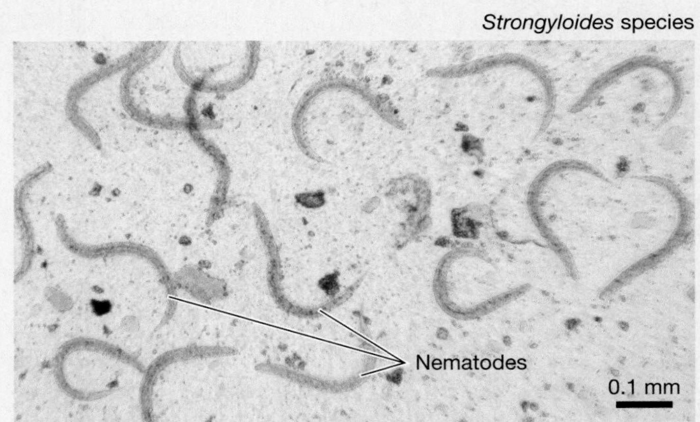

FIGURE 33.21 Most Nematodes Are Free Living, but Some Are Parasites.

worms, for example, infect about 40 million people in the United States alone, and *Onchocerca volvulus* causes an eye disease that infects around 20 million people in Africa and Latin America. Advanced infections by the roundworm *Wuchereria bancrofti* result in the blockage of lymphatic vessels, causing fluid accumulation and massive swelling—the condition known as elephantiasis.

Parasites actually represent a tiny fraction of the 25,000 nematode species that have been described to date, however. The vast majority of nematodes are free living. They are important ecologically because they are found in virtually every habitat known.

Nematodes are also fabulously abundant. Biologists have found 90,000 roundworms in a single rotting apple and have estimated that rich farm soils contain up to 9 billion roundworms per acre. Although they are not the most species-rich animal group, they are the most abundant. The simple nematode body plan has been extraordinarily successful.

Feeding Roundworms feed on a wide variety of materials, including bacteria, fungi, plant roots, small protists or animals, or detritus. In most cases, the structure of their mouthparts is specialized in a way that increases the efficiency of feeding on a particular type of organism or material.

Movement Roundworms move with the aid of their hydrostatic skeleton. Most roundworms live in soil or inside a host, so when contractions of their longitudinal muscles cause them to wriggle, the movements are resisted by a stiff substrate. As a result, the worm pushes off the substrate, and the body moves.

Reproduction Sexes are separate in most nematode species, and asexual reproduction is rare or unknown. Sexual reproduction begins with internal fertilization and culminates in egg laying and the direct development of offspring. Individuals go through a series of four molts over the course of their lifetime.

Arthropoda (Arthropods)

In terms of duration in the fossil record, species diversity, and abundance of individuals, arthropods are easily the most successful lineage of eukaryotes. They appear in the fossil record over 520 million years ago and have long dominated the animals observed in both marine and terrestrial environments. Well over a million living species have been described, and biologists estimate that millions or perhaps even tens of millions of arthropod species have yet to be discovered.

Morphologically, **arthropods** are distinguished by segmented bodies and a sophisticated, jointed exoskeleton. They have a highly reduced coelom but possess an extensive body cavity called a hemocoel, enclosed by the exoskeleton, and paired, jointed appendages. The body is organized into distinct head and trunk regions in all arthropods; in many species there is an additional grouping of segments into two distinct trunk regions, usually called an abdomen and a thorax. ● You should be able to indicate the origin of the elements of the arthropod body plan on Figure 33.19.

Metamorphosis is common in arthropods. Their larvae have segmented bodies but may lack a hardened exoskeleton. As in other ecdysozoans, larval and adult forms grow by molting.

At least some segments along the arthropod body produce paired, jointed appendages. Arthropod appendages have an array of functions: sensing aspects of the environment, exchanging gases, feeding, or locomotion via swimming, walking, running, jumping, or flying. Arthropod appendages provide the ability to sense environmental stimuli and make sophisticated movements in response. Most species also have sophisticated, image-forming compound eyes. A **compound eye** contains many lenses, each associated with a light-sensing, columnar structure. (Human and cephalopod eyes are **simple eyes**, meaning they have just one lens.) Most arthropods also have a pair of antennae on the head. **Antennae** are long, tentacle-like appendages that contain specialized receptor cells used to touch or smell.

Although the phylogeny of arthropods is still being worked out, most data sets agree that the phylum as a whole is monophyletic; that the myriapods (millipedes and centipedes), chelicerates (spiders), insects, and crustaceans represent four major subphyla within the phylum; and that crustaceans and insects are closely related. The phylum Arthropoda is so large and diverse that a detailed treatment would fill a book this size; space permits just a few notes about the four major lineages.

Arthropoda > Myriapods (Millipedes, Centipedes)

The **myriapods** have relatively simple bodies, with a head region and a long trunk featuring a series of short segments, each bearing one or two pairs of legs (**Figure 33.22**). If eyes are present, they consist of a few to many simple structures clustered on the sides of the head. The 11,600 species that have been described to date inhabit terrestrial environments all over the world.

Feeding Millipedes and centipedes have mouthparts that can bite and chew. These organisms live in downed, rotting logs and other types of dead plant material that litters the ground in forests and grasslands. Millipedes are detritivores. Centipedes, in contrast, use a pair of poison-containing fangs just behind the mouth to hunt an array of insects. Large centipedes can inject enough poison to debilitate a human.

Movement Myriapods walk or run on their many legs; a few species burrow. Some millipedes have over 190 trunk segments, each with two pairs of legs, for a total of over 750 legs. Centipedes usually have fewer than 30 segments, with one pair of legs per segment.

Reproduction Myriapod sexes are separate, and fertilization is internal. Males deposit sperm in packets that are picked up by the female or transferred to her by the male. After females lay eggs in the environment, the eggs hatch into juveniles that develop into adults via a series of molts.

Scolopendra species

1 cm

FIGURE 33.22 Myriapods Have a Pair of Legs on Each Body Segment.

Arthropoda > Chelicerata (Spiders, Ticks, Mites, Horseshoe Crabs, Daddy Longlegs, Scorpions)

Most of the 70,000 species of chelicerates are terrestrial, although the horseshoe crabs and sea spiders are marine. The lineage is considered a class in most of the traditional classification schemes and has several subgroups or subclasses. The most prominent of these chelicerate lineages is the arachnids (spiders, scorpions, mites, and ticks).

The chelicerate body consists of anterior and posterior regions (**Figure 33.23a**). The anterior region lacks antennae for sensing touch or odor but usually contains eyes. The group is named for appendages called **chelicerae,** found near the mouth. Depending on the species, the chelicerae are used in feeding, defense, copulation, movement, or sensory reception.

Feeding Spiders, scorpions, and daddy longlegs capture and sting insects or other prey. Although some of these species are active hunters, most spiders are sit-and-wait predators. They create sticky webs to capture prey, which fly or walk into the web and are subsequently trapped. The spider senses the vibrations of the struggling prey, pounces on it, and administers a toxic bite. Mites and ticks are ectoparasitic and use their piercing mouthparts to feed on host animals (**Figure 33.23b**). Horseshoe crabs eat a variety of protostomes as well as detritus. Most scorpions feed on insects; the largest scorpion species occasionally eat snakes and lizards.

Movement Like other arthropods, chelicerates move with the aid of muscles attached to an exoskeleton. They walk or crawl on their four pairs of jointed walking legs; some species are also capable of jumping. Horseshoe crabs and some other marine forms can swim slowly. Newly hatched spiders spin long, silken threads that serve as balloons and that can carry them on the wind more than 400 kilometers from the point of hatching.

Reproduction Sexual reproduction is the rule in chelicerates, and fertilization is internal in most groups. Courtship displays are extensive in many arthropod groups and may include both visual displays and the release of chemical odorants. In spiders, males use organs that are located on their legs to transfer sperm to females. These organs fit into the female reproductive tract in a "lock-and-key" fashion. Differences in the size and shape of male genitalia are often the only way to identify closely related species of spider.

(Continued on next page)

Arthropoda > Chelicerata (Spiders, Ticks, Mites, Horseshoe Crabs, Daddy Longlegs, Scorpions) *continued*

(a) Spider, showing general chelicerate features
Dolomedes fimbriatus

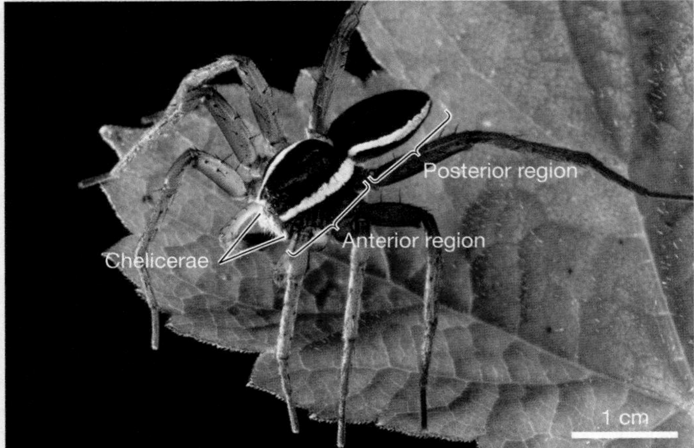

(b) Mites are ectoparasitic.
Dermatophagoides species

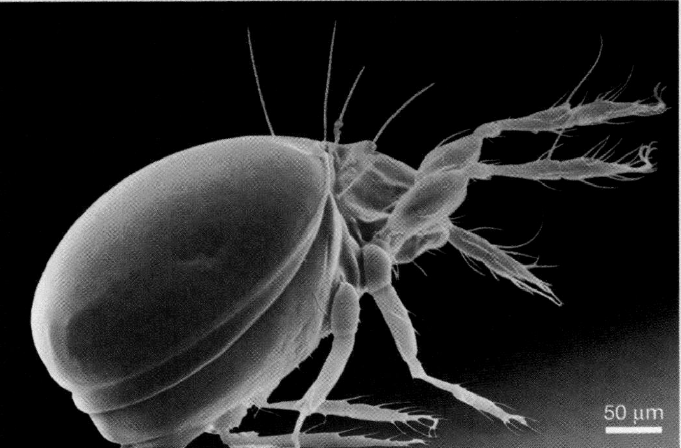

FIGURE 33.23 Chelicerates Have Two Body Regions.

EXERCISE Label the two major body regions and the chelicerae in part (b).

Development is direct, meaning that metamorphosis does not occur. In spiders, males may present a dead insect as a gift that the female eats as they mate; in some species, the male himself is eaten as sperm is being transferred. In scorpions, females retain fertilized eggs. After the eggs hatch, the young climb on the mother's back. They remain there until they are old enough to hunt for themselves.

Arthropoda > Insecta (Insects)

About 925,000 species of insects have been named thus far, but it is certain that many more exist. In terms of species diversity and numbers of individuals, insects dominate terrestrial environments. In addition, the larvae of some species are common in freshwater streams, ponds, and lakes. **Table 33.1** provides detailed notes about eight of the most prominent lineages, called orders, of insects.

Insects are distinguished by having three body regions—(1) the **head**, (2) **thorax**, and (3) **abdomen** (see Figure 33.7). Three pairs of walking legs are located on the ventral surface of the thorax. Most species have one or two pairs of wings, mounted on the dorsal (back) side of the thorax. Typically the head contains three sets of mouthparts that are derived from jointed appendages, a fourth pair of mouthparts, a pair of antennae that are used to touch and smell, and a pair of compound eyes.

Feeding Because most insect species have four sets of mouthparts (labrum, mandible, maxilla, and labium) that vary greatly in structure among species, insects are able to feed in every conceivable manner and on almost every type of food source available on land. In species with holometabolous metamorphosis, larvae have wormlike bodies; most are deposit feeders, though some are leaf eaters. Adults are predators or parasites on plant or animal tissues. Because so many insects make their living by feeding on plant tissues or fluids, the diversification of insects was closely correlated with the diversification of land plants. Insect predators usually eat other insects; insect parasites usually victimize other arthropods or mammals.

Movement Insects use their legs to walk, run, or swim, or they use their wings to fly. When insects walk or run, the sequence of movements usually results in three of their six legs maintaining contact with the ground at all times.

Reproduction Insect sexes are separate. Mating usually takes place through direct copulation, with the male inserting a sperm-transfer organ into the female. Most females lay eggs, but in a few species eggs are retained until hatching. Many species are also capable of reproducing asexually, through the production of unfertilized eggs via mitosis. In the vast majority of species, either incomplete or complete metamorphosis occurs, with complete metamorphosis being most common.

TABLE 33.1 **Prominent Orders of Insects**

Order	Common Name	Number of Known Species	Description
Coleoptera ("sheath-winged") 	Beetles	350,000	*Key traits:* Hardened forewings, called elytra, protect the membranous hindwings that power flight. During flight, the elytra are held to the side and act as stabilizers. *Feeding:* Adults are important predators and scavengers. Larvae are called grubs and often have chewing mouthparts. *Reproduction:* All have complete metamorphosis. *Notes:* The most species-rich lineage on the tree of life. Range in length from 0.25 mm to 10 cm. Scarab beetles were worshipped in ancient Egypt.
Lepidoptera ("scale-winged")  © Gustav W. Verderber	Butterflies, moths	180,000	*Key traits:* Wings are covered with tiny, often colorful scales. The forewings and hindwings hook together and move as a unit. *Feeding:* Larvae usually have chewing mouthparts and are herbivorous deposit feeders or food-mass feeders; adults often feed on nectar. *Reproduction:* All have complete metamorphosis. *Notes:* Some species migrate long distances.
Diptera ("two-winged") 	Flies (including mosquitoes, gnats, midges)	120,000	*Key traits:* Reduced hindwings, called halteres, act as stabilizers during flight. *Feeding:* Adults are usually liquid feeders; larvae are often parasitic. *Reproduction:* All have complete metamorphosis. *Notes:* First flies in fossil record appear 225 million years ago.
Hymenoptera ("membrane-winged") 	Ants, bees, wasps	115,000	*Key traits:* Membranous forewings and hindwings lock together via tiny hooks, thus acting as a single wing during flight. They have club-like antennae. *Feeding:* Most ants feed on plant material; most bees feed on nectar; wasps are predatory and often have parasitic larvae. *Reproduction:* All have complete metamorphosis. Males are haploid (they hatch from unfertilized eggs) and females are diploid (they hatch from fertilized eggs). Females deposit eggs via an ovipositor, which in some species is modified into a stinger used in defense. Larvae are often fed and protected by adults. *Notes:* Most species live in colonies, and many are "eusocial"—meaning that some individuals in the colony help raise the queen's offspring but never reproduce themselves.

(Continued on next page)

TABLE **33.1 Prominent Orders of Insects** *(continued)*

Order	Common Name	Number of Known Species	Description
Hemiptera ("different-winged") 	Bugs (including leaf hoppers, aphids, cicadas, scale insects)	85,000	*Key traits:* Have a thickened forewing with a membranous tip; also mouthparts that are modified for piercing and sucking. *Feeding:* Most suck plant juices, but some are predatory. *Reproduction:* All have complete metamorphosis. *Notes:* Range in length from 1 mm to 11 cm.
Orthoptera ("straight-winged") 	Grasshoppers, crickets	20,000	*Key traits:* Large, muscular hind legs power movement by jumping. *Feeding:* Most have chewing mouthparts and are leaf eaters. *Reproduction:* All have incomplete metamorphosis. In many species, males give distinctive songs to attract mates. *Notes:* Range in length from 5 mm to 11.5 cm.
Trichoptera ("hairy-winged") 	Caddisflies	12,000	*Key traits:* Wings are covered with minute hairs. *Feeding:* Adults lack mouthparts and do not feed; larvae are all aquatic and feed on detritus. Most larvae build a protective case of tiny stones and sticks. *Reproduction:* All have complete metamorphosis. *Notes:* Larvae and adults are important food sources for fish.
Odonata ("toothed") 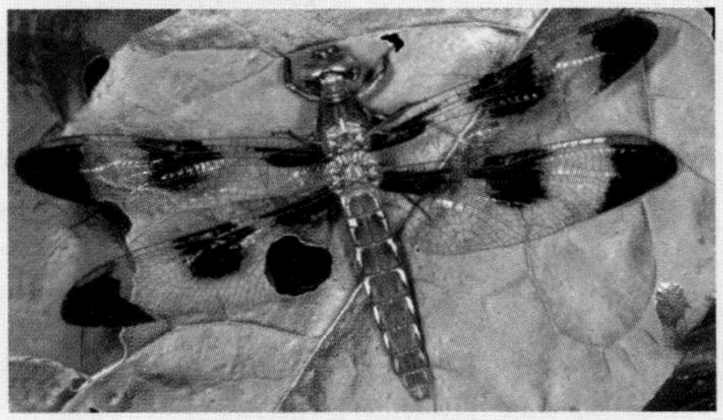	Dragonflies, damselflies	6,500	*Key traits:* Four membranous wings and long, slender abdomens. *Feeding:* Larvae and adults are predatory, often on flies. The "odon" in their name refers to strong, toothlike structures on their mandibles. *Reproduction:* All have complete metamorphosis. *Notes:* They hunt by sight and have reduced antennae (used for touch and smell) but enormous eyes with up to 28,000 lenses.

Arthropoda > Crustaceans (Shrimp, Lobster, Crabs, Barnacles, Isopods, Copepods)

The 67,000 species of **crustaceans** that have been identified to date live primarily in marine and freshwater environments. A few species of crab and some isopods are terrestrial, however. (Terrestrial isopods are known as sowbugs, pillbugs, or roly-polies). Crustaceans are common in surface waters, where they are important consumers. They are also important grazers and predators in shallow-water benthic environments.

The segmented body of most crustaceans is divided into two distinct regions: (1) the cephalothorax, which combines the head and thorax, and (2) the abdomen. Many crustaceans have a **carapace**—a platelike section of their exoskeleton that covers and protects the cephalothorax (**Figure 33.24a**). They are the only type of arthropod with two pairs of antennae, and they have sophisticated, compound eyes—usually mounted on stalks (**Figure 33.24b**). Crustaceans and trilobites were the first arthropods to appear in the fossil record.

Feeding Most crustaceans have 4–6 pairs of mouthparts that are derived from jointed appendages, and as a group they use every type of feeding strategy known. Barnacles (**Figure 33.24c**) and many shrimp are suspension feeders that use feathery structures located on head or body appendages to capture passing prey. Crabs and lobsters are active hunters, herbivores, and scavengers. Typically they have a pair of mouthparts called **mandibles** that bite or chew. Individuals capture and hold their food source with claws or other types of feeding appendages near their mouth and then use their mandibles to shred the food into small bits that can be ingested. As herbivores or detritivores, many species of crustaceans depend on algae for food.

Movement Crustacean limbs are highly diverse: species have many pairs of limbs and it is comman for a species to have more than one type of limb. Limb structures in crustaceans include paddle-shaped forms used in swimming, feathery structures used in capturing food that is suspended in water, and jointed appendages that make sophisticated walking or running movements possible. Barnacles are one of the few types of sessile crustaceans. Adult barnacles cement their heads to a rock or other hard substrate, secrete a protective shell of calcium carbonate, and use their legs to capture food particles and transfer them to the mouth.

Reproduction Most individual crustaceans are either male or female, and sexual reproduction is the norm. Fertilization is usually internal, and eggs are usually retained by the female until they hatch. Most crustaceans pass through several distinct larval stages; many species include a larval stage called a **nauplius**, which is usually planktonic. A nauplius has a single eye and appendages that develop into the two pairs of antennae and the mouthparts of the adult.

(a) Deep-sea lobster *Enoplometopus occidentalis*

Carapace

1 cm

(b) Red barnacle *Tetraclita species*

(c) Fiddler crab *Uca vocans*

FIGURE 33.24 Most Crustaceans Are Aquatic.

● **EXERCISE** Circle the lobster's cephalothorax (combined head and thorax).

Chapter Review

SUMMARY OF KEY CONCEPTS

Molecular phylogenies support the hypothesis that protostomes are a monophyletic group divided into two major subgroups: the Lophotrochozoa and the Ecdysozoa.

The protostomes comprise about 20 phyla and were originally identified because their embryos undergo early development in the same way. Only recently did biologists recognize the existence of the lophotrochozoans and ecdysozoans, however. Several phyla in the Lophotrochozoa have characteristic feeding structures called lophophores, and many have a distinctive type of larvae called trochophores. Species in the Ecdysozoa grow by molting—meaning that they shed their old external covering and grow a new, larger one.

You should be able to draw the phylogenetic relationships among the three major lineages of bilaterians, and label the origin of growth by molting, the trochophore larva, and the protostome and deuterostome patterns of development.

MB **Web Animation** at www.masteringbio.com
Protostome Diversity

Although the members of many protostome phyla have limbless, wormlike bodies and live in marine sediments, the most diverse and species-rich lineages—Mollusca and Arthropoda—have body plans with a series of distinctive, complex features. Molluscs and arthropods inhabit a wide range of environments.

All protostomes are bilaterally symmetric and triploblastic, but the nature of the body cavity is variable. The wormlike phyla of protostomes have well-developed coeloms that provide a hydrostatic skeleton used in locomotion. Flatworms lack a coelom, however, and both nematodes and rotifers evolved a pseudocoelom independently. The coelom is also much reduced or absent in the Mollusca and the Arthropoda lineages, although a body cavity is present. The mollusc's body is made up of a muscular foot, a visceral mass of organs, and a protective mantle. The arthropod body is segmented, divided into specialized regions such as the head, thorax, and abdomen of insects, and protected by an exoskeleton made of chitin.

You should be able to explain why flatworms lack a coelom, and why arthropods have a drastically reduced coelom.

Key events triggered the diversification of protostomes, including several lineages making the water-to-land transition, a diversification in appendages and mouthparts, and the evolution of metamorphosis in both marine and terrestrial forms.

The transition to living on land occurred several times independently during the evolution of protostomes. It was facilitated by watertight shells and exoskeletons and the ability to move to moist locations. Diversification in appendages and mouthparts gave protostomes the ability to move and find food in innovative ways. The evolution of a larval stage was significant because it allowed offspring in sessile species to disperse to new habitats and because it enabled juveniles and adults to find food in different ways.

You should be able to suggest hypotheses to explain why metamorphosis can be advantageous in both marine and terrestrial environments.

QUESTIONS

Test Your Knowledge

1. Why is it logical to observe that molluscs have a drastically reduced coelom or no coelom at all?
   a. They evolved from flatworms, which also lack a coelom.
   b. They are the most advanced of all the protostomes.
   c. Their bodies are encased by a mantle, which may or may not secrete a shell.
   d. They have a muscular foot and visceral mass that fulfill the functions of a coelom.
2. What major feature or features distinguish most of the wormlike phyla of protostomes from each other?
   a. Their mouthparts and feeding methods vary.
   b. Their modes of locomotion differ—they burrow, swim, crawl, or walk.
   c. Metamorphosis may be lacking, incomplete, or complete.
   d. They have a well-developed coelom and a tube-within-a-tube body plan.
3. What is the function of the arthropod exoskeleton?
   a. Because hard parts fossilize more readily than do soft tissues, the presence of an exoskeleton has given arthropods a good fossil record.
   b. It has no well-established function. (Trilobites had an exoskeleton, and they went extinct.)
   c. It provides protection and functions in locomotion.
   d. It makes growth by molting possible.
4. Why is it logical that Platyhelminthes have flattened bodies?
   a. They have simple bodies and evolved early in the diversification of protostomes.
   b. They lack a coelom, so their body cannot form a rounded tube-within-a-tube design.
   c. A flat body provides a surface area for gas exchange, which compensates for their lack of gas-exchange organs.
   d. They are sit-and-wait predators that hide from passing prey by flattening themselves against the substrate.
5. In number of species, number of individuals, and duration in the fossil record, which of the following phyla of wormlike animals has been the most successful? What feature is hypothesized to be responsible for this success?
   a. Annelida—a segmented body plan
   b. Platyhelminthes—bilateral symmetry but acoelomate condition
   c. Phoronida—a lophophore used as a feeding apparatus
   d. Echiura—a grooved proboscis used in suspension feeding
6. Which protostome phylum is distinguished by having body segments organized into distinct regions?
   a. Mollusca **c.** Annelida
   b. Arthropoda **d.** Nematoda

Test Your Knowledge answers: 1. d; 2. a; 3. c; 4. c; 5. a; 6. b

Test Your Understanding

Answers are available at www.masteringbio.com

1. Describe the traits that distinguish the Lophotrochozoa and Ecdysozoa. Describe the traits that are common to both groups.
2. Define a segmented body plan. Did segmentation evolve once or multiple times during the evolution of protostomes? Provide evidence to justify your answer.
3. The nature of the coelom changed drastically during the evolution of protostomes. Describe the structure and function of the coelom in nematodes, rotifers, and the wormlike phyla of protostomes other than flatworms.
4. Compare and contrast the role of metamorphosis in the life cycle of a sedentary, marine protostome such as a clam and a terrestrial, mobile protostome such as a butterfly.
5. Explain why the evolution of the exoskeleton and the jointed limb was such an important event in the evolution of arthropods.
6. The text claims that the diversification of protostomes was largely driven by diversification in mouthpart structure and feeding strategies. Considering only the wormlike phyla, give an example of this pattern.

Applying Concepts to New Situations

Answers are available at www.masteringbio.com

1. Recall that the phylum Platyhelminthes includes three major groups: the free-living turbellarians, the parasitic cestodes, and the parasitic trematodes. Because the parasitic forms are so simple morphologically, researchers suggest that they are derived from more complex, free-living forms. Draw a phylogeny of Platyhelminthes that would support the hypotheses that ancestral flatworms were morphologically complex, free-living organisms and that parasitism is a derived condition.
2. Brachiopoda is a phylum within the Lophotrochozoa. Even though they are not closely related to molluscs, brachiopods look and act like bivalve molluscs (clams or mussels). Specifically, brachiopods suspension feed, live inside two calcium carbonate shells that hinge together in some species, and attach to rocks or other hard surfaces on the ocean floor. How is it possible for brachiopods and bivalves to be so similar if they did not share a recent common ancestor?
3. The Mollusca includes a group of about 370 wormlike species called the Aplacophora. Although aplacophorans do not have a shell, their cuticle secretes calcium carbonate scales or spines. They lack a well-developed foot, and some species have a simple radula.
   - Predict where Aplacophora are found in the phylogenetic tree of Mollusca relative to bivalves, gastropods, chitons, and cephalopods. Explain your reasoning.
   - Predict where Aplacophora live and how they move. Explain your reasoning.
4. Consider the following: (a) The first arthropods to appear in the fossil record are marine crustaceans and trilobites, (b) terrestrial insects appeared relatively late in arthropod evolution, (c) the crustacean group called isopods includes both aquatic and terrestrial forms, (d) myriapods are strictly terrestrial, and (e) analyses of DNA sequence data suggest that crustaceans and insects are closely related. Based on these observations, do you agree or disagree with the hypothesis that arthropods made the water-to-land transition more than once? Explain your reasoning.

www.masteringbio.com is also your resource for • Answers to text, table, and figure caption questions and exercises • Answers to *Check Your Understanding* boxes • Online study guides and quizzes • Additional study tools including the *E-Book for Biological Science* 3rd ed., textbook art, animations, and videos.

Deuterostome Animals

34

In most habitats the "top predators"—meaning animals that prey on other animals and aren't preyed upon themselves—are deuterostomes.

KEY CONCEPTS

- The most species-rich deuterostome lineages are the echinoderms and the vertebrate groups called ray-finned fishes and tetrapods.
- Echinoderms and vertebrates have unique body plans. Echinoderms are radially symmetric as adults and have a water vascular system. All vertebrates have a skull and an extensive endoskeleton made of cartilage or bone.
- The diversification of echinoderms was triggered by the evolution of appendages called podia; the diversification of vertebrates was driven by the evolution of the jaw and limbs.
- Humans are a tiny twig on the tree of life. Chimpanzees and humans diverged from a common ancestor that lived in Africa 6–7 million years ago. Since then, at least 14 humanlike species have existed.

The **deuterostomes** include the largest-bodied and some of the most morphologically complex of all animals. They range from the sea stars that cling to dock pilings, to the fish that dart in and out of coral reefs, to the wildebeests that migrate across the Serengeti Plains of East Africa.

Biologists are drawn to deuterostomes in part because of their importance in the natural and human economies. Deuterostomes may not be as numerous as insects and other protostomes, but they act as key predators and herbivores in most marine and terrestrial habitats. If you diagram a food chain that traces the flow of energy from algae or plants up through several levels of consumers (see Chapter 30), deuterostomes are almost always at the top of the chain. In addition, humans rely on deuterostomes—particularly **vertebrates**, or animals with backbones—for food and power. Fish and domesticated livestock are key sources of protein in most cultures, and in the developing world agriculture is still based on the power generated by oxen, horses, water buffalo, or mules. In industrialized countries, millions of people bird-watch, plan vacations around seeing large mammals in national parks, or keep vertebrates as pets. Humans are deuterostomes—we have a special interest in our closest relatives, and a special fondness for them.

To introduce this key group, let's begin with an overview of the morphological traits that distinguish major lineages of deuterostomes and then delve into the phylogenetic relationships among the major vertebrate groups and the evolutionary innovations that led to their diversification. After summarizing the key characteristics of deuterostome lineages, the chapter closes with an introduction to human evolution.

Key Concept Important Information Practice It

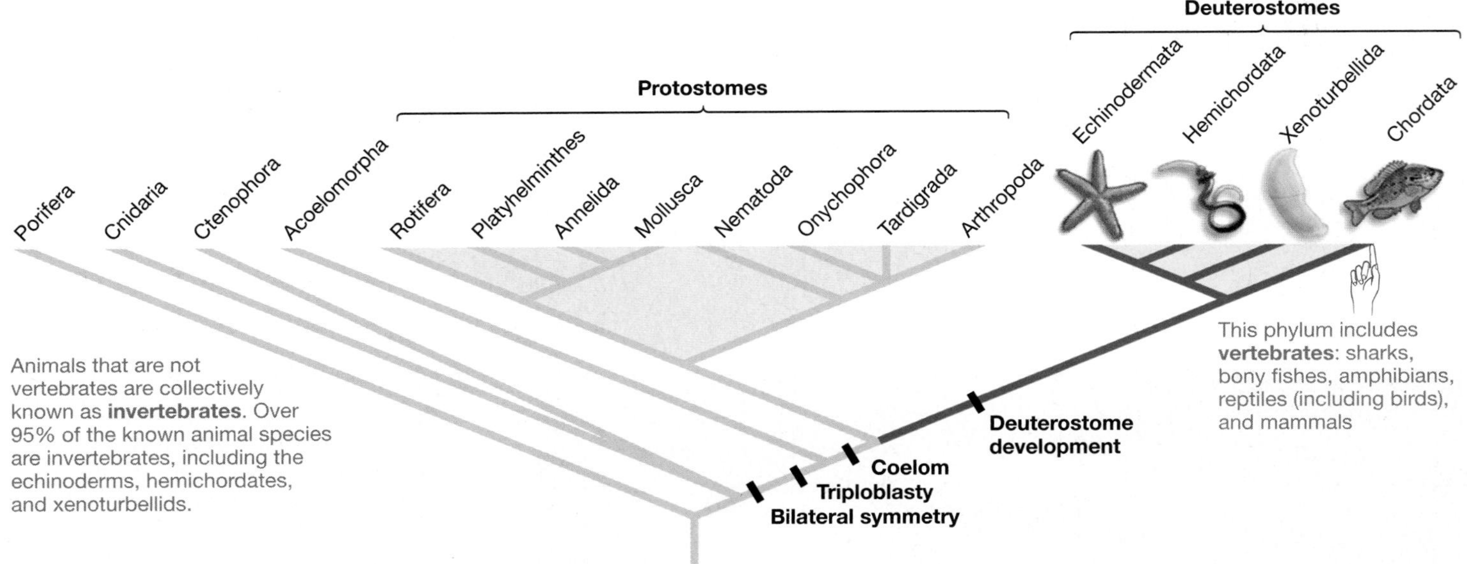

FIGURE 34.1 There Are Four Phyla of Deuterostomes. The deuterostomes comprise the phyla Echinodermata, Hemichordata, Xenoturbellida, and Chordata. The vertebrates are a subphylum of chordates.

● **QUESTION** Are invertebrates a monophyletic or a paraphyletic group?

34.1 An Overview of Deuterostome Evolution

Most biologists recognize just four phyla of deuterostomes: the Echinodermata, the Hemichordata, the Xenoturbellida, and the Chordata (**Figure 34.1**). The echinoderms include the sea stars and sea urchins. The hemichordates, or "acorn worms," are probably unfamiliar to you—they burrow in marine sands or muds and make their living by deposit feeding or suspension feeding. A lone genus with two wormlike species, called *Xenoturbella,* was recognized as a distinct phylum in 2006. The chordates include the vertebrates. The vertebrates, in turn, comprise the sharks, bony fishes, amphibians, reptiles (including birds), and mammals. Animals that are not vertebrates are collectively known as **invertebrates**. Echinoderms, hemichordates, and *Xenoturbella* are considered invertebrates, even though they are deuterostomes.

The deuterostomes were initially grouped together because they all undergo early embryonic development in a similar way. When a humpback whale, sea urchin, or human is just beginning to grow, cleavage is radial, the gut starts developing from posterior to anterior—with the anus forming first and the mouth second—and a coelom (if present) develops from outpocketings of mesoderm (see Chapter 32).

Although deuterostomes share important features of embryonic development, their adult body plans and their feeding methods, modes of locomotion, and means for reproduction are highly diverse. Let's explore who the deuterostome animals are and how they diversified, starting with the morphological traits that distinguish the echinoderms, chordates, and vertebrates.

What Is an Echinoderm?

All deuterostomes are considered bilaterians because they evolved from an ancestor that was bilaterally symmetric (see Chapter 32). But a remarkable event occurred early in the evolution of echinoderms: The origin of a unique type of radial symmetry. Adult **echinoderms** have bodies with five-sided radial symmetry, called pentaradial symmetry (**Figure 34.2a**), even though both their larvae and their ancestors are bilaterally symmetric (**Figure 34.2b**).

Recall from Chapter 32 that radially symmetric animals do not have well-developed head and posterior regions. As a result,

(a) Adult echinoderms are radially symmetric.

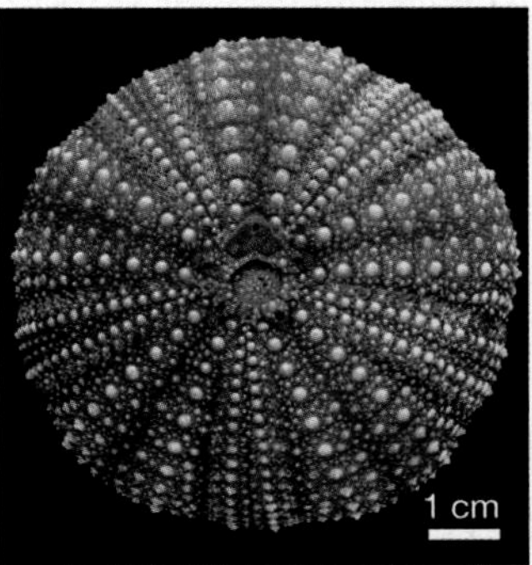

(b) Echinoderm larvae are bilaterally symmetric.

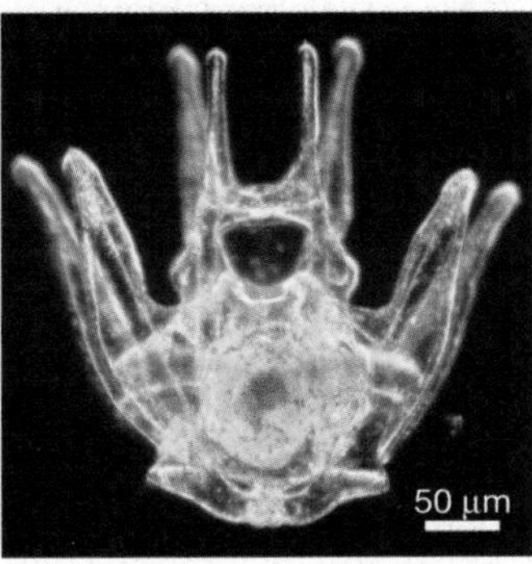

FIGURE 34.2 Body Symmetry Differs in Adult and Larval Echinoderms. **(a)** The skeleton of an adult and **(b)** the larva of sea urchin *Strongylocentrotus franciscanus.* Bilaterally symmetric larvae undergo metamorphosis and emerge as radially symmetric adults.

● **QUESTION** Why are echinoderms considered members of the lineage called Bilateria (bilaterally symmetric animals)?

(a) Echinoderms have a water vascular system.

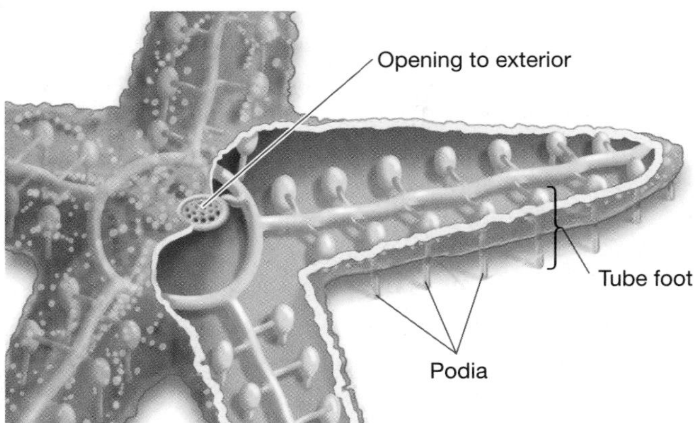

(b) Podia project from the underside of the body.

FIGURE 34.3 Echinoderms Have a Water Vascular System. (a) The water vascular system is a series of tubes and reservoirs that radiates throughout the body, forming a sophisticated hydrostatic skeleton. **(b)** Podia aid in movement because they extend from the body and can grab and release the substrate

EXERCISE Echinoderms have an endoskeleton just under their skin, which functions in protection. Label it in part (a).

they tend to interact with the environment in all directions at once instead of facing the environment in one direction. If adult echinoderms are capable of movement, they tend to move equally well in all directions instead of only headfirst.

The other remarkable event in echinoderm evolution was the origin of a unique morphological feature: a series of branching, fluid-filled tubes and chambers called the **water vascular system** (**Figure 34.3a**). One of the tubes is open to the exterior where it meets the body wall, so seawater can flow into and out of the system. Inside, fluids move via the beating of cilia that line the interior of the tubes and chambers. In effect, the water vascular system forms a sophisticated hydrostatic skeleton.

Figure 34.3a highlights a particularly important part of the system called tube feet. **Tube feet** are elongated, fluid-filled structures. **Podia** (literally, "feet") are sections of the tube feet that project outside the body (**Figure 34.3b**) and make contact with the substrate. As podia extend and contract in a coordinated way along the base of an echinoderm, they alternately grab and release from the substrate. As a result, the individual moves.

The other noteworthy feature of the echinoderm body is its **endoskeleton**, which is a hard, supportive structure located just inside a thin layer of epidermal tissue. (The structure in Figure 34.2a is an endoskeleton). As an individual is developing, cells secrete plates of calcium carbonate inside the skin. Depending on the species involved, the plates may remain independent and result in a flexible structure or fuse into a rigid case. Along with radial symmetry and the water vascular system, this type of endoskeleton is a synapomorphy—a trait that identifies echinoderms as a monophyletic group. If you understand this concept, you should be able to indicate the origin of pentaradial symmetry in adults, the water vascular system, and the echinoderm endoskeleton on Figure 34.1.

Two other phyla form a monophyletic group with echinoderms: Xenoturbellida and Hemichordata. The xenoturbellids ("strange-flatworms") have extremely simple, wormlike body plans. They don't have a gut, coelom, or brain, and they make their living absorbing nutrients in aquatic sediments. The hemichordates got their "half-chordates" name because they have an unusual feature found in chordates—openings into the throat called **pharyngeal gill slits**. The hemichordates are suspension feeders that live buried in muddy habitats on the ocean floor. As **Figure 34.4** shows, their pharyngeal gill slits function in feeding and gas exchange. Water enters the mouth, flows through structures where oxygen and food particles are extracted, and exits through the pharyngeal gill slits. They are not members of the phylum Chordata, however, because they lack several defining features shared by chordates. If you understand this point, you should be able to indicate two events on Figure 34.1: the origin of the pharyngeal gill slits observed in hemichordates and chordates, and the loss of pharyngeal gill slits in echinoderms and *Xenoturbella*.

What Is a Chordate?

The **chordates** are defined by the presence of four morphological features: (1) pharyngeal gill slits; (2) a stiff and supportive but flexible rod called a **notochord**, which runs the length of the body; (3) a bundle of nerve cells that runs the length of the body

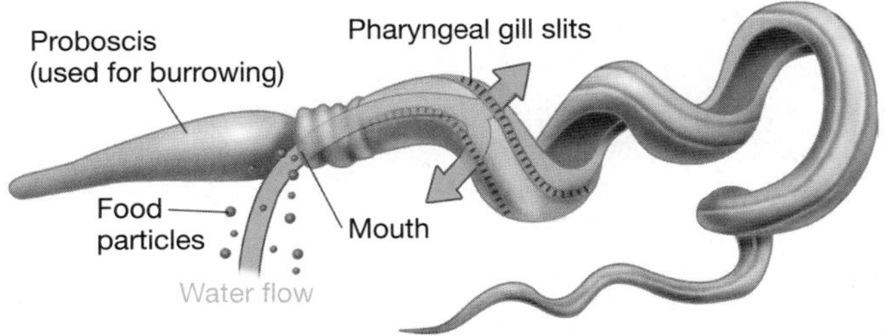

FIGURE 34.4 Hemichordates Suspension Feed Via Pharyngeal Gill Slits. The hemichordates, also known as acorn worms, burrow in marine muds and suspension feed as shown.

and forms a dorsal hollow **nerve cord**; and (4) a muscular, post-anal tail—meaning a tail that contains muscle and extends past the anus.

The phylum Chordata is made up of three major lineages, traditionally called subphyla: the (1) urochordates, (2) cephalochordates, and (3) vertebrates. All four defining characteristics of chordates—pharyngeal gill slits, notochords, dorsal hollow nerve cords, and tails—are found in these species.

- **Urochordates** are also called tunicates. As **Figure 34.5a** shows, pharyngeal gill slits are present in both larvae and adults and function in both feeding and gas exchange, much as they do in hemichordates. The notochord, dorsal hollow nerve cord, and tail are present only in the larvae, however.

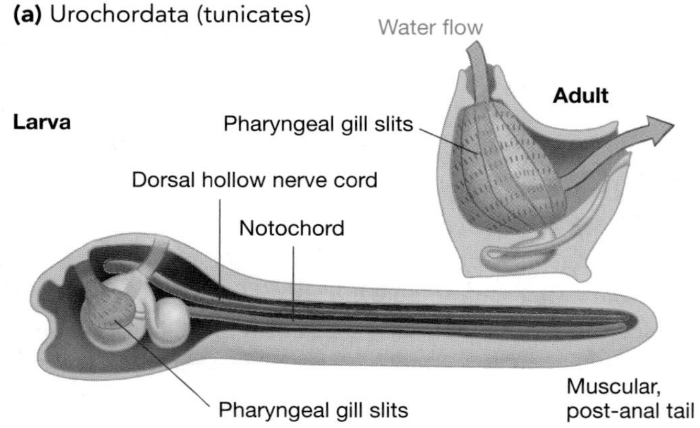

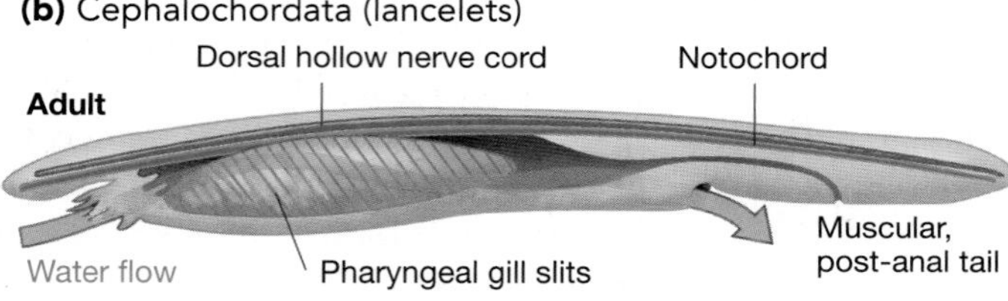

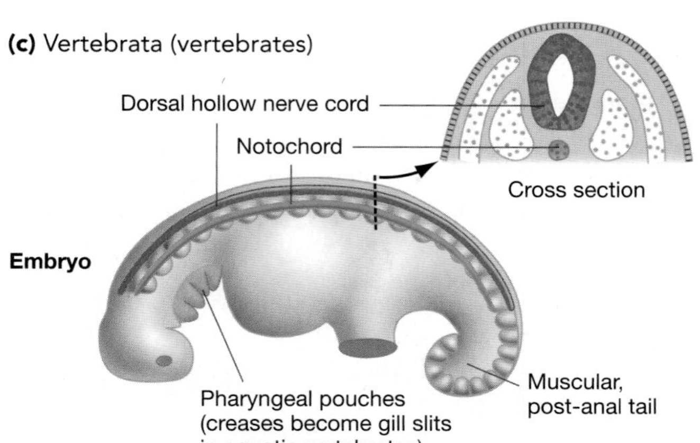

FIGURE 34.5 Four Traits Distinguish the Chordates. In chordates, either larvae or adults or both have notochords and muscular tails in addition to pharyngeal gill slits and dorsal hollow nerve cords. The three major chordate lineages are (a) tunicates, (b) lancelets, and (c) vertebrates.

QUESTION In tunicate larvae and in lancelets, what is the function of the notochord and tail?

Because the notochord stiffens the tail, muscular contractions on either side of a larva's tail wag it back and forth and result in swimming movements. As larvae swim or float in the upper water layers of the ocean, they drift to new habitats where food might be more abundant.

Urochordates underline one of the most fundamental conclusions about chordate evolution: The four features that distinguish the group enabled new types of feeding and movement. Pharyngeal gill slits function in suspension feeding in the adult. The notochord functions as a simple endoskeleton in the larva, while nerves organized into the dorsal cord stimulate muscles in the tail that make efficient swimming movements possible.

- **Cephalochordates** are also called lancelets or amphioxus; they are small, mobile suspension feeders that look a little like fish (**Figure 34.5b**). Adult cephalochordates live in ocean-bottom habitats, where they burrow in sand and suspension feed with the aid of their pharyngeal gill slits. The cephalochordates also have a notochord that stiffens their bodies, so that muscle contractions on either side result in fishlike movement when they swim during dispersal or mating.

- Vertebrates (**Figure 34.5c**) include the sharks, several lineages of fishes (**Box 34.1**, page 744), amphibians, reptiles (including birds), and mammals. In vertebrates, the dorsal hollow nerve cord is elaborated into the familiar spinal cord—a bundle of nerve cells that runs from the brain to the posterior of the body. Structures called pharyngeal pouches are present in all vertebrate embryos. In aquatic species, the creases between pouches open into gill slits and develop into part of the main gas-exchange organ—the **gills**. In terrestrial species, however, gill slits do not develop after the pharyngeal pouches form. A notochord also appears in all vertebrate embryos. Instead of functioning in body support and movement, though, it helps organize the body plan. You might recall from Chapter 22 that cells in the notochord secrete proteins that help induce the formation of **somites**, which are segmented blocks of tissue that form along the length of the body. Although the notochord itself disappears, cells in the somites later differentiate into the vertebrae, ribs, and skeletal muscles of the back, body wall, and limbs. In this way, the notochord is instrumental in the development of the trait that gave vertebrates their name.

What Is a Vertebrate?

The vertebrates are a monophyletic group distinguished by two traits: a column of cartilaginous or bony structures, called **vertebrae**, that form along the dorsal sides of most species, and a **cranium**, or skull—a bony, cartilaginous, or fibrous case that encloses the brain. The vertebral column is important because it protects the spinal cord. The cranium is important because it protects the brain along with sensory organs such as eyes.

The vertebrate brain develops as an outgrowth of the most anterior end of the dorsal hollow nerve cord and is important to the vertebrate lifestyle. Vertebrates are active predators and herbivores that can make rapid, directed movements with the aid of their endoskeleton. Coordinated movements are possible in part because vertebrates have large brains divided into three distinct regions: (1) a **forebrain**, housing the sense of smell; (2) a **midbrain**, associated with vision; and (3) a **hindbrain**, responsible for balance and hearing. In vertebrate groups that evolved more recently, the forebrain is a large structure called the **cerebrum**, and the hindbrain consists of enlarged regions called the **cerebellum** and **medulla oblongata**. (The structure and functions of the vertebrate brain are analyzed in detail in Chapter 45.) The evolution of a large brain, protected by a hard cranium, was a key innovation in vertebrate evolution.

Check Your Understanding

If you understand that...

- The deuterostomes include four phyla: the echinoderms, hemichordates, xenoturbellids, and chordates.
- Within the chordates, the vertebrates are the most species-rich and morphologically diverse subphylum.

You should be able to...

1) Draw a simple phylogenetic tree showing the relationships among the echinoderm, hemichordate, chordate, and vertebrate lineages.
2) Identify the key innovations that distinguish each of these four groups.
3) Explain why each of these innovations is important in feeding or movement.

34.2 An Overview of Vertebrate Evolution

Vertebrates have been the focus of intense research for well over 300 years, in part because they are large and conspicuous and in part because they include the humans. All this effort has paid off in an increasingly thorough understanding of how the vertebrates diversified. Let's consider what data from the fossil record have to say about key events in vertebrate evolution, then examine the current best estimate of the phylogenetic relationships among vertebrates.

The Vertebrate Fossil Record

Both echinoderms and vertebrates are present in the Burgess Shale deposits that formed during the Cambrian explosion 542–488 million years ago (see also Chapter 27). The vertebrate fossils in these rocks show that the earliest members of this lineage lived in the ocean about 540 million years ago, had streamlined, fishlike bodies, and appear to have had a skull made of cartilage. **Cartilage** is a stiff tissue that consists of scattered cells in a gel-like matrix of polysaccharides and protein fibers. Several groups of early vertebrates had endoskeletons made of cartilage, as do the sharks and rays living today.

Following the appearance of vertebrates, the fossil record documents a series of key innovations that occurred as this lineage diversified (**Figure 34.6**):

- Fossil vertebrates from the early part of the Silurian period, about 440 million years ago, are the first fossils to have bone. **Bone** is a tissue consisting of cells and blood vessels encased in a matrix made primarily of a calcium- and phosphate-containing compound called hydroxyapatite, along with a small amount of protein fibers. When bone first evolved, it did not occur in the endoskeleton.

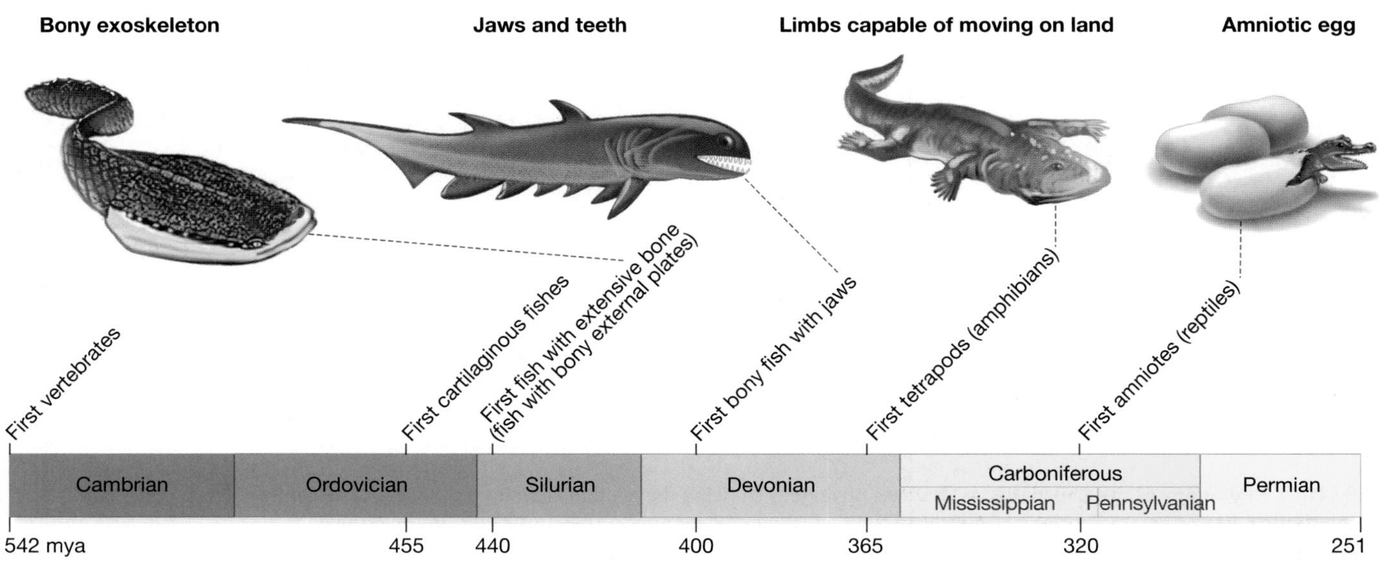

FIGURE 34.6 Timeline of the Vertebrate Fossil Record.

Instead, bone was deposited in scalelike plates that formed an **exoskeleton**—a hard, hollow structure that envelops the body. Based on the fossils' overall morphology, biologists infer that these animals swam with the aid of a notochord and that they breathed and fed by gulping water and filtering it through their pharyngeal gill slits. Presumably, the bony plates helped provide protection from predators.

- The first bony fish with jaws show up in the fossil record about 400 million years ago. The evolution of jaws was significant because it gave vertebrates the ability to bite, meaning that they were no longer limited to suspension feeding or deposit feeding. Instead, they could make a living as herbivores or predators. Soon after, jawbones with teeth appear in the fossil record. With jaws and teeth, vertebrates became armed and dangerous. The fossil record shows that a spectacular radiation of jawed fishes followed, filling marine and freshwater habitats.
- The next great event in the evolution of vertebrates was the transition to living on land. The first animals that had limbs and were capable of moving on land are dated to about 365 million years ago. These were the first of the **tetrapods** ("four-footed")—animals with four limbs (Figure 34.6).
- About 20 million years after the appearance of tetrapods in the fossil record, the first amniotes are present. The Amniota is a lineage of vertebrates that includes all tetrapods other than amphibians. The **amniotes** are named for a signature adaptation: the amniotic egg. An **amniotic egg** is an egg that

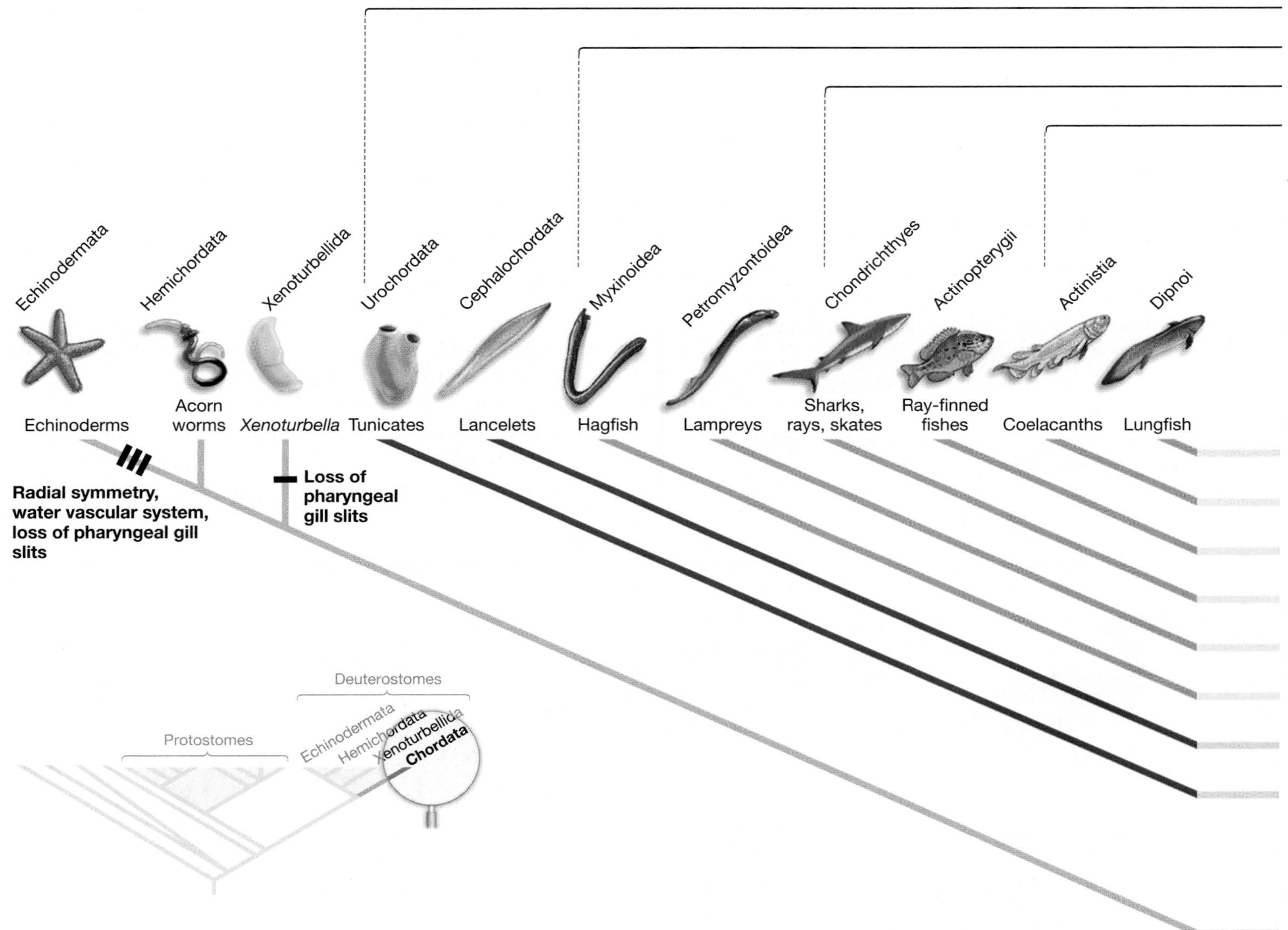

FIGURE 34.7 A Phylogeny of the Chordates. A chordate phylogeny showing the relationships among the major groups within the subphylum vertebrata. The short bars indicate where major innovations occurred as the lineages diversified.

● **QUESTION** Which group is more closely related to the amphibians: mammals or birds?

has a watertight shell or case enclosing a membrane-bound food supply, water supply, and waste repository. The evolution of the amniotic egg was significant because it gave vertebrates the ability to reproduce far from water. Amniotic eggs resist drying out, so vertebrates that produce amniotic eggs do not have to return to aquatic habitats to lay their eggs.

To summarize, the fossil record indicates that vertebrates evolved through a series of major steps, beginning about 540 million years ago with vertebrates whose endoskeleton consisted of a notochord. The earliest vertebrates gave rise to cartilaginous fishes (sharks and rays) and fish with bony skeletons and jaws. After the tetrapods emerged and amphibians resembling salamanders began to live on land, the evolution of the amniotic egg paved the way for the evolution of the first truly terrestrial vertebrates. The fossil record indicates that a radiation of reptiles followed, along with the animals that gave rise to mammals. Do phylogenetic trees estimated from analyses of DNA sequence data agree or conflict with these conclusions?

Evaluating Molecular Phylogenies

Figure 34.7 provides a phylogenetic tree that summarizes the relationships among vertebrates, based on DNA sequence data. The labeled bars on the tree indicate where major innovations occurred. Although the phylogeny of deuterostomes continues to be a topic of intense research, researchers are increasingly confident that the relationships described in Figure 34.7 are accurate. Box 34.1 discusses some of the implications of this tree, in terms of how lineages on the tree of life are named.

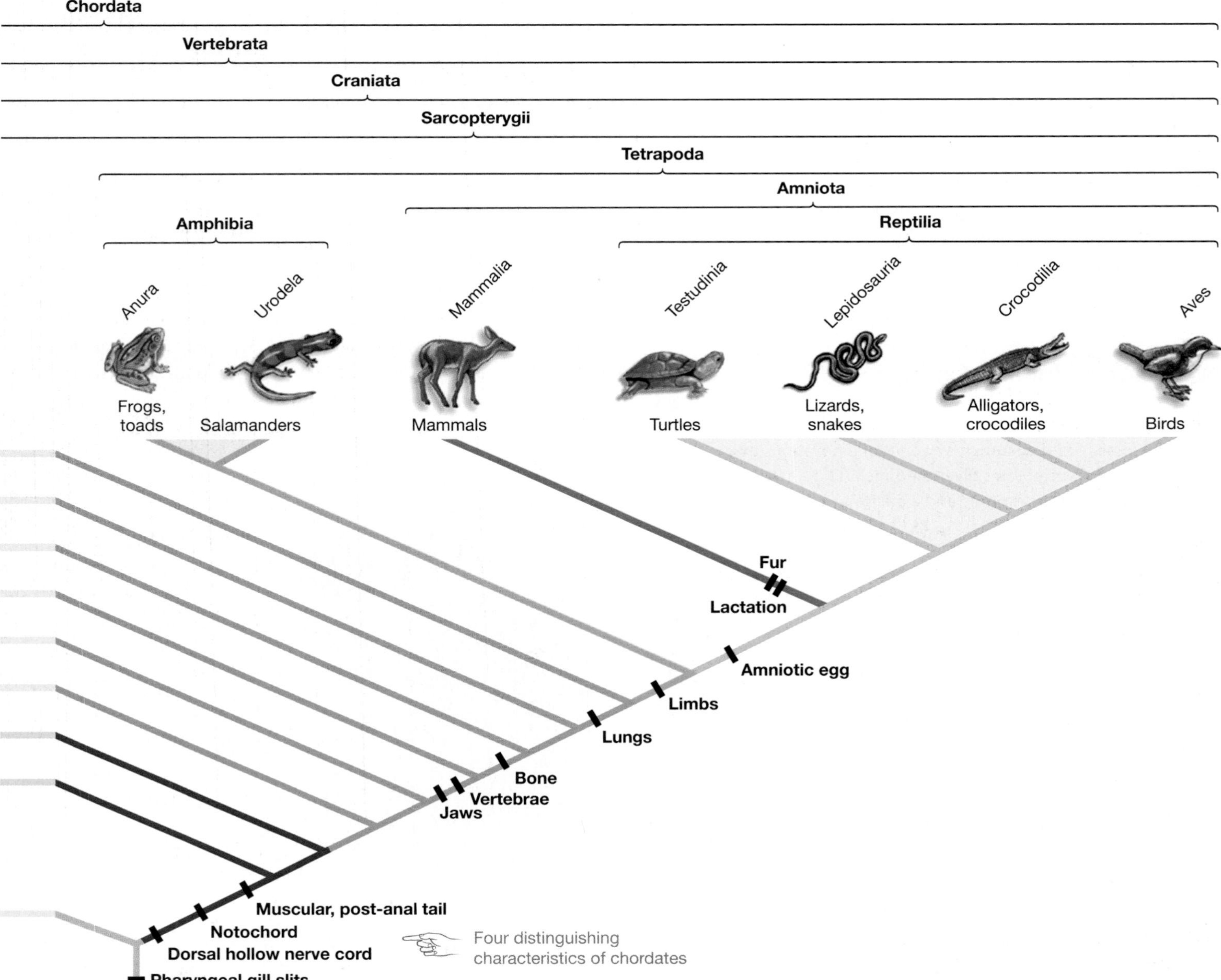

BOX 34.1 What Is a Fish?

In everyday English, people use the word *fish* to mean just about any vertebrate that lives in water and has a streamlined body and fins. But to a biologist, saying that an organism is a fish prompts a question: Which fish?

As the blue branches on Figure 34.7 show, there is no monophyletic group that includes all of the fishlike lineages. Instead, "fishy" organisms are a series of independent monophyletic groups. Taken together, they form what biologists call a **grade**—a sequence of lineages that are paraphyletic. You might recall from Chapter 30 that the seedless vascular plants also form a grade and are paraphyletic.

In naming schemes that recognize only monophyletic groups, there is no such thing as a fish. Instead, there are jawless fishes, cartilaginous fishes (sharks and rays), lobe-finned fishes, lungfishes, and ray-finned fishes. Each of the monophyletic fish groups will be explored in more detail in Section 34.5. ● If you understand the difference between a paraphyletic and monophyletic group, you should be able to draw what Figure 34.7 would look like if there actually were a monophyletic group called "The Fish."

The overall conclusion from this tree is that the branching sequence inferred from morphological and molecular data correlates with the sequence of groups in the fossil record. Reading up from the base of the tree, it is clear that the closest living relatives of the vertebrates are the cephalochordates. The most basal groups of chordates lack the skull and vertebral column that define the vertebrates, and the most ancient lineages of vertebrates lack jaws and bony skeletons. ● You should be able to indicate the origin of the cranium on Figure 34.7.

To understand what happened during the subsequent diversification of echinoderms and vertebrates, let's explore some of the major innovations involved in feeding, movement, and reproduction in more detail.

Check Your Understanding

If you understand that...

- The fossil record documents a series of innovations that occurred as vertebrates diversified, including a cartilaginous skeleton, bone, jaws, limbs, and a membrane-bound egg.

You should be able to...

1) Explain how each of these innovations allowed vertebrates to move, find food, or reproduce efficiently.
2) Indicate the origin of each of these key traits on a phylogenetic tree of vertebrates.

(MB) Web Animation at www.masteringbio.com
Deuterostome Diversity

34.3 What Themes Occur in the Diversification of Deuterostomes?

● In terms of numbers of species and range of habitats occupied, the most successful lineages of deuterostomes are the echinoderms, with 7000 named species, and the vertebrates, with 54,000 known species. Echinoderms are widespread and abundant in marine habitats. In some deepwater environments, echinoderms represent 95 percent of the total mass of organisms. Among vertebrates, the most species-rich and ecologically diverse lineages are the ray-finned fishes and the tetrapods (**Figure 34.8**). Ray-finned fishes occupy habitats ranging from deepwater environments, which are perpetually dark, to shallow ponds that dry up each year. Tetrapods include the large herbivores and predators in terrestrial environments all over the world.

Today there are about 7000 species of echinoderms, about 27,000 species of ray-finned fish, and about 27,000 species of tetrapods. To understand why these lineages have been so successful, it's important to recognize that they have unique body plans. Recall that echinoderms are radially symmetric and have a water vascular system, and that ray-finned fishes and tetrapods have a bony endoskeleton. In this light, the situation in deuterostomes appears to be similar to that in protostomes.

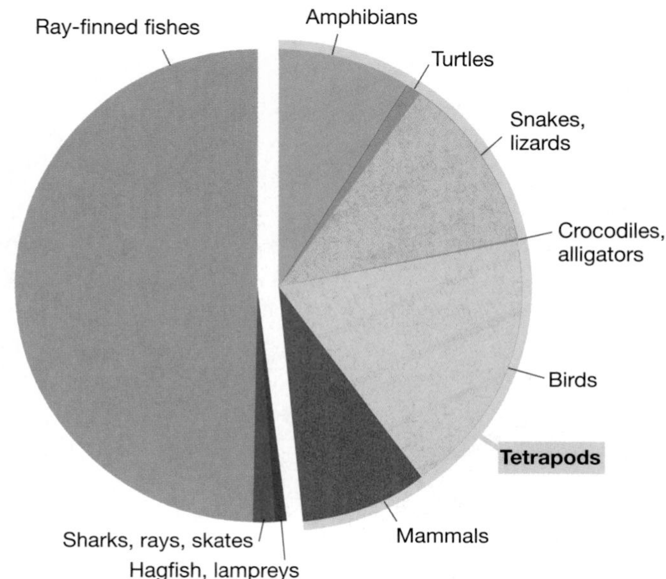

FIGURE 34.8 Relative Species Abundance Among Vertebrates.

Recall from Chapter 33 that the most evolutionarily successful protostome lineages are the arthropods and molluscs. A major concept in that chapter was that arthropods and molluscs have body plans that are unique among protostomes. Once their distinctive body plans evolved, evolution by natural selection led to extensive diversification based on novel methods for eating, moving, and reproducing.

Feeding

Animals eat to live, and it is logical to predict that each of the tens of thousands of echinoderm and vertebrate species eat different things in different ways. It is also logical to predict that echinoderms and vertebrates have traits that make diverse ways of feeding possible.

Both predictions are correct. Echinoderms have feeding strategies that are unique among marine animals. Many are based on the use of their water vascular system and podia. Ray-finned fishes and tetrapods, in contrast, depend on their jaws.

How Do Echinoderms Feed? Depending on the lineage and species in question, echinoderms make their living by suspension feeding, deposit feeding, or harvesting algae or other animals. In most cases, an echinoderm's podia play a key role in obtaining food. Many sea stars, for example, prey on bivalves. Clams and mussels respond to sea star attacks by contracting muscles that close their shells tight. But by clamping onto each shell with their podia and pulling, sea stars are often able to pry the shells apart a few millimeters (**Figure 34.9a**). Once a gap exists, the sea star extrudes its stomach from its body and forces the stomach through the opening between the bivalve's shells. Upon contacting the visceral mass of the bivalve, the stomach of the sea star secretes digestive enzymes. It then begins to absorb the small molecules released by enzyme action. Eventually, only the unhinged shells of the prey remain.

Podia are also involved in echinoderms that suspension feed (**Figure 34.9b**). In most cases, podia are extended out into the water. When food particles contact them, the podia flick the food down to cilia, which sweep the particles toward the animal's mouth. In deposit feeders, podia secrete mucus that is used to sop up food material on the substrate. The podia then roll the food-laden mucus into a ball and move it to the mouth.

The Vertebrate Jaw The most ancient groups of vertebrates have relatively simple mouthparts. For example, hagfish and lampreys lack jaws and cannot bite algae, plants, or animals. They have to make their living as ectoparasites or as deposit feeders that scavenge dead animals.

Vertebrates were not able to harvest food by biting until jaws evolved. The leading hypothesis for the origin of the jaw proposes that natural selection acted on mutations that affected the morphology of **gill arches**, which are curved regions of tissue between the gills. The jawless vertebrates have bars of cartilage that stiffen these gill arches. The gill-arch hypothesis proposes that mutation and natural selection increased the size of an arch and modified its orientation slightly, producing the first working jaw (**Figure 34.10**). Three lines of evidence, drawn from comparative anatomy and embryology, support the gill-arch hypothesis:

1. Both gill arches and jaws consist of flattened bars of bony or cartilaginous tissue that hinges and bends forward.
2. During development, the same population of cells gives rise to the muscles that move jaws and the muscles that move gill arches.
3. Unlike most other parts of the vertebrate skeleton, both jaws and gill arches are derived from specialized cells in the embryo called neural crest cells.

(a) Podia adhere to bivalve shells and pull them apart.

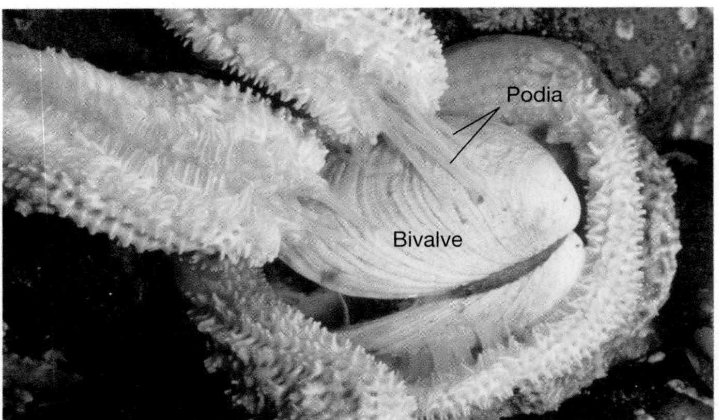

(b) Podia trap particles during suspension feeding.

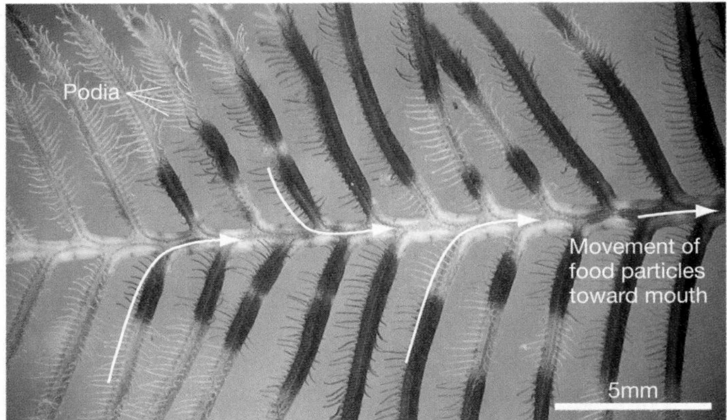

FIGURE 34.9 Echinoderms Use Their Podia in Feeding. (a) A sea star pries open a bivalve slightly and then everts its stomach into the bivalve to digest it. **(b)** Feather stars extend their podia when they suspension feed.

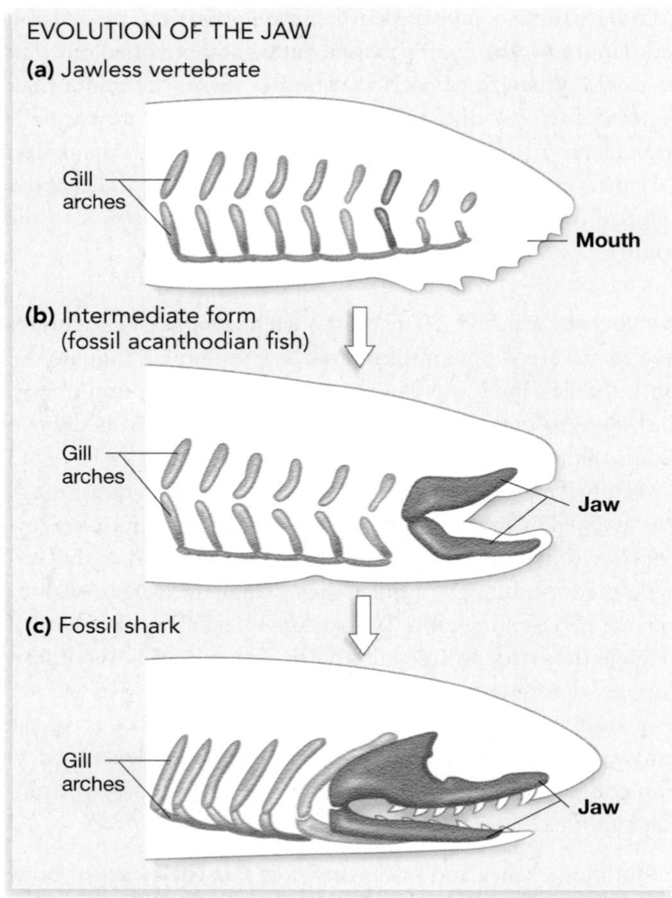

FIGURE 34.10 A Hypothesis for the Evolution of the Jaw. (a) Gill arches support the gills in jawless vertebrates. **(b)** In the fossil record, jawbones appear first in fossil fish called acanthodians. **(c)** Fossil sharks that appeared later had more elaborate jaws.

QUESTION The transition from gill arches to the jaws of acanthodian fishes is complex. Intermediate forms have yet to be found in the fossil record. Would intermediate stages in the evolution of the jaw have any function?

Taken together, these data support the hypothesis that jaws evolved from gill arches.

To explain why ray-finned fishes are so diverse in their feeding methods, biologists point to important modifications of the jaw. In most ray-finned fishes, for example, the jaw is protrusible—meaning it can be extended to nip or bite out at food. In addition, several particularly species-rich lineages of ray-finned fishes have a set of pharyngeal jaws. The **pharyngeal** ("throat") **jaw** consists of modified gill arches that function as a second set of jaws, located in the back of the mouth. Pharyngeal jaws are important because they make food processing particularly efficient. (For more on the structure and function of pharyngeal jaws, see Chapter 43.)

To summarize, the radiation of ray-finned fishes was triggered in large part by the evolution of the jaw, by modifications that made it possible to protrude the jaw, and by the origin of the pharyngeal jaw. The story of tetrapods is different, however. Although jaw structure varies somewhat among tetrapod groups, the adaptation that triggered their initial diversification involved the ability to get to food, not to bite it and process it.

Movement

The signature adaptations of echinoderms and tetrapods involve locomotion. We've already explored the water vascular system and tube feet of echinoderms; here let's focus on the tetrapod limb.

Most tetrapods live on land and use their limbs to move. But for vertebrates to succeed on land, they had to be able not only to move out of water but also to breathe air and avoid drying out. To understand how this was accomplished, consider the morphology and behavior of their closest living relatives, the lungfish (**Figure 34.11**). Most living species of lungfish inhabit shallow, oxygen-poor water. To supplement the oxygen taken in by their gills, they have lungs and breathe air. Some also have fleshy fins supported by bones and are capable of walking short distances along watery mudflats or the bottoms of ponds. In addition, some species can survive extended droughts by burrowing in mud.

Fossils provide strong links between lungfish and the earliest land-dwelling vertebrates. **Figure 34.12** shows three of the species involved. The first, an aquatic animal related to today's lungfishes, is from the Devonian period—about 382 million years ago. The second is one of the oldest tetrapods, or limbed vertebrates, found to date. This animal appears in the fossil record about 365 million years ago. The third is a more recent tetrapod fossil, dated to about 350 million years ago. The figure highlights the numbers and arrangement of bones in the fossil fish fin and the numbers and arrangement of bones in the limbs of early tetrapods. The color coding emphasizes that each fin or limb has a single bony element that is proximal (closest to the body) and then two bones that are

FIGURE 34.11 Lungfish Have Limb-like Fins. Some species of lungfish can walk or crawl short distances on their fleshy fins.

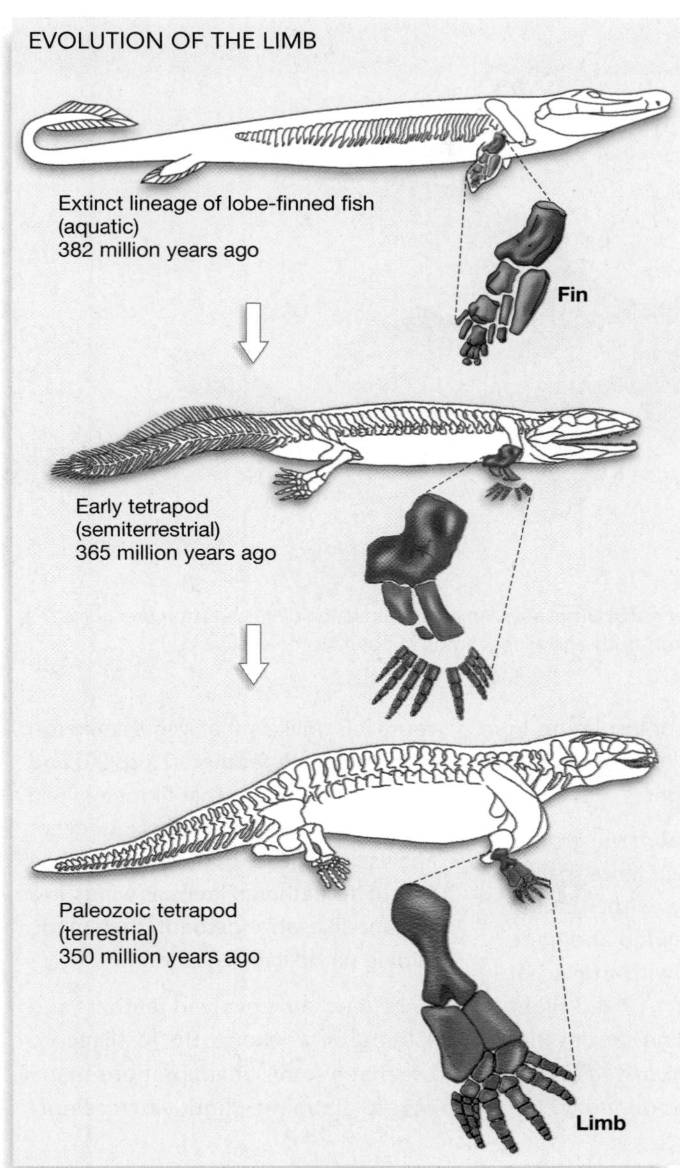

FIGURE 34.12 Fossil Evidence for a Fin-to-Limb Transition. The number and arrangement of bones in the fins and limbs of these fossil organisms agree with the general form of the modern tetrapod limb. The color coding indicates homologous elements.

distal (farther from the body) and arranged side by side, followed by a series of distal elements. Because the structures are similar, and because no other animal groups have limb bones in this arrangement, the evidence for homology is strong. Based on the lifestyle of living lungfish, biologists suggest that mutation and natural selection gradually transformed fins into limbs as the first tetrapods became more and more dependent on terrestrial habitats.

The hypothesis that tetrapod limbs evolved from fish fins has also been supported by molecular genetic evidence. Recent work has shown that several regulatory proteins involved in the development of zebrafish fins and the upper parts of mouse limbs are homologous. Specifically, the proteins produced by several different *Hox* genes are found at the same times and in the same locations in fins and limbs. These data suggest that these appendages are patterned by the same genes. As a result, the data support the hypothesis that tetrapod limbs evolved from fins.

Once the tetrapod limb evolved, natural selection elaborated it into structures that are used for running, gliding, crawling, burrowing, or swimming. In addition, wings and the ability to fly evolved independently in three lineages of tetrapods: the extinct flying reptiles called pterosaurs (pronounced *TARE-oh-sors*), birds, and bats. Tetrapods and insects are the only animals that have taken to the skies. **Box 34.2** explores how flight evolved in birds.

● To summarize, the evolution of the jaw gave tetrapods the potential to capture and process a wide array of foods. With limbs, they could move efficiently on land in search of food. What about the other major challenge of terrestrial life? How did tetrapods produce offspring that could survive out of water?

Reproduction

Among the various lineages of fish, a few species undergo only asexual reproduction. When sexual reproduction occurs, it may be based on internal or external fertilization and oviparity, viviparity, or ovoviviparity. In addition, parental care is extensive in some fish species, and it often involves guarding eggs from predators and fanning them to keep oxygen levels high. All fish lay their eggs or give birth in water, however. Tetrapods were the first vertebrates that were able to breed on land.

Three major evolutionary innovations gave tetrapods the ability to produce offspring successfully in terrestrial environments: (1) the amniotic egg, (2) the placenta, and (3) elaboration of parental care. Let's explore each of these innovations in turn.

What Is an Amniotic Egg? Amniotic eggs have shells that minimize water loss as the embryo develops inside. The first tetrapods, like today's amphibians (frogs, toads, and salamanders), lacked amniotic eggs. Although their eggs were encased by a membrane, the first tetrapods laid eggs that would dry out and die unless they were laid in water. This fact limited the range of habitats these animals could exploit. Like today's amphibians, the early tetrapods were largely restricted to living in or near marshes, lakes, or ponds.

In contrast, reptiles (including birds) and the egg-laying mammals produce amniotic eggs. The outer membrane of the amniotic egg is leathery in turtles, snakes, and lizards but hard—due to deposition of the calcium carbonate—in crocodiles, birds, and

BOX 34.2 The Evolution of Flight in Birds

In 2003 Xing Xu and colleagues announced the discovery of a spectacular fossilized dinosaur called *Microraptor gui.* As **Figure 34.13** shows, *M. gui* had feathers covering its wings, body, and legs. But as impressive as it is, *M. gui* was just one in an exciting series of feathered dinosaur species unearthed by Xu's group. Taken together, the newly discovered species answer several key questions about the evolution of birds, feathers, and flight:

FIGURE 34.13 Feathers Evolved in Dinosaurs. An artist's depiction of what *Microraptor gui,* a dinosaur that had feathers on its body and all four limbs, might have looked like in life.

- *Did birds evolve from dinosaurs?* On the basis of skeletal characteristics, all of these recently discovered fossil species clearly belong to a lineage of dinosaurs called the dromaeosaurs. The fossils definitively link the dromaeosaurs and the earliest known fossil birds.
- *How did feathers evolve?* The fossils support Xu's model that feathers evolved in a series of steps, beginning with simple projections from the skin and culminating with the complex structures observed in today's birds (**Figure 34.14**). *M. gui,* for example, had two distinct types of feathers but lacked the complex feathers found in contemporary birds. It is controversial, though, whether the original function of feathers was for courtship, other types of display, or insulation. In today's birds, feathers function in display, insulation, and flight.
- *Did birds begin flying from the ground up or from the trees down?* More specifically, did flight evolve with running species that began to jump and glide or make short flights, with the aid of feathers to provide lift? Or did flight evolve from tree-dwelling species that used feathers to glide from tree to tree, much as flying squirrels do today? Because it is unlikely that *M. gui* could run efficiently with feathered legs, Xu and colleagues propose that flight evolved from tree-dwelling gliders. A recent analysis suggests that *M. gui* used all four of its feathered limbs as wings, in a biplane-like arrangement that made gliding particularly efficient.

Once dinosaurs evolved feathers and took to the air as gliders, the fossil record shows that a series of adaptations made powered, flapping flight increasingly

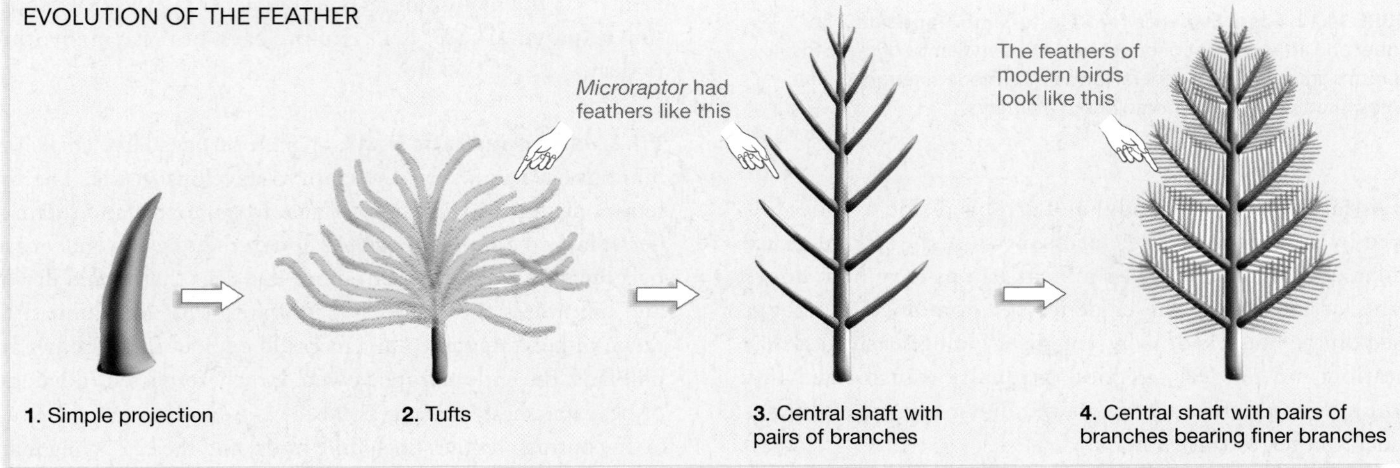

FIGURE 34.14 Feathers Evolved through Intermediate Stages. This model for the evolution of feathers is supported by the fossil record.

QUESTION Suggest a function for simple projections and tufts in the ancestors of birds, prior to the evolution of more complex feathers that made gliding or flapping flight possible.

efficient. Although most dinosaurs have a flat sternum (breastbone), the same structure in birds has an elongated projection called the keel, which provides a large surface area to which flight muscles attach (**Figure 34.15**; note that on a chicken or turkey, the flight muscles are called "breast meat"). Birds are also extraordinarily light for their size, primarily because they have a drastically reduced number of bones and because their larger bones are thin-walled and hollow—though strengthened by bony "struts" (see Figure 34.15). Birds are also capable of long periods of sustained activity year-round, because they are **endotherm**—meaning that they maintain a high body temperature by producing heat in their tissues. From dinosaurs that jumped and glided from tree to tree, birds have evolved into extraordinary flying machines.

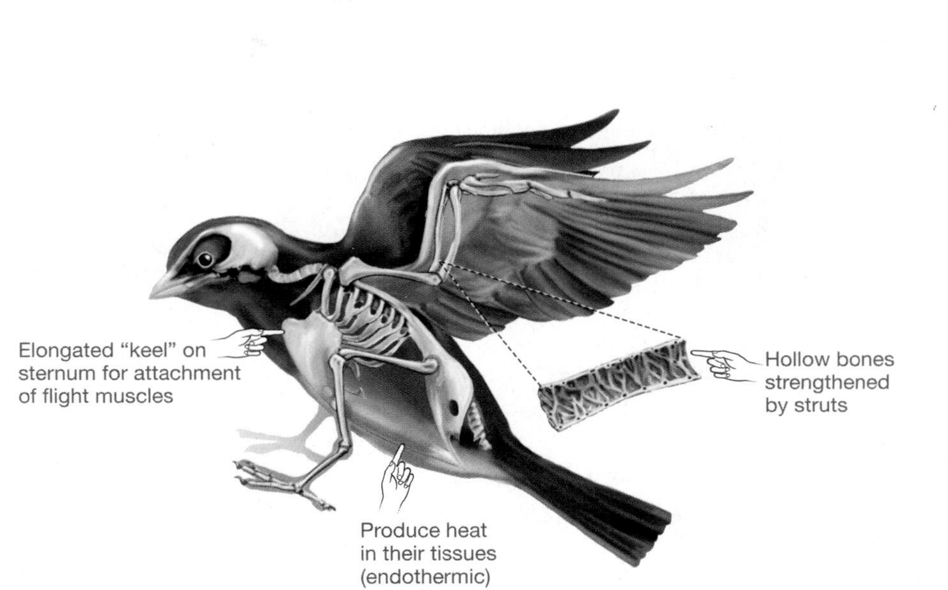

FIGURE 34.15 In Addition to Feathers, Birds Have Several Adaptations That Allow for Flight.

the egg-laying mammals. Besides having an outer shell or membrane that is largely watertight, an amniotic egg contains a membrane-bound supply of water in a protein-rich solution called **albumen** (**Figure 34.16**). The embryo itself is enveloped in a protective inner membrane known as the **amnion**. Inside an amniotic egg, the embryo is bathed in fluid. The egg itself is highly resistant to drying.

The evolution of the amniotic egg was a key event in the diversification of tetrapods because it allowed turtles, snakes, lizards, crocodiles, birds, and the egg-laying mammals to reproduce in any terrestrial environment—even habitats as dry as deserts. Members of the lineage called Amniota now occupy all types of terrestrial environments. But during the evolution of mammals, a second major innovation in reproduction occurred that eliminated the need for any type of egg laying: the placenta.

The Placenta Recall from Chapter 32 that egg-laying animals are said to be **oviparous**, while species that give birth are termed **viviparous**. In many viviparous animals, females produce an egg that contains a nutrient-rich yolk. Instead of laying the egg, however, the mother retains it inside her body. In these **ovoviviparous** species, the developing offspring depends on the resources in the egg yolk. In most mammals, however, the eggs that females produce lack yolk. After fertilization occurs and the egg is retained, the mother produces a placenta within her uterus. The **placenta** is an organ that is rich in blood vessels and that facilitates a flow of oxygen and

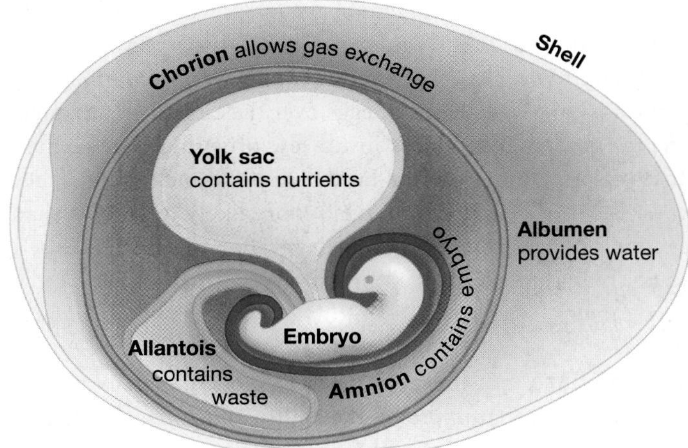

FIGURE 34.16 An Amniotic Egg. Amniotic eggs have membrane-bound sacs that hold nutrients, water, and waste and that allow gas exchange.

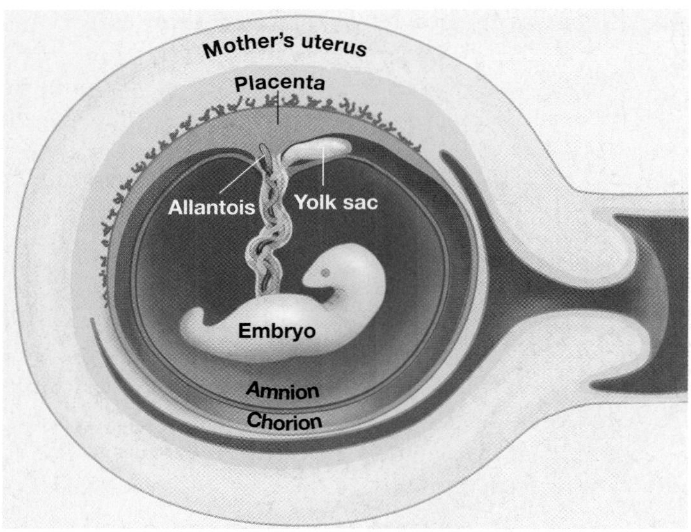

FIGURE 34.17 The Placenta Allows the Mother to Nourish the Fetus Internally.

QUESTION Compare the relative positions of the chorion and amnion here with those in the amniotic egg (Figure 34.16). Are they the same or different?

nutrients from the mother to the developing offspring (**Figure 34.17**). After a development period called **gestation**, the embryo emerges from the mother's body.

Why did viviparity and the placenta evolve? Biologists have formulated an answer to this question by pointing out that females have a finite amount of time and energy available to invest in reproduction. As a result, a female can produce a large number of small offspring or a small number of large offspring but not a large number of large offspring. Stated more formally, every female faces a **trade-off**—an inescapable compromise—between the quantity of offspring she can produce and their size. In some lineages, natural selection has favored traits that allow females to produce a small number of large, well-developed offspring. Viviparity and the placenta are two such traits. Compared with female insects or echinoderms, which routinely lay thousands or even millions of eggs over the course of a lifetime, a female mammal produces just a few offspring. But because those offspring are protected inside her body and fed until they are well developed, they are much more likely to survive than sea star or insect embryos are. Even after the birth of their young, many mammals continue to invest time and energy in rearing them.

Parental Care The term **parental care** encompasses any action by a parent that improves the ability of its offspring to survive, including incubating eggs to keep them warm during early development, keeping young warm and dry, supplying young with food, and protecting them from danger. In some insect and frog species, mothers carry around eggs or newly hatched young; in fishes, parents commonly guard eggs during development and fan them with oxygen-rich water.

The most extensive parental care observed among animals is provided by mammals and birds. In both groups, the mother and often the father continue to feed and care for individuals after birth or hatching—sometimes for many years (**Figure 34.18**). Female mammals also **lactate**—meaning that they produce a nutrient-rich fluid called milk and use it to feed their offspring after birth. With the combination of the placenta and lactation, placental mammals make the most extensive investment of time and energy in offspring known. Among large animals, the evolution of extensive parental care is hypothesized to be a major reason for the evolutionary success of mammals and birds.

(a) Mammal mothers feed and protect newborn young.

(b) Many bird species have extensive parental care.

FIGURE 34.18 Parental Care in Mammals and Birds.
(a) Female mammals feed and protect embryos inside their bodies until the young are well developed. Once the offspring is born, the mother feeds it milk until it is able to eat on its own. In some species, parents continue to feed and protect young for many years. **(b)** In birds, one or both parents may incubate the eggs, protect the nest, and feed the young after hatching occurs.

Check Your Understanding

If you understand that...

- Echinoderms and vertebrates have distinctive body plans. Subsequent diversification in each lineage was based on innovations that made it possible for species to feed, move, and reproduce in novel ways. Most echinoderms use their podia to move, but they feed in a wide variety of ways—including using their podia to pry open bivalves, suspension feed, or deposit feed.
- An array of key innovations occurred during the evolution of vertebrates: Jaws made it possible to bite and process food, limbs allowed tetrapods to move on land, and amniotic eggs could be laid on land.

You should be able to...

1) Summarize the leading hypotheses to explain how the jaw and limb evolved.

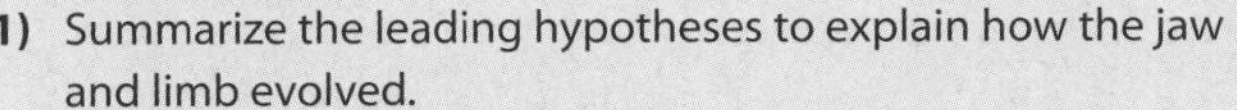

2) Diagram the structure of an amniotic egg.
3) Explain the role of increased parental care in the evolution of birds and mammals.

34.4 Key Lineages: Echinodermata

The echinoderms ("spiny-skins") were named for the spines or spikes observed in many species. They are bilaterally symmetric as larvae but undergo metamorphosis and develop into radially symmetric adults. As adults they all have a water vascular system and produce calcium carbonate plates in their skin to form an endoskeleton.

The echinoderms living today make up five major lineages, traditionally recognized as classes: (1) feather stars and sea lilies, (2) sea stars, (3) brittle stars and basket stars, (4) sea urchins and sand dollars, and (5) sea cucumbers (**Figure 34.19**). Most feather stars and sea lilies are sessile suspension feeders. Brittle stars and basket stars have five or more long arms that radiate out from a small central disk. They use these arms to suspension feed, deposit feed by sopping up material with mucus, or capture small prey animals. Sea cucumbers are sausage-shaped animals that suspension feed or deposit feed with the aid of modified tube feet called tentacles that are arranged in a whorl around their mouths. Sea stars, sea urchins, and sand dollars are described in detail in the text that follows.

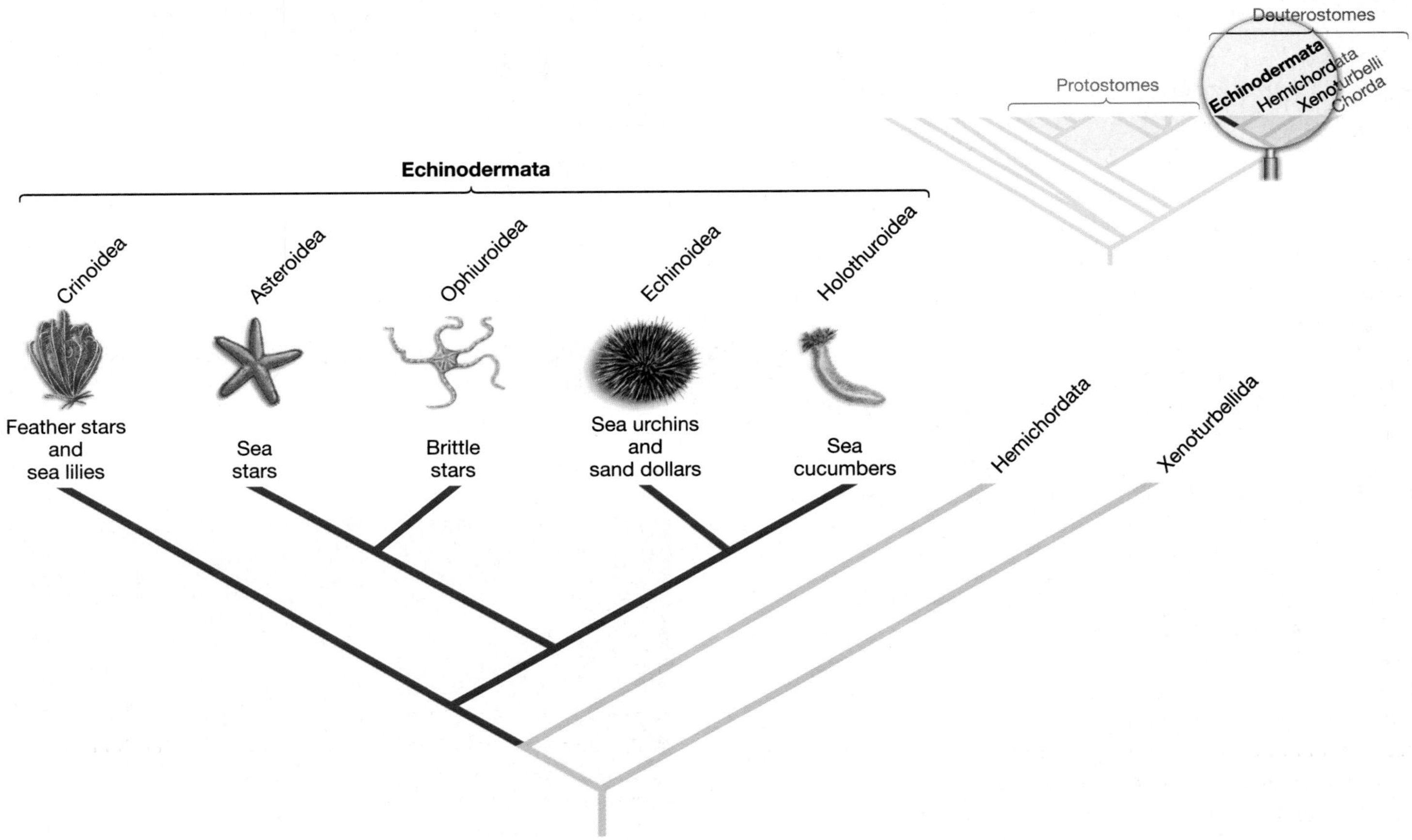

FIGURE 34.19 There Are Five Major Lineages of Echinoderms.

Echinodermata > Asteroidea (Sea Stars)

The 1700 known species of sea stars have bodies with five or more long arms—in some species up to 40—radiating from a central region that contains the mouth, stomach, and anus (**Figure 34.20**). Unlike brittle stars, though, the sea star's arms are not set off from the central region by clear, joint-like articulations. ● You should be able to indicate the origin of five arms, continuous with the central region, on Figure 34.19.

When fully grown, sea stars can range in size from less than 1 cm in diameter to 1 m across. They live on hard or soft substrates along the coasts of all the world's oceans. Although the spines that are characteristic of some echinoderms are reduced to knobs on the surface of most sea stars, the crown-of-thorns star and a few other species have prominent, upright, movable spines.

Feeding Sea stars are predators or scavengers. Some species pull bivalves apart with their tube feet and evert their stomach into the prey's visceral mass. Sponges, barnacles, and snails are also common prey. The crown-of-thorns sea star specializes in feeding on corals and is native to the Indian Ocean and western Pacific Ocean. Its population has skyrocketed recently—possibly because people are harvesting their major predator, a large snail called the triton, for its pretty shell. Large crown-of-thorns star populations have led to the destruction of large areas of coral reef.

Movement Sea stars crawl with the aid of their tube feet. Usually one of the five or more arms is used as the front or leading appendage as the animal moves.

Reproduction Sexual reproduction predominates in sea stars, and sexes are separate. At least one sea star arm is filled with reproductive organs that produce massive amounts of gametes—millions of eggs per female, in some species. Species that are native to habitats in the far north, where conditions are particularly harsh, care for their offspring by holding fertilized eggs on their body until the eggs hatch. Most sea stars are capable of regenerating arms that are lost in predator attacks or storms. Some species can reproduce asexually by dividing the body in two, with each of the two individuals then regenerating the missing half.

Pycnopodia helianthoides

FIGURE 34.20 Sea Stars May Have Many Arms.

Echinodermata > Echinoidea (Sea Urchins and Sand Dollars)

There are about 800 species of echinoids living today; most are sea urchins or sand dollars. Sea urchins have globe-shaped bodies and long spines and crawl along substrates (**Figure 34.21a**). Sand dollars are flattened and disk-shaped, have short spines, and burrow in soft sediments (**Figure 34.21b**). ● You should be able to indicate the origin of globular or disc-shaped bodies on Figure 34.19.

Feeding Sand dollars use their mucus-covered podia to collect food particles in sand or in other soft substrates. Most types of sea urchins are herbivores. In some areas of the world, urchins are extremely important grazers on kelp and other types of algae. In fact, when urchin populations are high, their grazing can prevent the formation of kelp forests. Most echinoids have a unique, jaw-like feeding structure in their mouths that is made up of five calcium carbonate teeth attached to muscles. In many species, this apparatus can extend and retract as the animal feeds.

Movement Using their podia, sea urchins crawl and sand dollars burrow. Sea urchins can also move their spines to aid in crawling along a substrate.

Reproduction Sexual reproduction predominates in sea urchins and sand dollars. Fertilization is external, and sexes are separate.

(a) Sea urchin

Echinus tylodes

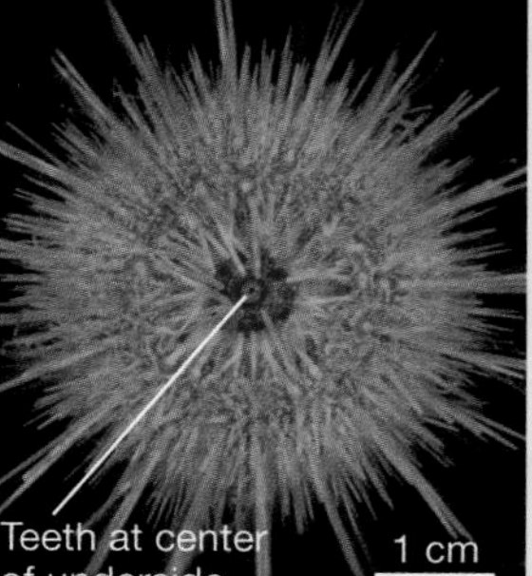

(b) Sand dollar

Dendraster excentricus

1 cm

FIGURE 34.21 Sea Urchins and Sand Dollars Are Closely Related.

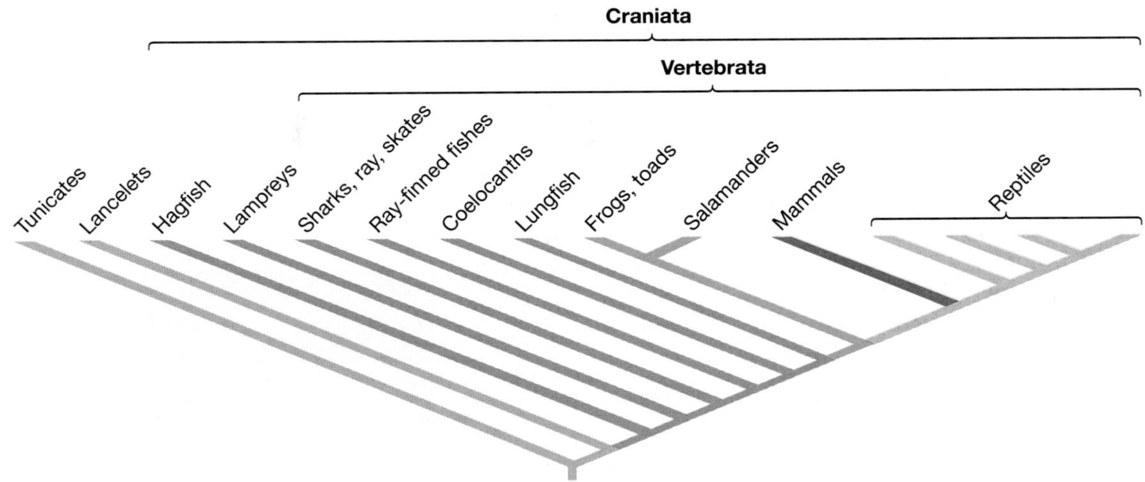

FIGURE 34.22 Craniates and Vertebrates Are Monophyletic Groups. All vertebrates are craniates, but not all craniates are vertebrates.

● **EXERCISE** Draw a brace at the top of the tree to indicate which groups belong to the Chordata.

34.5 Key Lineages: Chordata

The chordates comprise three major subgroups or subphyla: (1) the urochordates (also called tunicates, or sea squirts and salps), (2) the cephalochordates (lancelets), and (3) the craniates and vertebrates (**Figure 34.22**). There are about 1600 species of tunicates, 24 species of lancelets, and over 50,000 vertebrates. At some stage in their life cycle, all species in the phylum Chordata have a dorsal hollow nerve cord, a notochord, pharyngeal gill slits, and a muscular tail that extends past the anus.

Chordata > Urochordata (Tunicates)

The urochordates are also called tunicates, the two major sub-groups are known as the sea squirts (or ascidians) and salps. All of the approximately 300 species described to date live in the ocean. Sea squirts live on the ocean floor (**Figure 34.23a**), while salps live in open water (**Figure 34.23b**).

The distinguishing characters of urochordates include an exoskeleton-like coat of polysaccharide, called a tunic, that covers and supports the body; a U-shaped gut; and two openings, called siphons, where water enters and leaves an individual during feeding. ● You should be able to indicate the origin of the tunic, siphons, and the U-shaped gut on Figure 34.22.

Feeding Adult urochordates use their pharyngeal gill slits to suspension feed. The slits trap particles present in the water that enters one siphon and leaves through the other siphon.

Movement Larvae swim with the aid of the notochord, which stiffens the body and functions as a simple endoskeleton. Larvae are a dispersal stage and do not feed. Adults are sessile or float in currents.

Reproduction In most species, individuals produce both sperm and eggs. In some species, both sperm and eggs are shed into the water and fertilization is external; in other species, sperm are released into the water but eggs are retained, so that fertilization and early development are internal. Asexual reproduction by budding is also common in some groups.

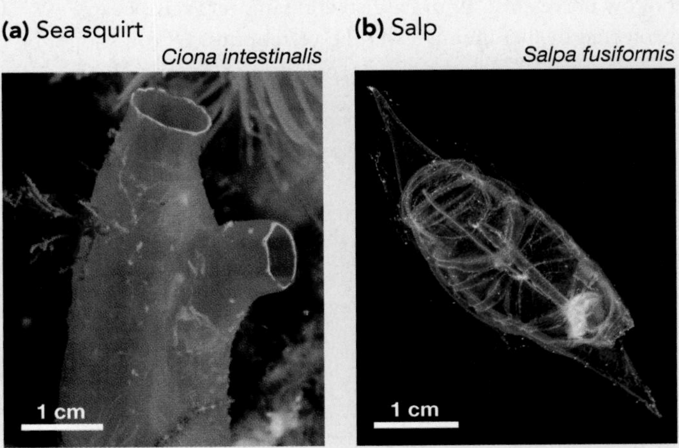

FIGURE 34.23 Sea Squirts and Salps Live in Different Habitats.

Chordata > Cephalochordata (Lancelets)

About two dozen species of cephalochordate have been described to date, all of them found in marine sands. Lancelets—also called amphioxus—have several characteristics that are intermediate between invertebrates and vertebrates. Chief among these is a notochord that is retained in adults, where it functions as an endoskeleton. You should be able to indicate the origin of the notochord that is retained in adults on Figure 34.22.

Feeding Adult cephalochordates feed by burrowing in sediment until only their head is sticking out into the water. They take water in through their mouth and trap food particles with the aid of their pharyngeal gill slits.

Movement Adults have large blocks of muscle arranged in a series along the length of the notochord. Lancelets are efficient swimmers because the flexible, rod-like notochord stiffens the body, making it wriggle when the blocks of muscle contract (**Figure 34.24**).

Reproduction Asexual reproduction is unknown, and individuals are either male or female. Gametes are released into the environment and fertilization is external.

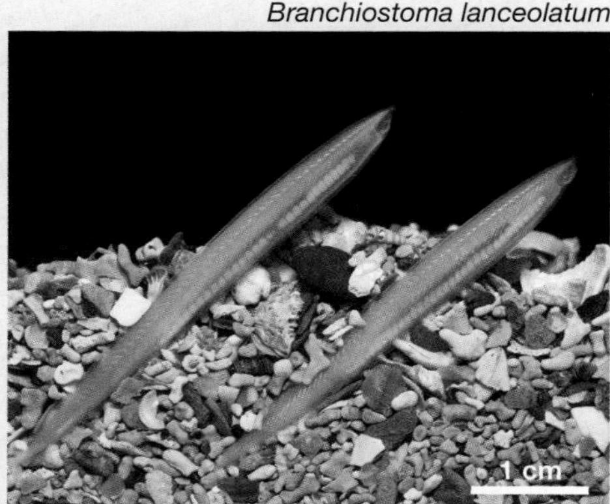

FIGURE 34.24 Lancelets Look like Fish but Are Not Vertebrates.

Chordata > Craniata > Myxinoidea (Hagfish) and Petromyzontoidea (Lampreys)

Although recent phylogenetic analyses indicate that hagfish and lampreys may belong to two independent lineages, some data suggest that they are a single group called the Agnatha ("not-jawed"). Because these animals are the only vertebrates that lack jaws, the 110 species in the two groups are still referred to as the jawless fishes. The hagfish and lampreys are the only surviving members of the earliest branches at the base of the vertebrates.

Hagfish and lampreys have long, slender bodies and are aquatic. Most species are less than a meter long when fully grown. Hagfish lack any sort of vertebral column, but lampreys have small pieces of cartilage along the length of their dorsal hollow nerve cord. Both hagfish and lampreys have brains protected by a cranium, as do the vertebrates. You should be able to indicate the origin of the cranium on Figure 34.22.

Feeding Hagfish are scavengers and predators (**Figure 34.25a**). They deposit feed on the carcasses of dead fish and whales, and some are thought to burrow through ooze at the bottom of the ocean, feeding on polychaetes and other buried prey. Lampreys, in contrast, are ectoparasites. They attach to the sides of fish or other hosts by suction, then use spines in their mouth and tongue to rasp a hole in the side of their victim (**Figure 34.25b**). Once the wound is open, they suck blood and other body fluids.

FIGURE 34.25 Hagfish and Lampreys Are Jawless Vertebrates.

(Continued on next page)

Chordata > Craniata > Myxinoidea (Hagfish) and Petromyzontoidea (Lampreys) *continued*

Movement Hagfish and lampreys have a well-developed notochord and swim by making undulating movements. Lampreys can also move themselves upstream, against the flow of water, by attaching their suckers to rocks and looping the rest of their body forward, like an inchworm. Although lampreys have fins that aid in locomotion, they do not have the paired lateral appendages—meaning fins or limbs that emerge from each side—found in vertebrates.

Reproduction Virtually nothing is known about hagfish mating or embryonic development. Some lampreys live in freshwater; others are **anadromous**—meaning they spend their adult life in the ocean, but swim up streams to breed. Fertilization is external, and adults die after breeding once. Lamprey eggs hatch into larvae that look and act like lancelets. The larvae burrow into sediments and suspension feed for several years before metamorphosing into free-swimming adults.

Chordata > Vertebrata > Chondrichthyes (Sharks, Rays, Skates)

The 970 species in this lineage are distinguished by their cartilaginous skeleton (*chondrus* is the Greek word for cartilage), the presence of jaws, and the existence of paired fins. Paired fins were an important evolutionary innovation because they stabilize the body during rapid swimming—keeping it from pitching up or down, yawing to one side or the other, or rolling. ● You should be able to indicate the origin of paired fins—which are also found in other groups of fishlike organisms—on Figure 34.22.

Most sharks, rays, and skates are marine, though a few species live in freshwater. Sharks have streamlined, torpedo-shaped bodies and an asymmetrical tail—the dorsal portion is longer than the ventral portion (**Figure 34.26a**) In contrast, the dorsal-ventral plane of the body in rays and skates is strongly flattened (**Figure 34.26b**).

Feeding A few species of ray and shark suspension feed on plankton, but most species in this lineage are predators. Skates and rays lie on the ocean floor and ambush passing animals; electric rays capture their prey by stunning them with electric discharges of up to 200 volts. Most sharks, in contrast, are active hunters that chase down prey in open water and bite them. The larger species of shark feed on large fish or marine mammals. Sharks are referred to as the "top predator" in many marine ecosystems, because they are at the top of the food chain—nothing eats them. Yet the largest of all sharks, the whale shark, is a suspension feeder. Whale sharks filter plankton out of water as it passes over their gills.

Movement Rays and skates swim by flapping their greatly enlarged pectoral fins. (Pectoral fins are located on the sides of an organism; dorsal fins are located on the dorsal surface.) Sharks swim by undulating their bodies from side to side and beating their large tails.

Reproduction Sharks use internal fertilization, and fertilized eggs may be shed into the water or retained until the young are hatched and well developed. In some viviparous species, embryos are attached to the mother by specialized tissues in a mammal-like placenta, where the exchange of gases, nutrients, and wastes takes place. Skates are oviparous, but rays are viviparous.

(a) Sharks are torpedo shaped. *Prionace glauca*

(b) Skates and rays are flat. *Taeniura melanospila*

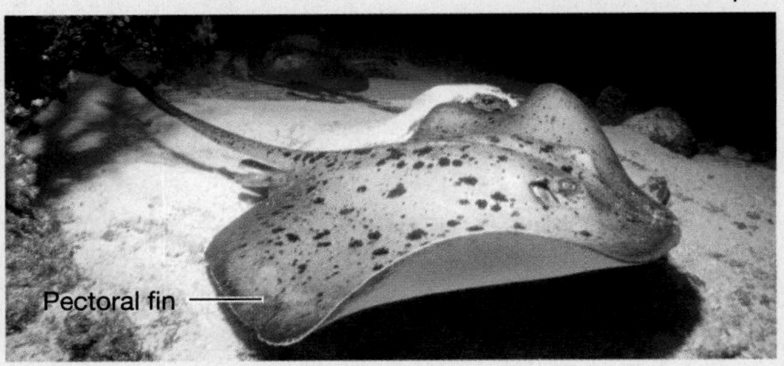

FIGURE 34.26 Sharks and Rays Have Cartilaginous Skeletons.

Chordata > Vertebrata > Actinopterygii (Ray-Finned Fishes)

Actinopterygii (pronounced *ack-tin-op-teh-RIJ-ee-i*) means "ray-finned." Logically enough, these fish have fins that are supported by long, bony rods arranged in a ray pattern. They are the most ancient living group of vertebrates that have a skeleton made of bone. Their bodies are covered with interlocking scales that provide a stiff but flexible covering, and they have a gas-filled **swim bladder**. The evolution of the swim bladder was an important innovation because it allowed ray-finned fishes to avoid sinking. Tissues are heavier than water, so the bodies of aquatic organisms tend to sink. Sharks and rays, for example, have to swim to avoid sinking. But ray-finned fishes have a bladder that changes in volume, depending on the individual's position. Gas is added to the bladder when a ray-finned fish swims down; gas is removed when the fish swims up. In this way, ray-finned fishes maintain neutral buoyancy in water of various depths and thus various pressures. ● You should be able to indicate the origin of rayed fins and the swim bladder on Figure 34.22.

The actinopterygians are the most successful vertebrate lineage based on number of species, duration in the fossil record, and extent of habitats occupied. Almost 27,000 species of ray-finned fishes have been identified thus far. In traditional classifications, Actinopterygii is considered a class.

The most important major lineage of ray-finned fishes is the Teleostei. About 96 percent of all living fish species, including familiar groups like the tuna, trout, cod, and goldfish, are teleosts (**Figure 34.27**).

Feeding Teleosts can suck food toward their mouths, grasp it with their protrusible jaws, and then process it with teeth on their jaws and with pharyngeal jaws in their throat. The size and shape of the mouth, the jaw teeth, and the pharyngeal jaw teeth all correlate with the type of food consumed. For example, most predatory teleosts have long, spear-shaped jaws armed with spiky teeth, as well as bladelike teeth on their pharyngeal jaws. Besides being major predators, ray-finned fishes are the most important large herbivores in both marine and freshwater environments.

Movement Ray-finned fishes swim by alternately contracting muscles on the left and right sides of their bodies from head to tail, resulting in rapid, side-to-side undulations. Their bodies are streamlined to reduce drag in water. Teleosts have a flexible, symmetrical tail, which reduces the need to use their pectoral (side) fins as steering and stabilizing devices during rapid swimming.

Reproduction Most ray-finned fish species rely on external fertilization and are oviparous; some species have internal fertilization with external development; still others have internal fertilization and are viviparous. Although it is common for fish eggs to be released in the water and left to develop on their own, parental care occurs in some species. Parents may carry fertilized eggs on their fins, in their mouth, or in specialized pouches to guard them until the eggs hatch. In freshwater teleosts, offspring develop directly; but marine species have larva that are very different from adult forms. As they develop, marine fish larvae undergo a metamorphosis to the juvenile form, which then grows into an adult.

FIGURE 34.27 Teleosts Are Ray-Finned Fishes That Have a Flexible Tail.

Chordata > Vertebrata > Sarcopterygii > Actinistia (Coelacanths) and Dipnoi (Lungfish)

Although coelacanths (pronounced *SEEL-uh-kanths*) and lungfish represent independent lineages, they are sometimes grouped together and called **lobe-finned fishes**. Lobe-finned fishes are common and diverse in the fossil record in the Devonian period, about 400 million years ago, but only eight species are living today. They are important, however, because they represent a crucial evolutionary link between the ray-finned fishes and the tetrapods. Instead of having fins supported by rays of bone, their fins are fleshy lobes supported by a linear—not radial—array of bones and muscles, similar to those observed in the limbs of tetrapods (**Figure 34.28**). ● You should be able to indicate the origin of a linear array of fin bones on Figure 34.22.

Coelacanths are marine and occupy habitats 150–700 m below the surface. In contrast, lungfish live in shallow, freshwater ponds and rivers (see Figure 34.11). As their name implies, lungfish have lungs and breathe air when oxygen levels in their habitats drop.

(Continued on next page)

Chordata > Vertebrata > Sarcopterygii > Actinistia (Coelacanths) and Dipnoi (Lungfish) *continued*

FIGURE 34.28 Coelacanths Are Lobe-Finned Fishes.

Some species burrow in mud and enter a quiescent, sleeplike state when their habitat dries up during each year's dry season.

Feeding Coelacanths prey on fish. Lungfish are **omnivorous** ("all-eating"), meaning that they eat algae and plant material as well as animals.

Movement Coelacanths swim by waving their pectoral and pelvic ("hip") fins in the same sequence that tetrapods use in walking with their limbs. Lungfish swim by waving their bodies, and they can use their fins to walk along pond bottoms.

Reproduction Sexual reproduction is the rule, with fertilization internal in coelacanths and external in lungfish. Coelacanths are ovoviviparous; lungfish lay eggs. Lungfish eggs hatch into larvae that resemble juvenile salamanders.

Chordata > Vertebrata > Amphibia (Frogs, Salamanders, Caecilians)

The 5500 species of **amphibians** living today form three distinct clades traditionally termed orders: (1) frogs and toads, (2) salamanders, and (3) caecilians (pronounced *suh-SILL-ee-uns*). Amphibians are found throughout the world and occupy ponds, lakes, or moist terrestrial environments (**Figure 34.29a**). Translated literally, their name means "both-sides-living." The name is appropriate because adults of most species of amphibian feed on land but lay their eggs in water. In many species of amphibians, gas exchange occurs exclusively or in part across their moist, mucus-covered skin. ● You should be able to indicate the origin of "skin-breathing" on Figure 34.22.

Feeding Adult amphibians are carnivores. Most frogs are sit-and-wait predators that use their long, extendible tongues to capture passing prey. Salamanders also have an extensible tongue, which some species use in feeding. Terrestrial caecilians prey on earthworms and other soil-dwelling animals; aquatic forms eat vertebrates and small fish.

Movement Most amphibians have four well-developed limbs. In water, frogs and toads move by kicking their hind legs to swim; on land they kick their hind legs out to jump or hop. Salamanders walk on land; in water they undulate their bodies to swim. Caecilians lack limbs and eyes; terrestrial forms burrow in moist soils (**Figure 34.29b**).

Reproduction Frogs are oviparous and have external fertilization, but salamanders and caecilians have internal fertilization. Most salamanders are oviparous, but many caecilians are viviparous. In some species of frogs, parents may guard or even carry eggs. In many frogs, young develop in the water and suspension feed on plant or algal material. Salamander larvae are carnivorous. Later the larvae undergo a dramatic metamorphosis into land-dwelling adults. For example, the fishlike tadpoles of frogs and toads develop limbs, and their gills are replaced with lungs.

(a) Frogs and other amphibians lay their eggs in water.

(b) Caecilians are legless amphibians.

FIGURE 34.29 Amphibians Are the Most Ancient Tetrapods.

Mammalia (Mammals)

Mammals are easily recognized by the presence of hair or fur, which serves to insulate the body. Like birds, mammals are endotherms that maintain high body temperatures by oxidizing large amounts of food and generating large amounts of heat. Instead of insulating themselves with feathers, though, mammals retain heat because the body surface is covered with layers of hair or fur. Endothermy evolved independently in birds and mammals. In both groups, endothermy is thought to be an adaptation that enables individuals to maintain high levels of activity— particularly at night or during cold weather.

In addition to being endothermic and having fur, mammals have **mammary glands**—a unique structure that makes lactation possible. The evolution of mammary glands gave mammals the ability to provide their young with particularly extensive parental care. Mammals are also the only vertebrates with facial muscles and lips and the only vertebrates that have a lower jaw formed from a single bone. In traditional classifications, Mammalia is designated as a class (**Figure 34.30**).

Mammals evolved when dinosaurs and other reptiles were the dominant large herbivores and predators in terrestrial and aquatic environments. The earliest mammals in the fossil record appear about 195 million years ago; most were small animals that were probably active only at night. Many of the 4800 species of mammal living today have good nocturnal vision and a strong sense of smell, as their ancestors presumably did. The adaptive radiation that gave rise to today's diversity of mammals did not take place until after the dinosaurs went extinct about 65 million years ago. After the dinosaurs were gone, the mammals diversified into lineages of small and large herbivores, small and large predators, or marine hunters—ecological roles that had once been filled by dinosaurs and the ocean-dwelling, extinct reptiles called mosasaurs.

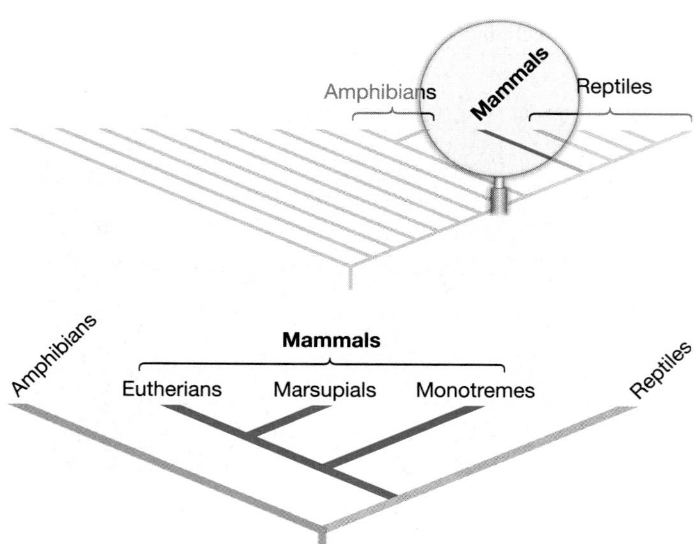

FIGURE 34.30 Mammals Are a Monophyletic Group.

Mammalia > Monotremata (Platypuses, Echidnas)

The **monotremes** are the most ancient lineage of mammals living today, and they are found only in Australia and New Zealand. They lay eggs and have metabolic rates—meaning a rate of using oxygen and oxidizing sugars for energy—that are lower than other mammals. Three species exist: one species of platypus and two species of echidna. ● You should be able to indicate the origin of fur and lactation on Figure 34.30.

Feeding Monotremes have a leathery beak or bill. The platypus feeds on insect larvae, molluscs, and other small animals in streams (**Figure 34.31a**). Echidnas feed on ants, termites, and earthworms (**Figure 34.31b**).

Movement Platypuses swim with the aid of their webbed feet. Echidnas walk on their four legs.

Reproduction Platypuses lay their eggs in a burrow, while echidnas keep their eggs in a pouch on their belly. The young hatch quickly, and the mother must continue keeping them warm and dry for another four months.

(a) Platypus *Ornithorhynchus anatinus*

(b) Echidna *Tachyglossus aculeatus*

FIGURE 34.31 Platypuses and Echidnas Are Egg-Laying Mammals.

Mammalia > Marsupiala (Marsupials)

The 275 known species of **marsupials** live in the Australian region and the Americas (**Figure 34.32**) and include the familiar opossums, kangaroos, wallabies, and koala. Although females have a placenta that nourishes embryos during development, the young are born after a short embryonic period and are poorly developed. They crawl from the opening of the female's reproductive tract to the female's nipples, where they suck milk. They stay attached to their mother until they grow large enough to move independently. ● You should be able to indicate the origin of the placenta and viviparity—traits that are also found in Eutherian mammals—on Figure 34.30.

Didelphis virginiana

FIGURE 34.32 Marsupials Give Birth after a Short Embryonic Period. Opossums are the only marsupials in North America.

Feeding Marsupials are herbivores, omnivores, or carnivores. In many cases, convergent evolution has resulted in marsupials that are extremely similar to placental species in overall morphology and way of life. For example, a recently extinct marsupial called the Tasmanian wolf was a long-legged, social hunter similar to the timber wolves of North America and northern Eurasia. A species of marsupial native to Australia specializes in eating ants and looks and acts much like the South American anteater, which is not a marsupial.

Movement Marsupials move by crawling, gliding, walking, running, or hopping.

Reproduction Marsupial young spend more time developing while attached to their mother's nipple than they do inside her body being fed via the placenta.

Mammalia > Eutheria (Placental Mammals)

The approximately 4300 species of **placental mammals, eutherians**, are distributed worldwide. They are far and away the most species-rich and morphologically diverse group of mammals.

Biologists group placental mammals into 18 lineages called orders. The six most species-rich orders are the rodents (rats, mice, squirrels; 1814 species), bats (986 species), insectivores (hedgehogs, moles, shrews; 390 species), artiodactyls (pigs, hippos, whales, deer, sheep, cattle; 293 species), carnivores (dogs, bears, cats, weasels, seals; 274 species), and primates (lemurs, monkeys, apes, humans; 235 species).

Hylobates lar

FIGURE 34.33 Eutherians Are the Most Species-Rich and Diverse Group of Mammals.

Feeding The size and structure of the teeth correlate closely with the diet of placental mammals. Herbivores have large, flat teeth for crushing leaves and other coarse plant material; predators have sharp teeth that are efficient at biting and tearing flesh. Omnivores, such as humans, usually have several distinct types of teeth. The structure of the digestive tract also correlates with the placental mammals' diet. In some plant-eaters, for example, the stomach hosts unicellular organisms that digest cellulose and other complex polysaccharides.

Movement In placental mammals, the structure of the limb correlates closely with the type of movement performed. Eutherians fly, glide, run, walk, swim, burrow, or swing from trees (**Figure 34.33**). Limbs are reduced or lost in aquatic groups such as whales and dolphins, which swim by undulating their bodies.

Reproduction Eutherians have internal fertilization and are viviparous. An extensive placenta develops from a combination of maternal and fetal tissues; and at birth, young are much better developed than in marsupials—some are able to walk or run minutes after emerging from the mother. ● You should be able to indicate the origin of delayed birth (extended development prior to birth) on Figure 34.30. All eutherians feed their offspring milk until the young have grown large enough to process solid food. A prolonged period of parental care, extending beyond the nursing stage, is common as offspring learn how to escape predators and find food on their own.

Reptilia (Turtles, Snakes and Lizards, Crocodiles, Birds)

The **reptiles** are a monophyletic group and represent one of the two major living lineages of amniotes—the other lineage consists of the extinct mammal-like reptiles and today's mammals. The major feature distinguishing the reptilian and mammalian lineages is the number and placement of openings in the side of the skull. Jaw muscles that make possible sophisticated biting and chewing movements pass through these openings and attach to bones on the upper part of the skull.

Several features adapt reptiles for life on land. Their skin is made watertight by a layer of scales made of the protein keratin, which is also a major component of mammalian hair. Reptiles breathe air through well-developed lungs and lay shelled, amniotic eggs that resist drying out. In turtles, the egg has a leathery shell; in other reptiles, the shell is made of stiff calcium carbonate. Fertilization is internal because the sperm and egg have to meet and form the zygote before the amniotic membrane and shell form. If fertilization were external, sperm would have to pass through the shell and amnion to reach the egg cell.

The reptiles include the dinosaurs, pterosaurs (flying reptiles), mosasaurs (marine reptiles), and other extinct lineages that flourished from about 250 million years ago until the mass extinction at the end of the Cretaceous Period, 65 million years ago. Today the Reptilia are represented by four major lineages, traditionally recognized as subclasses: (1) turtles, (2) snakes and lizards, (3) crocodiles and alligators, and (4) birds (**Figure 34.34**). Except for birds, all of these groups are **ectothermic** ("outside-heated")—meaning that individuals do not use internally generated heat to regulate their body temperature. It would be a mistake, however, to conclude that reptiles other than birds do not regulate their body temperature closely. Reptiles bask in sunlight, seek shade, and perform other behaviors to keep their body temperature at a preferred level.

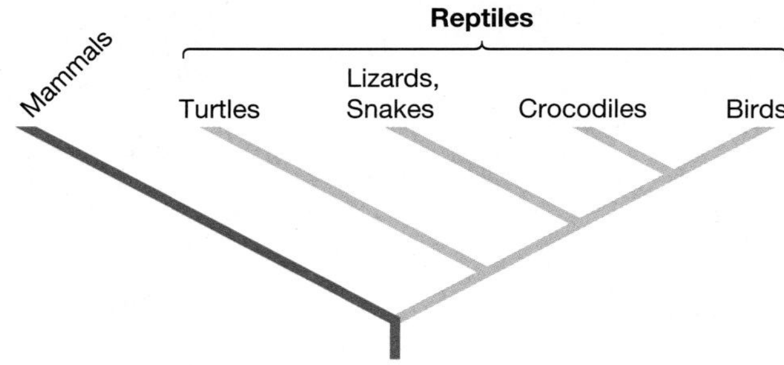

FIGURE 34.34 Reptiles Are a Monophyletic Group.

Reptilia > Testudinia (Turtles)

The 300 known species of turtles inhabit freshwater, marine, and terrestrial environments throughout the world. The testudines are distinguished by a shell composed of bony plates, covered with a material similar in composition to human fingernails, that fuse to the vertebrae and ribs (**Figure 34.35**). ● You should be able to indicate the origin of the turtle shell on Figure 34.34. The turtles' skulls are highly modified versions of the skulls of other reptiles. Turtles lack teeth, but their jawbone and lower skull form a bony beak.

Feeding Turtles are either carnivorous—feeding on whatever animals they can capture and swallow—or herbivorous. They may also scavenge dead material. Most marine turtles are carnivorous. Leatherback turtles, for example, feed primarily on jellyfish, and they are only mildly affected by the jellyfish's stinging cnidocytes (see Chapter 32). In contrast, species in the lineage of terrestrial turtles called the tortoises are plant-eaters.

Movement Turtles swim, walk, or burrow. Aquatic species usually have feet that are modified to function as flippers.

Reproduction All turtles are oviparous. Other than digging a nest prior to depositing eggs, parental care is lacking. The sex of a baby turtle is often not determined by sex chromosomes. Instead, in many species gender is determined by the temperature at which the egg develops. High temperatures produce mostly males, while low temperatures produce mostly females.

FIGURE 34.35 Turtles Have a Shell Consisting of Bony Plates.

Reptilia > Lepidosauria (Lizards, Snakes)

Most lizards and snakes are small reptiles with elongated bodies and scaly skin. Most lizards have well-developed jointed legs, but snakes are limbless (**Figure 34.36**). The hypothesis that snakes evolved from limbed ancestors is partially supported by the presence of vestigial hip and leg bones in boas and pythons. There are about 7000 species of lizards and snakes alive now. ● You should be able to indicate the origin of scaly skin on Figure 34.34.

Feeding Small lizards prey on insects. Although most of the larger lizard species are herbivores, the 3-meter-long monitor lizard from the island of Komodo is a predator that kills and eats deer. Snakes are carnivores; some subdue their prey by injecting poison through modified teeth called fangs. Snakes prey primarily on small mammals, amphibians, and invertebrates, which they swallow whole—usually headfirst.

Movement Lizards crawl or run on their four limbs. Snakes and limbless lizards burrow through soil, crawl over the ground, or climb trees by undulating their bodies.

Reproduction Although most lizards and snakes lay eggs, many are oviviviparous. Most species reproduce sexually, but asexual reproduction, via the production of eggs by mitosis, is known to occur in six groups of lizards and one snake lineage.

Morelia viridis

FIGURE 34.36 Snakes Are Limbless Predators.

Reptilia > Crocodilia (Crocodiles, Alligators)

Only 24 species of crocodile and alligator are known. Most are tropical and live in freshwater or marine environments. They have eyes located on the top of their heads and nostrils located at the top of their long snouts—adaptations that allow them to sit semi-submerged in water for long periods of time (**Figure 34.37**).

Feeding Crocodilians are predators. Their jaws are filled with conical teeth that are continually replaced as they fall out during feeding. Their usual method of killing small prey is by biting through the body wall. Large prey are usually subdued by drowning. Crocodilians eat amphibians, turtles, fish, birds, and mammals.

Movement Crocodiles and alligators walk or gallop on land. In water they swim with the aid of their large, muscular tails.

Reproduction Although crocodilians are oviparous, parental care is extensive. Eggs are laid in earth-covered nests that are guarded by the parents. When young inside the eggs begin to vocalize, parents dig them up and carry the newly hatched young inside their mouths to nearby water. Crocodilian young can hunt when newly hatched but stay near their mother for up to three years. ● You should be able to indicate the origin of extensive parental care—which also occurs in birds (and dinosaurs)—on Figure 34.34.

Alligator mississippiensis

FIGURE 34.37 Alligators Are Adapted for Aquatic Life.

Reptilia > Aves (Birds)

The fossil record provides conclusive evidence that birds descended from a lineage of dinosaurs that had a unique trait: **feathers**. In dinosaurs, feathers are hypothesized to have functioned as insulation and in courtship or aggressive displays. In birds, feathers insulate and are used for display but also furnish the lift, power, and steering required for flight. Birds have many other adaptations that make flight possible, including large breast muscles used to flap the wings. Bird bodies are lightweight because they have a reduced number of bones and organs and because their hollow bones are filled with air sacs linked to the lungs. Instead of teeth, birds have a horny beak. They are endotherms ("within-heating"), meaning that they have a high metabolic rate and use the heat produced, along with the insulation provided by feathers, to maintain a constant body temperature. The 9100 bird species alive today occupy virtually every habitat, including the open ocean (**Figure 34.38**). ● You should be able to indicate the origin of feathers, endothermy, and flight on Figure 34.34.

Feeding Plant-eating birds usually feed on nectar or seeds. Most birds are omnivores, although many are predators that capture insects, small mammals, fish, other birds, lizards, molluscs, or crustaceans. The size and shape of a bird's beak correlate closely with its diet. For example, predatory species such as falcons have sharp, hook-shaped beaks; finches and other seedeaters have short, stocky bills that can crack seeds and nuts; fish-eating species such as the great blue heron have spear-shaped beaks.

Movement Although flightlessness has evolved repeatedly during the evolution of birds, almost all species can fly. The size and shape of birds' wings correlate closely with the type of flying they do. Birds that glide or hover have long, thin wings; species that specialize in explosive takeoffs and short flights have short, stocky wings. Many seabirds are efficient swimmers, using their webbed feet to paddle or flapping their wings to "fly" under water. Ground-dwelling birds such as ostrich and pheasants can run long distances at high speed.

Reproduction Birds are oviparous but provide extensive parental care. In most species, one or both parents build a nest and incubate the eggs. After the eggs hatch, parents feed offspring until they are large enough to fly and find food on their own.

Diomedea melanophris

FIGURE 34.38 Birds Are Feathered Descendants of Dinosaurs.

34.6 Key Lineages: The Hominin Radiation

Although humans occupy a tiny twig on the tree of life, there has been a tremendous amount of research on human origins. This section introduces the lineage of mammals called the Primates, the fossil record of human ancestors, and data on the relationships among human populations living today.

The Primates

The lineage called Primates consists of two main groups: prosimians and anthropoids. The **prosimians** ("before-monkeys") consist of the lemurs, found in Madagascar, and the tarsiers, pottos, and lorises of Africa and south Asia. Most prosimians live in trees and are active at night (**Figure 34.39a**). The Anthropoidea or **anthropoids** ("human-like") include the New World monkeys found in Central and South America, the

(a) Prosimians are small, tree-dwelling primates.

Loris tardigradus

(b) New World monkeys are anthropoids.

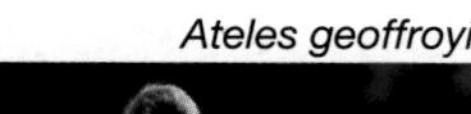

Ateles geoffroyi

FIGURE 34.39 There Are Two Main Lineages of Primates.
(a) Prosimians live in Africa, Madagascar, and south Asia. **(b)** Anthropoids include Old World monkeys, New World monkeys, and great apes.

Old World monkeys that live in Africa and tropical regions of Asia, the gibbons of the Asian tropics, and the Hominidae, or **great apes**—orangutans, gorillas, chimpanzees, and humans (**Figure 34.39b**). The phylogenetic tree in **Figure 34.40** shows the evolutionary relationships among these groups.

Primates are distinguished by having eyes located on the front of the face. Eyes that look forward provide better depth perception than do eyes on the sides of the face. Primates also tend to have hands and feet that are efficient at grasping, flattened nails instead of claws on the fingers and toes, brains that are large relative to overall body size, complex social behavior, and extensive parental care of offspring.

The lineage in Figure 34.40 that is composed of the great apes, including humans, is known as the Hominidae or **hominids**. From extensive comparisons of DNA sequence data, it is now clear that humans are most closely related to the chimpanzees and that our next nearest living relatives are the gorillas.

Compared with most types of primate, the great apes are relatively large bodied and have long arms, short legs, and no tail. Although all of the great ape species except for the orangutans live primarily on the ground, they have distinct ways of walking. When orangutans do come to the ground, they occasionally walk with their knuckles pressed to the ground. More commonly, though, they fist-walk—that is, they walk with the backs of their hands pressed to the ground. Gorillas and chimps, in contrast, only knuckle-walk. They also occasionally rise up on two legs—usually in the context of displaying aggression. Humans are the only great ape that is fully **bipedal** ("two-footed")—meaning they walk upright on two legs. Bipedalism is, in fact, the shared derived character that defines the group called hominins. The Homininae, or **hominins**, are a monophyletic group comprising *Homo sapiens* and more than a dozen extinct, bipedal relatives.

Fossil Humans

According to the fossil record, the common ancestor of chimps and humans lived in Africa about 7 million years ago. As a

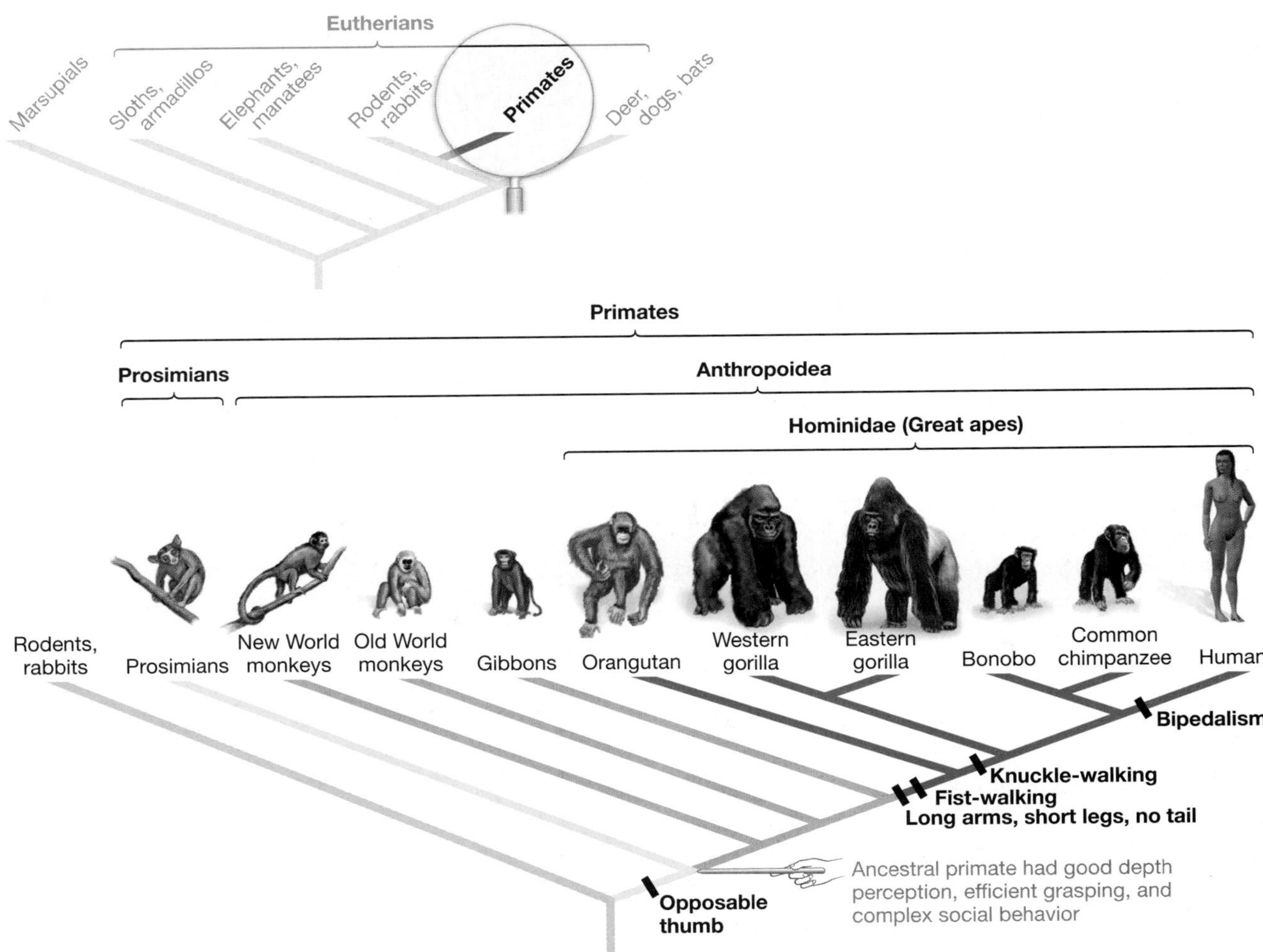

FIGURE 34.40 A Phylogeny of the Monkeys and Great Apes. Phylogenetic tree estimated from extensive DNA sequence data. According to the fossil record, humans and chimps shared a common ancestor 6 to 7 million years ago.

group, all the species on the branch leading to contemporary humans are considered hominins. The fossil record of hominins, though not nearly as complete as investigators would like, is rapidly improving. About 14 species have been found to date, and new fossils that inform the debate over the ancestry of humans are discovered every year. Although naming the hominin species and interpreting their characteristics remain intensely controversial, most researchers agree that they can be organized into four groups:

1. ***Australopithecus*** Four species of small apes called gracile australopithecines have been identified thus far (**Figure 34.41a**). The adjective *gracile*, or "slender," is appropriate because these organisms were slightly built. Adult males were about 1.5 meters tall and weighed about 36 kg. The genus name *Australopithecus* ("southern ape") was inspired by the earliest specimens, which came from South Africa. Several lines of evidence support the hypothesis that the gracile australopithecines were bipedal. For example, the hole in the back of their skulls where the spinal cord connects to the brain is oriented downward, just as it is in our species, *Homo sapiens*. In chimps, gorillas, and other vertebrates that walk on four feet, this hole is oriented backward.

2. ***Paranthropus*** Three species are grouped in the genus *Paranthropus* ("beside-human"). Like the gracile australopithecines, these robust australopithecines were bipedal. They were much stockier than the gracile forms, however—about the same height but an estimated 8–10 kilograms (20 pounds) heavier on average. In addition, their skulls were much broader and more robust (**Figure 34.41b**). All three species had massive cheek teeth and jaws, very large cheekbones, and a sagittal crest—a flange of bone at the top of the skull. Because muscles that work the jaw attach to the sagittal crest and cheekbones, researchers conclude that these organisms had tremendous biting power and made their living by crushing large seeds or coarse plant materials. One species is nicknamed "nutcracker man." The name *Paranthropus* was inspired by the hypothesis that the three known species are a monophyletic group that was a side branch during human evolution—an independent lineage that went extinct.

3. **Early *Homo*** Species in the genus *Homo* are called **humans**. As **Figure 34.41c** shows, species in this genus have flatter and narrower faces, smaller jaws and teeth, and larger braincases than the earlier hominins do. (The **braincase** is the portion of the skull that encloses the brain.) The appearance of early members of the genus *Homo* in the fossil record coincides closely with the appearance of tools made of worked stone—most of which are interpreted as handheld choppers or knives. Although the fossil record does not exclude the possibility that *Paranthropus* made tools, many researchers favor the hypothesis that extensive toolmaking was a diagnostic trait of early *Homo*.

(a) Gracile australopithecines (*Australopithecus africanus*)

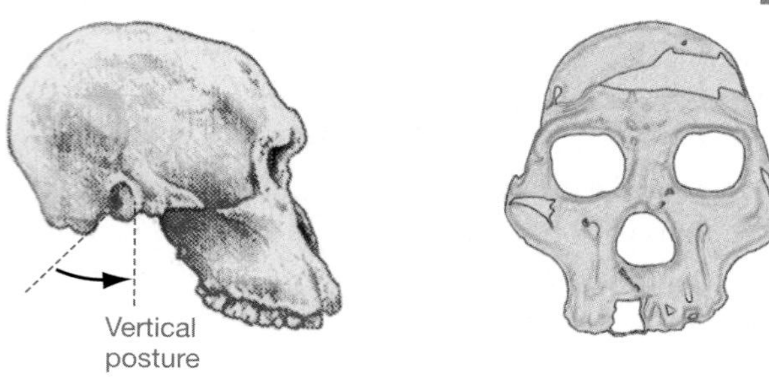

(b) Robust australopithecines (*Paranthropus robustus*)

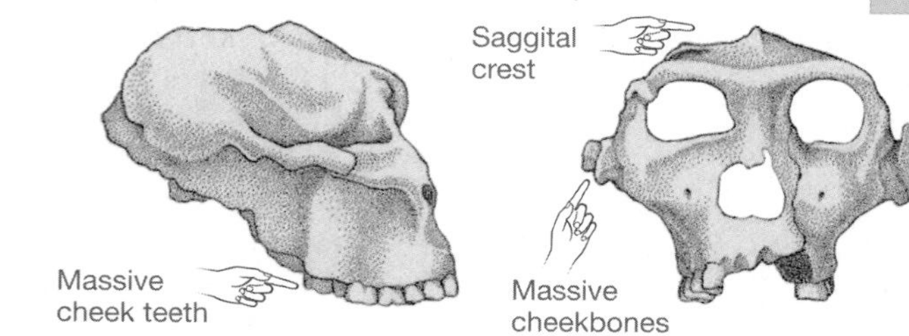

(c) Early *Homo* (*Homo habilis*)

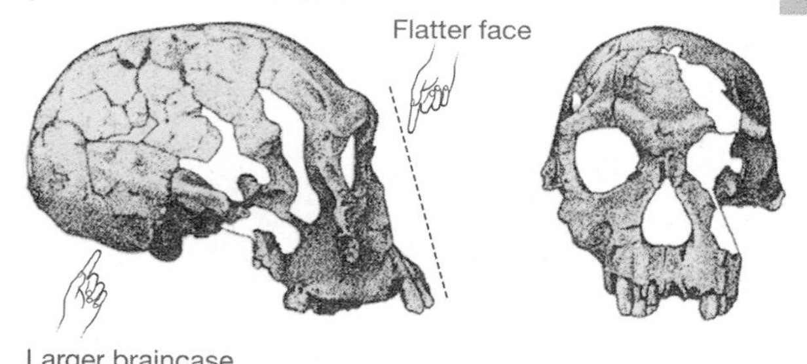

(d) Recent *Homo* (*Homo sapiens*, Cro-Magnon)

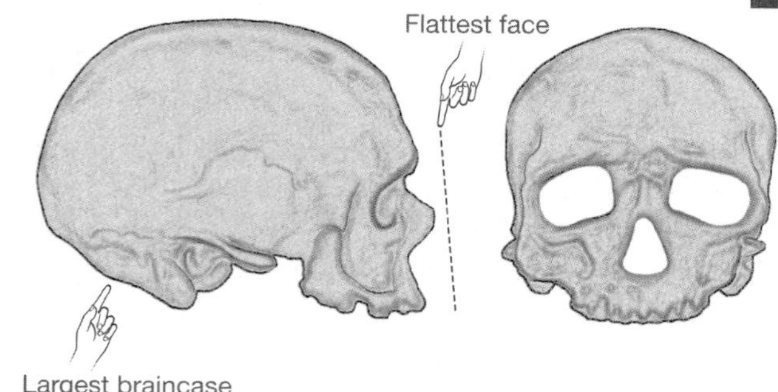

FIGURE 34.41 African Hominins Comprise Four Major Groups.

● **QUESTION** The skulls are arranged as they appear in the fossil record, from most ancient to most recent—(a) to (d). How did the forehead and brow ridge of hominins change through time?

TABLE **34.1 Characteristics of Selected Hominins**

Species	Location of Fossils	Estimated Average Braincase Volume (cm^3)	Estimated Average Body Size (kg)	Associated with Stone Tools?
Australopithecus afarensis	Africa	450	36	no
A. africanus	Africa	450	36	no
Paranthropus boisei	Africa	510	44	no?
Homo habilis	Africa	550	34	yes
H. ergaster	Africa	850	58	yes
H. erectus	Africa, Asia	1000	57	yes
H. heidelbergensis	Africa, Europe	1200	62	yes
H. neanderthalensis	Middle East, Europe, Asia	1500	76	yes
H. floresiensis	Flores (Indonesia)	380	28	yes
H. sapiens	Middle East, Europe, Asia	1350	53	yes

4. **Recent *Homo*** The recent species of *Homo* date from 1.2 million years ago to the present. As **Figure 34.41d** shows, these species have even flatter faces, smaller teeth, and larger braincases than the early *Homo* species do. The 30,000-year-old fossil in the figure, for example, is from a population of *Homo sapiens* (our species) called the **Cro-Magnons**. The Cro-Magnons were accomplished painters and sculptors who buried their dead in carefully prepared graves. There is also evidence that another species, the **Neanderthal** people (*Homo neanderthalensis*) made art and buried their dead in a ceremonial fashion. Perhaps the most striking recent *Homo*, though, is *H. floresiensis*. This species has been found only on the island of Flores in Indonesia, which was also home to a species of dwarfed elephants. *H. floresiensis* consisted of individuals that had braincases smaller than those of gracile australopithecines and were about a meter tall. Fossil finds suggest that the species inhabited the island from about 100,000 to 12,000 years before present and that dwarfed elephants were a major source of food.

Table 34.1 summarizes data on the geographic range, braincase volume, and body size of selected species within these four groups. **Figure 34.42** provides the time range of each species in the fossil record. Although researchers do not have a solid understanding of the phylogenetic relationships among the hominin species, several points are clear from the available data. First, the shared, derived character that defines the hominins is bipedalism. Second, several species from the lineage were present simultaneously during most of hominin evolution. For example, about 1.8 million years ago there may have been as many as five hominin species living in eastern and southern Africa. Because fossils from more than one species have been found in the same geographic location in rock strata of the same age, it is almost certain that different hominin species lived in physical contact. Finally, compared with the gracile and robust australopithecines and the great apes, species in the genus *Homo* have extremely large brains relative to their overall body size.

Why did humans evolve such gigantic brains? The leading hypothesis on this question is that early *Homo* began using

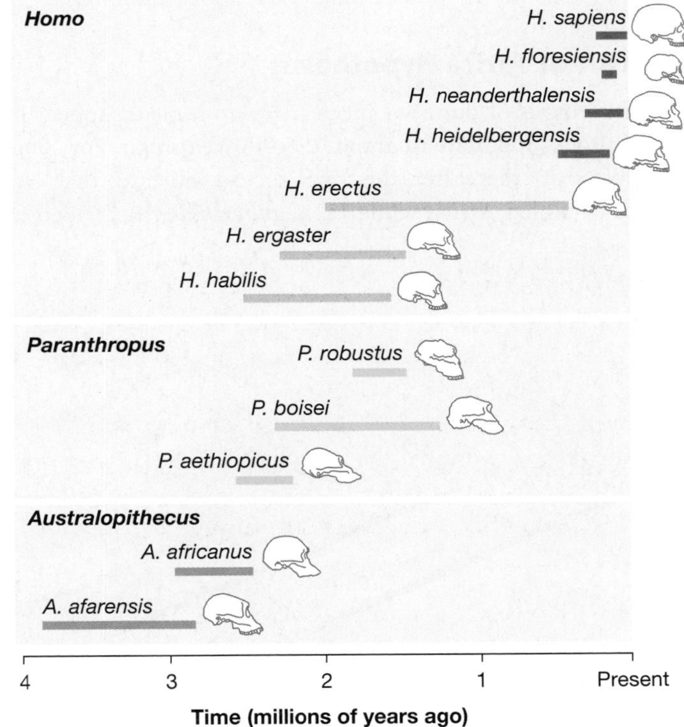

FIGURE 34.42 A Timeline of Human Evolution. Plot of the ages of selected fossil hominins.

● **QUESTION** How many species of hominin existed 2.2 million years ago, 1.8 million years ago, and 100,000 years ago?

symbolic spoken language along with initiating extensive tool use. The logic here is that increased toolmaking and language use triggered natural selection for the capacity to reason and communicate, which required a larger brain. To support this hypothesis, researchers point out that, relative to the brain areas of other hominins, the brain areas responsible for language were enlarged in the earliest *Homo* species. There is even stronger fossil evidence for extensive use of speech in *Homo neanderthalensis* and early *Homo sapiens*:

- The hyoid bone is a slender bone in the voice box of modern humans that holds muscles used in speech. In Neanderthals and early *Homo sapiens*, the hyoid is vastly different in size and shape from a chimpanzee's hyoid bone. Researchers recently found an intact hyoid bone associated with a 60,000-year-old Neanderthal individual and showed that it is virtually identical to the hyoid of modern humans.
- *Homo sapiens* colonized Australia by boat between 60,000 and 40,000 years ago. Researchers suggest that an expedition of that type could not be planned and carried out in the absence of symbolic speech.

To summarize, *Homo sapiens* is the sole survivor of an adaptive radiation that took place over the past 7 million years. From a common ancestor shared with chimpanzees, hominins evolved the ability to walk upright, make tools, and talk.

The Out-of-Africa Hypothesis

The first fossils of our own species, *Homo sapiens*, appear in African rocks that date to about 195,000 years ago. For some 130,000 years thereafter, the fossil record indicates that our species occupied Africa while *H. neanderthalensis* resided in Europe and the Middle East. Some evidence suggests that *H. erectus* may still have been present in Asia at that time. Then, in rocks dated between 60,000 and 30,000 years ago, *H. sapiens* fossils are found throughout Europe, Asia, Africa, and Australia. *H. erectus* had disappeared by this time, and *H. neanderthalensis* went extinct after coexisting with *H. sapiens* in Europe for perhaps a thousand years.

Phylogenies of *H. sapiens* estimated with DNA sequence data agree with the pattern in the fossil record. In phylogenetic trees that show the relationships among human populations living today, the first lineages to branch off lead to descendant populations that live in Africa today (**Figure 34.43**). Based on this observation, it is logical to infer that the ancestral population of modern humans also lived in Africa. The tree shows that lineages subsequently branched off to form three monophyletic groups. Because the populations within each of these clades live in a distinct area, the three lineages are thought to descend from three major waves of migration that occurred as a *Homo sapiens* population dispersed from east Africa to (1) north Africa, Europe, and central Asia, (2) northeast Asia and the Americas, and (3) southeast Asia and the South Pacific (**Figure 34.44**). To summarize, the data suggest that (1) modern humans originated in Africa; and (2) a population that left Africa split into three broad groups, which then spread throughout the world.

This scenario for the evolution of *H. sapiens* is called the **out-of-Africa hypothesis**. It contends that *H. sapiens* evolved independently of the earlier European and Asian species of *Homo*—meaning there was no interbreeding between *H. sapiens* and Neanderthals, *H. erectus*, or *H. floresiensis*. Stated another way, the out-of-Africa hypothesis proposes that *H. sapiens* evolved its distinctive traits in Africa and then dispersed throughout the world.

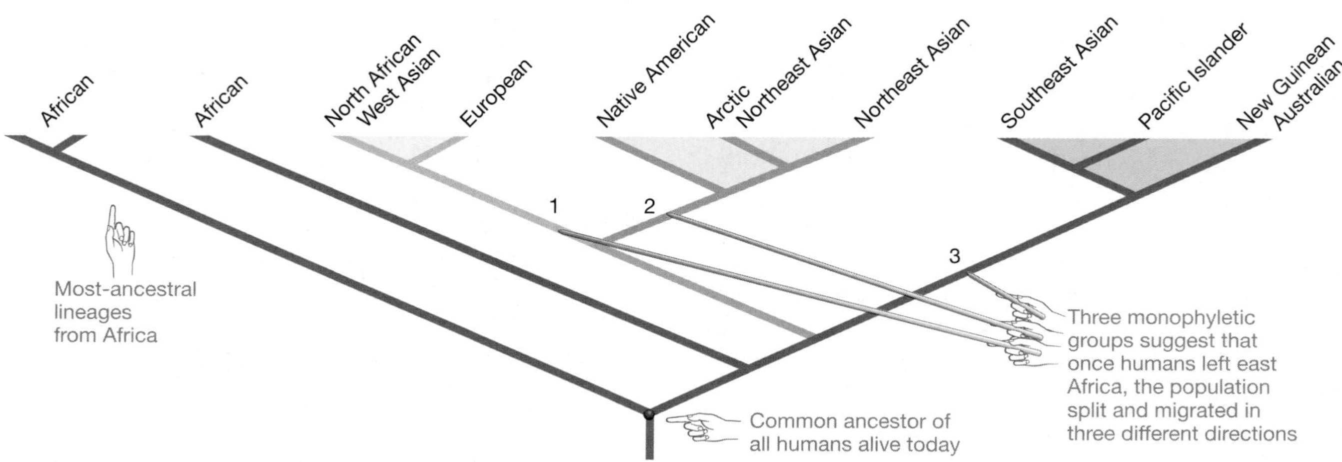

FIGURE 34.43 Phylogeny of Human Populations Living Today. Phylogeny of modern human populations, as estimated from DNA sequence data.

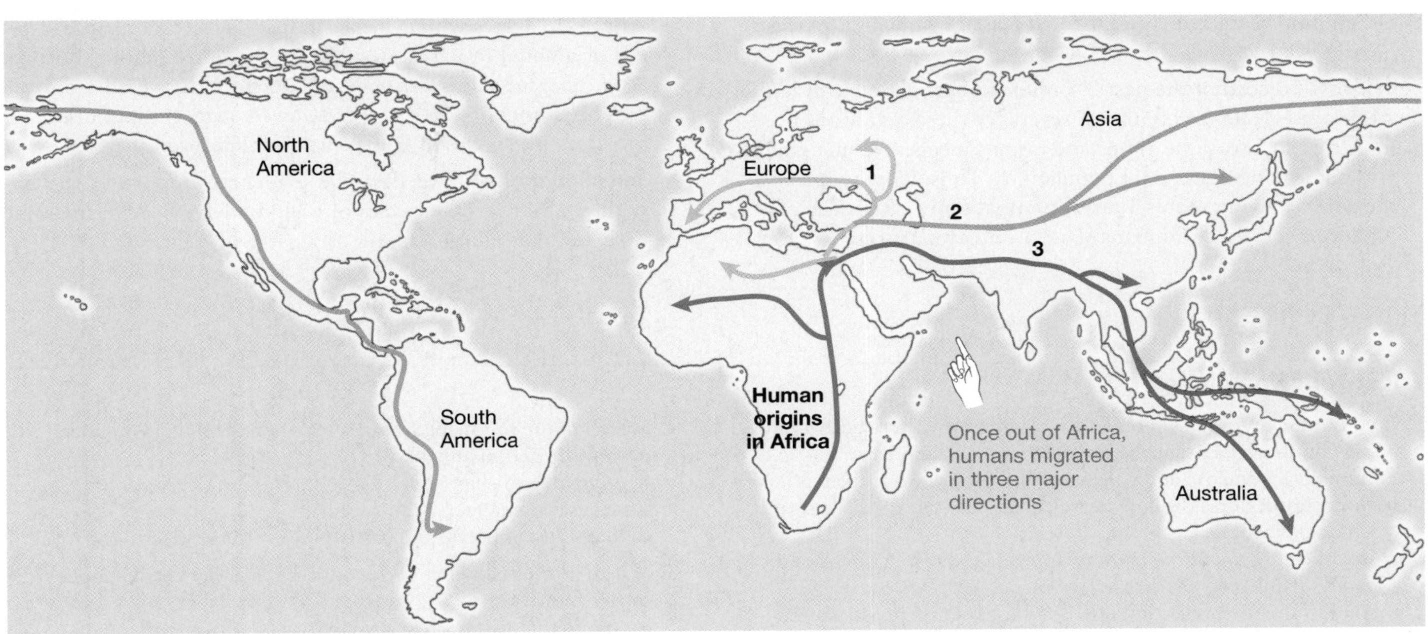

FIGURE 34.44 *Homo sapiens* Originated in Africa and Spread throughout the World. The phylogeny in Figure 34.43 supports the hypothesis that humans originated in Africa and spread out in three major migrations: to southeast Asia and the Pacific Islands, to Europe, and to northeast Asia and the New World.

Chapter Review

SUMMARY OF KEY CONCEPTS

The most species-rich deuterostome lineages are the echinoderms and the vertebrate groups called ray-finned fishes and tetrapods.

Echinoderms, ray-finned fishes, and tetrapods are also the most important large-bodied predators and herbivores in marine and terrestrial environments.

You should be able to describe what food chains in aquatic and terrestrial environments would look like if echinoderms, ray-finned fishes, and tetrapods did not exist.

Echinoderms and vertebrates have unique body plans. Echinoderms are radially symmetric as adults and have a water vascular system. All vertebrates have a skull and an extensive endoskeleton made of cartilage or bone.

Echinoderm larvae are bilaterally symmetric but undergo a metamorphosis into radially symmetric adults. Their water vascular system is composed of fluid-filled tubes and chambers and extends from the body wall in projections called podia. Podia can extend and retract in response to muscle contractions that move fluid inside the water vascular system.

Chordates are distinguished by the presence of a notochord, a dorsal hollow nerve cord, pharyngeal gill slits, and a muscular tail that extends past the anus. Vertebrates are distinguished by the presence of a cranium; most species also have vertebrae. In more derived groups of vertebrates, the body plan features an extensive endoskeleton composed of bone.

You should be able to sketch a sea star and a cephalochordate and label aspects of the body plan that qualify as synapomorphies.

(MB) **Web Animation** at www.masteringbio.com
Deuterostome Diversity

The diversification of echinoderms was triggered by the evolution of appendages called podia; the diversification of vertebrates was driven by the evolution of the jaw and limbs.

Most echinoderms move via their podia, and many species suspension feed, deposit feed, or act as predators with the aid of their podia.

Ray-finned fishes and tetrapods use their jaws to bite food and process it with teeth. Species in both groups move when muscles attached to their endoskeletons contract or relax. Tetrapods can move on land because their limbs enable walking, running, or flying. The evolution of the amniotic egg allowed tetrapods to lay eggs on land. Parental care was an important adaptation in some groups of ray-finned fishes and tetrapods—particularly mammals.

You should be able to explain why adaptations for efficient movement and feeding led to evolutionary success in terms of numbers and diversity of species in echinoderms and vertebrates.

Humans are a tiny twig on the tree of life. Chimpanzees and humans diverged from a common ancestor that lived in Africa

6–7 million years ago. Since then, at least 14 humanlike species have existed.

The fossil record of the past 3.5 million years contains at least 14 distinct species of hominins. Several of these organisms lived in Africa at the same time, and some lineages went extinct without leaving descendant populations. Thus, *Homo sapiens* is the sole surviving representative of an adaptive radiation. The phylogeny of living humans, based on comparisons of DNA sequences, agrees with evidence in the fossil record that *H. sapiens* originated in Africa and later spread throughout Europe, Asia, and the Americas. DNA sequences recovered from the fossilized bones of *H. neanderthalensis* suggest that *H. sapiens* replaced this species in Europe without interbreeding.

You should be able to describe evidence supporting the hypotheses that several species of hominin have lived at the same time and that *Homo sapiens* originated in Africa.

QUESTIONS

Test Your Knowledge

1. If you found an organism on a beach, what characteristics would allow you to declare that the organism is an echinoderm?
   - **a.** radially symmetric adults, spines, and presence of tube feet
   - **b.** notochord, dorsal hollow nerve cord, pharyngeal gill slits, and muscular tail
   - **c.** exoskeleton and three pairs of appendages; distinct head and body (trunk) regions
   - **d.** mouth that forms second (after the anus) during gastrulation
2. What is the diagnostic trait of vertebrates?
   - **a.** skull
   - **b.** jaws
   - **c.** endoskeleton constructed of bone
   - **d.** endoskeleton constructed of cartilage
3. Why are the pharyngeal jaws found in many ray-finned fishes important?
   - **a.** They allow the main jaw to be protrusible (extendible).
   - **b.** They make it possible for individuals to suck food toward their mouths.
   - **c.** They give rise to teeth that are found on the main jawbones.
   - **d.** They help process food.
4. Which of the following lineages make up the living Amniota?
   - **a.** reptiles and mammals
   - **b.** viviparous fishes
   - **c.** frogs, salamanders, and caecilians
   - **d.** hagfish, lampreys, and cartilaginous fishes (sharks and rays)
5. Which of the following does *not* occur in either cartilaginous fishes or ray-finned fishes?
   - **a.** internal fertilization and viviparity or ovoviviparity
   - **b.** external fertilization and oviparity
   - **c.** formation of a placenta
   - **d.** feeding of young
6. Most species of hominins are known only from Africa. Which species have been found in other parts of the world as well?
   - **a.** early *Homo*—*H. habilis* and *H. ergaster*
   - **b.** *H. erectus*, *H. neanderthalensis*, and *H. floresiensis*
   - **c.** gracile australopithecines
   - **d.** robust australopithecines

Test Your Knowledge answers: 1. a; 2. a; 3. d; 4. a; 5. d; 6. b

Test Your Understanding

Answers are available at www.masteringbio.com

1. Explain how the water vascular system of echinoderms functions as a type of hydrostatic skeleton.
2. List the four morphological traits that distinguish chordates. How are these traits involved in locomotion and feeding in larvae or adults?
3. Describe evidence that supports the hypothesis that jaws evolved from gill arches in fish.
4. Describe evidence that supports the hypothesis that the tetrapod limb evolved from the fins of lobe-finned fishes.
5. The text claims that "*Homo sapiens* is the sole survivor of an adaptive radiation that took place over the past 7 million years." Do you agree with this statement? Why or why not? Identify three major trends in the evolution of hominins, and suggest a hypothesis to explain each.
6. Explain how the evolution of the placenta and lactation in mammals improved the probability that their offspring would survive. Over the course of a lifetime, why are female mammals expected to produce fewer eggs than do female fish?

Applying Concepts to New Situations

Answers are available at www.masteringbio.com

1. Describe the conditions under which it might be possible for a new species of *Homo* to evolve from current populations of *H. sapiens*.
2. Compare and contrast adaptations that triggered the diversification of the three most species-rich animal lineages: molluscs, arthropods, and vertebrates.
3. Aquatic habitats occupy 73 percent of Earth's surface area. How does this fact relate to the success of ray-finned fishes? How does it relate to the success of coelacanths and other lobe-finned fishes?
4. Mammals and birds are endothermic. Did they inherit this trait from a common ancestor, or did endothermy evolve independently in these two lineages? Provide evidence to support your answer.

www.masteringbio.com is also your resource for • Answers to text, table, and figure caption questions and exercises • Answers to *Check Your Understanding* boxes • Online study guides and quizzes • Additional study tools including the *E-Book for Biological Science* 3rd ed., textbook art, animations, and videos.

Viruses

35

Photomicrograph created by treating seawater with a fluorescing compound that binds to nucleic acids. The smallest, most abundant dots are viruses. The larger, numerous spots are bacteria and archaea. The largest splotches are protists.

KEY CONCEPTS

- Viruses are tiny, noncellular parasites that infect virtually every type of cell known. They cannot perform metabolism on their own—meaning outside a parasitized cell—and are not considered to be alive. Different types of viruses infect particular species and types of cells.
- Although viruses are diverse morphologically, they can be classified as two general types: enveloped and nonenveloped.
- The viral infection cycle can be broken down into six steps: (1) entry into a host cell, (2) production of viral proteins, (3) replication of the viral genome, (4) assembly of a new generation of virus particles, (5) exit from the infected cell, and (6) transmission to a new host.
- In terms of diversity, the key feature of viruses is the nature of their genetic material. The genomes of viruses may consist of double-stranded DNA, single-stranded DNA, double-stranded RNA, or one of several types of single-stranded RNA.

A **virus** is an obligate, intracellular parasite. The adjective *obligate* is appropriate because viruses are completely dependent on host cells. The adjective *intracellular* is appropriate because viruses cannot replicate unless they enter a cell. The noun *parasite* is appropriate because viruses reproduce at the expense of their host cells.

Viruses can also be defined by what they are not. They are not cells and are not made up of cells, so they are not considered organisms. They cannot manufacture their own ATP or amino acids or nucleic acids, and they cannot produce proteins on their own. Viruses enter a **host cell**, take over its biosynthetic machinery, and use that machinery to manufacture a new generation of viruses. Outside of host cells, viruses cannot do anything—they simply exist.

Because they are not organisms, viruses are referred to as particles or agents and are not given scientific (genus + species) names. Most biologists would argue that viruses are not even alive. Yet viruses have a genome, they are superbly adapted to exploit the metabolic capabilities of their host cells, and they evolve. **Table 35.1** summarizes some characteristics of viruses.

The diversity and abundance of viruses almost defy description. Each type of virus infects a specific unicellular species or cell type in a multicellular species, and nearly all organisms examined thus far are parasitized by at least one kind of virus. The bacterium *Escherichia coli*, which resides in the human intestine, is afflicted by several dozen types of **bacteriophage** (literally, "bacteria-eater"). A bacteriophage is a virus that

Key Concept Important Information Practice It

SUMMARY TABLE **35.1 Characteristics of Viruses versus Characteristics of Organisms**

	Viruses	Organisms
Hereditary material	DNA or RNA; can be single stranded or double stranded	DNA; always double stranded
Plasma membrane present?	No	Yes
Can carry out transcription independently?	No—even if a viral polymerase is present, transcription of viral genome requires use of ATP and nucleotides provided by host cell	Yes
Can carry out translation independently?	No	Yes
Metabolic capabilities	Virtually none	Extensive—synthesis of ATP, reduced carbon compounds, vitamins, lipids, nucleic acids, etc.

infects bacteria. The surface waters of the world's oceans teem with bacteria and archaea, yet viruses outnumber them in this habitat by a factor of 10 to 1. A wine bottle filled with seawater taken from the ocean's surface contains about 10 billion virus particles—over one and a half times the world's population of humans.

35.1 Why Do Biologists Study Viruses?

Any study of life's diversity would be incomplete unless it included a look at the acellular parasites that exploit that diversity. But viruses are also important from a practical standpoint. To health-care workers, agronomists, and foresters, these parasites are a persistent—and sometimes catastrophic—source of misery and economic loss. The nature of viruses has been understood only since the 1940s, but they have been the focus of intense research ever since. Biologists study viruses because they cause illness and death. In the human body, virtually every system, tissue, and cell can be infected by one or more kinds of virus (**Figure 35.1**). Research on viruses is motivated by the desire to minimize the damage they can cause.

In addition, biologists study viruses with the goal of exploiting their ability to enter cells. Recall from Chapter 19 that viruses are being tested as possible therapeutic agents in the treatment of genetic diseases. If viruses can be engineered to carry normal copies of human genes into the cells of patients, it is possible that the gene products could cure symptoms.

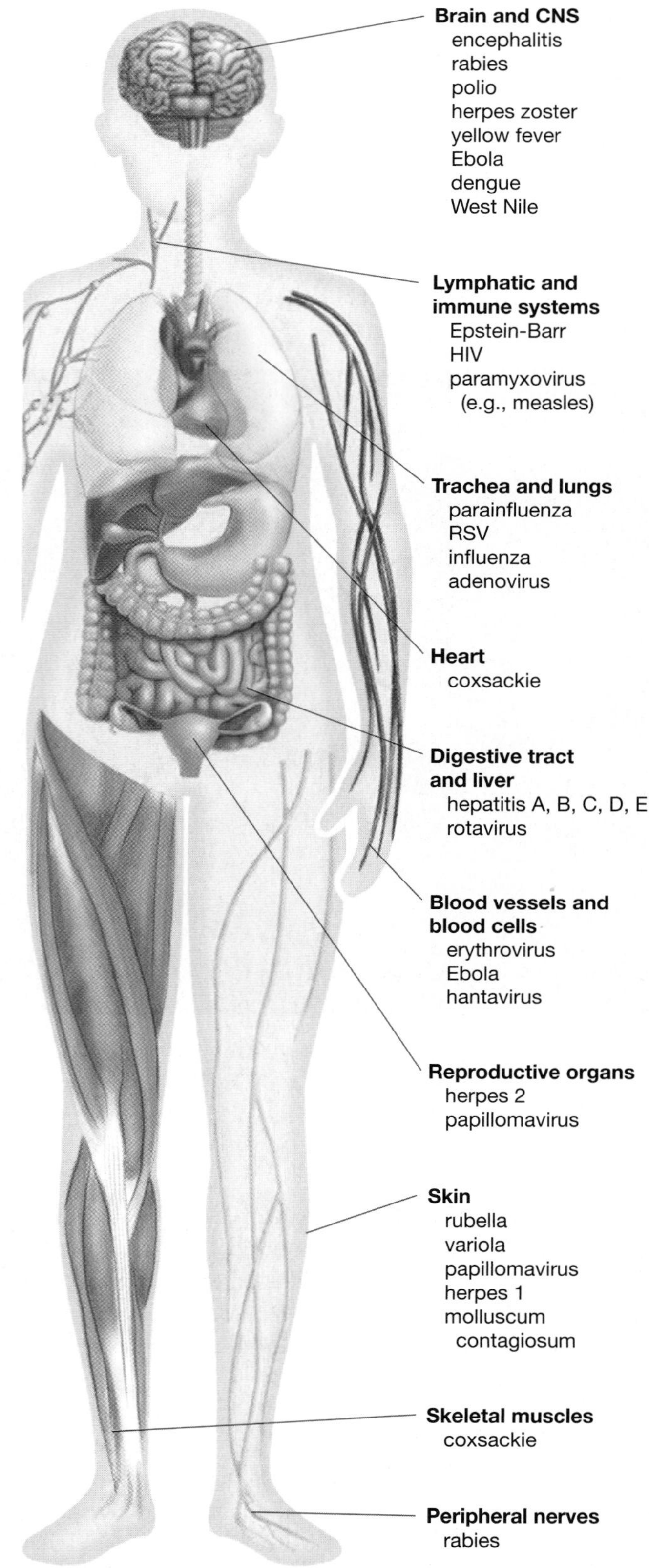

FIGURE 35.1 Human Organs and Systems That Are Parasitized by Viruses.

EXERCISE Choose two viruses that you are familiar with. Next to each, write the symptoms caused by an infection with this virus.

Recent Viral Epidemics in Humans

Physicians and researchers use the term **epidemic** (literally, "upon-people") to describe a disease that rapidly affects a large number of individuals over a widening area. Viruses have caused the most devastating epidemics in recent human history. During the eighteenth and nineteenth centuries, it was not unusual for Native American tribes to lose 90 percent of their members over the course of a few years to measles, smallpox, and other viral diseases spread by contact with European settlers. To appreciate the impact of these epidemics, think of 10 close friends and relatives—then remove nine.

An epidemic that is worldwide in scope is called a **pandemic**. The influenza outbreak of 1918–1919, called "Spanish flu," qualifies as the most devastating pandemic recorded to date. Influenza is a virus that infects the upper respiratory tract. The strain of influenza virus that emerged in 1918 infected people worldwide and was particularly **virulent**—meaning it tended to cause severe disease. Within hours of showing symptoms, the lungs of previously healthy people often became so heavily infected that affected individuals suffocated to death. Most victims were between the ages of 20 and 40. The viral outbreak occurred just as World War I was drawing to a close and killed far more people than did the conflict itself. For example, ten times as many Americans died of influenza than were killed in combat in the war. Worldwide, the Spanish flu is thought to have killed 20–50 million people.

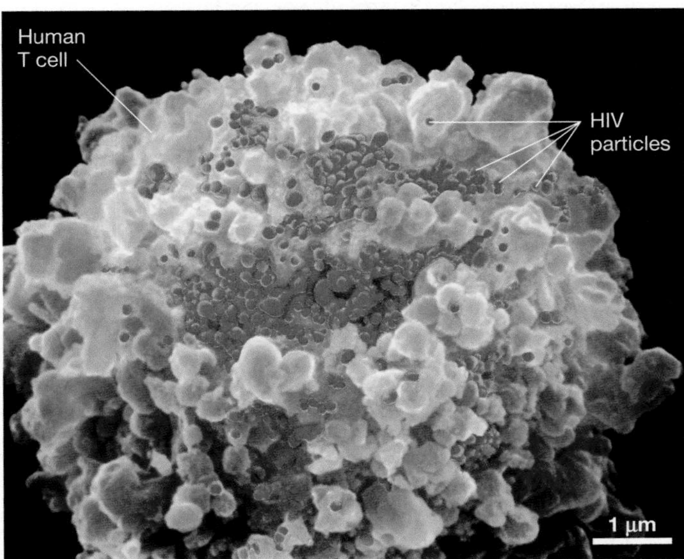

FIGURE 35.2 The Human Immunodeficiency Virus (HIV). Colorized scanning electron micrograph showing HIV particles emerging from an infected human T cell (a type of immune system cell).

Current Viral Epidemics in Humans: HIV

In terms of the total number of people affected, the measles and smallpox epidemics among native peoples of the Americas and the 1918 influenza outbreak are almost certain to be surpassed by the incidence of AIDS. **Acquired immune deficiency syndrome** (**AIDS**) is an affliction caused by the **human immunodeficiency virus** (**HIV**; **Figure 35.2**). HIV is now the most intensively studied of all viruses. Since the early 1980s, governments and private corporations from around the world have spent hundreds of millions of dollars on HIV research. More biologists are working on HIV than on any other type of virus. Given this virus's current and projected impact on human populations around the globe, the investment is justified.

How Does HIV Cause Disease? Like other viruses, HIV parasitizes specific types of cells. The cells most affected by HIV are called helper T cells and macrophages. These cells are components of the **immune system**, which is the body's defense system against disease. Chapter 49 explains just how crucial helper T cells and macrophages are to the immune system's response to invading bacteria and viruses. If an HIV particle succeeds in infecting one of these cells and reproducing inside, however, the cell dies as hundreds of new virus particles break out and infect more cells. Although the body continually replaces helper T cells and macrophages, the number produced does not keep pace with the number being destroyed by HIV. As a result, the total number of helper T cells in the bloodstream gradually declines as an HIV infection proceeds (**Figure 35.3**). When the T-cell count drops, the immune system's responses to invading bacteria and viruses become less and less effective. Eventually, too few helper T cells are left to fight off pathogens efficiently, and pathogenic bacteria and viruses

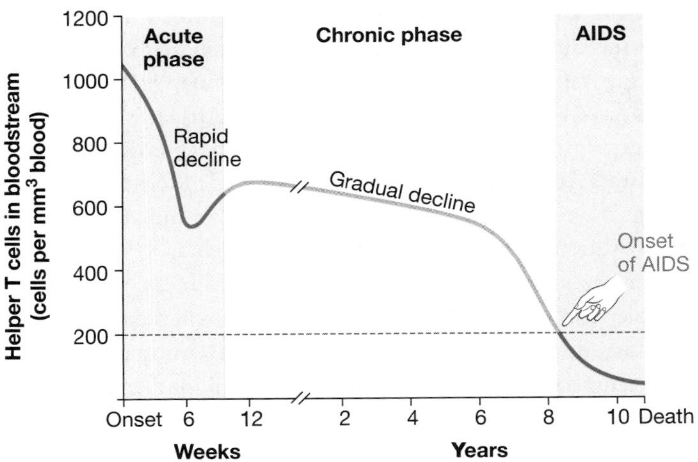

FIGURE 35.3 T-Cell Counts Decline during an HIV Infection. Graph of changes in the number of T cells that are present in the bloodstream over time, based on data from a typical patient infected with HIV. The acute phase of infection occurs immediately after infection and is sometimes associated with symptoms such as fever. Infected people usually show no disease symptoms in the chronic phase, even though their T-cell counts are in slow, steady decline. AIDS typically occurs when T-cell counts dip below 200/mm^3 of blood.

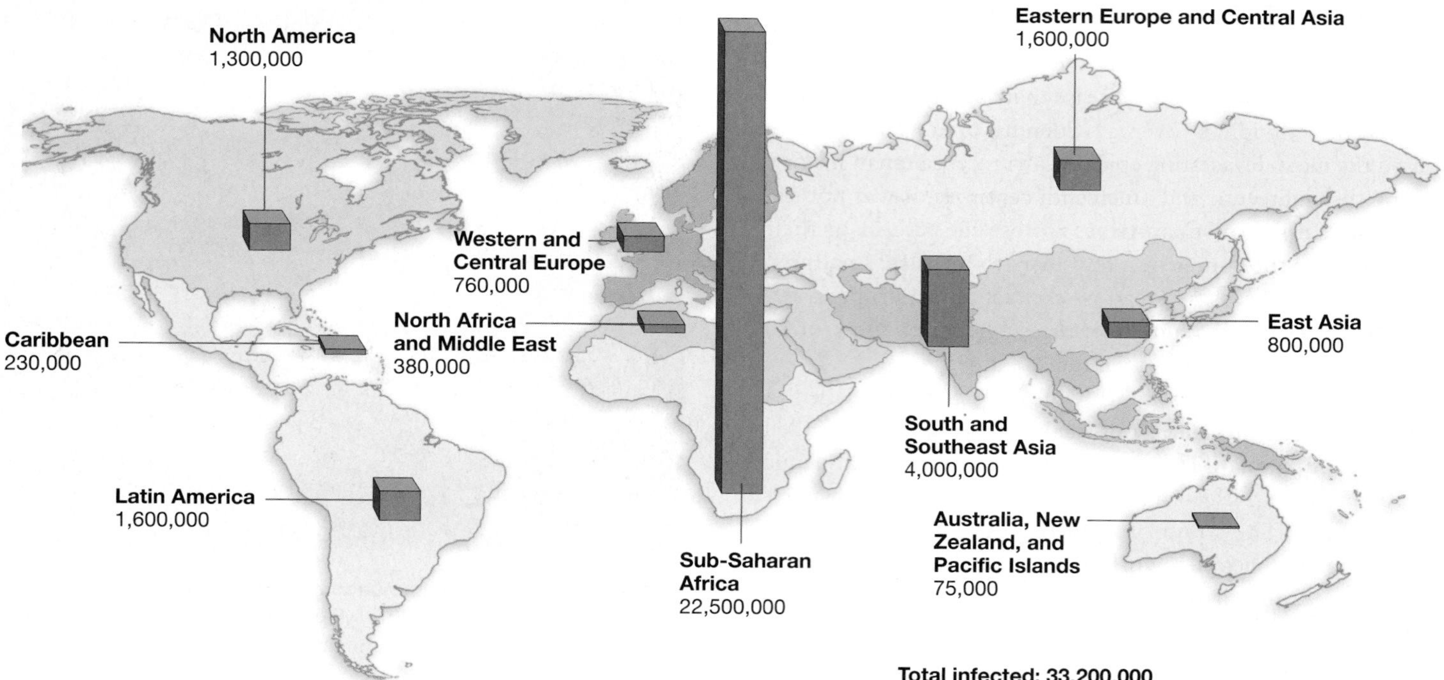

FIGURE 35.4 Geographic Distribution of HIV Infections. Data, compiled by the United Nations AIDS program, showing the numbers of people living with HIV in December 2007, by geographic area.

begin to multiply unchecked. In almost all cases, one or more of these infections proves fatal. HIV kills people indirectly—by making them susceptible to pneumonia, fungal infections, and unusual types of cancer.

What Is the Scope of the AIDS Epidemic? Researchers with the United Nations AIDS program estimate that AIDS has already killed 28 million people worldwide. HIV infection rates have been highest in east and central Africa, where one of the greatest public health crises in history is now occurring (**Figure 35.4**). In Botswana and elsewhere, blood-testing programs have confirmed that over 20 percent of individuals carry HIV. Although there may be a lag of as much as 8–12 years between the initial infection and the onset of illness, virtually all people who become infected with the virus will die of AIDS.

Currently, the UN estimates the total number of HIV-infected people worldwide at about 33.2 million. An additional 2.5 million people are infected each year, and the pandemic is growing. Researchers are particularly alarmed because the focus of the epidemic is shifting from its historical center of incidence—central and southern Africa—to south and east Asia. Infection rates are growing rapidly in some of the world's most populous countries—particularly India and China.

Most viral and bacterial diseases afflict the very young and the very old. But because HIV is a sexually transmitted disease, young adults are most likely to contract the virus and die. People who become infected in their late teens or twenties die of AIDS in their twenties or thirties. Tens of millions of people are being lost in the prime of their lives. Physicians, politicians, educators, and aid workers all use the same word to describe the epidemic's impact: staggering.

35.2 How Do Biologists Study Viruses?

Researchers who study viruses focus on two goals: (1) developing vaccines that help hosts fight off disease if they become infected and (2) developing antiviral drugs that prevent a virus from replicating efficiently inside the host. Both types of research begin with attempts to isolate the virus in question.

Isolating viruses takes researchers into the realm of "nanobiology," in which structures are measured in billionths of a meter. (One nanometer, abbreviated nm, is 10^{-9} meter.) Viruses range from about 20 to 300 nm in diameter. They are dwarfed by eukaryotic cells and even by bacterial cells (**Figure 35.5**). Millions of viruses can fit on the head of a pin.

If virus-infected cells can be grown in culture or harvested from a host individual, researchers can usually isolate the virus by passing the cells through a filter. The filters used to study viruses have pores that are large enough for viruses to pass through but are too small to admit cells. To test the hypothesis that the solution passing through the filter contains viruses that cause a specific disease, researchers expose uninfected host cells to this filtrate. If exposing host cells to the filtrate results in infection, then the virus-causation hypothesis is correct. In this

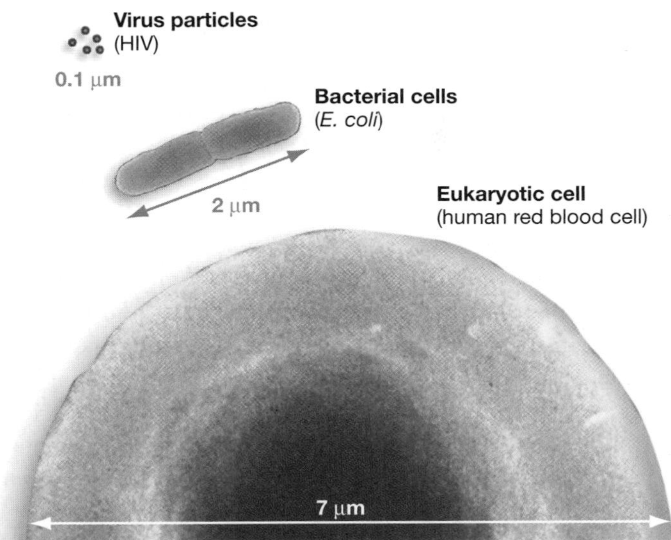

FIGURE 35.5 Viruses Are Tiny. Viruses are much smaller than bacterial cells or eukaryotic cells.

QUESTION What is the diameter of these virus particles in nanometers?

way, researchers can isolate a virus and confirm that it is the causative agent of infection. Recall from Chapter 28 that these steps are inspired by Koch's postulates, which established the criteria for linking a specific infectious agent with a specific disease.

Once biologists have isolated a virus, how do they study and characterize it? Let's begin with morphological traits, then consider how viral replication cycles vary.

Analyzing Morphological Traits

To see a virus, researchers usually rely on transmission electron microscopy (see **BioSkills 8**). Only the very largest viruses, such as the smallpox virus, are visible with a light microscope. Electron microscopy has revealed that viruses come in a wide variety of shapes, and many viruses can be identified by shape alone (**Figure 35.6**). In overall structure, however, viruses fall into just two general categories. Viruses can either be (1) enclosed by just a shell of protein called a **capsid** or (2) enclosed by both a capsid and a membrane-like **envelope**. Regarding their morphology, then, the important distinction among viruses is whether they are nonenveloped or enveloped.

Nonenveloped viruses consist of genetic material and possibly one or more enzymes inside a capsid—a protein coat. The nonenveloped virus illustrated in **Figure 35.7a** is an adenovirus. You undoubtedly have adenoviruses on your tonsils or in other parts of your upper respiratory passages right now. As the micrograph in Figure 35.6d shows, the morphology of nonenveloped viruses may be complex.

Enveloped viruses also have genetic material inside a capsid, but the capsid is surrounded by an envelope. The envelope consists of a phospholipid bilayer with a mixture of viral proteins and proteins derived from a membrane found in a host cell—specifically, the host cell in which the virus particle was manufactured (**Figure 35.7b**). Later sections in the chapter will detail how most of these viruses obtain their envelope from an infected host cell.

Once a virus has been isolated and its overall morphology characterized, researchers usually focus on understanding the nature of the virus's replication cycle and on attempts to develop a vaccine. A **vaccine** is a preparation that primes a host's immune system to respond to a specific type of virus. **Box 35.1** explains how vaccines work and why it has not been possible to develop a vaccine for certain viruses—particularly the flu viruses or HIV. Here let's focus on variation in virus replication cycles.

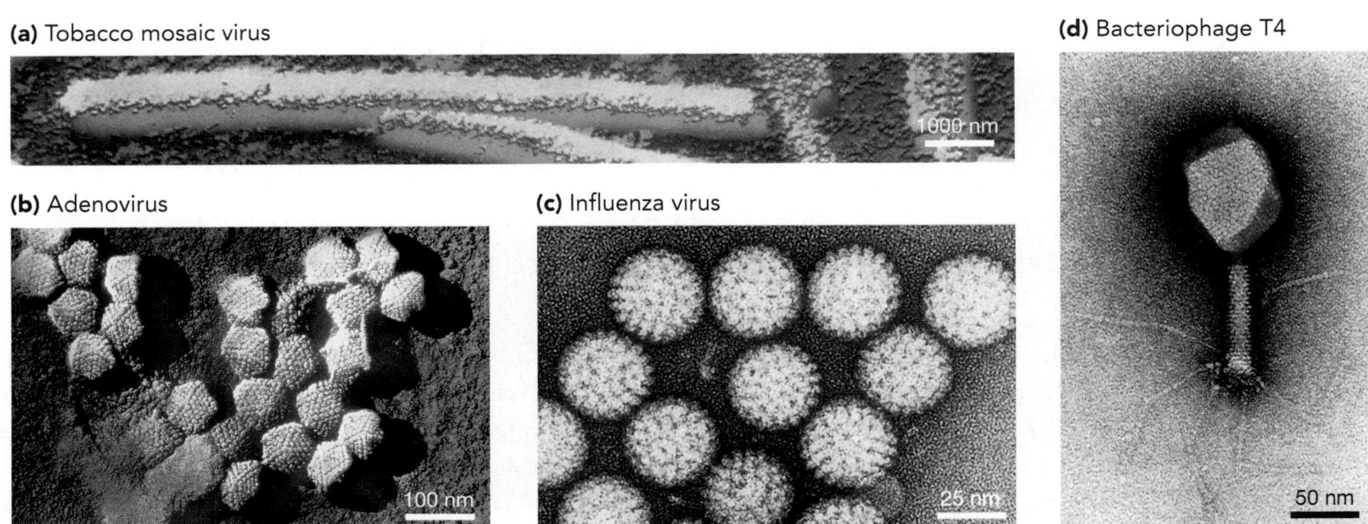

FIGURE 35.6 Viruses Vary in Size and Shape. Virus shapes include **(a)** rods, **(b)** polyhedrons, **(c)** spheres, and **(d)** complex shapes with "heads" and "tails."

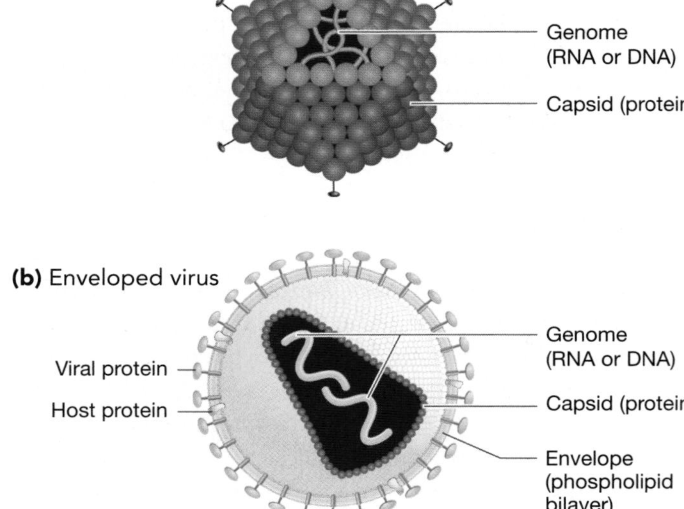

FIGURE 35.7 Viruses Are Nonenveloped or Enveloped. (a) A protein coat forms the exterior of nonenveloped viruses. **(b)** In enveloped viruses, the exterior is composed of a membranous sphere. Inside this envelope, the hereditary material is enclosed by a protein coat.

Analyzing Variation in Growth Cycles: Replicative and Latent Growth

Although it is likely that millions of types of virus exist, they all infect their host cells in one of two general ways: via replicative growth or in a dormant form referred to as latency in animal viruses or lysogeny in bacteriophages. All viruses undergo replicative growth; some can halt the replication cycle and enter a dormant state as well. Both types of viral infection begin when part or all of a virus particle enters the interior of a host cell.

Figure 35.8a shows a **lytic cycle**, or replicative growth. As illustrated, the viral genome enters the host cell and viral or host enzymes make copies of it, using nucleotides and ATP provided by the host. The host cell also manufactures viral proteins. When synthesis of the viral genome and viral proteins is complete, a new generation of virus particles assembles inside the host cell. A mature virus particle is called a **virion**. The replicative cycle is complete when the virions exit the cell—usually killing the host cell in the process.

Figure 35.8b diagrams a **lysogenic cycle**, or lysogeny, in a bacteriophage. Only certain types of viruses are capable of a latent infection. During lysogenic or latent growth, viral DNA

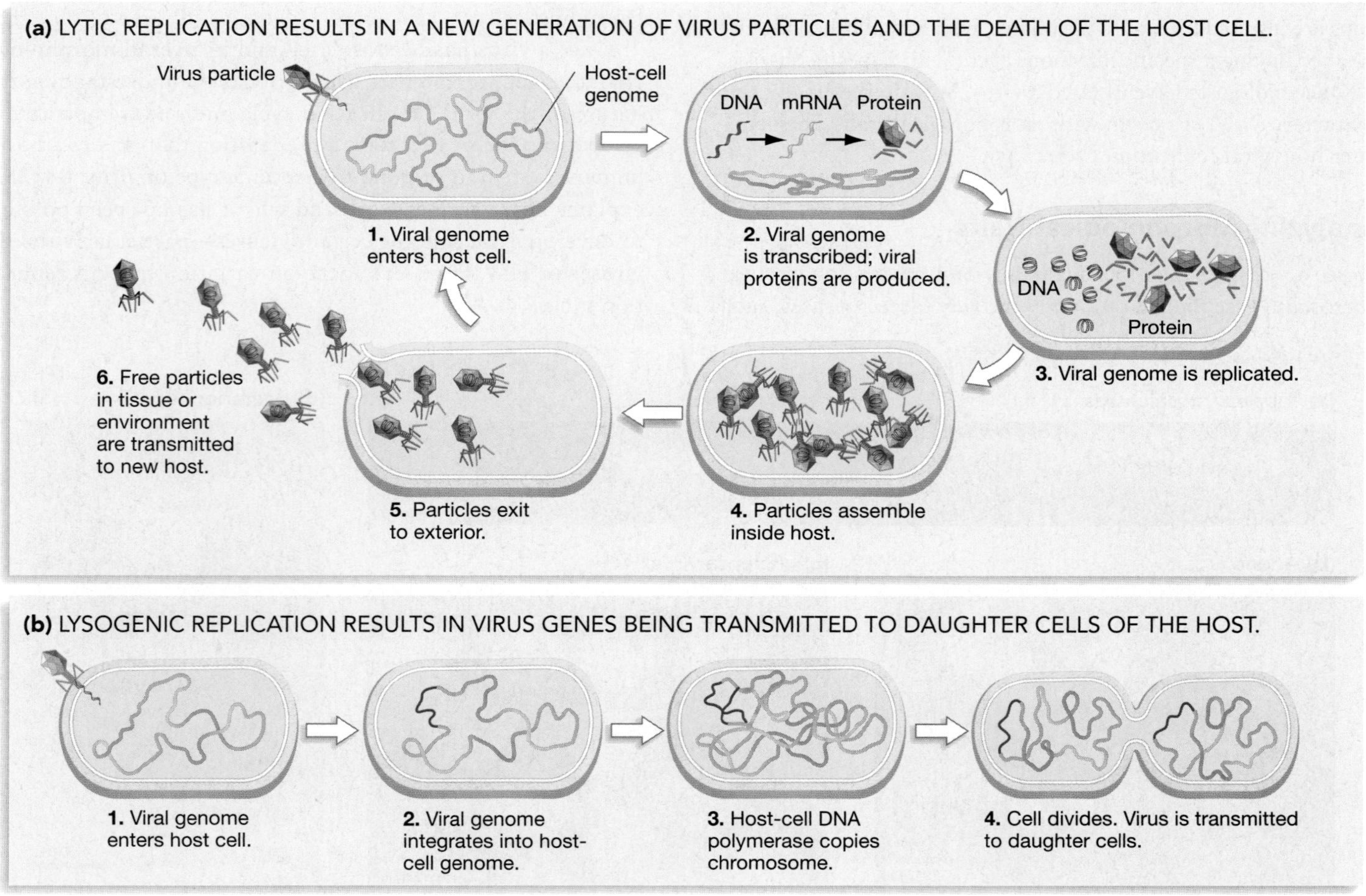

FIGURE 35.8 Viruses Replicate via Lytic or Lysogenic Cycles, or Both. (a) All viruses follow the same general lytic replication cycle. **(b)** Some viruses are also capable of lysogeny, meaning that their genome can become integrated into the host-cell chromosome.

● **EXERCISE** Compare and contrast a lysogenic virus to the transposable elements introduced in Chapter 20.

becomes incorporated into the host's chromosome. Often this integration occurs without serious damage to the host cell. Once the viral genome is in place, it is replicated by the host's DNA polymerase each time the cell divides. Copies of the viral genome are passed on to daughter cells just like one of the host's own genes. In the lysogenic or latent state, a virus is quiescent. This means that no new particles are being produced and no unrelated cells are being infected. The virus is transmitted from one generation to the next along with the host's genes. HIV and other retroviruses are also capable of being transmitted to daughter cells when they are latent. ● If you understand this concept, you should be able to explain the contrast in how viral genes are transmitted to offspring via lytic versus lysogenous growth.

It is not possible to treat a latent infection with drugs called **antivirals**, because the viruses are quiescent—they just sit there. But even lytic infections are notoriously difficult to treat because viruses use so many of the host cell's enzymes during the lytic replication cycle. Drugs that disrupt these enzymes usually harm the host much more than they harm the virus. To understand viral diversity and how antiviral drugs are developed, let's consider each phase of the lytic cycle in more detail.

Analyzing the Phases of the Replicative Cycle

● Six phases are common to replicative growth in virtually all viruses: (1) entry into a host cell, (2) transcription of the viral genome and production and processing of viral proteins, (3) replication of the viral genome, (4) assembly of a new generation of virions, (5) exit from the infected cell, and (6) transmission to a new host; corresponding steps are depicted in Figure 35.8a. Each virus has a particular way of entering a host cell and completing the subsequent phases of the cycle. Let's take a closer look.

BOX 35.1 How Are Vaccines Developed?

As Chapter 49 will show, one of the primary ways that an individual's immune system responds to parasite attack is by producing antibodies. An **antibody** ("against-body") is a protein that binds with high specificity to a particular site on another molecule. Any molecule that elicits the production of antibodies—and that antibodies stick to—is called an **antigen** ("against-produce").

Vaccines contain antigens. The antigens are usually components of a virus's exterior—the capsid from a nonenveloped virus or the envelope proteins from an enveloped virus. Exterior proteins are effective antigens because they are the part of the virus that can be attacked by antibodies, outside the cell.

To vaccinate a person or a domestic animal, viral antigens are injected or swallowed. Once inside the individual's blood, the antigens stimulate immune system cells called B cells (see Chapter 49), which produce antibodies to the antigen (**Figure 35.9**). Once antigens are coated with antibodies, the antigens are destroyed by other cells and components of the immune system.

(Continued on next page)

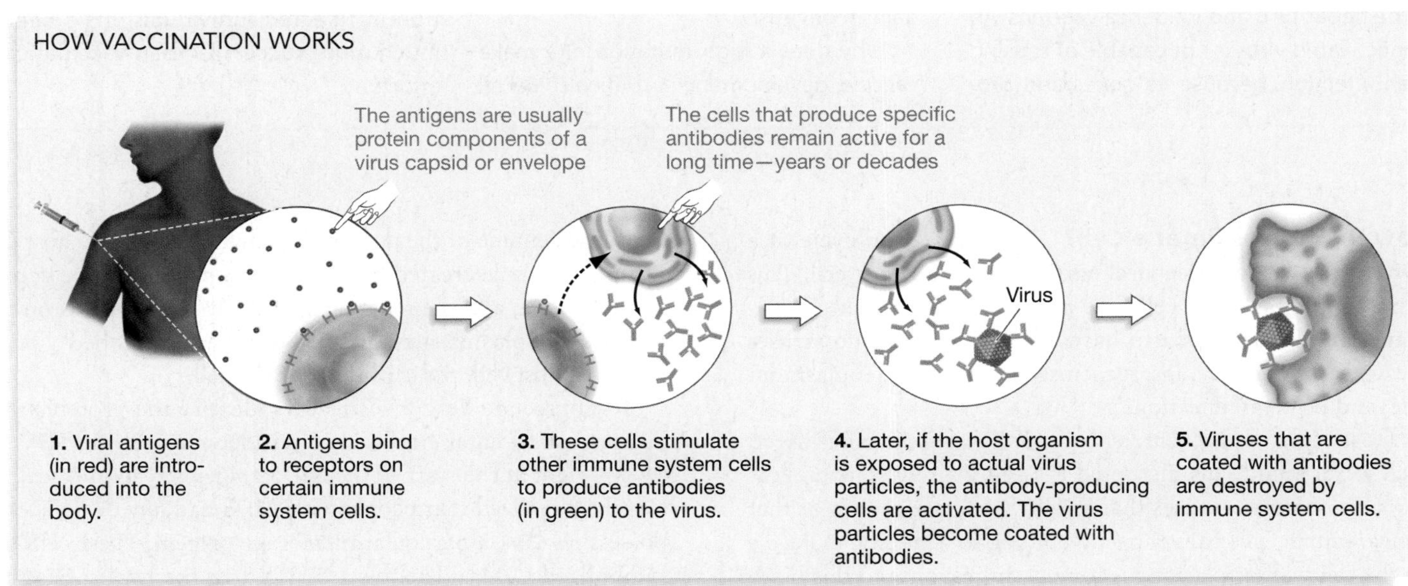

FIGURE 35.9 Vaccination Induces a Response from the Immune System. The immune system cells that respond to a vaccination are "immortalized," meaning that they stay active for a long time. Thus, the body can respond quickly to any future infections by the same virus.

Vaccinations cause fake or abortive infections. They function like fire drills or earthquake preparedness drills—they prepare the immune system to respond to a specific type of threat. Immune system cells that are stimulated to produce anti-bodies by a vaccination remain active in the vaccinated person for a long time—years or decades. If a vaccinated person is later exposed to the antigen in the form of active virus particles, the immune system can respond quickly and effectively enough to stop the infection before it threatens the individual's health.

In many cases, vaccines have been spectacularly successful at curbing or even eliminating viral diseases in humans and domestic animals. In humans alone, effective vaccines have been developed against smallpox, yellow fever, polio, measles, and some forms of hepatitis. These diseases used to kill or sicken hundreds of thousands of people each year.

There Are Three General Types of Vaccines

Successful vaccines consist of isolated viral proteins, inactivated viruses, or attenuated viruses. Isolated viral proteins are called subunit vaccines; familiar examples include the hepatitis B and influenza vaccines. An inactivated virus is not capable of causing an infection, because its genes and proteins have been damaged by chemical treatments—often exposure to formaldehyde—or irradiation with ultraviolet light. If you have been vaccinated for hepatitis A, you may have received an inactivated virus.

Attenuated viruses are also called "live" virus vaccines, because they consist of complete virus particles. The adjective attenuated means that the viruses lack virulence. Researchers can make a virus harmless to a host by culturing the virus on cells from species other than that host. In adapting to growth on the atypical cells, viral strains usually lose the ability to grow rapidly in their normal host cells. Although attenuated viruses still provoke a strong immune response, they are not capable of sustaining an infection in a vaccinated individual. The smallpox, polio, and measles vaccines consist of attenuated viruses.

Why Isn't There an Effective Vaccine for HIV and the Flu Viruses?

Researchers have not succeeded in developing vaccines against HIV and have been only moderately successful in developing a vaccine against flu viruses. The reason is that flu viruses and HIV have exceptionally high mutation rates. HIV actually has the highest mutation rate observed in any organism or virus.

Why does a high mutation rate make vaccine development so difficult? Recall from Chapter 16 that a mutation is defined as a change in the genetic material—DNA in organisms; DNA or RNA in viruses. Because the enzymes that copy the genes found in flu viruses and HIV are exceptionally error prone, many of the mature virus particles that are produced contain mutations in the genes for the virus's envelope proteins. When these genes are transcribed and translated, the resulting envelope proteins are likely to have an altered structure. Due to the high mutation rates of flu viruses and HIV, the antigens presented by these viruses constantly change through time. Stated another way, new strains of these viruses—with novel antigens—are constantly evolving. The antibodies produced against the envelope proteins of earlier strains do not work against strains that appear later, because the antigens presented by the strains are different. A vaccination that protected an individual against certain strains of flu virus or HIV would not help against other strains.

To date, it has not been possible to design antigens from the flu viruses or HIV that protect vaccinated individuals effectively. In the fight against HIV, condom use and other methods to prevent infection, as well as drugs that inhibit viral enzymes and thus stop or slow viral replication in infected individuals, have been much more successful than vaccination efforts.

How Do Viruses Enter a Cell? The replication cycle of a virus begins when a free viral particle enters a target cell. This is no simple task. All cells are protected by a plasma membrane, and many cells also have a cell wall. How do viruses breach these defenses, insert themselves into the cytoplasm inside, and begin an infection?

Most plant viruses enter host cells after a sucking insect, such as an aphid, has disrupted the cell wall with its mouthparts. In contrast, viruses that parasitize bacterial cells or that attack animal cells gain entry by binding to a specific molecule on the cell wall or plasma membrane. In response to this binding event, viruses that attack bacteria release an enzyme called lysozyme that degrades the bacterial cell wall. (Lysozyme is also found in human tears, where it acts as an antibiotic in the eye.) The genome of the nonenveloped virus enters the host cell through the hole created by lysozyme, a portion of the capsid seals the hole, and the remainder of the capsid remains on the cell wall or plasma membrane. But when enveloped viruses bind to a host cell, the capsid enters the cell.

To appreciate how investigators identify the proteins that viruses use to enter host cells, consider research on HIV. In 1981—right at the start of the AIDS epidemic—biomedical researchers realized that people with AIDS had few or no T cells possessing **CD4**, a particular membrane protein. These cells are symbolized $CD4^+$. This discovery led to the hypothesis that CD4 functions as the "doorknob" that HIV uses to enter host cells. The doorknob hypothesis predicts that if CD4 is blocked, then HIV will not be able to enter host cells.

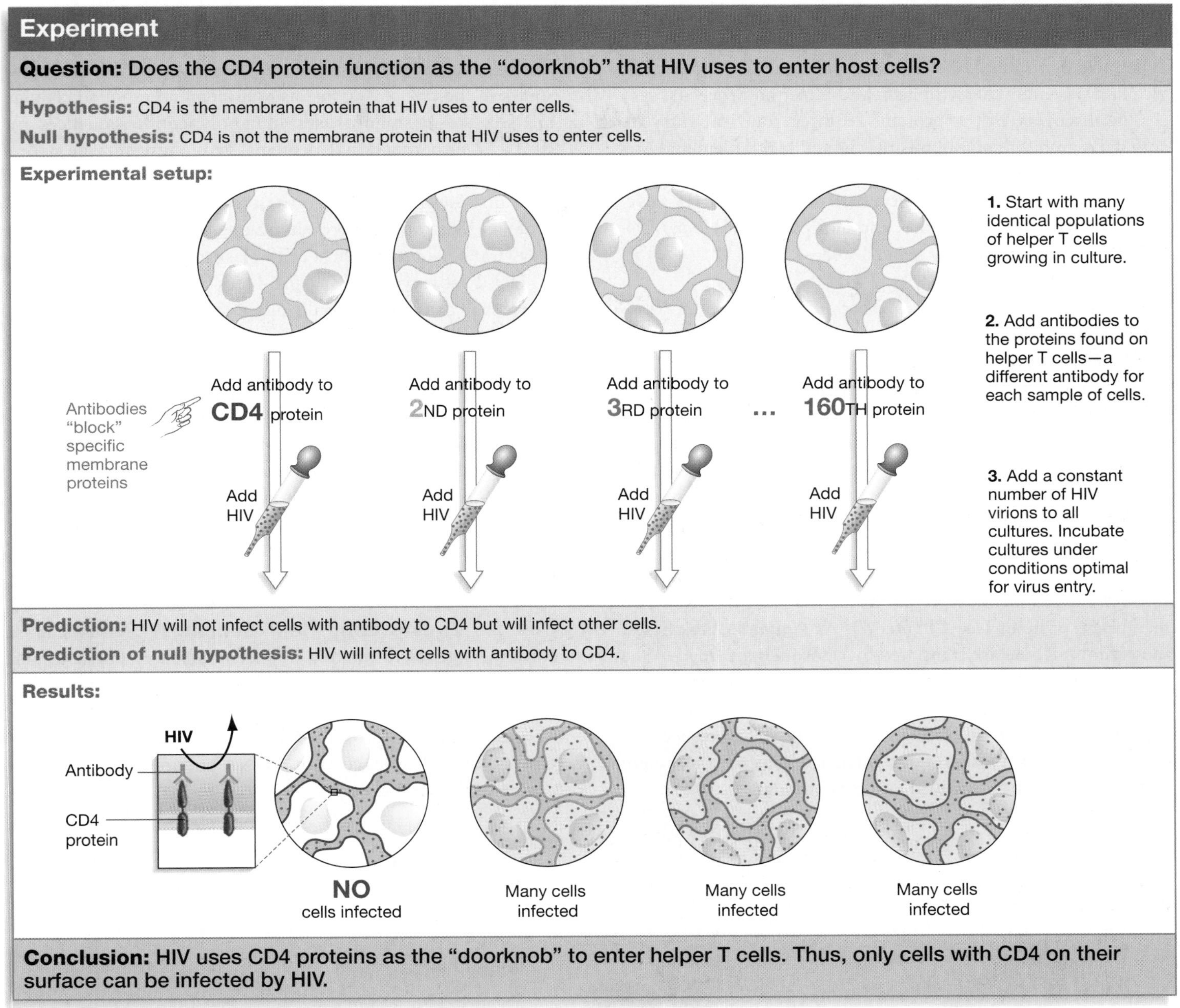

Experiment

Question: Does the CD4 protein function as the "doorknob" that HIV uses to enter host cells?

Hypothesis: CD4 is the membrane protein that HIV uses to enter cells.

Null hypothesis: CD4 is not the membrane protein that HIV uses to enter cells.

Experimental setup:

1. Start with many identical populations of helper T cells growing in culture.

2. Add antibodies to the proteins found on helper T cells—a different antibody for each sample of cells.

Antibodies "block" specific membrane proteins

Add antibody to CD4 protein | Add antibody to 2ND protein | Add antibody to 3RD protein | ... | Add antibody to 160TH protein

Add HIV | Add HIV | Add HIV | Add HIV

3. Add a constant number of HIV virions to all cultures. Incubate cultures under conditions optimal for virus entry.

Prediction: HIV will not infect cells with antibody to CD4 but will infect other cells.

Prediction of null hypothesis: HIV will infect cells with antibody to CD4.

Results:

HIV

Antibody

CD4 protein

NO cells infected | Many cells infected | Many cells infected | Many cells infected

Conclusion: HIV uses CD4 proteins as the "doorknob" to enter helper T cells. Thus, only cells with CD4 on their surface can be infected by HIV.

FIGURE 35.10 Experiments Confirmed that CD4 Is the Receptor Used by HIV to Enter Host Cells. In this experiment, the antibodies added to each culture bound to a specific protein found on the surface of helper T cells. Antibody binding blocked the membrane protein, so the protein could not be used by HIV to gain entry to the cells.

Two teams used the same experimental strategy to test this hypothesis (**Figure 35.10**). They began by growing large populations of helper T cells in culture. Then they took a sample of the cells and added an antibody to one of the cell-surface proteins found on helper T cells, along with HIV particles. They repeated this experiment 160 times but used a different sample of cells and added a different antibody each time. The key point here is that each of the 160 antibodies bound to and effectively blocked a different cell-surface protein. If one of the antibodies used in the experiment happened to bind to the receptor used by HIV, that antibody would cover up the receptor. In this way, the antibody would prevent HIV from entering the cell and protect that cell from infection. This approach led both research teams to reach exactly the same result: Only antibodies to CD4 protected the cells from viral entry.

Later work confirmed that HIV particles can enter cells only if the virions bind to a second membrane protein, called a **co-receptor**, in addition to CD4. In most individuals, proteins called CXCR4 and CCR5 function as co-receptors. If the proteins in a virion's envelope bind to both CD4 and a co-receptor, the lipid bilayers of the particle's envelope and the plasma membrane of the helper T cell fuse (**Figure 35.11**). When fusion occurs, HIV has breached the cell boundary. The viral capsid enters the cytoplasm and infection proceeds.

The discovery of the proteins required for viral entry inspired a search for compounds that would block them and prevent HIV from entering cells. Drugs that act in this way are called fusion inhibitors. A molecule that blocks the HIV envelope protein is currently in use, and molecules that block CD4, CXCR4, or CCR5 are now being tested.

Producing Viral Proteins Viruses cannot make the ribosomes and tRNAs necessary for translating their own mRNAs into proteins. For a virus to make the proteins required to produce a new generation of virus particles, it must exploit the host cell's biosynthetic machinery. Production of viral proteins begins soon after a virus enters a cell and continues after the viral genome is replicated. Viral mRNAs and proteins are produced and processed in one of two ways, depending on whether the proteins end up in the outer envelope of a particle or in the capsid.

RNAs that code for a virus's envelope proteins follow a route through the cell identical to that of the RNAs of the cell's own membrane proteins (see Chapter 7). As **Figure 35.12a** shows, these viral mRNAs are translated by ribosomes attached to the endoplasmic reticulum (ER). Afterward the resulting proteins are transported to the Golgi apparatus, where carbohydrate groups are attached, producing glycoproteins. The finished glycoproteins are then inserted into the plasma membrane, where they are ready to be assembled into new virions.

In contrast, a different route is taken by RNAs that code for proteins that make up the capsid or inner core of a virus particle (see **Figure 35.12b**). These RNAs are translated by ribosomes in the cytoplasm, just as non-membrane-bound cellular mRNAs are. In some viruses, long polypeptide sequences are later cut into functional proteins by a viral enzyme called **protease**. This enzyme cleaves viral polypeptides at specific locations—a critical step in the production of finished viral proteins. The resulting protein fragments are assembled into a new viral core near the host cell's plasma membrane.

The discovery that HIV produces a protease triggered a search for drugs that would block the enzyme. This search got a huge boost when researchers succeeded in visualizing the three-dimensional structure of HIV's protease, using the X-ray crystallographic techniques introduced in **BioSkills 8**. The molecule has an opening in its interior adjacent to the active site (**Figure 35.13a**). Polypeptides fit into the opening and are cleaved at the active site. Based on these data, researchers immediately began searching for molecules that could fit into the opening and prevent protease from functioning by blocking the active site (**Figure 35.13b**). Several HIV protease inhibitors are now on the market.

How Do Viruses Copy Their Genomes? Viruses must copy their genes to make a new generation of particles and continue an infection. Most viruses contain a polymerase that copies the viral genome inside the infected host cell, using nucleotides and ATP provided by the host cell. Many DNA viruses, for example, contain a DNA polymerase that makes a copy of the viral genome.

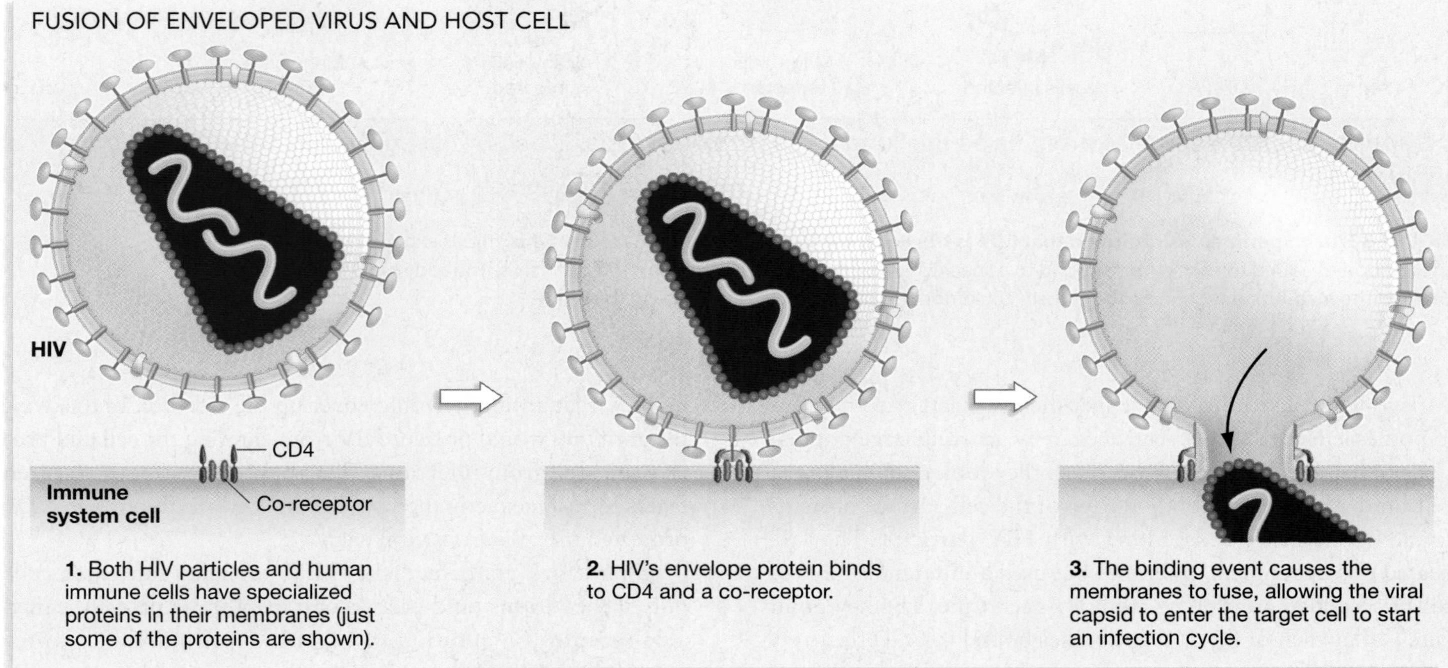

FIGURE 35.11 Enveloped Viruses Bind to Membrane Proteins of Target Cells.

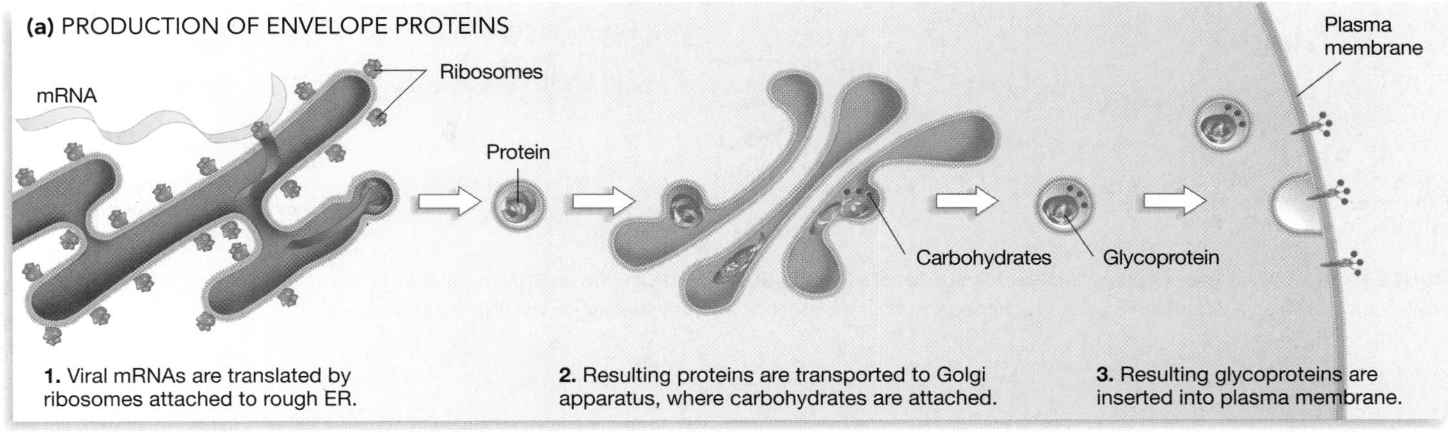

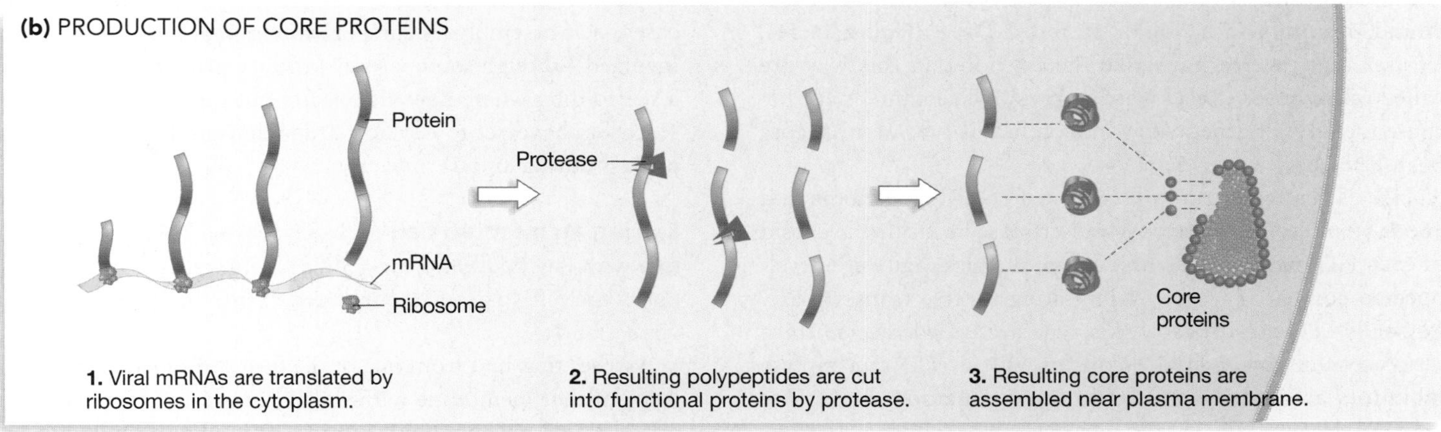

FIGURE 35.12 Production of Viral Proteins. (a) After being synthesized on the rough ER, envelope proteins are inserted into the plasma membrane. **(b)** After being synthesized by ribosomes in the cytoplasm and processed, core proteins assemble near the host cell's plasma membrane.

In some viruses, however, the genome consists of RNA. In most viruses that have an RNA genome, copies of the genome are synthesized by the viral enzyme **RNA replicase**, which is an RNA polymerase. RNA replicase synthesizes RNA from an RNA template, using ribonucleotides provided by the host cell.

In other RNA viruses, however, the genome is transcribed from RNA to DNA by a viral enzyme called **reverse transcriptase**. This enzyme is a DNA polymerase that makes a single-stranded **complementary DNA**, or **cDNA**, from a single-stranded RNA template. Reverse transcriptase then removes the RNA

(a) HIV's protease enzyme

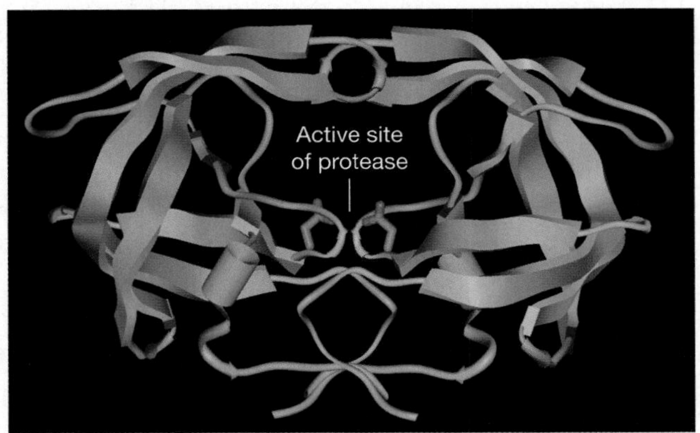

(b) Could a drug block the active site?

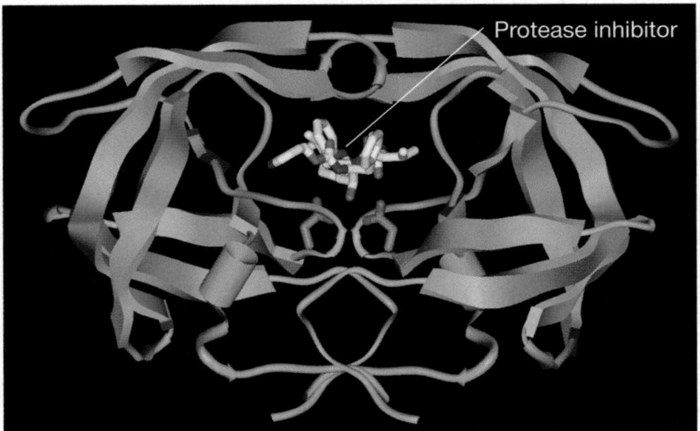

FIGURE 35.13 The Three-Dimensional Structure of Protease. (a) Ribbon diagram depicting the three-dimensional shape of HIV's protease enzyme. **(b)** Once protease's structure was solved, researchers began looking for compounds that would fit into the active site and prevent the enzyme from working.

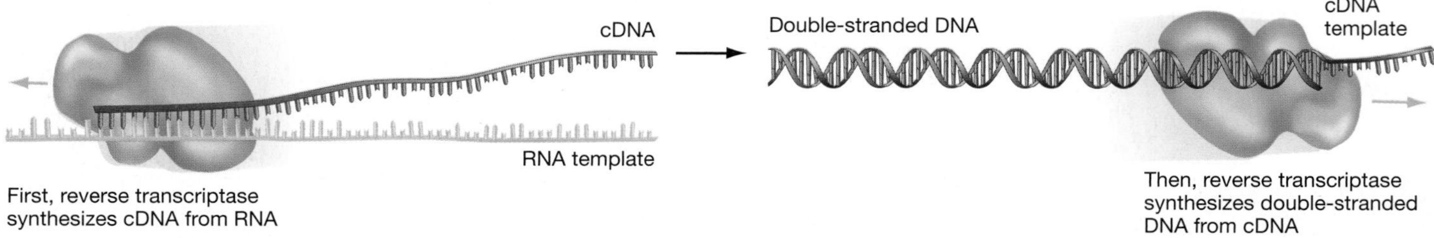

FIGURE 35.14 Reverse Transcriptase Catalyzes Synthesis of a Double-Stranded DNA from an RNA Template. The DNA produced by reverse transcriptase is called a cDNA because its base sequence is complementary to the RNA template.

strand and catalyzes the synthesis of a complementary DNA strand, resulting in a double-stranded DNA (**Figure 35.14**). Viruses that reverse-transcribe their genome in this way are called **retroviruses** ("backward viruses"). The name is apt because the flow of genetic information in this type of virus goes from RNA back to DNA.

HIV is a retrovirus. Two copies of the RNA genome and about 50 molecules of reverse transcriptase lie inside the capsid of each HIV particle. The first antiviral drugs that were developed to combat HIV act by inhibiting reverse transcriptase. Logically enough, drugs of this type are called reverse transcriptase inhibitors. **Box 35.2** explains why reverse transcriptase inhibitors are usually prescribed in combination with other antiviral drugs.

After reverse transcriptase makes a cDNA copy of the viral genome, another viral enzyme called integrase catalyzes the insertion of the double-stranded cDNA into a host-cell chromosome. Once it is integrated into the host's genome in this way, HIV may stay in a latent state for a period—its genes may "just sit there." More commonly, though, the viral genes are actively transcribed to mRNA, using the cell's RNA polymerase, and then translated into proteins by the host cell's ribosomes. In addition, some of the RNA transcripts serve as genomes for the next generation of HIV virions.

Assembly of New Virions Once the viral genome has been replicated and viral proteins are produced, they are transported to locations where a new generation of virions assembles inside the infected host cell. During assembly, the capsid forms around the viral genome. In many cases, copies of non-capsid proteins like DNA or RNA polymerases or reverse transcriptases are also packaged inside the capsid. And in enveloped viruses, envelope proteins become inserted into the host cell's plasma membrane.

In many cases, the details of the assembly process are not yet well understood. Bacteriophages and other nonenveloped viruses with complex morphologies appear to assemble in a step-by-step process resembling an assembly line. HIV and other enveloped viruses appear to assemble while attached to the inside surface of the host's plasma membrane. In most cases, self-assembly occurs—meaning that no enzymes are involved—though some viruses produce proteins that provide a scaffolding where new virions are put together. To date, researchers have yet to develop drugs that inhibit the assembly process during an HIV infection.

Exiting an Infected Cell Viruses leave a host cell in one of two ways: by budding from cellular membranes or by bursting out of the cell. In general, enveloped viruses bud; nonenveloped viruses burst.

Viruses that bud from one of the host cell's membranes take some of that membrane with them. As a result, they incorporate host-cell phospholipids into their envelope, along with envelope proteins encoded by the viral genome (**Figure 35.15a**). Most budding viruses infect host cells that lack a cell wall. In contrast, viruses that burst are able to infect host cells that have a cell wall. Bacteriophages produce lysozyme to create a hole in the cell wall of bacterial cells and then escape through the hole (**Figure 35.15b**).

How Are Viruses Transmitted to New Hosts? Once particles are released, the infection cycle is successful. Hundreds of newly assembled virions are now in extracellular space. What happens next?

If the host cell is part of a multicellular organism, the new generation of particles begins traveling through the body—often via the bloodstream or lymph system. There, they may be bound by antibodies produced by the immune system. In vertebrates, this binding marks the particles for destruction. But if a particle contacts an appropriate host cell before it encounters antibodies, then the particle will infect that cell. This starts the replication cycle anew.

What if the virus has infected a unicellular organism, or if the virus leaves a multicellular host entirely? For example, when people cough, sneeze, spit, or wipe a runny nose, they help rid their body of viruses and bacteria. But they also project the pathogens into the environment, sometimes directly onto an uninfected host. From the virus's point of view, this new host represents an unexploited habitat brimming with resources in

(a) Budding of enveloped viruses

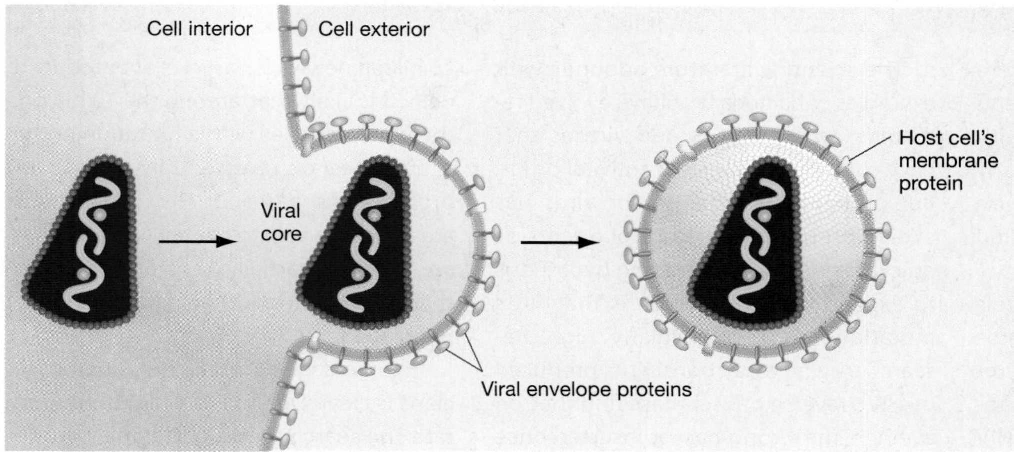

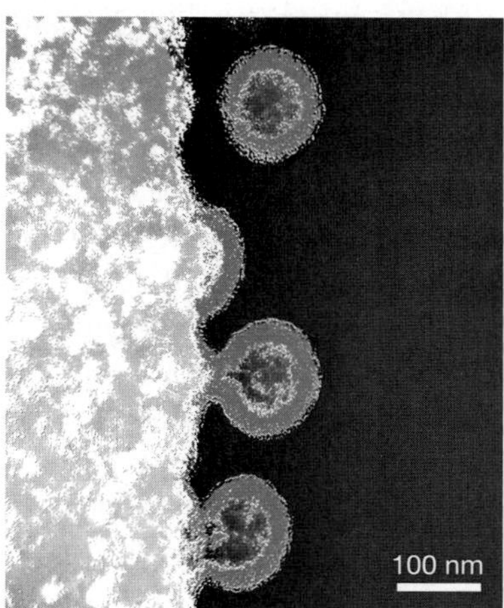

(b) Bursting of nonenveloped viruses

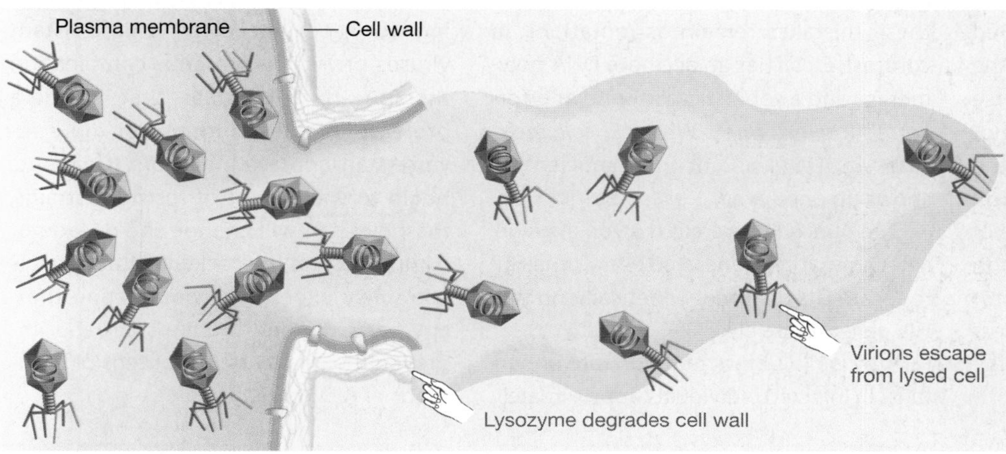

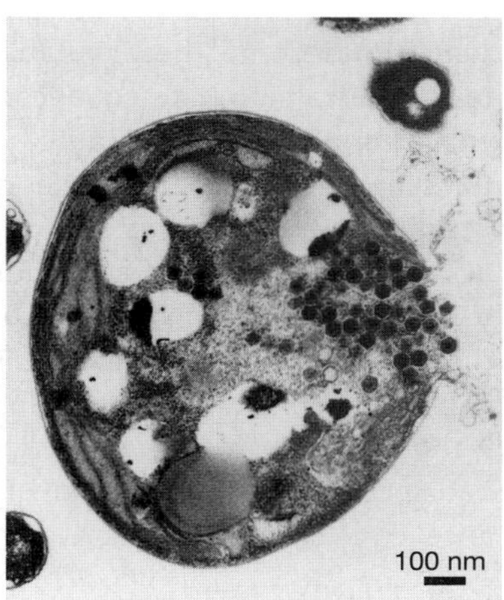

FIGURE 35.15 Viruses Leave Infected Cells by Budding or Bursting.

● **QUESTION** Both budding and bursting kill the host cell. Propose a hypothesis to explain why infection with a budding virus is fatal.

the form of target cells. The situation is analogous to that of a multicellular animal dispersing to a new habitat and colonizing it. Viruses that successfully colonize a new host replicate and increase in number. The alleles carried by these successful colonists increase in frequency in the total population. In this way, natural selection favors alleles that allow viruses to do two things: (1) replicate within a host and (2) be transmitted to new hosts.

For agricultural scientists, physicians, and public health officials, reducing the likelihood of **transmission**—the spread of pathogens from one individual to another— is often an effective way to combat a viral disease. For example, HIV particles are transmitted from person to person via body fluids such as blood, semen, or vaginal secretions. Faced with decades of disappointing results in drug and vaccine development, public health officials are aggressively promoting preventive medicine. Condom use reduces sexual transmission of HIV. Aggressive treatment of venereal diseases may also help; the lesions caused by chlamydia, genital warts, and gonorrhea encourage the transmission of HIV-contaminated blood during sexual intercourse. The most effective forms of preventive medicine are sexual abstinence or monogamy. In areas where blood supplies are routinely screened, HIV is rarely contracted by means other than unprotected sex with an infected person or needle sharing among drug abusers.

The effectiveness of preventive medicine underscores one of this chapter's fundamental messages: Viruses are a fact of life. Every organism is victimized by viruses; every organism has defenses against them. But the tree of life will never be free of these parasites. Mutation and natural selection guarantee that viral genomes will continually adapt to the de-

BOX 35.2 HIV Drug Cocktails and the Evolution of Drug Resistance

HIV reverse transcriptase inhibitors were dispensed widely in North America and Europe beginning in the mid-1990s, with spectacular results. After therapy, many patients no longer had detectable levels of HIV in their blood. The drugs knocked HIV populations down.

Within two years, however, HIV levels in many of the patients taking reverse transcriptase inhibitors began to rebound. To investigate why this was happening, researchers sequenced the HIV reverse transcriptase gene in these patients. The researchers found that a series of mutations had occurred in the reverse transcriptase gene. These mutations led to changed amino acid sequences in the enzyme's active site. Because reverse transcriptase inhibitor molecules did not fit as well into the altered version of the active site, the enzyme could function reasonably well even in the presence of the inhibitor molecules. Almost as soon as a new reverse transcriptase inhibitor went into widespread use, researchers had documented the evolution of resistance to the drug.

The scientific literature abounds with examples of bacteria that have evolved resistance to antibiotics and viruses that have evolved resistance to antiviral drugs. But perhaps no organism or virus has evolved resistance to control agents as quickly as HIV has. The leading hypothesis to explain this observation is that HIV's mutation rate is particularly high. Researchers who assay transcripts produced by HIV's reverse transcriptase find that, on average, the wrong base is inserted once every 8000 nucleotides. HIV's genome does not code for any of the error-correcting enzymes introduced in Chapter 14, so these mistakes remain as mutations. In contrast, *E. coli* has an accurate DNA polymerase and a sophisticated suite of error-correcting enzymes. When *E. coli* replicates its DNA, an incorrect nucleotide shows up once every 1 *billion* bases.

The punch line here is that, on average, a new mutant is generated every time HIV replicates its genome. Genetically, no two HIV particles are alike.

Why is HIV's high mutation rate important? In infected individuals, approximately 10 billion new viral particles are produced daily. It is likely that, among the 10 billion, there are particles with a mutation in the active sites of reverse transcriptase or protease. As a result, HIV populations are almost certain to contain variants that are at least partially resistant to drugs that cripple most other particles in the population.

The message to researchers and physicians is clear: Due to HIV's high mutation rate, the search for drug therapies promises to be an "arms race"—a constant battle between novel drugs and novel, resistant strains of the virus. In attempting to keep ahead of drug-resistant viruses, physicians prescribe combination therapy—drug "cocktails" that include a protease inhibitor and one or more reverse transcriptase inhibitors. If patients begin to show signs of resistant strains, the physician will change the dosage or composition of the cocktail. Although it is extremely expensive, combination therapy has extended the life span and improved the quality of life for tens of thousands of AIDS patients.

fenses offered by their hosts, regardless of whether those defenses are devised by an immune system or by biomedical researchers. Viruses are a constant threat for every organism alive.

Check Your Understanding

If you understand that...

- All organisms and cell types are parasitized by some type of virus.
- After infecting a cell, viruses may replicate or become latent.

You should be able to...

1) Diagram a lysogenic replication cycle.
2) Diagram the six phases of a lytic replication cycle: entry, protein synthesis, genome replication, viral assembly, exit, and transmission to a new host.
3) Explain how HIV performs each step in its replicative cycle.

MB Web Animation at www.masteringbio.com
The HIV Replicative Cycle

35.3 What Themes Occur in the Diversification of Viruses?

If viruses can infect virtually every type of organism and cell known, how can biologists possibly identify themes that help organize viral diversity? The answer is that, in addition to being identified as enveloped or nonenveloped, viruses can be categorized by the nature of their hereditary material—in essence, the type of molecule their genes are made of. The single most important aspect of viral diversity is the variation that exists in their genetic material.

Two other points are critical to recognize about viral diversity: (1) Biologists do not have a solid understanding of how viruses originate, but (2) it is certain that viruses will continue to diversify. Let's analyze the diversity of molecules that make up viral genes, then consider hypotheses to explain where viruses come from and recent data on new or "emerging" viruses.

The Nature of the Viral Genetic Material

DNA is the hereditary material in all cells. As cells synthesize the molecules they need to function, information flows from

SUMMARY TABLE **35.2 The Diversity of Viral Genomes**

Key: ss = single stranded; ds = double stranded; (+) = positive sense (genome sequence is the same as viral mRNA); (−) = negative sense (genome sequence is complementary to viral mRNA)

Genome	Example(s)	Host	Result of Infection	Notes
(+)ssRNA	TMV	Tobacco plants	Tobacco mosaic disease (leaf wilting)	TMV was the first RNA virus to be discovered.
(−)ssRNA	Influenza	Many mammal and bird species	Influenza	The negative-sense ssRNA viruses transcribe their genomes to mRNA via RNA replicase.
dsRNA	Phytovirus	Rice, corn, and other crop species	Dwarfing	Double-stranded RNA viruses are transmitted from plant to plant by insects. Many can also replicate in their insect hosts.
ssRNA that requires reverse transcription for replication	Rous sarcoma virus	Chickens	Sarcoma (cancer of connective tissue)	Rous sarcoma virus was identified as a cancer-decades before any virus causing agent in 1911, was seen.
ssDNA—can be (+), (−), or (+) and (−)	$\varphi \times 174$	Bacteria	Death of host cell	The genome for $\varphi \times 174$ is circular and was the first complete genome ever sequenced.
dsDNA	Baculovirus Smallpox Bacteriophage	Insects Humans Bacteria	Death Smallpox Death	These are the largest viruses in terms of genome size and overall size.

DNA to mRNA to proteins (Chapter 15). Although all cells follow this pattern, which is called the central dogma of molecular biology, some viruses break it. This conclusion traces back to work done in the 1950s, when biologists were able to separate the protein and nucleic acid components of a particle known as the tobacco mosaic virus, or TMV. Surprisingly, the nucleic acid portion of this virus consisted of RNA, not DNA. Later experiments demonstrated that the RNA of TMV, by itself, could infect plant tissues and cause disease. This was a confusing result because it showed that, in this virus at least, RNA—not DNA—functions as the genetic material.

Subsequent research revealed an amazing diversity of viral genome types. In some groups of viruses, such as the agents that cause measles and flu, the genome consists of RNA. In others, such as the particles that cause herpes and smallpox, the genome is composed of DNA. Further, the RNA and DNA genomes of viruses can be either single stranded or double stranded. The single-stranded genomes can also be classified as "positive sense" or "negative sense" or "ambisense." In a **positive-sense virus**, the genome contains the same sequences as the mRNA required to produce viral proteins. In a **negative-sense virus**, the base sequences in the genome are complementary to those in viral mRNAs. In an **ambisense virus**, some sections of the genome are positive sense while others are negative sense. **Table 35.2** summarizes the diversity of viral genome types. If you understand the concept of diversity in viral genomes, you should be able to name two types of viral genomes that are not based on double-stranded DNA. For each of these genomes, you should be able to diagram the steps in information flow from gene to protein.

Finally, the number of genes found in viruses varies widely. The tymoviruses that infect plants contain as few as three genes, but the genome of smallpox can code for up to 353 proteins.

Where Did Viruses Come From?

No one knows where viruses originated, but many biologists suggest that they may be derived from the plasmids and transposable elements introduced in Chapter 19 and Chapter 20. Viruses, plasmids, and transposable elements are all acellular, mobile genetic elements that replicate with the aid of a host cell. Simple viruses are actually indistinguishable from plasmids except for one feature: They encode proteins that form a capsid and allow the genes to exist outside of a cell.

Some biologists hypothesize that simple viruses, plasmids, and transposable elements represent "escaped gene sets." This hypothesis states that mobile genetic elements are descended from clusters of genes that physically escaped from bacterial or eukaryotic chromosomes long ago. According to this hypothesis, the escaped gene sets took on a mobile, parasitic existence because they happened to encode the information needed to replicate themselves at the expense of the genomes that once held them. In the case of viruses, the hypothesis is that the escaped genes included the instructions for making a protein capsid and possibly envelope proteins. According to the escaped-gene hypothesis, it is likely that each of the distinct types of RNA viruses and some of the DNA viruses represent distinct "escape events."

The same researchers contend that DNA viruses with large genomes may have originated in a very different manner, however. Here the hypothesis is that the large DNA viruses trace their ancestry back to free-living bacteria that once took up residence inside eukaryotic cells. The idea is that these organisms degenerated into viruses by gradually losing the genes required to synthesize ATP, nucleic acids, amino acids, and other compounds. Although this idea sounds speculative, it cannot be dismissed lightly. Chapter 28 introduced species in the genus *Chlamydia*, which are bacteria that live as parasites inside animal cells. And Chapter 29 provided evidence that the organelles called mitochondria and chloroplasts, which reside inside eukaryotic cells, originated as intracellular symbionts. Investigators contend that, instead of evolving into intracellular symbionts that aid their host cell, DNA viruses became parasites capable of destroying the host.

To date, neither of these hypotheses has been tested rigorously. To support the escaped-genes hypothesis, researchers would probably have to discover a brand-new virus that originated in this way, or viruses that had so recently derived from intact bacterial or eukaryotic genes that the viral DNA sequence still strongly resembled the DNA sequence of those genes. To support the degeneration hypothesis, researchers point to mimivirus, which still contains some genes involved in protein synthesis. A third alternative is also being discussed—the possibility that viruses coevolved with the first, RNA-based, forms of life (see Chapter 4). Currently, there is no widely accepted view of where viruses came from. Because viruses are so diverse, it is logical to predict that all three hypotheses may be valid.

Emerging Viruses, Emerging Diseases

Although it is not known how the various types of virus originated, it is certain that viruses will continue to diversify. With alarming regularity, the front pages of newspapers carry accounts of deadly viruses that are infecting humans for the first time. In 1993 a hantavirus that normally infects mice suddenly afflicted dozens of people in the southwestern United States. Nearly half of the people who developed hantavirus pulmonary syndrome died. Still higher fatality rates were recorded in 1995 when the Ebola virus, a variant of a monkey virus, caused a wave of infections in the Democratic Republic of Congo. By the time the outbreak subsided, over 200 cases had been reported; 80 percent were fatal. During a 1976 outbreak of Ebola, 90 percent of cases were fatal. Numerous reports of avian flu infecting and killing humans caused worldwide alarm in 2005–2006.

Hantavirus pulmonary syndrome, Ebola, and avian flu are examples of **emerging diseases:** new illnesses that suddenly affect significant numbers of individuals in a host population. In these cases, the causative agents were considered emerging viruses because they had switched from their traditional host species to a new host—humans. Many investigators consider HIV to be an emerging virus because it originated in chimps and was first transmitted to humans in the early to mid-twentieth century (see **Box 35.3**).

Physicians become alarmed when they see a large number of patients with identical and unusual disease symptoms in the same geographic area over a short period of time. The physicians report these cases to public health officials, who take on two urgent tasks: (1) identifying the agent that is causing the new illness and (2) determining how the disease is being transmitted.

Several strategies can be used to identify a pathogen. In the case of the outbreak of hantavirus pulmonary syndrome, officials recognized strong similarities between symptoms in the U.S. cases and symptoms caused by a hantavirus called Hantaan virus native to northeast Asia. The Hantaan virus rarely causes disease in humans; its normal host is rodents. To determine whether a Hantaan-like virus was responsible for the U.S. outbreak, researchers began capturing mice in the homes and workplaces of afflicted people. About a third of the captured rodents tested positive for the presence of a Hantaan-like virus. (The test that was done is explained in Chapter 49.) DNA sequencing studies confirmed that the virus was a previously undescribed type of hantavirus. Further, the sequences found in the mice matched those found in infected patients. Based on these results, officials were confident that a rodent-borne hantavirus was causing the wave of infections.

The next step in the research program, identifying how the agent is being transmitted, is equally critical. If a virus that normally parasitizes a different species suddenly begins infecting humans, if it can be transmitted efficiently from person to person, and if it causes serious illness, then the outbreak has the potential to become an epidemic. But if transmission takes place only between the normal host and humans, as is the case with rabies, then the number of cases will probably remain low. To date, for example, avian flu has not been transmitted efficiently from person to person—only inefficiently from birds to people.

Determining how a virus is transmitted takes old-fashioned detective work. By interviewing patients about their activities, researchers called epidemiologists decide whether each patient could have acquired the virus independently. Were the individuals infected with hantavirus in contact with

BOX 35.3 Where Did HIV Originate?

The key to discovering where HIV comes from lies in understanding the evolutionary history, or phylogeny, of the virus. Researchers have reconstructed this history by comparing the composition of HIV genes with sequences from closely related viruses that parasitize chimps, monkeys, and other mammals (**Figure 35.16**). The relationships among these gene sequences support several conclusions:

- HIV belongs to a group of viruses called the lentiviruses, which infect a wide range of mammals, including house cats, horses, goats, and primates. (*Lenti* is a Latin root that means "slow"; here it refers to the long period observed between the start of an infection by these viruses and the onset of the diseases they cause.)
- Many of HIV's closest relatives also have *immunodeficiency* in their names. Like HIV, these agents parasitize cells that are part of the immune system. Several of them cause diseases with symptoms reminiscent of AIDS. For unknown reasons, though, HIV's closest relatives don't appear to cause disease in their hosts. These viruses infect monkeys and chimpanzees and are called simian immunodeficiency viruses (SIVs).
- There are two distinct types of human immunodeficiency viruses, called HIV-1 and HIV-2. Although both can cause AIDS, HIV-1 is far more virulent and is the better studied of the two. It serves as the focus of this chapter.
- HIV-1's closest relatives are immunodeficiency viruses isolated from chimpanzees that live in central Africa. In contrast, HIV-2's closest relatives are immunodeficiency viruses that parasitize monkeys called sooty mangabeys. In central Africa, where HIV-1 infection rates first reached epidemic proportions, contact between chimpanzees and humans is extensive. Chimps are hunted for food and kept as pets. Similarly, sooty mangabeys are hunted and kept as pets in west Africa, where HIV-2 infection rates are highest. To make sense of these observations, biologists suggest that HIV-1 is a descendant of viruses that infect chimps, and HIV-2 is a descendant of viruses that infect sooty mangabeys. The "jumps" between host species probably occurred when a human cut up a monkey or chimpanzee for use as food.
- Several strains of HIV-1 exist. A virus **strain** consists of populations that have similar characteristics. The most important HIV-1 strains are called O for *out-group* (meaning, the most basal group relative to other strains), N for *new*, and M for *main*. Because few mechanisms exist for the virus to jump from humans back to chimps or monkeys, it is likely that each of these strains represents an independent origin of HIV from a chimp SIV strain. This is a key point. The existence of these distinct strains suggests that HIV-1 has jumped from chimps to humans several times and may do so again in the future.

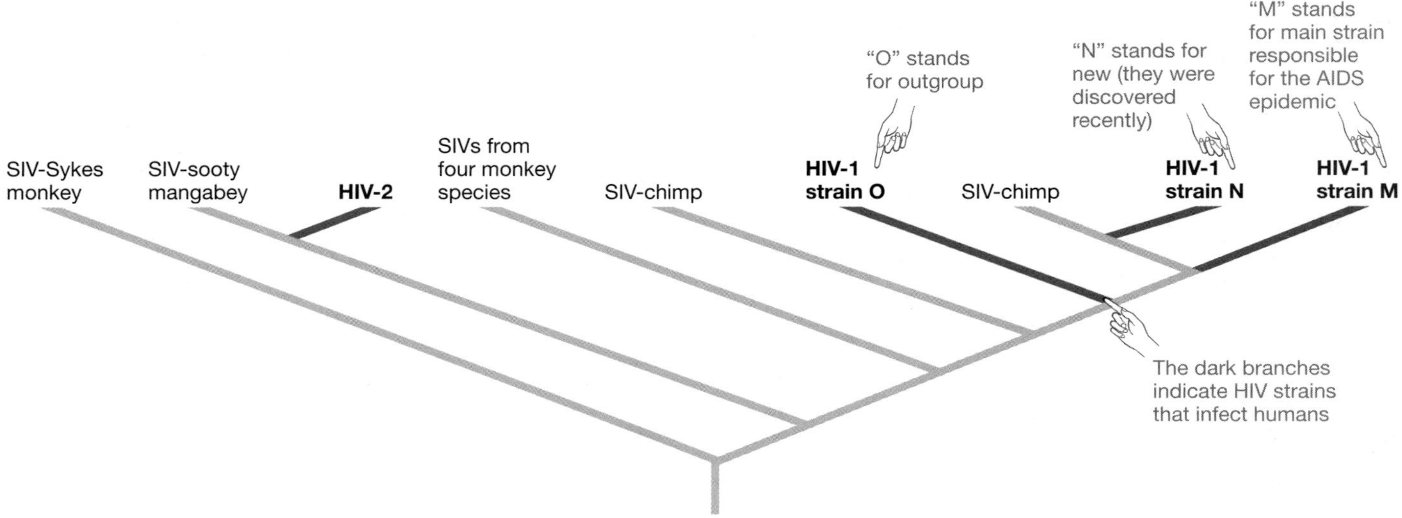

FIGURE 35.16 Phylogeny of HIV Strains and Types. Phylogenetic tree showing the evolutionary relationships among some of the immunodeficiency viruses that infect primates—including chimpanzees, humans, and several species of monkeys.

● **EXERCISE** On the appropriate branches, indicate where an SIV jumped to humans (draw and label bars across the appropriate branches).

mice? If so, were they bitten? Did they handle rodent feces or urine during routine cleaning chores or come into contact with contaminated food? Was the illness showing up in health-care workers, implying that it was being transmitted from human to human?

In the hantavirus outbreak, public health officials concluded that no human-to-human transmission was taking place. Health-care workers did not become ill; because patients had not had extensive contact with one another, it was likely that each had acquired the virus independently. The outbreak also coincided with a short-term, weather-related explosion in the local mouse population. The most likely scenario was that people had acquired the pathogen by inhaling dust or handling food that contained remnants of mouse feces or urine. In short, hantavirus did not have the potential to cause an epidemic. The best medicine was preventive: Homeowners were advised to trap mice in living areas, wear dust masks while cleaning any area where mice might have lived, and store food in covered jars.

The Ebola virus, in contrast, was clearly transmitted from person to person. Many doctors and nurses were stricken after tending to patients with the virus. Infections that originate in hospitals usually spread when carried from patient to patient on the hands of caregivers. But by carefully observing the procedures that were being followed by hospital staff, researchers concluded that the Ebola virus was being transmitted only through direct contact with body fluids (blood, urine, feces, or sputum). The outbreak was brought under control when hospital workers raised their standards of hygiene and insisted on the immediate disposal or disinfection of all contaminated bedding, utensils, and equipment. The situation with the avian flu outbreak of 2005–2006 was similar—person-to-person transmission was extremely rare and based only on extensive, direct contact with body fluids. Had the Ebola virus or avian flu virus been transmitted by casual contact, such as touch or inhalation—as is the common cold virus—a massive epidemic could have ensued.

Check Your Understanding

If you understand that...

Among viruses, several different types of molecules serve as the genetic material.

You should be able to...

1) Explain how the negative sense, single-stranded RNA genome of influenza viruses, which are complementary in sequence to viral mRNA, is used to produce a new generation of flu virions.
2) State whether a mutation that allowed avian flu virus to spread via airborne particles coughed out by infected individuals would make the virus more or less dangerous to humans? Explain your logic.

35.4 Key Lineages of Viruses

Because scientists are almost certain that viruses originated multiple times throughout the history of life, there is no such thing as the phylogeny of all viruses. Stated another way, there is no single phylogenetic tree that represents the evolutionary history of viruses as there is for the organisms discussed in previous chapters. Instead, researchers focus on comparing base sequences in the genetic material of small, closely related groups of viruses and using these data to reconstruct the phylogenies of particular lineages. Phylogenetic trees for viruses are usually inferred from comparisons of nucleic acid sequence data, using techniques introduced in Chapter 27.

The phylogenetic tree of the simian immunodeficiency viruses and human immunodeficiency viruses in Figure 35.16 is a good example of how researchers construct and interpret the phylogenies of particular groups of viruses. These types of phylogenies have been important in understanding viral diversity and the sources of emerging viruses. For example, phylogenetic analyses allowed biologists to recognize that there are two distinct types of HIV. The phylogenetic data diagrammed in Figure 35.16 indicate that HIV-1 and HIV-2 are distinct types of virus that originated from different host species (see Box 35.3). These data correlate with the observation that HIV-1 and HIV-2 differ in important ways. For instance, although both types of HIV infect immune system cells and are usually transmitted through sexual contact, HIV-2 is much less virulent and much less easily transmitted from person to person than is HIV-1.

To organize the diversity of viruses on a larger scale, researchers group them into seven general categories based on the nature of their genetic material. Within these broad groupings, biologists also identify a total of about 70 virus families that are distinguished by (1) the structure of the virion (often whether it is enveloped or nonenveloped), and (2) the nature of the host species.

Although they do not have formal scientific names, viruses within families are grouped into distinct genera for convenience. Within genera, biologists identify and name types of virus, such as HIV, the measles virus, and smallpox. Within each of these viral types, populations with distinct characteristics may be identified and named as strains. The strain is the lowest, or most specific, level of taxonomy for viruses. The O, N, and M strains of HIV-1, which are highlighted in Figure 35.16, are examples of separate virus strains. In the case of HIV-1, the named strains resulted from independent "jumps" of a simian immunodeficiency virus to a new host: humans. Each of these strains has distinguishing characteristics. The M strain, for example, was named *main* because it is responsible for most of the HIV infections known to date.

To get a sense of viral diversity, let's survey a few of the major groups that can be identified by the nature of their genetic material.

Double-Stranded DNA (dsDNA) Viruses

The double-stranded DNA viruses are a large group, composed of some 21 families and 65 genera. Smallpox (**Figure 35.17**) is perhaps the most familiar of these viruses. Although smallpox had been responsible for millions of deaths throughout human history, it was eradicated by vaccination programs. Smallpox is currently extinct in the wild; the only remaining samples of the virus are stored in research labs.

Genetic material As their name implies, the genes of these viruses consist of a single molecule of double-stranded DNA. The molecule may be linear or circular.

Host species These viruses parasitize hosts from throughout the tree of life, with the notable exception of land plants. They include virus families called the T-even and λ bacteriophages, some of which infect *E. coli*. In addition, the pox viruses, herpesviruses, and adenoviruses—some types of which parasitize humans—have genomes consisting of double-stranded DNA.

Replication cycle In most double-stranded DNA viruses that infect eukaryotes, viral genes have to enter the nucleus to be replicated. The viral genes are replicated only during S phase, when the host cell's chromosomes are being replicated. As a result, these types of viruses can sustain an infection only in cells that are actively dividing, such as the cells lining the respiratory tract or urogenital canal. Some of these viruses are capable of inducing replication in infected cells.

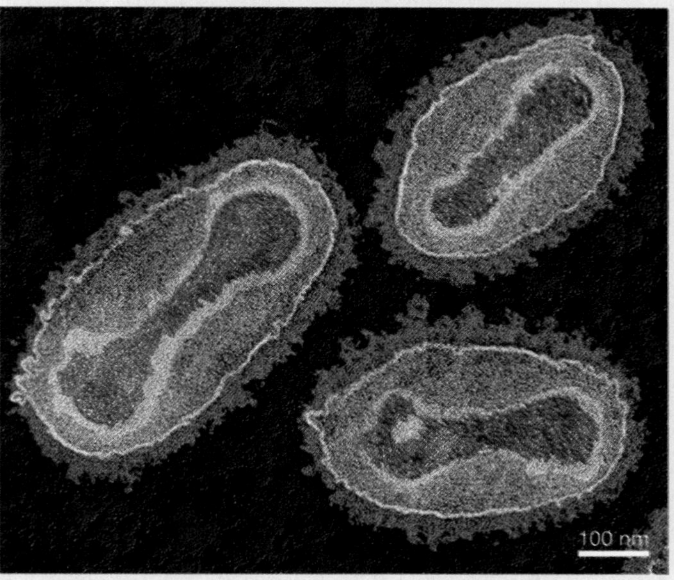

FIGURE 35.17 Smallpox Is a Double-Stranded DNA Virus.

RNA Reverse-Transcribing Viruses (Retroviruses)

The genomes of the RNA reverse-transcribing viruses are composed of single-stranded RNA. There is only one family, called the retroviruses.

Genetic material Virus particles have two copies of their single-stranded RNA genome, so they are diploid.

Host species Species in this group are known to parasitize only vertebrates—specifically birds, fish, or mammals. HIV is the most familiar virus in this group. The Rous sarcoma virus, the mouse mammary tumor virus, and the murine (mouse) leukemia virus are other retroviruses that have also been studied intensively. Rous sarcoma virus was the first virus shown to be associated with the development of cancer (in chickens); the mouse viruses were the first viruses known to increase the risk of cancer in mammals (**Figure 35.18**). In most cases, viruses that are associated with cancer development carry genes that contribute to uncontrolled growth of the cells they infect.

Replication cycle Retroviruses contain the enzyme reverse transcriptase inside their capsid. (A typical HIV particle contains about 50 reverse transcriptase molecules.) When the virus's RNA genome and reverse transcriptase enter a host cell's cytoplasm, the enzyme catalyzes the synthesis of a single-stranded cDNA from the original RNA. Reverse transcriptase then makes this cDNA double stranded. The double-stranded DNA version of the genome enters the nucleus with a viral protein called integrase. Integrase catalyzes the integration of the viral genes into a host chromosome. The virus may remain quiescent for a period, but eventually the genes are transcribed to RNA to begin lytic growth and the production of a new generation of virus particles.

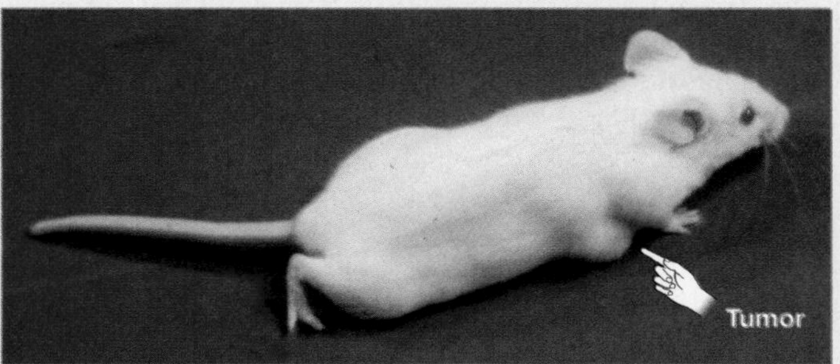

FIGURE 35.18 Some Retroviruses, Such as the Mouse Mammary Tumor Virus, Are Associated with Cancer.

Double-Stranded RNA (dsRNA) Viruses

There are 7 families of double-stranded RNA viruses and a total of 22 genera. Most of the viruses in this group are nonenveloped.

Genetic material In some families, virus particles typically have a genome consisting of 10–12 double-stranded RNA molecules; in other families, the genome is composed of just 1–3 RNA molecules.

Host species A wide variety of hosts, including fungi, land plants, insects, vertebrates, and bacteria, are victimized by viruses with double-stranded RNA genomes. Particularly prominent are viruses that cause disease in rice, corn, sugar cane, and other crops. **Figure 35.19** shows rice plants that are being attacked by a double-stranded RNA virus. Infections are also common in *Penicillium*, a filamentous fungus that produces the antibiotic penicillin. Reovirus and rotavirus infections, the leading cause of infant diarrhea in humans, are responsible for over 110 million cases and 440,000 deaths each year.

Replication cycle Once inside the cytoplasm of a host cell, the double-stranded genome of these viruses serves as a template for the synthesis of single-stranded RNAs, which are then translated into viral proteins. The proteins form the capsids for a new generation of virus particles. Copies of the genome are created when a viral enzyme makes the original single-stranded RNA double stranded.

FIGURE 35.19 Double-Stranded RNA Viruses, Such as the Ragged Stunt Virus, Parasitize a Wide Array of Organisms—Here, Rice.

Negative-Sense Single-Stranded RNA ([−]ssRNA) Viruses

There are 7 families and 30 genera in this group. Most members of this group are enveloped, but some negative-sense single-stranded RNA viruses lack an envelope.

Genetic material The sequence of bases in a negative-sense RNA virus is opposite in polarity to the sequence in a viral mRNA. Stated another way, the single-stranded virus genome is complementary to the viral mRNA. Depending on the family, the genome may consist of a single RNA molecule or up to eight separate RNA molecules.

Host species A wide variety of plants and animals are parasitized by viruses that have negative-sense single-stranded RNA genomes. If you have ever suffered from the flu, the mumps, or the measles, then you are painfully familiar with these viruses (**Figure 35.20**). The Ebola, Hantaan, and rabies viruses also belong to this group.

Replication cycle When the genome of a negative-sense single-stranded RNA virus enters a host cell, a viral RNA polymerase uses that genome as a template to make new viral mRNA. These viral mRNAs are then translated to form viral proteins. The viral mRNAs also serve as a template for the synthesis of new copies of the negative-sense single-stranded RNA genome.

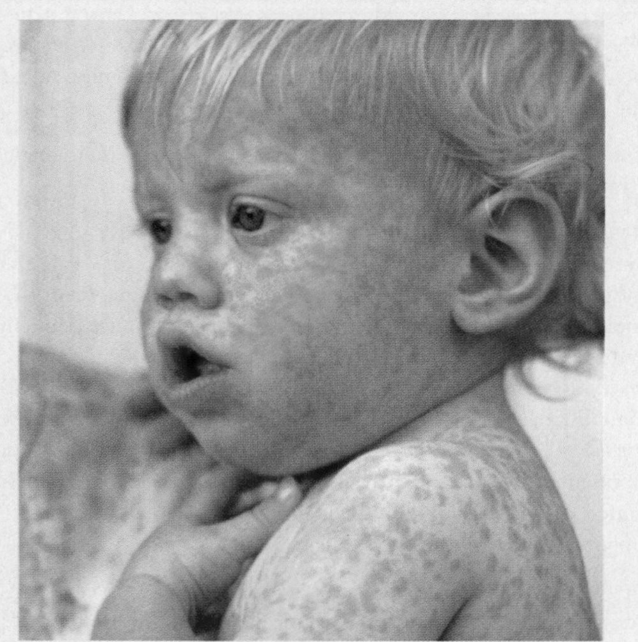

FIGURE 35.20 Negative-Sense Single-Stranded RNA Viruses, Such as the Measles Virus, Cause Some Common Childhood Diseases.

Positive-Sense Single-Stranded RNA ([+]ssRNA) Viruses

This is the largest group of viruses known, with 81 genera organized into 21 families.

Genetic material The sequence of bases in a positive-sense RNA virus is the same as that of a viral mRNA. Stated another way, the genome does not need to be transcribed in order for proteins to be produced. Depending on the species, the genome consists of one to three RNA molecules.

Host species Most of the commercially important plant viruses belong to this group. Because they kill groups of cells in the host plant and turn patches of leaf or stem white, they are often named mottle viruses, spotted viruses, chlorotic (meaning, lacking chlorophyll) viruses, necrotic (meaning, killed cells surrounded by intact tissue) viruses, or mosaic viruses. **Figure 35.21** shows a healthy cowpea leaf and a cowpea leaf that has been attacked by a positive-sense single-stranded RNA virus. Some species in this group of viruses specialize in parasitizing bacteria, fungi, or animals, however. A variety of human maladies, including the common cold, polio, and hepatitis A, C, and E, are caused by positive-sense RNA viruses.

Replication cycle When the genome of these viruses enters a host cell, the single-stranded RNA is immediately translated into viral proteins. These proteins include enzymes that make copies of the genome. The new generation of virus particles is assembled in a complex structure that is associated with host cell membranes.

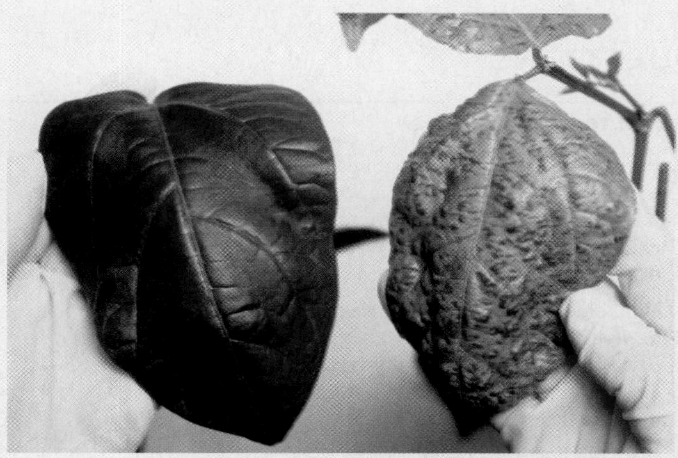

FIGURE 35.21 Positive-Sense Single-Stranded RNA Viruses, Such as the Cowpea Mosaic Virus, Cause Important Plant Diseases.

Chapter Review

SUMMARY OF KEY CONCEPTS

- **Viruses are tiny, noncellular parasites that infect virtually every type of cell known. They cannot perform metabolism on their own—meaning outside a parasitized cell—and are not considered to be alive. Different types of viruses infect particular species and types of cells.**

Viruses cause illness and death in plants, fungi, bacteria, and archaea, as well as in humans and other animals. Vaccination is frequently effective in preventing virus epidemics. Unfortunately, it is difficult or impossible to design a vaccine that can prepare the immune system for viruses that mutate very rapidly, such as HIV and the flu viruses.

You should be able to explain why, despite the diversity and abundance of viruses, most cells are healthy.

- **Although viruses are diverse morphologically, they can be classified as two general types: enveloped and nonenveloped.**

Both nonenveloped and enveloped viruses have a capsid made of protein. The capsid usually encloses viral enzymes as well as the viral genome. Nonenveloped viruses consist of a naked capsid, but in enveloped viruses, the capsid is surrounded by a membranous envelope. Nonenveloped viruses exit a cell by lysis; enveloped viruses exit a cell by budding.

You should be able to make sketches comparing the general morphologies of a nonenveloped and enveloped virus.

- **The viral infection cycle can be broken down into six steps: (1) entry into a host cell, (2) production of viral proteins, (3) replication of the viral genome, (4) assembly of a new generation of virus particles, (5) exit from the infected cell, and (6) transmission to a new host.**

A viral infection begins when the contents of a virus particle enter a host cell. For example, when HIV binds to a transmembrane protein called CD4 and a co-receptor, the virus's membrane-like envelope fuses with the host cell's plasma membrane, and the viral genome and proteins enter the cytoplasm. In the second phase of the replication cycle, viral proteins are produced. The viral genome is replicated in the third phase, and the viral proteins and genome then assemble into complete particles (the fourth phase of the infection cycle). The new generation of complete particles then buds or bursts from the cell, usually killing the host cell in the process. Once they are released, particles can infect new cells in the same multicellular organism or be transmitted to a new host.

You should be able to explain how at least three of the six phases of a viral replicative cycle can be stopped, using HIV as an example.

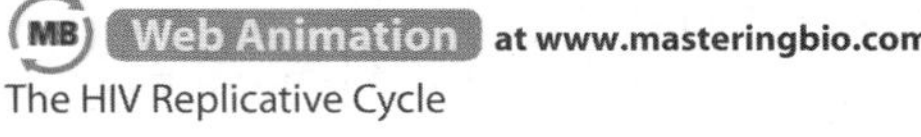

The HIV Replicative Cycle

In terms of diversity, the key feature of viruses is the nature of their genetic material. The genomes of viruses may consist of double-stranded DNA, single-stranded DNA, double-stranded RNA, or one of several types of single-stranded RNA.

Viral genomes are small and can consist of either RNA or DNA, but not both. Viral genomes do not code for ribosomes, and most viral genomes do not code for the enzymes needed to translate their own proteins or to perform other types of biosynthesis. When a virus infects a host cell, it uses that cell's enzymes, nutrients, ATP, and ribosomes to manufacture a new generation of virus particles. This observation explains why viral diseases are difficult to treat with drugs: Molecules that incapacitate enzymes needed by the virus are likely to damage host cells as well.

You should be able to explain how a viral genome that consists of single-stranded DNA is copied by a viral DNA polymerase.

QUESTIONS

Test Your Knowledge

1. How do viruses that infect animals enter an animal's cells?
   a. The viruses pass through a wound.
   b. The viruses bind to a membrane protein.
   c. The viruses puncture the cell wall.
   d. The viruses lyse the cell.
2. What does reverse transcriptase do?
   a. It synthesizes proteins from mRNA.
   b. It synthesizes tRNAs from DNA.
   c. It synthesizes DNA from RNA.
   d. It synthesizes RNA from DNA.
3. What do host cells provide for viruses?
   a. nucleotides and amino acids
   b. ribosomes
   c. ATP
   d. all of the above
4. When do most enveloped virus particles acquire their envelope?
   a. during entry into the host cell
   b. during budding from the host cell
   c. as they burst from the host cell
   d. as they integrate into the host cell's chromosome
5. What reaction does protease catalyze?
   a. polymerization of amino acids into peptides
   b. cutting of long peptide chains into functional proteins
   c. folding of long peptide chains into functional proteins
   d. assembly of viral particles
6. Why is it difficult to design a vaccine for viruses with high mutation rates, such as HIV and flu viruses?
   a. The vaccines tend to be unstable and deteriorate over time.
   b. So many protein fragments are presented by these viruses that the immune system overreacts.
   c. They have no protein fragments that can be recognized by a host cell.
   d. New mutations constantly change viral proteins.

Test Your Knowledge answers: 1. b; 2. c; 3. d; 4. b; 5. b; 6. d

Test Your Understanding

Answers are available at www.masteringbio.com

1. The outer surface of a virus consists of either a membrane-like envelope or a protein capsid. Which type of outer surface does HIV have? Which type does adenovirus have? How does the outer surface of a virus correlate with its mode of exiting a host cell, and why?
2. Compare the morphological complexity of HIV with that of bacteriophage T4. Which virus would you predict has the larger genome? Explain the logic behind your hypothesis.
3. Compare and contrast lytic growth with growth during the latent state. Is it possible for viral populations to increase if virions remain in the latent state?
4. Explain why viral diseases are more difficult to treat than diseases caused by bacteria.
5. Draw the lytic cycle of a nonenveloped virus with a positive-sense single-stranded RNA genome that infects cells in the roots of rice plants. Describe the modes of action of two drugs that could be developed to treat plants infected with this virus.
6. What type of data convinced researchers that HIV originated when a simian immunodeficiency virus "jumped" to humans? Do you agree with this conclusion? Why or why not?

Applying Concepts to New Situations

Answers are available at www.masteringbio.com

1. Suppose you could isolate a virus that parasitizes the pathogen *Staphylococcus aureus*—a bacterium that causes acne, boils, and a variety of other afflictions in humans. How could you test whether this virus might serve as a safe and effective antibiotic?
2. If you were in charge of the government's budget devoted to stemming the AIDS epidemic, would you devote most of the resources to drug development, vaccine development, or preventive medicine? Defend your answer.
3. Bacteria fight viral infections with restriction endonucleases (bacterial enzymes; introduced in Chapter 19). Restriction endonucleases cut up, or break, viral DNA at specific sequences. The enzymes do not cut a bacterium's own DNA, because the bases in the bacterial genome are protected from the enzyme by methylation (the addition of a CH_3 group). Generate a hypothesis to explain why members of the Eukarya do not have restriction endonucleases.
4. Consider these two contrasting definitions of life:
   a. An entity is alive if it is capable of replicating itself via the directed chemical transformation of its environment.
   b. An entity is alive if it is an integrated system for the storage, maintenance, replication, and use of genetic information.

   According to these definitions, are viruses alive? Explain.

www.masteringbio.com is also your resource for • Answers to text, table, and figure caption questions and exercises • Answers to *Check Your Understanding* boxes • Online study guides and quizzes • Additional study tools including the *E-Book for Biological Science* 3rd ed., textbook art, animations, and videos.

An Introduction to Ecology

50

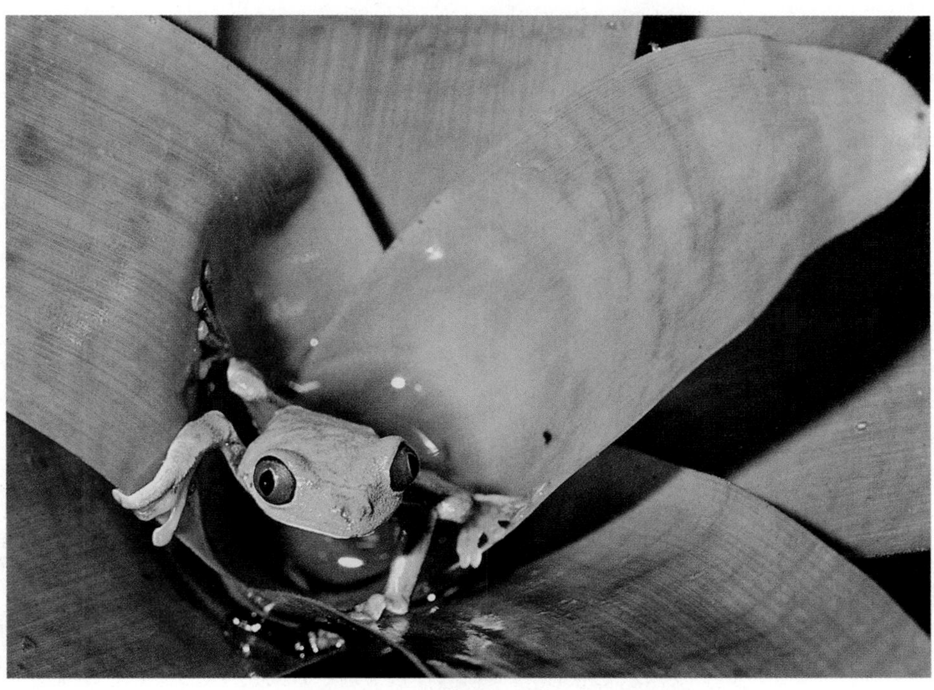

In addition to providing habitat for a frog, the puddle of water inside this plant provides an environment for an array of microscopic species. The plant itself is growing in the branch of a rain forest tree, and is just one of hundreds of plant, fungal, bacterial, and insect species that live in association with the tree.

KEY CONCEPTS

- Ecology focuses on how organisms interact with their environment. Because its goal is to understand the distribution and abundance of organisms, ecology provides a scientific foundation for the conservation of species and natural areas.
- Physical structure—particularly water depth—is the primary factor that limits the distribution and abundance of aquatic species. Climate—specifically, both the average value and annual variation in temperature and in moisture—is the primary factor that limits the distribution and abundance of terrestrial species.
- Climate varies with latitude, elevation, and other factors. Climate is changing rapidly around the globe.
- In addition to abiotic aspects of the environment such as physical structure and climate, species distribution is constrained by historical and biotic factors.

Ecology is the study of how organisms interact with their environment. Except for the episode 65 million years ago, when a mountain-sized asteroid struck Earth, environments are changing more rapidly right now than they have at any time in the past 3.5 billion years. As human impacts on the planet accelerate, ecology is becoming an increasingly prominent field in biological science. Efforts to maintain human health and welfare depend on our ability to understand and predict the consequences of the rapid environmental changes occurring all around us.

One of ecology's central goals is to understand the distribution and abundance of organisms. In many cases an ecologist's job is to identify factors that dictate why certain species live where they do, and how many individuals can live there. Some biologists ask why orangutans are restricted to forests in Borneo and why their numbers are declining so rapidly. Other biologists create mathematical models to predict how quickly the human immunodeficiency virus will increase in India, or how long it will take for the current human population of 6.6 billion to double. The distribution and abundance of these species is dictated by their ecology.

This chapter has two goals: to explore how biologists go about studying ecology, and to analyze the major types of environments that organisms occupy. It is a springboard to subsequent chapters in this unit, which analyze how organisms interact with their environment and the consequences of environmental change. In terms of understanding the biological problems that humans currently face, this may be the most important unit in the book.

50.1 Areas of Ecological Study

To understand why organisms live where they do and in what numbers, biologists break ecology into several levels of analysis. This is a common strategy in biological science. Cell biologists study how cells work at increasingly complex levels of organization—from individual molecules to complex multicellular organisms—while physiologists analyze processes at the level of ions and molecules as well as whole cells, tissues and organs, or a complete system. In ecology, researchers work at four main levels: (1) organisms, (2) populations, (3) communities, and (4) ecosystems. Let's examine each.

Organismal Ecology

At the finest level of organization, ecology focuses on how individuals interact with their environment. Researchers who study organismal ecology explore the morphological, physiological, and behavioral adaptations that allow individuals to live successfully in a particular area. The study of behavior is an important aspect of organismal ecology; it focuses on how organisms respond to particular stimuli from their environment (see Chapter 51). The environmental stimuli that trigger behavior may consist of changes in temperature or moisture, an escape response from prey, or a rival challenging for a mate. Organismal ecology also focuses on the physiological adaptations that allow individuals to thrive in heat, drought, cold, or other demanding physical conditions.

As an example, consider the sockeye salmon. After spending four or five years feeding and growing in the ocean, these salmon travel hundreds or thousands of kilometers to return to the stream where they hatched (**Figure 50.1a**). Females create nests in the gravel stream bottom and lay eggs. Nearby males compete for the chance to fertilize eggs as they are laid. When breeding is finished, all of the adults die.

At the level of organismal ecology, biologists want to know how these individuals interact with their physical surroundings and with other organisms in and around the stream. Which females get the best nesting sites and lay the most eggs, and which males are most successful in fertilizing eggs? How do individuals cope with the transition from living in salt water to living in freshwater?

Population Ecology

A **population** is a group of individuals of the same species that lives in the same area at the same time. When biologists study population ecology, they focus on how the numbers of individuals in a population change over time. Some of the tools that

(a) Organismal ecology

How do individuals interact with each other and their physical environment?

Salmon migrate from saltwater to freshwater environments to breed

(b) Population ecology

How and why does population size change over time?

Each female salmon produces thousands of eggs. Only a few will survive to adulthood. On average, only two will return to the stream of their birth to breed

(c) Community ecology

How do species interact, and what are the consequences?

Salmon are prey as well as predators

(d) Ecosystem ecology

How do energy and nutrients cycle through the environment?

Salmon die and then decompose, releasing nutrients that are used by bacteria, archaea, plants, protists, young salmon, and other organisms

FIGURE 50.1 Biologists Study Ecology at Four Main Levels.

population ecologists have developed to analyze and predict changes in population size are now being used to evaluate the fate of endangered species. For example, mathematical models of population growth have been used to predict the future of particular salmon populations (**Figure 50.1b**). Many salmon populations have declined precipitously as their habitats have become dammed or polluted, and salmon are an important source of food for both people and wildlife. If the factors that affect population size can be described accurately enough, mathematical models can assess the impact of proposed dams, changes in weather patterns, altered harvest levels, or specific types of protection efforts.

Community Ecology

A biological **community** consists of the species that interact with each other within a particular area. Researchers who study community ecology ask questions about the nature of the interactions between species and the consequences of those interactions. The work might concentrate on predation, parasitism, or competition. At this level of organization in ecology, biologists also analyze how groups of species respond to disturbances such as fires, floods, and volcanic eruptions. Because human activities are driving species to extinction and disturbing communities on a massive scale, community ecologists are being called on to generate hypotheses and data on how human impacts can be lessened.

As an example of the types of questions asked at the level of community ecology, consider the interactions among salmon and other species in the marine and stream communities where they live. When they are at sea, salmon eat smaller fish and are themselves hunted and eaten by orcas, sea lions, humans, and other mammals; when they return to freshwater to breed, they are preyed on by bears and bald eagles (**Figure 50.1c**). In both marine and freshwater habitats, salmon are subject to parasitism and disease. They are also heavily affected by disturbances—particularly changes in their food supply due to overfishing and the damming and degradation of breeding streams.

Ecosystem Ecology

Ecosystem ecology is an extension of community ecology. An **ecosystem** consists of all the organisms in a particular region along with nonliving components. These physical, or **abiotic** (literally, "not-living"), components include air, water, and soil. At the ecosystem level, biologists study how nutrients and energy move among organisms and between organisms and the surrounding atmosphere and soil or water. Because humans are adding massive amounts of nutrients to ecosystems all over the world and affecting energy flows and climate through global warming, this work has direct public policy implications. Ecosystem ecologists are responsible for assessing the impact of pollution and increased temperature on the distribution and abundance of species—in fact, on Earth's ability to support life.

Salmon are interesting to study at the ecosystem level because they form a link between marine and freshwater ecosystems. They harvest nutrients in the ocean and then, when they die and decompose, transport those molecules to streams (**Figure 50.1d**). In this way, salmon transport chemical energy and nutrients from one habitat to another. Because salmon are sensitive to pollution and to changes in water temperature, human-induced changes in marine and freshwater ecosystems have a large impact on their populations.

How Do Ecology and Conservation Efforts Interact?

The four levels of ecological study are synthesized and applied in conservation biology. **Conservation biology** is the effort to study, preserve, and restore threatened populations, communities, and ecosystems. Ecologists study how interactions between organisms and their environments result in a particular species being found in a particular area at a particular population size; conservation biologists apply these data to preserve species and restore environments. Conservation biologists are like physicians, except that instead of drawing on results from research in molecular biology and physiology, they draw on research in ecology and evolution. Instead of prescribing drugs for sick people, they prescribe remedies for threatened species and manage land to produce a diversity of species, clean air, pure water, and productive soils.

50.2 Types of Aquatic Ecosystems

If ecology is the study of how species interact with their environment, what makes up the environment? The short answer is that an organism's environment has both physical and biological components. The abiotic or physical components include temperature, precipitation, sunlight, and wind. The **biotic** ("living") components consist of other members of the organism's own species as well as individuals of other species.

Chapters 51–53 focus on biological interactions—how individuals interact with offspring, mates, competitors, predators, prey, and parasites. This chapter focuses on how the physical environment affects organisms. The discussion offers two payoffs: (1) Once you understand key aspects of the physical environment, you'll be able to think intelligently about the consequences of altering those conditions through global warming; and (2) knowledge of the physical environment will help you make sense of a fundamental observation—that no species lives everywhere. Organisms have a restricted set of physical conditions in which they can survive and thrive. Why?

Let's answer this question by analyzing the physical attributes of aquatic ecosystems in this section and focusing on the physical structure of terrestrial ecosystems in Section 50.3. Section 50.4 considers how global warming is affecting both types of environments.

What Physical Factors Play a Key Role in Aquatic Ecosystems?

In aquatic ecosystems, water depth and the rate of water movement qualify as the key physical factors that shape the environment. Water depth dictates how much light reaches the organisms that live in a particular region. Water movement presents a physical challenge: It can literally sweep organisms off their feet.

Water absorbs and scatters light, so the amount and types of wavelengths available to organisms change dramatically as water depth increases. As **Figure 50.2a** shows, ocean water specifically removes light in the blue and red regions of the visible spectrum. This is important because wavelengths in the blue and red regions are required for photosynthesis in many species (see Chapter 10). The total amount of light available to organisms also diminishes rapidly with increasing depth. In pure seawater, the total amount of light available at a depth of 10 m is less than 40 percent of what it is at the surface, and virtually no light reaches depths greater than 40 m in pure seawater (**Figure 50.2b**). In seawater that contains organisms or debris, light penetration is dramatically less than that observed in pure seawater. Light has a major influence on **productivity**—the total amount of carbon fixed by photosynthesis per unit area per year. Different species are found at different depths in aquatic environments, simply because the physical environment changes so radically.

The type and amount of water movement or flow is the other major influence in aquatic environments. Organisms that live in fast-flowing streams have to cope with the physical force of the water, which constantly threatens to move them downstream. Marine organisms that live in intertidal regions are exposed to the air periodically each day, as well as to violent wave action during storms.

Different water depths and types of water movement define the array of aquatic environments that are available to organisms. The boxes that follow summarize the types of freshwater and marine environments that organisms occupy.

(a) Only certain wavelengths of light are available under water.

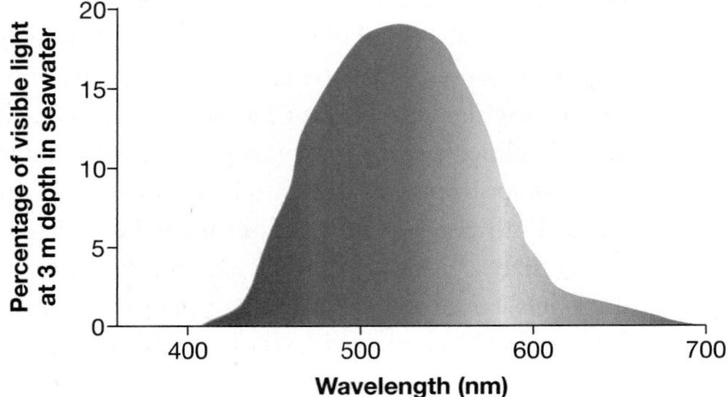

(b) Intensity of light declines with water depth.

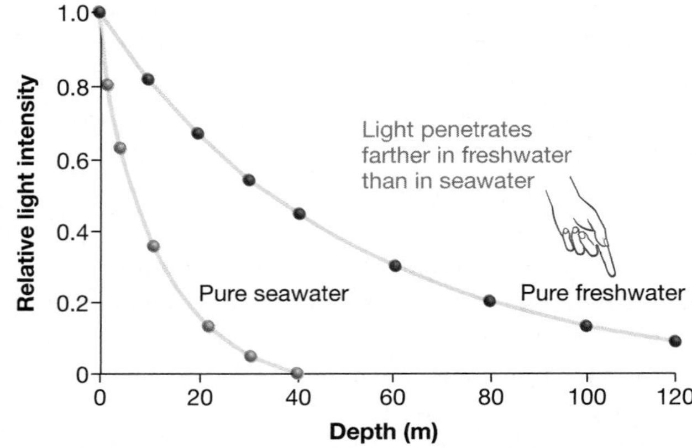

FIGURE 50.2 Availability of Light Changes Dramatically with Increasing Water Depth. (a) Graph showing the wavelengths of light available at a depth of 3 meters in seawater near an ocean coastline. **(b)** Graphs showing how rapidly the amount of light declines with depth of pure freshwater and pure seawater. When water contains organisms and organic debris, light availability declines even faster.

EXERCISE In green algae, photosynthesis is driven most efficiently by wavelengths of about 425 nm and 680 nm. Mark and label these wavelengths in part (a).

QUESTION What is the maximum depth you'd expect to find any photosynthetic organisms in freshwater? In salt water?

Freshwater Environments > Lakes and Ponds

Lakes and ponds are distinguished from each other by size. Ponds are small; lakes are large enough that the water in them can be mixed by wind and wave action. Most lakes and ponds occur in northern latitudes—they formed in depressions that were created by the scouring action of glaciers thousands of years ago. In the tropics, most lakes consist of old river channels. Elsewhere, many of the lakes and ponds that exist now were dug recently by people.

Water Depth Biologists describe the structure of lakes and ponds by naming five zones (**Figure 50.3**).

- The **littoral** ("seashore") **zone** consists of the shallow waters along the shore, where flowering plants are rooted.
- The **limnetic** ("lake") **zone** is offshore and comprises water that receives enough light to support photosynthesis.
- The **benthic** ("depths") **zone** is made up of the substrate.
- Regions of the littoral, limnetic, and benthic zones that receive sunlight are part of the **photic zone**.
- Portions of a lake or pond that do not receive sunlight make up the **aphotic zone**.

(Continued on next page)

Water Flow Water movement in lakes and ponds is driven by wind and temperature. The littoral and limnetic zones are typically much warmer and better oxygenated than the benthic zone, simply because they receive so much more solar radiation and are in contact with oxygen in the atmosphere. The benthic zone, in contrast, is relatively nutrient rich because dead and decomposing bodies sink and accumulate there. But as the temperature of the surface changes throughout the year, water from different depths can mix (see **Box 50.1**). Mixing driven by wind and changes in temperature allows well-oxygenated water from the surface to reach the benthic zone and nutrient-rich water from the benthic zone to enter the littoral and limnetic zones.

Organisms Cyanobacteria, algae, and other microscopic organisms, collectively called **plankton**, live in the photic zone, as do the fish that eat them. Animals (invertebrates and fish) that consume dead organic matter, or **detritus**, are common in the benthic zone.

● You should be able to predict which zone of lakes and ponds produces the most grams of organic material per m^2 per year, and explain your logic.

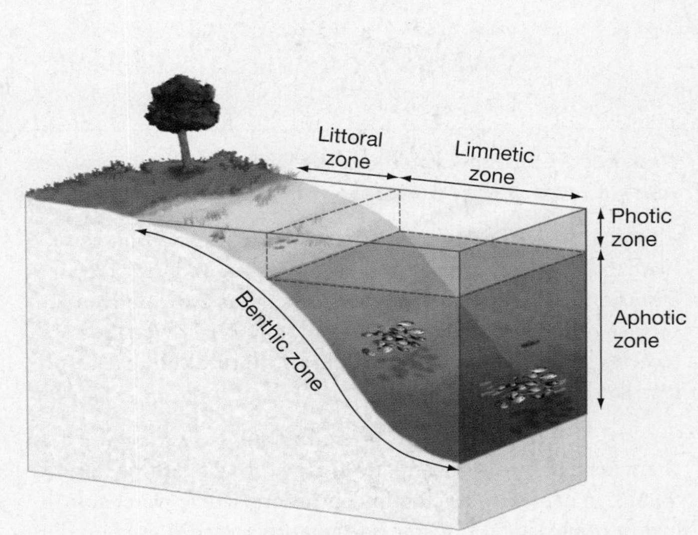

FIGURE 50.3 Lakes Have Distinctive Zones Defined by Water Depth and Distance from Shore.

BOX 50.1 Thermoclines and Lake Turnover

The northern and temperate regions of the world host many lakes. Each year, these bodies of water undergo remarkable changes known as the spring and fall **turnovers**.

The spring and fall lake turnovers occur in response to changes in air temperature. To see how this happens, examine the temperature profiles in **Figure 50.4**. In winter, water at the surface is locked up in ice at a temperature of 0°C. The water just under the ice is slightly warmer and is relatively oxygen rich, because it was exposed to the atmosphere as winter set in. The water at the bottom of the lake, in contrast, is at 4–5°C. A gradient in temperature such as this is called a **thermocline** ("heat-slope"); thermal stratification is said to occur. The water at the bottom is also oxygen poor, because organisms that decompose falling organic material use up available oxygen.

When spring arrives, the ice begins to melt. The temperature of the water rises until it reaches 4°C. This is important, because the density of liquid water is highest at 4°C. The water at the surface of the lake is now heavier than the water below it. As a result, it sinks. The water at the bottom of the lake is displaced and comes to the surface, completing the spring turnover. During the spring turnover, water at the bottom of the lake carries sediments and nutrients from the benthic zone up to the limnetic zone. This flush of nutrients triggers a rapid increase in the growth of algae and bacteria that biologists call the spring bloom.

When temperatures cool in the fall, the water at the surface reaches a temperature of 4°C and sinks, displacing water at the bottom and creating the fall turnover. The fall turnover is important because it brings oxygen-rich water from the surface down to the benthic layer, and because it again brings nutrients from the benthic zone up to the limnetic zone.

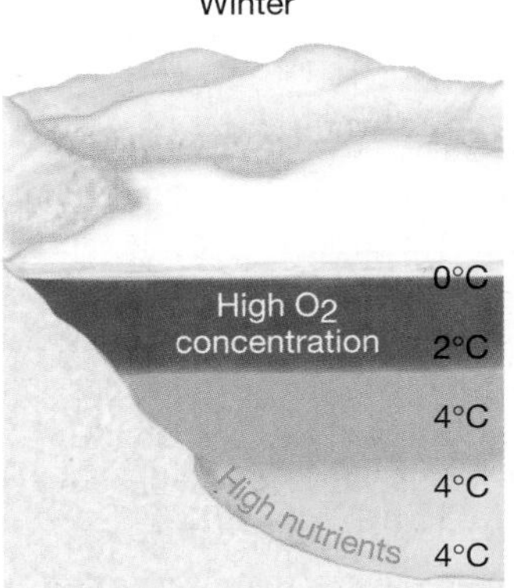

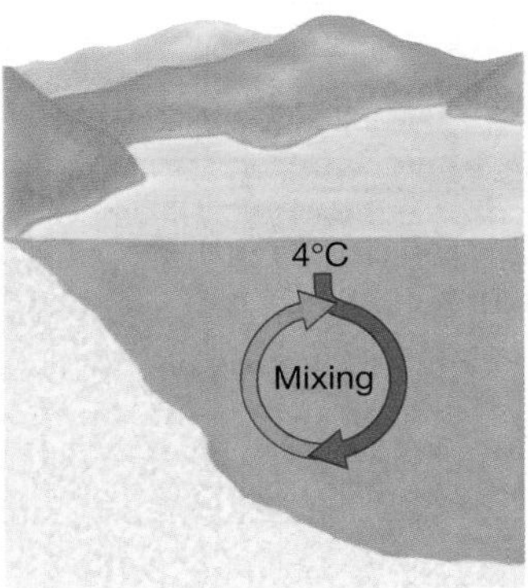

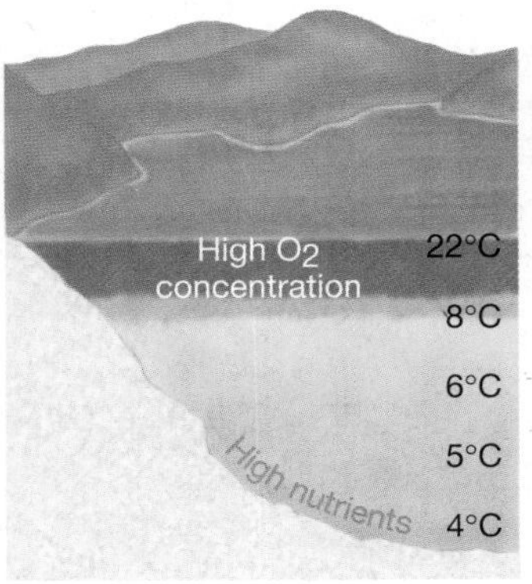

FIGURE 50.4 In Temperate Regions, Lakes Turn Over Each Spring and Fall.

● **QUESTION** In lakes like the one pictured here, there is a dramatic burst of photosynthesis each spring and fall. Why?

Freshwater Environments > Wetlands

Wetlands are shallow-water habitats where the soil is saturated with water for at least part of the year.

Water Depth Wetlands are distinct from lakes and ponds for two reasons: They have only shallow water, and they have **emergent vegetation**—meaning plants that grow above the surface of the water. All or most of the water in wetlands receives sunlight, and emergent plants capture sunlight before it strikes the water.

Water Flow Freshwater marshes and swamps are wetland types characterized by a slow but steady flow of water. **Bogs** (**Figure 50.5a**), in contrast, develop in depressions where water flow is low or nonexistent. If water is stagnant, oxygen is used up during the decomposition of dead organic matter faster than it enters via diffusion from the atmosphere. As a result, bog water is oxygen poor or even anoxic. Once the oxygen in the water is depleted, decomposition slows. Organic acids and other acids build up, lowering the pH of the water. At low pH, nitrogen becomes unavailable to plants (see Chapter 38).

Organisms The combination of acidity, lack of available nitrogen, and anoxic conditions makes bogs extremely unproductive habitats. Marshes and swamps, in contrast, offer ample supplies of oxygenated water and sunlight and are extraordinarily productive. **Marshes** lack trees and typically feature grasses (**Figure 50.5b**); **swamps** are dominated by trees and shrubs (**Figure 50.5c**). Because their physical environments are so different, there is little overlap in the types of species found in bogs, marshes, and swamps.

● You should be able to explain why carnivorous plants, which capture and digest insects, are relatively common in bogs but rare in marshes and swamps.

(a) Bogs are stagnant and acidic.

(b) Marshes have nonwoody plants.

(c) Swamps have trees and shrubs.

FIGURE 50.5 Wetland Types Are Distinguished by Water Flow and Vegetation.

Freshwater Environments > Streams

Streams are bodies of water that move constantly in one direction. Creeks are small streams; rivers are large. In terms of the environment that is available to organisms, the major physical variables in streams are the speed of the current and the availability of oxygen and nutrients.

Water Depth Most streams are shallow enough that sunlight reaches the bottom. Availability of sunlight is usually not a limiting factor for organisms.

Water Flow The structure of a typical stream varies along its length (**Figure 50.6**). Where it originates at a mountain glacier, lake, or spring, a stream tends to be cold, narrow, and fast. As it descends toward a lake, ocean, or larger river, a stream accepts water from tributaries and becomes larger, warmer, and slower.

Oxygen levels tend to be high in fast-moving streams because water droplets are exposed to the atmosphere when moving water splashes over rocks or other obstacles. Oxygen from the atmosphere diffuses into the droplets. In contrast, slow-moving streams that lack riffles or rapids tend to become relatively oxygen poor. Also, cold water holds more oxygen than warm water does (see Chapter 44).

(Continued on next page)

Freshwater Environments > Streams *continued*

Organisms It is rare to find photosynthetic organisms in small, fast-moving streams; nutrient levels tend to be low and most of the organic matter present consists of leaves and other materials that fall into the water from outside the stream. Fish, insect larvae, molluscs, and other animals have adaptations that allow them to maintain their positions in the fast-moving portions of streams. As streams widen and slow down, conditions become more favorable for the growth of algae and plants, and the amount of organic matter and nutrients increases. As a result, the same stream often contains completely different types of organisms near its source and near its end, or mouth.

● You should be able to explain why, in terms of their ability to perform cellular respiration, fish species found in cold, fast-moving streams tend to be much more active than fish species found in warm, slow-moving streams.

FIGURE 50.6 Distinctive Environments Appear along a Stream's Length.

Freshwater/Marine Environments > Estuaries

Estuaries form where rivers meet the ocean—meaning that freshwater mixes with salt water (**Figure 50.7**). In essence, an estuary includes slightly saline marshes as well as the body of water that moves in and out of these environments. Salinity varies with changes in river flows and with proximity to the ocean. Salinity has dramatic effects on osmosis and water balance (see Chapters 37 and 42); species that live in estuaries have adaptations that allow them to cope with variations in salinity.

Water Depth Most estuaries are shallow enough that sunlight reaches the substrate. Water depth may fluctuate dramatically, however, in response to tides, storms, and floods.

Water Flow Water flow in estuaries fluctuates daily and seasonally due to tides, storms, and floods. The fluctuation is important because it alters salinity, which in turn affects which types of organisms are present.

Organisms Because the water is shallow and sunlit, and because nutrients are constantly replenished by incoming river water, estuaries are among the most productive environments on Earth. They are often packed with young fish, which feed on abundant vegetation and plankton while hiding from predators.

● You should be able to compare and contrast the characteristics of estuaries and marshes, and explain why few of the same species are found in both.

FIGURE 50.7 Estuaries Are Highly Productive Environments. This photo was taken at low tide.

● **EXERCISE** Label freshwater environments, saltwater environments, and areas where freshwater and salt water are mixing.

Marine Environments > The Ocean

The world's oceans form a continuous body of salt water and are remarkably uniform in chemical composition. Regions within an ocean vary markedly in their physical characteristics, however, with profound effects on the organisms found there.

Water Depth Biologists describe the structure of an ocean by naming six regions (**Figure 50.8**).

- The **intertidal** ("between tides") **zone** consists of a rocky, sandy, or muddy beach that is exposed to the air at low tide but submerged at high tide.
- The **neritic zone** extends from the intertidal zone to depths of about 200 m. Its outermost edge is defined by the end of the **continental shelf**—the gently sloping, submerged portion of a continental plate.
- The **oceanic zone** is the "open ocean"—the deepwater region beyond the continental shelf.
- The bottom of the ocean is the **benthic zone**.
- The intertidal and sunlit regions of the neritic, oceanic, and benthic zones make up a **photic zone**.
- Areas that do not receive sunlight are in an **aphotic zone**.

Water Flow Water movement in the ocean is dominated by different processes at different depths. In the intertidal zone, tides and wave action are the major influences. In the neritic zone, currents that bring nutrient-rich water from the benthic zone of the deep ocean toward shore have a heavy impact. More specifically, nutrient-rich water is carried toward the surface when it hits the steep slope of the continental plate. Throughout the ocean, large-scale currents circulate water in the oceanic zone in response to prevailing winds and the Earth's rotation.

Organisms Each zone in the ocean is populated by distinct species that are adapted to the physical conditions present. Organisms that live in the intertidal zone must be able to withstand physical pounding from waves and desiccation at low tide. Productivity is high, however, due to the availability of sunlight and nutrients contributed by estuaries as well as by currents that sweep in nutrient-laden sediments from offshore areas.

Productivity is also high on the outer edge of the neritic zone, due to nutrients contributed by upwellings at the edge of the continental plate. Almost all of the world's major marine fisheries exploit organisms that live in the neritic zone. In the tropics, shallow portions of the neritic zone may host **coral reefs**. Because the water is warm and sunlight penetrates to the ocean floor in these habitats, coral reefs are among the most productive environments in the world (see Chapter 54).

If coral reefs are the rain forests of the ocean, then the oceanic zone is the desert. Sunlight is abundant in the photic zone of the open ocean, but nutrients are extremely scarce. When the photosynthetic organisms and the animals that feed on them die, their bodies drift downward out of the photic zone and are lost. In the open ocean, there is no mechanism for bringing nutrients back up from the bottom, as there is in the neritic zone or in most lakes (see Box 50.1). The aphotic zone of the open ocean is also unproductive because light is absent and photosynthesis is impossible. Most organisms present in the aphotic zone survive on the rain of dead bodies from the photic zone.

● You should be able to predict the following characteristics of fish that live in the aphotic zone: what they eat, whether they have eyes, whether they are capable of swimming, and whether they are rare or abundant.

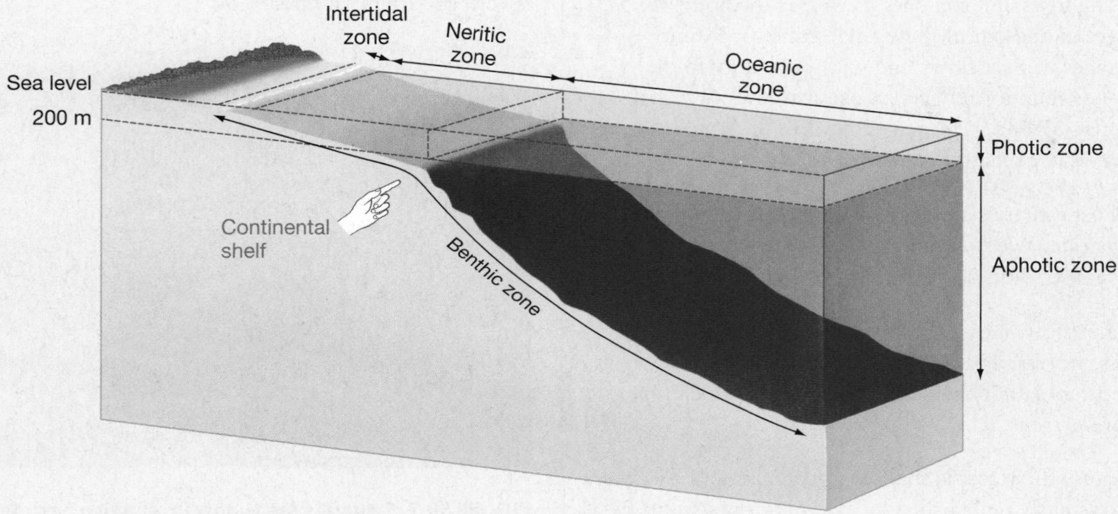

FIGURE 50.8 Oceans Have Distinctive Zones Defined by Water Depth and Distance from Shore.

Check Your Understanding

If you understand that...

- Aquatic environments are distinguished by the depth of water and the rate at which it moves.

You should be able to...

1) Explain why coastal environments are the most productive areas in the ocean.
2) Explain why bogs and marshes differ in productivity.

50.3 Types of Terrestrial Ecosystems

If you could walk from the equator in South America to the North Pole, you would notice startling changes in the organisms around you. Lush tropical forests with broad-leaved evergreen trees would give way to seasonally dry forests and then to deserts. The deserts would yield to the vast grasslands of central North America, which terminate at the boreal forests of the subarctic. If you pressed on, you would reach the end of the trees and the beginning of the most northerly community—the arctic tundra.

Broad-leaved evergreen forests, deserts, and grasslands are **biomes**: major groupings of plant and animal communities defined by a dominant vegetation type. **Figure 50.9** shows the global distribution of the most common types of biomes. Many distinct communities exist within each of these broad regions. For example, the Finger Lakes region of the northeastern United States is part of the temperate forest biome. However, the cool, wet, north-facing slopes in the area host plant communities dominated by eastern hemlock trees; warmer and drier south-facing slopes feature communities with red oak trees; and mild slopes and flat areas are covered with forests consisting of sugar maples and white oaks (**Figure 50.10**). The characteristics of these forest communities are similar enough to place them in the same biome, even though the specific species present vary with local conditions.

Each of the biomes found around the world is associated with a distinctive set of abiotic conditions. Just as water depth and movement have an overriding influence on aquatic ecosystems, the type of biome present in a terrestrial region depends on **climate**—the prevailing, long-term weather conditions found in an area. **Weather** consists of the specific short-term atmospheric conditions of temperature, moisture, sunlight, and wind.

- Temperature is critical because the enzymes that make life possible work at optimal efficiency only in a narrow range of temperatures (see Chapter 3). Temperature also affects the availability of moisture: Water freezes at low temperatures and evaporates rapidly at high temperatures.
- Moisture is significant because it is required for life, and because terrestrial organisms constantly lose water to the environment through evaporation or transpiration. To stay alive, they must reduce water loss and replace lost water.
- Sunlight is essential because it is required for photosynthesis.
- Wind is important because it exacerbates the effects of temperature and moisture. Wind increases heat loss due to evaporation and convection, and it increases water loss due to evaporation and transpiration. It also has a direct physical impact on organisms such as birds, flying insects, and plants by pushing them around.

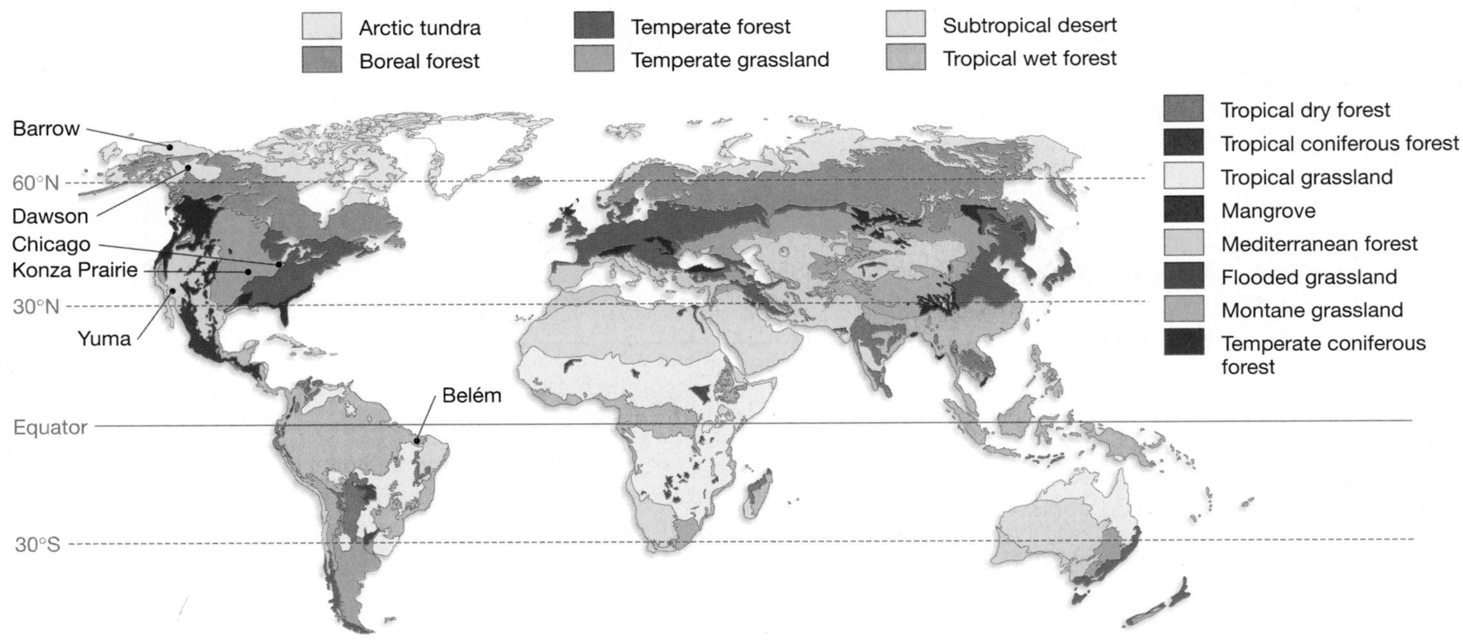

FIGURE 50.9 Distinct Biomes Are Found throughout the World. A recent classification identified the 14 major biomes shown here.

(a) Temperate forest biome:
Eastern hemlock community

(b) Temperate forest biome:
Red oak community

(c) Temperate forest biome:
Sugar maple community

FIGURE 50.10 Biomes Contain Distinct Communities. Within even a small area, the same biome can contain several distinct plant and animal communities.

QUESTION These communities are found in the northeastern United States and southeast Canada. How would temperate forest communities in New Zealand or northern France compare?

Of the four components of climate, temperature and moisture are far and away the most important to plants. More specifically, the nature of the biome that develops in a particular region is governed by (1) average annual temperature and precipitation, and (2) annual variation in temperature and precipitation. Each biome contains species that are adapted to a particular temperature and moisture regime.

The amount and variability of heat and precipitation structure terrestrial environments, much as water depth and flow rate structure aquatic environments. On land, photosynthesis and plant growth are maximized when temperatures are warm and conditions are wet; conversely, photosynthesis cannot occur efficiently at low temperatures or under drought stress. Biologists are particularly concerned with **net primary productivity** (**NPP**), which is the total amount of carbon that is fixed per year minus the amount of fixed carbon oxidized during cellular respiration. Fixed carbon that is consumed in cellular respiration provides energy for the organism but is not used for growth—that is, production of **biomass**. NPP is key because it represents the organic matter that is available as food for other organisms. In terrestrial environments NPP is often estimated by measuring **aboveground biomass**—the total mass of living plants, excluding roots.

To help you understand how temperature and precipitation influence biomes, let's take a detailed look at the six biomes that represent Earth's most extensive vegetation types, ranging from the wet tropics to the arctic. In each case, you'll be analyzing temperature and precipitation data from a specific location, typical of the biome in question.

Check Your Understanding

If you understand that...

- In terrestrial environments, distinctive biomes develop based on two aspects of temperature and moisture: average value and variability over the year.

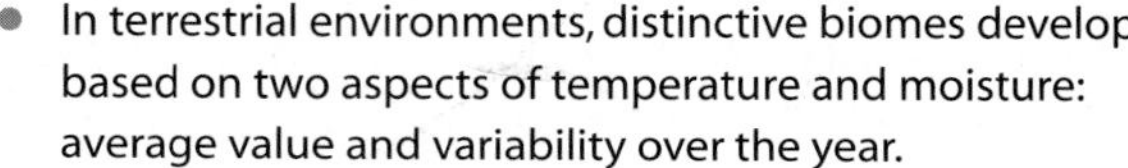

You should be able to...

Predict whether tropical dry forests, which have high year-round temperatures but a distinct dry season, are more or less productive than tropical wet forests.

Terrestrial Biomes > Tropical wet forest

Tropical wet forests—also called tropical rain forests—are found in equatorial regions around the world. Plants in this biome have broad leaves as opposed to narrow, needle-like leaves, and are evergreen. Older leaves are shed throughout the year, but there is no complete, seasonal loss of leaves.

Temperature The data in **Figure 50.11** are from Belém, Brazil, where mean monthly temperatures never drop below 25°C and never exceed 30°C. Compared to other biomes, tropical wet forests show almost no seasonal variation in temperature. This is important because temperatures are high enough to support growth throughout the year.

Precipitation Even in the driest month of the year, November, this region receives over 5 cm (2 in.) of rainfall—considerably more than the *annual* rainfall of many deserts.

Vegetation Favorable year-round growing conditions produce riotous growth, leading to extremely high productivity and aboveground biomass (**Figure 50.12**). Tropical wet forests are also

(Continued on next page)

Terrestrial Biomes > Tropical wet forest *continued*

renowned for their species diversity. It is not unusual to find over 200 tree species in a single 10 m × 100 m study plot. And based on counts of the insects and spiders collected from single trees, some biologists contend that the world's tropical wet forests may hold up to 30 million species of arthropods alone.

The diversity of plant sizes and growth forms in wet forest communities produces extraordinary structural diversity. In tropical wet forests, a few extremely large trees tower over a layer of large trees that form a distinctive **canopy** (the uppermost layers of branches). From the canopy to the ground, there is a complex assortment of vines, **epiphytes** (plants that grow entirely on other plants), small trees, shrubs, and herbs. This diversity of growth forms presents a wide array of habitat types for animals.

● You should be able to explain which aspects of structural diversity in a rain forest are lacking in the biome where you live.

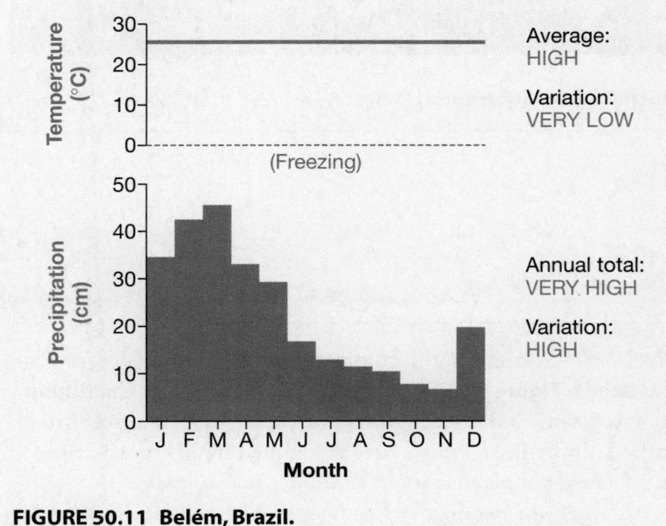

FIGURE 50.11 Belém, Brazil.

FIGURE 50.12 Tropical Wet Forest.

Terrestrial Biomes > Subtropical deserts

Subtropical deserts are found throughout the world in two distinctive locations: 30 degrees latitude, or distance from the equator, both north and south. Most of the world's great deserts—including the Sahara, Gobi, Sonoran, and Australian outback—lie on or about 30°N or 30°S latitude.

Temperature The data in **Figure 50.13** are from Yuma, Arizona, in the Sonoran Desert of southwestern North America. Mean monthly temperatures in subtropical deserts vary more than in tropical wet climates, but in Yuma temperatures still never fall below freezing. (Freezing nighttime temperatures are common elsewhere in the Sonora and in other subtropical deserts.)

Precipitation The most striking feature of the subtropical desert climate is low precipitation. The average annual precipitation in Yuma is just 7.5 cm (3 in.).

Vegetation The scarcity of water in deserts has profound implications. Conditions are rarely adequate to support photosynthesis, so the productivity of desert communities is a tiny fraction of average values for tropical forest communities. Further, as the photo in **Figure 50.14** shows, individual plants are widely spaced—a pattern that may reflect intense competition for water.

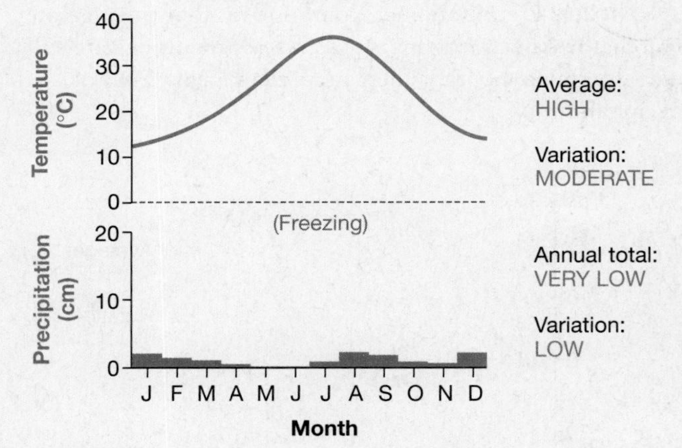

FIGURE 50.13 Sonoran Desert.

(Continued on next page)

Desert species adapt to the extreme temperatures and aridity in one of two ways: Growing at a low rate year-round, or breaking dormancy and growing rapidly in response to any rainfall. Cacti can grow year-round because they possess adaptations to cope with hot, dry conditions: small leaves or no leaves; a thick, waxy coating on leaves and stems; and the CAM pathway for photosynthesis (Chapter 10). The desert shrub called ocotillo, in contrast, is usually dormant. Individuals sprout leaves within days of a rainfall and drop them a week or two later when soils dry out again. The seeds of annual plants can also lie dormant for many years, then germinate after a rain.

● You should be able to be able to state hypotheses explaining why desert species are not found in rain forests, and why rain forest species are not found in deserts.

FIGURE 50.14 Subtropical Desert.

Terrestrial Biomes > Temperate grasslands

Temperate grassland communities are found throughout central North America and the heartland of Eurasia (see Figure 50.9). In North America they are commonly called prairies; in central Eurasia they are known as steppes.

Temperature A region of the world is called **temperate** if it has pronounced annual fluctuations in temperature—typically hot summers and cold winters—but not the temperature extremes recorded in the tropics or arctic. Temperature variation is important because it dictates a well-defined growing season. In the temperate zone, plant growth is possible only in spring, summer, and fall months when moisture and warmth are adequate. **Figure 50.15** presents climate data for a typical grassland community—in this case, the Konza Prairie near Manhattan, Kansas.

Precipitation Precipitation at Konza is over four times greater than that in Yuma, Arizona. Nonetheless, conditions are still quite dry; in no month is there more than 5 cm (2 in.) of precipitation.

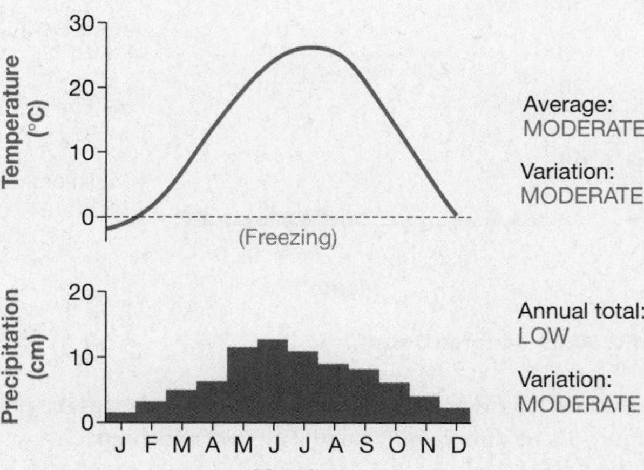

FIGURE 50.15 Konza Prairie.

Vegetation Grasses are the dominant life-form in temperate grasslands (**Figure 50.16**) for one of two reasons: (1) Conditions are too dry to enable tree growth, or (2) encroaching trees are burned out by fires. Prairie fires are ignited by lightning strikes or by native people managing land for game animals.

Although the productivity of temperate grasslands is generally lower than that of forest communities, grassland soils are often highly fertile. The subsurface is packed with roots and rhizomes, which add organic material to the soil as they die and decay. Further, grassland soils retain nutrients, because rainfall is low enough to keep key ions from dissolving and leaching out of the soil. It is no accident, then, that the grasslands of North America and Eurasia are the breadbaskets of those continents. The conditions that give rise to natural grasslands are ideal for growing wheat, corn, and other cultivated grasses.

● You should be able to explain why fires are much more common in grasslands than in deserts, even though deserts are drier.

FIGURE 50.16 Temperate Grassland.

In temperate areas with relatively high precipitation, grasslands give way to forests. Temperate forests are the most common biome found in eastern North America, western Europe, east Asia, Chile, and New Zealand.

Temperature As **Figure 50.17** shows, temperate forests experience a period in which mean monthly temperatures fall below freezing and plant growth stops. The data in this graph are from Chicago, Illinois, which was forested before being settled.

Precipitation Compared to grassland climates, precipitation in temperate forests is moderately high and relatively constant throughout the year. Chicago, for example, has an annual precipitation of 85 cm (34 in.); during most months, precipitation exceeds 5 cm (2 in.).

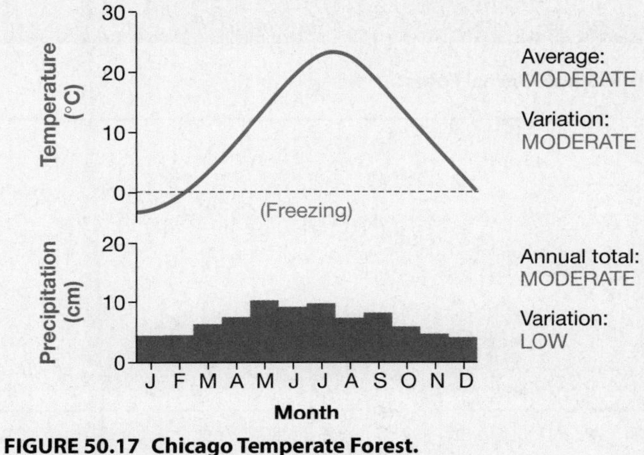

FIGURE 50.17 Chicago Temperate Forest.

FIGURE 50.18 Temperate Forest.

Vegetation In North America and Europe, temperate forests are dominated by deciduous species, which drop their leaves in autumn and grow new leaves in spring (**Figure 50.18**). Needle-leaved evergreens are also common. But in the temperate forests of New Zealand and Chile , broad-leaved evergreens predominate.

Most temperate forests have productivity levels that are lower than those of tropical forests yet higher than those of deserts or grasslands. The level of diversity is also moderate. A temperate forest in southeastern North America may have more than 20 tree species; similar forests to the north may have fewer than 10.

● You should be able to describe the characteristics (annual temperature, rainfall, vegetation) of the type of biome that is found in regions with climate characteristics intermediate between those of grasslands and temperate forests.

Terrestrial Biomes > Boreal forests

The boreal forest, or **taiga**, stretches across most of Canada, Alaska, Russia, and northern Europe. Because these regions are just south of the Arctic Circle, they are referred to as subarctic.

Temperature The data in **Figure 50.19** are from Dawson, in the Yukon Territory of Canada. The region is characterized by very cold winters and cool, short summers. Temperature variation is extreme; in the course of a year, subarctic areas may be subject to temperature ranges of more than 70°C.

Precipitation Annual precipitation in boreal forests is low, but temperatures are so cold that evaporation is minimal; as a result, moisture is usually abundant enough to support tree growth.

Vegetation Boreal forests are dominated by highly cold-tolerant conifers, including pines, spruce, fir, and larch trees. Except for larches, these species are evergreen. Two hypotheses have been offered to explain why evergreens predominate in cold environ-

(Continued on next page)

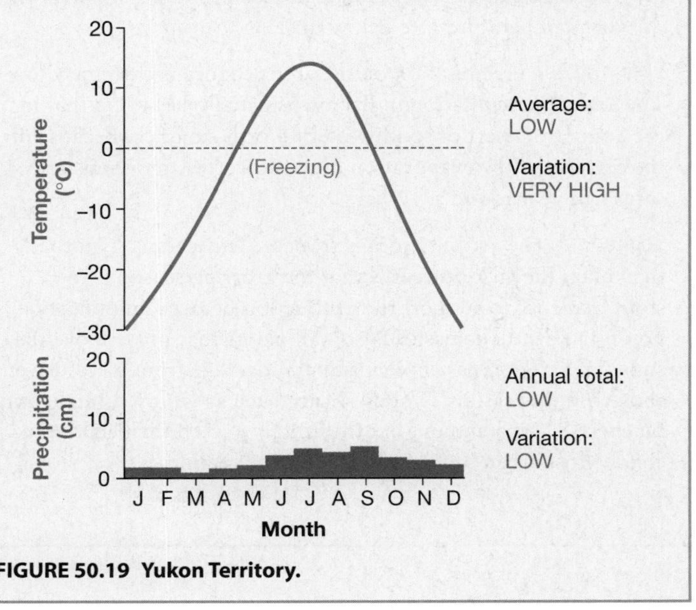

FIGURE 50.19 Yukon Territory.

Terrestrial Biomes > Boreal forests *continued*

ments, even though they do not photosynthesize in winter (**Figure 50.20**). The first is that evergreens can begin photosynthesizing early in the spring, even before the snow melts, when sunshine is intense enough to warm their needles. The second hypothesis is based on the observation that boreal forest soils tend to be acidic and contain little available nitrogen. Because leaves are nitrogen rich, species that must produce an entirely new set of leaves each year might be at a disadvantage. To date, however, these hypotheses have not been tested rigorously.

Based on these observations, it is not surprising that the productivity of boreal forests is low. Aboveground biomass is high, however, because slow-growing tree species may be long lived and gradually accumulate large standing biomass. Boreal forests also have exceptionally low species diversity. The boreal forests of Alaska, for example, typically contain seven or fewer tree species.

● You should be able to predict how the global distribution of boreal forests will change in response to global warming.

FIGURE 50.20 Boreal Forest.

Terrestrial Biomes > Arctic tundra

The arctic **tundra** biome, which lies poleward from the subarctic, is found throughout the arctic regions of the Northern Hemisphere and in regions of Antarctica that are not covered in ice.

Temperature Tundra develops in climatic conditions like those in Barrow, on the northern coast of Alaska (**Figure 50.21**). The growing season is 6–8 weeks long at most; for the remainder of the year, temperatures are below freezing.

Precipitation Precipitation on the arctic tundra is extremely low. The annual precipitation in Barrow is actually less than that in the Sonoran Desert of southwestern North America. Because of the extremely low evaporation rates, however, arctic soils are saturated year-round.

Vegetation The arctic tundra is treeless. The leading hypothesis to explain the lack of trees is that the growing season is too short and cool to support the production of large amounts of non-photosynthetic tissue. Also, tall plants that poke above the snow in winter experience substantial damage from wind-driven snow and ice crystals. Woody shrubs such as willows, birch, and blueberries are common, but they rarely exceed the height of a child. Most arctic tundra species hug the ground.

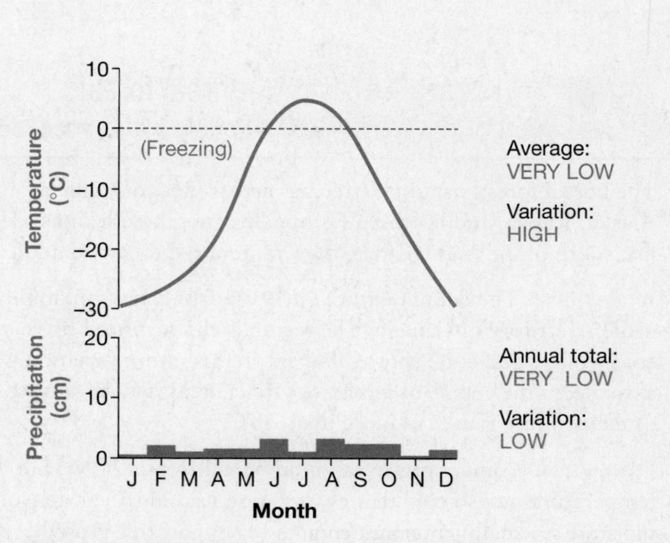

FIGURE 50.21 Barrow, Alaska.

(Continued on next page)

Terrestrial Biomes > Boreal forests *continued*

Arctic tundra has low species diversity, low productivity, and low aboveground biomass. Most tundra soils are in the perennially frozen state known as **permafrost**. The low temperatures inhibit both the release of nutrients from decaying organic matter and the uptake of nutrients into live roots. Unlike desert biomes, however, the ground surface in tundra communities is completely covered with plants or lichens (**Figure 50.22**). Animal diversity also tends to be low in the arctic tundra, although insect abundance—particularly of biting flies—can be staggeringly high.

● You should be able to predict how arctic tundras compare and contrast to alpine tundras, which are found at high elevation.

FIGURE 50.22 Arctic Tundra.

50.4 The Role of Climate and the Consequences of Climate Change

Each type of aquatic environment and terrestrial biome hosts species that are adapted to the abiotic conditions present at that location. Global warming—due to rising concentrations of atmospheric CO_2 (see Chapter 54)—is having a profound impact on these abiotic factors. Runoff from melting glaciers and pole ice is changing water depths along coasts, and warming ocean waters are disturbing long-established patterns in the direction and intensity of ocean currents. On land, higher air temperatures are altering growing seasons and water availability in biomes around the globe.

To understand how global warming is affecting the distribution and abundance of organisms, biologists begin by studying why climate varies in predictable ways around the planet. Then they ask how changes in these climate patterns are affecting the organisms present, with the aim of predicting the consequences of continued change. Let's do the same.

Global Patterns in Climate

Simple questions often have fascinating answers. This turns out to be true for questions about why climate varies around the globe. For example, why are some parts of the world warmer and wetter than others? Why do seasons exist?

Why Are the Tropics Warm and the Poles Cold? In general, areas of the world are warm if they receive a large amount of sunlight per unit area; they are cold if they receive a small amount of sunlight per unit area. Over the course of a year, regions at or near the equator receive much more sunlight per unit area—and thus much more energy in the form of heat—than regions that are closer to the poles. At the equator, the Sun is often directly overhead. As a result, sunlight strikes Earth there at or close to a 90° angle. At these angles, Earth receives a maximum amount of solar radiation per unit area (**Figure 50.23**). But because Earth's surface slopes away from the equator, the Sun strikes the surface at lower and lower angles moving toward the poles. When sunlight arrives at a low angle, much less energy is received per unit area. The pattern of decreasing average temperature with increasing latitude is a result of Earth's spherical shape.

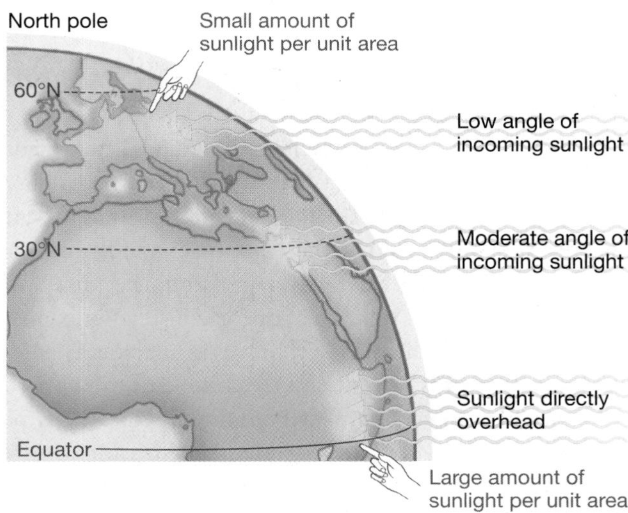

FIGURE 50.23 Solar Radiation per Unit Area Declines with Increasing Latitude. Over the course of a year, the Sun is frequently almost directly overhead at the equator. As a result, equatorial regions receive a large amount of solar radiation per unit area. At latitudes greater than 23.5°, the Sun is never directly overhead. As a result, high-latitude regions receive less solar radiation per unit area than do low-latitude regions.

(a) Circulation cells exist at the equator ...

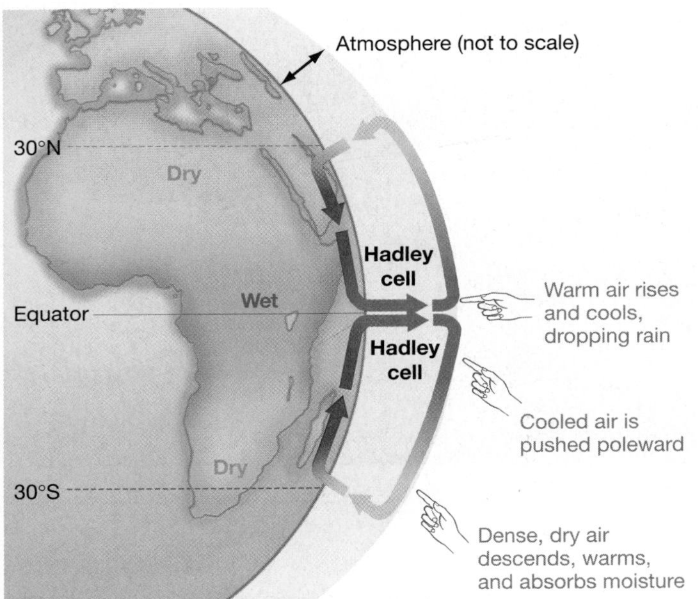

(b) ... and at higher latitudes.

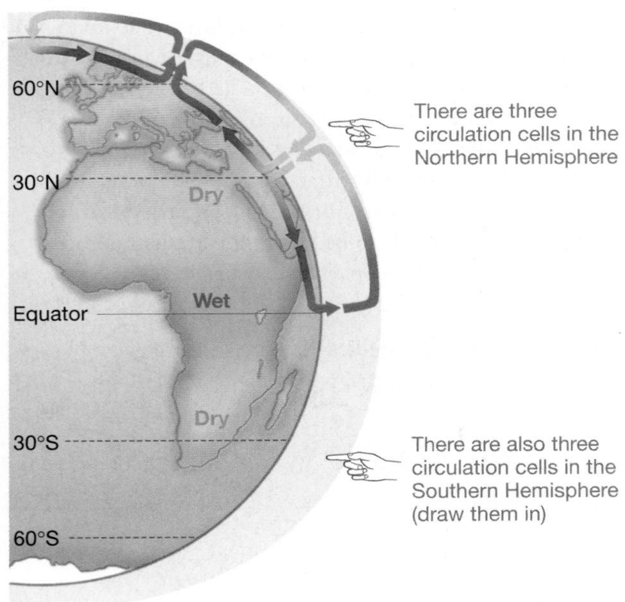

FIGURE 50.24 Global Air Circulation Patterns Affect Rainfall. Hadley cells explain why the tropics are much wetter than regions near 30° latitude north and south. Regions with rising air, such as those at the equator, tend to be wetter than regions with descending air.

● **QUESTION** Is air at the North Pole wet or dry? Is your answer consistent with the data in Figure 50.21?

Why Are the Tropics Wet? One of the most striking weather patterns on Earth involves precipitation. When average annual rainfall is mapped for regions around the globe, it is clear that areas along the equator receive the most moisture, while locations about 30 degrees latitude north and south of the equator are among the driest on the planet. Why do these patterns occur?

A major cycle in global air circulation, called a **Hadley cell**, is responsible for making the Amazon River basin wet and the Sahara Desert dry. Hadley cells are named after George Hadley, who in 1735 conceived of the idea of enormous air circulation patterns. As **Figure 50.24a** indicates, air that is heated by the strong sunlight along the equator expands and rises. Warm air can hold a great deal of moisture, because warm water molecules tend to stay in vapor form instead of condensing into droplets. As the air rises, however, it radiates heat to space. It also expands into the larger volume of the upper atmosphere, which lowers its density and temperature—a phenomenon known as adiabatic cooling. As rising air cools, its ability to hold water declines. When water vapor cools, water condenses. The result? High levels of precipitation occur along the equator.

As more air is heated along the equator, the cooler, "older" air above Earth's surface is pushed poleward. When the air mass has cooled enough, its density increases and it begins to sink. As it sinks, it absorbs more and more solar radiation reflected from Earth's surface and begins to warm. As the air warms, it also gains water-holding capacity. Thus, the air approaching the Earth "holds on" to its water and little rain occurs where it returns to the surface; the area is bathed in warm, dry air. The result is a band of deserts in the vicinity 30° latitude north and south. As this air moves toward each pole, it continues to pick up more moisture.

Similar air circulation cells occur between 30° and 60° of latitude, and between 60° latitude and the poles (**Figure 50.24b**). ● If you understand how Hadley cells work, you should be able to add diagrams to Figure 50.24b showing the three cells that occur in the Southern Hemisphere.

What Causes Seasonality in Weather? The striking global patterns in temperature and moisture that we've just reviewed are complicated by the phenomenon of seasons: regular, annual fluctuations in temperature, precipitation, or both. As **Figure 50.25** shows, seasonality occurs because Earth is tilted on its axis by 23.5 degrees. As a result of this incline, the Northern Hemisphere is tilted toward the Sun in June and faces the Sun most directly. Thus the Northern Hemisphere presents its least-acute angle to the Sun and receives the largest amount of solar radiation per unit area in June. The Southern Hemisphere, in contrast, is tilted away from the Sun in June, presents a steep angle for incoming sunlight, and receives its smallest quantity of solar radiation per unit area. As a result, in June it is summer in the Northern Hemisphere, and winter in the Southern Hemisphere. Conversely, in De-

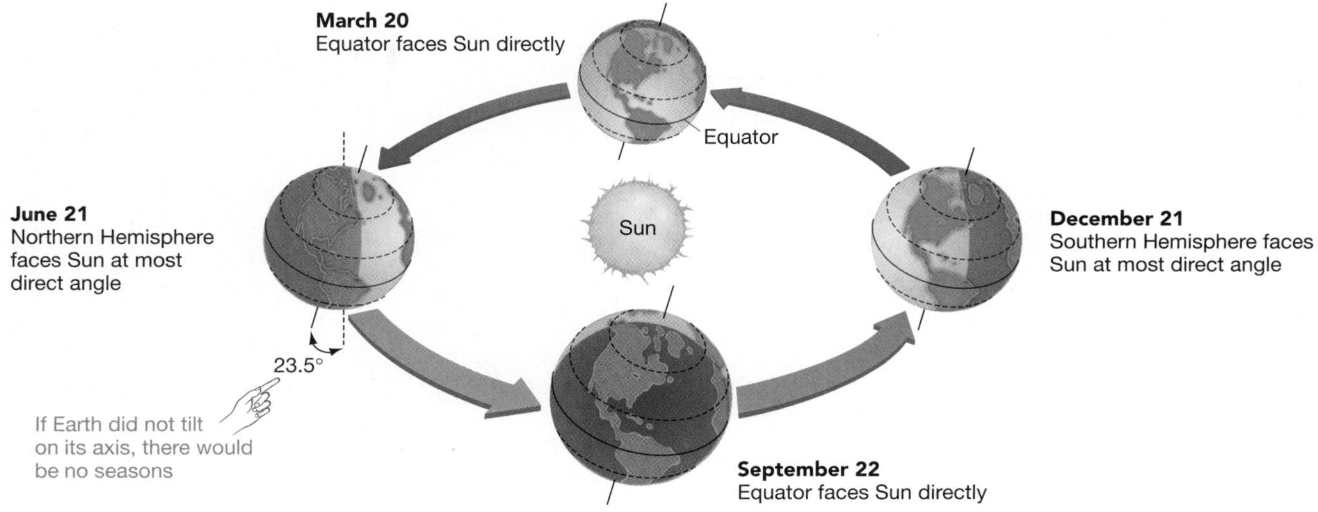

FIGURE 50.25 Earth's Orbit and Tilt Create Seasons at High Latitudes. Seasons occur due to annual variation in the amount of solar radiation that different parts of the Earth receive.

cember it is summer in the Southern Hemisphere, and winter in the Northern Hemisphere. In March and September the equator faces the Sun most directly, so the tropics receive the most solar radiation then. If Earth did not tilt on its axis, there would be no seasons.

Mountains and Oceans: Regional Effects on Climate

The broad patterns of climate that are dictated by global heating patterns, Hadley cells, and seasonality are overlain by regional effects. The most important of these are due to the presence of mountain ranges and proximity to an ocean.

Figure 50.26 shows how mountain ranges affect regional climate, using the Cascade Mountain range along the Pacific Coast of North America as an example. In this area of the world, the prevailing winds are from the west. These winds bring moisture-laden air from the Pacific Ocean onto the continent. As the air masses begin to rise over the mountains, the air cools and releases large volumes of water as rain. As a result, the area along the Pacific Coast from northern California to southeast Alaska hosts some of the only high-latitude rain forests in the world. One area along the Pacific Coast of Washington State gets an average of 340 cm (134 in.) of precipitation each year. (Mountain ranges also occur along the Pacific Coast of Southern California and Mexico; but these areas are arid because they are dominated by dry air that is part of the Hadley circulation.)

Once cooled air has passed the crest of a mountain range, the air is relatively dry because much of its moisture content has already been released. Areas that receive this dry air are said to be in a **rain shadow**. One of the only high-latitude deserts in the world exists to the east of the Cascade Mountains. It is created by a rain shadow and averages less than 25 cm (10 in.) of moisture annually. Similar rain shadows create the Great Basin desert of western North America, the Atacama Desert of South America, and the dry conditions of Asia's Tibetan Plateau.

While the presence of mountain ranges tends to produce extremes in precipitation, the presence of an ocean has a moderating influence on temperature. To understand why this is so, recall from Chapter 2 that water has an extremely high **specific heat**, meaning that it has a large capacity for storing heat energy. Because water molecules form hydrogen bonds with each other, it takes a great deal of heat energy to boil water or melt ice—in fact, to change the temperature of water at all. The hydrogen bonds have to be broken before the water molecules

FIGURE 50.26 Mountain Ranges Create Rain Shadows. When moisture-laden air rises over a mountain range, the air cools enough to condense water vapor. As a result, moisture falls to the ground as precipitation. On the other side of the mountain, little water is left to provide precipitation.

QUESTION What conditions would create the wettest spots on Earth? The driest spots?

themselves can begin to absorb heat and move faster, which we measure as increased temperature.

Because water has a high specific heat, it can absorb a great deal of heat from the atmosphere in summer, when the water temperature is cooler than the air temperature. As a result, the ocean moderates summer temperatures on nearby landmasses. Similarly, the ocean releases heat to the atmosphere in winter, when the water temperature is warmer than the air temperature. As a result, islands and coastal areas have much more moderate climates than do nearby areas inland.

MB Web Animation at www.masteringbio.com
Tropical Atmospheric Circulation

How Will Global Warming Affect Ecosystems?

Climate—particularly the amount and variability of heat and precipitation—has a dramatic effect on terrestrial ecosystems. Climate also has a significant impact on aquatic ecosystems. Heat and precipitation influence ocean currents and lake turnover, as well as water levels in lakes, streams, and oceans.

It is now well established that CO_2 pollution is causing climate change (see Chapter 54). Climate has changed frequently and radically throughout the history of life, but recent increases in average temperatures around the globe, combined with a projected increase of up to 5.8°C over the next 100 years, represents one of the most rapid periods of climate change since life began.

Biologists use three tools to predict how global warming will affect aquatic and terrestrial ecosystems.

1. *Simulation studies* are based on computer models of weather patterns in local regions. By increasing average temperatures in these models, researchers can predict how wind and rainfall patterns, storm frequency and intensity, and other aspects of weather and climate will change. In effect, simulation studies predict how temperature and precipitation profiles, like those in Figures 50.11, 13, 15, 17, 19, and 21, will change.

2. *Observational studies* are based on long-term monitoring at fixed sites around the globe. Researchers are documenting changes in key physical variables—water depth, water flow, temperature, and precipitation—as well as changes in the distribution and abundance of organisms. By extrapolation, they can predict how change may continue at these sites.

3. *Experiments* are designed to simulate changed climate conditions, and to record responses by the organisms present.

Experiments That Manipulate Temperature Even though the scientific community did not become alert to the reality of global warming until the 1980s, a large number of multiyear experiments have already been done to test how increased temperature affects organisms in nature. Because arctic regions are projected to experience particularly dramatic changes in temperature and precipitation, many of these experiments have been done in the tundra biome.

FIGURE 50.27 Experimental Increases in Temperature Change Species Composition in Artic Tundra. The transparent, open-topped chambers shown here are an effective way to increase average annual temperature. Inside the chambers, species composition changes relative to otherwise similar sites nearby.

QUESTION Why does a clear chamber like this increase average temperature inside?

Figure 50.27 shows how the most successful of these experiments are done. Transparent, open-topped chambers are placed at random locations around a study site, and the characteristics of the vegetation inside the chambers are recorded annually. The same measurements are made in randomly assigned study plots that lack chambers.

Across experiments, the average annual temperature inside the open-topped chambers increases by 1–3°C. This increase is consistent with conservative projections for how temperatures will change over the next century. In a recent analysis of data from 11 such study sites scattered throughout the arctic in North America and Eurasia, several strong patterns emerged: Overall, species diversity decreases. Compared with control plots, grasses and shrubs increase inside the chambers, and mosses and lichens decrease. These results support simulation and observational studies predicting that arctic tundra environments are giving way to boreal forest. Biomes are changing right before our eyes.

Experiments That Manipulate Precipitation Recent simulation and observational studies have converged on a remarkable conclusion: Increases in average global temperature are increasing variability in temperature and precipitation. Stated another way, global warming is making climates more extreme. For example, the amount of precipitation and the wind speeds associated with monsoon rains, cyclones, hurricanes, and other types of storm events is increasing, even though their frequency may be unchanged.

Experiment

Question: How does increasing variation in rainfall—without changing total amount—affect grassland biomes?

Hypothesis: Increasing variability in rainfall affects soil moisture, net primary productivity (NPP), and species diversity.

Null hypothesis: Increasing variability in rainfall has no effect on grassland biomes.

Experimental setup:

Precipitation (mm)
60
40
20
0
Jun Jul Aug Sep Oct
Control plots: Normal rainfall

Precipitation (mm)
100
80
60
40
20
0
Jun Jul Aug Sep Oct
Experimental plots: Extend the interval between rainstorms by collecting rainwater, then applying all at once

Record soil moisture, aboveground biomass (as an index of NPP), and species diversity, over a 4-year period.

Prediction: Soil moisture, NPP, and/or species diversity will differ between experimental plots and control plots.

Prediction of null hypothesis: Soil moisture, NPP, and species diversity will not differ between experimental plots and control plots.

Results:

NPP declined
Average NPP (g/m^2)
700
600
500
400
300
200
Control plots Experimental plots

Species diversity increased
Species diversity index
1.8
1.6
1.4
1.2
1.0
Control plots Experimental plots

Conclusion: Altering variability in rainfall has dramatic effects on soil moisture (data not shown), NPP, and species diversity.

FIGURE 50.28 Experimental Increases in Rainfall Variability Change Species Composition in Temperate Grasslands.

To test how increased variation in climate will affect a temperate grassland ecosystem, researchers set up the experiment described in **Figure 50.28**. The goal was to keep the average annual rainfall unchanged during the growing season, but increase variability. The biologists increased the severity of storms by capturing rain that fell over study plots during some storms and then applying that water to the same plots during later storms.

The results from the first four years of the experiment are striking: Average soil moisture and overall NPP declined in the experimental plots, compared with similar plots where rainfall was not manipulated. Overall species diversity is increasing in the high-variation plots, however, because the grass that produces most of the biomass at this site is doing poorly in the experimental plots—opening up room for other species to grow. The experiment is ongoing, however, because it is possible that long-term changes will differ from short-term changes, due to continued alterations in the species present.

Experiments like this are documenting that climate change causes dramatic alterations in the composition and distribution of biomes. Your world is changing.

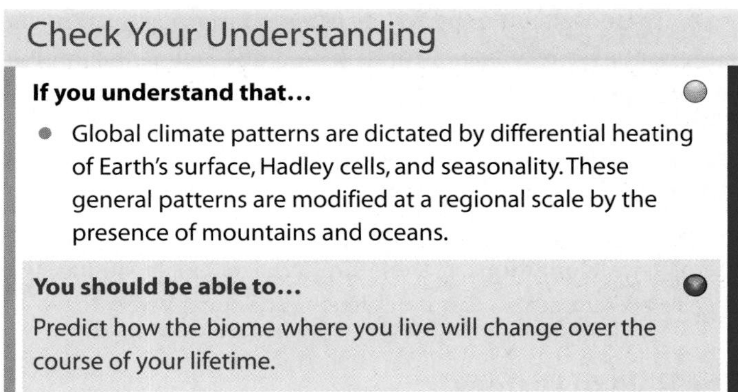
Check Your Understanding

If you understand that...

- Global climate patterns are dictated by differential heating of Earth's surface, Hadley cells, and seasonality. These general patterns are modified at a regional scale by the presence of mountains and oceans.

You should be able to...

Predict how the biome where you live will change over the course of your lifetime.

50.5 Biogeography: Why Are Organisms Found Where They Are?

Understanding why organisms are found where they are is one of ecology's most basic tasks. It is a fundamental aspect of predicting the consequences of climate change, and is required for designing nature preserves in a rational and effective way.

The study of how organisms are distributed geographically is called **biogeography**. You were introduced to this subject in Chapter 26, which analyzed how continental drift has contributed to the current distributions of ostriches and their relatives. Biogeography will come up repeatedly in this unit as well: Chapter 53 analyzes general patterns in how species numbers vary around the world, and Chapter 54 presents data showing that the geographic ranges of many species are changing rapidly in response to global warming. Here, let's focus on how interactions with the abiotic and biotic environments affect where a particular species is able to live.

The most fundamental observation about the **range**, or geographic distribution, of species is simple: No one species can survive the full range of environmental conditions present on Earth. *Thermus aquaticus* cells thrive in hot springs with temperatures above 70°C. These archaea can live in this environment because they have enzymes that do not denature at near-boiling temperatures. The same cells would die instantly, however, if they were transplanted to the frigid waters near polar ice. But vast numbers of other archaea are present in cold seawater. These cells have enzymes that can work efficiently at near-freezing temperatures. No organism thrives in both hot springs and cold ocean water, however. This observation is explained by the phenomenon called fitness trade-offs, introduced in Chapters 24 and 41: No enzyme can function well in both extremely high temperatures and extremely low temperatures.

Because of fitness trade-offs, organisms tend to be adapted to a limited set of physical conditions. Although some species have much broader geographic ranges than other species do—humans and the bacteria that live in and on them may have the widest distribution of any species at present—no organism can live everywhere.

To understand a species' distribution thoroughly, though, it is essential to examine historical and biotic factors in addition to the physical conditions present. For example, the temperature and moisture regimes of large areas of southern California, in North America, are almost identical to those found in the Mediterranean region of Europe and north Africa. Both regions are considered part of the same biome. But even though the physical conditions at these sites are extremely similar, the species present are almost completely different. Why?

The Role of History

To explain why certain species occur in a particular region while others don't, the first factor to consider is history—specifically, the history of dispersal. **Dispersal** refers to the movement of an individual from the place of its birth, hatching, or origin to the location where it lives and breeds as an adult. If a particular species is missing from an area, a physical barrier to dispersal—such as a river, a glacier, a mountain range, or an ocean—may be present. For example, the plant species found along North America's southwest coast are different from those found in the Mediterranean because the Pacific and Atlantic Oceans are effective barriers against the dispersal of seeds.

In many cases, human activity circumvents physical barriers to dispersal. For example, public health officials closely monitor annual flu outbreaks in south China—where most new strains originate—because they recognize that new strains disperse all over the planet in a matter of weeks or even days, via the respiratory passages of airplane passengers. Prior to the invention of air travel, disease epidemics in humans were much more likely to be confined to relatively small geographic areas.

Similarly, humans have transported thousands of seeds, birds, insects, and other species across physical barriers to new locations—sometimes purposefully and sometimes by accident. One accidental introduction has had disastrous consequences for the arid shrublands and grasslands of the western United States. In 1889 a native plant of Eurasia called cheatgrass was accidentally introduced to western North America in a shipment of crop seeds. Cheatgrass is an annual plant with seeds that germinate over winter, so that plants grow and set seed in early spring. As a result, cheatgrass seedlings get a "jump" on many North American plant species, which initiate growth later. Early growth allows cheatgrass to use more of the available water and nutrients. Within 30 years, cheatgrass had taken over tens of thousands of square miles of habitat (**Figure 50.29**).

If an **exotic** species —one that is not native—is introduced into a new area, spreads rapidly, and eliminates native species, it is said to be an **invasive species**. In North America alone, dozens of invasive species have had devastating effects on native plants and animals. Examples include kudzu, purple loosestrife, reed canary grass, garlic mustard, Russian thistle (also known as tumbleweed), European starlings, African honeybees, and zebra mussels. Similar lists could be compiled for other regions of the world.

Regarding the distribution of organisms, the story of invasive species is straightforward: They and other introduced species did not exist in certain parts of the world until recently, simply because they had never been dispersed there. But data show that only about 10 percent of the species introduced to an area actually become common, and a smaller percentage than that increases in population enough to be considered invasive. Why is it that some species introduced into an area do not thrive? Conversely, why are invasives so successful?

Biotic Factors

The distribution of a species is often limited by biotic factors—meaning interactions with other organisms. As an example, recall the data presented in Chapter 26 on the current and former

FIGURE 50.29 Cheatgrass Has Invaded Extensive Areas of North America.

distributions of hermit warblers and Townsend's warblers along the Pacific coast of North America. Both hermit warblers and Townsend's warblers live in evergreen forests. Experiments have shown that male Townsend's warblers directly attack male hermit warblers and evict them from breeding territories, and historical data indicate that the geographic range of Townsend's warblers has been expanding steadily at the expense of hermit warblers. The results support the hypothesis that a biotic factor—competition with another species—is limiting the range of hermit warblers.

Competition is not the only biotic factor to consider, however. In Africa, the range of domestic cattle is limited by the distribution of tsetse (pronounced *TSEE-tsee*) flies. Tsetse flies transmit a parasite that causes the disease trypanosomiasis, which is fatal in cattle (**Figure 50.30a**). Similarly, most of the songbirds native to Hawaii are limited to alpine habitats—above the range of the mosquitoes that transmit avian malaria. And the moth pictured in **Figure 50.30b** lays its eggs only in the flowers of yucca plants. As a result, this species does not exist outside the range of yucca plants.

(a) Distribution of cattle is limited by distribution of tsetse flies.

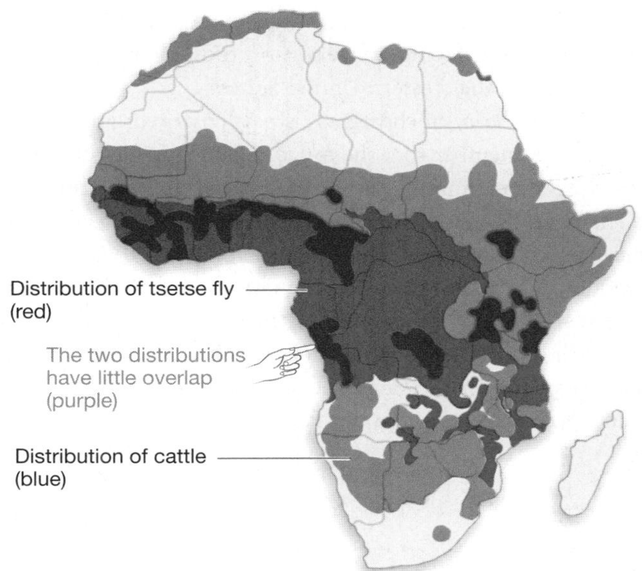

(b) Yucca moth distribution is limited by yucca plant distribution.

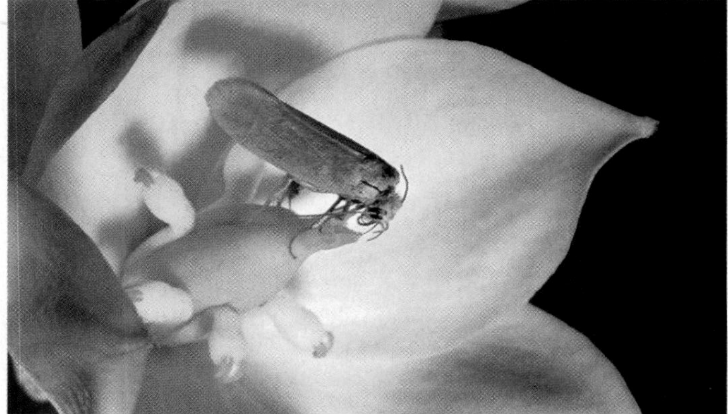

FIGURE 50.30 Distributions Can Be Limited by Biotic Factors.
(a) Because tsetse flies carry a disease that is fatal to cattle, tsetse flies and cattle rarely coexist. **(b)** Because yucca moths lay their eggs only in the flowers of yucca plants, the moths cannot live where yucca plants are absent.

Abiotic Factors

In many cases, introduced species may not become established in a region because of abiotic factors—particularly temperature and moisture. An area may simply be too cold or too dry for an organism to survive. For example, even though the invasive plant kudzu grows prolifically throughout the southeastern United States, it is killed when the soil freezes to a depth of a foot or more. As a result, it has not expanded its range farther north and west.

It is often difficult to separate the effects of biotic and abiotic factors on a species' range, however. To drive this point home, consider how cheatgrass became established in North America. Cheatgrass does not grow in wet grasslands because it does not compete well against the tall species that thrive there. But it has been able to invade two important types of biomes in North America: dry temperate grasslands and the arid, shrub-dominated habitats known as sage-steppe. In each case, the reasons for the success of cheatgrass were different.

Cheatgrass has been able to invade sage-steppe habitats because it is not affected by fire. Fire is an abiotic factor that can kill the shrubs that dominate sage-steppe (**Figure 50.31**). In fact, because cheatgrass grows in dense beds and dies back in

FIGURE 50.31 Sage-Steppe Is an Arid Biome Dominated by Shrubs. Large areas of western North America are covered with sage-steppe vegetation.

● **QUESTION** If cheatgrass were not present, what would happen if a fire started here?

early spring, it actually leads to an increase in the frequency and intensity of wildfires. Normally, fires are rare in sage-steppe biomes, because plants are widely spaced. This distribution usually prevents fires from spreading and affecting a large area. But when this biome begins to dry in late summer, beds of dead cheatgrass offer a continuous, extremely flammable source of fuel. If a fire does come through, it does not affect the cheatgrass for two reasons: (1) Cheatgrass is an annual, so there is no living tissue exposed once the growing season is over, and (2) cheatgrass seeds sprout readily in soils that have been depleted of organic matter by fire. Because sagebrush and other shrubs are perennials, their tissues are heavily damaged or killed by hot fires. By promoting the frequency and intensity of fire in sage-steppe biomes, the presence of cheatgrass creates an environment that favors the spread of cheatgrass.

How has this species been able to invade dry grassland biomes? Prior to the arrival of cheatgrass, grasslands in arid areas of the American West were dominated by the types of bunchgrasses shown in **Figure 50.32**. Bunchgrasses do not form a continuous sod; instead, they grow in compact bunches. The spaces between these bunches are occupied by what biologists call "black-soil crust." This material is a community of soil-dwelling cyanobacteria, archaea, and other microorganisms that grow when moisture is available. Many of these species also fix nitrogen (see Chapter 38). The nitrogen becomes available to the surrounding plants when the microbial cells die and decompose.

When European settlers arrived and began grazing cattle in this biome, however, both the black soils and bunchgrasses were affected. Cattle grazed on the bunchgrasses until the grasses died, and their hooves compacted and disrupted the black soils. Neither component of the ecosystem was allowed to recover, in part because intensive grazing continued and in part because cheatgrass arrived. Cheatgrass shades out black-soil crust organisms. When black-soil crust is gone, the amount of nitrogen available to bunchgrasses is reduced. Cheatgrass also competes successfully with bunchgrasses for water and nutrients, because it completes its growth early in the spring—before the native bunchgrasses have broken dormancy and begun to grow. And as in sage-steppe habitats, the expansion of cheatgrass increased the frequency and severity of fires, which further degraded the integrity of black-soil crust communities.

To summarize, changes in abiotic and biotic factors allowed cheatgrass to invade native biomes in North America and thrive as an invasive species. Currently, managing cheatgrass is the major problem facing conservation biologists and rangeland managers in the western United States.

Although the story of cheatgrass is particularly dramatic, it is important to remember that the range of every species on Earth is limited by a combination of historical, abiotic, and biotic factors. The range of a particular species depends on dispersal ability, the capacity to survive climatic conditions, and the ability to find food and avoid being eaten. Climate and other abiotic factors were the focus of this chapter; the next several chapters focus on the biotic aspects of ecology. How do interactions among organisms of the same species and interactions among different species affect geographic range and population size?

FIGURE 50.32 Arid Grasslands Feature Bunchgrasses and Black-Soil Crust. As their name implies, bunchgrasses grow in compact bunches instead of forming a sod. The black-soil crust that forms between bunchgrasses is composed of cyanobacteria, archaea, and other microscopic organisms that add organic matter and nutrients to the soil.

● **QUESTION** Suggest a hypothesis to explain why grasses with the bunching growth habit tend to be found in dry habitats, as compared to grasses that form sods.

Chapter Review

SUMMARY OF KEY CONCEPTS

Ecology focuses on how organisms interact with their environment. Because its goal is to understand the distribution and abundance of organisms, ecology provides a scientific foundation for the conservation of species and natural areas.

Biologists study ecology at four main levels: (1) organisms, (2) populations, (3) communities, and (4) ecosystems. The goal of organismal biology is to understand how individuals respond to stimuli from the environment, including members of their own species and other species. Population ecology focuses on how and why populations grow or decline. Community ecology is the study of how two or more species interact, and ecosystem ecology analyzes interactions between communities of organisms and their abiotic environment—particularly the flow of energy and nutrients. In efforts to preserve endangered species and communities, conservation biologists apply analytical methods and results from all four levels.

You should be able to explain how data from ecology's four levels would inform efforts to preserve an endangered species such as polar bears, pandas, tigers, or blue whales.

Physical structure—particularly water depth—is the primary factor that limits the distribution and abundance of aquatic species. Climate—specifically, both the average value and annual variation in temperature and in moisture—is the primary factor that limits the distribution and abundance of terrestrial species.

Water depth is a crucial factor in aquatic ecosystems because the wavelengths of sunlight that are required for photosynthesis do not penetrate water efficiently. Thus, only certain zones in a lake or ocean can support photosynthesis, and organisms in other aquatic zones must depend on food that rains out of the sunlit areas near the surface.

Distinct climatic regimes are associated with distinct types of terrestrial vegetation, or biomes. Biomes represent the major types of environments available to terrestrial organisms. Because photosynthesis is most efficient when temperatures are warm and water supplies are ample, the productivity and degree of seasonality in biomes varies with temperature and moisture.

You should be able to design an artificial stream and lake system that would maximize the diversity of aquatic ecosystems present. You should also be able to comment on how increased temperatures alone (not considering changes in precipitation) might affect the biomes described in Figures 50.11 through 50.22.

Climate varies with latitude, elevation, and other factors. Climate is changing rapidly around the globe.

An array of abiotic factors affect organisms, including the chemistry of water in aquatic habitats; the nature of soils in terrestrial habitats; and climate, which is a combination of temperature, moisture, sunlight, and wind. Climate varies around the globe because sunlight is distributed asymmetrically, with equatorial regions receiving on average more solar radiation than do regions toward the poles. Hadley cells create bands of wet and dry habitats, and the tilt of Earth's axis causes seasonality in the amount of sunlight that non-equatorial regions receive. The presence of mountains can create local areas of wet or dry habitats, and proximity to an ocean moderates temperatures in nearby terrestrial habitats. Global warming is causing temperature profiles and precipitation patterns to change in ecosystems throughout the planet.

You should be able to explain how climate would be expected to vary around the Earth if the planet were cylindrical and spun on an axis that was at a right angle to the Sun.

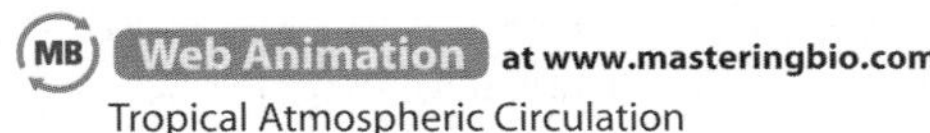

Tropical Atmospheric Circulation

In addition to abiotic aspects of the environment such as physical structure and climate, a species' distribution is constrained by historical and biotic factors.

The range, or geographic distribution, of a particular species depends on dispersal ability, the capacity to survive climatic conditions, and the ability to find food and avoid being eaten. For each species, a unique combination of historical, abiotic factors, and biotic factors dictates where individuals live and the size of the population.

You should be able to describe conditions that would maximize the likelihood that an introduced species becomes invasive.

QUESTIONS

Test Your Knowledge

1. Where do rain shadows exist?
   - **a.** the part of a mountain that receives prevailing winds and heavy rain
   - **b.** the region beyond a mountain range that receives dry air masses
   - **c.** the region along the equator where precipitation is abundant
   - **d.** the region near 30°N and 30°S latitude that receives hot, dry air masses

2. A region receives less than 5 cm (2 in.) of precipitation annually and has temperatures that never drop below freezing. Which type of biome is present?
   - **a.** subtropical desert
   - **b.** temperate grassland
   - **c.** boreal forest
   - **d.** arctic tundra

3. What is the predominant type of vegetation in a tropical wet forest?
   a. shrubs and bunchgrasses
   b. herbs, grasses, and vines
   c. broad-leaved deciduous trees
   d. broad-leaved evergreen trees

4. The littoral zone of a lake is most similar to which of the following environments?
   a. oceanic zone
   b. intertidal zone
   c. neritic zone
   d. marsh

5. Typically, where are oxygen levels highest and nutrient levels lowest in a stream?
   a. near its source
   b. near its mouth, or end
   c. where it flows through a swamp or marsh
   d. where it forms an estuary

6. Which of the following is most important in the success of an invasive species?
   a. historical factors
   b. abiotic factors
   c. biotic factors
   d. Historical, abiotic, and biotic factors are equally important in limiting invasive species.

Test Your Knowledge answers: 1. b; 2. a; 3. d; 4. b; 5. a; 6. c

Test Your Understanding

Answers are available at www.masteringbio.com

1. Name the four main levels of study in ecology. Write a question about the ecology of humans, trout, or oak trees that is relevant to each level.
2. In June, does the Northern Hemisphere receive more or less solar radiation than the equator does? Explain your answer.
3. Diagram the Hadley cell that exists between the equator and 30° latitude south. Indicate the direction of air flow, regions where air is warming or cooling, and regions where air is gaining or losing moisture. Use your diagram to explain why the Australian Outback is a subtropical desert.
4. Contrast the productivities of the intertidal, neritic, and oceanic zones of marine environments. Explain why large differences in productivity exist.
5. Compare and contrast the biomes found at increasing elevation on a mountain with the biomes found at increasing latitude in Figure 50.9.
6. Explain why productivity is much lower in bogs than it is in marshes or lakes that receive the same amount of solar radiation.

Applying Concepts to New Situations

Answers are available at www.masteringbio.com

1. Mars has an even more pronounced tilt than Earth does. Does Mars experience seasons? Why or why not?
2. Mountaintops are closer to the Sun than the surrounding area is. Why are mountaintops cold?
3. The southwest corners of some Caribbean islands are extremely dry, even though other areas of the islands receive substantial rainfall. Explain this observation. Base your answer on a hypothesis (one that you propose) about the direction of the prevailing winds in this region of the world.
4. Edinburgh, Scotland, and Moscow, Russia, are at the same latitude, yet these two cities have very different climates—Moscow has much colder winters. Explain why.

www.masteringbio.com is also your resource for • Answers to text, table, and figure caption questions and exercises • Answers to *Check Your Understanding* boxes • Online study guides and quizzes • Additional study tools including the *E-Book for Biological Science* 3rd ed., textbook art, animations, and videos.

Behavior

51

Peregrine falcons hunt ducks and other large birds.

KEY CONCEPTS

- After describing a behavior, biologists seek to explain both its proximate and ultimate causes—meaning, how it happens at the genetic and physiological levels and how it affects the individual's fitness.
- In a single species, behavior may range from highly stereotyped, invariable responses to highly flexible, conditional responses and from unlearned to learned responses.
- The types of learning that individuals do, the way they communicate, and the way they orient and navigate all correlate closely with their habitat and with the challenges they face in trying to survive and reproduce.
- When individuals behave altruistically, they are usually helping close relatives or individuals that help them in return.

To biologists, **behavior** is action—specifically, the response to a stimulus. Pond-dwelling bacteria swim toward drops of blood that fall into the water. Time-lapse movies of growing sunflowers document the steady movement of their flowering heads throughout the day, as they turn to face the shifting Sun. Filaments of the fungus *Arthrobotrys* form loops, then release a molecule that attracts roundworms. When a roundworm touches one of these loops, the fungal filaments swell—ensnaring the worm. *Arthrobotrys* cells then proceed to grow into the worm's body and digest it.

Among animals, action may be frequent and even spectacular. Peregrine falcons reach speeds of up to 320 km/hr (200 mph) when they dive in pursuit of ducks or other flying prey. Some Arctic terns migrate over 32,000 km (20,000 miles) each year. Deep inside a hive, in complete darkness, honeybees communicate the position and nature of food sources a mile away.

Ecology is the study of how organisms interact with their physical and biological environments, and behavioral biology is the study of how organisms respond to particular stimuli from those environments. What does an *Anolis* lizard do when it gets too hot? How does that same individual respond when a member of the same species approaches it? Or when a house cat approaches?

Although all organisms respond in some way to signals from their environment, most behavioral research is performed on vertebrates, arthropods, or molluscs. With their sophisticated nervous systems and skeletal-muscular systems, these animals can sense, process, and respond rapidly to a wide array of environmental stimuli.

Although most behavioral studies start by describing what animals do in response to a particular stimulus, the eventual goal is to explain the proximate and ultimate causes of behavior.

Proximate, or mechanistic, **causation** explains *how* actions occur in terms of the neurological, hormonal, and skeletal-muscular mechanisms involved. **Ultimate**, or evolutionary, **causation** explains *why* actions occur—based on their evolutionary consequences and history. Is a particular behavior adaptive, meaning that it increases an individual's fitness? If so, how does it help individuals produce offspring in a particular environment?

To illustrate the proximate and ultimate levels of causation, consider the spiny lobster shown in **Figure 51.1**. These animals spend the day hiding in cracks or holes in coral reefs. At night they emerge and wander away from the reef in search of clams, mussels, crabs, or other sources of food. Before dawn, they return to one of several dens they use on a regular basis. How do they find their way back? This question is proximate in nature—it asks how an individual does what it does. Although research is continuing, recent experiments have provided strong support for the hypothesis that spiny lobsters have receptors in their brains that detect changes in Earth's magnetic field, and that they navigate by using information on how the orientation and strength of the field changes as they move about. At the ultimate level of causation, biologists want to know why the behavior occurs. The leading hypothesis to explain homing behavior is that the ability to navigate allows spiny lobsters to search for food over a wide area under cover of darkness, then return to a safe refuge before sharks and other large predators that hunt by sight can find them.

● It is important to recognize that efforts to explain behavior at the proximate and ultimate levels are complementary. To understand what an organism is doing, biologists want to know how the behavior happens and why. Accomplishing this task requires them to draw on techniques and data from genetics, physiology, and evolutionary biology. But even though behavioral biologists work at the interface of many fields, their work is essentially ecological in nature. Questions about behavior are almost always aimed at understanding how individuals cope with changing physical conditions or how they interact with individuals of their own or other species—in short, how they interact with their environment.

FIGURE 51.1 Spiny Lobsters Can Find Their Way Back to Their Home Burrow. Spiny lobsters wander widely in search of food at night, then return to their hiding places in coral reefs before dawn.

● **QUESTION** Compared to other lobsters, spiny lobsters have extremely long antennae for their body size. Generate a hypothesis to explain why. (Your hypothesis will be at the ultimate level of causation.)

The goals of this chapter are to explore some key topics in behavioral biology and illustrate the spirit of the field. To begin, let's consider one way to categorize and organize the enormous variety of behavioral responses observed in animals.

51.1 Types of Behavior: An Overview

Some types of behavior are performed nearly the same way every time. When people yawn, the action is so similar that every instance takes about 6 seconds. Other types of behavior are highly flexible. When people smile, the action is variable. The same person may smile in dozens of ways, and different people tend to have their own individual style and frequency of smiling. In addition, some types of behavior are readily modified by learning, while other types of behavior are not. Formally, **learning** is defined as a change in behavior that results from a specific experience in the life of an individual. Human infants learn their native language through imitation and trial and error, but they begin to yawn and smile before they have a chance to learn these actions from their parents.

Figure 51.2 offers one way of summarizing these general observations about behavior. Notice that the graph has two axes.

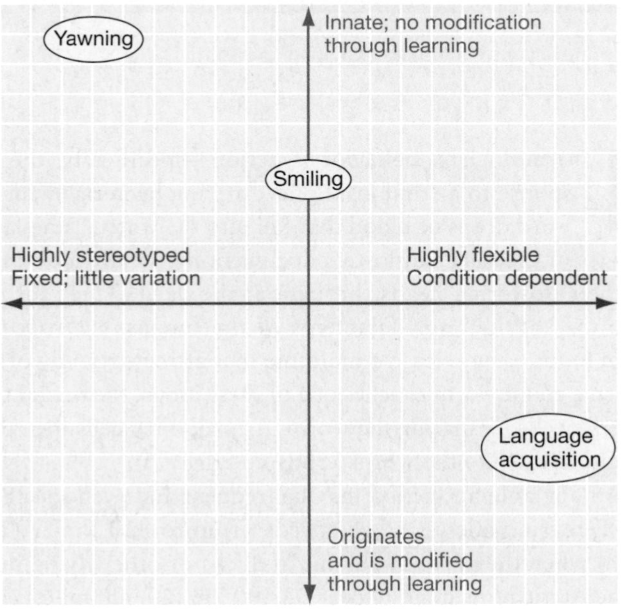

FIGURE 51.2 Behavioral Traits Vary in Flexibility and Dependence on Learning. These axes represent one way to organize the diversity of behavioral traits. Each type of behavior observed in a species could be plotted at a particular point on the graph.

The horizontal axis maps how flexible a behavior is in terms of its expression within and between individuals of the same species. This axis forms a continuum ranging from fixed, invariant, behaviors (often referred to as stereotyped behaviors) to highly flexible and variable responses. The vertical axis plots how much learning is involved in the nature of the response. This axis ranges from behaviors that require little or no learning for normal expression to responses that are highly dependent on learning. In humans, yawning would map in the upper left-hand corner of this graph. Language acquisition would be plotted in the lower right-hand corner.

When researchers first began to explore animal behavior experimentally, they focused on responses that are highly stereotyped and largely unlearned. This general research strategy is common in biological science. When biologists start to analyze a question—ranging from how DNA is transcribed to how plasma membranes work—they start by studying the simplest possible system. Once simple systems are understood, investigators can delve into more complex situations and questions. Behavioral biology started the same way. Early work focused on inflexible behavioral responses—those that map to the upper left-hand corner of the graph in Figure 51.2. Let's look at some conclusions from these pioneering studies of stereotyped responses, then consider examples of work on more flexible types of behavior.

Innate Behavior

Highly inflexible behavior patterns are called **fixed action patterns**, or **FAPs**. FAPs are stereotyped—meaning that they are performed in the same way every time—and are usually triggered by simple stimuli called **releasers** or **sign stimuli**. As an example, consider what happens in a darkened environment when a kangaroo rat hears the sound of a rattlesnake's rattle, or when a person hears a piercing scream. In both cases, a simple sign stimulus releases a FAP—a rapid jumping movement backward, away from the direction of the stimulus. In both cases, the FAP may result in survival. But kangaroo rats and humans don't learn to do this jump-back response, and it is performed in virtually the same way in response to the same stimulus. FAPs are examples of what biologists call **innate behavior**—behavior that is inherited and shows little variation based on learning or the individual's condition. (**Box 51.1** explores the role of genetics in behavior in more detail.)

It is common to observe innate behavior in response to (1) situations that have a high impact on fitness and demand a

BOX 51.1 A Closer Look at Behavior Genetics

Biologists explore the genetic bases of behavior to understand the proximate mechanisms responsible for specific behavioral traits. Are certain alleles associated with certain types of behavior? If so, how do the products of these alleles change what an individual does in response to certain stimuli?

To explore how biologists are answering these questions, consider the first gene shown to affect feeding behavior. When Marla Sokolowski was working as an undergraduate research assistant, she noticed that some of the fruit-fly larvae she was studying tended to move after feeding at a particular location, while others tended to stay put. By rearing these "rovers" and "sitters" to adulthood, breeding them, and studying the offspring, she was able to determine that the trait is inherited. Using information from the *Drosophila melanogaster* genome sequence (see Chapter 20) and genetic mapping techniques introduced in Chapter 19, Sokolowski found and cloned the gene responsible for the rover-sitter difference. She named it *foraging* (*for*).

Follow-up work showed that the protein product of *for* is involved in a signal transduction cascade (see Chapters 8 and 47)—meaning it is involved in the response to a signal. Rovers and sitters behave differently because they have different alleles of the *for* gene. The difference in behavior is apparent only in the presence of food, however.

By altering population density experimentally and documenting the reproductive success of rovers and sitters, Sokolowski has shown that the rover allele is favored when population density is high, but that the sitter allele reaches high frequency in low-density populations. These results are exciting because they link a proximate mechanism with an ultimate outcome—in this case, the presence of certain alleles is responsible for a difference in fitness in specific types of habitats. Genes that are homologous to *for* have now been found in honeybees and the roundworm *Caenorhabditis elegans*. Currently, researchers are focused on finding out which signaling pathway uses the *for* gene product and why different alleles lead to different types of feeding behavior.

Although the *for* gene presents a compelling story, biologists recognize that it is rare for behavioral traits to be so heavily affected by a single gene product. In reality, most aspects of behavior are influenced by the action of many genes. (Think of the number of genes that are likely to be involved in forming the human brain and thus in aspects of human personality and decision making.) To use a term that was introduced in Chapter 13, most behavioral traits are polygenic. Instead of representing discrete states such as roving and sitting, polygenic traits exhibit quantitative variation—meaning that individuals differ by degree and that many different phenotypes are observed when the population is considered as a whole.

reflex-like, unlearned response, or (2) situations where learning is not possible. Web-weaving in spiders and nest-building in birds are largely innate, because offspring have little or no opportunity to watch adults perform these behaviors and learn from them. In addition, the quality of a web or a nest has a high impact on fitness in spiders and birds—individuals need to get it right the first time.

Conditional Strategies and Decision Making

Although all species studied to date show some degree of innate behavior, it is much more common for an individual's behavior to change in response to learning and to show flexibility in response to changing environmental conditions. Animals are not stimulus-response machines. Instead, they take in information from the environment and, based on that information, make decisions about what to do. Animals make choices.

It is generally assumed that animals do not make conscious decisions. In most cases, though, biologists know relatively little about the proximate mechanisms involved in flexible responses and decision making. Fortunately, flexibility in behavior is well understood at the ultimate level. The guiding principle here is what biologists call cost-benefit analysis. Animals appear to take in information about their environment and weigh the costs and benefits of responding in various ways. Costs and benefits are measured in terms of their impact on fitness—the ability to produce offspring. As the following three examples show, individuals usually choose the option that maximizes their fitness.

What Decisions Do White-Fronted Bee-Eaters Make When Foraging? Given a choice, what should an animal eat? Biologists answered this question by assuming that individuals should forage in a way that maximizes the amount of usable energy they take in, given the costs of finding and ingesting their food and the risk of being eaten while they're at it. This claim—that animals should maximize their feeding efficiency—is called **optimal foraging**.

To test whether optimal foraging actually occurs, a researcher began studying feeding behavior in a bird called the white-fronted bee-eater (**Figure 51.3a**). Mated pairs of this species dig tunnels in riverbanks and raise their young inside. Appropriate riverbanks are rare, so many pairs build tunnels in the same location, forming a colony. To feed their young, pairs have to fly away from the colony, capture insects, and bring the prey back to the nest. Each pair defends a specific area where only they find food.

The key observation here is that some individuals forage a few meters from the colony, while some have to look for food hundreds of meters away. It takes only a few seconds to make a round trip to the closest feeding territories versus several minutes to the farthest territories. Thus, the fitness cost of each feeding trip varies widely among individuals.

(a) White-fronted bee-eaters are native to East Africa.

(b) Foraging behavior depends on distance traveled.

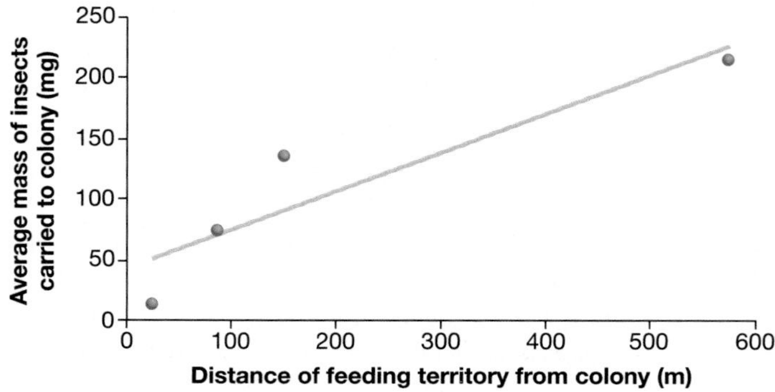

FIGURE 51.3 White-Fronted Bee-Eaters Vary Their Foraging Behavior Depending on Conditions.

If optimal foraging occurs, how do individuals maximize their benefits, given the cost they have to pay in time and energy? Individuals that have to find food far from the colony stay away longer on each trip. And as the data in **Figure 51.3b** show, individuals foraging far from the colony bring back a much larger mass of insects on each trip, on average, than do individuals that fly a short distance each time.

White-fronted bee-eaters appear to make decisions that maximize the energy they deliver to their offspring, given their costs in finding food. This behavior is highly flexible and condition dependent. If the same individual had a territory close to the colony instead of far away, it would make different decisions. Its behavior would change, but in each case it would be acting in a way that maximizes its fitness.

How Do Female Barn Swallows Choose Mates? Given a choice, who should an individual choose for a mate? Chapter 25 presented data indicating that females are usually the gender that is pickiest about mate choice, and that females choose males that contribute good alleles and/or resources to their offspring.

To see this type of decision making in action, consider an experiment on mate choice by female barn swallows native to Denmark. Although both male and female barn swallows help to build the nest and feed the young, the species exhibits a significant amount of sexual dimorphism—males are slightly

FIGURE 51.4 Barn Swallows Are Sexually Dimorphic.

larger and more brightly colored, and their outer tail feathers are about 15 percent longer than the same feathers in females (**Figure 51.4**). Research on other bird species has shown that individuals with particularly long tails, bright colors, and energetic courtship displays are usually the healthiest and best fed in the population (see Chapter 25). Recent work has shown that barn swallows with particularly long tails are more efficient in flight and thus more successful in finding food. If a female barn swallow chooses to mate with a long-tailed male, then she should be picking an individual that will pass high-fitness alleles on to her offspring *and* be able to help her rear them efficiently.

A researcher tested this prediction by altering tail length experimentally—in this case, by cutting off tail feathers and re-gluing them. The operation is painless and does not appear to affect flying ability. As the "Experimental Setup" section of **Figure 51.5** shows, the biologist caught a large series of males and randomly assigned them to one of four groups: shortened tails, "sham-operated" unchanged tails, untouched tails, and elongated tails. When he documented how long it took males in each group to find a mate, a strong trend emerged: Long-tailed males mated earliest—meaning that females preferred

Experiment

Question: Does tail length in barn swallows affect female choice of mates?

Hypothesis: Females prefer to mate with the longest-tailed males.

Null hypothesis: Tail length in males has no influence on female choice of mates.

Experimental setup:

Capture many males at the same site and randomly assign them to 1 of 4 groups:

Short tail: Outermost tail feathers cut

Control 1: Outermost tail feathers cut and glued back in place

Control 2: Males caught and handled, but no change in tail feathers

Elongated tail: Outermost feathers cut and elongated with feathers from other males, glued in place

Release males and document which males are first to mate, have a second nest that season, and raise the most offspring.

Results:

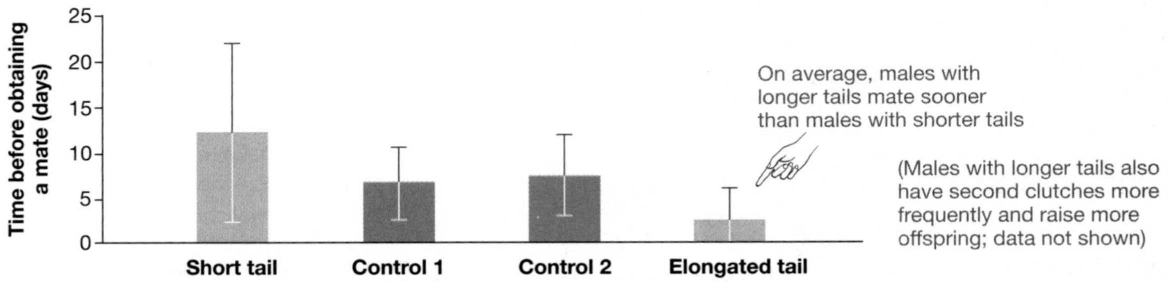

Time before obtaining a mate (days)

0, 5, 10, 15, 20, 25

Short tail, Control 1, Control 2, Elongated tail

On average, males with longer tails mate sooner than males with shorter tails

(Males with longer tails also have second clutches more frequently and raise more offspring; data not shown)

Conclusion: Females prefer to mate with long-tailed males.

FIGURE 51.5 Barn Swallow Females Prefer to Mate with Long-Tailed Males.

● QUESTION Why did the investigator bother to artificially lengthen and shorten tails? Why didn't he just record natural variation in tail length and observe mating success?

● QUESTION Why did he employ two controls?

them. This finding is important because pairs that nest early have time to complete a second nest later in the season and thus produce the most offspring.

Data on mate choice in barn swallows reinforce a central theme of this chapter: Animals make decisions in a way that maximizes their fitness.

Why Do Some Bluehead Wrasses Undergo a Sex Change? Many species of coral-reef fish have a distinctive mating system. Males defend territories that contain nesting sites and feeding areas. A group of females lives inside the boundaries of this territory. When these females lay eggs, the male fertilizes the eggs. Thus, a single male monopolizes all the matings in that territory. Invariably, this male is the largest individual in the group. This is logical because the male guards the territory, and because fights between fish are usually won by the biggest contestant.

When the territory-holding male dies, however, something unusual happens. The largest female in the group changes sex. Her reproductive organs change from egg production to sperm production, and her coloration changes (**Figure 51.6**). She becomes the dominant male and begins fertilizing all of the eggs laid in the territory.

Why does she do this? An idea called the size-advantage hypothesis states that if a group of fish is living in a territory dominated by a single male, and if that male dies, then the largest female should switch from female to male. The hypothesis is based on the observation that fish have indeterminate growth—meaning they continue growing throughout their lives.

To understand the logic behind the size-advantage hypothesis, suppose that a small female fish can lay 10 eggs a year but a large female of the same species can lay 20 eggs a year. If six small females and two large females live in a harem, the male that owns the territory fertilizes 100 eggs each year. If the male dies, the largest female can increase the number of offspring she produces each year from 20 to 80 by changing sex and taking over the role of dominant male. The switch is costly in terms of time and energy, but the benefit is large. There is no fitness advantage for smaller females to switch sex, though, because they would be defeated in fights and have 0 offspring per year instead of 10.

Like foraging in white-fronted bee-eaters, sex change in bluehead wrasses is flexible and condition dependent. If conditions don't change, females remain female. If conditions change, the same individual will stay female if she is small but become a male if she is large. In many cases animals have alleles that make a wide range of behavior possible; what an individual actually does is based on decisions that change, depending on conditions.

What other types of factors lead to flexible behavior? In many species, learning is crucial. Chapter 45 showed that at the proximate level, learning occurs because individual neurons and connections between neurons are modified in response to experience. For example, recall that in the sea slug *Aplysia*, the production of the neurotransmitter serotonin changes in response to experience. Changes in serotonin release lead to changes in the behavior of sea slugs in response to attacks by predators. Learning is usually based on changes in neurons.

Learning is a fundamentally important phenomenon in animal behavior. Let's take a closer look at learning from both proximate and ultimate perspectives. How does the process occur, and why?

Male, used to be female

Female

FIGURE 51.6 Many Species of Coral-Reef Fish Can Change Sex, Depending on Conditions. Bluehead wrasses: a group of females surround a male. Bluehead wrasses start life as females but can change sex and become male.

Check Your Understanding

If you understand that...

- Within a given species, behavioral traits may range from highly stereotyped to highly flexible and from innate to learned.
- Fixed action patterns (FAPs) are highly stereotyped, innate responses released by simple stimuli.
- Foraging, mate choice, and sex change are examples of flexible decision-making. Animals usually make decisions that maximize their fitness, given their current environmental conditions.

You should be able to...

1) Think of an example of a FAP and a condition-dependent behavior in a species that you are familiar with.
2) Pose hypotheses to explain why each type of behavior you just listed is adaptive.

51.2 Learning

Learning occurs when behavior changes in response to specific life experiences. Learning is particularly important in species that have large brains and a lifestyle dominated by complex social interactions—including humans, dolphins, chimps, and crows. In species such as these, FAPs and other types of inflexible, stereotyped behaviors are relatively rare. Instead, each individual is capable of a wide range of behavior. What an individual actually does is condition dependent or learned.

An enormous amount of research has been done on learning. This section touches on only a few highlights, to provide examples of what is known and illustrate how biologists go about studying it.

Simple Types of Learning: Classical Conditioning and Imprinting

Learning can be simple. For example, dogs salivate in response to the sight or smell of food. Ivan Pavlov discovered that if he rang a bell each time food was presented to a dog, the dog would learn to associate the sound of the bell with food. When stimulated by a bell in the absence of food, the dogs Pavlov was training would begin to salivate. This type of learning is called classical conditioning. In **classical conditioning**, individuals are trained by experience to give the same response to more than one stimulus—even a stimulus that has nothing to do with the normal response.

Another type of simple learning takes place in newly hatched ducks, geese, and certain other species of birds. Upon hatching, ducklings and goslings adopt as their mother the first moving thing they see. This type of learning is called **imprinting**. In nature, the "thing" that creates the imprint is the mother duck or goose. But Konrad Lorenz found that if he incubated eggs artificially until they hatched, he could be the first thing that the offspring saw. In response, young greylag geese would imprint on whatever boots he was wearing at the time (**Figure 51.7**). In contrast, mallard ducklings would imprint on him only if he crawled on all fours and quacked continuously.

FIGURE 51.7 Imprinting Occurs in Some Species of Birds and Lasts for Life. Goslings imprint on the first moving object they see—in this case, researcher Konrad Lorenz's boots.

Why does imprinting occur? At the ultimate level of causation, the leading hypothesis is that offspring must quickly learn to recognize and respond to their mother in order to survive. This explanation is most relevant to species that nest on the ground, such as ducks and geese. Ducklings and goslings are easy prey for foxes, raccoons, and other predators, so mother ducks and geese lead their offspring away from the nest to the safety of nearby water almost immediately. In these bird species, the ability to recognize a moving object and follow it immediately should increase the fitness of offspring.

Lorenz was also able to establish that imprinting occurs only in the early life of the animal, during a short interval called the **critical period** or **sensitive period**. In addition, his research showed that imprinting lasts for life and may establish not only the identity of the offspring's mother but also its species identity.

The key characteristics of imprinting—that it is fast and irreversible and occurs during a critical period—are not typical of most types of learning. Yet recent research has shown that imprinting and human language acquisition share some common characteristics. You may have personal experience with how quickly children learn a foreign language relative to adults. This observation suggests that there might be something like a critical period for language learning. The commonplace observation that babies learn the language of their parents is also important. The babbling noises that babies make are thought to be imitations of words and sounds uttered by their parents. Children who are born deaf cannot hear these sounds. Learning a spoken language may be almost impossible for these individuals, even with intensive training later in life. In contrast, people who become deaf as adults experience relatively slight changes in their speech. These data suggest that auditory feedback is important during the learning process and that language acquisition should be presented as shifted to the left in the graph depicted in Figure 51.2. Let's explore the topic of language learning further by considering research on how birds learn to sing.

More Complex Types of Learning: Birdsong

Birds sing to attract mates and to mark territories where they find food and raise offspring. In most cases, bird songs are unique to each species. Do young birds learn their species-specific song or suite of songs from their parents, or are songs innate? To answer these questions, biologists perform experiments to control the sound stimuli that birds hear over the course of their lives. The experimental protocols involve taking eggs from the nest, hatching them, and rearing individuals in the laboratory in a controlled auditory environment.

The results of these experiments show that, depending on the species, song-learning behavior falls at various locations on the learning continuum in Figure 51.2. For example, if young chickens or phoebes—birds that have simple vocalizations—are raised in isolation from other members of their species and never hear their species-specific calls, they still produce the correct vocalizations as adults. Song-learning behavior is innate in these species and may be highly stereotyped.

In contrast, white-crowned sparrows that are raised in isolation do not sing a normal song unless they hear a tape recording of their species-specific song during the early months of life. In this species—which sings a complex song—singing is learned during a critical period. Yet if researchers play the song of a song sparrow or other closely related species during the critical period for language acquisition, white-crowned sparrow chicks do not learn it. Instead they learn an abnormal song differing from that of their own species and from that of the alien introduced species. In this species, the ability to learn songs is restricted to songs of their own species (**Figure 51.8**).

In addition, the critical period for song learning in white-crowned sparrows is augmented by practice that occurs when the individual is older. White-crowned sparrows don't begin to sing until they are almost one year old, during the spring. Beginning singers produce a disorganized warbling called sub-song. With practice, this sub-song becomes more and more organized and progressively closer to adult song. Eventually an adult song develops and is "crystallized," meaning that it remains unchanged for the rest of the individual's life. For this improvement to occur, though, the individual must hear itself sing. Individuals that are prevented from hearing their own practice never sing a normal song. Yet if the same individual is deafened after it has developed adult song, it continues to sing a perfectly normal song. These data suggest that individuals have two critical periods for song learning during their lives: one as a nestling, when they must hear the song of their own species, and a second as a 1-year-old, when they must hear themselves practice.

To summarize, singing in white-crowned sparrows is heavily influenced by learning. But in this species, learning is constrained to certain periods of life and occurs only in response to certain types of stimuli.

If chickens are at one end of the learning continuum and if white-crowned sparrows are in the middle, then species such as mynahs, mockingbirds, lyrebirds, and some types of parrots occupy the other end. Individuals of these species continue to learn new songs throughout their lives by imitating sounds in their environment. They have relatively few age-related or species-specific limitations on song learning. "Polly-wanna-cracker" is not a species-specific call in parrots, and the ability to use this call does not depend on hearing it during a critical period early in life.

One of the great challenges that remains for research on song learning is to explain why so much variation exists in song-learning ability and in the size of song repertoires. Why is it adaptive for chickens to make a few innate calls while mynahs sing dozens of learned songs? Research continues.

Can Animals Think?

Research on classical conditioning, imprinting, and song learning has some important take-home messages: Several types of learning exist, and the ability to learn varies widely among species. It is clear that simple types of learning are widespread in animals and that at least some species can learn a great deal by imitation and practice. But can organisms other than humans think?

Cognition is defined as the recognition and manipulation of facts about the world, combined with the ability to form concepts and gain insights. Unlike the results of classical conditioning, imprinting, and song learning, the results of cognition

FIGURE 51.8 White-Crowned Sparrows Have a Critical Period for Song Learning Early in Life. To sing a normal song, white-crowned sparrows have to hear the song of their own species while they are chicks. They also must hear themselves practice their own song when they begin singing at age 1 year.

cannot be observed directly. Instead, to infer that animals are thinking, researchers must design experimental situations that require animals to manipulate facts or information and demonstrate an ability to form novel associations or insights. As an example of work on animal cognition, consider recent experiments on tool use by New Caledonian crows.

New Caledonian crows routinely make and use tools to find food in the wild. Each of these tools is held in the beak and used to remove insect larvae (grubs) or other types of food from plant stems or from crevices (**Figure 51.9a**). Biologists have documented that the crows manufacture the following general types of tools:

- Straight or curved sticks are broken off and cleaned before use. Several other bird species—including one of the Galápagos finches introduced in Chapter 41—are known to modify sticks or plant stems to "fish" for insects. Chimpanzees also make these types of tools. Recent laboratory experiments have shown that when New Caledonian crows are presented with an array of straight sticks along with a piece of food in a clear plastic tube, the crows select a stick whose width or length is appropriate for the tube's width and the food's distance from the end of the tube. The crows then insert the stick into the tube to fish out the food.
- Hooks are made by breaking complex leaves into pieces (**Figure 51.9b**).
- Spearing tools are constructed from a complex series of cuts and rips, made with the beak, along the edge of leaves from pandanus plants (**Figure 51.9c**).

(a) Straight sticks are used to fish or pry food from crevices.

(b) Hooks are made by breaking parts off leaves.

(c) Several cuts and tears are used to make spearing tools.

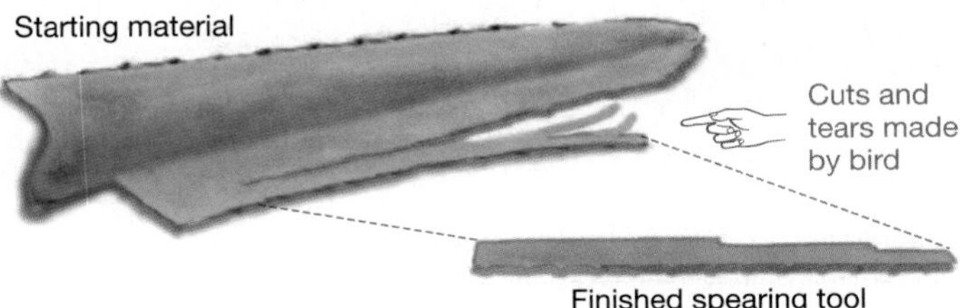

FIGURE 51.9 New Caledonian Crows Make and Use Several Types of Tools.

These observations support the hypothesis that crows can think. More specifically, crows appear to understand facts about the size and shape of raw materials and the location of food. Crows recognize that if they choose or manipulate materials in a certain way, they can use the resulting structure as a tool to acquire the food. The idea is that natural selection has favored the evolution of cognition in this species, because crows with the ability to manufacture and manipulate tools have higher fitness than do crows that lack this ability.

As **Figure 51.10** shows, the crows-can-think hypothesis was supported by recent experiments with two New Caledonian crows in captivity. The experimental setup consisted of a small bucket, containing food, in the bottom of a clear tube that was open at the top, with the handle of the bucket pointing up. In an initial experiment, a male crow and a female crow were given a choice of straight or bent wires to use in retrieving the food from the inside of the tube. The goal of the experiment was to test the hypothesis that the crows can distinguish the two shapes and associate the bent shape with the ability to lift the bucket and acquire the food. During the fifth trial with the apparatus and the choice of two wires, the male happened to remove the hooked wire from the area. Apparently in response, the female picked up the straight wire in her beak, bent it, and successfully used it to lift the bucket and obtain the food.

To confirm the validity of this observation, the researchers gave the pair a straight wire and let them use it until they either successfully retrieved the food or dropped the wire into the tube and were unable to retrieve it. Over the course of 17 trials, the male or female dropped the wire into the tube and was unable to retrieve it 7 times. Once the male was able to lift the bucket out and obtain the food using only the unmodified, straight wire. But 9 times, the female bent the straight wire and used it successfully to obtain the food.

These observations are nothing short of astonishing. They imply that the female was able to associate the shape of a wire

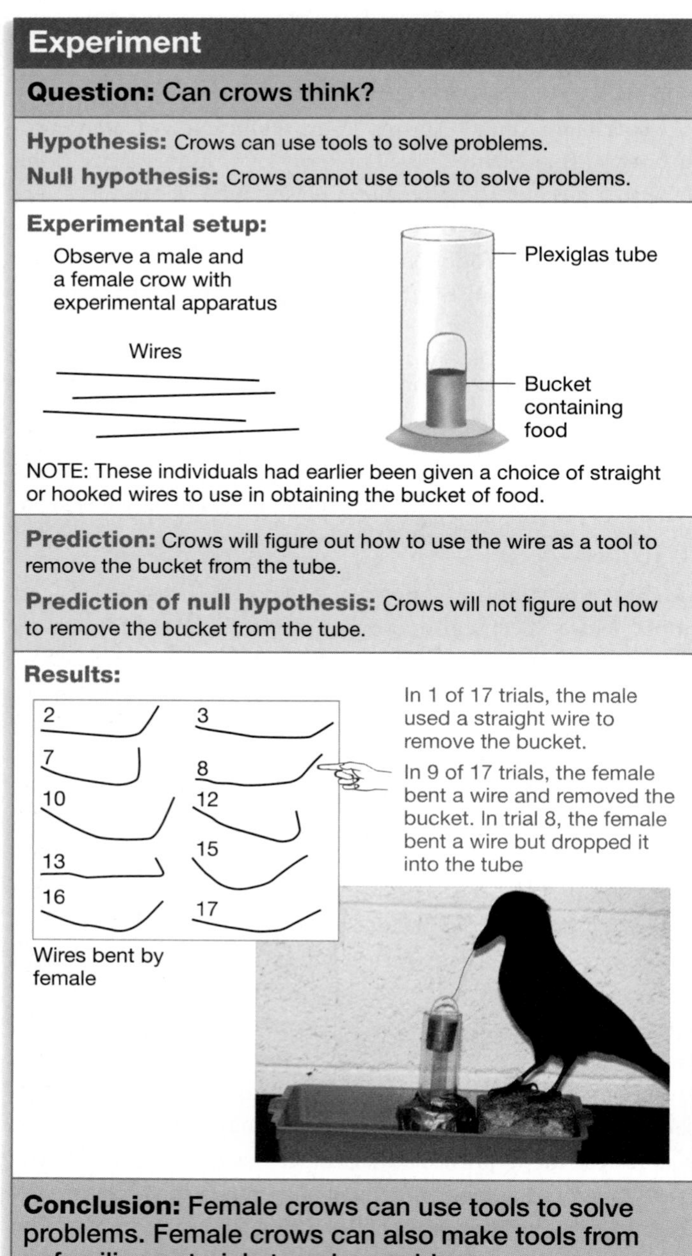

FIGURE 51.10 Experimental Evidence that New Caledonian Crows Understand How to Make Tools. Prior to these experiments, the crows involved had never seen the experimental apparatus or wires of the type available to them. Bending of materials to form hooks has not been observed in the wild.

with the ability to lift the bucket successfully, and that she also had the insight to understand that exerting a force on the wire would bend it into the shape required to be successful. In more colloquial terms, she appeared to understand what sort of tool she needed and how to make one. She had no model to imitate when she made her hooks, however, and she had had no prior experience with bending wires. Instead, she appeared to be thinking.

What Is the Adaptive Significance of Learning?

Biologists view learning as an adaptation that allows individuals to change their behavior in response to a changing and unpredictable environment. Under this hypothesis, the ability to learn varies among species because some species live in environments that are much more unpredictable than others.

In addition, the type of learning that occurs in a given species is correlated with the type of environmental unpredictability it encounters. Norway rats, for example are extremely adept at learning to navigate mazes (**Figure 51.11**). These animals live in sewers, the walls of homes, or other locations where the ability to learn travel routes is essential for survival. Norway rats are also exceptionally good at learning to avoid foods that contain poisons. In nature they feed on a wide variety of fruits, seeds, and insects. The availability of these foods changes with the season, and many potential food items are actually toxic. Thus, it is logical to observe that rats will taste only small quantities of novel foods and that they quickly learn to avoid any foods that induce illness.

Scrub jays, in contrast, are not particularly adept at negotiating mazes or learning to avoid foods. Instead they are proficient at remembering where they have cached (stored) seeds. In this species, spatial learning and a good memory are essential for survival. In addition, scrub jays learn how to prevent theft. This species lives in social groups, and it is not unusual for some members of the group to steal food that was cached by other jays in their group. Experiments have shown that if an individual that has stolen food in the past caches food while being watched by other jays, the individual will come back later, by itself, and move the stored food to a new location.

FIGURE 51.11 Different Species Are Adept at Different Types of Learning. There is often a strong correlation between the type of learning that occurs in a particular species and the problems the species must solve to survive and reproduce.

QUESTION Domestic dogs have been trained by people for thousands of years. Domestic dogs are adept at learning via classical conditioning. What is the relationship between these observations?

To summarize, the ability to learn and the type of learning that occurs vary within and among species. Learning is an adaptation that helps organisms cope with challenges from their environment.

Check Your Understanding

If you understand that...

- The same species of animal may learn in a variety of ways, ranging from simple forms such as imprinting and classical conditioning to imitation to cognition.
- It is common to observe that an individual is capable of learning only certain types of things—such as its species-specific song—only at certain times in its life.

You should be able to...

1) Define learning and give examples of at least two distinct types of learning.
2) Discuss the adaptive value of cognition.
3) Offer a hypothesis to explain why cognition appears to be rare among animals.

51.3 How Animals Act: Hormonal and Neural Control

Most behavior is modified at least slightly by some type of learning, and most animals have behavioral responses that range from highly stereotyped to highly flexible. In many cases, the adaptive value of learning various types of behavior has been documented. But we have yet to explore a question about proximate causation in detail. To understand how biologists go about addressing questions at the proximate level, consider the following: When behavior is flexible and condition dependent, it means that an individual has the potential to behave in a variety of ways. How are changes in behavior implemented and controlled? To answer this question, let's explore how lizards reproduce and how moths escape from bats.

Sexual Activity in *Anolis* Lizards

Anolis carolinensis lives in the woodlands of the southeastern United States. After spending the winter under a log or rock, males emerge in January and establish breeding territories (**Figure 51.12a**). Females become active a month later, and the breeding season begins in April. By May, females are laying an egg every 10–14 days. By the time the breeding season is complete three months later, the eggs produced by a female will total twice her body mass. **Figure 51.12b** maps this series of events, along with the corresponding changes observed in the male and female reproductive systems.

What causes these dramatic seasonal changes in behavior? The proximate answer is sex hormones—testosterone in males and estradiol in females. Testosterone is produced in the testes of males, and estradiol in the ovaries of females. The evidence for effects of these hormones is direct. Testosterone injections induce sexual behavior in castrated males, while estradiol injections induce sexual activity in females whose ovaries have been removed.

This result leaves some key questions unanswered, however. *Anolis* lizards are most successful if they reproduce early in the spring. In springtime their food supplies are increasing, and snakes and other predators are not yet hunting lizards to feed their own young. If testosterone and estradiol levels induced sexual activity in *Anolis* lizards at the wrong time of year, it would be a disaster for the individuals involved. In addition, it is critical for all *Anolis* individuals in a population to start their sexual activity at the same time, so that females can find males that are ready to breed, and vice versa. What environmental cues trigger the production of sex hormones in early spring? How do male and female *Anolis* synchronize their sexual behavior so that they are ready to produce gametes at the same time?

(a) Courtship display of a male *Anolis* lizard

(b) Changes in sexual organs through the year

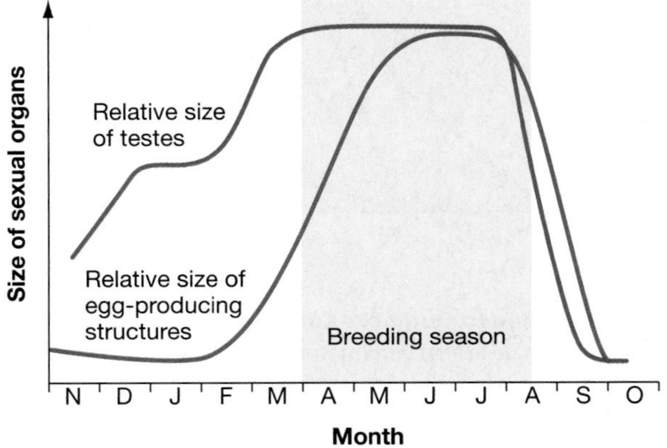

FIGURE 51.12 Sexual Behavior in *Anolis* Lizards.

To answer these questions, a biologist brought a large group of sexually inactive adult lizards into the laboratory during the winter and divided them into five treatment groups. The physical environment was exactly the same in all treatments. Each individual received identical food, and in all treatments high "daytime" temperatures were followed by slightly lower "nighttime" settings. The biologist also continued to monitor the condition of lizards that remained in natural habitats nearby.

To test the hypothesis that changes in day length signal the arrival of spring and trigger initial changes in sex hormones, the biologist exposed the five treatment groups in the laboratory to artificial lighting that simulated the long days and short nights of spring. To test the hypothesis that social interactions among individuals are responsible for synchronizing sexual behavior, the social setting was varied among treatment groups: (1) single isolated females; (2) groups of females only; (3) pairs of lizards: single females, each with a single male; (4) single females, each with a group of castrated (nonbreeding) males; or (5) single females, each with a group of uncastrated (breeding) males.

Each week, the researcher examined the ovaries of females in each group. He also monitored the ovaries of females in nearby natural habitats, since those females were not exposed to springlike conditions. As **Figure 51.13** shows, the differences in the animals' reproductive systems were dramatic. Females that were exposed to springlike conditions began producing eggs; females in the field that were not exposed to springlike conditions did not. But in addition, females that were exposed to breeding males began producing eggs much earlier than did the females placed in the other treatment groups. These results support the hypothesis that two types of stimulation are necessary to produce the hormonal changes that lead to sexual behavior. Females need to experience springlike light and temperatures *and* exposure to breeding males.

What aspect of male behavior causes the difference in female egg production? When males court females, they bob up and down and extend a brightly colored patch of skin called a dewlap (Figure 5.12a). To test the hypothesis that this visual stimulation triggers changes in estradiol production, the investigator repeated the previous experiment but added a twist: He placed some females with males that had intact dewlaps, and other females with males whose dewlaps had been surgically removed. The result? Females grouped with dewlap-less males were slow to produce eggs—just as slow, in fact, as the females in the first experiment that had been grouped with castrated males. These latter females had not been courted at all. The result suggests that the dewlap is a key visual signal. The experiments succeeded in identifying the environmental cues that trigger hormone production and the onset of sexual behavior.

Escape Behavior in Noctuid Moths

Many bats hunt night-flying insects. To find insects in the dark, the bats emit high-pitched pulses of sound and listen for the echo. This process, called **echolocation**, is remarkably acute. If you toss a small stone into the air, a foraging bat will pursue it but then turn away at the last instant. Bats are able to tell the difference between the echos from a stone and an insect.

As **Figure 51.14** shows, bats emit a relatively slow series of pulsed sound when they are searching for prey, switch to a faster series when they are approaching a flying insect, and finish with an extremely rapid sequence as they close in. In this way they get progressively detailed information about the prey's location. The "shouting" involved in echolocation is

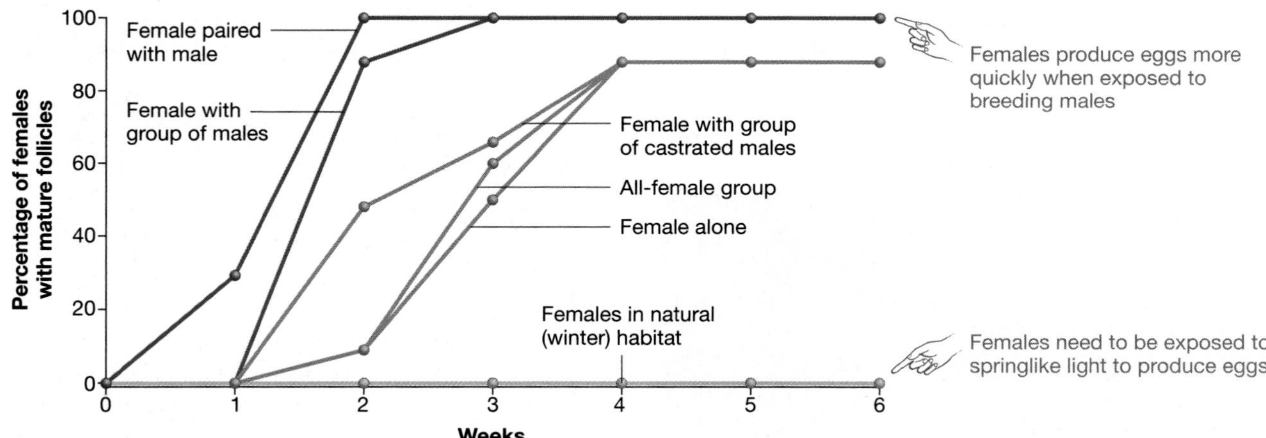

FIGURE 51.13 Exposure to Springlike Conditions and Breeding Males Stimulates Female Lizards to Produce Eggs. The percentage of female lizards with mature follicles, plotted over time. Five treatment groups were exposed to springlike light and temperature; data are also plotted for females left in a natural (winter) habitat. Each data point represents the average value for a group of 6–10 females.

QUESTION Some critics contend that the females that remained outside do not represent a legitimate control in this experiment. What conditions would represent a better control?

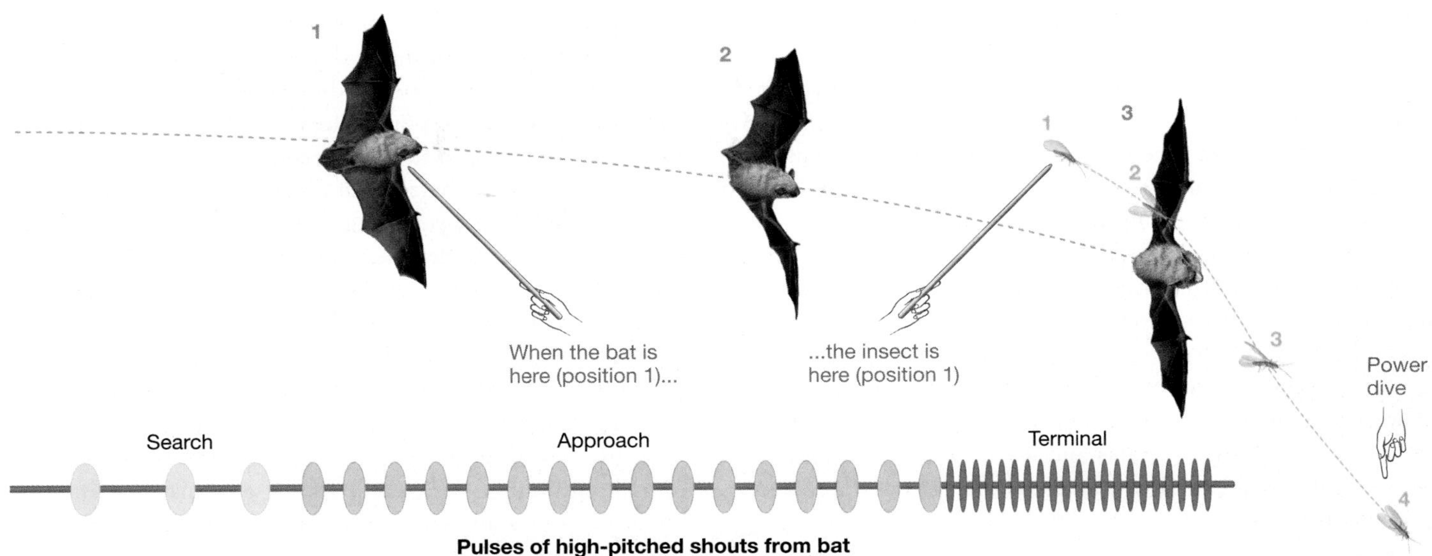

FIGURE 51.14 Bats Hunt By Echolocation. This drawing shows how the pulses of high-pitched sounds change as a bat locates and closes in on a flying insect. To read this diagram, put one finger on the bat at its position 1 and another finger on the insect's position 1. Then change your fingers to the bat's and insect's position 2, then position 3, noting how the bat's calls change as the positions change.

● **QUESTION** Did the bat catch the insect?

energetically expensive, so it is reasonable to observe that bats switch to rapid, high-cost sound production only when it is likely to pay off in a kill.

Humans can't hear the sounds that bats make, but moths and other insects that are preyed on by bats can. When a noctuid moth hears a bat in the distance, it turns and flies away. But if the bat is very close, the moth makes a power dive—it drops from the air like a stone. The adaptive significance of the power-dive behavior is obvious. But at the proximate level, how do moths detect and avoid bats?

Kenneth Roeder did pioneering work on this question. He showed that on either side of their bodies, moths have ears consisting of a tympanic membrane attached to two sensory neurons. When the membrane vibrates in response to bat sounds, the sensory neurons generate action potentials (see Chapter 46). One of the neurons is sensitive to high-pitched sounds of low to moderate loudness, while the other sends action potentials only in response to high-pitched sounds of high intensity. Roeder hypothesized that when the sensory neurons send an extremely rapid sequence of action potentials to the moth brain in response to increasingly loud sounds from an onrushing bat, the moth brain responds by shutting down the neurons that control the flight muscles. The result is chaotic flight or a power dive—desperate behavior that might save the moth's life. In this case, researchers have identified and characterized the neurons responsible for a particularly dramatic example of behavior.

Although a great deal remains to be learned about the hormonal and neural basis for behavior, biologists are making good progress. As techniques advance for measuring hormone concentrations as well as for finding and recording action potentials from individual neurons, the proximate basis for complex behaviors should become increasingly well understood.

51.4 Communication

The song of a bird, the bobbing dewlap of a male *Anolis carolinensis*, and the sentences in this text all have the same overall goal: communication. In biology, **communication** is defined as any process in which a signal from one individual modifies the behavior of a recipient individual. A **signal** is any information-containing behavior. Communication is a crucial component of animal behavior, because it creates a stimulus that elicits a response.

By definition, communication is a social process. For communication to occur, it is not enough that a signal is sent; the signal must be received and acted on. A lizard's bobbing dewlap does not qualify as communication unless another individual sees it and responds to the message. In addition, some biologists maintain that for an event to qualify as communication, the signal that is sent must be intended as a signal. According to these investigators, predicting a person's behavior by watching their body language does not qualify as communication. Similarly, if students unintentionally fall asleep in class and the instructor changes his behavior to wake them up, the exchange does not qualify as communication.

Modes of Communication

The information that organisms communicate may be encoded and delivered in a variety of ways. Communication can be acoustic, as in the song of crickets or birds. It can be visual, as in the color patterns of birds or fish. It can be olfactory, as in scent marking by mice and wolves. It can be tactile, as in communication in some spiders.

One of the most general observations about communication is that the type of signal used by an organism correlates with its habitat. For example, light is quickly dispelled in aquatic habitats, but sound is not. Based on this observation, it is logical to observe that humpback whales rely on songs for long-distance communication. In some cases, humpback whale songs can travel hundreds of kilometers. Groups of whales use acoustic communication to keep together as they move from summer feeding areas to winter breeding grounds, and individual males sing to attract mates and warn rivals. Correlations between habitat and mode of communication are also observed in visual, olfactory, and tactile communication. Animals that are active during the day and that live in open or treeless habitats tend to rely on visual communication during courtship and territorial displays. Bats, wolves, and other animals that are active at night communicate via sound or scent; ants and termites that live underground rely on olfactory and tactile communication.

It is also common to observe several modes of communication being used in conjunction. For example, male red-winged blackbirds establish a breeding territory in the spring by giving a display that combines auditory and visual signals. The birds' display is based on revealing the bright red patches of feathers near their shoulders and giving a loud call. To test the hypothesis that both components of the display are important, researchers have manipulated both acoustic and visual elements. In one experiment, biologists used a speaker to play recordings of red-winged blackbird songs in existing territories (**Figure 51.15a**). In response, the territory-owning males flew toward the speaker and appeared to hunt for it. In a second experiment, researchers made a crude model of a male red-winged blackbird by sewing red patches on a black sock stuffed with rags. When the model was placed in existing territories, territory owners approached the model and inspected it (**Figure 51.15b**). But if the model was presented along with a recorded blackbird song, the resident males attacked the model (**Figure 51.15c**). These experiments support the hypothesis that both auditory and visual information are important in territorial displays of red-winged blackbirds. More specifically, the data indicate that while each type of stimulus alone induces a response, the combination of both types of stimuli provokes a much more powerful change in behavior.

Each mode of communication has advantages and disadvantages. Although acoustic communication can be extremely effective in some habitats, songs and calls are short lived. A red-winged blackbird's red feathers remain in place for at least 6 months, but each call lasts less than 3 seconds. Thus, calls and songs have to be repeated to be effective. Frequent repetition of calls and songs requires a large expenditure of time and energy. In addition, acoustic communication has been shown to attract predators. It is no surprise that when a hawk or falcon approaches a marsh inhabited by red-winged blackbirds, things get very quiet. Communication systems have been honed by natural selection to maximize their benefits and minimize their costs.

FIGURE 51.15 Experimental Evidence that Visual and Auditory Stimuli Are Important in Territorial Behavior. Territory-owning red-winged blackbirds react differently to **(a)** auditory stimuli, **(b)** visual stimuli, and **(c)** auditory plus visual stimuli. Territory owners inspect each stimulus presented in (a) and (b) but attack the combination of the two in (c).

EXERCISE Researchers have clipped the red feathers from male red-winged blackbirds and released the birds into territories. Predict how territory owners respond to these individuals.

Research on animal communication has gone far beyond describing types of communication and documenting correlations between modes of communication and habitat characteristics, however. To appreciate the array of questions that biologists ask about communication and how they go about answering them, let's consider two research programs on animal signaling—one classical and one more recent. The classical work revealed a remarkable form of communication in honeybees; the more recent work has focused on the question of how and why signalers lie.

A Case History: The Honeybee Dance

Honeybees are highly social animals that live in hives. Inside the hive, a queen bee lays eggs that are cared for by individuals called workers. Besides caring for young and building and maintaining the hive, workers obtain food for themselves and other members of the colony by gathering nectar and pollen from flowering plants.

Biologists noticed that bees appear to recruit to food sources, meaning that if a new source is discovered by one or a few individuals, many more bees begin showing up over time. This observation inspired the hypothesis that successful food-finders have some way of communicating the location of food to other individuals. Karl von Frisch suspected that successful food-finders communicate information about food sources when they interact with other workers inside the hive. Because beehives are completely dark, von Frisch hypothesized that communication was tactile in nature.

In the 1930s Von Frisch began studying bee communication by observing bees that built hives inside the glass-walled chambers he had constructed. He found that if he placed a feeder containing sugar water near one of these observation hives, a few of the workers began moving in a circular pattern on the vertical, interior walls of the hive. Von Frisch called these movements the "round dance" (**Figure 51.16a**). Other bee workers appeared to follow the progress of the dance by touching the displaying individual as it danced, and to respond by flying away from the hive in search of the food source.

To investigate the function of these movements further, von Frisch placed feeders containing sugar water at progressively greater distances from the hive. Using this technique, he was able to get bees to visit feeders at a distance of several kilometers from the hive. By catching bees at the feeders and dabbing them with paint, he could individually mark successful food-finders. Follow-up observations at the hive confirmed that marked foragers danced when they returned to the hive. Follow-up observations at the feeders confirmed that marked foragers returned with greater and greater numbers of unmarked bees. To explain these data, von Frisch proposed that the round dance contained information about the location of food and that workers got information from the dance by touching the dancer and following the dancer's movements.

When von Frisch placed feeders at longer distances from the hive, however, he found that successful food-finders no longer did the round dance. Instead, they performed a new type of display. He named these movements the "waggle dance," because they combined circular movements like those of the round dance with short, straight runs (**Figure 51.16b**). During these runs, the dancer vigorously moved her abdomen from side to side.

These observations supported the hypothesis that both the round dance and waggle dance communicated information about food sources. But a key observation allowed von Frisch to push this result further. He noticed that the orientation of the waggle part of the dance varied and that the variation correlated with the direction of the food source from the hive. In addition, he observed that the length of the straight, "waggling" run was

(a) The round dance

(b) The waggle dance

FIGURE 51.16 Honeybees Perform Two Types of Dances. (a) During the round dance, successful food-finders move in a circle. **(b)** During the waggle dance, successful foragers move in a circle but then make straight runs through the circle. During the straight part of the dance, the dancer waggles her abdomen.

proportional to the distance the foragers had to fly to reach the feeder. By varying the location of the food source and observing the orientation of the waggle dance given by marked workers, von Frisch was able to confirm that dancing bees were communicating the position of the food relative to the current position of the Sun. For example, if he placed food directly away from the Sun's current position, marked bees would give the waggle portion of their dance directly downward (**Figure 51.17a**). But if the food was 90 degrees to the right of the Sun, bees would waggle 90 degrees to the right of vertical (**Figure 51.17b**).

Based on these experiments, von Frisch concluded that the difference between the round dance and the waggle dance was a matter of distance. The round dance is used to indicate the presence of food within 80 to 100 m of the hive. The waggle dance is used to indicate the direction and distance to food that is over 100 m from the hive.

These results are nothing short of astonishing. Honeybees do not have large brains, yet they are capable of symbolic language. What's more, they are able to interpret the angle of the waggle dance performed on a vertical surface and to respond by flying horizontally along the corresponding angle. Further work has confirmed that the dance language of bees includes several modes of communication. In addition to the tactile information in the movements themselves, bees make sounds during the dance and give off scents that indicate the nature of the food source.

When Is Communication Honest or Deceitful?

As far as is known, honeybees are always honest. Stated another way, honeybee dances consistently "tell the truth" about the location of food sources. This observation is logical, because the honeybees that occupy a hive are closely related and cooperate closely in the rearing of offspring. As a result, it is advantageous for an individual to convey information accurately. If a food-finder provided inaccurate or misleading information to its hivemates, fewer offspring would be reared and thus the food-finder's fitness would be reduced.

In many cases, however, natural selection has favored the evolution of deceitful communication, or what is commonly called lying. Recent research on deceitful communication has highlighted just how complex interactions between signalers and receivers can be. A few examples will help drive this point home.

Deceiving Individuals of Another Species

Figure 51.18 illustrates a few of the hundreds of examples of deceitful communication that have been documented among members of different species. Many of the best-studied instances involve predation:

- The anglerfish in Figure 51.18a has an appendage that dangles near its mouth and looks remarkably like a minnow. If another fish approaches this "lure" and attempts to eat it, the anglerfish attacks.

(a) Straight runs down the wall of the hive indicate that food is opposite the direction of the Sun.

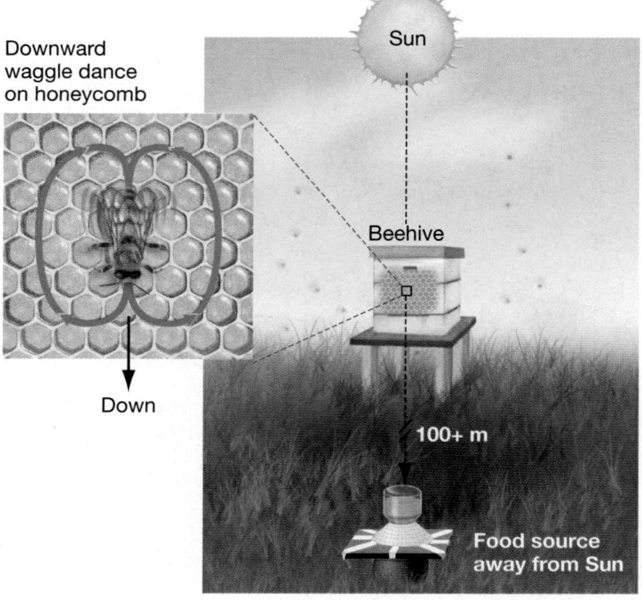

(b) Straight runs to the right indicate that food is 90° to the right of the Sun.

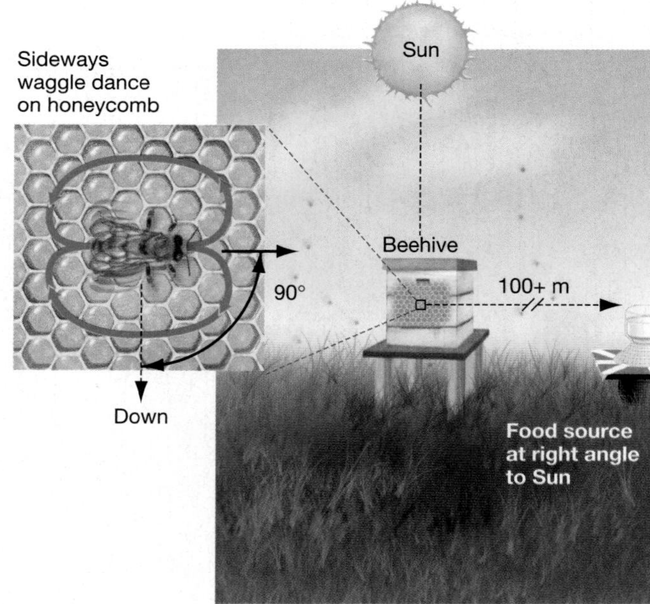

FIGURE 51.17 The Direction of the Waggle Dance Indicates the Location of Food Relative to the Sun. Waggle dances are done only when the food is at least 100 m from the hive.

EXERCISE The length of the straight run in the waggle dance indicates the relative distance of the food. Diagram a waggle dance that indicates food in the direction of the Sun and twice as far away as the food sources indicated by the dances drawn here.

(a) Anglerfish use a "lure" to attract prey.

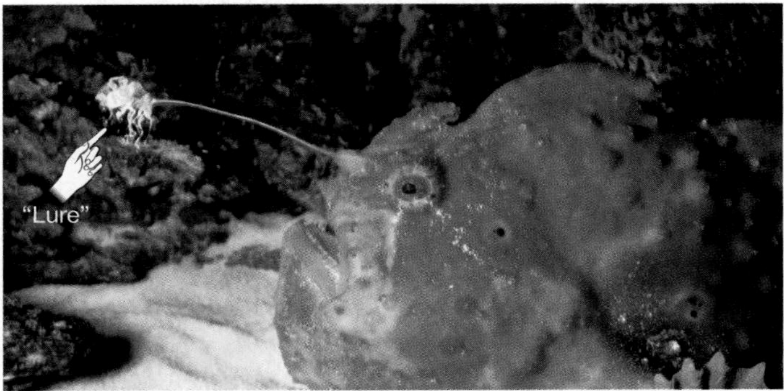

(b) Female *Photuris* fireflies flash the courtship signal of another species, then eat males that respond.

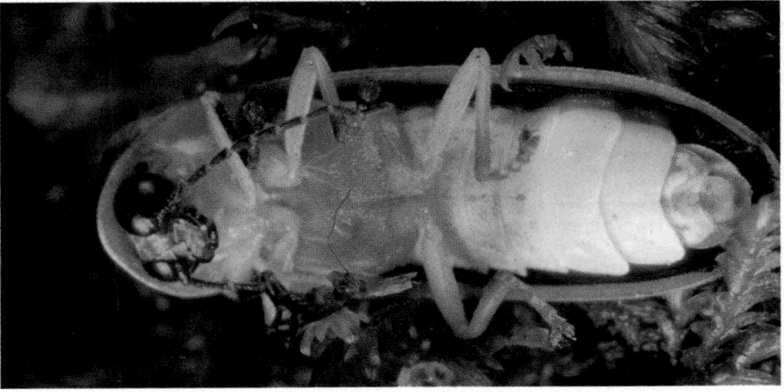

(c) This butterfly looks like a bad-tasting species but actually tastes good.

FIGURE 51.18 Deceitful Communication Is Common in Nature.

- Male and female fireflies flash a species-specific signal to each other during courtship. Predatory *Photuris* fireflies can mimic the pattern of flashes given by females of several other species. A *Photuris* female attracts a male of different species with the appropriate set of flashes, and then attacks and eats him (Figure 51.18b).
- The butterfly shown in Figure 51.18c is highly palatable to birds and other predators. It is rarely attacked, though, because it strongly resembles butterfly species that contain toxins and are highly unpalatable to predators. This phenomenon, known as mimicry, is discussed in more detail in Chapter 53.

In each of these examples, individuals increase their fitness by providing inaccurate or misleading information to members of a different species. Deceitful communication is also known in plants, where it functions in pollination. For example, several orchid species have flowers that look and smell like female wasps and that accomplish pollination by "fooling" male wasps into attempting copulation.

Deceiving Individuals of the Same Species In some cases, natural selection has favored the evolution of traits or actions that deceive members of an organism's own species. When mantis shrimps molt, for example, they lack any external covering and cannot use their large claws. Thus, they are unable to defend the cavities where they live. But if another mantis shrimp approaches with the intent of evicting the individual and taking over the cavity, the molting individual bluffs. It raises its claws in the normal aggressive display and may even lunge at the intruder.

Perhaps the best-studied type of deceit in nonhuman animals, however, involves the mating system of bluegill sunfish. Male bluegills set up nesting territories in the shallow water along lake edges, fertilize the eggs laid in their nests, care for the developing embryos by fanning them with oxygen-rich water, and protect newly hatched offspring from large predators. Some males cheat on this system, however, by mimicking females. To understand how this happens, examine the female, normal male, and female-mimic male in **Figure 51.19**. Female mimic males look like females but have well-developed testes

FIGURE 51.19 In Bluegill Sunfish, Some Males Look and Act Like Females. Female-mimic males have fully functional testes and produce large numbers of sperm.

and produce large volumes of sperm. Mimics also act like females during courtship movements with territory-owning males. They even adopt the usual egg-laying posture. Then when normal females approach the nest and begin courtship, the female-mimic males join in. The territory-owning male tolerates the mimic, apparently thinking that he is successfully courting two females at the same time. But when the actual female begins to lay eggs, the mimic responds by releasing sperm and fertilizing some of them, and then darting away. In this way, the female-mimic male fathers offspring but does not help care for them.

When Does Deception Work? In analyzing deceitful communication in bluegill sunfish and other species, researchers point out that in most cases, lying works only when it is relatively rare. The logic behind this hypothesis is that if deceit becomes extremely common, then natural selection will strongly favor individuals that can detect and avoid or punish liars. But if liars are rare, then natural selection will favor individuals that are occasionally fooled but are more commonly rewarded by responding to signals in a normal way. In mantis shrimps, for example, molting occurs only once or twice a year. Thus, the vast majority of contests over hiding places involve individuals that have an intact exoskeleton and claws that qualify as a deadly weapon. As a result, intruders usually benefit by interpreting threat displays as honest and responding to the stimulus by retreating. If molting were frequent and bluffing behavior common, then intruders would benefit by challenging territory owners, which are likely to be defenseless. Research on deceit and honesty in communication has been extraordinarily productive and continues at a rapid pace.

Check Your Understanding

If you understand that...

- Communication is an exchange of information between individuals.
- In most instances, the mode of communication that animals use maximizes the probability that the information will be transferred efficiently.
- Communication can be honest or deceitful, depending on the intent of the signaler.

You should be able to...

1) Explain the adaptive significance of different modes of communication (e.g. auditory, tactile, visual) and give examples of each.
2) Explain why natural selection favors individuals that can detect deceitful communication and either avoid liars or punish them.

51.5 Migration and Navigation

In response to many stimuli, organisms move. Ants follow odor trails; bees fly toward food sources. A movement that results in a change of position, like the flight paths of hunting bats, escaping moths, and foraging bees, is called **orientation**. The simplest type of orientation is termed **taxis**, and it involves positioning the body, or part of the body, toward or away from a stimulus. A moth attracted to a porch light is an example of a positive **phototaxis**—orientation toward light. A female cricket approaching a calling male is an example of a positive **phonotaxis**—orientation toward sound. A person who retreats from the sound of a siren blaring is an example of negative phonotaxis.

Orienting movements cover fairly short distances. But in some cases, bats and other species travel thousands of miles in search of places to feed and breed. Why?

Why Do Animals Move with a Change of Seasons?

In ecology, **migration** is defined as the long-distance movement of a population associated with a change of seasons. A few examples will bring home the simple point that migratory movements can be spectacular in their extent and the navigation challenge they pose:

- Arctic terns nest along the Atlantic coast of North America in warm months, fly south along the coast of Africa to wintering grounds off Antarctica, and then fly back north along the eastern coast of South America. Chicks follow their parents south but make the return trip on their own. The journey totals over 32,000 km (20,000 miles).
- Many of the monarch butterflies native to North America spend the winter in the mountains of central Mexico or southwest California. Tagged individuals are known to have flown over 3000 km (1870 miles) to get to a wintering area. In spring, monarchs begin the return trip north. They do not live to complete the trip, however. Instead, mating takes place along the way, and females lay eggs en route. Although adults die on the journey, offspring continue to head north. After several generational cycles of reproduction and death in the original habitats in northern North America have passed, individuals again migrate south to overwinter. Instead of a single generation making the entire migratory cycle, then, the round trip takes several generations.
- Salmon that hatch in rivers along the Pacific Coast of North America and northern Asia migrate to the ocean when they are a few months to several years old, depending on the species. After spending several years feeding and growing in the North Pacific Ocean, they return to the stream where they hatched. There they mate and die.

In most cases, it is relatively straightforward to generate hypotheses for why migration exists at the ultimate level. Arctic terns feed on fish that are available in different parts of the world at different seasons. Thus, individuals that migrate achieve higher reproductive success than do individuals that do not migrate. Monarch butterflies may achieve higher reproductive success by migrating to wintering areas and new breeding areas than by trying to overwinter in northern North America. Salmon eggs and young are safer and thrive better in freshwater habitats than they do in the ocean, but adult salmon can find much more food in saltwater habitats than in freshwater.

At the proximate level, however, explaining migratory movements is often extremely difficult. How do all of these animals find their way? What cues guide them on these immense journeys? The surprising answer is that in many or even most cases, biologists don't really know.

How Do Animals Find Their Way?

To organize research into how animals find their way during migratory movements, biologists distinguish three categories of navigation: (1) **Piloting** is the use of familiar landmarks, (2) **compass orientation** is movement that is oriented in a specific direction, and (3) **true navigation** is the ability to locate a specific place on Earth's surface. A thought experiment will help you understand these categories. Suppose you grew up on the shores of Lake Superior and were transported to the middle of this enormous lake. Because you were blindfolded and made to spin in place en route, you are disoriented and have no idea where you are. But you are given a magnetic compass. The compass tells you where north, south, east, and west are. Unfortunately this information will not help you find home, because you have no idea where you are in relation to home. You are stuck because you have no mechanism of true navigation. Now suppose that a helicopter flying overhead drops you a map. The map has an X marking your current position and an H marking your home. The map also has a scale and a compass symbol. From this information, you determine that home is 100 miles to the west. The map has solved your navigational problem. The compass has solved your orientation problem. Now that you know you live to the west, you can use the compass to find west and paddle home. As you approach the shore, you recognize familiar landmarks. Piloting gets you home.

Organisms other than humans do not have magnetic compasses or printed maps. How do they navigate? Although biologists know little about the nature of a map sense in animals, piloting and compass orientation are increasingly well understood.

Piloting A substantial amount of data suggests that at least some migratory animals use piloting to find their way. In some species of migratory birds and mammals, offspring follow their parents south in the fall and north in the spring. Young appear to memorize the route. Piloting even plays a role in species with more sophisticated navigational abilities. Homing pigeons, for example, can find their way home even when they are released in strange terrain a long distance away. But if these birds are equipped with frosted spectacles so that they see the world as a foggy haze, they return to within a mile or so of their home but do not actually find it. Based on these experiments, researchers have concluded that homing pigeons use piloting to navigate the final stages of a journey home.

Compass Orientation How do animals perform compass orientation? To date, most research on compass orientation has been done on migratory birds. To determine where north is, these animals appear to use the Sun during the day and the stars at night. Let's consider each type of compass orientation separately, because they work differently.

The Sun is difficult to use as a compass reference, because its position changes during the day. It rises in the east, is due south at noon (in the Northern Hemisphere), and sets in the west. To use the Sun as a compass reference, then, an animal must have an internal clock that defines morning, noon, and evening. Fortunately, most animals have such a clock. The **circadian clock** that exists in organisms maintains a 24-hour rhythm of chemical activity. The clock is set by the light–dark transitions of day and night. It tells individuals enough about the time of day that they can use the Sun's position to find magnetic north.

The situation is actually simpler on clear nights, because migratory birds in the Northern Hemisphere can use the North Star to find magnetic north and select a direction for migration. But what if the weather is cloudy and neither the Sun nor stars are visible? Under these conditions, migratory birds appear to orient using Earth's magnetic field. Exactly how they detect magnetism is under debate. One hypothesis contends that the birds can detect magnetism by their visual system, through an unknown molecular mechanism. An alternative hypothesis maintains that individuals have small particles of magnetic iron—the mineral called magnetite—in their bodies. Changes in the positions of magnetic particles, in response to Earth's magnetic field, could then be detected and provide reliable information for compass orientation.

Although research on mechanisms of compass orientation continues, one important point is clear: Birds and perhaps other organisms have multiple mechanisms of finding a compass direction. At least some species can use a Sun compass, a star compass, and a magnetic compass. Which system they use depends on the weather and other circumstances.

MB Web Animation at www.masteringbio.com
Homing Behavior in Digger Wasps

51.6 The Evolution of Self-Sacrificing Behavior

The types of behavior reviewed thus far—including FAPs, learned behaviors, honest and deceitful communication, and migration—all have a key common element: They help individuals respond to environmental stimuli in a way that increases their fitness. There is a type of behavior that appears to contradict this pattern, however: altruism. **Altruism** is behavior that has a fitness cost to the individual exhibiting the behavior and a fitness benefit to the recipient of the behavior. It is the formal term for self-sacrificing behavior. Altruism decreases an individual's ability to produce offspring but helps others produce more offspring.

The existence of altruistic behavior appears to be paradoxical, because if certain alleles make an individual more likely to be altruistic, those alleles should be selected against. Why does it occur?

Kin Selection

Even though it is not possible for behavior to evolve for the good of the species, self-sacrificing behavior does occur in nature. For example, black-tailed prairie dogs perform a behavior called alarm calling (**Figure 51.20**). These burrowing mammals live in large communities, called towns, throughout the Great Plains region of North America. When a badger, coyote, hawk, or other predator approaches a town, some prairie dogs give alarm calls that alert other prairie dogs to run to mounds and scan for the threat. Giving these calls is risky. In several species of ground squirrels and prairie dogs, researchers have shown that alarm-callers draw attention to themselves by calling and are in much greater danger of being attacked than non-callers are.

How can natural selection favor the evolution of self-sacrificing behavior? William D. Hamilton answered this question by creating a mathematical model to assess how an allele that contributes to altruistic behavior could increase in frequency in a population. To model the fate of altruistic alleles, Hamilton represented the fitness cost of the altruistic act to the actor as C and the fitness benefit to the recipient as B. Both C and B are measured in units of offspring produced. His model showed that the allele could spread if

$$Br > C$$

where r is the coefficient of relatedness. The **coefficient of relatedness** is a measure of how closely the actor and beneficiary are related. Specifically, r measures the fraction of alleles in the actor and beneficiary that are identical by descent—that is, inherited from the same ancestor (see **Box 51.2**).

This result is called **Hamilton's rule.** It is important because it shows that individuals can pass their alleles on to the next generation not only by having their own offspring but also by helping close relatives produce more offspring. Stated another way, an altruist might produce slightly fewer offspring because of the sacrifices the individual makes to help others. But if those sacrifices help close relatives produce many more offspring than the relatives could produce on their own, then copies of the altruist's alleles will increase in frequency.

According to Hamilton's rule, if the fitness benefits of altruistic behavior are high for the recipients, if the recipients are close relatives, and if the fitness costs to the altruist are low, then alleles associated with altruistic behavior will be favored by natural selection and will spread throughout the population. The keys are that (1) close relatives are very likely to have copies of the altruistic allele, and (2) help from close relatives allows these relatives to produce more offspring than they could produce without help. Biologists use the term **kin selection** to refer to natural selection that acts through benefits to relatives. ● If you understand this concept, you should be able to give an example of a species for which kin selection should be important, and explain why. You should also be able to explain whether kin selection occurs in plants.

Does Hamilton's rule work? Do animals really favor relatives when they act altruistically? To test the kin-selection hypothesis, a researcher studied which of the inhabitants of a black-tailed prairie dog town were most likely to give alarm calls. Within a large prairie dog town, individuals live in small groups called coteries that share the same underground burrow. Members of each coterie defend a territory inside the town. By tagging offspring that were born over several generations, the researcher identified the genetic relationships among individuals in the town. More specifically, the researcher determined, within each coterie, to which of the following three categories each prairie dog belonged: (1) an individual with no close genetic relatives in its coterie; (2) an individual with no offspring in the coterie but at least one sibling, cousin, uncle, aunt, niece, or nephew; and (3) an individual with at least one offspring or grandoffspring in the coterie.

FIGURE 51.20 Black-Tailed Prairie Dogs Are Highly Social. Prairie dogs live with their immediate and extended family within large groups called towns. The individual with the upright posture has spotted an intruder and may give an alarm call.

BOX 51.2 Calculating the Coefficient of Relatedness

The coefficient of relatedness, r, varies between 0.0 and 1.0. If two individuals have no identical alleles that were inherited from the same ancestor, then their r value is 0.0. Because every allele in pairs of identical twins is identical, their coefficient of relatedness is 1.0.

What about other relationships? **Figure 51.21a** shows how r is calculated between half-siblings. (To review what the boxes, circles, and lines in a pedigree mean, see Figure 13.21.) Half-siblings share one parent. Thus, r represents the probability that half-siblings share alleles as a result of inheriting alleles from their common parent. It is critical to realize that in each parent-to-offspring link of descent, the probability of any particular allele being transmitted is 1/2. This is so because meiosis distributes alleles from the parent's diploid genome to their haploid gametes randomly. Thus, half the gametes produced by a parent get one of the alleles present at each gene, and half the gametes produced get the other allele. Half-siblings are connected by two such parent-to-offspring links. The overall probability of two half-siblings sharing the same allele by descent is $1/2 \times 1/2 = 1/4$ (To review rules for combining probabilities, see **BioSkills 9**.)

To think about this calculation in another way, focus on the red arrows in Figure 51.21a. The left arrow represents the probability that the mother transmits a particular allele to her son. The right arrow represents the probability that the mother transmits the same allele to her daughter. Both probabilities are 1/2. Thus, the probability that the mother transmitted the same allele to both her son and daughter is $1/2 \times 1/2 = 1/4$.

Figure 51.21b shows how r is calculated between full siblings. The challenge here is to calculate the probability of two individuals sharing the same allele as a result of inheriting it through their mother or through their father. The probability that full siblings share alleles as a result of inheriting them from one parent is 1/4. Thus, the probability that full siblings share alleles inherited from either their mother or their father is $1/4 + 1/4 = 1/2$. (To review rules for combining probabilities, see BioSkills 9.) ● If you understand how to calculate r, you should be able to draw a pedigree with 3 generations and calculate r between any 2 individuals on the diagram.

(a) What is the r between half-siblings?

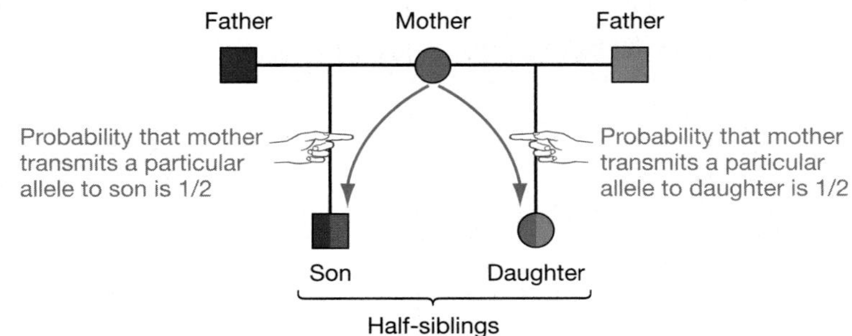

What is the probability that half-siblings inherit the same allele from their common parent?

Answer: r between half-siblings $= 1/2 \times 1/2 = 1/4$

(b) What is the r between full siblings?

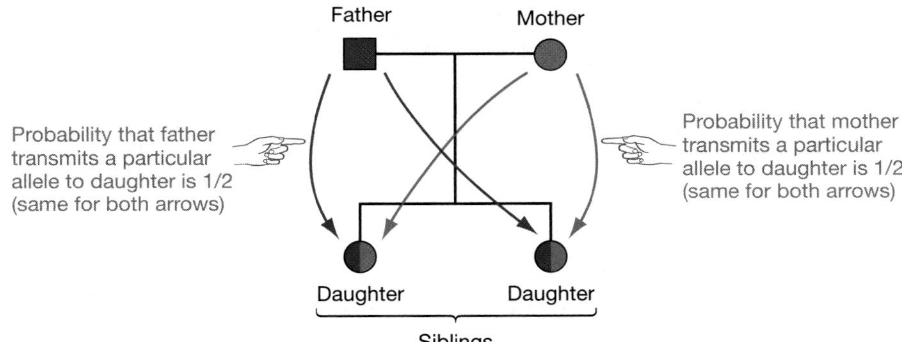

What is the probability that full siblings inherit the same allele from their father or their mother?

Answer: Probability that they inherit same allele from father $= 1/2 \times 1/2 = 1/4$
Probability that they inherit same allele from mother $= 1/2 \times 1/2 = 1/4$
Overall probability that they inherit the same allele $= 1/4 + 1/4 = 1/2$
r between full siblings $= 1/2$

FIGURE 51.21 The Coefficient of Relatedness Is Calculated from Information in Pedigrees.

● **QUESTION** According to Hamilton's rule, who should help each other more: half-siblings or full siblings?

● **QUESTION** What is the r between first cousins?

The kin-selection hypothesis predicts that individuals who do not have close genetic relatives nearby will rarely give an alarm call. To evaluate this prediction, the biologist recorded the identity of callers during 698 experiments. In these studies, a stuffed badger was dragged through the colony on a sled. Were prairie dogs with close relatives nearby more likely to call, or did kinship have nothing to do with the probability of alarm calling? The data in **Figure 51.22** illustrate that black-tailed prairie dogs are much more likely to call if they live in a coterie that includes close relatives. This same pattern—of preferentially dispensing help to kin—has been observed in many other species of social mammals and birds. Most cases of self-sacrificing behavior that have been analyzed to date are consistent with Hamilton's rule and are hypothesized to be the result of kin selection.

Reciprocal Altruism

During long-term studies of highly social animals, such as lions, chimpanzees, and vampire bats, biologists have observed nonrelatives helping each other. Chimps and other primates, for example, occasionally spend considerable time grooming unrelated members of their social group—cleaning their fur and removing ticks and other parasites from their skin. In vampire bats, individuals that have been successful in finding food are known to regurgitate blood meals to non-kin that have not been successful and that are in danger of starving.

How can self-sacrificing behavior like this evolve if kin selection is not acting? The leading hypothesis to explain altruism among nonrelatives is called **reciprocal altruism**, which is an exchange of fitness benefits that are separated in time. Data that have been collected so far support the reciprocal-altruism hypothesis in at least some instances. Among vervet monkeys, for example, individuals are most likely to groom unrelated individuals that have groomed or helped them in the past. Similarly, vampire bats are most likely to donate blood meals to non-kin that have previously shared food with them. Reciprocal altruism is also widely invoked as an explanation for the helpful and cooperative behavior commonly observed among unrelated humans.

To summarize, altruistic behavior increases the fitness of individuals by favoring kin or by increasing the likelihood of receiving help in the future from non-kin. Altruism is a flexible, condition-dependent behavior that occurs in a surprisingly wide array of species that live in social groups.

Check Your Understanding

If you understand that...

- Hamilton's rule states that alleles for altruistic behavior increase in frequency if the fitness cost of the behavior for the actor is low, the fitness benefit to the recipient is high, and the actor and recipient are closely related.
- Reciprocal altruism is based on an exchange of fitness benefits, separated in time.

You should be able to...

1) Use Hamilton's rule to describe situations in which a black-tailed prairie dog is likely and unlikely to give an alarm call.
2) Explain why reciprocal altruism has been observed only in species where individuals have good memories and live in small, long-lived social groups.

Experiment

Question: Do black-tailed prairie dogs prefer to help relatives when they give an alarm call?

Kin-selection hypothesis: Individuals give an alarm call only when close relatives are near.

Null hypothesis: The presence of relatives has no influence on the probability of alarm calling.

Experimental setup:

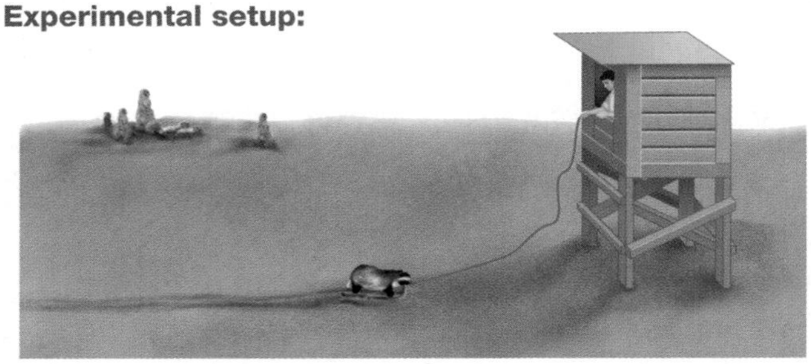

1. Determine relationships among individuals in prairie dog coterie.

2. Drag stuffed badger across territory of coterie.

3. From observation tower, record which members of coterie give an alarm call.

4. Repeat experiment 698 times. Each individual prairie dog coterie is tested 6-9 times over 3-year period.

Prediction of kin-selection hypothesis: Individuals in coteries that contain a close genetic relative are more likely to give an alarm call than are individuals in coteries that do not contain a close genetic relative.

Prediction of null hypothesis: The presence of relatives in coteries will not influence the probability of alarm calling.

Results:

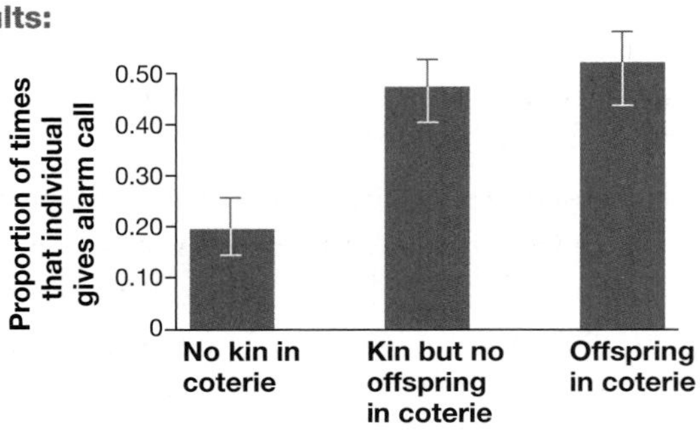

Proportion of times that individual gives alarm call

0.50
0.40
0.30
0.20
0.10
0

No kin in coterie
Kin but no offspring in coterie
Offspring in coterie

Conclusion: Alarm calling usually benefits relatives.

FIGURE 51.22 Experimental Evidence that Black-Tailed Prairie Dogs Are More Likely to Give Alarm Calls If Relatives Are Nearby.

Chapter Review

SUMMARY OF KEY CONCEPTS

After describing a behavior, biologists seek to explain both its proximate and ultimate causes—meaning how it happens at the genetic and physiological levels and how it affects the individual's fitness.

At the proximate level, experiments and observations focus on understanding how specific gene products, neuron activity, and hormonal signals cause behavior. At the ultimate level, researchers seek to understand the adaptive significance of behavior, or how it enables individuals to survive and reproduce. By combining proximate and ultimate viewpoints and studying the genetic basis of behavior, biologists can seek a comprehensive understanding of how and why animals do what they do.

You should be able to give an explanation at the proximate and ultimate levels for biological phenomena other than behavior—for example, cellular respiration, meiosis, water transport in plants, or the action potential in animal neurons.

In a single species, behavior may range from highly stereotyped, invariable responses to highly flexible, conditional responses and from unlearned to learned responses.

FAPs are examples of highly stereotyped, unlearned behaviors that are released by extremely simple stimuli. Language acquisition and tool making are examples of behavior that are highly dependent on learning, cognition, and the environmental conditions experienced by an individual. Most of the behavior observed in animals is condition dependent, meaning an individual has the genetic, neuronal, and hormonal mechanisms in place for behaving in a variety of ways, and the actual behavior observed depends on environmental conditions. Foraging in white-fronted bee-eaters, mate choice in barn swallow, sex change in coral-reef fish, and the onset of sexual behavior in *Anolis* lizards are examples of flexible, condition-dependent behaviors triggered by changing environmental stimuli.

You should be able to explain why migration in some birds is considered condition dependent, based on the observation that some species do not migrate if people supply food in feeders.

The types of learning that individuals do, the way they communicate, and the way they orient and navigate all correlate closely with their habitat and with the challenges they face in trying to survive and reproduce.

Animals usually behave in a way that increases their ability to survive and reproduce in their current environment. Deceitful communication can be adaptive—as when anglerfish fool prey into attacking a "lure" attached to the anglerfish's face or when molting mantis shrimps try to bluff their way into retaining ownership of a hiding place.

You should be able to predict the consequences for deceitful communication in bluegill sunfish if most large, territory-owning males were fished out of a lake.

(MB) Web Animation at www.masteringbio.com
Homing Behavior in Digger Wasps

When individuals behave altruistically, they are usually helping close relatives or individuals that help them in return.

Although it is common to observe self-sacrificing behavior in animals, most altruistic behavior is directed toward close relatives. When this is the case, alleles that lead to self-sacrificing can increase in frequency due to kin selection. In addition, some animals that live in close-knit social groups engage in reciprocal altruism—meaning they exchange help over time.

You should be able to describe the characteristics of a species for which altruistic behavior is expected to be extremely common.

QUESTIONS

Test Your Knowledge

1. What do proximate explanations of behavior focus on?
   a. how displays and other types of behavior have changed through time, or evolved
   b. the functional aspect of a behavior, or its "adaptive significance"
   c. genetic, neurological, and hormonal mechanisms of behavior
   d. psychological interpretations of behavior—especially motivation
2. What evidence suggests that there is a critical period for song learning in some bird species?
   a. In mynahs and mockingbirds, individuals continue to sing new songs throughout their lives.
   b. Individuals do not learn to sing normally unless they are allowed to hear themselves practice.
   c. Individuals that never hear their species-specific song can still sing normally as adults.
   d. Birds that are deafened soon after hatching never sing normally, but birds that are deafened several months after hatching sing normally.
3. What is the difference between orienting behavior and piloting?
   a. Orienting movements are fast responses to visual or auditory stimuli; piloting is finding the way using landmarks.
   b. Orienting is the ability to follow a compass during navigation; piloting is the ability to use map information during navigation.
   c. Orienting is used during long-distance movements such as migration; piloting is used in short-distance movements involved in homing.
   d. Orienting is the ability to use a Sun or star compass; piloting is the ability to use a magnetic compass.
4. Which of the following statements about the waggle dance of the honeybee is not correct?
   a. The length of a waggling run is proportional to the distance from the hive to a food source.
   b. Sounds and scents produced by the dancer provide information about the nature of the food source.

c. The dancer uses no elements of the round dance.
d. The orientation of the waggling run provides information about the direction of the food from the hive, relative to the Sun's position.

5. Why are biologists convinced that the sex hormone testosterone is required for normal sexual activity in male *Anolis* lizards?
   a. Male *Anolis* lizards with larger testes court females more vigorously than do males with smaller testes.
   b. The testosterone molecule is not found in female *Anolis* lizards.
   c. Male *Anolis* lizards whose gonads had been removed did not develop dewlaps.
   d. Male *Anolis* lizards whose gonads had been removed did not court females.

6. What does Hamilton's rule specify?
   a. why animals do things "for the good of the species"
   b. why reciprocal altruism can lead to fitness gains for unrelated individuals
   c. how alleles that favor self-sacrificing acts increase in frequency via kin selection
   d. the conditions under which more complex behaviors evolve from simpler behaviors

Test Your Knowledge answers: 1. c; 2. d; 3. a; 4. c; 5. d; 6. c

Test Your Understanding

Answers are available at www.masteringbio.com

1. Make a graph like that in Figure 51.2, with axes that plot the degree to which a particular behavior is modified by learning and the degree to which behavior is stereotyped and inflexible. Plot the following behaviors on the graph: tool use in New Caledonian crows; exploration of mazes by rats; singing by chickens, by white-crowned sparrows, and by mynahs; and blushing in humans.
2. Discuss the proximate and ultimate causes of the following behaviors: (a) homing behavior by spiny lobsters, (b) sexual behavior by *Anolis* lizards, and (c) the "jump-back" response of kangaroo rats.
3. Propose a hypothesis to explain the adaptive significance of tool use in New Caledonian crows. Recall that a female crow bent an unfamiliar material into a hooked tool. Explain why this observation supports the hypothesis that individuals of this species can think.
4. What environmental stimuli cause changes in hormone levels that lead to egg laying in *Anolis* lizards? Based on the data presented on sexual behavior in *Anolis*, propose a hypothesis to explain the proximate basis of sex switching in coral-reef fish.
5. For an animal to navigate, it must have a "map" and a "compass." Explain why both types of information are needed. Describe three types of compasses that have been identified in migratory or homing species.
6. Compare and contrast kin selection and reciprocal altruism. Be sure to identify the conditions under which self-sacrificing behavior is expected to evolve under kin selection versus reciprocal altruism.

Applying Concepts to New Situations

Answers are available at www.masteringbio.com

1. Mated pairs of cranes give territorial displays called unison calls. Cranes that are raised in isolation from other cranes give unison calls normally, and the display is performed the same way every time. Your friend argues that the unison call of cranes is a FAP, because it is stereotyped and not influenced by learning. Another friend argues that it is a condition-dependent behavior, because only mated pairs do it and because the tendency to give unison calls varies among pairs and with time of year. (Some pairs are more aggressive than others and give unison calls more frequently in territorial contexts.) Who's right?
2. Most tropical habitats are highly seasonal. But instead of alternating warm and cold seasons, there are alternating wet and dry seasons. Most animal species breed during the wet months. If you were studying a species of *Anolis* native to the tropics, what environmental cue would you simulate in the lab to bring them into breeding condition? How would you simulate this cue? Further, think about the sensory organs that lizards use to receive this cue. Are they the same as or different than the receptors that *Anolis carolinensis* uses to recognize that spring has arrived in the southeastern United States?
3. What is the significance of the observation that the product of the *for* gene in fruit flies is involved in a signal transduction pathway? Propose a hypothesis to explain why fruit-fly larvae with the rover genotype and phenotype have higher fitness when population density is high and individuals are crowded.
4. A biologist once remarked that he'd be willing to lay down his life to save two brothers or eight cousins. Explain what he meant. Based on the theory of kin selection and reciprocal altruism, predict the conditions under which people are expected to be nice to one another.

www.masteringbio.com is also your resource for • Answers to text, table, and figure caption questions and exercises • Answers to *Check Your Understanding* boxes • Online study guides and quizzes • Additional study tools including the *E-Book for Biological Science* 3rd ed., textbook art, animations, and videos.

Population Ecology

52

A flock of white pelicans swimming in a channel of the Mississippi River. This chapter explores how and why growth rates in populations change through time.

KEY CONCEPTS

- Life tables summarize how likely it is that individuals of each age class in a population will survive and reproduce.
- The growth rate of a population can be calculated from life-table data or from the direct observation of changes in population size over time.
- Researchers observe a wide variety of patterns when they track changes in population size over time, ranging from no growth, to regular cycles, to continued growth independent of population size.
- Data from population ecology studies help biologists evaluate prospects for endangered species and design effective management strategies.

If you asked a biologist to name two of the most pressing global issues facing your generation, she might say global warming and extinction of species. If you were asked to name two of the most pressing issues facing the region where you live, you might say traffic and the price of housing. All four of these problems have a common cause: recent and dramatic increases in the size of the human population.

A **population** is a group of individuals of the same species that live in the same area at the same time. In both ecology and evolutionary biology, populations are the basic unit of analysis. Evolutionary biologists study how the characteristics of populations change through time; ecologists study how populations interact with their environment. Increasingly, ecologists are studying how species are being affected by growing human populations.

Population ecology is the study of how and why the number of individuals in a population changes over time. Biologists also analyze changes in the ages of individuals in a population, the proportion of males to females, and geographic distribution. With the explosion of human populations across the globe, the massive destruction of natural habitats, and the resulting threats to species throughout the tree of life, population ecology has become a vital field in biological science. The mathematical and analytical tools introduced in this chapter help biologists predict changes in population size and design management strategies to save threatened species.

Biologists ask a wide array of questions in population ecology. How are individuals distributed in space? How old are they, and how likely are they to reproduce or die? Is the population growing, declining, or staying the same through time, and why?

To answer these questions, let's consider some of the basic tools that biologists use to study populations, then follow up with examples of how biologists study changes in the population size of humans and other species over time. The chapter concludes by asking how all of these elements fit together in efforts to limit human population growth and save endangered species.

52.1 Demography

The number of individuals that are present in a population depends on four processes: birth, death, immigration, and emigration. Populations grow due to births—here meaning any form of reproduction—and **immigration**, which occurs when individuals enter a population by moving from another population. Populations decline due to deaths and **emigration**, which occurs when individuals leave a population to join another population. Analyzing birth rates, death rates, immigration rates, and emigration rates is fundamental to **demography**: the study of factors that determine the size and structure of populations through time.

To make detailed predictions about the future of a population, however, biologists must understand the makeup of the population in more detail. If a population consists primarily of young individuals with a high survival rate and reproductive rate, the population size should increase over time. But if a population comprises chiefly old individuals with low reproductive rates and low survival rates, then it is almost certain to decline over time. To predict the future of a population, biologists need to know how many individuals of each age are alive, how likely individuals of different ages are to survive to the following year, how many offspring are produced by females of different ages, and how many individuals of different ages immigrate and emigrate each **generation**—the average time between a mother's first offspring and her daughter's first offspring.

Demographic data provide an important tool for biologists charged with designing management programs for endangered species. To understand the nature of these data and how they are used in conservation programs, let's examine a classical tool for describing the demography of a population.

Life Tables

Formal demographic analyses of populations are based on a type of data set called a life table. A **life table** summarizes the probability that an individual will survive and reproduce in any given time interval over the course of its lifetime. Life tables were invented almost 2000 years ago; in ancient Rome they were used to predict food needs. In modern times, life tables have been the domain of life-insurance companies, which have a strong financial interest in predicting the likelihood of a person dying at a given age. More recently, biologists have employed life tables to study the demographics of endangered species and other nonhuman populations.

To understand how researchers use life tables, consider the lizard *Lacerta vivipara* (**Figure 52.1**). *L. vivipara* is a common resident of open, grassy habitats in western Europe.

FIGURE 52.1 *Lacerta vivipara* Are Native to Europe. Translated literally, *vivipara* means "live birth." Females in most populations bear live young, though females in *L. vivipara* populations from northern Spain and southwest France lay eggs.

As their name suggests, most populations of the species are ovoviviparous (see Chapter 32) and give birth to live young. Researchers set out to estimate the life table of a low-elevation population in the Netherlands, with the goal of comparing the results to data that other researchers had collected from *L. vivipara* populations in the mountains of Austria and France and in lowland habitats in Britain and Belgium. The researchers wanted to know whether populations that live in different environments vary in basic demographic features.

To complete the study, the researchers visited their study site daily during the seven months that these lizards are active during the year. Each day the researchers captured and marked as many individuals as possible. Because this program of daily monitoring continued for seven years, the biologists were able to document the number of young produced by each female in each year of her life. If a marked individual was not recaptured in a subsequent year, they assumed that it had died sometime during the previous year. These data allowed researchers to calculate the number of individuals that survived each year in each particular age group and how many offspring each female produced. What did the data reveal?

Survivorship **Survivorship** is a key component of a life table and is defined as the proportion of offspring produced that survive, on average, to a particular age. For example, suppose 1000 *L. vivipara* are born in a particular year. These individuals represent a **cohort**—a group of the same age that can be followed through time. How many individuals would survive to age 1, age 2, age 3, and so on?

As **Table 52.1** shows, the biologists calculated that in the population living in the Netherlands, survivorship from birth to age 1 was 0.424. If 1000 females were born in a particular year in this population, on average 424 would still be alive one year later. Survivorship from birth to age 2 was 0.308, meaning an average of 308 female lizards would survive for two years.

TABLE **52.1 Life Table for *Lacerta vivipara* in the Netherlands**

Age	Number Alive	Survivorship	Fecundity	Survivorship × Fecundity = Average Number of Offspring Produced per Female of Age *x*
0	1000	1.000	0.00	0.00
1	424	0.424	0.08	0.03
2	308	0.308	2.94	0.91
3	158	0.158	4.13	0.65
4	57	0.057	4.88	0.28
5	10	0.010	6.50	0.07
6	7	0.007	6.50	0.05
7	2	0.002	6.50	0.01

Data are from H. Strijbosch and R.C.M. Creemers, 1988. Comparative population demography of sympatric populations of *Lacerta vivipara* and *Lacerta agilis*. *Oecologia* 76: 20–26. With kind permission of Springer Science and Business Media.

To recognize general patterns in survivorship and make comparisons among populations or species, biologists plot the logarithm of the number of survivors versus age. The resulting graph is called a **survivorship curve**. Studies on a wide variety of species indicate that three general types of survivorship curves exist (**Figure 52.2a**). Humans have what biologists call a type I survivorship curve. In this pattern, survivorship throughout life is high, and most individuals approach the species' maximum life span. Songbirds, in contrast, experience relatively constant mortality throughout their lives once they have left the nest, resulting in a type II survivorship curve. Many plants have type III curves due to extremely high death rates for seeds and seedlings. **Figure 52.2b** provides a graph for you to plot survivorship of *L. vivipara*.

Survivorship curves are important in conservation work, because they pinpoint the stage of life when endangered species have particularly low survivorship. Biologists can then put measures into place to protect individuals during that period. For example, when they are very young, trees or herbaceous plants can be protected from drought by germinating them in a nursery and from predation by putting a wire cage over seedlings.

Fecundity The number of female offspring produced by each female in the population is termed **fecundity**. (In most cases, biologists keep track only of females when calculating life-table data, because only females produce offspring.) Researchers documented the reproductive output of the same *L. vivipara* lizard females year after year and thus were able to calculate a quantity called **age-specific fecundity**, which is defined as the average number of female offspring produced by a female in age class *x*. An **age class** is a group of individuals of a specific age—for instance, all female lizards between 4 and 5 years old. As **Box 52.1** shows, data on survivorship and fecundity allow researchers to calculate the growth rate of a population. How do data on survivorship and fecundity in the Netherlands populations compare with populations in different types of habitats?

(a) Three general types of survivorship curves

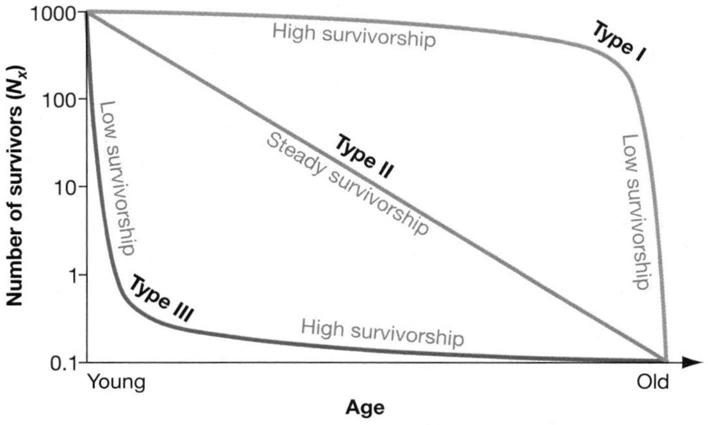

(b) Exercise: Survivorship curve for *Lacerta vivipara*

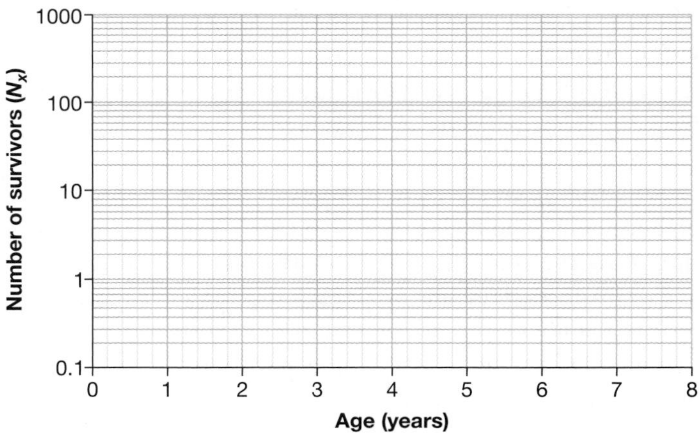

FIGURE 52.2 Survivorship Curves Identify When Mortality Rates Are Low, Steady, or High.

● **EXERCISE** Fill in the graph in part **(b)** with data on survivorship of *Lacerta vivipara* given in Table 52.1. Compare the shape of the curve to the generalized graphs in part (a).

BOX 52.1 Using Life Tables to Calculate Population Growth Rates

If immigration and emigration are not occurring, the data in a life table can be used to calculate a population's growth rate. This is logical, because survivorship and fecundity are ways to express death rates and birth rates—the other two factors influencing population size. To see how a population's growth rate can be estimated from life-table data, let's look at each component in a life table more carefully.

Survivorship is symbolized as l_x, where x represents the age class being considered. Survivorship for age class x is calculated by dividing the number of individuals in that age class (N_x) by the number of individuals that existed as offspring (N_0):

$$l_x = \frac{N_x}{N_0} \quad \textbf{(Eq. 52.1)}$$

Age-specific fecundity is symbolized m_x, where x again represents the age class being considered. Age-specific fecundity is calculated as the total number of offspring produced by females of a particular age, divided by the total number of females of that age class present. It represents the average number of offspring produced by a female of age x.

Documenting age-specific survivorship and fecundity allows researchers to calculate the net reproductive rate, R_0, of a population:

$$R_0 = \sum_{i=0}^{x} l_x m_x \quad \textbf{(Eq. 52.2)}$$

The **net reproductive rate** represents the growth rate of a population per generation. The logic behind the equation for R_0 is that the growth rate of a population per generation equals the average number of female offspring that each female produces over the course of her lifetime. A female's average lifetime reproduction, in turn, is a function of survivorship and fecundity at each age class. In Table 52.1, R_0 is the sum of the survivorship × fecundity values in the right column.

If R_0 is greater than 1, then the population is increasing in size. If R_0 is less than 1, then the population is declining. ● If you understand these concepts, you should be able to use the data in Table 52.1 to calculate R_0 in the Netherlands population of *L. vivipara*;[1] state how many female offspring an average *L. vivipara* female produces over the course of her lifetime,[2] and describe whether the population is growing, stable, or declining.[3] Answers are given below.

[1] R_0 = 2.0; [2] 2.00; [3] The population is growing rapidly.

The Role of Life History

The life-table data in Table 52.1 are interesting because they contrast with results from other populations of *L. vivipara*. For example, notice that almost no 1-year-old female *L. vivipara* lizards in the Netherlands reproduces. But in Brittany, France, 50 percent of 1-year-old female lizards reproduce. Both situations contrast sharply with data on females in the mountains of Austria. In populations that live at high elevation, females do not begin breeding until they are 4 years old. In addition, most females in the Austrian population live much longer than do individuals in either lowland population—in the Netherlands or France. When the three sites are compared, it is clear that fecundity is high but survivorship is low in Brittany. In contrast, fecundity is low but survivorship is high in Austria. The population in the Netherlands is intermediate in both characteristics.

Why isn't it possible for *L. vivipara* females to have both high fecundity and high survival? Biologists answer this question using the concept of **fitness trade-offs** (see Chapters 24 and 41). Fitness trade-offs occur because every individual has a restricted amount of time and energy at its disposal—meaning that its resources are limited. If a female lizard devotes a great deal of energy to feeding a large number of offspring as they develop, it is not possible for her to devote that same energy to her immune system, growth, nutrient stores, or other traits that increase survival. (Ask yourself whether a woman who has 10 children would be expected to live a longer or shorter time than a woman who has 2.) A female can maximize fecundity, maximize survival, or strike a balance between the two. An organism's **life history** consists of how an individual allocates resources to growth, reproduction, and activities or structures that are related to survival.

Life history trade-offs are universal. To drive this point home, **Figure 52.3** graphs the average number of eggs laid—a measure of fecundity—in 10 species of birds versus the probability that a female of that species survives to the following year. There are no points in the upper right corner of the graph because it is not possible to have high fecundity and high survivorship. There are no points in the lower left corner of the graph because species with low survivorship and low fecundity have low population growth and go extinct.

In almost all cases, biologists find that life history is shaped by natural selection in a way that maximizes an individual's fitness in its environment. For example, Chapter 41 provided experimental evidence that there is a fitness trade-off between egg size and egg number in side-blotched lizards, and that females in the study population strike a balance in egg size and number that is optimum for their habitat.

To make sense of the life-history data from *L. vivipara* populations, biologist contend that females who live a long time but mature late and have few offspring each year have high fitness in cold, high-elevation habitats. In contrast, females who have short lives but mature early and have large numbers of offspring each year do better in warm, low-elevation habitats. ● If you understand the concept of fitness trade-offs, you

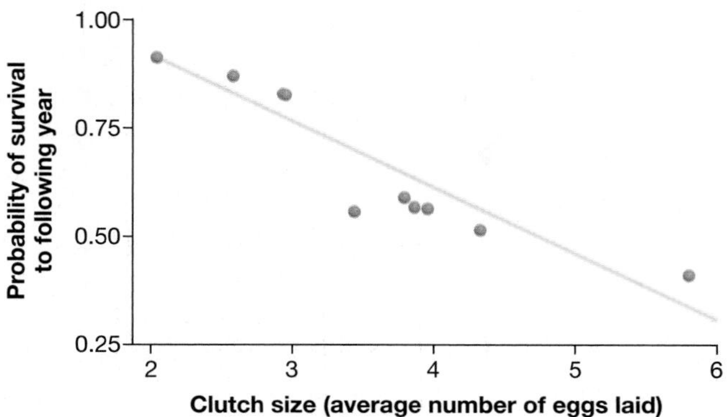

FIGURE 52.3 There Is a Trade-off between Survival and Reproduction. Graph of survivorship versus fecundity. Each data point represents a different species of bird. A statistical analysis called regression was used to create a "line of best fit" to these data points.

QUESTION To study survival-reproduction trade-offs within species of birds, researchers randomly add or subtract eggs to clutches. Predict what happens to survivorship in females that rear additional versus a reduced number of young.

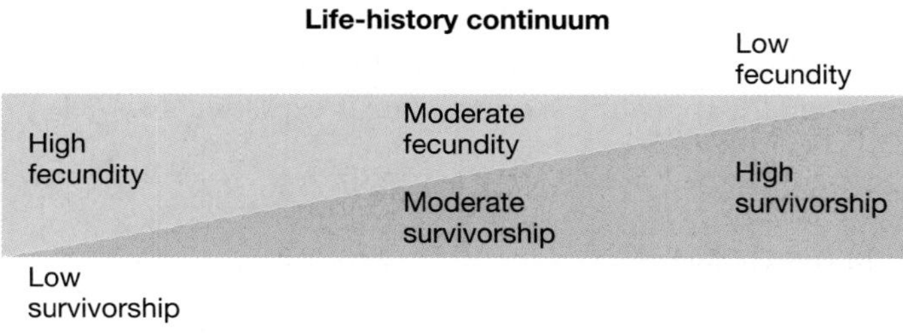

FIGURE 52.4 Life-History Traits Form a Continuum. Every organism can be placed somewhere on this life-history continuum. The placement of a species is most meaningful when it is considered relative to closely related species.

EXERCISE Place *Arabidopsis thaliana* and the coconut palm on the continuum. For plants, list the types of traits you would expect to see on the left, middle, and right of the continuum. Include growth habit (herbaceous, shrub, or tree) as well as relative disease- and predator-fighting ability, seed size, seed number, body size, and investment in roots versus shoots.

should be able to (1) predict how survivorship and fecundity should compare in *L. vivipara* in the warmest part of their range (southwest France and northern Spain), and (2) comment on why females in these populations lay many eggs instead of giving birth to a few live young.

In general, organisms with high fecundity tend to grow quickly, reach sexual maturity at a young age, and produce many small eggs or seeds. The mustard plant *Arabidopsis thaliana*, for example, germinates and grows to sexual maturity in just 4 to 6 weeks. In this species, individuals usually live only a few months but may produce as many as 10,000 tiny seeds. In contrast, organisms with high survivorship tend to grow slowly and invest resources in traits that reduce damage from enemies and increase their own ability to compete for water, sunlight, or food. A coconut palm, for example, may take a decade to reach sexual maturity but live 60–70 years and produce offspring each year. Coconut palms invest resources in making stout stems that allow them to grow tall, a relatively extensive root system, large fruits (coconuts), molecules that make herbivores sick, and enzymes that reduce infections from disease-causing fungi. These traits increase the survivorship of both seedlings and adults but decrease fecundity.

An *Arabidopsis thaliana* plant and a coconut palm represent two ends of a broad continuum of life-history characteristics (**Figure 52.4**). Even within the species *L. vivipara*, analyses of life tables have shown that there is considerable variation in life-history traits.

Research has also documented that within populations, life-history traits can change if conditions change. For example, suppose that a series of wet years produced an abundance of food for lizards. Females might survive better in response, or they might begin breeding at an earlier age or produce more offspring at a particular age. Such changes might directly affect a population's growth rate. How do biologists analyze growth rates in populations? And what factors make growth rates increase or decrease?

52.2 Population Growth

The most fundamental questions that biologists ask about populations involve growth or decline in numbers of individuals. For conservationists, analyzing and predicting changes in population size is fundamental to managing threatened species.

Recall that four processes affect a population's size: Births and immigration add individuals to the population; deaths and emigration remove them. It follows that a population's overall growth rate is a function of birth rates, death rates, immigration rates, and emigration rates. In this section we consider only the impact of births and deaths on population growth.

A population's growth rate is the change in the number of individuals in the population (ΔN) per unit time (Δt). If no immigration or emigration is occurring, then the growth rate is equal to the number of individuals (N) in the population times the difference between the birth rate per individual (b) and death rate per individual (d). The difference between the birth rate and death rate per individual is called the **per-capita rate of increase** and is symbolized r. (*Per capita* means "for each individual.") If the per-capita birth rate is greater than the per-capita death rate, then r is positive and the population is growing. But if the per-capita death rate begins to exceed the per-capita birth rate, then r becomes negative and the population declines. Within populations, r varies through time and can be positive, negative, or 0.

When conditions are optimal for a particular species—meaning birth rates per individual are as high as possible and death rates per individual are as low as possible—then r reaches a maximal value called the **intrinsic rate of increase,** r_{max}. When this occurs, a population's growth rate is expressed as

$$\Delta N/\Delta t = r_{max}N$$

In species such as *Arabidopsis* and fruit flies, which breed at a young age and produce many offspring each year, r_{max} is high. In contrast, r_{max} is low in species such as giant pandas and coconut palms, which take years to mature and produce few offspring each year. Stated another way, r_{max} is a function of a species' life-history traits.

Each species has a characteristic r_{max} that does not change. But at any specific time, each population of that species has an instantaneous growth rate, or per-capita rate of increase, symbolized by r, which is less than or equal to r_{max}. The instantaneous growth rate of a population at a particular time is likely to be much lower than r_{max}. A population's r is also likely to be different from r values of other populations of the same species, and to change over time.

Exponential Growth

The graphs in **Figure 52.5** plot changes in population size for various values of r, under the condition known as exponential growth. **Exponential population growth** occurs when r does not change over time. The key point about exponential growth is that the growth rate does not depend on the number of individuals in the population. Biologists say that this type of population growth is **density independent.**

It's important to emphasize that exponential growth adds an increasing number of individuals as the total number of individuals, N, gets larger. As an extreme example, an r of 0.02 per year in a population of 1 billion adds over 20 million individuals per year. The same growth rate in a population of 100 adds just over 2 individuals per year. Even if r is constant, the number of individuals added to a population is a function of N.

In nature, exponential growth is observed when a few individuals found a new population in a new habitat or when a population has been devastated by a storm or some other type of catastrophe and then begins to recover, starting with a few surviving individuals. But it is not possible for exponential growth to continue indefinitely. If *Arabidopsis* populations grew exponentially for a long period of time, they would eventually fill all available habitat. When **population density**—the number of individuals per unit area—gets very high, we would expect the population's per-capita birth rate to decrease and the per-capita death rate to increase, causing r to decline. Stated another way, growth in the natural world is most often **density dependent.**

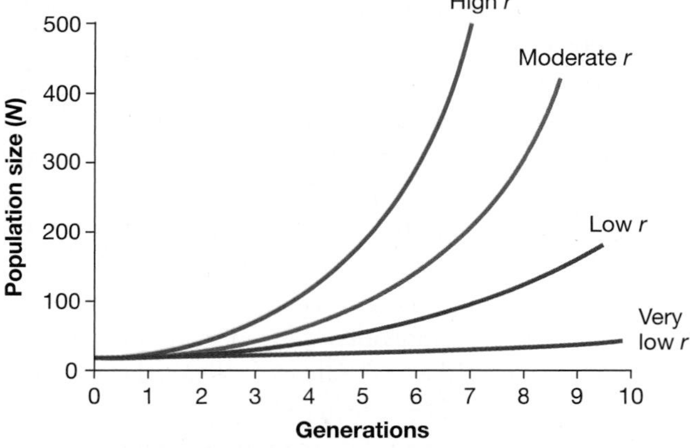

FIGURE 52.5 Exponential Growth Is Independent of Population Size. When the per-capita growth rate r does not change over time, exponential growth occurs. Population size may increase slowly or rapidly, depending on the size of r.

QUESTION Consider a species on the left-hand side of the life-history continuum in Figure 52.4 (high fecundity, low survivorship) versus a species on the right-hand side (low fecundity, high survivorship). Under ideal conditions, which species has higher r? Explain your reasoning.

Logistic Growth

To analyze what happens when growth is density dependent, biologists use a parameter called carrying capacity. **Carrying capacity, *K*,** is defined as the maximum number of individuals in a population that can be supported in a particular habitat over a sustained period of time. The carrying capacity of a habitat depends on a large number of factors: food, space, water, soil quality, resting or nesting sites, and the intensity of disease and predation. Carrying capacity can change from year to year, depending on conditions.

If a population of size N is below the carrying capacity K, then the population should continue to grow. More specifically, a population's growth rate is proportional to $(K - N)/K$:

$$\frac{\Delta N}{\Delta t} = r_{max}N\left(\frac{K - N}{K}\right)$$

This expression is called the logistic growth equation. In the expression $(K - N)/K$, the numerator defines the number of additional individuals that can be accommodated in a habitat with carrying capacity K; dividing $K - N$ by K turns this number of individuals into a proportion. Thus, $(K - N)/K$ describes the proportion of "unused resources and space" in the habitat. It can be thought of as the environment's resistance to growth. When N is small, then $(K - N)/K$ is close to 1 and the growth rate should be high. But as N gets larger, $(K - N)/K$ gets smaller. When N is at carrying capacity, then $(K - N)/K$ is equal to 0 and growth stops. Thus, as a population approaches a habitat's carrying capacity, its growth rate should slow.

The logistic growth equation describes **logistic population growth,** or changes in growth rate that occur as a function of population size. Just as exponential growth is density independent, logistic growth is density dependent. **Box 52.2** explores population growth models and how they are applied in more detail.

BOX 52.2 Developing and Applying Population Growth Equations: A Closer Look

To explore how biologists model changes in population size in more detail, let's consider data on whooping cranes, which are large, wetland-breeding birds native to North America (**Figure 52.6**). Whooping cranes may live more than 20 years in the wild, but it is common for females to be six or seven years old before they breed for the first time. Female cranes lay just two eggs per year and usually rear just one chick. Based on these life-history characteristics, we would expect the growth rate of whooping crane populations to be extremely low.

Conservationists monitor this species closely, because hunting and habitat destruction reduced the total number of individuals in the world to about 20 in the mid-1940s. Since then, intensive conservation efforts have resulted in a current total population of about 518. That number includes a group of 237 individuals that breeds in Wood Buffalo National Park in the Northwest Territories of Canada.

In addition to the Wood Buffalo population, two new populations of cranes have been established recently by releasing offspring raised in captivity. One of these groups lives near the Atlantic coast of Florida year-round; the other group migrates from breeding areas in northern Wisconsin to wintering areas along Florida's Gulf coast.

FIGURE 52.6 Whooping Cranes Have Been the Focus of Intensive Conservation Efforts.

According to the biologists who are managing whooping crane recovery, the species will no longer be endangered when the two newly established populations have at least 25 breeding pairs—meaning there would be about 125 individuals in each—and when neither population needs to be supplemented with captive-bred young in order to be self-sustaining. How long will this take?

Discrete Growth

Whooping cranes breed once per year, so the simplest way to express a crane population's growth rate is to compare the number of individuals at the start of one breeding season to the number at the start of the following year's breeding season.

When populations breed during discrete seasons, their growth rate can be calculated as for whooping cranes. To create a general expression for how populations grow over a discrete time interval, biologists use N to symbolize population size, with a subscript to indicate the year or generation. N_0 is the population size at time zero (the starting point), and N_1 the population size one breeding interval later. In equation form, the growth rate is given as

$$N_1/N_0 = \lambda \qquad \textbf{(Eq. 52.3)}$$

The parameter λ (lambda) is called the **finite rate of increase**. (In mathematics, a *finite rate* refers to an observed rate over a given period of time. A *parameter* is a variable or constant term that affects the shape of a function but does not affect its general nature.) The current total at Wood Buffalo is 237. Suppose that biologists count 248 cranes on the breeding grounds next year. The population growth rate could be calculated as $248/237 = 1.046 = \lambda$. Stated another way, the population will have grown at the rate of 4.6 percent per year. Rearranging the expression in Equation 52.3 gives

$$N_1 = N_0\lambda \qquad \textbf{(Eq. 52.4)}$$

Stated more generally, the size of the population at the end of year t will be given by

$$N_t = N_0\lambda^t \qquad \textbf{(Eq. 52.5)}$$

This equation summarizes how populations grow when breeding takes place seasonally. The size of the population at time t is equal to the starting size, times the finite rate of increase multiplied by itself t times. In a sense, λ works like the interest rate at a bank. For species that breed once per year, the "interest" on the population is compounded annually. A savings account with a 5 percent annual interest rate increases by a factor of 1.05 per year. If a population is growing, then its λ is greater than one. The population is stable when λ is 1.0 and declining when λ is less than 1.0.

If a population's age structure is stable, meaning that the proportion of females in each age class is not changing over time, then its finite rate of increase also has a simple relationship to its net reproductive rate: $\lambda = R_0/g$, where g is the generation time. In essence, dividing the net reproductive rate by generation time transforms it into a discrete rate.

Continuous Growth

The parameters λ and r—the finite rate of increase and the per-capita growth rate—have a simple mathematical relationship. The best way to understand their relationship is to recall that λ expresses a population's growth rate over a discrete interval of time. In contrast, r gives the population's per-capita growth rate at any particular instant. This is why r is also called the instantaneous rate of increase. The relationship between the two parameters is given by

$$\lambda = e^r \qquad \textbf{(Eq. 52.6)}$$

where e is the base of the natural logarithm, or about 2.72. (For help with using logarithms, see **BioSkills 5**. Also, note that the relationship between any finite rate and any instantaneous rate is given by finite rate $= e^{\text{instantaneous rate}}$.)

(Continued on next page)

(continued)

Substituting Equation 52.6 into Equation 52.5 gives

$$N_t = N_0 e^{rt} \qquad \textbf{(Eq. 52.7)}$$

This expression summarizes how populations grow when they breed continuously, as do humans and bacteria, instead of at defined intervals. For species that breed continuously, the "interest" on the population is compounded continuously. When the growth rates λ and r are equivalent, however, the differences between discrete and continuous growth are negligible. Because r represents the growth rate at any given time, and because r and λ are so closely related, biologists routinely calculate r for species that breed seasonally. In the whooping crane example, $\lambda = 1.046$ per year $= e^r$. To solve for r, take the natural logarithm of both sides. (**BioSkills 5** explains how to do this on your calculator.) In this case, $r = 0.045$ per year.

The instantaneous rate of increase, r, is also directly related to the net reproductive rate, R_0, introduced in Box 52.1. In most cases, r is calculated as $\ln R_0/g$, where g is the generation time. Thus, r can be calculated from life-table data. It is a more useful measure of growth rate than R_0, because r is independent of generation time.

To summarize, biologists have developed several ways of calculating and expressing a population's growth rate. Growth rate expressed as λ has the advantage of being easy to understand, and R_0 has the advantage of being calculated directly from life-table data. Although r is slightly more difficult conceptually, it is the most useful expression for growth rate, because it is independent of generation time and is relevant for species that breed either seasonally or continuously.

Applying the Models

To get a better feel for r and for Equation 52.7, consider the following series of questions about whooping cranes. The key to answering these questions is to realize that Equation 52.7 has just four parameters. Given three of these parameters, you can calculate the fourth. ● If you understand this concept, you should be able to solve the following four problems:

1. If 20 individuals were alive in 1941 and 518 existed in 2007, what is r? Here $N_t = 518$, $N_0 = 20$, and $t = 66$ years. Substitute these values into Equation 52.7 and solve for r. Then check your answer at the end of this box.[1]
2. In the most recent report issued by the biologists working on the Wood Buffalo crane recovery program, it is estimated that the flock should be able to sustain an r of 0.046 for the foreseeable future. If the flock currently contains 237 individuals, how long will it take that population to double? Here $N_0 = 237$ and $N_t = 2 \times 237 = 474$. In this case, you solve Equation 52.7 for t. Then check your answer.[2]
3. In 2002 a pair of birds in the flock that lives in Florida year-round successfully raised offspring (nine years after the first cranes were introduced there). In 2003 another pair of cranes in this population bred successfully. If the number of breeding pairs continues to double each year, how long will it take to reach the goal of 25 breeding pairs? (Note that $\lambda = 2.0$ if a population is doubling each year.) Given that $N_0 = 2$ and $N_t = 25$, solve for t again, and check your answer.[3]
4. The whooping crane flock that migrates between Wisconsin and Florida was founded in 2001. If its development is like that of the resident flock in Florida, the first successful breeding attempt will occur in 2010—nine years after the initial introduction. Suppose that the instantaneous growth rate for the number of breeding pairs in this population will be 0.05. In what year will the breeding population reach 25 pairs? This is the year that whooping cranes should come off the endangered species list. Here $N_0 = 1$, $N_t = 25$, and $r = 0.05$. Solve for t, and add this number of years to 2010; then check your answer.[4]

[1] $r = 0.049$; [2] $t = 15$ years; [3] $t = 3.64$ years; [4] 2074

Figure 52.7a illustrates density-dependent growth for a hypothetical population. Notice that the graph has three sections. Initially, growth is exponential—meaning that r is constant. With time, N increases to the point where competition for resources or other density-dependent factors begins to occur. As a result, the growth rate begins to decline and eventually reaches 0. When the population is at the habitat's carrying capacity, the graph of population size versus time is flat.

Figure 52.7b shows data on logistic growth from laboratory populations of two species of protists in the group called ciliates (see Chapter 29): *Paramecium aurelia* and *P. caudatum*. In this experiment, an investigator placed 20 individuals from one of the *Paramecium* species into 5 mL of a solution. He created many replicates of these 5-mL environments for each species separately. He kept conditions as constant as possible in each replicate by adding the same number of bacterial cells every day for food, washing the solution every second day to remove wastes, and maintaining the pH at 8.0. As the graphs show, both species exhibited logistic growth in this environment. The carrying capacity differed in the two species, however. The maximum density of *P. aurelia* averaged 448 individuals per mL, but that for *P. caudatum* averaged just 128 individuals per mL.

In both of the *Paramecium* species in this experiment, exponential growth could be sustained for only about 5 days. Exponential growth is almost always short lived. When population size increases, a population's growth rate has to decline. Why? What factors cause growth rates to change?

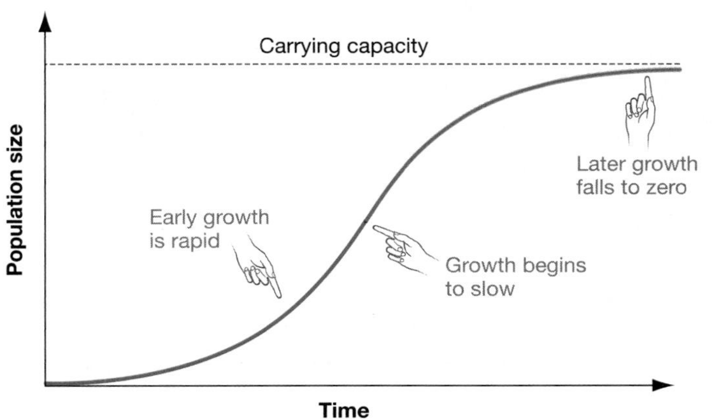

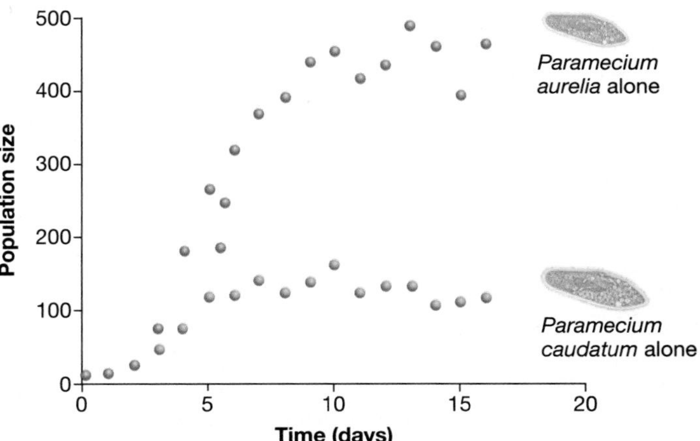

FIGURE 52.7 Logistic Growth Is Dependent on Population Size. (a) A curve illustrating the pattern predicted by the logistic growth equation. This pattern often occurs when a small number of individuals colonizes an unoccupied habitat. Initially, r is high because competition for resources is low to nonexistent. Carrying capacity depends on the quality of the habitat and can vary over time. **(b)** Data from laboratory experiments with two species of *Paramecium*.

EXERCISE In part (b), draw dashed horizontal lines indicating the carrying capacity for each species.

What Limits Growth Rates and Population Sizes?

Population sizes change as a result of two general types of factors: (1) density-independent and (2) density-dependent factors. Density-independent factors alter birth rates and death rates irrespective of the number of individuals in the population. Density-independent changes are usually triggered by changes in the abiotic environment—variation in weather patterns or catastrophic events such as cold snaps, hurricanes, volcanic eruptions, or drought. In contrast, density-dependent factors are usually biotic and change in intensity as a function of population size. For example, predation rates on deer may increase when deer population sizes are high, or competition for food and starvation may become more common. When trees are crowded, they have less water, nutrients, and sunlight at their disposal and make fewer seeds. Death rates may increase and birth rates decrease when populations are at high density.

To get a better understanding of how biologists study the density-dependent factors that affect population size, consider the data in **Figure 52.8**:

- The graph in Figure 52.8a presents results of an experimental study of a coral-reef fish called the bridled goby. Each data point represents an identical artificial reef, constructed by a researcher from the rubble of real coral reefs. The initial density, plotted along the horizontal axis, represents the marked bridled gobies that were released on each artificial reef at the start of the experiment. The proportion surviving, plotted on the vertical axis, represents the introduced individuals that were still living at that reef 2.5 months later. This graph shows a strong density-dependent relationship in survivorship.
- The graph in Figure 52.8b is from a long-term study of song sparrows on Mandarte Island, British Columbia. Each data point represents a different year. The density of females, plotted along the horizontal axis, is the number of females that bred on the island; the clutch size, plotted on the vertical axis, is the average number of eggs laid by each female. This graph indicates a strong density-dependent relationship in fecundity.

Density-dependent changes in survivorship and fecundity cause logistic population growth. In this way, density-dependent factors define a particular habitat's carrying capacity. If gobies get crowded, they die or emigrate. If song sparrows are crowded, the average number of eggs and offspring that they produce declines.

It's important to recognize that K varies among species and populations, however, and that this variation affects both growth rates and population sizes. K varies because for any particular species, some habitats are better than other habitats due to differences in food availability, predator abundance, and other density-dependent factors. Stated another way, K varies in space. It also varies with time, as conditions in some years are better than in others. In addition, the same region may have a very different carrying capacity for different species. The same area will tend to support many more individuals of a small-bodied species than of a large-bodied species, simply because large individuals demand more space and resources. These simple observations help explain the variation in total population size that exists among species and among populations of the same species.

(a) Survival declines at high population density.

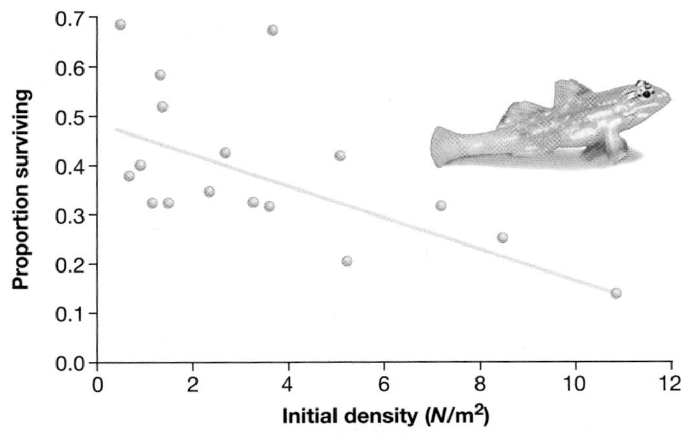

(b) Fecundity declines at high population density.

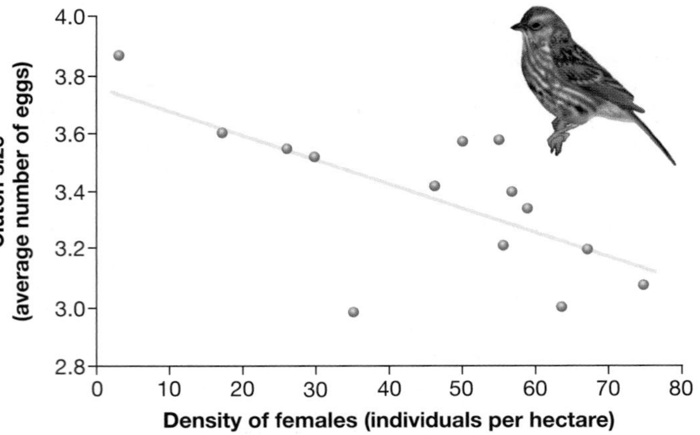

FIGURE 52.8 Density-Dependent Growth Results from Changes in Survivorship and Fecundity. (a) When bridled gobies are introduced at different population densities to artificial reefs, survivorship is highest at low density. **(b)** In the song sparrow population on Mandarte Island, there is a strong negative correlation between population density in a particular year and the average number of eggs (clutch size) produced by females.

QUESTION When song sparrow populations are at high density and extra food is provided to females experimentally, average clutch size is much higher than expected from the data in part (b). Based on these data, state a hypothesis to explain why average clutch size declines at high density.

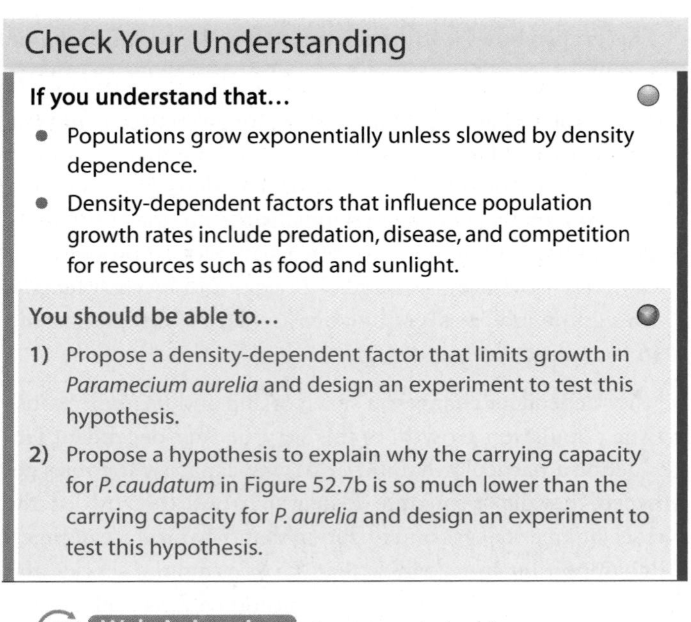
Check Your Understanding

If you understand that...

- Populations grow exponentially unless slowed by density dependence.
- Density-dependent factors that influence population growth rates include predation, disease, and competition for resources such as food and sunlight.

You should be able to...

1) Propose a density-dependent factor that limits growth in *Paramecium aurelia* and design an experiment to test this hypothesis.
2) Propose a hypothesis to explain why the carrying capacity for *P. caudatum* in Figure 52.7b is so much lower than the carrying capacity for *P. aurelia* and design an experiment to test this hypothesis.

MB Web Animation at www.masteringbio.com
Modeling Population Growth

52.3 Population Dynamics

The tools introduced in the previous two sections provide a foundation for exploring how biologists study **population dynamics**—changes in populations through time. Research on population dynamics has uncovered a wide array of patterns in natural populations in addition to exponential and logistic growth.

As an initial example of how population sizes change through time, consider data on the longest-running experiment in the history of biological science: the Park Grass study in Rothamsted, United Kingdom. In 1856 researchers established a series of 0.1–0.2-ha (about 1/3-acre) plots in a hay meadow that had uniform soil characteristics and vegetation. Since then, some of the plots have received regular fertilizer treatments or lime (a mix of calcium and magnesium oxides), and some have been left untreated as controls. Each year researchers recorded which plant species were present in each plot, as an index of population size.

The weather was the same in all plots, the fertilizing and liming regimes were constant in treated plots, and presumably all plots were exposed to the same predators and diseases. Even so, several dramatic changes in population size have taken place. When the data from 1920 to 1979 were analyzed for the control plots, four major patterns emerged among the 43 species present. Ten of the species showed a clear spike in abundance, followed by a decline over the course of the 60-year period (**Figure 52.9a**). Seven species increased in abundance and then reached a plateau (**Figure 52.9b**). Five species declined steadily, and 21 showed no discernible change (**Figure 52.9c**).

In several cases, patterns of change over time appeared to correlate with life-history traits. For example, species that maintained a constant population size throughout the 60-year interval have longer life spans and a lower r_{max} than do other species at the site. To interpret this observation, biologists suggest that these species allocate more energy and resources to competitive ability than to reproduction. As a result, these species are able to maintain abundance over a longer period. In contrast, species that spiked and declined or that increased and maintained high abundance had a higher r_{max}, an earlier age at first reproduction, and shorter life spans than did species

(a) Some species increased, then declined.

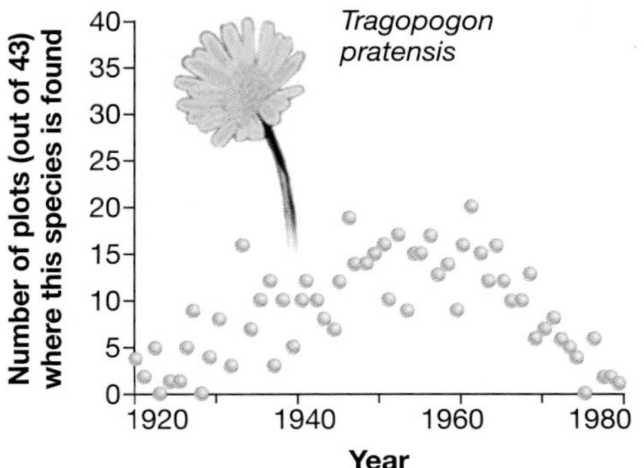

(b) Some species increased and maintained high population size.

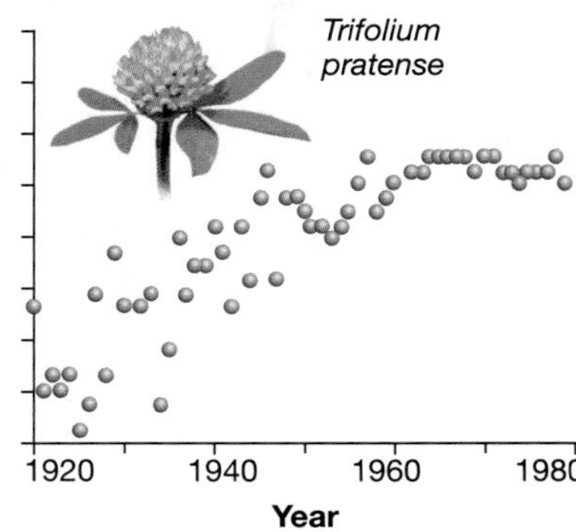

(c) Some species maintained high population size.

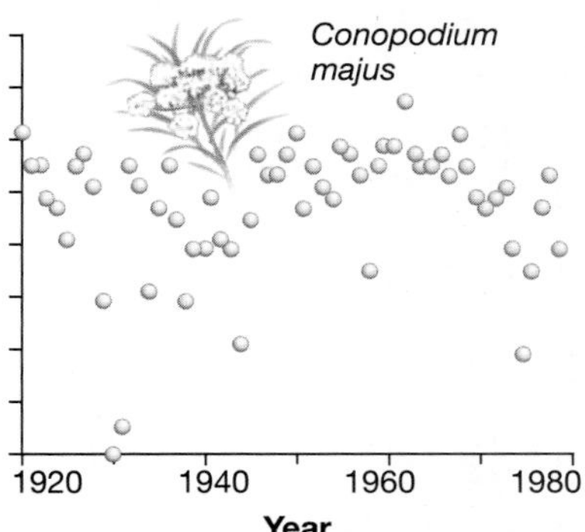

FIGURE 52.9 Distinct Patterns of Long-Term Changes in the Populations of Meadow Species. Graphs showing some of the distinct patterns in population dynamics from 1920 to 1979 among species in the Park Grass experiment, which began in 1856.

that maintained a constant population size. Researchers are still working on the question of why some of these species remained constant over time and why others declined.

How Do Metapopulations Change through Time?

If you browse through an identification guide for trees or birds or butterflies, you'll likely find range maps indicating that most species occupy a broad area. In reality, however, the habitat preferences of many species are restricted, and individuals occupy only isolated patches within that broad area. The meadows occupied by Glanville fritillaries—an endangered species of butterfly native to the Åland islands off the coast of Finland (**Figure 52.10**)—are a well-studied example of this pattern.

If individuals from a species occupy many small patches of habitat, so that they form many independent populations, they are said to represent a **metapopulation**, or population of populations. Glanville fritillaries exist naturally as metapopulations. But because humans are reducing large, contiguous areas of forest and grasslands to isolated patches or reserves, more and more species are being forced into a metapopulation structure.

Recent research on Glanville fritillaries by Ilkka Hanski and colleagues illustrates the consequences of metapopulation structure for endangered species. The work began with a survey of meadow habitats on the islands. Because fritillary caterpillars feed on just two types of host plant, *Plantago lanceolata* and *Veronica spicata*, the team was able to pinpoint potential butterfly habitats. They estimated the fritillary population size within each patch by counting the number of larval webs in each. Of the 1502 meadows that contained the host plants, 536 had Glanville fritillaries. The patches ranged from 6 m^2 to 3 ha in area. Most had only a single breeding pair of adults and one larval group, but the largest population contained hundreds of pairs of breeding adults and 3450 larvae.

FIGURE 52.10 Glanville Fritillaries Live in Patches of Meadow. The Glanville fritillary is extinct in much of its former range; it is now found only on the Åland islands near Finland.

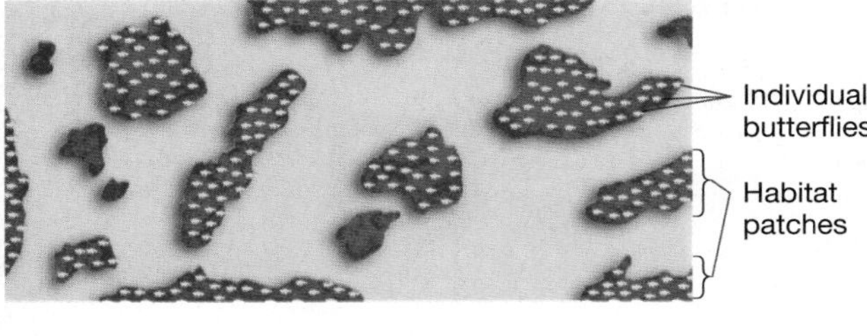

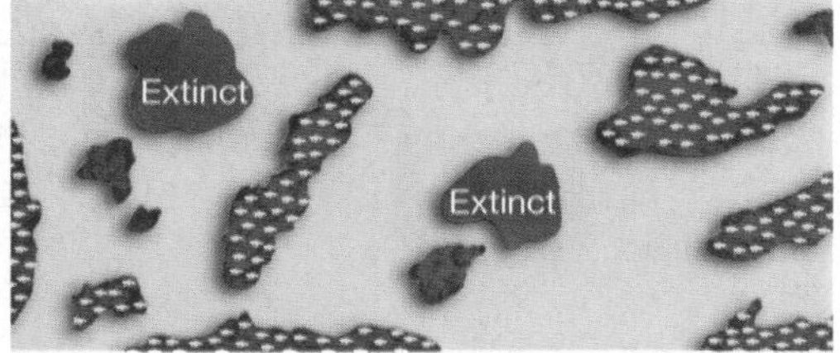

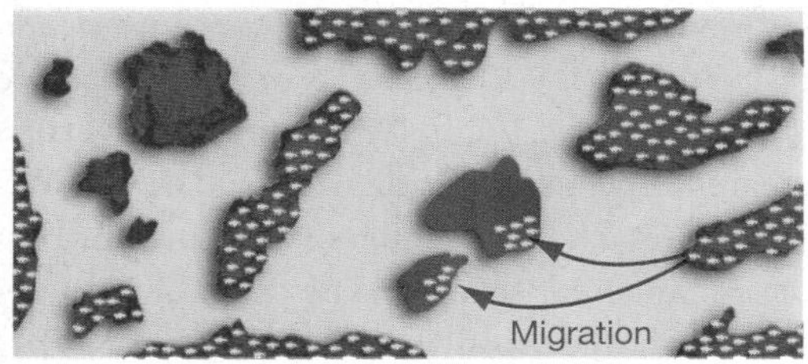

FIGURE 52.11 Metapopulation Dynamics Depend on Extinction and Recolonization. The overall population size of a metapopulation stays relatively stable even if subpopulations go extinct. These populations may be restored by migration, or unoccupied habitats might be colonized.

Figure 52.11 illustrates how a metapopulation like this is expected to change over the years. Given enough time, each population within the larger metapopulation is expected to go extinct. The cause could be a catastrophe, such as a storm or an oil spill; it could also be a disease outbreak or a sudden influx of predators. Migration from nearby populations can reestablish populations in these empty habitat fragments, however. In this way, there is a balance between extinction and recolonization. Even though subpopulations blink on and off over time, the overall population is maintained at a stable number of individuals.

Does this metapopulation model correctly describe the population dynamics of fritillaries? To answer this question, Hanski's group conducted a mark-recapture study. (**Box 52.3** on page 1188 provides details on how a mark-recapture study is done.) Of the 1731 butterflies that the biologists marked and released, 741 were recaptured over the course of the summer study period. Of the recaptured individuals, 9 percent were found in a previously unoccupied patch. This migration rate is high enough to suggest that patches where a population has gone extinct will eventually be recolonized. Hanski and co-workers repeated the survey two years after their initial census. Just as the metapopulation model predicts, some populations had gone extinct and others had been created. On average, butterflies were lost from 200 patches each year; 114 unoccupied patches were colonized and newly occupied each year. The overall population size was relatively stable even though constituent populations came and went.

To summarize, the history and future of a species divided into metapopulations is driven by the birth and death of populations, just as the dynamics of a single population are driven by the birth and death of individuals. As the final section in this chapter will show, these dynamics have important implications for saving endangered species.

Why Do Some Populations Cycle?

One of the most striking patterns in population dynamics has also been one of the most difficult to understand. **Population cycles** are regular fluctuations in size that some animal populations exhibit. **Figure 52.12** shows a classic case of population cycling: snowshoe hare and lynx populations in northern Canada. Snowshoe hares are herbivores and are the primary prey of lynx.

Most hypotheses to explain population cycles hinge on some sort of density-dependent factor. The idea is that predation, disease, or food shortages intensify dramatically at high population density and cause population numbers to crash. It has been extremely difficult, however, to document a causal

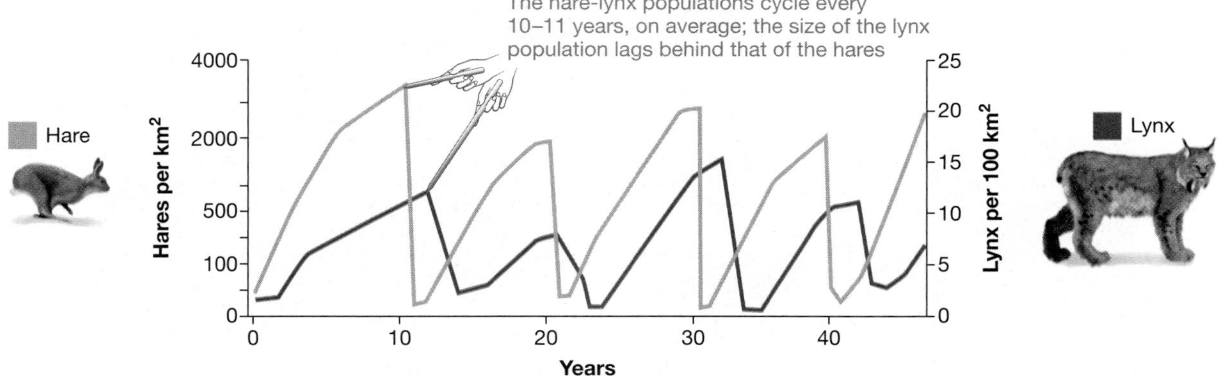

FIGURE 52.12 Hare and Lynx Populations Cycle in Synchrony. To make it easier to read, the left-hand vertical axis, which indicates hare density, is not linear.

relationship between population cycles and factors such as predation intensity, disease outbreaks, or food availability.

Although the hare-lynx cycle was documented in the early 1900s, it was not well understood. Two hypotheses were proposed to explain these population dynamics:

1. Hares use up all their food when their populations reach high density and starve; in response, lynx also starve.
2. Lynx populations reach high density in response to increases in hare density. At high density, lynx eat so many hares that the prey population crashes.

Stated another way, either hares control lynx population size or lynx control hare population size.

To test these hypotheses rigorously, researchers set up a series of 1-km^2 study plots in boreal forest habitats that were as identical as possible (**Figure 52.13**). Three plots were left as unmanipulated controls. One plot was ringed with an electrified fence with a mesh that excluded lynx but allowed hares to pass freely. Two plots received additional food for hares year-round. One plot had a predator-exclusion fence and was also supplemented with food for hares year-round. The biologists then monitored the size of the hare and lynx populations over an 11-year period, or enough time for a complete cycle in the two populations.

As the data in Figure 52.13 show, plots with predators excluded showed higher hare populations than did control plots. This result supports the hypothesis that predation by

Experiment

Question: What factors control the hare-lynx population cycle?

Hypothesis: Predation, food availability, or a combination of those two factors controls the hare-lynx cycle.

Null hypothesis: The hare-lynx cycle isn't driven by predation, food availability, or a combination of those two factors.

Experimental setup:

Document hare population in seven study plots (similar boreal forest habitats, each 1 km^2) over 11 years (duration of one cycle).

3 plots: Unmanipulated controls

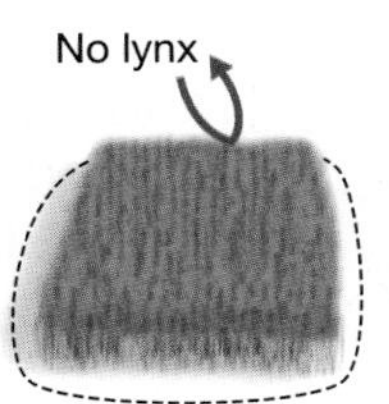

1 plot: Electrified fence excludes lynx but allows free access by hares.

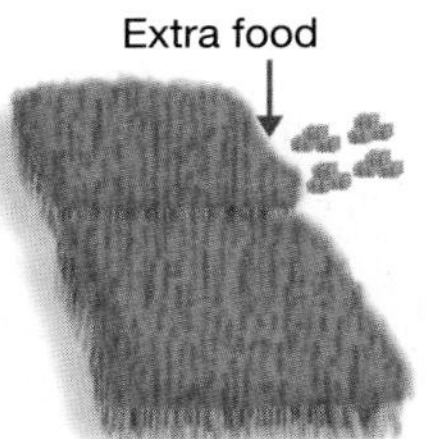

2 plots: Supply extra food for hares.

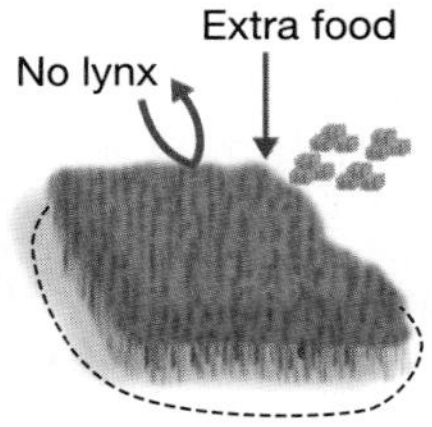

1 plot: Electrified fence excludes lynx but allows free access by hares; supply extra food for hares.

Prediction: Hare populations in at least one type of manipulated plot will be higher than the average population in control plots.

Prediction of null hypothesis: Hare populations in all of the plots will be the same.

Results:

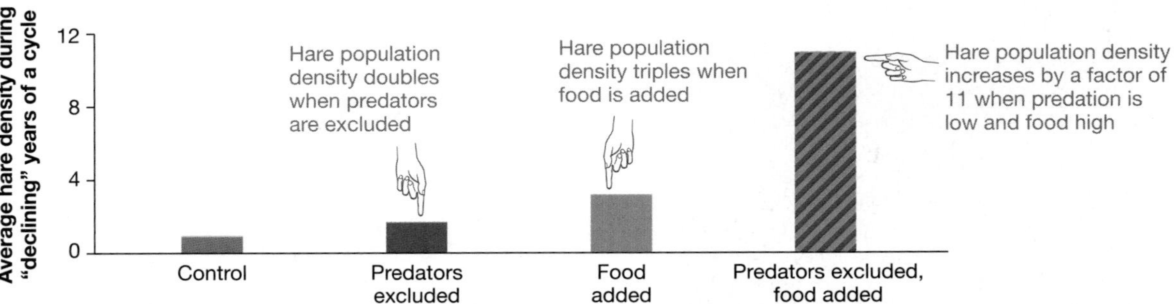

Conclusion: Hare populations are limited by both predation and food availability. When predation and food limitation occur together, they have a greater effect than either factor does independently.

FIGURE 52.13 Experimental Evidence that Predation and Food Availability Drive the Hare-Lynx Cycle.

QUESTION State a hypothesis to explain why lack of predation and increased food, in combination, have such a large effect on hare population size.

lynx reduces hare populations. Plots with supplemental food had even higher peak populations, however, and the plot with supplemental food and predators excluded had a hare population over 11 times as large as the average in control plots.

These data support the hypotheses that hare populations are limited by availability of food as well as by predation, and that food availability and predation intensity interact—meaning the combined effect of food and predation is much larger than their impact in isolation. Research continues to figure out why this is so. In the meantime, the study illustrates the power of a well-designed experiment to solve long-standing controversies in biology.

How Does Age Structure Affect Population Growth?

As the life-table data presented in Section 52.1 showed, age has a dramatic effect on the probability that an individual will reproduce and survive to the following year. But the previous analysis of life-table data left out a critical point about the ages of individuals in a population: A population's **age structure**—meaning the proportion of individuals that are at each possible age—has a dramatic influence on the population's growth over time. To see how changes in age structure can affect population dynamics, let's consider two case histories for which biologists have documented changes in age structure—one example concerns a flowering plant, and the other focuses on our own species.

Age Structure in a Woodland Herb The common primrose, pictured in **Figure 52.14a**, grows in the woodlands of Western Europe. Primroses grow on the forest floor and can germinate, mature, and produce offspring only in the relatively high-light environment created when a tall tree falls and opens a gap in the forest canopy. These sunny spaces are short lived, however, because mature trees and saplings in and around the gap grow and fill the space. As they do, light levels in the gap decline and the environment becomes increasingly unsuitable for herbs such as primrose.

Biologists recently set out to document how primrose populations respond to this ephemeral woodland environment. The researchers hypothesized that populations in new gaps and populations in large gaps—meaning those with the highest light availability—would experience high rates of reproduction and an age structure characterized by large numbers of juveniles. The researchers also hypothesized that as light levels declined in a particular gap due to the growth of surrounding trees, primrose reproduction would decline. As a result, the age structure of populations in older gaps should be characterized by a large proportion of individuals in adult stages, with fewer juveniles.

To test these hypotheses, the investigators selected eight common primrose populations growing in a range of light levels, from high to low. The team established 1-m^2 study plots inside each gap and studied approximately 350 individuals per population. Each individual was assigned to one of five ages: seedlings, juveniles, or adults in one of three size classes. As **Figure 52.14b** shows, the data supported the hypothesis that in gaps with high light levels, populations have a larger proportion of juveniles than do gaps with low light levels. In addition, the proportion of juvenile individuals declined over time in all gaps as light levels declined. These data have three important messages about the dynamics of primrose populations:

1. Populations that are dominated by juveniles should experience rapid growth. Population size should then decline with time, however, due to a density-independent factor: shading by trees.

(a) Common primroses live in sunlit gaps in forests.

(b) Age structure of primrose population varies with age of gap.

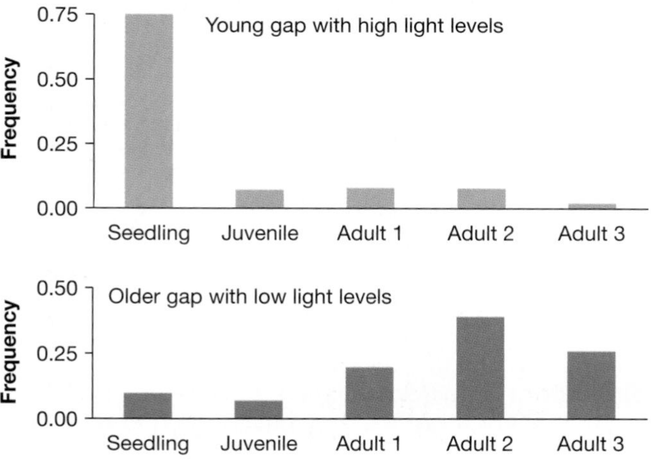

FIGURE 52.14 The Age Structure of Common Primrose Populations Changes over Time. In the graphs, "Adult 1," "Adult 2," and "Adult 3" individuals are of increasingly large size and thus age.

EXERCISE Create a third graph showing the age distribution of a gap that has completely filled in and closed, meaning that light levels are now as low as they were before the gap existed.

2. The long-term trajectory of the overall primrose population in an area may depend primarily on the frequency and severity of windstorms that knock down trees and create sunlit gaps. If so, it means that the dynamics of primrose populations are governed by an abiotic, density-independent factor.

3. In a large tract of forest, the primrose population will consist of a large number of subpopulations, each found in canopy gaps. Some of these subpopulations are likely to be very small, and each subpopulation will arise, grow, and then go extinct over time. Recall that biologists say that a group of small, isolated populations such as this represents a metapopulation, a population of populations. Primroses have metapopulation structure.

Age Structure in Human Populations As another example of how age structure affects population dynamics, consider humans. In countries where industrial and technological development occurred many decades ago, survivorship has been high and fecundity low for several decades. As a result, these populations are not growing quickly. When researchers create a graph with horizontal bars representing the numbers of males and females of each age group for such a population, an age pyramid like that in **Figure 52.15a** results. The most striking pattern in these data is that there are similar numbers of people in most age classes. The evenness occurs because the same number of infants are being born each year and because most survive to old age. In this way, analyzing an age pyramid can give biologists important information about a population's history.

Studying age distributions can also help researchers predict a population's future. For example, the populations of developed nations are not expected to grow especially quickly, because only modest numbers of individuals will reach reproductive age in the near future. The white lines and lighter bars in Figure 52.15a show the projected age structure of a developed nation (Sweden) in 2050. Modest changes in the age distribution have occurred since 2000 because survivorship has increased while fecundity has remained the same. The projections highlight a major public policy concern in countries such as this: how to care for an increasingly aged population.

In contrast, the age distribution is bottom-heavy in less-developed nations. As **Figure 52.15b** shows, these populations are dominated by the very young. This type of age distribution occurs when populations have undergone rapid growth—with more children being born than exist in older age classes. Due to dramatic improvements in health care, most people in the younger age classes now survive to reproductive age.

For Figure 52.15b, the projected age distribution in 2050 in Honduras is based on continued high survivorship and a pattern that has occurred repeatedly around the world: When infant mortality drops in a country, fecundity rates begin to decline within a generation or two. The white lines and light bars in Figure 52.15b are relatively even for ages 0–55—meaning that rapid growth has stopped—because fecundity is expected to be relatively low and survivorship high. The projected data also illustrate a major public policy concern in less-developed countries—providing education and jobs for an enormous group of young people who will be reaching adulthood during your lifetime.

The data in Figure 52.15b make another important point: Because of recent and rapid growth in developing countries, overall population size will increase dramatically in these nations over the course of your lifetime. A large part of the increase will be due to increased survivorship. But the number

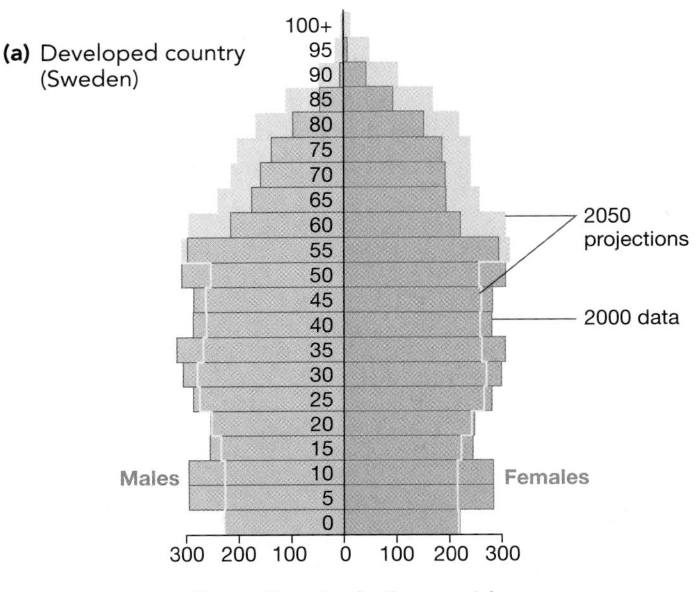

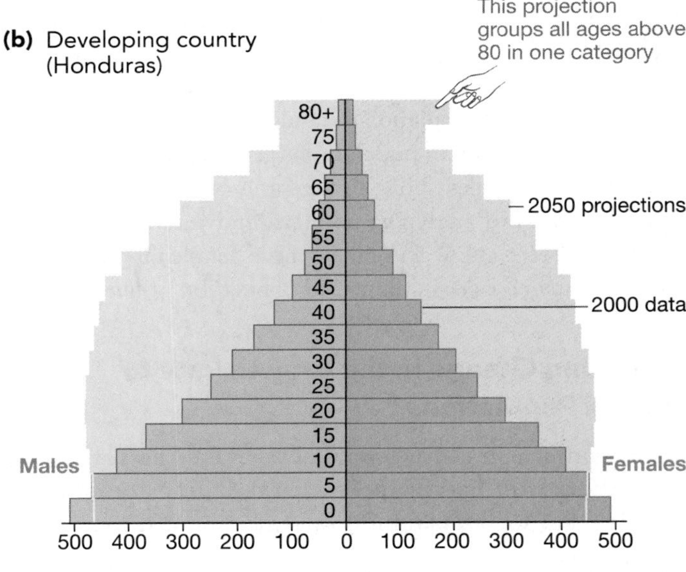

FIGURE 52.15 The Age Structure of Human Populations Varies Dramatically. Age distributions showing the number of males and females in 5-year age increments from 0 to 100.

BOX 52.3 Mark-Recapture Studies

To estimate the population size of sedentary (largely immobile) organisms, researchers sample appropriate habitats by counting the individuals that occur along transects or inside rectangular plots—called quadrats—set up at random locations in the habitat. These counts can be extrapolated to the entire habitat area to estimate the total population size. In addition, they can be compared to later censuses to document trends over time.

In contrast, estimating the total population size of organisms that are mobile and that do not congregate into herds or flocks or schools is much more of a challenge. In species such as the Glanville fritillary, the population inside sample quadrats or along transects changes constantly as individuals move in and out. Further, it can be difficult to track whether a particular individual has already been counted.

If individuals can be captured and then tagged in some way, however, the total population size of a mobile species can be estimated by using an approach called mark-recapture. To begin a mark-recapture study, researchers catch individuals and mark them with leg bands, ear tags, or some other method of identification. After the marked individuals are released, they are allowed to mix with the unmarked animals in the population for a period of time. Then a second trapping effort is conducted, and the percentage of marked individuals that were captured is recorded.

To estimate the total population from these data, researchers make a key assumption: The percentage of marked and recaptured individuals is equal to the percentage of marked individuals in the entire population. This assumption should be valid when no bias exists regarding which individuals are caught in each sample attempt. It is important that individuals do not learn to avoid traps after being caught once and that they do not emigrate or die as a result of being trapped.

The relationship between marked and unmarked individuals can be concisely expressed algebraically:

$$\frac{m_2}{n_2} = \frac{m_1}{N}$$

In this equation, m_2 is the number of marked animals in the second sample (the recapture), n_2 is the total number of animals (marked and unmarked) in the second sample, m_1 is the number of marked animals in the first sample, and N is the total population size. Having measured m_2, n_2, and m_1, the researcher can estimate N. ● If you understand how to estimate population size in a mark-recapture study, you should be able to do so in this example: Suppose researchers marked 255 animals and later were able to trap a total of 162 individuals in the population, of which 78 were marked. What is the estimate for total population size? Solve for N (give the nearest integer) and check your answer.[1]

[1] $N = 530$

of offspring being born each year is also expected to stay high, even though fecundity is predicted to decline significantly. This result seems paradoxical but is based on a key observation: There are now so many young women in these populations that the overall number of births will stay high even though the average number of children per female is much less than it was a generation ago. To capture this point, biologists say that these populations have momentum or inertia. Combined with high survivorship, population pyramids like these make continuing increases in overall population size inevitable.

As these examples show, understanding age structure is a key component of analyzing population dynamics in humans and other species. Now let's take a more detailed look at recent and projected changes in the world population of humans.

Analyzing Change in the Growth Rate of Human Populations

Studies on changes in whooping crane and the Park Grass populations over 60-year periods have been extraordinarily valuable. But six decades turns out to be short in comparison to data available on human population dynamics.

Archeological and anthropological data have been used to estimate the size of human populations over the past 250 years (**Figure 52.16**). Although the shape of the curve is superficially similar to the examples of exponential growth in Figure 52.5, the growth rate for humans has actually increased over time since 1750, leading to a steeply rising curve.

It is difficult to overemphasize just how dramatically the human population has increased over the past 250 years. **Table 52.2** presents data that will allow you to calculate how long it has taken this population to add each billion people, and to determine how many times the population has doubled—from a starting point in the year A.D. 1420—and how many years the

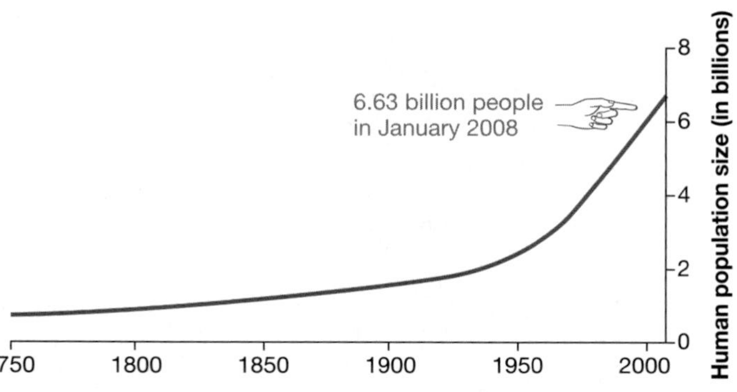

FIGURE 52.16 The Human Population Has Been Growing Rapidly. Graph based on estimates from historical and census data.

TABLE 52.2 **Milestones in Human Population Growth**

Date	Human Population (in hundreds of millions)
1420	375
1720	750
1800	1000
1875	1500
1927	2000
1961	3000
1974	4000
1987	5000
1999	6000
Number of years required to add 1 billion people:	
Number of years required to double population:	

EXERCISE Using the data provided, fill in the correct number of years in the blank cells in the table. To calculate the time required for the human population to reach 1 billion, assume that *Homo sapiens* originated 150,000 years ago.

population took to double each time. To put your calculations on doubling time in perspective, consider that until your grandparents' generation, no individual had lived long enough to see the human population double. But many members of the generation born in the early 1920s have lived to see the population *triple*. And to drive home the impact of continued population doubling, consider the following: If you were given a penny on January 1, then $0.02 on January 2, $0.04 on January 3, and so on, you would be handed $10,737,418 on January 31.

As this book goes to press, the world population is estimated at 6.6 billion. About 77 million additional people—equivalent to the current population of Egypt, or more than double the state of California—are being added each year.

It is not possible to overstate the consequences of recent and current increases in human population. In addition to being the primary cause of the habitat losses and species extinctions analyzed in Chapter 55, overpopulation is linked by researchers to declines in living standards, mass movements of people, political instability, and acute shortages of water, fuel, and other basic resources in many parts of the world.

The one encouraging trend in the data is that the growth rate of the human population has already peaked and begun to decline. The highest growth rates occurred between 1965 and 1970, when populations increased at an average of 2.04 percent per year. Between 1990 and 1995, the overall growth rate in human populations averaged 1.46 percent per year; the current growth rate is 1.2 percent annually. In humans, r may be undergoing the first long-term decline in history. The question is: Will the human population stabilize or begin to decline in time to prevent global—and potentially irreversible—damage?

Will Human Population Size Peak in Your Lifetime? The United Nations Population Division makes regular projections for how human population size will change between now and 2050. For most readers of this book, these projections describe what the world will look like as you reach your early 60s.

The UN projections are based on four scenarios, which hinge on different values for fertility rates—the average number of children that each woman has during her lifetime. Currently, the worldwide average is 2.7. This represents an enormous reduction from the fertility rates during the 1950s, which averaged 5.0 children per woman. **Figure 52.17** shows how total population size is expected to change by 2050 if average fertility continues to decline and averages 2.5 (high), 2.1 (medium), or 1.7 (low) children per woman. The middle number, 2.1, is the **replacement rate**, where each woman produces exactly enough offspring to replace herself and her offspring's father. When this fertility rate is sustained for a generation, $r = 0$ and there is **zero population growth** (**ZPG**).

A glance at Figure 52.17 should convince you that the four scenarios are starkly different. If average fertility around the world stays at its present level, world population will be closing in on 12 billion about the time you might consider retiring. This is close to double the current population. To realize what this might mean, imagine what traffic or home prices in your city would be like if double the number of people were using the roads and needing housing.

The "high-fertility" projection assumes that fertility rates continue to drop but average 2.5 children per woman over this interval. This model predicts that world population in 2050 will reach nearly 11 billion—a 75 percent increase over today's population—with no signs of peaking. The low-fertility projection, in contrast, predicts that the total human population in 2050 will be about 7.4 billion and will already have peaked.

The graphs make an important point: The future of the human population hinges on fertility rates—on how many children each of the women living today decides to have. Those decisions, in turn, depend on a wide array of factors,

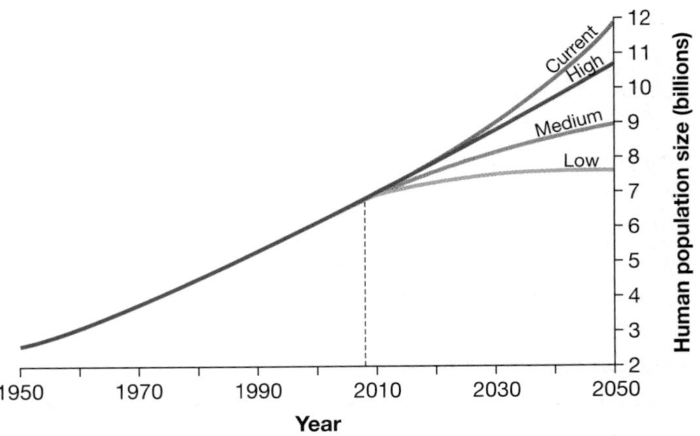

FIGURE 52.17 Projections for Human Population Growth to 2050. The "Current," "High," "Medium," and "Low" labels refer to fertility rates.

including how free women are to choose their family size and how much access women have to education. When women are allowed to become educated, they tend to delay when they start having children and have a smaller overall family size. Access to education and reliable birth control methods (see Chapter 48), in addition to overall economic development, access to quality health care, and other issues—including the course of the AIDS epidemic (see Chapter 35)—will all play a large role in determining how world population changes over your lifetime.

To summarize, humans may be approaching the end of a period of rapid growth that lasted well over 500 years. How quickly growth rates decline and how large the population can eventually become will be decided primarily by changes in fertility rates.

Check Your Understanding

If you understand that...

- Biologists use observational data—such as the life history of species in the Park Grass experiment—or experiments—such as manipulating food and predation in snowshoe hare populations—to test hypotheses about changes in the size of populations over time.
- In many cases, understanding population dynamics depends on understanding changes in a population's age structure.

You should be able to...

1) Explain why the age structure of human populations differs between developed and developing nations.
2) Explain why changes in fertility rates have such a dramatic impact on projections for the total human population in 2050.

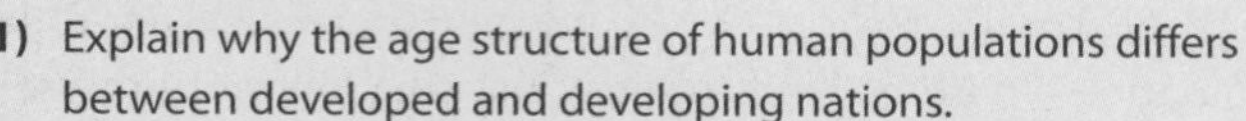

Human Population Growth and Regulation

52.4 How Can Population Ecology Help Endangered Species?

Conservationists draw heavily on concepts and techniques from population ecology when designing programs to save species threatened with extinction. As habitat destruction and invasive species push species throughout the world into decline, the study of population growth rates and population dynamics has taken on an increasingly applied tone.

To illustrate one of the ways that biologists apply the theory and results introduced in this chapter, let's analyze how an understanding of life-table data and geographic structure can help direct conservation action.

Using Life-Table Data to Make Population Projections

Demographic data such as age-specific survivorship and fecundity are important for saving endangered species and for other applied problems. To understand why, suppose that you were in charge of reintroducing a population of lizards to a nature reserve, and that research conducted when this species occupied the site previously documented the survivorship and fecundity data. Your initial plan is to take 1000 newly hatched females from a captive breeding center and release them into the habitat (along with enough males for breeding to occur). Is this population likely to become established and grow, or will you need to keep introducing offspring that have been raised in captivity?

To answer this question, you need to use the life-table data in **Figure 52.18a** and calculate (1) how many adults will survive to each age class each year and (2) how many offspring will be produced by each adult age class each year over the course of several years. **Figure 52.18b** starts these calculations for you. The fate of the original 1000 females is indicated in red. In the second year, just 330 of these individuals have survived; 40 are left as 3-year-olds.

How many offspring did this generation produce? As just stated, 330 of the original 1000 females remain after one year. As the first purple number in Figure 52.18b indicates, these survivors have an average of 3.0 female offspring apiece, so they contribute 990 new juveniles to the population. In the second year, the 200 females that are left have an average of 4.0 female offspring each and contribute 800 juveniles. In their third year, the 40 surviving females average 5.0 offspring and contribute 200 juveniles.

Figure 52.18c extends the calculations by showing what happens as the offspring of the original females begin to breed. Their offspring are shown in green. By adding subsequent generations and continuing the analysis, you should be able to predict whether the population will stay the same, decline, or increase over time. In this way, life-table data can be used to predict the future of populations.

Part of the value of a population projection based on life-table data—like the one begun in Figure 52.18—is that it allows biologists to alter values for survivorship and fecundity at particular ages and assess the consequences. For example, suppose that a predatory snake began preying on juvenile lizards. According to the model in Figure 52.18, what would be the impact of a change in juvenile mortality rate? Analyses like this allow biologists to determine which aspects of survivorship and fecundity are especially sensitive for particular species. The studies done to date support some general conclusions:

- Whooping cranes, sea turtles, spotted owls, and many other endangered species have high juvenile mortality, low adult mortality, and low fecundity. In these species, the fate of a population is extremely sensitive to increases in adult mortality. Based on this insight, conservationists have

(a) Life table

Age (x)	Survivorship (l_x)	Fecundity (m_x)
0 (birth)	----	0.0
1	0.33	3.0
2	0.2	4.0
3	0.04	5.0

(b) Fate of first-generation females

Year	0 (newborns)	1-year-olds	2-year-olds	3-year-olds	Total population size (N)
1st	1000 (just introduced)				1000
2nd	990 (= 330 × 3.0)	330 (= 1000 × 0.33)			1320 (= 990 + 330)
3rd	800 (= 200 × 4.0)		200 (= 1000 × 0.20)		
4th	200 (= 40 × 5.0)			40 (= 1000 × 0.04)	
5th					

(c) Fate of first- and second-generation females

Year	0 (newborns)	1-year-olds	2-year-olds	3-year-olds	Total population size (sum across all rows)
1st	1000				1000
2nd	990	330			1320
3rd	800 + 981 (981 = 327 × 3.0)	327 (= 990 × 0.33)	200		2308 (= 800 + 981 + 327 + 200)
4th	200 + 792 (792 = 198 × 4.0)		198 (= 990 × 0.20)	40	
5th	195 (195 = 39 × 5.0)			39 (= 990 × 0.04)	

FIGURE 52.18 Life-Table Data Can Be Used to Project the Future of a Population. (a) Life table providing age-specific survivorship and fecundity for a hypothetical population of lizards. **(b)** Predicted fate of 1000 one-year-old females introduced into a habitat just before the breeding season. The number of individuals in this cohort is shown in red; the number of offspring they produce each year is indicated in purple. **(c)** Extension of the data in part (b), indicating how many of the offspring produced by the original females in their first year survived in subsequent years, shown in purple, and how many offspring they produced in each subsequent year, shown in green.

EXERCISE Assume that all 4-year-old females die after producing three young. Fill in the 4th and 5th years in the table.

recently begun an intensive campaign to reduce the loss of adult female sea turtles in fishing nets. Previously, most conservation action had focused on protecting eggs and nesting sites.

- In humans and other species with high survivorship in most age classes, rates of population growth are extremely sensitive to changes in age-specific fecundity. Because of this, programs to control human population growth focus on two issues: lowering fertility rates through the use of birth control, and delaying the age of first reproduction by improving women's access to education.

In some or even most cases, however, the population projections made from life-table data may be too simplistic to be useful. For example, conservationists may need to expand the basic demographic models to account for occasional disturbances such as fires or storms or disease outbreaks. And what about species where overall population dynamics are dictated by a metapopulation structure?

Preserving Metapopulations

Decades of work have convinced biologists that the conclusions about metapopulation dynamics introduced in the previous section are correct, and that an increasing number of species are being forced into a metapopulation structure. Habitat destruction caused by suburbanization and other human activities leaves small populations isolated in pockets of intact habitat.

Work on Glanville fritillaries and other species has shown that a small, isolated population—even one within a nature preserve—is unlikely to survive over the long term. In Glanville fritillaries, data collected by Hanski's group have identified the attributes of subpopulations that are most likely to persist: They are large, occupy larger geographical areas, and are closer to neighboring populations (and hence more likely to be colonized). The data also indicate that fritillary populations with low genetic diversity—probably due to inbreeding—are more likely to go extinct than are populations of the same size that have higher genetic diversity.

Results like these have important messages for conservation biologists:

- Areas that are being protected for threatened species should be substantial enough in area to maintain large populations that are unlikely to go extinct in the near future.
- When it is not possible to preserve large tracts of land, an alternative is to establish systems of smaller tracts that are connected by corridors of habitat, so that migration between patches is possible.
- If the species that is threatened exists as a metapopulation, it is crucial to preserve at least some patches of unoccupied habitat to provide future homes for immigrants. If a population is lost from a preserve, the habitat should continue to be protected so it can be colonized in the future.

These results also have a more general message: Although traditional population growth models such as the expressions for exponential and logistic growth are simple and elegant, the factors they ignore—immigration and emigration—are crucial to understanding the dynamics of most populations. Because metapopulation structure is common, biologists must use more sophisticated models to predict the fates of populations.

Population Viability Analysis

A **population viability analysis**, or **PVA**, is a model that estimates the likelihood that a population will avoid extinction for a given time period. In most cases, a PVA attempts to combine basic demographic models for a species with data on geographic structure and the rate and severity of habitat disturbance. Typically, a population is considered viable if the analysis predicts that it has a 95 percent probability of surviving for at least 100 years. Natural resource managers are currently using PVA to assess the effects of logging, suburbanization, and other land management practices on sensitive species and to evaluate the merits of alternative recovery plans for endangered species.

A recent study—conducted on an endangered marsupial called Leadbeater's possum—illustrates the value of a carefully constructed PVA. Leadbeater's possum (**Figure 52.19a**) inhabits old-growth forests in southeastern Australia and relies on dead trees for nest sites. The goal of the PVA was to assess the effects that logging these habitats might have on the viability of the species. The question was, if logging reduced the ability of possums to migrate to new habitats, what would the impact on the population be?

(a) Leadbeater's possum

(b) Population viability analysis (PVA)

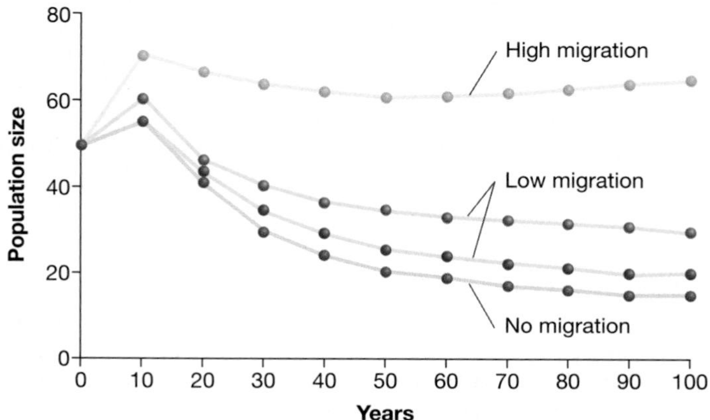

FIGURE 52.19 A PVA Can Predict the Consequences of Habitat Loss and Fragmentation. (a) Leadbeater's possum, a marsupial that breeds in old-growth forests of southeast Australia. **(b)** Results of a PVA: the fate of possum populations for four different rates of migration among habitat patches. The rates range from a case in which patches are isolated and no migration occurs to a case in which 5 percent of the individuals migrate each year. The no-migration scenario simulates what might happen if extensive deforestation occurred and created isolated patches of old-growth forest.

Life-table data are difficult to obtain for Leadbeater's possum, because this species is rare, lives in trees, and is active at night. Enough data were available from field studies, however, to allow for an estimate of age-specific survival and average fecundity per female per year. Biologists used these data to make population projections while varying the spatial configuration of habitat patches. The analysis allowed the researchers to assess the impact of migration and to simulate the effects of logging, fires, storms, and other disturbances.

Figure 52.19b illustrates the researchers' results. Each data point represents a population in a particular year. The four curves describe the changes in population size predicted to occur over time, based on different assumptions about the migration rates between groups of possums in patches of old-growth forest. When migration is high, the overall population size is predicted to stabilize at approximately 65 individuals. When no migration among fragments occurs, the final population size is predicted to be fewer than 20 individuals. Based on these results, the biologists concluded that extensive timber harvesting would pose a serious threat to the species. The PVA showed that reducing the size and number of remaining old-growth fragments would reduce the possibility for migration and lead to rapid population declines.

Like all analyses based on simulations, though, a PVA makes many assumptions about future events and is only as accurate as the data entered into it. In particular, the usefulness of PVAs has been challenged because data on age-specific fecundity and survivorship and other basic demographic information are lacking or poorly documented in many endangered species.

A group of biologists recently defended the approach, however, by analyzing 21 long-term studies that have been completed on threatened populations of animals. To do their analysis, the researchers split each of the 21 data sets in two. Data from the first half of each study was used to run a PVA on each of the 21 species. After comparing the predictions of the PVAs to the actual data from the second half of each study, the researchers found that the correspondence was extremely close. This result strengthens confidence in PVA as a good predictive method for land managers.

Chapter Review

SUMMARY OF KEY CONCEPTS

A population is a group of individuals of the same species that occupy the same area at the same time. Population ecology is the study of how and why the number of individuals in a population changes over time. Population size changes in response to changes in birth, death, immigration, and emigration rates. Biologists use a variety of mathematical and analytical tools to study population ecology.

Life tables summarize how likely it is that individuals of each age class in a population will survive and reproduce.

Life tables are a basic tool in demography—the study of patterns in births and deaths in populations. When life tables of different species or populations are compared, it is clear that individuals have distinct ways of allocating energy and resources to activities that promote survival versus activities that promote reproduction. Because the resources available to an individual are always limited, any increase in allocation of resources to survival and competitive ability necessitates a decrease in the resources allocated to reproduction. This trade-off between survival and reproduction is the most fundamental aspect of a species' life history.

You should be able to explain why a life table is relevant only to a particular population at a particular time, and predict how the first life tables—constructed in ancient Rome—differ from the current life table for the same population.

The growth rate of a population can be calculated from life-table data or from directly observing changes in population size over time.

One of the most basic characteristics of a population is its growth rate. Exponential growth occurs when the per-capita growth rate, r, does not change over time. Eventually, however, growing populations approach the carrying capacity of their environment.

Laboratory and field studies have confirmed the existence of exponential and logistic growth in species ranging from whooping cranes to *Paramecium*. Field research has documented that as population density increases, survivorship and fecundity may decrease—leading to increased death rates and lowered birth rates and thus a reduction in growth rate. Density dependence in population growth is due to competition for resources, disease, predation, or other factors that increase in intensity when population size is high.

You should be able to draw a logistic growth curve for a hypothetical human population, describe how r changes along its length, and suggest factors responsible for each change in r.

(MB) **Web Animation** at www.masteringbio.com
Modeling Population Growth

A wide variety of patterns are observed when researchers track changes in population size over time, ranging from no growth, to regular cycles, to continued growth independent of population size.

Experiments have shown that density-dependent factors drive the regular population cycles observed in certain species. In the case of snowshoe hare and lynx populations, both predator-prey interactions and food availability affect these species' synchronized population cycles.

Changes in population size through time may occur due to changes in the age structure of populations—specifically the number of individuals in various age classes. A population with few juveniles and many adults past reproductive age, like the human populations of developed nations, may be declining or stable in size. In contrast, a population with a large proportion of juveniles is likely to increase rapidly in size. Human populations in the developing nations currently have this type of age distribution.

The total human population has been increasing rapidly since about 1750 and is currently over 6.6 billion. In countries where survivorship has increased due to medical advances, fecundity has decreased. Based on various scenarios for average female fertility, the total human population in 2050 is expected to be between 7 and 11 billion.

You should be able to explain why human populations are projected to continue rapid growth, even though fertility rates are declining.

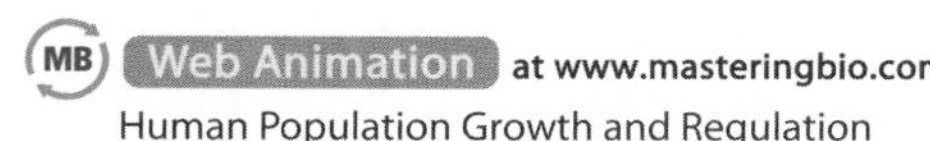

Human Population Growth and Regulation

Data from population ecology studies help biologists evaluate prospects for endangered species and design effective management strategies.

Human activities are isolating populations into metapopulations occupying small, fragmented habitats. The dynamics of a metapopulation are driven by the birth and death of populations, just as the dynamics of a population are driven by the birth and death of individuals. Migration among habitat patches is essential for the stability of a metapopulation, so conservationists are trying to preserve unoccupied patches of habitat and establish corridors that link habitat fragments.

Demographic data are the basis of population projections that are fundamental to population viability analysis, or PVA. Most PVAs attempt to model the effects that different management strategies might have on populations of endangered species. A PVA estimates the probability that a population will persist for a certain number of years under a prescribed set of demographic and habitat conditions.

You should be able to explain why a small, isolated population is virtually doomed to extinction, but why population size may be stable in a species consisting of a large number of small, isolated populations.

QUESTIONS

Test Your Knowledge

1. What is the defining feature of exponential growth?
   a. The population is growing very quickly.
   b. The growth rate is constant.
   c. The growth rate increases rapidly over time.
   d. The growth rate is very high.
2. What four factors define population growth?
   a. age-specific birth rates, age-specific death rates, age structure, and metapopulation structure
   b. survivorship, age-specific mortality, fecundity, death rate
   c. mark-recapture, census, quadrat sampling, transects
   d. births, deaths, immigration, emigration
3. In what populations does exponential growth tend to occur?
   a. in populations that colonize new habitats
   b. in populations that experience intense competition
   c. in populations that experience high rates of predation
   d. in metapopulations
4. Which of the following is *not* a reason that population growth declines as population size approaches the carrying capacity?
   a. Climate becomes unfavorable.
   b. Competition for resources increases.
   c. Predation rates increase.
   d. Disease rates increase.
5. If most individuals in a population are young, why is the population likely to grow rapidly in the future?
   a. Death rates will be low.
   b. The population has a skewed age distribution.
   c. Immigration and emigration can be ignored.
   d. Many individuals will begin to reproduce soon.
6. Why have population biologists become particularly interested in the dynamics of metapopulations?
   a. because humans exist as a metapopulation
   b. because whooping cranes exist as a metapopulation
   c. because many populations are becoming restricted to small islands of habitat
   d. because metapopulations explain why populations occupying large, contiguous areas are vulnerable to extinction

Test Your Knowledge answers: 1. b; 2. d; 3. a; 4. a; 5. d; 6. c

Test Your Understanding

Answers are available at www.masteringbio.com

1. Explain Equations 52.3 and 52.5 in words.
2. Draw type I, II, and III survivorship curves on a graph with labeled axes. Explain why the growth rate of species with a type I survivorship curve depends primarily on fertility rates. Explain why the growth rate of species with a type III survivorship curve is extremely sensitive to changes in adult survivorship.
3. Offer a hypothesis to explain why humans have undergone near-exponential growth for over 500 years. Why can't exponential growth continue indefinitely? Describe two documented examples of density-dependent factors that influence population growth in natural populations.
4. Compare and contrast the dynamics of a population that resides in a large contiguous habitat to the dynamics of a metapopulation. Be sure to consider density-dependent and density-independent factors on birth rates and death rates, and the importance of immigration and emigration. Assume that the total amount of area occupied is the same in both populations.
5. Make a rough sketch of the age distribution in developing versus developed countries, and explain the significance of the differences. How is AIDS, which is a sexually transmitted disease, affecting the age distribution in countries hard hit by the epidemic?
6. Compare the pros and cons of using R_0, λ, and r to express growth rates. What is the difference between r (the per-capita rate of increase) and r_{max} (the maximum or intrinsic growth rate)?

Applying Concepts to New Situations

Answers are available at www.masteringbio.com

1. When wild plant and animal populations are logged, fished, or hunted, only the oldest or largest individuals tend to be taken. What impact does harvesting have on a population's age structure? How might harvesting affect the population's life-table and growth rate?
2. Design a system of nature preserves for an endangered species of beetle whose larvae feed on only one species of sunflower. The sunflowers tend to be found in small patches that are scattered throughout dry grassland habitats. Explain the rationale behind your proposal.
3. The population of the United States is projected to increase dramatically over the next 50 years, even though fertility rates are only slightly above replacement level and the age distribution is largely stable. How is this possible?
4. In most species the sex ratio is at or near 1.00, meaning that there is an approximately equal number of males and females. In China, however, there is a strong preference for male children. According to the 2000 census there, the sex ratio of newborns is almost 1.17, meaning that close to 117 boys are born for every 100 girls. Based on these data, researchers project that there will soon be about 50 million more men than women of marriageable age in China. Discuss how this skewed sex ratio might affect the population growth rate in China.

www.masteringbio.com is also your resource for • Answers to text, table, and figure caption questions and exercises • Answers to *Check Your Understanding* boxes • Online study guides and quizzes • Additional study tools including the *E-Book for Biological Science* 3rd ed., textbook art, animations, and videos.

53 Community Ecology

KEY CONCEPTS

- Interactions among species, such as competition and consumption, have two main outcomes: (1) They affect the distribution and abundance of the interacting species, and (2) they are agents of natural selection and thus affect the evolution of the interacting species. The nature of interactions between species frequently changes over time.
- The assemblage of species found in a biological community changes over time and is primarily a function of climate and chance historical events.
- Species diversity is high in the tropics and lower toward the poles. The mechanism responsible for this pattern is still being investigated.

This chapter explores how species that live in coral reefs and other communities interact with each other, and how communities change over time.

Chapter 52 focused on the dynamics of populations—how and why they grow or decline and how they change over time and through space. That chapter considered each population as an isolated entity. But in reality, populations of different species form complex assemblages called communities. A biological **community** consists of interacting species, usually living within a defined area.

The concept of a biological community was introduced in Chapter 50. Each of the biomes analyzed in that chapter represents a broad grouping of plant communities. The goal of that chapter was to describe the general characteristics of selected biomes and analyze broad correlations between a region's climate and the type of biome present.

The task in this chapter is to analyze the dynamics of biological communities—how they develop and change over time. Biologists want to know how communities work. How do species inside communities interact with each other, and what are the long-term consequences? What happens to communities when they are disturbed by fires or flood, and why is the number of species higher in some areas than others? Let's delve in.

53.1 Species Interactions

The species in a community interact almost constantly. Members of different species eat one another, pollinate each other, exchange nutrients, compete for resources, and provide habitats for each other. As a result, the fate of a particular population may be tightly linked to the other species that share its habitat.

To study species interactions, biologists focus on analyzing the effects on the fitness of the individuals involved. Recall from Chapter 24 that **fitness** is defined as the ability to survive and produce offspring. Does the relationship between two species provide a fitness benefit to members of one species (a + interaction) but hurt members of the other species (a − interaction)? Or does the association have no effect on the fitness of a participant (a "0" interaction)? Three broad categories of interaction are analyzed in this chapter: the −/− relationship known as competition, the +/− interactions called consumption and parasitism, and the +/+ association termed mutualism. A fourth category of interaction, **commensalism**, is defined as a +/0 association. An example is the birds that follow moving army ants in the tropics. As the ants march along the forest floor, they hunt insects and small vertebrates. As they do, birds follow and pick off prey species that fly or jump up out of the way of the ants (**Figure 53.1**). The birds are commensals that benefit from the association (+) but have no measurable impact on the ants (0).

In addition to introducing the array of tools that biologists use to study competition, consumption, and mutualism, this section focuses on three key themes:

1. Species interactions may affect the distribution and abundance of a particular species. Recall from Chapter 52 that most of the density-dependent factors that produce logistic growth are based on species interactions—predation, disease, or competition for space and resources. In addition, changes in species interactions often explain short-term changes in population size and distribution.
2. Species act as agents of natural selection when they interact. Deer are fast and agile in response to natural selection exerted by their major predators, wolves and cougars. The speed and agility of deer, in turn, promote natural selection that favors wolves and cougars that are fast and that have superior eyesight and senses of smell. To capture this point, biologists say that species interactions resemble an arms race. In humans, an arms race is said to occur when one nation develops a new weapon, which prompts a rival country to develop a defensive weapon, which pushes the original country to manufacture an even more powerful weapon, and so on. In biology, a coevolutionary arms race occurs between predators and prey, parasites and hosts, and other types of interacting species. "Coevolutionary" indicates that species influence each others' evolution, leading to reciprocal adaptation. In this way, changes in species interactions lead to long-term changes in the characteristics of populations, a phenomenon called **coevolution**, in addition to having short-term impacts on population size.
3. The outcome of interactions among species is dynamic and conditional. Consider the relationship between army ants and birds that follow them, which is usually commensal. If bird attacks start to force other insects into the path of the ants, then both ants and birds benefit and the relationship becomes mutualistic. But if birds begin to steal prey that would otherwise be taken by ants, then the relationship becomes parasitic. The outcome of the interaction may depend on the number and types of prey, birds, and ants present and may change over time.

FIGURE 53.1 Commensals Gain a Fitness Advantage but Don't Affect the Species They Depend On. Birds that associate with army ants are commensals. They have no measurable fitness effect on the ants but gain from the association by capturing insects that try to fly or climb out of the way of the ants.

Competition

Competition is a −/− interaction that occurs when different individuals use the same resources and when those resources are limiting—meaning that lack of access to those resources prevents individuals from surviving better and having more offspring.

Ever since ecological studies began, researchers have focused on competition as an important interaction within and between species. The attention is justified in part by the central place that competition holds in the theory of evolution by natural selection. Darwin pointed out that individuals within a population compete for the resources that are required for growth and reproduction. Further, some individuals are more successful in this competition and leave more offspring than others do. If the traits that lead to success are heritable, then the frequency of alleles in the population changes and evolution by natural selection occurs.

Competition that occurs between members of the same species is called **intraspecific** (literally, "within species") **competition**. Intraspecific competition for space, sunlight, food, and other resources intensifies as a population's density increases. As a result, intraspecific competition is a major cause of density-dependent growth (see Chapter 52).

(a) Consumptive competition

These trees are competing for water and nutrients.

(b) Preemptive competition

Space preempted by these barnacles is unavailable to competitors.

(c) Overgrowth competition

The large fern has overgrown other individuals and is shading them.

(d) Chemical competition

Few plants are growing under these *Salvia* shrubs.

(e) Territorial competition

Grizzly bears drive off black bears.

(f) Encounter competition

Spotted hyenas and vultures fight over a kill.

FIGURE 53.2 Many Types of Competition Exist.

Interspecific ("between species") **competition** occurs when individuals from different species use the same limiting resources. As **Figure 53.2** shows, the mechanisms involved vary widely.

1. *Consumptive competition* occurs when individuals consume the same resources (Figure 53.2a).
2. *Preemptive competition* exists when one species makes space unavailable to other species (Figure 53.2b).
3. *Overgrowth competition* happens when one species grows above another (Figure 53.2c).
4. *Chemical competition* takes place when one species produces toxins that negatively affect another species (Figure 53.2d).
5. *Territorial competition* arises when a mobile species protects its feeding or breeding territory against other species (Figure 53.2e).
6. *Encounter competition* occurs when two species interfere directly for access to specific resources (Figure 53.2f).

Using the Niche Concept to Analyze Competition Early work on competition focused on the concept of the niche. A **niche** can be thought of as the range of resources that the species is able to use or the range of conditions it can tolerate. G. Evelyn Hutchinson proposed that a species' niche could be envisioned by plotting these habitat requirements along a series of axes. **Figure 53.3a**, for example, represents one niche axis for a hypothetical species. In this case, the habitat requirement plotted is the size of seeds eaten by members of this population, which might be a function of mouth or tooth size. Other niche axes could represent other types of foods used or the temperatures, humidity, and other environmental conditions tolerated by the species.

Interspecific competition occurs when the niches of two species overlap. The two species plotted in **Figure 53.3b**, for example, compete for seeds of intermediate size. When competition occurs, each individual will get fewer seeds on average and fitness will decline. This is why competition is considered a −/− interaction.

What Happens When One Species Is a Better Competitor? G. F. Gause claimed it is not possible for species with the same niche to coexist. This hypothesis is called the **competitive exclusion principle**, and it was inspired by a series of experiments Gause did with similar species of the unicellular pond-dweller *Paramecium*. When Gause placed small populations of *P. caudatum* and *P. aurelia* in separate laboratory cultures, both species exhibited logistic growth (see Chapter 52). But Gause showed that when the two species are put in the same

(a) One species eats seeds of a certain size range.

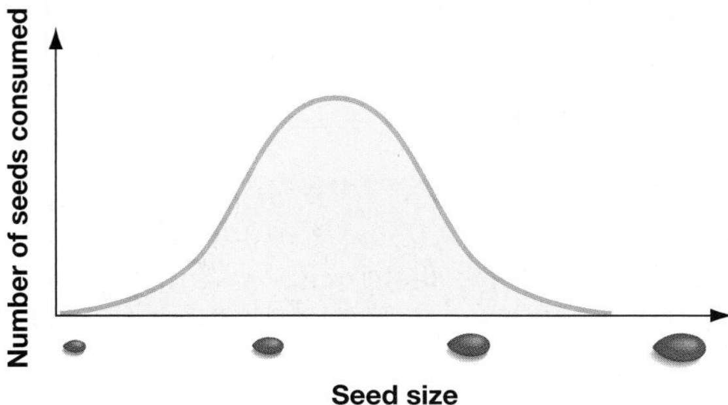

(b) Partial niche overlap: competition for seeds of intermediate size

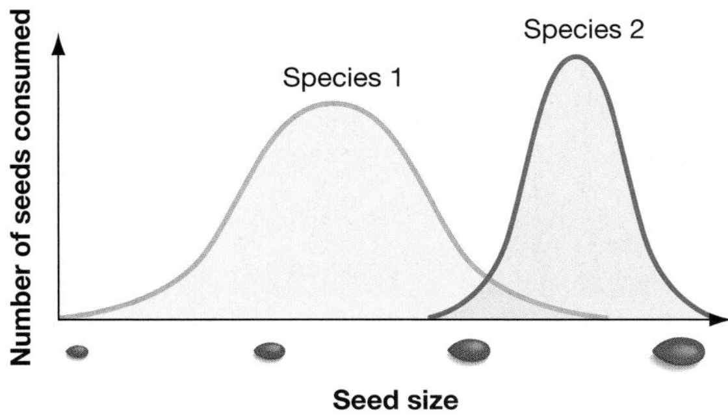

FIGURE 53.3 Niche Overlap Leads to Competition. (a) Graph describing one aspect of a species' fundamental niche, meaning the range of resources that it can use or range of conditions it can tolerate. **(b)** Competition occurs when the niches of different species overlap. In this case, both species use seeds of intermediate size.

EXERCISE Mark the horizontal axis to indicate the range of seed sizes where competition occurs.

(a) Competitive exclusion in two species of *Paramecium*

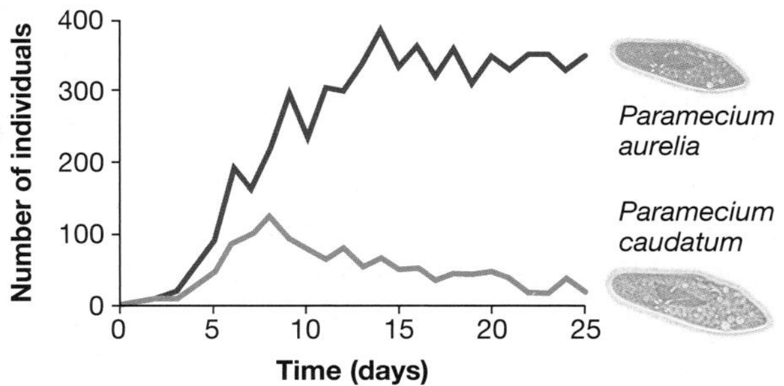

(b) Competitive exclusion occurs when competition is asymmetric ...

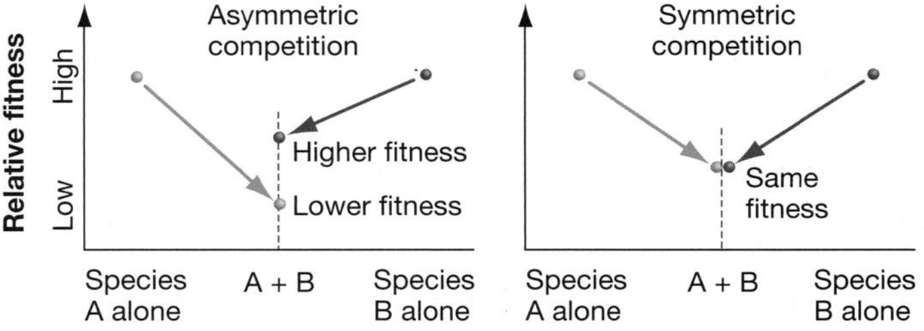

(c) ... and niches overlap completely.

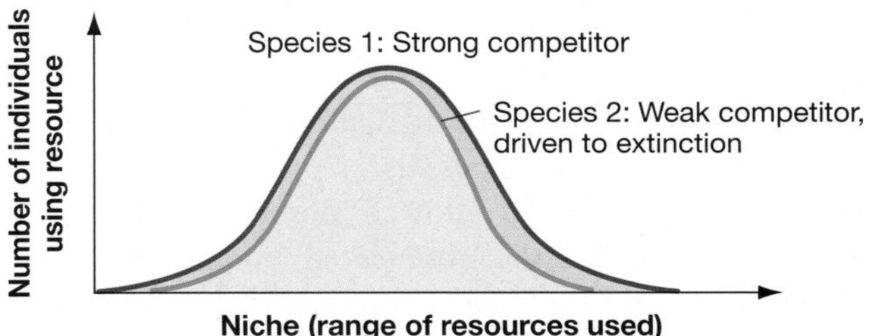

(d) When competition is asymmetric and niches do not overlap completely, weaker competitors use nonoverlapping resources.

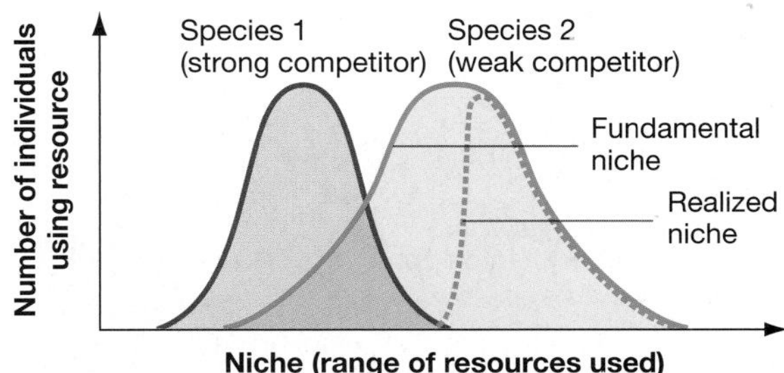

FIGURE 53.4 Niche Overlap Leads to Competitive Exclusion or Restricted Habitat Use. (a–c) If two species have completely overlapping niches, there is no refuge for the weaker competitor. It may be driven to extinction. **(d)** A realized niche is a subset of a species' fundamental niche.

culture together, only the *P. aurelia* population exhibits a logistic growth pattern. *P. caudatum*, in contrast, is driven to extinction (**Figure 53.4a**).

Gause's result is a product of asymmetric competition. When **asymmetric competition** occurs, one species suffers a much greater fitness decline than the other species does (**Figure 53.4b**). Under **symmetric competition**, however, each of the interacting species experiences a roughly equal decrease in fitness.

If asymmetric competition occurs and the two species have completely overlapping niches, as diagrammed in **Figure 53.4c**, then the stronger competitor is likely to drive the weaker competitor to extinction. But if the niches do not overlap completely, then the species that is the weaker competitor should be able to retreat to an area of non-overlap. In cases like this, an important distinction arises between a species' **fundamental niche**, which is the combination of resources or areas used or conditions tolerated in the absence of competitors, and its **realized niche**, which is the portion of resources or areas used or conditions tolerated when competition occurs (**Figure 53.4d**).

In Gause's laboratory cultures, competition between *P. aurelia* and *P. caudatum* was asymmetric and competitive exclusion occurred. How can researchers study competition in the field, under natural conditions?

Experimental Studies of Competition Joseph Connell began a classic study of competition in the late 1950s, after observing an interesting pattern in an intertidal rocky shore in Scotland. He noticed that there were two species of barnacles with distinctive distributions. Barnacle larvae are mobile, but adults live attached to rocks. The adults of one species, *Chthamalus stellatus*, occurred in an upper intertidal zone, while the adults of the other species, *Semibalanus balanoides* (formerly named *Balanus balanoides*), were restricted to a lower intertidal zone (**Figure 53.5**). The upper intertidal zone is a more severe environment for barnacles, because it is exposed to the air for longer periods at low tide each day. The young of both species were found together in the lower intertidal zone, however.

To explain these observations, Connell hypothesized that adult *Chthamalus* were competitively excluded from the lower intertidal zone. The alternative hypothesis is that adult *Chthamalus* are absent from the lower intertidal zone because they do not thrive in the physical conditions there.

Connell tested these hypotheses by performing the experiment shown in **Figure 53.6**. He began by removing a number of rocks that had been colonized by *Chthamalus* from the upper intertidal zone and transplanting them into the lower intertidal zone. He screwed the rocks into place and allowed *Semibalanus* larvae to colonize them. Once the spring colonization period was over, Connell divided each rock into two groups. In one half, he removed all *Semibalanus* that were in contact with or next to a *Chthamalus*.

This experimental design allowed Connell to document *Chthamalus* survival in the absence of competition with *Semibalanus* and compare it with survival during competition. This is a common experimental strategy in competition studies: One of the competitors is removed, and the response by the remaining species is observed.

Connell's results support the hypothesis of competitive exclusion. In the unmodified areas, *Semibalanus* killed many of the young *Chthamalus* by growing against them and lifting them off the substrate. As the graph in Figure 53.6 shows, *Chthamalus* survival was much higher when all of the *Semibalanus* were removed.

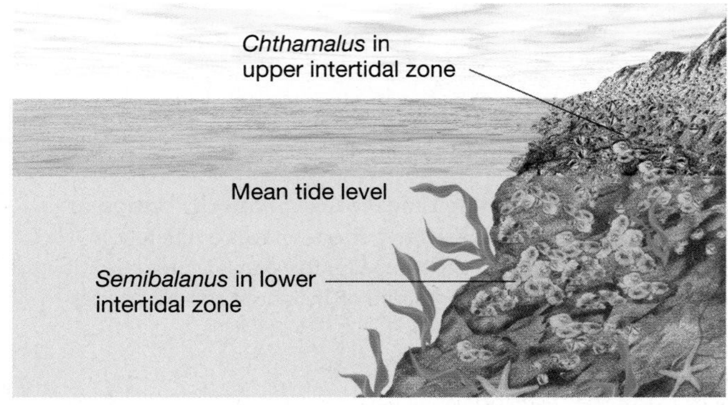

FIGURE 53.5 Barnacle Species Are Distributed in Distinct Zones. In natural habitats, adult *Chthamalus* and *Semibalanus* barnacles do not coexist.

Experiment

Question: Why is the distribution of adult *Chthamalus* restricted to the upper intertidal zone?

Hypothesis: Adult *Chthamalus* are competitively excluded from the lower intertidal zone.

Alternative hypothesis: Adult *Chthamalus* do not thrive in the physical conditions of the lower intertidal zone.

Experimental setup:

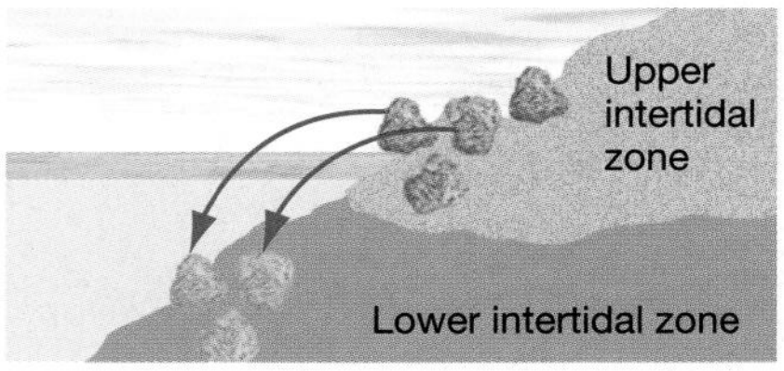

Upper intertidal zone
Lower intertidal zone

1. Transplant rocks containing young *Chthamalus* to lower intertidal zone.

2. Let *Semibalanus* colonize the rocks.

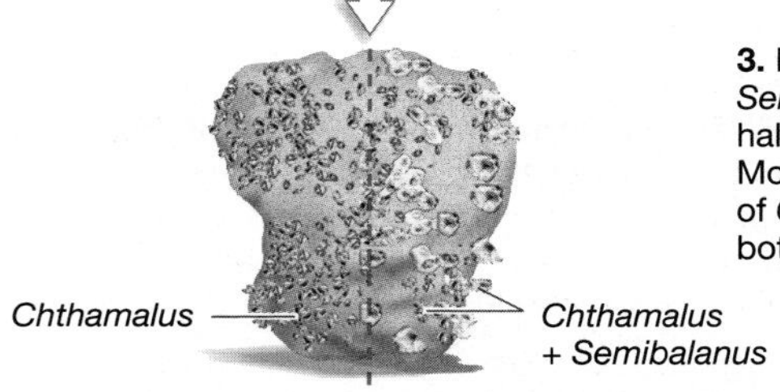

3. Remove *Semibalanus* from half of each rock. Monitor survival of *Chthamalus* on both sides.

Chthamalus
Chthamalus + *Semibalanus*

Prediction: *Chthamalus* will survive better in the absence of *Semibalanus*.

Prediction of alternative hypothesis: *Chthamalus* survival will be low and the same in the presence or absence of *Semibalanus*.

Results:

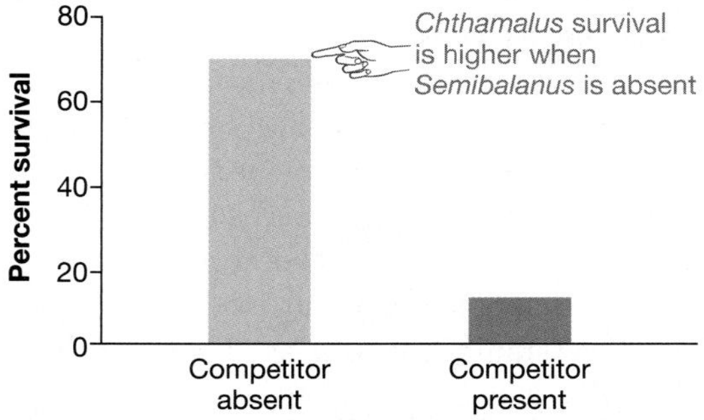

Percent survival
80
60
40
20
0
Competitor absent
Competitor present
Chthamalus survival is higher when *Semibalanus* is absent

Conclusion: *Semibalanus* is competitively excluding *Chthamalus* from the lower intertidal zone.

FIGURE 53.6 Experimental Evidence for Competitive Exclusion.

● QUESTION Why was it important to carry out both treatments on the same rock? Why not use separate rocks?

● QUESTION Connell also did the reciprocal removal experiment—removing *Chthamalus* from experimental rocks. Predict the outcome of this experiment.

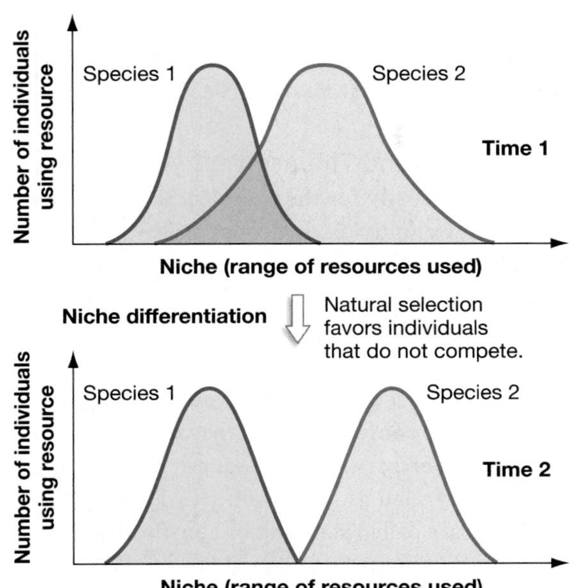

FIGURE 53.7 Over Time, Competition Can Lead to Niche Differentiation.

● **QUESTION** Biologists call the outcome of this process resource partitioning. Explain why.

Mechanisms of Coexistence: Fitness Trade-offs and Niche Differentiation Why haven't *Semibalanus* and other superior competitors taken over the world? Biologists answer this question by invoking the concept of fitness trade-offs (see Chapter 24). The key here is that the ability to compete for a particular resource—like space on rocks or edible seeds of a certain size—is only one aspect of an organism's niche. If individuals are extremely good at competing for a particular resource, then they are probably less good at enduring drought conditions, warding off disease, or preventing predation.

In the case of *Semibalanus* and *Chthalamus* growing in the intertidal, the fitness trade-off is rapid growth and success in competing for space versus the ability to endure the harsh physical conditions of the upper intertidal. *Semibalanus* are fast-growing and large; *Chthalamus* grow slowly but can survive long exposures to the air and to intense sun and heat. Neither species can do both things well. Fitness trade-offs limit the ability of superior competitors to spread.

It's also important to realize that because competition is a −/− interaction, there is strong natural selection on both species to avoid it. **Figure 53.7** shows the predicted outcome: An evolutionary change in traits reduces the amount of niche overlap, and thus the amount of competition. This change in resource use is called **niche differentiation** or resource partitioning; the change in species' traits is called **character displacement.** The fundamental idea is that competition exerts natural selection, and that the characteristics of species change in a way that reduces competition.

Peter and Rosemary Grant recently documented character displacement occurring in Galápagos finches. You might recall, from Chapter 24, that the Grants observed dramatic increases in average beak size and body size of a *Geospiza fortis* population during a drought in 1977. The changes occurred because the major food source available during the drought was fruits from a plant called *Tribulus cistoides*, and because only the largest-beaked individuals were able to crack these fruits and eat the seeds inside. In 1982, however, individuals from a species called *Geospiza magnirostris* arrived on the island and began breeding. *G. magnirostris* are about twice the size of *G. fortis* and use *T. cistoides* fruits as their primary food source (**Figure 53.8**).

A severe drought recurred in 2003 and 2004; most finches died of starvation. When the Grants measured the surviving *G. fortis* they made a remarkable observation: In contrast to the 1977 drought, only the *smallest*-beaked individuals survived. Data on feeding behavior indicated that *G. magnirostris* were outcompeting *G. fortis* for *T. cistoides* fruits. Only *G. fortis* that could eat extremely small seeds efficiently could survive.

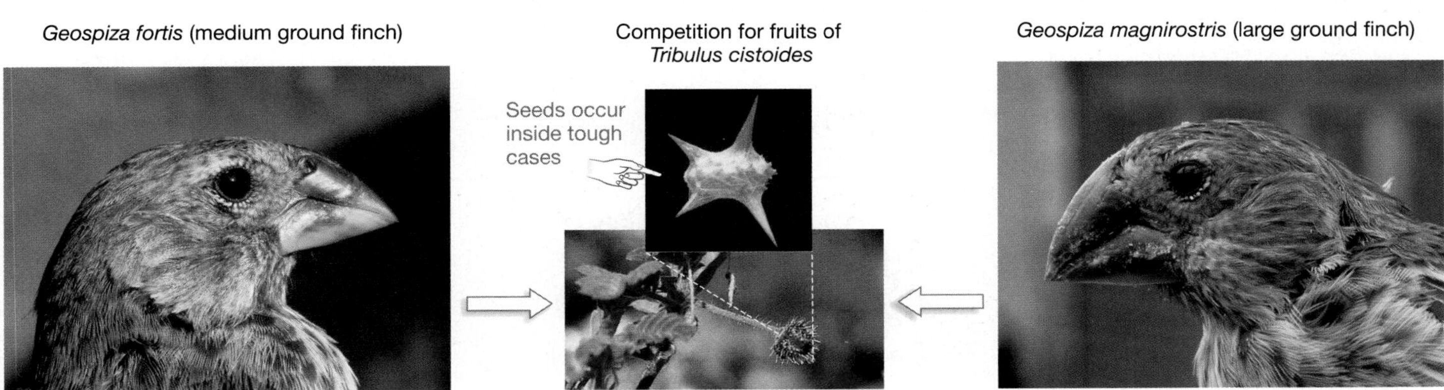

FIGURE 53.8 A Natural Experiment in Niche Differentiation. During drought years, *Geospiza fortis* have to compete for *Tribulus cistoides* fruits with *Geospiza magnirostris* if they are present.

Under similar environmental conditions, then, the presence of competition altered the course of evolution. This is a dramatic example of character displacement. Large ground finches and medium ground finches are partitioning a resource to avoid competition.

Consumption

Consumption occurs when one organism eats another. There are three major types of consumption:

1. **Herbivory** takes place when **herbivores** ("plant-eaters") consume plant tissues. Bark beetles mine cambium; cicadas suck sap; caterpillars chew leaves.
2. **Parasitism** occurs when a **parasite** consumes relatively small amounts of tissue or nutrients from another individual, called the **host**. Parasitism often occurs over a long period of time. It is not necessarily fatal, and parasites are usually small relative to their host. An array of worms and unicellular protists parasitizes humans; ticks parasitize cattle and other large mammals. Not all parasitism involves consumption, however—social parasites in birds and insects lay their eggs in other species' nests and induce them to raise the young.
3. **Predation** occurs when a predator kills and consumes all or most of another individual. The consumed individual is called the prey. Woodpeckers eat bark beetles; finches eat seeds; ladybird beetles devour aphids; wasps kill caterpillars.

To illustrate how biologists analyze the impact of consumption, let's consider a series of questions about predators, herbivores, parasites, and their victims.

How Do Prey Defend Themselves? With respect to fitness, predation is costly for the prey species and beneficial for the predator. Prey individuals do not passively give up their lives to increase the fitness of their predators, however. Instead, prey may hide or run, fly, or swim away when they sense the presence of a predator. Analyses have shown that many species find safety in numbers—schooling and flocking behavior is an effective way to reduce the risk of predation, in part because predators become confused when they dive or swim into a mass of prey. Other prey species sequester or spray toxins or employ weaponry such as sharp spines or kicking hooves. The traits just listed are called **standing** or **constitutive defenses**, because they are always present (**Figure 53.9a**).

Some of the best-studied constitutive defenses involve the phenomenon called mimicry. **Mimicry** occurs when one species closely resembles another species. **Figure 53.9b** illustrates two major types of mimicry. The wasp on the far left is brightly colored and dangerous, because of its stinger. It looks like other dangerous insects—particularly other wasps and bees. When harmful prey species resemble each other, **Müllerian mimicry** is said to occur. To explain the existence of **Müllerian** mimics, biologists propose that the existence of similar-looking dangerous prey in the same habitat increases the likelihood that predators will learn to avoid them. In this way, Müllerian mimicry

(a) Constitutive defenses of animals vary.

Camouflage: blending into the background

Schooling: safety in numbers

Weaponry: fighting back

(b) Mimicry can protect both dangerous and harmless species.

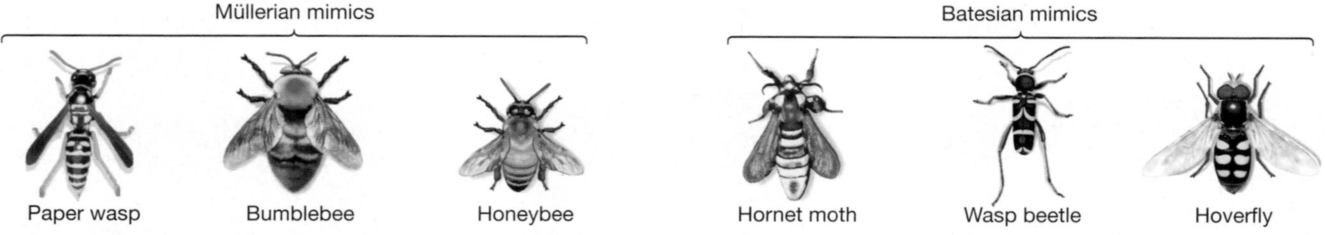

FIGURE 53.9 Constitutive Defenses Are Always Present. (a) Prey have an array of adaptations to reduce the likelihood of predation. **(b)** Müllerian mimicry occurs among dangerous prey species; Batesian mimicry occurs between a dangerous prey species and harmless prey species.

(a) Prey and predator

Blue mussels

Crabs

(b) Correlation between predation rate and prey defense

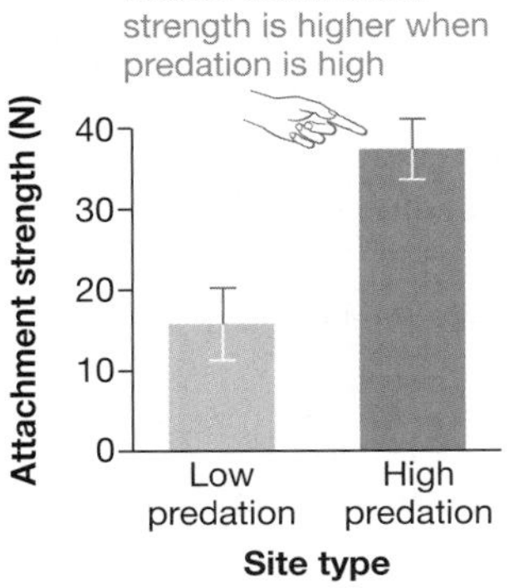

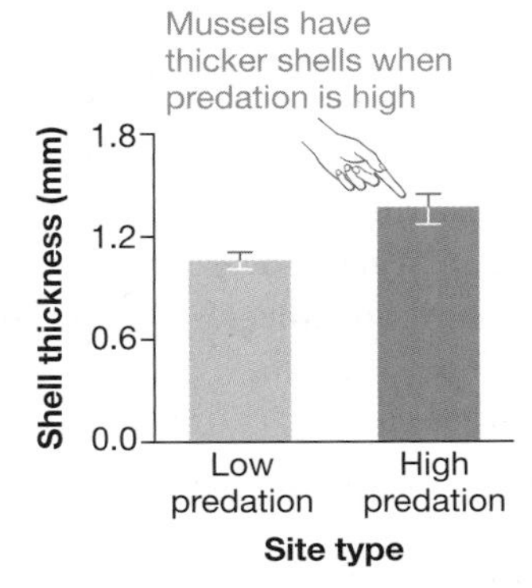

FIGURE 53.10 Inducible Defenses Are Produced Only when Prey Are Threatened. (a) In shallow-water environments in Maine, blue mussels (left) are preyed on by crabs (right). **(b)** To measure how strongly mussels are attached to the substrate, researchers drilled a hole in their shells and measured the force, in newtons (N), required to pull them off. They also measured how large mussel shells were relative to an individual's size by dividing shell weight by soft-tissue weight.

should reduce the likelihood of dangerous individuals being attacked. But wasps also act as a model for harmless species of moth, beetle, and fly that resemble wasps. To explain these mimics, biologists propose that predators avoid the harmless mimics because they mistake them for a dangerous wasp. This resemblance is known as **Batesian mimicry**.

The key point here is that prey have adaptations that reduce their likelihood of becoming victims. These adaptations are responses to natural selection exerted by predators. Constitutive defenses are expensive, however, in terms of the energy and resources that must be devoted to producing and maintaining them. Based on this observation, it should not be surprising to learn that many prey species have **inducible defenses**—meaning defensive traits that are produced only in response to the presence of a predator. Induced defenses then decline if predators leave the habitat. Inducible defenses are efficient energetically, but they are slow—it takes time to produce them.

To see how inducible defenses work, consider recent research on the blue mussels that live in an estuary along the coast of Maine (**Figure 53.10a**). Biologists had documented that predation on mussels by crabs was high in an area of the estuary with relatively slow tidal currents (a "low-flow" area) but low in an area of the estuary with relatively rapid tidal currents (a "high-flow" area). The researchers hypothesized that if blue mussels possess inducible defenses, then heavily defended prey individuals should occur in the low-flow area, where predation pressure is higher, but not in the high-flow area, where water movement reduces the number of crabs present.

To evaluate this hypothesis, the biologists measured mussel shell characteristics in the two areas. They found that mussels in the high-predation area were more strongly attached to their base or substrate and had thicker shells than did mussels in the low-predation area (**Figure 53.10b**). These traits make the mussels more difficult to remove from the substrate and harder to crush and so function as effective antipredator defenses.

The data in Figure 53.10b are correlational in nature, however, so they are open to interpretation. A critic of the inducible defense explanation could offer a reasonable alternative hypothesis—for example, that only constitutive defenses exist and that crabs have eliminated weakly attached mussels with thin shells from the low-flow areas. The observed differences could also be due to differences in light, temperature, or other abiotic factors that might affect mussel traits but have nothing to do with predation.

To test the inducible-defense hypothesis more rigorously, biologists carried out the experiment diagrammed in **Figure 53.11**. The tank on the left allowed the researchers to measure shell growth in mussels that were "downstream" from crabs. As predicted by the induced-defense hypothesis, the mussels exposed to a crab in this way developed significantly tougher shells than did mussels that were not exposed to a crab. These results suggest that, even without direct contact, mussels can sense the presence of crabs and increase their investment in defenses. In a similar experiment, the investigators compared mussels that were exposed to water running through broken mussel shells versus intact but empty mussel shells. They recorded a significant increase in shell thickness in the tank downstream from the broken shells. This result supports the hypothesis that mussels can detect the presence of predators from molecules released by broken shells.

Results like these underscore several themes of this section: Species interactions are dynamic and result in coevolution—in this case, adaptations and counteradaptations in an evolutionary arms race between predators and prey.

Are Animal Predators Efficient Enough to Reduce Prey Populations? Research on animal defense systems supports the hypothesis that species interactions have a strong impact on the evolution of predator and prey populations. Can predators also affect the short-term distribution and abundance of prey populations? Prey are typically smaller than predators, have larger litter or clutch sizes, and tend to begin reproduction at a younger age. As a result, they have a much larger intrinsic growth rate, r_{max}—that is, the maximum growth rate that a population can achieve under ideal

Experiment

Question: Are mussel defenses induced by the presence of crabs?

Hypothesis: Mussels increase investment in defense in the presence of crabs.

Null hypothesis: Mussels do not increase investment in defense in the presence of crabs.

Experimental setup:

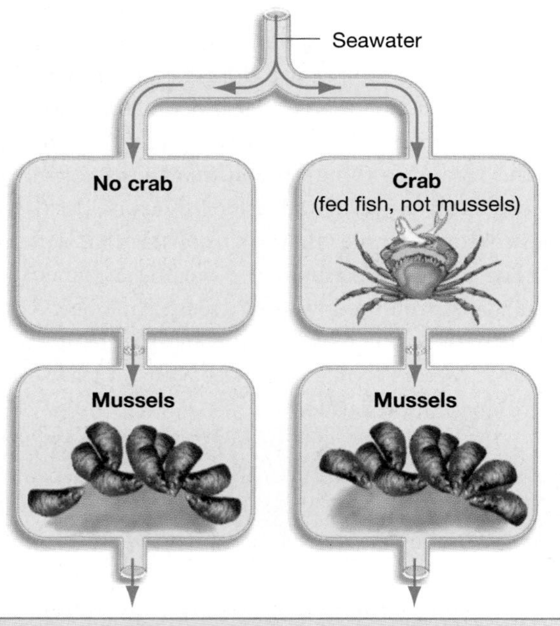

Prediction: Mussels downstream of the crab tank will have thicker shells than mussels downstream of the empty tank.

Prediction of null hypothesis: Mussels in the two tanks will have shells of equal thickness.

Results:

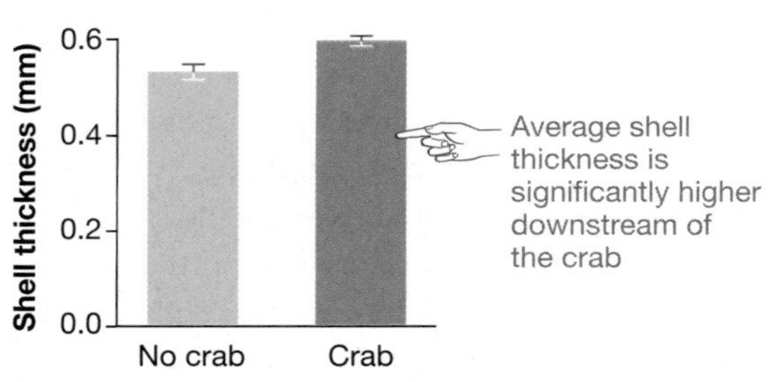

Conclusion: Mussels increase investment in defense when they detect crabs. Shell thickness is an inducible defense.

FIGURE 53.11 Experimental Evidence for Inducible Defenses.

QUESTION Why did the researchers feed the crabs fish instead of mussels?

conditions (see Chapter 52). If prey reproduce rapidly and are also well defended, it is not clear whether predators should be able to kill enough of them to reduce the prey population significantly—particularly if predators tend to take old or sick members of a population.

In several cases, data from predator-removal programs—in which wolves, cougars, coyotes, or other predators are actively killed by human hunters—support the hypothesis that predators actually do reduce the size of prey populations. For example, a wolf control program in Alaska during the 1970s decreased predator abundance to 55–80 percent below pre-control density. Concurrently, the population of moose, on which wolves prey, tripled. This observation suggests that wolf predators had reduced this moose population far below the number that could be supported by the available space and food.

Other types of experiments have also supported the hypothesis that predators play a role in density-dependent growth of prey populations. Recall from Chapter 52 that some populations go through regular cycles and that the regular population cycles of snowshoe hare and lynx appears to be driven, at least in part, by density-dependent increases in predation. Taken together, the data available to date indicate that in many instances, predators are efficient enough to reduce prey populations.

Why Don't Herbivores Eat Everything—Why Is the World Green? If predators affect the size of populations in prey that can run or fly or swim away, then consumers should have a devastating impact on plants and on mussels, anemones, sponges, and other sessile (nonmoving) animals.

In some cases, this prediction turns out to be correct. For example, consider the results of a recent **meta-analysis**—a study of studies, meaning an analysis of a large number of data sets on a particular question. Biologists who compiled the results of more than 100 studies on herbivory found that the median percentage of mass removed from aquatic algae by herbivores was 79 percent. Herbivores eat the vast majority of algal food available in aquatic biomes. The figure dropped to just 30 percent for aquatic plants, however, and only 18 percent for terrestrial plants.

Why don't herbivores don't eat more of the food available on land? Stated another way, why is the world green? Biologists routinely consider three possible answers to this question. Herbivores could be kept in check by predation or disease; plant tissues could offer poor or incomplete nutrition; or plants could defend themselves effectively against attack:

1. The **top-down control hypothesis** states that herbivore populations are limited by predation and disease. The "top-down" name is inspired by the food chains introduced in Chapter 30 and explored in detail in Chapter 54. It's appropriate because predators and parasites remove herbivores that eat plants. In a recent test of this idea, researchers monitored herbivory on islands created by a 1986 dam project in Venezuela. On some small islands in the new lake, predators disappeared. There are now many more herbivores on these islands than on similar sites nearby where predators were present, and a much higher percentage of primary production is being eaten. Predator-free islands, for example, had

just 25 percent of the small trees found on similar islands that contained predators.

2. The **poor-nutrition hypothesis** contends that plants are a poor food source in terms of the nutrients they provide for herbivores. More specifically, plant tissues have less than 10 percent of the nitrogen found in animal tissues, by weight. If the growth and reproduction of herbivores are limited by the availability of nitrogen, then their populations will be low and the impact of herbivory relatively slight. Herbivores could eat more plant material to gain nitrogen, but at a cost—they would be exposed to predation and expend energy processing the food. To evaluate the poor-nutrition hypothesis, a recent meta-analysis examined 185 studies of how insect herbivores had responded when the nitrogen concentration of plants was increased experimentally via fertilization. In over half of the cases, herbivores showed a significant increase in growth rate or reproduction when the plants that they feed on were fertilized. Based on this result, the researchers concluded that nitrogen limitation is an important factor in a large fraction of plant-herbivore interactions.

3. The **plant-defense hypothesis** holds that plants defend themselves effectively enough to limit herbivory. Most plant tissues are defended by weapons such as thorns, prickles, or hairs, or by potent poisons such as nicotine, caffeine, and cocaine. In addition, no animal species can, without help from protists or bacteria, digest cellulose or lignin, which are components of wood. Manufacturing defensive compounds seems like the perfect solution to the problem of herbivory. In practice, however, plants face a complex challenge in defending themselves. Consider data on interactions between cottonwood trees and two of their herbivores: beavers and leaf beetles. Cottonwoods resprout after they are cut down by a beaver (**Figure 53.12a**), and resprouted trees contain high concentrations of a defensive compound that deters further attack by beavers (**Figure 53.12b**). This is an induced defense. But larvae of a particular leaf beetle species eat this defensive compound readily. In fact, the larvae store enough of it in their bodies to act as a defensive compound against their own major predator—ants. The data in **Figure 53.12c** show that larvae that grow on resprouted cottonwood trees survive longer than do larvae that grow on normal cottonwoods when the larvae are placed on an ant mound. This is an example of an indirect effect in species interactions. The response by cottonwoods to herbivory by beavers benefits another herbivore—leaf beetles. The net result is that there is no perfect, one-size-fits-all defensive strategy. Natural selection should favor plants that evolve an ever-changing suite of compounds to deter the ever-changing array of herbivores they face.

Taken together, the data reviewed here suggest that there is no single answer to the question of why herbivores don't eat a greater fraction of the available plant food. Data have supported all three of the hypotheses we have examined. Top-down control, nitrogen limitation, and effective defense are all important factors in limiting the impact of herbivory, although the particular mix of factors will vary from plant species to plant species and from habitat to habitat.

Adaptation and Arms Races Over the long term, how do species that interact via consumption affect each other's evolution? When predators and prey or herbivores and plants interact over time, coevolutionary arms races result: Consumers evolve traits that increase their efficiency; in response, prey

(a) Cottonwood tree felled by beavers

(b) Resprouted trees have more defensive compounds.

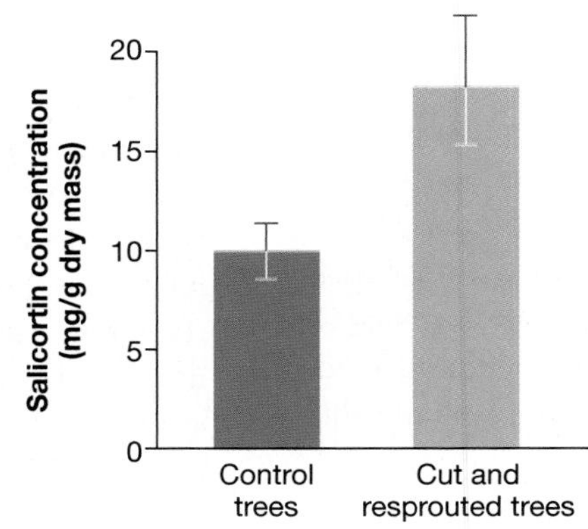

(c) Survival of beetle larvae placed on ant mound

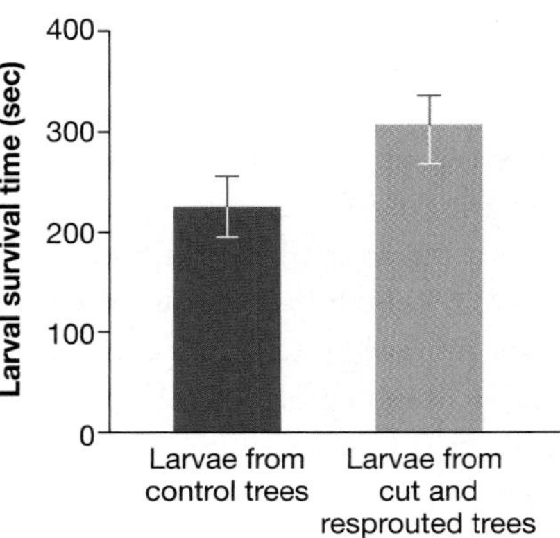

FIGURE 53.12 Defensive Compounds in Cottonwood Trees Discourage Beavers but Favor Beetles.

QUESTION Beetle larvae survive better on resprouted versus control trees, so they should damage resprouted trees more. If natural selection has maximized cottonwood fitness, do beavers or beetle larvae present the greater overall threat to them?

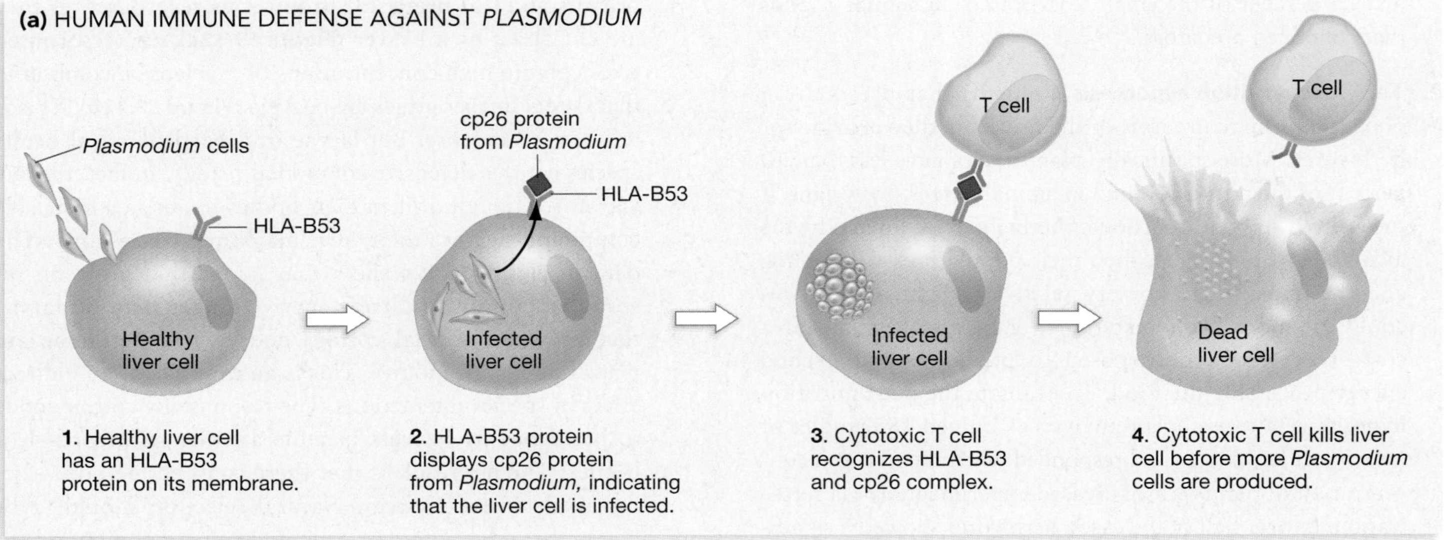

evolve traits that make them unpalatable or elusive, which leads to selection on consumers for traits that counter the prey adaptation, and so on.

To see a coevolutionary arms race in action, consider interactions between humans and the most serious of all human parasites—species in the genus *Plasmodium*. Recall from Chapter 29 that *Plasmodium* are unicellular protists that cause malaria in a wide array of vertebrates. Malaria kills at least a million people a year, most of them preschool-age children. Recent data suggest that humans and the causative agent of malaria—four species in the genus *Plasmodium*—are locked in a coevolutionary arms race.

Plasmodium has several distinct cell types, each of which infects liver, red blood cell, or other types of cells in mosquitoes or humans (see Chapter 29). Natural selection favors alleles that allow *Plasmodium* cells to infect their hosts efficiently and multiply inside. Cells in the human immune system do not sit passively by as *Plasmodium* destroys liver cells and red blood cells. Instead, immune system cells (introduced in Chapter 49) seek out and destroy *Plasmodium* cells. Natural selection favors alleles that allow mosquitoes and humans to resist infection.

In West Africa, for example, there is a strong association between an allele called *HLA-B53* and protection against malaria. In liver cells that are infected by *Plasmodium*, HLA-B53 proteins on the surface of the liver cell display a parasite protein called cp26 (**Figure 53.13a**). The display is a signal that immune system cells can read. It means "I'm an infected cell. Kill me before they kill all of us." Immune system cells destroy the liver cell before the parasite cells inside can multiply. In this way, people who have at least one copy of the *HLA-B53* allele are better able to beat back malarial infections. People who have the *HLA-B53* allele appear to be winning the arms race against malaria.

(b) *Plasmodium* strains have different versions of the cp protein.

***Plasmodium* strain**	**Infection rate**	**Interpretation**
cp26	Low	HLA-B53 binds to these proteins. Immune response is effective.
cp29	Low	
cp26 and cp29 strains together	High	Immune response fails when these strains infect the same person.
cp27	High	HLA-B53 does not bind to these proteins. Immune response is not as effective.
cp28	Average	

FIGURE 53.13 Interactions between the Human Immune System and Plasmodium. (a) If HLA-B53 binds to a particular *Plasmodium* protein, then infected cells are recognized and destroyed. **(b)** Some strains of *Plasmodium* appear to avoid detection by the immune system better than others do. Also, certain strains defeat the immune response if they infect the same person at the same time.

Follow-up research has shown that the arms race is far from won, however. *Plasmodium* populations in West Africa now have a variety of alleles for the protein recognized by HLA-B53. Some of these variants bind to HLA-B53 and trigger an immune response in the host, but others escape detection. Furthermore, many people in West Africa are infected with several different strains of *Plasmodium*, some of which are better at evading the immune system response. In some cases, the recognition step by HLA-B53 breaks down when certain strains are found together (**Figure 53.13b**). To make sense of these observations, researchers suggest that natural selection has favored the evolution of *Plasmodium* strains with weapons that counter HLA-B53.

An arms race continues to rage between *Plasmodium* and humans. Certain human proteins act as antimalarial weapons. But as predicted by coevolutionary theory, *Plasmodium* has evolved effective responses and continues to evolve.

(MB) **Web Animation** at www.masteringbio.com
Life Cycle of a Malaria Parasite

Can Parasites Manipulate Their Hosts? To thrive, parasites do not just have to invade tissues and grow while evading defensive responses by their host. They also have to be transmitted to new hosts. To a parasite, an uninfected host represents uncolonized habitat, teeming with resources. What have biologists learned about how parasites are transmitted to new hosts?

To answer this question, consider species of land snails that are parasitized by flatworms—specifically, by flukes in the genus *Leucochloridium*. Researchers who studied this association discovered something unusual. When the flukes have matured and are ready to be transmitted to their next host, a bird, they burrow into the snail's tentacles and wriggle. In addition, infected snails become attracted to light, even though uninfected snails avoid sunlit areas and prefer dark, shady environments (**Figure 53.14**). When infected snails move out of the shade into the open and glide about with wriggling tentacles, they are more easily spotted and consumed by birds. To interpret these observations, biologists suggest that the worms manipulate the behavior of the snail, and that the change in snail behavior makes the parasite more likely to be transmitted to a new host.

Studies of how parasites are transmitted to new hosts reinforce a general message: Extensive coevolution occurs among species that interact via consumption. Experiments on mutualistic interactions carry the same message.

Mutualism

Mutualisms are +/+ interactions that involve a wide variety of organisms and rewards. Many species of bees, for example, visit flowers to harvest nectar and pollen. Bees benefit because nectar is used as a food source for adult bees and the pollen is fed to larvae. Flowering plants also benefit because, in the process of visiting flowers, foraging bees carry pollen from one plant to another and accomplish pollination. Chapters 30 and 40 detailed some of the adaptations found in flowering plants that increase the efficiency of pollination. Other chapters highlighted an array of other mutualisms:

- One of the most important of all mutualisms occurs between fungi and plant roots. Chapter 31 reviewed experimental evidence indicating that fungi receive sugars and other carbon-containing compounds in exchange for nitrogen or phosphorus needed by the plant partner.
- Arguably the most critical of all mutualisms involves bacteria that fix nitrogen and certain species of plants. As Chapters 28 and 38 pointed out, the partnership is based on host plants providing sugars and protection and the bacteria supplying nitrogen in return.
- Chapter 32 cited mutualisms between (1) rancher ants and the aphids that the ants protect in exchange for sugar-rich honeydew, and (2) farmer ants and the fungi that they cultivate for food.

FIGURE 53.14 A Parasite That Manipulates Host Behavior. The behavior of snails that are infected with flukes in the genus *Leucochloridium* is dramatically different from the behavior of uninfected snails.

● **EXERCISE** Design an experiment to test the hypothesis that infected snails are more likely than uninfected snails to be eaten by birds.

Figure 53.15 illustrates some other interesting mutualisms. Figure 53.15a shows ants in the genus *Crematogaster*, which live in acacia trees native to Africa and the New World tropics. The ants live in bulbs at the base of acacia thorns and feed on small structures that grow from tree branches. These ants protect the tree by attacking and biting herbivores and by cutting vegetation from the ground below the host tree. Figure 53.15b illustrates cleaner shrimp in action. These shrimp pick external parasites from the jaws and gills of fish. In this mutualism, one species receives dinner while the other obtains medical attention.

As these examples show, the rewards from mutualistic interactions range from the transportation of gametes to food, housing, medical help, and protection. It is important to note, however, that even though mutualisms benefit both species, the interaction does not involve individuals from different species being altruistic or "nice" to each other. Expanding on a point that Charles Darwin introduced in 1862, Judith Bronstein described mutualisms as "a kind of reciprocal parasitism; that is, each partner is out to do the best it can by obtaining what it needs from its mutualist at the lowest possible cost to itself." Her point is that the benefits received in a mutualism are a byproduct of each individual pursuing its own self-interest, by maximizing its ability to survive and reproduce.

(a) Mutualism between ants and acacia trees

(b) Mutualism between cleaner shrimp and fish

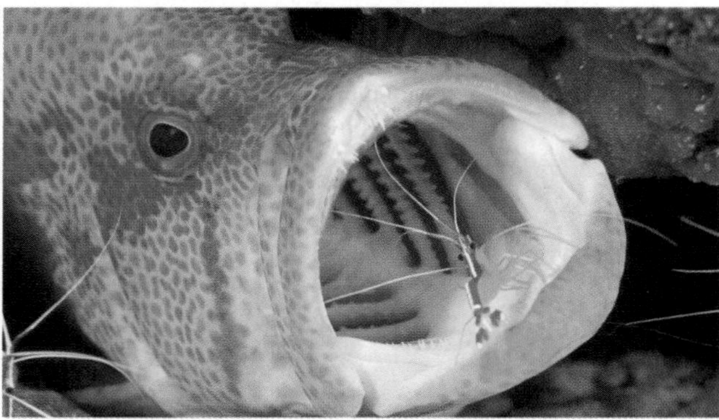

FIGURE 53.15 Mutualisms Take Many Forms. (a) In certain species of acacia tree, ants in the genus *Crematogaster* live in large bulbs at the base of spines and attack herbivores that threaten the tree. The ants eat nutrient-rich tissue produced at the tips of leaves or on branches. **(b)** Cleaner shrimp remove and eat parasites that take up residence on the gills of fish.

● QUESTION What is the cost of these associations to the acacia tree, the *Crematogaster* the cleaner shrimp, and the host fish?

In this light, it is not surprising that some species "cheat" on mutualistic systems. For example, deceit pollination occurs when certain species of plants produce a showy flower but no nectar reward. Pollinators have to be deceived to make a visit and carry out pollination. Evolutionary studies show that deceit pollinators evolved from ancestral species that did provide a reward. Over time, a +/+ interaction evolved into a +/− interaction.

A recent experimental study of mutualism provides another good example of the dynamic nature of these interactions. This study focused on ants and treehoppers. Ants are insects that live in colonies; treehoppers are small, herbivorous insects that feed by sucking sugar out of the phloem of plants. Treehoppers excrete the sugary solution honeydew from their posteriors. The honeydew, in turn, is harvested for food by ants.

It is clear that ants benefit from this association. But do the treehoppers? Biologists hypothesized that the ants might protect the treehoppers from their major predator, jumping spiders. These spiders feed heavily on juvenile treehoppers.

To test the hypothesis that ants protect treehoppers, the researchers studied ant-treehopper interactions over a three-year period. As **Figure 53.16** shows, the researchers marked out a 1000-m^2 study plot. Each year they removed the ants from one group of the treehopper host plants inside the plot but left the others alone to serve as a control. Then they compared the growth and survival of treehoppers on plants with and without ants. Recall that this is a common research strategy for studying species interactions. To assess the fitness costs or benefits of the interaction, researchers remove one of the participants experimentally and document the effect on the other

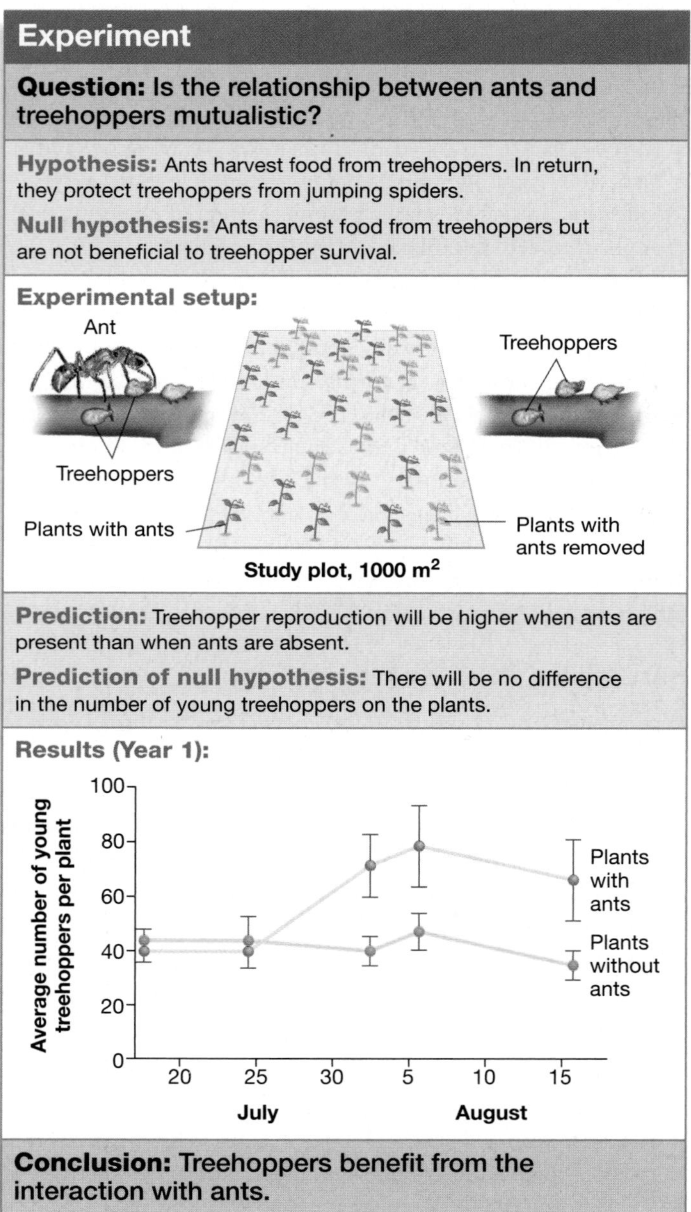

FIGURE 53.16 Experimental Evidence that the Treehopper-Ant Interaction Is Mutualistic.

SUMMARY TABLE 53.1 **Species Interactions**

Type of Interaction	Fitness Effects	Short-Term Impact: Distribution and Abundance	Long-Term Impact: Coevolution
Competition	−/−	Reduces population size of both species; if competition is asymmetric, competitive exclusion reduces range of one species.	Niche differentiation via selection to reduce competition
Consumption	+/−	Impact on prey population depends on prey density and effectiveness of defenses.	Strong selection on prey for effective defense; strong selection on consumer for traits that overcome defenses
Parasitism	+/−	Impact on host population depends on parasite density and effectiveness of defenses.	Strong selection on host for effective defense; strong selection on parasite for traits that overcome defenses
Mutualism	+/+	Population size and range of both species are dependent on each other.	Strong selection on both species to maximize fitness benefits and minimize fitness costs of relationship
Commensalism	+/0	Population size and range of commensal may depend on size and distribution of host.	Strong selection on commensal to increase fitness benefits in relationship; no selection on host

participant's survival and reproduction, compared with the survival and reproduction of control individuals that experience a normal interaction.

In both the first and third years of the study, the number of treehopper young on host plants increased in the treatment with ants but showed a significant decline in the treatment with no ants. This result supports the hypothesis that treehoppers benefit from the interaction with ants because the ants protect the treehoppers from predation by jumping spiders.

In the second year of the study, however, the researchers found a very different pattern. There was no difference in offspring survival, adult survival, or overall population size between treehopper populations with ants and those without ants. Why? The researchers were able to answer this question because they also measured the abundance of spiders that prey on treehoppers in each of the three years. Their census data showed that in the second year of the study, spider populations were very low.

Based on these results, the investigators concluded that the benefits of the ant-treehopper interaction depend entirely on predator abundance. Treehoppers benefit from their interaction with ants in years when predators are abundant but are unaffected in years when predators are scarce. If producing honeydew is costly to treehoppers, then the +/+ mutualism changes to a +/− interaction when spiders are rare.

Mutualism is like parasitism, competition, and other types of species interactions in an important respect: The outcome of the interaction depends on current conditions. Because the costs and benefits of species interactions are fluid, an interaction between the same two species may range from parasitism to mutualism to competition. **Table 53.1** summarizes the fitness effects, short-term impacts on population size, and long-term evolutionary aspects of species interactions.

Check Your Understanding

If you understand that...

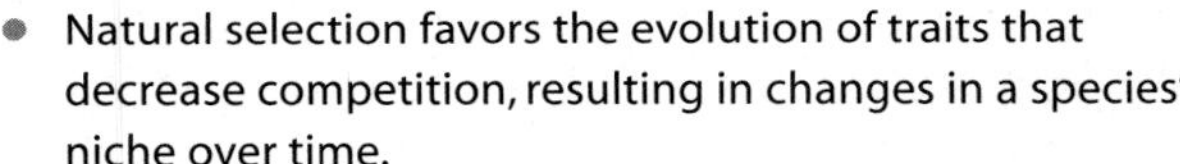

- Natural selection favors the evolution of traits that decrease competition, resulting in changes in a species' niche over time.

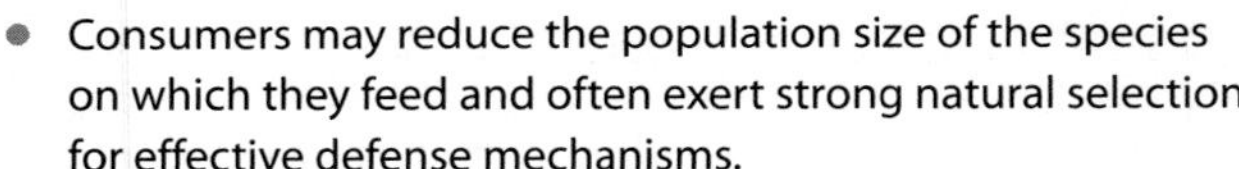

- Consumers may reduce the population size of the species on which they feed and often exert strong natural selection for effective defense mechanisms.
- Mutualisms benefit the species involved and can lead to highly coevolved associations, such as mycorrhizal fungi and symbiotic nitrogen-fixing bacteria.

You should be able to...

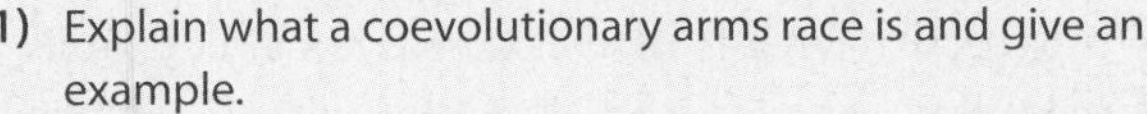

1) Explain what a coevolutionary arms race is and give an example.
2) Give specific examples of how competition, consumption, and mutualism each affect population size and long-term evolution.

53.2 Community Structure

Biologists have made important progress in understanding the nature of species interactions and their consequences. In terms of understanding the structure and function of biological communities, however, research on species interactions has a limitation: It usually focuses on just two species at a time. But biological communities contain many thousands of species. To understand how communities work, biologists broaden the scope of research and explore how combinations of many species interact.

The question that biologists initially asked about communities concerned structure: Do biological communities have a tightly prescribed organization and composition, or are they merely loose assemblages of species? If communities are highly structured entities, then their makeup should be predictable. For example, if a community is destroyed by a disturbance and then allowed to recover, the diversity and abundance of species at that site should be identical when recovery is complete. But if communities can be made up of many different combinations of species, depending on which arrive earlier or later, then community composition will be difficult to predict. The diversity and abundance of species found in a region would vary substantially before a disturbance and after recovery.

How Predictable Are Communities?

Beginning with a paper published in 1936, Frederick Clements promoted the view that biological communities are stable, integrated, and orderly entities with a highly predictable composition. His hypothesis was that species interactions are so extensive and coevolution is so important that the groups of species called communities have become highly integrated and interdependent units in nature. Stated another way, the species within a community cannot live without each other.

To drive his idea home, Clements likened the development of a plant community to the development of an individual organism. He argued that communities develop by passing through a series of predictable stages dictated by extensive interactions among species and that this development culminates in a stable final stage known as a **climax community**. According to Clements, the nature of the climax community is determined by the area's climate and does not change over time. Further, he held that if a fire or other disturbance destroys the climax community, it will reconstitute itself by repeating its predictable developmental stages.

Henry Gleason, in contrast, contended that the community found in a particular area is neither stable nor predictable. He claimed that plant and animal communities are ephemeral associations of species that just happen to share similar climatic requirements. According to Gleason, it is largely a matter of chance whether a similar community develops in the same area after a disturbance occurs. Gleason downplayed the role of biotic factors, such as species interactions, in structuring communities. To him, abiotic factors and history—for example, which seeds and juvenile animals happened to arrive after a disturbance—were the key elements in determining which species are found at a particular location.

Which viewpoint is more accurate? Let's consider observational and experimental data.

Mapping Current and Past Species' Distributions If communities are predictable assemblages, then the ranges of species that make up a particular community should be congruent. Stated another way, the same group of species should almost always be found growing together.

When biologists began documenting the ranges of tree species along elevational gradients, however, they found that species came and went independently of each other. As you go up a mountain, for example, you might find white oak trees growing with hickories and chestnuts. As you continue to gain elevation, chestnuts might disappear but white oaks and hickories remain. Farther upslope, hickories might drop out of the species mix while white pines and red pines start to appear.

Data on the historical composition of plant communities supported these observations. Studies of fossil pollen documented that the distribution of plant species and communities at specific locations throughout North America has changed radically since the end of the last ice age about 11,000 years ago. An important pattern emerged from these data: Species do not come and go in the fossil record in tightly integrated units. Instead, the ranges of individual species tended to change independently of one another. For example, at a study site in Mirror Lake, New Hampshire, the percentage of pollen present from pines and oaks rose and fell in tandem over the past 14,000 years—just as predicted for a tightly integrated, pine-oak forest community. In contrast, the percentage of pollen from spruce, fir, cedar, beech, and most other tree species changed independently of each other. In general, studies of fossil pollen suggest that the composition of most plant communities has been dynamic and contingent on historical events, rather than static.

Experimental Tests To explore the predictability of community structure experimentally, biologists constructed 12 identical ponds (**Figure 53.17**). They filled the ponds at the same time with water that contained enough chlorine to kill any preexisting organisms. If community structure is predictable, then each pond should develop the same community of species once the chlorine vaporized and made the water habitable. If community structure is unpredictable, then each pond should develop a different community.

To test these predictions, the researchers sampled water from the ponds repeatedly for one year. They measured temperature, chemical makeup, and other physical characteristics of the water and recorded the diversity and abundance of each planktonic species by examining the samples under the microscope. They found a total of 61 species in all of the ponds but discovered that individual ponds each had only 31 to 39 species. This observation is important. Each pond contained just half to two-thirds of the total number of species that lived in the experimental area and that were available for colonization. A number of species occurred in most or all of the 12 ponds, but each pond had a unique species assemblage. Why?

To explain their results, the researchers contended that some species are particularly good at dispersing and are likely to colonize all or most of the available habitats. Other species disperse more slowly and tend to reach only one or a few of the

Experiment

Question: Are communities predictable or unpredictable?

Communities-are-predictable hypothesis: The group of species present at a particular site is highly predictable.

Communities-are-unpredictable hypothesis: The group of species present at a particular site is highly unpredictable.

Experimental setup:

1. Construct 12 identical ponds. Fill at the same time and sterilize water so that there are no preexisting organisms.

2. Examine water samples from each pond. Identify each plankton species present in each sample.

Communities-are-predictable prediction: Identical plankton communities will develop in all 12 ponds.

Communities-are-unpredictable prediction: Different plankton communities will develop in different ponds.

Results (after 1 year):

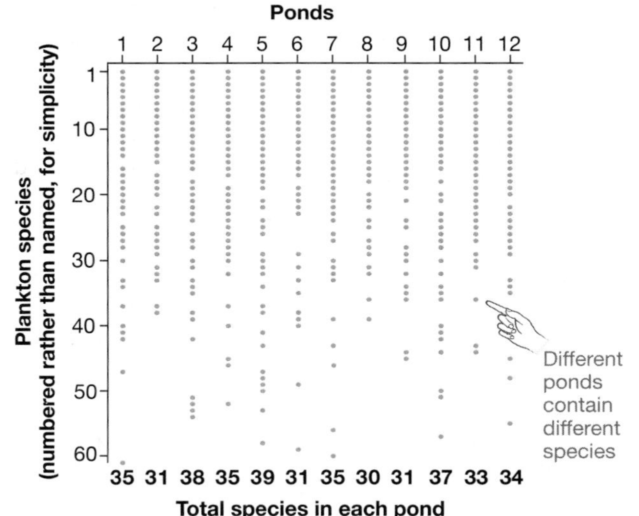

Conclusion: Although about half of the species present appear in all or most ponds, each pond has a unique composition. Both hypotheses are partially correct.

FIGURE 53.17 Experimental Evidence that Identical Communities Do Not Develop in Identical Habitats.

● QUESTION How many species are found in all 12 ponds?

available habitats. Further, the investigators proposed that the arrival of certain competitors or predators early in the colonization process greatly affects which species are able to invade successfully later. As a result, the specific details of community assembly and composition are somewhat contingent and difficult to predict. At least to some degree, communities are a product of chance and history.

The overall message of research on community structure is that Clements's position was too extreme and that Gleason's view is closer to being correct. ● Although both biotic interactions and climate are important in determining which species exist at a certain site, chance and history also play a large role.

How Do Keystone Species Structure Communities?

Even though communities are not predictable assemblages dictated by obligatory species interactions, the presence of certain consumers can have an enormous impact on the species present. Experiments have shown that, in some cases, the structure of an entire community can change dramatically if a single species of predator or herbivore is removed from a community or added to it.

As an example of this research, consider an experiment that Robert Paine conducted in intertidal habitats of the Pacific Northwest of North America. In this environment, the sea star *Pisaster ochraceous* is an important predator (**Figure 53.18a**). When Paine removed *Pisaster* from experimental areas, what had been diverse communities of algae and invertebrates became overgrown with solid stands of the California mussel *Mytilus californianus* (**Figure 53.18b**). Although *M. californianus* is a dominant competitor, its populations had been held in

(a) Predator: *Pisaster ochraceous*

(b) Prey: *Mytilus californianus*

FIGURE 53.18 A Keystone Predator and Its Prey

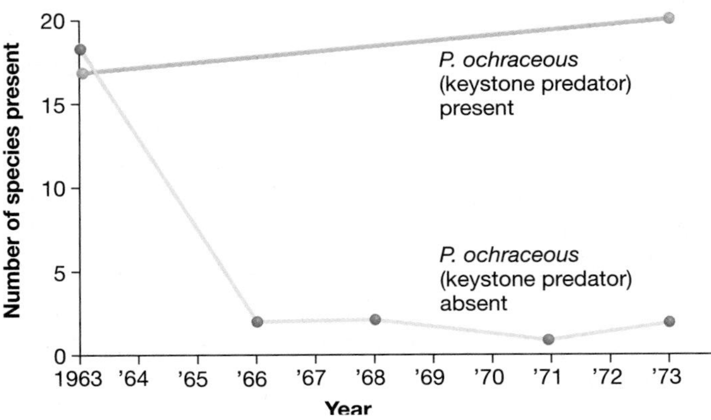

FIGURE 53.19 Keystone Predation Alters Community Structure in a Rocky Intertidal Habitat. Each data point represents the average number of species present in several study plots.

check by sea-star predation. When this predator was gone, the species richness and structural complexity of the habitat changed radically (**Figure 53.19**).

To capture the effect that a predator such as *Pisaster* can have on a community, Paine coined the term "keystone species." A **keystone species** has a much greater impact on the surrounding species than its abundance and total biomass would suggest. For example, wolves were introduced to the Greater Yellowstone Ecosystem of western North America in 1995. Although their population has grown to only about 150 individuals, their presence has led to far-reaching changes in plant and animal communities. Wolves feed on elk and do not tolerate the presence of coyotes. As elk declined, populations of favorite elk foods like aspen, willow, and cottonwood have increased, triggering an increase in beavers, which compete with elk for these plants. As coyotes declined, mouse populations increased and led to larger populations of hawks.

Sea stars, wolves, and other keystone species vary in abundance in space or over time. As a result, the composition and structure of their communities vary. Communities are ever-changing entities whose composition is difficult to predict.

53.3 Community Dynamics

Once biologists had a basic understanding of how species interact and how communities are structured, they turned to questions about how communities change through time. Like cells, individuals, and species, communities can be described in one word: dynamic.

Disturbance and Change in Ecological Communities

Community composition and structure may change radically in response to changes in abiotic and biotic conditions. Biologists have become particularly interested in how communities respond to disturbance. A **disturbance** is any event that removes some individuals or biomass from a community. (Recall from Chapter 50 that biomass is the total mass of living organisms.)

Forest fires, windstorms, floods, the fall of a large canopy tree, disease epidemics, and short-term explosions in herbivore numbers all qualify as disturbances. These events are important because they alter light levels, nutrients, unoccupied space, or some other aspect of resource availability.

Biologists have come to realize that the impact of disturbance is a function of three factors: (1) the type of disturbance, (2) its frequency, and (3) its severity—for example, the speed and duration of a flood, or the intensity of heat during a fire. Most communities experience a characteristic type of disturbance, and in most cases, disturbances occur with a predictable frequency and severity. To capture this point, biologists refer to a community's **disturbance regime**. For example, fires kill all or most of the existing trees in a boreal forest every 100 to 300 years, on average. In contrast, numerous small-scale tree falls, usually caused by windstorms, occur in temperate and tropical forests every few years.

How Do Researchers Determine a Community's Disturbance Regime? Ecologists use two strategies to determine a community's natural pattern of disturbance. The first approach is based on inferring long-term patterns from data obtained in a short-term analysis. For example, an observational study might document that 1 percent of all boreal forest on Earth burns in a given year. Assuming that fires occur randomly, researchers project that any particular piece of boreal forest has a 1 in 100 chance of burning each year. According to this reasoning, fires will recur in that particular area every 100 years, on average.

This extrapolation approach is straightforward to implement, but it has important drawbacks. In boreal forests, for example, fires do not occur randomly in either space or time. They are more likely in some areas than in others, and they tend to occur in particularly dry years. Unless sampling is extensive, it is difficult to avoid errors caused by extrapolating from particularly disturbance-prone or disturbance-free areas or years.

The second approach to determining disturbance regimes is based on reconstructing the history of a particular site. Flooding frequency, for example, can be estimated by analyzing sediments, because floods deposit distinctive groups of sediment particles. Researchers estimate the frequency and impact of storms by finding wind-killed trees and determining their date of death. (Investigators do so by comparing patterns in the growth rings of the dead trees with those of living individuals nearby; see Chapter 36.) The disturbances that have been most extensively studied using historical techniques, however, are forest fires.

Forest fires often leave a layer of burned organic matter and charcoal on the surface of the ground. As a result, researchers can dig a soil pit, find charcoal layers, and use radioisotope dating to establish when the fires occurred. It is also possible to date the death of trees killed by fire by comparing their growth rings with those of living trees. Further, trees that are not killed by fire are often scarred. When a fire burns close enough to kill a patch of cambium tissue, a scar forms. These fire scars occur most often at the tree's base, where dead leaves and twigs accumulate and furnish fuel. Fire scars can be dated by analyzing growth rings.

Why Is It Important to Understand Disturbance Regimes?

To appreciate why biologists are so interested in understanding disturbance regimes, consider a recent study on the fire history of giant sequoia groves in California (**Figure 53.20a**). Giant sequoias grow in small, isolated groves on the western side of the Sierra Nevada range. Individuals live more than a thousand years, and many have been scarred repeatedly by fires. A biologist obtained samples of cross sections through the bases of 90 giant sequoias in five different groves. As **Figure 53.20b** shows, the cross sections contained numerous rings that had been scarred by fire. To determine the date of each disturbance, the researcher counted tree rings back from the present. He found that, in most of the groves, 10 to 53 fires had occurred each century for the past 1530 to 2000 years (**Figure 53.20c**). The data indicated that each tree had been burned an average of 64 times.

This study established that fires are extremely frequent in the community examined. Because not enough time would pass for large amounts of fuel to accumulate between fires, they were probably of low severity. Partly because of this work, the biologists responsible for managing sequoia groves now set controlled fires or let low-intensity natural fires burn instead of suppressing them immediately. Similarly, studies of disturbance regimes along the Colorado River in southwestern North America inspired land managers to release a huge pulse of water from the reservoirs behind dams on the waterway recently. The flood that resulted was designed to mimic a natural disturbance. According to follow-up studies, the artificial flood appears to have benefited the plant and animal communities downstream.

Succession: The Development of Communities after Disturbance

Severe disturbances remove all or most of the organisms from an area. The recovery that follows is called **succession**. The name was inspired by the observation that certain species succeed others over time. **Primary succession** occurs when a disturbance removes the soil and its organisms as well as organisms that live above the surface. Glaciers, floods, volcanic eruptions, and landslides often initiate primary succession. **Secondary succession** occurs when a disturbance removes some or all of the organisms from an area but leaves the soil intact. Fire and logging are examples of disturbances that initiate secondary succession.

(a) Giant sequoias after a fire

(b) Fire scars in the growth rings

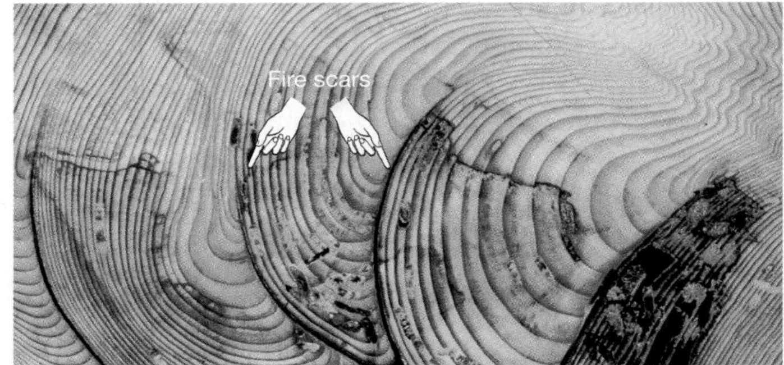

(c) Reconstructing history from fire scars

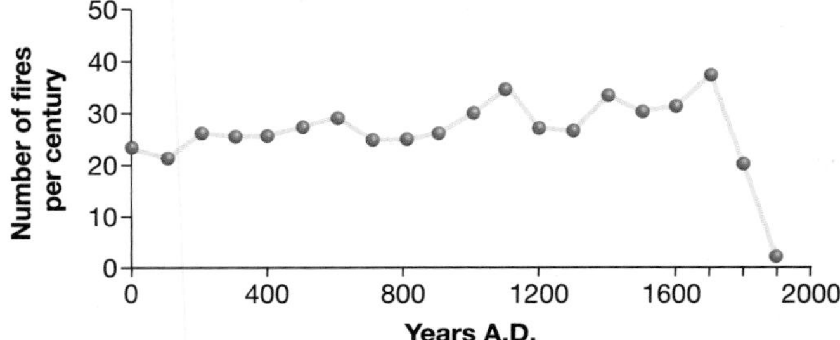

FIGURE 53.20 History of Disturbance in a Fire-Prone Community. Because trees form one ring (light band/dark band) every year, researchers can count the rings to determine how often fires have occurred during the last 2000 years.

QUESTION About how many fires occurred during the last century?

As **Figure 53.21** shows, a sequence of plant communities develops as succession proceeds. Early successional communities are dominated by species that are short lived and small and that disperse their seeds over long distances. Late successional communities are dominated by species that tend to be long lived, large, and good competitors for resources such as light and nutrients. The specific sequence of species that appears over time is called a successional pathway. What determines the pattern and rate of species replacement during succession at a particular time and place?

Theoretical Considerations Biologists focus on three factors to predict the outcome of succession in a community: (1) the particular traits of the species involved, (2) how species interact, and (3) historical and environmental circumstances such as the size of the area involved and weather conditions. Before going on to consider a detailed case history, we need to explore these factors in more detail.

Species traits, such as dispersal capability and the ability to withstand extreme dryness, are particularly important early in succession. As common sense would predict, recently disturbed sites tend to be colonized by plants and animals with good dispersal ability. When these organisms arrive, however, they often have to endure harsh environmental conditions. These **pioneering species** tend to have "weedy" life histories. (A **weed** is a plant that is adapted for growth in disturbed soils.) Early successional species devote most of their energy to reproduction and little to competitive ability. They have small seeds, rapid growth, and a short life span, and they begin reproducing at an early age. As a result, they have a high reproductive rate. But in addition, they can tolerate severe abiotic conditions, such as high light levels, poor nutrient availability, and drying.

Once colonization is under way, the course of succession tends to depend less on how species cope with aspects of the abiotic environment and more on how they interact with other species. This change occurs because plants that grow early in succession change abiotic conditions in a way that makes the conditions less severe. Because plants provide shade, they reduce temperatures and increase humidity. Their dead bodies also add organic material and nutrients to the soil. As abiotic conditions improve, biotic interactions become more important.

During succession, existing species can have one of three effects on subsequent species: (1) facilitation, (2) tolerance, or (3) inhibition. **Facilitation** takes place when the presence of an early arriving species makes conditions more favorable for the arrival of certain later species by providing shade or nutrients. **Tolerance** means that existing species do not affect the probability that subsequent species will become established. **Inhibition** occurs when the presence of one species inhibits the establishment of another. For example, a plant species that requires high light levels to germinate may be inhibited late in

FIGURE 53.21 Succession in Midlatitude Temperate Forests. Sequence of photos showing how succession leads to the development of a temperate forest from a disturbed state (in this case, an abandoned agricultural field).

succession by the presence of mature trees that prevent sunlight from reaching the forest floor.

In addition to species traits and species interactions, the pattern and rate of succession depend on the historical and environmental context in which they occur. For example, researchers found that the communities that developed after forest fires disturbed Yellowstone National Park in 1988 depended on the size of the burned patch and how hot the fire had been at that location. Succession is also affected by the particular weather or climate conditions that occur during the process. Variation in weather and climate causes different successional pathways to occur in the same place at different times.

Analyzing species traits, species interactions, and the historical/environmental context provides a useful structure for understanding why particular successional pathways occur. To see this theoretical framework in action, let's examine data on the course of primary succession that has occurred in Glacier Bay, Alaska.

A Case History: Glacier Bay, Alaska An extraordinarily rapid and extensive glacial recession is occurring at Glacier Bay (**Figure 53.22**, bottom). In just 200 years, glaciers that once filled the bay have retreated approximately 100 km, exposing extensive tracts of barren glacial sediments to colonization. Because of this event, Glacier Bay has become an important site for studying succession.

The top of Figure 53.22 shows the plant communities found in the area. The oldest sites are dense forests of Sitka spruce and western hemlock. Areas that have been deglaciated for about 100 years are inhabited by scattered spruce trees and dense thickets of a shrub called Sitka alder. Sites that have been deglaciated for 45 to 80 years are also covered with dense alder thickets, but the emergent trees are primarily cottonwood. Locations that have been ice free for 20 years or less do not have a continuous plant cover. Instead, they host scattered individuals of willow and a small shrub called *Dryas*.

These observations inspired a hypothesis for the pattern of succession in Glacier Bay: With time, the youngest communities of *Dryas* and willow succeed to alder thickets, which subsequently become dense spruce-hemlock forests. Stated another way, there is a single successional pathway throughout the bay. A recent study has challenged this hypothesis, however. Researchers who reconstructed the history of each community by studying tree rings found that three distinct successional pathways have occurred (**Figure 53.23**): (1) In the lower part of the bay, soon after the ice retreated, Sitka spruce began growing and quickly formed dense forests. Western hemlock arrived after spruce and is now common in the understory. (2) At middle-aged sites in the upper part of the bay, alder thickets were dominant for several decades, and spruce is just beginning to become common. These forests will probably never be as dense as the ones in the lower bay, however, and there is no sign that western hemlock has begun to establish itself. (3) In contrast, the youngest sites in the uppermost part of the bay may be following a third pathway. Alder thickets became dominant fairly early, but spruce trees are scarce. Instead, cottonwood trees are abundant.

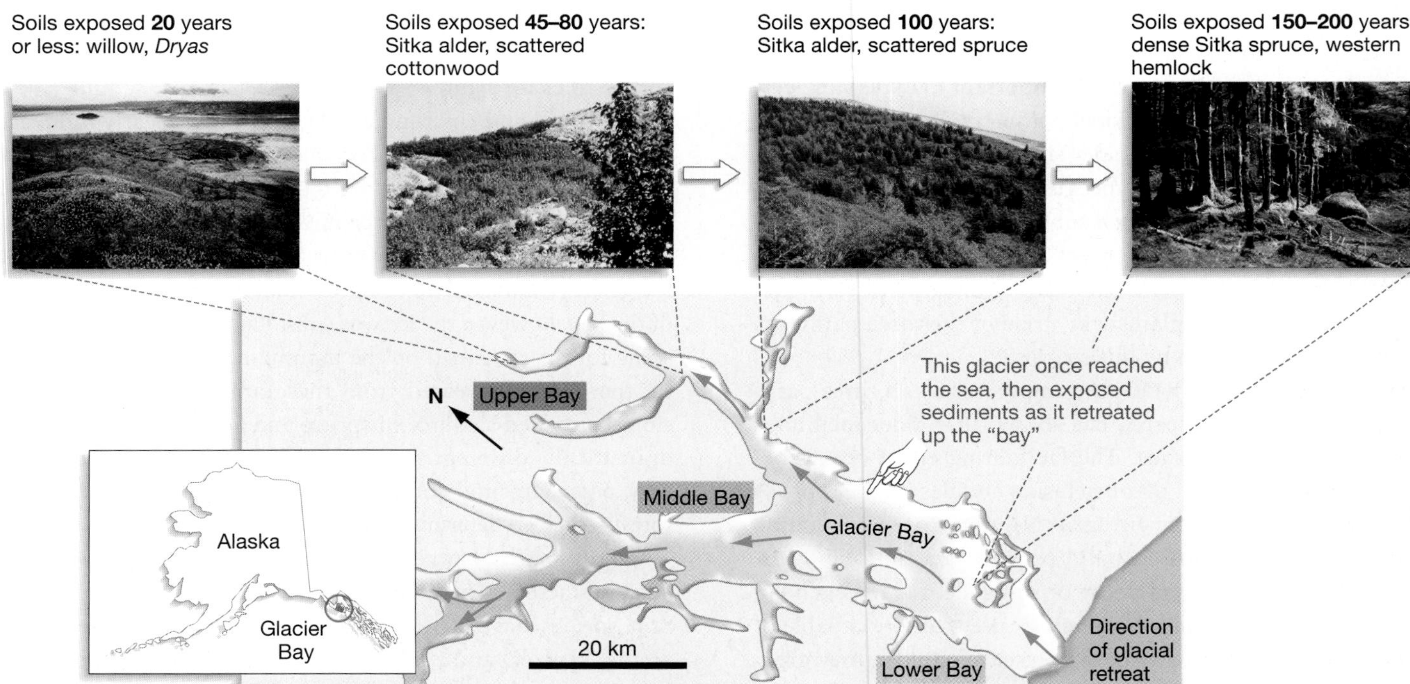

FIGURE 53.22 Evidence That a Single Successional Pathway Occurs in Glacier Bay.

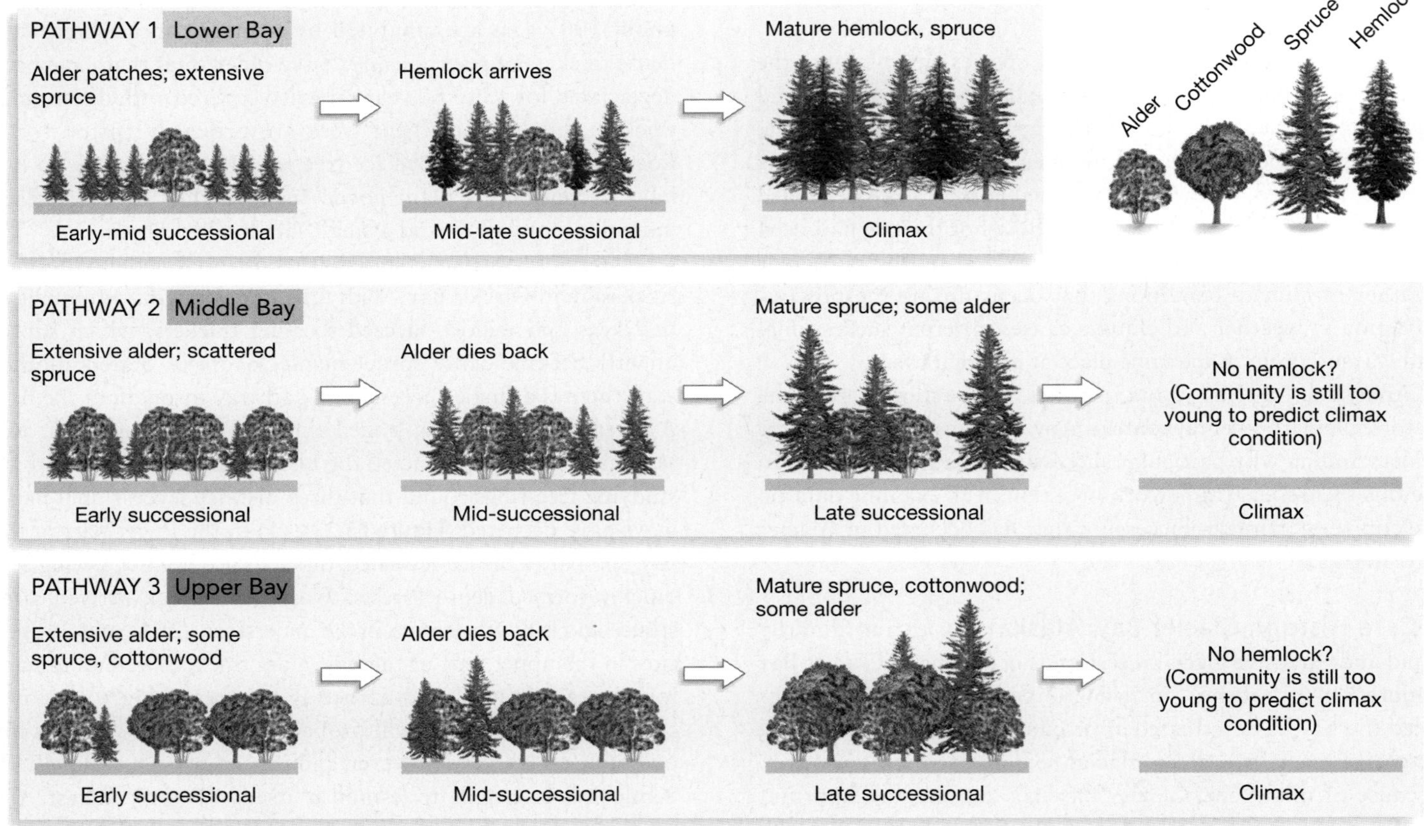

FIGURE 53.23 Evidence for Multiple Successional Pathways in Glacier Bay.

● **QUESTION** Suppose that a forest fire wiped out all of the plant communities throughout Glacier Bay. Would you predict that the successional pathways described here would reoccur?

These data present a challenge: How do species traits, species interactions, and dispersal patterns interact to generate the three observed successional pathways?

Species traits may be especially important in explaining certain details about the successional pathways. Western hemlock, for example, is abundant at older sites but largely absent from young ones. This is logical, because its seeds germinate and grow only in soils containing a substantial amount of organic matter and because the young trees can tolerate deep shade but not bright sunlight. Western hemlock's intolerance of early successional conditions explains why none of the three pathways began with colonization by this species.

Species interactions have been important in all three pathways. For example, research has shown that alder facilitates the growth of Sitka spruce. The facilitating effect occurs because symbiotic bacteria that live inside nodules on the roots of alder convert atmospheric nitrogen (N_2) to nitrogen-containing molecules that alder use to build proteins and nucleic acids. When alder leaves fall and decay or roots die, the nitrogen becomes available to spruce. Although spruce trees are capable of invading and growing without the presence of alder, they grow faster when alder stands have added nitrogen to the soil.

Competition is another important species interaction. For example, shading by alder reduces the growth of spruce until spruce trees are tall enough to protrude above the alder thicket. Once the spruce trees breach the alder canopy, however, alder dies out, because it is unable to compete with spruce trees for light.

Historical and environmental context also clearly influences succession at Glacier Bay. For example, geologists have found evidence that the ice was more than 1100 m thick in the upper part of Glacier Bay during the mid-1700s. Because forests grow to an elevation of only 700 m or 800 m in this part of Alaska, the glacier eliminated all of the existing forests. In the lower part of the bay, however, the ice was substantially thinner. As a result, some forests remained on the mountain slopes beyond the ice. As the glacier retreated from these areas, the forests on the slopes provided a source of spruce and hemlock seeds and set a dramatically different successional pathway in motion. In this way, environmental context—in this case, distance to existing forests—helped determine how the community developed.

To summarize, successional pathways are determined by an array of factors. These factors include the adaptations that certain species have to their abiotic environment, interactions among species, and the history of the site. Species traits and species interactions tend to make succession predictable, while history and chance events contribute a degree of unpredictability to succession.

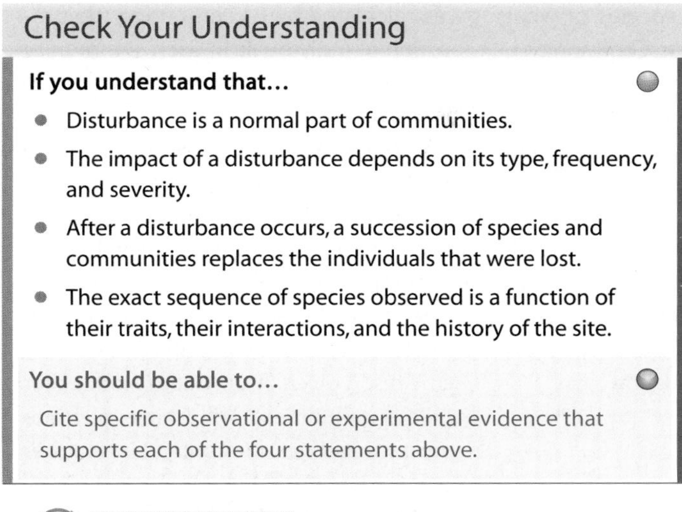

Check Your Understanding

If you understand that...

- Disturbance is a normal part of communities.
- The impact of a disturbance depends on its type, frequency, and severity.
- After a disturbance occurs, a succession of species and communities replaces the individuals that were lost.
- The exact sequence of species observed is a function of their traits, their interactions, and the history of the site.

You should be able to...

Cite specific observational or experimental evidence that supports each of the four statements above.

(MB) Web Animation at www.masteringbio.com
Primary Succession

53.4 Species Richness in Ecological Communities

The diversity of species present is a key feature of biological communities, and it can be quantified in two ways. **Species richness** is a simple count of how many species are present in a given community. **Species diversity**, in contrast, is a weighted measure that incorporates a species' relative abundance as well as its presence or absence (see **Box 53.1**). But at scales above relatively small study plots, it is rare to have data on relative abundance. As a result, ecologists sometimes use species richness and species diversity interchangeably. To introduce research on richness and diversity, let's focus on a simple question. Why are some communities more species rich than others?

Predicting Species Richness: The Theory of Island Biogeography

When researchers first began counting how many species are present in various areas, a strong pattern emerged: Larger patches of habitat contain more species than do smaller patches of habitat. The observation is logical because large areas should contain more types of niches and thus support higher numbers of species. But early work on species richness highlighted another pattern—one that was harder to explain. Islands in the ocean have smaller numbers of species than do areas of the same size on continents.

Robert MacArthur and Edward O. Wilson tackled this question by assuming that speciation occurs so slowly that the number of species present on an island is a product of just two events: immigration and extinction. The rates of both of these processes, they contended, should vary with the number of species present on an island. Immigration rates should decline as the number of species on the island increases, because individuals that arrive are more likely to represent a species that is already present. But extinction rates should increase as species richness increases, because niche overlap and competition for resources will be more intense. The result is an equilibrium—a balance between the arrival of new species and the extinction of existing ones (**Figure 53.24a**). If species richness changes due to a hurricane or fire or some other disturbance, then continued immigration and extinction should restore the equilibrium value.

MacArthur and Wilson also realized that immigration and extinction rates should vary as a function of island size and how far the island was to a continent or other source of immigrants. Immigration rates should be higher on large islands that are close to mainlands, because immigrants are more likely to find large islands that are close to shore than small ones that are far

(a) Species richness depends on the number of existing species.

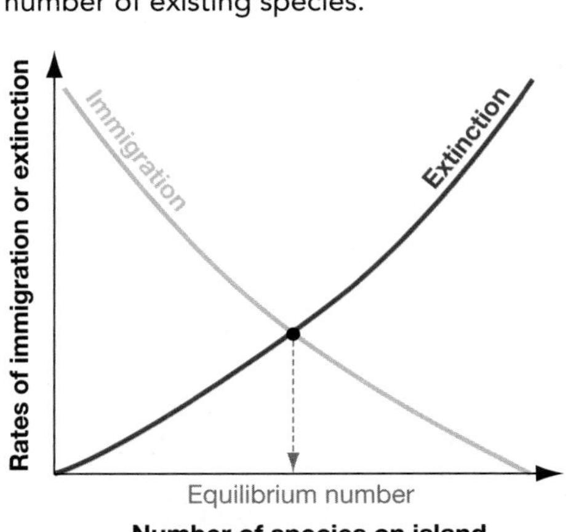

(b) Species richness depends on island size.

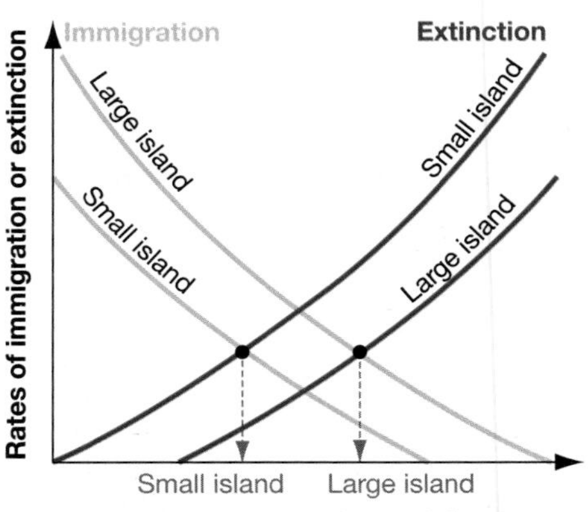

(c) Species richness depends on remoteness of the island.

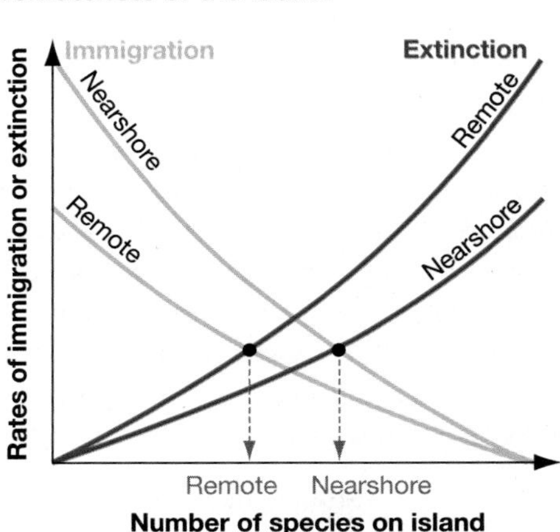

FIGURE 53.24 Species Richness Varies as a Function of Island Characteristics. If species richness on islands is dictated only by immigration and extinction rates, then species richness will vary as a function of (a) number of existing species, (b) island size, and (c) remoteness of the island from large landmasses.

QUESTION Suppose that the mainland habitats closest to an island were wiped out by suburbanization. How would this affect the curves in part (c)?

BOX 53.1 Measuring Species Diversity

To measure species diversity, a biologist could simply count the number of species present in a community. The problem is that simple counts provide an incomplete picture of diversity. The relative abundance of species is also an important component of diversity. To grasp this point, consider the composition of the three hypothetical communities shown in **Figure 53.25.** These communities are nearly identical in species composition but differ greatly in the relative abundance of each species.

These data can be used to compare two measures of species diversity. Species richness is simply the number of species found in a community. In this case, communities 1 and 2 have equal species richness and community 3 is lower in richness by one species. It is important to note, however, that communities 2 and 3 have similar relative abundances of each species, or what biologists call high *evenness*. Community 1, in contrast, is highly uneven. Fifty-five percent of the individuals in community 1 belong to species A, and other species are relatively rare. An uneven community has lower effective diversity than its species richness would indicate.

To take evenness into account, other diversity indices have been developed. A simple example is the Shannon index, given by the following equation:

$$H' = -\sum_{i=1}^{S} p_i \ln p_i$$

In this equation, p_i is the proportion of individuals in the community that belong to species i. The index is summed over all of the species in the study. The Shannon index for the three hypothetical communities, shown at the bottom of Figure 53.25, was calculated by (1) computing the proportion of individuals in each community that belong to each species, (2) taking the natural logarithm of each of these proportions (see **BioSkills 5** for an introduction to logarithms), (3) multiplying each natural logarithm times the proportion for each species, and (4) summing the total across the six species in the community. ● If you understand the equation, you should be able to do these calculations and get the same result given in the figure. Notice that while communities 1 and 2 have the same species richness, community 2 has higher diversity because of its greater evenness. Community 3 has lower species richness than community 1 but higher diversity.

● If you understand how to use and interpret the Shannon index, you should be able to calculate species richness and the Shannon diversity index for 3 communities that "double" Communities 1–3 in Figure 53.25. For example, one of the three new communities should be identical to Community 1, except that there are 2 species with 10 individuals present like species "A", 2 species with 1 individual present like species "B", 2 species with 1 individual present like species "C", 2 species with 3 individuals present like species "D", and so on. There are a total of 12 species in two of the communities, and 10 in the third. ● You should also be able to compare and contrast your results with the richness and diversity values given in the figure.

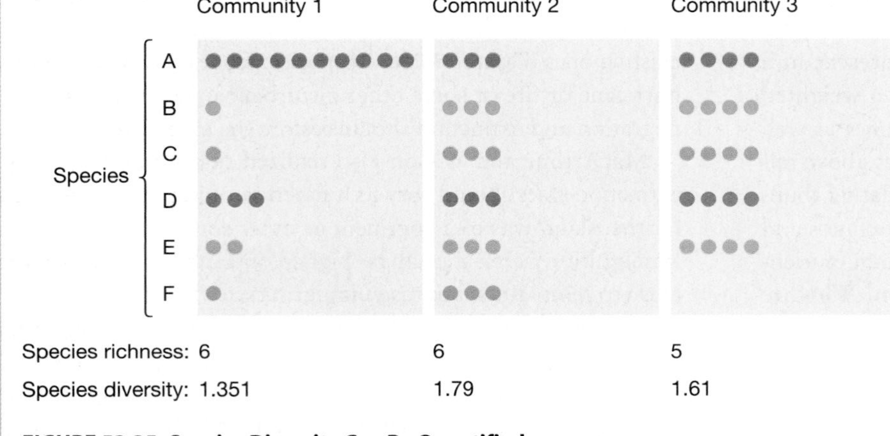

FIGURE 53.25 Species Diversity Can Be Quantified.

from shore. Extinction rates should be highest on small islands that are far from shore, because fewer resources are available to support large populations and because fewer individuals arrive to keep the population going. As a result, species richness should be higher on larger islands than smaller islands (**Figure 53.24b**), and on nearshore islands versus remote islands (**Figure 53.24c**).

This model, called the theory of island biogeography, is important for several reasons:

- It is relevant to a wide variety of island-like habitats such as alpine meadows, lakes and ponds, and caves.
- It made specific predictions that could be tested. For example, researchers have measured species richness on tiny islands, removed all of the species present, and then measured whether the same number of species recolonized the island and reached an equilibrium number. Predictions about immigration and extinction rates have also been measured by observing islands over time.
- It is relevant to the metapopulation dynamics introduced in Chapter 52 and can help inform decisions about the design of natural preserves. In general, the most-species rich reserves should be ones that are (1) relatively large, and (2) located close to other relatively large habitat areas.

Global Patterns in Species Richness

Biologists have long understood that large habitat areas tend to be species-rich, and the theory of island biography has been

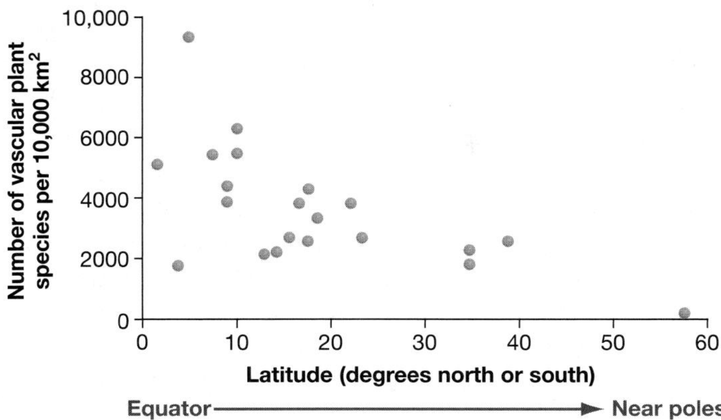

FIGURE 53.26 There Is a Strong Latitudinal Gradient in Species Diversity. One example of a latitudinal gradient in species diversity.

successful in framing thinking about how species richness should vary among island-like habitats. But researchers have had a much more difficult time explaining what may be the most striking pattern in species richness.

Biologists who began cataloguing the flora and fauna of the tropics in the mid-1800s quickly recognized that these habitats contain many more species than do temperate or subarctic environments. Data compiled in the intervening years have confirmed the existence of a strong latitudinal gradient in species diversity—for communities as a whole as well as for many taxonomic groups. In birds, mammals, fish, reptiles, many aquatic and terrestrial invertebrates, orchids, and trees, for example, species diversity declines as latitude increases (**Figure 53.26**). Although this pattern is not universal—a number of marine groups, as well as shorebirds, show a positive relationship between latitude and diversity—it is widespread. Why does it occur?

To explain why species diversity might decline with increasing latitude, biologists have to consider two fundamental principles. First, the causal mechanism must be abiotic, because latitude is a physical phenomenon produced by Earth's shape. The explanation must be a physical factor that varies predictably with latitude and that could produce changes in species diversity. Second, the species diversity of a particular area is the sum of four processes: speciation, extinction, immigration (colonization), and emigration (dispersal). Thus, the latitudinal gradient must be caused by an abiotic factor that affects the rate of speciation, extinction, immigration, or emigration in a way that would lead to more species in the tropics and fewer near the poles.

Over 30 hypotheses have been proposed to explain the latitudinal gradient. One of these maintains that high productivity in the tropics promotes high diversity by increasing speciation rates and decreasing extinction rates. (Recall from Chapter 50 that productivity is the total amount of photosynthesis per unit area per year.) The idea is that increased biomass production supports more herbivores and thus more predators and parasites and scavengers. In addition, speciation rates should increase when niche differentiation occurs within populations of herbivores, predators, parasites, and scavengers. Although this high-productivity hypothesis is supported by the global-scale correlation between productivity and diversity, experimental studies challenge it. For example, researchers who add fertilizer to aquatic or terrestrial communities routinely observe significant increases in productivity but decreases in diversity.

In light of results such as these, researchers have concluded that productivity alone is probably not a sufficient explanation for the higher species diversity in the tropics. Research continues, however, and investigators have recently begun to focus on an abiotic factor that is correlated with productivity: temperature. Biologists who analyzed data on gastropods and other marine invertebrates have documented a strong correlation between the temperature of marine waters and species diversity. This observation has inspired the energy hypothesis: High temperatures increase species diversity by increasing productivity and the likelihood that organisms can tolerate the physical conditions in a region.

A third hypothesis is that tropical regions have had more time for speciation to occur than other regions have. Temperate and arctic latitudes were repeatedly scoured by ice sheets over the last 2 million years, but tropical regions were not. Recent data suggest, however, that tropical forests were dramatically reduced in size by widespread drying trends during the ice ages. Existing forests may be much younger than originally thought. If so, then the contrast in the age of northern and southern habitats may not be enough to explain the dramatic difference in species diversity.

A fourth hypothesis was inspired by the observation that species diversity is much higher in mid-successional communities than in pioneer or mature communities. The **intermediate disturbance hypothesis** holds that regions with a moderate type, frequency, and severity of disturbance should have high species richness and diversity. The logic here is that, with intermediate levels of disturbance, communities will contain pioneering species as well as species better adapted to late-successional conditions. For example, recent studies have confirmed that tree falls and canopy gaps occur regularly in tropical forests and that fires occur in these biomes occasionally. As yet, however, there are no convincing data showing that intermediate levels of disturbance are more likely to occur in the tropics than they are at higher latitudes.

Each of the factors discussed here may influence diversity, but no single one offers a convincing explanation for the global diversity gradient. Although the high temperatures present in the tropics seem to have an especially important role in promoting species richness, there is no simple answer to the question of why the tropics are so much more species rich than temperate and high-latitude habitats.

Chapter Review

SUMMARY OF KEY CONCEPTS

Interactions among species, such as competition and consumption, have two main outcomes: (1) They affect the distribution and abundance of the interacting species, and (2) they are agents of natural selection and thus affect the evolution of the interacting species. The nature of interactions between species frequently changes over time.

A community is an assemblage of interacting species.

To categorize the different types of interactions that occur among species, biologists consider whether each participant experiences a net fitness cost or benefit from the interaction. These costs and benefits depend on the conditions that prevail at a particular time and place and may change through time.

Competition is a −/− interaction that occurs when the niches of two species overlap—meaning they use the same resources. Competition may result in the complete exclusion of one species. It may also result in niche differentiation, in which competing species evolve traits that allow them to exploit different resources or live in different areas.

Consumption is a +/− interaction that occurs when consumers eat prey, which resist through standing defenses or inducible defenses. Predators are efficient enough to reduce the size of many prey populations. Levels of herbivory are relatively low in terrestrial ecosystems, however, because predation and disease limit herbivore populations, because plants provide little nitrogen, and because many plants contain toxic compounds or other types of defenses. Parasites generally spend all or part of their life cycle in or on their host (or hosts) and usually have traits that allow them to escape host defenses. In turn, hosts have evolved counteradaptations that help fight off parasites.

Mutualism is a +/+ interaction that provides participating individuals with food, shelter, transport of gametes, or defense against predators. For each species involved, the costs and benefits of a mutualism may vary over time and from place to place.

You should be able to give an example of how competition can evolve into commensalism, and how a mutualistic relationship can evolve into a parasitic one.

MB Web Animation at www.masteringbio.com
Life Cycle of a Malaria Parasite

The assemblage of species found in a biological community changes over time and is primarily a function of climate and chance historical events.

Ecologists have debated whether communities are fixed, predictable entities or simply places where the distributions of various species overlap. Historical and experimental evidence support the view that communities are dynamic rather than static and that their composition is neither entirely predictable nor stable over time.

In addition to climate, the composition of a community is influenced by disturbance. In extreme cases, disturbance may remove all organisms and all soil from a large area. Each community has a characteristic disturbance regime—meaning a type, severity, and frequency of disturbance that it experiences. Three types of factors influence the pattern of succession that occurs after a disturbance. First, the historical and environmental context of the site affects which species are available to join the resulting communities. The dispersal ability of different species also affects their availability. Second, the physiological traits of any given species influence the kinds of abiotic environmental conditions it can tolerate and dictate when it can successfully join a community. Third, interactions among species influence if and when a species appears during succession.

You should be able to to explain why climate makes the three successional pathways documented in Glacier Bay similar, and why chance historical events make them different.

MB Web Animation at www.masteringbio.com
Primary Succession

Species diversity is high in the tropics and lower toward the poles. The mechanism responsible for this pattern is still being investigated.

One of the most widely studied patterns in community ecology is the latitudinal gradient in species diversity. Within most specific taxonomic groups and within communities as a whole, species diversity declines from the equator to the poles. Recent research suggests that the high species richness observed in the tropics results from a combination of factors—the most important of which may be high temperatures that increase productivity and provide relatively benign abiotic conditions.

You should be able to suggest a hypothesis to explain an exception to the latitudinal gradient rule—specifically, that most species of shorebirds live at high latitudes, not the tropics. (Note that many shorebird species nest near the abundant lakes found in arctic tundra).

QUESTIONS

Test Your Knowledge

1. What is competitive exclusion?
   a. the evolution of traits that reduce niche overlap and competition
   b. interactions that allow species to occupy their fundamental niche
   c. the degree to which the niches of two species overlap
   d. the claim that species with the same niche cannot coexist

2. What is niche differentiation?
   a. the evolution of traits that reduce niche overlap and competition
   b. interactions that allow species to occupy their fundamental niche
   c. the degree to which the niches of two species overlap
   d. the claim that species with the same niche cannot coexist

3. Why is the phrase "coevolutionary arms race" an appropriate way to characterize the long-term effects of species interactions?
   a. Both plants and animals have evolved weapons for defense that are so effective that many plants are not eaten and predators cannot reduce prey populations to extinction.
   b. Adaptations that give one species a fitness advantage in an interaction are likely to be countered by adaptations in the other species that eliminate this advantage.
   c. In all species interactions except for mutualism, at least one species loses (suffers decreased fitness).
   d. Even mutualistic interactions can become parasitic if conditions change. As a result, interacting species are always "at war."

4. Why are inducible defenses advantageous?
   a. They are always present—thus, an individual is always able to defend itself.
   b. They make it impossible for a consumer to launch surprise attacks.
   c. They result from a coevolutionary arms race.
   d. They make efficient use of resources, because they are produced only when needed.

5. Which of the following is *not* correlated with species diversity?
   a. latitude
   b. productivity
   c. longitude
   d. island size

6. What is net primary productivity?
   a. an individual's lifetime reproductive success (lifetime fitness)
   b. an individual's average annual reproductive success
   c. the total amount of photosynthesis that occurs in an area of a given size per year
   d. the amount of energy that is stored in standing biomass per year

Test Your Knowledge answers: 1. d; 2. a; 3. b; 4. d; 5. c; 6. d

Test Your Understanding

Answers are available at www.masteringbio.com

1. The text claims that species interactions are conditional and dynamic. Do you agree with this statement? Why or why not? Cite specific examples to support your answer.
2. State three hypotheses that have been proposed to explain the low level of herbivory in terrestrial plant communities. Are these hypotheses mutually exclusive? (In other words, can more than one be correct?) Explain why or why not.
3. Biologists have tested the hypotheses that communities are highly predictable versus highly unpredictable. State the predictions that these hypotheses make with respect to (a) the presence and impact of keystone species, (b) changes in the distribution of the species in a particular community over time, and (c) the communities that should develop at sites where abiotic conditions are identical. Which hypothesis appears to be more accurate?
4. What is a disturbance? List five examples of disturbance. Compare and contrast their effects. For each type of disturbance, compare and contrast the consequences of high-frequency and low-frequency disturbances and high and low severity of disturbances.
5. Summarize the life-history attributes of early successional versus late successional species. Why are these attributes considered adaptations?
6. Describe the latitudinal gradient in species diversity that exists for most taxonomic groups. Discuss the pros and cons of one hypothesis to explain this pattern.

Applying Concepts to New Situations

Answers are available at www.masteringbio.com

1. Some insects harvest nectar by chewing through the wall of the structure that holds the nectar As a result, they obtain a nectar reward, but pollination does not occur. Suppose that you observed a certain bee species obtaining nectar in this way from a particular orchid species. Over time, how might you expect the characteristics of the orchid population to change in response to this bee behavior?
2. Using this chapter's information on fire regimes in giant sequoia groves, propose a management plan for Sequoia National Park. Explain the logic behind your plan.
3. Suppose that a two-acre lawn on your college's campus is allowed to undergo succession. Describe how species traits, species interactions, and the site's history might affect the community that develops.
4. Design an experiment to test the hypothesis that increasing species richness increases a community's productivity and ability to resist disturbance and recover from disturbance.

www.masteringbio.com is also your resource for • Answers to text, table, and figure caption questions and exercises • Answers to *Check Your Understanding* boxes • Online study guides and quizzes • Additional study tools including the *E-Book for Biological Science* 3rd ed., textbook art, animations, and videos.

54 Ecosystems

KEY CONCEPTS

- An ecosystem has four components: (1) the abiotic environment, (2) primary producers, (3) consumers, and (4) decomposers. These components are linked by the movement of energy and nutrients.
- The productivity of terrestrial ecosystems is limited by warmth and moisture, while nutrient availability is the key constraint in aquatic ecosystems. As energy flows from producers to consumers and decomposers, much of it is lost.
- To analyze nutrient cycles, biologists focus on the nature of the reservoirs where elements reside and on how quickly elements move between reservoirs.
- Humans are causing large, global changes in the abiotic environment. The burning of fossil fuels has led to rapid global warming. Extensive fertilization is increasing productivity and causing pollution.

The Qori Kalis glacier in the Andes Mountains of Peru in the year 2006. In 1978 the glacier extended past the edge of the lake in the lower left-hand corner of this photograph. Global warming is causing dramatic reductions in the size of glaciers and ice fields in many locations throughout the world.

An **ecosystem** consists of the organisms that live in an area together with their physical, or abiotic, environment. With the explosive growth of human populations, dramatic changes are occurring in both the biotic and abiotic components of ecosystems around the globe. Extinctions and other effects on the biotic elements of ecosystems are explored in Chapter 55; this chapter analyzes changes in the chemical and physical characteristics of the environment. Human impacts on the abiotic environment include global warming, acid rain, a hole in the atmosphere's ozone layer, and nitrate pollution (see Chapter 28). Ecosystem studies are taking a prominent place in biological science.

How does an ecosystem differ from a community? In most cases, ecosystems are composed of multiple communities along with their chemical and physical environments. For example, biologists who study lakes recognize a number of communities within a lake. There are distinct communities of interacting species along the lake bottom, at the surface, and in the shallow water of the littoral zone. Those different communities are studied as a unit called the lake ecosystem, because energy and matter flow among them.

Adjacent ecosystems—such as a river, a lake, and the surrounding forest—can be distinguished because their energy and nutrient flows are self-contained. Occasionally, though, energy and nutrients can also be transferred between ecosystems. This happens when nutrients dissolved in rainwater run off of agricultural ecosystems into nearby streams, or when an eagle kills a fish and carries it into a forest ecosystem.

This chapter introduces ecosystem studies with a look at how energy flows among the components of an ecosystem; it continues by exploring how carbon, nitrogen, and other key elements cycle through organisms, sediments, the oceans, and the atmosphere; and it ends by analyzing how humans are affecting the abiotic environment. Ecosystem ecology has taken on an increasingly applied focus, because humans are adding massive amounts of energy to ecosystems, disrupting nutrient cycles, and changing the chemistry of lakes, oceans, and the atmosphere. Understanding ecosystem ecology is fundamental to managing the future of our planet.

54.1 Energy Flow and Trophic Structure

If you imagine that an ecosystem is like an economy, then energy is its currency. Ecosystems have four components: (1) the abiotic environment, (2) primary producers, (3) consumers, and (4) decomposers. The four components are linked by a flow of energy (**Figure 54.1**).

A **primary producer** is an **autotroph** (literally, "self-feeder")—meaning, an organism that can synthesize its own food from inorganic sources. In most ecosystems, producers use solar energy to manufacture their own photosynthesis. But in some ecosystems, such as hydrothermal vents and iron-rich rocks deep below Earth face, bacteria use the chemical energy contained in inorg compounds such as hydrogen (H_2), methane (CH_4), or hyd gen sulfide (H_2S) to make food (see Chapter 28).

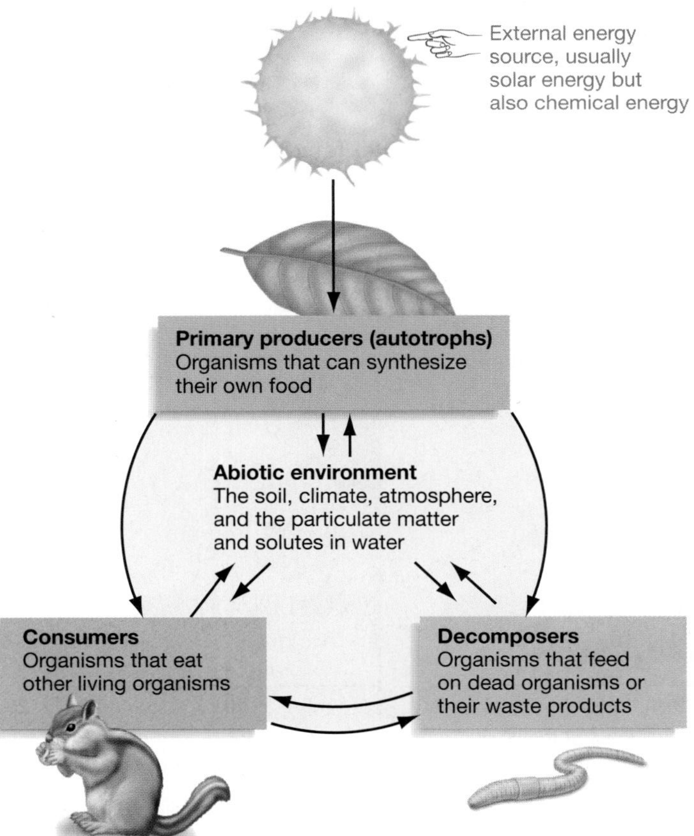

FIGURE 54.1 The Four Components of an Ecosystem Interact. The arrows represent energy. A similar diagram could be drawn to represent the flow of nutrients.

Primary producers form the basis of ecosystems because they transform the energy in sunlight or inorganic compounds into the chemical energy stored in sugars. Primary producers use this chemical energy in two ways: Most supports maintenance or respiratory costs, but some makes growth and reproduction possible. The energy that is invested in new tissue is called **net primary productivity** (**NPP**).

Net primary productivity represents the amount of energy that is available to the second and third components of an ecosystem: consumers and decomposers. **Consumers** eat living organisms. **Herbivores** are consumers that eat plants; **carnivores** are consumers that eat animals. **Decomposers**, or **detritivores**, obtain energy by feeding on the dead remains of other organisms or waste products.

The fourth and final component of an ecosystem is the abiotic environment, which includes soil, climate, the atmosphere, the Sun, and the particulate matter and solutes in water. The four components are linked because energy moves from the Sun or inorganic compounds to consumers, decomposers, and the abiotic environment.

Understanding NPP is a primary focus of ecosystem ecology because it dictates the amount of energy available to consumers and decomposers, and because global warming is altering NPP worldwide. Let's consider two of the most fundamental questions about net primary productivity: How does it vary among the world's ecosystems, and what happens to it?

Global Patterns in Productivity

Figure 54.2 summarizes data on NPP from around the globe. A quick look at the color key should convince you that the terrestrial ecosystems with highest productivity are located in the wet tropics. Notice that, with the exception of the world's major deserts, NPP on land declines from the equator toward the poles. Productivity patterns in marine ecosystems are different, however. Marine productivity is highest along coastlines, and it can be as high near the poles as it is in tropics. The oceanic zones, introduced in Chapter 50, have extremely low NPP. Typically, a square meter of open ocean produces a maximum of 35 g (1.2 oz) of organic matter each year. In terms of productivity per m^2, the open ocean is a desert.

Figure 54.3 presents the NPP data a different way—organized by ecosystem instead of by geography. Figure 54.3a provides data on average NPP per square meter per year for

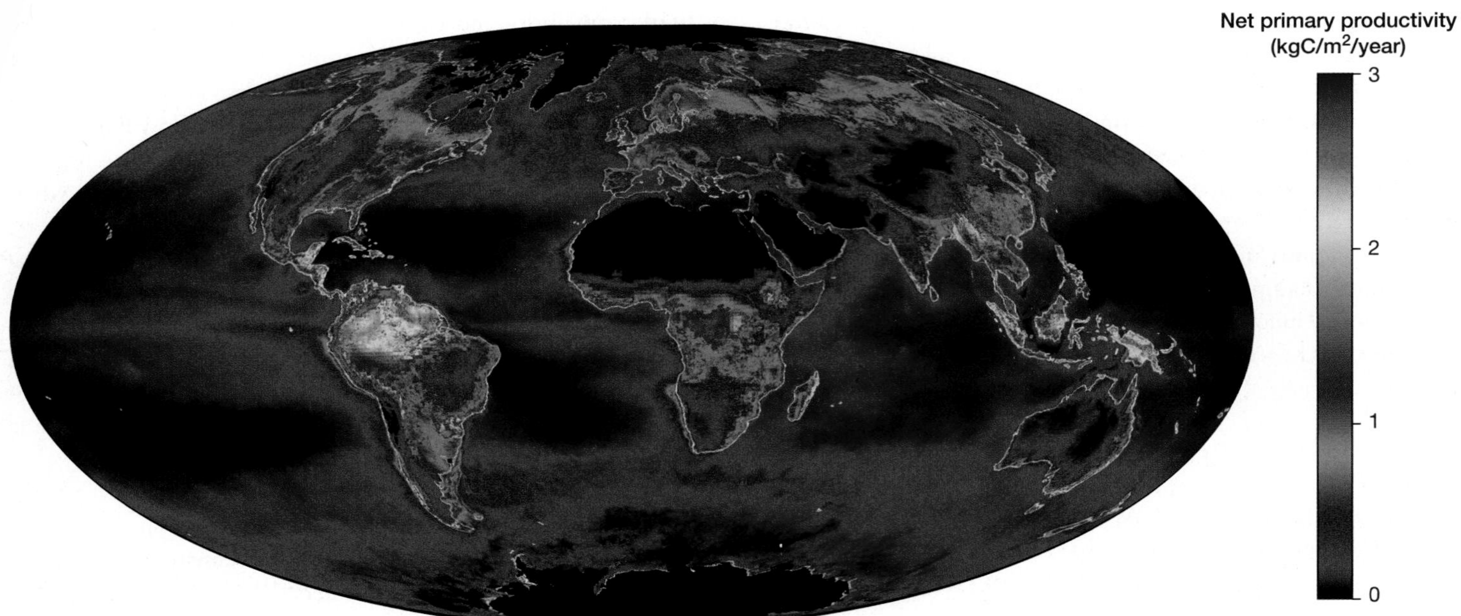

FIGURE 54.2 Net Primary Productivity Varies among Regions. The terrestrial ecosystems with the highest primary productivity are found in the tropics, where warm temperatures and high moisture encourage high rates of photosynthesis. Tundras and deserts have the lowest productivity. The highest productivity in the oceans occurs in nutrient-rich coastal areas.

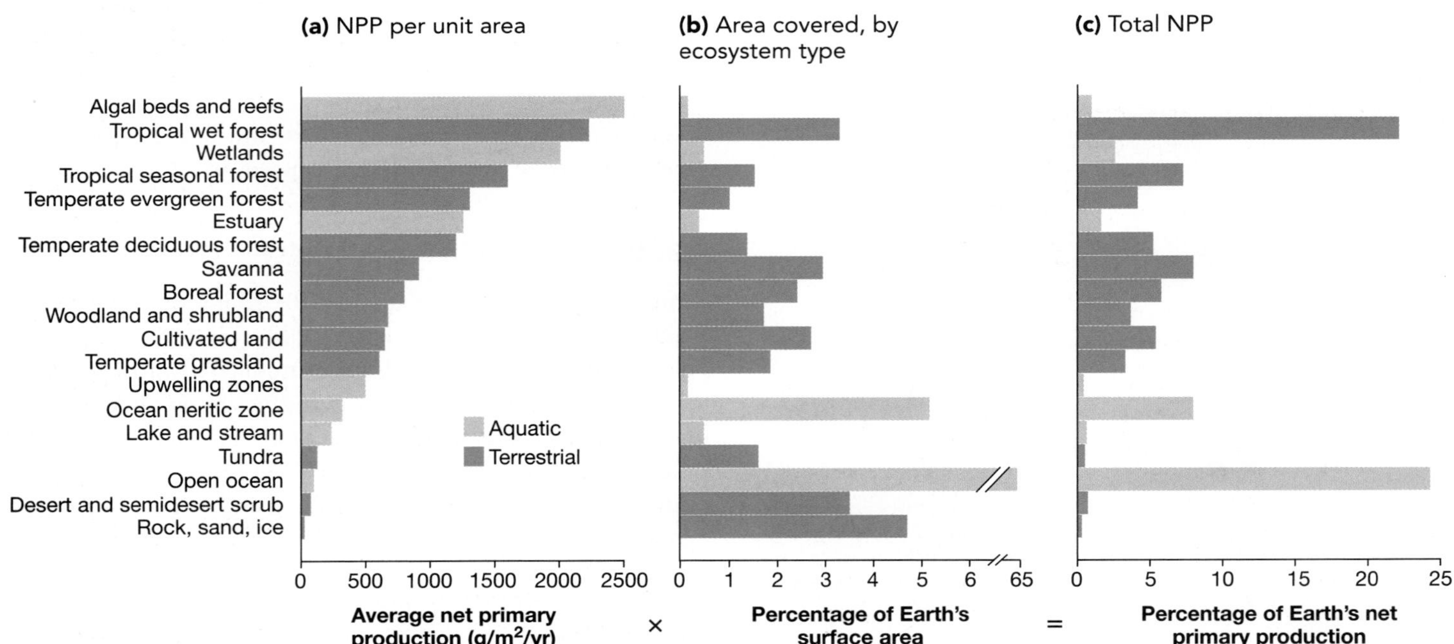

FIGURE 54.3 Net Primary Productivity Varies among Ecosystems. (a) Among biomes, average annual NPP per square meter varies by over three orders of magnitude. **(b)** Most of Earth's surface is covered by open ocean. The most common terrestrial habitat consists of unvegetated rock, sand, or ice. **(c)** Even though it has a low NPP per square meter, the open ocean is so vast that it is responsible for over 25 percent of Earth's total NPP.

EXERCISE Using information in Chapter 50, explain the patterns in part (a).

each biome; Figure 54.3b documents the total area that is covered by each type of ecosystem; and Figure 54.3c presents the percentage of the world's total productivity—the result of multiplying the data in part (a) by the data in part (b). Note that tropical wet forests and tropical seasonal forests, which have a dry season, cover less than 5 percent of Earth's surface but together account for over 30 percent of total NPP. Among aquatic ecosystems, the most productive habitats are algal beds and coral reefs, wetlands, and estuaries. Most of the total NPP provided by aquatic ecosystems derives from the open ocean, however. Even though NPP per square meter is low in these regions, the open ocean is so extensive that its total production is high.

These patterns raise an interesting question: What limits NPP in terrestrial and marine ecosystems? The short answer is that productivity is limited by any factor that limits the rate of photosynthesis—specifically, temperature and the availabilities of water, sunlight, and nutrients. Different limiting factors prevail in different environments, however. Let's take a closer look.

What Limits Productivity? The data in Figure 54.2 and Figure 54.3 document that terrestrial productivity is lowest in deserts and arctic regions. This observation suggests that the overall productivity of terrestrial ecosystems is limited by a combination of temperature and availability of water and sunlight.

To explain why the productivity of marine habitats is so much higher along coastlines than in deepwater regions, biologists focus on nutrient limitation. As Chapter 50 pointed out, the neritic and intertidal zones along coasts receive nutrients from two major sources: (1) rivers that carry and deposit nutrients from terrestrial ecosystems, and (2) nearshore ocean currents that bring nutrients from the cold, deep water of the oceanic zone back up to the surface. Both of these sources are absent in the surface waters of the open ocean. In addition, nutrients found in organisms near the surface of the open ocean—where sunlight is abundant—constantly fall to dark, deeper waters in the form of dead cells and are lost.

Analyses have confirmed that trace elements such as zinc, iron, and magnesium are particularly rare in the open ocean. These atoms are important because they are required as enzyme cofactors. Iron, for example, is essential to the proteins that are involved in electron transport chains (see Chapters 9 and 10). On the basis of these observations, biologists have proposed that the productivity of open-ocean ecosystems could be increased dramatically by fertilizing them with iron. The results of one iron-fertilization experiment are shown in **Figure 54.4**. The data show large increases in the concentration of chlorophyll *a* in surface waters over a two-week interval. Consistent with this result, recent research on ocean waters that are naturally enriched by nearby iron-containing rocks indicates that these regions have exceptionally high NPP. Results like these provide strong support for the hypothesis that NPP in marine ecosystems is limited primarily by the availability of nutrients and that iron is particularly important in the open ocean.

Experiment

Question: Is net primary productivity (NPP) in the open ocean limited by nutrients?

Hypothesis: NPP in the open ocean is limited by availability of iron.

Null hypothesis: NPP in the open ocean is not limited by availability of iron.

Experimental setup:

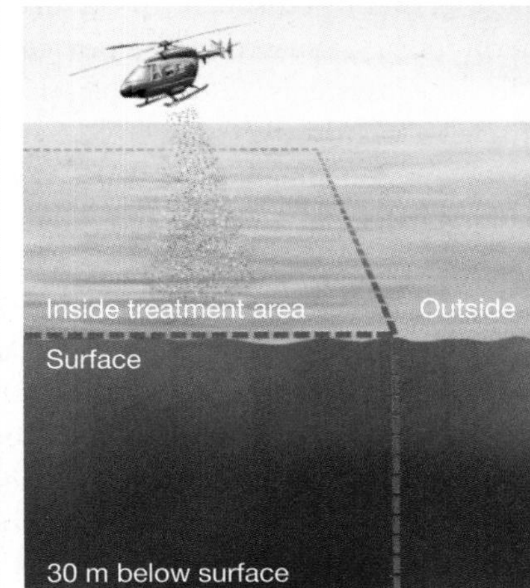

1. Add 350 kg iron (as $FeSO_4$) to a treatment area—a patch of ocean 8 km × 10 km.

2. Take water samples for a two-week period outside and inside the treatment area, at surface and at a depth of 30 m, and record amount of chlorophyll *a* present (as indicator of NPP).

Prediction: Amount of chlorophyll *a* near the surface inside the treatment area will increase relative to amounts outside the treatment area or at 30 m below the surface.

Prediction of null hypothesis: Amount of chlorophyll *a* will be the same in all measurements.

Results:

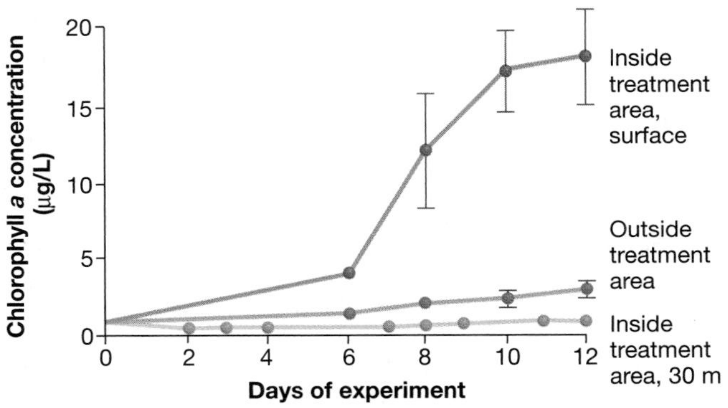

Conclusion: NPP in the open ocean is limited by the scarcity of nutrients—specifically, iron.

FIGURE 54.4 Fertilization with Iron Increases NPP in the Open Ocean.

QUESTION Most of the increased chlorophyll *a* was present in a single species of diatom (see Chapter 29). What happens to the biomass present in diatoms after they die?

How Does Energy Flow through an Ecosystem?

What happens to NPP? Net primary productivity results in **biomass**—organic material that non-photosynthetic organisms can eat. In every terrestrial and marine environment in the world, the chemical energy in primary producers eventually moves to one of two types of organisms: primary consumers or primary decomposers.

A **primary consumer** is an herbivore—an organism that eats plants or algae or other photosynthetic cells. As Chapter 53 pointed out, however, only a small percentage of all plant tissue is consumed by herbivores. Tissues that are not consumed eventually die. In forest ecosystems, for example, dead animals and dead plant tissues ("plant litter") are collectively known as **detritus**. Detritus is consumed by a variety of **primary decomposers**, including bacteria, archaea, and fungi.

Trophic Levels, Food Chains, and Food Webs Biomass represents chemical energy. Every time an herbivore eats a leaf or a fungus absorbs molecules from decaying wood, chemical energy flows from primary producers to primary consumers or decomposers. To describe these energy flows, biologists identify distinct feeding levels in an ecosystem. Organisms that obtain their energy from the same type of source are said to occupy the same **trophic** ("feeding") **level**.

A **food chain** connects the trophic levels in a particular ecosystem. In doing so, it describes how energy moves from one trophic level to another. As **Figure 54.5** shows, primary consumers are a key link in what biologists call the **grazing food chain**: the collection of organisms that eat plants, along with the organisms that eat herbivores. Consumers that eat herbivores are called **secondary consumers**. Organisms at the top trophic level in a food chain are not killed and eaten by any other organisms but enter the **decomposer food chain** when they die—along with biomass from dead primary producers and other consumers. A decomposer food chain starts with primary decomposers and includes primary consumers that usually specialize in eating primary decomposers. At higher trophic levels, though, the grazing and decomposer food chains merge. For example, the robin in Figure 54.5 functions as a secondary consumer in the grazing food chain and a tertiary consumer in the decomposer food chain.

In addition to feeding at several trophic levels, most organisms eat more than one type of food. As a result, food chains are almost always embedded in more complex **food webs**. The food web shown in **Figure 54.6** is a more complete description of the trophic relationships among the organisms in an ecosystem.

Food webs are a compact way of summarizing energy flows and documenting the complex trophic interactions that occur in communities. The keystone species introduced in Chapter 53, for example, are usually at the top of food chains and food webs. Removing or adding them has dramatic effects on a wide array of species at lower trophic levels. ● If you understand this concept, you should be able to predict how removing the top predator in Figure 54.5 would affect each of the other species in the food web.

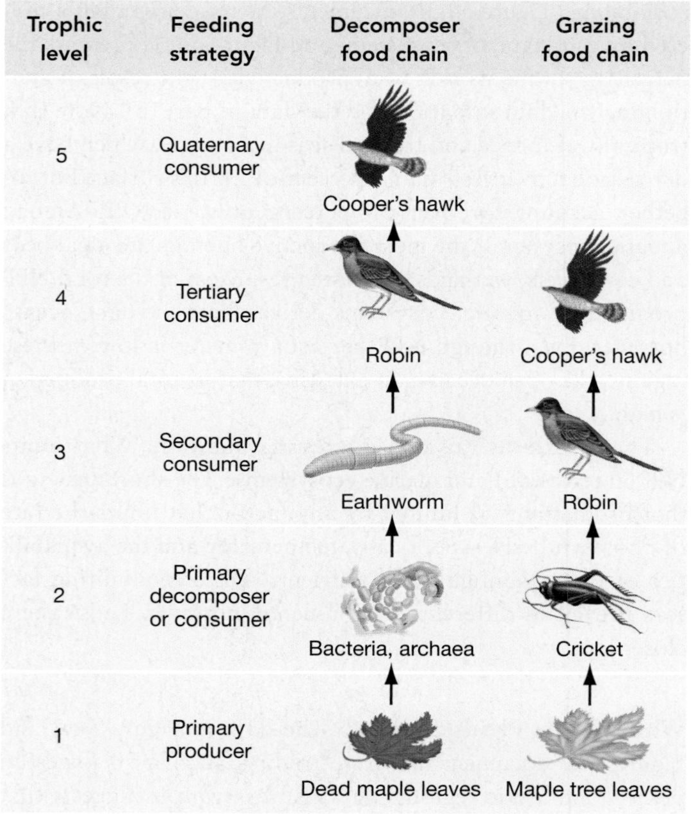

Trophic level	Feeding strategy	Decomposer food chain	Grazing food chain
5	Quaternary consumer	Cooper's hawk	
4	Tertiary consumer	Robin	Cooper's hawk
3	Secondary consumer	Earthworm	Robin
2	Primary decomposer or consumer	Bacteria, archaea	Cricket
1	Primary producer	Dead maple leaves	Maple tree leaves

FIGURE 54.5 Trophic Levels Identify Steps in Energy Transfer. Examples from a temperate-forest ecosystem are shown; many other species exist at each level in this ecosystem.

● **EXERCISE** Label the "top predator" in these food chains—a species that is not eaten by any other species.

Why Is Energy Lost at Each Trophic Level? All ecosystems share a characteristic pattern: The total biomass produced each year declines from lower trophic levels up to the higher levels. Although biomass production at each trophic level varies widely among ecosystems, the data in **Figure 54.7** are representative. The most striking observation is that biomass production at upper trophic levels is often only 10 percent of the production at the next lowest level.

To understand why this pattern occurs, consider the interface between primary producers and primary consumers. Much of the NPP produced each year is unavailable to herbivores, because it resides in indigestible substances such as the lignin in wood or because it is protected by noxious defensive compounds. NPP that is not eaten by an herbivore enters the decomposer food web. Even if the material is ingested by an herbivore,

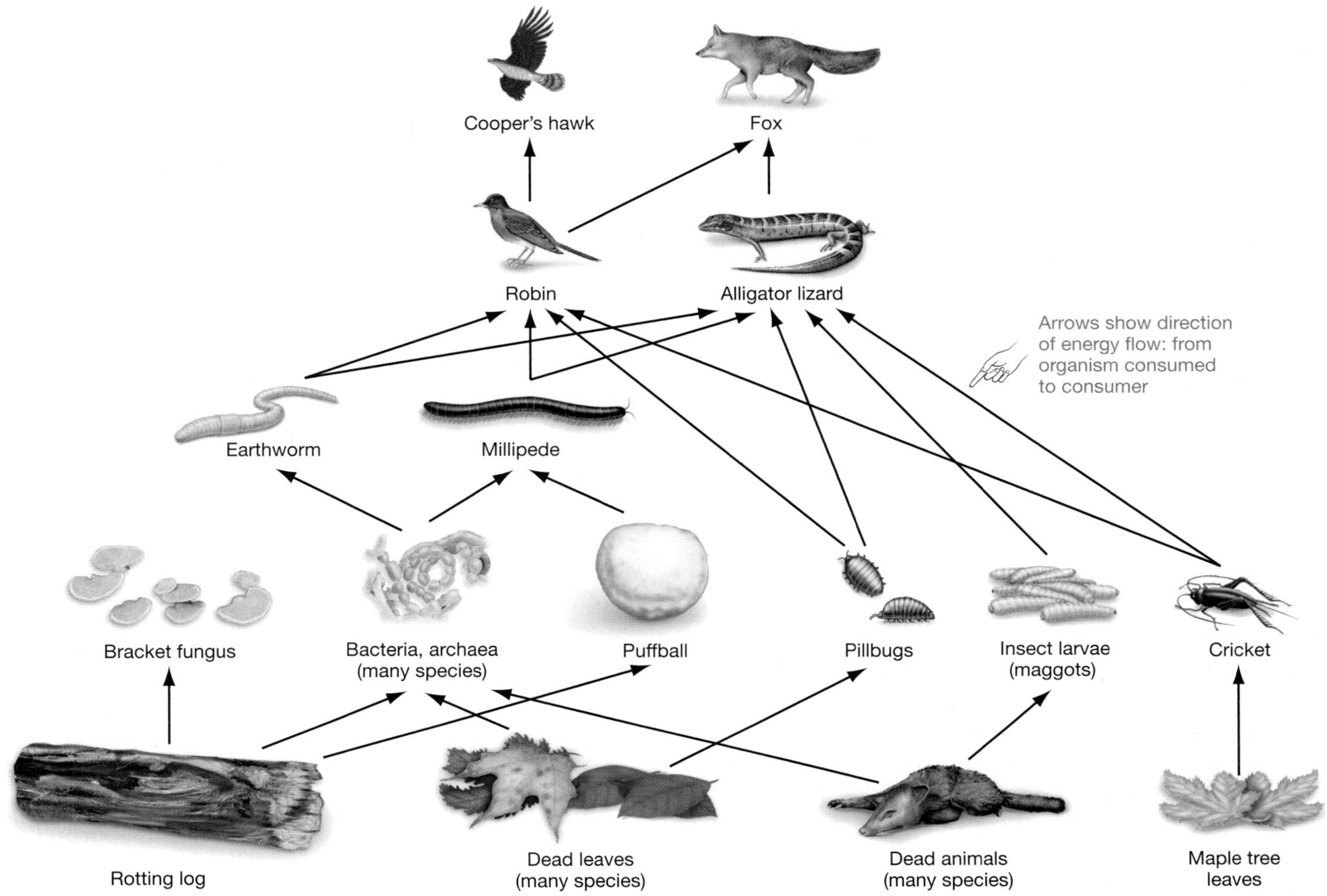

Cooper's hawk
Fox
Robin
Alligator lizard
Arrows show direction of energy flow: from organism consumed to consumer
Earthworm
Millipede
Bracket fungus
Bacteria, archaea (many species)
Puffball
Pillbugs
Insect larvae (maggots)
Cricket
Rotting log
Dead leaves (many species)
Dead animals (many species)
Maple tree leaves

FIGURE 54.6 Food Webs Describe Trophic Relationships in an Ecosystem. Food webs offer a more comprehensive analysis of feeding relationships than food chains do. This food web shows only a fraction of the total feeding relationships and species of a temperate deciduous forest. More complete food webs include species in the grazing food chain, along with data on the average amount of energy transferred between each link in the web.

Production of biomass (g/m2/year)
Tertiary consumers
3
10%
Secondary consumers
30
15%
Efficiency of energy transfer
Primary consumers and decomposers
200
20%
Primary producers
1000

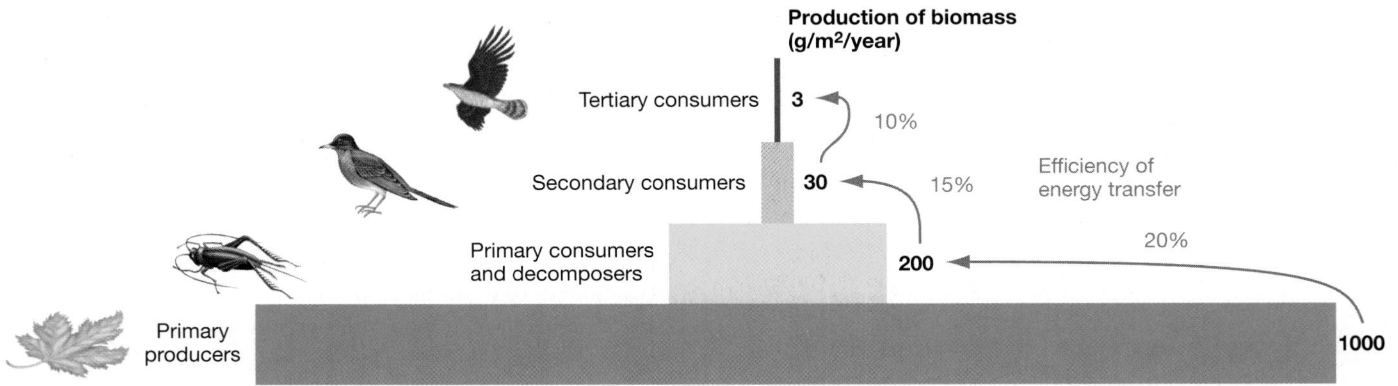

FIGURE 54.7 Productivity Declines at Higher Trophic Levels. In all ecosystems, annual production of biomass is highest at the lowest (bottom) trophic level and declines at higher levels. This pattern is the pyramid of productivity.

most of the energy stored in the chemical bonds of carbon compounds is lost as heat as it is metabolized or used to keep the consumer alive—not to grow or reproduce. At the next higher trophic level, many herbivores are never consumed by secondary consumers, because these herbivores defend themselves effectively and die of other causes. Of the energy that is successfully ingested by secondary consumers, some is again lost as heat or used up in efforts to stay alive and capture prey.

The general point here is simple: The amount of biomass produced at the second trophic level must be less than productivity at the first trophic level; productivity at the third trophic level must be less than that at the second. This pattern holds true for the entire food chain and produces a pyramid of productivity. Production of biomass is highest at the lowest trophic level. ● If you understand this concept, you should be able to explain why keystone species have relatively low biomass. You should also be able to explain why biologists consider vegetarian diets in humans much more efficient than diets that include large quantities of meat.

What Limits the Length of Food Chains? It is interesting to note that none of the food chains, food webs, or trophic structures presented in this chapter thus far have more than four or five trophic levels. When a biologist reviewed the characteristics of food chains that had been documented by researchers working in a wide variety of ecosystems, he found that the maximum number of links ranged from 1 to 6. Each link joins two trophic levels. In this study, terrestrial and lake ecosystems had 3.7 links per food chain, on average, while streams had an average of 3.2 links. Why don't ecosystems have 8 or 9 or 10 trophic levels? Why does the overall average number of levels seem to be about 3.5? Several competing hypotheses have been offered to explain this observation. Let's consider three:

- *Hypothesis 1: Energy Transfer* As energy is transferred up a food chain, a large fraction of that energy is lost. By the time energy reaches the top trophic level, there may not be enough left to support an additional suite of consumers. Suppose that the efficiency of energy transfer between each pair of trophic levels is 10 percent. If the initial trophic level in a hypothetical ecosystem produces 10,000 kcal per day, then the second, third, and fourth levels will produce 1000, 100, and 10 kcal per day, respectively. Can any organism obtain enough energy to survive at a fifth trophic level?

  The hypothesis that food-chain length is limited by productivity leads to a strong prediction: There should be more trophic levels in ecosystems with higher productivity or higher efficiency of energy transfer than in ecosystems with lower productivity or lower efficiency of energy transfer. When a researcher analyzed data on the productivity of four aquatic and four terrestrial ecosystems, however, he found that low-productivity ecosystems were as likely to contain four trophic levels as high-productivity ones were. To date, research has not supported the prediction that the most productive ecosystems have the longest food chains.

- *Hypothesis 2: Stability* Stuart Pimm proposed an alternative hypothesis to the energy-transfer explanation for why food chains are short. His idea was that long food chains are easily disrupted by droughts, floods, or other disturbances and thus tend to be eliminated. Using mathematical models, he demonstrated that long food chains took longer to return to their previous state following a disturbance than did short food chains. He then proposed that long food chains are unlikely to persist in a variable environment.

  Pimm's hypothesis also predicts that the length of food chains should increase with the stability of the environment. A biologist tested this prediction by comparing the animal communities that develop inside tree holes in Australia and Great Britain. After water accumulates in depressions or in the holes in tree branches or roots, leaf litter from the trees falls in and forms the basis of a food web. The researcher found that tree-hole communities in Australia and Great Britain were similar in many respects, except that annual leaf fall was much more variable in the British communities than in the Australian ones. As predicted by the stability hypothesis, the British ecosystems had only two trophic levels while Australian habitats supported three.

  Other researchers have challenged the assumptions behind Pimm's theoretical analysis, however, and support for the stability hypothesis remains tentative. Experimental and observational tests of this hypothesis are continuing.

- *Hypothesis 3: Environmental Complexity* Biologists have also hypothesized that food-chain length is a function of an ecosystem's physical structure. Specifically, researchers have proposed that tundra, grasslands, lake and sea bottoms, streambeds, and intertidal zones offer a largely flat, or two-dimensional, surface to the organisms living there. In contrast, forests and open-water environments in lakes and rivers offer three-dimensional volumes. This hypothesis predicts that three-dimensional ecosystems should have longer food chains than two-dimensional sites do.

  To test this hypothesis, researchers examined 113 publications that described food chains in a wide variety of aquatic and terrestrial environments. These data supported the prediction that food webs are significantly longer in three-dimensional ecosystems. As a result, the investigators concluded that dimensionality does influence food-web structure. As yet, however, no one knows why. The mechanism for this pattern remains to be determined.

To summarize, there is unlikely to be a single, simple answer to the question of what limits food-chain length. Inefficiencies of energy transfer, environmental stability, and environmental complexity may all influence the number of trophic levels that can be supported in a given ecosystem. Multiple causation is not unusual in biological science. In genetics, for example, gene expression is rarely triggered by a single regulatory site in DNA and just one regulatory protein. Instead, the amount and timing of gene expression result from interactions among dozens of regulatory DNA sequences and proteins. Recognizing multiple causation is productive, because it can suggest exciting new experiments and analyses. For example, if food-chain length does have multiple causes, then habitats that are particularly productive, stable, and highly complex in physical structure should have the longest food chains. Research continues on this fundamental question in ecosystem ecology.

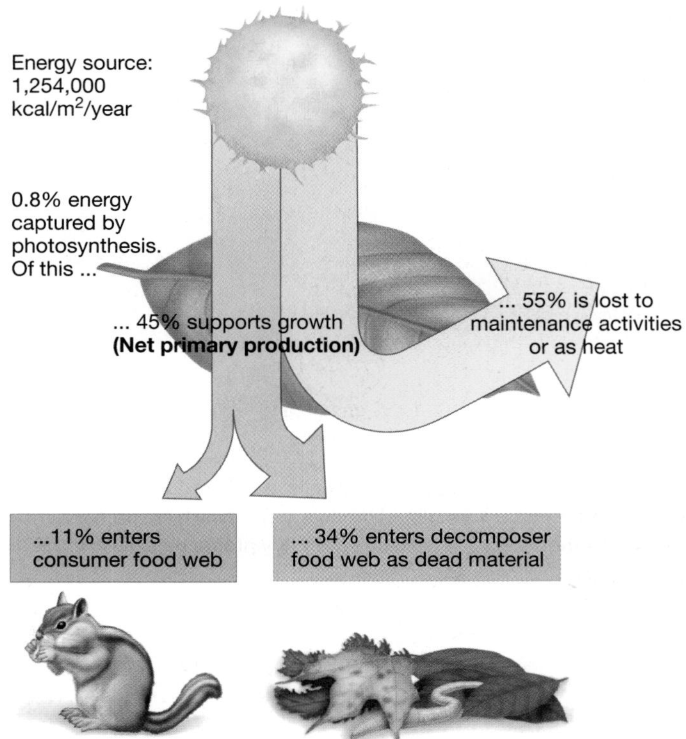

FIGURE 54.8 Energy Flow through the Hubbard Brook Forest Ecosystem. Energy from sunlight is transformed to chemical energy by photosynthesis. Photosynthetic products fuel new plant growth, which eventually enters either the consumer or decomposer food web.

EXERCISE Calculate the amount of energy transferred to primary consumers and decomposers each year, in kcal/m^2.

QUESTION If NPP in this ecosystem increases due to global warming, predict the impact on primary consumers and decomposers.

Analyzing Energy Flow: A Case History

Research on energy flow in ecosystems has uncovered a series of important general principles. In both terrestrial and marine ecosystems, NPP varies with region and biome. On land, the most important constraints on productivity are warmth and moisture; in aquatic habitats, nutrient availability is key. The chemical energy in biomass flows into consumer or decomposer food webs, where it is passed up a series of trophic levels. In terrestrial ecosystems, the total biomass declines with each trophic level.

To see these general principles in action, let's analyze energy flow in a specific ecosystem. For almost five decades, a team of researchers has been studying how energy and nutrients flow through a temperate forest ecosystem in the northeastern United States: the Hubbard Brook Experimental Forest in New Hampshire. Hubbard Brook may qualify as the most intensively studied ecosystem in the world.

As **Figure 54.8** shows, energy flow in this ecosystem begins when plants capture the energy in solar radiation via photosynthesis. At Hubbard Brook, the amount of energy entering the ecosystem from sunlight varies throughout the year and, to a lesser extent, from year to year. From June 1, 1969, to May 31, 1970, which was a typical year, 1,254,000 kilocalories (kcal) of solar radiation per square meter (m^2) reached the forest. If this amount of energy were available in the form of electricity, it would easily power two 150-watt lightbulbs that burned continuously for one year.

By documenting rates of photosynthesis in a large sample of forest plants, the biologists calculated that the plants used 10,400 kcal/m^2 of energy in photosynthesis. This value represents **gross primary productivity**, which is the total amount of photosynthesis in a given area and time period. The team also calculated **gross photosynthetic efficiency**, or the efficiency with which plants use the total amount of energy available to them, as the ratio of gross photosynthesis to solar radiation in kcal/m^2. At Hubbard Brook, efficiency was $10{,}400 \div 1{,}254{,}000 = 0.8$ percent. This value is typical of other ecosystems as well.

Why is efficiency so low? The answer has several components. Plants in temperate biomes cannot photosynthesize at all in winter, although sunlight is available. If conditions get dry during the summer, stomata close to conserve water and photosynthesis stops due to a lack of CO_2. Even when conditions are ideal, the pigments that drive photosynthesis can respond to only a fraction of the wavelengths available and thus a fraction of the total energy received.

Given that only a tiny fraction of incoming sunlight is converted to chemical energy and that only a fraction of this gross primary productivity is used to build biomass, let's consider the next question: What happens to the NPP at Hubbard Brook?

How Does Energy Flow through the Hubbard Brook Ecosystem? Between 1969 and 1971 the amount of energy that entered the consumer food web at Hubbard Brook varied from about 1 percent of net primary productivity per year to more than 40 percent. Leaves, seeds, and fruits were the most commonly consumed plant parts; wood was rarely eaten.

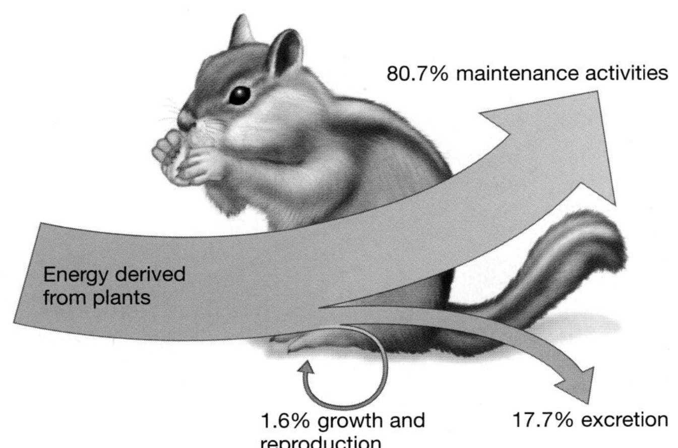

FIGURE 54.9 How Do Consumers Use Primary Production? A small fraction of the energy consumed by chipmunks, which are primary consumers (herbivores), is used for secondary production. Most of the energy is used for cellular respiration.

QUESTION Aphids are ectothermic, largely sedentary, and feed on high-quality foods containing almost no indigestible molecules. Describe how the percentages given here would compare in aphids.

Figure 54.9 illustrates what happened to the energy obtained by consumers, using chipmunks as an example consumer. Chipmunks are small rodents that are seed predators. On average, these primary consumers harvest 31 kcal/m^2 of energy each year. Of that total, 17.7 percent is unused and excreted and 80.7 percent is lost to respiration and other maintenance processes. Just 1.6 percent of the yearly 31 kcal/m^2 goes into the production of new chipmunk tissue by growth and reproduction. The production of new tissue by primary consumers is called **secondary production**.

Compared to endotherms such as chipmunks, secondary production is much higher in ectothermic consumers. Caterpillars, for example, transform about 5.4 percent of the energy they ingest into new tissue. Because ectotherms rely principally on heat gained from the environment and do not oxidize sugars to keep warm, they devote much less energy to cellular respiration than endotherms do. Even in the case of the ectotherms, though, it is clear that only a tiny fraction of the available solar radiation is involved in secondary production.

What about biomass that enters the decomposer food web? At Hubbard Brook, about 75 percent of total net primary productivity is not eaten while alive. This energy enters the decomposer food web and is passed up subsequent trophic levels, just as it is in the consumer food web. But in addition, the researchers at Hubbard Brook found that large amounts of energy leave the decomposer food web in the form of detritus that washes into streams. This transfer of energy from the forest ecosystem to the aquatic ecosystem is important, because it is a major source of energy for aquatic organisms. At Hubbard Brook, photosynthesis by aquatic algae and plants introduced only about 10 kcal/m^2/yr of energy. In contrast, each year 6039 kilocalories washed into each square meter of streambed from the surrounding forest.

Long-term studies have documented the flow of energy into, through, and out of Hubbard Brook and other types of ecosystems. Although the specific numbers found at Hubbard Brook are unique to that site and the time interval of that study, the general patterns have turned out to be typical of ecosystems around the globe.

Check Your Understanding

If you understand that... 

- In an ecosystem, energy flows from sunlight or inorganic compounds with high potential energy to producers, and from there to consumers and decomposers.
- In most ecosystems, NPP is limited by conditions that limit the rate of photosynthesis: temperature and/or the availability of sunlight, water, and nutrients.
- Productivity diminishes at each subsequent trophic level in the ecosystem.

You should be able to... 

1) Summarize variation in annual NPP among ecosystems and briefly explain why some ecosystems are particularly high or low.
2) Give two reasons that the amount of biomass present in terrestrial ecosystems diminishes with increasing trophic levels.
3) Evaluate at least one hypothesis to explain why most food chains have only two or three links.
4) Predict what will happen to consumer and decomposer food chains in ecosystems where global warming leads to increases or decreases in NPP.

54.2 Biogeochemical Cycles

Energy is not the only quantity that is transferred when one organism eats another. The organisms that are eaten also contain carbon (C), nitrogen (N), phosphorus (P), calcium (Ca), and other elements that act as nutrients. Atoms are constantly reused as they cycle through trophic levels and air, water, and soil. The path that an element takes as it moves from abiotic systems through organisms and back again is referred to as its **biogeochemical** ("life-Earth-chemical") **cycle**.

Because humans are now disturbing biogeochemical cycles on a massive scale, research on this aspect of ecosystem ecology is exploding. To get a basic understanding of how a particular biogeochemical cycle works, researchers focus on three fundamental questions:

1. What are the nature and size of the **reservoirs**, or areas where elements are stored for a period of time? In the case of carbon, the biomass of living organisms is an important reservoir, as are sediments and soils. Another significant carbon reservoir is buried in the form of coal and oil.

2. How fast does the element move between reservoirs, and what factors influence these rates? The global photosynthetic rate, for example, measures the rate of carbon flow from CO_2 in the atmosphere to living biomass. When fossil fuels burn, carbon that was buried in coal or petroleum moves into the atmosphere as carbon dioxide (CO_2).

3. How does a biogeochemical cycle interact with other cycles? For example, researchers are trying to understand how changes in the nitrogen cycle affect carbon.

Before taking up these issues, though, let's begin with a general overview of what nutrient cycling is.

Biogeochemical Cycles in Ecosystems

Figure 54.10 shows a simplified version of a generalized terrestrial nutrient cycle. In this case, the cycle starts when nutrients are taken up from the soil by plants and assimilated into plant tissue. If the plant is eaten, the nutrients pass to the animal members of the ecosystem's consumer food web; if the plant dies, the nutrients enter the decomposer food web. Once consumed by an animal, nutrients are excreted in fecal matter or urine, taken up by a parasite or predator, or added to the dead biomass reservoir when the animal dies.

The nutrients in plant litter, animal excretions, and dead animal bodies are used by bacteria, archaea, roundworms, fungi, and other primary decomposers. The combination of microscopic decomposers and the ions and molecules they release combine to form the soil organic matter. Soil organic matter is a complex mixture of partially and completely decomposed detritus. Completely decayed organic material is called **humus**, because it is rich in a family of carbon-containing molecules called humic acids. (Chapter 38 described other components of the soil.) Eventually, the nutrients in soil organic matter are converted to an inorganic form. For example, cellular respiration by soil-dwelling bacteria and archaea converts the nitrogen present in amino acids that are found in detritus to ammonium (NH_4^+) or nitrate (NO_3^-) ions. Once this step is accomplished, the nutrients are available for uptake by plants.

A key feature of this process is that nutrients are reused. Reuse is not complete, however. Nutrients leave the ecosystem whenever plant or animal biomass leaves. For example, plant biomass is removed if an herbivore enters the ecosystem, eats a plant, and migrates out of the ecosystem before excreting nutrients or dying. Nutrients also leave ecosystems when flowing water or wind removes particles or inorganic ions and deposits them somewhere else. Soil erosion has a huge impact on nutrient cycles: It removes nutrients rapidly and in potentially large quantities.

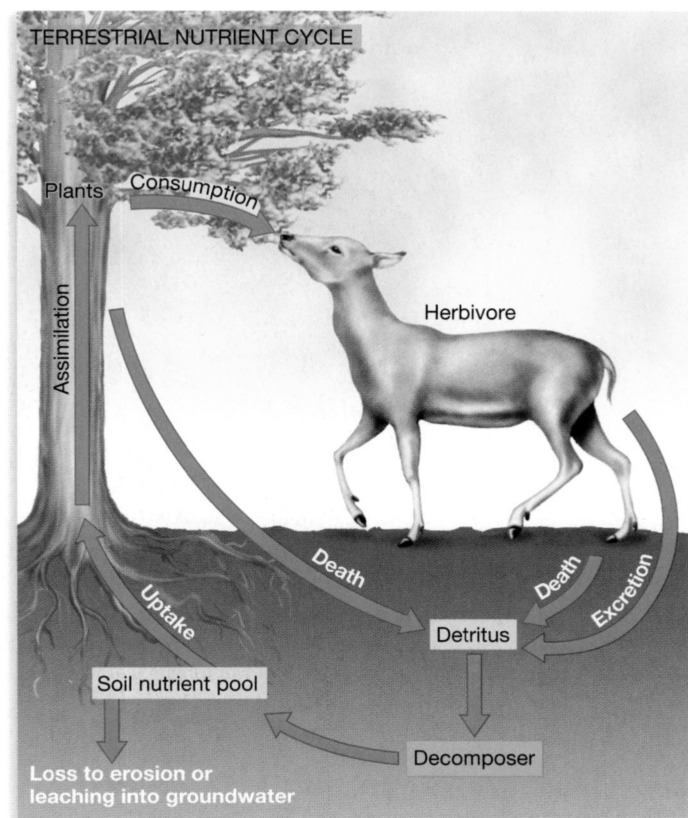

FIGURE 54.10 Generalized Terrestrial Nutrient Cycle. Nutrients cycle from organism to organism in an ecosystem via assimilation, consumption, and decomposition. Nutrients are exported from ecosystems when water or organisms leave the area.

What Factors Control the Rate of Nutrient Cycling? Of the many links in a nutrient cycle, the decomposition of detritus most often limits the overall rate at which nutrients move through an ecosystem. Until decomposition occurs, nutrients stay tied up in intact tissues. The decomposition rate, in turn, is influenced by two types of factors: abiotic conditions such as temperature and precipitation, and the quality of the detritus as a nutrient source for the fungi, bacteria, and archaea that accomplish decomposition.

To appreciate the importance of abiotic conditions on the decomposition rate, consider the difference in detritus accumulation between boreal forests and tropical wet forests. Chapter 50 indicated that boreal forests are found in areas where temperature is low. As a result, soils in these ecosystems are cold and wet. Tropical wet forests, in contrast, occur in areas where temperatures and rainfall are high. Soils tend to remain moist and warm all year long.

(a) Boreal forest: Accumulation of detritus and organic matter

(b) Tropical wet forest: Almost no organic accumulation

FIGURE 54.11 Temperature and Moisture Affect Decomposition Rates. (a) In boreal forests, decomposition rates are limited by cold soil temperatures. Organic matter builds up, because the input of detritus into the soil exceeds the decomposition rate. **(b)** In tropical wet forests, warm temperatures allow decomposition to proceed rapidly, so organic matter does not build up.

Figure 54.11 illustrates typical soils from boreal forests and tropical wet forests. Notice that the uppermost part of the soil in a boreal forest consists of partially decomposed detritus and organic matter. There is no such layer at the top of the soil in a tropical forest. The contrast occurs because the cold and wet conditions in boreal forests limit the metabolic rates of decomposers. As a result, decomposition fails to keep up with the input of detritus, and organic matter accumulates there. In the tropics, conditions are so favorable for fungi, bacteria, and archaea that decomposition keeps pace with detrital inputs. Nutrients cycle slowly through boreal forests but rapidly through wet tropical forests.

The quality of detritus also exerts a powerful influence on the decomposition rate and thus on nutrient availability. For example, decomposers can be hampered by the presence of large compounds in detritus that are difficult to digest, such as lignin. The presence of lignin, which is a primary constituent of wood, is one reason that wood takes much longer to decompose than leaves do. As Chapter 31 pointed out, only basidiomycete fungi, ascomycete fungi, and a few bacteria have the enzymes required to completely degrade lignin. The growth of decomposers is also inhibited if detritus is low in nitrogen.

What Factors Influence the Rate of Nutrient Loss?

Nutrient availability has a profound effect on productivity, so the rate of nutrient loss is an important characteristic of an ecosystem. Several of the major impacts that humans have on ecosystems—such as farming, logging, burning, and soil erosion—accelerate nutrient loss.

To test the effect of vegetation removal on nutrient export, researchers initiated a large-scale experiment at the Hubbard Brook Experimental Forest. They chose two similar **watersheds**—areas drained by a single stream—for study (**Figure 54.12**). They then cut all vegetation, including the trees, from the forests in one of the two watersheds. In the following three years, this clear-cut area was treated with an herbicide to prevent vegetation from regrowing. As a result, the experimental watershed was devegetated. An untreated watershed served as a control.

Before removing the vegetation, the researchers had documented that 90 percent of the nutrients in the ecosystem was in soil organic matter, and an additional 9.5 percent was in plant biomass. After the vegetation was removed, the team monitored the concentrations of nutrients in the streams exiting the two watersheds. The graph in Figure 54.12 documents the amount of dissolved substances that subsequently washed out of the stream in each watershed over the course of four years. Losses from the devegetated site were typically 10 times as high as they were from the control site.

Based on these data, the researchers concluded that devegetation has a huge impact on nutrient export. Instead of being held in the ecosystem and recycled, nutrients wash out in the absence of vegetation. The loss occurs because nutrients in the soil are either dissolved in water or attached to small particles of sand or clay (see Chapter 38). If plant roots no longer hold the soil particles in place and if they no longer take up and recycle aqueous nutrients, then the molecules and ions wash out of the soil and are lost to the ecosystem. Consequently, soil quality may decline over time if the area is kept

(a) Hubbard Brook Experimental Forest

Experiment

(b) Question: How does the presence of vegetation affect the rate of nutrient export in a temperate-forest ecosystem?

Hypothesis: Presence of vegetation lowers the rate of nutrient export because it increases soil stability and recycling of nutrients.

Null hypothesis: Presence of vegetation has no effect on the rate of nutrient export.

Experimental setup:

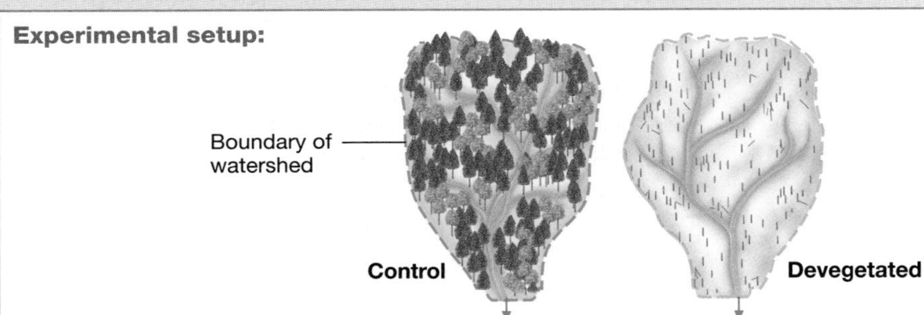

1. Choose two similar watersheds. Document nutrient levels in soil organic matter, plants, and streams.

2. Devegetate one watershed, and leave the other intact.

3. Monitor the amount of dissolved substances in streams.

Prediction: Amount of dissolved substances in stream in devegetated watershed will be much higher than amount of dissolved substances in stream in control watershed.

Prediction of null hypothesis: No difference will be observed in amount of dissolved substances in the two streams.

Results:

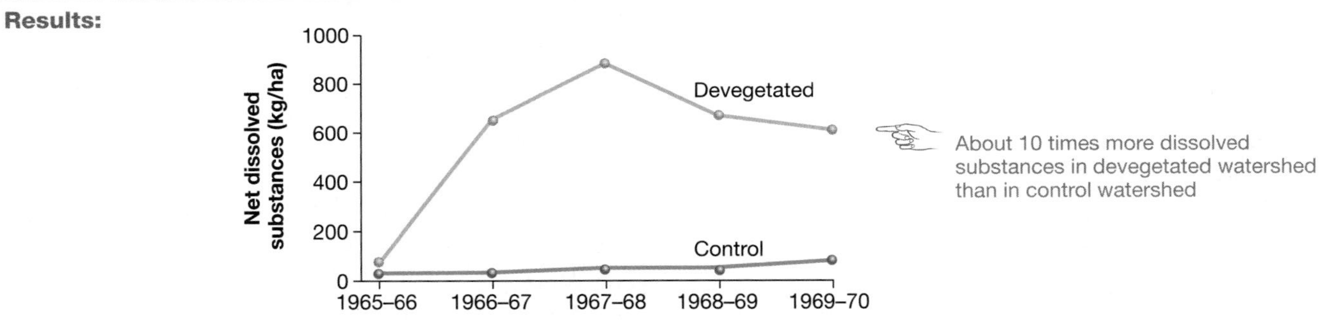

Conclusion: Presence of vegetation limits nutrient loss. Removing vegetation leads to large increases in nutrient export.

FIGURE 54.12 Experimental Evidence that Deforestation Increases the Rate of Nutrient Loss from Ecosystems.

QUESTION In effect, this experiment removed one of the arrows in Figure 54.10. Which one?

in a devegetated state. Long-term devegetation of this type has occurred in formerly forested areas of the Middle East, North Africa, and elsewhere due to intensive farming and grazing. As a result, the productivity of these regions is a tiny fraction of what it once was.

Global Biogeochemical Cycles

When nutrients leave one ecosystem, they enter another. In this way, the movement of ions and molecules among ecosystems links local biogeochemical cycles into one massive global system. Local and global cycles interact when water, organisms, or wind move nutrients. As an introduction to how these global biogeochemical cycles work, let's consider the global water, carbon, and nitrogen cycles. These cycles have recently been heavily modified by human activities—with serious ecological consequences.

The Global Water Cycle A simplified version of the **global water cycle** appears in **Figure 54.13**. The diagram shows the estimated amount of water that moves between major components of the cycle over the course of a year.

To analyze this cycle, begin with evaporation of water out of the ocean and the subsequent precipitation of water back into the water. For the marine component of the cycle, evaporation exceeds precipitation—meaning that, over the oceans, there is a net gain of water to the atmosphere. When this water vapor moves over the continents, it is joined by a small amount of water that evaporates from lakes and streams and a large volume of water that is transpired by plants. The total volume of water in the atmosphere over land is balanced by the amount of rain and other forms of precipitation that occurs on the continents. The cycle is completed by the water that moves from the land to the oceans via streams and **groundwater**—water that is found in soil.

Humans are affecting the water cycle in complex ways. Perhaps the simplest and most direct impacts concern rates of groundwater replenishment. Asphalt and concrete surfaces added through suburbanization reduce the amount of precipitation that percolates from the surface to enter deep soil layers. The conversion of grasslands and forests into agricultural fields also increases the amount of water that runs off Earth's surface into streams instead of penetrating to groundwater layers. Intact fields and forests lose less water than croplands do because extensive root systems that hold water are in place year-round. Most importantly, the dramatic increases in irrigated agriculture over the past three decades have resulted in massive quantities of water being removed from groundwater storage and brought to the surface.

In combination, these impacts have resulted in alarming drops in groundwater storage. The **water table** is the upper limit below the surface that the ground is saturated with stored water, and it is dropping on every continent. Between 1986 and 2006, the water table north of Beijing, China, experienced drops averaging almost 61 m (200 ft). Throughout India, water tables are falling at a rate of between 1 m and 3 m per year. Similar rates are being documented in Yemen, parts of Mexico, and states in the southern Great Plains region of the United States. Lack of water is exacerbating political tensions in several areas of the world, including the Middle East. Humans are mining water in many parts of the world—meaning that populations have already grown beyond carrying capacity for water.

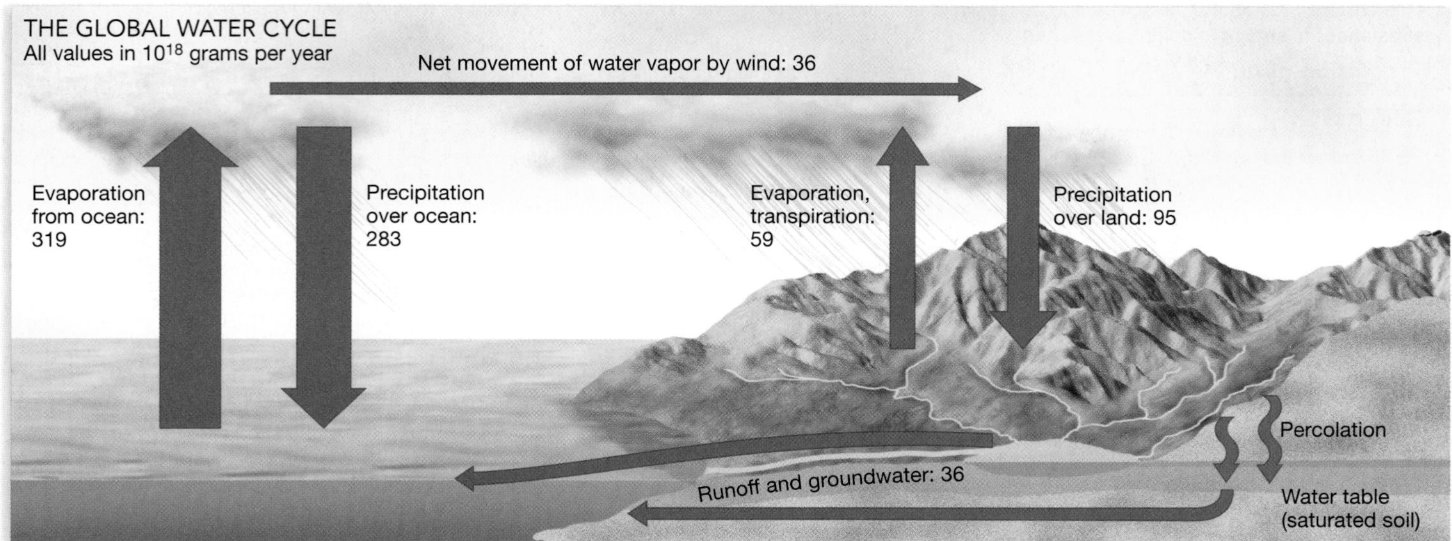

FIGURE 54.13 The Global Water Cycle.

QUESTION Predict how the amount of water evaporated from the ocean is changing in response to global warming. Discuss one possible consequence of this change.

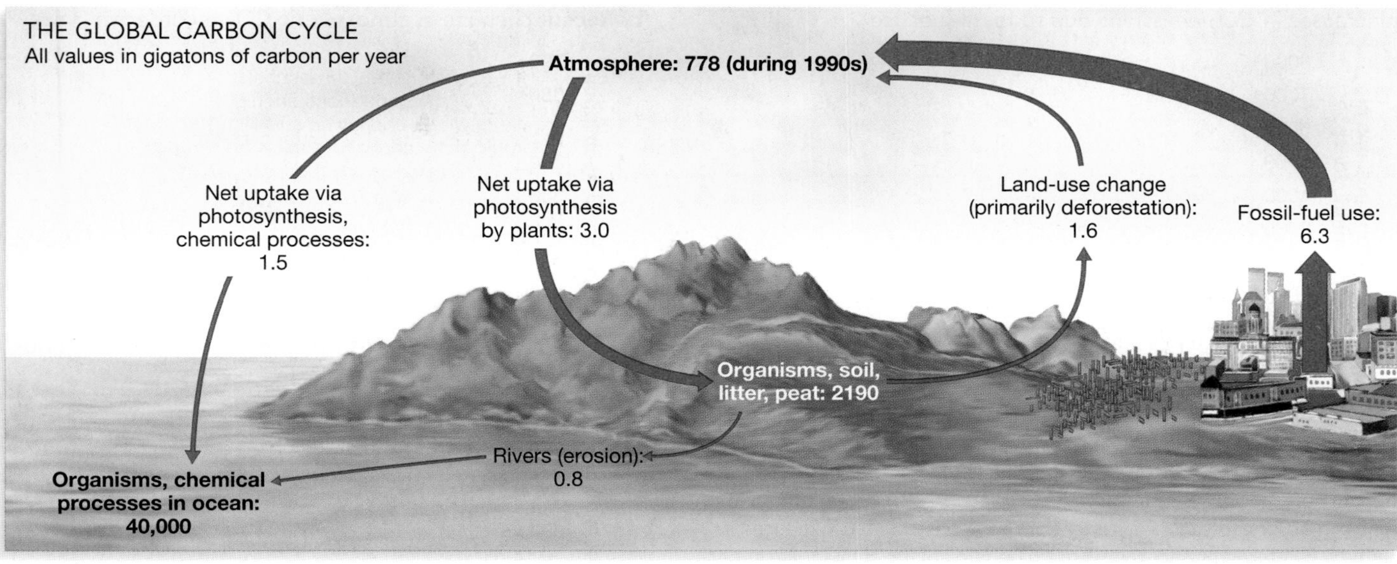

FIGURE 54.14 The Global Carbon Cycle. The arrows indicate how carbon moves into and out of ecosystems. Deforestation and the use of fossil fuels are adding 7.9 gigatons of carbon to the atmosphere each year. Of that 7.9 gigatons of carbon produced by human activities, 2 gigatons are fixed by photosynthesis in terrestrial ecosystems and 2 gigatons are fixed by physical and chemical processes in the oceans. The remainder—3.9 gigatons—is added to the atmosphere.

The Global Carbon Cycle As **Figure 54.14** shows, the **global carbon cycle** documents the movement of carbon among terrestrial ecosystems, the oceans, and the atmosphere. Of these three reservoirs, the ocean is by far the largest. The atmospheric reservoir is also important despite its relatively small size, because carbon moves into and out of it rapidly.

The arrows in Figure 54.14 emphasize that carbon frequently moves into and out of the atmospheric pool through organisms. In both terrestrial and aquatic ecosystems, for example, photosynthesis is responsible for taking carbon out of the atmosphere and incorporating it into tissue. Cellular respiration, in contrast, releases carbon that has been incorporated into living organisms to the atmosphere, in the form of carbon dioxide.

How have humans changed the carbon cycle? The fossil fuels found on Earth, which are derived from carbon-rich sediments, are estimated to contain a total of 5000–10,000 gigatons of carbon. (A gigaton, symbolized Gt, is a billion tons). This is one-eighth to one-fourth the size of the oceanic reservoir and 2.5 to 5 times the size of the terrestrial reservoir. In effect, burning fossil fuels moves carbon from an inactive geological reservoir to an active reservoir—the atmosphere. When you burn gasoline, you are releasing carbon atoms that have been locked up in coal or petroleum reservoirs for hundreds of millions of years.

Land-use changes have also altered the global carbon cycle. Deforestation, for example, reduces an area's net primary productivity. It also releases CO_2 when fire is used for clear-cutting or when dead limbs, twigs, and stumps are left to decompose. Researchers who used data on the amount of land cleared for agriculture and forestry to estimate the amount of carbon released found that, at a global scale, a net movement of carbon to the atmosphere has been occurring from terrestrial ecosystems for at least the past 100 years. The change from net carbon storage to net carbon release is the result of expanding human populations (see Chapter 52).

Figure 54.15a highlights the dramatic increase in carbon released from fossil-fuel burning over the past century; **Figure 54.15b** shows the consequences of this accelerated carbon release. In just 45 years, CO_2 in the atmosphere at Mauna Loa Observatory on the island of Hawaii has increased from about 315 to over 375 parts per million (ppm)—meaning milligrams of CO_2 per kilogram of air. The same trend has been observed at sites around the globe. As the data in **Figure 54.15c** show, CO_2 concentrations in the atmosphere have now risen to levels far above the 280 ppm that was typical prior to 1860.

These changes in the global carbon cycle are important because carbon dioxide functions as a **greenhouse gas**: It traps heat that has been radiated from Earth and keeps it from being lost to space, similar to the way the glass of a greenhouse traps heat. More specifically, carbon dioxide is one of several gases in the atmosphere that absorb and reflect the infrared wavelengths radiating from Earth's surface. Increases in amounts of greenhouse gases are warming Earth's climate by increasing the atmosphere's heat-trapping potential.

at www.masteringbio.com

The Global Carbon Cycle

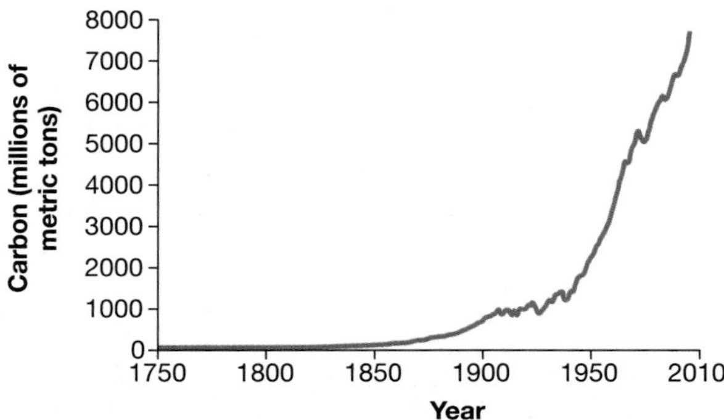

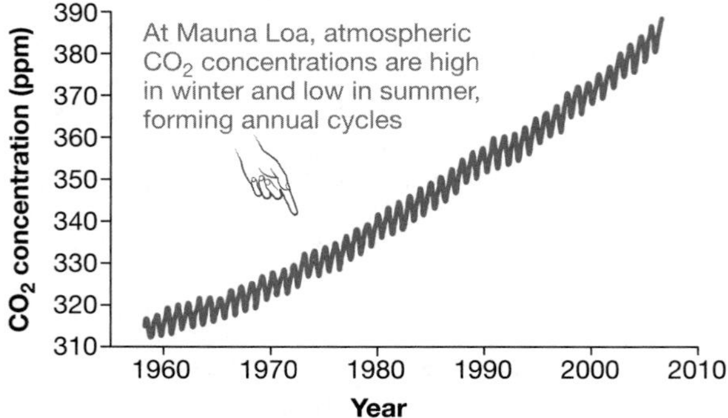

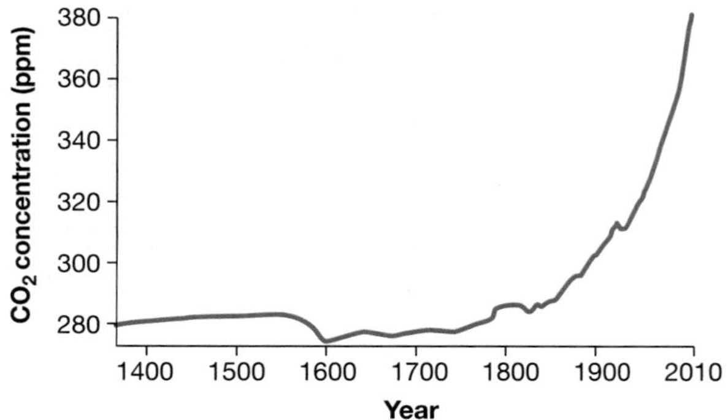

FIGURE 54.15 Humans Are Causing Increases in Atmospheric Carbon Dioxide. (a) Rates of carbon flow from fossil-fuel burning have increased as human populations have increased (see Chapter 52). **(b)** Because the Mauna Loa Observatory in Hawaii is far from large-scale human influences, it should accurately represent the average condition of the atmosphere in the Northern Hemisphere. **(c)** For centuries, average CO_2 concentrations in the atmosphere were about 280 parts per million (ppm).

QUESTION Why are atmospheric CO_2 concentrations low in the Northern Hemisphere in summer and high in winter? What pattern would you expect in the Southern Hemisphere?

The Global Nitrogen Cycle **Figure 54.16** illustrates the **global nitrogen cycle**. A key aspect of this biogeochemical cycle is that plants are able to use nitrogen only in the form of ammonium or nitrate ions (NH_4^+ or NO_3^-). As a result, the vast pool of molecular nitrogen (N_2) that is in the air blanketing Earth—N_2 makes up 78 percent of the atmosphere—is unavailable to plants. Nitrogen is added to ecosystems in a usable form only when it is reduced, or "fixed," meaning when it is converted from N_2 to NH_3. Nitrogen fixation results from lightning-driven reactions in the atmosphere and from enzyme-catalyzed reactions in bacteria that live in the soil and oceans. (Chapter 38 explained how bacteria fix nitrogen.)

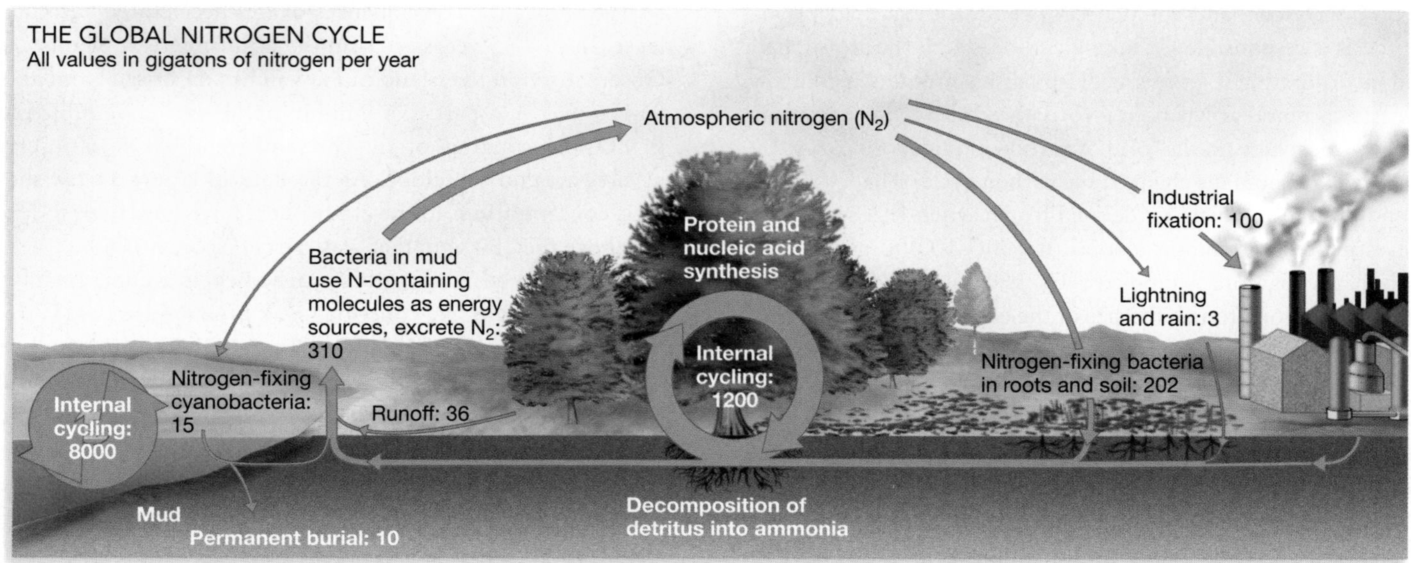

FIGURE 54.16 The Global Nitrogen Cycle. Nitrogen enters ecosystems as ammonia or nitrate via fixation from atmospheric nitrogen. It is exported in runoff and as nitrogen gas given off by bacteria that use nitrogen-containing compounds as an electron acceptor.

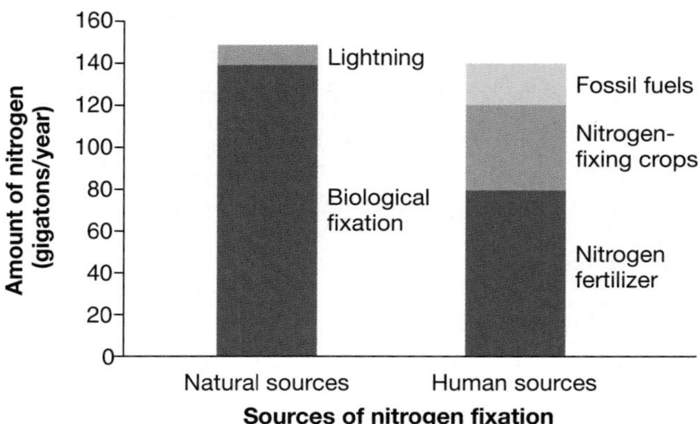

FIGURE 54.17 Humans Are Adding Large Amounts of Nitrogen to Ecosystems. Human activities now fix almost as much nitrogen each year as natural sources do. Thus, human activities have almost doubled the total amount of nitrogen available to organisms.

The nitrogen cycle has been profoundly altered by human activities. The amount of nitrogen fixation from human sources is now approximately equal to the amount of nitrogen fixation from natural sources (**Figure 54.17**). There are three major sources of this human-fixed nitrogen: (1) industrially produced fertilizers, (2) the cultivation of crops, such as soybeans and peas that harbor nitrogen-fixing bacteria, and (3) the production of nitric oxide during the combustion of fossil fuels.

Adding nitrogen to terrestrial ecosystems usually increases productivity. In some cases, then, a massive increase in nitrogen availability is beneficial. But in other situations, it is not. Chapter 28, for example, detailed how excessive applications of nitrogen-containing fertilizers on farmlands have produced nitrogen-laced runoff, which has had disastrous impacts—the formation of oxygen-free "dead zones"—in numerous aquatic ecosystems. Researchers have also documented that nitrogen inputs can lead to a significant loss of biodiversity in terrestrial ecosystems. For example, when biologists added nitrogen to plots of native grasslands in midwestern North America, a few competitively dominant species tended to take over. As those species grew rapidly, they displaced other species that did not respond to nitrogen inputs as strongly. In the grassland ecosystem, increased nitrogen boosted productivity but decreased species diversity, by altering the balance of competitive interactions. The same result has occurred in study plots within the Park Grass experiment, introduced in Chapter 52. The plots that have been fertilized with nitrogen since 1856 contain many fewer species than unfertilized plots do.

As the data in this section make clear, several of the most pressing environmental problems facing our species result from recent and massive alterations in biogeochemical cycles. Even local changes in biogeochemical cycles tend to have large-scale consequences, because nutrients are transported among ecosystems. If you are in your late teens or early twenties as you read this text, you are part of a generation that is expected to experience the most traumatic episode of environmental change in human history. The trauma has two sources: the massive loss of species documented in Chapter 55 and the profound changes in global biogeochemistry recorded here. Let's take a closer look at the consequences of altering biogeochemical cycles.

BOX 54.1 What Is Your "Ecological Footprint"?

The Ecological Footprint is an online tool that helps individuals evaluate their resource consumption. Once you've completed a questionnaire, the program calculates your resource-use "footprint"—meaning, the amount of productive land and water area required to support your lifestyle. The calculation is based on information about four aspects of resource use:

- *Food consumption* Because so much energy is lost between trophic levels, people who eat only primary producers have much smaller footprints than do individuals who eat a great deal of meat and dairy products. Eating foods that are transported long distances also inflates a footprint.
- *Consumer goods* Purchasing manufactured goods and generating large amounts of solid waste increases a footprint; recycling reduces it.
- *Shelter* Footprints are maximized by people who live alone in large homes and do not limit their use of electricity and running water.
- *Transportation* As you might predict, footprints increase when individuals drive gas-guzzling cars long distances by themselves, and when people fly a great deal. Footprints are reduced by fuel-efficient vehicles, car-pooling, bicycling, walking, and public transportation.

Many students (and professors!) who live in industrialized nations are shocked to find that their lifestyle demands over 20 acres of productive land—even though the world average is 2.2 acres/person.

In addition to raising personal awareness, the footprint makes another important point: It is not possible for Earth's resources to support a "westernized" lifestyle for more than a tiny fraction of the total human population. Most researchers agree that the human population has already overshot carrying capacity for high-resource-use lifestyles.

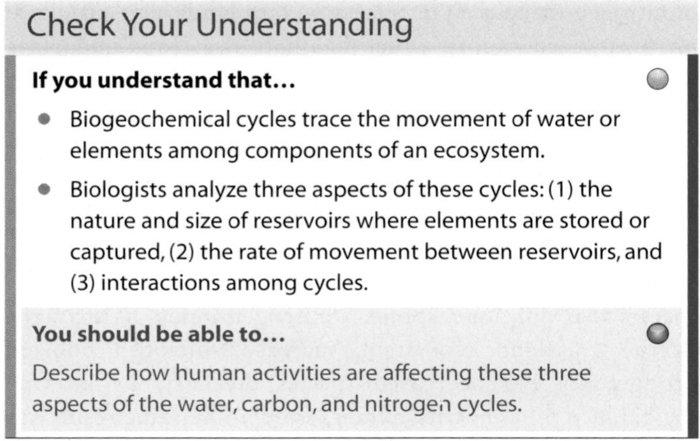

Check Your Understanding

If you understand that...

- Biogeochemical cycles trace the movement of water or elements among components of an ecosystem.
- Biologists analyze three aspects of these cycles: (1) the nature and size of reservoirs where elements are stored or captured, (2) the rate of movement between reservoirs, and (3) interactions among cycles.

You should be able to...

Describe how human activities are affecting these three aspects of the water, carbon, and nitrogen cycles.

54.3 Human Impacts on Ecosystems

Two factors are responsible for the human impacts on ecosystems being documented by biologists. The first is the increase in human population size, analyzed in Chapter 52. The second is an increase in resources used by people. Figure 54.15a documented increases in fossil-fuel use over about the past 150 years; **Figure 54.18** provides data from 2006 on the average annual oil consumption per person around the world. The key point is that residents of industrialized countries, though relatively few in number, have a disproportionately large impact on biogeochemical cycles because they use so much energy, water, food, and other resources (**Box 54.1** on page 1237).

Chapter 55 will assess how changes in human population size and resource use are affecting species extinctions. Here let's focus on the consequences for biogeochemical cycles. Explosive human population growth and intensive resource consumption are causing the average global temperature to rise and, in many biomes, NPP to increase. How are ecosystems responding?

Global Warming

Recall that carbon dioxide concentration in the atmosphere has been increasing throughout the twentieth century and that most analyses indicate that the increase is due to the burning of fossil fuels and clear-cutting of land, particularly forests, for homes and agriculture. Whether the increase in CO_2 is producing global warming—an increase in Earth's surface temperature, averaged over the globe, has been controversial until recently, however.

In 1988 an international group of scientists, called the Intergovernmental Panel on Climate Change (IPCC), was formed to evaluate the consequences of rising CO_2. The group has since produced a series of reports summarizing the state of scientific knowledge on the issue. The 1998 IPCC report concluded that current evidence suggests a "discernible human influence on climate." This was the IPCC's first statement supporting the hypothesis that rising CO_2 concentrations due to human activities are having a measurable impact on climate. The IPCC's next major report, released in 2001, stated that "There is new

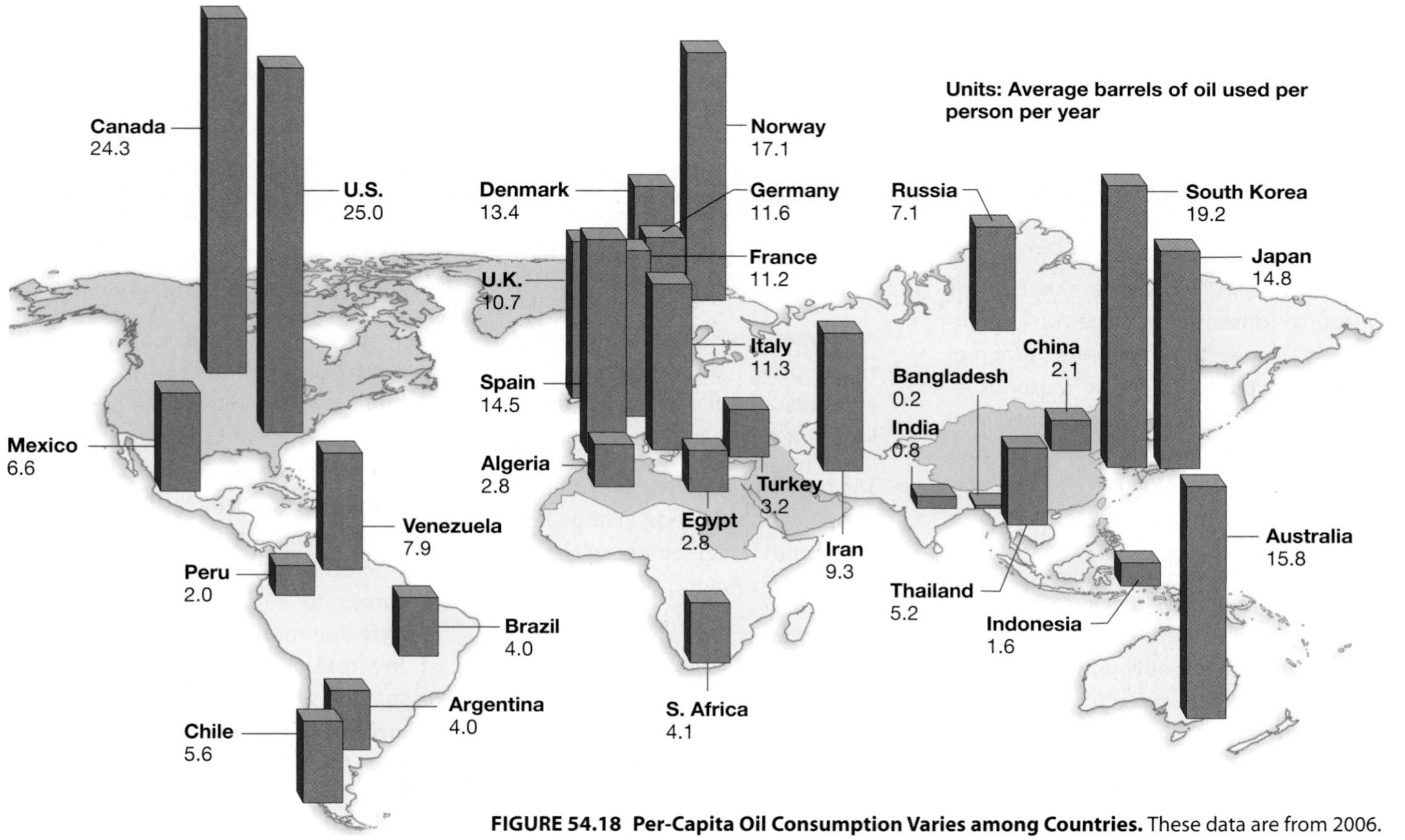

FIGURE 54.18 Per-Capita Oil Consumption Varies among Countries. These data are from 2006.

and stronger evidence that most of the warming observed over the last 50 years is attributable to human activities." In their most recent report, issued in 2007, the IPCC declared that evidence for global warming is unequivocal and that it is "very likely" due to human-induced changes in greenhouse gases.

How much will average temperatures rise in our lifetimes? Predicting the future state of a system as complex and variable as Earth's climate is extremely difficult. Global climate models are the primary tools that scientists use to make these projections. A global climate model is based on a large series of equations that describe how the concentrations of various gases in the atmosphere, solar radiation, transpiration rates, and other parameters interact to affect climate. The models currently being used suggest that average global temperature will undergo additional increases of 1.1–6.4°C (2.0–11.5°F) by the year 2100. The low number is based on models that assume no further increase in greenhouses gases over present levels; the high number is based on models that assume continued intensive use of fossil fuels and increases in greenhouse gases.

How will ecosystems respond to this increase? Answering this question is equally difficult, because ecosystems respond to warming in ways that increase or decrease CO_2 concentrations and thus exacerbate or mitigate warming. Positive feedbacks, for example, occur when warmer and drier climate conditions lead to more fires, which in turn release more CO_2, which leads to even more warming. Researchers recently documented that a form of positive feedback is already occurring in arctic tundras. Traditionally, tundras sequester carbon in the form of soil organic matter, because decomposition rates are extremely low there. During a series of warm summers in the 1980s, however, researchers found that decomposition rates increased sufficiently to release carbon from stored soil organic matter and to produce a net flow of carbon to the atmosphere.

Negative feedbacks, in contrast, arise when warmer conditions lead to increased rates of photosynthesis and hence to an increase in the uptake of CO_2. For example, experiments have shown that the growth rates of several tree species and some agricultural crops increase in direct response to increasing atmospheric CO_2. Because CO_2 is required for photosynthesis, it can act as a fertilizer.

Will ecosystem responses increase or decrease global warming? Currently, it is not clear whether positive or negative feedbacks will predominate. Researchers are working to answer this question by using computer models, experiments like the two highlighted in Chapter 50, and analyses of how ecosystems responded to past climate changes.

Even though global temperatures have risen only slightly in comparison with projections for the next 50–100 years, biologists have already documented dramatic impacts on organisms.

- The geographic ranges of many organisms—including the malaria parasite—are changing. Figure 54.19a shows recent data on the ranges of small crustaceans, called copepods, in the North Atlantic. Copepods are important predators in marine plankton and are eaten by many fish. The maps show changes that occurred in a 40-year period, beginning in 1960, in the number of copepod species typical of marine waters off southern Europe versus the number of copepod species typical of northern European waters. The key observation is that southern species have moved north, while the range of cold-water species has declined.
- The timing of events is changing in many seasonal environments. The graph in **Figure 54.19b** shows changes in the date of first flowering for a plant native to the Midwestern United States. Over the past 60 years, the first day of the year when flowers are produced by *Baptisia leucantha* plants has moved up by about 10 days. To interpret this observation, biologists hypothesize that the climate has warmed enough to promote growth and flowering earlier in the year. Similarly, recent studies have confirmed that migratory

(a) Cold-water copepods are declining in the North Atlantic.

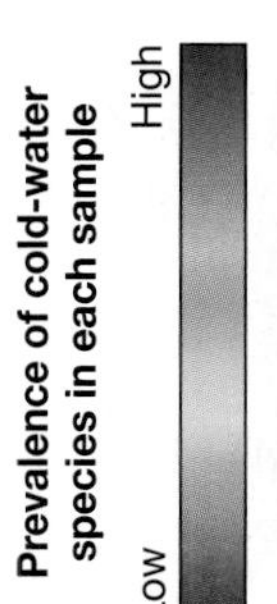

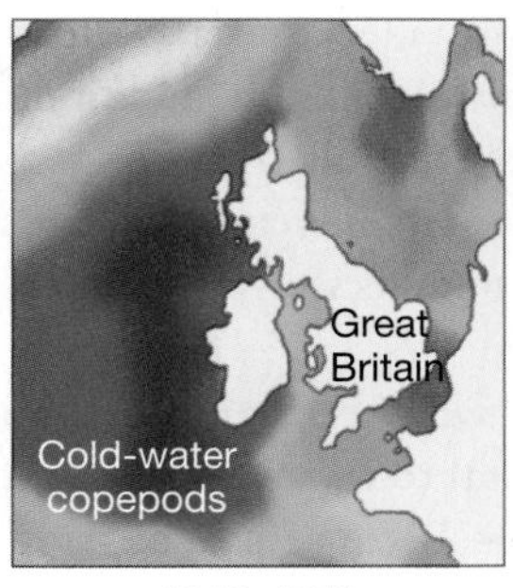

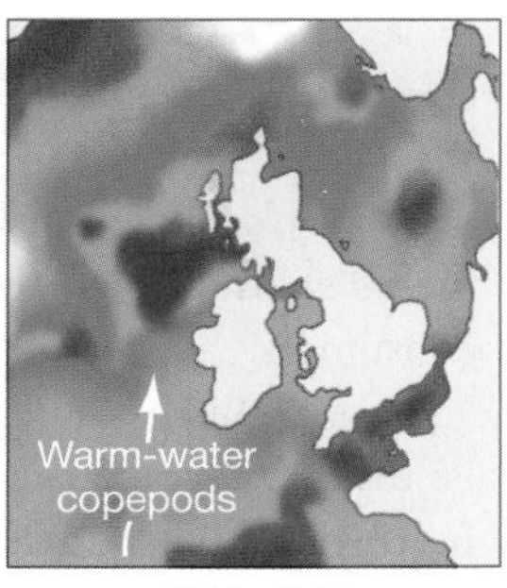

(b) Flowering times for some species in midwestern North America are earlier in the year.

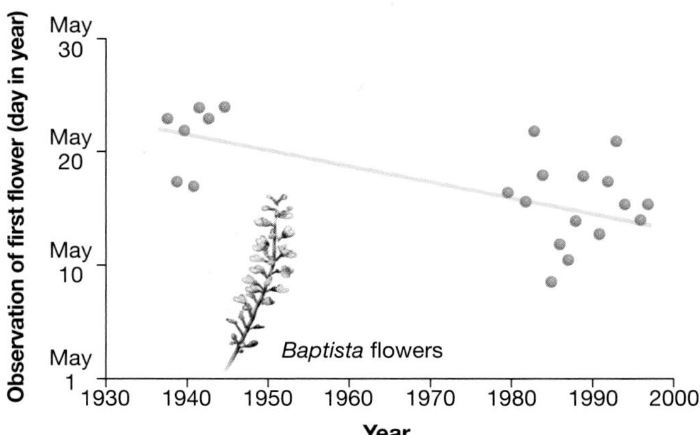

FIGURE 54.19 Global Warming Is Causing Changes in Geographic Ranges and Behavior. (a) Most species of copepod are restricted to waters of a certain temperature. **(b)** In a data set from Wisconsin, 19 of the 55 events tracked over a 61-year period occurred significantly earlier. Only one event occurred later; the remaining 35 showed no change.

bird species are arriving on their breeding grounds in Europe earlier and that fungi are fruiting earlier in some seasonal environments.

- There is strong evidence that most of the 122 frog species that have gone extinct recently succumbed to a parasitic fungus. Data published in 2006 suggest that global warming has increased the frequency and severity of these fungal infections.
- Coral reefs are among the most productive and species-rich ecosystems in the world. Their integrity is threatened, however, by increases in water temperature that cause reef-building corals to expel their photosynthetic algae. When this "coral bleaching" continues due to sustained exposure to warm water, corals begin to die of starvation.
- In some populations, changing temperatures are already causing allele frequencies to change—meaning that some species are evolving in response to global warming. In the fruit fly *Drosophila subobscura*, for example, alleles that increase fitness in hot habitats have increased in frequency in Europe, South America, and North America independently.

Global warming has already caused significant changes in geographic ranges, behavior, and allele frequencies. If the climate models published to date are correct, it will cause many more changes over the course of your lifetime.

Productivity Changes

Several of the changes that humans are inducing in biogeochemical cycles have the same effect: They alter NPP. For example, experiments have documented that warming temperatures, the addition of nitrogen and other nutrients, and rising CO_2 levels increase NPP in some terrestrial ecosystems.

Figure 54.20 shows how average annual NPP has changed globally, in (1) terrestrial environments between 1982 and 1999 (Figure 54.20a) and (2) marine environments from 1999 to 2004 (Figure 54.20b). These are the most recent intervals for which data are available. On land, global NPP increased by over 6 percent during the 17-year study interval. In the oceans, though, the 1999–2004 data show a convincing correlation between increased surface water temperatures and *decreased* productivity.

The overall increase in terrestrial productivity is thought to be due to rising temperatures, increased rainfall in the tropics, and CO_2 fertilization—all factors that increase the rate of photosynthesis in plants. The leading explanation for the drop in marine productivity is more involved. You might recall from Chapter 50 that lakes can become stratified because water is most dense at 4°C and much less dense at higher temperatures. Seawater also becomes stratified. For example, water in the benthic zone is at 4°C year-round in large regions of the ocean (**Figure 54.21a**). When the temperature of surface water rises due to global warming, the water at the surface becomes even less dense than benthic water. This is important because water in the benthic zone is nutrient rich, due to the rain of decomposing organic material from the surface. When surface water becomes lighter, water currents are much less likely to be strong enough to overcome the density difference and bring nutrient-rich, 4°C water all the way up to the surface, where nutrients can spur the growth of photosynthetic bacteria and algae (**Figure 54.21b**). If currents don't become stronger, then global warming causes surface waters to become more nutrient poor.

It's important to note, though, that these global changes are underlain by considerable local variation. On land, productivity during the 1982–1999 interval tended to increase near the equator and at 60°N latitude but to decrease in the arctic and an array of more localized areas. In the ocean, productivity during the 1999–2004 interval dropped dramatically in large areas of the pelagic zone but increased in a number of other regions.

What are the consequences of these local and global changes in NPP? The short answer is, "It depends."

In local areas of the ocean, the consequences of dramatically increased productivity have usually been negative. For example, Chapter 28 detailed how nitrate ions—derived from fertilizers applied to cornfields in midwestern North America—are washing into the Gulf of Mexico. Increased nitrate concentrations have stimulated the growth of planktonic organisms, whose subsequent decomposition has used up available oxygen and triggered the formation of an anoxic "dead zone." Biologists also hypothesize that increases in nitrate and other fertilizers of human origin have contributed to the frequency or intensity of harmful algal blooms in the ocean—large populations of dinoflagellates that release toxins into the surrounding water (see Chapter 29).

Biologists are concerned that if overall NPP in the ocean continues to decline in response to global warming, positive feedback will occur: Atmospheric CO_2 will rise due to less atmospheric CO_2 being fixed by photosynthesis. In contrast, increased NPP on land provides a negative feedback on CO_2 concentrations and global warming. In addition, lower NPP in the ocean means that the productivity of the world's fisheries may decline, but biomass increases in terrestrial environments should boost the numbers of primary and higher-level consumers.

Overall, it is not clear whether changes in productivity will be beneficial or detrimental to ecosystems. The answer should be forthcoming, however. Because global temperatures continue to rise and large-scale additions of nitrogen, phosphorus, and carbon dioxide are ongoing, humans are implementing a global-scale experiment on the effects of altering NPP. Some of the most urgent problems facing your generation are rooted in ecosystem ecology.

(a) Average annual change in terrestrial NPP, 1982–1999

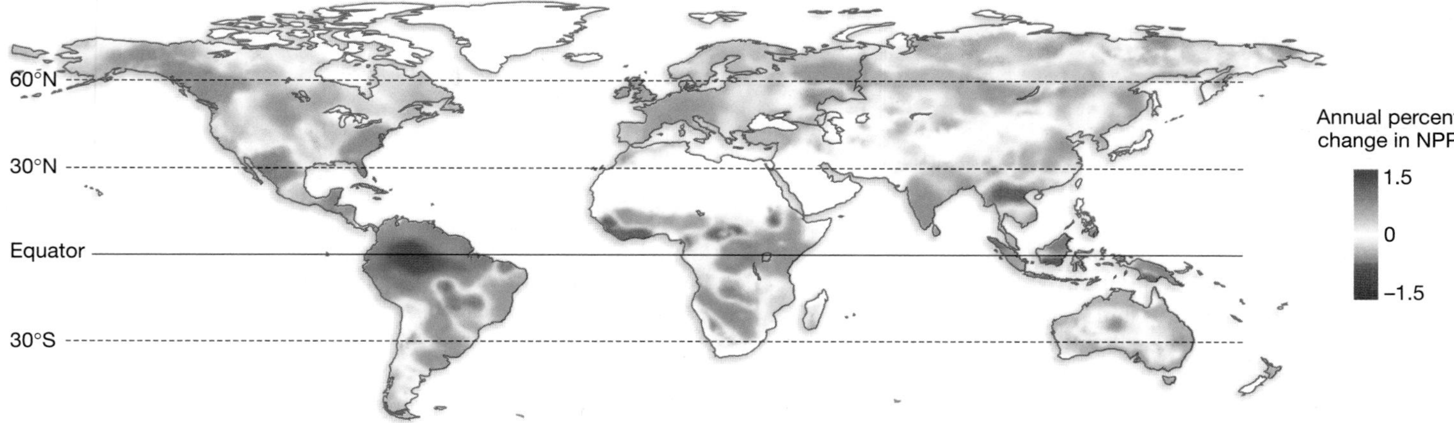

(b) Average annual change in marine NPP, 1999–2004

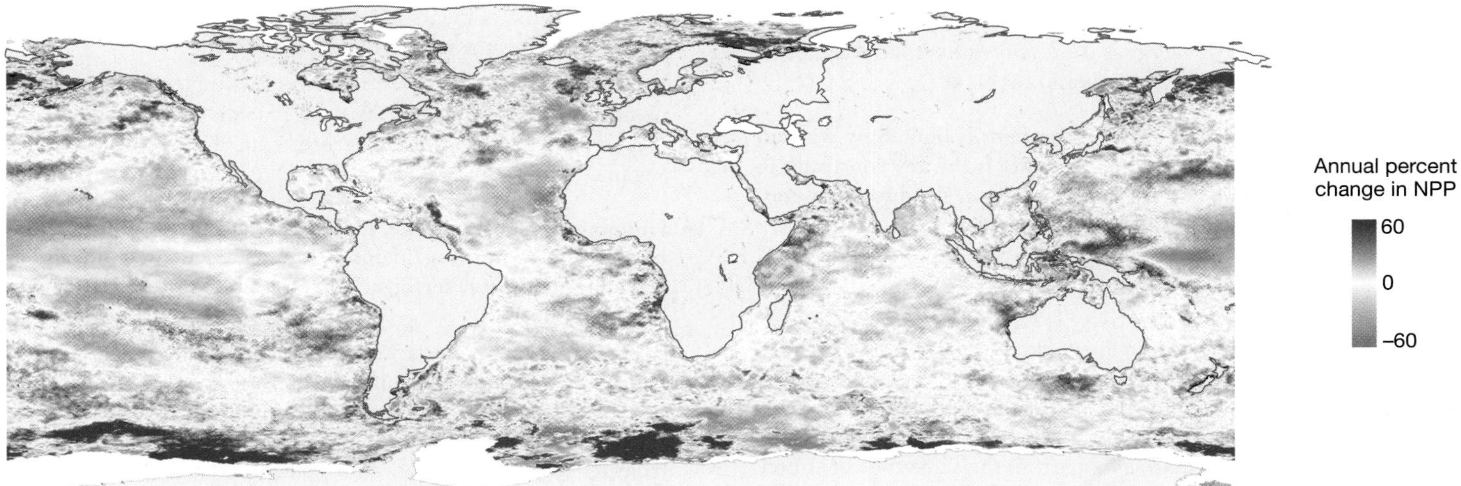

FIGURE 54.20 Consequences of Global Warming: Recent Changes in Terrestrial and Marine NPP.

(a) Much of the ocean is stratified by density and temperature.

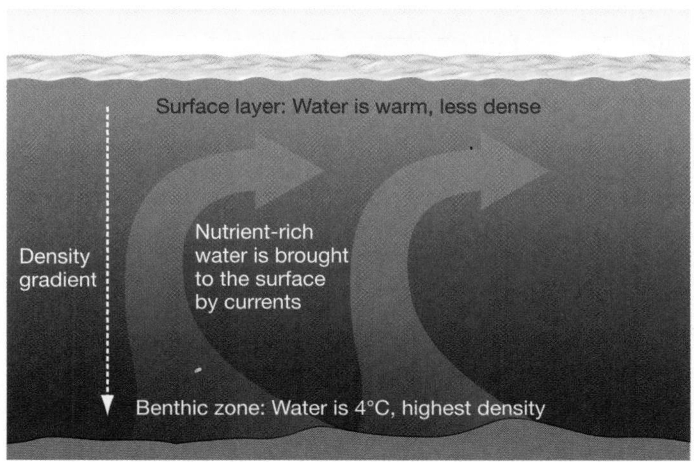

(b) Global warming increases the density gradient, making it less likely for layers to mix.

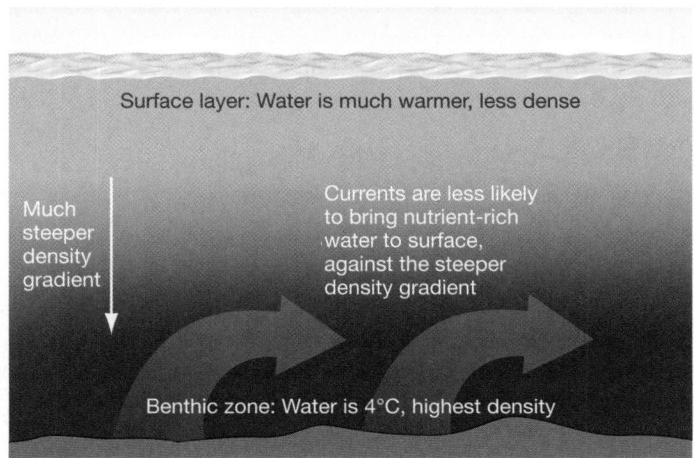

FIGURE 54.21 Temperature Stratification in the Ocean May Affect Productivity.

Chapter Review

SUMMARY OF KEY CONCEPTS

An ecosystem has four components: (1) the abiotic environment, (2) primary producers, (3) consumers, and (4) decomposers. These components are linked by the movement of energy and nutrients.

An ecosystem consists of one or more communities of interacting species and their abiotic environment. As energy flows through ecosystems and as nutrients cycle through them, energy and nutrients are exchanged between biotic and abiotic components of the ecosystem. Energy flows into ecosystems as a result of photosynthesis or the respiration of inorganic molecules with high potential energy. Chemical energy from producers enters food webs via primary consumers or decomposers.

You should be able to draw the relationships among the four components of an ecosystem where no photosynthesis occurs—the outside source of energy is iron- and sulfur-containing compounds released from magma (liquid rock).

The productivity of terrestrial ecosystems is limited by warmth and moisture, while nutrient availability is the key constraint in aquatic ecosystems. As energy flows from producers to consumers and decomposers, much of it is lost.

Among terrestrial ecosystems, net primary productivity (NPP) is highest in tropical wet forests and tropical dry forests. Among aquatic ecosystems, productivity is highest in coral reefs, wetlands, and estuaries. Although productivity is extremely low in the oceanic zone, the area covered by this ecosystem is so extensive that the open ocean accounts for the highest percentage of Earth's overall productivity.

Organisms that acquire energy from the same type of source are said to occupy the same trophic level. Most ecosystems have at least three trophic levels: primary producers, herbivores or primary decomposers, and carnivores. Because energy transfer from one trophic level to the next is inefficient, ecosystems have a pyramid of productivity: Biomass production is highest at the lowest trophic level and lower at each higher trophic level.

The feeding relationships among species in a particular ecosystem are described by a food chain or food web. Food chains rarely exceed five or six trophic levels. The energy-flow, dynamic-stability, and environmental-complexity hypotheses have been proposed to explain why food webs and food chains are not longer.

You should be able to describe the nature of a food chain where a minimum of energy would be lost.

To analyze nutrient cycles, biologists focus on the nature of the reservoirs where elements reside and on how quickly elements move between reservoirs.

Nutrients move through ecosystems in biogeochemical cycles. The rate of nutrient cycling is strongly affected by the rate of decomposition of detritus. The decomposition rate, in turn, is affected by abiotic environmental conditions such as temperature and by the quality of the detritus. Nutrients are also lost from ecosystems. Experiments have shown that the loss of vegetation greatly increases the rate of nutrient loss.

You should be able to describe conditions under which decomposition rates in terrestrial and marine environments are extremely low—creating a carbon sink due to a buildup of organic matter.

Humans are causing large, global changes in the abiotic environment. The burning of fossil fuels has led to rapid global warming. Extensive fertilization is increasing productivity and causing pollution.

Average global temperatures are increasing rapidly because land-use changes and burning of fossil fuels have increased the flow of carbon in the form of CO_2 into the atmosphere, and because carbon dioxide acts as a greenhouse gas. Nitrogen fixation from fertilizer production and from the planting of nitrogen-fixing crop species is now approximately equal to the amount of nitrogen fixation from natural sources. These increases have led to increased productivity, but also to pollution and to loss of biodiversity.

You should be able to propose three steps to reduce global warming and explain the logic behind each.

Web Animation at www.masteringbio.com
The Global Carbon Cycle

QUESTIONS

Test Your Knowledge

1. What is the difference between a community or group of communities and an ecosystem?
   a. An ecosystem comprises a community and the abiotic environment.
   b. An ecosystem is a type of community.
   c. A biome includes only the plant community or communities present in an environment.
   d. An ecosystem includes only the abiotic aspects of a particular environment.

2. Which of the following ecosystems would you expect to have the highest primary production?
   a. subtropical desert
   b. temperate grassland
   c. boreal forest
   d. tropical dry forest

3. Most of the net primary productivity that is consumed is used for what purpose?
   a. respiration by primary consumers
   b. respiration by secondary consumers
   c. growth by primary consumers
   d. growth by secondary consumers

4. According to the dynamic-stability hypothesis for food-chain length, food chains will be shorter in which type of environment?
   a. cold
   b. constant
   c. variable
   d. low in nutrient availability

5. Which of the following is normally the longest-lived reservoir for carbon?
   a. atmosphere (CO_2)
   b. marine plankton (primary producers *and* consumers)
   c. petroleum
   d. wood

6. Devegetation has what effect on ecosystem dynamics?
   a. It increases belowground biomass.
   b. It increases nutrient export.
   c. It increases rates of groundwater recharge (penetration of precipitation to the water table).
   d. It increases the pool of soil organic matter.

Test Your Knowledge answers: 1. a; 2. d; 3. a; 4. c; 5. c; 6. b

Test Your Understanding

Answers are available at www.masteringbio.com

1. Draw a pyramid of productivity for a temperate-forest ecosystem, and explain its shape.
2. Explain the difference between gross primary productivity and net primary productivity. Which is larger, and why?
3. Explain why decomposition rates are higher in some ecosystems than in others, and give examples. How does the decomposer food web regulate nutrient availability in an ecosystem?
4. Compare and contrast the energy-flow, dynamic-stability, and environmental-complexity hypotheses for food-chain length.
5. Draw a diagram of the global water, carbon, or nitrogen cycle. Label major reservoirs and flows. Compare and contrast the life span of water or an element in each reservoir, and evaluate factors that affect the rate of movement between reservoirs.
6. Why are the open oceans nutrient poor? Why are neritic zones and intertidal habitats relatively nutrient rich?

Applying Concepts to New Situations

Answers are available at www.masteringbio.com

1. Suppose you had a small set of experimental ponds at your disposal and an array of pond-dwelling algae, plants, and animals. How could you use radioactive isotopes of carbon or phosphorus to study energy flows or nutrient cycling in these experimental ecosystems?
2. Some researchers are concerned that fertilizing the open oceans with iron would lead to overfertilization of neritic zones, because larger amounts of nutrients would be carried into coastal regions by ocean currents. Outline an experiment and a computer simulation study to test this hypothesis. Evaluate which approach—the experiment or the computer simulation—would likely be more effective in addressing the hypothesis.
3. Suppose that herbivores were removed from a temperate deciduous forest ecosystem. Predict what would happen to the rate of nitrogen cycling. Explain the logic behind your prediction.
4. Explain why human-caused changes to the global carbon cycle are affecting Earth's climate. State why you think that these changes are beneficial or detrimental.

www.masteringbio.com is also your resource for • Answers to text, table, and figure caption questions and exercises • Answers to *Check Your Understanding boxes* • Online study guides and quizzes • Additional study tools including the *E-Book for Biological Science* 3rd ed., textbook art, animations, and videos.

55 Biodiversity and Conservation Biology

KEY CONCEPTS

- Biodiversity can be analyzed at the genetic, species, and ecosystem levels.
- If recent rates of extinction due to human population growth and habitat destruction continue, a mass extinction will occur.
- Humans depend on biodiversity for the products that wild species provide and for ecosystem services that protect the quality of the abiotic environment.

Wangari Maathai, 2004 Nobel Peace Prize Winner, founded the Green Belt Movement that has sponsored the planting of over 1 billion trees, beginning in Kenya and expanding internationally. She is shown here planting a tree.

Most biologists choose their profession for two reasons: They love organisms, and they love answering questions about organisms. But even as the extent of our knowledge about life explodes and the science of biology expands, the number of species decreases. For people who have devoted their lives to the study of life, the irony is cruel. In response, many students cite a third reason for becoming a biologist: Learning how to preserve biodiversity. That is the focus of this chapter.

Let's start by getting a better understanding of what biodiversity is, and how biologists quantify and analyze the diversity of life. Section 55.2 follows up by exploring data on where biodiversity is highest.

The second half of the chapter focuses on efforts to preserve biodiversity. Section 55.3 delves into data on the extent and causes of extinction, and Section 52.4 focuses on efforts to prevent extinction—beginning with the question of why we should bother. As biologists, what reasons can we provide for preserving species? Are these reasons compelling enough to influence real estate developers, farmers, fishermen, loggers, miners, and others whose lives are directly affected by conservation efforts?

Biologists who are concerned about conservation realize that to be effective, they have to do more than collect and analyze data on threatened species and communities. They must also be comfortable in the economic and public policy arenas. A conservation biologist is similar to a physician managing the recovery of an acutely ill patient. Success demands more than the clinical skills of making a correct diagnosis and prescribing medication; it requires an empathy with the person affected, the ability to manage the cost and implementation of care, and a devotion to collaborating with other professionals to achieve a cure. Conservation biology is a demanding and synthetic field.

55.1 What Is Biodiversity?

Perhaps the simplest way to think about biodiversity is in terms of the tree of life—the phylogenetic tree of all organisms, introduced in Chapter 1 and explored in detail in Chapters 28 through 35. If biologists are eventually able to estimate the complete tree of life, using the techniques reviewed in Chapter 27, the branches would represent all of the lineages of organisms living today and the tips would represent all of the species.

When biodiversity increases, branches and tips are added and the tree of life gets fuller and bushier. When extinctions occur, tips and perhaps branches are removed; the tree of life becomes thinner and sparser.

Coordinated, multinational efforts are under way to estimate the phylogenetic tree of major branches on the tree of life, including the fungi, brown algae, and land plants. These and other efforts to estimate phylogenies are producing a better and better picture of the relationships among lineages and species, and a clearer understanding of how extinction is affecting biodiversity. But even though the tree of life is a compelling way to document biodiversity, it does not tell the entire story.

Biodiversity Can Be Measured and Analyzed at Several Levels

To get a complete understanding of the diversity of life, biologists recognize and analyze biodiversity on three levels:

1. **Genetic diversity** is the total genetic information contained within all individuals of a species and is measured as the number and relative frequency of all alleles present in a species. Because no two members of the same species are genetically identical, each species is the repository of an immense array of alleles. At this level, biodiversity is everywhere around you, from the variety of apples at the grocery store to the variation in coloration and singing ability of sparrows in your backyard. **Box 55.1** details recent technical innovations that are allowing biologists to document genetic diversity at a large scale.

2. **Species diversity** is based on the variety of species on Earth. Recall from Chapter 53 that, in practice, species diversity is measured by quantifying the number and relative frequency of species in a particular region. There is an additional aspect of species diversity, however, that biologists call taxonomic diversity and document by estimating phylogenies. Besides understanding the number and relative frequency of species present, researchers also want to know their

BOX 55.1 Environmental Sequencing: A New Approach to Quantifying Genetic Diversity

When the first techniques for analyzing genes and proteins became available in the 1960s, biologists immediately began quantifying the level of genetic diversity within species. The basic approach was to collect tissue samples from many individuals within a population or species, isolate proteins or DNA from the samples, and document how many different alleles existed at a particular gene or suite of genes. With some notable exceptions, the overall message of this work is that most species contain a remarkably high level of allelic diversity. This was an important result, because it indicated that changes in natural selection or other evolutionary processes could cause rapid changes in the frequencies of existing alleles and thus a rapid evolutionary response to environmental change. It also demonstrated that losing a species to extinction means that a large number of unique alleles are lost.

Recent work on genetic diversity has focused on studying entire communities or ecosystems rather than individual populations or species. Research on bacteria from the Sargasso Sea furnishes a good example of this approach. The Sargasso Sea is located in the Caribbean region, near the islands of the Bahamas. Because it is exceptionally nutrient poor and species poor, it is considered one of the ocean's great deserts in terms of biodiversity. The Sargasso Sea was an attractive ecosystem to study for just this reason: Biologists routinely start by studying simple systems before progressing to more complex situations.

To inventory the complete array of bacterial genes present, a research team collected cells from different water depths and locations. The team isolated DNA from the samples and sequenced it using techniques introduced in Chapter 20. They called the work **environmental sequencing**, because their goal was to analyze all of the genes present in an ecosystem. After analyzing over 1 billion base pairs, the team concluded that at least 1800 bacterial species were present, of which 148 were previously undiscovered. They also identified more than 1.2 million alleles that had never before been characterized.

The results suggest that an enormous amount of genetic diversity exists even in Earth's simplest ecosystems. To grasp why this is important, consider that the researchers discovered more than 780 new bacterial alleles that code for proteins similar to the rhodopsin molecules found in your eyes. Do these proteins function as light receptors, like rhodopsin does? Or are these molecules absorbing light energy and using it to pump protons out of the cells, establishing a proton gradient that can drive synthesis of ATP and keep the organisms alive? No one knows—yet. New knowledge inspires new questions.

evolutionary relationships. For example, some lineages on the tree of life are extremely species rich. Prominent examples include the African cichlid fish, introduced in Chapter 27, and the 35,000 species of orchids. Other lineages, in contrast, are represented by a single species (**Figure 55.1**). Some biologists argue that it is particularly important to preserve populations from species-poor lineages, because they are the last living representatives of their lineages. If those populations go extinct, an entire lineage is lost forever.

3. **Ecosystem diversity** is the variety of biotic communities in a region along with abiotic components, such as soil, water, and nutrients. Ecosystem diversity is more difficult to define and measure than genetic diversity or species diversity, because ecosystems do not have sharp boundaries. Recall from Chapter 54 that ecosystems are complex and dynamic assemblages of organisms that interact with each other and their nonliving environment. Attempts to measure ecosystem diversity focus on capturing the array of biotic communities in a region, along with variation in the physical conditions present. Areas around estuaries, for example, tend to have high ecosystem diversity due to the combination of stream, wetland, intertidal, neritic, and upland habitats (see Chapter 50). When an estuary is dredged or filled, biodiversity is affected at all three levels: genetic and species diversity change due to the different numbers and types of individuals present, and ecosystem diversity is altered due to the change in abiotic conditions.

Biodiversity can be recognized and quantified on several distinct levels, but it is also dynamic. Mutations that create new alleles increase genetic diversity; natural selection, genetic drift, and gene flow may eliminate certain alleles or change their frequency, leading to an increase or decrease in overall genetic diversity. Speciation increases species diversity; extinction decreases it. Changes in climate or other physical conditions can result in the formation of new ecosystems, as can the evolution of new species that interact in novel ways. Disturbances such as volcanic eruptions, human activities, and glaciation can destroy ecosystems.

Biodiversity is not static. It has been changing since life on Earth began.

(a) Red panda

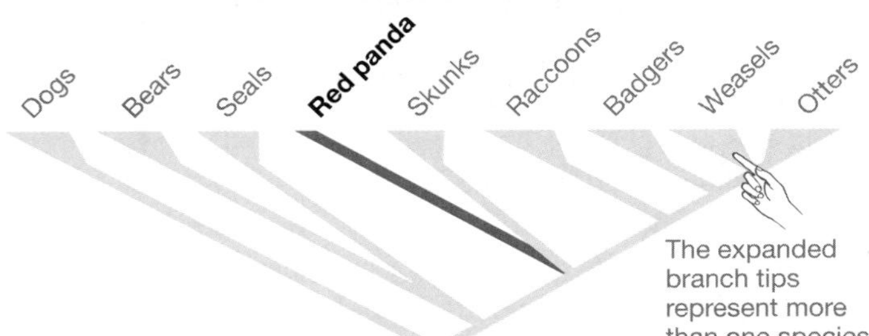

(b) Yangtze river dolphin

FIGURE 55.1 Phylogenetically Distinct Species May Be High-Priority Targets for Conservation. The red panda and the Yangtze river dolphin have few close relatives and represent distinct branches on the tree of life. If conservation measures attempt to preserve taxonomic diversity, these species would be a high priority for preservation. Unfortunately, preservation efforts for the Yangtze river dolphin are too little too late—this species is now functionally extinct (too few individuals remain to breed and sustain the population).

How Many Species Are Living Today?

One of the simplest questions about biodiversity is also one of the most difficult to answer. How many species are there on Earth? The answer is not known. Given the massive effort that it would take to document every form of life on Earth, the answer will probably never be known.

Biologists are well aware, though, that the approximately 1.5 million species cataloged to date represent a tiny fraction of the number actually present. Chapter 31, for example, noted that an average of 6 fungal species live on each well-studied plant species. Because over 300,000 plant species have been described, a conservative estimate based on the 6:1 fungi:plant ratio suggests that 1.65 million species of fungi exist, even though only 80,000 have been described thus far. The issue is even more acute in poorly studied lineages like bacteria and archaea, where researchers routinely discover dozens or hundreds of previously unknown species with each new direct sequencing or environmental sequencing study (see Chapter 28 and Box 55.1). For example, a recent direct sequencing study estimated that over 500 species of bacteria live in the human mouth, although only about half have been described and named. Even among well-studied groups like birds and mammals, new species are discovered almost every year.

Given that only a fraction of the organisms alive have been discovered to date, how can biologists go about estimating the total number of species on Earth? Two general approaches have been used. One is based on intensive surveys of species-rich groups at small sites. A second is based on attempts to identify all of the species present in a particular region.

Taxon-Specific Surveys Terry Erwin and J. C. Scott published a classic study of species diversity in a single taxon. They began by estimating the number of insect species that live in the canopy of a single tropical tree (**Figure 55.2**). After using an insecticidal fog to knock down insects from the top of a *Luehea seemannii* tree, they identified over 900 species of beetles among the individuals that fell. Most of these species were new to science.

FIGURE 55.2 Estimating Species Richness via Intensive Local Sampling. A small sample of the insects collected from a single tree in the Amazonian rain forest.

To use these data as an indicator of global arthropod species diversity, Erwin and Scott used the following train of logic: Based on earlier work with insects on this tree species, they estimated that 160 of the 900 beetle species live only on *L. seemannii*. Worldwide, beetles represent about 40 percent of all known arthropods. Thus, it was reasonable to suggest that 400 species of arthropods live only in the canopy of *L. seemannii*. By adding an estimate of arthropods specializing on the trunk and roots of this tree, Erwin projected that it is host to 600 specialist arthropods. If each of the 50,000 species of tropical tree harbors the same number of arthropod specialists, then the world total of arthropod species exceeds 30 million species. Based on such studies, biologists estimate that at least 10 million, and possibly as many as 100 million, species of all types exist today.

More recent single-taxon surveys have sampled larger geographic areas. For example, Philippe Bouchet and colleagues conducted a massive survey of marine molluscs in coral-reef habitats along the west coast of New Caledonia, a tropical island in the southwest Pacific Ocean. The team spent more than a year collecting molluscs at 42 sites over a total area of almost 300 square kilometers. The survey represented the most thorough sampling effort ever made to determine the species diversity of molluscs, and produced more than 127,000 individuals representing 2738 species. These numbers far exceed the mollusc diversity recorded for any comparable-sized area. The species total was 2 to 3 times the total number of mollusc species that had been reported for similar habitats in the region.

In reporting their findings, the biologists emphasized that 20 percent of the species found were represented by a single specimen. This observation suggests that many species are exceedingly rare and thus likely to be missed by less-intensive sampling efforts. And when the investigators compared their data with a survey in progress at a second site, different in reef structure but only 200 kilometers away from the original site, they found that only 36 percent of species were shared between the two sites. These results support the hypothesis that the global biodiversity of molluscs—currently thought to be about 93,000 species—is an underestimate.

All-Taxa Surveys The first effort to find and catalog all of the species present in a large area is now under way. The location is the Great Smoky Mountains National Park in the southeastern United States (**Figure 55.3**). A consortium of biologists, volunteers, and research organizations initiated this all-taxa survey in 1999. To date, the survey has discovered over 650 species that are new to science and over 4650 species that had never before been found in the park. When the inventory is complete, in 2015, biologists will have a much better database to use in estimating the extent of global biodiversity.

(a) Great Smoky Mountains National Park

(b) All-taxon survey workers assess organisms in a sample.

FIGURE 55.3 The First All-Taxon Survey Is Now Under Way. The first attempt to document every species living in a prescribed area is under way at Great Smoky Mountains National Park, along the border of Tennessee and North Carolina.

Check Your Understanding

If you understand that...

- Biologists document biodiversity at the genetic, species, and ecosystem levels.

You should be able to...

Outline a study that would quantify levels of genetic, species, and ecosystem diversity on your campus.

55.2 Where Is Biodiversity Highest?

If documenting the extent of genetic, species, and ecosystem diversity is the most fundamental goal of research on biodiversity, then the second-most important aim is to understand its geographic distribution. Chapter 53 introduced the most prominent geographic pattern in the distribution of biodiversity: In most taxonomic groups, species richness is highest in the tropics and declines toward the poles. Tropical rain forests are particularly species rich. Even though they represent just 7 percent of Earth's land area, they are thought to contain at least 50 percent of all species present. As Chapter 53 noted, understanding why tropical forests are so species rich and why a latitudinal gradient occurs is one of the most active areas of research in community ecology and biodiversity.

Biologists have also been working to understand the distribution of species richness at finer spatial scales, however. For example, the map in **Figure 55.4a** was constructed by dividing the world's landmasses into a grid of cells 1° latitude by 1° longitude. In the tropics, this translates to rectangles approximately 111 km by 111 km (69 mi by 69 mi); because lines of longitude converge at the poles, the rectangles are about 111 km by 55 km at 60° latitude. Using published data, researchers then plotted how many species of birds breed in each cell in the grid. These data are consistent with the latitudinal gradient in species richness, but indicate that some areas of the tropics are much more species rich than others. Biologists use the term **hotspot** to capture this point. In terms of bird species richness, the Andes mountains, the Amazon River basin, portions of East Africa, and southwest China are important hotspots.

Researchers have also been keenly interested in understanding which regions of the world have a high proportion of **endemic species**—meaning, species that are found in an area and nowhere else. **Figure 55.4b** maps the location of endemic species of breeding birds, based on the same cells and data as the species richness map in Figure 55.4a.

Mapping species richness hotspots and centers of endemism can inspire interesting questions—principally, efforts to understand *why* certain regions contain many species or a high proportion of endemic taxa. But in addition, biologists are studying the geographic distribution of biodiversity as a way of focusing conservation efforts. In the year 2000, for example, a team set out to identify regions of the world that meet two criteria: (1) They contain at least 1500 endemic plant species, and (2) at least 70 percent of their traditional or primary vegetation has been lost. The goal of the study was to identify regions of the world that are in most urgent need of conservation action—areas where efforts to preserve habitat would have the highest return on investment. The idea was that efforts to protect a diversity of primary producers would guarantee the protection of a wide array of primary, secondary, tertiary, and quaternary consumers.

The 25 regions that met the two "conservation hotspot" criteria are shown in **Figure 55.4c**. Although these areas represent just 1.4 percent of the Earth's land area, they contain 44 percent of all known plant species and 35 percent of all known vertebrate species. It's also important to note that key regions with high plant species diversity—specifically the rain forests found in the Amazon, the Congo River basin, and New Guinea—are not included as conservation hotspots, because they have not yet lost 70 percent of their primary vegetation. Effectively protecting these areas along with the hotspots defined on the map would provide protection for 60 percent of all land plants in just 5 percent of the Earth's land surface.

(a) Hotspots in terms of species richness of birds

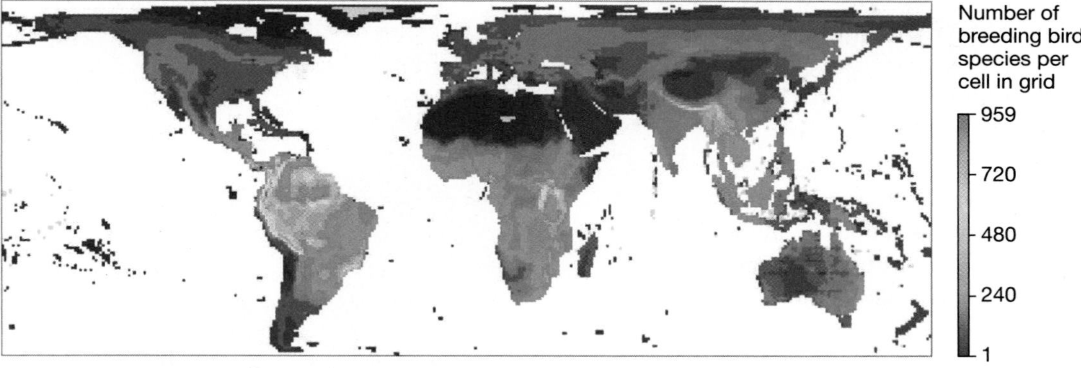

(b) Hotspots in terms of endemic species of birds

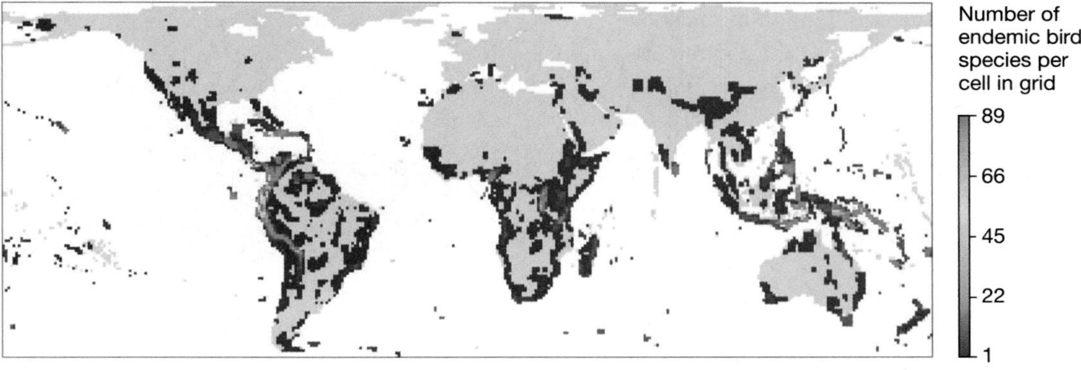

(c) Hotspots in terms of high proportion of endemic plants *and* high threat

FIGURE 55.4 Conservation "Hotspots" Can Be Defined in Different Ways. The maps in **(a)** and **(b)** are based on a grid that is 1° longitude by 1° latitude. The areas shown in red in **(c)** have lost 70 percent of their original vegetation and contain a high diversity of vascular plant species.

QUESTION Suppose you were going to set up natural areas focused on preserving bird species. Which areas on the three maps are in agreement in terms of conservation priorities? Which are in conflict?

This analysis serves as your introduction to the applied aspects of biodiversity studies—the field known as conservation biology. To understand the importance of the data in Figure 55.4 thoroughly, we need to explore two things in depth: why so many species are threatened with annihilation, and why we should do something about it. Let's consider each issue in turn.

55.3 Threats to Biodiversity

No species lasts forever. Climate change, disease, competition from newly arrived species, and habitat alteration are all natural processes. They have been happening since life began. Extinction, like death, is a fact of life.

If extinction is natural, why are biologists so concerned about habitat and species conservation? The answer is rate. Today species are vanishing faster than at virtually any other time in Earth's history. Modern rates of extinction are 100 to 1000 times greater than the average, or "background," rate recorded in the fossil record over the past 550 million years.

Either directly or indirectly, current extinctions are being caused by the demands of a rapidly growing human population, which is currently increasing by about 77 million people per year. If present trends in human population growth and species extinctions continue, a mass extinction on the scale of events described in Chapter 27 will occur. Based on the data in hand, the vast majority of biologists agree with the claim that the sixth mass extinction in the history of multicellular life is now under way. Between the time you are reading this and the time your great-grandchildren are grown, human impacts on the planet promise to equal or exceed those of the gigantic asteroid that smashed into Earth 65 million years ago. It's important to note, however, that human impacts on biodiversity are not a new phenomenon.

FIGURE 55.5 On Easter Island, Social Decline Coincided with the Exhaustion of Natural Resources. Easter Island was one of the last Pacific Islands to be colonized by humans. The prosperity that supported the artwork shown in the inset was based on resources that were overexploited.

Humans Have Affected Biodiversity throughout History

Historically, humans have a poor record of conserving resources and protecting species. For example, recent research on fossil birds found on islands in the South Pacific suggests that about 2000 bird species were wiped out as people colonized this area about A.D. 1200. Many of these extinctions occurred due to predation by humans or rats, pigs, and other animals introduced by humans.

Humans even have a poor record of preserving species they depend on for survival. To drive this point home, consider data on the easternmost island in the South Pacific—Easter Island (**Figure 55.5**). When European explorers arrived on Easter Island in 1722, about 1000 people lived there. The island was treeless and dotted with gigantic stone statues. Researchers who analyzed buried pollen samples taken from swamps on the island discovered that it had once been covered with lush forest dominated by palm trees. A rapid decline in tree pollen coincided with the arrival of the first human settlers.

Fossil digs confirmed that the fauna of the island underwent drastic changes at the same time. In the oldest human garbage piles excavated by biologists, bones from dolphins, seabirds, and land birds are abundant. But these species dropped out of the fossil record by around A.D. 1300—about the time that loss of forests, or deforestation, was complete.

Jared Diamond interpreted these data as evidence of an ecological disaster. In his view, people arrived to find a lush tropical island brimming with natural resources. The human population flourished and people had the leisure time and resources to carve the gigantic statues and roll them into place on beds of palm logs. But after deforestation was complete and local extinctions had begun, the system collapsed. Without palm trunks to make canoes, Easter Island natives did less dolphin hunting and less fishing. Soil erosion may have cut into the productivity of banana and sweet potato plantations. The population crashed, and the great statues fell.

Current Threats to Biodiversity

Most extinctions that have occurred over the past 1000 years took place on islands and were caused by overhunting or the introduction of **exotic species**—nonnative competitors, diseases, or predators. Starting in the twentieth century, though, the situation began to change. **Endangered species**, which are almost certain to go extinct unless effective conservation programs are put in place, are now more likely to live on continents than islands. In addition, habitat destruction has replaced overhunting and species introductions as the primary threat to such species.

Figure 55.6 shows the results of a recent analysis on causes of endangerment for 488 endangered species native to Canada. Several patterns jump out of the data:

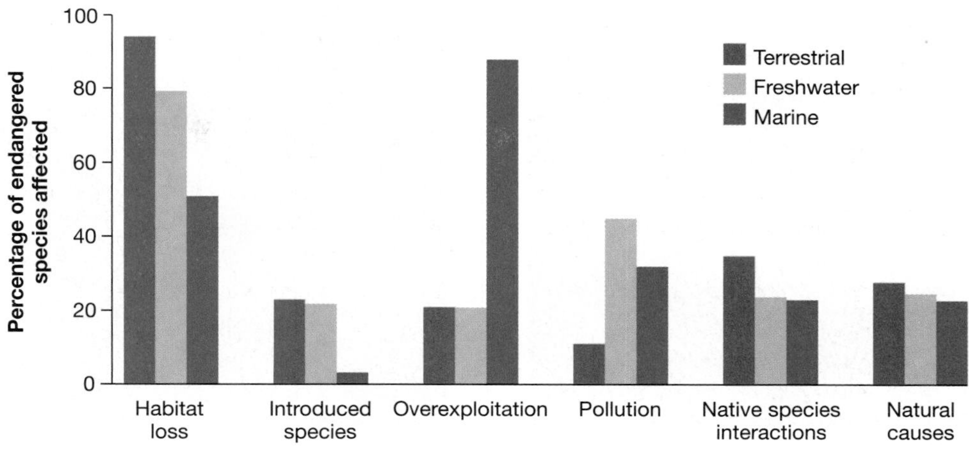

FIGURE 55.6 Endangerment Is Caused by Multiple Factors. These data are from Canada.

QUESTION Why do the data for terrestrial, freshwater, and marine species each add up to more than 100 percent?

- Virtually all of the endangered species are affected by more than one factor. This is important, because it means that conservation biologists may have to solve more than one problem for any given species to recover.
- Habitat loss is the single most important factor in the decline of these species. It is a significant issue for over 90 percent of endangered species in terrestrial environments.
- Overharvesting is the dominant problem for marine species, while pollution plays a large role for freshwater species.
- Factors beyond human control can be important. These include predation or competition with native species, natural disturbances such as fires or droughts, or the fact that some species have narrow niches and have historically been rare. Stated another way, background extinctions will continue to occur.

Do these conclusions hold up for other parts of the world? Although invasive species are a more significant problem in some regions (**Figure 55.7**; also see Chapter 50), most analyses completed to date are broadly consistent with the Canadian data. For example, overexploitation is a grave concern for marine fisheries worldwide. Two-thirds of harvestable species are now considered fully exploited or depleted, and a recent study of the global fishing industry concluded that 90 percent of large bottom-dwelling and open-water fish have been removed from the world's oceans. The species affected include tuna, swordfish, marlin, cod, halibut, and flounder. Overhunting has also emerged recently as a dire threat to

(a) Invasive species increase **competition**.

Purple loosestrife is crowding out native organisms in North American marshes.

(b) Invasive species introduce **disease**.

An introduced fungus has virtually wiped out the American chestnut.

(c) Invasive species increase **predation**.

The brown tree snake has extinguished dozens of bird species on Guam.

FIGURE 55.7 Invasive Species Are Destructive. In many parts of the world, invasive species are a leading cause of endangerment. [Photo (b) © Gary Braasch.]

FIGURE 55.8 Overhunting Continues to Be a Problem in Some Regions. The recent growth of the "bushmeat" trade has devastated many populations of primates and other mammals in Africa. Overharvesting is also a major cause of endangerment in marine species worldwide (see Figure 55.6).

FIGURE 55.9 Suburbanization Is a Major Cause of Habitat Loss in North America.

many African mammal populations (**Figure 55.8**). Biologists view habitat changes driven by global warming—already implicated in the extinction of dozens or hundreds of amphibians—a serious enough threat to deserve close monitoring. Due to the numbers of species and ecosystems affected, human alteration of natural habitat is now the dominant cause of biodiversity decline worldwide, as it is in Canada.

Habitat Destruction Humans cause **habitat destruction** by logging and burning forests, damming rivers, dredging or filling estuaries and wetlands, plowing prairies, grazing livestock, excavating minerals, and building housing developments, golf courses, shopping centers, office complexes, airports, and roads (**Figure 55.9**).

On a global scale, one of the most important types of habitat destruction is deforestation—especially the conversion of primary forests to agricultural fields and human settlements. To appreciate the extent of deforestation in some areas, consider satellite images of wet tropical rain forest in Rondônia, Brazil, that were made in 1975 and in 2001 (**Figure 55.10**). Analyses of satellite photos such as these have shown that as many as 3 million hectares, an area about 10 percent larger than the state of Maryland, was deforested in the Amazon each year during the 1990s. (A hectare, abbreviated ha, is 100 meters by 100 meters, approximately the size of two football fields.)

More recent analyses suggest that the global rate of deforestation has now slowed compared to peak rates in the 1990s. The latest report from the United Nations Food and Agriculture Organization, released in 2005, states that the world's

(a) The devastation of deforestation

(b) Satellite view of deforestation in Rondônia, Brazil

FIGURE 55.10 Rates of Deforestation in Tropical Wet Forests Are Approaching 1 Percent per Year. The satellite photos in part (b) show the extent of deforestation in Rondônia, Brazil, in less than 30 years. The darker green areas in the photo on the right are intact forest; the lighter areas have been burned or logged and converted to agricultural fields and pastures.

forests experienced a net loss of about 7.3 million ha/year between 2000 and 2005. This is down from the average net loss of 8.9 million ha/year during the previous decade. Although most conservationists greeted this report as good news, they also urged caution. Net losses dropped because of extensive forest planting and regrowth in China, North America, and Europe, even though over 8 million ha/year of primary forest were lost in South America and Africa each year from 2000 to 2005. Forest loss in South America and Africa is particularly important because it affects biodiversity hotspots.

Habitat Fragmentation In addition to destroying natural areas outright, human activities fragment large, contiguous areas of natural habitats into small, isolated fragments. **Habitat fragmentation** concerns biologists for several reasons.

1. It can reduce habitats to a size that is too small to support some species. This is especially true for keystone predators such as mountain lions, grizzly bears, and bluefin tuna, which need vast natural spaces in which to feed, find mates, and reproduce successfully.
2. By creating islands of habitat in a sea of human-dominated landscapes, fragmentation reduces the ability of individuals to disperse from one habitat to another. In effect, habitat fragmentation is forcing many species into the metapopulation structure introduced in Chapter 52. The small, isolated populations that make up a metapopulation are much more likely than large populations to be wiped out by catastrophic events such as storms, disease outbreaks, or fires. They can also suffer from inbreeding depression and random loss of alleles due to genetic drift (see **Box 55.2**).
3. Fragmentation creates large amounts of "edge" habitat. The edges of intact habitat are subject to invasion by weedy species and are exposed to more intense sunlight and wind, creating difficult conditions for plants.

The decline in habitat quality caused by fragmentation is being documented in a long-term experiment in a tropical wet forest. In an area near Manaus, Brazil, that was slated for clear-cutting, a research group set up 66 square, 1-hectare experimental plots that remained uncut. Thirty-nine of these study plots were located in fragments designed to contain 1, 10, or 100 hectares of intact forest. Twenty-seven of the plots were set up nearby, in continuous wet forest. As the "Experimental setup" section of **Figure 55.11** shows, the distribution of the study plots allowed the research team to monitor changes inside forest fragments of different sizes and to compare these changes with conditions in unfragmented forest.

When the research group surveyed the plots at least 10 years after the initial cut, they recorded two predominant effects: a rapid loss of species diversity, especially from the smaller fragments, and a startling drop in **biomass**, or

Experiment

Question: How does fragmentation affect the quality of tropical wet forest habitats?

Hypothesis: Fragmentation reduces the quality of wet forest habitats.

Null hypothesis: Fragmentation has no effect on the quality of wet forest habitats.

Experimental setup:

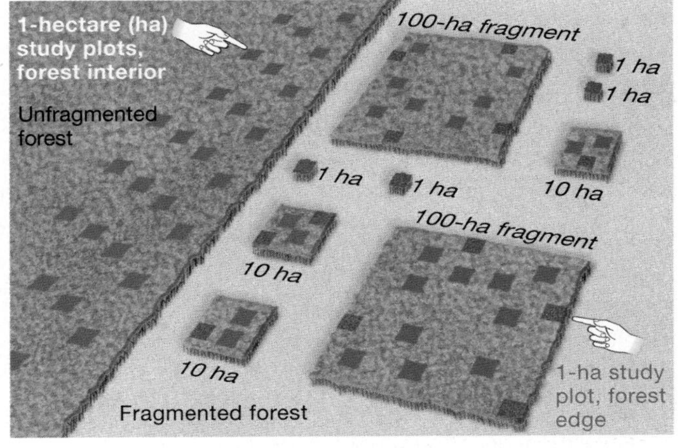

Prediction: Species diversity and biomass will decline in forest fragments compared with those of the forest interior, particularly along edges of fragments.

Prediction of null hypothesis: Species diversity and biomass will be the same inside forest fragments and along edges as in the forest interior.

Results:

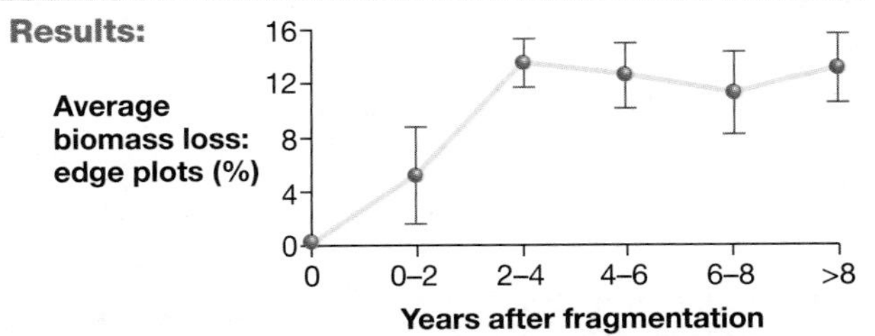

Conclusion: Biomass declines sharply along edges of forest fragments. To interpret this result, hypothesize that large trees near forest edges tend to die.

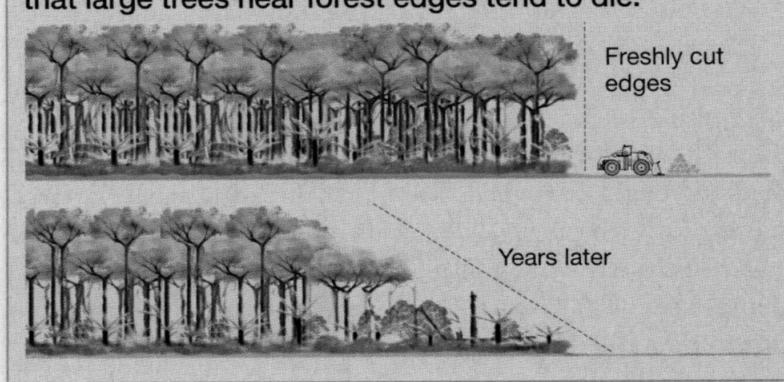

FIGURE 55.11 Experimental Evidence for Edge Effects in Fragmented Forests. Researchers tracked 66 study plots among four 1-hectare fragments, three 10-hectare fragments, and two 100-hectare fragments.

BOX 55.2 An Introduction to Conservation Genetics

Analyses presented in Chapters 24 and 25 highlighted the four evolutionary forces—natural selection, genetic drift, gene flow, and mutation—that affect allele frequencies and cause evolution. In addition, inbreeding and other forms of nonrandom mating change genotype frequencies. Conservation biologists have been keenly interested in how these processes are affecting small, endangered populations.

Several decades of observational and experimental work on conservation genetics supports the following conclusions:

- When populations are fragmented into a metapopulation structure, gene flow between isolated groups is reduced or eliminated. You might recall from Chapter 25 that the major impact of gene flow between populations is to homogenize them genetically. When gene flow is reduced, then, isolated populations begin to differentiate.
- Genetic drift is much more pronounced in small populations than large populations. Chapter 25 presented theory and experiments on the major impact of drift: Because it leads to the random loss or fixation of alleles, drift reduces genetic variation in populations. Loss of genetic variation has now been documented in a wide array of endangered species and other small populations. This is a concern because if the environment is changed by the evolution of a new disease or through global warming, the population may lack alleles that confer high fitness in the new environment.
- Small populations become inbred, and inbreeding often leads to lowered fitness—the phenomenon known as inbreeding depression.

In general, then, small populations that are isolated from each other are expected to be much less healthy, genetically, than robust populations that occupy large, contiguous areas.

When biologists document fitness declines in small, isolated populations, one option is to increase gene flow experimentally by importing individuals of the same species from a different population. For example, **Figure 55.12** shows data on annual population growth rate (λ; see Chapter 52) in a small population of bighorn sheep isolated on a refuge in northwest Montana. From the time the population was founded in 1922 until gene flow was experimentally increased, the average population size was about 42 individuals and average growth rate declined. After introducing 5 new individuals in 1985 and 10 additional sheep in 1990–1994, though, λ has stabilized and population size has increased. Introducing new alleles in this way can counteract the effects of genetic drift and inbreeding.

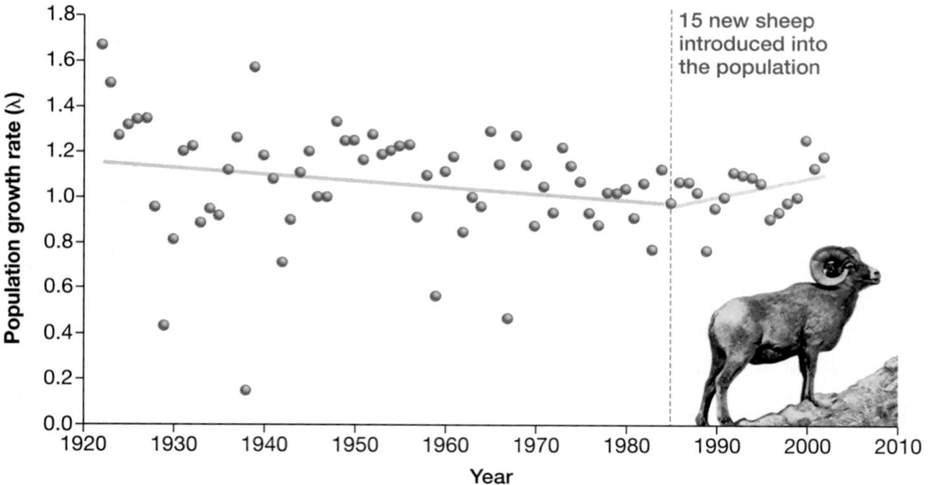

FIGURE 55.12 Experimental Evidence that Gene Flow Can Alleviate Genetic Problems in Small Populations. The y-axis plots the finite rate of increase (λ) in an isolated population of bighorn sheep, before and after unrelated individuals were introduced.

EXERCISE Draw a horizontal line indicating the value of λ where population size is constant (no growth).

the total amount of fixed carbon, in the study plots located near the edges of logged fragments. In tropical wet forests, most of the biomass is concentrated in large trees. The edges of the experimental plots contained many downed and dying trees. Based on this observation, the researchers inferred that the decrease in biomass occurred because large trees near the edges of fragments died from exposure to high winds and dry conditions. Follow-up work showed that in edge habitats and in the fragmented patches, large-seeded, slow-growing trees typical of undisturbed forests are being replaced by early successional, "weedy" species (see Chapter 53). A large number of bird and understory plant species have disappeared as well.

The take-home message of this experiment and other research is clear: When habitats are fragmented, the quality *and* quantity of habitat decline drastically. In addition to losing over 8 million hectares of primary forest in South America and Africa each year, we are losing large amounts of high-quality habitat to fragmentation. Just how many extinctions are all of these problems projected to cause?

MB Web Animation at www.masteringbio.com
Habitat Fragmentation

How Can Biologists Predict Future Extinction Rates?

If the most fundamental question about biodiversity is how many species are alive, then the most basic question in conservation biology is how rapidly species are going extinct. Both questions have been difficult to answer.

Biologists use two approaches to estimate current extinction rates and predict how they might change in the near future. The first is based on direct counts of species that are known to have gone extinct recently or are in imminent danger of going extinct. The second is based on predicting the consequences of habitat loss, using existing data on the relationship between the size of habitats and the number of species present.

Estimates Based on Direct Counts The best information on current extinction rates comes from studies on the best-studied of all major lineages on the tree of life: birds. Based on data in the fossil record, background extinctions in lineages like birds are estimated to occur at the rate of 1 extinction per million species alive per year. Of the approximately 10,000 bird species known, then, one species should go extinct every 100 years due to normal or background events.

Recent analyses have focused on species that were wiped out over the past 1000 years, coincident with the expansion of Polynesian people throughout the Pacific and Europeans throughout the world. These data suggest that birds have been going extinct at a rate 100 times background levels—meaning that a bird species has gone extinct every year, on average. Although the success of conservation programs targeted at birds has lowered this rate in the last few decades, estimates based on the numbers of species that are currently endangered suggest that the rate could reach 1000–1500 extinctions per million species alive per year by the end of this century. If this prediction holds, you may live long enough to see 10 or more species of bird go extinct each year.

Similar trends are occurring in listings of other well-studied groups. For example, a research group looked at endemic species of birds, mammals, reptiles, amphibians, and conifers, and determined that 794 of the species are in imminent danger of extinction. This is three times the number of all species from these groups known to have gone extinct since the year 1500. Instead of a handful of species going extinct each year, we may be losing dozens or hundreds. The message from these analyses is that the current rate of extinction is already many times higher than background or normal levels, and it is poised to increase dramatically in the near future.

Species–Area Relationships A second strategy for predicting the future of Earth's biodiversity focuses on the consequences of the most urgent problem: habitat destruction. Given reasonable projections of habitat loss, biologists can estimate rates of extinction based on well-documented **species–area relationships**.

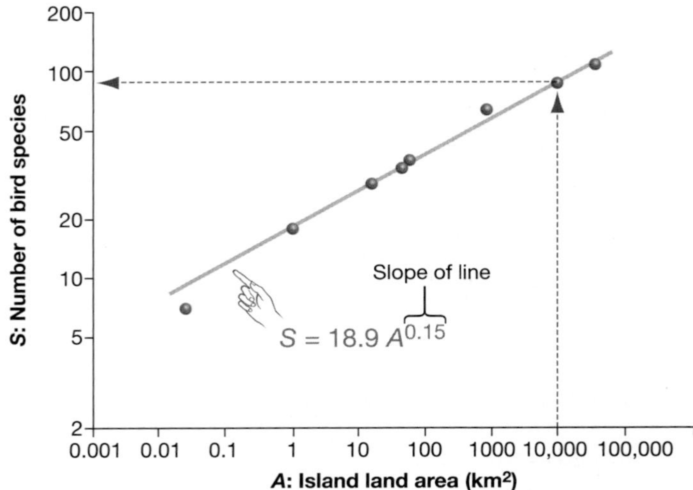

FIGURE 55.13 Species–Area Plots Quantify the Relationship between Species Richness and Habitat Area. Graph plotting the sizes of islands in the South Pacific (horizontal axis) versus the number of bird species present on each island (vertical axis). Notice the log-log scale. The dashed arrows show the number of bird species that are expected to live on an island with an area of 10,000 km^2.

To understand how this work is done, consider that recent estimates of habitat destruction in the world's tropical wet forests suggest that about 10 percent of the original area covered by these biomes is being lost each decade. It is not unreasonable, then, to project that 90 percent of the world's tropical forest habitat will be lost over the next century.

To predict how many species will be driven to extinction as a result, biologists use species–area curves like the one shown in **Figure 55.13**. This graph was generated by a biologist who analyzed the number of bird species found on islands in the Bismarck Archipelago near New Guinea, in the South Pacific. The graph shows the number of species on islands of various sizes. Notice that both axes on the graph are logarithmic. The solid line drawn through the points is described by the function $S = (18.9)A^{0.15}$, where S is the number of species and A is the area. In this island group, the function indicates that each tenfold increase in island area increases the number of bird species present by about 40 percent; a tenfold decline in area reduces bird species by about 30 percent.

These data turn out to be typical for other habitats and taxonomic groups as well. When biologists have plotted species–area relationships for plants, butterflies, mammals, or birds from islands or continental habitats around the globe, the relationship is consistently described by a function of the form $S = cA^z$. The c term is a constant that scales the data. Its value is high in species-rich areas, such as coral reefs or tropical wet forests, and low in species-poor areas, such as arctic tundras. The exponent z represents the slope of the line on a log-log plot. Thus, z describes how rapidly species numbers change with area. Typically, z is about 0.25.

To understand how biologists use these analyses to project extinction rates, study Figure 55.13 again. Ask yourself: If 90 percent of the habitats were destroyed—for example, if *A* were reduced from 10,000 km^2 to 1000 km^2—what percentage of species would disappear? According to the graph, the number of bird species in the Bismarck Archipelago would drop from about 75 to about 53. Thus, the answer is roughly 30 percent. This is an important conclusion. If z is higher than 0.25 for taxa that inhabit tropical wet forests and other threatened habitats, the prediction is that more than 30 percent of all species will be wiped out in the next 100 years.

Check Your Understanding

If you understand that...

- Until recently, most human-induced extinctions have occurred on islands and were due to overhunting or the impact of introduced species.
- Although invasive species remain a major threat to biodiversity, the majority of current problems are on continents and are due to habitat destruction and fragmentation.
- Current and projected extinction rates can be estimated from direct counts and from an analysis of species–area relationships.

You should be able to...

1) Explain why fragmentation reduces habitat quality and leads to genetic problems.
2) Use the species–area relationship to compare the proportion of species lost when 90 percent of a 100,000 km^2 habitat is destroyed in a region where $c = 19$ and $z = 0.20$ versus the same-sized region where $c = 19$ and $z = 0.25$.

55.4 Preserving Biodiversity

The conservation of biodiversity is among the most urgent but least-recognized issues facing us today. Unlike ozone holes, acid rain, phosphorus pollution, global warming, and other environmental problems that humans have recently begun to address, extinction is irreversible. The only solution to the biodiversity crisis is to prevent the loss of alleles, species, and ecosystems.

Solving any global problem requires a common goal to be defined. In the case of preserving biodiversity, the objective is to sustain diverse communities in natural landscapes while supporting the extraction of resources required to maintain the health and well-being of the human population.

In almost every case, the underlying causes of the biodiversity crisis are political and economic pressures that encourage short-term overexploitation of land and other resources and discourage **sustainability**—the managed use of resources at a rate only as fast as the rate at which they are replaced. What reasons can biologists give for preserving biodiversity, in addition to the direct economic benefits listed in **Box 55.3**? From a biological point of view, why does biodiversity matter?

Why Is Biodiversity Important?

The benefits of biodiversity extend beyond the direct use of diverse genes and species by humans to include **ecosystem services**—processes that increase the quality of the abiotic environment. Recall from Chapter 10 that green plants and other photosynthetic organisms produce the oxygen we breathe, and from Chapter 30 that the presence of land plants builds soil, reduces soil erosion, moderates local temperature and wind conditions, and increases the volume of water retained in lakes, streams, and soils. Species from throughout the tree of life are involved in cycling nitrogen, carbon, and other nutrients through ecosystems (see Chapters 28–31 and 54). But do ecosystems improve the abiotic environment more when biodiversity is high?

The effort to answer this question began in the mid-1980s and has blossomed into one of the most active research frontiers in community ecology and conservation biology.

Biodiversity Increases Productivity **Figure 55.14** summarizes a classic experiment on how species richness affects one of the most basic aspects of an ecosystem function: the production of biomass. As the "Experimental Setup" portion of the figure shows, David Tilman and colleagues classified 32 grassland plant species into five functional categories. The functional groups differ by the timing of their growing season and by whether they allocate most of their resources to manufacturing woody stems or seeds. The researchers planted plots measuring 13 m × 13 m with a mixture of between 0 and 32 randomly chosen species, representing from 0 to 5 functional categories. After 2 years of plant growth, the researchers harvested and weighed the aboveground tissues from the plots. The data provided an index of **net primary productivity (NPP)**—the total amount of photosynthesis per unit area per year that ends up in biomass.

Tilman's group compared the productivity of plots with different numbers of species and functional groups, and found that both the number and type of species present had important effects. Plots with more species and with a wider diversity of functional groups were more productive. Total biomass leveled off as species richness and functional diversity increased, however. This observation suggests that increasing species diversity improves ecosystem function only up to a point. The data support the hypothesis that at least some species in ecosystems are redundant.

Follow-up experiments in an array of ecosystems not only supported the conclusion that species richness has a positive impact on NPP, but showed that several causal mechanisms may be at work.

Experiment

Question: Do high species richness and high functional diversity of species increase aspects of ecosystem function such as net primary productivity (NPP)?

Hypothesis: NPP increases with increasing species richness and with increasing functional diversity of species.

Null hypothesis:

Experimental setup:

Plant a total of 289 experimental plots, each with up to 32 species and up to 5 functional groups:

Cool-season grasses: Grow in spring

Warm-season grasses: Grow in summer

Legumes: Fix nitrogen

Woody plants: Trees, shrubs

Forbs: Lots of seeds

Examples of experimental plots:

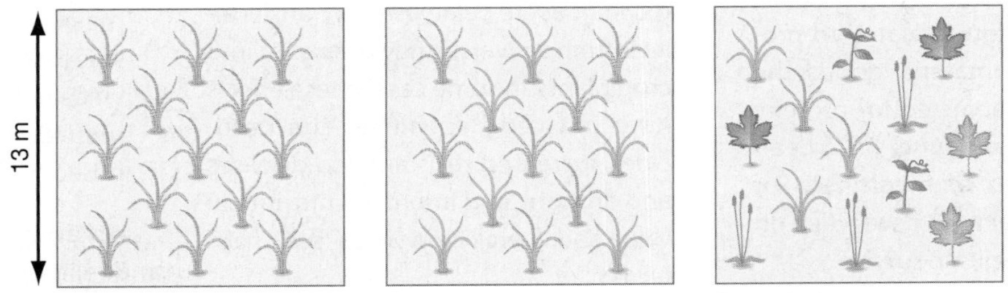

13 m

1 species
1 functional group

2 species
1 functional group

6 species
4 functional groups

Prediction:

Prediction of null hypothesis:

Results:

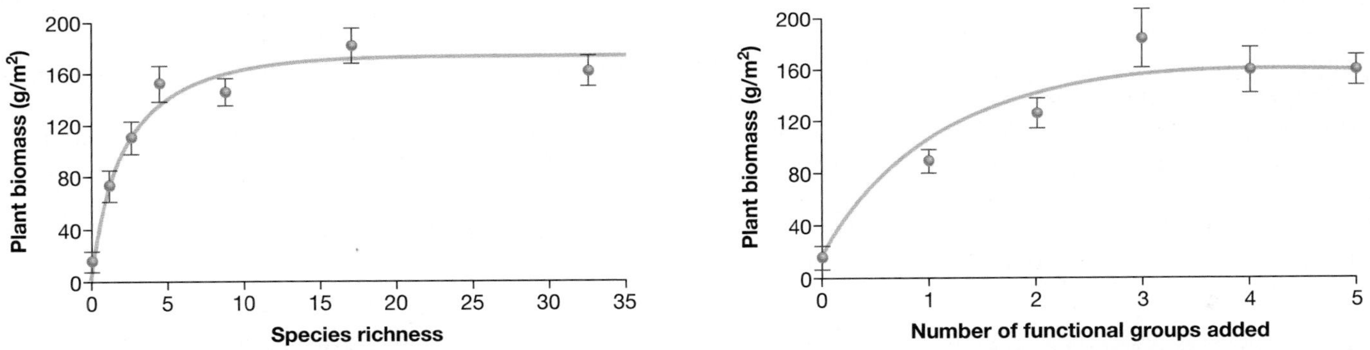

Plant biomass (g/m²)

Species richness

Plant biomass (g/m²)

Number of functional groups added

Conclusion: In this plant community, NPP increases with increasing species richness and increasing functional diversity of plants, at least up to a point.

FIGURE 55.14 Evidence that the Productivity of Ecosystems Depends on the Number and Types of Species Present.

EXERCISE Write in the null hypothesis and both sets of predictions.

- *Resource use efficiency* Diverse assemblages of plant species make more efficient use of the sunlight, water, and other resources available and thus lead to greater overall productivity. For example, some prairie plants extract water near the soil surface, while others use water available a meter or more below the surface. When species diversity is high, more overall water is used and more photosynthesis can occur.
- *Facilitation* Certain species or functional groups facilitate the growth of other species by providing them with nutrients, partial shade, or other benefits. In prairies, the presence of onion-family plants may discourage herbivores, or decaying roots from nitrogen-fixing species may fertilize other species.
- *Sampling effects* In many habitats it is common to observe that one or two species are extremely productive. If the number of species in a study plot is low, it is likely that the "big producer" will be missing and NPP will be low. But if the number of species in a study plot is high, then it is likely

BOX 55.3 Economic Benefits of Biodiversity

The first large-scale, highly organized human societies began to form about 10,000 years ago, when people at several locations around the world began to domesticate wild plants and farm their food. Since that time, wild species have provided the raw material to fuel the continued development of human societies. Vast tracts of forest have been felled for building materials and fuel; wild plants have been selectively bred to yield food and fibers; animals have been domesticated to provide food, labor, and material goods; the oceans have been harvested for protein; and plants, animals, and fungi have been processed as sources of medicines. For thousands of years, humans have relied on a diversity of wild species to survive.

The direct use of biodiversity continues today:

- Research programs collectively known as bioprospecting focus on assessing bacteria, archaea, plants, fungi, and frogs as novel sources of drugs or ingredients in consumer products. Bioprospecting has benefited from the recent explosion of genetic and phylogenetic information, because biologists can now search genomes from a wide array of species to find alleles with desired functions. Recent advances in biotechnology facilitated the development of a new painkiller from the paralyzing sting of tropical cone snails, and a blood anticoagulant from the saliva of vampire bats.
- Agricultural scientists are preserving diverse strains of crop plants in seed banks and continue to use wild relatives of domesticated species in breeding programs aimed at improving crop traits. In addition, genetic engineering techniques are being used to transfer alleles from a diverse array of species into crop plants. In some cases these efforts have reduced dependence on pesticides, increased resistance to diseases and drought, and improved nutritional value and overall crop yields (see Chapter 19).
- The production of almonds, apples, cherries, chocolate, alfalfa, and an array of other crops depends on the presence of wild pollinators. In the United States alone, insect-pollinated crops produce \$40 billion worth of products annually.
- Strategies for cleaning up oil spills, abandoned mines, and contaminated industrial sites are incorporating bioremediation—the use of bacteria, archaea, and plants to metabolize pollutants and render them harmless (see Chapters 28 and 30).
- Recreation based on visiting wild places, or ecotourism, is a major industry internationally and is growing rapidly. In South Africa, for example, the number of tourists visiting wildlife preserves increased from 454,428 in 1986 to over 6 million in 1999. In 2004, ecotourism grew three times faster than the tourism industry as a whole.
- High ecosystem diversity—particularly the presence of forests or grasslands on steep slopes and the presence of wetlands in low-lying areas—dramatically reduces flood damage and the danger posed by mudslides.

An array of biologists, philosophers, and religious leaders argue that, in addition to making money from biodiversity, humans have an ethical obligation to preserve species and ecosystems. Their position is that organisms have intrinsic worth, and that humans diminish the world by extinguishing species and destroying ecosystems. If this is so, then extinction is a moral issue as well as a biological and economic problem. One of the most important reasons to preserve biodiversity may simply be that it is the right thing to do.

the big producer is present and NPP will be high. Simply due to sampling, then, high-species plots will tend to outproduce low-species plots.

The three mechanisms listed are not mutually exclusive—several can operate at the same time. Further, long-term experiments like the one diagrammed in Figure 55.14 have shown that the reason for the pattern can change over time. As plants mature, resource use efficiency and facilitation may become more important than sampling effects.

Does Biodiversity Lead to Stability? When biologists refer to the stability of a community, they mean its ability to (1) withstand a disturbance without changing, (2) recover to former levels of productivity or species richness after a disturbance, and (3) maintain productivity and other aspects of ecosystem function as conditions change over time. **Resistance** is a measure of how much a community is affected by a disturbance. **Resilience** is a measure of how quickly a community recovers following a disturbance.

To test the hypothesis that diversity increases resistance and resilience, the team that did the work highlighted in Figure 55.14 followed up on a natural experiment. In 1987–1988 a severe drought hit their study site. When the drought ended, the team asked whether species richness affected the response to the disturbance. The results are shown in **Figure 55.15**. The graph documents the change in total biomass that occurred from the year prior to the drought until the height of the drought. This quantity reflects how resistant the community is to disturbance. A completely resistant community would show no change. As predicted, drought resistance appeared to be higher in more diverse communities than in less diverse ones.

Experiment

Question: Are more species-rich communities more stable than less-diverse communities?

Hypothesis: Resistance to disturbance should increase with increasing species richness.

Null hypothesis: There is no relationship between species richness and resistance.

Experimental setup:

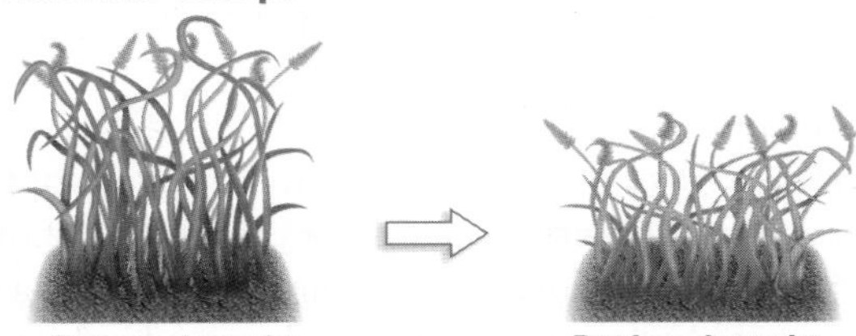

Before drought **During drought**

Compare biomass of experimental plots before drought and at peak of drought. (This was a natural experiment—severe drought just happened to occur during study.)

Prediction: Plots that were more species rich before the drought will be more resistant to change.

Prediction of null hypothesis: All the plots will have similar resistance regardless of species richness before the drought.

Results:

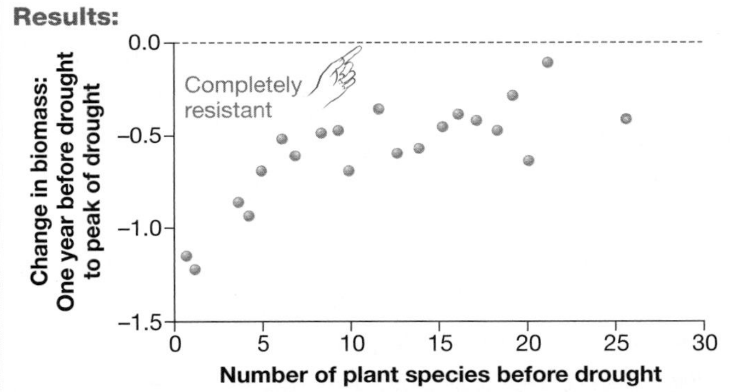

Conclusion: Resistance to disturbance increases with increasing species richness.

FIGURE 55.15 Evidence that Species-Rich Communities Resist Disturbance. Data support the hypothesis that diverse communities are more stable, meaning they change less during a disturbance, than less-diverse communities are.

● **QUESTION** The 0.0 value on the vertical graph axis is labeled "Completely resistant." Why?

In addition, the group analyzed the change in biomass in each plot four years after the drought versus biomass prior to the drought. This analysis focused on how resilient the community is. A completely resilient community would recover quickly from the disturbance and have the same biomass at both times. The data indicated that most plots containing five or fewer species showed a significant lowering of biomass after the disturbance, indicating that they had not recovered. But in all of the plots that contained more than five species, biomass after the drought was the same as biomass prior to the disturbance.

More recent experiments on California grasslands suggests that species-rich ecosystems may also be less prone to invasion by exotic species. Taken together, these results lead biologists to be increasingly confident that species richness has a strongly positive effect on how ecosystems function. In North American grasslands, at least, communities that are more diverse appear to be more productive, more resistant to disturbance and invasion, and more resilient than communities that are less diverse. Increased richness increases the services provided by ecosystems. By implication, biologists can infer that if ecosystems are simplified by extinctions, then productivity and other attributes might decrease.

Designing Effective Protected Areas

When biologists first recognized the growing threats to biodiversity worldwide, they joined with government agencies, economists, community leaders, private landowners, and others to set aside protected areas. To assess the progress of this effort, the World Conservation Union sponsors the World Parks Congress every decade. In 1992 the Congress met in Caracas, Venezuela, and established a goal of setting aside 10 percent of Earth's land surface in protected areas. In 2003 the Congress met in Durban, South Africa, and announced that this goal had been surpassed: Protected areas covered 11.5 percent of Earth's terrestrial surface.

How efficiently is the existing network of protected areas protecting biodiversity? Researchers are attempting to answer this question via a geographic approach called the **Gap Analysis Program** (**GAP**). A GAP analysis tries to identify gaps between geographic areas that are particularly rich in biodiversity and areas that are actually managed for the preservation of biodiversity. One GAP analysis combined data sets on the distribution of mammals, birds, amphibians, and freshwater turtles with a map of world protected areas. The analysis revealed that many species' ranges occur completely outside any protected areas. It also pinpointed regions in Mexico, Madagascar, and elsewhere where the gap between species' ranges and protected areas is particularly high.

To date, most GAP analyses suggest that, because relatively few species richness hotspots are included in existing protected areas, the 11.5 percent of Earth's surface area that is now being managed for biodiversity will not be enough to conserve many species. Efforts to fill these gaps are now focused on preserving species richness hotspots and centers of endemism like those illustrated in Figure 55.4.

In addition to working on where reserves are set up, biologists are focusing more attention on how reserves are designed. For example, it is often impossible to preserve large areas of high-quality habitat. In most cases, biologists have to accept the fact that habitats are fragmented and species exist as metapopulations. Given this reality, what is the best way to prevent small, isolated populations from going extinct?

The leading hypothesis in reserve design was to make sure that strips of undeveloped habitat, called **wildlife corridors**, connected populations that would otherwise be isolated. In some cases wildlife corridors were as simple as a walkway under a major highway, so that animals could move from one side to the other without being killed. By facilitating the movement of individuals, the goals of corridors were to (1) allow areas to be recolonized if a species was lost, and (2) introduce new alleles that would counteract the deleterious effects of genetic drift and inbreeding, detailed in Box 55.2.

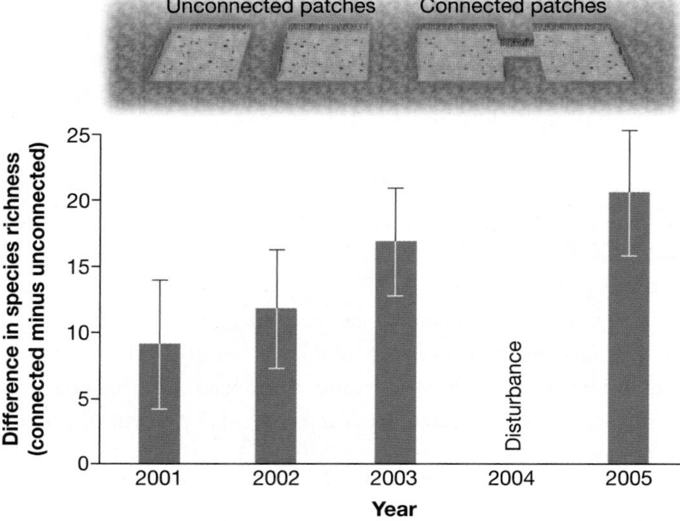

FIGURE 55.16 Experimental Evidence that Wildlife Corridors Work. Species richness could not be assessed in 2004 because the study plots were burned as part of a program to restore natural habitats.

● **EXERCISE** Draw and label bars showing what this graph would look like if corridors had no effect on species richness.

Do corridors work? A recent experiment suggests that the answer is yes. A research team established a series of restored natural areas in the middle of a large, species-poor pine plantation. This was the equivalent of restoring patches of grassland habitat in the middle of an enormous cornfield. Some of the restoration sites were connected by corridors, while others were isolated. By monitoring the composition of species inside each habitat patch over time, the group was able to show that connected patches are steadily gaining more species over time compared to unconnected patches (**Figure 55.16**). This is strong evidence that wildlife corridors can increase overall species richness in a metapopulation.

Experiments like these bring us to the forefront of work on biodiversity and conservation biology. Your generation is facing the most serious global environmental crisis in the history of our species. As someone with a background in biology, you have the intellectual tools to help solve it. In addition to studying life, biologists have to save it.

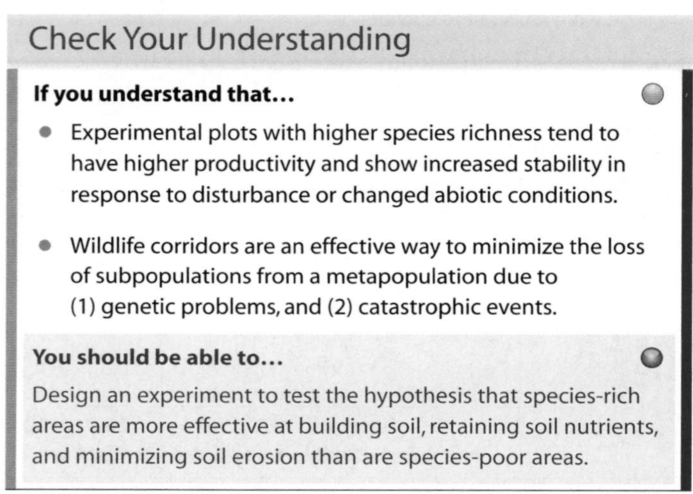

Check Your Understanding

If you understand that...

- Experimental plots with higher species richness tend to have higher productivity and show increased stability in response to disturbance or changed abiotic conditions.
- Wildlife corridors are an effective way to minimize the loss of subpopulations from a metapopulation due to (1) genetic problems, and (2) catastrophic events.

You should be able to...

Design an experiment to test the hypothesis that species-rich areas are more effective at building soil, retaining soil nutrients, and minimizing soil erosion than are species-poor areas.

Chapter Review

SUMMARY OF KEY CONCEPTS

● **Biodiversity can be analyzed at the genetic, species, and ecosystem levels.**

Genetic diversity is well characterized within some species, but biologists are only beginning to use genome sequencing techniques to explore the extent of genetic diversity in ecosystems. Research on the total number of species alive today is also at a preliminary stage, with estimates ranging from 10 to 100 million. To date, the message from efforts to characterize biodiversity is that it is much more extensive than expected and that a great deal remains to be learned.

You should be able to describe conditions that maximize genetic, species, and ecosystem diversity. ●

● **If recent rates of extinction due to human population growth and habitat destruction continue, a mass extinction will occur.**

Historically, most human-caused extinctions have occurred on islands because of direct exploitation or the introduction of exotic herbivores and predators. Habitat destruction is currently the leading cause of extinctions, however. Experiments in the Brazilian Amazon have shown that habitat loss leads not only to a rapid decline in biodiversity but also to a decline in the quality of the remaining habitats due to fragmentation.

To estimate how many species will go extinct in the near future, biologists combine data on current rates of habitat loss—usually estimated from satellite images taken over time—with data on the average number of species found in habitats of a given size. These species–area analyses suggest that if 90 percent of habitats are destroyed as expected during the next century, then over 30 percent of all species will become extinct. According to data on the rates of extinction in birds and other well-studied groups, it is likely that 60 percent of all species will be wiped out within 500 years.

You should be able to explain why species that have a metapopulation structure, due to habitat fragmentation, may be at higher risk of extinction than a species of similar abundance that exists in a large, contiguous habitat.

MB **Web Animation** at www.masteringbio.com
Habitat Fragmentation

Humans depend on biodiversity for the products that wild species provide and for ecosystem services that protect the quality of the abiotic environment. Species diversity is important for maintaining the productivity of natural ecosystems and their ability to build and hold soil, moderate local climates, retain and cycle nutrients, retain surface water and recharge groundwater, prevent flooding, and produce oxygen. At the ecosystem level, experiments have shown that high species richness increases aspects of ecosystem function such as productivity, resistance to disturbance, and ability to recover from disturbance. Humans also gain direct economic benefits from fishing, forestry, agriculture, tourism, and other activities that depend on biodiversity.

You should be able to outline a plan for restoring an abandoned airstrip that would maximize its ability to deliver ecosystem services.

QUESTIONS

Test Your Knowledge

1. What does a species–area plot show?
   a. The overall distribution, or area, occupied by a species.
   b. The relationship between the body size of a species and the amount of territory or home range it requires.
   c. The number of species found, on average, in tropical versus northern areas.
   d. The number of species found, on average, in a habitat of a given size.
2. What does a GAP analysis do?
   a. It compares the current distributions of species with the locations of preserved habitats.
   b. It quantifies gross *a*boveground *p*roductivity.
   c. It uses data on the rates at which lists of threatened species are growing to project the rate of future extinctions.
   d. It uses genome sequencing techniques to quantify genetic (allelic) diversity in an ecosystem.
3. What is the difference between species richness and species diversity?
   a. Species diversity incorporates data on species interactions.
   b. Species diversity takes relative abundance into account.
   c. Species diversity is weighted by taxonomic (phylogenetic) diversity.
   d. Species diversity is adjusted for conservation priority (endangered status of species present).
4. What is a biodiversity "hotspot?"
   a. an area where an all-taxon survey is underway
   b. an area where an environmental sequencing study has been completed
   c. a habitat with high NPP
   d. an area with high species richness
5. Why do small populations become inbred?
   a. They are often part of a metapopulation structure.
   b. Genetic drift becomes a prominent evolutionary force.
   c. Over time, all individuals become increasingly related.
   d. Natural selection does not operate efficiently in small populations.
6. What is the primary cause of endangerment in marine environments?
   a. overexploitation
   b. pollution
   c. global warming
   d. invasive species

Test Your Knowledge answers: 1. d; 2. a; 3. b; 4. d; 5. c; 6. a

Test Your Understanding

Answers are available at www.masteringbio.com

1. The primary cause of endangerment on continents has changed from direct exploitation to habitat loss. Explain why this change occurred.
2. Compare and contrast biodiversity at the genetic, species, and ecosystem levels. How do biologists analyze the extent of biodiversity at each of these levels?
3. Biologists claim that the all-taxa survey now under way at the Great Smoky Mountains National Park in the United States will improve their ability to estimate the total number of species living today. Discuss the benefits and limitations that this data set will provide in understanding the extent of global biodiversity.
4. How are species–area curves used to relate rates of habitat destruction to projected extinction rates?
5. Discuss evidence to support the hypotheses that species richness increases ecosystem functions such as productivity, resistance to disturbance, and resilience. Describe the logic behind the resource use efficiency, facilitation, and sampling effect hypotheses to explain why species richness increases productivity, and thus the ecosystem services that benefit humans.
6. Explain why the fragmentation of habitats reduces their ability to support biodiversity. Explain why the construction of wildlife corridors can help maintain biodiversity in a fragmented landscape.

Applying Concepts to New Situations

Answers are available at www.masteringbio.com

1. Projections for an impending mass extinction are contingent on the continuation of present trends in human population growth and in habitat destruction. Do you believe that these trends will continue? In general, are you optimistic or pessimistic about the future of biodiversity? Explain your logic.
2. You are helping design a series of reserves in a tropical country. List the steps you would recommend for gathering data and creating a plan that would protect a large number of species in a small amount of land.
3. The maps that follow chronicle the loss of old-growth forest (>200 years old) that occurred in Warwickshire, England, and in the United States. In your opinion, under what conditions is it ethical for conservationists who live in these countries to lobby government officials in Brazil, Indonesia, and other tropical countries to slow the rate of loss of old-growth forest?

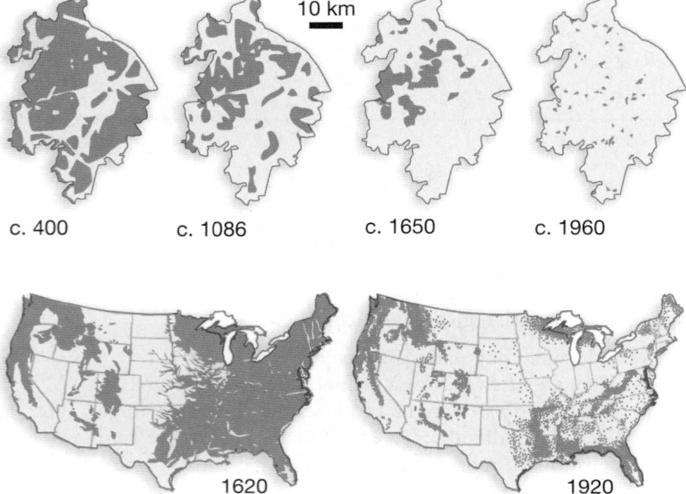

4. Make a list of characteristics that would render a species particularly vulnerable to extinction by humans. Make a list of characteristics that would render a species particularly resistant to pressure from humans. Try to think of an example of each type of species.

www.masteringbio.com is also your resource for • Answers to text, table, and figure caption questions and exercises • Answers to *Check Your Understanding* boxes • Online study guides and quizzes • Additional study tools including the *E-Book for Biological Science* 3rd ed., textbook art, animations, and videos.

BioSkills 1

Reading Graphs

Graphs are the most common way to report data, for a simple reason. Compared to reading raw numerical values in the form of a table or list, a graph makes it much easier to understand what the data mean.

Learning how to read and interpret graphs is one of the most basic skills you'll need to acquire as a biology student. As when learning piano or soccer or anything else, you need to understand a few key ideas to get started and then have a chance to practice—a *lot*—with some guidance and feedback.

Getting Started

To start reading a graph, ask yourself three questions:

1. *What do the axes represent?* As **Figure BS1.1a** shows, the horizontal axis of a graph is also called the *x*-axis or the abscissa. The vertical axis of a graph is also called the *y*-axis or the ordinate. Each axis represents a quantity that varies over a range of values. These values are indicated by the ticks and labels on the axis. In our example, the *x*-axis represents the width of broccoli stalks, in centimeters, while the *y*-axis represents the number of individuals in the sample. Note that the units being represented on each axis should *always* be clearly labeled or implied.

   To create a graph, researchers plot the independent variable on the *x*-axis and the dependent variable on the *y*-axis. The *independent* and *dependent* terms are appropriate because the values on the *y*-axis depend on the *x*-axis values. In our example, the researchers wanted to show the distribution of flowering stalk sizes in a large sample of individuals. Thus, the number of individuals present depended on the size of the flowering stalk. (This is similar to the way you would plot a graph showing the heights of students in your class.)

   In many graphs in biology, the independent variable is either time or the various treatments used in an experiment. In these cases, the *y*-axis records how some quantity changes as a function of time or as the outcome of the treatments applied to the experimental cells or organisms. The value on the *y*-axis depends on the value on the *x*-axis, but not vice versa.

(a) Read the axes—what's being plotted?

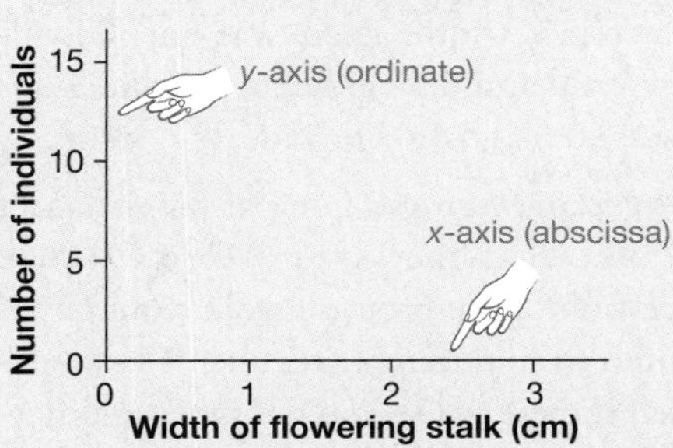

(b) Look at the bars or data points—what do they represent?

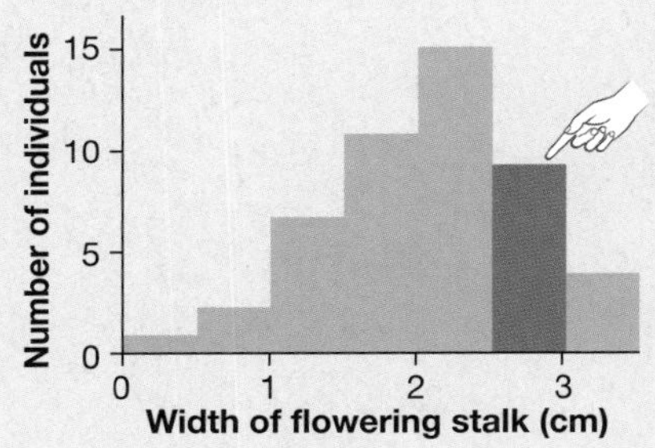

(c) What's the punchline?

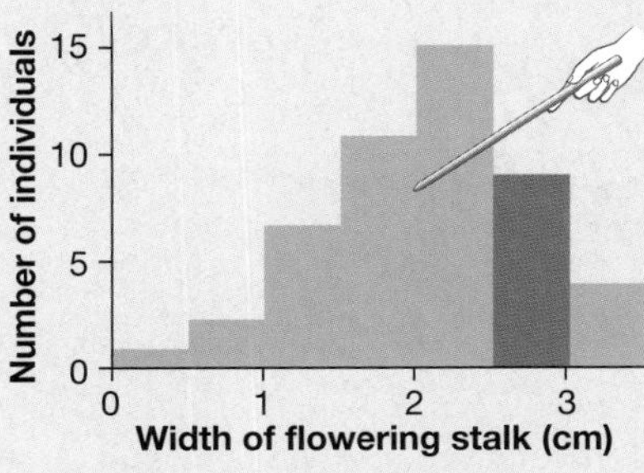

FIGURE BS1.1 Histograms Are a Common Way to Graph Data.

EXERCISE Draw what this histogram would look like if almost all the individuals in the sample had flowering stalks between 3.0 and 3.5 cm wide.

EXERCISE Suppose you measured the width of flowering stalks in a large sample of individuals over time—measuring the same individuals every week for 10 weeks. Draw a scatter plot predicting how the average width of flowering stalks might change over time.

2. *What do the bars or the data points represent?* Most of the graphs in this text are histograms like the one shown here, or bar charts—which are similar to histograms, but plot data that have discrete values instead of a continuous range of values—or scatter plots, where individual data points are plotted. Once you've read the axes, you need to figure out what each bar in the bar chart or each data point in the scatter plot represents and how the researchers determined it. In our broccoli stalk example, the bars in **Figure BS1.1b** represent the number of individuals in the sample with flowering stalks of a certain width. There was one individual with a stalk between 0 and 0.5 cm wide, two individuals with a stalk between 0.5 and 1.0 cm wide, and so on.
3. *What is the overall trend or message?* Look at the data as a whole, and figure out what they mean. **Figure BS1.1c** suggests an interpretation of the broccoli stalk example. If a bar chart plots values from different treatments in an experiment, ask yourself if these values are the same or different. If a scatter plot shows how some quantity changes over time, ask yourself if that quantity is increasing, decreasing, or staying the same.

Getting Practice

Working with this text will give you lots of practice with reading graphs—they appear in almost every chapter. In many cases we've put a little pointing hand, like your instructor's hand at the whiteboard, with a label that suggests an interpretation or draws your attention to an important point on the graph. In other cases, you should be able to figure out what the data mean on your own or with the help of other students or your instructor.

Make an effort to develop this skill. One of the most common complaints from professors who teach upper-level courses is, "My students can't read graphs!" When you become competent and confident at interpreting graphs, your performance will improve in this course, in subsequent courses, and on professional and graduate school admission tests.

Reading a Phylogenetic Tree

Phylogenetic trees show the evolutionary relationships among species, just as a genealogy shows the relationships among people in your family. They are unusual diagrams, however, and it can take practice to interpret them correctly.

To understand how evolutionary trees work, consider **Figure BS2.1**. Notice that a phylogenetic tree consists of branches, nodes, and tips. Branches represent populations through time. Nodes (also called forks) occur where an ancestral group splits into two or more descendant groups (see point B in Figure BS2.1). If more than two descendant groups emerge from a node, the node is called a polytomy (see node C). Tips (also called terminal nodes) are the tree's endpoints, which represent groups living today or a dead end—a branch ending in extinction. The names at the tips can represent species or larger groups such as mammals or conifers. Recall from Chapter 1 that a taxon (plural: *taxa*) is any named group of organisms. A taxon could be a single species, such as *Homo sapiens*, or a large group of species, such as Primates. Groups that occupy adjacent branches on the tree are called sister taxa.

The phylogenetic trees used in this text are all rooted—meaning the bottom, or most basal, node on the tree is the most ancient. To determine where the root on a tree occurs, biologists include one or more out-group species when they are collecting data to estimate a particular phylogeny. An out-group is a taxonomic group that is known to have diverged prior to the rest of the taxa in the study. In Figure BS2.1, "Taxon 1" is an out-group to the monophyletic group consisting of taxa 2–6. A monophyletic group consists of an ancestral species and all of its descendants. The root of a tree is placed between the out-group and the monophyletic group being studied. This position in Figure BS2.1 is node A.

Understanding monophyletic groups is fundamental to reading and estimating phylogenetic trees. Monophyletic groups may also be called lineages or clades and can be identified using the "one-snip test": If you cut any branch on a phylogenetic tree, all of the branches and tips that fall off represent a monophyletic group. Using the one-snip test, you should be able to convince yourself that the monophyletic groups on a tree are nested. In Figure BS2.1, for example, the monophyletic group comprising node A and taxa 1–6 contains a monophyletic group consisting of node B and taxa 2–6, which includes the monophyletic group represented by node C and taxa 4–6.

To put all these new terms and concepts to work, consider the phylogenetic tree in **Figure BS2.2**, which shows the relationships

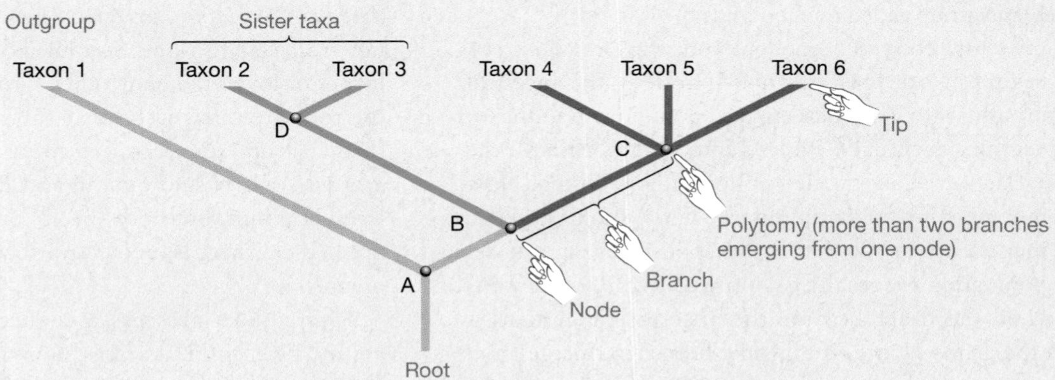

FIGURE BS2.1 Phylogenetic Trees Have Roots, Branches, Nodes, and Tips.
EXERCISE Circle all nine monophyletic groups present.

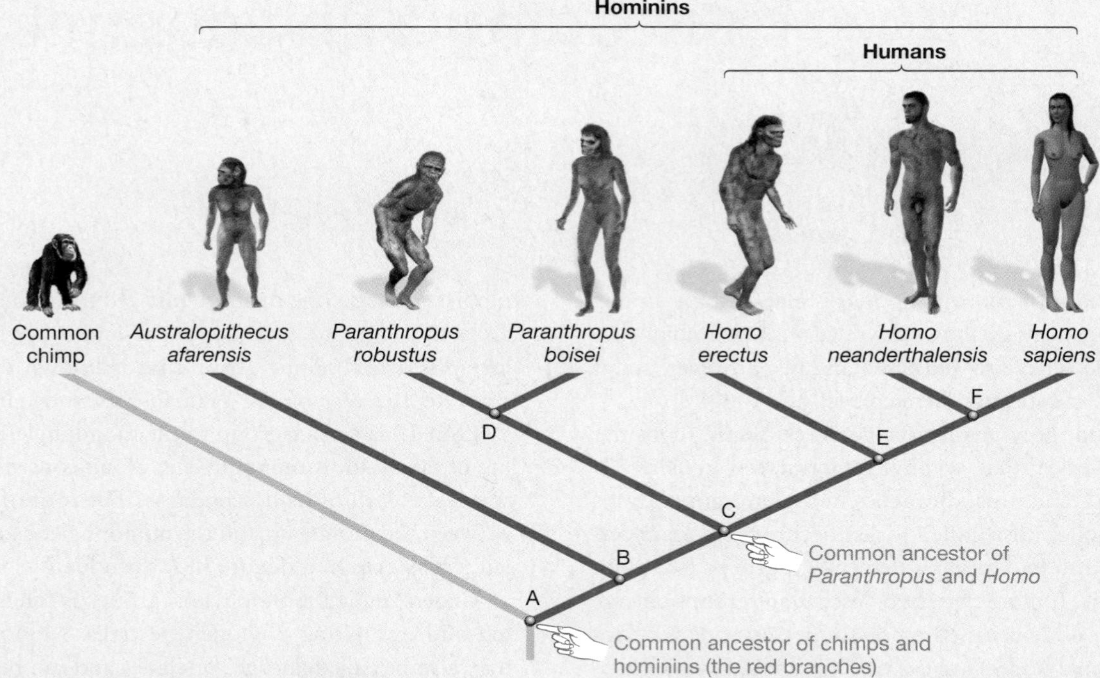

FIGURE BS2.2 An Example of a Phylogenetic Tree. A phylogenetic tree showing the relationships of species in the monophyletic group called hominins.

EXERCISE All of the hominins walked on two legs—unlike chimps and all of the other primates. Add a mark on the phylogeny to show where upright posture evolved, and label it "origin of walking on two legs." Circle and label a pair of sister species. Circle the monophyletic group called hominins. Label an out-group to the monophyletic group called humans (species in the genus *Homo*).

between common chimpanzees and six human and humanlike species that lived over the past 5–6 million years. Chimps functioned as an out-group in the analysis that led to this tree, so the root was placed at node A. The branches marked in red identify a monophyletic group called the hominins.

To practice how to read a tree, put your finger at the tree's root and begin to work your way up. At node A, the ancestral population split into two descendant populations. One of these populations eventually evolved into today's chimps; the other gave rise to the six species of hominins pictured. Now continue moving your finger up the tree until you hit node C. It should make sense to you that at this splitting event, one descendant population eventually gave rise to two *Paranthropus* species, while the other became the ancestor of humans—species in the genus *Homo*. If multiple branches emerge from a node, creating a polytomy, it would be because the populations involved split from one another so quickly that it is not possible to tell which split off earlier or later.

As you study Figure BS2.2, you should consider a couple of important points. First, there are many equivalent ways of drawing this tree. For example, this version shows *Homo sapiens* on the far right. But the tree would be identical if the two branches emerging from node E were rotated 180°, so that the species appeared in the order *Homo sapiens*, *Homo neanderthalensis*, *Homo erectus*. Trees are read from root to tips, not from left to right. Second, no species on any tree is any higher or lower than any other. Groups that branch off close to the root are referred to as basal, and groups that branch off further from the root are more derived. *Australopithecus afarensis* is a basal hominin and *Homo erectus* is a more derived hominin, but both species are at tips—*Homo erectus* is not higher. There is, in fact, no such thing as a higher or lower organism.

Figure BS2.3 presents a chance to test your tree-reading ability. Five of the six trees shown in this diagram are identical in terms of the evolutionary relationships they represent. One differs. You should be able to answer the question posed in the figure's caption and then, on tree (a), circle three nested monophyletic groups and label a node, a branch, a tip, and the out-group.

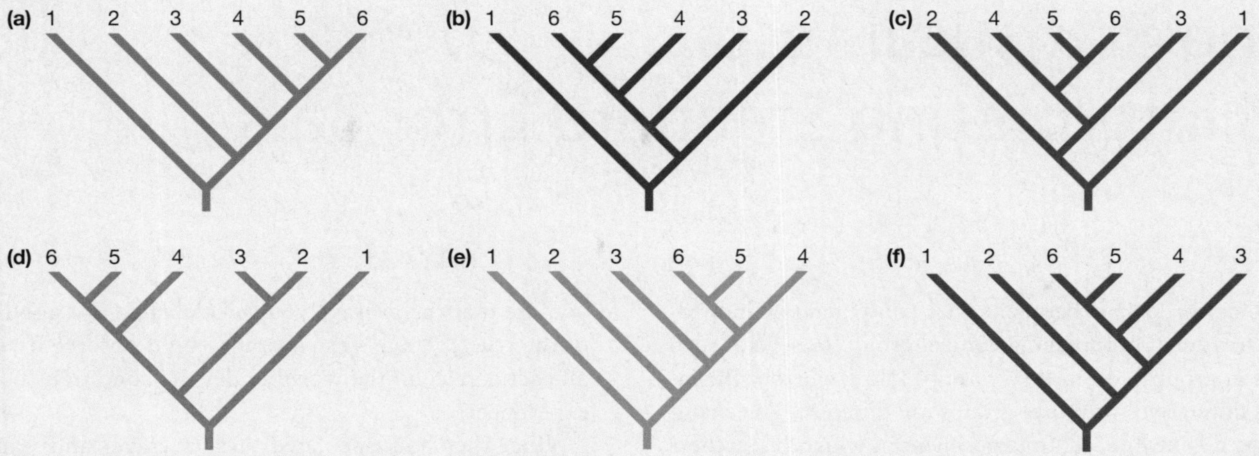

FIGURE BS2.3 Alternative Ways of Drawing the Same Tree.

● **QUESTION** Five of these six trees describe exactly the same relationships among taxa 1 through 6. Identify the tree that is different from the other five.

Answer: The unlike tree is (d).

BioSkills 3

Using Statistical Tests and Interpreting Standard Error Bars

When biologists do an experiment, they collect data on individuals in a treatment group and a control group (or several such comparison groups). Then they want to know whether the individuals in the two (or more) groups are different. For example, Chapter 1 introduced an experiment in which researchers fed three types of fruits to two different types of fruit-eating organisms—cactus mice and birds called thrashers—and measured how much of each type of fruit the animals ate.

Figure BS3.1 graphs the data for cactus mice and thrashers. The treatments are plotted on the x-axes as hackberry (H), non-pungent chilies (NP), and pungent chilies (P). The heights of the bars on the graphs indicate the average percentage of each type of fruit consumed by the 5 mice and 10 thrashers tested. The thin "I-beams" on each bar indicate the standard error of each average. The standard error is a quantity that indicates the uncertainty in the calculation of an average. For example, if two of the mice ate all of the hackberry offered, two ate none, and one ate half, then your estimate of how much an average mouse consumes would be 50 percent. The standard error of that average would be very large, though, because the amounts that the mice ate ranged from 0 to 100 percent. In contrast, if two of the mice ate 49 percent of the hackberry, two ate 51 percent, and one ate half, the average would still be 50 percent but the standard error would be very small. In effect, the standard error quantifies how confident you are that the average you've calculated is a good estimate of the true average—the average you'd observe if you tested all cactus mice in the world under the conditions used in the experiment.

Once they had calculated these averages and standard errors, the biologists wanted to know the answers to two questions: Do cactus mice eat all three types of fruit equally? And, do thrashers eat all three types of fruit equally?

After looking at the data, you might conclude that cactus mice ate different amounts of the fruits offered—specifically, that they ate much more hackberry than non-pungent chilies or pungent chilies. You might be less sure about suggesting that thrashers ate less hackberry and more of the non-pungent and pungent chilies. But how could you come to conclusions like this rigorously, instead of subjectively?

The answer is to use a statistical test. The first step in a statistical test is to specify the null hypothesis, which is that there is no difference among the groups. The second step is to calculate a test statistic, which is a number that characterizes the size of the difference among the groups. In this case, the test statistic compares the actual differences in amount of fruit consumed to the differences predicted by the null hypothesis. The third step is to determine the probability of getting a test statistic as large as the one calculated just by chance. The answer comes from a reference distribution—a mathematical function

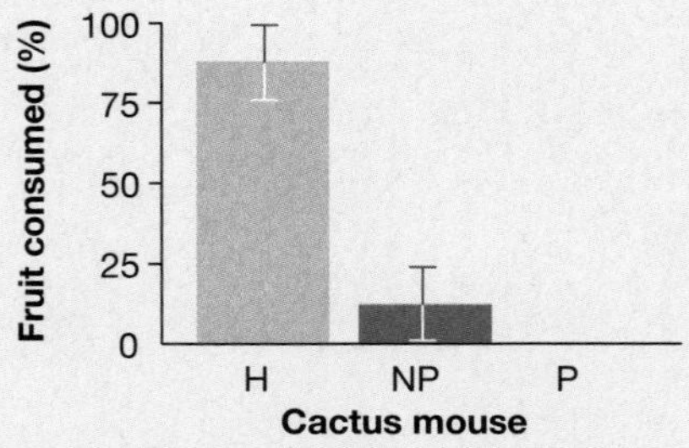

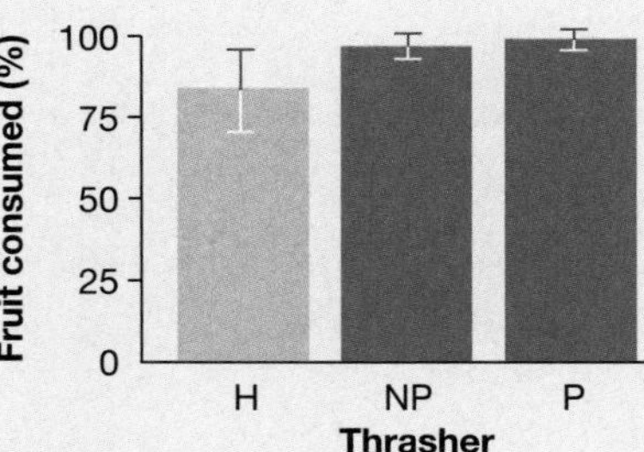

FIGURE BS3.1 Standard Error Bars Indicate the Uncertainty in an Average.

that specifies the probability of getting various values of the test statistic if the null hypothesis is correct. (If you take a statistics course, you'll learn which reference distributions are relevant to different types of data.)

You are very likely to see small differences among treatment groups just by chance—even if no differences actually exist. If you flipped a coin 10 times, for example, you are unlikely to get exactly five heads and five tails, even if the coin is fair. A reference distribution tells you how likely you are to get each of the possible outcomes of the 10 flips if the coin is fair, just by chance.

In this case, the reference distribution indicated that if the null hypothesis of no actual fruit preference is correct, you would see differences as large as those observed for mice only 0.8 percent of the time just by chance. For thrashers, though, you would expect to see differences as large as those observed in the experiment 22 percent of the time, just by chance, if the null hypothesis is correct. By convention, biologists consider a difference among treatment groups to be statistically significant if you have less than a 5 percent probability of observing it just by chance. Based on this convention, the researchers were able to claim that the null hypothesis is not correct for cactus mice—they really do prefer hackberries over chilies. But the null hypothesis is correct for thrashers—the data indicate that they show no preference among the three fruits.

It is likely that you'll be doing actual statistical tests early in your undergraduate career. To use this text, though, you only need to be aware of what statistical testing does. And you should take care to inspect the standard error bars on graphs in this book. As a *very* rough rule of thumb, averages often turn out to be significantly different, according to an appropriate statistical test, if two times the standard errors do not overlap.

BIOSKILLS 4

Reading Chemical Structures

If you haven't had much chemistry yet, learning basic biological chemistry can be a challenge. One of the stumbling blocks is simply being able to read chemical structures efficiently and understand what they mean. This skill will come much easier once you have a little notation under your belt and you understand some basic symbols.

Atoms are the basic building blocks of everything in the universe, just as cells are the basic building blocks of your body. Every atom has a 1- or 2-letter symbol. The following table shows the symbols for most of the atoms you'll encounter in this book. You should memorize these. The table also offers details on how the atoms form bonds as well as how they are represented in visual models.

When atoms attach to each other by covalent bonding, a molecule forms. Biologists have a couple of different ways of representing molecules—you'll see each of these in the book and in class. A molecular formula like those in **Figure BS4.1a** simply lists the atoms present in a molecule, with subscripts indicating how many of each atom are present. If the formula has no subscript, only one of that type of atom is present. A methane (natural gas) molecule, for example, can be written as CH_4 because it consists of one carbon atom and four hydrogen atoms. Structural formulas like those in **Figure BS4.1b** show which atoms in the molecule are bonded to each other, with each bond indicated by a dash. The structural formula for methane indicates that each of the four hydrogen atoms forms one covalent bond with carbon, and that carbon makes a total of four covalent bonds. Note that single covalent bonds are symbolized by a single dash; double bonds are indicated by two dashes.

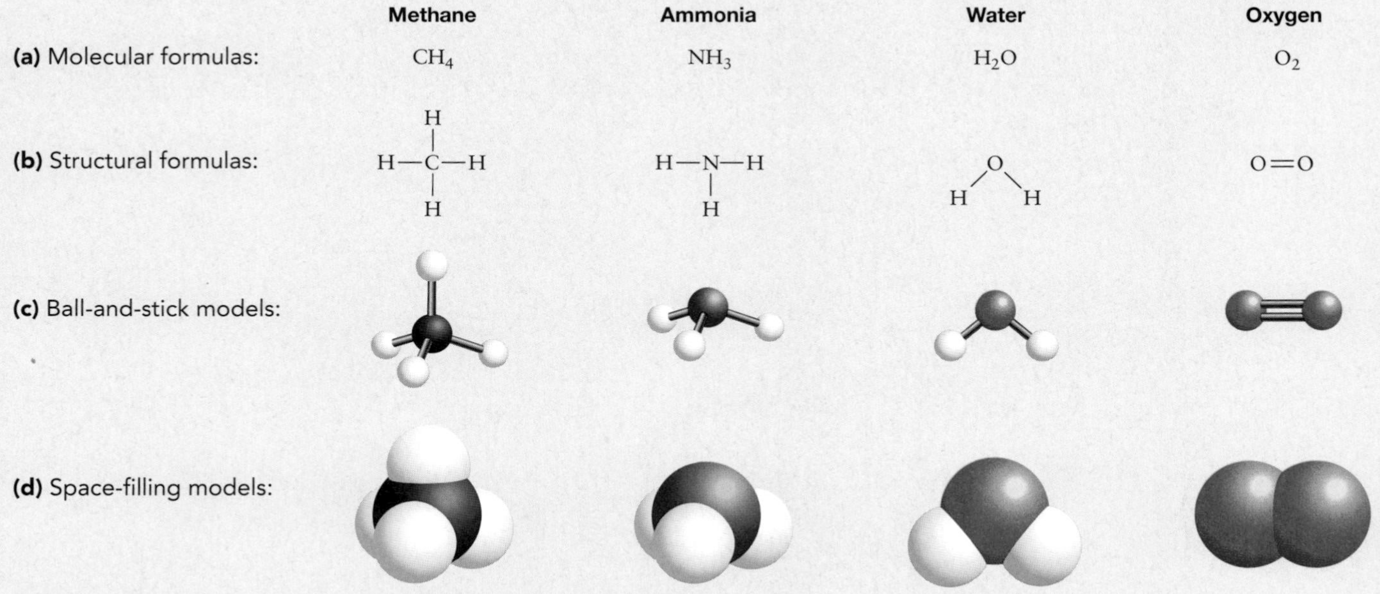

	Methane	Ammonia	Water	Oxygen
(a) Molecular formulas:	CH_4	NH_3	H_2O	O_2
(b) Structural formulas:				O=O
(c) Ball-and-stick models:				
(d) Space-filling models:				

FIGURE BS4.1 Molecules Can Be Represented in Several Different Ways.

EXERCISE Carbon dioxide consists of a carbon atom that forms a double bond with each of two oxygen atoms, for a total of four bonds. It is a linear molecule. Write carbon dioxide's molecular formula, then draw its structural formula, a ball-and-stick model, and a space-filling model.

Even simple molecules have distinctive shapes, because different atoms make covalent bonds at different angles. Ball-and-stick and space-filling models show the geometry of the bonds accurately. In a ball-and-stick model, a stick is used to represent each covalent bond (see **Figure BS4.1c**). In space-filling models, the atoms are simply stuck onto each other in their proper places (see **Figure BS4.1d**).

Atom	Symbol	Number of Bonds It Can Form	Standard Color Code*
Hydrogen	H	1	white
Carbon	C	4	black
Nitrogen	N	3	blue
Oxygen	O	2	red
Sodium	Na	1	(not used in this text)
Magnesium	Mg	2	(not used in this text)
Phosphorus	P	5	orange or purple
Sulfur	S	2	yellow
Calcium	Ca	2	(not used in this text)

*In ball-and-stick or space-filling models.

To learn more about a molecule when you look at a chemical structure, ask yourself three questions:

1. *Is the molecule polar—meaning that some parts are more negatively or positively charged than others?* Molecules that contain nitrogen or oxygen atoms are often polar, because these atoms have such high electronegativity (see Chapter 2). This trait is important because polar molecules dissolve in water.
2. *Does the structural formula show atoms that might participate in chemical reactions?* For example, are there charged atoms or amino ($—NH_2$) or carboxyl (—COOH) groups that might act as a base or an acid?
3. *In ball-and-stick and especially space-filling models of large molecules, are there interesting aspects of overall shape?* For example, is there a groove where a protein might bind to DNA, or a cleft where a substrate might undergo a reaction in an enzyme?

BIOSKILLS 5

Using Logarithms

You have probably been introduced to logarithms and logarithmic notation in algebra courses, and you will encounter logarithms at several points in this course. Logarithms are a way of working with powers—meaning, numbers that are multiplied by themselves one or more times. Scientists use exponential notation to represent powers. For example,

$$a^x = y$$

means that if you multiply a by itself x times, you get y. In exponential notation, a is called the base and x is called the exponent. The entire expression is called an exponential function.

What if you know y and a, and you want to know x? This is where logarithms come in.

$$x = \log_a y$$

This equation reads, x is equal to the logarithm of y to the base a. Logarithms are a way of solving exponential functions. They are important because so many processes in biology (and chemistry and physics, for that matter) are exponential in nature. To understand what's going on, you have to describe the process with an exponential function and then use logarithms to work with that function.

Although a base can be any number, most scientists use just two bases when they employ logarithmic notation: 10 and e. Logarithms to the base 10 are so common that they are usually symbolized in the form $\log y$ instead of $\log_{10} y$. A logarithm to the base e is called a natural logarithm and is symbolized ln (pronounced *EL-EN*) instead of log. You write "the natural logarithm of y" as $\ln y$. The base e is an irrational number (like π) that is approximately equal to 2.718. Like 10, e is just a number. But both 10 and e have qualities that make them convenient to use in biology (and chemistry, and physics).

Most scientific calculators have keys that allow you to solve problems involving base 10 and base e. For example, if you know y, they'll tell you what $\log y$ or $\ln y$ are—meaning that they'll solve for x in our example above. They'll also allow you to find a number when you know its logarithm to base 10 or base e. Stated another way, they'll tell you what y is if you know x, and y is equal to e^x or 10^x. This is called taking an antilog. In most cases, you'll use the inverse or second function button on your calculator to find an antilog (above the log or ln key).

To get some practice with your calculator, consider the equation

$$10^2 = 100$$

If you enter 100 in your calculator and then press the log key, the screen should say 2. The logarithm tells you what the exponent is. Now press the antilog key while 2 is on the screen. The calculator screen should return to 100. The antilog solves the exponential function, given the base and the exponent.

If your background in algebra isn't strong, you'll want to get more practice working with logarithms—you'll see them frequently during your undergraduate career. Remember that once you understand the basic notation, there's nothing mysterious about logarithms. They are simply a way of solving exponential functions, which describe what happens when something is multiplied by itself a number of times—like cells that divide and then divide again and then again.

Using logarithms will also come up when you are studying something that can have a huge range of values, like the concentration of hydrogen ions in a solution or the intensity of sound that the human ear can detect. In cases like this, it's convenient to express the numbers involved as exponents. Using exponents makes a large range of numbers smaller and more tractable. For example, instead of saying that hydrogen ion concentration in a solution can range from 1 to 10^{-14}, the pH scale allows you to simply say that it ranges from 1 to 14. Instead of giving the actual value, you're expressing it as an exponent. It just simplifies things.

Making Concept Maps

A concept map is a graphical device for organizing and expressing what you know about a topic. It has two main elements: (1) concepts that are identified by words or short phrases and placed in a box or circle, and (2) labeled arrows that physically link two concepts and explain the relationship between them. The concepts are arranged hierarchically on a page, with the most general concepts at the top and the most specific ideas at the bottom.

The combination of a concept, a linking word, and a second concept is called a proposition. Good concept maps also have cross-links—meaning, labeled arrows that connect different elements in the hierarchy, as you read down the page.

Concept maps were initially developed by Joseph Novak in 1972 and have proven to be an effective studying and learning tool. They can be particularly valuable if constructed by a group, or when different individuals exchange and critique concept maps they have created independently. Although concept maps vary widely in quality and can be graded using objective criteria, there are many equally valid ways of making a high-quality concept map on a particular topic.

When you are asked to make a concept map in this text, you will usually be given at least a partial list of concepts to use. As an example, suppose you were asked to create a concept map on experimental design and were given the following concepts: results, predictions, control treatment, experimental treatment, controlled (identical) conditions, conclusions, experiment, hypothesis to be tested, null hypothesis. One possible concept map is shown in **Figure BS6.1**.

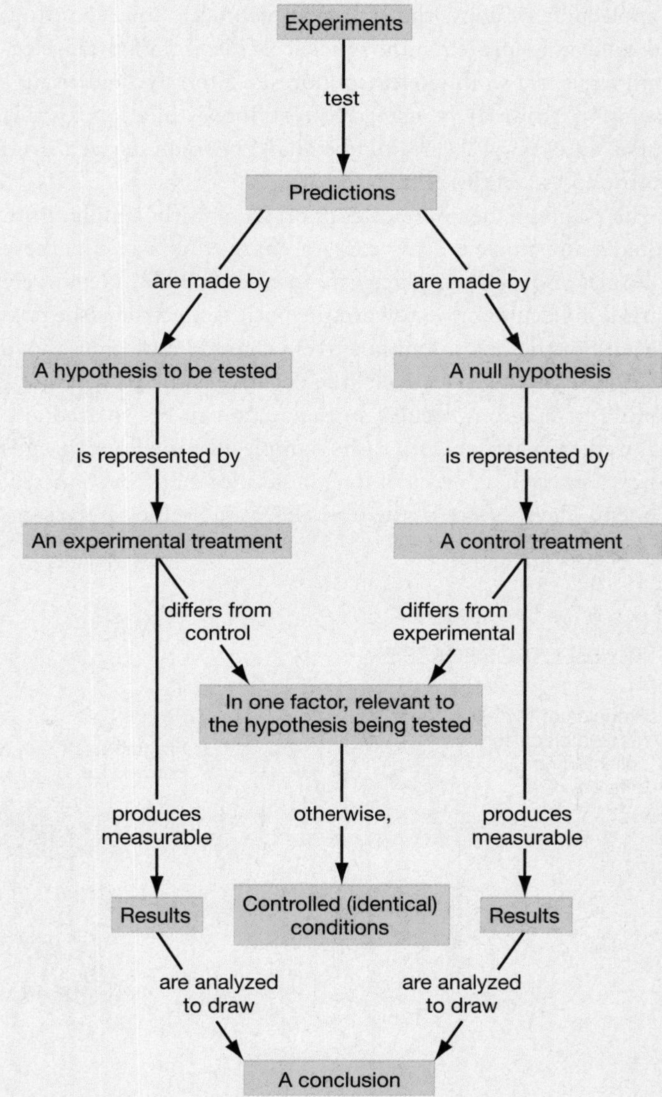

FIGURE BS6.1 A Concept Map on Principles of Experimental Design.

Dr. Doug Luckie, who uses concept maps extensively in his teaching at Michigan State University, points out that good concept maps have four qualities.

- They exhibit an organized hierarchy, indicating how each concept on the map relates to larger and smaller concepts.
- The concept words are specific—not vague.
- The propositions are accurate.
- There is cross-linking between different elements in the hierarchy of concepts.

As you practice making concept maps, go through these criteria and use them to evaluate your own work, as well as the work of fellow students. For more about concept maps, see Dr. Luckie's online handout "Concept Maps: What the Heck Are These" (http://ctools.msu.edu/ctools/How-to-Cmap.pdf) and web-based concept-mapping software (https://www.msu.edu/~luckie/ctoolsbasic/).

Using Electrophoresis to Separate Molecules

In molecular biology, the standard technique for separating and analyzing proteins and nucleic acids is called gel electrophoresis or, simply, electrophoresis (literally "electricity-moving"). You may be using electrophoresis in a lab for this course, and you will certainly be analyzing data derived from electrophoresis in this text.

The principle behind electrophoresis is fairly simple. Both proteins and nucleic acids carry a charge. As a result, these molecules move when placed in an electric field. Negatively charged molecules move toward the positive electrode (the positive end of the field), and positively charged molecules move toward the negative electrode (the negative end). To separate a mixture of macromolecules so that each can be isolated and analyzed, researchers place the sample in a gelatinous substance. The "gel" consists of long molecules that form a matrix of fibers. The presence of the fibers keeps molecules in the sample from moving around randomly, but the gelatinous matrix also has pores through which the molecules can pass. When an electrical field is applied across the gel, the molecules in the well move through the gel toward an electrode. Molecules that are smaller or more highly charged for their size move faster than do larger or less highly charged molecules. As they move, the molecules separate by size and by charge.

Figure BS7.1 shows the electrophoresis setup used in an experiment investigating how RNA molecules polymerize, described in Chapter 4. Step 1 shows how investigators loaded samples of macromolecules, taken on different days during the experiment, into cavities or "wells" at the top of the gel slab. In this and many other cases, the researchers also filled a well with a sample containing fragments of known size, called a size standard or "ladder." In step 2 the researchers immersed the gel in a solution that conducts electricity and applied a voltage

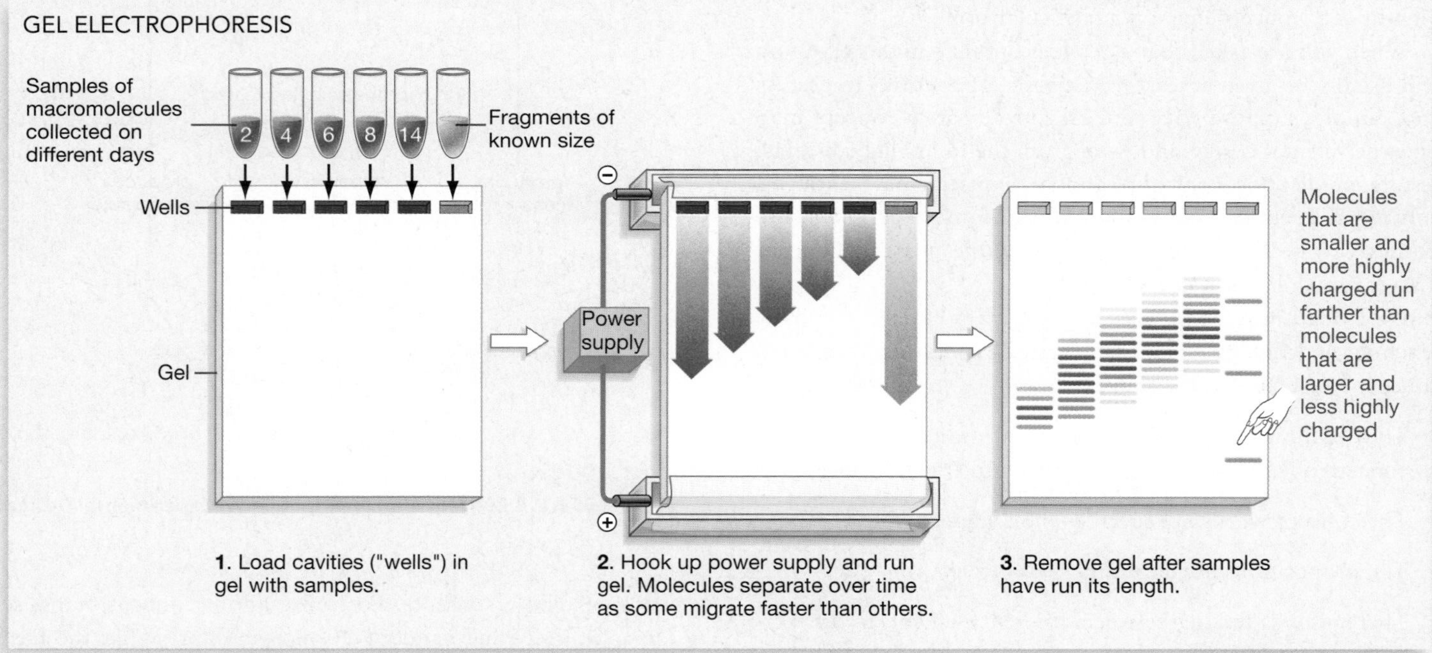

FIGURE BS7.1 Macromolecules Can Be Separated via Gel Electrophoresis.

QUESTION DNA and RNA run toward the positive electrode. Why are these molecules negatively charged?

across the gel. After the samples had run down the gel for some time (step 3), they removed the electric field. By then, molecules of different size and charge had separated from one another. In this case, small RNA molecules had reached the bottom of the gel. Above them were larger RNA molecules, which had run more slowly.

Once molecules have been separated in this way, they have to be detected. Often, proteins or nucleic acids can be stained or dyed. In this case, however, the researchers had attached a radioactive atom to the monomers used in the experiment, so they could visualize the resulting polymers by laying X-ray film over the gel. Because radioactive emissions expose film, a black dot appears wherever a radioactive atom is located in the gel. This technique for visualizing macromolecules is called autoradiography. If the samples are loaded into a rectangular well, as is commonly done, molecules of a particular size and charge form a band in an autoradiograph.

The autoradiograph that resulted from the polymerization experiment is shown in **Figure BS7.2**. The samples, taken on days 2, 4, 6, 8, and 14 of the experiment, are labeled along the bottom. The far right lane contains macromolecules of known size; this lane is used to estimate the size of the molecules in the experimental samples. The bands that appear in each sample lane represent the different polymers that had formed. Darker bands contain more radioactive marker, indicating the presence of many radioactive molecules. Lighter bands contain fewer molecules.

To read a gel, you look for (1) the presence or absence of bands in some lanes—meaning, some experimental samples—versus others, and (2) contrasts in the darkness of the bands present—meaning, differences in the amount of molecule present. For example, several conclusions can be drawn from the data in Figure BS7.2. First, a variety of polymers formed at each stage. After the second day, for example, polymers from 12 to 18 monomers long had formed on the clay particles used in this experiment. Second, the overall length of polymers produced increased with time. At the end of the fourteenth day, most of the RNA molecules were between 20 and 40 monomers long.

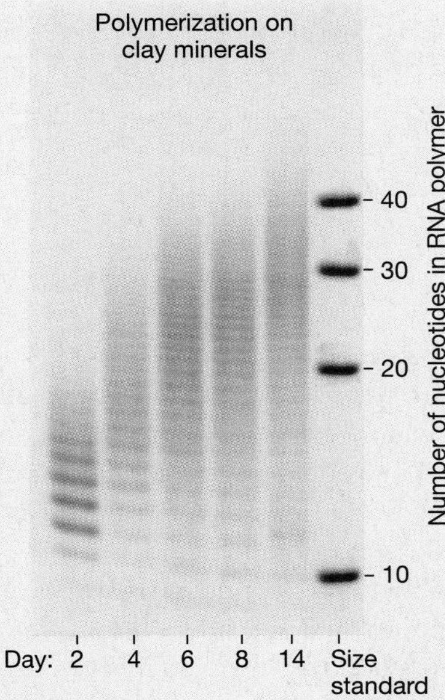

FIGURE BS7.2 Autoradiography Is a Technique for Visualizing Macromolecules. The molecules in a gel can be visualized in a number of ways. In this case, the RNA molecules in the gel exposed an X-ray film because they had radioactive atoms attached. When developed, the film is called an autoradiograph.

● **EXERCISE** Add labels to the photograph that read: "Top of gel—large molecules" and "Bottom of gel—small molecules."

BIOSKILLS 8

Observing Microscopic Structures and Processes

With the unaided eye, it is not possible to see a lot of biology. Biologists have to use microscopes to study small multicellular organisms, individual cells, and the contents of cells. And to understand what individual macromolecules or multimolecular machines like ribosomes look like, researchers use data from a technique called X-ray crystallography.

You'll probably use dissecting microscopes and compound light microscopes to view specimens during your labs for this course, and throughout this text you'll be seeing images generated from other types of microscopy and from X-ray crystallographic data. One of the fundamental skills you'll be acquiring as an introductory student, then, is a basic understanding of how these techniques work. The key is to recognize that each approach for visualizing microscopic structures has strengths and weaknesses. As a result, each technique is appropriate for studying certain types or aspects of cells or molecules.

Trends in Microscopy: Increasing Magnification and Clarity

If you use a dissecting microscope during labs, you'll recognize that it works by magnifying light that bounces off a whole specimen—often a live organism. You'll be able to view the specimen in three dimensions, which is why these instruments are sometimes called stereomicroscopes, but the maximum magnification is only about 20 to 40 times normal size (20× to 40×).

To view smaller objects, you'll probably use a compound microscope. Compound microscopes magnify light that is passed through a specimen. The instruments used in introductory labs are usually capable of 400× magnifications; the most sophisticated compound microscopes available can achieve magnifications of about 2000×. This is enough to view individual bacterial or eukaryotic cells and see large structures inside cells, like condensed chromosomes (see Chapter 11). To prepare a specimen for viewing under a compound light microscope, the tissues or cells are usually sliced to create a thin enough section for light to pass through efficiently. The section is then dyed to increase contrast and make structures visible. In many cases, different types of dyes are used to highlight different types of structures.

Until the 1950s, the compound microscope was the biologist's only tool for viewing cells directly. But the invention of the electron microscope provided a new way to view specimens. Two basic types of electron microscopy are now available: one that allows researchers to examine cross sections of cells at extremely high magnification, and one that offers a view of surfaces at somewhat lower magnification.

Transmission Electron Microscopy (TEM)

The transmission electron microscope is an extraordinarily effective tool for viewing cell structure at high magnification. TEM forms an image from electrons that pass through a specimen, just as a light microscope forms an image from light rays that pass through a specimen.

Biologists who want to view a cell under a transmission electron microscope begin by "fixing" the cell, meaning that they treat it with a chemical agent that stabilizes the cell's structure and contents while disturbing them as little as possible. Then they permeate the cell with an epoxy plastic that stiffens the structure. Once this epoxy hardens, the cell can be cut into extremely thin sections with a glass or diamond knife. Finally, the sectioned specimens are impregnated with a metal—often lead. (The reason for this last step is explained shortly.)

Figure BS8.1a outlines how the transmission electron microscope works. A beam of electrons is produced by a tungsten filament at the top of a column and directed downward. (All of the air is pumped out of the column, so that the electron beam isn't scattered by collisions with air molecules.) The electron beam passes through a series of lenses and through the specimen. The lenses are actually electromagnets, which alter the path of the beam much like a glass lens in a dissecting or compound microscope bends light. The lenses magnify and focus the image on a screen at the bottom of the column. There the electrons strike a coating of fluorescent crystals, which emit visible light in response—just like a television screen. When the microscopist moves the screen out of the way and allows the electrons to expose a sheet of black-and-white film, the result is a micrograph—a photograph of an image produced by microscopy.

The image itself is created by electrons that pass through the specimen. If no specimen were in place, all the electrons would pass through and the screen (and micrograph) would be uniformly bright. Unfortunately, cell materials by themselves would also appear fairly uniform and bright. This is because an atom's ability to deflect an electron depends on its mass. In

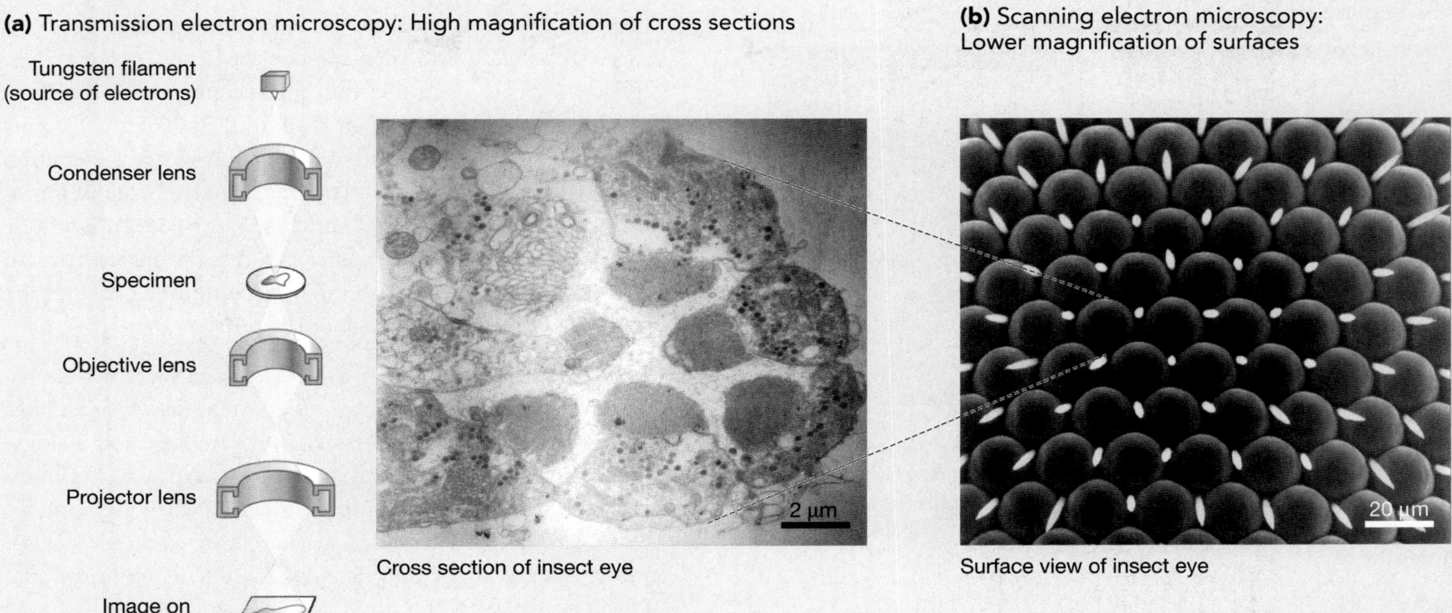

FIGURE BS8.1 There Are Two Basic Types of Electron Microscopy.

turn, an atom's mass is a function of its atomic number. The hydrogen, carbon, oxygen, and nitrogen atoms that dominate biological molecules have low atomic numbers. This is why cell biologists must saturate cell sections with lead solutions. Lead has a high atomic number and scatters electrons effectively. Different macromolecules take up lead atoms in different amounts, so the metal acts as a "stain" that produces contrast. With TEM, areas of dense metal scatter the electron beam most, producing dark areas in micrographs.

The advantage of TEM is that it can magnify objects up to 250,000×—meaning that intracellular structures are clearly visible. The downsides are that researchers are restricted to observing dead, sectioned material, and they must take care that the preparation process does not distort the specimen.

Scanning Electron Microscopy (SEM)

The scanning electron microscope is the most useful tool biologists have for looking at the surfaces of cells. Materials are prepared for scanning electron microscopy by coating their surfaces with a layer of metal atoms. To create an image of this surface, the microscope scans the surface with a narrow beam of electrons. Electrons that are reflected back from the surface or that are emitted by the metal atoms in response to the beam then strike a detector. The signal from the detector controls a second electron beam, which scans a TV-like screen and forms an image magnified up to 50,000 times the object's size. Because SEM records shadows and highlights, it provides images with a three-dimensional appearance (**Figure BS8.1b**). It cannot magnify objects nearly as much as TEM can, however.

Studying Live Cells and Real-Time Processes

Until the 1960s, it was not possible for biologists to get clear, high-magnification images of living cells. But a series of innovations over the past several decades has made it possible to observe organelles and subcellular structures in action.

The development of video microscopy, where the image from a light microscope is captured by a video camera instead of by an eye or a film camera, proved revolutionary. It allowed specimens to be viewed at higher magnification, because video cameras are more sensitive to small differences in contrast than are the human eye or still cameras. It also made it easier to keep live specimens functioning normally, because the increased light sensitivity of video cameras allows them to be used with low illumination, so specimens don't overheat. And when it became possible to digitize video images, researchers began using computers to remove out-of-focus background material and increase image clarity.

Another important innovation was the use of a fluorescent molecule called green fluorescent protein, or GFP, which allows researchers to tag specific molecules or structures and follow their movement over time. GFP is naturally synthesized in jellyfish that fluoresce, or emit light. By affixing GFP molecules

(a) Conventional fluorescence image of single cell

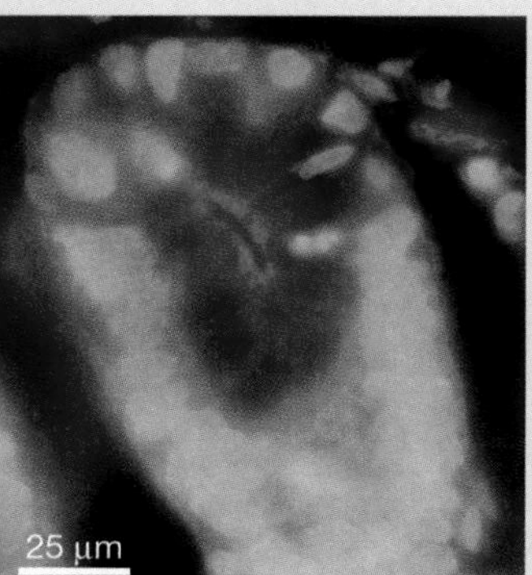

(b) Confocal fluorescence image of same cell

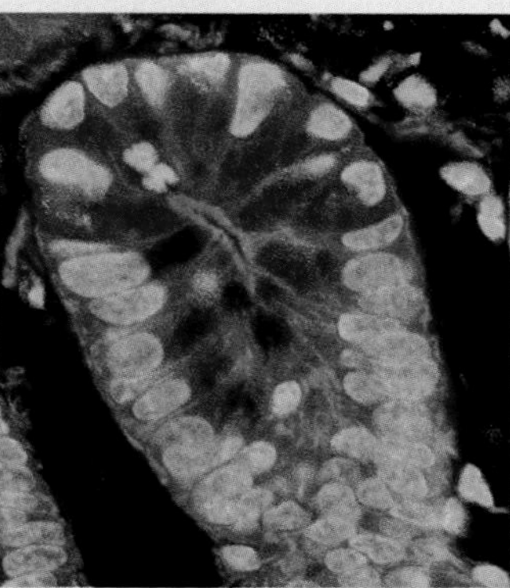

FIGURE BS8.2 Confocal Microscopy Provides Sharp Images of Living Cells. (a) The conventional image of this mouse intestinal cell is blurred, because it results from light emitted by the entire cell. **(b)** The confocal image is sharp, because it results from light emitted at a single plane inside the cell.

to another protein and then inserting it into a cell, investigators can follow the protein's fate over time and even videotape its movement. For example, researchers have videotaped GFP-tagged proteins being transported from the rough ER through the Golgi apparatus and out to the plasma membrane. This is cell biology: the movie.

Visualizing Structures in 3-D

The world is three-dimensional. To understand how microscopic structures and macromolecules work, it is essential to understand their shape and spatial relationships. Consider three techniques currently being used to reconstruct the 3-D structure of cells, organelles, and macromolecules.

- Confocal microscopy is carried out by mounting cells that have been treated with one or more fluorescing tags on a microscope slide and then focusing a beam of ultraviolet light at a specific depth within the specimen. The fluorescing tag emits visible light in response. A detector for this light is then set up at exactly the position where the emitted light comes into focus. The result is a sharp image of a precise plane in the cell being studied (**Figure BS8.2**). By altering the focal plane, a researcher can record images from an array of depths in the specimen; a computer can then be used to generate a 3-D image of the cell.
- Electron tomography uses a transmission electron microscope to generate a 3-D image of an organelle or other subcellular structure. The specimen is rotated around a single axis, with the researcher taking many "snapshots." The individual images are then pieced together with a computer. This technique has provided a much more accurate view of mitochondrial structure than was possible using traditional TEM (see Chapter 7).
- X-ray crystallography, or X-ray diffraction analysis, is the most widely used technique for reconstructing the 3-D structure of molecules. As its name implies, the procedure is based on bombarding crystals of a molecule with X-rays. X-rays are scattered in precise ways when they interact with the electrons surrounding the atoms in a crystal, producing a diffraction pattern that can be recorded on X-ray film or other types of detectors (**Figure BS8.3**). By varying the orientation of the X-ray beam as it strikes a crystal and documenting the diffraction patterns that result, researchers can construct a map representing the density of electrons in the crystal. By relating these electron-density maps to information about the primary structure of the nucleic acid or protein, a 3-D model of the molecule can be built. Virtually all of the molecular models used in this book were built from X-ray crystallographic data.

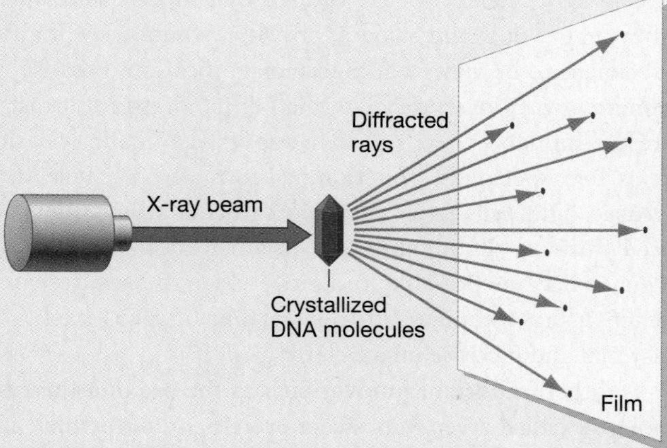

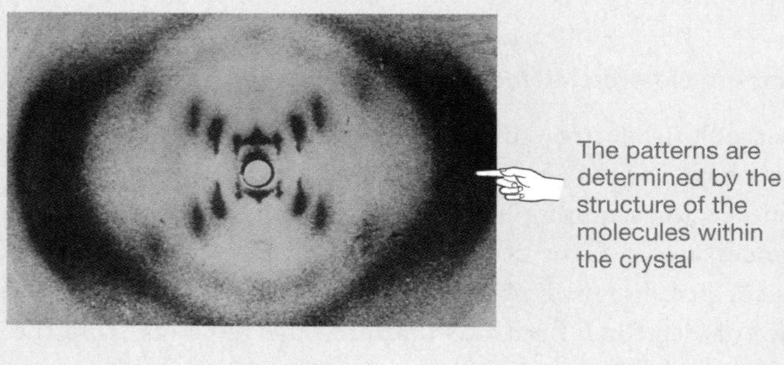

FIGURE BS8.3 X-Ray Crystallography. When crystallized molecules are bombarded with X-rays, the radiation is scattered in distinctive patterns. The photograph at the right shows an X-ray film that recorded the pattern of scattered radiation from DNA molecules.

BioSkills 9

Combining Probabilities

In several cases in this text, you'll need to combine probabilities from different events in order to solve a problem. One of the most common applications is in genetics problems. For example, Punnett squares work because they are based on two fundamental rules of probability, called the "both-and rule" and the "either-or" rule. Each rule pertains to a distinct situation.

The both-and rule applies when you want to know the probability that two or more independent events occur together. Let's use the rolling of two dice as an example. What is the probability of rolling two sixes? These two events are independent, because the probability of rolling a six on one die has no effect on the probability of rolling a six on the other die. (In the same way, the probability of getting a gamete with allele *R* from one parent has no effect on the probability of getting a gamete with allele *R* from the other parent. Gametes fuse randomly.) The probability of rolling a six on the first die is 1/6. The probability of rolling a six on the second die is also 1/6. The probability of rolling a six on both die, then, is $1/6 \times 1/6 = 1/36$. In other words, if you rolled two die 36 times, on average you would expect to roll two sixes once. It should make sense that the both-and rule is also called the multiplication rule or product rule. In the case of a cross between two parents heterozygous at the *R* gene, the probability of getting allele *R* from the father is 1/2 and the probability of getting *R* from the mother is 1/2. Thus, the probability of getting both alleles and creating an offspring with genotype *RR* is $1/2 \times 1/2 = 1/4$.

The either-or rule, in contrast, applies when you want to know the probability of an event happening when there are several ways for the event to occur. In this case, the probability that the event will occur is the sum of the probabilities of each way that it can occur. For example, suppose you wanted to know the probability of rolling either a one or a six when you toss a die. The probability of drawing each is 1/6, so the probability of getting one or the other is $1/6 + 1/6 = 1/3$. (The either-or rule is also called the addition rule or sum rule.) If you rolled a die three times, on average you'd expect to get a one or a six once. In the case of a cross between two parents heterozygous at the *R* gene, the probability of getting an *R* allele from the father and an *r* allele from the mother is $1/2 \times 1/2 = 1/4$. Similarly, the probability of getting an *r* allele from the father and an *R* allele from the mother is $1/2 \times 1/2 = 1/4$. Thus, the combined probability of getting the *Rr* genotype in either of the two ways is $1/4 + 1/4 = 1/2$.

Questions

1. Suppose that four students each toss a coin. What is the probability of four "tails"?
2. After a single roll of a die, what is the probability of getting either a two, a three, or a six?

Answers: 1. $1/2 \times 1/2 \times 1/2 \times 1/2 = 1/16$; 2. $1/6 + 1/6 + 1/6 = 1/2$

Glossary

abiotic Not alive (e.g., air, water, and soil). Compare with **biotic.**

ABO blood types Genetically determined classes of human blood that are distinguished by the presence or absence of specific glycolipids on the surface of red blood cells. Also called *ABO blood groups.*

aboveground biomass The total mass of living plants in an area, excluding roots.

abscisic acid (ABA) A plant hormone that inhibits cell elongation and stimulates leaf shedding and dormancy.

abscission In plants, the normal (often seasonal) shedding of leaves, fruits, or flowers.

abscission zone The region at the base of a petiole that thins and breaks during dropping of leaves.

absorption In animals, the uptake of ions and small molecules derived from food across the lining of the intestine and into the bloodstream.

absorption spectrum The amount of light of different wavelengths absorbed by a pigment. Usually depicted as a graph of light absorbed versus wavelength. Compare with **action spectrum.**

acclimation, acclimatization Gradual physiological adjustment of an organism to new environmental conditions that occur naturally or as part of a laboratory experiment.

acetylation Addition of an acetyl group (CH_3COO^-) to a molecule.

acetylcholine (ACh) A neurotransmitter, released by nerve cells at neuromuscular junctions, that triggers contraction of muscle cells. Also used as a neurotransmitter between neurons.

acetyl CoA A molecule produced by oxidation of pyruvate (the final product of glycolysis) in a reaction catalyzed by pyruvate dehydrogenase. Can enter the Krebs cycle and also is used as a carbon source in the synthesis of fatty acids, steroids, and other compounds.

acid Any compound that gives up protons or accepts electrons during a chemical reaction or that releases hydrogen ions when dissolved in water.

acid-growth hypothesis The hypothesis that auxin triggers elongation of plant cells by inducing the synthesis of proton pumps whose activity makes the cell wall more acidic, leading to expansion of the cell wall and an influx of water.

acoelomate An animal that lacks an internal body cavity (coelom). Compare with **coelomate** and **pseudocoelomate.**

acquired immune deficiency syndrome (AIDS) A human disease characterized by death of immune system cells (in particular helper T cells and macrophages) and subsequent vulnerability to other infections. Caused by the human immunodeficiency virus (HIV).

acquired immune response See **acquired immunity.**

acquired immunity Immunity to a particular pathogen or other antigen conferred by antibodies and activated B and T cells following exposure to the antigen. Is characterized by specificity, diversity, memory, and self-nonself recognition. Compare with **innate immunity.**

acrosomal reaction A set of events occurring in a sperm cell upon encountering an egg cell, including release of acrosomal enzymes and formation of the acrosomal process, which helps the sperm cell reach the egg.

acrosome A caplike structure, located on the head of a sperm cell, that contains enzymes capable of dissolving the outer coverings of an egg.

ACTH See **adrenocorticotropic hormone.**

actin A globular protein that can be polymerized to form filaments. Actin filaments are part of the cytoskeleton and constitute the thin filaments in skeletal muscle cells.

actin filament A long fiber, about 7 nm in diameter, composed of two intertwined strands of polymerized actin protein; one of the three types of cytoskeletal fibers. Involved in cell movement. Also called a *microfilament.* Compare with **intermediate filament** and **microtubule.**

action potential A rapid, temporary change in electrical potential across a membrane, from negative to positive and back to negative. Occurs in cells, such as neurons and muscle cells, that have an excitable membrane.

action spectrum The relative effectiveness of different wavelengths of light in driving a light-dependent process such as photosynthesis. Usually depicted as a graph of some measure of the process versus wavelength. Compare with **absorption spectrum.**

activation energy The amount of energy required to initiate a chemical reaction; specifically, the energy required to reach the transition state.

active site The portion of an enzyme molecule where substrates (reactant molecules) bind and react.

active transport The movement of ions or molecules across a plasma membrane or organelle membrane against an electrochemical gradient. Requires energy (e.g., from hydrolysis of ATP) and assistance of a transport protein (e.g., pump).

adaptation Any heritable trait that increases the fitness of an individual with that trait, compared with individuals without that trait, in a particular environment.

adaptive radiation Rapid evolutionary diversification within one lineage, producing numerous descendant species with a wide range of adaptive forms.

adenosine diphosphate (ADP) A molecule consisting of adenine, a sugar, and two phosphate groups. Addition of a third phosphate group produces adenosine triphosphate (ATP).

adenosine triphosphate (ATP) A molecule consisting of adenine, a sugar, and three phosphate groups that can be hydrolyzed to release energy. Universally used by cells to store and transfer energy.

adenylyl cyclase An enzyme that can catalyze the formation of cyclic AMP (cAMP) from ATP. Involved in controlling transcription of various operons in prokaryotes and in some eukaryotic signal-transduction pathways.

adhesion The tendency of certain dissimilar molecules to cling together due to attractive forces. Compare with **cohesion.**

adipocyte A fat cell.

adipose tissue A type of connective tissue whose cells store fats.

ADP See **adenosine diphosphate.**

adrenal glands Two small endocrine glands that sit above each kidney. The outer portion (cortex) secretes several steroid hormones; the inner portion (medulla) secretes epinephrine and norepinephrine.

adrenaline See **epinephrine.**

adrenocorticotropic hormone (ACTH) A peptide hormone, produced and secreted by the anterior pituitary, that stimulates release of steroid hormones (e.g., cortisol, aldosterone) from the adrenal cortex.

adventitious root A root that develops from a plant's shoot system instead of from the plant's root system.

aerobic Referring to any metabolic process, cell, or organism that uses oxygen as an electron acceptor. Compare with **anaerobic.**

afferent division The part of the nervous system, consisting mainly of sensory neurons, that transmits information about the internal and external environment to the central nervous system. Compare with **efferent division.**

agar A gelatinous mix of polysaccharides that is commonly used to grow cells on solid media.

age class All the individuals of a specific age in a population.

age structure The proportion of individuals in a population that are of each possible age.

age-specific fecundity The average number of female offspring produced by a female in a certain age class.

agglutination Clumping together of cells, typically caused by antibodies.

aggregate fruit A fruit (e.g., raspberry) that develops from a single flower that has many separate carpels. Compare with **multiple** and **simple fruit.**

AIDS See **acquired immune deficiency syndrome.**

albumen A solution of water and protein (particularly albumins), found in amniotic eggs, that nourishes the growing embryo. Also called *egg white.*

albumin A class of large proteins found in plants and animals, particularly in the albumen of eggs and in blood plasma.

alcohol fermentation Catabolic pathway in which pyruvate produced by glycolysis is converted to ethanol in the absence of oxygen.

aldosterone A hormone produced in the adrenal cortex that stimulates the kidney to conserve salt and water and promotes retention of sodium.

alimentary canal See **digestive tract.**

allele A particular version of a gene.

allopatric speciation The divergence of populations into different species by physical isolation of populations in different geographic areas. Compare with **sympatric speciation.**

allopatry Condition in which two or more populations live in different geographic areas. Compare with **sympatry.**

allopolyploidy (adjective: allopolyploid) The state of having more than two full sets of chromosomes (polyploidy) due to hybridization between different species. Compare with **autopolyploidy.**

allosteric regulation Regulation of a protein's function by binding of a regulatory molecule, usually to a specific site distinct from the active site, causing a change in the protein's shape.

alpha(α)-helix A protein secondary structure in which the polypeptide backbone coils into a spiral shape stabilized by hydrogen bonds between atoms.

alternation of generations A life cycle involving alternation of a multicellular haploid stage (gametophyte) with a multicellular diploid stage (sporophyte). Occurs in most plants and some protists.

alternative splicing In eukaryotes, the splicing of primary RNA transcripts from a single gene in different ways to produce different mature mRNAs and thus different polypeptides.

altruism Any behavior that has a cost to the individual (such as lowered survival or reproduction) and a benefit to the recipient. See **reciprocal altruism.**

alveolus (plural: alveoli) One of the tiny air-filled sacs of a mammalian lung.

ambisense virus A virus whose genome contains both positive-sense and negative-sense sequences.

amino acid A small organic molecule with a central carbon atom bonded to an amino group ($-NH_3$), a carboxyl group ($-COOH$), a hydrogen atom, and a side group. Proteins are polymers of 20 common amino acids.

aminoacyl tRNA A transfer RNA molecule that is covalently bound to an amino acid.

aminoacyl tRNA synthetase An enzyme that catalyzes the addition of a particular amino acid to its corresponding tRNA molecule.

ammonia (NH_3) A small molecule, produced by the breakdown of proteins and nucleic acids, that is very toxic to cells. Is a strong base that gains a proton to form the ammonium ion (NH_4^+).

amnion The membrane within an amniotic egg that surrounds the embryo and encloses it in a protective pool of fluid (amniotic fluid).

amniotes A major lineage of vertebrates (Amniota) that reproduce with amniotic eggs. Includes all reptiles (including birds) and mammals.

amniotic egg An egg that has a watertight shell or case enclosing a membrane-bound water supply (the amnion), food supply (yolk sac), and waste sac (allantois).

amoeba Any unicellular protist that lacks a cell wall, is extremely flexible in shape, and moves by means of pseudopodia.

amoeboid motion See **cell crawling.**

amphibians A lineage of vertebrates many of whom breathe through their skin and feed on land but lay their eggs in water; represent the earliest tetrapods. Include frogs, salamanders, and caecilians.

amphipathic Containing hydrophilic and hydrophobic elements.

amylase Any enzyme that can break down starch by catalyzing hydrolysis of the glycosidic linkages between the glucose residues.

amyloplasts Dense, starch-storing organelles that settle to the bottom of plant cells and that may be used as gravity detectors.

anabolic pathway Any set of chemical reactions that synthesizes larger molecules from smaller ones. Generally requires an input of energy. Compare with **catabolic pathway.**

anadromous Having a life cycle in which adults live in the ocean (or large lakes) but migrate up freshwater streams to breed and lay eggs.

anaerobic Referring to any metabolic process, cell, or organism that uses an electron acceptor other than oxygen, such as nitrate or sulfate. Compare with **aerobic.**

anaphase A stage in mitosis or meiosis during which chromosomes are moved to opposite ends of the cell.

anatomy The study of the physical structure of organisms.

androgens A class of steroid hormones that generally promote male-like traits (although females have some androgens, too). Secreted primarily by the gonads and adrenal glands.

aneuploidy (adjective: aneuploid) The state of having an abnormal number of copies of a certain chromosome.

angiosperm A flowering vascular plant that produces seeds within mature ovaries (fruits). The angiosperms form a single lineage. Compare with **gymnosperm.**

animal A member of a major lineage of eukaryotes (Animalia) whose members typically have a complex, large, multicellular body, eat other organisms, and are mobile.

animal model Any disease that occurs in a nonhuman animal and has many parallels to a similar disease of humans. Studied by medical researchers in hope that findings may apply to human disease.

anion A negatively charged ion.

annelids Members of the phylum Annelida (segmented worms). Distinguished by a segmented body and a coelom that functions as a hydrostatic skeleton. Annelids belong to the lophotrochozoan branch of the protostomes.

annual Referring to a plant whose life cycle normally lasts only one growing season—less than one year. Compare with **perennial.**

anoxygenic Referring to any process or reaction that does not produce oxygen. Photosynthesis in purple sulfur and purple nonsulfur bacteria, which does not involve photosystem II, is anoxygenic. Compare with **oxygenic.**

antenna (plural: antennae) A long appendage that is used to touch or smell.

antenna complex Part of a photosystem, containing an array of chlorophyll molecules and accessory pigments, that receives energy from light and directs the energy to a central reaction center during photosynthesis.

anterior Toward an animal's head and away from its tail. The opposite of posterior.

anterior pituitary The part of the pituitary gland containing endocrine cells that produce and release a variety of peptide hormones in response to other hormones from the hypothalamus. Compare with **posterior pituitary.**

anther The pollen-producing structure at the end of a stamen in flowering plants (angiosperms).

antheridium (plural: antheridia) The sperm-producing structure in most land plants except angiosperms.

anthropoids One of the two major lineages of primates, including apes, humans, and all monkeys. Compare with **prosimians.**

antibiotic Any substance, such as penicillin, that can kill or inhibit the growth of bacteria.

antibody An immunoglobulin protein, produced by B cells, that can bind to a specific part of an antigen, tagging it for attack by the immune system. All antibody molecules have a similar Y-shaped structure and, in their monomer form, consist of two identical light chains and two identical heavy chains.

anticodon The sequence of three bases (triplet) in a transfer RNA molecule that can bind to a mRNA codon with a complementary sequence.

antidiuretic hormone (ADH) A peptide hormone, secreted from the posterior pituitary gland, that stimulates water retention by the kidney. Also called *vasopressin.*

antigen Any foreign molecule, often a protein, that can stimulate a specific response by the immune system.

antigen presentation Process by which small peptides, derived from ingested particulate antigens (e.g., bacteria) or intracellular antigens (e.g., viruses in infected cell) are complexed with MHC proteins and transported to the cell surface where they are displayed and can be recognized by T cells.

antiparallel Describing the opposite orientation of the strands in a DNA double helix with one strand running in the $5' \rightarrow 3'$ direction and the other in the $3' \rightarrow 5'$ direction.

antiporter A carrier protein that allows an ion to diffuse down an electrochemical gradient, using the energy of that process to transport a different substance in the opposite direction *against* its concentration gradient. Compare with **symporter.**

antiviral Any drug or other agent that can kill or inhibit the transmission or replication of viruses.

anus In a multicellular animal, the end of the digestive tract where wastes are expelled.

aorta In terrestrial vertebrates, the major artery carrying oxygenated blood away from the heart.

aphotic zone Deep water receiving no sunlight. Compare with **photic zone.**

apical Toward the top. In plants, at the tip of a branch. In animals, on the side of an epithelial layer that faces the environment and not other body tissues. Compare with **basal.**

apical bud A bud at the tip of a stem, where growth occurs to lengthen the stem.

apical dominance Inhibition of lateral bud growth by the apical meristem at the tip of a plant branch.

apical meristem A group of undifferentiated plant cells, at the tip of a stem or root, that is responsible for primary growth. Compare with **lateral meristem.**

apomixis The formation of mature seeds without fertilization occurring; a type of asexual reproduction.

apoplast In plant roots, a continuous pathway through which water can flow, consisting of the porous cell walls of adjacent cells and the intervening extracellular space. Compare with **symplast.**

apoptosis Series of genetically controlled changes that lead to death of a cell. Occurs frequently during embryological development and later may occur in response to infections or cell damage. Also called *programmed cell death.*

aquaporin A type of channel protein through which water can move by osmosis across a plasma membrane.

arbuscular mycorrhizal fungi (AMF) Fungi whose hyphae enter the root cells of their host plants.

Archaea One of the three taxonomic domains of life consisting of unicellular prokaryotes distinguished by cell walls made of certain polysaccharides not found in bacterial or eukaryotic cell walls, plasma membranes composed of unique isoprene-containing phospholipids, and ribosomes and RNA polymerase similar to those of eukaryotes. Compare with **Bacteria** and **Eukarya.**

archegonium (plural: archegonia) The egg-producing structure in most land plants except angiosperms.

arteriole One of the many tiny vessels that carry blood from arteries to capillaries.

artery Any thick-walled blood vessel that carries blood (oxygenated or not) under relatively high pressure away from the heart to organs throughout the body. Compare with **vein.**

arthropods Members of the phylum Arthropoda. Distinguished by a segmented body; a hard, jointed exoskeleton; paired appendages; and an extensive body cavity called a hemocoel. Arthropods belong to the ecdysozoan branch of the protostomes.

articulation A movable point of contact between two bones of a skeleton. See **joint.**

artificial selection Deliberate manipulation by humans, as in animal and plant breeding, of the genetic composition of a population by allowing only individuals with desirable traits to reproduce.

ascocarp A large, cup-shaped reproductive structure produced by some ascomycete fungi. Contains many microscopic asci, which produce spores.

ascomycete See **sac fungus.**

ascus (plural: asci) Specialized spore-producing cell found at the ends of hyphae in sac fungi (ascomycetes).

asexual reproduction Any form of reproduction resulting in offspring that are genetically identical to the parent. Includes binary fission, budding, and parthenogenesis. Compare with **sexual reproduction.**

asymmetric competition Ecological competition between two species in which one species suffers a much greater fitness decline than the other. Compare with **symmetric competition.**

atomic mass unit (amu) A unit of mass equal to 1/12 the mass of one carbon-12 atom; about the mass of 1 proton or 1 neutron. Also called *dalton.*

atomic number The number of protons in the nucleus of an atom, giving the atom its identity as a particular chemical element.

ATP See **adenosine triphosphate.**

ATP synthase A large membrane-bound protein complex in chloroplasts, mitochondria, and some bacteria that uses the energy of protons flowing through it to synthesize ATP. Also called F_oF_1 *complex.*

atrioventricular (AV) node A region of the heart between the right atrium and right ventricle where electrical signals from the atrium are slowed briefly before spreading to the ventricle. This delay allows the ventricle to fill with blood before contracting. Compare with **sinoatrial (SA) node.**

atrium (plural: atria) A thin-walled chamber of the heart that receives blood from veins and pumps it to a neighboring chamber (the ventricle).

autocrine Relating to a chemical signal that affects the same cell that produced and released it.

autoimmunity A pathological condition in which the immune system attacks self cells or tissues of an individual's own body.

autonomic nervous system The part of the peripheral nervous system that controls internal organs and involuntary processes, such as stomach contraction, hormone release, and heart rate. Includes parasympathetic and sympathetic nerves. Compare with **somatic nervous system.**

autophagy The process by which damaged organelles are surrounded by a membrane and delivered to a lysosome to be destroyed.

autopolyploidy (adjective: autopolyploid) The state of having more than two full sets of chromosomes (polyploidy) due to a mutation that doubled the chromosome number.

autoradiography A technique for detecting radioactively labeled molecules separated by gel electrophoresis by placing an unexposed film over the gel. A black dot appears on the film wherever a radioactive atom is present in the gel. Also can be used to locate labeled molecules in fixed tissue samples.

autosomal inheritance The inheritance patterns that occur when genes are located on autosomes rather than on sex chromosomes.

autosome Any chromosome that does not carry genes involved in determining the sex of an individual.

autotroph Any organism that can synthesize reduced organic compounds from simple inorganic sources such as CO_2 or CH_4. Most plants and some bacteria and archaea are autotrophs. Also called *primary producer.* Compare with **heterotroph.**

auxin Indoleacetic acid, a plant hormone that stimulates phototropism and some other responses.

avirulence (*avr*) loci Genes in pathogens encoding proteins that trigger a defense response in plants. Compare with **resistance (R) loci.**

axillary bud A bud that forms in the angle between a leaf and a stem and may develop into a lateral (side) branch. Also called *lateral bud.*

axon A long projection of a neuron that can propagate an action potential and transmit it to another neuron.

axon hillock The site in a neuron where an axon joins the cell body and where action potentials are first triggered.

axoneme A structure found in eukaryotic cilia and flagella and responsible for their motion; composed of two central microtubules surrounded by nine doublet microtubules (9 + 2 arrangement).

background extinction The average rate of low-level extinction that has occurred continuously throughout much of evolutionary history. Compare with **mass extinction.**

Bacteria One of the three taxonomic domains of life consisting of unicellular prokaryotes distinguished by cell walls composed largely of peptidoglycan, plasma membranes similar to those of eukaryotic cells, and ribosomes and RNA polymerase that differ from those in archaeans or eukaryotes. Compare with **Archaea** and **Eukarya.**

bacterial artificial chromosome (BAC) An artificial version of a bacterial chromosome that can be used as a cloning vector to produce many copies of large DNA fragments.

bacteriophage Any virus that infects bacteria.

baculum A bone inside the penis usually present in mammals with a penis that lacks erectile tissue.

bark The protective outer layer of woody plants, composed of cork cells, cork cambium, and secondary phloem.

baroreceptors Specialized nerve cells in the walls of the heart and certain major arteries that detect changes in blood pressure and trigger appropriate responses by the brain.

basal Toward the base. In plants, at the base of a branch where it joins the stem. In animals, on the side of an epithelial layer that abuts underlying body tissues. Compare with **apical.**

basal body A structure of nine pairs of microtubules arranged in a circle at the base of eukaryotic cilia and flagella where they attach to the cell. Structurally similar to a centriole.

basal lamina A thick, collagen-rich extracellular matrix that underlies most epithelial tissues (e.g., skin) in animals.

basal metabolic rate (BMR) The total energy consumption by an organism at rest in a comfortable environment. For aerobes, often measured as the amount of oxygen consumed per hour.

basal transcription complex A large multiprotein structure that assembles near the promoter of eukaryotic genes and initiates transcription. Composed of basal transcription factors, TATA-binding protein, coactivators, and RNA polymerase.

basal transcription factor General term for proteins, present in all cell types, that bind to eukaryotic promoters and help initiate transcription. Compare with **regulatory transcription factor.**

base Any compound that acquires protons or gives up electrons during a chemical reaction or accepts hydrogen ions when dissolved in water.

basidiomycete See **club fungus.**

basidium (plural: basidia) Specialized spore-producing cell at the ends of hyphae in club fungi.

basilar membrane The membrane in the vertebrate cochlea on which the bottom portion of hair cells sit.

basolateral Toward the bottom and sides. In animals, the side of an epithelial layer that faces other body tissues and not the environment.

Batesian mimicry A type of mimicry in which a harmless or palatable species resembles a dangerous or poisonous species. Compare with **Müllerian mimicry.**

B cell A type of leukocyte that matures in the bone marrow and, with T cells, is responsible for acquired immunity. Produces antibodies and also functions in antigen presentation. Also called *B lymphocyte.*

B-cell receptor (BCR) An immunoglobulin protein (antibody) embedded in the plasma membrane of mature B cells and to which antigens bind.

behavior Any action by an organism.

benign tumor A mass of abnormal tissue that grows slowly or not at all, does not disrupt surrounding tissues, and does not spread to other organs. Benign tumors are not cancers. Compare with **malignant tumor.**

benthic Living at the bottom of an aquatic environment.

benthic zone The area along the bottom of an aquatic environment.

beta(β)-pleated sheet A protein secondary structure in which the polypeptide backbone folds into a sheetlike shape stabilized by hydrogen bonding.

bilateral symmetry An animal body pattern in which there is one plane of symmetry dividing the body into a left side and a right side. Typically, the body is long and narrow, with a distinct head end and tail end. Compare with **radial symmetry.**

bilaterian A member of a major lineage of animals (Bilateria) that are bilaterally symmetrical at some point in their life cycle, have three embryonic germ layers, and have a coelom. All protostomes and deuterostomes are bilaterians.

bile A complex solution produced by the liver, stored in the gall bladder, and secreted into the intestine. Contains steroid derivatives called bile salts that are responsible for emulsification of fats during digestion.

binary fission See **fission** (1).

binomial nomenclature A system of naming species by using two-part Latinized names composed of a genus name and a species name (e.g., *Homo sapiens,* humans; *Canis familiaris,* dogs). Always italicized, with genus name capitalized.

biodiversity The diversity of life considered at three levels: genetic diversity (variety of alleles in a population, species, or group of species); species diversity (variety and relative abundance of species present in a certain area); and ecosystem diversity (variety of communities and abiotic components in a region).

biogeochemical cycle The pattern of circulation of an element or molecule among living organisms and the environment.

biogeography The study of how species and populations are distributed geographically.

bioinformatics The field of study concerned with managing, analyzing, and interpreting biological information, particularly DNA sequences.

biological species concept The definition of a species as a population or group of populations that are reproductively isolated from other groups. Members of a species have the potential to interbreed in nature to produce viable, fertile offspring but cannot produce viable, fertile hybrid offspring with members of other species. Compare with **morphological** and **phylogenetic species concept.**

bioluminescence The emission of light by a living organism.

biomass The total mass of all organisms in a given population or geographical area; usually expressed as total dry weight.

biome A large terrestrial ecosystem characterized by a distinct type of vegetation and climate.

bioremediation The use of living organisms, usually bacteria or archaea, to degrade environmental pollutants.

biotechnology The application of biological techniques and discoveries to medicine, industry, and agriculture.

biotic Living, or produced by a living organism. Compare with **abiotic.**

bipedal Walking primarily on two legs.

bipolar cell A cell in the vertebrate retina that receives information from one or more photoreceptors and passes it to other bipolar cells or ganglion cells.

bivalves A lineage of molluscs that have two shells, such as clams and mussels.

bladder A mammalian organ that holds urine until it can be excreted.

blade The wide, flat part of a plant leaf.

blastocoel Fluid-filled cavity in the blastula of many animal species.

blastocyst Specialized type of blastula in mammals. A spherical structure composed of trophoblast cells on the exterior and a cluster of cells (the inner cell mass), which fills part of the interior space.

blastomere A small cell created by cleavage divisions in early animal embryos.

blastopore A small opening (pore) in the surface of an early vertebrate embryo, through which cells move during gastrulation.

blastula In vertebrate development, a hollow ball of cells (blastomere cells) that is formed by cleavage of a zygote and immediately undergoes gastrulation. See **blastocyst.**

blood A type of connective tissue consisting of red blood cells and leukocytes suspended in a fluid portion called plasma.

blood pressure See **diastolic blood pressure** and **systolic blood pressure.**

body mass index (BMI) A mathematical relationship of weight and height used to assess obesity in humans. Calculated as weight (kg) divided by the square of height (m^2).

body plan The basic architecture of an animal's body, including the number and arrangement of limbs, body segments, and major tissue layers.

bog A wetland that has no or almost no water flow, resulting in very low oxygen levels and acidic conditions.

Bohr shift The rightward shift of the oxygen-hemoglobin dissociation curve that occurs with decreasing pH. Results in hemoglobin being more likely to release oxygen in the acidic environment of exercising muscle.

bone A type of vertebrate connective tissue consisting of living cells and blood vessels within a hard extracellular matrix composed of calcium phosphate ($CaPO_4$) and small amounts of calcium carbonate ($CaCO_3$) and protein fibers.

bone marrow The soft tissue filling the inside of long bones containing stem cells that develop into red blood cells and leukocytes throughout life.

Bowman's capsule The hollow, double-walled cup-shaped portion of a nephron that surrounds a glomerulus in the vertebrate kidney.

braincase See **cranium.**

brain stem The most posterior portion of the vertebrate brain, connecting to the spinal cord and responsible for autonomic body functions such as heart rate, respiration, and digestion.

branch (1) A part of a phylogenetic tree that represents populations through time. (2) Any extension of a plant's shoot system.

bronchiole One of the small tubes in mammalian lungs that carry air from the bronchi to the alveoli.

bronchus (plural: bronchi) In mammals, one of a pair of large tubes that lead from the trachea to each lung.

brown adipose tissue A specialized form of adipose tissue whose cells contain a high density of mitochondria as well as stored fats and that can produce extra body heat. Found in some mammals.

bryophytes Members of several phyla of green plants that lack vascular tissue including liverworts, hornworts, and mosses. Also called *nonvascular plants.*

budding Asexual reproduction via outgrowth from the parent that eventually break free as an independent individual; occurs in yeasts and some invertebrates.

buffer A substance that, in solution, acts to minimize changes in the pH of that solution when acid or base is added.

bulbourethral glands In male mammals, small paired glands at the base of the urethra that secrete an alkaline mucus (part of semen), which lubricates the tip of the penis and neutralizes acids in the urethra during copulation. In humans, also called *Cowper's glands.*

bulk flow The directional movement of a substantial volume of fluid due to pressure differences, such as movement of water through plant phloem and movement of blood in animals.

bundle-sheath cell A type of cell found around the vascular tissue (veins) of plant leaves.

C_3 photosynthesis The most common form of photosynthesis in which atmospheric CO_2 is used to form 3-phosphoglycerate, a three-carbon sugar.

C_4 photosynthesis A variant type of photosynthesis in which atmospheric CO_2 is first fixed into four-carbon sugars, rather than the three-carbon sugars of classic C_3 photosynthesis. Enhances photosynthetic efficiency in hot, dry environments, by reducing loss of oxygen due to photorespiration.

cadherin Any of a class of cell-surface proteins involved in cell adhesion and important for coordinating movements of cells during embryological development.

callus In plants, a mass of undifferentiated cells that can generate roots and other tissues necessary to create a mature plant.

Calorie A unit of energy often used to measure the energy content of food. Also called *kilocalorie.*

Calvin cycle In photosynthesis, the set of light-independent reactions that use NADPH and ATP formed in the light-dependent reactions to drive the fixation of atmospheric CO_2 and reduction of the fixed carbon, ultimately producing sugars. Also called *carbon fixation* and *light-independent reactions.*

calyx All of the sepals of a flower.

CAM See **crassulacean acid metabolism.**

cambium (plural: cambia) See **lateral meristem.**

Cambrian explosion The rapid diversification of animal body types that began about 543 million years ago and continued for approximately 40 million years.

camera eye The type of eye in vertebrates and cephalopods, consisting of a hollow chamber with a hole at one end (through which light enters) and a sheet of light-sensitive cells against the opposite wall.

cAMP See **cyclic AMP.**

cancer General term for any tumor whose cells grow in an uncontrolled fashion, invade nearby tissues, and spread to other sites in the body.

canopy The uppermost layers of branches in a forest (i.e., those fully exposed to the Sun).

5′ cap A chemical grouping, consisting of 7-methylguanylate and three phosphate groups, that is added to the 5′ end of newly transcribed messenger RNA molecules.

CAP binding site A DNA sequence upstream of certain prokaryotic operons to which catabolite activator protein can bind, increasing gene transcription.

capillarity The tendency of water to move up a narrow tube due to surface tension, adhesion, and cohesion.

capillary One of the numerous small, thin-walled blood vessels that permeate all tissues and organs, and allow exchange of gases and other molecules between blood and body cells.

capillary bed A thick network of capillaries.

capsid A shell of protein enclosing the genome of a virus particle.

carapace In crustaceans, a large platelike section of the exoskeleton that covers and protects the cephalothorax (e.g., a crab's "shell").

carbohydrate Any of a class of molecules that contain a carbonyl group, several hydroxyl groups, and several to many carbon-hydrogen bonds. See **monosaccharide** and **polysaccharide.**

carbon cycle, global The worldwide movement of carbon among terrestrial ecosystems, the oceans, and the atmosphere.

carbon fixation See **Calvin cycle.**

carbonic anhydrase An enzyme that catalyzes the formation of carbonic acid (H_2CO_3) from carbon dioxide and water.

cardiac cycle One complete heartbeat cycle, including systole and diastole.

cardiac muscle The muscle tissue of the vertebrate heart. Consists of long branched fibers that are electrically connected and that initiate their own contractions; not under voluntary control. Compare with **skeletal** and **smooth muscle.**

carnivore (adjective: carnivorous) An animal whose diet consists predominantly of meat. Most members of the mammalian taxon Carnivora are carnivores. Some plants are carnivorous, trapping and killing small animals, then absorbing nutrients from the prey's body. Compare with **herbivore** and **omnivore.**

carotenoid Any of a class of accessory pigments, found in chloroplasts, that absorb wavelengths of light not absorbed by chlorophyll; typically appear yellow, orange, or red. Includes carotenes and xanthophylls.

carpel The female reproductive organ in a flower. Consists of the stigma, to which pollen grains adhere; the style, through which pollen grains move; and the ovary, which houses the ovule. Compare with **stamen.**

carrier A heterozygous individual carrying a normal allele and a recessive allele for an inherited trait; does not display the phenotype of the trait but can pass the recessive gene to offspring.

carrier protein A membrane protein that facilitates diffusion of a small molecule (e.g., glucose) across the plasma membrane by a process involving a reversible change in the shape of the protein. Also called *carrier* or *transporter.*

carrying capacity (K) The maximum population size of a certain species that a given habitat can support.

cartilage A type of vertebrate connective tissue that consists of relatively few cells scattered in a stiff matrix of polysaccharides and protein fibers.

Casparian strip In plant roots, a waxy layer containing suberin, a water-repellent substance, that prevents movement of water through the walls of endodermal cells, thus blocking the apoplastic pathway.

catabolic pathway Any set of chemical reactions that breaks down larger, complex molecules into smaller ones, releasing energy in the process. Compare with **anabolic pathway.**

catabolite activator protein (CAP) A protein that can bind to the CAP binding site upstream of certain prokaryotic operons, facilitating binding of RNA polymerase and stimulating gene expression.

catabolite repression A type of positive transcriptional control in which the end product of a catabolic pathway inhibits further transcription of the gene encoding an enzyme early in the pathway.

catalysis (verb: catalyze) Acceleration of the rate of a chemical reaction due to a decrease in the free energy of the transition state, called the activation energy.

catalyst Any substance that increases the rate of a chemical reaction without itself undergoing any permanent chemical change.

catecholamines A class of small compounds, derived from the amino acid tyrosine, that are used as hormones or neurotransmitters. Include epinephrine, norepinephrine, and dopamine.

cation A positively charged ion.

cation exchange In botany, the release (displacement) of cations, such as magnesium and calcium from soil particles, by protons in acidic soil water. The released cations are available for uptake by plants.

CD4 A membrane protein on the surface of some T cells in humans. $CD4^+$ T cells can give rise to helper T cells.

CD8 A membrane protein on the surface of some T cells in humans. $CD8^+$ T cells can give rise to cytotoxic T cells.

Cdk See **cyclin-dependent kinase.**

cDNA See **complementary DNA.**

cDNA library A set of cDNAs from a particular cell type or stage of development. Each cDNA is carried by a plasmid or other cloning vector and can be separated from other cDNAs. Compare with **genomic library.**

cecum A blind sac between the small intestine and the colon. Is enlarged in some species (e.g., rabbits) that use it as a fermentation vat for digestion of cellulose.

cell A highly organized compartment bounded by a thin, flexible structure (plasma membrane) and containing concentrated chemicals in an aqueous (watery) solution. The basic structural and functional unit of all organisms.

cell body The part of a neuron that contains the nucleus and where incoming signals are integrated. Also called the *soma.*

cell crawling A form of cellular movement involving actin filaments in which the cell produces bulges (pseudopodia) that stick to the substrate and pull the cell forward. Also called *amoeboid motion.*

cell cycle Ordered sequence of events in which a eukaryotic cell replicates its chromosomes, evenly partitions the chromosomes to two daughter cells, and then undergoes division of the cytoplasm.

cell-cycle checkpoint Any of several points in the cell cycle at which progression of a cell through the cycle can be regulated.

cell division Creation of new cells by division of pre-existing cells.

cell-mediated (immune) response The type of immune response that involves generation of cytotoxic T cells from $CD8^+$ T cells. Defends against pathogen-infected cells, cancer cells, and transplanted cells. Compare with **humoral (immune) response.**

cell membrane See **plasma membrane.**

cell plate A double layer of new plasma membrane that appears in the middle of a dividing plant cell; ultimately divides the cytoplasm into two separate cells.

cell sap An aqueous solution found in the vacuoles of plant cells.

cell theory The theory that all organisms are made of cells and that all cells come from preexisting cells.

cellular respiration A common pathway for production of ATP, involving transfer of electrons from compounds with high potential energy (often NADH and $FADH_2$) to an electron transport chain and ultimately to an electron acceptor (often oxygen).

cellulose A structural polysaccharide composed of β-glucose monomers joined by β-1,4-glycosidic linkages. Found in the cell wall of algae, plants, bacteria, fungi, and some other groups.

cell wall A protective layer located outside the plasma membrane and usually composed of polysaccharides. Found in algae, plants, bacteria, fungi, and some other groups.

Cenozoic era The most recent period of geologic time, beginning 65 million years ago. during which mammals became the dominant vertebrates and angiosperms became the dominant plants. Also called *Age of Mammals.*

central dogma The long-accepted hypothesis that information in cells flows in one direction: DNA codes for RNA, which codes for proteins. Exceptions are now known (e.g., retroviruses).

central nervous system (CNS) The brain and spinal cord of vertebrate animals. Compare with **peripheral nervous system (PNS).**

centriole One of two small cylindrical structures, structurally similar to a basal body, found together within the centrosome near the nucleus of a eukaryotic cell.

centromere Constricted region of a replicated chromosome where the two sister chromatids are joined and the kinetochore is located.

centrosome Structure in animal and fungal cells, containing two centrioles, that serves as a microtubule-organizing center for the cell's cytoskeleton and for the mitotic spindle during cell division.

cephalization The formation of a distinct anterior region (the head) where sense organs and a mouth are clustered.

cephalochordates One of the three major chordate lineages (Cephalochordata), comprising small, mobile organisms that live in marine sands; also called *lancelets* or *amphioxi.* Compare with **urochordates** and **vertebrates.**

cephalopods A lineage of molluscs including the squid, octopuses, and nautiluses. Distinguished by large brains, excellent vision, tentacles, and a reduced or absent shell.

cerebellum Posterior section of the vertebrate brain that is involved in coordination of complex muscle movements, such as those required for locomotion and maintaining balance.

cerebrum The most anterior section of the vertebrate brain. Divided into left and right hemispheres and four lobes: parietal lobe, involved in complex decision making (in humans); occipital lobe, receives and interprets visual information; parietal lobe, involved in integrating sensory and motor functions; and temporal lobe, functions in memory, speech (in humans), and interpreting auditory information.

cervix The narrow passageway between the vagina and the uterus of female mammals.

chaetae (singular: chaeta) Bristle-like extensions found in some annelids.

channel protein A membrane protein that facilitates diffusion of a specific ion or small molecule across a plasma membrane through a central pore in the protein. Includes water channels (aquaporins) and ion channels. Also called simply *channel*.

character displacement The tendency for the traits of similar species that occupy overlapping ranges to change in a way that reduces interspecific competition.

chelicerae A pair of clawlike appendages found around the mouth of certain arthropods called chelicerates (spiders, mites, and allies).

chemical bond An attractive force binding two atoms together. Covalent bonds, ionic bonds, and hydrogen bonds are types of chemical bonds.

chemical carcinogen Any chemical that can cause cancer.

chemical energy The potential energy stored in covalent bonds between atoms.

chemical equilibrium A dynamic but stable state of a reversible chemical reaction in which the forward reaction and reverse reactions proceed at the same rate, so that the concentrations of reactants and products remain constant.

chemical evolution The theory that simple chemical compounds in the ancient atmosphere and ocean combined by spontaneous chemical reactions to form larger, more complex substances, eventually leading to the origin of life and the start of biological evolution.

chemical reaction Any process in which one compound or element is combined with others or is broken down; involves the making and/or breaking of chemical bonds.

chemiosmosis An energetic coupling mechanism whereby energy stored in an electrochemical proton gradient (proton-motive force) is used to drive an energy-requiring process such as production of ATP.

chemokine Any of several chemical signals that attract leukocytes to a site of tissue injury or infection.

chemolithotroph An organism that produces ATP by oxidizing inorganic molecules with high potential energy such as ammonia (NH_3) or methane (CH_4). Also called *lithotroph*. Compare with **chemoorganotroph.**

chemoorganotroph An organism that produces ATP by oxidizing organic molecules with high potential energy such as sugars. Also called *organotroph*. Compare with **chemolithotroph.**

chemoreceptor A sensory cell or organ specialized for detection of specific molecules or classes of molecules.

chemotaxis Movement toward or away from a certain chemical.

chiasma (plural: chiasmata) The X-shaped structure formed during meiosis by crossing over between non-sister chromatids in a pair of homologous chromosomes.

chitin A structural polysaccharide composed of *N*-acetylglucosamine monomers joined end to end by β-1,4-glycosidic linkages. Found in cell walls of fungi and many algae, and in external skeletons of insects and crustaceans.

chitons A lineage of marine mollusc that have a protective shell formed of eight calcium carbonate plates.

chlorophyll Any of several closely related green pigments, found in chloroplasts and photosynthetic protists, that absorb light during photosynthesis.

chloroplast A chlorophyll-containing organelle, bounded by a double membrane, in which photosynthesis occurs; found in plants and photosynthetic protists. Also the location of amino acid, fatty acid, purine, and pyrimidine synthesis.

choanocyte A specialized flagellated feeding cell found in choanoflagellates (protists that are the closest living relatives of animals) and sponges (the oldest animal phylum).

cholecystokinin A peptide hormone secreted by cells in the lining of the small intestine. Stimulates the secretion of digestive enzymes from the pancreas and of bile from the liver and gallbladder.

chordates Members of the phylum Chordata, deuterostomes distinguished by a dorsal hollow nerve cord, pharyngeal gill slits, a notochord, a dorsal hollow nerve cord, and a post-anal tail. Include vertebrates, cephalochordata, and urochordata.

chromatid One of the two identical strands composing a replicated chromosome that is connected at the centromere to the other strand.

chromatin The complex of DNA and proteins, mainly histones, that compose eukaryotic chromosomes. Can be highly compact (heterochromatin) or loosely coiled (euchromatin).

chromatin remodeling The process by which the DNA in chromatin is unwound from its associated proteins to allow transcription or replication. May involve chemical modification of histone proteins or reshaping of the chromatin by large multi-protein complexes in an ATP-requiring process.

chromosome Gene-carrying structure consisting of a single long molecule of DNA and associated proteins (e.g., histones). Most prokaryotic cells contain a single, circular chromosome; eukaryotic cells contain multiple noncircular (linear) chromosomes located in the nucleus.

chromosome inversion See **inversion.**

chromosome painting A technique for producing high-resolution karyotypes by "painting" chromosomes with fluorescent tags that bind to particular regions of certain chromosomes. Also called *spectral karyotyping*.

chromosome theory of inheritance The principle that genes are located on chromosomes and that patterns of inheritance are determined by the behavior of chromosomes during meiosis.

chromosome translocation See **translocation** (2).

chylomicron A ball of protein-coated lipids used to transport the lipids through the bloodstream.

chytrid A member of a paraphyletic group of fungi (Chytridiomycota) that produce motile spores and gametes (both have flagella similar to those in sperm cells). Also called *chytridiomycete*.

cilium (plural: cilia) One of many short, filamentous projections of some eukaryotic cells containing a core of microtubules. Used to move the cell and/or to move fluid or particles along a stationary cell. See **axoneme.**

circadian clock An internal mechanism found in most organisms that regulates many body processes (sleep-wake cycles, hormonal patterns, etc.) in a roughly 24-hour cycle.

circulatory system The system in animals responsible for moving oxygen, carbon dioxide, and other materials (hormones, nutrients, wastes) around the body.

cisternae (singular: cisterna) Flattened, membrane-bound compartments that make up the Golgi apparatus.

clade See **monophyletic group.**

cladistic approach A method for constructing a phylogenetic tree that is based on identifying the unique traits of each monophyletic group. Compare with **phenetic approach.**

class In Linnaeus' system, a taxonomic category above the order level and below the phylum level.

Class I MHC protein An MHC protein that is present on the plasma membrane of virtually all nucleated cells and functions in presenting antigen to $CD8^+$ T cells.

Class II MHC protein An MHC protein that is present only on the plasma membrane of dendritic cells, macrophages, and B cells and functions in presenting antigen to $CD4^+$ T cells.

classical conditioning A type of learning in which an animal learns to associate two stimuli, so that a response originally given to just one stimulus can be evoked by the second stimulus as well.

cleavage In animal development, the series of rapid mitotic cell divisions, with little cell growth, that produces successively smaller cells and transforms a zygote into a multicellular blastula, or blastocyst in mammals.

cleavage furrow A pinching-in of the plasma membrane that occurs as cytokinesis begins in animal cells and deepens until the cytoplasm is divided.

climate The prevailing long-term weather conditions in a particular region.

climax community The stable, final community that develops from ecological succession.

clitoris A small rod of erectile tissue in the external genitalia of female mammals. Is formed from the same embryonic tissue as the male penis and has a similar function in sexual arousal.

cloaca An opening to the outside used by the excretory and reproductive systems in many nonmammalian vertebrate but few vertebrates.

clonal expansion Rapid cell division by a T cell or B cell in response to its binding the specific antigen recognized by receptors on the cell surface. Produces a large population of descendant cells all specific for that particular antigen.

clonal-selection theory The dominant explanation of the development of acquired immunity in vertebrates. According to the theory, the immune system retains a vast pool of inactive lymphocytes, each with a unique receptor for a unique antigen. Lymphocytes that encounter their antigens are stimulated to divide (selected and cloned), producing daughter cells that combat infection and confer immunity.

clone (1) An individual that is genetically identical to another individual. (2) A lineage of genetically identical individuals or cells. (3) As a verb, to make one or more genetic replicas of a cell or individual.

cloning vector A plasmid or other agent used to transfer recombinant genes into cultured host cells. Also called simply *vector.*

closed circulatory system A circulatory system in which the circulating fluid (blood) is confined to blood vessels and flows in a continuous circuit. Compare with **open circulatory system.**

club fungus A member of a lineage of fungi (Basidiomycota) that produce complex reproductive structures (e.g., mushrooms, puffballs) containing specialized spore-forming cells (basidia). Also called *basidiomycete.*

cnidocyte A specialized stinging cell found in cnidarians (e.g., jellyfish, corals, and anemones) that is used in capturing prey.

coactivator Any regulatory protein that help initiate transcription by bringing together the necessary transcription factors but does not itself bind to DNA.

cochlea The organ of hearing in the inner ear of mammals, birds, and crocodilians. A coiled, fluid-filled tube containing specialized pressure-sensing neurons (hair cells) that detect sounds of different pitches.

coding strand See **non-template strand.**

codominance An inheritance pattern in which heterozygotes exhibit both of the traits seen in either kind of homozygous individual.

codon A sequence of three nucleotides in DNA or RNA that codes for a certain amino acid or that initiates or terminates protein synthesis.

coefficient of relatedness (*r*) A measure of how closely two individuals are related. Calculated as the probability that an allele in two individuals is inherited from the same ancestor.

coelom An internal, usually fluid-filled, body cavity that is lined with mesoderm.

coelomate An animal that has a true coelom. Compare with **acoelomate** and **pseudocoelomate.**

coenocytic Containing many nuclei and a continuous cytoplasm through a filamentous body, without the body being divided into distinct cells. Some fungi are coenocytic.

coenzyme A small organic molecule that is a required cofactor for an enzyme-catalyzed reaction. Often donates or receives electrons or functional groups during the reaction.

coenzyme A (CoA) A nonprotein molecule that is required for many cellular reactions involving transfer of acetyl groups ($-COCH_3$).

coenzyme Q A nonprotein molecule that shuttles electrons between membrane-bound complexes in the mitochondrial electron transport chain. Also called *ubiquinone* or *Q*.

coevolution A pattern of evolution in which two interacting species reciprocally influence each other's adaptations over time.

cofactor A metal ion or small organic compound that is required for an enzyme to function normally. May be bound tightly to an enzyme or associate with it transiently during catalysis.

cognition The mental processes involved in recognition and manipulation of facts about the world, particularly to form novel associations or insights.

cohesion The tendency of certain like molecules (e.g., water molecules) to cling together due to attractive forces. Compare with **adhesion.**

cohesion-tension theory The theory that water movement upward through plant vascular tissues is due to loss of water from leaves (transpiration), which pulls a cohesive column of water upward.

cohort A group of individuals that are the same age and can be followed through time.

coleoptile A modified leaf that covers and protects the stems and leaves of young grasses.

collagen A fibrous, pliable, cable-like glycoprotein that is a major component of the extracellular matrix of animal cells. Various subtypes differ in their tissue distribution.

collecting duct In the vertebrate kidney, a large straight tube that receives filtrate from the distal tubules of several nephrons. Involved in the regulated reabsorption of water.

collenchyma cell In plants, an elongated cell with cell walls thickened at the corners that provides support to growing plant parts; usually found in strands along leaf veins and stalks. Compare with **parenchyma cell** and **sclerenchyma cell.**

colon The portion of the large intestine where feces are formed by compaction of wastes and reabsorption of water.

colony An assemblage of individuals. May refer to an assemblage of semi-independent cells or to a breeding population of multicellular organisms.

commensalism (adjective: commensal) A symbiotic relationship in which one organism (the commensal) benefits and the other (the host) is not harmed. Compare with **mutualism** and **parasitism.**

communication In ecology, any process in which a signal from one individual modifies the behavior of another individual.

community All of the species that interact with each other in a certain area.

companion cell In plants, a cell in the phloem that is connected via numerous plasmodesmata to adjacent sieve-tube members. Companion cells provide materials to maintain sieve-tube members and function in the loading and unloading of sugars into sieve-tube members.

compass orientation A type of navigation in which movement occurs in a specific direction.

competition In ecology, the interaction of two species or two individuals trying to use the same limited resource (e.g., water, food, living space). May occur between individuals of the same species (intraspecific competition) or different species (interspecific competition).

competitive exclusion principle The principle that two species cannot coexist in the same ecological niche in the same area because one species will out-compete the other.

competitive inhibition Inhibition of an enzyme's ability to catalyze a chemical reaction via a nonreactant molecule that competes with the substrate(s) for access to the active site.

complementary base pairing The association between specific nitrogenous bases of nucleic acids stabilized by hydrogen bonding. Adenine pairs only with thymine (in DNA) or uracil (in RNA), and guanine pairs only with cytosine.

complementary DNA (cDNA) DNA produced in the laboratory using an RNA transcript as a template and reverse transcriptase; corresponds to a gene but lacks introns. Also produced naturally by retroviruses.

complementary strand A newly synthesized strand of RNA or DNA that has a base sequence complementary to that of the template strand.

complement system A set of proteins that circulate in the bloodstream and can form holes in the plasma membrane of bacteria, leading to their destruction.

complete metamorphosis See **holometabolous metamorphosis.**

compound eye An eye formed of many independent light-sensing columns (ommatidia); occurs in arthropods. Compare with **simple eye.**

concentration gradient Difference across space (e.g., across a membrane) in the concentration of a dissolved substance.

condensation reaction A chemical reaction in which two molecules are joined covalently with the removal of an $-OH$ from one and an $-H$ from another to form water. Also called a *dehydration reaction.* Compare with **hydrolysis.**

conduction (1) Direct transfer of heat between two objects that are in physical contact. Compare with **convection.** (2) Transmission of an electrical impulse along the axon of a nerve cell.

cone In conifers (e.g., pines, cedars, and spruces), the reproductive structure in which microspores and macrospores are formed.

cone cell A photoreceptor cell with a cone-shaped outer portion that is particularly sensitive to bright light of a certain color. Also called simply *cone.* Compare with **rod cell.**

conjugation The process by which DNA is exchanged between unicellular individuals. Occurs in bacteria, archaea, and some protists.

connective tissue An animal tissue consisting of scattered cells in a liquid, jellylike, or solid extracellular matrix. Includes bone, cartilage, tendons, ligaments, and blood.

conservation biology The effort to study, preserve, and restore threatened populations, communities, and ecosystems.

constant (C) region The portion of an antibody's light chains or heavy chains that has the same amino acid sequence in the antibodies produced by every B cell of an individual. Compare with **variable (V) region.**

constitutive Always occurring; always present. Commonly used to describe enzymes and other proteins that are synthesized continuously or mutants in which one or more genetic loci are constantly expressed due to defects in gene control.

constitutive defense A defensive trait that is always manifested even in the absence of a predator or pathogen. Also called *standing defense.* Compare with **inducible defense.**

consumer See **heterotroph.**

continuous strand See **leading strand.**

control In a scientific experiment, a group of organisms or samples that do not receive the experimental treatment but are otherwise identical to the group that does.

convection Transfer of heat by movement of large volumes of a gas or liquid. Compare with **conduction.**

convergent evolution The independent evolution of analogous traits in distantly related organisms due to adaptation to similar environments and a similar way of life.

cooperative binding The tendency of the protein subunits of hemoglobin to affect each other's oxygen binding such that each bound oxygen molecule increases the likelihood of further oxygen binding.

copulation The act of transferring sperm from a male directly into a female's reproductive tract.

coral reef A large assemblage of colonial marine corals that usually serves as shallow-water, sunlit habitat for many other species as well.

co-receptor Any membrane protein that acts with some other membrane protein in a cell interaction or cell response.

cork cambium One of two types of lateral meristem, consisting of a ring of undifferentiated plant cells found just under the cork layer of woody plants; produces new cork cells on its outer side. Compare with **vascular cambium.**

cork cell A waxy cell in the protective outermost layer of a woody plant.

corm A rounded, thick underground stem that can produce new plants via asexual reproduction.

cornea The transparent sheet of connective tissue at the very front of the eye in vertebrates and some other animals. Protects the eye and helps focus light.

corolla All of the petals of a flower.

corona The cluster of cilia at the anterior end of a rotifer.

corpus callosum A thick band of neurons that connects the two hemispheres of the cerebrum in the mammalian brain.

corpus luteum A yellowish structure in an ovary that secretes progesterone. Is formed from a follicle that has recently ovulated.

cortex (1) The outermost region of an organ, such as the kidney or adrenal gland. (2) In plants, a layer of ground tissue found outside the vascular bundles and pith of a plant stem.

cortical granules Small enzyme-filled vesicles in the cortex of an egg cell. Involved in formation of the fertilization envelope after fertilization.

corticosterone A steroid hormone, produced and secreted by the adrenal cortex, that increases blood glucose and prepares the body for stress. The major glucocorticoid hormone in most reptiles, birds, and many mammals.

corticotropin-releasing hormone (CRH) A peptide hormone, produced and secreted by the hypothalamus, that stimulates the anterior pituitary to release ACTH.

cortisol A steroid hormone, produced and secreted by the adrenal cortex, that increases blood glucose and prepares the body for stress. The major glucocorticoid hormone in some mammals. Also called *hydrocortisone.*

cotransporter A transmembrane protein that facilitates diffusion of an ion down its previously established electrochemical gradient and uses the energy of that process to transport some other substance, in the same or opposite direction, *against* its concentration gradient. See **antiporter** and **symporter.**

cotyledon The first leaf, or seed leaf, of a plant embryo. Used for storing and digesting nutrients and/or for early photosynthesis.

countercurrent exchanger In animals, any anatomical arrangement that allows the maximum transfer of heat or a soluble substance from one fluid to another. The two fluids must be flowing in opposite directions and have a heat or concentration gradient between them.

covalent bond A type of chemical bond in which two atoms share one or more pairs of electrons. Compare with **hydrogen bond** and **ionic bond.**

cranium A bony, cartilaginous, or fibrous case that encloses and protects the brain of vertebrates. Forms part of the skull. Also called *braincase.*

crassulacean acid metabolism (CAM) A variant type of photosynthesis in which CO_2 is stored in organic acids at night when stomata are open and then released to feed the Calvin cycle during the day when stomata are closed. Helps reduce water loss and oxygen loss by photorespiration in hot, dry environments.

cristae (singular: crista) Sac-like invaginations of the inner membrane of a mitochondrion. Location of the electron transport chain and ATP synthase.

critical period See **sensitive period.**

Cro-Magnon A prehistoric European population of modern humans (*Homo sapiens*) known from fossils, paintings, sculptures, and other artifacts.

crossing over The exchange of segments of non-sister chromatids between a pair of homologous chromosomes that occurs during meiosis I.

cross-pollination Pollination of a flower by pollen from another individual, rather than by self-fertilization. Also called *crossing.*

crustaceans A lineage of arthropods that includes shrimp, lobster, and crabs. Many have a carapace (a platelike portion of the exoskeleton) and mandibles for biting or chewing.

culture In cell biology, a collection of cells or a tissue growing under controlled conditions, usually in suspension or on the surface of a dish on solid growth medium.

cup fungus See **sac fungus.**

Cushing's disease A human endocrine disorder caused by loss of feedback inhibition of cortisol on ACTH secretion. Characterized by high ACTH and cortisol levels and wasting of body protein reserves.

cuticle A protective coating secreted by the outermost layer of cells of an animal or a plant.

cyanobacteria A lineage of photosynthetic bacteria formerly known as blue-green algae. Likely the first life-forms to carry out oxygenic photosynthesis.

cyclic AMP (cAMP) Cyclic adenosine monophosphate; a small molecule, derived from ATP, that is widely used by cells in signal transduction and transcriptional control.

cyclic photophosphorylation Path of electron flow during the light-dependent reactions of photosynthesis in which photosystem I transfers excited electrons back to the electron transport chain of photosystem II, rather than to $NADP^+$. Also called *cyclic electron flow.* Compare with **Z scheme.**

cyclin One of several regulatory proteins whose concentrations fluctuate cyclically throughout the cell cycle.

cyclin-dependent kinase (Cdk) Any of several related protein kinases that are active only when bound to a cyclin. Involved in control of the cell cycle.

cytochrome *c* (cyt *c*) A soluble iron-containing protein that shuttles electrons between membrane-bound complexes in the mitochondrial electron transport chain.

cytokines A diverse group of autocrine signaling proteins, secreted largely by cells of the immune system, whose effects include stimulating leukocyte production, tissue repair, and fever. Generally function to regulate the intensity and duration of an immune response.

cytokinesis Division of the cytoplasm to form two daughter cells. Typically occurs immediately after division of the nucleus by mitosis or meiosis.

cytokinins A class of plant hormones that stimulate cell division and retard aging.

cytoplasm All of the contents of a cell, excluding the nucleus, and bounded by the plasma membrane.

cytoplasmic determinant A regulatory transcription factor or signaling molecule that is distributed unevenly in the cytoplasm of the egg cells of many animals and that directs early pattern formation in an embryo.

cytoplasmic streaming The directed flow of cytosol and organelles that facilitates distribution of materials within some large plant and fungal cells. Occurs along actin filaments and is powered by myosin.

cytoskeleton In eukaryotic cells, a network of protein fibers in the cytoplasm that are involved in cell shape, support, locomotion, and transport of materials within the cell. Prokaryotic cells have a similar but much less extensive network of fibers.

cytosol The fluid portion of the cytoplasm.

cytotoxic T cell An effector T cell that destroys infected cells and cancer cells. Is descended from an activated $CD8^+$ T cell that has interacted with antigen on an infected cell or cancer cell. Also called *cytotoxic T lymphocyte (CTL)* and *killer T cell.* Compare with **helper T cell.**

dalton (Da) See **atomic mass unit.**

Darwinian fitness See **fitness.**

day-neutral plant A plant whose flowering time is not affected by the relative length of day and night (the photoperiod). Compare with **long-day** and **short-day plant.**

dead space Portions of the air passages that are not involved in gas exchange with the blood, such as the trachea and bronchi.

deciduous Describing a plant that sheds leaves or other structures at regular intervals (e.g., each fall)

decomposer See **detritivore.**

decomposer food chain An ecological network of detritus, decomposers that eat detritus, and predators and parasites of the decomposers.

definitive host The host species in which a parasite reproduces sexually. Compare with **intermediate host.**

dehydration reaction See **condensation reaction.**

deleterious In genetics, referring to any mutation, allele, or trait that reduces an individual's fitness.

demography The study of factors that determine the size and structure of populations through time.

denaturation (verb: denature) For a macromolecule, loss of its three-dimensional structure and biological activity due to breakage of hydrogen bonds and disulfide bonds, usually caused by treatment with excess heat or extreme pH conditions.

dendrite A short extension from a neuron's cell body that receives signals from other neurons.

dendritic cell A type of leukocyte that ingests and digests foreign antigens, moves to a lymph node, and presents the antigens displayed on its membrane to $CD4^+$ T cells.

density dependent In population ecology, referring to any characteristic that varies depending on population density.

density independent In population ecology, referring to any characteristic that does not vary with population density.

deoxyribonucleic acid (DNA) A nucleic acid composed of deoxyribonucleotides that carries the genetic information of a cell. Generally occurs as two intertwined strands, but these can be separated. See **double helix.**

deoxyribonucleoside triphosphate (dNTP) A monomer that can be polymerized to form DNA. Consists of deoxyribose, a base (A, T, G, or C), and three phosphate groups; similar to a nucleotide, but with two more phosphate groups.

deoxyribonucleotide See **nucleotide.**

depolarization Change in membrane potential from its resting negative state to a less negative or a positive state; a normal phase in an action potential. Compare with **hyperpolarization.**

deposit feeder An animal that eats its way through a food-containing substrate.

dermal tissue system The tissue forming the outer layer of an organism. In plants, also called *epidermis;* in animals, forms two distinct layers: *dermis* and *epidermis.*

descent with modification The phrase used by Darwin to describe his hypothesis of evolution by natural selection.

desmosome A type of cell-cell attachment structure, consisting of cadherin proteins, that bind the cytoskeletons of adjacent animal cells together. Found where cells are strongly attached to each other. Compare with **gap junction** and **tight junction.**

determination In embryogenesis, progressive changes in a cell that commit it to a particular cell fate. Once a cell is fully determined, it can differentiate only into a particular cell type (e.g., liver cell, brain cell).

detritivore An organism whose diet consists mainly of dead organic matter (detritus). Various bacteria, fungi, and protists are detritivores. Also called *decomposer.*

detritus A layer of dead organic matter that accumulates at ground level or on seafloors and lake bottoms.

deuterostomes A major lineage of animals that share a pattern of embryological development, including radial cleavage, formation of the anus earlier than the mouth, and formation of the coelom by pinching off of layers of mesoderm from the gut. Includes echinoderms and chordates. Compare with **protostomes.**

developmental homology A similarity in embryonic form, or in the fate of embryonic tissues, that is due to inheritance from a common ancestor.

diabetes insipidus A human disease caused by defects in the kidney's system for conserving water. Characterized by production of large amounts of dilute urine.

diabetes mellitus A human disease caused by defects in insulin production (type I) or the response of cells to insulin (type II). Characterized by abnormally high blood glucose levels and large amounts of glucose-containing urine.

diaphragm An elastic, sheetlike structure. In mammals, the muscular sheet of tissue that separates the chest and abdominal cavities. Contracts and moves downward during inhalation, expanding the chest cavity.

diastole The portion of the heartbeat cycle during which the atria or ventricles of the heart are relaxed. Compare with **systole.**

diastolic blood pressure The force exerted by blood against artery walls during relaxation of the heart's left ventricle. Compare with **systolic blood pressure.**

dicot Any plant that has two cotyledons (embryonic leaves) upon germination. The dicots do not form a monophyletic group. Also called *dicotyledonous plant.* Compare with **eudicot** and **monocot.**

dideoxy sequencing A laboratory technique for determining the exact nucleotide sequence of DNA. Relies on the use of dideoxynucleotide triphosphates (ddNTPs), which terminate DNA replication.

diencephalon The part of the mammalian brain that relays sensory information to the cerebellum and functions in maintaining homeostasis.

differential centrifugation Procedure for separating cellular components according to their size and density by spinning a cell homogenate in a series of centrifuge runs. After each run, the supernatant is removed from the deposited material (pellet) and spun again at progressively higher speeds.

differential gene expression Expression of different sets of genes in cells with the same genome. Responsible for creating different cell types.

differentiation The process by which a relatively unspecialized cell becomes a distinct specialized cell type (e.g., liver cell, brain cell) usually by changes in gene expression. Also call *cell differentiation.*

diffusion Spontaneous movement of a substance from a region of high concentration to one of low concentration (i.e., down a concentration gradient).

digestion The physical and chemical breakdown of food into molecules that can be absorbed into the body of an animal.

digestive tract The long tube that begins at the mouth and ends at the anus. Also called *alimentary canal, gastrointestinal(GI) tract,* or the *gut.*

dihybrid cross A mating between two parents that are heterozygous for both of the two genes being studied.

dimer An association of two molecules, which may be identical (homodimer) or different (heterodimer).

dioecious Describing an angiosperm species that has male and female reproductive structures on separate plants. Compare with **monoecious.**

diploblast (adjective: diploblastic) An animal whose body develops from two basic embryonic cell layers—ectoderm and endoderm. Compare with **triploblast.**

diploid (1) Having two sets of chromosome s ($2n$). (2) A cell or an individual organism with two sets of chromosomes, one set inherited from the maternal parent and one set from the paternal parent. Compare with **haploid.**

directional selection A pattern of natural selection that favors one extreme phenotype with the result that the average phenotype of a population changes in one direction. Generally reduces overall genetic variation in a population.

direct sequencing A technique for identifying and studying microorganisms that cannot be grown in culture. Involves detecting and amplifying copies of certain specific genes in their DNA, sequencing these genes, and then comparing the sequences with the known sequences from other organisms.

disaccharide A carbohydrate consisting of two monosaccharides (sugar residues) linked together.

discontinuous strand See **lagging strand.**

discrete trait An inherited trait that exhibits distinct phenotypic forms rather than the continuous variation characteristic of a quantitative traits such as body height.

dispersal The movement of individuals from their place of origin (birth, hatching) to a new location.

disruptive selection A pattern of natural selection that favors extreme phenotypes at both ends of the range of phenotypic variation. Maintains overall genetic variation in a population. Compare with **stabilizing selection.**

distal tubule In the vertebrate kidney, the convoluted portion of a nephron into which filtrate moves from the loop of Henle. Involved in the regulated reabsorption of sodium and water. Compare with **proximal tubule.**

disturbance In ecology, any event that disrupts a community, usually causing loss of some individuals or biomass from it.

disturbance regime The characteristic disturbances that affect a given ecological community.

disulfide bond A covalent bond between two sulfur atoms, typically in the side groups of some amino acids (e.g., cysteine). Often contributes to tertiary structure of proteins.

DNA See **deoxyribonucleic acid.**

DNA cloning Any of several techniques for producing many identical copies of a particular gene or other DNA sequence.

DNA fingerprinting Any of several methods for identifying individuals by unique features of their genomes. Commonly involves using PCR to produce many copies of certain simple sequence repeats (microsatellites) and then analyzing their lengths.

DNA footprinting A technique used to identify stretches of DNA that are bound by particular regulatory proteins.

DNA library See **cDNA library** and **genomic library.**

DNA ligase An enzyme that joins pieces of DNA by catalyzing formation of a phosphodiester bond between the pieces.

DNA microarray A set of single-stranded DNA fragments, representing thousands of different genes, that are permanently fixed to a small glass slide. Can be used to determine which genes are expressed in different cell types, under different conditions, or at different developmental stages.

DNA polymerase Any enzyme that catalyzes synthesis of DNA from deoxyribonucleotides.

domain (1) A section of a protein that has a distinctive tertiary structure and function. (2) A taxonomic category, based on similarities in basic cellular biochemistry, above the kingdom level. The three recognized domains are Bacteria, Archaea, and Eukarya.

dominant Referring to an allele that determines the phenotype of a heterozygous individual. Compare with **recessive.**

dopamine A catecholamine neurotransmitter that functions mainly in a part of the mammalian brain involved with muscle control. Also functions as a hypothalamic inhibitory hormone that inhibits release of prolactin from the interior pituitary; also called *prolactin-inhibiting hormone (PIH).*

dormancy A temporary state of greatly reduced, or no, metabolic activity and growth in plants or plant parts (e.g., seeds, spores, bulbs, and buds).

dorsal Toward an animal's back and away from its belly. The opposite of ventral.

double fertilization An unusual form of reproduction seen in flowering plants, in which one sperm nucleus fuses with an egg to form a zygote and the other sperm nucleus fuses with two polar nuclei to form the triploid endosperm.

double helix The secondary structure of DNA, consisting of two antiparallel DNA strands wound around each other.

downstream In genetics, the direction in which RNA polymerase moves along a DNA strand. Compare with **upstream.**

Down syndrome A human developmental disorder caused by trisomy of chromosome 21.

dyad symmetry A type of symmetry in which an object can be superimposed on itself if rotated 180°. Occurs in some regulatory sequences of DNA. Also called *two-fold rotational symmetry.*

dynein Any one of a class of motor proteins that use the chemical energy of ATP to "walk" along an adjacent microtubule. Dyneins are responsible for bending of cilia and flagella, play a role in chromosome movement during mitosis, and can transport certain organelles.

early endosome A small membrane-bound vesicle, formed by endocytosis, that is an early stage in the formation of a lysosome.

ecdysone An insect hormone that triggers either molting (to a larger larval form) or metamorphosis (to the adult form), depending on the level of juvenile hormone.

ecdysozoans A major lineage of protostomes (Ecdysozoa) that grow by shedding their external skeletons (molting) and expanding their bodies. Includes arthropods, insects, crustaceans, nematodes, and centipedes. Compare with **lophotrochozoans.**

echinoderms A major lineage of deuterostomes (Echinodermata) distinguished by adult bodies with five-sided radial symmetry, a water vascular system, and tube feet. Includes sea urchins, sand dollars, and sea stars.

echolocation The use of echoes from vocalizations to obtain information about locations of objects in the environment.

ecology The study of how organisms interact with each other and with their surrounding environment.

ecosystem All the organisms that live in a geographic area, together with the nonliving (abiotic) components that affect or exchange materials with the organisms; a community and its physical environment.

ecosystem diversity The variety of biotic components in a region along with abiotic components, such as soil, water, and nutrients.

ecosystem services Alterations of the physical components of an ecosystem by living organisms, especially beneficial changes in the quality of the atmosphere, soil, water, etc.

ectoderm The outermost of the three basic cell layers in most animal embryos; gives rise to the outer covering and nervous system. Compare with **endoderm** and **mesoderm.**

ectomycorrhizal fungi (EMF) Fungi whose hyphae form a dense network that covers their host plant's roots but do not enter the root cells.

ectoparasite A parasite that lives on the outer surface of the host's body.

ectotherm An animal that does not use internally generated heat to regulate its body temperature. Compare with **endotherm.**

effector Any cell, organ, or structure with which an animal can respond to external or internal stimuli. Usually functions, along with a sensor and integrator, as part of a homeostatic system.

efferent division The part of the nervous system, consisting primarily of motor neurons, that carries commands from the central nervous system to the body.

egg A mature female gamete and any associated external layers (such as a shell). Larger and less mobile than the male gamete. In animals, also called *ovum.*

ejaculation The release of semen from the copulatory organ of a male animal.

ejaculatory duct A short duct connecting the vas deferens to the urethra, through which sperm move during ejaculation.

elastic Referring to a structure (e.g., lungs) with the ability to stretch and then spring back to its original shape.

electrical potential Potential energy created by a separation of electric charges between two points. Also called *voltage.*

electric current A flow of electrical charge past a point. Also called *current.*

electrocardiogram (EKG) A recording of the electrical activity of the heart, as measured through electrodes on the skin.

electrochemical gradient The combined effect of an ion's concentration gradient and electrical (charge) gradient across a membrane that affects the diffusion of ions across the membrane.

electrolyte Any compound that dissociates into ions when dissolved in water. In nutrition, refers to the major ions necessary for normal cell function.

electromagnetic spectrum The entire range of wavelengths of radiation extending from short wavelengths (high energy) to long wavelengths (low energy). Includes gamma rays, x-rays, ultraviolet, visible light, infrared, microwaves, and radio waves (from short to long wavelengths).

electron acceptor A reactant that gains an electron and is reduced in a reduction-oxidation reaction.

electron carrier Any molecule that readily accepts electrons from and donates electrons to other molecules.

electron donor A reactant that loses an electron and is oxidized in a reduction-oxidation reaction.

electronegativity A measure of the ability of an atom to attract electrons toward itself from an atom to which it is bonded.

electron microscope See **scanning electron microscope** and **transmission electron microscope.**

electron shell A group of orbitals of electrons with similar energies. Electron shells are arranged in roughly concentric layers around the nucleus of an atom, with electrons in outer shells having more energy than those in inner shells. Electrons in the outermost shell, the valence shell, often are involved in chemical bonding.

electron transport chain (ETC) Any set of membrane-bound protein complexes and smaller soluble electron carriers involved in a coordinated series of redox reactions in which the potential energy of electrons transferred from reduced donors is successively decreased and used to pump protons from one side of a membrane to the other.

electroreceptor A sensory cell or organ specialized to detect electric fields.

element A substance, consisting of atoms with a specific number of protons, that cannot be separated into or broken down to any other substance. Elements preserve their identity in chemical reactions.

elongation (1) The process by which messenger RNA lengthens during transcription. (2) The process by which a polypeptide chain lengthens during translation.

elongation factors Proteins involved in the elongation phase of translation, assisting ribosomes in the synthesis of the growing peptide chain.

embryo A young developing organism; the stage after fertilization and zygote formation.

embryo sac The female gametophyte in flowering plants that exhibit alternation of generations.

embryogenesis The process by which a single-celled zygote becomes a multicellular embryo.

embryophyte A plant that nourishes its embryos inside its own body. All land plants are embryophytes.

emergent vegetation Any plants in an aquatic habitat that extend above the surface of the water.

emerging disease Any infectious disease, often a viral disease, that suddenly afflicts significant numbers of humans for the first time; often due to changes in the host species for a pathogen or host population movements.

emigration The migration of individuals away from one population to other populations. Compare with **immigration.**

emulsification (verb: emulsify) The dispersion of fat into an aqueous solution. Usually requires the aid of an amphipathic substance such as a detergent or bile salts, which can break large fat globules into microscopic fat droplets.

endangered species A species whose numbers have decreased so much that it is in danger of extinction throughout all or part of its range.

endemic species A species that lives in one geographic area and nowhere else.

endergonic Referring to a chemical reaction that requires an input of energy to occur and for which the Gibbs free-energy change (ΔG) > 0. Compare with **exergonic.**

endocrine Relating to a chemical signal (hormone) that is released into the bloodstream by a producing cell and acts on a distant target cell.

endocrine gland A gland that secretes hormones directly into the bloodstream or interstitial fluid instead of into ducts. Compare with **exocrine gland.**

endocrine system All of the glands and tissues that produce and secrete hormones into the bloodstream.

endocytosis General term for any pinching off of the plasma membrane that results in the uptake of material from outside the cell. Includes phagocytosis, pinocytosis, and receptor-mediated endocytosis. Compare with **exocytosis.**

endoderm The innermost of the three basic cell layers in most animal embryos; gives rise to the digestive tract and organs that connect to it (liver, lungs, etc.). Compare with **ectoderm** and **mesoderm.**

endodermis In plant roots, a cylindrical layer of cells that separates the cortex from the vascular tissue.

endomembrane system A system of organelles in eukaryotic cells that performs most protein and lipid synthesis. Includes the endoplasmic reticulum (ER), Golgi apparatus, and lysosomes.

endoparasite A parasite that lives inside the host's body.

endophyte (adjective: endophytic) A fungus that lives inside the aboveground parts of a plant in a symbiotic relationship. Compare with **epiphyte.**

endoplasmic reticulum (ER) A network of interconnected membranous sacs and tubules found inside eukaryotic cells. See **rough** and **smooth endoplasmic reticulum.**

endoskeleton Bony and/or cartilaginous structures within the body that provide support. Examples are the spicules of sponges, the plates in echinoderms, and the bony skeleton of vertebrates. Compare with **exoskeleton.**

endosome See **early** and **late endosome.**

endosperm A triploid ($3n$) tissue in the seed of a flowering plant (angiosperm) that serves as food for the plant embryo. Functionally analogous to the yolk in some animal eggs.

endosymbiont An organism that lives in a symbiotic relationship inside the body of its host.

endosymbiosis theory The theory that mitochondria and chloroplasts evolved from prokaryotes that were engulfed by host cells and took up a symbiotic existence within those cells, a process termed primary endosymbiosis. In some eukaryotes, chloroplasts originated by secondary endosymbiosis, that is, by engulfing a chloroplast-containing protist and retaining its chloroplasts.

endotherm An animal whose primary source of body heat is internally generated heat. Compare with **ectotherm.**

endothermic Referring to a chemical reaction that absorbs heat. Compare with **exothermic.**

energetic coupling In cellular metabolism, the mechanism by which energy released from an exergonic reaction (commonly, hydrolysis of ATP) is used to drive an endergonic reaction.

energy The capacity to do work or to supply heat. May be stored (potential energy) or available in the form of motion (kinetic energy).

enhancer A regulatory sequence in eukaryotic DNA that may be located far from the gene it controls or within introns of the gene. Binding of specific proteins to an enhancer enhances the transcription of certain genes.

enrichment culture A method of detecting and obtaining cells with specific characteristics by placing a sample, containing many types of cells, under a specific set of conditions (e.g., temperature, salt concentration, available nutrients) and isolating those cells that grow rapidly in response.

entropy (*S*) A quantitative measure of the amount of disorder of any system, such as a group of molecules.

envelope, viral A membrane-like covering that encloses some viruses and their capsid coats, shielding them from attack by the host's immune system.

environmental sequencing The inventory of all the genes in a community or ecosystem by sequencing, analyzing, and comparing the genomes of the component organisms.

enzyme A protein catalyst used by living organisms to speed up and control biological reactions.

epicotyl In some embryonic plants, a portion of the embryonic stem that extends above the cotyledons.

epidemic The spread of an infectious disease throughout a population in a short time period. Compare with **pandemic.**

epidermis The outermost layer of cells of any multicellular organism.

epididymis A coiled tube wrapped around the testis in reptiles, birds, and mammals. The site of the final stages of sperm maturation and storage.

epigenetic inheritance Pattern of inheritance involving differences in phenotype that are not due to changes in the nucleotide sequence of genes.

epinephrine A catecholamine hormone, produced and secreted by the adrenal medulla, that triggers rapid responses relating to the fight-or-flight response. Also called *adrenaline.*

epiphyte (adjective: epiphytic) A nonparasitic plant that grows on trees or other solid objects and is not rooted in soil.

epithelium (plural: epithelia) An animal tissue consisting of sheet-like layers of tightly packed cells that lines an organ, a duct, or a body surface. Also called *epithelial tissue.*

epitope A small region of a particular antigen to which an antibody, B-cell receptor, or T-cell receptor binds.

equilibrium curve See **oxygen-hemoglobin equilibrium curve.**

equilibrium potential The membrane potential at which there is no net movement of a particular ion into or out of a cell.

ER signal sequence A short amino acid sequence that marks a polypeptide for transport to the endoplasmic reticulum where synthesis of the polypeptide chain is completed and the signal sequence removed. See **signal recognition particle.**

erythrocyte See **red blood cell.**

erythropoietin (EPO) A peptide hormone, released by the kidney in response to low blood oxygen levels, that stimulates the bone marrow to produce more red blood cells.

esophagus The muscular tube that connects the mouth to the stomach.

essential amino acid An amino acid that an animal cannot synthesize and must obtain from the diet. May refer specifically to one of the eight essential amino acids of adult humans: isoleucine, leucine, lysine, methionine, phenylalanine, threonine, tryptophan, and valine.

essential nutrient Any chemical element, ion, or compound that is required for normal growth, reproduction, and maintenance of a living organism and that cannot be synthesized by the organism.

ester linkage The covalent bond formed by a condensation reaction between a carboxyl group ($-COOH$) and a hydroxyl group ($-OH$). Ester linkages join fatty acids to glycerol to form a fat or phospholipid.

estradiol The major estrogen produced by the ovaries of female mammals. Stimulates development of the female reproductive tract, growth of ovarian follicles, and growth of breast tissue.

estrogens A class of steroid hormones, including estradiol, estrone, and estriol, that generally promote female-like traits. Secreted by the gonads, fat tissue, and some other organs.

estrous cycle A female reproductive cycle, seen in all mammals except Old World monkeys and apes (including humans), in which the uterine lining is reabsorbed rather than shed in the absence of pregnancy, and the female is sexually receptive only briefly during mid-cycle (estrus). Compare with **menstrual cycle.**

estuary An environment of brackish (partly salty) water where a river meets the ocean.

ethylene A gaseous plant hormone that induces fruits to ripen, flowers to fade, and leaves to drop.

eudicot A member of a monophyletic group (lineage) of angiosperms that includes complex flowering plants and trees (e.g., roses, daisies, maples). All eudicots have two cotyledons, but not all dicots are members of this lineage. Compare with **dicot** and **monocot.**

Eukarya One of the three taxonomic domains of life consisting of unicellular organisms (most protists, yeast) and multicellular organisms (fungi, plants, animals) distinguished by a membrane-bound cell nucleus, numerous organelles, and an extensive cytoskeleton. Compare with **Archaea** and **Bacteria.**

eukaryote A member of the domain Eukarya; an organism whose cells contain a nucleus, numerous membrane-bound organelles, and an extensive cytoskeleton. May be unicellular or multicellular. Compare with **prokaryote.**

eutherians A lineage of mammals (Eutheria) whose young develop in the uterus and are not housed in an abdominal pouch. Also called *placental mammals.*

evaporation The energy-absorbing phase change from a liquid state to a gaseous state. Many organisms evaporate water as a means of heat loss.

evo-devo Research field focused on how changes in developmentally important genes have led to the evolution of new phenotypes.

evolution (1) The theory that all organisms on Earth are related by common ancestry and that they have changed over time, predominantly via natural selection. (2) Any change in the genetic characteristics of a population over time, especially, a change in allele frequencies.

excitable membrane A plasma membrane that is capable of generating an action potential. Neurons, muscle cells, and some other cells have excitable membranes.

excitatory postsynaptic potential (EPSP) A change in membrane potential, usually depolarization, at a neuron dendrite that makes an action potential more likely.

exergonic Referring to a chemical reaction that can occur spontaneously, releasing heat and/or increasing entropy, and for which the Gibbs free-energy change (ΔG) < 0. Compare with **endergonic.**

exocrine gland A gland that secretes some substance through a duct into a space other than the circulatory system, such as the digestive tract or the skin sufrace. Compare with **endocrine gland.**

exocytosis Secretion of intracellular molecules (e.g., hormones, collagen), contained within membrane-bounded vesicles, to the outside of the cell by fusion of vesicles to the plasma membrane. Compare with **endocytosis.**

exon A region of a eukaryotic gene that is translated into a peptide or protein. Compare with **intron.**

exoskeleton A hard covering secreted on the outside of the body, used for body support, protection, and muscle attachment. Examples are the shell of molluscs and the outer covering (cuticle) of arthropods. Compare with **endoskeleton.**

exothermic Referring to a chemical reaction that releases heat. Compare with **endothermic.**

exotic species A nonnative species that is introduced into a new area. Exotic species often are competitors, pathogens, or predators of native species.

expansins A class of plant proteins that actively increase the length of the cell wall when the pH of the wall falls below 4.5.

exponential population growth The accelerating increase in the size of a population that occurs when the growth rate is constant and density independent. Compare with **logistic population growth.**

extensor A muscle that pulls two bones farther apart from each other, as in the extension of a limb or the spine. Compare with **flexor.**

extinct Said of a species that has died out.

extracellular digestion Digestion that takes place outside of an organism, as occurs in many fungi that make and secrete digestive enzymes.

extracellular matrix (ECM) A complex meshwork of proteins (e.g., collagen, fibronectin) and polysaccharides secreted by animal cells and in which they are embedded.

extremophile A bacterium or archaean that thrives in an "extreme" environment (e.g., high-salt, high-temperature, low-temperature, or low-pressure).

F_1 generation First filial generation. The first generation of offspring produced from a mating (i.e., the offspring of the parental generation).

facilitated diffusion Movement of a substance across a plasma membrane down its concentration gradient with the assistance of transmembrane carrier proteins or channel proteins.

facilitation In ecological succession, the phenomenon in which early-arriving species make conditions more favorable for later-arriving species. Compare with **inhibition** and **tolerance.**

facultative aerobe Any organism that can perform aerobic respiration when oxygen is available to serve as an electron acceptor but can switch to fermentation when it is not.

FAD/FADH$_2$ Oxidized and reduced forms, respectively, of flavin adenine dinucleotide. A nonprotein electron carrier that functions in the Krebs cycle and oxidative phosphorylation.

fallopian tube A narrow tube connecting the uterus to the ovary in humans, through which the egg travels after ovulation. Site of fertilization and cleavage. In nonhuman animals, called *oviduct.*

family In Linnaeus' system, a taxonomic category above genus and below order. In animals, family names usually end in the suffix *-idae*; Canidae (dogs, wolves and foxes) and Felidae (cats) are both in the order Carnivora.

fat A lipid consisting of three fatty acid molecules joined by ester linkages to a glycerol molecule. Also called *triacylglycerol* or *triglyceride.*

fatty acid A lipid consisting of a hydrocarbon chain bonded to a carboxyl group (–COOH) at one end. Used by many organisms to store chemical energy; a major component of animal and plant fats.

fauna All the animals characteristic of a particular region, period, or environment.

feather A specialized skin outgrowth, composed of β-keratin, present in all birds and only in birds. Used for flight, insulation, display, and other purposes.

feces The waste products of digestion.

fecundity The average number of female offspring produced by a single female in the course of her lifetime.

feedback inhibition A type of metabolic control in which high concentrations of the product of a metabolic pathway inhibit one of the enzymes early in the pathway. A form of negative feedback.

fermentation Any of several metabolic pathways that make ATP by transferring electrons from a reduced compound such as glucose to a final electron acceptor other than oxygen. Allows glycolysis to proceed in the absence of oxygen.

ferredoxin In photosynthetic organisms, an iron- and sulfur-containing protein in the electron transport chain of photosystem I. Can transfer electrons to the enzyme NADP$^+$ reductase, which catalyzes formation of NADPH.

fertilization Fusion of the nuclei of two haploid gametes to form a zygote with a diploid nucleus.

fertilization envelope A physical barrier that forms around a fertilized egg in amphibians and some other animals. Formed by an influx of water under the vitelline membrane.

fetal alcohol syndrome A condition, marked by hyperactivity, severe learning disabilities, and depression, thought to be caused by exposure of an individual to high blood alcohol concentrations during embryonic development.

fetus In live-bearing animals, the unborn offspring after the embryonic stage, which usually are developed sufficiently to be recognizable as belonging to a certain species. In humans, from 9 weeks after fertilization until birth.

fiber In plants, a type of elongated sclerenchyma cell that provides support to vascular tissue. Compare with **sclereid.**

fibronectin An abundant protein in the extracellular matrix that binds to other ECM components and to integrins in plasma membranes; helps anchor cells in place. Numerous subtypes are found in different tissues.

Fick's law of diffusion A mathematical relationship that describes the rates of gas exchange in animal respiratory systems.

fight-or-flight response Rapid physiological changes that prepare the body for emergencies. Includes increased heart rate, increased blood pressure, and decreased digestion.

filament Any thin, threadlike structure, particularly (1) the threadlike extensions of a fish's gills or (2) the slender stalk that bears the anthers in a flower.

filter feeder See **suspension feeder.**

filtrate Any fluid produced by filtration, in particular the fluid (pre-urine) in the nephrons of vertebrate kidneys.

filtration A process of removing large components from a fluid by forcing it through a filter. Occurs in a renal corpuscle of the vertebrate kidney, allowing water and small solutes to pass from the blood into the nephron.

finite rate of increase (λ) The rate of increase of a population over a given period of time. Calculated as the ending population size divided by the starting population size. Compare with **intrinsic rate of increase.**

first law of thermodynamics The principle of physics that energy is conserved in any process. Energy can be transferred and converted into different forms, but it cannot be created or destroyed.

fission (1) A form of asexual reproduction in which a prokaryotic cell divides to produce two genetically similar daughter cells by a process similar to mitosis of eukaryotic cells. Also called *binary fission.* (2) A form of asexual reproduction in which an animal splits into two or more individuals of approximately equal size; common among invertebrates.

fitness The relative ability of an individual to produce viable offspring compared with other individuals in the same population. Also called *Darwinian fitness.*

fitness trade-off See **trade-off.**

fixed action pattern (FAP) Highly stereotyped behavior pattern that occurs in a certain invariant way in a certain species. A form of innate behavior.

flaccid Limp as a result of low internal pressure (e.g., a wilted plant leaf). Compare with **turgid.**

flagellum (plural: flagella) A long, cellular projection that undulates (in eukaryotes) or rotates (in prokaryotes) to move the cell through an aqueous environment. See **axoneme.**

flatworms Members of the phylum Platyhelminthes. Distinguished by a broad, flat, unsegmented body that lacks a coelom. Flatworms belong to the lophotrochozoan branch of the protostomes.

flavin adenine dinucleotide See **FAD/FADH$_2$.**

flexor A muscle that pulls two bones closer together, as in the flexing of a limb or the spine. Compare with **extensor.**

floral meristem A group of undifferentiated plant cells that can give rise to the four organs making up a flower.

florigen In plants, a protein hormone that is synthesized in leaves and transported to the shoot apical meristem where it stimulates flowering.

flower In angiosperms, the part of a plant that contains reproductive structures. Typically includes a calyx, a corolla, and one or more stamens and/or carpels. See **perfect** and **inperfect flower.**

fluid feeder An animal that feeds by sucking or mopping up liquids such as nectar, plant sap, or blood.

fluid-mosaic model The widely accepted hypothesis that the plasma membrane and organelle membranes consist of proteins embedded in a fluid phospholipid bilayer.

fluorescence The spontaneous emission of light from an excited electron falling back to its normal (ground) state.

follicle An egg cell and its surrounding ring of supportive cells in a mammalian ovary.

follicle-stimulating hormone (FSH) A peptide hormone, produced and secreted by the anterior pituitary, that stimulates (in females) growth of eggs and follicles in the ovaries or (in males) sperm production in the testes.

follicular phase The first major phase of a menstrual cycle during which follicles grow and estrogen levels increase; ends with ovulation.

food Any nutrient-containing material that can be consumed and digested by animals.

food chain A relatively simple pathway of energy flow through a few species, each at a different trophic level, in an ecosystem. Might include, for example, a primary producer, a primary consumer, a secondary consumer, and a decomposer. Compare with **food web.**

food vacuole A membrane-bound organelle containing food ingested by a cell via phagocytosis.

food web Any complex pathway along which energy moves among many species at different trophic levels of an ecosystem.

foot One of the three main parts of the mollusc body; a muscular appendage, used for movement and/or burrowing into sediment.

forebrain One of the three main regions of the vertebrate brain; includes the cerebrum, thalamus, and hypothalamus. Compare with **hindbrain** and **midbrain.**

fossil Any trace of an organism that existed in the past. Includes tracks, burrows, fossilized bones, casts, etc.

fossil record All of the fossils that have been found anywhere on Earth and that have been formally described in the scientific literature.

founder effect A change in allele frequencies that often occurs when a new population is established from a small group of individuals (founder event) due to sampling error (i.e., the small group is not a representative sample of the source population).

free energy See **Gibbs free-energy change.**

free radical Any substance containing one or more atoms with an unpaired electron. Unstable and highly reactive.

frequency The number of wave crests per second traveling past a stationary point. Determines the pitch of sound and the color of light.

fronds The large leaves of ferns.

fruit In flowering plants (angiosperms), a mature, ripened plant ovary (or group of ovaries), along with the seeds it contains and any adjacent fused parts. See **aggregate, multiple,** and **simple fruit.**

fruiting body A structure formed in some prokaryotes, fungi, and protists for spore dispersal; usually consists of a base, a stalk, and a mass of spores at the top.

functional genomics The study of how a genome works, that is, when and where specific genes are expressed and how their products interact to produce a functional organism.

functional group A small group of atoms bonded together in a precise configuration and exhibiting particular chemical properties that it imparts to any organic molecule in which it occurs.

fundamental niche The ecological space that a species occupies in its habitat in the absence of competitors. Compare with **realized niche.**

fungi A lineage of eukaryotes that typically have a filamentous body (mycelium) and obtain nutrients by absorption.

fungicide Any substance that can kill fungi or slow their growth.

G_1 phase The phase of the cell cycle that constitutes the first part of interphase before DNA synthesis (S phase).

G_2 phase The phase of the cell cycle between synthesis of DNA (S phase) and mitosis (M phase); the last part of interphase.

gall A tumorlike growth that forms on plants infected with certain bacteria or parasites.

gallbladder A small pouch that stores bile from the liver and releases it as needed into the small intestine during digestion of fats.

gametangium (plural: gametangia) (1) The gamete-forming structure found in all land plants except angiosperms. Contains a sperm-producing antheridium and an egg-producing archegonium. (2) The gamete-forming structure of some chytrid fungi.

gamete A haploid reproductive cell that can fuse with another haploid cell to form a zygote. Most multicellular eukaryotes have two distinct forms of gametes: egg cells (ova) and sperm cells.

gametogenesis The production of gametes (eggs or sperm).

gametophyte In organisms undergoing alternation of generations, the multicellular haploid form that arises from a single haploid spore and produces gametes. A female gametophyte is commonly called an *embryo sac;* a male gametophyte, a *pollen grain.* Compare with **sporophyte.**

ganglion cell A neuron in the vertebrate retina that collects visual information from one or several bipolar cells and sends it to the brain via the optic nerve.

gap junction A type of cell-cell attachment structure that directly connects the cytoplasms of adjacent animal cells, allowing passage of water, ions, and small molecules between the cells. Compare with **desmosome** and **tight junction.**

gastrin A hormone produced by cells in the stomach lining in response to the arrival of food or to a neural signal from the brain. Stimulates other stomach cells to release hydrochloric acid.

gastrointestinal (GI) tract See **digestive tract.**

gastropods A lineage of mollusc distinguished by a large muscular foot and a unique feeding structure, the radula. Include slugs and snails.

gastrulation The process by which some cells on the outside of a young embryo move to the interior of the embryo, resulting in the three distinct germ layers (endoderm, mesoderm, and ectoderm).

gated channel A channel protein that opens and closes in response to a specific stimulus, such as the binding of a particular molecule or a change in the electrical charge on the outside of the membrane.

gel electrophoresis A technique for separating molecules on the basis of size and electric charge, which affect their differing rates of movement through a gelatinous substance in an electric field.

gemma (plural: gemmae) A small reproductive structure that is produced in some liverworts during the gametophyte phase and can grow into mature gametophyte.

gene A section of DNA (or RNA, for some viruses) that encodes information for building one or more related polypeptides or functional RNA molecules along with the regulatory sequences required for its transcription.

gene duplication The formation of an additional copy of a gene, typically by misalignment of chromosomes during crossing over. Thought to be an important evolutionary process in creating new genes.

gene expression Overall process by which the information encoded in genes is converted into an active product, most commonly a protein. Includes transcription and translation of a gene and in some cases protein activation.

gene family A set of genetic loci whose DNA sequences are extremely similar. Thought to have arisen by duplication of a single ancestral gene and subsequent mutations in the duplicated sequences.

gene flow The movement of alleles between populations; occurs when individuals leave one population, join another, and breed.

gene-for-gene hypothesis The hypothesis that there is a one-to-one correspondence between the resistance (*R*) loci of plants and the avirulence (*avr*) loci of pathogenic fungi; particularly, that *R* genes produce receptors and *avr* genes produce molecules that bind to those receptors.

gene pool All of the alleles of all of the genes in a certain population.

generation The average time between a mother's first offspring and her daughter's first offspring.

gene therapy The treatment of an inherited disease by introducing normal alleles.

genetic bottleneck A reduction in allelic diversity resulting from a sudden reduction in the size of a large population (population bottleneck) due to a random event.

genetic code The set of all 64 codons and the particular amino acids that each specifies.

genetic correlation A type of evolutionary constraint in which selection on one trait causes a change in another trait as well; may occur when the same gene(s) affect both traits.

genetic diversity The diversity of alleles in a population, species, or group of species.

genetic drift Any change in allele frequencies due to random events. Causes allele frequencies to drift up and down randomly over time, and eventually can lead to the fixation or loss of alleles.

genetic homology Similarities in DNA sequences or amino acid sequences that are due to inheritance from a common ancestor.

genetic map An ordered list of genes on a chromosome that indicates their relative distances from each other. Also called a *linkage map* or *meiotic map*. Compare with **physical map.**

genetic marker A genetic locus that can be identified and traced in populations by laboratory techniques or by a distinctive visible phenotype.

genetic model A set of hypotheses that explain how a certain trait is inherited.

genetic recombination A change in the combination of genes or alleles on a given chromosome or in a given individual. Also called *recombination.*

genetics The field of study concerned with the inheritance of traits.

genetic screen Any of several techniques for identifying individuals with a particular type of mutation. Also called a *screen.*

genetic variation (1) The number and relative frequency of alleles present in a particular population. (2) The proportion of phenotypic variation in a trait that is due to genetic rather than environmental influences in a certain population in a certain environment.

genitalia External copulatory organs.

genome All of the hereditary information in an organism, including not only genes but also other non-gene stretches of DNA.

genomic library A set of DNA segments representing the entire genome of a particular organism. Each segment is carried by a plasmid or other cloning vector and can be separated from other segments. Compare with **cDNA library.**

genomics The field of study concerned with sequencing, interpreting, and comparing whole genomes from different organisms.

genotype All of the alleles of every gene present in a given individual. May refer specifically to the alleles of a particular set of genes under study. Compare with **phenotype.**

genus (plural: genera) In Linnaeus' system, a taxonomic category of closely related species. Always italicized and capitalized to indicate that it is a recognized scientific genus.

geologic time scale The sequence of eons, epochs, and periods used to describe the geologic history of Earth.

geometric isomer A molecule that shares the same molecular formula as another molecule but differs in the arrangement of atoms or groups on either side of a double bond or ring structure. Compare with **optical isomer** and **structural isomer.**

germ cell In animals, any cell that can potentially give rise to gametes. Also called *germ-line cells.*

germination The process by which a seed becomes a young plant.

germ layer In animals, one of the three basic types of tissue formed during gastrulation; gives rise to all other tissues. See **endoderm, mesoderm,** and **ectoderm.**

germ theory of disease The theory that infectious diseases are caused by bacteria, viruses, and other microorganisms.

gestation The duration of embryonic development from fertilization to birth in those species that have live birth.

gibberellins A class of hormones found in plants and fungi that stimulate growth. Gibberellic acid is one of the major gibberellins.

Gibbs free-energy change (ΔG) A measure of the change in potential energy and entropy that occurs in a given chemical reaction. $\Delta G < 0$ for spontaneous reactions and >0 for nonspontaneous reactions.

gill Any organ in aquatic animals that exchanges gases and other dissolved substances between the blood and the surrounding water. Typically, a filamentous outgrowth of a body surface.

gill arch In aquatic vertebrates, curved region of tissue between the gills. Gills are suspended from the gill arches.

gill filament In fish, one of the many long, thin structures that extend from gill arches into the water and across which gas exchange occurs.

gill lamella (plural: gill lamellae) One of hundreds to thousands of sheetlike structures, each containing a capillary bed, that makes up a gill filament.

gland An organ whose primary function is to secrete some substance, either into the blood (endocrine gland) or into some other space such as the gut or skin (exocrine gland).

glia Collective term for several types of cells in nervous tissue that are not neurons and do not conduct electrical signals but provide support, nourishment, and electrical insulation and perform other functions. Also called *glial cells.*

global warming A sustained increase in Earth's average surface temperature.

glomerulus (plural: glomeruli) (1) In the vertebrate kidney, a ball-like cluster of capillaries, surrounded by Bowman's capsule, at the beginning of a nephron. (2) In the brain, a ball-shaped cluster of neurons in the olfactory bulb.

glucagon A peptide hormone produced by the pancreas in response to low blood glucose. Raises blood glucose by triggering breakdown of glycogen and stimulating gluconeogenesis. Compare with **insulin.**

glucocorticoids A class of steroid hormones, produced and secreted by the adrenal cortex, that increase blood glucose and prepare the body for stress. Include cortisol and corticosterone. Compare with **mineralocorticoids.**

gluconeogenesis Synthesis of glucose from non-carbohydrate sources (e.g., proteins and fatty acids). Occurs in the liver in response to low insulin levels and high glucagon levels.

glucose Six-carbon monosaccharide whose oxidation in cellular respiration is the major source of ATP in animal cells.

glyceraldehyde-3-phosphate (G3P) The phosphorylated three-carbon compound formed as the result of carbon fixation in the first step of the Calvin cycle.

glycerol A three-carbon molecule that forms the "backbone" of phospholipids and most fats.

glycogen A highly branched storage polysaccharide composed of α-glucose monomers joined by 1,4- and 1,6-glycosidic linkages. The major form of stored carbohydrate in animals.

glycolipid Any lipid molecule that is covalently bonded to a carbohydrate group.

glycolysis A series of 10 chemical reactions that oxidize glucose to produce pyruvate and ATP. Used by all organisms as part of fermentation or cellular respiration.

glycoprotein Any protein with one or more covalently bonded carbohydrate groups.

glycosidic linkage The covalent bond formed by a condensation reaction between two sugar monomers; joins the residues of a polysaccharide.

glycosylation Addition of a carbohydrate group to a molecule.

glyoxisome Specialized type of peroxisome found in plant cells and packed with enzymes for processing the products of photosynthesis.

goblet cell A cell in the stomach lining that secretes mucus.

goiter A pronounced swelling of the thyroid gland; usually caused by a deficiency of iodine in the diet.

Golgi apparatus A eukaryotic organelle, consisting of stacks of flattened membranous sacs (cisternae), that functions in processing and sorting proteins and lipids destined to be secreted or directed to other organelles. Also called *Golgi complex.*

gonad An organ that produces reproductive cells, such as a testis or an ovary.

gonadotropin-releasing hormone (GnRH) A peptide hormone, produced and secreted by the hypothalamus, that stimulates release of FSH and LH from the anterior pituitary.

G protein Any of various peripheral membrane proteins that bind GTP and function in signal transduction. Binding of a signal to its receptor triggers activation of the G protein, leading to production of a second messenger or initiation of a phosphorylation cascade.

grade In taxonomy, a group of species that share a position in an inferred evolutionary sequence of lineages but that are not a monophyletic group. Also called a *paraphyletic group.*

Gram-negative Describing bacteria that look pink when treated with a Gram stain. These bacteria have a cell wall composed of a thin layer of peptidoglycan and an outer phospholipid layer.

Gram-positive Describing bacteria that look purple when treated with a Gram stain. These bacteria have cell walls composed of a thick layer of peptidoglycan.

Gram stain A dye that distinguishes the two general types of cell walls found in bacteria. Used to routinely classify bacteria as Gram-negative or Gram-positive.

granum (plural: grana) In chloroplasts, a stack of flattened, membrane-bound vesicles (thylakoids) where the light reactions of photosynthesis occur.

gravitropism The growth or movement of a plant in a particular direction in response to gravity.

grazing food chain The ecological network of herbivores and the predators and parasites that consume them.

great apes See **hominids.**

green algae A paraphyletic group of photosynthetic organisms that contain chloroplasts similar to those in green plants. Often classified as protists, green algae are the closest living relatives of land plants and form a monophyletic group with them.

greenhouse gas An atmospheric gas that absorbs and reflects infrared radiation, so that heat radiated from Earth is retained in the atmosphere instead of being lost to space.

green plant A member of a lineage of eukaryotes that includes green algae and land plants.

gross photosynthetic productivity The efficiency with which all the plants in a given area use the light energy available to them to produce sugars.

gross primary productivity In an ecosystem, the total amount of carbon fixed by photosynthesis, including that used for cellular respiration, over a given time period. Compare with **net primary productivity.**

ground meristem The middle layer of a young plant embryo. Gives rise to the ground tissue system.

ground tissue system In plants, all the tissues beneath the outer protective layers of epidermis and cork except for vascular tissue. Also called simply *ground tissue.*

groundwater Any water below the land surface.

growth factor Any of a large number of signaling molecules that are secreted by certain cells and that stimulate other cells to divide or to differentiate.

growth hormone (GH) A peptide hormone, produced and secreted by the mammalian anterior pituitary, that promotes lengthening of the long bones in children and muscle growth, tissue repair, and lactation in adults. Also called *somatotropin.*

GTP See **guanosine triphosphate.**

guanosine triphosphate (GTP) A molecule consisting of guanine, a sugar, and three phosphate groups. Can be hydrolyzed to release free energy. Commonly used in many cellular reactions; also functions in signal transduction in association with G proteins.

guard cell One of two specialized, crescent-shaped cells forming the border of a plant stoma. Guard cells can change shape to open or close the stoma. See also **stoma.**

gustation The perception of taste.

guttation Excretion of water droplets from plant leaves in the early morning, caused by root pressure.

gymnosperm A vascular plant that makes seeds but does not produce flowers. The gymnosperms include four lineages of green plants (cycads, ginkgoes, conifers, and gnetophytes). Compare with **angiosperm.**

habitat destruction Human-caused destruction of a natural habitat with replacement by an urban, suburban, or agricultural landscape.

habitat fragmentation The breakup of a large region of a habitat into many smaller regions, separated from others by a different type of habitat.

Hadley cell An atmospheric cycle of large-scale air movement in which warm equatorial air rises, moves north or south, and then descends at approximately 30°N or 30°S latitude.

hair cell A pressure-detecting sensory cell, found in the cochlea, that has tiny "hairs" (stereocilia) jutting from its surface.

hairpin A secondary structure in RNA consisting of a stable loop formed by hydrogen bonding between purine and pyrimidine bases on the same strand.

halophile A bacterium or archaean that thrives in high-salt environments.

halophyte A plant that thrives in salty habitats.

Hamilton's rule The proposition that an allele for altruistic behavior will be favored by natural selection only if $Br > C$, where B = the fitness benefit to the recipient, C = the fitness cost to the actor, and r = the coefficient of relatedness between recipient and actor.

haploid (1) Having one set of chromosomes ($1n$). (2) A cell or an individual organism with one set of chromosomes. Compare with **diploid.**

haploid number The number of distinct chromosome sets in a cell. Symbolized as n.

haplotype The set of alleles found on a single chromosome.

Hardy-Weinberg principle A principle of population genetics stating that genotype frequencies in a large population do not change from generation to generation in the absence of evolutionary processes (e.g., mutation, migration, genetic drift, random mating, and selection).

heart A muscular pump that circulates blood throughout the body.

heart murmur A distinctive sound caused by backflow of blood through a defective heart valve.

heartwood The older xylem in the center of an older stem or root, containing protective compounds and no longer functioning in water transport.

heat Thermal energy that is transferred from an object at higher temperature to one at lower temperature.

heat of vaporization The energy required to vaporize 1 gram of a liquid into a gas.

heavy chain The larger of the two types of polypeptide chains in an antibody molecule; composed of a variable (V) region, which contributes to the antigen-binding site, and a constant (C) region. Differences in heavy-chain constant regions determine the different classes of immunoglobulins (IgA, IgE, etc.). Compare with **light chain.**

helicase An enzyme that catalyzes the breaking of hydrogen bonds between nucleotides of DNA, "unzipping" a double-stranded DNA molecule.

helix-turn-helix motif A motif seen in many repressor proteins in prokaryotes, consisting of two α-helices connected by a short stretch of amino acids that form a turn.

helper T cell An effector T cell that secretes cytokines and in other ways promotes the activation of other lymphocytes. Is descended from an activated $CD4^+$ T cell that has interacted with antigen presented by dendritic cells, macrophages, or B cells.

hemimetabolous metamorphosis A type of metamorphosis in which the animal increases in size from one stage to the next, but does not dramatically change its body form. Also called *incomplete metamorphosis.*

hemocoel A body cavity, present in arthropods and some molluscs, containing a pool of circulatory fluid (hemolymph) bathing the internal organs.

hemoglobin An oxygen-binding protein consisting of four polypeptide subunits, each containing an oxygen-binding heme group. The major oxygen carrier in mammalian blood.

hemolymph The circulatory fluid of animals with open circulatory systems (e.g., insects) in which the fluid is not confined to blood vessels.

hemophilia A human disease, caused by an X-linked recessive allele, that is characterized by defects in the blood-clotting system.

herbaceous Referring to a plant that is not woody.

herbivore (adjective: herbivorous) An animal that eats primarily plants and rarely or never eats meat. Compare with **carnivore** and **omnivore.**

herbivory The practice of eating plant tissues.

heredity The transmission of traits from parents to offspring via genetic information.

heritable Referring to traits that can be transmitted from one generation to the next.

hermaphrodite An organism that produces both male and female gametes.

heterokaryotic Describing a cell or fungal mycelium containing two or more nuclei that are genetically distinct.

heterospory (adjective: heterosporous) In seed plants, the production of two distinct types of spore-producing structures and thus two distinct types of spores: microspores, which become the male gametophyte, and megaspores, which become the female gametophyte. Compare with **homospory.**

heterotherm An animal whose body temperature varies markedly with environmental conditions. Compare with **homeotherm.**

heterotroph Any organism that cannot synthesize reduced organic compounds from inorganic sources and that must obtain them by eating other organisms. Some bacteria, some archaea, and virtually all fungi and animals are heterotrophs. Also called *consumer.* Compare with **autotroph.**

heterozygote advantage A pattern of natural selection that favors heterozygous individuals compared with homozygotes. Tends to maintain genetic variation in a population. Also called *heterozygote superiority.*

heterozygous Having two different alleles of a certain gene.

hexose A monosaccharide (simple sugar) containing six carbon atoms.

hibernation An energy-conserving physiological state, marked by a decrease in metabolic rate, body temperature, and activity, that lasts for a prolonged period (weeks to months). Occurs in some animals in response to winter cold and scarcity of food. Compare with **torpor.**

hindbrain One of the three main regions of the vertebrate brain; includes the cerebellum and medulla oblongata. Compare with **forebrain** and **midbrain.**

histamine A molecule released from mast cells during an inflammatory response that causes blood vessels to dilate and become more permeable.

histone One of several positively charged (basic) proteins associated with DNA in the chromatin of eukaryotic cells.

histone acetyl transferase (HAT) In eukaryotes, one of a class of enzymes that loosen chromatin structure by adding acetyl groups to histone proteins.

histone deacetylase (HDAC) In eukaryotes, one of a class of enzymes that recondense chromatin by removing acetyl groups from histone proteins.

HIV See human immunodeficiency virus (HIV).

holoenzyme A multipart enzyme consisting of a core enzyme (containing the active site for catalysis) along with other required proteins.

holometabolous metamorphosis A type of metamorphosis in which the animal completely changes its form. Also called *complete metamorphosis*.

homeosis Replacement of one body part by another normally found elsewhere in the body as the result of mutation in certain developmentally important genes (homeotic genes).

homeostasis (adjective: homeostatic) The array of relatively stable chemical and physical conditions in an animal's cells, tissues, and organs. May be achieved by the body's passively matching the conditions of a stable external environment (conformational homeostasis) or by active physiological processes (regulatory homeostasis) triggered by variations in the external or internal environment.

homeotherm An animal that has a constant or relatively constant body temperature. Compare with **heterotherm.**

homeotic gene Any gene that specifies a particular location within an embryo, leading to the development of structures appropriate for that location. Mutations in homeotic genes cause the development of extra body parts or body parts in the wrong places.

hominids Members of the family Hominidae, which includes humans and extinct related forms; chimpanzees, gorillas, and orangutans. Distinguished by large body size, no tail, and an exceptionally large brain. Also called *great apes*.

hominins Humans and extinct related forms; species in the lineage that branched off from chimpanzees and eventually led to humans.

homologous chromosomes In a diploid organism, chromosomes that are similar in size, shape, and gene content. Also called *homologs*.

homology (adjective: homologous) Similarity among organisms of different species due to their inheritance from a common ancestor. Features that exhibit such similarity (e.g., DNA sequences, proteins, body parts) are said to be homologous. Compare with **homoplasy.**

homoplasy Similarity among organisms of different species due to convergent evolution. Compare with **homology.**

homospory (adjective: homosporous) In seedless vascular plants, the production of just one type of spore. Compare with **heterospory.**

homozygous Having two identical alleles of a certain gene.

hormone Any of numerous different signaling molecules that circulate throughout the body in blood or other body fluids and can trigger characteristic responses in distant target cells at very low concentrations.

hormone-response element A specific sequence in DNA to which a steroid hormone-receptor complex can bind and affect gene transcription.

host An individual or a species in or on which a parasite lives.

host cell A cell that has been invaded by an organism such as a parasite or a virus.

***Hox* genes** A class of homeotic genes found in several animal phyla, including vertebrates, that are expressed in a distinctive pattern along the anterior-posterior axis in early embryos and control formation of segment-specific structures.

human Any member of the genus *Homo*, which includes modern humans (*Homo sapiens*) and several extinct species.

human chorionic gonadotropin (hCG) A glycoprotein hormone produced by the human placenta from about week 3 to week 14 of pregnancy. Maintains the corpus luteum, which produces hormones that preserve the uterine lining.

Human Genome Project The multinational research project that sequenced the human genome.

human immunodeficiency virus (HIV) A retrovirus that causes AIDS (acquired immune deficiency syndrome) in humans.

humoral (immune) response The type of immune response that involves generation of antibody-secreting plasma cells from activated B cells. Defends against extracellular pathogens. Compare with **cell-mediated (immune) response.**

humus The completely decayed organic matter in soils.

Huntington's disease A degenerative brain disease of humans caused by an autosomal dominant allele.

hybrid The offspring of parents from two different strains, populations, or species.

hybrid zone A geographic area where interbreeding occurs between two species, sometimes producing fertile hybrid offspring.

hydrocarbon An organic molecule that contains only hydrogen and carbon atoms.

hydrogen bond A weak interaction between two molecules or different parts of the same molecule resulting from the attraction between a hydrogen atom with a partial positive charge and another atom (usually O or N) with a partial negative charge. Compare with **covalent bond** and **ionic bond.**

hydrogen ion (H^+) A single proton with a charge of 1+; typically, one that is dissolved in solution or that is being transferred from one atom to another in a chemical reaction.

hydrolysis A chemical reaction in which a molecule is split into smaller molecules by reacting with water. In biology, most hydrolysis reactions involve the splitting of polymers into monomers. Compare with **condensation reaction.**

hydrophilic Interacting readily with water. Hydrophilic compounds are typically polar compounds containing charged or electronegative atoms. Compare with **hydrophobic.**

hydrophobic Not interacting readily with water. Hydrophobic compounds are typically nonpolar compounds that lack charged or electronegative atoms and often contain many C–C and C–H bonds. Compare with **hydrophilic.**

hydroponic growth Growth of plants in liquid cultures instead of soil.

hydrostatic skeleton A system of body support involving fluid-filled compartments that can change in shape but cannot easily be compressed.

hydroxide ion (OH^-) An oxygen atom and a hydrogen atom joined by a single covalent bond and carrying a negative charge; formed by dissociation of water.

hyperpolarization Change in membrane potential from its resting negative state to an even more negative state; a normal phase in an action potential. Compare with **depolarization.**

hypersensitive response In plants, the rapid death of a cell that has been infected by a pathogen, thereby reducing the potential for infection to spread throughout a plant. Compare with **systemic acquired resistance.**

hypertension Abnormally high blood pressure.

hypertonic Comparative term designating a solution that has a greater solute concentration, and therefore a lower water concentration, than another solution. Compare with **hypotonic** and **isotonic.**

hypha (plural: hyphae) One of the strands of a fungal mycelium (the meshlike body of a fungus). Also found in some protists.

hypocotyl The stem of a very young plant; the region between the cotyledon (embryonic leaf) and the radicle (embryonic root).

hypothalamic-pituitary axis The functional interaction of the hypothalamus and the pituitary gland, which are anatomically distinct but work together to regulate most of the other endocrine glands in the body.

hypothalamus A part of the brain that functions in maintaining the body's internal physiological state by regulating the autonomic nervous system, endocrine system, body temperature, water balance, and appetite.

hypothesis A proposed explanation for a phenomenon or for a set of observations.

hypotonic Comparative term designating a solution that has a lower solute concentration, and therefore a higher water concentration, than another solution. Compare with **hypertonic** and **isotonic.**

immigration The migration of individuals into a particular population from other populations. Compare with **emigration.**

immune system In vertebrates, the system whose primary function is to defend the body against pathogens. Includes several types of cells (e.g., lymphocytes and macrophages) and several organs where they develop or reside (e.g., lymph nodes and thymus).

immunity (adjective: immune) State of being protected against infection by disease-causing pathogens either by relatively nonspecific mechanisms (innate immunity) or by specific mechanisms triggered by exposure to a particular antigen (acquired immunity).

immunization The conferring of immunity to a particular disease by artificial means.

immunoglobulin (Ig) Any of the class of proteins that function as antibodies.

immunological memory The ability of the immune system to "remember" an antigen and mount a rapid, effective response to a pathogen encountered years or decades earlier.

imperfect flower A flower that contains male parts (stamens) *or* female parts (carpels) but not both. Compare with **perfect flower.**

implantation The process by which an embryo buries itself in the uterine wall and forms a placenta. Occurs in mammals and a few other vertebrates.

imprinting A type of rapid, irreversible learning in which a young animal learns to recognize the individual caring for it. Occurs in birds and mammals.

inbreeding Mating between closely related individuals. Increases homozygosity of a population and often leads to a decline in the average fitness (inbreeding depression).

incomplete dominance An inheritance pattern in which the heterozygote phenotype is a blend or combination of both homozygote phenotypes.

incomplete metamorphosis See **hemimetabolous metamorphosis.**

independent assortment, principle of The concept that each pair of hereditary elements (alleles of the same gene) behaves independently of other genes during meiosis. One of Mendel's two principles of genetics.

indeterminate growth A pattern of growth in which an individual continues to increase its overall body size throughout its life.

indicator plate A laboratory technique for detecting mutant cells by growing them on agar plates containing a compound that when metabolized by wild-type cells yields a colored product.

induced fit Change in the shape of the active site of an enzyme, as the result of the initial weak binding of a substrate, so that it binds substrate more tightly.

inducer A small molecule that triggers transcription of a specific gene, often by binding to and inactivating a repressor protein.

inducible defense A defensive trait that is manifested only in response to the presence of a predator or pathogen. Compare with **constitutive defense.**

induction (1) The process by which one embryonic cell, or group of cells, alters the differentiation of neighboring cells. (2) Positive control of gene expression by a regulatory protein that binds to DNA and triggers transcription of a specific gene(s).

infection thread An invagination of the membrane of a root hair through which beneficial nitrogen-fixing bacteria enter the roots of their host plants (legumes).

inflammatory response An aspect of the innate immune response, seen in most cases of infection or tissue injury, in which the affected tissue becomes swollen, red, warm, and painful.

inhibition In ecological succession, the phenomenon in which early-arriving species make conditions less favorable for the establishment of certain later-arriving species. Compare with **facilitation** and **tolerance.**

inhibitory postsynaptic potential (IPSP) A change in membrane potential, usually hyperpolarization, at a neuron dendrite that makes an action potential less likely.

initiation (1) In an enzyme-catalyzed reaction, the stage during which enzymes orient reactants precisely as they bind at specific locations within the enzyme's active site. (2) In DNA transcription, the stage during which RNA polymerase and other proteins assemble at the promoter sequence. (3) In RNA translation, the stage during which a complex consisting of a ribosome, a mRNA molecule, and an aminoacyl tRNA corresponding to the start codon is formed.

initiation factors A class of proteins that assist ribosomes in binding to a messenger RNA molecule to begin translation.

innate behavior Behavior that is inherited genetically, does not have to be learned, and is typical of a species.

innate immune response See **innate immunity.**

innate immunity A set of nonspecific defenses against pathogens that exist before exposure to an antigen and involves mast cells, neutrophils, and macrophages; typically results in an inflammatory response. Compare with **acquired immunity.**

inner cell mass (ICM) A cluster of cells in the interior of a mammalian blastocyst that undergo gastrulation and eventually develop into the embryo.

inner ear The innermost portion of the mammalian ear, consisting of a fluid-filled system of tubes that includes the cochlea (which receives sound vibrations from the middle ear) and the semicircular canals (which function in balance).

insects The largest lineage of arthropods distinguished by having three body regions (a head, thorax, and abdomen), walking legs, and in most species one or two pairs of wings.

in situ hybridization A technique for detecting specific DNAs and mRNAs in cells and tissues by use of labeled probes. Can be used to determine where and when particular genes are expressed in embryos.

insulin A peptide hormone produced by the pancreas in response to high levels of glucose (or amino acids) in blood. Enables cells to absorb glucose and coordinates synthesis of fats, proteins, and glycogen. Compare with **glucagon.**

integral membrane protein Any membrane protein that spans the entire lipid bilayer. Also called *transmembrane protein.* Compare with **peripheral membrane protein.**

integration In the nervous system, processing of information from many sources.

integrator A component of an animal's nervous system that functions as part of a homeostatic system by evaluating sensory information and triggering appropriate responses. See **effector** and **sensor.**

integrin Any of a class of cell-surface proteins that bind to fibronectins and other proteins in the extracellular matrix, thus holding cells in place.

intercalated disc A specialized junction between adjacent heart muscle cells that contains gap junctions, allowing electrical signals to pass between the cells.

intermediate disturbance hypothesis The hypothesis that moderate ecological disturbance is associated with higher species diversity than either low or high disturbance.

intermediate filament A long fiber, about 10 nm in diameter, composed of one of various proteins (e.g., keratins, lamins); one of the three types of cytoskeletal fibers. Form networks that help maintain cell shape and hold the nucleus in place. Compare with **actin filament** and **microtubule.**

intermediate host The host species in which a parasite reproduces asexually. Compare with **definitive host.**

interneuron A neuron that passes signals from one neuron to another. Compare with **motor neuron** and **sensory neuron.**

internode The section of a plant stem between two nodes (sites where leaves attach).

interphase The portion of the cell cycle between one mitotic (M) phase and the next. Includes the G_1 phase, S phase, and G_2 phase.

interspecific competition Competition between members of different species for the same limited resource. Compare with **intraspecific competition.**

interstitial fluid The plasma-like fluid found in the region (interstitial space) between cells.

intertidal zone The region between the low-tide and high-tide marks on a seashore.

intraspecific competition Competition between members of the same species for the same limited resource. Compare with **interspecific competition.**

intrinsic rate of increase (r_{max}) The rate at which a population will grow under optimal conditions (i.e., when birthrates are as high as possible and death rates are as low as possible). Compare with **finite rate of increase.**

intron A region of a eukaryotic gene that is transcribed into RNA but is later removed, so it is not translated into a peptide or protein. Compare with **exon.**

invasive species An exotic (nonnative) species that, upon introduction to a new area, spreads rapidly and competes successfully with native species.

inversion A mutation in which a segment of a chromosome breaks from the rest of the chromosome, flips, and rejoins with the opposite orientation as before.

invertebrates A paraphyletic group composed of animals without a backbone; includes about 95 percent of all animal species. Compare with **vertebrates.**

ion An atom or a molecule that has lost or gained electrons and thus carries an electric charge, either positive (cation) or negative (anion), respectively.

ion channel A type of channel protein that allows certain ions to diffuse across a plasma membrane down an electochemical gradient.

ionic bond A chemical bond that is formed when an electron is completely transferred from one atom to another so that the atoms remain associated due to their opposite electric charges. Compare with **covalent bond** and **hydrogen bond.**

iris A ring of pigmented muscle just below the cornea in the vertebrate eye that contracts or expands to control the amount of light entering the eye through the pupil.

islets of Langerhans Clusters of cells in the pancreas that secrete insulin and glucagon directly into the blood.

isomer A molecule that has the same molecular formula as another molecule but differs from it in three-dimensional structure.

isotonic Comparative term designating a solution that has the same solute concentration and water concentration than another solution. Compare with **hypertonic** and **hypotonic.**

isotope Any of several forms of an element that have the same number of protons but differ in the number of neutrons.

joint A place where two components (bones, cartilages, etc.) of a skeleton meet. May be movable (an articulated joint) or immovable (e.g., skull sutures).

juvenile hormone An insect hormone that prevents larvae from metamorphosing into adults.

karyogamy Fusion of two haploid nuclei to form a diploid nucleus. Occurs in many fungi, and in animals and plants during fertilization of gametes.

karyotype The distinctive appearance of all of the chromosomes in an individual, including the number of chromosomes and their length and banding patterns (after staining with dyes).

keystone species A species that has an exceptionally great impact on the other species in its ecosystem relative to its abundance.

kidney In terrestrial vertebrates, one of a paired organ situated at the back of the abdominal cavity that filters the blood, produces urine, and secretes several hormones.

kilocalorie (kcal) A unit of energy often used to measure the energy content of food. Also called *Calorie.*

kinesin Any one of a class of motor proteins that use the chemical energy of ATP to transport vesicles, particles, or chromosomes along microtubules.

kinetic energy The energy of motion. Compare with **potential energy.**

kinetochore A protein structure at the centromere where spindle fibers attach to the sister chromatids of a replicated chromosome. Contains motor proteins that move a chromosome along a microtubule.

kingdom In Linnaeus' system, a taxonomic category above the phylum level and below the domain level.

kinocilium (plural: kinocilia) A single cilium that juts from the surface of many hair cells and functions in detection of sound or pressure.

kin selection A form of natural selection that favors traits that increase survival or reproduction of an individual's kin at the expense of the individual.

Klinefelter syndrome A syndrome seen in humans who have an XXY karyotype. People with this syndrome have male sex organs, may have some female traits, and are sterile.

knock-out mutant A mutant allele that does not function at all, or an organism homozygous for such a mutation. Also called *null mutant* or *loss-of-function mutant.*

Koch's postulates Four criteria used to determine whether a suspected infectious agent causes a particular disease.

Krebs cycle A series of chemical reactions, occurring in mitochondria, in which acetyl CoA is oxidized to CO_2, producing ATP and reduced compounds for the electron transport chain. Also called *citric acid cycle.*

labia major (plural: labium majus) One of two outer folds of skin that protect the labia minora, clitoris, and vaginal opening of female mammals.

labia minora (plural: labium minus) One of two inner folds of skin that protect the opening of the urethra and vagina.

labor The strong muscular contractions of the uterus that expel the fetus during birth.

lactation (verb: lactate) Production of milk from mammary glands of mammals.

lacteal A small lymphatic vessel extending into the center of a villus in the small intestine. Receives chylomicrons containing fat absorbed from food.

lactic acid fermentation Catabolic pathway in which pyruvate produced by glycolysis is converted to lactic acid in the absence of oxygen.

lagging strand In DNA replication, the strand of new DNA that is synthesized discontinuously in a series of short pieces that are later joined together. Also called *discontinuous strand.* Compare with **leading strand.**

lamellae (singular: lamella) Any set of parallel platelike structures (e.g., the crescent-shaped flaps on the gill filaments of fish gills that serve to increase surface area for gas exchange).

large intestine The distal portion of the digestive tract consisting of the cecum, colon, and rectum. Its primary function is to compact the wastes delivered from the small intestine and absorb enough water to form feces.

larva (plural: larvae) An immature stage of a species in which the immature and adult stages have different body forms.

late endosome A membrane-bound vesicle that arises from an early endosome and develops into a lysosome.

lateral bud A bud that forms in the angle between a leaf and a stem and may develop into a lateral (side) branch. Also called *axillary bud.*

lateral gene transfer Transfer of DNA between two different species, especially distantly related species. Commonly occurs among bacteria and archaea via plasmid exchange; also can occur in eukaryotes via viruses and some other mechanisms.

lateral meristem A layer of undifferentiated plant cells found in older stems and roots that is responsible for secondary growth. Also called *cambium* or *secondary meristem.* Compare with **apical meristem.**

lateral root A plant root extending from another, older root.

leaching Loss of nutrients from soil via percolating water.

leading strand In DNA replication, the strand of new DNA that is synthesized in one continuous piece, with nucleotides added to the 3′ end of the growing molecule. Also called *continuous strand.* Compare with **lagging strand.**

leaf The main photosynthetic organ of vascular plants.

leak channel Potassium channel that allows potassium ions to leak out of a neuron in its resting state.

learning An enduring change in an individual's behavior that results from specific experience(s).

leghemoglobin An iron-containing protein similar to hemoglobin. Found in root nodules of legume plants where it binds oxygen, preventing it from poisoning a bacterial enzyme needed for nitrogen fixation.

legumes Members of the pea plant family that form symbiotic associations with nitrogen-fixing bacteria in their roots.

lens A transparent, crystalline structure that focuses incoming light onto a retina or other light-sensing apparatus of an eye.

leptin A hormone produced and secreted by fat cells (adipocytes) that acts to stabilize fat tissue mass in part by inhibiting appetite and increasing energy expenditure.

leukocytes Several types of blood cells, including neutrophils, macrophages, and lymphocytes, that circulate in blood and lymph and function in defense against pathogens. Also called *white blood cells.*

lichen A symbiotic association of a fungus and a photosynthetic alga.

life cycle The sequence of developmental events and phases that occurs during the life span of an organism, from fertilization to offspring production.

life history The sequence of events in an individual's life from birth to reproduction to death, including how an individual allocates resources to growth, reproduction, and activities or structures that are related to survival.

life table A data set that summarizes the probability that an individual in a certain population will survive and reproduce in any given year over the course of its lifetime.

ligand Any molecule that binds to a specific site on a receptor molecule.

ligand-gated channel An ion channel that opens or closes in response to binding by a certain molecule. Compare with **voltage-gated channel.**

light chain The smaller of the two types of polypeptide chains in an antibody molecule; composed of a variable (V) region, which contributes to the antigen-binding site, and a constant (C) region. Compare with **heavy chain.**

light-dependent reactions In photosynthesis, the set of reactions, occurring in photosystem I and II, that use the energy of sunlight to split water, producing ATP, NADPH, and oxygen.

light-independent reactions See **Calvin cycle.**

lignin A substance found in the secondary cell walls of some plants that is exceptionally stiff and strong. Most abundant in woody plant parts.

limiting nutrient Any essential nutrient whose scarcity in the environment significantly reduces growth and reproduction of organisms.

limnetic zone Open water (not near shore) that receives enough light to support photosynthesis.

lineage See **monophyletic group.**

LINEs (long interspersed nuclear elements) The most abundant class of transposable elements in human genomes; can create copies of itself and insert them elsewhere in the genome. Compare with **SINEs.**

linkage In genetics, a physical association between two genes because they are on the same chromosome; the inheritance patterns resulting from this association.

linkage map See **genetic map.**

lipase Any enzyme that can break down fat molecules into fatty acids and monoglycerides.

lipid Any organic subtance that does not dissolve in water, but dissolves well in nonpolar organic solvents. Lipids include fats, oils, phospholipids, and waxes.

lipid bilayer The basic structural element of all cellular membranes consisting of a two-layer sheet of phospholipid molecules whose hydrophobic tails are oriented toward the inside and hydrophilic heads, toward the outside. Also called *phospholipid bilayer.*

littoral zone Shallow water near shore that receives enough sunlight to support photosynthesis. May be marine or freshwater; often flowering plants are present.

liver A large, complex organ of vertebrates that performs many functions including storage of glycogen, processing and conversion of food and wastes, and production of bile.

locomotion Movement of an organism under its own power.

locus (plural: loci) A gene's physical location on a chromosome.

logistic population growth The density-dependent decrease in growth rate as population size approaches the carrying capacity. Compare with **exponential population growth.**

long-day plant A plant that blooms in response to short nights (usually in late spring or early summer in the northern hemisphere). Compare with **day-neutral** and **short-day plant.**

long interspersed nuclear elements See **LINEs.**

loop of Henle In the vertebrate kidney, a long U-shaped loop in a nephron that extends into the medulla. Functions as a countercurrent exchanger to set up an osmotic gradient that allows reabsorption of water from a subsequent portion of the nephron.

loose connective tissue A type of connective tissue consisting of fibrous proteins in a soft matrix. Often functions as padding for organs.

lophophore A specialized feeding structure found in some lophotrochozoans and used in filter feeding.

lophotrochozoans A major lineage of protostomes (Lophotrochozoa) that grow by extending the size of their skeletons rather than by molting. Many phyla have a specialized feeding structure (lophophore) and/or ciliated larvae (trochophore). Includes rotifers, flatworms, segmented worms, and molluscs. Compare with **ecdysozoans.**

loss-of-function mutant See **knock-out mutant.**

lumen The interior space of any hollow structure (e.g., the rough ER) or organ (e.g., the stomach).

lung Any respiratory organ used for gas exchange between blood and air.

luteal phase The second major phase of a menstrual cycle, after ovulation, when the progesterone levels are high and the body is preparing for a possible pregnancy.

luteinizing hormone (LH) A peptide hormone, produced and secreted by the anterior pituitary, that stimulates estrogen production, ovulation, and formation of the corpus luteum in females and testosterone production in males.

lymph The mixture of fluid and white blood cells that circulates through the ducts and lymph nodes of the lymphatic system in vertebrates.

lymphatic system In vertebrates, a body-wide network of thin-walled ducts (or vessels) and lymph nodes, separate from the circulatory system. Collects excess fluid from body tissues and returns it to the blood; also functions as part of the immune system.

lymph node One of numerous small oval structures through which lymph moves in the lymphatic system. Filter the lymph and screen it for pathogens and other antigens. Major sites of lymphocyte activation.

lymphocytes Two types of leukocyte—B cells and T cells—that circulate through the bloodstream and lymphatic system and that are responsible for the development of acquired immunity.

lysogenic cycle A type of viral replication in which a viral genome enters a host cell, is inserted into the host's chromosome, and is replicated whenever the host cell divides. When activated, the viral DNA enters the lytic cycle, leading to production of new virus particles. Also called *lysogeny* or *latent growth*. Compare with **lytic cycle.**

lysosome A small organelle in an animal cell containing acids and enzymes that catalyze hydrolysis reactions and can digest large molecules. Compare with **vacuole.**

lysozyme An enzyme that functions in innate immunity by digesting bacterial cell walls. Occurs in saliva, tears, mucus, and egg white.

lytic cycle A type of viral replication in which a viral genome enters a host cell, new virus particles (virions) are made using host enzymes and eventually burst out of the cell, killing it. Also called *replicative growth*. Compare with **lysogenic cycle.**

macromolecule Any very large organic molecule, usually made up of smaller molecules (monomers) joined together into a polymer. The main biological macromolecules are proteins, nucleic acids, and polysaccharides.

macronutrient Any element (e.g., carbon, oxygen, nitrogen) that is required in large quantities for normal growth, reproduction, and maintenance of a living organism. Compare with **micronutrient.**

macrophage A type of leukocyte, capable of moving through body tissues, that engulfs and digests pathogens and other foreign particles; also secretes cytokines and presents foreign antigens to $CD4^+$ T cells.

MADS box A DNA sequence that codes for a DNA-binding motif in proteins; present in floral organ identity genes in plants. Functionally similar sequences are found in some fungal and animal genes.

major histocompatibility protein See **MHC protein.**

malaria A human disease caused by four species of the protist *Plasmodium* and passed to humans by mosquitoes.

malignant tumor A tumor that is actively growing and disrupting local tissues and/or is spreading to other organs. Cancer consists of one or more malignant tumors. Compare with **benign tumor.**

Malpighian tubules A major excretory organ of insects, consisting of blind-ended tubes that extend from the gut into the hemocoel. Filter hemolymph to form pre-urine and then send it to the hindgut for further processing.

mammals One of the two lineages of amniotes (vertebrates that produce amniotic eggs) distinguished by hair (or fur) and mammary glands. Includes the monotremes (platypuses), marsupials, and eutherians (placental mammals).

mammary glands Specialized exocrine glands that produce and secrete milk for nursing offspring. A diagnostic feature of mammals.

mandibles Any mouthpart used in chewing. In vertebrates, the lower jaw. In insects, crustaceans, and myriapods, the first pair of mouthparts.

mantle One of the three main parts of the mollusc body; the thick outer tissue that protects the visceral mass and may secrete a calcium carbonate shell.

Marfan syndrome A human syndrome involving increased height, long limbs and fingers, an abnormally shaped chest, and heart disorders. Probably caused by mutation in one pleiotropic gene.

marsh A wetland that lacks trees and usually has a slow but steady rate of water flow.

marsupials A lineage of mammals (Marsupiala) that nourish their young in an abdominal pouch after a very short period of development in the uterus.

mass extinction The extinction of a large number of diverse evolutionary groups during a relatively short period of geologic time (about 1 million years). May occur due to sudden and extraordinary environmental changes. Compare with **background extinction.**

mass feeder An animal that takes chunks of food into its mouth.

mass number The total number of protons and neutrons in an atom.

mast cell A type of leukocyte that is stationary (embedded in tissue) and helps trigger the inflammatory response to infection or injury, including secretion of histamine. Particularly important in allergic responses and defense against parasites.

maternal chromosome A chromosome inherited from the mother.

mechanoreceptor A sensory cell or organ specialized for detecting distortions caused by touch or pressure. One example is hair cells in the cochlea.

medium A liquid or solid in which cells can grow in vitro.

medulla The innermost part of an organ (e.g., kidney or adrenal gland).

medulla oblongata In vertebrates, a region of the brain stem that along with the cerebellum forms the hindbrain.

medusa (plural: medusae) The free-floating stage in the life cycle of some cnidarians (e.g., jellyfish). Compare with **polyp.**

megapascal (MPa) A unit of pressure (force per unit area), equivalent to 1 million pascals (Pa).

megaspore In seed plants, a haploid (n) spore that is produced in a megasporangium by meiosis of a diploid ($2n$) megasporocyte; develops into a female gametophyte. Compare with **microspore.**

meiosis In sexually reproducing organisms, a special two-stage type of cell division in which one diploid ($2n$) parent cell produces four haploid (n) reproductive cells (gametes); results in halving of the chromosome number. Also called *reduction division*.

meiosis I The first cell division of meiosis, in which synapsis and crossing over occur, and homologous chromosomes are separated from each other, producing daughter cells with half as many chromosomes (each composed of two sister chromatids) as the parent cell.

meiosis II The second cell division of meiosis, in which sister chromatids are separated from each other. Similar to mitosis.

meiotic map See **genetic map.**

membrane potential A difference in electric charge across a cell membrane; a form of potential energy. Also called *membrane voltage*.

memory Retention of learned information.

memory cells A type of lymphocyte responsible for maintenance of immunity for years or decades after an infection. Descended from a B cell or T cell activated during a previous infection.

meniscus (plural: menisci) The concave boundary layer formed at most air-water interfaces due to surface tension.

menstrual cycle A female reproductive cycle seen in Old World monkeys and apes (including humans) in which the uterine lining is shed (menstruation) if no pregnancy occurs. Compare with **estrous cycle.**

menstruation The periodic shedding of the uterine lining through the vagina that occurs in females of Old World monkeys and apes, including humans.

meristem (adjective: meristematic) In plants, a group of undifferentiated cells that can develop into various adult tissues throughout the life of a plant.

mesoderm The middle of the three basic cell layers in most animal embryos; gives rise to muscles, bones, blood, and some internal organs (kidney, spleen, etc.). Compare with **ectoderm** and **endoderm.**

mesoglea A gelatinous material, containing scattered ectodermal cells, that is located between the ectoderm and endoderm of cnidarians (e.g., jellyfish, corals, and anemones).

mesophyll cell A type of cell, found near the surfaces of plant leaves, that is specialized for the light-dependent reactions of photosynthesis.

Mesozoic era The period of geologic time, from 250 million to 65 million years ago, during which gymnosperms were the dominant plants and dinosaurs the dominant vertebrates. Ended with extinction of the dinosaurs. Also called *Age of Reptiles.*

messenger RNA (mRNA) An RNA molecule that carries encoded information, transcribed from DNA, that specifies the amino acid sequence of a polypeptide.

meta-analysis A comparative analysis of the results of many smaller, previously published studies.

metabolic pathway An ordered series of chemical reactions that build up or break down a particular molecule. Often, each reaction is catalyzed by a different enzyme.

metabolic rate The total energy use by all the cells of an individual. For aerobic organisms, often measured as the amount of oxygen consumed per hour.

metabolic water The water that is produced as a by-product of cellular respiration.

metabolism All the chemical reactions occurring in a living cell or organism.

metallothioneins Small plant proteins that bind to and prevent excess metal ions from acting as toxins.

metamorphosis Transition from one developmental stage to another, such as from the larval to the adult form of an animal.

metaphase A stage in mitosis or meiosis during which chromosomes line up in the middle of the cell.

metaphase plate The plane along which chromosomes line up during metaphase of mitosis or meiosis; not an actual structure.

metastasis The spread of cancerous cells from their site of origin to distant sites in the body where they may establish additional tumors.

methanogen A prokaryote that produces methane (CH_4) as a by-product of cellular respiration.

methanotroph An organism that uses methane (CH_4) as its primary electron donor and source of carbon.

methylation The addition of a methyl ($-CH_3$) group to a molecule.

MHC protein One of a large set of mammalian cell-surface glycoproteins involved in marking cells as self and in antigen presentation to T cells. Also called *MHC molecule.* See **Class I** and **Class II MHC protein.**

microbe Any microscopic organism, including bacteria, archaea, and various tiny eukaryotes.

microbiology The field of study concerned with microscopic organisms.

microfilament See **actin filament.**

micrograph A photograph of an image produced by a microscope.

micronutrient Any element (e.g., iron, molybdenum, magnesium) that is required in very small quantities for normal growth, reproduction, and maintenance of a living organism. Compare with **macronutrient.**

micropyle The tiny pore in a plant ovule through which the pollen tube reaches the embryo sac.

microRNA (miRNA) A small, single-stranded RNA associated with proteins in an RNA-induced silencing complex. Can bind to complementary sequences in mRNA molecules, allowing the associated proteins to degrade the bound mRNA or inhibit its translation. See **RNA interference.**

microsatellite A noncoding stretch of eukaryotic DNA consisting of a repeating sequence 1- to 5-base pair long. Also called *simple sequence repeat.*

microspore In seed plants, a haploid (n) spore that is produced in a microsporangium by meiosis of a diploid ($2n$) microsporocyte; develops into a male gametophyte. Compare with **megaspore.**

microtubule A long, tubular fiber, about 25 nm in diameter, formed by polymerization of tubulin protein dimers; one of the three types of cytoskeletal fibers. Involved in cell movement and transport of materials within the cell. Compare with **actin filament** and **intermediate filament.**

microtubule-organizing center (MTOC) General term for any structure (e.g., centrosome and basal body) that organizes microtubules in cells.

microvilli (singular:microvillus) Tiny protrusions from the surface of an epithelial cell that increase the surface area for absorption of substances.

midbrain One of the three main regions of the vertebrate brain; includes sensory integrating and relay centers. Compare with **forebrain** and **hindbrain.**

middle ear The air-filled middle portion of the mammalian ear, which contains three small bones (ossicles) that transmit and amplify sound from the tympanic membrane to the inner ear. Is connected to the throat via the eustachian tube.

middle lamella The layer of gelatinous pectins between the primary cell walls of adjacent plant cells. Helps hold the cells together.

migration (1) In ecology, a cyclical movement of large numbers of organisms from one geographic location or habitat to another. (2) In population genetics, movement of individuals from one population to another.

millivolt (mV) A unit of voltage equal to 1/1000 of a volt.

mimicry A phenomenon in which one species has evolved (or learns) to look or sound like another species. See **Batesian mimicry** and **Müllerian mimicry.**

mineralocorticoids A class of steroid hormones, produced and secreted by the adrenal cortex, that regulate electrolyte levels and the overall volume of body fluids. Aldosterone is the principal one in humans. Compare with **glucocorticoids.**

minisatellite A noncoding stretch of eukaryotic DNA consisting of a repeating sequence that is 6 to 500 base pairs long. Also called *variable number tandem repeat (VNTR).*

mismatch repair The process by which mismatched base pairs in DNA are fixed.

missense mutation A point mutation (change in a single base pair) that causes a change in the amino acid sequence of a protein. Also called *replacement mutation.*

mitochondrial matrix Central compartment of a mitochondrion, which is lined by the inner membrane; contains the enzymes and substrates of the Krebs cycle and mitochondrial DNA.

mitochondrion (plural: mitochondria) A eukaryotic organelle that is bounded by a double membrane and is the site of aerobic respiration.

mitosis In eukaryotic cells, the process of nuclear division that results in two daughter nuclei genetically identical to the parent nucleus. Subsequent cytokinesis (division of the cytoplasm) yields two daughter cells.

mitosis-promoting factor (MPF) A complex of a cyclin and cyclin-dependent kinase that phosphorylates a number of specific proteins needed to initiate mitosis in eukaryotic cells.

mitotic (M) phase The phase of the cell cycle during which cell division occurs. Includes mitosis and cytokinesis.

mitotic spindle Temporary structure, composed largely of microtubules, that is involved in the movement of chromosomes to the equatorial plate and then to opposite sides of the cell during mitosis and meiosis.

model organism An organism selected for intensive scientific study based on features that make it easy to work with (e.g., body size, life span), in the hope that findings will apply to other species.

molarity A common unit of solute concentration equal to the number of moles of a dissolved solute in 1 liter of solution.

mole The amount of a substance that contains 6.022×10^{23} of its elemental entities (e.g., atoms, ions, or molecules). This number of molecules of a compound will have a mass equal to the molecular weight of that compound expressed in grams.

molecular chaperone A protein that facilitates the three-dimensional folding of newly synthesized proteins, usually by an ATP-dependent mechanism.

molecular clock The hypothesis that certain types of mutations tend to reach fixation in populations at a steady rate over large spans of time. As a result, comparisons of DNA sequences can be used to infer the timing of evolutionary divergences.

molecular formula A notation that indicates only the numbers and types of atoms in a molecule, such as H_2O for the water molecule. Compare with **structural formula.**

molecular weight The sum of the mass numbers of all of the atoms in a molecule; roughly, the total number of protons and neutrons in the molecule.

molecule A combination of two or more atoms held together by covalent bonds.

molluscs Members of the phylum Mollusca. Distinguished by a body plan with three main parts: a muscular foot, a visceral mass, and a mantle. Include bivalves (clams, oysters), gastropods (snails, slugs), chitons, and cephalopods (squid, octopuses). Molluscs belong to the lophotrochozoan branch of the protostomes.

molting A method of body growth, used by ecdysozoans, that involves the shedding of an external protective cuticle or skeleton, expansion of the soft body, and growth of a new external layer.

monocot Any plant that has a single cotyledon (embryonic leaf) upon germination. Monocots form a monophyletic group. Also called a monocotyledonous plant. Compare with **dicot.**

monoecious Describing an angiosperm species that has both male and female reproductive structures on each plant. Compare with **dioecious.**

monohybrid cross A mating between two parents that are both heterozygous for a given gene.

monomer A small molecule that can covalently bind to other similar molecules to form a larger macromolecule. Compare with **polymer.**

monophyletic group An evolutionary unit that includes an ancestral population and all of its descendants but no others. Also called a *clade* or *lineage.* Compare with **paraphyletic group.**

monosaccharide A small carbohydrate, such as glucose, that has the molecular formula $(CH_2O)_n$ and cannot be hydrolyzed to form any smaller carbohydrates. Also called *simple sugar.* Compare with **disaccharide** and **polysaccharide.**

monosomy Having only one copy of a particular type of chromosome.

monotremes A lineage of mammals (Monotremata) that lay eggs and then nourish the young with milk. Includes just three living species: the platypus and two species of echidna.

morphological species concept The definition of a species as a population or group of populations that have measurably different anatomical features from other groups. Also called *morphospecies concept.* Compare with **biological** and **phylogenetic species concept.**

morphology The shape and appearance of an organism's body and its component parts.

motif In molecular biology, a domain (a section of a protein with a distinctive tertiary structure) found in many different proteins and often having specific functional properties.

motor neuron A nerve cell that carries signals from the central nervous system (brain and spinal cord) to an effector, such as a muscle or gland. Compare with **interneuron** and **sensory neuron.**

motor protein A class of proteins whose major function is to convert the chemical energy of ATP into motion. Includes dynein, kinesin, and myosin.

MPF See **mitosis-promoting factor.**

mRNA. See **messenger RNA.**

mucigel A slimy substance secreted by plant root caps that eases passage of the growing root through the soil.

mucosal-associated lymphoid tissue (MALT) Collective term for lymphocytes and other leukocytes associated with skin cells and with mucus-secreting epithelial tissues in the gut and respiratory tract. Plays important role in preventing entry of pathogens into the body.

mucus (adjective: mucous) A slimy mixture of glycoproteins (called mucins) and water that is secreted in many animal organs for lubrication.

Müllerian inhibitory substance A peptide hormone secreted by the embryonic testis that causes regression (withering away) of the female reproductive ducts.

Müllerian mimicry A type of mimicry in which two (or more) harmful species resemble each other. Compare with **Batesian mimicry.**

multicellularity The state of being composed of many cells that adhere to each other and do not all express the same genes with the result that some cells have specialized functions.

multiple fruit A fruit (e.g., pineapple) that develops from many separate flowers and thus many carpels. Compare with **aggregate** and **simple fruit.**

multiple sclerosis (MS) A human autoimmune disease caused by the immune system attacking the myelin sheaths that insulate nerve axons.

muscle fiber A single muscle cell.

muscle tissue An animal tissue consisting of bundles of long, thin contractile cells (muscle fibers).

mutagen Any physical or chemical agent that increases the rate of mutation.

mutant An individual that carries a mutation, particularly a new or rare mutation.

mutation Any change in the hereditary material of an organism (DNA in most organisms, RNA in some viruses).

mutualism (adjective: mutualistic) A symbiotic relationship between two organisms (mutualists) that benefits both. Compare with **commensalism** and **parasitism.**

mycelium (plural: mycelia) A mass of underground filaments (hyphae) that form the body of a fungus. Also found in some protists and bacteria.

mycorrhiza (plural: mycorrhizae) A mutualistic association between certain fungi and most vascular plants, sometimes visible as nodules or nets in or around plant roots.

myelin sheath Multiple layers of myelin, derived from the cell membranes of certain glial cells, that is wrapped around the axon of a neuron, providing electrical insulation.

myofibril Long, slender structure composed of contractile proteins organized into repeating units (sarcomeres) in vertebrate heart muscle and striated muscle.

myosin Any one of a class of motor proteins that use the chemical energy of ATP to move along actin filaments in muscle contraction, cytokinesis, and vesicle transport.

myriapods A lineage of arthropods with long segmented trunks, each segment bearing one or two pairs of legs. Includes millipedes and centipedes.

NAD^+/NADH Oxidized and reduced forms, respectively, of nicotinamide adenine dinucleotide. A nonprotein electron carrier that functions in many of the redox reactions of metabolism.

$NADP^+$/NADPH Oxidized and reduced forms, respectively, of nicotinamide adenine dinucleotide phosphate. A nonprotein electron carrier that is reduced during the light-dependent reactions in photosynthesis and extensively used in biosynthetic reactions.

natural experiment A situation in which groups to be compared are created by an unplanned, natural change in conditions rather than by manipulation of conditions by researchers.

natural selection The process by which individuals with certain heritable traits tend to produce more surviving offspring than do individuals without those traits, often leading to a change in the genetic makeup of the population. A major mechanism of evolution.

nauplius A distinct planktonic larval stage seen in many crustaceans.

Neanderthal A recently extinct European species of hominid, *Homo neanderthalensis*, closely related to but distinct from modern humans.

nectar The sugary fluid produced by flowers to attract and reward pollinating animals.

negative feedback A self-limiting, corrective response in which a deviation in some variable (e.g., body temperature, blood pH, concentration of some compound) triggers responses aimed at returning the variable to normal. Compare with **positive feedback.**

negative pressure ventilation Ventilation of the lungs that is accomplished by "pulling" air into the lungs by expansion of the rib cage. Compare with **positive pressure ventilation.**

negative-sense virus A virus whose genome contains sequences complementary to those in the mRNA required to produce viral proteins. Compare with **ambisense virus** and **positive-sense virus.**

nematodes. See **roundworms.**

nephron One of the tiny tubes within the vertebrate kidney that filter blood and concentrate salts to produce urine. Also called *renal tubule.*

neritic zone Shallow marine waters beyond the intertidal zone, extending down to about 200 meters, where the continental shelf ends.

nerve A long, tough strand of nervous tissue typically containing thousands of axons wrapped in connective tissue; carries impulses between the central nervous system and some other part of the body.

nerve cord A bundle of nerves extending from the brain along the dorsal (back) side of a chordate animal, with cerebrospinal fluid inside a hollow central channel. One of the defining features of chordates.

nervous tissue An animal tissue consisting of nerve cells (neurons) and various supporting cells.

net primary productivity (NPP) In an ecosystem, the total amount of carbon fixed by photosynthesis over a given time period minus the amount oxidized during cellular respiration. Compare with **gross primary productivity.**

net reproductive rate (R_0) The growth rate of a population per generation; equivalent to the average number of female offspring that each female produces over her lifetime.

neural Relating to nerve cells (neurons) and the nervous system.

neural tube A folded tube of ectoderm that forms along the dorsal side of a young vertebrate embryo and that will give rise to the brain and spinal cord.

neuroendocrine Referring to nerve cells (neurons) that release hormones into the blood or to such hormones themselves.

neuron A cell that is specialized for the transmission of nerve impulses. Typically has dendrites, a cell body, and a long axon that forms synapses with other neurons. Also called *nerve cell.*

neurosecretory cell A nerve cell (neuron) that produces and secretes hormones into the bloodstream. Principally found in the hypothalamus. Also called *neuroendocrine cell.*

neurotoxin Any substance that specifically destroys or blocks the normal functioning of neurons.

neurotransmitter A molecule that transmits electrical signals from one neuron to another or from a neuron to a muscle or gland. Examples are acetylcholine, dopamine, serotonin, and norepinephrine.

neutral In genetics, referring to any mutation or mutant allele that has no effect on an individual's fitness.

neutrophil A type of leukocyte, capable of moving through body tissues, that engulfs and digests pathogens and other foreign particles; also secretes various compounds that attack bacteria and fungi.

niche The particular set of habitat requirements of a certain species and the role that species plays in its ecosystem.

niche differentiation The change in resource use by competing species that occurs as the result of character displacement.

nicotinamide adenine dinucleotide See **NAD^+/NADH.**

nicotinamide adenine dinucleotide phosphate See **$NADP^+$/NADPH.**

nitrogen cycle, global The movement of nitrogen among terrestrial ecosystems, the oceans, and the atmosphere.

nitrogen fixation The incorporation of atmospheric nitrogen (N_2) into forms such as ammonia (NH_3) or nitrate (NO_3^-), which can be used to make many organic compounds. Occurs in only a few lineages of bacteria and archaea.

nociceptor A sensory cell or organ specialized to detect tissue damage, usually producing the sensation of pain.

node (1) In animals, any small thickening (e.g., a lymph node). (2) In plants, the part of a stem where leaves or leaf buds are attached. (3) In a phylogenetic tree, the point where two branches diverge, representing the point in time when an ancestral group split into two or more descendant groups. Also called *fork.*

node of Ranvier One of the periodic unmyelinated sections of a neuron's axon at which an action potential can be regenerated.

Nod factors Molecules produced by nitrogen-fixing bacteria that help them recognize and bind to roots of legumes.

nodule Lumplike structure on roots of legume plants that contain symbiotic nitrogen-fixing bacteria.

noncyclic electron flow See **Z scheme.**

nondisjunction An error that can occur during meiosis or mitosis in which one daughter cell receives two copies of a particular chromosome, and the other daughter cell receives none.

nonpolar covalent bond A covalent bond in which electrons are equally shared between two atoms of the same or similar electronegativity. Compare with **polar covalent bond.**

non-sister chromatids The chromatids of a particular type of chromosome (after replication) with respect to the chromatids of its homologous chromosome. Crossing over occurs between non-sister chromatids. Compare with **sister chromatids.**

non-template strand The strand of DNA that is not transcribed during synthesis of RNA. Its sequence corresponds to that of the mRNA produced from the other strand. Also called *coding strand.*

nonvascular plants See **bryophytes.**

norepinephrine A catecholamine used as a neurotransmitter in the sympathetic nervous system. Also is produced by the adrenal medulla and functions as a hormone that triggers rapid responses relating to the fight-or-flight response.

Northern blotting A technique for identifying specific RNAs separated by gel electrophoresis by transferring them to filter paper and hybridizing with a labeled DNA probe complementary to the RNA of interest. Compare with **Southern blotting** and **Western blotting.**

notochord A long, gelatinous, supportive rod down the back of a chordate embryo, below the developing spinal cord. Replaced by vertebrae in most adult vertebrates. A defining feature of chordates.

nuclear envelope The double-layered membrane enclosing the nucleus of a eukaryotic cell.

nuclear lamina A lattice-like sheet of fibrous nuclear lamins, which are one type of intermediate filaments. Lines the inner membrane of the nuclear envelope, stiffening the envelope and helping organize the chromosomes.

nuclear localization signal (NLS) A short amino acid sequence that marks a protein for delivery to the nucleus.

nuclear pore An opening in the nuclear envelope that connects the inside of the nucleus with the cytoplasm and through which molecules such as mRNA and some proteins can pass.

nuclear pore complex A large complex of dozens of proteins lining a nuclear pore, defining its shape and transporting substances through the pore.

nuclease Any enzyme that can break down RNA or DNA molecules.

nucleic acid A macromolecule composed of nucleotide monomers. Generally used by cells to store or transmit hereditary information. Includes ribonucleic acid and deoxyribonucleic acid.

nucleic acid hybridization Base pairing between a single-stranded nucleic acid and a complementary sequence in a different nucleic acid (e.g., a labeled probe). Is used experimentally in Southern blotting and Northern blotting.

nucleoid In prokaryotic cells, a dense, centrally located region that contains DNA but is not surrounded by a membrane.

nucleolus In eukaryotic cells, specialized structure in the nucleus where ribosomal RNA processing occurs and ribosomal subunits are assembled.

nucleosome A repeating, bead-like unit of eukaryotic chromatin, consisting of about 200 nucleotides of DNA wrapped twice around eight histone proteins.

nucleotide A molecule consisting of a five-carbon sugar (ribose or deoxyribose), a phosphate group, and one of several nitrogen-containing bases. DNA and RNA are polymers of nucleotides containing deoxyribose (deoxyribonucleotides) and ribose (ribonucleotides), respectively. Equivalent to a nucleoside plus one phosphate group.

nucleotide excision repair The process of removing a damaged region in one strand of DNA and correctly replacing it using the undamaged strand as a template.

nucleus (1) The center of an atom, containing protons and neutrons. (2) In eukaryotic cells, the large organelle containing the chromosomes and surrounded by a double membrane. (3) A discrete clump of neuron cell bodies in the brain, usually sharing a distinct function.

null hypothesis A hypothesis that specifies what the results of an experiment will be if the main hypothesis being tested is wrong. Often states that there will be no difference between experimental groups.

null mutant See **knock-out mutant.**

nutrient A substance that an organism requires for normal growth, maintenance, or reproduction.

nutritional balance A state in which an organism is taking in enough nutrients to maintain normal health and activity.

nymph The juvenile form of an animal that undergoes hemimetabolous (incomplete) metamorphosis; resembles a miniature version of the adult form.

oceanic zone The waters of the open ocean beyond the continental shelf.

oil A fat that is liquid at room temperature.

Okazaki fragment Short segment of DNA produced during replication of the 5′ to 3′ template strand. Many Okazaki fragments make up the lagging strand in newly synthesized DNA.

olfaction The perception of odors.

olfactory bulb A bulb-shaped projection of the brain just above the nose. Receives and interprets odor information from the nose.

oligodendrocyte A type of glial cell that wraps around axons of some neurons in the central nervous system, forming a myelin sheath that provides electrical insulation. Compare with **Schwann cell.**

oligopeptide A chain composed of fewer than 50 amino acids linked together by peptide bonds. Often referred to simply as *peptide.*

ommatidium (plural: ommatidia) A light-sensing column in an arthropod's compound eye.

omnivore (adjective: omnivorous) An animal whose diet regularly includes both meat and plants. Compare with **carnivore** and **herbivore.**

oncogene Any gene whose protein product stimulates cell division at all times and thus promotes cancer development. Often is a mutated form of a gene involved in regulating the cell cycle. See **proto-oncogene.**

one-gene, one-enzyme hypothesis The hypothesis that each gene is responsible for making one (and only one) protein, in most cases an enzyme that catalyzes a specific reaction. Many exceptions to this hypothesis are now known.

oocyte A cell in the ovary that can undergoes meiosis to produce an ovum.

oogenesis The production of egg cells (ova).

oogonia (singular: oogonia) The diploid cells in an ovary that can divide by mitosis to create more oogonia and primary oocytes, which can undergo meiosis.

open circulatory system A circulatory system in which the circulating fluid (hemolymph) is not confined to blood vessels. Compare with **closed circulatory system.**

open reading frame (ORF) Any DNA sequence, ranging in length from several hundred to thousands of base pairs long, that is flanked by a start codon and a stop codon. ORFs identified by computer analysis of DNA may be functional genes, especially if they have other features characteristic of genes (e.g., promoter sequence).

operator In prokaryotic DNA, a binding site for a repressor protein; located near the start of an operon.

operculum The stiff flap of tissue that covers the gills of teleost fishes.

operon A region of prokaryotic DNA that codes for a series of functionally related genes and is transcribed from a single promoter into a polycistronic mRNA.

opsin A transmembrane protein that is covalently linked to retinal, the light-detecting pigment in rod and cone cells.

optical isomer A molecule that shares the same molecular formula as another molecule but differs in the arrangement of atoms or groups around a carbon atom; left-handed or right-handed form of a molecule. Compare with **geometric** and **structural isomer.**

optimal foraging The concept that animals forage in a way that maximizes the amount of usable energy they take in, given the costs of finding and ingesting their food and the risk of being eaten while they're at it.

orbital The spherical region around an atomic nucleus in which an electron is present most of the time.

order In Linnaeus' system, a taxonomic category above the family level and below the class level.

ORF See **open reading frame.**

organ A group of tissues organized into a functional and structural unit.

organelle Any discrete, membrane-bound structure within a cell (e.g., mitochondrion) that has a characteristic structure and functions.

organic For a compound, containing carbon and hydrogen and usually containing carbon-carbon bonds. Organic compounds are widely used by living organisms.

organism Any living entity that contains one or more cells.

organogenesis A stage of embryonic development, just after gastrulation in vertebrate embryos, during which major organs develop from the three embryonic germ layers.

orientation A deliberate movement that results in a change in position relative to some external cue, such as toward the Sun or away from a sound.

origin of replication The site on a chromosome at which DNA replication begins.

osmoconformer An animal that does not actively regulate the osmolarity of its tissues but conforms to the osmolarity of the surrounding environment.

osmolarity The concentration of dissolved substances in a solution, measured in moles per liter.

osmoregulation The process by which a living organism controls the concentration of water and salts in its body.

osmoregulator An animal that actively regulates the osmolarity of its tissues.

osmosis Diffusion of water across a selectively permeable membrane from a region of high water concentration (low solute concentration) to a region of low water concentration (high solute concentration).

osmotic potential See **solute potential.**

ossicle One of three small bones, in the middle ear of mammals, that transmit and amplify sound from the tympanic membrane to the inner ear.

ouabain A plant toxin that poisons the sodium-potassium pumps of animals.

outcrossing Reproduction by fusion of the gametes of different individuals, rather than self-fertilization. Typically refers to plants.

outer ear The outermost portion of the mammalian ear, consisting of the pinna (ear flap) and the ear canal. Funnels sound to the tympanic membrane.

outgroup A taxon that is closely related to a particular monophyletic group but is not part of it.

out-of-Africa hypothesis The hypothesis that modern humans (*Homo sapiens*) evolved in Africa and spread to other continents, replacing other *Homo* species without interbreeding with them.

oval window A membrane separating the fluid-filled cochlea from the air-filled middle ear through which sound vibrations pass from the middle ear to the inner ear in mammals.

ovary The egg-producing organ of a female animal, or the seed-producing structure in the female part of a flower.

oviduct See **fallopian tube.**

oviparous Producing eggs that are laid outside the body where they develop and hatch. Compare with **ovoviviparous** and **viviparous.**

ovoviviparous Producing eggs that are retained inside the body until they are ready to hatch. Compare with **oviparous** and **viviparous.**

ovulation The release of an ovum from an ovary of a female vertebrate. In humans, an ovarian follicle releases an egg at the end of the follicular phase of the menstrual cycle.

ovule In flowering plants, the structure inside an ovary that contains the female gametophyte and eventually (if fertilized) becomes a seed.

ovum (plural: ova) See **egg.**

oxidation The loss of electrons from an atom during a redox reaction, either by donation of an electron to another atom or by the shared electrons in covalent bonds moving farther from the atomic nucleus.

oxidative phosphorylation Production of ATP molecules from the redox reactions of an electron transport chain.

oxygen-hemoglobin equilibrium curve The graphical depiction of the percentage of hemoglobin in the blood that will bind to oxygen at various partial pressures of oxygen.

oxygenic Referring to any process or reaction that produces oxygen. Photosynthesis in plants, algae, and cyanobacteria, which involves photosystem II, is oxygenic. Compare with **anoxygenic.**

oxytocin A peptide hormone, secreted by the posterior pituitary, that triggers labor and milk production in females and that stimulates pair bonding, parental care, and affiliative behavior in both sexes.

p53 A tumor-suppressor protein (molecular weight of 53 kilodaltons) that responds to DNA damage by stopping the cell cycle and/or triggering apoptosis. Encoded by the *p53* gene.

pacemaker cell A specialized cardiac muscle cell in the sinoatrial (SA) node of the vertebrate heart that has an inherent rhythm and can generate an electrical impulse that spreads to other heart cells.

paleontology The study of organisms that lived in the distant past.

Paleozoic era The period of geologic time, from 543 million to 250 million years ago, during which fungi, land plants, and animals first appeared and diversified. Began with the Cambrian explosion and ended with the extinction of many invertebrates and vertebrates.

pancreas A large gland in vertebrates that has both exocrine and endocrine functions. Secretes digestive enzymes into a duct connected to the intestine and several hormones (notably, insulin and glucagon) into the bloodstream.

pandemic The spread of an infectious disease in a short time period over a wide geographic area and affecting a very high proportion of the population. Compare with **epidemic.**

parabiosis An experimental technique for determining whether a certain physiological phenomenon is regulated by a hormone, by surgically uniting two individuals so that hormones can pass between them.

parabronchus (plural: parabronchi) One of the many tiny parallel air tubes that run through a bird's lung where gas exchange occurs.

paracrine Relating to a chemical signal that is released by one cell and affects neighboring cells.

paraphyletic group An evolutionary unit that includes an ancestral population and *some* but not all of its descendants. Paraphyletic groups are not meaningful units in evolution. Compare with **monophyletic group.**

parapodia (singular: parapodium) Appendages found in some annelids from which bristle-like structures (chaetae) extend.

parasite An organism that lives on or in a host species and that damages its host.

parasitism (adjective: parasitic) A symbiotic term relationship between two organisms that is beneficial to one organism (the parasite) but detrimental to the other (the host). Compare with **commensalism** and **mutualism.**

parasitoid An organism that has a parasitic larval stage and a free-living adult stage. Most parasitoids are insects that lay eggs in the bodies of other insects.

parasympathetic nervous system The part of the autonomic nervous system that stimulates functions for conserving or restoring energy, such as reduced heart rate and increased digestion. Compare with **sympathetic nervous system.**

parathyroid glands Four small glands, located near or embedded in the thyroid gland of vertebrates, that secrete parathyroid hormone.

parathyroid hormone (PTH) A peptide hormone, secreted from the parathyroid glands, that increases blood calcium by promoting Ca^{2+} release from bones, Ca^{2+} uptake in the intestines, and Ca^{2+} reabsorption in the kidneys.

parenchyma cell In plants, a general type of cell with a relatively thin primary cell wall. These cells, found in leaves, the centers of stems and roots, and fruits, are involved in photosynthesis, starch storage, and new growth. Compare with **collenchyma cell** and **sclerenchyma cell.**

parental care Any action by which an animal expends energy or assumes risks to benefit its offspring (e.g., nest-building, feeding of young, defense).

parental generation The adult organisms used in the first experimental cross in a formal breeding experiment.

parietal cell A cell in the stomach lining that secretes hydrochloric acid.

parsimony The logical principle that the most likely explanation of a phenomenon is the most economical or simplest. When applied to comparison of alternative phylogenetic trees, it suggests that the one requiring the fewest evolutionary changes is most likely to be correct.

parthenogenesis Development of offspring from unfertilized eggs; a type of asexual reproduction.

partial pressure The pressure of one particular gas in a mixture; the contribution of that gas to the overall pressure.

pascal (Pa) A unit of pressure (force per unit area).

passive transport Diffusion of a substance across a plasma membrane or organelle membrane. When this occurs with the assistance of membrane proteins, it is called facilitated diffusion.

patch clamping A technique for studying the electrical currents that flow through individual ion channels by sucking a tiny patch of membrane to the hollow tip of a microelectrode.

paternal chromosome A chromosome inherited from the father.

pathogen (adjective: pathogenic) Any entity capable of causing disease, such as a microbe, virus, or prion.

pattern formation The series of events that determines the spatial organization of an embryo, including alignment of the major body axes and orientation of the limbs.

pattern-recognition receptor One of a class of membrane proteins on leukocytes that bind to molecules on the surface of many bacteria. Part of the innate immune response.

PCR See **polymerase chain reaction.**

peat Semidecayed organic matter that accumulates in moist, low-oxygen environments such as bogs.

pectin A gelatinous polysaccharide found in the primary cell wall and middle lamella of plant cells. Attracts and holds water, forming a gel that helps keep the cell wall moist.

pedigree A family tree of parents and offspring, showing inheritance of particular traits of interest.

pellet Any solid material that collects at the bottom of a test tube below a layer of liquid (the supernatant) during centrifugation.

penis The copulatory organ of male mammals, used to insert sperm into a female.

pentose A monosaccharide (simple sugar) containing five carbon atoms.

PEP carboxylase An enzyme that catalyzes addition of CO_2 to phosphoenol pyruvate, a three-carbon compound, forming a four-carbon organic acid. Found in mesophyll cells of plants that perform C_4 photosynthesis.

pepsin A protein-digesting enzyme present in the stomach.

pepsinogen The precursor of the digestive enzyme pepsin. Is secreted from cells in the stomach lining and converted to pepsin by the acidic environment of the stomach lumen.

peptide See **oligopeptide.**

peptide bond The covalent bond (C–N) formed by a condensation reaction between two amino acids; links the residues in peptides and proteins.

peptidoglycan A complex structural polysaccharide found in bacterial cell walls.

per capita rate of increase (*r*) The growth rate of a population, expressed per individual. Calculated as the per capita birthrate minus the per capita death rate. Also called *per capita growth rate.*

perennial Describing a plant whose life cycle normally lasts for **more than one year.** Compare with **annual.**

perfect flower A flower that contains both male parts (stamens) and female parts (carpels). Compare with **imperfect flower.**

perforation In plants, a small hole in the primary and secondary cell walls of vessel elements that allow passage of water.

pericarp The part of a fruit, formed from the ovary wall, that surrounds the seeds and protects them. Corresponds to the flesh of most edible fruits and the hard shells of most nuts.

pericycle In plant roots, a layer of cells that give rise to lateral roots.

periderm The outermost portion of bark consisting of cork cambium, cork cells, and a third small layer of cells (phelloderm).

peripheral membrane protein Any membrane protein that does not span the entire lipid bilayer and associates with only one side of the bilayer. Compare with **integral membrane protein.**

peripheral nervous system (PNS) All the components of the nervous system that are outside the central nervous system (the brain and spinal cord). Includes the somatic nervous system and the autonomic nervous system.

peristalsis Rhythmic waves of muscular contraction that push food along the digestive tract.

permafrost A permanently frozen layer of icy soil found in most tundra and some taiga.

permeability The tendency of a structure, such as a membrane, to allow a given substance to diffuse across it.

peroxisome An organelle found in most eukaryotic cells that contains enzymes for oxidizing fatty acids and other compounds including many toxins, rendering them harmless. See **glyoxisome.**

petal One of the leaflike organs arranged around the reproductive organs of a flower. Often colored and scented to attract pollinators.

petiole The stalk of a leaf.

pH A measure of the concentration of protons in a solution and thus of acidity or alkalinity. Defined as the negative of the base-10 logarithm of the proton concentration: $pH = -\log[H^+]$.

phagocytosis Uptake by a cell of small particles or cells by pinching off the plasma membrane to form small membrane-bound vesicles; one type of endocytosis.

pharyngeal gill slits A set of parallel openings from the throat through the neck to the outside. A diagnostic trait of chordates.

phelloderm In the stems of woody plants, a thin layer of cells located between the outer cork cells and inner cork cambium.

pharyngeal jaw A secondary jaw in the back of the mouth, found in some fishes. Derived from modified gill arches.

phenetic approach A method for constructing a phylogenetic tree by computing a statistic that summarizes the overall similarity among populations, based on the available data. Compare with **cladistic approach.**

phenotype The detectable physical and physiological traits of an individual, which are determined its genetic makeup. Also the specific trait associated with a particular allele. Compare with **genotype.**

phenotypic plasticity Within-species variation in phenotype that is due to differences in environmental conditions. Occurs more commonly in plants than animals.

pheophytin In photosystem II, a molecule that accepts excited electrons from a reaction center chlorophyll and passes them to an electron transport chain.

pheromone A chemical signal, released by one individual into the external environment, that can trigger responses in a different individual.

phloem A plant vascular tissue that conducts sugars; contains sieve-tube members and companion cells. Primary phloem develops from the procambium of apical meristems; secondary phloem, from the vascular cambium of lateral meristems. Compare with **xylem.**

phonotaxis Orientation toward or away from sound.

phosphodiester linkage Chemical linkage between adjacent **nucleotide residues** in DNA and RNA. Forms when the phosphate group of one nucleotide condenses with the hydroxyl group on the sugar of another nucleotide. Also known as *phosphodiester bond.*

phosphofructokinase The enzyme that catalyzes synthesis of fructose-1,6-bisphosphate from fructose-6-phosphate, a key reaction (step 3) in glycolysis.

phospholipid A class of lipid having a hydrophilic head (a phosphate group) and a hydrophobic tail (one or more fatty acids). Major components of the plasma membrane and organelle membranes.

phosphorylase An enzyme that breaks down glycogen by catalyzing hydrolysis of the α-glycosidic linkages between the glucose residues.

phosphorylation (verb: phosphorylate) The addition of a phosphate group to a molecule.

phosphorylation cascade A series of enzyme-catalyzed phosphorylation reactions commonly used in signal transduction pathways to amplify and convey a signal inward from the plasma membrane.

photic zone In an aquatic habitat, water that is shallow enough to receive some sunlight (whether or not it is enough to support photosynthesis). Compare with **aphotic zone.**

photon A discrete packet of light energy; a particle of light.

photoperiodism Any response by an organism to the relative lengths of day and night (i.e., photoperiod).

photophosporylation Production of ATP molecules using the energy released as light-excited electrons flow through an electron transport chain during photosynthesis. Involves generation of a proton-motive force during electron transport and its use to drive ATP synthesis.

photoreceptor A molecule, a cell, or an organ that is specialized to detect light.

photorespiration A series of light-driven chemical reactions that consumes oxygen and releases carbon dioxide, basically reversing photosynthesis. Usually occurs when there are high O_2 and low CO_2 concentrations inside plant cells, often in bright, hot, dry environments when stomata must be kept closed.

photoreversibility A change in conformation that occurs in certain plant pigments when they are exposed to their preferred wavelengths of light and that triggers responses by the plant.

photosynthesis The complex biological process that converts the energy of light into chemical energy stored in glucose and other organic molecules. Occurs in plants, algae, and some bacteria.

photosystem One of two types of units, consisting of a central reaction center surrounded by antenna complexes, that is responsible for the light-dependent reactions of photosynthesis.

photosystem I A photosystem that contains a pair of P700 chlorophyll molecules and uses absorbed light energy to produce NADPH.

photosystem II A photosystem that contains a pair of P680 chlorophyll molecules and uses absorbed light energy to split water into protons and oxygen and to produce ATP.

phototaxis Orientation toward or away from light.

phototroph An organism that produces ATP through photosynthesis.

phototropins A class of plant photoreceptors that detect blue light and initiate phototropic responses.

phototropism Growth or movement of an organism in a particular direction in response to light.

phylogenetic species concept The definition of a species as the smallest monophyletic group in a phylogenetic tree. Compare with **biological** and **morphological species concept.**

phylogenetic tree A diagram that depicts the evolutionary history of a group of species and the relationships among them.

phylogeny The evolutionary history of a group of organisms.

phylum (plural: phyla) In Linnaeus' system, a taxonomic category above the class level and below the kingdom level. In plants, sometimes called a *division.*

physical map A map of a chromosome that shows the number of base pairs between various genetic markers. Compare with **genetic map.**

physiology The study of how an organism's body functions.

phytoalexin Any small compound produced by a plant to combat an infection (usually a fungal infection).

phytochrome A specialized plant photoreceptor that exists in two shapes depending on the ratio of red to far-red light and is involved in the timing certain physiological processes, such as flowering, stem elongation, and germination.

phytoplankton Small drifting aquatic organisms (plankton) that are photosynthetic.

phytoremediation The use of plants to clean contaminated soils.

pigment Any molecule that absorbs certain wavelengths of visible light and reflects or transmits other wavelengths.

piloting A type of navigation in which animals use familiar landmarks to find their way.

pinocytosis Uptake by a cell of extracellular fluid by pinching off the plasma membrane to form small membrane-bound vesicles; one type of endocytosis.

pioneering species Those species that appear first in recently disturbed areas.

pit In plants, a small hole in the secondary cell walls of tracheids that allow passage of water.

pitch The sensation produced by a particular frequency of sound. Low frequencies are perceived as low pitches; high frequencies, as high pitches.

pituitary gland A small gland directly under the brain that is physically and functionally connected to the hypothalamus. Produces and secretes an array of hormones that affect many other glands and organs.

placenta A structure that forms in the pregnant uterus from maternal and fetal tissues. Exchanges nutrients and wastes between mother and fetus, anchors the fetus to the uterine wall, and produces some hormones. Occurs in most mammals and in a few other vertebrates.

placental mammals See **eutherians.**

plankton Any small organism that drifts near the surface of oceans or lakes and swims little if at all.

Plantae The monophyletic group that includes red, green, and glaucophyte algae and land plants.

plant-defense hypothesis The hypothesis that rates of herbivory are limited by plant defenses such as toxins and spines.

plantlet A small plant, particularly one that forms on a parent plant via asexual reproduction and drops, becoming an independent individual.

plasma cell An effector B cell, which produces large quantities of antibodies. Is descended from an activated B cell that has interacted with antigen.

plasma membrane A membrane that surrounds a cell, separating it from the external environment and selectively regulating passage of molecules and ions into and out of the cell. Also called *cell membrane.*

plasmid A small, usually circular, supercoiled DNA molecule independent of the cell's main chromosome(s) in prokaryotes and some eukaryotes.

plasmodesmata (singular: plasmodesma) Physical connection between two plant cells, consisting of gaps in the cell walls through which the two cells' plasma membranes, cytoplasm, and smooth ER can connect directly. Functionally similar to gap junctions in animal cells.

plasmogamy Fusion of the cytoplasm of two individuals. Occurs in many fungi.

plastid One of a family of plant organelles, bounded by a double membrane, that includes chloroplasts; chromoplasts, which house pigment-containing vacuoles; and leucoplasts, which store oils, starch, or proteins.

plastocyanin A small protein that shuttles electrons from photosystem II to photosystem I during photosynthesis.

plastoquinone (PQ) A nonprotein electron carrier in the chloroplast electron transport chain. Receives excited electrons from pheophytin and passes them to more electronegative molecules in the chain. Also carries protons to the lumen side of the thylakoid membrane, generating a proton-motive force.

platelet A small membrane-bound cell fragment in vertebrate blood that functions in blood clotting. Derived from large cells in the bone marrow.

pleiotropy (adjective: pleiotropic) The ability of a single gene to affect more than one phenotypic trait.

ploidy The number of complete chromosome sets present. *Haploid* refers to a ploidy of 1; *diploid,* a ploidy of 2; *triploid,* a ploidy of 3; and *tetraploid,* a ploidy of 4.

podium (plural: podia) See **tube foot.**

point mutation A mutation that results in a change in a single nucleotide pair in a DNA molecule.

polar (1) Asymmetrical or unidirectional. (2) Carrying a partial positive charge on one side of a molecule and a partial negative charge on the other. Polar molecules are generally hydrophilic.

polar bodies The tiny, nonfunctional cells produced during meiosis of a primary oocyte, due to most of the cytoplasm going to the ovum.

polar covalent bond A covalent bond in which electrons are shared unequally between atoms differing in electronegativity, resulting in the more electronegative atom having a partial negative charge and the other atom, a partial positive charge. Compare with **nonpolar covalent bond.**

polar nuclei In flowering plants, the nuclei in the female gametophyte that fuse with one sperm nucleus to produce the endosperm. Most species have two.

pollen grain In seed plants, a male gametophyte enclosed within a protective coat.

pollen tube In flowering plants, a structure that grows out of a pollen grain after it reaches the stigma, extends down the style, and through which two sperm cells are delivered to the ovule.

pollination The process by which pollen reaches the carpel of a flower (in flowering plants) or reaches the ovule directly (in conifers and their relatives).

poly(A) tail In eukaryotes, a sequence of 100–250 adenine nucleotides added to the 3′ end of newly transcribed messenger RNA molecules.

polycistronic mRNA An mRNA molecule that contains more than one protein-coding segment, each with its own start and stop codons and each coding for a different protein. Common in prokaryotes.

polygenic inheritance The inheritance patterns that result when many genes influence one trait.

polymer Any long molecule composed of small repeating units (monomers) bonded together. The main biological polymers are proteins, nucleic acids, and polysaccharides.

polymerase chain reaction (PCR) A laboratory technique for rapidly generating millions of identical copies of a specific stretch of DNA by incubating the original DNA sequence of interest with primers, nucleotides, and DNA polymerase.

polymerization (verb: polymerize) The process by which many identical or similar small molecules (monomers) are covalently bonded to form a large molecule (polymer).

polymorphism (adjective: polymorphic) (1) The occurrence of more than one allele at a certain genetic locus in a population. (2) The occurrence of more than two distinct phenotypes of a trait in a population.

polyp The immotile (sessile) stage in the life cycle of some cnidarians (e.g., jellyfish). Compare with **medusa.**

polypeptide A chain of 50 or more amino acids linked together by peptide bonds. Compare with **oligopeptide** and **protein.**

polyploidy (adjective: polyploid) The state of having more than two full sets of chromosomes.

polyribosome A structure consisting of one messenger RNA molecule along with many attached ribosomes and their growing peptide strands.

polysaccharide A linear or branched polymer consisting of many monosaccharides joined by glycosidic linkages. Carbohydrate polymers with relatively few residues often are called *oligosaccharides.*

polyspermy Fertilization of an egg by multiple sperm.

poor-nutrition hypothesis The hypothesis that herbivore populations are limited by the poor nutritional content of plants, especially low nitrogen.

population A group of individuals of the same species living in the same geographic area at the same time.

population cycle A regular fluctuation in size exhibited by certain populations.

population density The number of individuals of a population per unit area.

population dynamics Changes in the size and other characteristics of populations through time.

population ecology The study of how and why the number of individuals in a population changes over time.

population viability analysis (PVA) A method of estimating the likelihood that a population will avoid extinction for a given time period.

positive feedback A physiological mechanism in which a change in some variable stimulates a response that increases the change. Relatively rare in organisms but is important in generation of the action potential. Compare with **negative feedback.**

positive pressure ventilation Ventilation of the lungs that is accomplished by "pushing" air into the lungs by positive pressure in the mouth. Compare with **negative pressure ventilation.**

positive-sense virus A virus whose genome contains the same sequences as the mRNA required to produce viral proteins. Compare with **ambisense virus** and **negative-sense virus.**

posterior Toward an animal's tail and away from its head. The opposite of anterior.

posterior pituitary The part of the pituitary gland that contains the ends of hypothalamic neurosecretory cells and from which oxytocin and antidiuretic hormone are secreted. Compare with **anterior pituitary.**

postsynaptic neuron A neuron that receives signals, usually via neurotransmitters, from another neuron at a synapse. Muscle and gland cells also may receive signals from presynaptic neurons.

post-translational control Regulation of gene expression by modification of proteins (e.g., addition of a phosphate group or sugar residues) after translation.

postzygotic isolation Reproductive isolation resulting from mechanisms that operate after mating of individuals of two different species occurs. The most common mechanisms are the death of hybrid embryos or reduced fitness of hybrids.

potential energy Energy stored in matter as a result of its position or molecular arrangement. Compare with **kinetic energy.**

prebiotic soup A hypothetical solution of sugars, amino acids, nitrogenous bases, and other building blocks of larger molecules that may have formed in shallow waters or deep-ocean vents of ancient Earth and given rise to larger biological molecules.

Precambrian era The interval between the formation of the Earth, about 4.6 billion years ago, and the appearance of most animal groups about 543 million years ago. Unicellular organisms were dominant for most of this era, and oxygen was virtually absent for the first 2 billion years.

predation The killing and eating of one organism (the prey) by another (the predator).

predator Any organism that kills other organisms for food.

prediction A measurable or observable result of an experiment based on a particular hypothesis. A correct prediction provides support for the hypothesis being tested.

pressure-flow hypothesis The hypothesis that sugar movement through phloem tissue is due to differences in the turgor pressure of phloem sap.

pressure potential (ψ_P) A component of the potential energy of water caused by physical pressures on a solution. In plant cells, it equals the wall pressure plus turgor pressure. Compare with **solute potential (ψ_S).**

presynaptic neuron A neuron that transmits signals, usually by releasing neurotransmitters, to another neuron or to an effector cell at a synapse.

pre-urine See **filtrate.**

prezygotic isolation Reproductive isolation resulting from any one of several mechanisms that prevent individuals of two different species from mating.

primary cell wall The outermost layer of a plant cell wall, made of cellulose fibers and gelatinous polysaccharides, that defines the shape of the cell and withstands the turgor pressure of the plasma membrane.

primary consumer An herbivore; an organism that eats plants, algae, or other primary producers. Compare with **secondary consumer.**

primary decomposer A decomposer (detritivore) that consumes detritus from plants.

primary growth In plants, an increase in the length of stems and roots due to the activity of apical meristems. Compare with **secondary growth.**

primary immune response An acquired immune response to a pathogen that the immune system has not encountered before. Compare with **secondary immune response.**

primary oocyte The large diploid cell in an ovarian follicle that can initiate meiosis to produce a haploid ovum.

primary producer Any organism that creates its own food by photosynthesis or from reduced inorganic compounds and that is a food source for other species in its ecosystem. Also called *autotroph.*

primary RNA transcript In eukaryotes, a newly transcribed messenger RNA molecule that has not yet been processed (i.e., it has not received a 5′ cap or poly(A) tail, and still contains introns).

primary spermatocyte A diploid cell in the testis that can initiate meiosis I to produce two secondary spermatocytes.

primary structure The sequence of amino acids in a peptide or protein; also the sequence of nucleotides in a nucleic acid. Compare with **secondary, tertiary,** and **quaternary structure.**

primary succession The gradual colonization of a habitat of bare rock or gravel, usually after an environmental disturbance that removes all soil and previous organisms. Compare with **secondary succession.**

primase An enzyme that synthesizes a short stretch of RNA to use as a primer during DNA replication.

primates The lineage of mammals that includes prosimians (lemurs, lorises, etc.), monkeys, and great apes (including humans).

primer A short, single-stranded RNA molecule that base pairs with the 5′ end of a DNA template strand and is elongated by DNA polymerase during DNA replication.

prion An infectious form of a protein that is thought to cause disease by inducing the normal form to assume an abnormal three-dimensional structure. Cause of spongiform encephalopathies, such as mad cow disease.

probe A radioactively or chemically labeled single-stranded fragment of a known DNA or RNA sequence that can bind to and thus detect its complementary sequence in a sample containing many different sequences.

proboscis A long, narrow feeding appendage through which food can be obtained.

procambium A group of cells in the center of a young plant embryo that gives rise to the vascular tissue.

product Any of the final materials formed in a chemical reaction.

productivity The total amount of carbon fixed by photosynthesis per unit area per year.

progesterone A steroid hormone produced in the ovaries and secreted by the corpus luteum after ovulation; causes the uterine lining to thicken.

programmed cell death See **apoptosis.**

prokaryote A member of the domain Bacteria or Archaea; a unicellular organism lacking a nucleus and containing relatively few organelles or cytoskeletal components. Compare with **eukaryote.**

prolactin A peptide hormone, produced and secreted by the anterior pituitary, that promotes milk production in female mammals and has a variety of effects on parental behavior and seasonal reproduction in other vertebrates.

prometaphase A stage in mitosis or meiosis during which the nuclear envelope breaks down and spindle fibers attach to chromatids.

promoter A short nucleotide sequence in DNA that binds RNA polymerase, enabling transcription to begin. In prokaryotic DNA, a single promoter often is associated with several contiguous genes. In eukaryotic DNA, each gene generally has its own promoter.

promoter-proximal elements In eukaryotes, regulatory sequences in DNA that are close to a promoter and that can bind regulatory transcription factors.

proofreading The process by which a DNA polymerase recognizes and removes a wrong base added during DNA replication and then continues synthesis.

prophase The first stage in mitosis or meiosis during which chromosomes become visible and the mitotic spindle forms. Synapsis and crossing over occur during prophase of meiosis I.

prosimians One of the two major lineages of primates, including lemurs, tarsiers, pottos, and lorises. Compare with **anthropoids.**

prostate gland A gland in male mammals that surrounds the base of the urethra and secretes a fluid that is a component of semen.

protease An enzyme that can degrade proteins by cleaving the peptide bonds between amino acid residues.

protein A macromolecule consisting of one or more polypeptide chains composed of 50 or more amino acids linked together. Each protein has a unique sequence of amino acids and, in its native state, a characteristic three-dimensional shape.

protein kinase An enzyme that catalyzes the addition of a phosphate group to another protein, typically activating or inactivating the substrate protein.

proteome The complete set of proteins produced by a particular cell type.

proteomics The systematic study of the interactions, localization, functions, regulation, and other features of the full protein set (proteome) in a particular cell type.

protist Any eukaryote that is not a green plant, animal, or fungus. Protists are a diverse paraphyletic group. Most are unicellular, but some are multicellular or form aggregations called colonies.

protoderm The exterior layer of a young plant embryo that gives rise to the epidermis.

proton-motive force The combined effect of a proton gradient and an electric potential gradient across a membrane, which can drive protons across the membrane. Used by mitochondria and chloroplasts to power ATP synthesis via the mechanism of chemiosmosis.

proton pump A membrane protein that can hydrolyze ATP to power active transport of protons (H^+ ions) across a plasma membrane against an electrochemical gradient. Also called *H^+ ATPase.*

proto-oncogene Any gene that normally encourages cell division in a regulated manner, typically by triggering specific phases in the cell cycle. Mutation may convert it into an oncogene.

protostomes A major lineage of animals that share a pattern of embryological development, including spiral cleavage, formation of the mouth earlier than the anus, and formation of the coelom by splitting of a block of mesoderm. Includes arthropods, mollusks, and annelids. Compare with **deuterostomes.**

proximal tubule In the vertebrate kidney, the convoluted section of a nephron into which filtrate moves from Bowman's capsule. Involved in the largely unregulated reabsorption of electrolytes, nutrients, and water. Compare with **distal tubule.**

proximate causation In biology, the immediate, mechanistic cause of a phenomenon (how it happens), as opposed to why it evolved. Also called *proximate explanation.* Compare with **ultimate causation.**

pseudocoelomate An animal with an internal fluid-filled body cavity (coelom) that is lined with endoderm and mesoderm layers.

pseudogene A DNA sequence that closely resembles a functional gene but is not transcribed. Thought to have arisen by duplication of the functional gene followed by inactivation due to a mutation.

pseudopodium (plural: pseudopodia) A temporary bulge-like extension of certain cells used in cell crawling and ingestion of food.

puberty The various physical and emotional changes that an immature animal undergoes leading to reproductive maturity. Also the period when such changes occur.

pulmonary artery A short, thick-walled artery that carries oxygen-poor blood from the heart to the lungs.

pulmonary circulation The part of the circulatory system that sends oxygen-poor blood to the lungs. Is separate from the rest of the circulatory system (the systemic circulation) in mammals and birds.

pulmonary vein A short, thin-walled vein that carries oxygen-rich blood from the lungs to the heart.

pulse-chase experiment A type of experiment in which a population of cells or molecules at a particular moment in time is marked by means of a labeled molecule and then their fate is followed over time.

pump Any membrane protein that can hydrolyze ATP to power active transport of a specific ion or small molecule across a plasma membrane against its electrochemical gradient. See **proton pump.**

Punnett square A diagram that depicts the genotypes and phenotypes that should appear in offspring of a certain cross.

pupa (plural: pupae) A metamorphosing insect that is enclosed in a protective case.

pupation A developmental stage of many insects, in which the body metamorphoses from the larval form to the adult form while enclosed in a protective case.

pupil The hole in the center of the iris through which light enters a vertebrate or cephalopod eye.

pure line In animal or plant breeding, a strain of individuals that produce offspring identical to themselves when self-pollinated or crossed to another member of the same population. Pure lines are homozygous for most, if not all, genetic loci.

purines A class of small, nitrogen-containing, double-ringed bases (guanine, adenine) found in nucleotides. Compare with **pyrimidines.**

pyrimidines A class of small, nitrogen-containing, single-ringed bases (cytosine, uracil, thymine) found in nucleotides. Compare with **purines.**

pyruvate dehydrogenase A large enzyme complex, located in the inner mitochondrial membrane, that is responsible for conversion of pyruvate to acetyl CoA during cellular respiration.

quantitative trait A heritable feature that exhibits phenotypic variation along a smooth, continuous scale of measurement (e.g., human height), rather than the distinct forms characteristic of discrete traits.

quaternary structure The overall three-dimensional shape of a protein containing two or more polypeptide chains (subunits); determined by the number, relative positions, and interactions of the subunits. Compare with **primary, secondary,** and **tertiary structure.**

radial cleavage The pattern of embryonic cleavage seen in protostomes, in which cells divide at right angles to each other to form tiers. Compare with **spiral cleavage.**

radial symmetry An animal body pattern in which there are least two planes of symmetry. Typically, the body is in the form of a cylinder or disk, with body parts radiating from a central hub. Compare with **bilateral symmetry.**

radiation Transfer of heat between two bodies that are not in direct physical contact. More generally, the emission of electromagnetic energy of any wavelength.

radicle The root of a plant embryo.

radula A rasping feeding appendage in gastropods (snails, slugs).

rain shadow The dry region on the side of a mountain range away from the prevailing wind.

range The geographic distribution of a species.

Ras protein A type of G protein that is activated by binding of signaling molecules to receptor tyrosine kinases and then initiates a phosphorylation cascade, culminating in a cell response.

Rb protein A tumor-suppressor protein that helps regulate progression of a cell from the G_1 phase to the S phase of the cell cycle. Defects in Rb protein are found in many types of cancer.

reactant Any of the starting materials in a chemical reaction.

reaction center Centrally located component of a photosystem containing proteins and a pair of specialized chlorophyll molecules. Is surrounded by antenna complexes and receives excited electrons from them.

reactive oxygen intermediates (ROIs) Highly reactive oxygen-containing compounds that are used in plant and animal cells to kill infected cells and for other purposes.

reading frame The division of a sequence of DNA or RNA into a particular series of three-nucleotide codons. There are three possible reading frames for any sequence.

realized niche The ecological niche that a species occupies in the presence of competitors. Compare with **fundamental niche.**

receptor-mediated endocytosis Uptake by a cell of certain extracellular macromolecules, bound to specific receptors in the plasma membrane, by pinching off the membrane to form small membrane-bound vesicles.

receptor tyrosine kinase (RTK) Any of a class of cell-surface signal receptors that undergo phosphorylation after binding a signaling molecule. The activated, phosphorylated receptor then triggers a signal-transduction pathway inside the cell.

recessive Referring to an allele whose phenotypic effect is observed only in homozygous individuals. Compare with **dominant.**

reciprocal altruism Altruistic behavior that is exchanged between a pair of individuals at different points in time (i.e., sometimes individual A helps individual B, and sometimes B helps A).

reciprocal cross A breeding experiment in which the mother's and father's phenotypes are the reverse of that examined in a previous breeding experiment.

recombinant Possessing a new combination of alleles. May refer to a single chromosome or DNA molecule, or to an entire organism.

recombinant DNA technology A variety of techniques for isolating specific DNA fragments and introducing them into different regions of DNA and/or a different host organism.

recombination See **genetic recombination.**

rectal gland A salt-excreting gland in the digestive system of sharks, skates, and rays.

rectum The last portion of the digestive tract where feces are held until they are expelled.

red blood cell A hemoglobin-containing cell that circulates in the blood and delivers oxygen from the lungs to the tissues. Also called *erythrocyte.*

redox reaction Any chemical reaction that involves the transfer of one or more electrons from one reactant to another. Also called *reduction-oxidation reaction.*

reduction An atom's gain of electrons during a redox reaction, either by acceptance of an electron from another atom or by the electrons in covalent bonds moving closer to the atomic nucleus.

reduction-oxidation reaction See **redox reaction.**

reflex An involuntary response to environmental stimulation. May involve the brain (e.g., conditioned reflex) or not (e.g., spinal reflex).

refractory No longer responding to stimuli that previously elicited a response. For example, the tendency of voltage-gated sodium channels to remain closed immediately after an action potential.

regeneration Growth of a new body part to replace a lost body part.

regulatory cascade In embryonic development, a progressive series of interactions among genes and/or cytoplasmic determinants that organizes the body plan of the embryo.

regulatory sequence, DNA Any segment of DNA that is involved in controlling transcription of a specific gene by binding certain proteins.

regulatory site A site on an enzyme to which a regulatory molecule can bind and affect the enzyme's activity; separate from the active site where catalysis occurs.

regulatory transcription factor General term for proteins that bind to DNA regulatory sequences (eukaryotic enhancers, silencers, and promoter-proximal elements), but not to the promoter itself, leading to an increase or decrease in transcription of specific genes. Compare with **basal transcription factor.**

reinforcement In evolutionary biology, the natural selection for traits that prevent interbreeding between recently diverged species.

release factors Proteins that can trigger termination of RNA translation when a ribosome reaches a stop codon.

releaser See **sign stimulus.**

renal corpuscle In the vertebrate kidney, the ball-like structure at the beginning of a nephron, consisting of a glomerulus and the surrounding Bowman's capsule. Acts as a filtration device.

replacement mutation See **missense mutation.**

replacement rate The number of offspring each female must produce over her entire life to "replace" herself and her mate, resulting in zero population growth. The actual number is slightly more than 2 because some offspring die before reproducing.

replica plating A method of identifying bacterial colonies that have certain mutations by transferring cells from each colony on a master plate to a second (replica) plate and observing their growth when exposed to different conditions.

replication fork The Y-shaped site at which a double-stranded molecule of DNA is separated into two single strands for replication.

repolarization Return to a normal membrane potential after it has changed; a normal phase in an action potential.

repressor Any regulatory protein that inhibits transcription.

reproduction The ability of an organism to make an exact or nearly exact copy of itself.

reptiles One of the two lineages of amniotes (vertebrates that produce amniotic eggs) distinguished by adaptations for reproduction on land. Includes turtles, snakes and lizards, crocodiles and alligators, and birds. Except for birds, all are ectotherms.

reservoir In biogeochemical cycles, a location in the environment where elements are stored for a time.

residues In a polymer, the individual units derived from the monomers that covalently bind to form the polymer. Proteins contain amino acid residues; nucleic acids, nucleotide residues; and polysaccharides, sugar residues.

resilience, community A measure of how quickly a community recovers following a disturbance.

resistance, community A measure of how much a community is affected by a disturbance.

resistance (*R*) loci Genes in plants encoding proteins involved in sensing the presence of pathogens and mounting a defensive response. Compare with **avirulence (*avr*) loci.**

respiration See **cellular respiration.**

respiratory distress syndrome A syndrome in which premature infants can suffocate due to insufficient surfactant in their lungs.

respiratory system The collection of cells, tissues, and organs responsible for gas exchange between an animal and its environment.

resting potential The membrane potential of a cell in its resting, or normal, state.

restriction endonucleases Bacterial enzymes that cut DNA at a specific base-pair sequence (restriction site). Also called *restriction enzymes.*

restriction site A specific short sequence in a DNA molecule that is recognized and cleaved by a particular restriction endonuclease. Also called *restriction endonuclease recognition site.*

retina A thin layer of light-sensitive cells (rods and cones) and neurons at the back of a camera-type eye, such as that of cephalopods and vertebrates.

retinal A light-absorbing pigment, derived from vitamin A, that is linked to the protein opsin in rods and cones of the vertebrate eye.

retrovirus A virus with an RNA genome that reproduces by transcribing its RNA into a DNA sequence and then inserting that DNA into the host's genome for replication.

reverse transcriptase A enzyme of retroviruses (RNA viruses) that can synthesize double-stranded DNA from a single-stranded RNA template.

Rh factor A protein present on the surface of red blood cells in some but not all humans. Clumping may occur when blood from an Rh^+ individual is mixed with that from an Rh^- individual.

rhizobia (singular: rhizobium) Members of the bacterial genus *Rhizobia;* nitrogen-fixing bacteria that live in root nodules of members of the pea family (legumes).

rhizoid The hairlike structure that anchors a bryophyte (nonvascular plant) to the substrate.

rhizome A modified stem that runs horizontally underground and produces new plants at the nodes (a form of asexual reproduction). Compare with **stolon.**

rhodopsin A transmembrane complex that is instrumental in detection of light by rods and cones of the vertebrate eye. Is composed of the transmembrane protein opsin covalently linked to retinal, a light-absorbing pigment.

ribonucleic acid (RNA) A nucleic acid composed of ribonucleotides that usually is single stranded and functions as structural components of ribosomes (rRNA), transporters of amino acids (tRNA), and translators of the message of the DNA code (mRNA).

ribonucleotide See **nucleotide.**

ribosomal RNA (rRNA) A RNA molecule that forms part of the structure of a ribosome.

ribosome A large complex structure that synthesizes proteins by using the genetic information encoded in messenger RNA strands. Consists of two subunits, each composed of ribosomal RNA and proteins.

ribosome binding site In a bacterial mRNA molecule, the sequence just upstream of the start codon to which a ribosome binds to initiate translation. Also called the *Shine-Dalgarno sequence.*

ribozyme Any RNA molecule that can act as a catalyst, that is, speed up a chemical reaction.

ribulose bisphosphate (RuBP) A five-carbon compound that combines with CO_2 in the first step of the Calvin cycle during photosynthesis.

RNA See **ribonucleic acid.**

RNA interference (RNAi) Degradation of an mRNA molecule or inhibition of its translation following its binding by a short RNA (microRNA) whose sequence is complementary to a portion of the mRNA.

RNA polymerase One of a class of enzymes that catalyze synthesis of RNA from ribonucleotides using a DNA template. Also called *RNA pol.*

RNA processing In eukaryotes, the changes that a primary RNA transcript undergoes in the nucleus to become a mature mRNA molecule, which is exported to the cytoplasm. Includes the addition of a 5′ cap and poly(A) tail and splicing to remove introns.

RNA replicase A viral enzyme that can synthesize RNA from an RNA template.

rod cell A photoreceptor cell with a rod-shaped outer portion that is particularly sensitive to dim light, but not used to distinguish colors. Also called simply *rod*. Compare with **cone cell.**

root (1) An underground part of a plant that anchors the plant and absorbs water and nutrients. (2) In a phylogenetic tree, the bottom, most ancient node.

root apical meristem (RAM) A group of undifferentiated plant cells at the tip of a plant root that can differentiate into mature root tissue.

root cap A small group of cells that covers and protects the tip of a plant root. Senses gravity and determines the direction of root growth.

root hair A long, thin outgrowth of the epidermal cells of plant roots, providing increased surface area for absorption of water and nutrients.

root pressure Positive (upward) pressure of xylem sap in the vascular tissue of roots. Is generated during the night as a result of the accumulation of ions from the soil and subsequent osmotic movement of water into the xylem.

root system The belowground part of a plant.

rotifers Members of the phylum Rotifera. Distinguished by a cluster of cilia (corona) at the anterior end and a pseudocoelom. Rotifers belong to the lophotrochozoan branch of the protostomes.

rough endoplasmic reticulum (rough ER) The portion of the endoplasmic reticulum that is dotted with ribosomes. Involved in synthesis of plasma membrane proteins, secreted proteins, and proteins localized to the ER, Golgi apparatus, and lysosomes. Compare with **smooth endoplasmic reticulum.**

roundworms Members of the phylum Nematoda. Distinguished by an unsegmented body with a pseudocoelom and no appendages. Roundworms belong to the ecdysozoan branch of the protostomes. Also called *nematodes.*

rRNA See **ribosomal RNA.**

rubisco The enzyme that catalyzes the first step of the Calvin cycle during photosynthesis: the addition of a molecule of CO_2 to ribulose bisphosphate. Also called *ribulose 1,5-bisphosphate carboxylase/oxygenase.*

ruminants A group of hoofed mammals (e.g., cattle, sheep, deer) that have a four-chambered stomach specialized for digestion of plant cellulose. Ruminants regurgitate the cud, a mixture of partially digested food and cellulose-digesting bacteria, from the largest chamber (the rumen) for further chewing.

sac fungus A member of a monophyletic lineage of fungi (Ascomycota) that produce large, often cup-shaped reproductive structures that contain asci. Also called *cup fungus* and *ascomycete.*

salicylic acid A compound produced by plants that may play a role in systemic acquired resistance (SAR) against pathogens; a component of aspirin.

salivary glands Vertebrate glands that secrete saliva (a mixture of water, mucus-forming glycoproteins, and digestive enzymes) into the mouth.

sampling error The accidental selection of a nonrepresentative sample from some larger population, due to chance.

saprophyte An organism that feeds primarily on dead plant material.

sapwood The younger xylem in the outer layer of wood of a stem or root, functioning primarily in water transport.

sarcomere The repeating contractile unit of a skeletal muscle cell; the portion of a myofibril located between adjacent Z disks.

sarcoplasmic reticulum Sheets of smooth endoplasmic reticulum in a muscle cell. Contains high concentrations of calcium, which can be released into the cytoplasm to trigger contraction.

saturated Referring to fats and fatty acids in which all the carbon-carbon bonds are single bonds. Such fats have relatively high melting points. Compare with **unsaturated.**

scanning electron microscope (SEM) A microscope that produces images of the surfaces of objects by reflecting electrons from a specimen coated with a layer of metal atoms. Compare with **transmission electron microscope.**

scarify To scrape, rasp, cut, or otherwise damage the coat of a seed. Necessary in some species to trigger germination.

Schwann cell A type of glial cell that wraps around axons of some neurons outside the brain and spinal cord, forming a myelin sheath that provides electrical insulation. Compare with **oligodendrocyte.**

sclereid In plants, a type of sclerenchyma cell that usually functions in protection, such as in seed coats and nutshells. Compare with **fiber.**

sclerenchyma cell In plants, a cell that has a thick secondary cell wall and provides support; typically contains the tough structural polymer lignin and usually is dead at maturity. Includes fibers and sclereids. Compare with **collenchyma cell** and **parenchyma cell.**

screen See **genetic screen.**

scrotum A sac of skin, containing the testes, suspended just outside the abdominal body cavity of many male mammals.

secondary cell wall The inner layer of a plant cell wall formed by certain cells as they mature. Provides support or protection.

secondary consumer A carnivore; an organism that eats herbivores. Compare with **primary consumer.**

secondary growth In plants, an increase in the width of stems and roots due to the activity of lateral meristems. Compare with **primary growth.**

secondary immune response The acquired immune response to a pathogen that the immune system has encountered before. Compare with **primary immune response.**

secondary production The total amount of new body tissue produced by animals that eat plants. May involve growth and/or reproduction.

secondary spermatocyte A cell produced by meiosis I of a primary spermatocyte in the testis. Can undergo meiosis II to produce spermatids.

secondary structure In proteins, localized folding of a polypeptide chain into regular structures (e.g., α-helix and β-pleated sheet) stabilized by hydrogen bonding between atoms of the backbone. In nucleic acids, elements of structure (e.g., helices and hairpins) stabilized by hydrogen bonding and other interactions between complementary bases. Compare with **primary, tertiary,** and **quaternary structure.**

secondary succession Gradual colonization of a habitat after an environmental disturbance (e.g., fire, windstorm, logging) that removes some or all previous organisms but leaves the soil intact. Compare with **primary succession.**

second law of thermodynamics The principle of physics that the entropy of the universe or any closed system increases during any spontaneous process.

second-male advantage The reproductive advantage of a male who mates with a female last, after other males have mated with her.

second messenger A nonprotein signaling molecule produced or activated inside a cell in response to stimulation at the cell surface. Commonly used to relay the message of a hormone or other extracellular signaling molecule.

secretin A peptide hormone produced by cells in the small intestine in response to the arrival of food from the stomach. Stimulates secretion of bicarbonate (HCO_3^-) from the pancreas.

sedimentary rock A type of rock formed by gradual accumulation of sediment, as in riverbeds and on the ocean floor. Most fossils are found in sedimentary rocks.

seed A plant reproductive structure consisting of an embryo, associated nutritive tissue (endosperm), and an outer protective layer (seed coat). In angiosperms, develops from the fertilized ovule of a flower.

seed coat A protective layer around a seed that encases both the embryo and the endosperm.

seed plants Members of several phyla of green plants that have vascular tissue and make seeds. Include **angiosperms** and **gymnosperms.**

seedless vascular plants Members of several phyla of green plants that have vascular tissue but do not make seeds. Include horsetails, ferns, lycophytes, and whisk ferns.

segment A well-defined region of the body along the anterior-posterior body axis, containing similar structures as other, nearby segments.

segmentation Division of the body or a part of it into a series of similar structures; exemplified by the body segments of insects and worms and by the somites of vertebrates.

segmentation genes A group of genes that affect body segmentation in embryonic development. Includes gap genes, pair-rule genes, and segment polarity genes.

segmented worms See **annelids.**

segregation, principle of The concept that each pair of hereditary elements (alleles of the same gene) separate from each other during the formation of offspring (i.e., during meiosis). One of Mendel's two principles of genetics.

selective adhesion The tendency of cells of one tissue type to adhere to other cells of the same type.

selectively permeable membrane Any membrane across which some solutes can move more readily than others.

selective permeability The property of a membrane that allows some substances to diffuse across it much more readily than other substances.

self Property of a molecule or cell such that immune system cells do not attack it, due to certain molecular similarities to other body cells.

self-fertilization In plants, the fusion of two gametes from the same individual to form a diploid offspring. Also called *selfing.*

self-incompatible Incapable of self-fertilization.

semen The combination of sperm and accessory fluids that is released by male mammals and reptiles during ejaculation.

semiconservative replication The mechanism of replication used by cells to copy DNA. Results in each daughter DNA molecule containing one old strand and one new strand.

seminal vesicles In male mammals, paired reproductive glands that secrete a sugar-containing fluid into semen, which provides energy for sperm movement. In other vertebrates and invertebrates, often stores sperm.

senescence The process of aging.

sensitive period A short time period in a young animal's life during which learning of certain critical behaviors can occur. Also called the *critical period.*

sensor Any cell, organ, or structure with which an animal can sense some aspect of the external or internal environment. Usually functions, along with an integrator and effector, as part of a homeostatic system.

sensory neuron A nerve cell that carries signals from sensory receptors to the central nervous system. Compare with **interneuron** and **motor neuron.**

sepal One of the protective leaflike organs enclosing a flower bud and later supporting the blooming flower.

septum (plural: septa) Any wall-like structure. In fungi, septa divide the filaments (hyphae) of mycelia into cell-like compartments.

serotonin A neurotransmitter involved in many brain functions, including sleep, pleasure, and mood.

serum The liquid that remains when cells and clot material are removed from clotted blood. Contains water, dissolved gases, growth factors, nutrients, and other soluble substances. Compare with **plasma.**

sessile Permanently attached to a substrate; not capable of moving to another location.

set point A normal or target value for a regulated internal variable, such as body heat or blood pH.

severe combined immunodeficiency disease (SCID) A human disease characterized by an extremely high vulnerability to infectious disease. Caused by a genetic defect in the immune system.

sex chromosome Any chromosome carrying genes involved in determining the sex of an individual. Compare with **autosome.**

sex-linked inheritance Inheritance patterns observed in genes carried on sex chromosomes, so females and males have different numbers of alleles of a gene and may pass its trait only to one sex of offspring. Also called *sex-linkage.*

sexual dimorphism Any trait that differs between males and females.

sexual reproduction Any form of reproduction in which genes from two parents are combined via fusion of gametes, producing offspring that are genetically distinct from both parents. Compare with **asexual reproduction.**

sexual selection A pattern of natural selection that favors individuals with traits that increase their ability to obtain mates. Acts more strongly on males than females.

shell A hard protective outer structure. In protists, also called a *test.*

Shine-Dalgarno sequence See **ribosome binding sequence.**

shoot apical meristem (SAM) A group of undifferentiated plant cells at the tip of a plant stem that can differentiate into mature shoot tissues.

shoot system The aboveground part of a plant comprising stems, leaves, and flowers (in angiosperms).

short-day plant A plant that blooms in response to long nights (usually in late summer or fall in the northern hemisphere). Compare with **day-neutral** and **long-day plant.**

short interspersed nuclear elements See SINES.

shotgun sequencing A method of sequencing genomes that is based on breaking the genome into small pieces, sequencing each piece separately, and then figuring out how the pieces are connected.

sieve plate In plants, a pore-containing structure at one end of a sieve-tube member in phloem.

sieve-tube member In plants, an elongated sugar-conducting cell in phloem that has sieve plates at both ends, allowing sap to flow to adjacent cells.

signal In behavioral ecology, any information-containing behavior.

signal receptor Any cellular protein that binds to a particular signaling molecule (e.g., a hormone or neurotransmitter) and triggers a response by the cell. Receptors for water-soluble signals are transmembrane proteins in the plasma membrane; those for many lipid-soluble signals (e.g., steroid hormones) are located inside the cell.

signal recognition particle (SRP) A RNA-protein complex that binds to the ER signal sequence in a polypeptide as it emerges from a ribosome and transports the ribosome-polypeptide complex to the ER membrane where synthesis of the polypeptide is completed.

signal transducers and activators of transcription (STATs) In mammals, a group of regulatory transcription factors that, upon phosphorylation, can activate transcription of certain genes.

signal transduction The process by which a stimulus (e.g., a hormone, a neurotransmitter, or sensory information) outside a cell is amplified and converted into a response by the cell. Usually involves a specific sequence of molecular events, or signal transduction pathway.

signal transduction cascade See **phosphorylation cascade.**

sign stimulus A simple stimulus that elicits an invariant, stereotyped behavioral response (fixed action pattern) from an animal. Also called a *releaser.*

silencer A regulatory sequence in eukaryotic DNA to which repressor proteins can bind, inhibiting transcription of certain genes.

silent mutation A mutation that does not detectably affect the phenotype of the organism.

simple eye An eye with only one light-collecting apparatus (e.g., one lens), as in vertebrates. Compare with **compound eye.**

simple fruit A fruit (e.g., apricot) that develops from a single flower that has a single carpel or several fused carpels. Compare with **aggregate** and **multiple fruit.**

simple sequence repeat See **microsatellite.**

SINEs (short interspersed nuclear elements) The second most abundant class of transposable elements in human genomes; can create copies of itself and insert them elsewhere in the genome. Compare with **LINEs.**

single nucleotide polymorphism (SNP) A site on a chromosome where individuals in a population have different nucleotides. Can be used as a genetic marker to help track the inheritance of nearby genes.

single-strand DNA-binding proteins (SSBPs) A class of proteins that attach to separated strands of DNA during replication or transcription, preventing them from re-forming a double helix.

sink Any tissue, site, or location where an element or a molecule is consumed or taken out of circulation (e.g., in plants, a tissue where sugar exits the phloem). Compare with **source.**

sinoatrial (SA) node A cluster of cardiac muscle cells, in the right atrium of the vertebrate heart, that initiates the heartbeat and determines the heart rate. Compare with **atrioventricular (AV) node.**

siphon A tubelike appendage of many molluscs, that is often used for feeding or propulsion.

sister chromatids The paired strands of a recently replicated chromosome, which are connected at the centromere and eventually separate during anaphase of mitosis and meiosis II. Compare with **non-sister chromatids.**

sister groups Closely related taxa, which occupy adjacent branches in a phylogenetic tree. Also called *sister taxa.*

skeletal muscle The muscle tissue attached to the bones of the vertebrate skeleton. Consists of long, unbranched muscle fibers with a characteristic striped (striated) appearance; controlled voluntarily. Also called *striated muscle.* Compare with **cardiac** and **smooth muscle.**

sliding-filament model The hypothesis that thin (actin) filaments and thick (myosin) filaments slide past each other, thereby shortening the sarcomere. Shortening of all the sarcomeres in a myofibril results in contraction of the entire myofibril.

slug (1) A member of a certain lineage of terrestrial gastropods, closely related to snails but lacking a shell. (2) A mobile aggregation of cells of a cellular slime mold.

small intestine The portion of the digestive tract between the stomach and the large intestine. The site of the final stages of digestion and of most nutrient absorption.

small nuclear ribonucleoproteins See **snRNPs.**

smooth endoplasmic reticulum (smooth ER) The portion of the endoplasmic reticulum that does not have ribosomes attached to it. Involved in synthesis and secretion of lipids. Compare with **rough endoplasmic reticulum.**

smooth muscle The unstriated muscle tissue that lines the intestine, blood vessels, and some other organs. Consists of tapered, unbranched cells that can sustain long contractions. Not voluntarily controlled. Compare with **cardiac** and **skeletal muscle.**

snRNPs (small nuclear ribonucleoproteins) Complexes of proteins and small RNA molecules that function in splicing (removal of introns from primary RNA transcripts) as components of spliceosomes.

sodium-potassium pump A transmembrane protein that uses the energy of ATP to move sodium ions out of the cell and potassium ions in. Also called *Na^+/K^+-ATPase.*

solute Any substance that is dissolved in a liquid.

solute potential (ψ_S) A component of the potential energy of water caused by a difference in solute concentrations at two locations. Also called *osmotic potential.* Compare with **pressure potential (ψ_P).**

solution A liquid containing one or more dissolved solids or gases in a homogeneous mixture.

solvent Any liquid in which one or more solids or gases can dissolve.

soma See **cell body.**

somatic cell Any type of cell in a multicellular organism except eggs, sperm, and their precursor cells. Also called *body cells.*

somatic nervous system The part of the peripheral nervous system (outside the brain and spinal cord) that controls skeletal muscles and is under voluntary control. Compare with **autonomic nervous system.**

somatostatin A hormone secreted by the pancreas and hypothalamus that inhibits the release of several other hormones.

somites Paired blocks of mesoderm on both sides of the developing spinal cord in a vertebrate embryo. Give rise to muscle tissue, vertebrae, ribs, limbs, etc.

soredium (plural: soredia) In lichens, a small reproductive structure that consists of fungal hyphae surrounding a cluster of green algae.

source Any tissue, site, or location where a substance is produced or enters circulation (e.g., in plants, the tissue where sugar enters the phloem). Compare with **sink.**

Southern blotting A technique for identifying specific DNA fragments separated by gel electrophoresis by transferring them to filter paper, separating the strands, and hybridizing with a labeled probe complementary to the fragment of interest. Compare with **Northern** and **Western blotting.**

speciation The evolution of two or more distinct species from a single ancestral species.

species A distinct, identifiable group of populations that is thought to be evolutionarily independent of other populations and whose members can interbreed. Generally distinct from other species in appearance, behavior, habitat, ecology, genetic characteristics, etc.

species–area relationship The mathematical relationship between the area of a certain habitat and the number of species that it can support.

species diversity The variety and relative abundance of the species present in a given ecological community.

species richness The number of species present in a given ecological community.

specific heat The amount of energy required to raise the temperature of 1 gram of a substance by 1°C; a measure of the capacity of a substance to absorb energy.

spectral karyotyping See **chromosome painting.**

spectrophotometer An instrument used to measure the wavelengths of light that are absorbed by a particular pigment.

sperm A mature male gamete; smaller and more mobile than the female gamete.

spermatid An immature sperm cell.

spermatogenesis The production of sperm. Occurs continuously in a testis.

spermatogonia (singular: spermatogonium) The diploid cells in a testis that can give rise to primary spermatocytes.

spermatophore A gelatinous package of sperm cells that is produced by males of species that have internal fertilization without copulation.

sphincter A muscular valve that can close off a tube, as in a blood vessel or a part of the digestive tract.

spicule Stiff spike of silica or calcium carbonate found in the body of many sponges.

spindle fibers Groups of microtubules that attach to chromosomes and push and pull them during mitosis and meiosis.

spiracle In insects, a small opening that connects air-filled tracheae to the external environment, allowing for gas exchange.

spiral cleavage The pattern of embryonic cleavage seen in deuterostomes, in which cells divide at oblique angles to form a spiral coil of cells. Compare with **radial cleavage.**

spleen A dark red organ, found near the stomach of most vertebrates, that filters blood, stores extra red blood cells in case of emergency, and plays a role in immunity.

spliceosome In eukaryotes, a large, complex assembly of snRNPs (small nuclear ribonucleoproteins) that catalyzes removal of introns from primary RNA transcripts.

splicing The process by which introns are removed from primary RNA transcripts and the remaining exons are connected together.

sporangium (plural: sporangia) A spore-producing structure found in seed plants, some protists, and some fungi (e.g., chytrids).

spore (1) In bacteria, a dormant form that generally is resistant to extreme conditions. (2) In eukaryotes, a single cell produced by mitosis or meiosis (not by fusion of gametes) that is capable of developing into an adult organism.

sporophyte In organisms undergoing alternation of generations, the multicellular diploid form that arises from two fused gametes and produces haploid spores. Compare with **gametophyte.**

sporopollenin A watertight material that encases spores and pollen of modern land plants.

stabilizing selection A pattern of natural selection that favors phenotypes near the middle of the range of phenotypic variation. Reduces overall genetic variation in a population. Compare with **disruptive selection.**

stamen The male reproductive structure of a flower. Consists of an anther, in which pollen grains are produced, and a filament, which supports the anther. Compare with **carpel.**

standing defense See **constitutive defense.**

stapes The last of three small bones (ossicles) in the middle ear of vertebrates. Receives vibrations from the tympanic membrane and by vibrating against the oval window passes them to the cochlea.

starch A mixture of two storage polysaccharides, amylose and amylopectin, both formed from α-glucose monomers. Amylopectin is branched, and amylose is unbranched. The major form of stored carbohydrate in plants.

start codon The AUG triplet in mRNA at which protein synthesis begins; codes for the amino acid methionine.

statocyst A sensory organ of many arthropods that detects the animal's orientation in space (i.e., whether the animal is flipped upside down).

statolith A tiny stone or dense particle found in specialized gravity-sensing organs in some animals such as lobsters.

statolith hypothesis The hypothesis that amyloplasts (dense, starch-storing plant organelles) serve as statoliths in gravity detection by plants.

STATs See **signal transducers and activators of transcription.**

stem cell Any relatively undifferentiated cell that can divide to produce daughter cells identical to itself or more specialized daughter cells, which differentiate further into specific cell types.

stereocilium (plural: stereocilia) One of many stiff outgrowths from the surface of a hair cell that are involved in detection of sound by terrestrial vertebrates or of waterborne vibrations by fishes.

steroid A class of lipid with a characteristic four-ring structure.

steroid-hormone receptor One of a family of intracellular receptors that bind to various steroid hormones, forming a hormone-receptor complex that acts as a regulatory transcription factor and activates transcription of specific target genes.

sticky end The short, single-stranded ends of a DNA molecule cut by a restriction endonuclease. Tend to form hydrogen bonds with other sticky ends that have complementary sequences.

stigma The moist tip at the end of a flower carpel to which pollen grains adhere.

stolon A modified stem that runs horizontally over the soil surface and produces new plants at the nodes (a form of asexual reproduction). Compare with **rhizome.**

stoma (plural: stomata) Generally, a pore or opening. In plants, a microscopic pore on the surface of a leaf or stem through which gas exchange occurs.

stomach A tough, muscular pouch in the vertebrate digestive tract between the esophagus and small intestine. Physically breaks up food and begins digestion of proteins.

stop codon One of three mRNA triplets (UAG, UGA, or UAA) that cause termination of protein synthesis. Also called a *termination codon.*

strain A population of genetically similar or identical individuals.

stream A body of water that moves constantly in one direction.

striated muscle See **skeletal muscle.**

stroma The fluid matrix of a chloroplast in which the thylakoids are embedded. Site where the Calvin cycle reactions occur.

structural formula A two-dimensional notation in which the chemical symbols for the constituent atoms are joined by straight lines representing single (—), double (=), or triple (≡) covalent bonds. Compare with **molecular formula.**

structural gene A stretch of DNA that codes for a functional protein or functional RNA molecule, not including any regulatory sequences (e.g., a promoter, enhancer).

structural homology Similarities in organismal structures (e.g., limbs, shells, flowers) that are due to inheritance from a common ancestor.

structural isomer A molecule that shares the same molecular formula as another molecule but differs in the order in which covalently bonded atoms are attached. Compare with **geometric** and **optical isomer.**

style The slender stalk of a flower carpel connecting the stigma and the ovary.

subspecies A population that has distinctive traits and some genetic differences relative to other populations of the same species but that is not distinct enough to be classified as a separate species.

substrate (1) A reactant that interacts with an enzyme in a chemical reaction. (2) A surface on which a cell or organism sits.

substrate-level phosphorylation Production of ATP by transfer of a phosphate group from an intermediate substrate directly to ADP. Occurs in glycolysis and in the Krebs cycle.

succession In ecology, the gradual colonization of a habitat after an environmental disturbance (e.g., fire, flood), usually by a series of species. See **primary** and **secondary succession.**

sucrose A disaccharide formed from glucose and fructose. One of the two main products of photosynthesis.

sulfate reducer A prokaryote that produces hydrogen sulfide (H_2S) as a by-product of cellular respiration.

summation The additive effect of different postsynaptic potentials at a nerve or muscle cell, such that several subthreshold stimulations can cause an action potential.

supernatant The liquid above a layer of solid particles (the pellet) in a tube after centrifugation.

surface tension The cohesive force that causes molecules at the surface of a liquid to stick together, thereby resisting deformation of the liquid's surface and minimizing its surface area.

surfactant A mixture of phospholipids and proteins produced by lung cells that reduces surface tension, allowing the lungs to expand more.

survivorship curve A graph depicting the percentage of a population that survives to different ages.

suspension culture A population of cells grown in a flask containing a liquid nutrient medium. The flask is rotated continuously to keep the cells suspended and the nutrients mixed.

suspension feeder Any organism that obtains food by filtering small particles or small organisms out of water or air. Also called *filter feeder.*

sustainability The planned use of environmental resources at a rate no faster than the rate at which they are naturally replaced.

sustainable agriculture Agricultural techniques that are designed to maintain long-term soil quality and productivity.

swim bladder A gas-filled organ of many ray-finned fishes that regulates buoyancy.

swamp A wetland that has a steady rate of water flow and is dominated by trees and shrubs.

symbiosis (adjective: symbiotic) Any close and prolonged physical relationship between individuals of two different species. See **commensalism, mutualism,** and **parasitism.**

symmetric competition Ecological competition between two species in which both suffer similar declines in fitness. Compare with **asymmetric competition.**

sympathetic nervous system The part of the autonomic nervous system that stimulates fight-or-flight responses, such as increased heart rate, increased blood pressure, and decreased digestion. Compare with **parasympathetic nervous system.**

sympatric speciation The divergence of populations living within the same geographic area into different species as the result of their genetic (not physical) isolation. Compare with **allopatric speciation.**

sympatry Condition in which two or more populations live in the same geographic area, or close enough to permit interbreeding. Compare with **allopatry.**

symplast In plant roots, a continuous pathway through which water can flow through the cytoplasm of adjacent cells that are connected by plasmodesmata. Compare with **apoplast.**

symporter A carrier protein that allows an ion to diffuse down an electrochemical gradient, using the energy of that process to transport a different substance in the same direction *against* its concentration gradient. Compare with **antiporter.**

synapomorphy A shared, derived trait found in two or more taxa that is present in their most recent common ancestor but is missing in more distant ancestors. Useful for inferring evolutionary relationships.

synapse The interface between two neurons or between a neuron and an effector cell.

synapsis The physical pairing of two homologous chromosomes during prophase I of meiosis. Crossing over occurs during synapsis.

synaptic cleft The space between two communicating nerve cells (or between a neuron and effector cell) at a synapse, across which neurotransmitters diffuse.

synaptic plasticity Long-term changes in the responsiveness or physical structure of a synapse that can occur after particular stimulation patterns. Thought to be the basis of learning and memory.

synaptic vesicle A small neurotransmitter-containing vesicle at the end of an axon that releases neurotransmitter into the synaptic cleft by exocytosis.

synaptonemal complex A network of proteins that holds non-sister chromatids together during synapsis in meiosis I.

synthesis (S) phase The phase of the cell cycle during which DNA is synthesized and chromosomes are replicated.

system In biology, a more complex organization resulting from the combination of various components, such as a group of organs that work together to perform a physiological function.

systemic acquired resistance (SAR) A slow, widespread response of plants to a localized infection that protects healthy tissue from invasion by pathogens. Compare with **hypersensitive response.**

systemic circulation The part of the circulatory system that sends oxygen-rich blood from the lungs out to the rest of the body. Is separate from the pulmonary circulation in mammals and birds.

systemin A peptide hormone, produced by plant cells damaged by herbivores, that initiates a protective response in undamaged cells.

systole The portion of the heartbeat cycle during which the heart muscles are contracting. Compare with **diastole.**

systolic blood pressure The force exerted by blood against artery walls during contraction of the heart's left ventricle. Compare with **diastolic blood pressure.**

taiga A vast forest biome throughout subarctic regions, consisting primarily of short conifer trees. Characterized by intensely cold winters, short summers, and high annual variation in temperature.

taproot A large vertical main root of a plant.

taste bud Sensory structure, found chiefly in the mammalian tongue, containing spindle-shaped cells that respond to chemical stimuli.

TATA-binding protein (TBP) A protein that binds to the TATA box in eukaryotic promoters and is a component of the basal transcription complex.

TATA box A short DNA sequence in many eukaryotic promoters about 30 base pairs upstream from the transcription start site.

taxis Movement toward or away from some external cue.

taxon (plural: taxa) Any named group of organisms at any level of a classification system.

taxonomy The branch of biology concerned with the classification and naming of organisms.

TBP See **TATA-binding protein.**

T cell A type of leukocyte that matures in the thymus and, with B cells, is responsible for acquired immunity. Involved in activation of B cells ($CD4^+$ helper T cells) and destruction of infected cells ($CD8^+$ cytotoxic T cells). Also called *T lymphocytes.*

T-cell receptor (TCR) A transmembrane protein found on T cells that can bind to antigens displayed on the surfaces of other cells. Composed of two polypeptides called the alpha chain and beta chain. See **antigen presentation.**

tectorial membrane A membrane in the vertebrate cochlea that takes part in the transduction of sound by bending the stereocilia of hair cells in response to sonic vibrations.

telomerase An enzyme that replicates the ends of chromosome (telomeres) by catalyzing DNA synthesis from an RNA template that is part of the enzyme.

telomere The region at the end of a linear chromosome.

telophase The final stage in mitosis or meiosis during which sister chromatids (replicated chromosomes in meiosis I) separate and new nuclear envelopes begin to form around each set of daughter chromosomes.

temperate Having a climate with pronounced annual fluctuations in temperature (i.e., warm summers and cold winters) but typically neither as hot as the tropics nor as cold as the poles.

temperature A measurement of thermal energy present in an object or substance, reflecting how much the constituent molecules are moving.

template strand (1) The strand of DNA that is transcribed by RNA polymerase to create RNA. (2) An original strand of RNA used to make a complementary strand of RNA.

tendon A band of tough, fibrous connective tissue that connects a muscle to a bone.

tentacle A long, thin, muscular appendage of gastropod molluscs.

termination (1) In enzyme-catalyzed reactions, the final stage in which the enzyme returns to its original conformation and products are released. (2) In DNA transcription, the dissociation of RNA polymerase from DNA when it reaches a termination signal sequence. (3) In RNA translation, the dissociation of a ribosome from mRNA when it reaches a stop codon.

territory An area that is actively defended by an animal from others of its species.

tertiary structure The overall three-dimensional shape of a single polypeptide chain, resulting from multiple interactions among the amino acid side chains and the peptide backbone. Compare with **primary, secondary,** and **quaternary structure.**

test A hard protective outer structure seen in some protists. Also called a *shell.*

testcross The breeding of an individual of unknown genotype with an individual having only recessive alleles for the traits of interest in order to infer the unknown genotype from the phenotypic ratios seen in offspring.

testis (plural: testes) The sperm-producing organ of a male animal.

testosterone A steroid hormone, produced and secreted by the testes, that stimulates sperm production and various male traits and reproductive behaviors.

tetrapod Any member of the taxon Tetrapoda, which includes all vertebrates with two pairs of limbs (amphibians, mammals, birds, and other reptiles).

theory A proposed explanation for a broad class of phenomena or observations.

thermal energy The kinetic energy of molecular motion.

thermocline A gradient (cline) in environmental temperature across a large geographic area.

thermophile A bacterium or archaean that thrives in very hot environments.

thermoreceptor A sensory cell or an organ specialized for detection of changes in temperature.

thermoregulation Regulation of body temperature.

thick filament A filament composed of bundles of the motor protein myosin; anchored to the center of the sarcomere. Compare with **thin filament.**

thigmotropism Growth or movement of an organism in response to contact with a solid object.

thin filament A filament composed of two coiled chains of actin and associated regulatory proteins; anchored at the Z disk of the sarcomere. Compare with **thick filament.**

thorn A modified plant stem shaped as a sharp protective structure. Helps protect a plant against feeding by herbivores.

threshold potential The membrane potential that will trigger an action potential in a neuron or other excitable cell. Also called simply *threshold.*

thylakoid A flattened, membrane-bound vesicle inside a plant chloroplast that functions in converting light energy to chemical energy. A stack of thylakoids is a granum.

thymus An organ, located in the anterior chest or neck of vertebrates, in which immature T cells generated in the bone marrow undergo maturation.

thyroid gland A gland in the neck that releases thyroid hormone (which increases metabolic rate) and calcitonin (which lowers blood calcium).

thyroid-stimulating hormone (TSH) A peptide hormone, produced and secreted by the anterior pituitary, that stimulates release of thyroid hormones from the thyroid gland.

thyroxine (T_4) A peptide hormone containing four iodine atoms that is produced and secreted by the thyroid gland. Acts primarily to increase cellular metabolism. In mammals, T_4 is converted to the more active hormone triiodothyronine (T_3) in the liver.

tight junction A type of cell-cell attachment structure that links the plasma membranes of adjacent animal cells, forming a barrier that restricts movement of substances in the space between the cells. Most abundant in epithelia (e.g., the intestinal lining). Compare with **desmosome** and **gap junction.**

tip The end of a branch on a phylogenetic tree. Represents a specific species or larger taxon that has not (yet) produced descendants—either a group living today or a group that ended in extinction. Also called *terminal node.*

Ti plasmid A plasmid carried by *Agrobacterium* (a bacterium that infects plants) that can integrate into a plant cell's chromosomes and induce formation of a gall.

tissue A group of similar cells that function as a unit, such as muscle tissue or epithelial tissue.

tolerance In ecological succession, the phenomenon in which early-arriving species do not affect the probability that subsequent species will become established. Compare with **facilitation** and **inhibition.**

tonoplast The membrane surrounding a plant vacuole.

top-down control hypothesis The hypothesis that the size of herbivore populations is limited by predation or disease rather than by limited or toxic nutritional resources.

topoisomerase An enzyme that cuts and rejoins DNA downstream of the replication fork, to ease the twisting that would otherwise occur as the DNA "unzips."

torpor An energy-conserving physiological state, marked by a decrease in metabolic rate, body temperature, and activity, that lasts for a short period (overnight to a few days or weeks). Occurs in some small mammals when the ambient temperature drops significantly. Compare with **hibernation.**

totipotent Capable of dividing and developing to form a complete, mature organism.

trachea (plural: tracheae) (1) In insects, one of the small air-filled tubes that extend throughout the body and function in gas exchange. (2) In terrestrial vertebrates, the airway connecting the larynx to the bronchi. Also called *windpipe.*

tracheid In vascular plants, a long, thin water-conducting cell that has gaps in its secondary cell wall, allowing water movement between adjacent cells. Compare with **vessel element.**

trade-off In evolutionary biology, an inescapable compromise between two traits that cannot be optimized simultaneously. Also called *fitness trade-off.*

trait Any heritable characteristic of an individual.

transcription The process by which RNA is made from a DNA template.

transcriptional control Regulation of gene expression by various mechanisms that change the rate at which genes are transcribed to form messenger RNA. In negative control, binding of a regulatory protein to DNA represses transcription; in positive control binding of a regulatory protein to DNA promotes transcription.

transcriptome The complete set of genes transcribed in a particular cell.

transduction Conversion of information from one mode to another. For example, the process by which a stimulus outside a cell is converted into a response by the cell.

transfer cell In land plants, a cell that transfers nutrients from a parent plant to a developing plant seed.

transfer RNA (tRNA) One of a class of RNA molecules that have an anticodon at one end and an amino acid binding site at the other. Each tRNA picks up a specific amino acid and binds to the corresponding codon in messenger RNA during translation.

transformation (1) Incorporation of external DNA into the genome. Occurs naturally in some bacteria; can be induced in the laboratory by certain processes. (2) Conversion of a normal cell to a cancerous one.

transgenic Referring to an individual plant or animal whose genome contains DNA introduced from another individual, either from the same or a different species.

transitional form A fossil species or population with traits that are intermediate between older and younger species.

transition state A high-energy intermediate state of the reactants during a chemical reaction that must be achieved for the reaction to proceed. Compare with **activation energy.**

transition state facilitation The second stage (after initiation) in enzyme-catalyzed reactions, in which the enzyme enables formation of the transition state.

translation The process by which proteins and peptides are synthesized from messenger RNA.

translational control Regulation of gene expression by various mechanisms that alter the life span of messenger RNA or the efficiency of translation.

translocation (1) In plants, the movement of sugars and other organic nutrients through the phloem by bulk flow. (2) A type of mutation in which a piece of a chromosome moves to a nonhomologous chromosome. (3) The process by which a ribosome moves down a messenger RNA molecule during translation.

transmembrane protein Any membrane protein that spans the entire lipid bilayer. Also called *integral membrane protein.*

transmission The passage or transfer (1) of a disease from one individual to another or (2) of electrical impulses from one neuron to another.

transmission electron microscope (TEM) A microscope that forms an image from electrons that pass through a specimen. Compare with **scanning electron microscope.**

transpiration Water loss from aboveground plant parts. Occurs primarily through stomata.

transporter See **carrier protein.**

transport protein Collective term for any membrane protein that enables a specific ion or small molecule to cross a plasma membrane. Includes carrier proteins and channel proteins, which carry out passive transport (facilitated diffusion), and pumps, which carry out active transport.

transposable elements Any of several kinds of DNA sequences that are capable of moving themselves, or copies of themselves, to other locations in the genome. Include LINEs and SINEs.

tree of life A diagram depicting the genealogical relationships of all living organisms on Earth, with a single ancestral species at the base.

triacylglycerol See **fat.**

trichome A hairlike appendage that grows from epidermal cells of some plants. Trichomes exhibit a variety of shapes, sizes, and functions depending on species.

triglyceride See **fat.**

triiodothyronine (T_3) A peptide hormone containing three iodine atoms that is produced and secreted by the thyroid gland. Acts primarily to increase cellular metabolism. In mammals, T_3 has a stronger effect than does the related hormone thyroxine (T_4).

triose A monosaccharide (simple sugar) containing three carbon atoms.

triplet code A code in which a "word" of three letters encodes one piece of information. The genetic code is a triplet code because a codon is three nucleotides long and encodes one amino acid.

triploblast (adjective: triploblastic) An animal whose body develops from three basic embryonic cell layers: ectoderm, mesoderm, and endoderm. Compare with **diploblast.**

trisomy The state of having three copies of one particular type of chromosome.

tRNA See **transfer RNA.**

trochophore A larva with a ring of cilia around its middle that is found in some lophotrochozoans.

trophic level A feeding level in an ecosystem.

tropomyosin A regulatory protein present in thin (actin) filaments that blocks the myosin-binding sites on these filaments, thereby preventing muscle contraction.

troponin A regulatory protein, present in thin (actin) filaments, that can move tropomyosin off the myosin-binding sites on these filaments, thereby triggering muscle contraction. Activated by high intracellular calcium.

true navigation The type of navigation by which an animal can reach a specific point on Earth's surface.

trypsin A protein-digesting enzyme present in the small intestine that activates several other protein-digesting enzymes.

trypsinogen The precursor of protein-digesting enzyme trypsin. Secreted by the pancreas and activated by the intestinal enzyme enterokinase.

T tubules Membranous tubes that extend into the interior of muscle cells. Propagate action potentials throughout a muscle cell and trigger release of calcium from the sarcoplasmic reticulum.

tube foot One of the many small, mobile, fluid-filled extensions of the water vascular system of echinoderms; the part extending outside the body is called a podium. Used in locomotion and feeding.

tuber A modified plant rhizome that functions in storage of carbohydrates.

tumor A mass of cells formed by uncontrolled cell division. Can be benign or malignant.

tumor suppressor A gene (e.g., *p53* and *Rb*) or the protein it encodes that prevents cell division, particularly when the cell has DNA damage. Mutated forms are associated with cancer.

tundra The treeless biome in polar and alpine regions, characterized by short, slow-growing vegetation, permafrost, and a climate of long, intensely cold winters and very short summers.

turgid Swollen and firm as a result of high internal pressure (e.g., a plant cell containing enough water for the cytoplasm to press against the cell wall). Compare with **flaccid.**

turgor pressure The outward pressure exerted by the fluid contents of a plant cell against its cell wall.

Turner syndrome A human genetic disorder caused by the presence of only one X chromosome and no Y chromosome ("XO"). Individuals with this condition are female but sterile.

turnover In lake ecology, the complete mixing of upper and lower layers of water that occurs each spring and fall in temperate-zone lakes.

twofold rotational symmetry See **dyad symmetry.**

tympanic membrane The membrane separating the middle ear from the outer ear in terrestrial vertebrates, or similar structures in insects. Also called the *eardrum.*

ubiquinone See **coenzyme Q.**

ultimate causation In biology, the reason that a trait or phenomenon is thought to have evolved; the adaptive advantage of that trait. Also called *ultimate explanation.* Compare with **proximate causation.**

umami The taste of glutamate, responsible for the "meaty" taste of most proteins and of monosodium glutamate.

umbilical cord The cord that connects a developing mammalian embryo or fetus to the placenta and through which the embryo or fetus receives oxygen and nutrients.

unequal crossover An error in crossing over during meiosis I in which the two non-sister chromatids match up at different sites. Results in gene duplication in one chromatid and gene loss in the other.

unsaturated Referring to fats and fatty acids in which at least one carbon-carbon bond is a double bond. Double bonds produce kinks in the fatty acid chains and decrease the compound's melting point. Compare with **saturated.**

upstream In genetics, opposite to the direction in which RNA polymerase moves along a DNA strand. Compare with **downstream.**

urea A water-soluble excretory product of mammals and sharks. Used to remove from the body excess nitrogen derived from the breakdown of amino acids. Compare with **uric acid.**

ureter In vertebrates, a tube that transports urine from one kidney to the bladder.

urethra The tube that drains urine from the bladder to the outside environment. In male vertebrates, also used for passage of sperm during ejaculation.

uric acid A whitish excretory product of birds, reptiles, and terrestrial arthropods. Used to remove from the body excess nitrogen derived from the breakdown of amino acids. Compare with **urea.**

urochordates One of the three major chordate lineages (Urochordata), comprising sessile, filter-feeding animals that have a polysaccharide exoskeleton (tunic) and two siphons through which water enters and leaves; also called tunicates or sea squirts. Compare with **cephalochordates** and **vertebrates.**

uterus The organ in which developing embryos are housed in those vertebrates that give live birth. Common in most mammals and in some lizards, sharks, and other vertebrates.

vaccination The introduction into an individual of weakened, killed, or altered pathogens to stimulate development of acquired immunity against those pathogens.

vaccine A preparation designed to stimulate an immune response against a particular pathogen without causing illness. Vaccines consist of inactivated (killed) pathogens, live but weakened (attenuated) pathogens, or portions of a viral capsid (subunit vaccine).

vacuole A large organelle in plant and fungal cells that usually is used for bulk storage of water, pigments, oils, or other substances. Some vacuoles contain enzymes and have a digestive function similar to lysosomes in animal cells.

vagina The birth canal of female mammals; a muscular tube that extends from the uterus through the pelvis to the exterior.

valence The number of unpaired electrons in the outermost electron shell of an atom; determines how many covalent bonds the atom can form.

valence electron An electron in the outermost electron shell, the valence shell, of an atom. Valence electrons tend to be involved in chemical bonding.

valves In circulatory systems, flaps of tissue that prevent backward flow of blood, particularly in veins and between the chambers of the heart.

van der Waals interactions A weak electrical attraction between two hydrophobic side chains. Often contributes to tertiary structure in proteins.

variable number tandem repeat See **minisatellite.**

variable (V) region The portion of an antibody's light chains or heavy chains that has a highly variable amino acid sequence and forms part of the antigen-binding site. Compare with **constant (C) region.**

vasa recta In the vertebrate kidney, a network of blood vessels that runs alongside the loop of Henle of a nephron. Functions in reabsorption of water and solutes from the filtrate.

vascular bundle A cluster of xylem and phloem strands in a plant stem.

vascular cambium One of two types of lateral meristem, consisting of a ring of undifferentiated plant cells inside the cork cambium of woody plants; produces secondary xylem (wood) and secondary phloem. Compare with **cork cambium.**

vascular tissue system In plants, any tissue that is involved in conducting water or solutes from one part of a plant to another. Also called simply *vascular tissue.* See **phloem** and **xylem.**

vas deferens A pair of muscular tubes that store and transport semen from the epididymis to the ejaculatory duct. In nonhuman animals, called the *ductus deferens.*

vector A biting insect or other organism that transfers pathogens from one species to another. See also **cloning vector.**

vegetative organs The nonreproductive parts of a plant including roots, leaves, and stems.

vein Any blood vessel that carries blood (oxygenated or not) under relatively low pressure from the tissues toward the heart. Compare with **artery.**

vena cava (plural: vena cavae) A large vein that returns oxygen-poor blood to the heart.

ventral Toward an animal's belly and away from its back. The opposite of dorsal.

ventricle (1) A thick-walled chamber of the heart that receives blood from an atrium and pumps it to the body or to the lungs. (2) One of several small fluid-filled chambers in the vertebrate brain.

vertebra (plural: vertebrae) One of the cartilaginous or bony elements that form the spine of vertebrate animals.

vertebrates One of the three major chordate lineages (Vertebrata), comprising animals with a dorsal column of cartilaginous or bony structures (vertebrae) and a skull enclosing the brain. Includes fishes, amphibians, mammals, reptiles, and birds. Compare with **cephalochordates** and **urochordates.**

vessel element In vascular plants, a short, wide water-conducting cell that has gaps through both the primary and secondary cell walls, allowing unimpeded passage of water between adjacent cells. Compare with **tracheid.**

vestigial trait Any rudimentary structure of unknown or minimal function that is homologous to functioning structures in other species. Vestigial traits are thought to reflect evolutionary history.

vicariance The physical splitting of a population into smaller, isolated populations by a geographic barrier.

villi (singular: villus) Small, fingerlike projections (1) of the lining of the small intestine or (2) of the fetal portion of the placenta adjacent to maternal arteries. Function to increase the surface area available for absorption of nutrients and gas exchange (in the placenta).

virion A single mature virus particle.

virulence The ability of a pathogen or parasite to cause disease and death.

virulent Referring to pathogens that can cause severe disease in susceptible hosts.

virus A tiny intracellular parasite that uses host cell enzymes to replicate; consists of a DNA or RNA genome enclosed within a protein shell (capsid). In enveloped viruses, the capsid is surrounded by a phospholipid bilayer derived from the host cell plasma membrane, whereas nonenveloped viruses lack this protective covering.

visceral mass One of the three main parts of the mollusc body; contains most of the internal organs and external gill.

visible light The range of wavelengths of electromagnetic radiation that humans can see, from about 400 to 700 nanometers.

vitamin An organic micronutrient that usually functions as a coenzyme.

vitelline envelope A fibrous sheet of glycoproteins that surrounds mature egg cells in many vertebrates. Surrounded by a thick gelatinous matrix (the jelly layer) in some species. In mammals, called the *zona pellucida.*

viviparous Producing live young (instead of eggs) that develop within the body of the mother before birth. Compare with **oviparous** and **ovoviviparous.**

volt (V) A unit of electrical potential (voltage).

voltage Potential energy created by a separation of electric charges between two points. Also called *electrical potential.*

voltage clamping A technique for imposing a constant membrane potential on a cell. Widely used to investigate ion channels.

voltage-gated channel An ion channel that opens or closes in response to changes in membrane voltage. Compare with **ligand-gated channel.**

wall pressure The inward pressure exerted by a cell wall against the fluid contents of a plant cell.

water cycle, global The movement of water among terrestrial ecosystems, the oceans, and the atmosphere.

water potential (ψ) The potential energy of water in a certain environment compared with the potential energy of pure water at room temperature and atmospheric pressure. In living organisms, ψ equals the solute potential (ψ_S) plus the pressure potential (ψ_P).

water potential gradient A difference in water potential in one region compared with that in another region. Determines the direction that water moves, always from regions of higher water potential to regions of lower water potential.

watershed The area drained by a single stream or river.

water table The upper limit of the underground layer of soil that is saturated with water.

water vascular system In echinoderms, a system of fluid-filled tubes and chambers that functions as a hydrostatic skeleton.

Watson-Crick pairing See **complementary base-pairing.**

wavelength The distance between two successive crests in any regular wave, such as light waves, sound waves, or waves in water.

wax A class of lipid with extremely long hydrocarbon tails, usually combinations of long-chain alcohols with fatty acids. Harder and less greasy than fats.

weather The specific short-term atmospheric conditions of temperature, moisture, sunlight, and wind in a certain area.

weathering The gradual wearing down of large rocks by rain, running water, and wind; one of the processes that transform rocks into soil.

weed Any plant that is adapted for growth in disturbed soils.

Western blotting A technique for identifying specific proteins separated by gel electrophoresis by transferring them to filter paper and exposing them to a labeled antibody that binds to the protein of interest. Compare with **Northern blotting** and **Southern blotting.**

wetland A shallow-water habitat where the soil is saturated with water for at least part of the year.

white blood cells See **leukocytes.**

wild type The most common phenotype seen in a population; especially the most common phenotype in wild populations compared with inbred strains of the same species.

wildlife corridor Strips of wildlife habitat connecting populations that otherwise would be isolated by man-made development.

wilt To lose turgor pressure in a plant tissue.

wobble hypothesis The hypothesis that some tRNA molecules can pair with more than one mRNA codon, tolerating some variation in the third base, as long as the first and second bases are correctly matched.

wood Xylem resulting from secondary growth. Also called *secondary xylem.*

xeroderma pigmentosum A human disease characterized by extreme sensitivity to ultraviolet light. Caused by an autosomal recessive allele that results in a defective DNA repair system.

X-linked inheritance Inheritance patterns for genes located on the mammalian X chromosome. Also called *X-linkage.*

X-ray crystallography A technique for determining the three-dimensional structure of large molecules, including proteins and nucleic acids, by analysis of the diffraction patterns produced by X-rays beamed at crystals of the molecule.

xylem A plant vascular tissue that conducts water and ions; contains tracheids and/or vessel elements. Primary xylem develops from the procambium of apical meristems; secondary xylem, or wood, from the vascular cambium of lateral meristems. Compare with **phloem.**

xylem sap The watery fluid found in the xylem of plants.

yeast Any fungus growing as a single-celled form. Also, a specific lineage of ascomycetes.

Y-linked inheritance Inheritance patterns for genes located on the mammalian Y chromosome. Also called *Y-linkage.*

yolk The nutrient-rich cytoplasm inside an egg cell; used as food for the growing embryo.

Z disk The structure that forms each end of a sarcomere. Contains a protein that binds tightly to actin, thereby anchoring thin filaments.

zero population growth (ZPG) A state of stable population size due to fertility staying at the replacement rate for at least one generation.

zona pellucida The gelatinous layer around a mammalian egg cell. In other vertebrates, called the *vitelline envelope.*

zone of (cellular) division In plant roots, a group of apical meristematic cells just behind the root cap where cells are actively dividing.

zone of (cellular) elongation In plant roots, a group of young cells, located behind the apical meristem, that are increasing in length.

zone of (cellular) maturation In plant roots, a group of plant cells, located several centimeters behind the root cap, that are differentiating into mature tissues.

Z scheme Path of electron flow in which electrons pass from photosystem II to photosystem I and ultimately to $NADP^+$ during the light-dependent reactions of photosynthesis. Also called *noncyclic electron flow.*

zygomycete A member of a paraphyletic group of fungi (Zygomycota) characterized by a durable, thick-walled zygosporangium that forms from yoked hyphae of different mating types.

zygosporangium (plural: zygosporangia) The spore-producing structure in fungi that are members of the Zygomycota.

zygote The diploid cell formed by the union of two haploid gametes; a fertilized egg. Capable of undergoing embryological development to form an adult.

Credits

IMAGE CREDITS

Frontmatter

viT Natalie B. Fobes Photography **viB** David Quillin **x** Jeff Rotman/Nature Picture Library **xv** Anthony Bannister/Gallo Images/Corbis; **xix** Lee W. Wilcox **xxiii** Robert Fried/robertfriedphotography.com

Chapter 1

Opener Jeff Rotman/Nature Picture Library. **1.1a** Burndy Library/Omikron/Photo Researchers, Inc. **1.1bM** Walker/Photo Researchers, Inc. **1.3a** Kelly Buono/Dr. Richard Amasino. **1.3c** Bruce Forster/Getty Images Inc. **1.4aL** University of Florida. **1.4aR** Michael Lustbader/Photo Researchers, Inc. **1.4bLR** Scott P. Carroll. **1.4cL** Tom Murray. **1.4cR** Andrew A. Forbes. **1.6a** Samuel F. Conti and Thomas D. Brock. **1.6b** Kwangshin Kim/Photo Researchers, Inc. **1.7/1** Dr. David Phillips/Visuals Unlimited. **1.7/2** Dennis Kunkel/Dennis Kunkel Microscopy, Inc. **1.7/3** Kolar, Richard/Animals Animals/Earth Scenes **1.7/4** Biophoto Associates/Photo Researchers, Inc. **1.7/5** Darwin Dale/Photo Researchers, Inc. **1.10b** Michael Hughes/Aurora & Quanta Productions Inc. **1.11a** Joshua J. Tewksbury/Joshua J. Tewksbury. **1.11b** William Weber/Visuals Unlimited. **1.11c** Robert Dobbs/Joshua J. Tewksbury

Chapter 2

Opener Colin Monteath/Hedgehog House/Minden Pictures. **2.1** Mitchell Layton/PCN Photography. **2.6c** Albert Copley/Visuals Unlimited. **2.11** Beth Plowes–Proteapix. **2.14** Dietmar Nill/Picture Press/Photolibrary.com. **2.15c** Geostock/Getty Images Inc. **2.19bL** David Glick/Getty Images Inc. Stone Allstock. **2.19bR** Sergio Bartelsman/eStock

Chapter 3

Opener The Scripps Research Institute **3.4b** Martin Bough/Fundamental Photographs, NYC **3.10a-c** Clare Sansom. **3.11a** Microworks/Phototake NYC. **3.11b** Walter Reinhart/Phototake NYC. **3.13BLMR** Clare Sansom. **3.14a-b** Clare Sansom. **3.20aLR** Thomas A. Steitz. **3.23aLR, 3.23bLR** Thomas A. Steitz

Chapter 4

4.5 A. Barrington Brown/Science Source Photo Researchers, Inc. **4.11** Reprinted with permission from Science 292: 1319-1325 Fig 4B (2001) by Wendy K. Johnston, Peter J. Unrau, Michael S. Lawrence, Margaret E. Glasner, David P. Bartel "RNA-Catalyzed RNA Polymerization: Accurate and General RNA-Templated Primer Extension." Copyright 2004 American Association for the Advancement of Science

Chapter 5

Opener Dr. Jeremy Burgess Photo Researchers, Inc. **5.6a** Biophoto Associates/Photo Researchers, Inc. **5.6b** Dr. Jacob S. Ishay. **5.6c** Dr. Manfred Jericho Terry J. Beveridge.

Chapter 6

Opener Kit Pogliano . **6.1a** Alec D. Bangham, M.D., F.R.S. Alec D. Bangham, M.D., F.R.S. **6.1b** Fred Hossler Visuals Unlimited. **6.7aL** James J. Cheetham James J. Cheetham. **6.11a** Dorling Kindersley Dorling Kindersley Media Library. **6.11b** Clive Streeter Dorling Kindersley Media Library. **6.11c** Phil Degginger Color-Pic, Inc. **6.24b** Timothy A. Cross. **6.26L** Andrew Syred Getty Images Inc.-Stone Allstock. **6.26M** Dr. David Phillips Visuals Unlimited. **6.26R** Joseph F. Hoffman Joseph F. Hoffman/Yale University School of Medicine

Chapter 7

Opener Albert Tousson Phototake NYC. **7.1** Dr. T.J. Beveridge Visuals Unlimited. **7.2** Stanley C. Holt/Biological Photo Service **7.3** Gopal Murti/ Visuals Unlimited **7.5** Wanner/Eye of Science Photo Researchers, Inc. **7.7** Fawcett Photo Researchers, Inc. **7.8** Omikron Photo Researchers, Inc. **7.9** Dr. Don Fawcett Photo Researchers, Inc. **7.10** Biophoto Associates Photo Researchers, Inc. **7.12** Fawcett/Friend Photo Researchers, Inc. **7.12** Dr. Don Fawcett Photo Researchers, Inc. **7.13** Dr. Gopal Murti Visuals Unlimited. **7.16** E.H. Newcomb & W.P. Wergin/Biological Photo Service Biological Photo Service. **7.17** T. Kanaseki & Donald Fawcett Visuals Unlimited. **7.18** E.H. Newcomb & W.P. Wergin Biological Photo Service. **7.19** E. H. Newcomb & S.E. Frederick Biological Photo Service. **7.20a** Don Fawcett/S. Ito & A. Like/Photo Researchers, Inc. **7.20b** Don W. Fawcett/Photo Researchers, Inc. **7.20c** Biophoto Associates Photo Researchers, Inc. **7.20d** Dr. Dennis Kunkel Visuals Unlimited. **7.23** Don W. Fawcett Photo Researchers, Inc. **7.24a** Don W. Fawcett Photo Researchers, Inc. **7.27a** James D. Jamieson, M.D. James D. Jamieson, M.D. **7.27b** James D. Jamieson, M.D. **7.27c** James D. Jamieson, M.D. **7.31a-c** Dr. Victor Small. **7.33** Conly L. Rieder Biological Photo Service. **7.34a-b** American Society for Cell Biology. **7.35a** John E. Heuser, M.D. **7.36L** Dennis Kunkel Phototake NYC. **7.36R** Dennis Kunkel Phototake NYC. **7.37a** Don W. Faucett Photo Researchers, Inc. **7.38** Charles J. Brokaw The Rockefeller University Press

Chapter 8

Opener E.H. Newcomb & W.P. Wergin Biological Photo Service. **8.2** Henry M. Walker, Samuel R. & Marie-Louise Rosenthal, Profs **8.3** Biophoto Associates Photo Researchers, Inc. **8.4** Barry F. King Biological Photo Service. **8.6** C. T. Huang, Karen Xu, Gordon McFeters, and Philip S. Stewart Philip S. Stewart. **8.7** SPL Photo Researchers, Inc. **8.9** Photo Researchers, Inc. Photo Researchers, Inc. **8.10aL** Don W. Fawcett Photo Researchers, Inc. **8.10aR** Dr. Don Fawcett Photo Researchers, Inc. **8.11a** Dr. Don Fawcett/Gida Matoltsy Photo Researchers, Inc. **8.14a** E. H. Newcomb & W.P. Wergin Biological Photo Service. **8.14b** Dr. Don Fawcett Photo Researchers, Inc.

Chapter 9

Opener Richard Megna Fundamental Photographs, NYC. **9.6b** Darren McCollester/Reuters Corbis/Bettmann. **9.6TL** Peter Anderson Dorling Kindersley Media Library. **9.6TR** Dorling Kindersley Dorling Kindersley Media Library. **9.16** Terry Frey. **9.25a** Dr. Yasuo Kagawa

Chapter 10

Opener Murray Cooper/NPL Minden Pictures. **10.2aM** John Durham/Science Photo Library Photo Researchers, Inc. **10.2aR(detail)** John Durham/Science Photo Library Photo Researchers, Inc. **10.2bR** Dr. J. Burgess Photo Researchers, Inc. **10.3bL** Biophoto Associates Photo Researchers, Inc. **10.3bM** Dr. George Chapman Visuals Unlimited **10.3bR** Richard Green Photo Researchers, Inc. **10.3T** E. H. Newcomb & P. K. Hepler Biological Photo Service **10.5b** Sinclair Stammers/Science Photo Library Photo Researchers, Inc. **10.10** David Newman Visuals Unlimited. **10.18L** James A. Bassham James A. Bassham. **10.18R** James A. Bassham. **10.21a** Dr. Jeremy Burgess/Science Photo Library Photo Researchers, Inc. **10.24a** Adam Hart-Davis Photo Researchers, Inc. **10.24b** David Muench Corbis/Bettmann

Chapter 11

Opener Collection CNRI Phototake NYC. **11.2a** Conly L. Rieder, Ph.D. **11.2b** Photo Researchers, Inc. **11.3a** From: Paulson, J.R. and Laemmli, U.K. Cell 12 (1977) 817-828. **11.9a** David M. Phillips/Visuals Unlimited **11.9bR** Calentine/Visuals Unlimited. **11.10** Micrographs by Conly L. Rieder, Division of Molecular Medicine, Wadsworth Center, Albany, New York 12201-0509

Chapter 12

Opener David Phillips/The Population Council Photo Researchers, Inc. **12.2a** Hesed M. Padilla-Nash. **12.6c** Wessex Reg. Genetics Centre Wellcome Trust Medical Photographic Library. **12.8a-d** David A. Jones. **12.12** Doug Sokell Visuals Unlimited

Chapter 13

Opener Brian Johnston Brian Johnston. **13.9a** Robert Calentine Visuals Unlimited. **13.9bLR** Carolina Biological Supply Company Phototake NYC. **13.17a** Robert Calentine Visuals Unlimited

Chapter 14

Opener Dr. Gopal Murti/Science Photo Library Photo Researchers, Inc. **14.1** Belanger et al., 2004. Pyruvate oxidase is a determinant of Avery's rough morphology. J. Bacteriology 186:8164-8174. American Society of Microbiology. **14.4b** Oliver Meckes/Max-Planck-Institut-Tubingen/Photo Researchers, Inc. **14.10a** Dr. Gopal Murti/Science Photo Library Photo Researchers, Inc.

Chapter 15

Opener Halaska, Jacob Photolibrary.com

Chapter 16

Opener Joseph G. Gall. **16.05** Fred Stevens/Mykoweb. **16.06** Bert W. O'-Malley, M.D. Bert W. O'Malley, M.D. **16.09** Oscar Miller/Science Photo Library/Photo Researchers, Inc. **16.15a** Venkitaraman Ramakrishnan. **16.16** E.V. Kiseleva/Donald Fawcett/Visuals Unlimited. **16.19b** Bill Longcore /Photo Researchers, Inc. **16.21b** Bill Longcore /Photo Researchers, Inc.

Chapter 17

Opener EM Unit, VLA/Science Photo Library/Photo Researchers, Inc. **applying concepts** Michael Gabridge/Visuals Unlimited

Chapter 18

Opener Dr. Timothy J. Richmond Nature Magazine DC. **18.02a** Ada Olins/Don Fawcett/Photo Researchers, Inc. **18.04** Victoria E. Foe **18.04** Barbara Hamkalo

Chapter 19

Opener Tsien Laboratory **19.14** Baylor Coll. of Medicine/Peter Arnold, Inc. **19.17a** Brad Mogen/Visuals Unlimited

Chapter 20

Opener Sanger Institute Welcome Trust Medical Photographic Library. **20.07** and **20.07b** Reprinted by Permission from Macmillan Publishers Ltd. From Fig 4 in "Nature Medicine", 11 October 1, 2005 "Genetic Fingerprinting" by Alec J. Jeffreys pp1035-1039. The first application of DNA fingerprinting—an immigration case. **20.12** Patrick O. Brown, M.D.

Chapter 21

Opener Dr. Yorgos Nikas/Photo Researchers, Inc. **21.01** Chin-Sang. **21.02a-b** Elsevier Science Ltd. **21.04** Roslin Institute Empics. **21.05a** Richard Hutchings/Photo Researchers, Inc. **21.05b** Institute for Laboratory Animal Research ILAR/The National Academies. **21.06a** F. Rudolf Turner. **21.06b** Gary Grumbling. **21.07** Prof. Dr. Christiane Nusslein-Volhard. **21.09** Wolfgang Driever. **21.10a** Stephen Paddock. **21.10b** Stephen J. Small. **21.10c** Stephen Paddock. **21.12L** David Scharf/Photo Researchers, Inc. **21.12M** David Scharf/Peter Arnold, Inc. **21.12R** (c)Science VU/Dr. F.R. Turner/Visuals Unlimited. **21.14** Anthony Bannister/NHPA/Photo Researchers, Inc. **21.15a-b** Natural History Magazine

Chapter 22

Opener Yorgos Nikas/Photo Researchers, Inc. **22.02** Holger Jastrow. **22.04.** Gregory Ochocki/Photo Researchers, Inc. **22.07a** Michael Whitaker/Science Photo Library/Photo Researchers, Inc. **22.07bRML** Victor D. Vacquier. **22.14a** Kathryn W. Tosney. **22.14b** Gary C. Schoenwolf

Chapter 23

Opener Dr. John Runions/Oxford Brookes University **23.01** Holt Studios International/Photo Researchers, Inc. **23.03a** Ray Evert. **23.03c** Cabisco/Visuals Unlimited **23.04b** Biodisc/Visuals Unlimited **23.08b** Ken Wagner/Phototake NYC **23.10a-c** Neelima Sinha **23.12** John L.

Chapter 24

Opener Kim Taylor/NPL/Minden Pictures **24.02a** Bettmann/Corbis **24.02b** National Portrait Gallery, London **24.03a** Ken Lucas/Visuals Unlimited **24.03b** T. A. Wiewandt/DRK Photo **24.03c** Knut Finstermeier/Dr. Svante Paabo **24.04aR** Robert Lubeck/Animals Animals Earth Scenes **24.05aL** CMCD/Getty Images Inc. **24.05aR** Vincent Zuber/Custom Medical Stock Photo, Inc. **24.05bL** Mary Beth Angelo/Photo Researchers, Inc. **24.05bR** Custom Medical Stock Photo, Inc. **24.06aBL** George D. Lepp/Photo Researchers, Inc. **24.06aBR** Mickey Gibson Animals Animals/Earth Scenes **24.06aTL** Marie Read Animals Animals/Earth Scenes **24.06aTR** Tui De Roy Bruce Coleman Inc. **24.08L** Michael K. Richardson Springer-Verlag GmbH & Co KG **24.08M** Ronan O'Rahilly Springer-Verlag GmbH & Co KG **24.08R** National Museum of Health and Medicine **24.12** Frans Lanting/Minden Pictures **24.17aTL** From Fig 1C in Abzanhov et al. 2004. Science 305: 1462-1465 **24.17aTR** From Fig 1C in Abzanhov et al. 2004. Science 305: 1462-1465 **24.17aBL,BR** The Royal Society of London Reproduced by permission from K. Petren et al., A phylogeny of Darwin's finches based on microsatellite DNA length variation. Proceedings of the Royal Society of London, Series B, 266:321-329 (1999), p. 327, fig. 3. Copyright (c) 1999 Royal Society of London. Image courtesy of Kenneth Petren, University of Cincinnati. **24.17bL** From Figure 3D in Abzanhov et al. 2004. Science 305: 1462-1465 **24.17bR** From Fig 3E in Abzhanov et al. 2004. Science 305: 1462-1465

Chapter 25

Opener Phil Savoie/Minden Pictures **25.07a** Max Westby **25.08a** James Hughes/Visuals Unlimited **25.12a** Cyril Laubscher/Dorling Kindersley Media Library **25.13** Otorohanga Zoological Society Inc. **25.14a** Marc Moritsch/National Geographic Image Collection **25.15aT** Linda Mitchell Photography **25.15bB** R. & A. Simpson VIREO/The Academy of Natural Sciences **25.15bT** B. Schorre VIREO/The Academy of Natural Sciences **25.15cB** Van Os, Joseph/Getty Images Inc. **25.15cT** Jeremy Woodhouse/Getty Images, Inc

Chapter 26

Opener Wayne Lankinen/DRK Photo **26.09** Jordan Rofkar **26.12LMR** Jason Rick Loren H. Rieseberg

Chapter 27

Opener Pete Oxford/Minden Pictures **27.06a** William L. Crepet/American Journal of Botany **27.06b** Martin Land/Science Photo Library/Photo Researchers, Inc. **27.06c** Monte Hieb & Harrison Hieb/Monte Hieb/Geocraft **27.06d** John Gerlach/DRK Photo **27.09b** Nature Magazine **27.09b** Shuhai Xiao/Nature Magazine **27.09cB** Ken Lucas/Visuals Unlimited **27.09cT** Simon Conway Morris Simon Conway Morris **27.09dB** Ed Reschke/Peter Arnold, Inc. **27.09dT** B.J. Miller Biological Photo Service **27.12a** left, middle Jonathan B. Losos and Kevin de Quieroz/Prof. Jonathan B. Losos **27.13a** Colin Keates/Dorling Kindersley Media Library **27.13b** Tim Fitzharris/Minden Pictures **27.13c** Don P. Northup www.africancichlidphotos.com **27.13d** Joe Tucciarone/Photo Researchers, Inc. **27.15** David Hardy/Photo Researchers, Inc. **27.16bLR** Glen A. Izett/U.S. Geological Survey, Denver **27.16cR** Peter H. Schultz

Chapter 28

Opener Beth Donidow/Visuals Unlimited **28.03** Science VU/Visuals Unlimited **28.04** J. Robert Waaland/Biological Photo Service **28.06** Richard L. Carlton/Visuals Unlimited **28.08** Yul Roh American Association for the Advancement of Science **28.10** Kwangshin Kim/Photo Researchers, Inc. **28.10b** Todd Bannor/Custom Medical Stock Photo, Inc. **28.10c** Carolina Biological/Visuals Unlimited **28.13aL** CNRI/Science Photo Library/Photo Researchers, Inc. **28.13aR** Dr. Heide Schulz American Association for the Advancement of Science **28.13bL** Gary Gaugler/Visuals Unlimited **28.13bR** David M. Phillips/Visuals Unlimited **28.13cL** Linda Stannard University of Cape Town/Science Photo Library/Photo Researchers, Inc. **28.13c** Richard W. Castenholz **28.14a** R. Hubert, Microbiology Program, Dept. of Animal Science, Iowa State University **28.18** J. C. Revy/Phototake NYC **28.19** Centers for Disease Control & Prevention **28.20** S. Amano, S. Miyadoh and T. Shomura (The Society for Actinomycetes Japan) **28.21** R. Calentine/Visuals Unlimited **28.22** National Cancer Institute/SPL/Photo Researchers, Inc. **28.23a** Yves. V. Brun **28.23b** Prof. Hans Reichenbach **28.26** Kenneth M. Stedman **28.26** Corale Brierley/Visuals Unlimited **28.27** Marli Miller **28.27** Bonnie K. Baxter **28.29** Roland Birke/Tierbild Okapia/Photo Researchers, Inc.

Chapter 29

Opener Steve Gschmeissner/Photo Researchers, Inc. **29.02a** Norman T. Nicoll/Natural Visions **29.02b** Gregory Ochocki/Photo Researchers, Inc. **29.02c** Tom and Therisa Stack/Tom Stack & Associates, Inc. **29.02r** John Anderson/Animals Animals Earth Scenes **29.03a** Michael Abbey/Photo Researchers, Inc. **29.03b** Wim van Egmond/Visuals Unlimited **29.03c** Cabisco/Visuals Unlimited **29.03d** Sherman Thomson/Visuals Unlimited **29.03e** National Institute for Environmental Studies, Japan http://www.nies.go.jp/biology/mcc/home.htm **29.03f** Dr. Sc. Yuuji Tsuki **29.04** Sanford Berry/Visuals Unlimited **29.04** J. Robert Waaland/Biological Photo Service **29.07** Biophoto Associates/Photo Researchers, Inc. **29.13a** Andrew Syred/Science Photo Library/Photo Researchers, Inc. **29.13b** David M. Phillips/Visuals Unlimited **29.13c** Andrew Syred/Science Photo Library/Photo Researchers, Inc. **29.14a** Biophoto Associates/Photo Researchers, Inc. **29.14b** Bruce Coleman Inc. **29.15** M.I. Walker/Photo Researchers, Inc. **29.18** D. Fleetham/Animals Animals Earth Scenes **29.23** M. Abbey/Photo Researchers, Inc. **29.24** Carolina Biological/Visuals Unlimited **29.26** Michael Abbey/Visuals Unlimited **29.28** Dr. Philip J. Spagnuolo **29.30** Wim van Egmond/Visuals Unlimited **29.32** Phototake NYC **29.33** Andrew Syred/SPL/Photo Researchers, Inc. **29.34** Joyce Photographics/Photo Researchers, Inc. **29.36** Peter Parks/Image Quest 3-D **29.38** Robert De Goursey/Visuals Unlimited **29.40** Scott Camazine/Photo Researchers, Inc.

Chapter 30

Opener Darrell Gulin/DRK Photo **30.01a** Tom Myers/Photo Researchers, Inc. **30.01b** D. P. Burnside/Photo Researchers, Inc. **30.03a** Runk/Schoenberger/Grant Heilman Photography, Inc. **30.05** Holt Studios Int./Photo

Researchers, Inc. **30.06L** Linda Graham **30.06R** Lee W. Wilcox **30.07a/1** K.G. Vock/Okapia/Photo Researchers, Inc. **30.07a/2-3** Lee W. Wilcox **30.07b-1** Alvin E. Staffan/Photo Researchers, Inc. **30.07b/2** Biophoto Associates/Photo Researchers, Inc. **30.07b/3** Milton Rand/Tom Stack & Associates, Inc. **30.07b/4** Rod Planck/Photo Researchers, Inc. **30.07c-bM** Larry Lefever/Grant Heilman Photography, Inc. **30.07cBR** Stephen J. Krasemann/DRK Photo **30.07cTL** Walter H. Hodge/Peter Arnold, Inc. **30.07cTM** Degginger, E R/Animals Animals Earth Scenes **30.07cTR** ©ENLIGHTENED IMAGES/Animals Animals Earth Scenes **30.08/1** Paul K. Strother **30.08/2** Robert & Linda Mitchell Photography **30.08/3** Thomas A. Wiewandt/DRK Photo **30.08/4** David L. Dilcher **30.10** Lee W. Wilcox **30.13a-b** Lee W. Wilcox **30.14** David T. Webb **30.18a-bLR** Lee W. Wilcox **30.23a** James L. Castner **30.23b** Gerald C. Kelley/Photo Researchers, Inc. **30.23c** Gerry Ellis/GLOBIO **30.25a** John Cancalosi/Peter Arnold, Inc. **30.25b** Holt Studios Int.(Nigel Cattlin)/Photo Researchers, Inc. **30.27b/1** Robert & Linda Mitchell Photography **30.27b/2** Runk/Schoenberger/Grant Heilman Photography, Inc. **30.27b/3** Don Farrall/Getty Images, Inc. **30.27b/4** Kim Heacox Photography/DRK Photo **30.27T/2** Ed Reschke/Peter Arnold, Inc. **30.27T/3** Pat O'Hara/Corbis-Bettmann **30.27T/4** Robert & Linda Mitchell Photography **30.29a** Carolina Biological Supply Company/Phototake NYC **30.20bBL** C.C. Lockwood/Animals Animals Earth Scenes **30.20bTL** Runk/Schoenberger/Grant Heilman Photography, Inc. **30.20bR** Bill Banaszewski/Visuals Unlimited **30.29a-c** Lee W. Wilcox **30.31** Wim van Egmond **30.32** Andrew J. Martinez/Photo Researchers, Inc. **30.33** John Clayton/NIWA National Institute of Water and Atmospheric Research **30.35a-b** Robert and Beth Plowes Photograhy/Proteapix **30.36** Ed Reschke/Peter Arnold, Inc. **30.37** Visuals Unlimited **30.39** Lee W. Wilcox **30.40** Karen S. Renzaglia **30.41** Michael Clayton **30.42aL** David T. Webb **30.42aR** David Cavagnaro/DRK Photo **30.42b** Lee W. Wilcox **30.44** David T. Webb **30.45a** Biophoto Associates/Photo Researchers, Inc. **30.45b** Martin Land/Science Photo Library/Photo Researchers, Inc. **30.46** NICK GREEN/Photolibrary.com **30.47a-b** Lee W. Wilcox **30.49a** Brian Johnston **30.49bLR** Lee W. Wilcox. 30.48 Peter Chadwick/Dorling Kindersley Media Library

Chapter 31

Opener Steve Austin/CORBIS **31.01L** Mycorrhizal Applications, Inc. **31.01R** Jim Deacon Institute of Cell and Molecular Biology, The University of Edinburgh **31.03** John Cang Photography **31.04a** David M. Dennis /Tom Stack & Associates, Inc. **31.04b** Poehlmann/Phototake NYC **31.05a** George Musil/Visuals Unlimited **31.05b** Tony Brain/Science Photo Library/Photo Researchers, Inc. **31.06b** Biophoto Associates/Photo Researchers, Inc. **31.11a** Stan Flegler Visuals Unlimited **31.11b** David T. Webb **31.12** Michael P. Gadomski/Photo Researchers, Inc. **31.15** James A. Wubah **31.16** J. I. Ronny Larsson, Professor of Zoology **31.17** Biophoto Associates/Photo Researchers, Inc. **31.18** Jim Deacon Institute of Cell and Molecular Biology, The University of Edinburgh **31.19** Jeff Lepore/Photo Researchers, Inc. **31.20b** Pat O'Hara/DRK Photo **31.21** Stephen Sharnoff/Visuals Unlimited

Chapter 32

Opener David B. Fleetham/SeaPics.com **32.01** Perennou/Nuridsa /Photo Researchers, Inc. **32.01b** Mark W. Moffett/Minden Pictures **32.02a** Harry Rogers/Photo Researchers, Inc. **32.02b** Glenn Oliver/Visuals Unlimited **32.02cCM** Fallows/SeaPics.com **32.03a** Walker England/Photo Researchers, Inc. **32.03b** Ken Lucas/Visuals Unlimited **32.03c** David J. Wrobel/Visuals Unlimited **32.04** Roland Birke /Tierbild Okapia/Photo Researchers, Inc. **32.05a** Jeffrey L. Rotman/CORBIS **32.05b** Pitchal Frederic/Corbis Sygma **32.05c** J. Alex Halderman **32.09bB** L. Newman & A. Flowers/Photo Researchers, Inc. **32.09bT** Roger Steene/Image Quest 3-D **32.13a** J.P. Ferrero/Jacana/Photo Researchers, Inc. **32.12a** Flip Nicklin/Minden Pictures **32.12a** Flip Nicklin/Minden Pictures **32.12b** inset Marc Chamberlain/SeaPics.com **32.12b** Hiroya Minakuchi/Minden Pictures **32.13b** Steve Earley/Animals Animals Earth Scenes **32.14a** Pete Oxford/Minden Pictures **32.14b** Heidi & Hans-Jurgen Koch/Minden Pictures **32.15a** Daniel Sambraus/Photo Researchers, Inc. **32.15b** Anthony Bannister/NHPA/Photo Researchers, Inc. **32.15b** Josef Ramsauer & Prof. Dr. Robert Patzner **32.16a** Satoshi Kuribayashi/DRK Photo **32.16a** Satoshi Kuribayashi/OSF/DRK Photo **32.16b** Jim Brandenburg/Minden Pictures **32.17a** Andrew Syred/Photo Researchers, Inc. **32.17b** Eye of Science/Photo Researchers, Inc. **32.18a** Peter Parks/Image Quest 3-D **32.19a** Roland Seitre/Peter Arnold, Inc. **32.19b** Jeff Foott/DRK Photo **32.19c** Chamberlain, MC/DRK Photo **32.19d** Jim Greenfield/Image Quest 3-D **32.18b** Armin Maywald/Foto Natura/Minden Pictures **32.20bLBMBR** Proceedings of the National Academy of Sciences **32.22a** Doug Perrine/DRK Photo **32.22b** Dan Suzio/Photo Researchers, Inc. **32.21a** Karl-Heinz Marschner Holger de Groot/Biologis **32.21b** D. Parer & E. Parer-Cook/Auscape/Minden Pictures **32.23aTM** Phototake NYC **32.23aB** Sinclair Stammers/Science Photo Library/Photo Researchers, Inc. **32.23bT** Jose B. Ruiz/Nature Picture Library **32.B** Jose B. Ruiz/Nature Picture Library **32.26** Andrew J. Martinez/Photo Researchers, Inc. **32.27a-b** Holger de Groot/Biologis **32.28** Andrew J. Martinez/Photo Researchers, Inc. **32.29** Matthew D. Hooge

Chapter 33

Opener Barbara Strnadova/Photo Researchers, Inc. **33.01a** Jean Claude Revy/Phototake NYC **33.01b** Sinclair Stammers/Photo Researchers, Inc. **33.04a** Peter Parks/Image Quest 3-D **33.05** Bill Beatty/Visuals Unlimited **33.04b** Carmel McDougall **33.09a** Mark W. Moffett/Minden Pictures **33.09b** Phototake NYC **33.09c** Dennis Kunkel Phototake NYC **33.10b** M.C. Chamberlain/DRK Photo **33.10c** Stephen Dalton/Photo Researchers, Inc. **33.12** Jan Van Arkel/Minden Pictures **33.13a** Newman & Flowers/Photo Researchers, Inc. **33.13b** Manfred Kage/Peter Arnold, Inc. **33.13c** Oliver Meckes/Photo Researchers, Inc. **33.14a** Peter Batson/Image Quest 3-D **33.14b** Alexa Bely **33.14c** Martin Dohrn/Photo Researchers, Inc. **33.15a** Degginger, E R/Animals Animals Earth Scenes **32.16b** Norbert Wu/Peter Arnold, Inc. **33.16a** Roger Steen/Image Quest Marine **33.17** Kjell B. Sandved/Photo Researchers, Inc. **33.18** Fred Bavendam/Peter Arnold, Inc. **33.20a** Michael Fogden/DRK Photo **33.20b** Diane Nelson/Visuals Unlimited **33.21** Luis M. de la Maza/Phototake NYC. **33.22** Mark Smith/Photo Researchers, Inc. **33.23a** Gerard Blondeau/Phototake NYC **T33.1/1** Jay Cossey **T33.1/2** Gustav W. Verderber **T33.1/3** Joachim Lippi **T33.1/4** Trounce Wikimedia Commons **T33.1/5** J. Gall/Photo Researchers, Inc. **T33.1/6** Nature's Images/Photo Researchers, Inc. **T33.1/7** Dr. J. K. Lindsey **T33.1/8** Scott Camazine/Photo Researchers, Inc. **33.24a** E. R. Degginger/Photo Researchers, Inc. **33.24b** M C Chamberlain/DRK Photo

Chapter 34

Opener Paul Nicklen/National Geographic Image Collection **34.02b** Thurston Lacalli **34.02a** Kaj R. Svensson/Photo Researchers, Inc. **34.03b** John D. Cunningham/Visuals Unlimited **34.09a** Gerald & Buff Corsi/Visuals Unlimited **34.09b** Greg Rouse/Scripps Institution of Oceanography Explorations **34.11** Rudie Kuiter/OceanwideImages.com **34.18a** James D. Watt/SeaPics.com **34.18b** S. Boon Fu/Peter Arnold, Inc. **34.20** James D. Watt/Image Quest 3-D **34.21R** Tom Stack/Tom Stack & Associates, Inc. **34.21** Southestern Regional Taxonomic Center (SERTC)/South Carolina Department of Natural Resources **34.23a** Sue Daly/Nature Picture Library **34.23b** David Wrobel/SeaPics.com **34.25a** Tom McHugh/Photo Researchers, Inc. **34.25b** Tom Stack/Tom Stack & Associates, Inc. **34.24** Heather Angel/Natural Visions **34.26a** Fred McConnaughey/Photo Researchers, Inc. **34.26b** Jeff Jaskolski/SeaPics.com **34.27** Amar & Isabelle Guillen/SeaPics.com **34.28** Peter Scoones/Getty Images Inc **34.29a** Michael Fogden/DRK Photo **34.29b** R.Andrew Odum/Peter Arnold, Inc. **34.31** Tom McHugh/Photo Researchers, Inc. **34.32** Tom Vezo/Peter Arnold, Inc. **34.33** Gerry Ellis/Minden Pictures **34.35** Gerald and Buff Corsi/Visuals Unlimited **34.36** George Grall/National Geographic Image Collection **34.37** Doug Perrine/DRK Photo **34.38** Gerald and Buff Corsi/Visuals Unlimited **34.39** M.S. Mayilvahnan **34.39b** Tom McHugh/Photo Researchers, Inc.

Chapter 35

Opener Jed Fuhrman **35.02** NIBSC/Science Photo Library/Photo Researchers, Inc. **35.06a** Omikron/Photo Researchers, Inc. **35.06b** Biophoto Associates/Photo Researchers, Inc. **35.06c** K.G. Murti/Visuals Unlimited **35.06d** Oliver Meckes/E.O.S./Max-Planck-Institut-Tubingen/Photo Researchers, Inc. **35.13a-b** Abbott Laboratories **35.15aR** NIBSC/Science Photo Library/Photo Researchers, Inc. **35.15bR** James L. Van Etten **35.17** Hans Gelderblom/Eye of Science/Meckes/Ottawa/Photo Researchers, Inc. **35.18** Philip Leder **35.19** Nigel Cattlin/Holt Studios/Photo Researchers, Inc. **35.20** Lowell Georgia/Photo Researchers, Inc. **35.21** David Parker/SPL/Photo Researchers, Inc.

Chapter 36

Opener Tom McHugh/Photo Researchers, Inc. **36.04a** Tim Hauf Photography/Visuals Unlimited **36.04bLMR** John E. Weaver University of Nebraska-Lincoln Libraries **36.05** Phil Stoffer U.S. Geological Survey **36.06a** Wally Eberhart/Visuals Unlimited **36.06b** Wolfgang Poelzer/Water Frame/Peter Arnold, Inc. **36.08a-b** Gerald D. Carr **36.08c** Frans Lanting/Minden Pictures **36.09R** Alan Majchrowicz/Peter Arnold, Inc. **36.10e** Will Cook/Charles W. Cook **36.10a** Ken Wagner/Phototake NYC **36.10b** Kenneth W. Fink/Photo Researchers, Inc. **36.10c** Lee W. Wilcox **36.10d** Bill Beatty/Visuals Unlimited **36.12a-d** Lee W. Wilcox **36.14a** Lee W. Wilcox **36.14b** M. Harvey/DRK Photo **36.14d-f** Lee W. Wilcox **36.14c** Doug Wechsler/Animals Animals Earth Scenes **36.14e** TORSTEN BREHM/Nature Picture Library **36.15b** Walker/Photo Researchers, Inc. **36.15a** Ed Reschke/ Peter Arnold, Inc. **36.18b** Lee W. Wilcox **36.18b** Ed Reschke/Peter Arnold,

Inc. **36.20RL** Lee W. Wilcox **36.22b** Lee W. Wilcox **36.21** Andrew Syred/Science Photo Library/Photo Researchers, Inc. **36.22a** Dr. Brad Mogen/Visuals Unlimited/Getty Images **36.23LR** Lee W. Wilcox **36-24a-c** Lee W. Wilcox **36.25a** Biophoto Associates/Science Source/Photo Researchers, Inc. **36.25b** Lee W. Wilcox **36.26a** Bruce Iverson/Bruce Iverson, Photomicrography **36.26b** G. Shih and R. Kessel/Visuals Unlimited **36.26c** Richared Kessel & Dr. Gene Shih Visuals Unlimited **36.27LMR** Lee W. Wilcox **36.28a-b** Michael Clayton **36.29c** Adam Hart-Davis/Science Photo Library/Photo Researchers, Inc. **36.29b** James W. Richardson/Visuals Unlimited **36.29a** Lee W. Wilcox;

Chapter 37

Opener David Nunuk/Science Photo Library/Photo Researchers, Inc. **37.02** Dr. Ulf Mehlig/Wikipedia Commons **37.02R** Alan Watson "Peter Arnold, Inc." **37.05a-b** David T. Webb **37.07LR** Lee W. Wilcox **37.08a** Lee W. Wilcox **37.09** Larry Lefever/Grant Heilman Photography, Inc. **37.11M** G. Shih and R. Kessel/Visuals Unlimited **37.11B** Lee W. Wilcox **37.11T** Ken Wagner/Phototake NYC **37.16a** John D. Cunningham/Visuals Unlimited **37.16a** Lee W. Wilcox **37.16b** Lee W. Wilcox **37.19R** Biophoto Associates/Photo Researchers, Inc. **36.18L** Jean Claude Revy/Phototake NYC **37.21** Martin H. Zimmerman/ Harvard Forest **37.24** Michael R. Sussman/American Society of Plant Biologists

Chapter 38

Opener Angelo Cavalli/Getty Images **38.03** Emanuel Epstein **38.03a** Photo Researchers, Inc. **38.02b** Nigel Cattlin/Photo Researchers, Inc. **38.02c** Photo Researchers, Inc. **38.06a** The Institute of Texan Cultures **38.06b** Thony Belizaire/Getty Images, Inc. **38.09R** Dennis Drenner/Visuals Unlimited **38.13b** Eduardo Blumwald **38.14** Hugh Spencer/Photo Researchers, Inc. **38.14** Andrew Syred/SPL/Photo Researchers, Inc. **38.15** E.H. Newcomb & S.R. Tandon/Biological Photo Service **38.17a** Richard Thom/Visuals Unlimited **38.17b** Gerry Ellis/GLOBIO **38.19** Frank Greenaway/Dorling Kindersley Media Library **38.18** Carlos Munoz-Yague Eurelios/Photographic Press Agency **38.20** Ken W. Davis /Tom Stack & Associates, Inc.

Chapter 39

Opener Lee W. Wilcox **39.04a-b** Malcolm B. Wilkins **39.12LR** Malcolm B. Wilkins **39.16a-b** Runk/Schoenberger/Grant Heilman Photography, Inc. **39.17a** American Society of Plant Biologists **39.20** Thomas Bjorkman **39.22L** Carolina Biological Supply Company/Phototake NYC **39.21** Donald Specker/Animals Animals Earth Scenes **39.23a-b** Lee W. Wilcox **39.27** Malcolm B. Wilkins **39.30b** American Society of Plant Biologists **39.32** Adel Kader **39.39** Nigel Cattlin/Holt Studios International /Photo Researchers, Inc.

Chapter 40

Opener Brian Johnston **40.01aL** Wayne P. Armstrong **40.01aR** Mitsuhiko Imamori Nature Production **40.01bL** Jerome Wexler/Photo Researchers, Inc. **40.01bR** Danny Ellinger Minden Pictures **40.03a** Dan Suzio/Photo Researchers, Inc. **40.03b-c** Jerome Wexler/Photo Researchers, Inc. **40.04bB** Rod Planck/Photo Researchers, Inc. **40.04bTL** John Gerlach/DRK Photo **40.04bTR** Tom & Therisa Stack/Tom Stack & Associates, Inc. **40.07a** D. Cavagnaro/DRK Photo **40.06a-b** Leonard Lessin/Photo Researchers, Inc. **40.07bTB** Lee W. Wilcox **40.13** "Candace Galen, University of Missouri. American Journal of Botany. 2003; 90:724-729. Sunny-side up: flower heliotropism as a source of parental environmental effects on pollen quality and performance in the snow buttercup, Ranunculus adoneus (Ranunculaceae) by Candace Galen, and Maureen L. Stanton. Reproduced by permission of the publisher. **40.18** Lee W. Wilcox **40.17TL** R.J. Erwin "Photo Researchers, Inc." **40.17TM** Mark Stouffer Animals Animals/Earth Scenes **40.17TR** Gerard Lacz/Animals Animals Earth Scenes **40.17ML** Tom Edwards/Visuals Unlimited **40.17MM** David Stuckel/Visuals Unlimited **40.17MR** Gregory K. Scott/Photo Researchers, Inc **40.17BL** Fritz Prenzel/Peter Arnold, Inc. **40.17BM** David M. Schleser/Nature's Images, Inc/Photo Researchers, Inc. **40.17BR** Sylvan Wittwer/Visuals Unlimited

Chapter 41

Opener Rob Nunnington/Minden Pictures **41.02** The Royal Society of London **41.03a** Educational Images/Custom Medical Stock Photo **41.03bT** Carolina Biological Supply Company/Phototake NYC **41.03bB** Carolina Biological Supply Company/Phototake NYC **41.03c** Carolina Biological Supply Company/Phototake NYC **41.04** Ed Reschke/Peter Arnold, Inc. **41.05a** Innerspace Imaging/Photo Researchers, Inc. **41.05b** Ed Reschke/Peter Arnold, Inc. **41.05c** G.W. Willis/Animals Animals Earth Scenes **41.06a** Nina Zanetti, Pearson Benjamin Cummings **41.10** Tom Stewart/CORBIS **41.11** Natalie Fobes/Getty Images Inc. **41.13a** Fred Hossler/Visuals Unlimited **41.13b** Oliver Meckes & Nicole Ottawa/Photo Researchers, Inc. **41.13c** P.M. Motta, A. Caggiati, G. Macchiarelli/Science Photo Library/Photo Researchers, Inc. **41.18a** Kent, Breck P./Animals Animals Earth Scenes **41.18b** Mendez, Raymond A/Animals Animals Earth Scenes **41.18c** Wayne Lynch/DRK Photo **41.19b** Editrice Kurtis S.r.l. **41.19a** Ed Reschke/Peter Arnold, Inc.;

Chapter 42

Opener Frans Lanting/Minden Pictures **42.13a** Fred Hossler/Visuals Unlimited

Chapter 43

Opener Jonathan Blair/Corbis Bettmann **43.03** Mark Smith/Photo Researchers, Inc. **43.04bLMR** Karel F. Liem **43.09a** Carolina Biological Supply Company/Phototake NYC **43.11a** Ed Reschke Peter Arnold, Inc.

Chapter 44

Opener Tony Freeman/PhotoEdit Inc. **44.06a** Walter E. Harvey/Photo Researchers, Inc. **44.06b** John D. Cunningham/Visuals Unlimited **44.22a** Lennart Nilsson Albert Bonniers Forlag AB **44.22b** Carolina Biological Supply/Visuals Unlimited

Chapter 45

Opener Marcus E. Raichle **45.02** Ronald F. Mervis, PhD, Neurostructural Research Labs and The Center for Aging and Brain Repair, University of South Florida College of Medicine **45.12a** C. Raines/Visuals Unlimited **45.14** Dennis Kunkel/Visuals Unlimited **45.16a** Oliver Meckes & Nicole Ottawa/Photo Researchers, Inc. **45.23a** Norbert Wu/Peter Arnold, Inc. **45.24b** Elsevier Science Ltd.

Chapter 46

Opener Stanley Breeden/DRK Photo **46.03a** Carole M. Hackney **46.07a** David Scharf/Peter Arnold, Inc. **46.10aR** Don Fawcett/T. Kwwabara/Photo Researchers, Inc. **46.13RL** David Quillin **46.14a** Michael and Patricia Fogden **46.14bLR** John A. Pearce **46.19B** James E. Dennis/Phototake NYC **46.19T** James E. Dennis/Phototake NYC **T46.01L** Brian Eyden/Science Photo Library/Photo Researchers, Inc. **T46.01M** Innerspace Imaging/Science Photo Library/Photo Researchers, Inc. **T46.01R** Brian Eyden/Science Photo Library/Photo Researchers, Inc.

Chapter 47

Opener Ralph A. Clevenger/Corbis **47.08R** Bernard Castelein/Nature Picture Library **47.08M** Duncan McEwan/Nature Picture Library **47.08L** Stephen Dalton/Photo Researchers, Inc. **47.09R** Perennou Nuridsany/Photo Researchers, Inc. **47.09M** Hans Pfletschinger/Peter Arnold Inc. **47.09L** BIOS/Heras Joël/Peter Arnold Inc. **47.11** Jordan Rehm/D. Scott Weigle

Chapter 48

Opener Leszczynski, Zigmund/Animals Animals/Earth Scenes **48.01a** Nova Scientific Corp/Oxford Scientific Films Ltd. **48.01b** Tom Adams/Visuals Unlimited **48.01c** Charles J. Cole **48.02a** Oxford Scientific Films/Animals Animals Earth Scenes **48.04TB** Reprinted by permission from Catherine S.C. Price et al., Sperm competition between Drosophila males involves both development and incapacitation. Nature 400:449-452 (1999), figs. 2 and 3. Copyright (c) 1999 Macmillan Magazines Limited. Image courtesy of Jerry A. Coyne, University of Chicago/Nature Magazine **48.05a** G. & C. Merker/B. Tomberlin/Visuals Unlimited **48.06a** Brent D. Opell Brent D. Opell **48.10B** From Nature "A stimulatory phalloid organ in a weaver bird" M. Winterbottom, T. Burke & T.R. Birkhead/Nature/ 6 May 1999, 399: 28 Fig 1b. Photo courtesy T.R. Birkhead **48.10T** From Nature "A stimulatory phalloid organ in a weaver bird" M. Winterbottom, T. Burke & T.R. Birkhead/Nature/6 May 1999, 399: 28 Fig 1a. Photo courtesy T.R. Birkhead **48.15a-c** Mitsuaki Iwago/Minden Pictures **48.16a** Claude Edelmann /Petit Format/Photo Researchers, Inc. **48.16b** Lennart Nilsson Albert Bonniers Forlag AB **48.16c** Petit Format/Nestle/Science Source/Photo Researchers, Inc. **48.18aLR** American Association for the Advancement of Science

Chapter 49

Opener Dennis Kunkel/CNRI Phototake NYC **49.01B** Fred Hossler/Visuals Unlimited **49.02a** Photo Researchers Inc **49.02b-c** Steve Gschmeissner/Photo Researchers, Inc. **49.05b** Steve Gschmeissner/SPL/ Photo Researchers, Inc. **49.05a** David M. Phillips/Visuals Unlimited

Chapter 50

Opener Stephen J. Krasemann/DRK Photo **50.01a-d** Natalie B. Fobes Photography **50.05b** Gerry Ellis/Minden Pictures **50.05c** Tim Fitzharris/Minden Pictures **50.05a** Gerry Ellis/Minden Pictures **50.06** George Gerster/ Photo Researchers, Inc. **50.07** Michael Collier **50.10a-c** Edward Kinsman/Photo Researchers, Inc. **50.12** Hans Reinhard/OKAPIA/Photo Researchers, Inc. **50.14** C.K. Lorenz/Photo Researchers, Inc. **50.16** © Judd Patterson **50.18** Tom Edwards/Animals Animals/Earth Scenes **50.20** Francis Lepine/Animals Animals Earth Scenes **50.22** Tom Bean/DRK Photo **50.27** Sandy Scheltema/Courtesy of The Age **50.29 inset** William H. Mullins/Photo Researchers, Inc. **50.29** Raymond Gehman/Corbis Bettmann **50.30b** Michael Fogden/DRK Photo **50.31** Ric Ergenbright/Corbis **50.32** Mark Newman/Photo Researchers, Inc. **50.32** Scott T. Smith Corbis Bettmann

Chapter 51

Opener Thomas Mangelsen/Minden Pictures **51.01** A. Flowers & L. Newman/Photo Researchers, Inc. **51.03a** © Callie de Wet **51.04** © Gary Woodburn/http://pbase.com/woody **51.06** Fred McConnaughey/Photo Researchers, Inc. **51.07** Nina Leen/Getty Images/Time Life Pictures **51.09a** Gavin Hunt **51.10** Photo courtesy of the Behavioural Ecology Research Group, Department of Zoology, Oxford University, Alexander Weir **51.11** Matt Meadows/Peter Arnold, Inc. **51.12a** Manuel Leal, Department of Biological Sciences, Union College **51.18a** Ken Lucas/Visuals Unlimited **51.18b** Paul A. Zahl/Photo Researchers, Inc. **51.18c** Ken Preston-Mafham/Premaphotos Wildlife **51.19** Bryan D. Neff **51.20** Tom Vezo/Peter Arnold, Inc.

Chapter 52

Opener National Geographic Image Collection **52.01** Robert Maier Animals Animals/Earth Scenes **52.06** Mark Trabue/USDA/NRCS **52.10LR** Marko Nieminen Ilkka Hanski **52.14a** Paal Hermansen /NHPA/Photo Researchers, Inc. **52.19a** BIOS (F. Mercay) Peter Arnold, Inc.

Chapter 53

Opener Ergenbright Scott Tuason/Image Quest Marine **53.01** PREMAPHOTOS/Nature Picture Library **53-01** Christian Ziegler **53.02a** G.R.Dick Roberts Photo Library/The Natural Sciences Image Library (NSIL) **53.02b** Fred Bavendam/Minden Pictures **53.02c** Tom Till/DRK Photo **53.02d** D. Cavagnaro/Visuals Unlimited **53.02e** Stephen J. Krasemann/DRK Photo **53.02f** Hal Beral/Visuals Unlimited **53.08LR** Jeff Podos **53.08MB** B.R. Grant **53.08MT** From Fig 1D in Science 14 July 2006 Vol. 313. no. 5784, pp. 224 – 226. Evolution of Character Displacement in Darwin's Finches Peter R. Grant* and B. Rosemary Grant. [Photos are by the authors] **53.09aL** Dave B. Fleetham/Visuals Unlimited **53.09aM** Chris Newbert/Minden Pictures **53.09aR** J. Sneesby/B. Wilkins/Getty Images Inc **53.10aL** Andrew J. Martinez Photo Researchers, Inc. **53.10aR** Zig Leszczynski/Animals Animals Earth Scenes **53.12a** Thomas G. Whitham/Northern Arizona University **53.15a** Mark Moffett/Minden Pictures **53.15b** Doug Perrine/SeaPics.com **53.18b** Thomas Kitchin/Tom Stack & Associates, Inc. **53.18a** Nancy Sefton/Photo Researchers, Inc. **53.20a** David Kjaer/Nature Picture Library **53.20b** Tony C. Caprio/Tony C. Caprio **53.21/1** Breck P. Kent/Animals Animals Earth Scenes **53.21/2** G. Carleton Ray/Photo Researchers, Inc. **53.21/3** Michael P. Gadomski/Animals Animals/Earth Scenes **53.21/4** James P. Jackson/Photo Researchers, Inc. **53.21/5** Bruce Heinemann/Getty Images, Inc. **53.21/6** Michael P. Gadomski/Photo Researchers, Inc. **53.22/1-4** Glacier Bay National Park Photo Glacier Bay National Park and Preserve **53.22/3** Christopher L. Fastie

Chapter 54

Opener Lonnie G. Thompson Byrd Polar Research Center The Ohio State University **54.02** MODIS NASA **54.11b** Randall J. Schaetzl, Michigan State University **54.11a** Richard Hartnup/Wikipedia Comons **54.12a** U.S. Forest Service, Northern Research Station

Chapter 55

Opener Micheline Pelletier/Corbis Bettmann **55.01b** Institute of Hydrobiology, Chinese Academy of Sciences **55.01a** Tom & Pat Lesson/DRK Photo **55.02** Edward S. Ross, California Academy of Sciences **55.03a** Carr Clifton/Minden Pictures **55.03b** Kevin W. Fitz All Species Photography & Sound **55.04b** from: Fig1C in Nature Vol 436 August 18, 2005 page 1016 "Global hotspots of species richness are not congruent with endemism or threat". C. David L. Orne et al. **55.04a** from Fig 1A in Nature Vol 436 August 18, 2005 page 1016 "Global hotspots of species richness are not congruent with endemism or threat". C. David L. Orne et al. **55.05B** Fred Bruemmer/DRK Photo **55.07a** John Mitchell/Photo Researchers, Inc. **55.07b** Gary Braasch Gary Braasch Photography **55.07c** David Dennis/Animals Animals Earth Scenes **55.09** Jim Wark/Airphoto **55.08** Martin Harvey/Alamy **55.10a** Jacques Jangoux/Peter Arnold, Inc. **55.10bLR** NASA/Goddard Space Flight Center

Backmatter

BS7.02 Reproduced by permission from J.P. Ferris et al., Synthesis of long prebiotic oligomers on mineral surfaces. Nature 381:59-61 (1996), Fig. 2. Copyright (c) 1996 Macmillan Magazines Limited. Image courtesy of James P. Ferris, Rensselaer Polytechnic Institute **BS8.01aL** Photo courtesy of Peter M. O'Day, Juan Bacigalupo, Joan E. Haab, and Cecilia Vergara. The Journal of Neuroscience, October 1, 2000. 20(19): 7193-7198, Fig. 1C. (c) 2004 by the Society for Neuroscience **BS8-01bR** Dennis Kunkel/Phototake NYC **BS8.02aTL** Michael W. Davidson/Florida State University/Molecular Expressions **BS8.02bTR** Michael W. Davidson/Florida State University/Molecular Expressions; **BS8.03B** Rosalind Franklin/Photo Researchers, Inc.

ILLUSTRATION CREDITS

1.12 Data from J.J. Tewksbury and G.P. Nabhan. Seed dispersal direct deterrence by capsaicin in chilies. 2001. *Nature* 412:403–404, Fig. 1a. Used with permission from Macmillan Publishers Ltd.

3.25a Data from N.N. Nawani and B.P. Kapadnis. 2001. *Journal of Applied Microbiology* 90:803–808, Fig. 3. Data also from N.N. Nawani et al. 2002. *Journal of Applied Microbiology* 93:965–975, Fig. 7. Blackwell Publishing. **3.25b** Data from T. Hansen et al. 2002. *FEMS Microbiology Letters* 216:249–253, Fig. 1. Blackwell Publishing.

6.25a PDB Document Object Identifier (DOI):10.2210/pdb1J4N/pdb. **6.25b** PDB DOI:10.2210/pdb1ORS/pdb and PDB DOI:10.2210/ pdb1ORQ/pdb.

7.4 PDB DOI:10.2210/pdb2AW4/pdb and PDB DOI:10.2210/ pdb2AVY/pdb.

8.10b After *Molecular Biology of the Cell,* 4th edition, Fig. 19.5. © 2002 by Bruce Alberts et al. With permission from Garland Science/Taylor & Francis, LLC.

9.2 PDB DOI:10.2210/pdb1ERK/pdb and PDB DOI:10.2210/ pdb2ERK/pdb. **9.15** PDB DOI:10.2210/pdb1PFK/pdb. **9.25** Modified from H. Wang and G. Oster. Energy transduction in the F_1 motor of ATP synthase. 1998. *Nature* 396:279–282, Fig. 1. With permission from Macmillan Publishers Ltd.

11.16 American Cancer Society's *Cancer Facts and Figures–2003*. Reprinted with permission.

16.2 Data from S. Murakami et al. 2002. *Science*. 296:1285–1290. Fig. 2a. Used with permission from AAAS. **16.15** PDB DOI:10.2210/pdb2J00/pdb and PDB DOI:10.2210/pdb2J01/pdb.

17.18b PDB DOI:10.2210/pdb2ORI/pdb.

18.2b After B. Dorigo et al. 2004. *Science* 306:1571–1573, Fig. 1. With permission from AAAS. **18.15** PDB DOI:10.2210/pdb1TSR/pdb.

20.1 Data from "Public collections of DNA and RNA reach 100 gigabases." 2005. EMBL-EBI Press Release, Hixton. **20.10** After *Molecular Cell Biology*, 5/e by Harvey Lodish, et al. © 1986, 1990, 1995, 2000, 2004 by W. H. Freeman and Company. Used with permission.

21.13 After S.B. Carroll. Homeotic genes and the evolution of arthropods and chordates (Review). 1995. Nature 376: 479–485. With permission from Macmillan Publishers Ltd.

24.4b *Pakicetus* based on J.G.M. Thewissen et al. 2001. *Nature* 413:277–281, Fig. 2. *Rhodocetus* based on P.D. Gingerich et al. 2001. *Science* 293:2239–2242, Fig. 3. **24.13** Data from P.T. Boag and P.R. Grant. 1981. *Science* 214:82–85. With permission from AAAS. **24.14** Modified from P.R. Grant and B.R. Grant. 2001. *Science* 296:707–711, Fig. 1. With permission from AAAS.

25.6 Data from S. Freeman and J. Herron, *Evolutionary Analysis,* 3e, Figs. 6.15a and 6.15c. © 2003 by Prentice-Hall, Inc. **25.7b** Data from W.E. Kerr and S. Wright. 1954. *Evolution* 8:172–177. **25.11** Data from M.O. Johnston. 1992. *Evolution* 46:688–702.

28.2 After G.L. Armstrong et al. 1999. *JAMA* 281:61–66, Fig. 1. © American Medical Association. All rights reserved. **28.5** Data from M.T. Madigan and J.M. Martinko. 2006. *Brock Biology of Microorganisms,* 11e, Fig. 5.9.

34.12 After E.B. Daeschler et al. 2006. *Nature* 440:757–763, Fig. 6. Also after N.H. Shubin et al. 2006. *Nature* 440:764–771, Fig. 4. With permission from Macmillan Publishers Ltd.

43.17 Data from L.O. Schulz et al. 2006. *Diabetes Care* 29:1866–1871. © 2006 American Diabetes Association. Used with permission from The American Diabetes Association.

48.4 Data from C.S.C. Price et al. Sperm competition between *Drosophila* males involves both displacement and incapacitation. 1999. *Nature* 400:449–452, Fig. 3c. Used with permission from Macmillan Publishers Ltd.

49.6a PDB DOI:10.2210/pdb1IGT/pdb. **49.6b** PDB DOI:10.2210/pdb1TCR/pdb. **49.7** PDB DOI:10.2210/pdb2VIR/pdb.

50.28 Data from A.K. Knapp et al. 2002. *Science* 298:2202–2205, Figs. 1, 2a, and 3b.

51.22 Data from J.L. Hoogland. 1983. *Animal Behavior* 31:472–479.

52.3 Modified from C.K. Ghalambor and T.E. Martin. 2001. *Science* 292:494–497, Fig. 1c. Used with permission from AAAS. **52.7b** Data from G.F. Gause. 1934. *The Struggle for Existence*. New York: Hafner Press. **52.9** Data from M. Dodd et al. 1995. *Journal of Ecology* 83:277–285. **52.14b** Data from T. Valverde and J. Silvertown. Variation in the demography of a woodland understorey herb (*Primula vulgaris*) along the forest regeneration cycle: Projection matrix analysis. 1998. *Journal of Ecology* 86:545–562. Blackwell Publishing.

53.4 Data from G.F. Gause. 1934. *The Struggle for Existence*. New York: Hafner Press. **53.18** Data from R.T. Paine. Intertidal community structure: Experimental studies on the relationship between a dominant competitor and its principal predator. 1974. *Oecologia* 15:93–120. With kind permission of Springer Science Business Media. **53.19** Data from R.T. Paine. 1974. *Oecologia* 15:93–120.

54.4 Data from A. Tsuda et al. 2003. *Science* 300:958–961. **54.18** Data from *BP Statistical Review of World Energy 2007*, pp. 6–21. **54.19a** Data from G. Beaugrand et al. 2002. *Science* 296:1692–1694. **54.19b** Data from N.L. Bradley et al. 1999. *Proceedings of the National Academy of Science* 96:9701–9704.

55.4a,b C.D.L. Orme et al. Global hotspots of species richness are not congruent with endemism or threat. 2005. *Nature* 436:1016–1019. Reprinted by permission from Macmillan Publishers Ltd. **55.4c** N. Myers et al. Biodiversity hotspots for conservation priorities. 2000. *Nature* 403:853–858, Fig. 1. Reprinted by permission from Macmillan Publishers Ltd. **55.6** O. Venter et al. 2006. *BioScience* 56:903–910, Fig. 2. Reproduced with permission of American Institute of Biological Sciences via Copyright Clearance Center. **55.12** Data from J.T. Hogg et al. 2006. *Proceedings of the Royal Society B: Biological Sciences* 273:1491–1499, Fig.1. © Proceedings of the Royal Society B: Biological Sciences. Additional unpublished from J.T. Hogg. **55.16** Data from E.I. Damschen et al. 2006. *Science* 313:1284–1286, Fig. 2b. Use with permission from AAAS.

Index

Boldface page numbers indicate a glossary entry; page numbers followed by an *f* indicate a figure; those followed by *t* indicate a table.

B

O

P

U

V

SUMMARY TABLES

BIOSKILLS

Dear Student:

As the author of ***Biological Science*, Third Edition,** I set out to write a book that would be an effective learning tool for you. I would appreciate hearing about your experience with this textbook and its multimedia support, and I invite your comments and suggestions for improvements!

Many thanks,

Scott Freeman

Grading System **A**=excellent/very useful, **B**=good/useful, **C**=moderately helpful, **D**=poor/very little help, **F**=not helpful

1. Please grade the following study aids in *Biological Science*, Third Edition:

(circle one)

Gold highlighting of Key Concepts and important information within the narrative — Grade: **A B C D F**

Comments ____

"If you understand... you should be able to..." in blue typeface within the narrative — Grade: **A B C D F**

Comments ____

Check Your Understanding boxes — Grade: **A B C D F**

Comments ____

Figure caption questions and exercises — Grade: **A B C D F**

Comments ____

Summary tables — Grade: **A B C D F**

Comments ____

Overall Comments ____

2. Did you use your *MasteringBiology*™ subscription? Yes [ ] No [ ] If yes, please grade the following online resources:

Web Animations — Grade: **A B C D F**

Comments ____

BioFlix animations and tutorials — Grade: **A B C D F**

Comments ____

Instructor-assigned tutorials — Grade: **A B C D F**

Comments ____

Chapter quizzes — Grade: **A B C D F**

Comments ____

Answer keys for textbook questions — Grade: **A B C D F**

Comments ____

3. Please list three things you liked most about the textbook and/or online resources:

Comments ______________________________

4. Please describe two or three topics that were difficult for you to understand, and please include page numbers.

Comments ______________________________

5. Do you have any other suggestions for improving future editions of ***Biological Science***?
Please provide page numbers where appropriate.

Comments ______________________________

School: ______________________________

Your Name (Optional): ______________________________

Email: ______________________________

Date: ______________________________

May Benjamin Cummings have permission to quote your comments in promotions for ***Biological Science***? ☐ Yes ☐ No

NO POSTAGE NECESSARY IF MAILED IN THE UNITED STATES

BUSINESS REPLY MAIL

FIRST-CLASS MAIL PERMIT NO. 275 SAN FRANCISCO CA

POSTAGE WILL BE PAID BY ADDRESSEE

BENJAMIN CUMMINGS
PEARSON EDUCATION
1301 SANSOME STREET
SAN FRANCISCO CA 94111-9328